ADVANCES IN ENERGY SCIENCE AND EQUIPMENT ENGINEERING II

T0239623

PROCEEDINGS OF THE 2ND INTERNATIONAL CONFERENCE ON ENERGY EQUIPMENT SCIENCE AND ENGINEERING (ICEESE 2016), 12–14 NOVEMBER 2016, GUANGZHOU, CHINA

Advances in Energy Science and Equipment Engineering II

Editors

Shiquan Zhou
China-EU Institute for Clean and Renewable Energy at Huazhong University of Science and Technology, Wuhan, China

Aragona Patty
Institute of Applied Industrial Technology Division, Portland Community College, Portland, OR, USA

Shiming Chen
College of Civil Engineering, Tongji University, Shanghai, China

VOLUME 1

CRC Press
Taylor & Francis Group
Boca Raton London New York

CRC Press is an imprint of the
Taylor & Francis Group, an **informa** business
A BALKEMA BOOK

Published by:
CRC Press/Balkema
P.O. Box 447, 2300 AK Leiden, The Netherlands
e-mail: Pub.NL@taylorandfrancis.com
www.crcpress.com – www.taylorandfrancis.com

First issued in paperback 2020

© 2017 by Taylor & Francis Group, LLC
CRC Press/Balkema is an imprint of the Taylor & Francis Group, an informa business

No claim to original U.S. Government works

Typeset by V Publishing Solutions Pvt Ltd., Chennai, India

ISBN: 978-1-138-71798-5 (set of 2 volumes)
ISBN 13: 978-0-367-73629-3 (vol 1)(pbk)
ISBN 13: 978-1-138-05682-4 (vol 1)(hbk)
ISBN 13: 978-1-138-05683-1 (vol 2)

This book contains information obtained from authentic and highly regarded sources. Reasonable efforts have been made to publish reliable data and information, but the author and publisher cannot assume responsibility for the validity of all materials or the consequences of their use. The authors and publishers have attempted to trace the copyright holders of all material reproduced in this publication and apologize to copyright holders if permission to publish in this form has not been obtained. If any copyright material has not been acknowledged please write and let us know so we may rectify in any future reprint.

Except as permitted under U.S. Copyright Law, no part of this book may be reprinted, reproduced, transmitted, or utilized in any form by any electronic, mechanical, or other means, now known or hereafter invented, including photocopying, microfilming, and recording, or in any information storage or retrieval system, without written permission from the publishers.

For permission to photocopy or use material electronically from this work, please access www.copyright.com (http://www.copyright.com/) or contact the Copyright Clearance Center, Inc. (CCC), 222 Rosewood Drive, Danvers, MA 01923, 978-750-8400. CCC is a not-for-profit organization that provides licenses and registration for a variety of users. For organizations that have been granted a photocopy license by the CCC, a separate system of payment has been arranged.

Trademark Notice: Product or corporate names may be trademarks or registered trademarks, and are used only for identification and explanation without intent to infringe.

Visit the Taylor & Francis Web site at
http://www.taylorandfrancis.com

and the CRC Press Web site at
http://www.crcpress.com

Table of contents

Mechanical engineering

Environmental and architectural engineering

VOLUME 2

Structural and materials science

Computer simulation & computer and electrical engineering

Advances in Energy Science and Equipment Engineering II – Zhou, Patty & Chen (Eds)
© 2017 Taylor & Francis Group, London, ISBN 978-1-138-71798-5

Preface

The previous First International Conference on Energy Equipment Science and Engineering (ICEESE 2015) successfully took place on May 30-31, 2015 in Guangzhou, China. All accepted papers were published by CRC Press/Balkema (Taylor & Francis Group) and have been indexed by Ei Compendex and Scopus and CPCI.

The 2016 2nd International Conference on Energy Equipment Science and Engineering (ICEESE 2016) was held on November 12–14, 2016 in Guangzhou, China. ICEESE 2016 is bringing together innovative academics and industrial experts in the field of energy equipment science and engineering to a common forum. The primary goal of the conference is to promote research and developmental activities in energy equipment science and engineering and another goal is to promote scientific information interchange between researchers, developers, engineers, students, and practitioners working all around the world.

The conference will be held every year to make it an ideal platform for people to share views and experiences in energy equipment science and engineering and related areas. We invite original papers describing an idea or concept, addressing issues and problems, or focusing empirically on potential or realistic fields.

ICEESE 2016 has received about 1000 papers from 8 countries and regions. The papers originate from both academia and industry concentrating on the international flavor of this event in the topics of energy equipment science and engineering. Based on the peer review reports, 434 papers were accepted to be presented in ICEESE 2016 by the editors. All the accepted papers will be presented on the conference, mainly by oral presentations in 5 sessions: Energy and Environmental Engineering, Mechanical Engineering, Environmental and Architectural Engineering, Structural and Materials Science, Computer Simulation & Computer and Electrical Engineering.

We express our thanks to all the members of 2016 2nd International Conference on Energy Equipment Science and Engineering (ICEESE 2016). Thanks are also given to CRC Press/Balkema (Taylor & Francis Group) for producing this volume.

We hope that ICEESE 2016 has been successful and enjoyable to all participants.

The Organizing Committee of ICEESE 2016

Advances in Energy Science and Equipment Engineering II – Zhou, Patty & Chen (Eds)
© 2017 Taylor & Francis Group, London, ISBN 978-1-138-71798-5

Organizing committees

CONFERENCE CHAIRS

Prof. Shiquan Zhou, *Huazhong University of Science & Technology*
Prof. Aragona Patty, *Portland Community College, United States*

TECHNICAL PROGRAM COMMITTEE

Prof. Zhandeng Dong, *Chinese Academy for Environmental Planning(CAEP), China*
Dr. Qing Fu, *Sun Yat-sen University, China*
Prof. Hongwei Wang, *Hebei University, China*
Dr. Bo Cai, *Research Institute of Petroleum Exploration and Development-Langfang Branch, PetroChina, China*
Prof. Tao Zhu, *China University of Mining & Technology (Beijing), China*
Prof. Zicheng Zhou, *Mechanical Engineering Dept., Shanghai Urban Construction College, China*
A. Prof. Jinyou Dai, *College of Petroleum Engineering, China University of Petroleum-Beijing, China*
Assoc. Prof. Eng. Krzysztof Witkowski, *University of Zielona Gora, Poland*
Prof. Mohd Khairol Anuar Mohd Ariffin, *Universiti Putra Malaysia, Malaysia*
Prof. Mingming Mao, *Shandong University of Technology, China*
A. Prof. Yu-Ming Fei, *Chilhlee University of Technology, Taiwan*
Prof. Dr. Aidy Ali, *Department of Mechanical Engineering, Universiti Pertahanan Nasional, Malaysia*
A. Prof. Gajendra Sharma, *Kathmandu University, Department of Computer Science and Engineering, Nepal*
Dr. Harish Kumar Sahoo, *International Institute of Information Technology (IIIT), Bhubaneswar, India*
Prof. Alaimo Andrea, *Faculty of Engineering and Architecture, Italy*
Prof. Govind Sharan Dangayach, *Malaviya National Institute of Technology, India*
Prof. Antonio Gil, *Public University of Navarra, Spain*
Prof. Dzintra Atstaja, *BA School of Business and Finance, Latvia*
Dr. Asimananda Khandual, *Biju Pattanaik University of Technology, India*
Dr. Fábio Robereto Chavarette, *Department of Mathematics, Brasil*
Prof. Heyong He, *Fudan University, China*
Prof. Leonid Getsov, *The Polzunov Central Boiler and Turbine Inst, Russia*
Prof. Loganina, *Penza State University of Architecture and Construction, Russia*
Prof. Maroš Soldán, *Slovak Technical University, Slovak Republic*
Dr. Saima Shabbir, *Department of Materials Science and Engineering, Institute of Space Technology, Pakistan*
Prof. Saffi Mohamed, *Mohamed V University in Rabat (ESTS), Morocco*
Prof. Ahmad Rezaeian, *Isfahan University of Technology, Iranian/Canadian*
Prof. Yingkui Yang, *Hubei University, China*

Energy and environmental engineering

Advances in Energy Science and Equipment Engineering II – Zhou, Patty & Chen (Eds)
© 2017 Taylor & Francis Group, London, ISBN 978-1-138-71798-5

A study on the usage of activated semi-coke for COD removal in waste water treatment

Liu Jing & Meili Du
Chemistry and Chemical Engineering College, Xi'an University of Science and Technology, Xi'an, Shaanxi, China

ABSTRACT: In this work, properties of active semi-coke were analyzed and the apparent morphology and the distribution of adsorption characteristics of active semi-coke were characterized by SEM and low temperature N_2 adsorption. It is found that active semi-coke has developed pores which are mainly formed due to the distribution of the micro hole and the middle pore, and belongs to the I type of adsorption. And then, active semi-coke was used to treat waste water which exhibited a COD level of 320 g/mL. The best treatment process was explored and the best treatment conditions were determined as follows: temperature—50°C, amount of active semi-coke added—7 g, pH—7.6, and adsorption time—60 minutes. Under these conditions, the removal rate of COD can reach up to 89.06%.

1 INTRODUCTION

According to statistics, there is an annual output of 20 billion tons of coke in Yulin and surrounding areas. With advanced technology applied ceaselessly, the concept of circular economy goes in to permeate, based on coke production mode purely in the past, and the coke industry has gradually expanded the chain downstream. In recent years, considerable attention has been devoted to performing waste water treatment by using activated semi-coke, which is made from Yulin coke. Activated semi-coke is a new material for adsorbents. It has been widely used in the city sewage treatment plants and in industrial waste water purification. It has broad market prospects.

2 MATERIALS AND METHODS

2.1 Materials and equipment

Materials: Waste water was collected from the campus sewage outlet. It contained a lot of suspended solids, with a pH of 6.0. The amount of COD present was 320 mg/L. Activated semi-coke was prepared under laboratory conditions.

Equipment: A HH-W21-600 thermostatic water bath, HH·W21 TU-1900 double beam ultraviolet visible spectrophotometer, SEM, and N_2-adsorption instrument are the various pieces of equipment used in this work.

2.2 Experimental section

Active semi-coke was added to 60 mL water. The sample was taken in a beaker and the pH value was adjusted. At a certain temperature, under rapid and slow stirring allowing the sample to stand for some time, the upper clear liquid was taken from the beaker and tested to determine the amount of COD. The effects of activated semi-coke dosage, temperature, pH, and time on treatment results were investigated. The extent of adsorption of activated semi-coke for COD removal in waste water was determined.

3 RESULTS AND DISCUSSION

3.1 Properties' analysis of activated semi-coke

Fig. 1 shows the SEM image of active semi-coke. From Figure 1, the active semi-coke surface can be seen clearly. It is rough and exhibits many voids. This confirms that activated semi-coke can be used as an adsorption material.

Fig. 2 shows the low temperature N_2-adsorption curve of active semi-coke. The adsorption isotherm of active semi-coke sample belongs to the I type of adsorption isotherm. It can be seen in the low pressure area adsorption. With an increase in pressure, there is a sharp rise in the $P/P^0 = 0.92$ region near the slow bend; when $P/P^0 > 0.20$, the adsorption is not a horizontal line, but is a larger slope, and also has a "tail" phenomenon. The results show that the adsorption of active semi-coke is I type and is mainly distributed in the micro hole and the middle hole. By analyzing the SEM image and the N_2 adsorption test results, it can be observed that active semi-coke can be used to reduce the COD levels in waste water.

Figure 1.   SEM image of activated semi-coke.

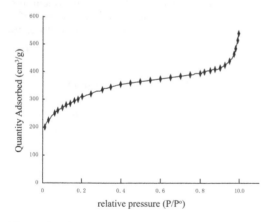

Figure 2.   Graph showing N$_2$-adsorption curve of activated semi-coke.

3.2   Waste water treatment

3.2.1   Influence of the active semi-coke dosage on COD removal

Fig. 3 shows the effect of the amount of activated semi-coke added on the COD of the water sample. It can be seen from the diagram that, when the amount of activated semi-coke added was increased to 6.5 g, the level of COD decreased from 320 g/mL to 100 g/mL; with a continuous increase

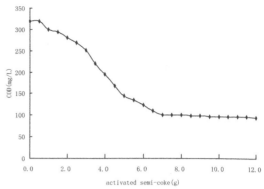

Figure 3.   Graph showing the influence of activated semi-coke dosage on COD levels.

in the amount of investment, the rate of reduction of COD levels slowed down until the dosage of 7 g, where the corresponding COD level was 98 g/mL after which the basic remained unchanged. This is because, an increase in the amount of activated semi-coke led to an increase in water absorbed from organic matter in the active site and pore numbers. Water samples of organic molecules continued to enter the pores of activated carbon on the surface and within the filling, or in combination with the active site, greatly reduced the concentration of organic compounds, and thereby the COD value decreased. When the dosage reached 7 g, after the feeding amount continues to increase, the unit quality in active semi-coke organic molecules reduces and the magnitude of the adsorption mass transfer driving force decreases, thereby leading to difficulty in filling of molecular organics into the channel, and in surface binding of active semi-coke to the active site, leading to the trend of an unchanged COD removal rate and adsorption capacity. Therefore, 7 g was selected as the dosage of activated semi-coke.

3.2.2   Influence of temperature on COD removal

Fig. 4 shows that by adding 7 g of active semi-coke to water and by adjusting the temperature from 25°C to 60°C, the COD levels of water was reduced sharply. The COD level was decreased from 129 g/mL to 68 g/mL. When the temperature was maintained at 60°C, activated semi-coke had a good effect for removal of COD. If the temperature continues to rise, the COD removal rate should continue to increase. The reason is that the high temperature is not only conducive to the adsorption equilibrium, but also is conducive to accelerate the decomposition of organic matter. But too high temperatures will cause trouble for the operation to progress, also will increase the cost, and not the economy. And so, for the treatment

4

of domestic sewage water, temperature can be adjusted to 50°C. Therefore, 50°C was selected as the suitable temperature.

3.2.3 *Influence of pH on COD removal*
The effect of pH on the removal rate and adsorption capacity is shown in Fig. 5. pH has a greater impact on active semi-coke-removed COD levels. Under strong acidic conditions, the effect of active semi-coke on the COD removal rate is low, the level of COD is 140 mg/g, and the removal rate was 56.31%; under less acidic conditions, COD levels decreased gradually, and at pH = 7.7, the level of COD reached its lowest of 49 mg/g, and the COD removal rate reached 84.70%. Under strong alkaline conditions, COD levels began to increase. Alkaline conditions are not conducive to the activated semi-coke adsorption of organic molecules. The presence of acidic organic compounds in water samples (such as humus, HA) under acidic conditions, the surface of active semi-coke consists of C-OH^{2+} ions; a small HA ions dissociates and

H^+ and A^- ions combined with active semi-coke adsorption block the surface with positive—OH^{2+} ions; with an increase in the pH, the H^+ concentration decreased, hindered and weakened, and active semi-coke adsorption of humic acids was more prone. When pH = 7.7, the blocking effect is reduced to a minimum, and the COD level reached the minimum. When pH > 7.80, the surface of active semi-coke particles is C-O$^-$ based. Organic molecular dissociation degree in A$^-$, between activated semi-coke humic acid particles and molecular electrostatic repulsion leads to a large amount of decrease in activated semi-coke adsorption of organic compounds in water samples, and a corresponding increase in COD levels. Therefore, pH = 7.7 was selected as the suitable pH.

3.2.4 *Influence of adsorption time on the COD removal rate*
Fig. 6 shows the effect of adsorption time on the removal rate of COD. The adsorption time is prolonged. The rate of COD removal decreased gradually at the initial time. After 60 min, COD levels were no longer reduced, and the residual COD concentration remained constant. This indicated that the adsorption process is closed to balance. At that time, the COD removal rate reached 89.06%, and the water residual COD concentration reduced from the original 320 mg/L to 33 mg/L. The possible reason for this result is that the adsorption channel is unobstructed at the early stage of adsorption, the porosity is larger, the pore velocity is smaller, the adhesion ability between the organic matter and organic matter is less, and a large amount of organic matter is very large. Therefore, it is easy to enter into the channel, and COD levels rapidly decline. As the adsorption continued, the level of active semi-coke channel-adsorbed-organic matter increased gradually and the occurrence of filling and plugging and porosity decreased, with a corresponding

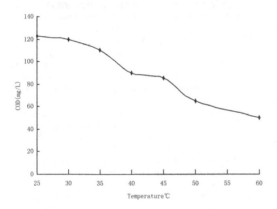

Figure 4. Graph showing the influence of temperature on COD.

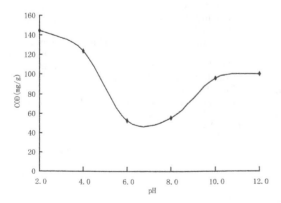

Figure 5. Graph showing the Influence of pH on the COD removal rate.

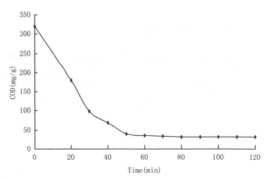

Figure 6. Graph showing the influence of adsorption time on the COD removal rate.

increase in the flow rate. The magnitude of adhesion force between the objects becomes smaller, which makes adsorption difficult, and then the adsorption amount of organic molecules is stable, and the concentration of COD remains constant. Therefore, 60 min was selected as the suitable adsorption time.

4 CONCLUSIONS

Activated semi-coke can be used as adsorption materials to reduce COD levels in waste water.

The adsorption isotherm of the active semi-coke sample belongs to the I type of adsorption isotherms. The adsorption of active semi-coke is I type and is mainly distributed in the micro hole and the middle hole.

When the amount of active semi-coke added was increased to 6.5 g, the level of COD reduced from 320 g/mL to 100 g/mL; with a continuous increase in the amount of activated semi-coke investment, the rate of COD removal increased, until the dosage of 7 g and a corresponding COD value of 98 g/mL, after which the basic remained constant.

When the temperature is 50°C, activated semi-coke exhibited a good effect for removal of COD.

pH has a greater impact for active semi-coke-removed COD. Under strong acidic conditions, the influence of active semi-coke on the COD removal rate is low. At pH = 7.7, the COD level is the lowest.

The adsorption time is prolonged and COD levels decreased gradually at the initial time. After 60 min, COD levels were no longer reduced, and the residual COD concentration remained constant.

Therefore, the best treatment conditions are as follows: temperature—50°C, amount of active semi-coke added—7 g, pH—7.6, and adsorption time—60 minutes. Under these conditions, the removal rate of COD can reach up to 89.06%.

ACKNOWLEDGMENTS

This work was financially supported by the project supported by the Natural Science Basic Research Plan in Shaanxi Province of China (2011 K07-08) and Scientific Research Cultivating Foundation of Xi'an University of Science and Technology (2010022).

REFERENCES

C.T.H, H. Teng, Influence of mesopore volume and adsorbate size on adsorption capacities of activated carbons in aqueous solutions [J]. Carbon, 2000, 38: 863–869.

Ehrburger P, Puussetn, Dziedzinl P. Active surface area of microporous carbons [J]. Carbon, 1992, 30(7): 1105–1109.

Houston G J, H A Oye. Reactivity testing of anode carbonmaterials [J]. Light Metals, 1985: 885–899.

Ikenaga NO, Ohgaim Y, M atsushimaH. Preparation of zinc ferrite in the presence of carbon material and its application to hot gas cleaning [J]. Fuel, 2004, 83: 661–669.

Ishii Chiaki, Kaneko Katsumi. Anomalous magnetism of activated carbon having ultra high surface area [J]. Tanso, 2000, 193: 218–222.

Lhuissier, J., L Bezamanifary. Use of under -calcined coke for the production of low reactivity anodes [J]. Light Metals, 2009: 979–983.

Li, X., J Xue, T Chen. Characterization of porous structure and its correlation to sodium expansion of graphite cathode materials using image analysis [J]. Light Metals, 2013: 1263–1267.

Liu Zhiguang. Analysis of the competitiveness of coal-based synthetic natural gas [J]. Sino-global energy, 2010, 15: 30–34.

Mac Dongalgj A F, Quinn D F. Carbon adsorbents for natural gas storage [J]. Fuel, 1997, 77(1): 61–65.

Maryam M, Josephine H. Effect of pyrolysis and CO_2 gasification pressure on surface area and pore size distribution of petroleum coke gasification pressure [J]. Energy & Fuels, 2011,25(11):1–21.

Mui K, E L K, Lau K S T. Production of activated carbons from waste tire-process design and economical analysis [J]. Waste Management, 2004, 24: 875–888.

Otowa T, Ymada M, Tanibaba R, et al. Gas separation technology In: Eds. Vansant E F, Dewolfs R. Amsterdam: Elsevier, 1990: 263–267.

Qingrong Qian, Motoi Machida, Hideki Tatsumoto. Preparation of activated carbons from cattle-manure compost by zinc chloride activation [J]. Bioresource Technology, 2007, 98: 353–360.

Qu Deyang, Shi Hang. Studies of activated carbons used indouble-layer capacitors [J]. Power Source, 1998, 74: 99–107.

Samanos, B C Dreyer. Impact of coke calcination level and anode baking temperature on anode properties [J]. Light Metals, 2001: 681–688.

Sharma D K. Extraction of coal through dilute alkaline hydrolytic treatment at low temmperature and atmospheric pressure [J]. Fuel Processing Technology, 1988, 19(1): 73–94.

Shouxin Liu, Jian Sun, Zhanhua Huang. Carbon spheres/activated carbon composite materials with high Cr (VI) adsorption capacity prepared by a hydrothermal method [J]. Journal of Hazardous Materials, 2010, 173(1–3): 377–383.

Sulaiman, D R Garg. Use of under calcined coke to produce baked anode for aluminium reduction lines [J]. Light Metals, 2012: 1147–1151.

Tran N K, Bhatia K S. Air reactivity of petroleum cokes-role of inaccessible porosity [J]. Ind Eng Chem, 2007, 46(10): 3265–3274.

Xing Wei, Zhang Mingjie, Yan Zifeng. Synthesis and activation mechanism of coke based super activated carbons. Acta Physics-Chimica Sinica, 2002, 18(4): 340–345.

Advances in Energy Science and Equipment Engineering II – Zhou, Patty & Chen (Eds)
© 2017 Taylor & Francis Group, London, ISBN 978-1-138-71798-5

Research on the spontaneous combustion index of cooking oil tar in fog discharge pipes

Z.W. Xie, Z.P. Jiang & S.F. Gong
Safety and Environment Institute, China Jiliang University, Hangzhou, Zhejiang, China

ABSTRACT: Cooking oil tar from a university faculty dining room, a university student canteen, a hotel, an enterprise staff canteen cooking system, moisture, ash, volatile oil, and ignition parameters under natural conditions are obtained by using the industry analysis method. On the basis of industrial analysis, a Thermogravimetric analysis experiment was carried out by means of a large material thermal analysis device developed by us. The TG-DTG definition was used to obtain grease samples under different heating rates. The experimental study provides the ignition index parameters for the design of an automatic fire protection system of fog discharge pipes.

1 INTRODUCTION

Cooking fire is a big problem, which may cause huge economic losses and have an adverse social impact (Xie Z.W, 2009). It is very important to prevent burning of fire in the fog discharge pipe of a cooking system. In order to reduce cooking oil fire in fog discharge pipes substantially (Xie Z.W, 2009), research on the combustion characteristics of cooking oil tar should be carried out firstly and then the fitting fire alarm and extinguishing system must be designed. The field work is very difficult due to the different shapes of the smoke channels. In order to achieve the measurement work as far as possible, some typical characteristics should be researched and summarized. The field sampling object is divided into the following two categories: the large dining room kitchen fume and large hotel kitchen fume.

2 TESTING METHOD FOR COOKING OIL TAR

2.1 Measured object selection

Various parameters of cooking oil tar obtained from a university faculty dining room, a university student canteen, a hotel, and an enterprise staff canteen cooking system are shown in Table 1. A university faculty cafeteria can accommodate 1000 people dining at the same time, where is adjacent to the kitchen and dining room, having been put into use about 5 years. The inner wall of the exhaust pipe in the kitchen stove has been on the upper part of the layer deposition about 2 months, and the black viscous material samples are scraped

from the pipe. A university faculty student dining hall can accommodate 2000 people dining at the same time, and the kitchen and dining room have almost been put to use for 6 years. The choice of the inner wall of the smoke collecting pipe is on the upper part of the kitchen stove, with a stratification time of 1 month, scraping black viscous material samples. A large hotel in the Chinese kitchen and Chinese restaurant are adjacent to the kitchen fumes having been put to use for 3 years, with black sticky material samples at the entrance of the fume purifier. A large enterprise dining hall can accommodate 800 people dining simultaneously, and the kitchen and the restaurant are next to it, which have been put to use for 10 years. At the corner of the fume hood, the layer has been deposited for about 4 months, and the surface layer is black and sticky.

2.2 Basic characteristic parameters

For fuels, the chemical composition and content of the fuel are determined by using the elemental analysis method; the main components are carbon, hydrogen, oxygen, nitrogen, sulfur, and also a litter of ash and water. For solid fuels, components can also be determined by performing industrial analysis. Industrial analysis includes moisture, ash, and volatile matter (Li X.H, 2007). In order to investigate the chemical constituents of grease stain and combustion characteristics, an experiment was carried out to explore the content of carbon, hydrogen, nitrogen, and sulfur by using the elemental analysis method. The measurement of levels of moisture and ash and degrees of volatility and heat was achieved by using the method of industry

Table 1. Sample collection and experiments.

Number	Sample	Temperature rise rate	Initial mass (g)	Number	Sample	Temperature rise rate	Initial mass (g)
1	Staff canteen	5°C/min	16.3	3	Large hotel	5°C/min	16.7
		10°C/min	16.7			10°C/min	16.8
		15°C/min	17.0			15°C/min	16.0
		20°C/min	14.9			20°C/min	15.2
		25°C/min	16.4			25°C/min	16.2
2	Dining hall	5°C/min	16.6	4	Enterprise canteen	5°C/min	16.5
		10°C/min	16.0			10°C/min	16.1
		15°C/min	16.1			15°C/min	17.1
		20°C/min	16.9			20°C/min	15.9
		25°C/min	16.4			25°C/min	15.4

Table 2. Elemental analysis results of cooking oil tar sample elements.

Sample	Cooking oil	Elemental analysis				
		$\omega(O)$ (%)	C_{ad} (%)	H_{ad} (%)	N_{ad} (%)	S_{ad} (%)
Staff canteen	J1	27.760	40.960	6.990	1.765	0.087
	J2	26.460	39.650	7.425	1.765	0.054
	J3	27.140	41.620	6.894	1.684	0.080
Dining hall	X1	28.038	41.370	7.060	1.783	0.088
	X2	26.725	40.047	7.499	1.783	0.055
	X3	27.411	42.036	6.963	1.701	0.081
Large hotel	H1	27.482	40.550	6.920	1.747	0.086
	H2	26.195	39.254	7.351	1.747	0.053
	H3	26.869	41.204	6.825	1.667	0.079
Enterprise canteen	Q1	25.880	41.180	6.960	1.654	0.091
	Q2	24.570	39.830	7.385	1.674	0.058
	Q3	25.220	41.850	6.804	1.593	0.087

Table 3. Analysis results with characteristics of coke residue of cooking oil tar sample elements.

Sample	Cooking oil	M_{ad} (%)	V_{ad} (%)	A_{ad} (%)	Qb_{ad} (J/g)	Characteristics of coke residue
Staff canteen	J1	13.20	71.71	8.13	22483.31	Honeycomb
	J2	12.60	71.70	8.01	27721.53	
	J3	13.40	69.33	8.32	26542.39	
Dining hall	X1	14.63	71.53	5.43	27258.04	
	X2	14.77	71.31	6.32	23369.77	
	X3	15.96	71.18	5.72	24568.34	
Large hotel	H1	18.10	71.35	2.73	28412.36	
	H2	17.30	70.42	5.33	27051.12	
	H3	18.50	73.02	3.11	22879.25	
Enterprise canteen	Q1	16.29	68.22	2.76	26954.43	
	Q2	15.58	69.38	4.77	24971.58	
	Q3	16.63	68.72	2.72	22553.87	

analysis. In this paper, the research method based on coal was used. The basic characteristic parameters of cooking oil tar are shown in Tables 2 and 3. Since the samples and experiment need to reflect a certain statistical significance, the average value is used to denote it.

3 MACRO-TG APPARATUS

A Thermogravimetric (TG) system built for performing pyrolysis studies of cooking oil tar is shown schematically in Figure 1. The main components of the system are as follows: a balance sensor which

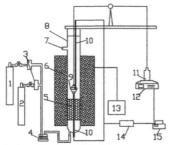

1. N$_2$ cylinder, 2. O$_2$ cylinder, 3. flow meter, 4. vapor arise, 5. porous ceramic, 6. furnace, 7. outlet and gas collector, 8. quartz reactor, 9. nickel crucible, 10. thermocouple, 11. 200 g weight, 12. electronic balance, 13. furnace controller, 14. temperature collector (USB2816), and 15. computer

Figure 1. Schematic diagram of the laboratory scale thermobalance (Macro-TG).

provides a means of measuring weight changes in all runs, a cylindrical quartz reactor with an inlet for the carrier gas (N2:O2 = 4:1) and an outlet for the volatiles and tars, a temperature controlled furnace, and a computer to continuously record the weight and temperature for samples and record the temperature for the furnace during the entire reaction period. The quartz reactor measured 350 mm in length with an internal diameter of 40 mm and is externally heated by a 3 kW temperature controlled furnace. The effective heating zone of the quartz reactor is about 250 mm. The sample basket is a 30 mL nickel crucible and approximately 15 g of the sample is placed in the crucible. And then, it is freely hung by using a suspension wire from one end of the level into the furnace tube. A type-K thermocouple located directly above the crucible is introduced to register the sample temperature during the pyrolysis process and another type-K thermocouple is located directly below the crucible to register the furnace temperature. The Macro-TG reactor is capable of pyrolyzing samples up to 1300°C at a prefixed heating rate (in the range of 0–100°C/min). Subsequently, approximately 15 g of sample, which is slightly placed on the electronic balance, is also suspended by using a suspension wire and then connected to another end of the level. The carrier gas is firstly introduced at a large rate of flow to purge residual impurities within the system, and then is established at 200 ml/min. When the electronic balance output is stable, the furnace power is turned on for pyrolysis to begin. As the sample begins to decompose, the balance output will change. According to the level law, the weight loss of the sample is equal to the weight increase of the electronic balance output. Thus, the change in the mass of the sample will be known (Xie Z.W, 2015). Simultaneously, the mass-change

to time and reaction temperature is recorded by using a personal computer at sampling rates of 1 s. In order to simulate the atmosphere of cooking fog discharge pipes, a vapor arise is installed in the gas system, which can provide wet air.

In order to eliminate the potential effect of gas convection by gas flowing, a section of a porous ceramic layer is placed on the bottom of the reactor. As gas is introduced into the quartz tube, it firstly passes through the porous ceramic layer and then is preheated to reaction temperature. A separate blank experiment is conducted to eliminate the system error and it is used for baseline correction during pyrolysis of the sample. By this means, magnitudes of buoyancy force and impact force of the carrier gas to the sample basket will reduce to a lesser extent.

4 THERMOGRAVIMETRIC ANALYSIS EXPERIMENT

In this study, experiments were conducted at the heating rates of 5, 10, 15, 20, and 25°C/min. Samples were heated from room temperature to 800°C at the respective heating rates and then held at 800°C for 10 min. When the experiments were completed, the furnace power was turned off but the carrier gas was allowed to flow until the reactor was cooled down to the room temperature. All experiments were carried out under atmospheric pressure.

In this section, the weight of the sample used was 15 g. Figure 2 gives the weight loss in relation to temperature during the pyrolysis of the sample. The TG curve is the Thermogravimetric curve when the simple weight changes with the temperature, while the DTG curve is the Derivative of the TG curve. It can be observed that TG curves are moved to the higher temperature when increasing the heating rate. DTG curves of all the samples bear a resemblance during the reaction and there are four peaks in each curve. Low heating rates affected the shape of the curves in the favor of narrower and sharper, but an increasing heating rate shifted the DTG peak to the right. Thermogravimetric curve characteristics are shown in Table 4. The T in the Table 4 means the time interval when the temperature changes from one point to another.

5 DETERMINATION OF IGNITION TEMPERATURE

Ignition temperature has different definitions, such as the temperature curve mutation method, DTG curve catastrophe point method (Wang F.J,

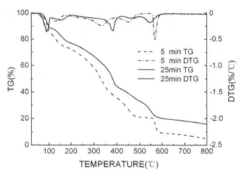

(a) Staff canteen

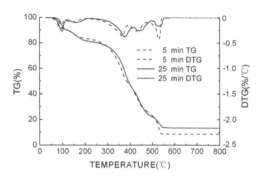

(b) Dining hall

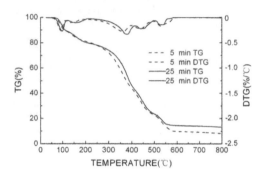

(c) Large hotel

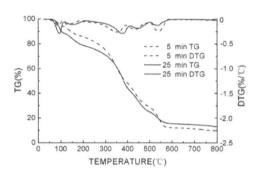

(d) Enterprise canteen

Figure 2.   Thermogravimetric curves of cooking oil tar.

Wang W.X, 2004), DTA curve method, TG curve boundary method, and the TG-DTG joint definition method. The TG-DTG joint definition method was used in this work. Figure 3 shows the results of the ignition temperature of the staff canteen under the different heating rates of 5, 10, 15, 20, and 25°C/min and Table 5 lists the ignition temperatures of four kinds of samples which were obtained from the staff canteen, dining hall, large hotel, and enterprise canteen under the different heating rates. The experimental results by comparison of four kinds of samples under the different heating rates are shown in Figure 4.

As it can be seen in Figure 4 and Table 5, when the heating rate increases, there is a corresponding increase in the ignition temperature. According to all results, a conclusion can be drawn that the ignition temperature under the different heating rates and different samples is entirely above 250°C.

6   CONCLUSION

Various parameters of cooking oil tar samples obtained from a university faculty dining room, a university student canteen, a hotel, an enterprise staff canteen cooking system, and the ignition temperatures at different rise rates were analyzed. Thermogravimetric analysis of cooking oil tar was conducted by using a specially designed laboratory-scale thermobalance (Macro-TG) with a sample loading of 15 g under a dynamic air atmosphere. The effects of heating rates (5, 10, 15, 20 and 25°C/min) were examined. It is relatively easy for the fume to get accumulated in the fog discharge pipe. The cooking oil tar will accumulate till it forms a certain thickness for several days and months in the corner. Once the fire soared into the flue, the cooking oil tar will be quickly ignited and the fire will spread rapidly. According to the research in this paper, the ignition temperature threshold was determined as 250°C while designing a fire prevention system.

Table 4. Thermogravimetric curve characteristics of cooking oil tar.

Sample	Temperature rise rate/(°C/min)	Temperature change /°C	T_{m1}	T_{m2}	T_{m3}	T_{m4}	Weight loss/%	Remain/%
Staff canteen	5	106~183	130.3				8.64	5.3
		183~376		336.1			36.00	
		376~475			448.5		16.55	
		475~621				562.8	13.64	
	25	113~200	154.4				11.00	15.6
		200~418		381.4			34.78	
		418~490			476.1		8.09	
		490~622				549.2	14.32	
Dining hall	5	111~197	138.8				7.3	8.7
		197~392		360.2			35.7	
		392~472			430.7		21.4	
		472~549				529.0	15.9	
	25	121~218	165.8				8.9	13.1
		218~416		373.6			39.4	
		416~497			444.9		19.7	
		497~559				533.8	8.7	
Large hotel	5	110~208	148.7				9.3	8.4
		208~399		360.2			36.8	
		399~497			441.2		21.2	
		497~594				537.6	12.8	
	25	119~223	165.8				9.5	13.5
		223~421		379.0			38.3	
		421~506			450.6		17.6	
		506~560				540.7	8.9	
Enterprise canteen	5	116~210	156.1				9.2	10.0
		210~418		376.0			39.8	
		418~502			451.4		15.9	
		502~594				536.0	16.5	
	25	107~185	136.2				9.8	13.4
		185~393		352.9			34.4	
		393~494			434.5		19.8	
		494~582				540.5	9.7	

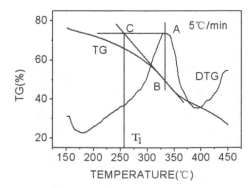

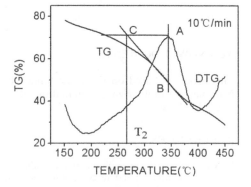

Figure 3. (a) Graph showing ignition temperatures of staff canteen at the heating rate of 5°C/min.

Figure 3. (b) Graph showing ignition temperatures of staff canteen at the heating rate of 10°C/min.

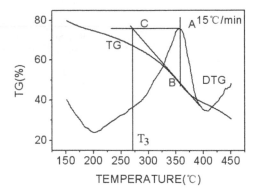

Figure 3. (c) Graph showing ignition temperatures of staff canteen at the heating rate of 15°C/min.

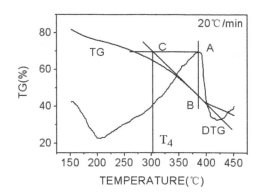

Figure 3. (d) Graph showing ignition temperatures of staff canteen at the heating rate of 20°C/min.

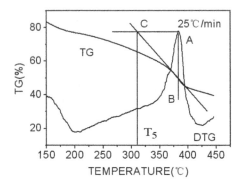

Figure 3. (e) Graph showing ignition temperatures of staff canteen at the heating rate of 25°C/min.

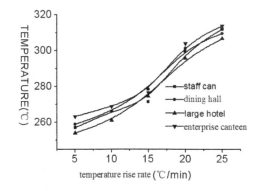

Figure 4. Graph showing ignition temperatures at different rise rates.

Table 5. Ignition temperatures at different rise rates.

Rise rate sample	5°C/min	10°C/min	15°C/min	20°C/min	25°C/min
Staff canteen	257.07	266.25	271.53	299.61	311.95
Dining hall	258.92	266.00	278.37	301.27	309.59
Large hotel	254.07	261.16	274.62	295.63	306.70
Enterprise canteen	263.30	268.84	276.60	303.95	313.78

ACKNOWLEDGMENTS

This article is based on the work funded by the National Natural Science Foundation of China (No. 51304179).

REFERENCES

Li X.H. (2007). The research on combustibility of sediment in ventilation and air-conditioning systems and how it affects building fire, *Hunan university.*

Wang F.J., Zhao C.S. (2004). Research on ignition and burnout characteristics of mixed fuels of petroleum coke and coal in CFB combustor, *power system engineering*, 20, 35–36.

Wang W.X., Wang F.J. (2004). Thermogravimetric analysis on the pyrolysis and combustion characteristics of mixed fuels of petroleum coke and coal, *Journal of fuel chemistry and technology*, 32, 522–526.

Xie Z.W., Jiang Z.P. (2015). Study on the characteristics of cooking fume fire based on fire dynamics simulation, *Advances in energy science and equipment engineering*, 11, 915–920.

Xie Z.W., Liang X.Y., Yuan Q., Qu F. (2009). Summarization of Research on Cooking Fume Fire. *Journal of Catastrophology.* 24, 89–96.

Xie Z.W., Wu C., Lang X.Y., Yuan Q. (2009). Fuzzy event tree analysis of cooking fume duct fire accident. *Fire safety science.* 18, 10–14.

Advances in Energy Science and Equipment Engineering II – Zhou, Patty & Chen (Eds)
© 2017 Taylor & Francis Group, London, ISBN 978-1-138-71798-5

Determination of tetracycline-resistant bacteria and resistance genes in soils from Tianjin

Y.K. Zhao, N. Wu, L. Jin & F. Yang
College of Engineering and Technology, Tianjin Agricultural University, Tianjin, China

X.M. Wang
Tianjin Key Laboratory of Aqua-Ecology and Aquaculture, College of Fisheries, Tianjin Agricultural University, Tianjin, China

F.H. Liang
College of Horticulture and Landscape, Tianjin Agricultural University, Tianjin, China

M. Zeng
College of Marine and Environment, Tianjin University of Science and Technology, Tianjin, China

ABSTRACT: Manure application could accelerate the spread of Antibiotic Resistant Bacteria (ARB) and Antibiotic Resistance Genes (ARGs) in soils. In this study, the prevalence of tetracycline resistant bacteria and resistance genes was investigated in manured agricultural soils in Tianjin, China. Anti-tetracycline bacteria were found in the range of 1.86×10^4 to 4.75×10^4 CFU/g dry soil, occupying 0.4% to 0.9% of total viable counts. Three tetracycline resistance genes (*tet*BP, *tet*O and *tet*X) were detected in all sampling sites, with relative abundances of 7.04×10^{-6} to 4.04×10^{-5}, 7.53×10^{-5} to 4.33×10^{-4} and 1.25×10^{-4} to 4.17×10^{-3} respectively. No significant correlations between cultivable tetracycline resistant bacteria and *tet* genes were found in this study. Only *tet*BP genes showed significant negative correlation with soil pH. The results highlight the important role of soils in resistance genes capture as environmental reservoir.

1 INTRODUCTION

Antibiotics are widely used to treat animal diseases and promote animal growth in livestock system. However, many antibiotics are poorly absorbed by animals and subsequently are excreted into the environment by urine or feces. The antibiotics tend to persist and accumulate in soils after repeated manure application. Residual antibiotics may exert selection pressure on environmental microorganisms, accelerating the spread of Antibiotic Resistant Bacteria (ARB) and Antibiotic Resistance Genes (ARGs). ARB and ARGs in soils have the potential to pose risks to human health, as susceptible pathogenic bacteria can become resistant by acquiring resistance genes in the environmental media (Chen C, 2014). Therefore, it is important to investigate the behavior of ARB and ARGs in agricultural soils.

The tetracycline group of antibiotics were selected as target substances, due to their broad usage in livestock industry. Tetracycline-resistant bacteria and tetracycline resistance (*tet*) genes have been found in various soil environments (Wu N, 2010, Wang F H, 2014), indicating the ubiquitous occurrence of tetracycline resistance. To date, more than 40 classes of *tet* genes have been described. The three main resistance mechanisms encoding by these *tet* genes are efflux pump proteins, Ribosomal Protection Proteins (RPPs) and inactivating enzymes (Wu N, 2010).

In this work, the prevalence of tetracycline resistance was investigated in agricultural soils to which organic manures had been applied from 4 districts in Tianjin, China. Culture-dependent method was used to assess the resistance rate of bacteria in soils exposed to different tetracycline concentrations. The quantitative PCR (qPCR) was used to quantify three *tet* genes. Meanwhile, the relationship between ARGs, ARB and environmental factors was also investigated.

2 MATERIALS AND METHODS

2.1 Soil samples

Soil samples were collected from agricultural fields to which organic manures had been applied. The samples were taken from Jixian, Xiqing, Wuqing and Dagang Districts (define as JX, XQ, WQ and DG respectively) in Tianjin, China, between March and May 2016. Samples were taken from 0 to 10 cm surface soils. Fresh samples were processed for ARB cultivation. The other samples were stored at -80°C before DNA extraction and chemical analysis. Soil pH was determined with a soil to water ratio of 1:2.5. Total Carbon (TC) was measured by a by a TOC analyzer (TOC-VCPH, SHIMADZU). Organic Matter (OM) was determined by the $K_2Cr_2O_7$ oxidation method. The physicochemical properties of soil samples are described in Wu et al. (in press).

2.2 Detection of ARB

Each soil sample was measured in duplicate to determine the numbers of ARB using colony forming units count method (Chen C, 2014) with minor modifications. Briefly, 5.0 g of wet soil was mixed with 45 mL sterile saline solution (0.85% NaCl) and shaken vigorously on an oscillator at 200 rpm for 20 min. 100 µL of ten-fold serial dilutions for each suspension were spread onto a broth agar plates containing selected concentrations of antibiotics (8 and 16 µg/mL of tetracycline). Broth agar plates with no antibiotics was used to determine the total cultivable bacteria numbers.

2.3 Detection of ARGs

Total DNA was extracted from 0.5 g of soil (fresh weight) by using FastDNA SPIN kit for Soil (MP Blomedicals, LLC., France). The concentration and quality of the extracted DNA was determined by spectrophotometer analysis (NanoDrop ND-1000, Theromo Fisher Scientific). PCR assays were used for broad-scale screening of the presence/absence of *tet* genes, and to get the standards for subsequent qPCR analysis. All PCR assays were conducted in a Peltier Thermal Cycler (Bio-Rad). Primers and annealing temperatures are described in Table 1.

After PCR amplification, gel slices of an agarose gel containing the PCR products were excised, and purified using EasyPure Quick Gel Extraction Kit (TransGen). The purified PCR products were ligated into p-GEM T easy vector (Promega) and then cloned into *Escherichia coli* DH5α (Tiangen). Clones containing target gene inserts were picked and sequenced. If the gene inserts were verified as the object resistance genes using the BLAST alignment tool, clones that had right gene inserts were chosen as the standards for the subsequent qPCR. Plasmids carrying target genes were extracted with Plasmid Kit (TaKaRa).

Table 1. Primers and PCR conditions used in this study.

Target genes	Sequences (5′—3′)	Amplicon size (bp)	Annealing temp (°C)	Reference
*tet*BP	aaaacttatt atattatagtg tggagtatcaa taatattcac	169	40	(Aminov et al., 2001)
*tet*O	acggaragttt attgtatacc tggcgtatct ataatgttgac	171	45	(Aminov et al., 2001)
*tet*X	caataattgg tggtggaccc ttcttaccttgg acatcccg	468	58	(Ng et al., 2001)
16S rRNA	cggtgaata cgttcycgg ggwtaccttgtt acgactt	123	56	(Suzuki et al., 2000)

Three target genes (*tet*BP, *tet*O and *tet*X) were quantified by qPCR using a SYBR-Green approach. 16S rRNA genes were quantified according to the TaqMan qPCR method (Suzuki M.T, 2000). 10-fold serial dilutions of a known copy number of plasmid carrying respective genes were generated to produce the standard curve. The qPCR mixtures (total volume, 20 µL) consisted of 10 µL of SYBR Premix Ex Taq (TaKaRa), 0.3 µL of each primer (20 µM), 1.6 µL of template DNA and 7.8 µL of ddH$_2$O. The qPCR procedure was conducted using an iCycler IQ5 Thermocycler (Bio-Rad). The protocol was as the following program: 45 s at 95°C, followed by 40 cycles of 10 s at 95°C, 30 s at the annealing temperatures. Product specificity was confirmed by melting curve analysis (55–95°C, 0.5°C per read, 30 s hold). Each reaction ran in triplicate. R^2 values were more than 0.99 for all calibration curves.

2.4 Data analysis

Averages and standard deviations of all data were determined using Microsoft Excel, 2007. All statistical analyses were performed using SPSS version 19.0, and significant differences were determined by the *t* test. Bivariate correlation analyses were conducted to obtain Spearman's coefficients.

3 RESULTS AND DISCUSSION

3.1 Occurrence of tetracycline-resistant bacteria

As shown in Figure 1, total concentrations of cultivable bacterial cells in soils ranged from 2.05×10^6

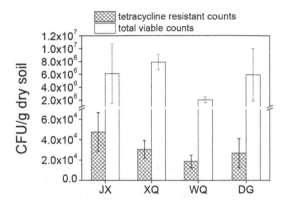

Figure 1. Tetracycline-resistant counts (resistant to 16 µg/mL of tetracycline) and total viable microbial counts in soil samples.

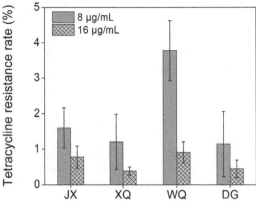

Figure 2. The tetracycline resistance rate of bacteria in soil samples exposed to different tetracycline concentrations (8 and 16 µg/mL of tetracycline).

to 7.86×10^6 CFU/g dry soil. Cultivable live anti-tetracycline bacteria, define as counts resistant to 16 µg/mL of tetracycline (Munir M, 2011), were observed in all soil samples, ranging from 1.86×10^4 to 4.75×10^4 CFU/g dry soil.

The relative abundances of tetracycline-resistant strains observed in this study (0.4% to 0.9%) were much lower than those in wastewater-irrigated soils, approximately 5% (Negreanu Y, 2012). These resistant bacteria may come directly from the indigenous soil environment. They may also result from the stress of antibiotics in soil which needs to be further studied.

Meanwhile, the tetracycline resistance rate of bacteria in soils exposed to different tetracycline concentrations are investigated (Fig. 2). With the increase of tetracycline concentration (from 8 to 16 µg/mL), the resistance rate of bacteria showed a decreasing trend (1.2%–3.8% at 8 µg/mL, 0.4%–0.9% at 16 µg/mL). No significant difference was observed in resistance rate under different tetracycline concentrations. When exposed to the same tetracycline concentration of 8 µg/mL, the resistance rate of bacteria in WQ is significantly higher than those in other sampling sites ($P < 0.05$). However, when exposed to 16 µg/mL tetracycline, there is no significant difference in resistance rate of bacteria among different samples.

3.2 Occurrence of tetracycline resistance genes

Two RPPs genes (tetBP and tetO), and one enzymatic modification gene (tetX) were quantified. Absolute gene copy numbers of tet genes in soil samples were presented in Figure 3. In all sampling sites, the absolute abundance of tetBP genes (4.78×10^5 to 2.10×10^6 copies/g dry soil) were lower than other two genes (5.12×10^6 to 6.20×10^8

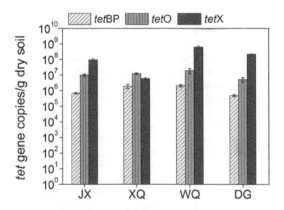

Figure 3. Detected levels of tet genes (tetBP, tetO and tetX) normalized to sample volume (copies per g dry soil) in soil samples.

copies/g dry soil). This result was in agreement with the observation in soils adjacent to swine feedlots (Wu et al., 2009). Moreover, the highest absolute abundances of three genes were all occurred in sample from WQ. This could be related to the high 16S rRNA gene copies in WQ (10^{11} copies/g dry soil).

To minimize variance caused by different extraction and analytical efficiencies and differences in background bacterial abundances, the absolute number of tet genes were normalized to that of ambient 16S rRNA genes (Fig. 4). The same general trends in gene abundance were seen in the normalized data relative to absolute data, since the total number of 16S rRNA gene copies was relatively consistent among different sites at 10^{10} copies/g dry soil, except for WQ (10^{11} copies/g

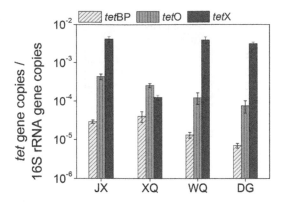

Figure 4. Detected levels of *tet* genes (*tet*BP, *tet*O and *tet*X) normalized to copies of ambient 16S rRNA gene present in soil samples.

dry soil). The relative abundances (target gene/16S rRNA genes) of *tet* genes showed significant variation over sampling sites, ranging from 10^{-6} to 10^{-3}.

The relative abundances of *tet*O ranged from 7.53×10^{-5} to 4.33×10^{-4}, comparable to those in soils near swine feedlots (Wu N, 2010) and in wastewater irrigated soils (Negreanu Y, 2012). The relative abundances of *tet*X were in the range of 1.25×10^{-4} to 4.17×10^{-3}. In all sampling sites (except for XQ), the relative abundances of *tet*X were above 1×10^{-3}, approximately 1–2 order of magnitude higher than those of other two genes. Compared to *tet*O and *tet*X, *tet*BP genes were lower in relative abundances (7.04×10^{-6} to 4.04×10^{-5}) in all samples, and we hypothesized that it could be attributed to the limited transferability of *tet*BP genes.

In this study, *tet* genes were detected with high abundance. The high prevalence of *tet* genes could be related to residual antibiotics in soils introduced by manure application or wastewater irrigation. Overall, the results highlight that soil plays an important role in resistance genes capture as the environmental reservoir.

3.3 *Relationship between ARGs, ARB and environmental factors*

According to the Spearman's correlation coefficients (Table 2), no significant correlations between cultivable tetracycline resistant bacteria and *tet* genes were found in this study. This is not surprising given that only a very small fraction of soil microbes were targeted in cultivation. A considerable part of *tet* genes quantified could possibly exist in uncultivable bacteria. In addition, the abundance of tetracycline resistant bacteria showed significant positive correlation with soil organic matter.

Table 2. Correlation analysis of relative abundances of ARGs, ARB and environmental factors.

	*tet*BP	*tet*O	*tet*X	pH	TC	OM	ARB
*tet*BP	1.0	0.8	−0.2	−1.0**	0.4	0.6	0.6
*tet*O	0.8	1.0	0.4	−0.8	−0.2	0.8	0.8
*tet*X	−0.2	0.4	1.0	0.2	−0.8	0.2	0.2
pH	−1.0**	−0.8	0.2	1.0	−0.4	−0.6	−0.6
TC	0.4	−0.2	−0.8	−0.4	1.0	−0.4	−0.4
OM	0.6	0.8	0.2	−0.6	−0.4	1.0	1.0**
ARB	0.6	0.8	0.2	−0.6	−0.4	1.0**	1.0

** represents statistical significance with $P < 0.01$; ARB represents tetracycline resistant bacteria.

Recent data have shown a significant negative relationship between ARGs and soil pH, since soil pH exerted a strong selection pressure on soil microbes and appeared to have a pervasive effect on the abundance of bacteria (Wang F H, 2014). In this study, the significant negative relationship was only found between *tet*BP genes and pH.

4 CONCLUSIONS

This study demonstrates the presence of high levels of tetracycline resistance in agricultural soils in Tianjin area, raising concerns that the land application of manures contributes to the environmental reservoir of resistance. No significant correlations between cultivable tetracycline resistant bacteria and *tet* genes were found in this study. Only *tet*BP genes showed significant negative correlation with soil pH. The relationship between ARGs and other environmental factors such as residual antibiotics, needs to be further studied. Overall, the results highlight the important role of soils in resistance genes capture as environmental reservoir.

ACKNOWLEDGEMENTS

This research was financially supported by Natural Science Foundation of Tianjin (No. 14JCYBJC43700, 15JCYBJC53700), National Undergraduate Training Programs for Innovation and Entrepreneurship (No. 201610061014), Scientific Research Foundation for the Returned Overseas Chinese Scholars, and National Natural Science Foundation of China (No. 51308392).

REFERENCES

Aminov, R. I., Garrigues-Jeanjean, N. & Mackie, R. I. 2001. Molecular ecology of tetracycline resistance:

Development and validation of primers for detection of tetracycline resistance genes encoding ribosomal protection proteins. *Applied and Environmental Microbiology* 67: 22–32.

Chen, C., Li, J., Chen, P., Ding, R., Zhang, P. & Li, X. 2014. Occurrence of antibiotics and antibiotic resistances in soils from wastewater irrigation areas in Beijing and Tianjin, China. *Environmental Pollution* 193: 94–101.

Munir, M., Wong, K. & Xagoraraki, I. 2011. Release of antibiotic resistant bacteria and genes in the effluent and biosolids of five wastewater utilities in Michigan. *Water Research* 45: 681–693.

Negreanu, Y., Pasternak, Z., Jurkevitch, E. & Cytryn, E. 2012. Impact of Treated Wastewater Irrigation on Antibiotic Resistance in Agricultural Soils. *Environmental Science & Technology* 46: 4800–4808.

Ng, L.K., Martin, I., Alfa, M. & Mulvey, M. 2001. Multiplex PCR for the detection of tetracycline resistant genes. *Molecular and Cellular Probes* 15: 209–215.

Suzuki, M.T., Taylor, L.T. & Delong, E.F. 2000. Quantitative analysis of small-subunit rRNA genes in mixed microbial populations via 5'-nuclease assays. *Applied and Environmental Microbiology* 66: 4605–4614.

Wang, F.H., Qiao, M., Lv, Z.E., Guo, G.X., Jia, Y., Su, Y.H. & Zhu, Y.G. 2014. Impact of reclaimed water irrigation on antibiotic resistance in public parks, Beijing, China. *Environmental Pollution* 184: 247–253.

Wu, N., Qiao, M. & Zhu, Y.G. 2009. Quantification of five tetracycline resistance genes in soil from a swine feedlot. *Asian Journal of Ecotoxicology* 4(5): 705–710 (in Chinese).

Wu, N., Qiao, M., Zhang, B., Cheng, W.D. & Zhu, Y.G. 2010. Abundance and diversity of tetracycline resistance genes in soils adjacent to representative swine feedlots in China. *Environmental Science & Technology* 44: 6933–6939.

Wu, N., Zhang, W.Y., Liu, H.F., Wang, X.B., Yang, F., Zeng, M., Chen, P.P. & Wang, X. 2016. Detection of sulfonamide-resistant bacteria and resistance genes in soils, Tianjin, China. *Key Engineering Materials* (in press).

Advances in Energy Science and Equipment Engineering II – Zhou, Patty & Chen (Eds)
© *2017 Taylor & Francis Group, London, ISBN 978-1-138-71798-5*

Differences of auxin transport inhibitor treatment on several development processes between different *Nicotiana* varieties seedlings

Lijuan Zhou
School of Life Sciences, Yunnan University, Kunming, Yunnan, China
Yunnan Reascend Tobacco Technology (GROUP) CO., LTD., Kunming, Yunnan, China

Xiaolong Zhang
Yunnan Reascend Tobacco Technology (GROUP) CO., LTD., Kunming, Yunnan, China
Yunnan Tianhedi Biotechnology CO., LTD., Kunming, Yunnan, China

Xiaoya Li
Yunnan Reascend Tobacco Technology (GROUP) CO., LTD., Kunming, Yunnan, China

Weijuan Liu
Yunnan Reascend Tobacco Technology (GROUP) CO., LTD., Kunming, Yunnan, China
Yunnan Tianhedi Biotechnology CO., LTD., Kunming, Yunnan, China

Jingqiang Wu & Yihong Pang
Yunnan Reascend Tobacco Technology (GROUP) CO., LTD., Kunming, Yunnan, China

ABSTRACT: Auxin is a crucial hormone that modulates many important developmental processes and responses to the environment including embryogenesis, leaf development, organogenesis, vascular patterning, senescence, and stress tolerance. In *Arabidopsis*, auxin signaling and transport enforce stomatal patterning as well as the size of their stem cell compartment, but whether this regulation also existed in different *Nicotiana* varieties remains unknown. We use two *Nicotiana* varieties as an example to analyse the function of the auxin transport inhibitor (NAP) on its stomatal and pavement cell development on the epidermis of the cotyledon. In this work, we reported that there was an inter-specific difference between K326 and Yun203 under exogenous auxin inhibitors (NPA) treatment in stomatal and pavement cell development. NPA treatment can induce patterning and morphogenes exhibited abnormality of stoma in two *Nicotiana* varieties, but the sensitivity of auxin regulation was different between two varieties.

1 INTRODUCTION

Nicotiana is an important commercial crop and the leaf structure is a critical factor to its quality. The *Nicotiana* leaf epidermis contains three cell types: Pavement Cells (PCs), stomata, and trichrome. Mature PCs have an interlocking jagsaw-puzzle-like shape, thereby resulting in coordinated multiple polarities and intercalated growth. Their main functions are protecting plants by the following methods, such as preventing moisture loss, resisting outside invasion, holding internal material, and maintaining temperature. Stomata are epidermal pores used for water and gas exchange between a plant and the atmosphere. Both the entry of carbon dioxide for photosynthesis and the evaporation of water that drives transpiration and temperature regulation are modulated by the activities of stomata (Dong and Bergmann, 2010). Auxin widely participates in plant development, such as by coordinating and patterning the placement of organs and cells; most of this research was based on the model plant, *Arabidopsis*, for its small and completely sequenced genome (Benjamins and Scheres, 2008; Normanly J, 2009; Tromas A, 2013; Xu T, 2014; Xu T, 2010). Auxin also regulates seed germination, cell division, polarity formation as well as stomata movement in *Nicotiana* (Petrášek J, 2002; Yan L I, 2010; Zhen-Hua, 2013). Recently Le and Zhou reported auxin transport and signaling regulated stomatal development and patterning in *Arabidopsis and Nicotiana* (Le J, 2014). In this study, we used the auxin transport inhibitor (NPA) (Geldner N, 2001) to study the different sensitivities of auxin in pavement cell and stoma development in different *Nicotiana* varieties (K326, Yun203).

2 MATERIALS AND METHODS

2.1 *Plants and growth conditions*

All experiments described in this study involve *Nicotiana tabacum* (K326, Yun203). *Nicotiana tabacum* was grown at 24°C in 1/2 MS agar plates in an incubator with 12-hr-light/12-hr-dark cycles, and specific stages of rosette leaves were used.

2.2 *Chemical treatment*

Seedlings were grown in three types of culture plates, all obtained from Sigma-Aldrich. NPA (Sigma 33371) was dissolved in DMSO and used at a final concentration of 20 μM and as a control, and the same volume of DMSO was added.

2.3 *Microscopy*

For observation by bright-field microscopy, seedlings were analyzed by using a BSK41 compound light microscope (Olympus Instruments).

2.4 *Quantification*

Epidermal cell types and pavement cell/stomatal shape were quantified with bright field images of cotyledons of 14-d-old seedlings by using ImageJ 1.46r software (Wayne Rasband, NIH, Bethesda, USA). Depending on the cotyledon size, up to three areas in the abaxial epidermis were selected, avoiding the margin and the area close to the petiole, and cell types or lobe length in these areas were quantified. The area size was 200 μm × 200 μm in cotyledons of seedlings. Four to five cotyledons of individual seedlings, respectively, were analyzed per genotype and condition. Statistical analyses were performed with IBM SPSS Statistics 21.0 software.

3 RESULTS

3.1 *Effect of NPA on seed germination rate of different Nicotiana varieties*

For both varieties, 20 uM NPA treatment had seldom effect on their seed germination, but NPA does prolong the time needed for seed germination in *Nicotiana*. As shown in Figure1, both kind of seeds germinated mainly in 6 days after planting in the MS medium. The most significantly inhibited for seed germination on 6dap, 9.99% and 14.6% were germinated with NPA treatment in Yun203 and K326, respectively. More than 85% of the seeds were germinated with or without NPA treatment in both varieties after 7dap. While, the seed germination rate of K326 in NPA treatment more

dramatically decreased, which means its seed is more sensitive to NPA than that of Yun203. Thus, auxin polar transport during seed germination is essential to *Nicotiana* seedlings, since the disrupts PIN recycling, and thus auxin transport, by inhibiting the ARF-GEF activity.

3.2 *Effect of NPA on the constituents of the epidermis cell number of both varieties*

We then measure the number of constituents of epidermis cells on the abaxial epidermis of the cotyledon in different *Nicotiana* varieties seedlings. As can be seen from Figure 2, NPA treatment resulted in more divisions in the epidermis of the cotyledon. The stomatal density (stomata number per $μm^2$ epidermis area) increased from 1.16 to 1.81 in K326 and induced a stomatal index increase from 1.02 to 1.39 in Yun203. As for the stomatal index, NPA induced a stomatal index decrease from 0.21 to 0.17 in K326 and induced a stomatal decrease from 0.17 to 0.16 in Yun203.

3.3 *Effect of NPA on stomatal and pavement cell shape of both varieties*

As can be seen from Figure 3, the stomatal size was inhibited by NPA treatment; the length of the stoma long axis decreased from 22.47 to 19.32 in K326 and decreased from 24.86 to 22.93 in Yun203. The PC shape was also changed as the lobe number and length decreased by 7.9% (K326)/7.3% (Yun203) and 64.08% (K326)/50.32% (Yun203) when compared to untreated cotyledons. These results demonstrated that auxin transport is essential for the division frequency regulation and the final cell constituents on the epidermis. Also, the change of the PC shape by inhibitor treatment proved that auxin transport involves in PC morphogenesis, while K326 is more sensitive to NPA than Yun203.

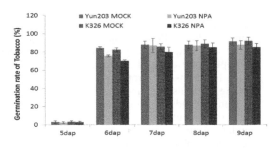

Figure 1. Chart showing the effect of the auxin polar transport inhibitor on the germination rate in K326 and Yun203 (n>100). Error bars show Standard Deviation (SD) (n>30).

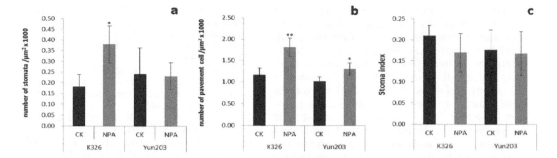

Figure 2. Charts showing change in the constituents of epidermis cell number. (a-c) Images of the cotyledon epidermis of 14-d-old *Nicotiana* (K326 and Yun203) with or without 20μM NPA treatment. (a and b) Stomata/pavement cell density of K326/Yun203 was increased in response to NPA treatment, error bars show Standard Deviation (SD) (n = 20). (c) Stomata index of K326/Yun203 was decreased in response to NPA treatment, error bars show Standard Deviation (SD) (n = 20). *Differences between treated plants and control plants are significant (P<0.05). **Differences between treated plants and control plants are highly significant (P<0.01), (n>100).

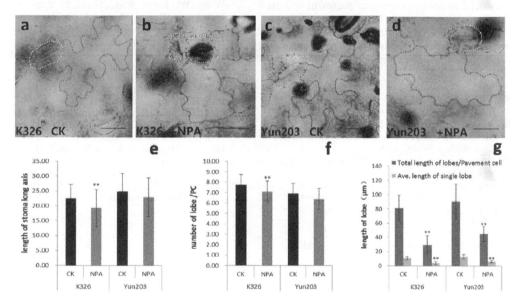

Figure 3. Images showing change in the size of stoma and shape of the pavement cell with NPA treatment. (a-d) Images of the cotyledon epidermis of 14-d-old *Nicotiana* (K326 and Yun203) with (a and c) or without (b and d) 20 μM NPA treatment. Bars = 20μm. Red dots indicate the typical shape of the pavement cell, while yellow dots show mature stomata or stomata clusters (b). (e) Size of stomata of 14-d-old *Nicotiana* (K326 and Yun203) with 20 μM NPA treatment. Error bars show Standard Deviation (SD) (n>60). (f) Number of lobes per pavement cell of 14-d-old *Nicotiana* (K326 and Yun203) with or without 20 μM NPA treatment. Error bars show Standard Deviation (SD) (n>90). (g) Length of single lobe and total length of lobes per pavement cell of 14-d-old *Nicotiana* (K326 and Yun203) with or without 20 μM NPA treatment. Error bars show Standard Deviation (SD) (n>90). **Differences between treated plants and control plants are highly significant (P<0.01).

4 DISCUSSION

As a hormone, in general, auxin modulates almost every aspect of plant growth and development. Auxin transport, as well as signaling, participates in many aspects of the development process such as root elongation, embryogenesis, floral development, leaf development, vascular patterning, and so on (Benjamins and Scheres, 2008; Sauer M, 2013; Vanneste and Friml, 2009). Our results also proved that auxin transport is essential for seed germination, epidermis development, and root gravity tropism in both *Nicotiana* seedling varieties. Auxin transport enforces the epidermis cell number as well as the shape of the epidermis cell in *Nicotiana* varieties. But under the treatment of same concentration of inhibitors such as NPA, both varieties exhibited a difference partly in response in pavement and stomatal development, more in detail, the sensitivity to NPA of epidermis cell density and shape differed. *K326* seems to be more resistant to NPA, and thus to auxin transport inhibition. The reason for this inter-specific difference needs to be further studied, for differed sensitivities to chemicals or different epidermis cell regulation mechanisms in these two varieties may help us to understand more details about the stomatal development regulation mechanism based on more data from more varieties throughout the plant kingdom. For the former, our results may remind researchers of the difference in sensitivity to chemical inhibitors in various varieties when they need to adopt exogenous treatment and draw their conclusion. For the latter, the possibility may exist, and if it is so, more research will be carried out on stomatal and pavement cell development regulation mechanism and comparison.

REFERENCES

Benjamins, R., and Scheres, B. (2008). Auxin: The Looping Star in Plant Development. *Annual Review of Plant Biology* 59, 443–465.

Dong, J., and Bergmann, D.C. (2010). Stomatal Patterning and Development. 91, 267–297.

Geldner, N., Friml, J., Stierhof, Y.D., Jurgens, G., and Palme, K. (2001). Auxin transport inhibitors block PIN1 cycling and vesicle trafficking. *Nature* 413, 425–428.

Le, J., Liu, X.G., Yang, K.Z., Chen, X.L., Zou, J.J., Wang, H.Z., Wang, M., Vanneste, S., Morita, M., and Tasaka, M. (2014). Auxin transport and activity regulate stomatal patterning and development. *Nature Communications* 5, 3090–3090.

Normanly, J., Slovin, J.P., and Cohen, J.D. (2009). "Auxin Biosynthesis and Metabolism."

Petrášek, J., Elčkner, M., Morris, D.A., and Zažímalová, E. (2002). Auxin efflux carrier activity and auxin accumulation regulate cell division and polarity in tobacco cells. *Planta* 216, 302–8.

Sauer, M., Robert, S., and Kleine-Vehn, J. (2013). Auxin: simply complicated. *J Exp Bot* 64, 2565–2577.

Tromas, A., Paque, S., Stierle, V., Quettier, A.L., Muller, P., Lechner, E., Genschik, P., and Perrot-Rechenmann, C. (2013). Auxin-binding protein 1 is a negative regulator of the SCF(TIR1/AFB) pathway. *Nat Commun* 4, 2496.

Vanneste, S., and Friml, J. (2009). Auxin: a trigger for change in plant development. *Cell* 136, 1005–1016.

Xu, T., Dai, N., Chen, J., Nagawa, S., Cao, M., Li, H., Zhou, Z., Chen, X., De Rycke, R., Rakusova, H., Wang, W., Jones, A.M., Friml, J., Patterson, S.E., Bleecker, A.B., and Yang, Z. (2014). Cell surface ABP1-TMK auxin-sensing complex activates ROP GTPase signaling. *Science* 343, 1025–1028.

Xu, T., Wen, M., Nagawa, S., Fu, Y., Chen, J.G., Wu, M.J., Perrot-Rechenmann, C., Friml, J., Jones, A.M., and Yang, Z. (2010). Cell surface—and rho GTPase-based auxin signaling controls cellular interdigitation in Arabidopsis. *Cell* 143, 99–110.

Yan, L.I., Hong-Yan, L.I., Wang, Q., Kang-Ning, L.I., Zhang, L.L., Zhao, X.M., Yu-Guang, D.U., and Hou, H.S. (2010). Role and Relationship of Oligochitosan, NO and Phytohormones in Stomatal Movement of Tobacco (Nicotiana tabacum L.cv.Samsun NN). *Plant Physiology Communications* 46, 575–578.

Zhen-Hua, L.I. (2013). Advances in Hormones During Tobacco Seed Germination. *Journal of Yunnan Agricultural University*.

Advances in Energy Science and Equipment Engineering II – Zhou, Patty & Chen (Eds)
© 2017 Taylor & Francis Group, London, ISBN 978-1-138-71798-5

Cellulose refining of biobutanol

Renjie Chen

College of Life Science and Technology, Beijing University of Chemical Technology, Beijing, P.R. China

ABSTRACT: As the main product of acetone, butanol is highly valued for its excellent fuel properties and attractive potential. In this paper, the application of cheap agricultural wastes, such as rice straw lignin cellulose is introduced as a renewable resource for the production of biofuel, such as butanol. By using the method of *Clostridium* fermentation, the study was carried out from several aspects, such as selection of the substrate, hydrolysis of hemicellulose, hydrolysis of the cellulose enzyme, production of butanol by fermentation of the hydrolysate, etc. This research work was carried out in order to find new technology and new ideas for the production of butanol, to reduce the energy dependence of the petrochemical gasoline transportation, to realize the sustainable development of bioenergy in biomass production.

1 INTRODUCTION

The molecular formula of butanol is $C_4H_{10}O$, a colorless, pungent liquid, which can form an explosive mixture of vapor and air. Butanol has a boiling point of 117.7 DEG C, melting point of 90.2 DEG C, flash point of 35~35.5 DEG C, relative density of 0.8109 and spontaneous combustion of 365 degrees centigrade; the refractive index 1.3993. At 20 DEG, it is in the aqueous solubility of 7.7%; water has a solubility of 20.1% in positive butanol. Butanol is not only a kind of organic chemical raw material, but also is a kind of new biofuel with great potential. As a fuel, the physical parameters are shown in Table 1.

The following are the advantages of using butanol as a biofuel: when compared with petroleum refining fuels, it does not produce SOX or NOX combustion and has significant environmental benefits. When compared with the traditional existing biofuels such as ethanol, butanol is more tolerant with water pollution in gasoline and it has a higher blending ratio with gasoline such that the vehicle need not be transformed. Steam is low, and so it can be transported through the pipeline. Butanol can also improve the vehicle mileage and fuel efficiency.

When the rice hull is used in animal feed, because of its peculiar shape, low density, and high ash content, it is often difficult to digest for animals. It is used in combustion for heat energy conversion, by using the special boiler; otherwise, a large amount of ash easily solidifies in the high temperature, thereby resulting in the problem of boiler blockage. Therefore, rice hulls are often treated as waste. However, it is convenient for us to collect them under this condition. Rice hulls appear in large-scale food processing enterprises, and the location is relatively concentrated. These cannot be easily degenerated if they have a good resistance to moisture. Finally, we decided to use the method of pre-hydrolysis to destroy the hemicellulose of rice husk, and then let the exposed cellulose increased and working face with cellulose enzyme and further enzymatic hydrolysis into fermentable sugars and used to fuel alcohol production experiments.

Clostridium production of butanol comprises a number of strains, including *acetobutylicum, beijerinckii, saccaroperbutylacetonicum, saccharoacetobutylicum, aurantibutyricum, pasteurianum, sporogenes* and *tetanomorphum*. Among these *Clostridiums, C. acetobutylicum, C. beijerinckii, C. saccharoacetobutylicum, C. saccharoperbutylacetonicum,* and so on are able to obtain higher

Table 1. Comparison of basic physical parameters of gasoline, butanol, ethanol, and methanol.

Parameter	Butanol	Ethanol	Methanol	Gasoline
Energy density (MJ/L)	29.2	19.6	16	32
Evaporation heat (MJ/kg)	0.43	0.92	1.2	0.36
Research octane number	96	107	106	91–99
Mecoctane number	78	89	92	81–89
Gas fuel ratio	11.2	9.0	6.4	14.6

butanol production rates. In these wild strains, the substrate utilization efficiency of different strains, the most suitable pH, and the optimum temperature, and the difference in the proportion of the product are so huge. Most *Clostridium* butanol can be used to achieve the metabolism of 5C and 6C sugar, starch, and other parts of *Clostridium* can be used as a substrate in the metabolism of syngas and glycerol. The fermentation process of butanol is generally considered as a two-phase process. The first fermentation process is the acid production process. In this process, the substrate is consumed to generate acetic acid, acid, and other small molecules, and thus cause the reduction of pH in the culture system, thereby inducing some specific gene response. At the same time, the rate of reduction of the enzyme activity increased and the fermentation period was moved into the second stage, namely production stage solvent. At this stage, small molecule acid is reacted by the enzyme, to product the reaction solvent (acetone, butanol, and ethanol). *Clostridium* metabolism produces butanol at its late phase, which has considerable toxicity for *Clostridium* fermentative metabolism that can result in changing of the bacterial cell membrane surface of phospholipid and fatty acid, thereby leading to corresponding changes in the structure and fluidity of membranes. The product inhibition of butanol is also shown in the transportation of the solvent, the polarity of the membrane, and the intracellular ATP equilibrium. In general, the butanol tolerance of *Clostridium* butanol is in the range of 1%–2%. Among the high producing strains screened out, the *acetobutylicum* ATCC824 C. mutant strain SA-1 and the mutant strain BA101 C. *beijerinckii* and NCIMB8052 C. *beijerinckii* were two strains with high butanol tolerance and high yield. The SA-1 evolutionary engineering methods are employed to change the microbial life system and improve the butanol concentration in the system, but BA101 is used by N-methyl-N9-nitro-nitroso guanidine mutagenesis screening. BA101 can ferment to produce more than 2% of butanol, and its production solvent stage of small molecule acid conversion rate significantly increased, and SA-1 butanol tolerance when compared to the original strain ATCC824 increased by 121%.

2   EXPERIMENTS

The methods used to refine biobutanol are divided into three stages: pretreatment, enzymatic hydrolysis, and fermentation.

2.1   *Pretreatment*

Crop straw is very rich in renewable biomass resources. Due to the filamentous crystalline structure of the cell wall and the embedded connection of lignin, it is difficult for the straw to be degraded. Straw pretreatment is an important part of the degradation process and even hydrolysis of straw cellulose. The focus of pretreatment is to reduce the biological conversion time, reduce the dosage of cellulase, and increase the yield of alcohol. The goals are to change the natural cellulose structure, destroy the connection between the cellulose, lignin, and hemicellulose, reduce cellulose crystallinity, remove lignin, and increase the loosening of the raw material to increase the effective contact of cellulose to the cellulase system, so as to increase the efficiency of the enzyme. Common pretreatment methods usually include the chemical method (acid, alkali, etc.), the physical and chemical method (steam explosion method, wet oxidation method, blasting method, etc.), biological method (enzyme treatment), etc. The main components of rice straw are cellulose, hemicellulose, and lignin. The crystal structure of the cellulose is compact, and it is formed by the combination of B-1,4-glycosidic bond (usually by 4000 ~ 8 000 glucose molecules in series, molecular weight of 200 ~ 2000 K), and is embedded by lignin and hemicellulose. The complex spatial structure can avoid the destruction of microorganisms and various physical and chemical factors, and also greatly reduce the molecular conversion of cellulose. Therefore, in order to make full use of the rice straw, we must choose a suitable pretreatment method, which can destroy the original structure of cellulose to decrease crystallinity, remove the lignin and hemicellulose, and increase the contact surface of the enzyme and cellulose in order to improve the efficiency of enzymatic hydrolysis and improve efficient utilization of the straw of rice. By referring to the literature, we obtained the different pretreatment methods of rice straw and determined the best pretreatment method of rice straw cellulose.

The chemical method mainly refers to the use of acid or alkali in the dissolution of lignin and hemicellulose, reduces the crystallinity of cellulose, while increasing the specific surface area of lignin, and thus improves the efficiency of enzymatic hydrolysis. The wetting process of NaOH solution is one of the earliest and most widely used pretreatment methods, and the treatment temperature and pressure are lower than other pretreatment methods. NaOH can open the cross-linked wood lignin and xylan ester bond that can partly dissolve lignin and hemicellulose of the raw material and reduce cellulose crystallinity, while increasing the lignin material's specific surface area so that it can get a higher enzymatic saccharification rate, that is a more effective preprocessing method. This experiment using NaOH solution pretreatment of rice straw was carried out to study the effect of the mass fraction of NaOH, solid content of rice straw, pretreatment time factors on the process for enzymatic hydrolysis of rice straw, through enzymolysis effect, further defined the NaOH pretreatment effect.

Rice harvest was carried out in 2013 from the Chinese Academy of Agricultural Sciences, Shunyi District, Beijing experimental field. The rice juice of the rice stem was obtained after cutting in sealed bags, and placed in the refrigerator freezer to save −20 C. The quality scores were 1%, 2%, and 3% NaOH solution, and the equipment included high temperature sterilization pot, high-speed centrifuge, analysis balance, conical flask, measuring cups, pH test paper, deionized water, etc.

Dry wheat straw was taken, with a small mill to crush.

Different mass fractions of NaOH, pretreatment temperatures and pretreatment times (a total of 15 groups) were set up.

After cooling to room temperature, a high speed centrifuge was used to carry out centrifugation and cleaning 3 to 5 times.

The product attained neutral pH after centrifugal filtration in the filtration apparatus, which was confirmed by using test paper pH.

Filtration of wheat straw was carried out in 105 under dry conditions, and was stored in a desiccator to spare.

Specific experimental groups are as follows:

Table 2. Sodium hydroxide pretreatment experiment data sheet.

Groups	NaOH concentration (%)	Temperature (°C)	Time (min)
1	1	100	40
2	3	100	40
3	2	120	40
4	2	80	120
5	1	80	80
6	3	100	120
7	2	100	80
8	1	100	120
9	3	120	80
10	1	120	80
11	2	80	40
12	2	120	120
13	3	80	80
14	0	0	0
15	0	0	0

Because cellulose's structure was destroyed by alkaline hydrolysis, the results of alkaline hydrolysis can be obtained by means of transmission electron microscopy to observe the degree of damage to cellulose structure and through the method of infrared spectroscopy. But in this experiment, through enzymatic hydrolysis, one can achieve efficient and optimal inverse conditions to alkaline hydrolysis conditions, and so it is little important to observe by using an electron microscope; this can be explained by the subsequent experimental procedures.

2.2 Enzymatic hydrolysis

Enzymatic hydrolysis of cellulose is a combined action of beta - 1, 4 - endoglucanase and beta 1,4 cut glucanase, and beta glucosidase synergy. First, the hydrolysis of random EG enzyme-cut amorphous regions of the cellulose chain was performed, and so crystalline cellulose appeared. More cellulose molecules create conditions for CBH enzymatic hydrolysis of cellulose; CBH enzyme hydrolysis products, such as cellobiose, are obtained by BG enzyme hydrolysis into glucose. The process of cellulase hydrolysis of cellulose can be simply expressed as follows: EG - CBH - BG. The present study shows that EG enzyme actually includes at least EG I, EG II, EG III, and EG V four, CBH includes at least CBH I and CBH II enzymes.

We used some experimental materials, such as potassium dihydrogen phosphate buffer solution, cellulase, and deionized water and instruments, such as shaker, NREL structure analyzer, pH meter, analysis balance, and conical flask.

These are numbered as without pretreatment and after NaOH pretreatment of straw powder of about 1.5 g (15 groups).

Potassium dihydrogen phosphate buffer solution (pH 4.5, 20 ml) was added to 15 groups of samples. The cellulase dosage was 0.4 ml per group. As shown in the table and after hydrolysis for 120 h, the supernatant can be obtained after centrifugation by using the NREL method for structural analysis.

The enzymatic hydrolysis efficiency of different sugars was calculated.

Specific experimental groups are shown in Table 3.

The optimal conditions of hydrolysis were determined by measuring the hydrolysis efficiency of the main products of glucose, xylose, and Arabia sugar. The results are shown as Tables 4–6:

From the above analysis, we can see that the third group is the most optimal solution group after enzymatic hydrolysis, the concentration of NaOH is 2%, alkali hydrolysis temperature is 120 DEG C, and hydrolysis time is 40 min; enzymatic hydrolysis of total sugar efficiency was for 20.92%, out of which glucose production efficiency was 16.18%, xylose production efficiency was 4.31%, and Arabic sugar production efficiency was 0.43%. In addition to the most optimal group, the three groups with higher sugar yields were fourth groups, fifth groups and tenth groups. Therefore, we can take the above four groups to carry on the fermentation processing.

Table 3. Enzymatic hydrolysis treatment table.

Groups	NaOH concentration (%)	Temperature (°C)	Time (min)	Enzymolysis dregs quality (g)	Water (ml)	Enzyme (ml)
1	1	100	40	1.5206	20	0.4
2	3	100	40	1.5115	20	0.4
3	2	120	40	1.5079	20	0.4
4	2	80	120	1.5094	20	0.4
5	1	80	80	1.5152	20	0.4
6	3	100	120	1.5024	20	0.4
7	2	100	80	1.5116	20	0.4
8	1	100	120	1.5230	20	0.4
9	3	120	80	1.5189	20	0.4
10	1	120	80	1.5003	20	0.4
11	2	80	40	1.5059	20	0.4
12	2	120	120	1.5063	20	0.4
13	3	80	80	1.5123	20	0.4
14	0	0	0	1.5118	20	0.4
15	0	0	0	1.5337	20	0.4

Table 4. Enzymatic hydrolysis yield of glucose.

Groups	NaOH concentration (%)	Temperature (°C)	Time (min)	Enzymolysis dregs quality (g)	Water (ml)	Enzyme (ml)	Efficiency of enzymatic hydrolysis of glucose
1	1	100	40	1.5206	20	0.4	5.11
2	3	100	40	1.5115	20	0.4	8.79
3	2	120	40	1.5079	20	0.4	16.18
4	2	80	120	1.5094	20	0.4	6.04
5	1	80	80	1.5152	20	0.4	9.45
6	3	100	120	1.5024	20	0.4	8.16
7	2	100	80	1.5116	20	0.4	8.91
8	1	100	120	1.5230	20	0.4	9.03
9	3	120	80	1.5189	20	0.4	0.75
10	1	120	80	1.5003	20	0.4	8.27
11	2	80	40	1.5059	20	0.4	4.24
12	2	120	120	1.5063	20	0.4	2.41
13	3	80	80	1.5123	20	0.4	5.79
14	0	0	0	1.5118	20	0.4	1.89
15	0	0	0	1.5337	20	0.4	1.68

2.3 Fermentation

At present, there are four kinds of microbes, namely *acetobutylicum C, beijerinckii C, saccharoperbutylacetonicum C* and *saccharobutylicum*. *C. acetobutylicum* and *C. beijerinckii* are widely studied and applied. Its brief introduction is as follows: anaerobic, Gram positive bacteria whose cells were fusiform, secondary end spores and spores were oval, colony projections, translucent gray white, margin irregularly circular. C. acetobutylicum grows well in CO_2, N_2 and anaerobic environments, in pH 4.5 and 7.0 and 35 degrees Celsius temperature and 37 degrees

C range. It is a widely available carbon source, including hexose and pentose monosaccharides, oligosaccharides and polysaccharides. Therefore, we can use starch, molasses, straw and other biomass resources to perform fermentation. The experimental strain used was ABE1201-1 *C. acetobutylicum* of *Clostridium* butanol.

Butanol fermentation was performed by using the synthetic culture medium, and the culture medium formula is as follows: 60 g/L of glucose, 2.2 g/L of ammonium acetate, 1 g/L of K_2HPO_4 - $3H_2O$, 1 g/L of KH_2PO_4, 0.2 g/l of $MgSO_4$ $7H_2O$, 0.01 g/l of $FeSO_4$, and 0.01 g/l of $MnSO_4$. The seed

Table 5. The efficiency of enzymatic hydrolysis of xylose.

Groups	NaOH concentration (%)	Temperature (°C)	Time (min)	Enzymolysis dregs quality (g)	Water (ml)	Enzyme (ml)	Efficiency of enzymatic hydrolysis of glucose
1	1	100	40	1.5206	20	0.4	1.31
2	3	100	40	1.5115	20	0.4	3.11
3	2	120	40	1.5079	20	0.4	4.31
4	2	80	120	1.5094	20	0.4	1.83
5	1	80	80	1.5152	20	0.4	2.98
6	3	100	120	1.5024	20	0.4	1.33
7	2	100	80	1.5116	20	0.4	2.57
8	1	100	120	1.5230	20	0.4	3.27
9	3	120	80	1.5189	20	0.4	0.37
10	1	120	80	1.5003	20	0.4	2.99
11	2	80	40	1.5059	20	0.4	1.54
12	2	120	120	1.5063	20	0.4	2.49
13	3	80	80	1.5123	20	0.4	1.44
14	0	0	0	1.5118	20	0.4	0.18
15	0	0	0	1.5337	20	0.4	0.26

Table 6. Enzymatic hydrolysis yield of Arabia sugar.

Groups	NaOH concentration (%)	Temperature (°C)	Time (min)	Enzymolysis dregs quality (g)	Water (ml)	Enzyme (ml)	Efficiency of enzymatic hydrolysis of glucose
1	1	100	40	1.5206	20	0.4	0.38
2	3	100	40	1.5115	20	0.4	0.39
3	2	120	40	1.5079	20	0.4	0.43
4	2	80	120	1.5094	20	0.4	0.39
5	1	80	80	1.5152	20	0.4	0.5
6	3	100	120	1.5024	20	0.4	0.21
7	2	100	80	1.5116	20	0.4	0.4
8	1	100	120	1.5230	20	0.4	0.49
9	3	120	80	1.5189	20	0.4	0.17
10	1	120	80	1.5003	20	0.4	0.56
11	2	80	40	1.5059	20	0.4	0.15
12	2	120	120	1.5063	20	0.4	0.24
13	3	80	80	1.5123	20	0.4	0.24
14	0	0	0	1.5118	20	0.4	0.26
15	0	0	0	1.5337	20	0.4	0.23

medium was cultured with 250 ml glucose bottle and the medium volume was 150 ml.

Production of culture medium

The culture medium was obtained by passing high pure nitrogen gas for 1 min, to ensure the formation of an anaerobic environment in the bottle to facilitate the growth of the cell.

Sterilization was carried out at a temperature of 116 degrees in the sterilizing pot for 25 min.

In the sterile stage of inoculation, the inoculation amount was 10%, and the temperature of the sample was maintained at 37 degrees Celsius under the conditions of static culture for 48 h.

Three groups with better hydrolysis effect were selected. The products were ethanol, acetone, acetic acid, butyl alcohol, and butyrate. The results are shown as Table 7: tenth groups were prepared twice.

The experimental data show that the concentration of butanol is far greater than that of ethanol, which is one of the reasons why we chose to produce butanol. The calculation shows that the ratio

Table 7. Concentration of main products.

Groups	Ethanol concentration (g/L)	Acetone concentration (g/L)	Acetic acid concentration (g/L)	Butanol concentration (g/L)	Butyrate concentration (g/L)
4	0.6803	0.1962	2.6002	1.717	1.938
5	0.6125	0.1447	0.3799	1.7353	1.9983
10-1	0.8531	0.2292	1.8505	3.3874	1.6975
10-2	0.8490	0.1983	1.8591	3.3256	1.7375

of butanol: ethanol: acetone is equal to 13:4:1, which is much higher than the previous test results i.e., butanol: ethanol: acetone = 6:3:1. This is one of the innovative points of this experiment.

3 FUTURE PROSPECTS OF BUTANOL

Calculating that the substrate cost ratio is about 60% of the production cost is a very important economic indicator in butanol fermentation. In butanol used in transport energy targets, including starch, soluble sugar and grain substrate, the fermentation of economic value could not be achieved, which is one of the key reasons for the butanol fuel not being catered to the market. Therefore, more economical and economical agricultural wastes, including the biomass, as well as the industrial wastes, will be more suitable for the economic fermentation of butanol. The production of ethanol by distillation with low value raw material has been proved to be a very economical way to produce ethanol. The current strain transformation method improves the technical level of the yield and conversion rate, is far from not of butanol fermentation of economic evaluation have relatively large impact, and substrate and product recovery cost is the butanol fermentation method to produce the most important economic index. Batch fermentation process to produce butanol economically and continuous fermentation are not competitive; the reason is that the batch fermentation process requires additional sterilization equipment and procedures, as well as pipeline valves and other equipment costs. However, the problem of continuous fermentation of bacteria is also quite a challenge to the production of butanol. As a result of the gradual establishment of butanol as a biological transport energy fuel, it will lead to a huge market share. The production of butanol by using the fermentation method is the most important cost factor in the production of butanol fermentation. It is the most effective method to control the production of butanol by using a relatively inexpensive substrate. However, it is still a challenging work to select an appropriate and cheap substrate, and the product recycling technology of energy saving and consumption reduction. It also requires the continuous efforts and practice of scientific research workers. Another very important aspect of the investigation is the environmental problems of the project. In recent decades, the global greenhouse gas emissions have caused drastic changes in the environment. And so, whether the development of biomass fuel alcohol or other butanol, the project must be put to environmental assessment first; active nature of the process in protecting the environment, rational use of resources, and the overall consideration of various factors, are bio energy development directions in the future.

4 DISCUSSION AND CONCLUSION

By using *C. acetobutylicum* ABE1201-1, alkaline hydrolysis and enzymatic hydrolysis were performed under the optimal conditions for an NaOH mass fraction of 2%; alkali hydrolysis temperature is 120 DEG C, hydrolysis time is 40 min, enzymatic hydrolysis of total sugar efficiency is 20.92%, out of which glucose production efficiency was 16.18%, xylose production efficiency was 4.31%, and arabinose production efficiency was 0.43%.

Fermentation produces butanol, ethanol, and acetone in the ratio 13:4:1, where the butanol yield is much higher than the previous experiment.

Glucose and xylose are used to produce butanol by alkaline hydrolysis pretreatment; other methods, such as hydrothermal method and acid method, only use glucose and not xylose.

The selected strains have the advantage that the proportion of butanol is higher than that of the traditional one, that is, the more the carbon source is converted into butanol, the lesser are the impurities.

In the past, the experimental fermentation produced butanol, ethanol, and acetone in the ratio 6:3:1 and the current experiment produced butanol, ethanol, and acetone in the ratio 13:4:1. It can be seen that butanol production is far higher.

It can be used in the process of evolutionary engineering to produce stable and high yield butanol-producing strains and increase the yield of butanol in the fermentation stage.

In this experiment, the method of batch fermentation is adopted. It is simple, but it takes a long time. There are a number of fermentation methods, such as continuous fermentation of the first stage, two stages of continuous fermentation and other methods, which are performed by continuously adding to the cell to maintain the stability of sugar concentration, and save time for fermentation. But the requirement of the experiment instruments is high.

REFERENCES

Ezeji TC, Milne C, Price ND, Blaschek HP. Achievements and perspectives to overcome the poor solvent resistance in acetone and butanol-producing microorganisms [J]. Appl Microbiol Biotechnol., 2010,85:1697–1712.

Ezeji TC, Qureshi N, Blaschek HP. Acetone Butanol Ethanol (ABE) production from concentrated substrate: reduction in substrate inhibition by fed-batch technique and product inhibition by gas stripping. [J] Appl Microbiol Biotechnol., 2004; 63:653–658.

Ezeji TC, Qureshi N, Blaschek HP. Production of Acetone Butanol (AB) from liquified corn starch, a commercial substrate, using Clostridium beijerickii coupled with product recovery by gas stripping. [J] J Ind MicrobiolBiotechnol., 2007; 34:771–777.

Lu CC. Butanol production from lignocellulosic feedstock by acetone-butanol-ethanol fermentation with integrated product recovery [D]. The Ohio state university, 2011.

Qureshi N, Blaschek HP. Recovery of butanol from fermentation broth by gas stripping. [J] Renew Energy., 2001; 22:557–564.

Qureshi N, Lai LL, Blaschek HP. Scale-up of a high productivity continuous biofilm reactor to produce butanol by adsorbed cells of Clostridium beijerinckii. [J] Food Bioprod Process., 2004; 82:164–173.

Servinsky, M.D., J.T. Kiel, N.F. Dupuy, C.J. Sund. Transcriptional analysis of different carbohydrate utilization by Clostridium acetobutylicum. Microbiology, 2010:156, 3478–3491.

Yu JL, Zhang T, Zhong J, Zhang X, Tan TW. Biorefinery of sweet sorghum stem [J]. Biotechnol. Adv., 2012, 30(4): 811–816.

Zhang T, Du N, Tan TW. Biobutanol production from sweet sorghum bagasse [J]. Journal of Biobased Materials and Bioenergy, 2011,5:1–6.

Advances in Energy Science and Equipment Engineering II – Zhou, Patty & Chen (Eds)
© *2017 Taylor & Francis Group, London, ISBN 978-1-138-71798-5*

A study on microbial enhanced oil recovery with the synergy of organic matter CX

Donghan Li & Yanhua Suo
Northeast Petrolium University, Daqing, China

Wei Li, Xiaolin Wu, Qing Luo, Zhaowei Hou, Rui Jin & Rui Wang
Exploration and Development Research Institute of Daqing Oilfield Company Ltd., Daqing, China

ABSTRACT: The technology of Microbial Enhanced Oil Recovery (also known as MEOR) has received much attention in recent years and has undergone rapid development because of its many advantages. Studies showed that the bacteria could degrade the light component of crude oil preferentially, thereby resulting in the worse rheological characteristics and increased difficulty in exploitation of crude oil. Focusing on the key problem, a new method using the synergy of organic matter in the process of microbial flooding that could promote the degradation efficiency of the heavy components of crude oil from bacteria by increasing the dissolution of heavy components is studied in this paper. With the synergy of additive organic compound, the relative proportion of saturated hydrocarbons has been changed and that of $\sum nC21^-/\sum nC22^+$ reached up to 39.06%, which leads to the forward transference of the peak value distribution curve of hydrocarbon components. Meanwhile, a degradation rate of 67.9% of heavy components of crude oil has been achieved and the rheological properties are altered obviously; the viscosity of crude oil reduced from 124 mP·s to 40 mP·s. The results of the physical simulation displacement test showed that the efficiency of MEOR could be improved by 9.72% by using the technology of additive agents and a pilot field test of MEOR has been carried out in ChaoYanggou reservoir in Daqing Oilfield. The field test has had a positive effect on increasing oil production and decreasing water cut, where accumulative enhanced oil production was more than 40 thousand tons.

1 INTRODUCTION

Microbial Enhanced Oil Recovery (also named MEOR) is a promising technology for tertiary oil recovery, which has advantages such as lower cost, more applicable in many types of oil reservoirs, easier for injection and production, no damage to the oil reservoir and no pollution to the environment, especially efficient in depleted or nearly depleted oil fields. Daqing Oilfield has studied this technology since 1965 and the initial research was carried out to judge water-flooded layers by analyzing the change of the microbial characteristics. In recent years, more experimental studies were focused on the mechanism of MEOR and the screening and evaluation of bacteria used in MEOR. Several microbial flooding tests were implemented in Daqing Oilfield and the analysis showed a higher viscosity of crude oil than that before microbial treatment, indicating the negative effects of the oilfield test.

In this paper, we mainly provide the method of controlling the degradation of light oil components

by bacteria and increase the dissolved proportion of heavy oil components (including asphalt) in water by adding organic matter screened from the amount of organic compounds. Thus, the bacteria for MEOR can easily utilize the dissolved oil components accordingly. This process will contribute to the reduction of viscosity of crude oil and alter the fluid performance. Therefore, the displacement efficiency and oil recovery can be improved.

2 MATERIALS AND METHODS

2.1 Materials

2.1.1 Bacterial strain
The bacterial strain (named Y-3) was separated from the produced water sample and preserved by the Exploration and Development Research Institute of Daqing Oilfield Company.

2.1.2 Culture medium
The ingredients in the culture medium for bacteria Y-3 comprised of the following: KH_2PO_4, 0.5 g/L;

K$_2$HPO$_4$, 4 g/L; NH$_4$Cl, 1 g/L; CaCl$_2$, 2 mg/L; MgCl$_2$·6H$_2$O, 2 mg/L; FeCl$_3$, 0.2 mg/L; and mineral water, 1 L. The mixed broth was sterilized at 121°C for 20 min.

2.1.3 Crude oil

Demulsified crude oil was sampled from the combined-treatment station.

2.2 Experimental methods

2.2.1 Analysis of the physiological–biochemical characteristic of bacteria

The physiological and biochemical identification and experimental procedures were carried out as indicated in the literature (Yang, 2013).

2.2.2 The measuring method of asphaltene

Sixty grams of crude oil were added to normal hexane (about 600 gram). The sample was filtered after the crude oil has been dissolved and was placed for 24 hours. The viscosity of crude oil was measured when the wax and resin had been evaporated from the solvent by using a water bath at a temperature of 65°C. The content of asphaltene could be quantified by weighing the quality of insoluble substance. After asphaltene had been isolated, the change in the viscosity of crude oil could be compared.

2.2.3 The measuring method of group composition of crude oil

The sample of crude oil was soaked in petroleum ether for 12 hours and then, gelatine and asphaltene were filtered out. The residue was separated by silica–aluminum oxide gel chromatography, by leaching with the mixture of normal hexane, dichloromethane and trichloromethane. The quality of saturated hydrocarbons, aromatic hydrocarbons and non-hydrocarbons were calculated, respectively.

2.2.4 Detection of the components of crude oil in the water

A certain amount of organic matter CX (screened from a variety of organic compounds) was added into the experimental sample and extracted with acetone/normal hexane after a certain time. The condition of chromatography was selected as follows: The size of the capillary-column used was 30 m × 0.25 mm × 0.25 μm. The temperature of the injection port was 280°C and column chromatography was carried out with programming temperature. The carrier gas used was high purity helium and filtered by using a deoxidation tube and then analyzed by GC/MS.

2.2.5 The measurement of microbial growth curve

Different concentrations of organic matter CX were added to the microbial culture for Y-3, respectively. The initial germ concentration was 10^3 cells per millilitres (cfu). At the temperature of 45°C, the culture was placed under shaking for 24° hours and then the germ concentration was measured again.

3 RESULTS AND ANALYSIS

3.1 The result of identification of the bacteria strain

A bacterium has been separated and screened from the water sample and identified as *Bacillus licheniformis*, whose size was 2.0 μm (length) × 0.8 μm (width) with the white-colony and bar-shape.

The strain was named as Y-3 and preserved in a laboratory in Daqing oil field. The evaluation showed that exhibited good degradation properties of crude oil and enhancing oil recovery. And so, it was used in this study (as shown in Figure 1 and Table 1).

3.2 Effect of the synergy of organic matter CX on the oil components dissolved in water

After adding organic matter CX to the culture medium for growth of bacteria Y-3, the content of the long-chain alkane including C$_{14}$–C$_{22}$ dissolved in the broth has been increased greatly after 7 days. With the delay of time, more and more components of crude oil such as some esters, arene, and cyclane have also increased.

When compared with the control experimental sample cultured without organic matter CX, the components of crude oil in the broth have little been detected, which verified the perfect solubilization of organic matter CX in the components of crude oil, as shown in Figures 2–4.

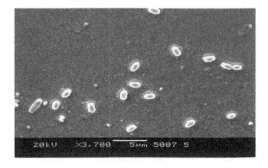

Figure 1. The shape of bacteria Y-3 observed by using an electron microscope.

Table 1. The physiological and biochemical characteristics of strain Y-3.

Item of test	Result of test
Gram's dye	Positive
Shape of cells	Rod-shaped
Diameter of cells	>1 μm
Formation of spore	+
Catalase	+
Oxidase	+
Anaerobic growth	+
MR test	+
VP test	+
Using citrate	+
Growth at 50°C	+
Growth on 50% NaCl	+
Growth with pH 5.7	+
Amylolysis	+
Gelatin hydrolysis	+
Nitrate reduction	+

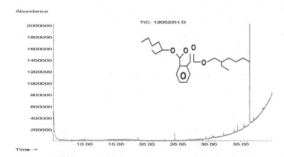

Figure 2. The oil components detected in the water sample by GC-MS (after 7 days, the control group).

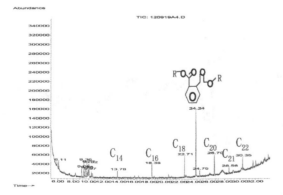

Figure 3. The oil components detected in the water sample by GC-MS (after 7 days, the experimental group).

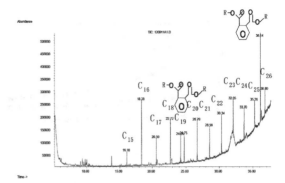

Figure 4. The oil component detected in the water sample by GC-MS (after 15 days, the experimental group).

3.3 Effect of degradation by bacteria on crude oil with the synergy of organic matter CX

An analysis of change in the group components in crude oil has been performed between the control sample and the experimental sample. The results showed that the asphaltene content has decreased and that of saturated hydrocarbon has increased with the synergy of different concentrations of organic matter CX. The relative proportion of saturated hydrocarbon has changed and that of $\sum nC21^-/\sum nC22^+$ reached up to 39.06%, which led to the forward transference of the peak value distribution curve of hydrocarbon components.

The asphaltene content in the culture medium has also decreased after 30 days, indicating the degradation of bacteria on the asphaltene content. The best concentration of organic matter CX was 200 mg/L and the degradation rate of asphaltene was 67.9%.

3.4 The result of physical simulation core flooding test

A core flooding test has been finished by simulating the conditions of the oil reservoir. Two groups of microbial water flooding tests with organic matter CX and without organic matter CX were designed.

The results showed that the oil recovery has been enhanced by 5% and reached up to 9.72% in the group with added organic matter CX, when compared to the oil recovery of 3.87% in the group without organic matter CX.

3.5 The change in the viscosity of crude oil after treatment

The rheological curve of crude oil considering viscosity and shear rate has been analyzed. In this study, two experimental groups and one control

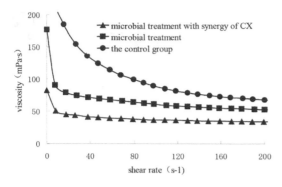

Figure 5. The change in rheological properties of crude oil before treatment and after treatment.

group have been designed and the results showed that the viscosity of crude oil has been reduced greatly after treatment.

At the temperature of 45°C, the viscosity of crude oil was 40 mP·s after microbial treatment with the synergy of organic matter CX, when compared with the 120 mP·s in the control group and the 69.2 mP·s in the experimental group treated only by bacteria. The above results show that the rheological properties have been improved greatly and the crude oil could be displaced more efficiently from the reservoir, as shown in Figure 5.

4 THE EFFECT OF THE PILOT FIELD TEST

Aimed at the difficult development situations and the huge oil resource potential in the Daqing low permeability oil field, a microbial flooding field test has been carried out in Chaoyanggou oil field, with permeability below $25 \times 10^{-3}\mu m^2$. The cultured bacteria solution with a certain concentration of organic matter CX was injected into the oil formation setup.

4.1 Bacteria may degrade heavy components of crude oil with the synergy of CX

The average viscosity of crude oil sampled from three producing wells declined from 94.3 mPa·s to 76.0 mPa·s, which showed that microbes injected exhibited a selective degradation of heavy components of crude oil in the formation. Meanwhile, the wax content decreases from 12.4% to 7.6% and the gum content decreased from 17.5% to 13.6%. The decrease in the wax content or gum content can also improve the properties of crude oil and restore good to normal production and lessen the application of other stimulation treatments, for example, hot water washing or acidizing. Moreover, the

solidifying point of the crude oil declined from 40°C to 35.7°C, thereby contributing to the flow capacity of the crude oil. The interfacial tension between oil and water also decreased by 6.5 mN/m from 46.3 mN/m to 39.8 mN/m, thereby suggesting that the microbial metabolism has produced active materials composed of bio-surfactants primarily under reservoir conditions with the synergy of CX.

4.2 MEOR contributes to the establishment of an effective driving system with the synergy of CX

Start-up pressure exists in low permeability reservoirs and only when the displacement pressure is higher than start-up pressure, the water flooding controlled reserves can be effectively developed. Producing well 60–124 in the test block had been closed for 3 years because of not producing liquids before microbial injection. After 3 months of microbial injection, well 60–124 was opened again and its liquid-producing and oil-producing capacities were maintained above 1.0t and 0.2t per day, respectively. The startup of the exhausted well indicates that the technology of MEOR can contribute to reduce resistance during water flooding, by driving bypassed oil in exhausted wells and building an effective driving system in oil layers.

The analytical results of the field test showed that the producing wells belong to the main channel sand and developed fracture zones had a quicker and better effect. Meanwhile, the wells with good relationship of connectivity between injection wells and production wells exhibited a better response. The daily oil production increased from 32.4 tons to 49.7 tons and the incremental oil amounted to 45.5 thousand tons. At the same time, the water cut has dropped to 3.5%, as shown in Figure 6.

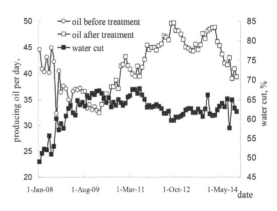

Figure 6. Graph showing the change in oil production and water cut before treatment and after treatment.

5 CONCLUSION

The organic matter screened in this study has a good effect of synergy with the bacteria used in MEOR. By adding the compound into the microbial culture, the component of crude oil including the long-chain alkane and asphaltene can be dispersed into the aqueous phase and degraded by bacteria subsequently. Therefore, the viscosity of crude oil will be reduced and the rheological properties will be improved apparently. In our study, the optimal concentration of CX is 100 mg/L and an exceeding amount of CX will have a negative impact on the growth of bacteria. The synergy of CX has been confirmed by performing physical simulation core flooding test and field test.

The mechanism of MEOR is very complicated and the process of microbial degradation on crude oil needs to be controlled. Selecting a variety of methods and chemical substances to promote the microbial flooding efficiency and enhance the oil recovery is of importance. Continuous research will be carried out considering different types of organic matter and the concentration of matter, as well as the chemotaxis of bacteria to crude oil.

REFERENCES

Liu Chang. (2015). The application of microbial oil recovery technology in EOR. *J. Chemical Intermediates.* 30(4):22–24.

Lu Shufeng. (2015). Research and application of microbial enhanced oil recovery technology. *J. Petrochemical Industry Technology.* 6,5–6.

Park Y J, Kim K R, Kim J H. (1999). Gas chromatographic organic acid profiling analysis of brandies and whiskeys for pattern recognition analysis. *J. J Agric Food Chem.* 47(6):2322–2326.

Wei dong-Wang. (2012). Laboratory research and field trials of microbial oil recovery technique. *J. Oil drilling & production technology.* 34(1):107–113.

Xiaolin-Wu, Jian jun-Le, Rui Wang. (2013). Progress in pilot tests of microbial enhanced oil recovery in Daqing oilfield. *J. Microbiology China.* 40(8):1478–1486.

Xiuzhu-Dong, Miaoying-Chai. (2001). Handbook of common bacterial system identification. *J. M.BeiJing: scientific publishing.* 34–56.

Xu Tao, Chen Chang, Liu Chunqiao, et al. (2009). A novel way to enhance the oil recovery ratio by Steptococcus sp. *J. J Basic Microb.* 49(5):477–481.

Yang Hua, Huang Jun, Zhao Yonggui, et al. (2013). Identification and characterization of *Thauera sp.* strain TN9. *J Appl Environ Biol.* 19 (2):318–323.

Ying-Guo. (2011). Exogenous microbial enhanced water flooding pilot tests in China. *J. Science & Technology Review.* 29(22):51–54.

Advances in Energy Science and Equipment Engineering II – Zhou, Patty & Chen (Eds)
© 2017 Taylor & Francis Group, London, ISBN 978-1-138-71798-5

A novel method for status diagnosis of power equipment based on feature correlation

Jian-cai Huang

School of Control and Computer Engineering, North China Electric Power University, Baoding, Hebei Province, China

ABSTRACT: The number of the fault samples of the power equipments is usually limited. In this situation, the traditional methods have low correct rate. In order to get the high correct rate, a novel method is proposed in this paper. The global feature correlation of the data samples of the power equipment is computed using the multivariate fitting. Then, the deviation levels of the samples under the different status are computed based on the global correlation. By comparing the deviation levels of the samples and that of the on-line data, the status of the power equipment is analyzed. The method is used to detect the leakage current of the transmission line insulator in the experiment. The results show that the method has the higher correct rate than SVM and ANN.

1 INTRODUCTION

The status detection of the power equipments can provide the run status data quickly, which is the basis of the maintenance (Guo et al. 2014). This technology has been applied on many power equipments, such as transformer (Li et al. 2014, Chang et al. 2013), circuit breaker (Teng et al. 2014) etc. But, during the diagnosis one of the outstanding problems is that some fault samples are severely insufficient, which influences the analysis for the power equipments.

The reason why the fault samples are very few is that: (1) The fault rate of the power equipments is low and the fault data is often captured difficultly. Besides, the enterprises don't pay attention to this work. These factors result in severely insufficient samples in some fields. (2) Many power equipments are very expensive, which is not suit to acquire samples with the destructive fault tests. Though some enterprises have some fault samples, they don't want to open the data. So, the status diagnosis task of the power equipments is usually under small sample size(<30), even under minimum samples size(≤5). From the above, studying a method suitable for small sample size and minimum sample size is important.

The main data-driven fault diagnosis methods for the power equipments judge the problems with the classifiers (Geng et al. 2015, Gholami et al. 2015), which includes two steps: (1) Collecting the samples and training the classifier; (2) Accepting the on-line data, inputting it into the trained classifier and getting the diagnosis result. The classifiers usually include Bayesian Classifier (BC), Artificial Neural Network (ANN) and Support Vector Machines (SVM). While, the BC method has to compute the conditional probability and needs a large number of samples, which is not suit for the small sample size (Tschiatschek et al. 2015, Lu et al. 2009). The ANN method also needs many samples to get high accuracy (Netto et al. 2015, Li 2008). Though the SVM method can solve the small sample problem to a certain extent, its performance is influenced by the sample characteristics. For example, we have used the insulator leakage current data of the transmission lines to train the SVM and tested the output. The results showed the following: when the number of the training samples was below 6, its correct rate was below 40%. That is, in minimum sample size the accuracy of the SVM method is very low.

During the analysis for the power equipment data, it is necessary to extract the feature vector whose elements are the features. Ahmad (2015), Yang (2014), Zheng (2014), Yang (2013), Rao (2007), Rao (2008), Rao (2009) consider that there is correlation between the features and that the feature correlation of the samples in the different class is different. In this paper, a novel method for the status diagnosis of the power equipments under small sample size and minimum sample size is proposed, which is based on the feature correlation. The multivariate fitting is used to describe the correlation. The feature correlation of the samples of the power equipment is computed which is used to judge the on-line data to diagnose the fault status. To verify the validity of the method proposed, it is applied on the insulator leakage current data collected in the artificial experiment.

2 STATUS DIAGNOSIS METHOD FOR THE POWER EQUIPMENTS BASED ON FEATURE CORRELATION

2.1 *Multivariate fitting for the features*

There is correlation between the elements in the feature vector of the power equipment samples. For the feature number is usually more than 1, the multivariate fitting is used to describe the correlation.

The feature vector extracted from the sample x is $\alpha = [\alpha_1, \alpha_2, ..., \alpha_p]$, in which α_i ($i = 1, 2,...,p$) is a element of the vector and p is the number of the elements. Then, the features have the correlation as the formula (1).

$$\alpha_t = F_i(\alpha_1, \alpha_2,...,\alpha_{i-1}, \alpha_{i+1}...,\alpha_p) + e_i \qquad (1)$$

That is, any feature α_i can be computed by the other p-1 features. In the formula (1), e_i is the prediction error. F_i is the fitting function, which estimates a value from the other p-1 values. The fitting function can be Linear (L), Linear + Interaction (LI), Quadratic + Interaction (QI) or pure Quadratic (Q) model types. For simpler, here the L model is taken for the fitting, which is shown in the formula (2).

$$F_i = b_0 + \Sigma^p_{j=1, j \neq i} b_j \alpha_j \qquad (2)$$

The purpose of solving the formula (2) is to compute the coefficients $b_0, b_1,...,b_{i-1}, b_{i+1}, ..., b_p$ by the feature values of the samples. In the formula (1), the smaller the prediction error e_i is, the better the fitting function is. The formula (3) is used to evaluate the prediction error.

$$SSE_i = \Sigma^q_{k=1}(\alpha_{ik} - F_{ik})^2 \qquad (3)$$

In the formula (3), α_{ik} is the ith feature value of the kth sample, F_{ik} is the fitting result of α_{ik} and q is the number of all of the samples. The ith feature has the same fitting function. The smaller SSE_i is, the better F_{ik} is. So, when the formula (3) gets the minimum value, the fitting function F_i is the wanted one, which is a multivariate fitting. The solution process is taken by the least square fitting here.

By the above method, the global correlation can be gotten. The samples of the different class have the different deviation levels from the global correlation, by which they can be distinguished. This method doesn't limit the number of the samples and it is suitable for the small sample size and the minimum sample size.

2.2 *Diagnosis method based on multivariate fitting*

The fault diagnosis of the power equipments is based on the samples. That is, the sample data are computed to find the pattern which is used to judge the new on-line data to diagnose the fault of the power equipment. So, the diagnosis process is divided to two stages.

Stage 1: Classification process of the sample data of the power equipment based on multivariate fitting.

Based on the multivariate fitting in 2.1, the following steps show the classification process.

1. Extracting the feature vector $\alpha = [\alpha_1, \alpha_2, ..., \alpha_p]$ of each sample collected from the power equipment, in which α_i is the kth feature of the sample and p is the length of the feature vector.
2. According to the method in 2.1, realizing the multivariate fitting.
3. Computing the deviation level of each sample from the global correlation using the formula (4)

$$e = \Sigma^p_{k=1}(\alpha_k - F_k)^2 \qquad (4)$$

In the formula (4), F_i is the fitting value of α_i.
4. The lower and upper bounds of the deviation level in the same class are selected.

Stage 2: Process of distinguishing the on-line data

1. Extracting the feature vector of the on-line data (new data).
2. According to the fitting result, computing all the feature values.
3. According to the formula (4), computing its deviation level from the global correlation.
4. Computing the distance between its deviation level and the lower and upper bounds of the different classes, the smallest one is the class to which the on-line data belongs, by which the status of the power equipment can be diagnosed.

3 EXPERIMENT VERIFICATION

3.1 *Experiment setting and sample data*

In order to verify the method proposed above, the leakage current of the transmission line insulator is collected in the artificial fog chamber. In the experiment, the insulator type is FXBW4-110/100. The coating method is selected to contaminate the insulator, using the diatomite to simulate the insoluble substance and the sodium chloride to simulate the soluble substance. The Equivalent Salt Deposit Density (ESDD) and the No Soluble Deposit Density (NSDD) of the contamination is $\rho_{ESDD} = 0.025$ mg/cm^2 and $\rho_{NSDD} = 0.125$mg/cm^2 respectively. In the fog chamber, the relative humidity is 90% and the temperature is 27°C. The boosting voltage method is used to collect the leakage current data.

During the experiment, the phenomenon of the insulator surface can be divided into three stages roughly. In the first stage, there is no discharge on the surface. In the second stage, the discharge occurs, which is more serious with the increasing of the voltage. When the flashover is happen, it is the third stage. The global waveform of the collected leakage current is shown in figure 1, in which the dotted lines separate the waveform into three parts according to the three stages.

3.2 *Method verification using leakage current*

Each data segment including 8000 data point from the waveform in figure 1 is taken as a sample. The sample set is gotten from the leakage current, in which 30 samples are in stage 1, 27 ones in stage 2 and 2 ones in stage 3. Each stage can be taken as a class. So, class 1 and class 2 have the small sample size and class 3 has the minimum sample size. The samples are used to verify the method proposed above. The steps are as the following.

1. Taking 20 samples in class1, 15 ones in class 2 and 1 sample in class 3.
2. Extracting the feature vector of each sample. Here, the feature vector is composed of the maximum absolute value of the leakage current and the amplitudes of 50Hz, 150Hz and 250Hz components.

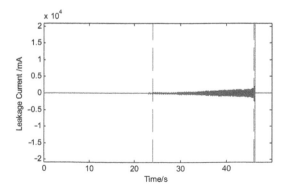

Figure 1. Leakage current waveform.

Table 1. Lower and upper bounds of different classes.

Class	Lower bound	Upper bound
1	0.0249	24.7674
2	123.7158	4246.0

Table 2. Test results of different methods.

Method	Test data number	Number of diagnosis error	Correct rate
Method proposed here	23	2	91.3%
SVM	23	4	82.6%
ANN	23	7	69.6%

3. According to the formulas (2) and (3), the coefficients of each feature are obtained.
4. Computing the lower and upper bounds of the fitting error of the samples in each class. The results are shown in the Table 1.

Because there is one sample taken in class 3, only one fitting error is gotten here which is 7.27×10^8.

The result shows that the fitting errors of the samples in the different classes are different.

The rest samples are taken to test the method according to 2.2, in which 10 samples are in class 1, 12 ones in class 2 and 1 sample in class 3. The test result is shown in Table 2.

From Table 2, the method proposed in this paper gets the higher correct rate than SVM and ANN.

4 CONCLUSION

There is correlation between the sample features of the power equipment. The multivariate fitting is used to describe the correlation, by which the global correlation is gotten. The lower and upper bounds of the deviation levels from the global correlation in the different classes are computing. By comparing the deviation level of the test data with the lower and upper bounds, the data is divided into different classes to diagnose the status of the power equipment. By comparing the results of the different methods, the proposed method in this paper has higher correct rate than SVM and ANN, which is more suitable for the small sample and minimum sample size.

ACKNOWLEDGEMENT

This work was supported by the Fundamental Research Funds for the Central Universities under Grant 2014MS136.

REFERENCES

Ahmad Saikhu & Deneng Eka Putra. (2015). Discriminant Analysis Implementation Based on Variable Predictive Models for Similarity Pattern Classification. *The 6th International Conference on Information & Communication Technology and Systems*, IV1–IV6.

Chang Jing, Huang Jiadong, Song Peng, & Han Tingjian. (2013). The Assessment of Transformer Status Based on Gray Theory. *International Conference on Sensors, Mechatronics and Automation*, 11–512: 1147–1152.

Geng Tianxiang, Qi Xin, Xiang Li, Tang Rui, & Wang Bin. (2015). A Novel Disturbance Locating Method for Power Systems Based on Sequenced Minimum Distance Classifier In: *International Conference on Electrical Engineering and Mechanical Automation*, 123–129.

Gholami, M., Gharehpetian, G. B., Mohammadi, M. (2015). Intelligent hierarchical structure of classifiers to assess static security of power system. *Journal of Intelligent & Fuzzy Systems*, 28(6): 2875–2880.

Guo, Xiangfu & Ming Zhong. (2014). Analysis on the Security Platform of Status Monitoring and Assessment Technology for Power Grid Equipment. *Proceedings of the 2014 International Conference on Information Technology and Career Education (ICITCE 2014)*, http:// www.shs-conferences.org/ articles/ shsconf/ pdf/ 2015/ 01/ shsconf_icitce2014_02016.pdf.

Li Enwen & Song Bin. (2014). Transformer Health Status Evaluation Model Based on Multi-feature Factors. *International Conference on Power System Technology*, 1417–1422.

Li Jianbo. (2008). Study on Power Transformer Fault Diagnosis Technology Based on Dissolved Gases Analysis. Jilin university.

Lu Jin-ling, Zhu Yongli, Ren Hui, & Meng Zhongqiang. (2009). The Research on Transient Stability Assessment Methods Based on Bayesian Network Classifier. *Asia-Pacific Power and Energy Engineering Conference*, 1776–1779.

Netto Ulisses Chemin, Grillo Diego de Castro, Lonel Ivan Donisete, Pellini Eduardo Lorenzetti, & Coury Denis Vinicius. (2015). An ANN based forecast for IED network management using the IEC61850 standard. *Electric Power Systems Research*, 130: 148–155.

Rao Raghuraj & S. Lakshminarayanan. (2008). Variable predictive model based classification algorithm for effective separation of protein structural classes. *Computational Biology and Chemistry*, 2008, 32: 302–306.

Rao Raghuraj & Samavedham Lakshminarayanan. (2007). VPMCD: Variable interaction modeling approach for class discrimination in biological systems. *FEBS Letters*, 591: 826–830.

Rao Raghuraj & Samavedham Lakshminarayanan. (2009). Variable predictive models—A new multivariate classification approach for pattern recognition applications. *Pattern Recognition*, 42: 7–16

Teng Yun, An Zhiyao, Yu Jibo, Wang Ji, & Zhang Yonggang. (2014). Study of Circuit Breaker Status Evaluating Model Based on Matter-Element Theory. *Applied Mechanics and Materials*, 670–671: 1458–1461.

Tschiatschek Sebastian & Pernkopf Franz. (2015). On Bayesian Network Classifiers with Reduced Precision Parameters. *IEEE Transactions on Pattern Analysis and Machine Intelligence*, 37(4): 774–785.

Yang Yu, Pan Haiyang, Ma Li, & Cheng Junsheng. (2014) A roller bearing fault diagnosis method based on the improved ITD and RRVPMCD. *Measurement*, 55: 255–264.

Yang Yu, Wang Huanhuan, Cheng Junsheng, & Zhang Kang. (2013). A fault diagnosis approach for roller bearing based on VPMCD under variable speed condition[J]. *Measurement*, 46: 2306–2312.

Zheng Jinde, Cheng Junsheng, Yang Yu, & Luo Songrong. (2014). A rolling bearing fault diagnosis method based on multi-scale fuzzy entropy and variable predictive model-based class discrimination. *Mechanism and Machine Theory*, 78: 187–200.

Advances in Energy Science and Equipment Engineering II – Zhou, Patty & Chen (Eds)
© 2017 Taylor & Francis Group, London, ISBN 978-1-138-71798-5

Utilization of comprehensive methods to evaluate the amount of geothermal resources in Dongqianhu area of Ningbo, Zhejiang province

Xiaojuan Cao
School of Earth Sciences and Resources, China University of Geosciences, Beijing, China
China Institute of Geoenvironmental Monitoring, Beijng, China

Binghua Li
Beijing Water Science and Technology Institute, Beijing, China

Ying Dong
China Institute of Geoenvironmental Monitoring, Beijing, China

ABSTRACT: In order to determine the location and to evaluate the amount of geothermal resources in Dongqianhu area, comprehensive methods including geology and hydrogeological survey, geophysical exploration and geochemical analysis were used. The results showed that Xi'ao was suitable to drill geothermal wells. The temperature and yield amount of groundwater from that well can reach 42°C and 252 m^3 per day, respectively, and thus it can belong to the medium low geothermal resource category. The total available heat energy of geothermal wells in 100 years was 32.74 MW.

1 INTRODUCTION

Geothermal resource is an important renewable energy source, which is located in the interior of earth. It is clean and environment-friendly, and can be widely used for bathing, heating and power generation purposes (Lund J.W, 2011). Utilization of geothermal resources is valuable for energy structure adjustment, energy conservation, and environment improvement (Zhu J.L, Malafeh, S, 2015). Shallow geothermal resources of cities at the prefecture level and above in China is 10^{11} kWh, which is equivalent to 0.23 billion tons of standard coal per year. Utilizing these geothermal resources can reduce 0.61 billion tons of CO_2 and 1.89 million tons of SO_2 emission per year (Lin W.J, 2012), thereby helpfully slowing down the trend of global warming.

2 GEOLOGY AND HYDROLOGY OF THE STUDY AREA

The study area, called Dongqianhu, is located on the southeast of Ningbo city, Zhejiang province. Its square is about 70 km². Most of its surface is covered by acidic pyroclastic rocks of Upper Jurassic, and the left side is covered by quaternary deposits. Figure 1 shows a lot of larger, extended further and deeper fractures mainly trending north-

north-east (NNE), which appeared successively in the middle late period of the Yanshan movement.

The storage condition of groundwater in the study area is influenced by the geological structure system. Formation of Dongqianhu area is largely controlled by Meihu fault. The surface of Meihu fault is made of smooth, soothing, undulate and thrusting scratches. The fault plane inclination is close to 90 degrees and trends to NE direction. The fault width is generally ranged from 10 m to 30 m, and it is filled with cataclastic rock and mylonite rock. Meanwhile there are gray-green and purple-gray rhyolite porphyry and rhyolitic breccia lava on the east of Dongqianhu area. These crushed rocks in the field are formed as results of the activity of Meihu fault. The lithology and filling fractures showed that

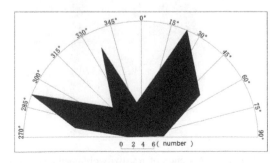

Figure 1. The rose diagram of water in bedrock fractures.

water-bearing properties of the fracture was good, while the flow rate of groundwater was slow.

3 METHODS

In this work, hydrogeological survey, geophysical exploration, and geochemical analysis methods were all used. Based on the survey of lithology and hydrogeology, conditions of groundwater recharge, runoff, and discharge were found, and its origin was also determined. Geophysical exploration methods, including high precision gravity measurement, audio geoeletric field measurement, heat release mercury, and radon measurement were applied to determine the development of the fault zone and the conditions of the thermal storage structure. The geochemical analysis method was used to identify the characteristics of deep underground geothermal resources.

4 RESULTS AND DISCUSSION

4.1 *Location of the geothermal storage structure*

Eleven geophysical sections were deployed in four positions. The possibility of geothermal storage structure existence in the botanical garden was very low. However, the possibility of existence in the south of Hengjie village was very high. Figure 2 shows the section of resistivity obtained by using the Controlled Source Audio frequency MagnetoTelluric method (CSAMT). Figure 3 shows four possible faults in the study area.

The measuring lines of CSMAT shows that resistivities of different layers were not the same,

and their arrangement was in the following order: deep layer > shallow layer > medium layer. Based on the data of CSMAT, it can be deduced that there were four possible faults in this work zone. The cut depth of fault 1 was between 1300 m and 14000 m, and those of fault 2, fault 3, and fault 4 were about 1200 m, 900 m, 500 m, respectively.

In addition, contents of Hg gas and radon gas were measured in the abnormal zone of CSAMT. Figure 4 showed concentrations of Hg gas and radon gas in that abnormal zone. Furthermore, data also indicated that the Xi'ao ditch was a fracture with obvious linear extension and its direction was NW 340 degree. Tensile fractures groups with North-North-West (NNW) direction distributed on the cutting excavation sections in the south and north of Shashan. Those fractures were located on the extension line of Xi'ao to Dongqianhu, indicating that the NNW trending fracture existed.

Figure 5 shows curves of low resistivity in different deep fracture zones of Xi'ao. Based on the interpretation of geophysical sections, it can be deduced that there was abnormal low resistivity, which was

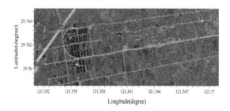

Figure 3. Horizontal distribution of possible faults obtained by CSAMT in the study area.
(*1. Green lines indicate measuring lines of CSAMT and 2. red lines indicate possible faults).

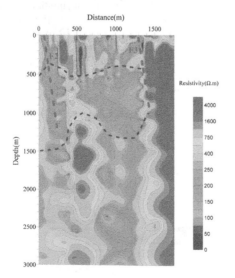

Figure 2. A section of resistivity obtained by CSAMT in one measuring line.

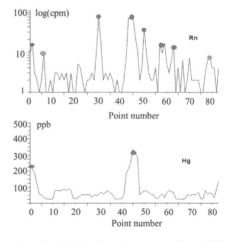

Figure 4. Graphs showing the concentrations of Hg gas and radon gas in the abnormal zone of CSAMT.
(*1. Red points indicate abnormal concentration; 2. green points indicate high concentration; 3. pink points indicate higher concentration).

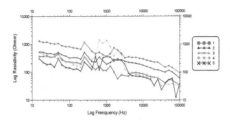

Figure 5. Curves of low resistivity in different deep fracture zones of Xi'ao.
(*1. Normal curve of low resistivity in the 50 m fracture zone, 2. abnormal curve of low resistivity, 3. normal curve of low resistivity in the 75 m fracture zone, 4. normal curve of low resistivity in the 125 m fracture zone, and 5. normal curve of low resistivity in the 175 m fracture zone).

caused by fracture structures. The characteristics of these fracture structures are as follows: smaller width with shallow downward extension on the south and slightly larger width with deep downward extension on the north. Moreover, the electrical characteristics and the change of resistivity value showed that fracture structures in the north may exhibit better water-bearing properties.

4.2 Chemical characteristics of groundwater in Dongqianhu area

During the process of migration, groundwater may interact with surrounding rocks under various temperatures, pressures, and other physical conditions, thereby dissolving different substances and leading to formation their special chemical components (Zhang Y.F, 2016). When comparing components of two shallow groundwater samples with those of two deep hot underground water samples in Dongqianhu area, it showed that there was a great difference between these various samples.

Moreover, groundwater classification by using the Chebotarev method was performed to classify those groundwater samples according to their main ionic components. Figure 6 demonstrates that all of these samples belong to Na-HCO$_3$. But deep groundwater samples contained a higher Na$^+$ content, thus representing their origin from deep underground. The main ionic components of shallow groundwater were present on behalf of typical shallow origins.

4.3 Temperature and amount of groundwater in geothermal wells

Two geothermal wells were drilled in Xi'ao area. The depth of hot groundwater in wells was between 850 m and 1000 m. Figures 7 and 8 show the results of the pumping test.

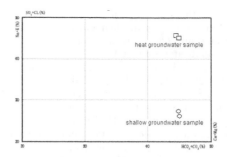

Figure 6. Diagram of groundwater classification by using the Chebotarev method.

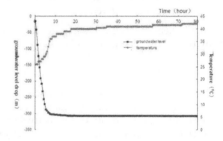

Figure 7. Graph showing the time–groundwater level–temperature trend during the pumping test.

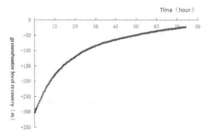

Figure 8. Graph showing groundwater level recovery trend of the pumping test.

According to these results, it can be seen that the temperature and yield of one well was 42°C and 252 m^3 per day and the temperature and yield of the other well was 37.5°C and 62.8 m^3 per day.

4.4 The total amount of heat emitted during geothermal well exploitation per year

The total amount of heat emitted during geothermal well exploitation per year was calculated by using the following equation:

$$Q_w = 365 Q \rho_w C_w (t_w - t_0) \tag{1}$$

Where Q_w is the total amount of heat exploited from the geothermal well per year, unit (J/a); Q is the groundwater yield of the geothermal well, unit (m^3/d); ρ_w is the density of water, unit (kg/m^3);

C_w is the specific heat of hot groundwater, unit (J/kg. °C); t_w is the temperature of hot groundwater, unit (°C); t_o is the reference temperature of groundwater, unit (°C).

The yield, temperature of the geothermal well was 252 m³/d and 42°C and the density and specific heat of hot groundwater were 992.8 kg/m³ and 4186.8 J/kg.°C. The reference temperature of groundwater was 15°C. Substitute these data in equation (1), and then the total amount of heat exploited from the geothermal well per year can be obtained, which is 1032.3×10^{10} J/a.

4.5 *Available heat storage capacity per unit area of the geothermal well*

Available heat storage capacity was calculated by using the following equation (2):

$$Q_r = KHC_r(t_r - t_0) \qquad (2)$$

Where Q_r is the available heat storage capacity, unit (kcal/m²); K is the recovery rate of the geothermal resource; H is the depth of the thermal reservoir, unit(m); C_r is the average heat storage capacity of the thermal reservoir, unit (kcal /m³°C); t_r is the average temperature of the thermal reservoir, unit (°C); t_0 is the temperature of the normal temperature formation zone, unit (°C).

In this work, the values of K, H, C_r, and t_0 can be obtained from the literature, which are 0.15, 170 m, 594 kcal/m³°C, and 15°C, respectively. t_r was calculated by using the SiO$_2$ geo-thermometer method, which was 75°C. By substituting these data in equation (2), the available heat storage capacity per unit area of geothermal well was obtained, which reached to 3.48×10^5 kcal /m².

4.6 *Evaluation of the amount of geothermal resources*

Generally, geothermal energy consists of two parts; one is the thermal energy accumulation in the rock and the other is the thermal energy storage in hot underground water. Based on the current economic and technical conditions, exploitation of hot underground water is the primary method of geothermal resource utilization. Therefore, the task should be focused on evaluating the total quantity of heat and recoverable amount of hot underground water.

According to the classification of Table 1, the geothermal resource in this study area was a typical low-medium temperature resource because its temperature was 41.5°C. It can be used for heating, bathing, and entertainment.

According to the classification of Table 2, reserves of geothermal resources were calculated for 100 years, and thus the total available heat from this geothermal well was 1032.3×10^{12} J, which was equivalent to 32.74 MW. The impacted area of the well was 7×10^5 m².

Table 1. Classification of geothermal resources according to their temperature.

Classification		Temperature (°C)
High geothermal resource		$t \geq 150$
Medium geothermal resource		$90 \leq t < 150$
Low geothermal resource	Hot water	$60 \leq t < 90$
	Hot-warm water	$40 \leq t < 60$
	Warm water	$25 \leq t < 40$

Table 2. Scale classification of geothermal resources.

Classification	High T (°C)		Low-medium T (°C)	
	E1* (MW)	Time (a)*	E2* (MW)	Time (a)
Large	>50	30	>50	100
Medium	10–50	30	10–50	100
Small	<10	30	<10	100

E1* means electric energy; Time (a)* means duration of geothermal resource available time; and E2* means thermal energy.

5 CONCLUSIONS

In this study, geology and hydrology in Dongqianhu area of Ningbo, Zhejiang province were investigated in detail with comprehensive methods. The location of Xi'ao had many tensile fractures and crush rocks, which were ideal for groundwater storage. A geothermal well was built in this area. Hot groundwater was mainly stored in the depth range of 850 m–1000 m. The yield and temperature of the geothermal well was 252 m³/d and 42°C. The geothermal resource in this area was belonging to typical low-medium temperature field. The total available heat from the geothermal well was 1032.3×10^{12} J, and the impacted area by exploitation of these wells was 7×10^5 m².

REFERENCES

Lin, W.J, Wu, Q.H, Wang, G.L. 2012. Shallow geothermal energy resource potential evaluation and environmental effect in China (In Chinese). *Journal of Arid Land Resources & Environment* 26(3):57–61.

Lund, J.W, Freeston, D.H, Boyd, T.L. 2011. Direct utilization of geothermal energy 2010 worldwide review. *Energies,* 40(3):159–180.

Malafeh, S. and Sharp, B. 2015. Role of royalties in sustainable geothermal energy development. *Energy Policy* 85:235–242.

Zhang, Y.F, Tan, H.B, Zhang, W.J, Dong, Q. 2016. Geochemical constraint on origin and evolution of solutes in geothermal springs in western Yunnan, China. *Chemie der Erde-Geochemistry* 76(1):63–75.

Zhu, J.L, Hu, K.Y, Lu, X.L, et al., 2015. A review of geothermal energy resources, development, and applications in China: Current status and prospects. *Energy* 93:466–483.

Advances in Energy Science and Equipment Engineering II – Zhou, Patty & Chen (Eds)
© 2017 Taylor & Francis Group, London, ISBN 978-1-138-71798-5

Effects of storage temperatures on the nutrition quality of Macadamia nuts during long-term storage

Li-Qing Du, Xing-Hao Tu, Qiong Fu, Xi-Xiang Shuai, Ming Zhang & Fei-Yue Ma
Key Laboratory of Tropical Fruit Biology, Ministry of Agriculture, South Subtropical Crop Research Institute,
Chinese Academy of Tropical Agricultural Science (CATAS), Zhanjiang, Guangdong, China

ABSTRACT: The effects of different storage temperatures (20 and 4°C) on the nutrition and quality in Macadamia nuts stored for long durations were studied in this paper. The investigation focused on the increase in acid, peroxide and carbonyl values, as well as contents of soluble protein and soluble total sugar. It was shown that PerOxide Values (POVs) and Acid Values (AVs) were increased by 0.90 mmol/kg and 2.82 mmol/kg, respectively. The nutritional quality of Macadamia nuts stored at room temperature for 4 months was corresponding to that of Macadamia nuts stored at 4°C for 12 months.

1 INTRODUCTION

The Macadamia nut (*Macadamia integrifolia*) is a native fruit of the rainforests of eastern Australia (Storey, 1954) and it is now grown in other parts of the world. It is rich in monosaturated fatty acids and is delicious (Pankaew, 2016A). Meanwhile, it is a rich source of lipids, proteins, and important micronutrients (Pankaew, 2016A, Piza, 2014). A study of the chemical composition of 22 Macadamia cultivars in Brazil performed by Maro et al. (Maro, 2012, Sandra, 2016) reported that most moisture contents reached 11%, lipids amounted to 65%, proteins contributed 20%, crude fiber comprised 30% and ash constituted 2% of the total nutrition content (Marisa, 2010, Garg, 2003, Neto, 2010). After harvesting, Macadamia nuts need to be stored. To maintain the quality, it is necessary to know their storage characteristics in order to optimize the storage operations and to obtain high quality Macadamia nuts. The storage temperature was an important factor.

The aims of the present work were to evaluate the effects of different storage temperatures (20 and 4°C) in Macadamia nuts over a period of 12 months. The Macadamia nut oxidation trends were estimated by the amounts of both primary and secondary oxidation products. Water loss rate, PerOxide Values (POVs) and Acid Values (AVs) of Macadamia nut samples were determined to assess the development of rancidity. In this sense, any methodology aimed at prolonging its shelf-life would be of great interest to the sector.

2 MATERIALS AND METHODS

2.1 Materials and reagents

The Macadamia nuts were collected from South Subtropical Crop Research Institute in August 2014 and these were known as Nanya No.1. The peel was removed from the material and placed in a refrigerator (4°C). The same size Macadamia nuts, which had no insects and no mechanical injury, were selected for the following experiment.

All reagents obtained from Tianjin Chemical Reagents Co. (Tianjin, China) was of analytical grade. Coomassie brilliant blue was purchased from Sigma–Aldrich (Steinheim, Germany).

2.2 Determination of physical and chemical indicators and nutrients

Five kilograms of Macadamia nuts with shell were placed for 12 month at 20 or 4°C to evaluate the effects of different storage temperatures. During the period, samples were stored in perforated polyethylene storage bags at a relative humidity of 45–55% to avoid dehydration and decay. Every month, 40 Macadamia nuts were used to perform the following determination of physical and chemical indicators.

In order to investigate the effects of storage temperature on nutrition and quality of Macadamia nuts, the water loss rate, PerOxide Values (POVs) and Acid Values (AVs) were used for evaluation and extraction Macadamia nut quality in the present study. POVs of Macadamia nuts were determined according to AOCS recommended practice Cd 8-53. The AVs of Macadamia nuts were determined according to AOCS recommended practice Cd 3d-63. The soluble total sugar and soluble protein levels of samples were determined according to anthrone colorimetry and Coomassie brilliant blue.

3 RESULTS AND DISCUSSION

3.1 *Effects on mass loss rate*

Generally, the mass loss rate was an important factor for evaluating the preservation of fruits and vegetables. Meanwhile, it was an important reason contributing to increased fruit wilting, deterioration, and decay. As shown in Fig. 1, the water loss rate of Macadamia nuts increased drastically with an increase in the storage time from 1 to 12 months. And it was observed that, 20°C had stronger effects on the mass loss rate of Macadamia nuts than 4°C. With an increase in the storage time, the water loss rate reaches the peak value (20.61%). The Macadamia nuts were seriously shriveled and browned and lost their commercial value. The possible reason was that respiration and transpiration in the Macadamia nuts, led to loss of tissue water and water-soluble nutrients. After 12 months, the mass loss rate reaches 6.31% at 4°C storage temperature. However, the appearance and color of Macadamia nuts were not significantly changed.

3.2 *Effects on POV*

Primary oxidation products (hydroperoxides) were determined by POV measurement during storage. The POVs of Macadamia nuts at two different temperatures are shown in Fig. 2. At 20°C, the POV

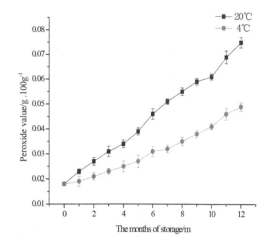

Figure 2. Graph showing the effects on POVs.

of Macadamia nuts exhibited identical increasing trends during storage. The POV rose rapidly and reached 1.2 mmol/kg in the 12th month. A significant difference ($p < 0.01$) in the POV was observed between the first month and the 12th month at two storage temperatures. When compared with samples stored at 20°C, the samples stored at 4°C showed lower POVs. The POVs ascended constantly with an extension of storage time at 4°C. After 12 months, the POVs of Macadamia nuts were below 6 mmol/kg (NY/T—2011).

3.3 *Effects on AV*

The Acid Value (AV), which is an important index of nuts, represents the rancidity of the sample. It is measured by the amount of free fatty acids in the sample. A sample of low Acid Value can be regarded as a good quality sample. The AV values of all the samples were determined for 12 months of storage at 20 and 4°C (Fig. 3). For samples stored at 20°C, the difference in AV values of the two storage temperatures was unnoticeable in the first 5 months. When compared with samples stored at 20°C, after 5 months, a tremendous rise in AVs of the Macadamia nuts was observed. In the 12th month of analysis, the AVs of samples stored at 20 and 4°C increased by 4.16 mmol/kg and 2.82 mmol/kg, respectively. The results showed that the CA, at 4°C, exhibited an effective inhibition in the form of rancidity in Macadamia nuts.

3.4 *Effects on soluble total sugar and soluble protein*

Soluble total sugars and soluble proteins are considered to be primary nutrients. In this work, soluble total sugar and soluble protein contents

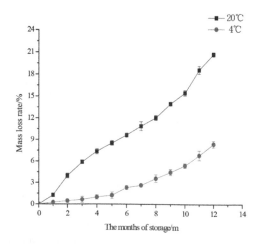

Figure 1. Graph showing the effects on the mass loss rate.

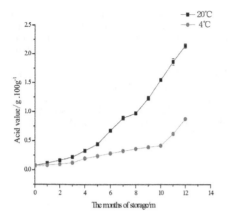

Figure 3. Graph showing the effects on AV.

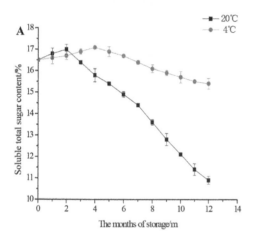

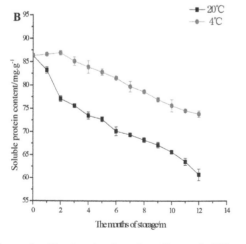

Figure 4. Graphs showing the effects of different storage temperatures on soluble total sugar content (A) and soluble protein content (B) of Macadamia nuts.

of Macadamia nuts were determined to be 204.61 mg/(100 g) and 1,706.86 mg/(100g) in the initial stage, respectively. With an increase in the storage time, soluble total sugar and soluble protein contents decreased, which led to a decline in the nutrition and quality of Macadamia nuts. The contents of the soluble total sugars and soluble proteins at 20 and 4°C storage temperatures are shown in Fig. 4A and B.

From Fig. 4A, it can be observed that, after 12 months, the contents of soluble total sugars in Macadamia nuts stored at 20 and 4°C were 8.81 mg/(100 g) and 179.84 mg/(100 g), respectively. It decreased by 22.4% and 12.1%, respectively. From Fig. 4B, it can be observed that, after 12 months, the contents of soluble proteins in Macadamia nuts stored at 20 and 4°C were 1118.01 mg/(100 g) and 1651.98 mg/(100 g), respectively. It decreased by 34.5% and 3.2%, respectively. Therefore, it can be concluded that hypothermia was more suitable for long-term storage and nutritional quality retention of Macadamia nuts.

4 CONCLUSION

In the present study, hypothermia exhibited a strong inhibitory effect on the rancidity of Macadamia nuts. The effects of different storage temperatures (20 and 4°C) on nutrition and quality in Macadamia nuts stored for a long term was studied in this paper. The investigation focused on the increase in acid, peroxide and carbonyl values, as well as contents of soluble proteins and soluble total sugars. It was shown that POV and AV were increased by 0.90 mmol/kg and 2.82 mmol/kg, respectively. The nutritional quality of Macadamia nuts stored at room temperature for 4 months was corresponding to that of Macadamia nuts stored at 4°C for 12 months. Macadamia nut oxidation stability strengthened with a decrease in the storage temperature. Hypothermia can effectively inhibit the formation of oxidation products of Macadamia nuts and prevent loss of tissue water and water-soluble nutrients. This is an important effect on the nutrition quality of Macadamia nuts.

ACKNOWLEDGMENTS

This work was financially supported by the funds from Basic Scientific Research Project of Nonprofit Central Research Institutions (1630062015011, 1630062015013 and 1630062016007), Modern Agricultural (Important Tropical Crops) Industrial Technology Project of Guangdong Province, Public Service Sectors (Agriculture) Research

Projects (201303077-2), and Key Laboratory of Tropical Fruit Biology, Ministry of Agriculture.

REFERENCES

Garg M. L., Blake, R. J., Wills, R. B. H.: J. Nutrition (2003) 133:1060–1063.

Marisa M. W.: Food Chem. (2010) 121:1103–1108.

Maro L. A. C., Pio R., Penoni E. dos S., et al.: Ci^encia Rural (2012) 42(12): 2166–2171.

Neto M. M., Nogueira, N. R.: (2010). In XXX Encontro Nacional de Engenharia de Produç–ao pp.1–10.

Pankaew P., Janjai S., Nilnont W., Phusampao C., Bala B.K.: Food and Bioproducts Processing (2016A) 100: 16–24.

Piza P. L., Moriya, L. M.: Revista Brasileira de Fruticultura (2014) 36(1):39–45.

Sandra L.B. N., Christianne E.C. R.: Trends in Food Science and Technology (2016) 54:148–154.

Storey W. B., Hamilton R. A.: Yearbook (1954) 38:63–67.

Advances in Energy Science and Equipment Engineering II – Zhou, Patty & Chen (Eds)
© 2017 Taylor & Francis Group, London, ISBN 978-1-138-71798-5

Experimental research on the bubble behavior of the sub-cooled pool boiling process under microgravity

Y.Y. Jie, H.Y. Yong, C.X. Qian & L.G. Yu
College of Aerospace Science and Engineering, National University of Defense Technology, Changsha, P.R. China

ABSTRACT: The effect of heat flux and surface energy on the bubble behavior of the sub-cooled pool boiling process under microgravity has been researched by utilizing the drop tower facility of National Microgravity Laboratory (NMLC) in Beijing. Two ceramic plates of dimensions $10 \times 25 \times 1.2$ mm^3 were used as heaters. Distilled water and HFE7500 were used as the working liquids. The number, size, and movement of bubbles under microgravity during the experiment were analyzed. The bubble regrowth of distilled water was observed. Bubbles of lower surface energy grow rapidly and the departure diameter is bigger than bubbles of high surface energy on the same solid plate. The model of pool boiling is mainly determined by heat flux, thus affecting the coefficient of heat transfer.

1 INTRODUCTION

Economic propellants with high performance are required for on-orbit refueling and deep space exploration missions (Juan, 2014). Due to the highly efficient thrust and clean production, cryogenic liquids like oxygen and hydrogen have been playing a significant role in aerospace missions and refueling projects since Apollo 11 was sent to the Moon by Saturn V in 1969 (Kutter, 2005). The extremely low storage temperature of cryogenic propellants induces many severe challenges—the mainly including mass loss and pressure increase when heat leakage occurs—in its future aerospace application. Approaches have been studied to reduce the boil-off of liquid (Guernsey, 2005) and control the pressure of cryogenic tanks (Panzarella, Plachta and Kassemi, 2014; Barsi and Kassemi, 2007). The temperature field of cryogenic propellants is heterogeneous and even can boil when heat leakage occurs, which possibly threatens tank safety. However, the heat transfer of pool boiling is more efficient than conduction and rare nature convection under microgravity due to its latent heat when liquid evaporates. The agitating effect induced by bubble growth, migration, and coalescence also contributes to pool boiling heat transfer. Research on bubble behavior and heat transfer of pool boiling under microgravity has become a significant study field (Straub, 2001; Straub and Zell, 2010). However, the amount of heat flux required to cause nucleate pool boiling and the effect of surface energy on the bubble behavior are not explained well. And few studies have been car-

ried out on both how boiling heat transfer can be applied to transport external heat out of the tank and the pool boiling bubble behavior of cryogenic propellants under microgravity due to the lack of microgravity experiments (Straub, 2006). The basic mechanism of boiling is not developed yet due to the complex progress when the phase changes (Straub, 2001) and the relative bubble behavior needs further research.

For the purpose of exploring the effect of heat flux, surface energy, liquid volume, and heater pose on pool boiling heat transfer and bubble dynamics under microgravity, eight rounds of pool boiling experiments were conducted by using the drop tower facility of National Microgravity Laboratory (NMLC) in Beijing. Bubble growth, regrowth, and migration can be observed frame by frame. In this paper, experimental results are introduced in part—the effect of heat flux and surface energy on the bubble behavior of the sub-cooled pool boiling process under microgravity.

2 EXPERIMENTAL APPARATUS

The duration of microgravity of the drop tower is about 3.6 seconds and gravitational acceleration is about 10^{-5} g$_0$.

A 502.4 mL tank manufactured by using acrylic resin (PMMA) was used as the boiling pool. Two ceramic plates were applied as the heaters. Pt100 thermistors (ranging from 0–100°C and from 0–200°C) were used and the one ranging from 0–200°C was attached to the back surface of the heater to measure the surface temperature. The left two were settled in the symmetry axis of two heaters. The nominal resistances of the heater was 26.2, 10.2, and 6.1 Ω (20°C), respectively. The heating voltage was supplied by the energy system of the drop tower. The pressure gauge (range: –100–200 kPa) was set in the tank top to measure the inner pressure of the tank. A video camera (25 Hz) was used to record the bubble behavior during the duration of each round experiment. An LED light is used as a light source.

The whole schematic of the experimental facility is shown in Fig. 1.

There were two procedures in each round of experiments—(1) heat the system to a boil and (2) drop the cabin after a short period of time. In the first procedure, the light was turned on and the heater began to work till bubbles were observed by the observation camera to confirm that pool boiling occurred before the drop. And then, wait for a certain period of time (from 5 seconds to 40 seconds) to ensure that the pool boiling was developing. Then, the experiment facility was dropped freely from the tower and experiments under microgravity began. The video camera was working during the whole experiment both at normal gravity and microgravity to record the experiment operations and the results.

Table 1. Different combinations of experimental arrangements.

Working liquid	Volume %	Power W	Heater pose
1 Distilled water	50	57	Sideling
2 Distilled water	50	57, 57	Side, side
3 Distilled water	50	57, 57	Side, forth
4 Distilled water	90	57, 57	Side, forth
5 HFE7500	50	57, 57	Side, forth
6 HFE7500	90	57, 57	Side, forth
7 HFE7500	90	22, 22	Athwart, forth
8 HFE7500	90	22, 94	Athwart, forth

Figure 1.   Illustration of the experimental facility.

3   RESULTS AND ANALYSIS

Eight rounds of pool boiling experiments were conducted successfully and the experimental data of temperature, pressure, pictures, and videos were acquired. The original pressure inside the tank was 107 kPa and the temperature was about 25–35°C. A constant voltage of 24 V was applied to the heater when it was working.

3.1   *Bubble behavior of pool boiling of different liquid and heat fluxes*

Part of the experiment results of rounds 4, 6, 7, and 8—the effect of heat flux and surface energy on the bubble behavior of sub-cooled pool boiling process are given below.

Bubble growth and migration were observed frame by frame. From Fig. 3, it can be observed that the bubble shape of different surface energy liquids is different on the same solid surface. HFE7500 corresponds to the low surface energy liquid and retains good spherical shape. Distilled water appears more like an ellipse as it corresponds to the medium surface energy. The bubble of distilled water at –10 s and –0.04 s changes little in shape but migrates toward the surface center and increases its diameter by absorbing other little bubbles that are nucleated around at 0 s. From photo analysis, it can be observed that the bubble of HFE7500 migrates more frequently and quickly than that of distilled water. The contact edge of the triple phase of distilled water is not so smooth and then evolves into a circle under microgravity. From photos of four groups, it can be observed that the bubble tends to migrate to the center of the heater surface and stay there under developing conditions. From Fig. 3(d), it can be seen that the bubble shake was observed due to its big size. The temperature of bulk liquid is lower than the saturation temperature and the temperature of the bottom part of the bubble is higher, concluding that the temperature gradient exists along the bubble surface from bottom to top. As a consequence, the top part condenses as the sub-cooling increases with bubble growth while the bottom part of the bubble evaporates, thereby resulting in the bubble shake.

The heat flux is another determinant of the bubble behavior, especially of bubble growth. Fig. 3 (b)–(d) show bubble growth and migration of different heat fluxes. Fig. 3(b) shows the results of round 7 of 22 W supply on the heater and has the smallest bubble right after the heater began working. Fig. 3(c) shows the results of experiment 6 of 57 W supply and the bubble size is medium. Fig. 3(d) represents the 94 W supply experiment of round 8 and shows the biggest bubble. The bubble

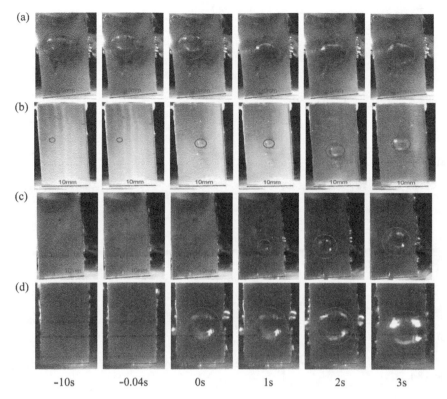

(a)

(b)

(c)

(d)

-10s   -0.04s   0s   1s   2s   3s

Figure 2. Nucleate boiling continues to develop under microgravity ((a), (b), (c), and (d) represent rounds 4, 7, 6, and 8).

of round 8 grows faster at the first second than those of the other two rounds, which have lower power supply.

The observed bubble number of HFE7500 at –0.04 s (about 10) is much more than that of the bubble at 0 s (about 1) on the heater surface.

3.2 *Experimental data of different liquid and heat fluxes*

Fig. 3 shows the experimental data of temperature, pressure, and microgravity signals of rounds 2, 5, 7, and 8. The data show that the duration of microgravity recorded is less than 3.6 seconds and is about 2 seconds.

From Fig. 3, it can be observed that the pressure in the tank remains almost constant ensuring that the saturation temperature is constant. In Fig. 3(a), it can be observed that the temperature of the heater surface, bulk liquid, and liquid increases sharply at the moments of A1, B1, and C1, which indicates that the heater worked as reported by Straub (2001). Then, the temperature increases slowly till the drop begins. The duration of pool boiling at normal gravity is about

40 seconds. At the moment of drop, the surface temperature decreases sharply and immediately increases to A5 and then temperature vibrates. In Fig. 3(b)–(d), it can be observed that the temperatures of the heater surface, liquid nearby, and bulk liquid change constantly without the sharp jump or drop. They all increase rapidly during the initial period of heating and then maintain almost the same temperature before the drop, which indicates that pool boiling has developed completely. The heater surface temperature decreases while the drop begins and slowly increases until the end of experiment, as shown in Fig. 3(b). In Fig. 3(c) and (d), it can be seen that the temperatures of the heater surface and liquid nearby change a little at the moment the drop begins and sharply increases after the drop. The temperature is controlled by heat absorbing and transferring, from which the heat transfer coefficient is directly derived. It can be concluded that the heat transfer increases in distilled water and changes in HFE7500 without certain predictions. In Fig. 3(b), it can be seen that the coefficient of heat transfer increases while it decreases in Fig. 3(c) and (d).

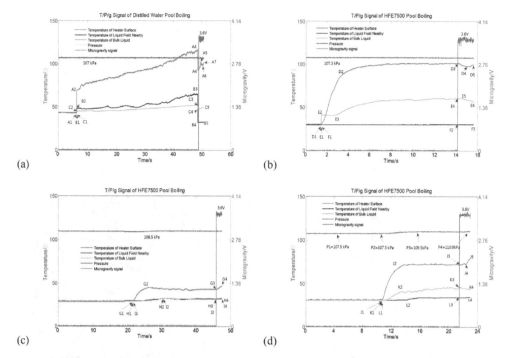

Figure 3. Experiment data of temperature, pressure, and microgravity signal ((a), (b), (c), and (d) represent rounds 2, 5, 7, and 8.).

What contradicts in the results is that the effect of the heat flux on the heater surface temperature is opposite when compared between Fig. 3(b) and (c) and between Fig. 3(c) and (d). It needs to be further validated.

4 CONCLUSIONS

A series of experiments focusing on the effect of heat flux and surface energy on the bubble behavior of the sub-cooled pool boiling process under microgravity have been conducted with the application of a drop tower.

The liquid with lower surface energy has a smaller contact angle on the same surface. It is easier for the bubble of this type liquid to retain its spherical shape and has less drag force to exhibit migration on the surface. The heat flux can mainly control the time of occurrence of the first bubble and the speed of bubble growth.

Pool boiling of those liquids with medium surface energy increases heat transfer at the initial period of boiling and then decreases. Pool boiling of the liquid with lower surface energy can decrease the coefficient of heat transfer depending on the amount of heat flux.

REFERENCES

Barsi, S. & Kassemi, M. 2007. *45th AIAA Aerospace Sciences Meeting and Exhibit in U.S.*: Validation of Tank Self-Pressurization Models in Normal Gravity. 2007. Reno, NV: AIAA.

Guernsey, C. S., Baker, R. S., Plachta, D. & Kittel, P. 2005. *AIAA/ASME/SAE/ASEE Joint Porpulsion Conference in U.S.*: Cryogenic propulsion with zero boil-off storage applied to outer planetary exploration. *10–13 July* 2005. Tucson, AZ: AIAA.

Juan, F., 2014. Research on the self-pressurization phenomenon and liquid mass gauge of cryogenic propellant storage. *Ph. D, National University of Defense Technology.*

Kutter, B., Zegler, F. & Lucas, S. 2005. Atlas centaur extensibility to long-duration in-space applications. *AIAA*, 2005–6738.

Panzarella, C., Plachta, D. & Kassemi, M. 2004. Pressure control of large cryogenic tanks in microgravity. *Cryogenics* 44, 475–483.

Straub, J. 2001. Boiling heat transfer and bubble dynamics in microgravity. *Adv. in Heat Transfer, 35*, 57–172.

Straub, J. & Zell, M. 2010, Transport-mechanisms in natural nucleate boiling in absence of external forces. *Heat Mass Transfer, 46*, 1147–1157.

Straub, J. 2006. Highs and lows of 30 years research of fluid physics in microgravity, a personal memory. *Microgravity Sci. Tec. XVIII-3/4*, 14–20.

Advances in Energy Science and Equipment Engineering II – Zhou, Patty & Chen (Eds)
© 2017 Taylor & Francis Group, London, ISBN 978-1-138-71798-5

Dynamic changes in NDVI and their relationship with precipitation in Ningxia, NW of China

Tao Wang

College of Geomatics, Xi'an University of Science and Technology, Xi'an, China

ABSTRACT: Dynamic changes in NDVI and their relationship with precipitation were analyzed on the basis of 250 m resolution and 16 d synthesized MODIS NDVI data, as well as daily precipitation data from 9 weather stations from 2000 to 2014 in Ningxia, Northwest China. The result showed the following: (1) annual average NDVI in the vegetation growing season had a linear growth trend with a rate of 0.052/10a, but vegetation NDVI tended to decrease over the last 4 years. (2) The average NDVI value had an increasing trend, accounting for 91.62% of the total area, which was mainly distributed in the central arid region, southern hills and mountains region. Only 8.38% of the total area had a decreasing trend, which was mainly in the northern Yellow River irrigated region and the central arid region of Ningxia. (3) The correlation coefficient between the average NDVI value and precipitation had a linear growth trend. Positive and significantly-positive correlation coefficients were mainly distributed in the southern hills and mountainous region of Ningxia, whereas the negative and significantly-negative correlation coefficients were mainly distributed in the northern Yellow River irrigated area of Ningxia, especially around Yinchuan city.

1 INTRODUCTION

Vegetation cover is an important indicator of global surface eco-environmental change (Hao et al., 2012). The Normalized Difference Vegetation Index (NDVI) is one of the important indexes for research on vegetation cover change and its relationship with climate change or human activity (Li et al., 2013). Ningxia, located in northwest China, belongs to arid and semi-arid regions, and it is doubly influenced by the Mid-latitude Westerlies of the Northern Hemisphere and the Southeast Asia Monsoon (Wu et al., 2008). Those two climate models are leading to different vegetation cover changes in Ningxia: dry climate and little precipitation are harmful to vegetation growth when under the control of the Mid-latitude Westerlies, whereas humid climate and more precipitation are beneficial to vegetation growth when under the control of the Southeast Asia Monsoon.

The resolution of NOAA and SPOT NDVI is 1 km and above, whereas MODIS NDVI has different resolutions of 1 km, 500 m and 250 m, and Landsat NDVI has a resolution of 30 m. Furthermore, the time series data of NOAA, SPOT and MODIS NDVI are better, and the former two types are longer than the latter but have low resolution. NOAA and SPOT NDVI data are usually applied in research on large-scale vegetation cover changes,

such as at global and national scales (Fensholt et al., 2012), whereas MODIS NDVI is more widely applied at middle and small scales (Wardlow et al., 2008). Due to its poor continuity, Landsat NDVI is applied to vegetation cover change less often, and it is more often used to verify low resolution NDVI data (Yang et al., 2012).

The research on vegetation cover changes in Ningxia is rich. For example, Li et al. (2011) suggested that vegetation in central Ningxia had a deteriorating trend, whereas the southern and northern regions improved overall, and vegetation cover change was highly correlated with precipitation from 1982 to 2006. The conclusion of Li et al. (2011) was in agreement with Du et al. (2015), who also expressed that vegetation cover has had a growth trend over the past 30 years, especially after 1990 when vegetation cover improved from a degraded state to sustainable recovery. Wang et al. (2013) described a decrease in desertification area, where vegetation improved and continually increased. The research described above mainly used NOAA and SPOT NDVI images as data sources with long time series but low-resolution. In this paper, mid-resolution MODIS NDVI images and precipitation data were used to analyze vegetation NDVI change and its relationship with precipitation to provide a scientific basis for eco-environmental policy development and implementation.

2 MATERIALS AND METHODS

2.1 Study area

The Ningxia Hui Autonomous Region (Ningxia) is located in Northwest China between 35°14′N–39°23′N and 104°17′E–107°39′E (Figure 1). Ningxia belongs to an arid and semi-arid climate, where the average temperature and total precipitation in 2013 were 9.5°C and 346 mm, respectively. Due to large differences in altitude, temperature and precipitation are quite different among the various regions. In the central mid-arid region and the northern Yellow River irrigation region, annual average temperature and total precipitation were 10.5°C and 166 mm, respectively, whereas they were 7.8°C and 676 mm, respectively, in the southern hills and mountains (BSNHAR. 2014). The northern region, which has less precipitation, is an important agricultural region that makes use of Yellow River water to develop agriculture, and the vegetation situation is better, whereas the central arid region, which has less precipitation, is poor in vegetation, and the vegetation cover in the southern hills and mountains is correspondingly better due to more precipitation (Du et al., 2015).

2.2 Data

The data used in this research include: (1) 250 m resolution and 16 d synthesized MODIS NDVI 13Q1-Level 3 images from 2000 to 2015, downloaded from http://ladsweb.nascom.nasa.gov; and (2) daily precipitation data of 9 weather stations in Ningxia from 2000 to 2014, downloaded from the China Meteorological Data Network (http://data.cma.cn).

Figure 1.   Study area and weather station distribution.

2.3 Linear trend method

Linear trend methods are mainly used in two aspects: one is the temporal analysis of NDVI time series data to reflect the trend in NDVI change over time, and the other is spatial analysis of NDVI to show the trend in NDVI change in the study area. The linear trend method is calculated as follows (Xu. 2002):

$$a = \frac{\sum_{i=1}^{n} x_i y_i - n x' y'}{\sum_{i=1}^{n} x_i^2 - n x'^2} \tag{1}$$

where y represents the MODIS NDVI images of the study area from 2000 to 2015, x represents the years from 2000 to 2015, a is a coefficient, x′ and y′ are the average values of x and y, respectively. Positive or negative of a-value reflect increasing or decreasing linear trends in NDVI, respectively.

2.4 Correlation coefficient

The correlation coefficient is mainly used to analyze the relationship between two independent variables, calculated as (Xu. 2002):

$$r = \frac{\sum_{i=1}^{n}(x_i - x')(y_i - y')}{\sqrt{\sum_{i=1}^{n}(x_i - x')^2}\sqrt{\sum_{i=1}^{n}(y_i - y')^2}} \tag{2}$$

where x and y are annual NDVI and precipitation, respectively. The value of r represents the distribution between [−1, 1], and r = 0 indicates no correlation between the two variables. An r value close to −1 indicates strong negative correlation, and a value close to 1 indicates a strong positive correlation.

3 RESULTS

3.1 Vegetation growing season

The vegetation growing season is a comprehensive reflection of climate change. Multiple images of the average NDVI change curve and its slope were used to determine the range of the vegetation growing season according to the slope turning points (Chen et al., 2010). The change process was in line with four polynomial fitting results ($R^2 = 0.9417$). Then, the first derivative was calculated to get the slope of the four polynomial fitting curves, which showed that, overall, the vegetation growing season of Ningxia started at the end of April (the 8th image, average NDVI value of 0.1634) and ended in late of September (the 17th image, average NDVI value of 0.3030). The vegetation growing season in Ningxia included the 8th to the 17th images, with a total of 10 images.

3.2 Temporal change of vegetation NDVI

Annual average NDVI values and linear change trends were obtained on the basis of different images from the vegetation growing season from 2000 to 2015. The annual vegetation NDVI value was distributed in the range of 0.21–0.34, which indicated a low vegetation cover level in Ningxia. An annual average NDVI value during the vegetation growing season showed a linear growth trend and significantly (P < 0.001) passed the 0.05 significance level test, which suggested that vegetation NDVI would increase linearly in the coming period. By the slope of the fitting curve, the growth rate of the vegetation NDVI average value was 0.052/10a from 2000 to 2015. It is noteworthy that the decline in NDVI was consecutive from 2012 to 2015. Thus, there is a need to strengthen the monitoring of vegetation and to keep abreast of regional eco-environmental change (Figure 2).

3.3 Spatial change of vegetation NDVI

The average NDVI in the vegetation growing season showed a growth trend from 2000 to 2015, which accounting for 91.62% of the total area of Ningxia, and it was mainly distributed in the central arid region and the southern hills and mountains area. Furthermore, the northern Yellow River irrigation region also had a larger areal distribution (Figure 3a).

The area with a decreased trend accounted for 8.38%, and it was mainly distributed in the northern Yellow River irrigation region and the west of the central arid region. The change rate of the average NDVI showed a decrease process from the south to the north of Ningxia during 2000–2015. The greatest linear growth rate was distributed in the

southern hills and mountains region, followed by the central arid region and the northern Yellow River irrigation region, especially for the NDVI around urban areas, with an obvious decreasing trend (Figure 3b).

3.4 Relationship between NDVI and precipitation

The spatial distribution of the correlation coefficient between average NDVI and precipitation showed that the coefficient was mainly positive, accounting for 93.64% of the Ningxia area, especially in the central arid region and the southern hills and mountainous region, and a large area distribution was also shown in the northern Yellow River irrigation region. Negative correlations accounted for 5.36% of the area, which was mainly distributed in the northern Yellow River irrigation region where human activity was intensive, such as Yinchuan (the capital of Ningxia Province) and its nearby area (Figure 4a).

When $\alpha = 0.05$, the critical value of the correlation coefficient was 0.5319, so the correlation coefficient was divided into 3 types, <–0.5319, –0.5319–0.5319, and >0.5319, where <–0.5319 and >0.5319 denote significant negative correlation and significant positive correlation, respectively. In the spatial distribution of the correlation coefficient, the proportions of significant and non-significant correlation coefficients were roughly equal (48.21% and 51.79%, respectively). The significant areas were mainly distributed in the southern hills and mountains region and the west area of the central arid region, whereas non-significant areas were mainly distributed in the northern Yellow River irrigation region, especially in urban areas, as well as the eastern part of the central arid region (Figure 4b).

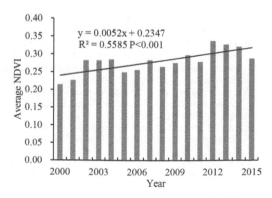

Figure 2. Trend of the annual average NDVI in the vegetation growing season from 2000 to 2015.

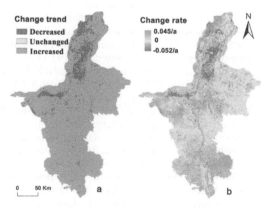

Figure 3. Spatial distribution of the change trend (a) and the change rate (b) in the vegetation growing season from 2000 to 2015.

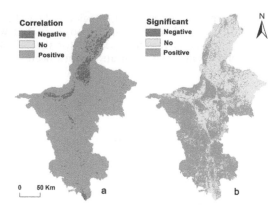

Figure 4. Spatial distribution of the correlation coefficient between average NDVI and precipitation from 2000 to 2015.

4 CONCLUSION AND DISCUSSION

MODIS NDVI data from 2000 to 2015 and daily precipitation data of 9 weather stations from 2000 to 2014 were used to analyze the dynamic change in NDVI and its relationship with precipitation. The results showed the following: (1) the vegetation growing season in Ningxia started late in April and ended in late September. (2) The annual average NDVI value in the vegetation growing season had a linear growth rate of 0.052/a, but it decreased in the last 4 years, indicating the need to strengthen vegetation monitoring and to keep abreast of regional eco-environmental change. (3) In the spatial data, the average NDVI value mainly had an increasing trend. The area with an increasing trend accounted for 91.62% and was mainly distributed in the central arid region and the southern hills and mountains region. The decreasing trend accounted for 8.38% of the area, which was mainly distributed in the northern Yellow River irrigation region and the western part of the central arid region. The spatial change rate decreased from south to north. (4) The correlation coefficient between the average NDVI value and precipitation showed a linear growth trend. The positive and significantly positive correlations were mainly distributed in the southern hills and mountains region, whereas the negative and non-significant negative correlations were mainly distributed in the northern Yellow River irrigation region, especially in Yinchuan and its nearby area.

MODIS NDVI provided a valid data source for eco-environmental monitoring. However, as it only depended on remote sensing data, the mechanism of eco-environmental change could not be explained, especially regarding eco-environmental change under the influence of human activity, which all needs to be combined with field survey data in order to explain the mechanism of eco-environmental change.

ACKNOWLEDGEMENTS

This work was supported by the Science Foundation of the Xi'an University of Science & Technology (2014007), and the Science Foundation of the State Key Laboratory of Soil Erosion and Dryland Farming on the Loess Plateau (A314021402-1616).

REFERENCES

Bureau of Statistics in Ningxia Hui Autonomous Region (BSNHAR). (2014). Statistics Yearbook in 2014. *China Statistic Press*, Beijing China, pp. 1–100. (in Chinese)

Chen, Q., Zhou, Q., Zhang, H. F., *et al.* (2010). Spatial dis parity of NDVI response in vegetation growing season to climate change in the Three-River Headwaters Region. *Ecology and Environmental Sciences.* 19(6), 1284–1289. (in Chinese)

Du, L. T., Song, N. P., Wang, L., *et al.* (2015). Impact of global warming on vegetation activity in Ningxia province from 1982 to 2013. *Journal of Natural Resources.* 30(12), 2095–2106. (in Chinese)

Fensholt, R. & S. R. Proud (2012). Evaluation of earth ob servation based global long term vegetation trends-comparing GIMMS and MODIS global NDVI time series. *Remote. Sens. Environ. 119*, 131–147.

Hao, F., Zhang, X., Ouyang, W., *et al.* (2012). Vegetation NDVI linked to temperature and precipitation in the upper catchments of Yellow River. *Environ. Model. Assess. 17*, 389–398.

Li, Z., Huffman, T., McConkey, B., *et al.* (2013). Monitor ing and modelling spatial and temporal patterns of grassland dynamics using time-series MODIS NDVI with climate and stocking data. *Remote. Sens. Environ. 138*, 232–244.

Wang, Z. J., Qiu, X. H., Tang, Z. H., *et al.* (2013). Analysis on the factors influencing the evolution of desertification in Ningxia of Chin from 1999 to 2009. *Journal of Desert Research. 33(2)*, 325–333. (in Chinese)

Wardlow, B. D., & S. L. Egbert (2008). Large-area crop mapping using time-series MODIS 250 m NDVI data: an assessment for the U.S. Central Great Plain. *Remote. Sens. Environ. 112*, 1096–1116.

Wu, G. H., Wang, N. A., Hu, S. X., *et al.* (2008). Physical geography. *Higher Education Press*, Beijing, China, pp. 50–100. (in Chinese)

Xu, J. H. (2002). Mathematical methods in contemporary geography (Second Edition). *Higher Education Press*, Beijing, China, pp. 37–43. (in Chinese)

Yang, Q., Qin, Z., Li, W., *et al.* (2012). Temporal and spa tial variations of vegetation cover in Hulun Buir grassland of Inner Mongolia, China. *Arid. Land. Res. Manag. 26*, 328–343.

Advances in Energy Science and Equipment Engineering II – Zhou, Patty & Chen (Eds)
© 2017 Taylor & Francis Group, London, ISBN 978-1-138-71798-5

Simulation analysis of photovoltaic inverter under the condition of different solar radiation

M.X. Chen, S.L. Ma, L. Huang & J.W. Wu
School of Automation Science and Electrical Engineering, Beihang University, Beijing, China

P. Wang
Changping Power Supply Company, Beijing, China

ABSTRACT: Grid-connected inverter is considered as an important component in Photovoltaic (PV) systems. In PV grid-connected systems, the major requirement for inverter is to achieve maximum output power from the source with good dynamic performance. This paper presents a control strategy for PV grid-connected inverter under the condition of different solar radiation levels. Two control objectives were set up: first, the DC-link voltage is maintained at a stable operating point. Second, the current loop reference is tracked according to the requirements of the grid-connected current control. A simulation model was built in MATLAB/SIMULINK in order to verify the validity and feasibility of the proposed control strategy.

Keyword: Photovoltaic cells; Double loop cascade control; PV inverter; MPPT

1 INTRODUCTION

The rapid increase in global energy consumption has accelerated the need for greener energy sources. There is a long list of clean energy resource types such as solar, wind, sea tide, geothermal, and biomass. Among them, PV from solar is much preferable due to its implementation simplicity with less maintenance (Borowy B., 1996; Zhou, 2011; ZHANG, 2012). Nowadays, PV systems have witnessed enormous demand for the sake of their enormous potential to be the easiest solution we have right now to substitute our diminishing fossil fuel energy sources (LU, 2009; WANG, 2008). The research of PV grid-connected system can be generally divided into two parts: the first part is the Maximum Power Point Tracking (MPPT) technology of photovoltaic array; the second part is mainly focusing on the control strategy of power electronic converter (WANG, 2010).

2 PHOTOVOLTAIC POWER GENERATION STRUCTURE

2.1 Photovoltaic power generation topological

PV grid-connected systems are one of the main research directions of solar power generation. At present, voltage source input and current source output control methods have been commonly used in PV grid-connected inverter. Its control objective is: the inverter output current with gird is maintained at same frequency and same phase (power factor), and the output voltage and current are required to meet the power quality requirements. The topology of grid—connected inverter is the key part of the inverter, which is related to the efficiency and cost of the inverter. Generally, the topology structure can be roughly divided into two types, single—stage and two—stage model.

Among the possible selections of different topologies for grid connection of distributed energy sources, the single-stage approach is one of the easiest and cost effective topology. Moreover, such systems can be used in a micro-grid as a part of an energy management system as shown in Figure.1.

As shown in figure 1, the DC-link capacitor is connected between the solar cell and the grid-connected inverter. The controller detects

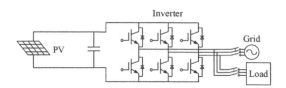

Figure 1. Typical configurations of single-stage PV grid—connected system.

and receives the output voltage and the output current of the photovoltaic cell module, as well as the grid-connected current of the inverter output. By adjusting the amplitude of the grid-connected current and controlling the output power of the PV module, the maximum power point tracking and grid-connected power generation are realized. Without the boost converter in single stage PV system, the transmission of the power is limited.

Two-stage PV system is divided into two parts which are DC/DC boost converter and the DC/AC inverter.

As shown in figure 2, the DC/DC boost converter connected to the PV panels step up the voltage of the DC-link to a proper level for the DC/AC inverter. The three-phase DC/AC inverter produces proper sinusoidal current in the grid with unity power factor based on the MPPT and Phase Locked Loop (PLL) algorithms.

DC/DC converter and DC/AC inverter are independent of each other in the two stage PV system, each with independent local control strategy and objectives. The design and realization of the control link is relatively convenient.

2.2 *PV inverter mathematical model*

Voltage source type grid-connected PV inverter system is one of the most commonly inverter which can be achieved on the AC side of the active and reactive power control. This paper selects three-phase voltage source type inverter as the research object.

Figure 3 shows the main circuit topology of three-phase PV inverter. U_{pv}, defined as the solar

cell battery, is connected with a boost circuit output side, namely PV inverter DC-link side and DC side of the inverter in parallel a large capacitor C, in order to stabilize the DC side voltage. The series connected Ls and Rs that compose the L-type filter attenuate the harmonic injected into the grid generated from the inverter with the SVPWM technique.

When the grid-connected inverter is stable, on the AC side of the inverter voltage and grid voltage running on the same frequency, the inverter steady-state equivalent circuit, without considering the equivalent resistance of the inverter loss and reactors, can be further the equivalent circuit of the inverter and simplified. Hence, there are:

$$\overline{U}_s = j\omega L_s \overline{I}_s + \overline{E}_s \qquad (1)$$

The mathematical model of the inverter in the three-phase static coordinate system is:

$$L\begin{bmatrix} \dfrac{di_{sa}}{dt} \\ \dfrac{di_{sb}}{dt} \\ \dfrac{di_{sc}}{dt} \end{bmatrix} = \begin{bmatrix} u_{sa} \\ u_{sb} \\ u_{sc} \end{bmatrix} - \begin{bmatrix} e_{sa} \\ e_{sb} \\ e_{sc} \end{bmatrix} - R\begin{bmatrix} i_{sa} \\ i_{sb} \\ i_{sc} \end{bmatrix} \qquad (2)$$

2.3 *Control strategy*

The control strategy of grid-connected inverter based on grid voltage orientation is shown in figure 4.

By sampling the PV panel voltage and the current, a proper control algorithm can be set. The DC bus voltage control algorithm gives the injected current reference. Moreover, a PLL

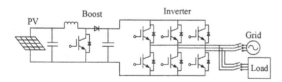

Figure 2. Typical configurations of two-stage PV grid—connected system.

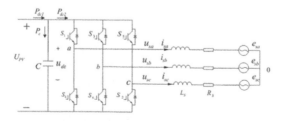

Figure 3. Main circuit of three-phase PV inverter.

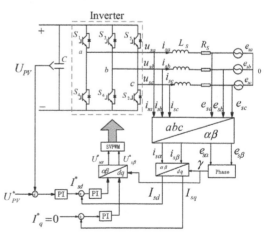

Figure 4. Control strategy of PV inverter.

algorithm synchronizes the injected current with the grid voltage. Wide discussions on PLL algorithms for single-phase inverters can be found in various papers.

Seen from Figure 4, the control loop is a double loop control system, that is, the DC voltage outer loop and the active and reactive current inner loop. In the DC voltage outer loop, the measured values are compared with the given value. The difference by PI adjustment device can realize the DC voltage control with zero steady-state error and DC voltage outer loop is stable and the adjustment of DC voltage. The DC voltage control can be achieved. Therefore, the current reference value of the inner loop of the active current is given by the output of the DC voltage outer loop, so that the control of the output active power of the grid-connected inverter is practical.

3 SIMULATION ANALYSIS

3.1 Simulation parameters

The proposed control strategy for a two-stage PV system is simulated in Matlab/Simulink simulation software platform. Single piece of photovoltaic panels parameters are shown in table 1, with five solar battery monomer series, parallel 20 groups, namely a total of 100 photovoltaic battery veneer form PV array power of the overall system simulation. The simulation parameters are shown in table 1. The simulation model is shown in Figure 5.

3.2 Simulation result

Figure 5 and Figure 6 show the inverter output three-phase current. As shown in Figure 5, inverter on startup after 0.08 s (4 cycles) stepped into steady state output, in 1000 W/m² radiation intensity, current amplitude stable at around 37.5 A; after 0.4 s,

Table 1. Simulation parameters.

Setting	Value
Uoc (V)	221.5
Isc (A)	102
Um (V)	175
Im (A)	99.2
Pm (kW)	17.4
Uoc (V)	300
Isc (A)	10
C (uF)	10
L (mH)	3
Lf (mH)	50
f_grid (Hz)	311
Um (V)	221.5

light intensity of solar radiation became 600 W/m², inverter after 0.04 s (two grid cycle) stepped into the stable state, with output current amplitude being stable at around 11 A; then, as shown in Figure 7, light intensity of radiation at 0.7 s from 600 W/m² suddenly increased to 800 W/m², the inverter output current kept in steady state output after a gird cycle, and the output current amplitude kept stable at 23 A.

As shown in Figure 8, the DC-link voltage was keeping on stable state at 0.14 s, and the amplitude of DC-link voltage was 700 V.

At 0.4 s, with light intensity fluctuating suddenly, the DC-link voltage recovered the 700 V in three grid cycle. At 0.7 s, with radiation intensity

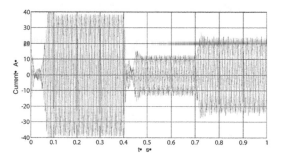

Figure 5. Inverter output three phase current.

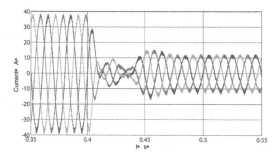

Figure 6. Inverter output three phase current (0.35 s~0.55 s).

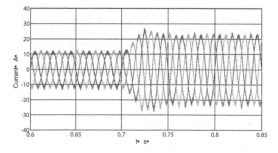

Figure 7. Inverter output three phase current (0.6 s~0.85 s).

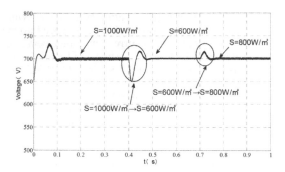

Figure 8.   DC-link voltage.

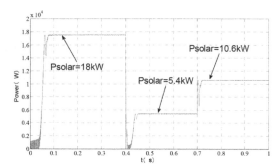

Figure 9.   PV solar output power.

grid-connected inverter is analyzed based on the grid voltage oriented vector control strategy. The inverter control structure diagram is proposed. The proposed control strategy for a two-stage PV system is simulated in Matlab/Simulink. The PV system is simulated under several conditions with different light intensity increasing and decreasing suddenly. The PV inverter output characteristics are discussed in detail.

Under the conditions of solar radiation intensity changing tempestuously and the changing in the output power of the solar cell, the PV inverter can quickly and accurately track the power output.

The DC-link voltage is maintained at 700 V and the fluctuations are within an acceptable range. The output current is consistent with the PV output characteristic curve.

ACKNOWLEDGMENT

This paper is Supported by the National Natural Science Foundation of China (51377007); Supported by the Specialized Research Fund for the Doctoral Program of Higher Education of China (20131102130006).

REFERENCES

Borowy B, Salameh Z. Methodology for the optimally sizing the combination of a battery bank and PV array in a wind/PV hybrid system. IEEE Trans Energy Convers 1996, 11(2): 367.

Haihua Zhou, Tanmoy Bhattacharya, Duong Tran, et al. Composite Energy Storage System Involving Battery and Ultracapacitor With Dynamic Energy Management in Microgrid Applications [J]. IEEE transactions on power electronics, 2011, 26(3): 923–930.

Lu Hongyi, HE Benteng. Application of the supercapacitor in a microgrid [J]. Automation of Electric Power Systems, 2009, 33(22): 89–91.

Wang Chengshan, WANG Shouxiang. Study on some key problems related to distributed generation systems [J]. Automation of Electric Power Systems, 2008, 32(20): 1–4.

Wang Chengshan, YANG Zhangang, WANG Shouxiang, et al. Analysis of structural characteristics and control approaches of experimental microgrid systems [J]. Automation of Electric Power Systems, 2010, 34(1): 99–105.

Zhang Ye, Guo Li, JIA Hongjie, et al. An energy management method of hybrid energy storage system based on smoothing control [J]. Automation of Electric Power Systems, 2012, 36(16): 36–41.

suddenly increasing, the DC-link voltage recovered the 700 V in two grid cycle, which indicated that the DC voltage loop control had strong ability of anti-disturbance.

As shown in Figure 9, output power of the solar cell was 18 kW according to the settings of solar radiation intensity and power output; under the radiation intensity, output power of the solar cell was 5.4 kW; in light intensity, solar cell output power was 10.6 kW.

From the above simulation results, it proved that the inverter control method and solar cell model is correct. Under the condition of light intensity occurring suddenly increasing or decreasing, the inverter can quickly and accurately track the power output. The DC-link voltage is keeping the stability of 700 V at mutation moment to ensure the stability of output power. The inverter output three-phase current under the process of light intensity changing still maintain unity power factor with grid.

4   CONCLUSIONS

The topological structures of grid-connected PV system are given in this paper. The two-stage PV

Advances in Energy Science and Equipment Engineering II – Zhou, Patty & Chen (Eds)
© 2017 Taylor & Francis Group, London, ISBN 978-1-138-71798-5

Understanding Chinese environmental pollution based on the dialectics of nature

Ping Gu & Wei Teng
School of Economics and Management, Jiangsu University of Science and Technology, P.R. China

ABSTRACT: More and more attention has been paid to the problem of environmental pollution by people. The various problems are as follows: how to correctly understand the roots of environmental pollution, how to think about the relationship between people and the environment at the theoretical level, and how to put forward effective ways to protect the environment. These require that we conduct in-depth mining of dialectics of nature. Based on the actual situation of environmental pollution, we research on the theory of dialectics of nature, which is conducive for us to correctly understand environmental pollution and explore ways to solve environmental problems.

1 INTRODUCTION

The environment is the foundation of human survival and development. It is not only directly related to the vital interests of the masses, but also related to the long-term development of the Chinese nation. With the continuous progress of the society, the ability of human beings changing their natural environment is continuing to improve. For the pursuit of immediate economic interests, the pollution that is caused due to use the environment improperly is shocking. In recent years, those environmental pollution incidents (such as haze, excessive levels of lead in blood, toxic campus, etc.) have aroused people's attention. These derived effects from environmental pollution have hystereses, but they are becoming more and more serious as time goes by. When they are perceived, it indicates that the pollution has developed to a very serious degree. These indirect pollution influences the production activities, reduces the quality of life, leads to the rise in the incidence of major diseases, and even leads to premature birth or deformity of the fetuses. At present, Chinese environmental problems have attracted the attention of the government. The eighteenth major report of the Communist Party of China first discusses the 'ecological civilization', and the central leaderships also propose many new ideas, for example 'green mountains and clear rivers are equivalent to wealth', 'promote green development, cycle development, low carbon development', 'structure beautiful China' and 'use strict legal system to protect the ecological environment'. These new ideas show that the Chinese government attaches great importance to environmental pollution. China is a socialist country, and so we should seriously study Engels's dialectics of nature. Reflecting the relationships between economic constructions with environmental pollution and analyzing the deep-seated causes of environmental pollution based on Engels's dialectics of nature are conducive to the harmonious coexistence of man and nature and conducive to construct a socialist modernization country orderly.

2 BASIC SITUATION OF CHINESE ENVIRONMENTAL POLLUTION

At present, environmental pollution mainly includes water pollution, air pollution, soil pollution, and noise pollution.

2.1 *Water pollution*

Water pollution refers that water is involved by some kinds of materials, which causes the chemical properties, physical properties and other aspects properties of water to be changed, thereby leading to the deterioration of water quality and threats to human and ecological environments (Li, 2014). The situation of water pollution in our country is little serious. June 2012 Wu Xiaoqing Vice Minister of Environmental Protection Ministry pointed out that the national surface water is slightly polluted in the mass, and out of the 26 lakes monitored, 53.8% of these lakes are eutrophic. <2013 China Land and Resources Bulletin > shows: in Chinese 4778 groundwater monitoring points, water quality of 43.9% monitored points is little poor, water quality of 15.7% monitored points is quite poor (Deng, 2014). In 2015, the Chinese State Council issued the <Water Pollution Prevention Action Plan>, which enhances control water pollution to the high of 'national water security'. This shows that the pollution capacity of the water environment bearing in our country is close to the limit and water pollution has affected the effective supply of the production and living water and ecosystem balance.

2.2 Air pollution

Air pollution usually refers to the phenomenon of the human activities or natural processes causing discharge of some substances into the atmosphere and therefore endangering human comfort, health, welfare, and environment (Xin, 2013). Soot is the main air contaminant in China. This has relationships with our country that are mainly relying on coal to develop the economy in the past period of time. The concentration of total suspended particles in the air is generally exceeded due to an excessive use of coal, and sulfur dioxide pollution has been at a dangerous level for a long time. In recent years, the automobile exhaust has intensified air pollution, which causes nitrogen oxides pollution to deteriorate continuously (Wang, 2011). In short, the Chinese air pollution situation is worrying.

2.3 Soil pollution

The phenomenon of the contaminants produced by human activities entering the soil and accumulating to a certain degree, thereby leading to the deterioration of soil quality and causing some indexes to exceed the national standard is called soil pollution (Wang, 2012). A variety of pollutants from life and production pollute the soil directly. The blind use of pesticides leads to the decline of soil quality (Zhu, 2010). Some toxic pollutants (such as mercury and chromium) may be enriched into the fruits of the crops due to the enrichment effect, which threatens human health. In the actual survey of 6 million 300 thousand square kilometers of land in our country, the rate at which soil pollution exceeded is 16.1%.

2.4 Noise pollution

Those sounds which disturb people's rest, study, and work are called as noise. When it causes adverse effects on people and the surrounding environment, noise evolves into noise pollution (Zhang, 2010). The creation and use of all kinds of machinery and equipment have brought prosperity and progress for human beings, but it produces more and more noise. The urban residents have been plagued by traffic noise due to an increase in the number of motor vehicles. Machine noise brings trouble to workers and surrounding residents. Constructing noise affects the lives of residents seriously. Social noise makes people less quiet at rest. In addition, noise not only makes people feel annoyed, but also causes damage to hearing, and even leads to a variety of fatal diseases.

3 BRIEF INTRODUCTION OF THE DIALECTICS OF NATURE

Engels summed up the highest achievements of natural science and carried out a philosophical revolution of natural view. He put forward the natural view of the dialectical materialist, revealed the dialectical law of natural development, emphasized the systematic connection of all things, and thought that man is a part of the natural environment (Peng, 2013).

3.1 Nature is in the process of constant movement and development

Engels inspected the previous research results of natural science and thought that the moving material is in a perpetual cycle: Those things that the evolution from the objects of the universe to the planets and the solar system, the historical changes from different layers of the earth to the atmosphere and the evolution from cells to humans show that all things of nature are in constant generation and disappearance.

3.2 The developing law of nature is dialectic

Dialectic is a science of the general laws about movement and development of nature, human society, and thought. The universal law of dialectics is summed up in the practice of nature and human society and it has objectivity. Dialectic provides an explanating method for the transitions in different research areas. And nature is a touchstone which tests the accuracy of dialectic. The things and phenomena in nature are both contradictory and interdependent. This dialectical thinking plays a very important role in viewing the relationship between human beings and environment correctly and solving the environmental problems.

3.3 Relationship between humans and the environment

Dialectics of nature regard man as an integral part of nature and object regarding nature as an enemy. It acknowledges the impact of the environment on human and society development, and also points out the positive effects of human initiative in front of the environment. Nature is the material premise of human beings' existence and development. It is the objective basis of human beings carrying out practical activities. Nature precedes human beings and society. Human beings and human society are the products of nature and human beings will not exist without nature.

4 ANALYSIS AND SOLUTIONS OF ENVIRONMENTAL PROBLEMS BASED ON THE DIALECTICS OF NATURE

4.1 Cause analysis of environmental problems

4.1.1 Understanding root
People make the natural environment provide services for them through labor, science, and technology. But it is restricted by the level of science and technology and the limitation of human knowledge, and so there is always a discrepancy between

the intended purpose and the outcome. Man does not have a correct understanding about the relationship between human beings and nature, thereby leading to exploit and abuse nature and damage of ecological balance.

4.1.2 Class root
Man is a class animal and just takes care of himself. Human beings are egoistic in the face of economic interests. The class nature of human greed leads them only to pursue the immediate economic benefits and high profits regardless of long-term social consequences and the impact on the environment. This is bound to destroy the ecological balance and increasing environmental pollution.

4.1.3 Social root
The capitalists only notice the immediate results when they produce and exchange goods. The environmental pollution that is gradually formed over time is completely ignored. The courses of development in many countries indicate that, in the early stage, the degree of environmental pollution is relatively small. With an increase in the national wealth, environmental pollution will become more and more serious. Our society is in the pursuit of economic benefit maximization and environmental protection is thrown to the winds by the capitalists who want to get more interests.

4.2 Main obstacles of treating environmental pollution

The problem of environmental pollution has become an important issue that cannot be avoided. It has been part of the agenda of many governments. Protecting the environment, controlling environmental pollution, and curbing the deterioration of the ecological environment have become important tasks of the government's social management. For China, protecting the environment is a basic national policy. Solving the nation's outstanding environmental problems is an important and arduous task for the Chinese government. There are three main obstacles that obstructed the governance of environmental pollution in China.

4.2.1 Weak awareness of environmental protection
At present, although people's environmental awareness has improved than before, it has not reached the satisfactory level. In particular, the awareness of participation in environmental protection is still at a low level. In the face of environmental pollution problems, people rely on the government overly; the lack of understanding of environmental law enforcement has brought a lot of pressure on environmental law enforcement.

4.2.2 Insufficient funding for protecting environment
Environmental protection system funding is insufficient and the law enforcement capacity construction costs of grassroots environmental protection agency are greatly affected after the reform of sewage charges. With strengthening of front-line law enforcement efforts of environmental monitoring, pollution, sewage charges, and pollution accident treatment, the requirements of the corresponding environmental funding will be higher. There is a huge contrast between the lack of law enforcement funding and the increasing workload of the protecting environment. In addition, due to the lack of R&D investment, people who deal with pollutants personally lack high-tech equipment to environmental pollution control.

4.2.3 Environmental regulation mechanism is not very strict
China has made great achievements in environmental protection; the situation of environmental pollution has been basically controlled. But also need to develop a more stringent environmental regulation mechanism. The policy of 'combine prevention and treatment, prevention is the main' indicates that the environmental protection policy should not only deal with the consequences, but also take measures to prevent new environmental problems. The loss caused by environmental pollution is controlled to a minimum by preventing the emergence of new pollution sources. Treating the consequences of environmental pollution is only a kind of remedial measures.

4.3 Solutions of environmental problems based on dialectics of nature

4.3.1 Understanding the laws of nature correctly
Grasping the laws of nature correctly helps us to understand the consequences of our own behaviors. We should not treat the environment as the invaders rule the different people. Man belongs to nature and exists in it and all dominant power that human beings own is to understand and use the laws of nature correctly. We should follow the laws of nature and protect the natural environment when we are in the pursuit of a better life.

4.3.2 Regulate the relationship between human beings and the natural environment
To regulate the relationship between human beings and the natural environment is the premise of achieving harmony between man and nature. Therefore, we need to re-understand and re-evaluate the relationship between human beings and the environment. Engels pointed out that we together with all that we have belong to

nature; human beings and nature are an indivisible whole, and human beings are the products of the long-term development of nature. We must criticize the idea of the opposition of human beings and nature, and human beings must be placed in nature. In the activities of reforming nature, we should not only pay attention to the immediate interests, but also be concerned about the long-term potential impact[3].

4.3.3 Exploitation must be eliminated in order to reduce the emission of pollutants

Reflecting on the development of the present industry, the capitalists only care about their own interests and exploitage the fragile environment. And in the case of blind pursuit of interests, a vicious circle of pollution treatment after first pollution has been formed[11]. Therefore, in order to eliminate the vicious circle, we must eliminate the contradictions of the modern industry. We cannot pursue the direct economic benefits and achieve industrialization at the cost of pollution.

4.3.4 Reform the social system and mode of production

The relationship between human beings and the natural environment is the foundation of the social system and mode of production. And the social system and mode of production reflect the relationship between human beings and the natural environment. Therefore, it is necessary to adjust the relationship between man and man, man and the natural environment, and also to carry out a comprehensive reform of the existing mode of production and the whole social system, which is connected with the mode of production. When people consciously grasp and use the laws of nature, people can have the ability to control and regulate the social impact of production activities consciously in order to ultimately achieve the reconciliation of human and nature, as well as the reconciliation of human beings itself.

4.3.5 Adhere to the concepts of scientific development and sustainable development

The United Nations Environment and Development Committee put forward the concept of sustainable development (Wang, 2012). Its essence is 'not only to meet the needs of the contemporary people, but also not to pose a hazard for future generations meeting their needs'. The Chinese government put forward the scientific development concept. It emphasizes people-oriented development and to promote the overall development of the economic society and people. It emphasizes that people use natural resources in a sustainable way, while pursuing development. The concepts of scientific development and sustainable development are conducive to improving the quality of human environment ethics and building the system of ecological civilization.

5 SUMMARY

In the situation of increasingly serious environmental pollution, re-thinking the causes of environmental problems and finding solutions become global tasks. This also makes dialectics of nature attract more and more attention. On one hand, its theory is comprehensive and constructive to make it become a theoretical building for the development of modern environmental philosophy; on the other hand, dialectics of nature provide a basic way to be familiar with the environment and realize the inner unity between man and nature. Reviewing dialectics of nature, relating dialectics of nature with policy and theory of the Chinese society, and exploring the contents and methods of the modern value of dialectics of nature are conducive to adjust the relationship between human beings and nature, thereby realizing the harmonious development of human beings and the natural environment.

REFERENCES

Deng shan. The characteristics of environmental pollution and the relevant legal principles [J]. *File*, 2014 (7):455. (in Chinese)

Li haoyang. Prevention and control measures of water pollution[J]. *Applied Chemical Industry*, 2014 (4): 729–742. (in Chinese)

Peng xiulan. The philosophical implication of Marx's view of ecological civilization [J]. *Journal of Tianzhong*, 2013 (6):28–31. (in Chinese)

Wang he. Study on the harm and control measures of urban air pollution[J]. *Heilongjiang Science and Technology Information*, 2011 (26):69. (in Chinese)

Wang yanjuan. The risk and pollution control of soil pollution [J]. *Private Science and Technology*, 2012 (10):178. (in Chinese)

Xin jia. The harm of air pollution and control measures [J]. *Science and Technology Entrepreneurs*, 2013 (17):166. (in Chinese)

Zhang xuejun, Wu aimin. The relationship between environment and development in the perspective of Marx's view of nature [J]. *Front*, 2010 (17):152–154. (in Chinese)

Zhu xiaoyu. Study on the content and distribution of copper in soil around the power plant [J]. *Environmental Science and Management*, 2010 (12):71–74. (in Chinese)

Advances in Energy Science and Equipment Engineering II – Zhou, Patty & Chen (Eds)
© 2017 Taylor & Francis Group, London, ISBN 978-1-138-71798-5

Suppression of spontaneous combustion based on "three zones" division and anti-fire control of a large-scale coal mining space

Z.H. Lu, Y.Z. Zhang & P.F. Shan
Energy School, Xi'an University of Science and Technology, Xi'an, China
Shenhua-NingXia Coal Ltd., Yin chuan, China

C.H. Zhao & X.J. Yue
Shenhua-NingXia Coal Ltd., Yin chuan, China

ABSTRACT: A "three-soft" coal seam has the characteristics of a long construction path, highly loose degrees of coal and a large amount of residual coal, which are very prone to spontaneous combustion during the period of withdrawal. The key points of fire prevention are able to reduce the temperature of the goaf and cut off the air leakage area effectively. In this paper, the research object is the spontaneous combustion problem of the 110205 retracement working face. After synthetically analyzing the cause for spontaneous combustion and locating the key area of the fire zone, we formatted a comprehensive fire suppression program including burying and backfilling the surface fractures, constructing the network of pressure balance ventilation, and injecting liquid N_2 and CO_2 into the goaf. These measures had created a perfect fire suppression effect including that the temperature reduced from 38°C to 18°C and the concentration of CO reduced from 0.012% to 0.005% at the retracement space. The problem of fire around the working face was controlled effectively.

1 OVERVIEW OF THE RETREATING WORKING FACE

The hosting geological condition of the Qingshuiying coal mine is very complex. Rock joints and fractures' spatial distribution are the main features of a typical "three soft" coal seam. The roof of the coal seam is composed of thick sandstone with apparent characteristics of strong permeability, good water conductivity, ease of weathering; the floor of the coal seam is composed of siltstone with characteristics of low intensity, poor stability, ease of expansion, and turning to mud when encountering water. The coal seam of the 110205 retracement working face is the secondary coal seam with a thickness of 4.8 m. And it belongs to the group I easily-spontaneous combustion coal seam category. The combustion period is 42 d. The slant length of retracement working face is 296 m and its height is 3.4 m. Due to the problems such as broken roof lithology and ease of cementation of floor when encountering water, in addition, it reserves the top coal and the bottom coal layers with a thickness of 0.6 m and 0.8 m to control integrity of the two parts to help retracement more safely. As the retracement passage needs a long time to finish and the looseness of coal rock

is very high, when the working face recovering to the position end supports, the phenomena of spontaneous combustion and toxic gases overflow appear. In this situation, an open flame starts appearing at the 98# hydraulic support segment, the CO concentration reaches 0.012% at the 141# hydraulic support segment, and the local goaf temperature reaches up to 38°C.

2 COMPREHENSIVE FIRE PREVENTION AND EXTINGUISHING COUNTERMEASURES

In view of the adverse factors of firefighting in the "three soft" retracement working face, the "floor-first and roof-afterward", "system-first and link-afterward", and "overall-first and partial-afterward" measures were formulated, which include filling the crack surface, isolating the leakage channel, adjusting the ventilation system, implementing average pressure ventilation, injecting liquid N_2 and CO_2 to lower the goaf temperature and oxygen concentration, and injecting polymer gel at the local fire source to embrace and isolate the fire source (Yu. 2000, Wang et al. 2013).

2.1 Division of goaf spontaneous combustion "three-zones"

In a comprehensive firefighting process, to suppress the activeness of spontaneous combustion of the goaf, it should be based on the "three zones" width to reasonably determine the space, time, and injection volume of liquid gas to ensure that the oxidation spontaneous combustion zone is always covered by the inert gases. At the end of the exploitation in the working face, set a sampling point every 50 m, 100 m, and 150 m in the return airway and inlet wind roadway respectively; during the construction period of the retracement passage, set sampling points every 60 m, 90 m, 120 m in the return airway and inlet wind roadway respectively, and analyze the oxygen concentration through the monitoring system. Take 10% ~ 18% oxygen contents as the basis of the division of the oxidation zone and use the coal spontaneous combustion test parameters and the actual underground to judge the necessary conditions for spontaneous combustion of the goaf:

$$(h > h_{min}) \cap (c > c_{min}) \cap (Q < Q_{max}) \qquad (1)$$

where h_{min} is the minimum float coal thickness of the goaf, m; c_{min} is the minimum oxygen concentration of the goaf, %; and Q_{max} is the maximum leakage intensity of the goaf, m³/s;

By substituting the result (equation 1) in Figure 1, with the result of oxygen concentration monitoring and the 'three tape' division of regional comprehensive measurement, the width of the heat dissipation area is determined to be 40 m ~ 60 m, the width of the oxidation temperature zone is determined to be 130 m ~ 200 m, and the width of the asphyxia zone is determined to be 200 m far from the working place.

2.2 Injection of liquid N_2 and CO_2

Liquid N_2 and CO_2 are able to absorb the heat spilling out of the retracement, thereby reducing the fire temperature and oxygen content. As the intensity of N_2 is similar to that of air, fire zone elevation has little effect on it, and it is good for the inertness, temperature declination and isolation of the retracement space; the intensity of CO_2 is relatively high which can sink into the bottom of the goaf rapidly after the gasification and diffusion in the fire zone (Song 2014). In order to assure the fire extinguishment's effectiveness, continuous injection of liquid N_2 and CO_2 is performed and it has a good control of the fire zone temperature; meanwhile, the oxygen content declines to minimize coal's spontaneous combustion.

The following equation is used to calculate the volume of the sealed fire zone:

$$V = V_1 + V_2 = LhL_1 + LHL_2\lambda \qquad (2)$$

where V is the total volume of the closed fire zone, m³; V_1 is the closed volume of the working face, m³; V_2 is the volume of the closed space of the mined-out area, m³; h is the return channel height, which is taken as 3.4 m; L_1 is the withdrawal channel width, which is taken as 3.4 m; L is the withdrawal channel length, which is taken as 296 m; H is the coal seam thickness, which is taken as 4.8 m; L_1 is the width of the oxidation temperate zone, which is taken as 200 m; λ is the rock loose coefficient of the mined-out area, which is taken as 0.8.

The equation to calculate the inert gas injection volume to the fire zone is as follows:

$$Q = K\frac{V(C_1 - C_2)}{C_2} \qquad (3)$$

where Q is the total amount of inert gas injected into the fire zone, m³; K is the reserve coefficient of the injected inert gas, which is taken as 1.6; C_1 is the average oxygen concentration of the fire zone before injecting inert gas, which is taken as 14%; and C_2 is the critical oxygen concentration of fire extinguishing, which is taken as 5%.

7.38×10^5 m³ inert gas should be injected to complete the extinguishment through theoretical calculation, by laying the nitrogen pipeline on the front position of brackets at the working face and drilling 15 holes at the top plate of the support tail beam. Each hole's diameter is 50 mm, depth is 4.3 m and angle of elevation is 35°. Liquid N_2 and CO_2 are continuously injected through these holes. The no. 1 liquid nitrogen tube was connected to the liquid nitrogen injection drilling holes 170#, 163#, and 158# at the rear side of the brackets, respectively; the no. 2 liquid nitrogen tube was connected to the liquid nitrogen injection drilling holes 107#, 104#, 103#, 98#, and 95# at the rear side of the brackets, respectively; the no. 3 liquid nitrogen

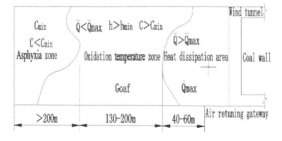

Figure 1. Schematic of the division of goaf spontaneous combustion "three zones".

tube was connected to the liquid nitrogen injection drilling holes 126#, 116#, 113#, and 131# at the rear side of the brackets, respectively; and the no. 4 liquid nitrogen tube was connected to the liquid nitrogen injection drilling holes 151#, 150#, and 140# at the rear side of the brackets, respectively. From 15th February to 4th March, 768.3 t nitrogen was injected cumulatively. From 27th February to 4th March, 548.8 t liquid CO_2 was injected cumulatively; the amount of liquid converting into volume was almost 8.89×10^5 m^3. Figure 2 shows the pipeline distribution of liquid N_2 and CO_2 injection. The liquid N_2 (CO_2) injection path is as follows: ground tankers → unloading hole → +1065 m horizon inset → air way nitrogen infusion pipeline (DN50) → high pressure hose → return-air way closed outside the reserve line → goaf.

2.3 *Average pressure ventilation*

The average pressure fireproof technology was adopted to adjust the ventilation system to alter the pressure distribution and reduce the pressure difference of the air leakage area, which contributes to the air leakage blocking, air and wind were controlled to prevent fire (Zhang & Li 2014). According to the ventilation barrier law:

$$M = \sqrt{\frac{\Delta h}{R}} \qquad (4)$$

where M is the amount of air leakage; Δh is the pressure difference between the start and the end of the leakage air path, Pa; R is the air resistance in the leakage air way, kg/m^3. From equation (4), it can be seen that if the closed resistance of the goaf gets infinite, the air leakage M approaches 0, which means that it approaches 0; it can also make the fire zone reach a balance status by reducing the inside and outside difference of the sealed mined-out area, the sandbag wall as well as the wall pillars.

1. Construct ventilation facilities by building nine permanent and three temporary airtight as well as four wind doors and one wind blocking wall on the main ventilation line. The specific positions are shown in Figure 3.
2. Set up a pressure regulating fan and windshield. Set up 2*75 kW and 2*30 kW pressure regulating fans at the +1065 m horizontal inset and transport downhill outer damper respectively of the working face. At the same time, set up regulation windscreens at the +786 m level shaft station of the total return air side to form an average pressure system, as shown in Figure 3.

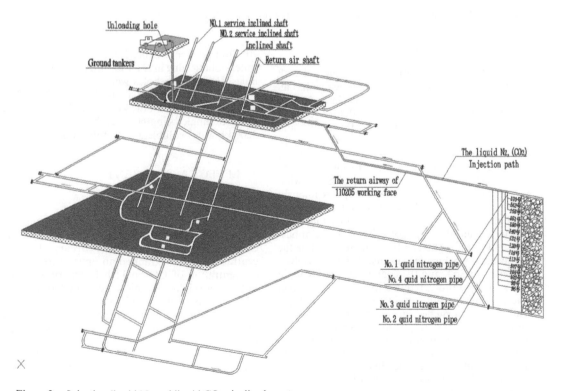

Figure 2.   Injection liquid N_2 and liquid CO_2 pipeline layout.

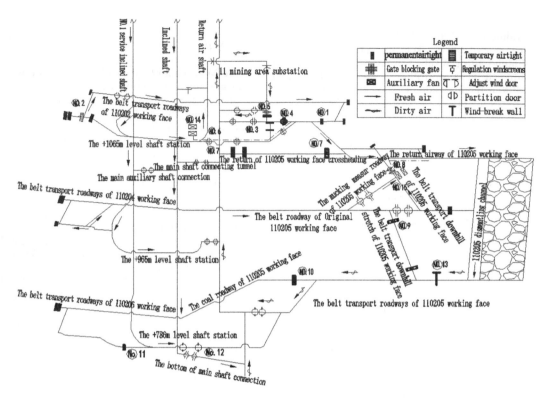

Figure 3. Chart of a working face, pressure-balanced ventilation system.

2.4 *Topical gel injection*

Gel is a kind of new extinguishing material with good permeable and sealing features, which has an obvious effect when applying it on limited space and extinguishes the fire topically. Before gelation, the liquidity of gel exhibits good flowability and its diffusion permeability is also good. Usually, 1 m³ gel can permeate into 4 ~ 5 m³ thick loose coal layers and combine with it firmly; this kind of property has an effect on plugging the wind as well as cutting off oxygen. Except extinguishing the fire directly, the gel can isolate the goaf through its stacking interaction and permeability, which shortens the length of spontaneous combustion and reduces the combustion space. After the retracement working face is unsealed, we dissolved the base Na_2SiO_3 and the coagulant NH_4HCO_3 in water in the ratio 2:1 respectively in the distinct where the temperature and CO density were high. And then, we switched on two bumps at the same time, delivered the proportioning solution to the mixer, and placed it in the borehole. The gelation time was controlled to be 5 min~10 min. In addition, we drilled seven holes between the 83# and 90# hydraulic support, with each holes' diameter being 15 mm, elevation angles being 35°, 60°, and

70°, and the depth being 2.3 m; and then, 1160 kg water-soluble glue in total was injected through the holes. Besides, ten holes were drilled between the stands 92#, 93#, 100#, 108#, 121#, 134#, 135#, 137#, and 141#, with each hole's diameter being 25 mm, elevation angle being 60°, depth being 2.3 m, and 3600 kg water-soluble glue in total was injected.

3 EFFECT ASSESSMENT

3.1 *Concentration of the CO gas*

During the fire prevention and extinguishing period, we arranged for experts to monitor gas concentration. In 4th ~ 10th February, the concentration of CO exhibited an upward trend. The concentration of CO on 11th February was seriously overweight, of which the CO concentration of the end part of the 112# stent segment rose up to 0.012% and its upper corner rose up to 0.0096%; on 15th February, after implementing the average pressure ventilation, the overall concentration of the iconic gas in the working face sealed upper part was higher than that in the lower part. This indicated that the relative position of the fire zone still tended to the working face's upper part and the fire zone did not narrow obviously; however, after injecting 159 t liquid nitrogen on 17th February, the CO concentration decreased rapidly and maintained in 0.0003% ~ 0.007%. With an increase in the amount of liquid N_2 and CO_2 being injected, there was an overall decline in the CO concentration. The change curves are shown in Figure 4, with the concentrations of CO at working face, upper corner, and the return air decreasing from 0.012%, 0.0096%, and 0.0085% to 0.0005%, 0.0004%, and 0.0003%, respectively.

3.2 Temperature changes

In the mid-shift on 12th February, the overall temperature of the working face exhibited a growing trend, among which the end parts of the 156# stent segment rose up to 36°C and the temperature of the return air was up to 22°C. Since the injection

of liquid N_2 and CO_2 on 16th February, the overall temperature in the goaf decreased obviously, which showed that gasified liquid N_2 and CO_2 made the fire zone easy to control. The working face was unsealed on 6th March, and its temperature was rebounded, which implied that the effectiveness of glue injection on isolating the partial fire zone was obvious, and the temperature of the working face remained constant at 16°C ~ 18°C. As shown in Figure 5, the temperature of the working face, upper corner and return air decreased from 21°C, 22°C, and 20°C to 18°C, 14°C, and 16°C, respectively.

4 CONCLUSION

1. In view of the fire characteristics of the retracement space, the measures of floor crack landfills and passage air isolation; closing underground construction and implementing average pressure ventilation; and liquid N_2, CO_2, and polymer gel injection on the partial fire source were practiced to tackle the conditions of spontaneous combustion in the retracement space, which extinguish the fire.
2. Through the comprehensive fire prevention and extinguishing countermeasures, the temperature of the fire zone decreased from 38 °C to 18 °C, and the CO concentration decreased from 0.012% to 0.0005%.

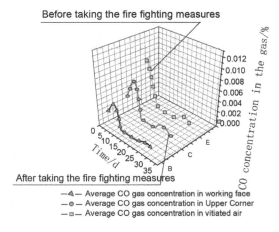

Figure 4. Curves showing changes in concentrations of CO.

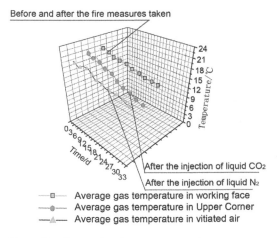

Figure 5. Curves showing temperature change.

REFERENCES

Song YM (2014). Research on fire prevention and extinguishment by carbon dioxide in goaf [J]. China Coal. 40(4):106–108.

Wang XB, Kang XR, Xu YH (2013). Application of fire prevention and extinguishment by liquid nitrogen during recovery of working face with great mining height in Yangchangwan Coal Mine [J]. China Coal. 39(3):102–104.

Wu YG (2011). Distribution law of gas and change rule of "three zones" in the goaf of Mechanized top-coal caving working face under the continuous nitrogen injection [J]. Journal of Chian Coal Society. 36(6):964–967.

Yu MG (2000). The latest development and application of coal mine fire prevention technology in China [J]. Mining Safety & Environmental Protection, 27(1): 21–23.

Zhang XH, Li W (2014). Study on Comprehesive Control Technology of Large Area Firing Zone in Above Seam Mining Goaf of Bulianta Mine [J]. Coal engineering. 46(2):52–54.

Advances in Energy Science and Equipment Engineering II – Zhou, Patty & Chen (Eds)
© 2017 Taylor & Francis Group, London, ISBN 978-1-138-71798-5

Flow field analysis of the transmission line surface in wind and sand environments based on Fluent

Guangru Hua & Wenhao Li
Department of Mechanical Engineering, North China Electric Power University, Baoding, China

ABSTRACT: The surface condition of the transmission line impacts the corona characteristics to a great degree. Research on the surface corrosion of the LGJ-400/50 Aluminium Conductor Steel Reinforced is of significance. An indoor closed wind tunnel system is utilized to simulate the actual wind and sand experiment. We established the test section and 3D model of the wire. The software Fluent is adopted to simulate the sand erosion law of wires in four different angles. Results indicate that the positive pressure in the windward side of the wire is maximal, the negative pressure in the area which has an angle of 45° with the leeward surface is maximal, the speed and direction of wind around the wire both change frequently, and the speed in the triangle area formed by the leeward side is zero, which is related to the distribution and vorticity of the flow field. Meanwhile, a Kaman vortex street is formed in the leeward side, which causes a reverse impact towards the lee side surface, and this provides a reference basis for determining the data collection position of the physical simulation experiment.

The LGJ-400/50 Aluminum Conductor Steel Reinforced is the major transmission line of the northwest 750 kv power transmission project (Hao et al. 2013, Yue et al. 2009).The wire surface condition deteriorates due to the sand's consistent impact, which will cause corona discharge (Hao et al. 2011, Tomotaka et al. 1988, Walker et al. 2000). Therefore, the research of the law on how sand affects the surface corrosion characteristics of the LGJ-400/50 Aluminum Conductor Steel Reinforced is of significance, which will improve the transmission capacity of the transmission line. Therefore, in this paper, the erosion of the overhead line under a sand environment is simulated. As it is related to the wire flow fields and the wire erosion in the sand environment, the software Fluent is used to simulate the wire flow field in four angles.

1 MODEL ESTABLISHMENT

1.1 Simulation system of the wind tunnel and the suspension of the wire

The experiment adopts an axial flow fan, cooperating with the expansion period, guide section, contraction section and test section, which form a closed cycle of a wind tunnel in a laboratory.

A 0.6 m LGJ-400/50 wire is taken as the experimental sample. There are four wire samples in order to simulate four different attack angles. As shown in Figure 1, the four wire samples are suspended evenly in an air duct of which the length, width, and height are respectively 10 m, 1.5 m and 1.1 m, and they are suspended at a height of 42 cm above the ground. Only the mid-section of the wire's longitudinal section is selected and the suspension diagram is shown in Figure 2.

1.2 Model building and ANSYS meshing

A wind tunnel test section can be equal to a rectangular duct, which is opened at both ends.

Figure 1. Picture of the wire suspension.

Wires are suspended evenly on the middle plane of the test section. Wires are idealized to a same-diameter cylinder. The test section and the wire's 3D finite-element model are shown in Figure 3.

With regard to ANSYS meshing, a tetrahedral grid is adopted, and "Fine" is chosen to refine

Figure 2. Schematic of the wire suspension angle.

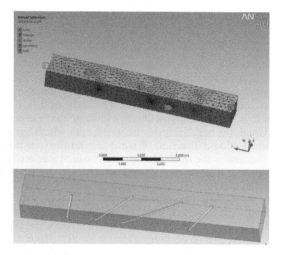

Figure 3. Schematic of a 3D model of the test section and Ansys meshing.

the grid; the superlative degree "High" is chosen as the smoothness degree. The grid is shown in Figure 3 and the density of the grid is high in the wire end and in the other complex stress gradient section.

1.3 Setting of model parameters

A standard κ-ε model is selected as the turbulence model. As sand erodes the wire, it is necessary to use the DPM model. Sand particles are defined as discrete phase incident particles, and air is defined as a continuous phase; the density of the discrete phase is 2650 kg/m³; the incident speed along the direction of the X axis is 21.23 m/s; the size of the particle 0.25 mm; the mass flow-rate is 0.12 kg/s; the hydraulic diameter is 1.2, which is the interaction region between the continuous phase and discrete phase. Inlet velocity and discrete phase velocity are both 21.23 m/s; the outlet is the pressure outlet; the inner boundary is the boundary of the fluid to be studied; the wall is the wire; the standard initialization method is chosen, and the iteration steps are set to 200. The residual curve is convergent to 1e–03.

2 ANALYSIS OF SIMULATED RESULTS

2.1 Analysis of the variation law of the wall pressure of the wire

A pressure nephogram is shown in Figure 4, which shows the sand erosion process. It can be seen that, in the direction of four attack angles, the pressure of the wire changes regularly along the circumferential direction. The pressure in the first half period of the test is positive. The pressure in the second half is negative, which can reach up to 640 Pa.

The pressure nephogram of a 90° attack angle is shown in Figure 5. It can be seen that the positive pressure on the upwind side is maximum. As the red zone shows, the maximum static pressure value is 201 Pa. As the green zone shows, the leeward side is influenced by a negative pressure which is equal to 219 Pa. The pressure is distributed symmetrically along the cross section and the maximum negative pressure appears in the region, which has an angle of 45° with the leeward side instead of the leeward side, and the maximum negative pressure is 387 Pa.

At a 60° attack angle, the pressure on the wire's cross section is shown in Figure 6, from which it can be seen that the pressure is distributed symmetrically along the circumferential direction. The positive pressure in the upwind side is the maximum, and the negative pressure in the region

Figure 4. Pressure nephogram.

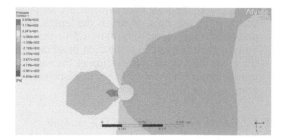

Figure 5. Pressure nephogram of a 90° attack angle.

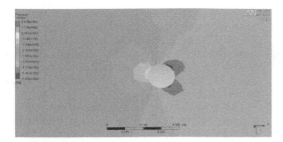

Figure 6. Pressure nephogram of a 60° attack angle.

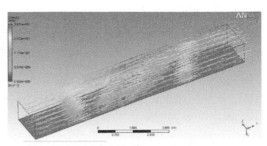

Figure 7. Global distribution of the velocity vector.

which has a 45° angle with the leeward side is the maximum.

2.2 *Analysis of the velocity variation around the wire*

Figure 7 shows a velocity vector diagram based on the finite element model. In the case that the inlet wind speed is 21.23 m/s, the speed of the wind around the wire can reach up to 25.7 m/s, and the maximum speed can reach up to 34.31 m/s as the red region indicates. Around the wire, the velocity direction changes frequently, and so the vector is denser. Taking the 90° attack angle as an example, the amplified velocity vector is shown in Figure 8. As the deep blue region shows, the speed in the triangle area formed in the leeward side is zero.

Figure 9 shows the distribution of the tangential velocity vector around the wire and the length of the arrow line represents the size of the vector. It can be seen that the tangential velocity is distributed uniformly in the case of a 90° attack angle; in the cases of 45° and 30°, the outer tangential velocity is greater than the inner tangential velocity; in the case of 60°, the tangential velocity has the maximum value, as the red area indicates. The above condition is related to the distribution of the flow field.

Figure 8. Diagram of the velocity vector around the wire of an attack angle of 90°.

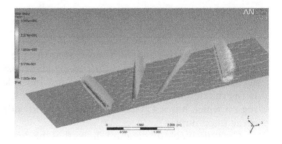

Figure 9. Distribution of tangential velocity.

2.3 *Analysis of the wire's flow field and vorticity's change rules*

The vorticity and flow field of the wire are shown in Figure 10. There is a red vorticity belt close to the leeward of wire which is long and narrow. The vorticity belt leads to air convolution, and further leads to a reverse impact on the surface of the leeward side. The process of reverse impact can be observed in Figure 11. Two symmetric vortexes are formed on the surface of leeward side, which leads to a back-flow impact.

The existence of a Karman vortex street is validated according to its forming condition. This experiment adopted the cylinder's Reynolds number Re to validate. If 47 < Re < 105, the

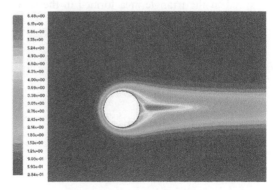

Figure 10. Cloud diagram of vorticity.

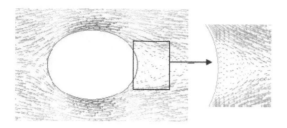

Figure 11. Vector diagram of a continuous phase flow field.

forming condition is satisfied. The equation of Re is given as follows:

$$R_e = \frac{\rho v l}{\mu} \qquad (1)$$

where ρ is the fluid density, v is the fluid velocity, l is the characteristic length, and μ is the viscosity coefficient.

The result shows that Re = 4200, the forming condition of the Karman vortex street is satisfied. When combined with the results of the above analysis, it provides the basis for determining the data collection position of physical simulation experiments.

3 CONCLUSIONS

The pressure cloud diagram shows that the pressure is distributed symmetrically along the cross section. The positive pressure in the upwind side is the maximum, and the negative pressure in the region which has a 45° angle with the leeward side is the maximum.

The velocity cloud diagram shows that the direction of velocity around the wire changes frequently, and the velocity vector is relatively dense; the wind speed around the wire reaches up to 25.7 m/s. The speed in the triangle area formed in the leeward side is zero. The tangential velocity is distributed uniformly in the case of a 90° attack angle; in the case of 45° and 30°, the outer tangential velocity is larger than the inner tangential velocity; in the case of 60°, the tangential velocity has the maximum value. The above condition is related to the distribution of the flow field.

The flow field vector diagram shows that the vorticity belt leads to air convolution, and further leads to a reverse impact on the surface of the leeward side. Two symmetric vortexes are formed on the surface of the leeward side, which leads to a back-flow impact. A Karman vortex street is formed on the leeward side, which provides the basis for determining the data collection position of physical simulation experiments.

ACKNOWLEDGMENTS

This work was financially supported by the Beijing Natural Science Foundation (2132038).

REFERENCES

Hao, P. & Wang, B. & Li, G. & Du, K.F. & Zeng, D.J & Zhang, X. 2013. Structural design of stiffened shells based on imperfection sensitivity analysis [J]. *Chinese Journal of Applied Mechanics* 30(3): 350–355.

Hao, Y.H. & Xing, Y.M. & Zhao, Y.R. & Yan, L. 2011. Erosion mechanism and evaluation method of steel structure coating eroded under sandstorm environment [J]. *Journal of Building Materials* 14(3): 345–350.

Tomotaka, S. & Yoshiyyuki, H. 1988. Aging effect of conductor surface conditions on corona characteristics [J]. *TEEE Transactions on Power Delivery* 3(4): 1903–1912.

Walker, C.I. & Bodkin, G.C. 2000. Empirical wear relationships for slurry pump part I side–liners [J]. *Wear* 242: 140–146.

Yue, G.W. & Lin, H.X. & Chang, X. 2009. *Research on scientific problems of sand-dust storm [M]*. Shandong: Shandong people's publishing house.

Advances in Energy Science and Equipment Engineering II – Zhou, Patty & Chen (Eds)
© *2017 Taylor & Francis Group, London, ISBN 978-1-138-71798-5*

An analysis of the domestic tourists' consumption structure and the behavioral characteristics in tourism resources destination—based on a domestic tourism sampling survey in Huangshan city in 2015

Yurong He & Jianhong Lu
Tourism College, Huangshan University, Anhui, China

ABSTRACT: According to the relative statistic data in Huangshan city in 2015, in the article, the consumption structure, tourism motivation, and the behavior characteristics of tourists are analyzed. The results show that the domestic tourist consumption structure is unreasonable. The overall level of tourism consumption is low, the proportion of basic tourism consumption is high, the tourists' stay time is short, etc. Based on this, suggestions on fostering new tourism formats, building public information systems, and accelerating the self-drive and self-service facilities to optimize the tourism consumption structure in Huangshan city are put forward.

1 INTRODUCTION

A tourism consumption structure is an indicator of the tourism development level of a country or region, and also an important index of the development level of the tourism group economy. It directly reflects the overall development of the local tourism industry (Wang, 2005). Huangshan city is the core of the international cultural tourism demonstration area in the southern Anhui province. There are unique tourism resources and cultural tourism resources in Huangshan city. It is a typical resource-based tourism destination. By the end of 2015, there were 55 natural A-level scenic areas, including 5 5-A-level scenic spots, 22 4-A-level scenic spots, 12 3-A-level scenic spots, and 18 2-A-level scenic spots in Huangshan city. The tourism industry is an important economy pillar and motive industry, which can completely improve and promote the development of the regional economy, culture and ecosystem. An accurate measurement of the development level of the tourism industry is necessary. According to the relevant data of tourist consumption behavior and structure characteristics of a sample survey made in the city of Huangshan, in this paper, the domestic tourism consumption structure and the characteristics of tourist consumption behavior are studied, in order to promote the transformation and upgrading of the Huangshan city tourism industry and to provide countermeasures and suggestions for the government.

2 METHODS

2.1 *Data and index*

Data are obtained from the sample survey of the domestic tourism statistics in Huangshan city in 2015. On the basis of the study of the tourism consumer behavior index, in this paper, 30 detailed effective data of the following three factors are studied: the demographic characteristics of tourists, tourist behavior characteristics, and tourism consumption formation. There are 2000 questionnaire samples and 1914 of the samples are valid.

2.2 *Analysis method*

On the basis of the sampling survey statistics, in this paper, Microsoft Excel is used to analyze the survey data and make the corresponding charts. Single factor analysis is used to inspect the difference between related samples and to analyze the significance by judging when the value of P is less than 0.05.

3 RESULTS

3.1 *The characteristics of tourists consumption structure*

The tourism consumption structure is the proportional relationship of the consumptions between the various types of tourism products and relevant tourism products. Tourism consumption includes food, accommodation, transportation, sightseeing, shopping, and entertainment. Accommodation, food, transportation, and sightseeing are known as the basic tourism consumptions. Tourism shopping, entertainment, and communication belong to the non-basic tourism consumptions and the non-basic tourism consumption is of great flexibility. According to relevant statistics, tourist shopping consumption accounts for over 60% in developed countries and regions. In the domestic tourism-developed areas, the tourist shopping consumption has accounted for about 40%. Domestic tourists have spent 1565 Yuan on average in Huangshan city in 2015. The basic tourism consumption ratio is up to 69.68%, and the non-basic tourism consumption is 30.32%. In the non-basic tourism consumption, the shopping proportion accounted for only 16.18%, which is far below the levels of developed countries and districts. As shown in Figure 1, the entertainment proportion accounted for 5.75%, communication proportion accounted for 4.11%, and other proportions accounted for 4.29%.

3.2 Analysis of the behavior characteristics influencing the tourist consumption structure

Tourism consumption behavior refers to the behavior and related activities of tourists when they gather information to make decisions and purchase, enjoy, assess, and deal with the products (Zhou, 2010). There are many factors which influence tourist consumption behavior and consumption structure. In this paper, the domestic tourism consumption of Huangshan city is analyzed from the perspective of demographic characteristics of tourists, tourism motivation, tourism information acquisition methods, organizational form of tourism, and tourist preference.

3.2.1 The tourist's demographic characteristics
From the demographic characteristics of tourists, in the investigated domestic tourists, males account for 54.03%, which is slightly more than that of females (45.97%). Tourists are mainly middle-aged. The young-aged tourists who are between 25 and 44 years of age account for a large proportion of 71.84%. From the level of education, tourists of college degrees account for 82.71%. Data prove that domestic tourists who have higher education levels and the middle-income group whose revenues are between 2500 and 4999 Yuan per month are the main customers of Huangshan city.

3.2.2 Obtaining travel information
Data prove that an average of 50.11% of domestic tourists who travel to Huangshan city obtain travel information through the Internet to learn about Huangshan city, 23.03% through the introduction of friends or relatives, and 12.61% tourists are advised by a travel agency, as shown in Figure 2. Tourists' consumption behavior was greatly influenced by the information. Internet network media represented by the Internet and mobile phones have become the mainstream channels for tourists to obtain travel information of Huangshan city.

3.2.3 Modes of travelling
With the analysis of ways of travelling, the proportion of self-driving tour and independent travel is 39.68%, and the proportion of the travel agency organization declines to only 18.43%, as shown in Figure 3. Data prove that self-service and travel self-driving tour have become mainstream forms of tourism to the domestic tourists who came to Huangshan city in 2015.

3.2.4 Tourists' purposes
Tourists' purposes are generally divided into sightseeing, leisure vacation, business activities, visiting friends and relatives, taking part in science and technology meetings, and religious worship. Tourists' purposes are the subjective condition of traveling, which determines the action of tourists and the choice of tourism content. Tourists' purposes of consumers are the key factors for the tourism enterprises to know the needs of the consumers and to design tourism

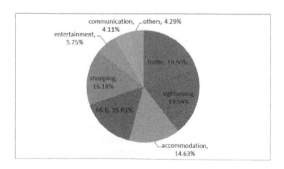

Figure 1.   Pie-chart of the consumption structure.

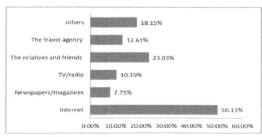

Figure 2.   Chart on sources of travel information.

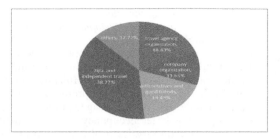

Figure 3. Chart showing modes of travelling.

Figure 5. Chart showing tourists' preferences.

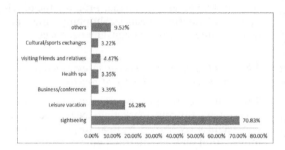

Figure 4. Chart showing tourists' purposes.

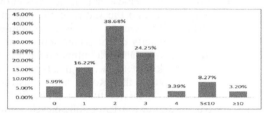

Figure 6. Graph showing tourists' stay.

3.2.6 *Tourists' stay time*

The length of tourists' stay time is a comprehensive reflection of the local tourism development status. The length of time provides space for the consumption of food, accommodation, transportation, traveling, shopping, and entertainment. Travelling budgets and different travelling purposes have a significant impact on tourists' stay time (Guo, 2011). From the consumption characteristics, the tourism consumption level and tourists' stay time has a significant positive correlation. The average stay time of tourists is 2.59 days in 2015, as shown in Figure 6. The average data prove that the tourists' stay time is short and the results also point out the main purpose of sightseeing tourism is directly related to the length.

products and projects to meet the demand of the target market. As shown in Figure 4, in 2015, tourists travelled for sightseeing (70.83%), leisure vacation (16.28%), business/conference (3.39%), health treatment (3.35%), visiting friends and relatives (4.47%), and cultural and sports exchanges (3.22%). Data prove that sightseeing is still the main purpose of the domestic tourists visiting Huangshan city.

3.2.5 *Tourists' preferences*

Tourists' preference refers to a kind of psychological tendency of tourists when choosing a target of tourist attractions. Due to the different education backgrounds and the quality of the tourists, tourists' preferences are different. Understanding tourists' travelling preferences and satisfying them is the key for tourism destination to attract tourists. Scenic spots with high resources endowment, such as the Huangshan scenic spot, XiDi, HongCun, and Tunxi Old Street attract more tourists. 75.60% of tourists visited the Huangshan scenic area, 73.57% of domestic tourists visited the Tunxi Old Street, and 56.55% of the domestic tourists visited HongCun. As shown in Figure 5, it shows that there is an imbalance of the "hot and cold" phenomenon in the process of tourism development in Huangshan city, and the integration and coordination of tourism resources are particularly important.

4 CONCLUSIONS

The above analysis of statistic results shows that Huangshan city is a resource-based tourism destination. The development of the tourism industry is still in a lower level. The tourism measures have not been integrated well. The consumption structure is unreasonable. The length of tourists' stay time is short. Self-driving tour has become a mainstream form of tourism. Sightseeing is still the main purpose of tourism. Accordingly, the following measures should be taken to improve the tourism consumption structure and to promote the development of the tourism industry.

4.1 Speeding up the construction of facilities of self-driving tour and independent travel to implement "swim fast brigade slow"

The rapid growth of the self-driving tour culture brings serious traffic problems to Huangshan city and scenic spots, which would also influence the efficiency and quality of tourism. Convenient transportation service systems should be built to meet the demand of self-driving tour and independent travel and achieve "swim fast brigade slow". The convenient tourism transportation service system includes five parts. One part is to set up signs of the scenic spot capacity at the major intersections of the scenic area to let tourists know the passenger flow and the status of the parking lot in advance. The second part is to announce the maximum capacity of the scenic spot in time. The third part is to establish a multistorey parking system in the parking-tense scenic area (spot). The fourth is to strengthen the "Road" camp construction and perfect its service function. The fifth is to construct a digital tourism public bicycle rental system like Hangzhou to increase the tourists' travel experience.

4.2 Perfecting tourism public information service system

It is necessary to improve the tourism public information service system and use information technology to promote and guide tourism consumption. Wise tourism services need to be provided for tourists to collect information, book tickets and hotels, and make travel arrangements through network, mobile applications and other tools, which can guide tourists to make scientific planning by avoiding hot spots, and to improve the quality of tour. The scenic spots need to change its traditional role to perfect the audio guide system and provide convenient service for self-driving travel and independent travel tourists.

4.3 Cultivating multiple forms of tourism operation to extend tourists stay

4.3.1 Realizing tourism industry transformation and upgrading through resources integration

In order to realize the integration of resources and create the new project, tourism enterprises and scenic spots must communicate effectively, Thereby building a comprehensive cultural tourism industry chain which can combine sightseeing, leisure, vacation, shopping, business, conferences, food, culture and health services with each other to lengthen the tourists' stay time and to drive the development of related industries in Huangshan city.

4.3.2 Strengthening the tourism industry and Hui culture fusion and developing a wide variety of night tourism products

Tourism enterprises and scenic spots can develop many kinds of night tourism products depending on the characteristics of ancient Huizhou culture at night. By developing tourism products at night, it can reach the goal of extending the time of travel and improving hotel occupancy around tourist attractions and the city. It can also attract the local residents to consume. Creating the tourism service industry chain in the night can bring economic income to the scenic spots and tourist attractions' surrounding areas, and further promote the development of catering, business, shopping, entertainment, leisure, and fitness consumer services.

4.4 Exploring diverse tourism propaganda forms to increase the publicity of tourism

The influence of information technology should be used on tourism marketing. New media and new technology should be used to provide timely information on tourism, travel guides, road traffic, weather forecast, etc. via WeChat and tourism festival activities in Huangshan city. It can also increase publicity of Huangshan city tourism brand and its influence. On the other hand, artificial consulting services are necessary in this Information Age. More consulting services in different languages should be offered at bus stations, railway stations, and airports. More tourism materials, such as brochures and roadmaps, should be provided freely to the tourists.

REFERENCES

Guo Jinhui & Guo Weifeng (2011). Tourists' Lengths of Stay from Survival Analysis: A Case of Mountain Wuyi Destination during National Day, J. Yibin University. 112,26–28.

Li BeiBei & Zheng Guozhang (2013). A Demonstrative Analysis of Consumptive Changes in Anhui Inbound Tourists. J. Shanxi, Normal University Natural Science Edition. 27, 112–117.

Wang Yuan & Huang Zhenfang (2005). An Analysis on Domestic Tourists Consumption Structure and the Related Factors of Behavior—Taking Nanjing City As an Example. J. Nanjing Normal University (Natural-Science). l28,123–126.

Wang Zhongfu & Wang Erda (2009). Cluster Analysis of the Domestic Tourism Consumption Pattern in Dalian City Tourism Destination, J. Dalian University of Technology (Social Sciences). 30,68–72.

Zhou Wenli & Li Shiping (2010). An Empirical Analysis on Consumption Structure of Urban and Rural Residents' Domestic Tours: Based on ELES Theory. J. Tourism Science. 24,29–37.

Advances in Energy Science and Equipment Engineering II – Zhou, Patty & Chen (Eds)
© 2017 Taylor & Francis Group, London, ISBN 978-1-138-71798-5

An analysis of the power grid development level based on cluster analysis

Jun Li, Dong Peng & Hui Li
State Power Economic Research Institute, China State Grid Corp, Beijing, China

Jianfeng Shi
Department of Economic Management, North China Electric Power University, Baoding, China

ABSTRACT: There is a strong correlation between the development level of the power grid and economic development. In order to provide better service to economic development, it is necessary to study the development level of the power grid. In this paper, cluster analysis was firstly adopted to explore the development level of power grids in twelve countries and analyze development characteristics of power grids in various countries. And then, twelve countries were divided into six categories according to the development characteristics of the power grid and the results of cluster analysis. Finally, some policy suggestions are proposed for power grids of China based on the characteristics of power grid development in developed countries.

1 INTRODUCTION

Since the reform and opening up, the power grid of China has entered into a rapid development stage. The total power generation of China reached 46037 KWh in 2011, which surpasses the United States and ranks first in the world (Liu 2012). China has ranked first in the world in the following three indicators by 2013: total electricity generation, the total electricity consumption, and total installed capacity. However, due to the huge population, China is positioned in a significantly lower level than the developed countries in terms of per capita power generation, per capita electricity consumption, and per capita installed capacity. The significant gap between China and the developed countries indicates that there is still much room for development of China's power grid in the future.

A power grid is the backbone of national economic development, and the development of the power grid provides services for economic development. After 1990s, the development of the world power grid gradually slowed down in the overall per capita electricity consumption (Hong 2013). Moreover, the development of the power grid in different countries also presents a huge difference. The developed countries are still clearly ahead of the developing countries in many indicators. Generally speaking, the development of the power grid needs to exceed the development of economy to a certain extent to provide a better service to economic development. In order to provide safe and economical electricity to guarantee economic development, it is necessary to explore the development characteristics and define the development level of the power grid in the developed and developing countries, and thereby provide policy suggestions for the development of China's power grid based on the experience of developed countries.

There are a few studies that have studied the development of power grids and some achievements have been obtained. Lawton F (1962) described the construction process of the power industry of the Soviet Union. Fumio A (2007) narrated the development process of the Japanese grid in a historical perspective. However, the two studies mentioned above did not depict and study the characteristics of power grid development. Nam M H (2007) studied the composition of the Korean power grid, and illustrated the importance of power construction in the development of the whole national economy. Mukhopadhyay S et al. (2006) considered that the electric power reform will play a positive role in promoting the development of India's electric power grid. Kukde P K (2004) stated the current situation of the Japanese power grid after the reform and listed some features of the power grid, but deeper research on the development trend of the power grid was not made. Based on the research of the development process of the British power grid, Elders I M et al. (2005) made some predictions about the development of the power grid. In addition, Hu Liexiang et al. (2011) used the logistic regression model to explore the development of the power grid and divided development of the power grid into four stages. In a word, the existing studies are mainly focused on the longitudinal study of a certain country and therefore, a study that focuses on the comparative analysis of development characteristics among different countries is needed.

Aiming at the gaps existing in the research on the development level of the power grid, in this paper cluster analysis is adopted to explore the development level of the power grid in twelve countries and analyze development characteristics of power grids in various countries, and then the development level is assessed by comparison with other countries. Finally, some policy suggestions are proposed for the power grid of China in the light of the characteristics of power grid development in developed countries.

2 METHODS

2.1 Selection of relevant countries

In this paper, twelve countries are selected as cluster analysis samples. Some developed countries, such as United States (USA), Britain (UK), France (FRA), Germany (DEU), Italy (ITA), Japan (JPN), Australia (AUS), and South Korea (KOR), as the overall level of power grid is higher than China (CHN) are included in the sample, three developing countries, such as Brazil (BRA), India (IND) and Russia (RUS), as the overall level of the power grid is similar to China are also contained in the sample. There are gaps in the development level of the power grid among developed countries, which is favorable to formulate periodical targets for the development of China's power grid. A comparison between development levels of power grids in developing countries is helpful to define the common characteristics of the power grid in developing countries and find the development level of China in developing countries.

2.2 Data sources

In this article, per capita GDP, per capita electricity generation, per capita electricity consumption, per capita installed capacity, transmission and distribution losses as well as the fossil fuel power generation ratio of the total generating capacity were selected as feature quantities to measure the development level of power grids in the twelve countries. The relevant data are obtained from the World Bank (WB), International Energy Agency (IEA) and the US Department of Energy (EIA). In order to eliminate the impact caused by population

difference, per capita GDP, per capita electricity consumption, per capita electricity consumption, and per capita installed capacity are selected as the characteristic quantities to weigh the development level of the power grid. To exclude the price factor, GDP per capita is calculated in constant 2005 dollars. In order to remove differences in the total amount, the transmission and distribution losses are presented by the proportion of its share of total electricity generation.

2.3 System clustering

System cluster analysis is one of the most widely used clustering methods in various research fields. Firstly, system clustering regards every sample as a class respectively. Secondly, system clustering calculates the distance between classes and selects a pair of samples with minimum distance in all samples to merge into a new class. Thirdly, system clustering calculates the distance between the new class and other classes and merges a pair of samples with minimum distance into a new class. Finally, the third step operation will be repeated until all samples are combined into one class (Fan & Guan 2008). System clustering adopts the sample distance to measure the similarity degree between samples. A Euclidean distance calculation method is used to calculate the sample distance in this paper, and the calculation method is as follows:

$$d(x_i, x_j) = \left[\sum_{k=1}^{p}(x_{ik} - x_{jk})^2\right]^{\frac{1}{2}} \quad (1)$$

where x_i represents the ith sample; x_{ik} represents the value of x_i in the kth index; p is the number of indexes, and $d(x_i, x_j)$ is the distance between two samples.

If $d_{ij} = d(x_i, x_j)$ and $D = (d_{ij})_{p \times p}$, a distance matrix can be formed as follows:

$$D = \begin{bmatrix} 0 & d_{12} & \cdots & d_{1p} \\ d_{21} & 0 & \cdots & d_{2p} \\ \vdots & \vdots & \vdots & \vdots \\ d_{p1} & d_{p2} & \cdots & 0 \end{bmatrix} \quad (2)$$

where $d_{ij} = d_{ji}$.

In this paper, we used system clustering of SPSS to handle the twelve samples. Through standardized processing, the sample data are placed in [0 1]. The Euclidean distance is used to calculate the distance between samples, and the distance between one class and another class is calculated by the method of class average.

3 RESULTS AND ANALYSIS

A total of twelve samples' data were processed by SPSS and the results showed that the there was no missing sample and all twelve samples were valid.

The proximity matrix derived from Euclidean distance between the various samples is shown in Table 1. From the table, it can be observed that Japan and Germany have the smallest Euclidean distance of $d_{3,10} = 0.287$ and India and the United States have the largest Euclidean distance of $d_{1,7} = 2.170$. The distance between developed countries and developing countries is greater than the distance between developed countries and developing countries themselves. Moreover, the distance between developed countries and developing countries themselves is different. United States and Italy have the maximum distance of 1.08 in developed countries. Russia and India have the largest distance of 1.037 in developing countries. However, the latter is slightly smaller than the former. When compared with other developed countries, United States has the largest distance with developing countries, through which it can be considered that United States has the highest development level of the power grid. When compared with other developing countries, India has the largest distance with developed countries, through which can be considered that India has the lowest development level of the power grid.

The distance between China and developed countries is lower than that of Brazil and India, but higher than that of Russia, and it can be considered that the power grid development level of China is higher than that of India and Brazil, but lower than that of Russia. In addition, taking into account the minimum distance between Italy and developing countries, it can be regarded that the power grid development level of Italy is the lowest in the developed countries category. Because of the minimum distance between Russia and developed countries, it is considered that the power grid development level of Russia is the highest in the developing countries category.

The clustering process of the sample is shown in Table 2. The coefficients in the table are expressed by distance. When the clustering objects are samples, the Euclidean distance between samples is calculated as a coefficient; otherwise, the distance between classes calculated by using a class average method is taken as the coefficient. Based on the principle of Euclidean distance between cluster objects from small to large, system clustering is performed. After the 11-times system clustering, all the samples are finally clustered into one class. Because of the smallest coefficient, Japan and Germany are merged into class 1 in the first clustering. Britain and Italy are merged into class 2 in the second clustering. Class 1 and class 2 are merged into class 3 in the third clustering, that is class 3 is composed of Japan, Germany, Britain, and Italy. In the fourth clustering, class 3 and United States are amalgamated into class 4. In the fifth clustering, China and Russia are amalgamated into class 5, which proves that the power grid development level of China is the closest to Russia. In the sixth clustering, class 3 and South Korea are amalgamated into class 5. Class 4 and class 6 are merged into class 7 in the seventh clustering. India and Brazil are merged into class 8 in the eighth clustering. All of the developed countries are amalgamated into class 9 in the ninth clustering. All of the developing countries are merged into class 10 in the tenth clustering. All of countries are amalgamated into class 11 in the last clustering. The minimum clustering coefficient was 0.287, which appeared in the first clustering; the maximum clustering coefficient was 1.281, which appeared in the last clustering.

Table 1. Proximity matrix.

Case	Euclidean distance											
	USA	UK	JPN	FRA	KOR	CHN	IND	BRA	RUS	DEU	AUS	ITA
1: USA	0.000	1.052	0.717	1.070	0.750	1.627	2.170	1.963	1.271	0.780	0.457	1.080
2: UK	1.052	0.000	0.483	0.796	0.738	0.900	1.352	1.218	0.793	0.411	0.776	0.334
3: JPN	0.717	0.483	0.000	0.930	0.463	1.022	1.639	1.553	0.840	0.287	0.354	0.495
4: FRA	1.070	0.796	0.930	0.000	0.920	1.281	1.750	1.216	1.018	0.676	1.079	0.768
5: KOR	0.750	0.738	0.463	0.920	0.000	0.955	1.670	1.522	0.717	0.525	0.496	0.697
6: CHN	1.627	0.900	1.022	1.281	0.955	0.000	0.939	1.028	0.543	1.029	1.257	0.754
7: IND	2.170	1.352	1.639	1.750	1.670	0.939	0.000	0.818	1.037	1.653	1.830	1.279
8: BRA	1.963	1.218	1.553	1.216	1.522	1.028	0.818	0.000	0.941	1.437	1.753	1.113
9: RUS	1.271	0.793	0.840	1.018	0.717	0.543	1.037	0.941	0.000	0.854	0.968	0.603
10: DEU	0.780	0.411	0.287	0.676	0.525	1.029	1.653	1.437	0.854	0.000	0.559	0.417
11: AUS	0.457	0.776	0.354	1.079	0.496	1.257	1.830	1.753	0.968	0.559	0.000	0.779
12: ITA	1.080	0.334	0.495	0.768	0.697	0.754	1.279	1.113	0.603	0.417	0.779	0.000

Table 2. Agglomeration schedule.

Stage	Cluster combined Cluster 1	Cluster 2	Coefficients	Next stage
1	3	10	0.287	3
2	2	12	0.334	3
3	2	3	0.451	6
4	1	11	0.457	7
5	6	9	0.543	10
6	2	5	0.606	7
7	1	2	0.734	9
8	7	8	0.818	10
9	1	4	0.891	11
10	6	7	0.986	11
11	1	6	1.281	0

Table 3. List of class numbers and clustering coefficients.

Class	1	2	3	4	5	6
Coefficient	1.28	0.99	0.89	0.82	0.73	0.61

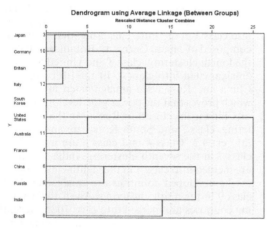

Figure 1. Schematic of the cluster tree.

levels of United States and Australia. As far as the developing countries are concerned, the power grid development levels of China and Russia are similar, and so are those of India and Brazil.

From the cluster tree, the power grid development level of twelve countries can be clustered into different classes and corresponding to different clustering coefficients. The specific class number and clustering coefficient are shown in Table 3.

In the light of clustering coefficients, the power grid development level of the twelve countries can be clustered into six classes, and the corresponding clustering coefficient is 0.61. The specific classification is as follows: Japan, Germany, Britain, Italy, and South Korea are taken as the first class; United States and Australia are taken as the second class; France is taken as the third class; China and Russia are taken as the fourth class; and Brazil and India are taken as the fifth and sixth categories, respectively. As far as the development level of the power grid is concerned, the second class of countries are the highest, followed by the third class of countries, the first class of countries, the fourth class of countries, the fifth class of countries, and the sixth class of countries is the lowest. It can be concluded that there is a strong correlation between the development level of the power grid and economic development, and the development level of the power grid is also higher in the developed countries.

From Fig. 1, we can see that the power grid development level of France is prominently unique and the features of power grid development in France are significantly different from other developed countries and developing countries; it could be attributed to the fact that the fossil fuel power generation ratio of the total generating capacity is much smaller than that of other countries. Because of the particularity of the development level, the French power grid can be separately classified as a class. The maximum distance between developing countries reached 0.986; correspondingly, the largest distance between developed countries is 0.891. Generally, the differences between developing countries are larger than that of developed countries. As far as the developed countries are concerned, the power grid development levels of Japan and Germany are similar, the power grid development levels of Britain and Italy are similar, and so are the power grid development

4 DISCUSSION AND SUGGESTIONS

In this paper, a system cluster method was adopted to analyze the development level of the selected twelve countries and divided the power grid development level of countries into different levels in the light of development characteristics of power grid in various countries.

From the above results, it can be concluded that there is a big gap between China and developed countries in the power grid development level, even less than Russia. In terms of per capita GDP, per capita electricity generation, per capita electricity consumption and per capita installed capacity, China is significantly smaller than developed countries. However, in terms of transmission and distribution losses as well as the fossil fuel power generation ratio of the total generating capacity, the difference between China and developed countries is much small, even ahead of some

82

developed countries. From the above analysis, it can be concluded that the low level of power grid development of China is determined by the low level of economic development and vast populations. The requirement of economic development on power grid construction should be considered in the development of the power grid; power grid construction should be ahead of economic development to some extent, so as to provide a better service to economy development.

ACKNOWLEDGMENTS

This study was financially supported by the National Grid Science and Technology Project of China State Grid Corp (Project Name: Research on the Method and Application of Reasonable Investment of Power Grid Considering the Difference of Development Needs) and the Key Projects of Basic Scientific Research Business of Central University (Project No. 2014ZD21).

REFERENCES

Elders, I. M., Ault, G. W., Galloway, S., & McDonald, J. R. (2005). Identification of long-term scenarios of electricity network development. In *2005 International Conference on Future Power Systems* (p. 6). IEEE.

Fan, Z. J., & Guan. W. (2008). A partition method of financial control in differentiated areas—Based on the application of system cluster analysis method. *Management World*, (4), 36–47.

Fumio, A. (2007). History of power systems development in Japan. In *Electric Power, 2007 IEEE Conference on the History of* (pp. 1–9). IEEE.

Hong, L. (2013). The stage research of power grid development and Its Enlightenment. *Doctoral dissertation, Zhejiang University*.

Hu, L. X., Xu, Q., Zhang, R., X., Zhao, R. X., Gan, D. Q., & Wu, H. (2011). Stage theory of power grid development. *Zhejiang Electric Power*, 30(12), 9–11.

Kukde, P. K., Sathe, D. A., & Kulkarni, S. V. (2004). Accelerated power development and reform programme in India. In *Power Engineering Society General Meeting, 2004. IEEE* (pp. 2346–2352). IEEE.

Lawton, F. L. (1962). Power Development in the USSR. Power Apparatus and Systems, Part III. Transactions of the American Institute of Electrical Engineers, 81(3), 385–397.

Liu, Z.Y. (2012). China Electric Power and Energy. *Energy comments, (5)*, 60–61.

Mukhopadhyay, S., Dube, S. K., & Soonee, S. K. (2006). Development of power market in India. *Power Engineering Society General Meeting* (Vol. 42, pp. 151–156). IEEE.

Nam, M. H. (2007). Early history of Korean electric light and power development. In *Electric Power, 2007 IEEE Conference on the History of* (pp. 192–200). IEEE.

Advances in Energy Science and Equipment Engineering II – Zhou, Patty & Chen (Eds)
© *2017 Taylor & Francis Group, London, ISBN 978-1-138-71798-5*

A study on farmers' behavior game evolution of the scale pig breeding ecological energy system

Bi-Bin Leng
School of Economics and Management, Jiangxi Science and Technology Normal University, Nanchang, Jiangxi, China
Center for Central China Economic Development Research, Nanchang University, Nanchang, Jiangxi, China

Qiao Hu
Center for Central China Economic Development Research, Nanchang University, Nanchang, Jiangxi, China

Wen-Bo Zhang & Jia-Ling Liu
School of Economics and Management, Jiangxi Science and Technology Normal University, Nanchang, Jiangxi, China

ABSTRACT: Large-scale pig breeding has been developed rapidly and its scale has also been enlarged in recent years, with which a new pollution problem occurs. Based on the scale pig breeding and the household bio-gas development system which has been constructed, we made use of evolutionary game mode to analyze farmers' behaviors of the scale pig breeding and household bio-gas development system. And then, we obtained farmers' behavior evolutionary path of the scale pig breeding and household bio-gas, and we found that they were influenced by income, investment, distribution of benefits, and the relative benefits. As a result, we obtained some relevant policy recommendations.

1 INTRODUCTION

As is known to all, most areas of China have suffered badly from haze for over 2 years, which warns us to protect the environment, as soon as possible. Against that backdrop, the 18th CPC National Congress put forward the proposal "Promotion of ecological civilization". The Central Economic Working Conference held in 9th. Dec. 2014, made it clear that the environment's bearing capacity has reached the limit at present. And so, we must conform to public's expectation for a pleasant ecological environment and promote a new way to develop green low-carbon economy. It shows the arduous and urgent nature of environmental protection.

In terms of studies on the pig breeding scale waste disposal, a large number of research scholars adhere to the principle that make the pollutants harmless and reduce quantity. With a view to the bio-gas development of pig manure, fertilizer effect of pig manure and so on, they put forward a mode which combines bio-gas resources development with the breeding cycle.

However, currently, with the continuous development of scale cultivation, breeding enterprises cannot dispose and utilize the vast pig manure brought about by the growing scale of pig breeding entirely. In 2013, we proposed and built the scale

pig breeding and household bio-gas resources cooperation development system. Through previous research, we found that the scale pig breeding ecological energy system is a mode of energy conservation and emission reduction. What is more, it is in accord with the requirement of the national strategy of energy conservation and emissions reduction. According to the scale pig breeding and household bio-gas resources development system study, we combine farmer's behavior theory with evolutionary game theory, through extensive investigation, and then we analyze and discuss the evolution process of farmers' behaviors in the cooperation development system. As a result, we concluded that the relevant policies promoted the cooperative development system to work stably.

2 FARMERS' BEHAVIOR EVOLUTION IN THE COOPERATIVE DEVELOPMENT SYSTEM

2.1 *Evolutionary game model about farmers' participation in the cooperative development system*

Let us consider the whole farmers' group taking part in the cooperation system as two sub-groups.

Table 1. Game matrix.

Farmers X	Farmers Y	
	Develop	No development
Participation	$b_1+(Q-C)\,\lambda_1+\beta(\lambda_1-\lambda_2)\,Q$, $b_2+(Q-C)\,\lambda_1+\beta(\lambda_1-\lambda_2)\,Q$	$b_1-C\,\lambda_1$, b_2
Non-participation	b_1, $b_2-C\,\lambda_2$	b_1, b_2

Owing to farmers have differences in resources, ability, and organizational status in this system, we call those relative feeble farmers as disadvantaged peasant household, which is denoted by X. On the contrary, and those of the advantage peasant household, were denoted by Y. Both X and Y have same strategy choices, which are "participation" or "non-participation". If both sides chose "non-participation", X and Y only get normal earnings, which are recorded as b_1 and b_2, respectively. In contrast, the two sides' earning can be recorded as $b_1+(Q-C)\lambda_1$ and $b_2+(Q-C)\lambda_2$, respectively where, Q is the farmers' total revenue obtained from the cooperation development system through participating in it, C ($C<Q$) is the farmers' total input in this system, λ_1 and λ_2 are quotient coefficients of groups X and Y for total revenue and total input, $\lambda_1+\lambda_2=1$, $0<\lambda_1\leq\lambda_2$.

When group X selects "participation" strategy, and group Y selects the "non-participation" strategy, the result is that the investment $C\lambda_1$ made by X for the cooperation is irrecoverable; when group Y selects "participation", and group X selects the "non-participation" strategy, the investment $C\lambda_2$ made by Y for the cooperation is irrecoverable. β ($\beta>0$, much smaller than C and Q) is the correlation coefficient of the relative gains and absolute gains. Finally, we obtained the matrix about farmers' participation and cooperation in the exploitation system.

2.2 Evolutionary game of farmers

We can assume that the ratio of the individual who chooses "participation" is P_x in group X, and so the ratio of "non-participation" is $1-P_x$. For the same reason, P_y is the proportion of individuals who select "participation" in group Y, and the proportion of choosing "non-participation" is $1-P_y$ in group Y.

In conclusion, we can obtain the expected revenue for strategies of "participation" and "non-participation" in group X, which is named as E_{x1}, E_{x2}. Finally, we can figure out the average profit E_x:

$$E_{x1}=[b_1+(Q-C)\lambda_1+\beta(\lambda_1-\lambda_2)Q]P_y \\ +(b_1-C\lambda_1)(1-P_y) \quad (1)$$

$$E_{x2}=b_1P_y+b_1(1-P_y) \quad (2)$$

$$\overline{E}_x=E_{x1}P_x+E_{x2}(1-P_x) \quad (3)$$

The same procedure can be easily adapted to group Y, which is named as E_{y1}, E_{y2}. And then, the average profit $\overline{E}_y$ is given as follows:

$$E_{y1}=[b_2+(Q-C)\,\lambda_2+\beta(\lambda_2-\lambda_1)\,Q]\,P_x \\ +(b_2-C\lambda_2)(1-P_x) \quad (4)$$

$$E_{y2}=b_2P_x+b_2(1-P_x) \quad (5)$$

$$\overline{E}_y=E_{y1}P_y+E_{y2}(1-P_y) \quad (6)$$

The replicator dynamics equation of farmers' group X participating in the cooperation development system in the proportion P_x is given as follows:

$$\frac{\partial P_x}{\partial t}=P_x(E_{x1}-\overline{E}_x) \\ =\{[b_1+(Q-C)\,\lambda_1+\beta(\lambda_1-\lambda_2)\,Q]\,P_y \\ +(b_1-C\lambda_1)(1-P_y)-E_{x1}P_x+E_{x2}(1-P_x)\}P_x \quad (7)$$

The replicator dynamics equation of farmers group Y participating in the cooperation development system in the proportion P_y is given as follows:

$$\frac{\partial P_y}{\partial t}=P_y(E_{y1}-\overline{E}_y) \\ =\{[b_2+(Q-C)\,\lambda_2+\beta(\lambda_2-\lambda_1)Q]P_x \\ +(b_2-C\lambda_2)-E_{y1}P_y+E_{y2}(1-P_y)\}P_y \quad (8)$$

Order $\dfrac{\partial P_x}{\partial t}=0$

The solution that can make the group X and group Y evolve into a stable state is as follows:

$E_1(0,0)\ E_2(1,0)\ E_3(0,1)\ E_4(1,1)$,

$$E_5\left(\frac{C\lambda_2}{Q\lambda_2+\beta Q(\lambda_2-\lambda_1)},\frac{C\lambda_1}{Q\lambda_1+\beta Q(\lambda_1-\lambda_2)}\right).$$

According to Lyapunov theorem of first approximation, if the Jacobian matrix eigenvalue of the replicator dynamics equations is negative, the equilibrium point is stable; if the eigenvalue is positive, then the equilibrium point will be unstable; if the determinant value of the corresponding matrix is negative, the equilibrium point is the saddle point of the evolutionary game. The calculation results are that the equilibrium point $E_5(0, 0)$ is asymptotically stable, $(1, 0)$ not stable, $(0, 1)$ not stable, and $(1, 1)$ gradually stable, the equilibrium point $E_5(\dfrac{C\lambda_2}{Q\lambda_2 + \beta Q(\lambda_2 - \lambda_1)}, \dfrac{C\lambda_1}{Q\lambda_1 + \beta Q(\lambda_1 - \lambda_2)})$ is a saddle point. In a plane coordinate system, E_1, E_2, E_3, and E_4 form evolutionary game areas, which are divided by saddle point $E_5(\dfrac{C\lambda_2}{Q\lambda_2 + \beta Q(\lambda_2 - \lambda_1)},$ $\dfrac{C\lambda_1}{Q\lambda_1 + \beta Q(\lambda_1 - \lambda_2)})$ into four parts; we mark $E_1E_5E_3$ as ①, mark $E_3E_5E_4$ as ②, mark $E_4E_5E_2$ as ③, mark $E_1E_5E_2$ as ④, and the dynamic evolution game eventually converging to the equilibrium is mainly determined by initial state and saddle points. If the initial state of the mode in areas ② and ③, it will become a stable state to participate in the cooperative development system eventually through the dynamic evolution game. If the initial game position starts at areas ① and ④, then the final game will be stable to maintain a state of not participating in the development system.

Therefore, enlarging the areas ② and ③ and decreasing the areas ① and ④ can make evolutionary game finally equalize within a (participation, participation) status. However, the four divided parts of the area depend on the saddle points $E_5(\dfrac{C\lambda_2}{Q\lambda_2 + \beta Q(\lambda_2 - \lambda_1)}, \dfrac{C\lambda_1}{Q\lambda_1 + \beta Q(\lambda_1 - \lambda_2)})$. If E_5 moves towards the lower left corner, it would be more possible for an evolutionary game to reach a gradual stable state.

3 ANALYSES ON THE EVOLUTIONARY PATH OF FARMERS' PARTICIPATION IN THE COOPERATIVE DEVELOPMENT SYSTEM

If Q increases when compared to the initial state, the x- and y-values of the saddle point E_5 would both reduce, and the saddle point would be more and more close to the origin. Accordingly, the "non-participation" region $E_1E_2E_5E_3$ is less than the initial state, the corresponding "participation" area $E_2E_5E_3E_4$ will expand, the probability of system coverage to E_4 will increase. "Participation" is a strategy to make evolution stable. Therefore, keeping other conditions constant, with an increase in the value of Q, it will stimulate the enthusiasm of farmers to take part in the cooperative system.

When $Q \to \infty$, the saddle point E_5 will get close to point $E_1(0, 0)$, the entire system evolution space has become a "participation" region, and all farmers would participate in the cooperative development system.

If C is bigger than its initial state, the x- and y-values of saddle point E_5 would both increase, the saddle point E_5 would remain close to $E_4(1, 1)$. Meanwhile, the "non-participation" region $E_1E_2E_5E_3$ is bigger than the initial state, the corresponding "participation" area $E_2E_5E_3E_4$ will reduce, the evolution of the system is more likely towards to $E_1(0, 0)$, and so "non-participation" is a strategy that makes evolution stable. Thus, in the case of other conditions remaining constant, if the cost of participating in the cooperative development system is very huge, it will hinder the development of the system.

If the status of λ_1 increases, the "non-participation" region $E_1E_2E_5E_3$ will contract and the corresponding "participation" area $E_2E_5E_3E_4$ will increase, and then system converges to $E_4(1, 1)$. "Participation" is the strategy that makes the evolutionary game stable.

If β increases, the "non-participation" region $E_1E_2E_5E_3$ will be larger, and the corresponding "participation" region $E_2E_5E_3E_4$ will reduce, then the system is more likely to evolve towards $E_1(0, 0)$. At this time, "non-participation" is the strategy that makes the evolutionary game stable.

Therefore, the evolutionary path of the farmer group in the cooperative development system is affected by the income, investment, benefit distribution, and relative gain. As long as the development of the cooperative development system can bring considerable benefits to farmers, it may encourage farmers to participate in the system actively. If it costs too much for farmers to participate in the cooperative development system, it will hinder the passion of farmers. When the differences between farmers are great, the behavior of farmers' cooperative participation is easy to realize. On one hand, reducing the cost can improve farmers' income and raise farmers' enthusiasm; on the other hand, we must pay attention to the complementary relationship between the parties. Therefore, we should attach importance to the effective combination of different farmers, such as the farmers retaining labors and not retaining labors, and the farmers requiring bio-gas slurry and not requiring bio-gas slurry. A fair distribution of benefits received from the cooperation development system, lowering β and improving λ_1, will expand "participation" areas $E_2E_5E_3E_4$ and promote the system to evolve towards $E_4(1, 1)$.

4 CONCLUSIONS

The pig breeding industry is an important industry of animal husbandry and agriculture; its scale model has been the main breeding mode in our country. What is more, it is also a pivotal part of the "agriculture-countryside-farmer" problem. The number and scale of pig breeding enterprises constantly expanding have brought much new environmental pollution. On the whole, farmers tend to choose a long-term process of repeated games because of their limited rationality in cooperation. And then, they adjust and evolve the strategy to take part in the cooperative development system. Through an evolutionary game, we obtain the following conclusions:

Firstly, the total earnings that farmers gain from the cooperative development system can motivate their positivity; secondly, the increasing cost will hinder the development of the system; thirdly, narrowing the investment and profit distribution gap between the disadvantaged farmers and advantaged farmers is beneficial to maintain a sustainable development of the cooperative system; fourthly, the absolute earnings distribution gap between farmers groups can influence the realization of cooperation equilibrium. When both sides have less difference, they are more sensitive to yield gap, which makes it difficult to form a stable and effective cooperation.

ACKNOWLEDGMENTS

The authors gratefully acknowledge the grant of project Scale Pig Breeding Ecological Energy System Stability Feedback Simulation Study (71501085) supported by the National Natural Science Foundation of China and the Project of Humanities and Social Sciences in Colleges and Universities by Jiangxi Province (GL1536).

REFERENCES

Bhattacharya S C, S Abdul P. *Low Greenhouse Gas Biomass Options for Cooking in the Developing Countries* [J]. Biomass and Bio energy, 2002, 22(4): 305–317.

Chen Y, Yang G H, Sweeney S, et al. *Household Biogas Use in Rural China: A Study of Opportunities and Constraints* [J]. *Renewable and Sustainable Energy Reviews*, 2010, 14(1): 545–549.

Jury C, Benetto E, Koster D, et al. *Life Cycle Assessment of Biogas Production by Mono Fermentation of Energy Crops and Injection into the Natural Gas Grid* [J]. *Biomass and Bio energy*, 2010, 34(1): 54–66.

Leng Bibing, Tu Guoping, Jia Renan. *Scale Pig Breeding and Household Biogas Development System Dynamic Stability Basing on SD Evolutionary Game Model* [J]. Systems engineering, 2014(3): 104–111.

Tu Guoping, Leng Bibing. *The exploration of scale pig breeding development model's cooperation with household biomass resources* [J]. Agricultural economy, 2013, (7): 101–104.

Wang X, Di C et al. *The Influence of Using Biogas Digesters on Family Energy Consumption and Its Economic Benefit in Rural Areas -Comparative Study between Lianshui and Guichi in China* [J]. Renewable and Sustainable Energy Reviews, 2007, 11(5): 1018–1024.

Yan L, MinQet al. *Energy Consumption and Bio-Energy Development in Rural Areas of China* [J]. Resources Science, 2005, 27(1): 8–14.

Zhang Jiaqiang. *The western household biogas development present situation and the evaluation of the potential* [J]. Agricultural technology economy, 2008, (5): 103–109.

Zhou bin, Chen Xingpeng, Wang Yuanliang. *Regional cu mulative carbon footprint measure system dynamics model simulation experimental research-Taking Gannan Tibetan autonomous prefecture for example* [J]. Scientific and technological progress and countermeasures, 2007, 27(23): 37–41.

Advances in Energy Science and Equipment Engineering II – Zhou, Patty & Chen (Eds)
© 2017 Taylor & Francis Group, London, ISBN 978-1-138-71798-5

Surface runoff harvesting in redirecting sustainable water conditions in the UPNM campus

M. Vikneswaran & Muadz Azhar
Faculty of Civil Engineering, Universiti Pertahanan Nasional Malaysia (UPNM), Kuala Lumpur, Malaysia

ABSTRACT: Literally, water is the most critical element that concerns everyone. Water is very important to face any existing probability of emergency in the future. Most importantly it is the capability of producing clean water in an emergency situation. This is to ensure the sustainability of water needs for the human beings to survive. To achieve that, this research work has been conducted to determine the parameters of surface runoff in the selected unpaved areas and then test it by using the proposed water filter media. After that, the same parameters were tested to determine the effectiveness of the water filter media. The aim of this study is to determine whether the proposed design of water filter media is capable of producing clean water to be used as an alternative water source, according to the Department of Environment. This study represents an attempt to determine the characteristics and parameters of surface runoff in the unpaved area in a university campus from January until May 2016. Special emphasis was placed on the water quality rather than water quantity. The results of these experiments were analyzed based on National Interim Water Quality Parameters. Overall, the analysis results show that the proposed design of water filter media is capable of making the water clean. However, continuous research must be performed to ensure better results. The study proves that after the samples were tested on the proposed design water filter media, it has a higher value of Water Quality Index (WQI) (80.90) when compared to that the quality index before it was treated (81.97).

1 INTRODUCTION

Water is one of the most important elements in human beings' daily life. Water can be obtained from the sources like rivers, rain, snow, etc. However, improper water management implications can result in water pollution, running out of water, and wastage of electricity. Water might be present everywhere, but one must never take it for granted. A water conservationist must include the policies, strategies, and activities to manage fresh water as a sustainable resource, to protect the water environment, and to satisfy the current and future demands of human beings. Population, household size, and growth affect the degree of water consumption. Water is one of the most essential elements of health and is so important that your body actually has a specific drought management system in place to prevent dehydration and ensure your survival. Factors such as climate change will increase the pressure on natural water resources, especially in manufacturing and agricultural irrigation.

The aim of water conservation efforts include the following:

Ensuring availability of water for future generations: this requires that the withdrawal of fresh water from an ecosystem does not exceed its natural replacement rate. In other words, it needs sufficient clean water for our children.

Energy conservation: water pumping, delivery and wastewater treatment facilities consume a significant amount of energy. In some regions of the world, over 15% of total electricity consumption is devoted to water management. Apparently, water is the best source of conservation of electricity.

Habitat conservation: minimizing human water use helps to preserve fresh fir habitats for local wildlife and migrating waterfowl, as well as reducing the need to build new dams and other water diversion infrastructures. Basically, the dam costs the highest number of the deaths in water habitats.

Fresh water, a renewable but restricted resource, is limited in many areas of the developing world because of unplanned withdrawal of waters from rivers and underground aquifers causing severe environmental complications like arsenic contamination (Rahman et al. 2003). In several countries, the total water being consumed has exceeded the annual quantity of renewal, thereby creating a no-sustainable condition (Choudhury & Vasudevan, 2003). In addition to that, rainwater runoff during rainfall from roofs and other sealed surfaces when heavy rain is leading to accumulated flooding in the urban areas of many countries such as Bangladesh, where the drainage system is not properly designed as well as the volume of rainwater runoff. As early as ancient days of civilization, rainwater harvesting, the method of collecting and using the precipitation from a catchment area is considered as an alternative option for water supply in Bangladesh (Yusuf, 1999). Unless it comes into connection with a surface or collection system, the quality of rainwater meets Environmental Protection Agency standards (Choudhury & Vasudevan, 2003) and the independent character of its harvesting system has made it appropriate for scattered settlement and individual operation (Rahman & Yusuf, 2000).

Surface runoff and water treatment represents a procedure used to make water more adequate for the desired end-user. These can meet utilization needs as drinking water, industrial processes, and restorative and numerous different employments. All water treatment procedures are performed to uproot existing segments in water and enhancing it for subsequent use. It may permit treated water to be released into the regular habitat without any unfavorable environmental effect. On the other hand, surface runoff harvesting is one of the systems that could be used as an alternative source to achieve a sustainable environment.

2 RESEARCH BACKGROUND

Non-point sources pollutants in storm water runoff especially in urban areas highly depend on human activities, rainfall characteristics, and the land use. Latest studies on the characteristics of storm water pollution may differ and various effects depend on different types of land (Ramesh and Kiran, 2006).

Non-point sources pollution is the primary cause of storm water pollution. Type, concentration, and characteristics of pollutants brought by the storm water runoff depend on the land surface, whether it is permeable or non-permeable. Urban storm water runoff often accommodates significant concentrations of dissolved and suspended impurities causing disadvantage on the quality of water. Other than that, urban storm water runoff is documented for impressive peak flows and high runoff volumes. Drown drainage rainwater harvesting is a critical practice to improve water productivity and to cope with climate change in the drier marginal environments. The accurate determination of the location and types of rainwater harvesting interference through a land suitability assessment is the key to successful implementation. (Al Anwar Shamiri, Feras M. Ziadat, 2011). It is a suitable option for improving agricultural production and enhancing

environmental protection in the water-lacking areas (Oweis and Hachum, 2006). Rainwater harvesting can be merely defined as "the process of concentrating precipitation through runoff and storing it for beneficial use" (Frasier, 1994), or it may be defined as the "collection of runoff for its productive use" (FAO, 1991). The insufficiency of water resources in the arid areas forced people to realize the importance and role of rainwater harvesting in increasing and securing additional water supplies. Currently, it is used as an important tool to address problems associated with climate change in environments (Al Anwar Shamiri, Feras M. Ziadat, 2011).

Rainwater harvesting is a mutual practice in the countries and areas, where the annual precipitation is high and pure drinking and usable water is uncommon. All over the world, the economic condition has prompted the low-income groups to harvest rainwater for household and vital uses. Some countries of the world in different regions have shown the popularity of this technique. Originating almost 5000 years ago in Iraq, rainwater harvesting is trained throughout the Middle East, the Indian subcontinent, in Mexico, Africa as well as in Australia and United States. As the population of the world is increasing, water usage for irrigation increased, thereby resulting in the crisis of water supply in different regions. Among the available sources for water supply, rainwater harvesting has become the most economical way to meet the water emergency (Boers and Ben-Asher, 1982).

There were plenty types of filter concepts that could be applied, such as sand filters, slow-sand filters, charcoal filters, and many others. Apart from filtering stated before, there was a simple storage system used to store the surface runoff. The materials were used in the filter as a medium of filtration filter including silica, sand, gravel, and pebbles.

3 RESEARCH METHODS

First, to achieve the objective of this research, the selected site of an unpaved area had been chosen to collect the sample of surface runoff. When the samples were collected, they were kept in the freezer in order to avoid the samples from being interrupted. After the samples were collected, the laboratory experiment analysis test was conducted to determine the parameters of both samples.

Next, the surface runoff storage and the water filter media system had been designed to fulfill the third objective. A good filtration process had been able to produce a good quality of water. For this study, a simple filtration design was proposed. Specifically, three layers of filters were designed which consists of sand, aggregates, and charcoal as the main media components in the filter.

To achieve the second objective, the quality of the surface runoff on an unpaved area is being determined by using two different parts of testing that has been conducted in the laboratory. There are seven different experiments being conducted testing pH, turbidity, total suspended solid, and Dissolved Oxygen (DO). As for the chemical part, Biochemical Oxygen Demand (BOD), ammonia nitrogen and Chemical Oxygen Demand (COD) tests have been performed.

In this study, a proposed design of surface run-off from the unpaved area was designed, as shown in Figure 1. Firstly, the filter design was proposed by using a simple filter, which consists of sand, aggregates, and charcoal. Secondly, a storage tank was proposed to have a pumping system so that the water can be used easily for any purposes such as watering of landscaping and toilet flushing, as shown in Figure 2.

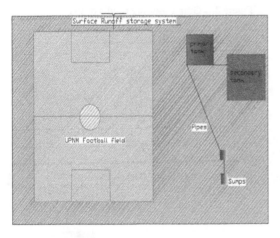

Figure 1. Proposed layout plan design of the surface runoff storage system.

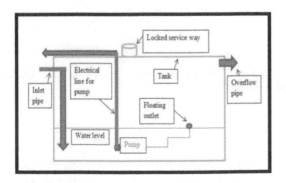

Figure 2. Design of the surface runoff storage tank.

4 RESULTS AND DISCUSSION

4.1 *Total suspended solids*

The results show the digression value of the total suspended solids. The results also indicated that the water filter media were managed to reduce the value of total suspended solids. Figure 3 shows the results of total suspended solids.

4.2 *Biochemical oxygen demand*

The BOD results for both surface runoff before and after testing on water filter media are illustrated in Figure 4. It shows that the value of BOD of the surface runoff before being tested on water filter media has a high value when compared to the unpaved area. It shows that the samples were cleaner after being tested with water filter media.

4.3 *Chemical oxygen demand*

Figure 5 gives the results of COD for the sample. The samples show that the COD value of the surface runoff after testing was greater than that before testing.

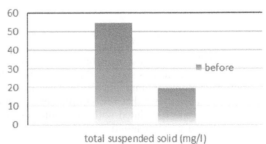

Figure 3. Graph showing results of total suspended solids.

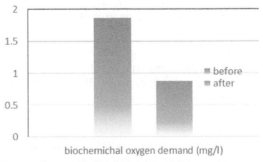

Figure 4. Graph showing results of Biochemical Oxygen Demand (BOD).

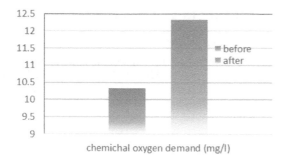

Figure 5. Graph showing results of Chemical Oxygen Demand (COD).

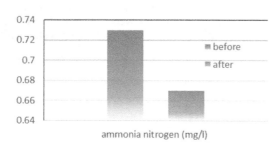

Figure 6. Graph showing results of ammonia nitrogen.

4.4 *Ammonia nitrogen*

The results for ammonia nitrogen can be seen in Figure 6. The results showed that before test, the removal volume of ammonia nitrogen was from the highest. This shows that the sample before being tested was more polluted with ammonia nitrogen than that after being tested.

4.5 *Dissolved oxygen*

Figure 7 provides the results for dissolved oxygen for each type of sample. The results of dissolved oxygen for all samples are between 5.00 mg/L and 7.00 mg/L, which belong to class II.

4.6 *pH*

Figure 8 shows the pH results before and after testing using water filter media. Both results were in class 1 category.

4.7 *Turbidity*

Figure 9 shows the decreasing value of turbidity of water samples after being tested with water filter media. This shows that the surface runoff after being tested with water filter media is cleaner.

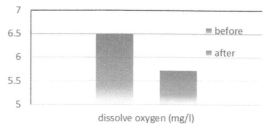

Figure 7. Graph showing results of dissolved oxygen.

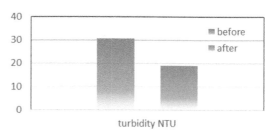

Figure 8. Graph showing results of pH.

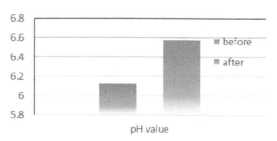

Figure 9. Graph showing results of turbidity.

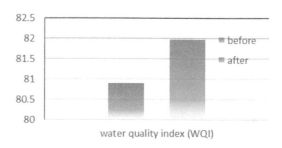

Figure 10. Graph showing results for Water Quality Index (WQI).

4.8 *Water quality index*

Figure 10 shows the WQI graph. From the Water Quality Index (WQI) table, the values for both before and after testing with water filter media are quite the same. It can be concluded that both are in class 2 and the water can be used for recreational purpose.

5 CONCLUSION

Based on overall results, from the Water quality Index (WQI) of both paved and unpaved areas, both surface runoff samples on paved and unpaved areas have a good quality of water, which are in category of class 2. This class is suitable for recreational use with body contact. Therefore, it can be concluded that the water filtration system can improve the surface runoff water quality and reduce pollution in the water with the use of sand, aggregates, and pebbles as a media in the filter. With comparisons with water class and uses, the parameters have met the prescribed standard limits. Thus, the filter media need to be redesigned in order to remove the ammonia nitrogen and to increase the pH value by using other suitable methods.

ACKNOWLEDGMENTS

This work was completed in UPNM Campus. The authors would like to express their special thanks to the environmental laboratory supervisor and technicians for providing access and guidance to the research facilities.

REFERENCES

Al Anwar Shamiri, Feras m Ziadat, (January 2011). An Investigation of the Potential Role for Rainwater Catchment Systems in Rural Water Supply in Botswana. Univ. of Botswana Research and publications committee. MSc thesis.

A.M. Shinde, (2015). Removal of Endosulfan from Water Using Wood. MSc thesis.

Boers and Ben-Asher, (1982). Available another Sources for Water Supply for the Water Emergency. MSc thesis.

Mohamad Zakuan bin Abdul Razak (2015). Study on surface runoff harvesting for sustainable UPNM campus.

Navotny and Chester, (1981). Document on Urban Storm Water Runoff its Peak Flows and High Runoff Volumes, and Drown Drainage System. MSc thesis.

Oweis and Prinz, (1994). Integrating Rainwater Harvesting Within a Sound Farming System. MSc thesis.

Ramesh, H. S. and Kiran, K.R. (2006). Studies on the Characteristic of Storm Water Pollution and Various Effect Depend on Different Type of Land. MSc thesis.

Retrieved from http://clear-ion.tradeindia.com/Exporters_Suppliers/Exporter2483.31736/Sand-Activated-carbon-Iron-Removal-Filter.html

Retrieved from http://www.purewaterproducts.com/articles/birm

United State Geological Survey (2000). Characteristic of Runoff, 30(11–16), 792–798.

Advances in Energy Science and Equipment Engineering II – Zhou, Patty & Chen (Eds)
© *2017 Taylor & Francis Group, London, ISBN 978-1-138-71798-5*

Leveling marine magnetic survey data using the wavelet multi-scale analysis

Qiang Liu, Xiaodong Yin, Gang Bian & Junsheng Zhao
Department of Hydrographic and Cartography, Dalian Naval Academy, Dalian, Liaoning Province, China

ABSTRACT: Based on the characteristics of marine magnetic survey, the leveling method of survey line systematic errors is discussed in this paper. The theory of frequency filtering in the study of leveling aeromagnetic data is referenced. Wavelet multi-scale analysis method is adopted to separate the low-frequency background field, the high-frequency local magnetic anomaly field and systematic error. Different levels of magnetic data is unified by the leveling method. The simulation experiment analysis shows that wavelet multi-scale analysis can effectively sift out the low-frequency magnetic anomaly data, and unify the level of the entire area magnetic field. The method that introduced in this paper is advised to be used in future marine magnetic survey.

1 INTRODUCTION

Marine magnetic survey is mainly carried out by surface ships with the magnetometer sensors towed using cables long enough to avoid the ship magnetic effect. The survey environment of each point on a survey line is similar, and the influence of systematic errors on each point is coincident, which is called survey line systematic error in marine survey. It is mainly affected by some common systematic factors on every survey point, such as the magnetic effect of the ship, geomagnetic diurnal variation, remnant systematic errors after corrections and surveying base value of each voyage. The systematic errors can be different on diverse survey lines. The influence of systematic error on the magnetic anomaly map is analyzed by Bian Gang (2014), who has proved that the position and local direction of the contour line on magnetic anomaly map can be affected and changed by the systematic errors.

In aeromagnetic survey, the adjustment of survey line systematic errors is called leveling processing. The main leveling methods are commonly based on the principle of step by step iteration and the least square, which minimize the crossover differences (CODs). Most laterally extensive airborne surveys operate under these acquisition methodologies, yet the corrections are often ineffective due to strong gradients in the anomaly field and the low flight altitude at which modern surveys may operate (James C. 2015). A new technique that can remove the leveling errors in airborne geophysical data is based on the assumption that the data are continuous from line to line (H. Huang 2008). A single flight line is selected as a reference to level and tie all survey lines to this continuously varying datum, which produces a marked improvement in the quality of the unleveled raw data. In addition, other technique called micro-leveling (Luo Yao 2012) has been proposed to level airborne geophysical data. It has proved that the effect of leveling is remarkable, which adopts respectively different frequency filters along the survey line direction and vertical survey line direction to separate systematic errors from the unleveled aeromagnetic data.

Currently, the application of filtering to level the survey line data has become a major trend in leveling processing (Mauring E. 2006 & Beiki M. 2010). Meanwhile, with an advantage in time-frequency separation, wavelet multi-scale analysis has also been widely used to process geophysical signal in recent years (Li Zongjie 1997 & Leblanc G E. 2001). In this paper, we try to use wavelet multi-scale analysis method to separate the high and low frequency information in marine magnetic survey line data to level survey line systematic errors, so as to unify the magnetic field level. A simulation experiment has been designed to verify the feasibility of this method.

2 METHOD PRINCIPLE AND PROCEDURE

2.1 Wavelet multi-scale analysis principle

In the space of geomagnetic field, the magnetic anomaly of any point can be expressed as a function

$f(t)$. It can be expanded to the Continuous Wavelet Transform (CWT). Its expression is

$$W_f(a,b) = \langle f, \psi_{a,b}(t) \rangle = |a|^{-\frac{1}{2}} \int_{-\infty}^{\infty} f(t) \bar{\psi}\left(\frac{t-b}{a}\right) d \quad (1)$$

$$\psi_{a,b}(t) = |a|^{-\frac{1}{2}} \psi\left(\frac{t-b}{a}\right), a \in R, \ a \neq 0, \ b \in R \quad (2)$$

In Equation (1) and (2), $\psi_{a,b}(t)$ is called wavelet function, which is obtained by conducting expansion and translation on $\psi(t)$ in different time scales. $W_f(a,b)$ is wavelet coefficient obtained by wavelet transform. R represents the real number field, $\psi(t)$ is the wavelet prototype or the mother wavelet, a is expansion parameter, b is translation parameter.

When the wavelet satisfies certain condition, there exists an inverse transform on the continuous wavelet. The formula is

$$f(t) = C_\psi^{-1} \int_{-\infty}^{\infty} \int_{-\infty}^{\infty} W_f(a,b) \psi_{a,b}(t) \frac{da}{a^2} db \quad (3)$$

In Equation (3), $C_\psi = \int_{-\infty}^{\infty} |\psi(\omega)|^2 / |\omega|^{d\omega < \infty}$ is the condition that the wavelet function should satisfy.

The low-frequency approximation signal and the high-frequency detail can be separated by changing the expansion scale parameter a in wavelet transform. The order of decomposition can be determined according to the need of different frequency information. To some extent, it has a band-pass filter function. We attempt to use the characteristics that wavelet analysis can separate different frequency information to extract the background field and the local geomagnetic anomaly information from the geomagnetic field function $f(t)$. In practical application, the wavelet transform is used in the discrete form, and the expression of the two-dimensional discrete wavelet (2-DDW) multi-scale analysis is

$$A_j f(x,y) = A_{j+1} f(x,y) + D_{j+1}^h f(x,y) + D_{j+1}^v f(x,y) + D_{j+1}^d f(x,y) \quad (4)$$

In Equation (4), $A_{j+1}f(x,y)$ is the low-frequency approximation of the magnetic anomaly signal, $D_{j+1}^h f(x,y)$, $D_{j+1}^v f(x,y)$, $D_{j+1}^d f(x,y)$ respectively represent the horizontal, vertical, diagonal high-frequency detail information of the magnetic anomaly in the space (Yang Wencai 2001). The wavelet multi-scale decomposition structure chart is shown in figure 1.

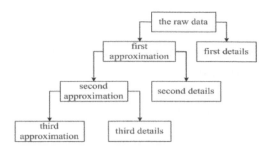

Figure 1. The third order wavelet multi-scale decomposition structure chart.

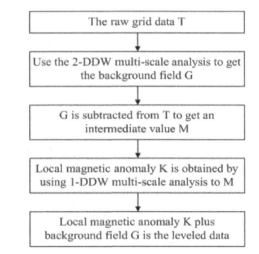

Figure 2. The flow chart of leveling method.

2.2 Concrete procedure

In this paper, wavelet multi-scale analysis method is applied to adjust the systematic errors among different lines in the same survey area, the main steps are as follows:

1. Gridding the raw data. In marine magnetic survey, the distribution of survey data is irregular with a large amount of data along the line direction and a small amount of data along the vertical survey line direction. The data in the survey area should be grided before the implementation of wavelet analysis.
2. Extracting magnetic background field data. We need to choose the suitable mother wavelet and determine the order of decomposition, then 2-DDW multi-scale analysis is performed to extract the low-frequency background field information.
3. The difference between the raw data and background field data obtained from step (2) is an intermediate amount containing local

geomagnetic anomalies and high-frequency leveling errors.

4. The 1-DDW multi-scale analysis is carried out on the intermediate amount in step (3) to extract lower frequency local geomagnetic anomalies than the leveling errors.

5. Adding the local geomagnetic anomaly data obtained in step (4) to the background field data will get the leveled data.

It is noteworthy that, it is essential to select the suitable mother wavelet when using the wavelet analysis method, for different mother wavelets selected may lead to different results. Currently, the suitable mother wavelet is selected based on the characteristics of the data as well as the comparisons of different mother wavelets. The main operating process is shown in figure 2.

3 SIMULATION ANALYSIS

To illustrate that the wavelet multi-scale analysis can effectively adjust the survey line systematic errors, we design a simulation test. The true value data in the test is the real magnetic anomaly data that has achieved the survey accuracy after several corrections. The main line direction is in the north and south, with a total of 20 lines. Line spacing is 0.5 km. The magnetic anomaly contour map of the true value in simulation is shown in figure 3. In order to simulate the condition that different lines have different systematic errors and the influence of survey line systematic error on the magnetic anomaly map, we add some random interference errors to every survey line data to make sure the systematic error is a constant on the same one line while they are different among different lines. The magnetic anomaly contour map containing random noises is shown in figure 4. We can see that the position and local direction of the contour line are changed by the systematic errors from the figure 4. This is close to the actual unadjusted magnetic anomaly con-

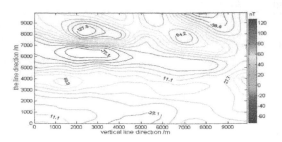

Figure 3.   The magnetic anomaly contour map of true value in the simulation.

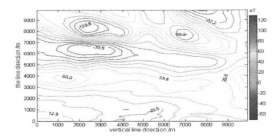

Figure 4.   The magnetic anomaly contour map of containing interference errors.

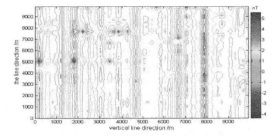

Figure 5.   The leveling error image.

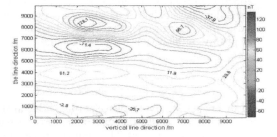

Figure 6.   The magnetic anomaly contour map after leveling.

tour map, which that illustrates simulation data is valid.

We select the sym6 mother wavelet and take 2 layers of low-frequency information as the background field data when performing 2-DDW multi-scale analysis on the raw data. Then, the high—frequency random interference errors are extracted from the intermediate amount (the difference between the raw data and background field data) by using 1-DDW multi-scale analysis, and the leveling error image is shown in figure 5. Finally, adding the local magnetic anomalies to the background field data is the leveled data, as shown in figure 6.

Contrasting the leveled magnetic anomaly map (Fig. 6) with the true value of magnetic anomaly map (Fig. 3), we can see the magnetic anomaly contour generally have almost the same trend in

Table 1. The statistical table of the leveling errors and simulation values in 20 lines.

Line	Leveling errors/nT	Interference errors/nT	Line	Leveling errors/nT	Interference errors/nT
1	−3.01	−2.78	11	3.09	3.95
2	3.87	4.86	12	−4.74	−2.27
3	2.69	2.56	13	3.33	2.51
4	4.85	4.89	14	−3.39	−3.16
5	−3.74	−3.69	15	−4.53	−2.98
6	−3.49	−3.05	16	3.99	4.06
7	2.90	1.55	17	−4.31	−3.43
8	3.49	1.53	18	3.39	3.47
9	−3.01	−1.14	19	4.00	1.69
10	−2.96	−0.51	20	−3.46	−3.95

these two maps, the changed position and local direction of the contour line caused by the interference errors are effectively improved, and the magnetic anomaly values are also closer. In order to illustrate the effect of the leveling method, we calculate the leveling errors in 20 lines and compare them with the interference errors added in corresponding lines (as shown in table 1). We can clearly see the leveling errors and the interference errors added are very close on most of the 20 lines, and the difference is only 0.04 nT in line 4 between this two errors, which proves that the leveling method of positive influence on error correction.

4 CONCLUSIONS

In this paper, we use the wavelet multi-scale analysis method to adjust the systematic error in the marine magnetic survey. The simulation analysis shows that the leveling method can effectively separate high-frequency systematic error from magnetic anomaly data and achieve unified magnetic field level in the whole test area.

It is worth mentioning that the high-frequency systematic errors separated by using our leveling method sometimes also contain some real and useful magnetic anomalies that are significant for the interpretation of the geomagnetic field data and detection of underwater target. Hence, they should not be eliminated from the survey data. Therefore, a more suitable leveling method that can separate systematic errors from meaningful geomagnetic data deserves further research and discussion.

REFERENCES

Alexander, Y. 2014. Principal component analysis for filtering and leveling of geophysical data. *J. Journal of Applied Geophysics*. 109, 266–280.

Beiki, M. & Bastani, M. & Pedersen, L. B. 2010. Leveling HEM and aeromagnetic data using differential polynomial fitting. *J. Geophysics*. 75(1): L13–L23.

Bian Gang & Xia Wei & Jin Shaohua 2014. Influence of Systematic Error on the Magnetic Anomaly Map in Marine Magnetic Survey. *J. The 7th International Congress on Image and Signal Processing*. pp. 753–757.

Fausto, F. 1998. Microlevelling procedures applied to regional aeromagnetic data: an example from the Transantarctic Mountains. *J. Geophysical Prospecting*. 46, 177–196.

Huang, H. 2008. Airborne geophysical data leveling based on line to line correlations. *J. Geophysics*. 73(3):F83–F89.

James, C. & D. Beamish 2015. Leveling aeromagnetic survey data without the need for tie-lines. *J. Geophysical and Prospecting*. 63, 451–460.

Leblanc, G. E. & Morris, W., A. 2001. Denoising of aeromagnetic data via the wavelet transform. *J. Geophysics*. 66(4): 1793–1804.

Li Zongjie & Yang Lin 1997. Applying wavelet transform in potential field data processing. *J. Geophysical Prospecting for Petroleum*. 36(3): 71–75.

Luo Yao & Wang Linfei & He Hui 2012. Micro-leveling processing of airborne geophysical data. *J. Geophysical and Geochemical Exploration*. 36(5):852–855.

Mauring, E. & Kihle, O. 2006. Leveling aero-geophysical data using a moving differential median filter. *J. Geophysics*. 71 (1): L5–L11.

Peng Lu 2012. A Study on Practical Methods for Aeromagnetic Data Leveling. *D. China University of Geosciences*.

Yang Wencai & Shi Zhiqun & Hou Zunze 2001. Discrete Wavelet Transform for Multiple Decomposition of Gravity Anomalies. *J. Chinese Journal of Geophysics*. 44(4): 535–540.

Yu Bo & Zhai Guojun & Liu Yanchun 2009. Data denoising based on Wavelet threshold in marine magnetic survey. *J. Science of Surveying and Mapping*. 34(5): 89–91.

Risk assessment of surface water quality in the Beijing mountainous area

Weiwei Zhang & Hong Li
Beijing Academy of Agriculture and Forestry Sciences, Beijing, China

ABSTRACT: Monitoring and assessing the surface water quality of river basins plays an important role in the control of surface water pollution, meeting increasing water demand, and ensure ecological environment protection. In this study, we analyzed and evaluated the surface water quality data of five major river basins in the Beijing mountainous area during 1980–2003 using a single factor water quality identification index. Our data indicated that total nitrogen, total phosphorus, and Hg were the major pollutants. However, other pollution indicators were also identified at some monitoring sites for specific years. The assessment of surface water quality in the Beijing mountainous basins provides a scientific basis for the management of the mountain basins, including water pollution control.

1 INTRODUCTION

With the rapid development of industrialization and urbanization, increasing amounts of domestic, industrial and agricultural wastes are transmitted to rivers by natural processes (such as precipitation and surface runoff), leading to the pollution and degradation of surface water quality, which threatens human health and damages the ecological environment. In recent years, the rapid development of urbanization in Beijing at very large social and economic scales has caused changes in land use and land cover. These changes, together with ineffective environmental protection measures, have led to increased water environment problems. These water environment problems create a sustainable developmental bottleneck in the strategic planning to build Beijing into a "world class city". The Beijing mountainous area is located upstream of the developed areas of the capital, and is an ecological environment protection and water conservation buffer. The water supplies, water environment, and ecosystem functioning of the Beijing mountainous area are very important. Thus, monitoring and assessing the surface water quality in the river basins of Beijing mountainous area play an indispensable role in the control of surface water pollution, meeting increased water demand, and ecological environment protection.

The rational evaluation of surface water quality is very important for the effective utilization of water resources at the basin scale. Currently, many methods are available to evaluate the water quality of river basins, such as the pollution index method (Nives, 1999; Pesce, 2000), the fuzzy evaluation method (Chang, 2001; Luo, 2002; Liou, 2005; Zou, 2006), the fuzzy matter element model method (Zhang, 2005), the neural network method (Shetty, 2003; Chaves, 2007; He, 2008), the gray fuzzy method (Karmakar, 2006; Karmakar, 2007), and the single factor water quality identification index method (Xu, 2005). All of these methods have advantages and disadvantages for assessing the water quality of river systems. In this study, the surface water quality data of the major river basins in the Beijing mountainous area during 1980–2003 were analyzed. The surface water quality assessment was performed using the single factor water quality identification index method. This can provide a scientific basis for the regulation of the water environment and agricultural development in the Beijing mountainous area.

2 MATERIALS AND METHODS

2.1 Data collection of surface water quality

There are 27 water quality monitoring sites (S1–S27) in the five major basins in the mountainous area of Beijing. Long-term monitoring data (1980–2003) for average annual surface water quality were obtained from the Beijing Water Authority. The data included seven water quality indicators: Permanganate index (COD_{Mn}), five-day biochemical requirement (BOD_5), total mercury (Hg), total cadmium (Cd), Total Lead (Pb), Total Phosphorus (TP), and Total Nitrogen (TN). Sampling took place in May and October each year for each water quality indicator. The data were evaluated using the basic entry analysis methods adopted from the national standard for surface water environmental quality. The average of the water quality concentrations in May and October served as the water quality value for that year. Water quality monitoring sites in the Beijing mountainous area are shown in Figure 1.

2.2 Surface water quality assessment method

Based on Xu's method, two numbers were used to create the single factor water quality identification index (P) (Equation 1). The first number (X) represents the water quality class of the water quality index—the larger the number, the more severe was the water pollution. The second number (Y) indicates the position of the monitoring value in its water quality class, using 0–9 as representation. The higher the figure was the more significant was pollution in that class.

$$P = XY \qquad (1)$$

When the water quality was in Class I–V, X was determined by comparing the monitoring data to the national standard for surface water environmental quality. When X = 1, 2, 3, 4, and 5, it indicated that the water quality was Class I, II, III, IV, and V, respectively. To determine Y, the position y (Equation 2) for the monitoring data in the X class was calculated, and numbers 0–9 were used for Y to show the range of y, as indicated in Table 1. The water quality monitoring data increased as the water quality class increased, so y was calculated using Equation 2:

$$y = \frac{\rho_i - \rho_{ikmin}}{\rho_{ik\,max} - \rho_{ikmin}} \qquad (2)$$

Table 1. The coding in the second place Y.

Y	0	1	2	3	4
y	$0 \le y$ < 0.1	$0.1 \le y$ < 0.2	$0.2 \le y$ < 0.3	$0.3 \le y$ < 0.4	$0.4 \le y$ < 0.5
Y	5	6	7	8	9

where ρ_i was the ith indicator in the monitoring data, $\rho_{ikmin} \le \rho_i \le \rho_{ikmax}$, ρ_{ikmin} and ρ_{ikmax} were the minimum and maximum values of the quality concentration for the ith indicator and X water quality class, respectively.

When the water quality was greater than the maximum of Class V, X was represented by 5 + n, where n was the excess ratio of the data to the maximum of Class V. Thus, when the water quality exceeded twice the Class V maximum X = 6. Similarly, when the water quality exceeded twice the Class V maximum, then X = 7, and so on. The value y was calculated using Equation 3 to show the position after n times exceeding the standard, and Y was determined using Table 1, and was represented by 0–9.

$$y = \frac{\rho_i - n\rho_{i5\,max}}{\rho_{i5\,max}} \qquad (3)$$

where $\rho_{i5\,max}$ is the maximum for Class V water for the ith indicator.

3 RESULTS AND DISCUSSION

3.1 Chaobai river basin

A total of 8 water quality monitoring sites were available in Chaobai River basin. Of these, the S1, S2, S3, and S4 monitoring sites are located upstream and downstream of the Miyun Reservoir on two tributaries of the Chaobai River—the Chao River and Bai River. Sites S5, S6, S7, and S8 are located below the confluence of the Chao and Bai Rivers. Site S5 is on the Yanqi River, an upstream tributary of the Huai River, and S6 and S7 are on the Huaisha River and Huaijiu Rivers, respectively (two tributaries of the Huairou Reservoir upstream of the Huai River), and S8 is located in the Huairou Reservoir. Based on the distribution of these sites, the Chaobai River basin was divided into two for the water quality evaluation: (1) the Miyun Reservoir basin (S1, S2, S3, and S4), and (2) the Huairou Reservoir basin (S5, S6, S7, and S8).

TN and TP were the major pollutants in the Miyun Reservoir basin; and TN, TP, and Hg were the major pollutants in Huaiyou Reservoir basin.

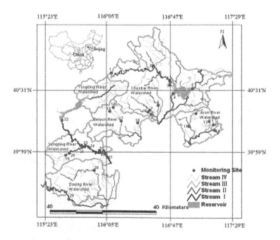

Figure 1. Water quality monitoring sites in the Beijing mountainous area.

The indicators of organic pollutants in both the Miyun Reservoir and the Huairou Reservoir basins did not exceed the limits for drinking water during the monitored period, and therefore the pollution risk was relatively small. The heavy metal indicator Cd did not exceed the limit for drinking water, and therefore the heavy metal pollution risk was also small. Although Pb did not exceed drinking water limits in the Miyun Reservoir basin, a pollution risk existed in the Huairou Reservoir basin for a single year (S5, 1992). Hg did not meet the requirements for surface drinking water standards at S1, S3, and S4 in the Miyun Reservoir basin for certain years, as well as at all the sites in the upstream areas of the Huaiyou Reservoir basin during 1987–1990. The pollution risk for TN was higher than that for TP at S1, S2, and S4 in the Miyun Reservoir basin during 1988–2003, while the risk of TP was higher than TN at S3 for some years (1992 and 1997), and the remaining years also showed a higher TN pollution risk compared with TP. The TN and TP pollution risks for the Miyun Reservoir at the Chao River entrance was greater than that at the Bai River entrance for most years, and the TN and TP risk of the Miyun Reservoir water inlet was lower than that for the water outlet for most years. In addition, the TN and TP levels at sites upstream of the Huai River exceeded the drinking water standards for most of the monitored years, thus exhibiting pollution risks, but the TN and TP concentrations within the Huairou Reservoir were smaller than that of the water inlet.

3.2 Jiyun river basin

The five monitoring sites (S9–S13) for the Jiyun River basin were located on tributaries of the Liuju River, a tributary of the Ji Canal, and thus there are no spatial upstream and downstream relationships. TN and Hg are the major pollutants in the Ji Canal basin. The values for TN were greater than the drinking water standard for all the monitored years, and there were other significant pollution indicators at various sites. In addition to the relatively high pollution risks for TN and Hg, the site at the West Valley Reservoir on the Zhenluoyingshi River showed substantial TP pollution, which exceeded the drinking water standard in many years. Values for COD_{Mn} also exceeded the drinking water standards in one year. In addition to Hg pollution in certain years, Pb pollution was also present for two years in the Cuo River tributary. Pollution risks were observed in the Hongsongyu Reservoir on the Hongsongyu River for TN, Hg, Pb, and TP in 1992. The COD_{Mn}, Hg, Cd, and Pb values were all greater than the drinking water standards at the site on the Tumenshi River for certain monitoring periods, indicating a high pol-

lution risk. The pollution risks for TN and TP was not ruled out, because data on TN and TP was not available. In addition, the pollution risk for TN and Hg was relatively high in the Haizi Reservoir, and Pb and TP risks were also evident in certain years.

3.3 Beiyun river basin

In the Bei Canal basin, monitoring sites S14 and S15 were located on the two influent rivers of the Ming Tombs Reservoir, and S16 was located within the Reservoir. The fourth site was on the Taoyukou Reservoir, which has no upstream and downstream relationship with the other three sites. The rivers associated with these four sites flow into the Wenyu River, a tributary of the North Canal.

TN, TP, and Hg were the main pollutants. The organic pollution indicators COD_{Mn} and BOD_5 for the two upstream tributaries of the Ming Tombs Reservoir were all lower than the drinking water standard, thus showing insignificant pollution risk. However, the level of organic pollution indicators for the Ming Tombs Reservoir exceeded the standard in the early 1980s. In addition, COD_{Mn} exceeded the standard for most years in the Taoyukou Reservoir, and thus a pollution risk was present. For the heavy metal indicators, Cd levels were lower than the standard, thus showing minor pollution risk. Except for the pollution risk at the Ming Tombs Reservoir site in 1989, Pb did not show any risk at monitoring sites in this basin. Hg exhibited pollution risks for some monitored years, and did not meet the drinking water standard. The data for TN and TP were only available at the sites of two Reservoirs; these data revealed concentrations that substantially exceeded the drinking water standard. The pollution risk in the Taoyukou Reservoir was greater than that of the Ming Tombs Reservoir after 2000.

3.4 Yongding river basin

Six of the eight monitoring sites (S20–S25) for the Yongding River basin were located on the main river, and the other two (S18 and S19) were located on its tributaries. TN, TP, and Hg were the main pollutants. The organic pollution indicator COD_{Mn} exceeded the drinking water standard at S19 in 1982, at S23 in 1994, while BOD_5 at S23 exceeded the standards in 1993. The levels at other sites were lower than the standard, and exhibited minor pollution risk. The level of COD_{Mn} in the downstream monitoring sites was generally lower than that of the upstream tributaries. Levels of COD_{Mn} showed significant spatial variation characteristics for upstream and downstream tributaries, which was not evident for BOD_5. The heavy metal pollution indicator, Cd, exceeded the drinking water standard at S21 in 1990, S23 in 1984 and 1987, and Pb exceeded the standard at S21 in 1983. All the sites in the basin revealed excessive Hg pollution in certain years, especially in the 1980s. No significant spatial variation characteristic for the time series data was observed for the concentrations of these three heavy metal indicators. More than three monitoring sites showed a level of TN and TP over the standard within the watershed in multiple years. Of these, the level of TN exceeded the standard at two sites in the main river for all monitored years and at the sites in the upstream tributary for most monitored years, so the pollution risk was substantial. TN levels at the sites in the upstream tributary were significantly higher than that in the downstream sites. In addition, the years with TN exceeding and within the standard alternated, and the spatial characteristics of the monitoring sites at the upstream and downstream tributaries changed with time.

3.5 Daqing river basin

Only 5 water quality indicators were available for the Daqing River basin. Values for TP and TN were not available. Hg was found to be the major pollutant in the area. The organic pollution indicators, COD_{Mn} and BOD_5 were all within the limits of the drinking water standard, and therefore the pollution risk was minor. For the heavy metal indicators, Cd and Pb levels were lower than the drinking water standard, thus exhibiting a small pollution risk. However, the pollution risk of Hg was present in certain years, because its levels were higher than the drinking water standard.

4 CONCLUSION

In this study, surface water quality data were analyzed and assessed for the five major basins in the Beijing mountainous area during 1980–2003 using the single factor water quality identification index method. The results provide a scientific basis for regulating the water environment and agricultural development in the Beijing mountainous area.

The main pollutants were TN, TP, and Hg, with other pollution indicators observed at different monitoring sites in particular years. From the characteristics of N/P migration and transformation, the accumulation of P is significantly more than that of N. The source of N pollution and its mechanism are well established, and therefore regulatory monitoring at the basin scale is relatively straightforward. However, understanding the source of P pollution is more complicated, and is always associated with other pollution (such as heavy metals), so it is difficult to manage and remediate. In addition, P becomes the determining factor for surface water pollution (such as eutrophication), and thus more attention needs to be given to the impact of P on the surface water environment of the Beijing mountainous area.

REFERENCES

Chang, N.B., Chen, H.W., Ning, S.K. 2001. Identification of river water quality using the Fuzzy Synthetic Evaluation aproach. *Journal of Environmental Management* 63(3): 293–305.

Chaves, P., Kojiri, T. 2007. Deriving reservoir operational strategies considering water quantity and quality objectives by stochastic fuzzy neural networks. *Advances in Water Resources* 30(5): 1329–1341.

He, L.M., He, Z.L. 2008. Water quality prediction of marine recreational beaches receiving watershed baseflow and stormwater runoff in southern California, USA. *Water Research* 42(10): 2563–2573.

Karmakar, S., Mujumdar, P.P. 2006. Grey fuzzy optimization model for water quality management of a river system. *Advances in Water Resources* 29(7): 1088–1105.

Karmakar, S., Mujumdar, P.P. 2007. A two-phase grey fuzzy optimization approach for water quality management of a river system. *Advances in Water Resources* 30(5): 1218–1235.

Liou, Y.T., Lo, S.L. 2005. A fuzzy index model for trophic status evaluation of reservoir waters. *Water Research* 39(7): 1415–1423.

Luo, H.J., Zhu, J.P., Jiang, H.H. 2002. The suggestion of select ing pollution items on water quality evaluation. *Environmental Monitoring in China* 18(4): 51–55.

Nives, S.G. 1999. Water quality evaluation by index in Dalama. *Water Research* 33(16): 3423–3440.

Pesce, S.F., Wunderlin, D.A. 2000. Use of water quality indices to verify the impact of Cordoba City on Suquia River. *Water Research* 34(11): 2915–2926.

Shetty, G.R., Malki, H., Chellam, S. 2003. Predicting contaminant removal during municipal drinking water nanofiltration using artificial neural networks. *Journal of Membrane Science* 212(1–2): 99–112.

Xu, Z.X. 2005. Single factor water quality identification index for environmental quality assessment of surface water. *Journal of Tongji University* 33(3): 321–325.

Zhang, X.Q., Liang, C. 2005. Application of fuzzy matter-element model based on coefficients of entropy in comprehensive evaluation of water quality. *Journal of Hydraulic Engineering* 36(9): 1057–1061.

Zou, Z.H., Yuan, Y., Sun, J.N. 2006. Entropy method for determination of weight of evaluating indicators in fuzzy synthetic evaluation for water quality assessment. *Journal of Environmental Sciences* 18(5): 1020–1023.

Advances in Energy Science and Equipment Engineering II – Zhou, Patty & Chen (Eds)
© *2017 Taylor & Francis Group, London, ISBN 978-1-138-71798-5*

Numerical simulation of an internal flow field in a biomass high efficiency hot blast stove

Guojian Li & Chunyuan Gu
Shanghai Institute of Applied Mathematics and Mechanics, Shanghai University, Shanghai, China
Shanghai Key Laboratory of Mechanics in Energy Engineering, Shanghai, China

ABSTRACT: A Hot Blast Stove (HBS) is one of the most critical devices for efficient utilization of biomass energy and reducing pollution from rural straw combustion. However, the problems of low Heat Transfer Efficiency (HET) and equipment damage due to local overheating severely restrict its application. The temperature and flow fields in the combustion process of a hot blast stove are studied by using a steady $k - \varepsilon$ model. Furthermore, the influence of inlet velocity and inlet position on the HTE and outlet temperature is analyzed; consequently, related parameters are optimized to improve the HTE. Simulation results show that at high temperature, the wall of the burner, the smoke outlet, and the lower parts of heat exchange tubes are the key components used to prevent overheating. Properly increasing the inlet velocity can significantly increase the HTE and the air speed in the lower regions of the exchange tube to improve their temperature. When the inlet mass flow rate is 0.97 kg/s, the outlet temperature is 451.3 K and HTE is 61.6%, which well matches the actual operating situation. When the inlet mass flow is increased to 1.4 kg/s and outlet temperature decreases to 444.5 K, the HTE is 84.4%, thereby indicating an increase of nearly 23%. The inlet position has a certain effect on the HTE of a HBS, and the existing inlet position is relatively reasonable.

1 INTRODUCTION

In the 21st century, biomass energy is a main type of new energies, which is one of the important directions of the national development strategy (Chen D, Hu X, & Li X, 2015). The research and development of biomass energy technology is spread in the world's governments and scientists. However, a large amount of biomass being abandoned or inefficiently burned still generates serious pollution (Steininger & Voraberger, 2003). Some evidences show that the burning of straw in rural areas is one of the serious problems (Wenting Liu, Ying Zhao, Shengyong Liu, & Fuying Yang, 2006) and causes serious air pollution, and consequently leads to a large number of traffic accidents. Therefore, at all levels, the government has banned the burning of straw. The violators would be heavily fined. If the case is very serious, they will be sentenced to crime according to related law. But the reason for the repeated burning is that, it is inconvenient to transport such large amounts of straw for farmers, and there is no place to store them. In order to solve the problems caused by the burning of straw and other biomass, storing the resources (Ping Luo, Guilan Wang, & Xiuhua Wang, 2011) and converting the waste biomass into "green coal", the high-efficiency biomass HBS has become

a research goal of many scientists and technicians (Yongming Sun, Zhenhong Yuan, & Zhenjun Sun, 2006).

The development of HBS as industrial equipment in foreign countries began in the 1940s. In 1970s, the United States, Japan, and other countries had achieved automation and after the 1980s, they developed HBSs toward the directions of the efficiency, high quality, energy conservation, environmental protection, and the economy (Yan, Shamim, Chou, Desideri, & Li, 2016). Meanwhile, the continuous development of new technology and new models highly improves the quality of HBSs (Wiklund, Helle, & Saxén, 2016). In the designing process, computer-aided study and simulation are widely used, which not only reduce the development cycle, but also ensure the reliability of the HBS (Borah AK, Singh PK, & Goswami P. 2013). The integration of furnace efficiency and waste heat recovery is put forward by Duleeka Sandamali Gunarathne, which can recover two-thirds of the available waste heat (Gunarathne, Mellin, Yang, Pettersson, & Ljunggren, 2016). Chuan Wang believes that replacing coal with biomass, can not only save energy, but also reduce a lot of carbon dioxide emissions (Gunarathne DS, Mellin P, Yang W, Lövgren J, Nilsson L, Yang W, Salman H, Hultgren A, & Larsson M. 2015).

A dynamic model for a HBS is developed by the finite difference approximation (Zetterholm, Ji, Sundelin, Martin, & Wang, 2015). The blast temperature increases while the flue gas temperature decreases, which allows for an increase of the blast temperature, thereby leading to improved energy efficiency for the hot stove system.

Biomass for the heating source, combined with solar radiation, has been used to achieve double heating of the drying air. The mathematical model of the hybrid solar biomass drying device was established and the drying uniformity of the drying chamber was predicted by Jackis Aukah. Through the CFD simulation, a uniform temperature and speed for the drying machine were obtained (Aukah, Muvengei, Ndiritu, & Onyango, 2015).

China's HBS, started relatively late, was in the early liberation period and began to imitate Soviet Union and Japan. With the rapid development of China's economy and the further requirements of the environment and food hygiene, China has begun to develop a HBS toward the direction of high efficiency, energy consumption, high quality, environmental protection, economy, and computer control, that falls far short when compared with the developed countries (Junlin Su, Zhenkun Wang, & Wang Wei, 2007).

At the end of 1998, cooperating with some manufacturers in Harbin, China's Agricultural Research and Design Institute successfully developed a highly efficient and indirect heating HBS of a small vertical cylinder type with coaling two returns. Agricultural Machinery Research Institute in Sichuan Province successfully developed a small and medium and tube-free HBS, which has a practical metal shell type and has been widely applied in the drying process of tea, grain, seeds, and so on (An L, & Fan Q, 2015; Wei Jin, Xuejun Zhang, Yun Liu, Jinshan Yan, Zenglu Shi, & Chaoxin Li, 2015).

In recent years, a small HBS has made great progress, which is suitable for China. A new type of straw HBS CL-40 has been developed, which opened a new age of biomass instead of using coal as a fuel. A novel combustion chamber of the HBS was designed, which is adapted to the combustion of graininess and rod-shaped fuel. It is effective to reduce the incomplete combustion of solid fuel and volatile gas. The temperature is controlled within a certain range and the ash and slag are effectively removed (Liguo Liu, Xuejun Zhang, Yun Liu, Jinshan Yan, Jie Sun, & Qianqian Xin, 2016).

Changzhou Renewable Energy Company, according to different temperature requirements from small and medium-sized enterprises, produced a series of HBSs, which can be used for heating, drying, baking. SY-1800RF vertical HBS, using straw as the raw material, is an efficient combustion furnace for drying tea and rice. It can generate 60~180°C hot air, and consume 54 kg energy fuel per hour. Its HTE can reach 60%~85%, and the cost is decreased by 50% when compared with using conventional coal as a fuel. However, in an actual application, heat exchange tubes are easily damaged. The preliminary study shows that heat transfer between the cold air and the coil tube is one of the important reasons for the damage, and the failure temperature of the furnace tube is 1255 °C (Nan Li, Zhiqiang Luo, & Jikang Li, 2015). However, since the high temperature will yield and distort the material, the temperature of flue gas outlet pipes is maintained in a certain range, and the low temperature (below 150 °C) will cause flue gas corrosion. In order to improve the service life, the low sulfur fuel is used, the combustion temperature is controlled, the anti-corrosion technology is adopted, and corrosion of flue gas is effectively prevented (Haipeng Su, 2012). It is necessary to analyze the temperature distribution and flow field distribution by simulation (Guo, Yan, Zhang, Liu, & Pei, 2014; Qi, Liu, Yao, & Li, 2015). However, hot air temperature and duration time are the key parameters of the evaluation of the HBS. Only under uniform temperature distribution and stable gas flow can ensure a normal and orderly function of the HBS (Weihan Chen, & Haibing Luo, 2005).

At present, the HBS is mainly based on the iron and steel industry, and biomass as the fuel of the drying furnace and heating furnace is ignored. However, the conditions of the biomass HBS and the blast furnace are different. It is necessary to analyze the cause of local damage by simulating its temperature and flow fields. Therefore, according to the actual structure of a SY-1800RF vertical HBS, a heat transfer model is established, the flow and temperature fields in the furnace under different conditions were studied, and the optimization scheme was proposed by using finite element software. Straw is used as fuel, which can not only protect the environment, but also deal with a large amount of straw accumulation in the most economical way. A large amount of heat is generated for baking, which turns waste into treasure. Hence, it is an urgent task to improve the HTE of the hot blast stove.

2 PHYSICAL MODEL AND MATHEMATICAL MODEL

2.1 *Physical model and conditional hypothesis*

Air is controlled in the furnace chamber. The flow rate, fluid density and pressure are different at different locations. The distribution of an air flow field in the HBS is analyzed by numerical simulation.

The physical model of the HBS is designed, as shown in Fig. 1. A HBS is a box structure (1810 × 750 × 1210 mm). A red rectangle area (280 × 250 mm) is the entrance, which is about 50 mm away from the right edge and 250 mm away from the top edge. A yellow rectangular area (450 × 400 mm) is the exit, 150 mm from the left edge and 400 mm from the bottom edge. The combustion furnace is simplified as a cylinder with the diameter of 600 mm. Inside the furnace chamber, thirty-six heat exchange tubes are uniformly distributed and the diameter of each tube is 92 mm.

According to the actual situation, the main assumptions of the finite element model are proposed.

1. The influence of the time factor on the heat transfer process is not considered, and the method of steady state analysis is used in the simulation analysis process (Ping Luo, Guilan Wang, & Xiuhua Wang, 2011).
2. The flowing medium in the HBS is ideal gas, with a low viscosity coefficient, which can ensure the flow in the furnace body to be fully developed.
3. At modeling time, it is considered that the heat exchange tube and each cylinder are tightly connected, and all the units are interconnected.

2.2 Mathematical model

The continuity equation and momentum equation of flow are given as follows:

$$\frac{\partial \rho}{\partial t} + \frac{\partial}{\partial x_i}(\rho u_i) = S_m. \tag{1}$$

$$\frac{\partial}{\partial t}(\rho u_i) + \frac{\partial}{\partial x_j}(\rho u_i u_j) = -\frac{\partial p}{\partial x_i} + \frac{\partial \tau_{ij}}{\partial c_j} + \rho g_i + F_i. \tag{2}$$

Figure 1. The physical model structure diagram.

where

$$\tau_{ij} = \left[\mu \left(\frac{\partial u_i}{\partial x_j} + \frac{\partial u_j}{\partial x_i} \right) \right] - \frac{2}{3} \mu \frac{\partial u_l}{\partial x_l} \delta_{ij}. \tag{3}$$

In order to keep the turbulent motion, it is necessary to continue to provide energy. Energy constantly spreads with the diffusion of the fluid under the effect of turbulence, and so only when the energy supply, energy diffusion and energy consumption are in a balance, turbulence can be in a constant state, and otherwise it will result in the disappearance of the turbulent kinetic energy with time.

According to the characteristics of the physical model, the standard $k - \varepsilon$ turbulent energy model is chosen to establish the mathematical model (Stamou & Katsiris, 2006), which is most widely used in engineering applications. This model introduces equations for the kinetic energy k and dissipation rate ε, which are defined as follows:

$$\frac{\partial(\rho k)}{\partial t} + \frac{\partial}{\partial x_i}(\rho u_i k) = \frac{\partial}{\partial x_j}\left[(\mu + \frac{\mu_t}{\sigma_k})\frac{\partial k}{\partial x_j}\right]$$
$$+ G_k + G_b - \rho \varepsilon - Y_M + S_k. \tag{4}$$

$$\frac{\partial(\rho \varepsilon)}{\partial t} + \frac{\partial}{\partial x_i}(\rho u_i \varepsilon) = \frac{\partial}{\partial x_j}\left[(\mu + \frac{\mu_t}{\sigma_\varepsilon})\frac{\partial \varepsilon}{\partial x_j}\right]$$
$$+ C_{1\varepsilon}\frac{\varepsilon}{k}(G_k + C_{3\varepsilon}G_b) - C_{2\varepsilon}\rho\frac{\varepsilon^2}{k} + S_\varepsilon. \tag{5}$$

where G_k and G_b are the turbulent kinetic energy, which are produced by the laminar velocity gradient and the buoyancy, respectively, Y_M is the fluctuation generated by excessive diffusion in the compressible turbulent flow, occurring in compressible turbulence, C_1, C_2, C_3, and σ_k are constants, S_k and S_ε are defined by the user. What is more, C_{1a}, C_{2a}, σ_k, and σ_ε are the parameters in the $k - \varepsilon$ model, and in the present study the following standard values are used: $C_{1a} = 1.44$, $C_{2a} = 1.92$, $\sigma_k = 1$, and $\sigma_\varepsilon = 1.3$. The value of μ_t is given as follows:

$$\mu_t = \rho C_\mu \frac{k^2}{\varepsilon}. \tag{6}$$

where C_u is 0.99.

2.3 Grid division and boundary condition setting

Tetrahedral meshes are used to simulate the flow field, and are encrypted in the boundary area. The standard wall equation is applied and the parameters of the $k - \varepsilon$ model are described in Section 1.2. The inlet air temperature is set at 305.15 K, air density ρ = 1.1576 kg/m³, specific heat c_p = 1005 J/(kg · K), the thermal conductivity k = 0.2688 $w/(m · K)$, and viscosity $\nu = 1.87 \times 10^{-5}$. Wall temperature, in accordance with the actual operation situation, is set at 770 K, the measured outlet temperature is 453 K, the HTE is about 60~85%, and the outlet flow rate Q_2 is 4500 m³/h.

Based on conservation of mass, the inlet mass flow rate is calculated to be 0.97 kg/s. The outlet is set to be a free-outlet boundary.

3 RESULTS ANALYSIS OF NUMERICAL SIMULATION

3.1 Analysis of the velocity field

After 3000 steps of iteration, an internal flow has been fully developed, and the iterative curve has achieved the convergence in the simulation.

Fig. 2 is the velocity contour (left) and the streamline (right) inside the HBS. Simulation results show that the inlet velocity v_1 is 11.97 m/s, outlet velocity v_2 is 6.21 m/s, and the process of the simulation has a certain energy loss.

As can be seen from Fig. 2, the air flow direction undergoes great changes inside the HBS, and the vortex is generated under the action of the inner tube of the heat exchanging pipes, which increase the velocity. Under the influence of temperature, the velocity will be greatly increased. Moreover, convective heat transfer will be extended with the increase of the velocity.

3.2 Analysis of the temperature field

Fig. 3 is the temperature contour of the whole model. Simulation shows that the outlet temperature T_2 is 451.3 K. It can be seen from Fig. 3 that, the temperature at the upper right region (entrance) is lower, while the temperature of the area near the flue gas pipe, combustion furnace, and the bottom of the heat exchange tubes, is higher. Therefore, these positions are prone to failure.

According to the simulation results, the HTE of the hot blast furnace can be calculated. The

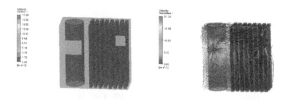

Figure 2. Velocity contour of the inlet and outlet (left) and streamline (right).

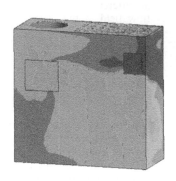

Figure 3. Temperature contour of the whole model.

calculation formula of heat dissipation by air through a HBS can be written as follows:

$$\Delta Q = Q_2 - Q_1 = MC_{p2}T_2 - MC_{p1}T_1. \tag{7}$$

where ΔQ is the flow difference between the inlet and outlet, Q_1 is the inlet flow rate, Q_2 is the outlet flow rate, and other parameters have been shown in Section 2.3.

Calculated from Eq. 7, heat dissipation is 541260 kJ/h.

In the actual process, the combustion furnace burns 50 kg biological particles per hour, and it can produce 21000 calories. [Note: one kilocalorie is equal to 4.184 kilojoules]. It can be obtained that the total amount of heat generated per hour is 878642 kJ.

The thermal efficiency formula of the HBS can be expressed as follows:

$$\eta = \frac{\Delta Q}{Q}. \tag{8}$$

The HTE of hot air is 61.60% according to Eq. 8.

From the above results, the outlet temperature is 451.3 K and the HTE of the HBS is 61.60%, which are corresponding with the actual measured temperature (453 K) and HTE (60%~85%). But the HTE is low, which may be not sufficient for heat transfer, and cause the heat exchange tubes overheat damage. And so, the operating parameters of the HBS may be not reasonable and need to be further optimized.

3.3 Influence of inlet velocity on flow field

Change the inlet velocity, while other conditions remain invariant, and explore the outlet temperature of the HBS, heat dissipation, and the HTE. The inlet mass flow rate is set at 0.2, 0.5, 0.97 1.4, 1.5, 2, 4, and 8.0 kg/s, the section velocity contours under different inlet velocities are shown in Fig. 4.

The section velocity contours of XZ directions can be seen from Fig. 4. With an increase in the inlet mass flow rate, the velocity near the inlet increases, which can cause more heat, reduce the temperature of this region, and improve the HTE.

The analysis results are shown in Fig. 5 and 6.

The outlet temperature increases with a decrease in the inlet mass flow rate, as shown in Fig. 5. When the mass flow rate changes from 0.2 kg/s to 2 kg/s, the outlet temperature is significantly reduced, and when it increases to 8 kg/s, the change of the outlet temperature slows down, but heat dissipation in the whole process increases linearly. The reason for this phenomenon is that the exchange time of the unit mass flow of air in the HBS is relatively short with an increase in the inlet mass flow, and heat dissipation by the unit mass flow of air is also decreased, and then the outlet temperature will be lower. However, due to the large inlet mass flow, the overall mass flow owns more heat, by which the HTE is improved for the HBS. Therefore, in

order to strike a balance, it is necessary to meet the requirements of the outlet temperature and then attain a higher efficiency. As the outlet temperature is controlled in the range of 445~455 K, the inlet mass flow is controlled in the range of 0.9~1.4 kg/s. The HTE of the HBS is shown in Fig. 9, which shows that the inlet mass flow rate is in the range of 0.2~1.4 kg/s. With an increase in the mass flow rate, the outlet temperature decreases, but the HTE is enhanced. Therefore, when the inlet mass is 1.4 kg/s, the HTE is the greatest, which is 84.4%.

3.4 Influence of inlet position on the flow field

Keep the inlet velocity and other conditions as constant, namely the inlet mass flow rate is 0.97 kg/s and the inner wall temperature is 770 K. The inlet position is arranged in the following five positions, which are shown in Fig. 7.

From Table 1, it can be seen that when the inlet position is in position 4 and position 2, although the outlet velocity has increased, the outlet temperature is reduced and leads to a decrease in heat

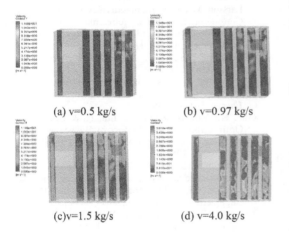

(a) v=0.5 kg/s   (b) v=0.97 kg/s

(c)v=1.5 kg/s   (d) v=4.0 kg/s

Figure 4.   The XZ section velocity contours in different inlet mass flow rates.

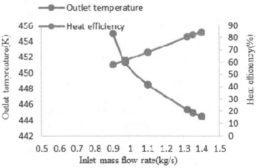

Figure 6.   Graph showing the variation curve of thermal efficiency.

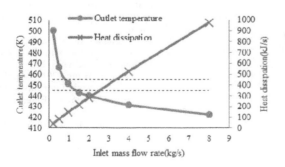

Figure 5.   Graph showing the outlet temperature curves and the change curve of heat away in different mass flow rates.

Figure 7.   The physical model of different inlet positions.

Table 1. Changes of parameters in different inlet positions.

Inlet position	Outlet velocity (m/s)	Outlet temperature (K)	Heat dissipation (kJ/s)
Position 1	6.21	451.3	150.4
Position 2	6.31	432.1	128.7
Position 3	6.02	438.0	135.1
Position 4	6.26	432.7	128.9
Position 5	6.03	442.2	139.8

dissipation. Considering the inlet velocity, the outlet temperature and the heat dissipation, position 1 is still more reasonable, where the HTE is the greatest.

4 CONCLUSIONS

1. The simulation results of the HTE of the HBS is consistent with the actual operational situation.
2. Properly increasing the inlet velocity can significantly increase the HTE and improve their temperature in the lower regions of the exchange tube. When the inlet mass flow rate increases from 0.97 kg/s to 1.4 kg/s, the HTE increases to 84.4% from 61.6%, thereby indicating a gap of nearly 20%.
3. The temperature is larger at the wall of the combustion furnace, near flue gas pipe and the bottom of the heat exchange tubes, where the stove is prone to damage and can be achieved by the application of the temperature resistant material, thickening, and improving the efficiency of the heat transfer process in the design. At the same time, the temperature of these dangerous locations is decreased by optimizing the inlet velocity.
4. The inlet position has a certain impact on the outlet velocity, outlet temperature and HTE. The HTE is the highest when the position 1 is selected as the entrance, and so the position 1 is relatively reasonable.
5. By performing simulation experiments, the HTE of the HBS is increased, and the goal of economic and environmental protection is achieved.

ACKNOWLEDGMENTS

This research is supported partly by the National Nature Science Funding of China (51274136), Shanghai Program for Innovative Research Team in Universities, Shanghai Leading Academic Discipline Project (S30106), Shanghai Key Laboratory of Mechanics in Energy Engineering, and the Shanghai Municipal Education Commission (Peak Discipline Construction Program).

REFERENCES

An, L., & Fan Q. (2015). Chengchang Zhang. Numerical Simulation and Optimization of Small Rapeseed Layered Bed Type Drying Equipment. J. Journal of Huazhong Agricultural University, 6, 125–129.

Aukah, J., Muvengei, M., Ndiritu, H., & Onyango, C. (2015). Simulation of Drying Uniformity inside Hybrid Solar Biomass Dryer using ANSYS CFX. J. Paper presented at the Proceedings of Sustainable Research and Innovation Conference.

Borah, A.K., Singh, P.K., & Goswami, P. (2013). Advances in Numerical Modeling of Heat Exchanger Related Fluid Flow and Heat Transfer. J. American Journal of Engineering Science and Technology Research, 1(9).

Chen, D., Hu, X., & Li, X. (2015). The way of energy utilization of biomass resources in iron and steel industry in China. J. Research on Iron and Steel, 4(43), 60–62.

Gunarathne, D. S., Mellin, P., Yang, W., Lövgren, J., Nilsson, L., Yang, W., Salman, H., Hultgren, A., & Larsson, M. (2015). Biomass as blast furnace injectant—Considering availability, pretreatment and deployment in the Swedish steel industry. J. Energy Conversion & Management, 102(6), 1137–1144.

Gunarathne, D.S., Mellin, P., Yang, W., Pettersson, M., & Ljunggren, R. (2016). Performance of an effectively integrated biomass multi-stage gasification system and a steel industry heat treatment furnace. J. Applied Energy, 170, 353–361.

Guo, H., Yan, B., Zhang, J., Liu, F., & Pei, Y. (2014). Optimization of the Number of Burner Nozzles in a Hot Blast Stove by the Way of Simulation. J. JOM, 66(7), 1241–1252.

Haipeng Su. (2012). The Flue Gas Acid Dew Point Temperature and Water Jacket Furnace Tail Heating Surface of Low Temperature Corrosion. J. Petroleum Engineering Construction, 38(2), 44–48.

Junlin Su, Zhenkun Wang, & Wang Wei. (2007). Study on The Hot Blast Furnace of Using Biomass Fuels. J. Energy-Saving Technologies, 25(2), 160–163.

Liguo Liu, Xuejun Zhang, Yun Liu, Jinshan Yan, Wei Jin, Jie Sun, & Qianqian Xin. (2016). Design and research of the combustion chamber of biomass briquette combustion stove. J. Journal of Agricultural Mechanization Research.

Nan Li, Zhiqiang Luo, & Jikang Li. (2015). Fracture Failure Analysis of Furnace Tube for Heat Exchange Converter. J. Heat Treatment of Metals, 6, 186–189.

Ping Luo, Guilan Wang, & Xiuhua Wang. (2011). Application of Biomass Hot Blast Stove in Greenhouse in Northeast China. J. Science-Technology Enterprise, (12X), 200–200.

Qi, F., Liu, Z., Yao, C., & Li, B. (2015). Numerical Study and Structural Optimization of a Top Combustion Hot Blast Stove. J. Advances in Mechanical Engineering, 7(2), 709675.

Stamou, A., & Katsiris, I. (2006). Verification of a CFD model for indoor airflow and heat transfer. J. Building and Environment, 41(9), 1171–1181.

Steininger, K. W., & Voraberger, H. (2003). Exploiting the medium term biomass energy potentials in Austria: a comparison of costs and macroeconomic impact. *J. Environmental and Resource Economics, 24*(4), 359–377.

Weihan Chen, & Haibing Luo. (2005). Numerical Simulation of Flow and Heat Transfer in Regenerative Reheating Furnace. *J.* Journal of Huazhong University of Science and Technology (Nature Science Edition), *33*(3), 17–19.

Wei Jin, Xuejun Zhang, Yun Liu, Jinshan Yan, Zenglu Shi, & Chaoxin Li. (2015). The Design and Research of Vertical Straw Energy-Saving Hot Blast Stove. *J.* Journal of Chinese Agricultural Mechanization, *36*(4), 249–252.

Wenting Liu, Ying Zhao, Shengyong Liu, & Fuying Yang. (2006). Design and Development of Biomass Briquette Hot Blast Stove. *J.* Journal of Henan Agricultural University, *40*(2), 201–204.

Wiklund, C. -M., Helle, M., & Saxén, H. (2016). Economic assessment of options for biomass pretreatment and use in the blast furnace. *J. Biomass and Bioenergy, 91*, 259–270.

Yan, J., Shamim, T., Chou, S., Desideri, U., & Li, H. (2016). Clean, efficient and affordable energy for a sustainable future. *J. Applied Energy*.

Yongming Sun, Zhenhong Yuan, & Zhenjun Sun. (2006). The status and future of bioenergy and biomass utilization in China. *J.* Renewable Energy, 78–82.

Zetterholm, J., Ji, X., Sundelin, B., Martin, P. M., & Wang, C. (2015). Model Development of a Blast Furnace Stove. *J. Energy Procedia, 75*, 1758–1765.

Advances in Energy Science and Equipment Engineering II – Zhou, Patty & Chen (Eds)
© *2017 Taylor & Francis Group, London, ISBN 978-1-138-71798-5*

Simulation research of the effect on gas drainage caused by the diameter of drilling in coalbeds

Jiajia Liu & Dan Wang

School of Safety Engineering, Heilongjiang University of Science and Technology, Harbin, China
State and Local Joint Engineering Laboratory for Gas Drainage and Ground Control of Deep Mines, Henan Polytechnic University, Jiaozuo, China
Hunan Key Lab of Coal Safety Mining Technology, Hunan University of Science and Technology, Xiangtan, China
School of Resources and Safety Engineering, China University of Mining and Technology, Beijing, China

Ang Liu

School of Resources and Safety Engineering, China University of Mining and Technology, Beijing, China

ABSTRACT: In order to reveal the mechanism of the effect of drilling diameter on gas drainage in coalbeds, a fluid–solid coupling model on gas-bearing coal was set under the effect of fluid–solid coupling. The model was set based on the effect of gas seepage and deformation of coal rock caused by the effective stress and the coal swelling induced by gas adsorption; meanwhile, the dynamic characteristics of porosity and permeability were considered. And then, the effect on gas drainage caused by the diameter of the hole drilled along the seam of 29031 working face in a certain mine was studied by the model. The results indicated that the diameter of drilling has some influences on the decrease of gas pressure in coal seams. Simultaneously, an exponential function between the diameter of drilling and the effective drainage radius is obtained. The influence radius of drainage and the effective drainage radius will gradually increase with an increasing diameter of drilling, but the increase rate will gradually reduce when the diameter of drilling exceeds 94 mm. Considering the actual situation and factors such as the engineering amount, the effect on gas drainage comprehensively, the most reasonable diameter of hole drilled along the coal seam of the 29031 working face is determined as 94 mm.

1 INTRODUCTION

In the process of coal mining, gas extraction in coal seams is the effective method to prevent disasters of coalbed methane leakage (Yu Qixiang, 1992). The sophisticated change of permeability in coal seams is a process of the intercoupling between the deformation of the coal skeleton and gas seepage during gas extraction (Kong Xiangyan, 2010). Moreover, the diameter of drilling affects gas extraction heavily for the reason that the oversized diameter of drilling will lead to a huge amount of work and the collapse of drilling; on the contrary, the undersized diameter of drilling will result in the unsatisfactory effect of gas extraction. Therefore, the reasonable diameter of drilling is crucial.

At present, scholars at home and abroad have carried out much research work on building the fluid–solid coupling models with regard to coal seams (LU Yiyu, 2014; Guo Ping, 2012; FEI Yuxiang,

2014; LANG Bing, 2013; YANG Xiaobin, 2014). The effect on gas extraction caused by the diameter of drilling is less studied by predecessors and there are different opinions on the only research results in selecting the diameter of drilling.

Therefore, in this paper, the fluid–solid coupling model in gas-bearing coal was established to describe the deformation of coal skeleton and the compressible gas under the consideration of the effects on gas seepage and the deformation of coal rock caused by the effective stress and the coal swelling induced by gas adsorption. Subsequently, the dynamic variable model on the permeability of gas-bearing coal was established by considering the dynamic characteristics of porosity and permeability, based on the definition of porosity and used the Kozeny–Carman equation. We studied the effect on gas extraction caused by the diameter of drilling based on the related physical properties' parameters of the 29031 working face in a certain coal mine by the numerical simulation software named COMSOL Multiphysics.

2 ESTABLISHED MODEL

2.1 Effective stress of coal rock

The effective stress of coal rock can be deduced in the consideration of the effective stress theory of Terzaghi, the effective stress, the stress of adsorption swelling, and the principles of elastic mechanics and physical chemistry of surfaces, and can be expressed as follows (Wu Shiyue, 2005):

$$\sigma_{ij}' = \sigma_{ij} - \frac{2aRT\rho_s\left[1-2\nu\ln(1+bp)\right]}{3V_m} \quad (1)$$

where ρ_s is the apparent density of coal, ν is the poisson ratio, a is the limited adsorbed amount per unit weight of coal, b is the adsorption constant of coal, R is the molar gas constant, T is the absolute temperature, and V_m is the molar volume.

2.2 Dynamic variation model of porosity

The fact is that the variation of porosity and permeability is dynamic. Then, the dynamic variation of the porosity model on gas-bearing coal in the stage of elasticity can be established by considering the definition of porosity, the deformation that is induced by compressing coal particles and the deformation is induced by coal particles' adsorption swelling, can be expressed as follows:

$$\phi=1-\frac{1-\phi_0}{1+\varepsilon_v}\left[1-\frac{\frac{3(1-2\nu)\Delta p}{E}}{+\frac{2a\rho_sRT(1-2\nu)\ln(1+bp)}{EV_m}}\right] \quad (2)$$

where ε_v is the volumetric strain, ϕ_0 is the initial porosity of gas-bearing coal, and E is the young modulus.

2.3 Dynamic variation model of permeability

The dynamic variation of the permeability model on gas-bearing coal in the stage of elasticity can be deduced by considering the definition of porosity and the equation of Kozeny–Carman, and can be expressed as follows (LIU Qingquan, 2015):

$$\frac{k}{k_0}=\left(\frac{1}{\phi_0}-\frac{1-\phi_0}{\phi_0(1+\varepsilon_v)}\left[1-\frac{\frac{3(1-2\nu)\Delta p}{E}}{+\frac{2a\rho_sRT(1-2\nu)\ln(1+bp)}{EV_m}}\right]\right)^3 \quad (3)$$

2.4 Control equation of the coal rock deformation field

The equation of the coal rock deformation field consists of the constitutive equation, the equilibrium differential equation, and the geometric equation.

2.4.1 The constitutive equation

For an isotropic elastomer, the constitutive equation about the elastic deformation under the effect of adsorption can be expressed as follows (LI Peicha, 2003):

$$\varepsilon_{ij}=\frac{\sigma_{ij}}{2G}-\left(\frac{1}{6G}-\frac{1}{9K}\right)\sigma_{kk}\delta_{ij}+\frac{1}{3}\delta_{ij}\left(\frac{\alpha p}{K}+\frac{\varepsilon_L p}{p+p_L}\right) \quad (4)$$

where G is the shear modulus, $G=E/2(1+\nu)$, σ_{kk} is the component of positive stress, $\sigma_{kk}=\sigma_{11}+\sigma_{22}+\sigma_{33}$, α is the Biot coefficient, $\alpha=1-K/K_m$, K is the bulk modulus of coal, $K=E/3(1-2\nu)$, K_m is the bulk modulus of the coal skeleton, ε_L is the volumetric strain of Langmuir, and p_L is the pressure constant of Langmuir.

2.4.2 The equilibrium differential equation

The equilibrium differential equation on mechanical properties in the deformation field of the coal skeleton can be obtained by the effective stress theory of Terzaghi, and can be expressed as follows:

$$\sigma_{ij,j}+F_i=0 \quad (5)$$

where $\sigma_{ij,j}$ is the stress tensor and F_i is the volume force.

2.4.3 The geometric equation

The following geometric equation of the gas-bearing coal can be obtained based on the continuous conditions on deformation:

$$\varepsilon_{ij}=\frac{1}{2}\left(\frac{\partial u_i}{\partial x_j}+\frac{\partial u_j}{\partial x_i}\right)=\frac{1}{2}\left(u_{i,j}+u_{j,i}\right) \quad (6)$$

where ε_{ij} is the component of strain and $u_{i,j}$ and $u_{j,i}$ are the components of displacement.

The equation of the coal rock deformation field can be deduced by combining Eq. (4), Eq. (5) and Eq. (6), and then can be expressed as follows:

$$Gu_{i,jj}+\frac{G}{1-2\nu}u_{j,ij}-\alpha p_i-K\frac{\varepsilon_L p}{p+p_L}p_i+F_i=0 \quad (7)$$

2.5 Control equation of coal rock methane's field

2.5.1 The continuous equation

Coalbed methane's seepage satisfied the law of conservation in mass and can be expressed as follows:

$$\frac{\partial m}{\partial t} + \nabla \cdot \left(\rho_g q_g \right) = Q_p \qquad (8)$$

where m is the gas content, t is time, ρ_g is the density of methane, q_g is the seepage velocity of methane, and Q_p are the sources and sinks.

2.5.2 The gas state equation

We have taken methane as an ideal gas and its density and pressure met the gas state equation; the equation can be expressed as follows:

$$\rho_g = \beta p \qquad (9)$$

where β is the compressibility factor of methane, $\beta = M_g / RT$, M_g is the molar mass of methane, and R is the ideal gas constant.

2.5.3 The equation of gas content in coal

Coalbed methane occurs in coal in the state of adsorption or free, and the gas content m in coal can be calculated by the following equation:

$$m = p \left(\beta\phi + \frac{ab\rho_0}{1+bp} \right) \qquad (10)$$

The seepage flow of coalbed methane meet the Darcy law under the function of pressure gradient, and can be expressed as follows:

$$q_g = -\frac{k_g}{\mu_g} \left(\nabla p + \rho_g g \nabla z \right) \qquad (11)$$

where k_g is the permeability of gas-bearing coal, μ_g is the dynamic viscosity of coalbed methane, g is the acceleration of gravity, the gravity term will be neglected in this paper because the gravity of coal bed methane is very small.

According to Peicha et al. [10], the variation of porosity in gas-bearing coal can be described as follows:

$$\frac{\partial \phi}{\partial t} = (1-\phi) \left(\frac{\partial \varepsilon_v}{\partial t} + \frac{1}{K_s} \frac{\partial p}{\partial t} \right) \qquad (12)$$

The control equation of the coal rock deformation field can be deduced by combining Eq. (8), Eq. (9), Eq. (10), Eq. (11) and Eq. (12), can be expressed as follows:

$$2 \left[\phi + \frac{p(1-\phi)}{K_s} + \frac{ab\rho_0}{(1+bp)^2} \right] \frac{\partial p}{\partial t} - \nabla \left[\frac{k_g}{\mu_g} \nabla p^2 \right]$$
$$+ 2(1-\phi) p \frac{\partial \varepsilon_v}{\partial t} = 0 \qquad (13)$$

The fluid–solid coupling model can be obtained by combining Eq. (2), Eq. (3), Eq. (7), and Eq. (13). And, it is the fluid–solid coupling model of gas extraction under the consideration of the effects on gas seepage and the deformation of coal rock caused by the effective stress and the coal swelling induced by gas adsorption.

3 ANALYSIS OF RESULTS

3.1 Occurrence features of methane

The 29031 working face starts from the track downhill of 29 in the west, adjoins the 211 mining area in the east, and reaches to the 2901 working face in the south; meanwhile, it is next to the 2901 working face in the north. The strike length of the working face is 658 m, its inclined length ranges from 126 to 136 m at an average of 132 m. The thickness of the coal seam ranges from 1 to 10 m at an average of 5.95 m. We obtained the primitive gas pressure as 1.16 MPa.

3.2 Geometric model

Aiming to obtain the appropriate diameter of drilling along the seam of the 29031 working face, we make a comparison on different groups with diameters of 65 mm, 94 mm, and 123 mm by combining the field diameters of drilling. A geometric model of 30 m × 50 m × 6 m is established with the strike length of 30 m, the inclined length of 50 m and the thickness of the coal seam 6 m. The related physical parameters of the model are listed in Table 1.

3.3 Definite conditions

While the initial time t = 0, the methane pressure distributed in the coal seam presented as the primary methane pressure of 1.16 MPa. The diameters of drillings were set as 65 mm, 94 mm, and 123 mm. The depth of the diameter of drillings were set as 48 m.

Table 1. Relevant parameters of the model.

Young modulus E/MPa	Poisson ratio v	Initial porosity ϕ_0/%	Initial permeability k_0/m^2
3000	0.33	0.0725	4.75e-17

Dynamic viscosity μ/Pa*s	Density of coal ρ/kg/m^3	Moisture M/%	Ash A/%
1.08e-5	1380	0.83	10.02

113

The model along the way on both sides of the working face towards the direction of the strike presented as a support by using stick. The inferior edge-appointed displacement in the direction of Z is 0 and the upper edge is free. And then, the overburden pressure is15.36 MPa.

3.4 *Simulation result analysis*

The effective gas drainage radius is the distance between the point with a coalbed methane pressure of 0.74 MPa within the gas drainage area and the center of gas drainage drilling; the influence radius is the distance between the point with a pressure declination of 10% and the center of gas drainage drilling. The distribution and iso-surface graphs of coalbed methane pressure under different diameters of drilling are shown as Fig. 1. Fig. 2 shows the decreased curves of coalbed methane pressure under different diameters of drilling.

From Fig. 1 and 2, it can be seen that the diameter of drilling has an influence on the decrease in gas pressure in the coal seam. The influence radius of drainage and the effective drainage radius would increase gradually with an increasing diameter of drilling. The diameter of 94 mm has a 0.27 m's increment when compared to 65 mm on the influence radius of drainage, but 123 mm only has a 0.09 m's increment when compared to 94 mm. The diameter of 94 mm has a 0.25 m's increment when compared to 65 mm on the effective drainage radius, but 123 mm only has a 0.03 m's increment when compared to 94 mm. Hence, the rate of the influence radius of drainage and the effective drainage radius will reduce gradually when the diameter is more than 94 mm. The specific effect area of gas extraction drilling is shown in Table 2.

The fitting curve equation between the diameter of drilling and the effective drainage radius can be expressed as $y = 2.13 - 1.15 \exp(-0.39x)$ and presented in Fig. 3; the correlation coefficient in

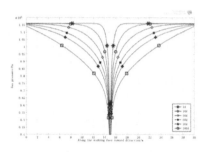

Figure 2. Decreased curves of coalbed methane pressure under different diameters of drilling is 94 mm.

Table 2. The effect area of gas extraction drilling.

Drilling diameter/m	Effect area/m	Effect area divided diameter	Effect drainage range/m	Effect drainage range divided diameter
0.065	9.84	151.43	1.35	20.81
0.094	10.11	107.55	1.60	16.94
0.123	10.20	82.94	1.77	14.40

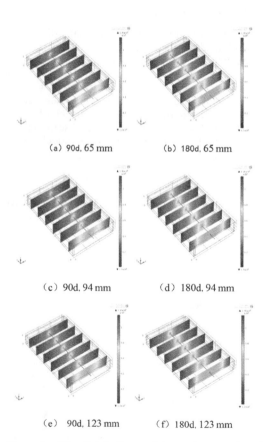

(a) 90d, 65 mm (b) 180d, 65 mm

(c) 90d, 94 mm (d) 180d, 94 mm

(e) 90d, 123 mm (f) 180d, 123 mm

Figure 1. The distribution and iso-surface graphs of coalbed methane pressure in different drilling diameters.

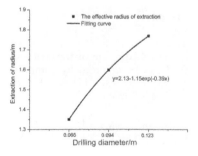

Figure 3. Variation curves of effective drainage radius in different diameters of drilling.

the equation is 0.99. From Fig. 3, with an increase in the diameter of drilling, the effective drainage area will gradually increase under the given drainage negative pressure and the extraction time. But the increase rate of the effective radius will reduce gradually when the diameter of drilling increases to a certain level. By combining the actual conditions in the 29031 working face, 94 mm is determined as the reasonable diameter of drilling.

4 CONCLUSION

The effect of gas extraction in the coal seam is more and more obvious with an increase in the gas extraction time under different diameters of drilling. Obviously, 180 d is a better gas extraction time when compared to 90 d. The diameter of drilling has an influence on the decrease of gas pressure in the coal seam. The influence radius of drainage and the effective drainage radius would increase gradually with an increasing diameter of drilling, and we obtain the fitting curve equation between the diameter of drilling and the effective drainage radius, it can be expressed as $y = 2.13 - 1.15\exp(-0.39x)$, the correlation coefficient in the equation is 0.99. With an increase in the diameter of drilling, the effective drainage area will gradually increase under the given drainage negative pressure and the extraction time. But the increase rate of the effective radius will reduce gradually when the diameter of drilling increases to a certain level. By combining the actual conditions in the 29031 working face, 94 mm is determined as the reasonable drilling diameter.

ACKNOWLEDGMENTS

This work was supported by the National Natural Science Foundation of China (51604101), funded by the Research Fund of State and Local Joint Engineering Laboratory for Gas Drainage & Ground Control of Deep Mines (Henan Polytechnic University) (G201608), and Open Research Fund Program of Hunan Province Key Laboratory of Safe Mining Techniques Of Coal Mines (Hunan University of Science and Technology) (201502).

REFERENCES

FEI Yuxiang, CAI Feng, et al. 2014. Gas-solid Coupled Model Based on Gas Seepage Characteristics of Drilling Drainage[J]. Safety in Coal Mines, (3):1–4.

Guo Ping, Cao Shugang, et al. 2012. Analysis of solid-gas coupling model and simulation of coal containing gas[J]. Journal of China Coal Society, 37(A2):330–335.

Kong, Xiangyan. 2010. Advanced Mechanics of Fluid in Porous Media[M]. Hefei: Press of University of Science and Technology of China: 661–679.

LANG Bing, YUAN Xinpeng, et al. 2013. Grouped pressure test to determine effective gas drainage radius[J]. Journal of Mining and Safety Engineering, 30(1): 132–135.

LI Peicha, KONG Xiangyan. 2003. Mathematical modeling of flow in saturated porous media on account of fluid-structure coupling effect[J]. Journal of Hydrody-namics, 30(10):419–426.

LIU Qingquan, CHENG Yuanping, et al. 2015. A Mathematical Model of Coupled Gas Flow and Coal Deformation with Gas Diffusion and Klinkenberg Effects[J]. Rock Mechanics and Rock Engineering, 48(3):1163–1180.

LU Yiyu, JIA Yajie, et al. 2014. Coupled fluid-solid model of coal bed methane and its application after slotting by high-pressure water jet[J]. Journal of Mining & Safety Engineering, (1):23–29.

Wu Shiyue, Zhao Wen. 2005. Parametric correlation between expansion deformation of coal mass and adsorption thermodynamics[J]. Journal of Northeastern University, 26(7):683–686.

YANG Xiaobin, TAO Zhenxiang, et al. 2014. Numerical simulation on fluid-solid coupling of gassy coal and rock[J]. Journal of Liaoning Technical University (Natural Science), (8):1009–1014.

Yu, Qixaing. 1992. Mine Gas Prevention and Control[M]. China University of mining and technology press: 229–230.

Advances in Energy Science and Equipment Engineering II – Zhou, Patty & Chen (Eds)
© 2017 Taylor & Francis Group, London, ISBN 978-1-138-71798-5

A study on the compensation mechanisms for cultivated land protection in China

Y.F. Yang & J.Q. Xie
School of Land Science and Technology, China University of Geosciences (Beijing), Beijing, China

ABSTRACT: In this paper, the existing protection and compensation conditions for domestic and foreign cultivated land based on the theoretical study of the compensation mechanism of cultivated land protection are analyzed. The proposals are given by exploring the problems in China: (1) establishing and improving the economic compensation mechanism on the basis of a deepened protection for cultivated lands by means of both management and control. (2) Adjusting and balancing the interest allocation relationship among relevant economic entities so as to form a complete restraining and encouraging mechanism for cultivated land protection. (3) Bringing 'land consolidation' into the compensation mechanism of cultivated land protection to form a constructive compensation mechanism for protecting cultivated lands, which promotes the protection by construction.

1 INTRODUCTION

Cultivated lands, which provide the material basis for both human survival and development, are special public resources, not only for their commodity value (economic value), but also for their ecological environment and social security value. In recent years, China has been insisting on implementing the most stringent cultivated land protection system and the most stringent land conservation system to ensure both development and safe baseline, in which it has been attaining remarkable success. (Yong Xinqin & Zhang Anlu, 2012).

While currently, with the accelerated process of urbanization and industrialization in China, construction land shortage and cultivated land protection's severe conditions contribute to the more and more prominent conflicts between cultivated land protection and economy development. And meanwhile, with the rapid development of farmland conversion and the growing challenge with both food security and ecological security, the protection for cultivated land is becoming increasingly difficult in inhibiting the sharp reduction of farmlands by simply raising the standards for land requisition compensation.

The trends are shown in Figure 1 and 2, which are the cultivated areas in 2009–2014 and the state-owned land of construction use supplied in 2009–2014.

In the process of the Chinese economic development, the protection and compensation of cultivated lands are important manifestations of urban and rural overall development, and therefore their

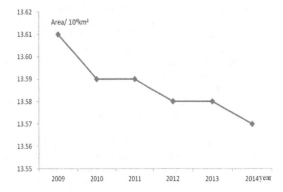

Figure 1. Graph showing the cultivated land area in 2009–2014.

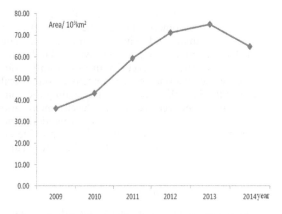

Figure 2. Graph showing the state-owned land of construction use supplied in 2009–2014.

function should be highlighted to make these institutionalized, stabilized, and continuous in playing a solid foundation role for the urban and rural overall development.

2 THORETICAL ANALYSIS FOR CULTIVATED LAND PROTECTION AND COMPENSATION MECHANISM

The protection for cultivated lands is to protect the sustainable productivity of basic agricultural products, which is necessary to meet the requirement of the Chinese population and economic development, to achieve harmony and unity in farmland quantity, quality, and environmental protection (Li Shiping & Li pin, 2005).

2.1 The externality of cultivated land protection

Externality, proposed by Marshall from Cambridge University in his "Principles of Economics", published in 1980, refers to the effect that production and consumption are behaviors of one economic agent (producers and consumers). Among which, the beneficial effect is called the external economy or positive externality, and on the contrary, the harmful effect is called external diseconomy or negative externality (external cost).

The protection for cultivated lands bears distinctive externality. The externality is shown in those regions which assumes more responsibilities for cultivated land protection over its counterparts who assume less responsibilities in the process of protecting the cultivated land, which is mainly seen in national food security, ecological environment, social security, etc. (Zhou Xiaoping et al., 2010).

2.2 Sustainable development theory

Chinese huge population and unbalanced economic development make the contradiction between human and cultivated land intense. The protection and usage of cultivated lands not only affect the food security in our country, but involve the sustainable development of our social economy. Therefore, based on reconciling the contradiction between economic development and cultivated land protection, the protection and compensation for cultivated lands will provide an inexhaustible driving force for urban and rural overall development.

2.3 Land development rights

The land development rights include farmland development rights, construction land development rights, and unused land development rights.

Under the condition of marketing economy, power relationship is the foundation of benefits transfer. A stable and accurate property right is the foundation of forming price (Zhou Jianchun, 2007). An imperfect property right setting will inevitably lead to the distortion of the price mechanism.

In the transition period of converting farmlands to construction lands, ownership transferred; meanwhile, land development rights, which are independent of ownership, are transferred to the government as well. However, judging from the point of view of composition of land acquisition price, only land compensation is included, while the compensation for land development rights is excluded. It seems that land development rights disappeared together with the loss of ownership in this transaction process (Zang Junmei et al., 2008). Hence, it is desired to establish the compensation mechanism for cultivated land protection in order to coordinate the imbalance of economic development.

3 ANALYSIS FOR THE CURRENT SITUATION OF BOTH DOMESTIC AND FOREIGN CULTIVATED LAND PROTECTION AND COMPENSATION

3.1 Analysis for the current situation of foreign cultivated land protection and compensation

The economic subsidy for cultivated lands is mixed with agricultural subsidies (especially in a mixture with the agricultural ecological subsidy and agriculture environmental protection plan). Initiated in the 1930s and finalized in December 1933 in the form of "Agricultural Adjustment Act", the agricultural subsidy for the United States aimed at setting the target price and providing the deficiency subsidy for agricultural products. The United States government adopted the amended Agricultural Law subsequently, in which the previous simple agricultural subsidy (business subsidy) is extended to the category of agricultural trade subsidy and resource conservation subsidy. The Farmland Protection Program has become an important part of the protection plan in the Agricultural law (Li jingyun, 2007; Alterman, 2006). The Farmland Protection Program provides financing to states or local governments and the private organizations to buy out the farmlands right on the edge of the city, and thus these farmlands can be used for agricultural production again (Yin hong, 2005). Farmlands can be efficiently protected in the means of compensation to farmland owners in the process of purchasing and transferring farmland development rights. Besides, the US government stimulates and rewards using arable lands for agricultural purposes by direct or indirect financial compensation, in which tax preference and relief policy is included. The German Ministry of Agriculture believes that agriculture bears the double function of food production and ecological maintenance and is of great significance especially to biological diversity and the protection and maintenance of water resources, soil, and climate. Thus, beginning in 1990s, the German Ministry of Agriculture targeted promulgating and enacting related agro-environmental policies and regulations. In the meantime, the German Ministry of Agriculture also encouraged farmers to engage in agricultural production in an environment-friendly way by ecological compensation and achieved good results (Li jingyun, 2007).The UK made great efforts in providing subsidies and implementing the protection policy in the framework of EU's Common Policy (CAP), its agricultural infrastructure construction and informational service to farmers (Yin hong, 2005).

In addition, other European Union countries, Japan, India, and Serbia are implementing the agricultural compensation policy to protect the quantity and quality of cultivated lands in direct or indirect ways by pursuing legislation.

3.2 Analysis for the current situation of domestic cultivated land protection and compensation

In recent years, some cities with better economic conditions like Chengdu city in Sichuan province, Foshan city and Dongguan city in Guangdong province, Haining city in Zhejiang province, in combination with local reality, preliminarily established the direct funding compensation mechanism to those farmers and collective economic organizations who took the responsibility of protecting cultivated lands. Farmers' enthusiasm to protect cultivated lands as subject liabilities are improved greatly, which is of great importance to promote urban and rural overall development.

Chengdu withdrew money from land-transferring fees, paid use for incremental construction lands and part of the farmland occupation tax to the local government. Compensation is given to those farmers who have contracted land management rights and have shouldered the responsibility for farmland protection, and to village collective organizations who shoulder the responsibility for not-contracted household farmland protection (the above subsidies object will not change because of the contracted land circulation). The compensation standard varies according to cultivated land gradation, 400 yuan/mu for the 1st level and 300 yuan/mu for the 2nd level annually. (Liao heping et al. 2010, Yang huizhen, 2009).

Foshan has formulated "Foshan prime farmland protection subsidy implementation measures" for fund collection, in which the crucial part should be played by districts and towns (mainly 80%), while the role of municipal subsidy is complementary (about 20%). Compensation is provided to those farmer collectives or other responsibility units, who have been undertaking the prime farmland protection assignment and signed letters of responsibility. Compensation is only targeted to the prime farmland designated in the Overall Land-Utilization Plan (Xu xiaomian, 2010). Dongguan city only provides subsidy to prime farmland protection areas and non-economic forest land areas, which is the excessive allocated part to village collective, excluding the part owned by residents committee, all state-owned and town-owned woodlands, enterprises, and troop stationed areas. The Municipal Finance Bureau will grant subsidies to the exceeding part based on a standard of 500 yuan/mu annually, if the percentage, 13.87%, which is to divide the farmland protection area in the city by the total area of the city, exceeds the average allocated proportion (Ma aihui et al., 2011, Zhou xiaoping et al., 2016, Niu haipeng & Zhang anlu, 2009).

Haining city compensates the farmers and collective economic organizations who take the responsibility for farmland protection by using money from the paid use of incremental construction lands and land-transferring fees (Chen xiuqin, 2011).

4 THE SHORTAGE FOR CURRENT CULTIVATED LAND PROTECTION AND COMPENSATION MECHANISM IN CHINA

At present, although in several experimental areas, the establishment of cultivated land protection mechanism brings some positive impact, it still has problems, difficulties, and shortage.

4.1 Simple content and form for cultivated land protection and compensation

At present, the protection and compensation for cultivated lands only targeted the productive property of cultivated lands, while the compensation for its contribution to ecology and environment is ignored. The cultivated land has the multi-function on economic output, social capacity, and ecosystem environment service (Xin kexia & Wang qinli, 2007) and is valuable to the economy, ecology, and society (Zhang xiaojun et al., 2006). The 12th 5-Year plan in China has proposed explicitly to build a green development, resource-saving, and environment-friendly society. Therefore, apart from financial compensation, cultivated land protection and compensation should also take ecological, environmental, and social compensation into consideration.

Most compensation is offered by giving money directly, while other forms of effective supplement is quite poor. Chengdu city allocates the responsibility and obligations among the Administrative Department of Land Resources, Department of Finance and Labor, and Social Security Department. The Administrative Department of Land Resources is in charge of the operation management and annual distribution plan for cultivated land protection fund, division of cultivated land protection category and supervision of the implementation status for cultivated land protection responsibility. The Department of Finance is in charge of monitoring the collection and usage for cultivated land protection fund. The Labor and Social Security Department is in charge of filing endowment insurance registration for farmers who undertook the responsibility of protecting cultivated lands. Apart from these tasks, the protection for cultivated lands still requires the supplements of agricultural technology.

Currently, most of the compensation for cultivated lands is limited to unified administrative divisions. Several experimental areas for cultivated land compensation only focus on the compensation for certain small unit administrative divisions. Distinctive regional differences still exist in current cultivated land protection plans.

4.2 Urgent improvement for scientific basis for cultivated land compensation

4.2.1 Unfulfilled transition from quantity to quality for cultivated land protection

The present compensation for cultivated land protection is limited to protect cultivated land quantity, rather than the unification of quality and quantity due to the defect in both system and technology. Currently, the cultivated land protection and compensation mechanism implemented in several experimental areas have not broken through the traditional ideas like "protecting for prospecting's sake". There should be explicit rules and requirements for cultivated land quality. Meanwhile, cultivated land quality quite lacks effective supervision and protection.

4.2.2 Inadequate information system

The current experimental area, such as Chengdu, lacks a three-dimensional, dynamic and real-time geographic information system and a supervision system for cultivated land protection and compensation, though it has already combined the results from "The Second National Land Investigation", signed the contract for cultivated land protection

and built five levels of cultivated land protection ledger based on district, town, village, group, and household.

4.3 Lack of efficient convergence to the land usage planning and the national land consolidation project

The cultivated land protection and compensation mechanism is not a simple protection plan or protecting for prospecting's sake. It should be combined with land usage planning and the national land consolidation project. In experimental areas, the compensation range and scope are based on general farmland usage and basic farmland usage delimited in the general plan of land utilization. Whether the compensation level is higher or only for basic farmlands embodies the compensation inclination to basic farmlands. However, though the establishment of the compensation mechanism improves farmers' enthusiasm greatly, it also brings problems of how to distribute the interest fairly and reasonably, such as the adjacent fields could get different compensation standards, which depends on whether they are basic farmlands, which places a higher requirement for scientific planning. At present, in experimental cities, the protection and compensation for cultivated lands lack the efficient convergence with land consolidation work. A scientific and holistic development is needed nationwide to accelerate.

5 TO PERFECT THE COMPENSATION MECHANISM OF CULTIVATED LAND PROTECTION

5.1 Broaden the compensation connotation, establish the integrated compensation mode with economic compensation, ecological compensation, and social compensation

The protection for cultivated land should be comprehensive. The ecological value, social value, and economic value should be evaluated comprehensively. Consideration should also be given to the quantity and quality of cultivated lands. In this way, the protection can be multi-faceted and comprehensive.

Economic compensation is to compensate the production function of cultivated lands for its comparative lower profits. Ecological compensation is provided to compensate the cultivated land for its ecological service function. The economic calculation for the ecological value and the evaluation for the ecological service function should be proceeded together with economic compensation. The ecological compensation mechanism for cultivated lands should be established, and in this way, the ecological value will be highlighted. Internalizing the ecological and environmental protection cost for cultivated lands is helpful in improving farmers' enthusiasm and environmental consciousness in the protected region of cultivated lands. Social compensation should be strengthened (intra-generational equity and inter-generational equity compensation for cultivated land protection). The fundamental principle such as "from the land, giving back to land," "who protects, who benefits" and "more usage, more compensation" should be established and improved. The compensation transfer payment system should be established. The problem of sharing benefits and costs within the region or beyond the region can be solved by public financial transfer payment. The interest relationship of different regions on cultivated land protection should be coordinated. Land development fund or intergenerational compensation fund should be established to balance the profits of land development within the generation and beyond the generation.

In the designing period, the cultivated land compensation mechanism should protect the quantity and quality for cultivated lands as a whole and offer compensation. The economic compensation is to secure the cultivated land from not being destroyed, not being used for non-agricultural construction, not being abandoned, and to ensure that the quantity is not reduced. Meantime, compensation should also be given to spontaneous, organized small-scale farmland consolidation, cultivated land water conservancy, cultivated land quantity increase, and cultivated land quality improvement.

5.2 Accelerate the basic work for farmland compensation and improve the efficiency of the system

Work involves improving the compensation by accounting standard and measurement technology, improving and strengthening the classification and grading for agricultural lands. Situations such as the reduction of cultivated land quality leading to the invisible reduction of cultivated land quantity should be avoided. The measurement standard and method for cultivated land quality should be built. Relevant database should also be built to form a good circulation system like dynamic monitoring, dynamic complement, and dynamic compensation.

At the same time, innovation should be accelerated. Cooperation with agricultural sectors should be strengthened with the idea of "Give man a fish, but also teach him how to fish". While protecting cultivated lands, the support system of science and technology and the system environment should be built to encourage farmers to protect the cultivated land quality. The first is to establish and improve the balanced fertilization system for main crops (such as soil testing fertilization). The second is to strengthen the research and extension for the technology of cultivated land protection, improve the agricultural industrial structure, and ensure fine and mass development of agricultural production. The third is to emphasize the training for farmers' agriculture technical level and professional skills, reward the technical innovation on the protection level for cultivated land quality in practice. All these are to ensure the stabilization and improvement for total cultivated land quantity and quality in urban and rural overall development, to ensure that all farmers can share the fruits of economic and social development.

5.3 Integrated planning to establish regional coordination and balance mechanism of economic compensation for rural cultivated land protection

After an investigation, the cultivated land decreased much faster in developed areas, but due to the limited amount of regional cultivated lands, the smooth execution of Cultivated Land Requisition-compensation balance is quite difficult, which brought an awkward situation of balancing the regional economy development and cultivated land protection. In agriculture-oriented areas, there is little upward pressure for cultivated land protection, but the level of economic development is low. The subject of cultivated land protection bears the responsibility of protecting the land, but loses the chance to get profits from land-transferring and corresponding economic compensation. From a global perspective, the regions with less responsibility on protecting cultivated lands and basic farmlands are the beneficiaries of cultivated land protection. These regions should provide proper compensation to their stressful peers with methods like financial transfer and share their responsibility in protecting cultivated lands and basic farmlands, so as to establish the regional coordination and balance mechanism for regional cultivated land economic compensation, as shown in Figure 3:

5.4 Combined with the national rural land consolidation opportunity, bring "promote the construction by compensation" into the cultivated land compensation mechanism, to strongly enhance protection measures for incentive control and stimulation

At present, the major measures for cultivated land protection include management and control protection, constructive protection, and incentive

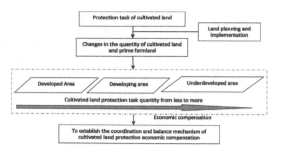

Figure 3. Flowchart showing the establishment of a regional economic compensation mechanism for regional coordination and balance.

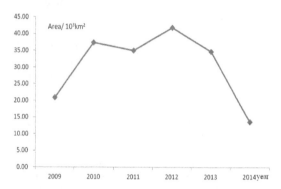

Figure 4. Graph showing the area of cultivated lands increased by land consolidation and improvement in 2009–2014.

protection. Among these, land consolidation is the primary means of constructive protection (increase the cultivated land quantity and improve the cultivated land quality).

However, in the current cultivated land protection work, constructive protection (land consolidation) is seldom introduced in the domain of cultivated land protection mechanism. While the land consolidation implementation model, which is "providing subsidies instead of investment, promoting construction with subsidies", not only can achieve the desired effect of increasing cultivated land quantity and improving cultivated land quality to realize the goal of protecting cultivated lands with "improving protection with construction", but also farmers' subject position in land consolidation and cultivated land protection is highlighted. Making farmers get more benefits from land consolidation and cultivated land protection, their enthusiasm in cultivated land construction and protection would be improved so that the double effect of "constructive protection and incentive protection" can be achieved. In recent years, the

Land and Resource System collected one hundred billion yuan every year to carry out land consolidation work in rural areas. The total area of cultivated lands that increased by land consolidation from 2009–2014 is 183440 km². The area of cultivated lands increased by land consolidation and improvement in 2009–2014 is shown in Figure 4.

Combined with national land consolidation planning as on one chessboard, the compensation mechanism of constructive cultivated lands should be improved to make it supplement and improve with both management and control protection and incentive protection in order to protect the cultivated land resource in our country.

6 SUMMARY

In short, the compensation mechanism should be guided with the basic principle of urban and rural overall development, adhere to the fundamental principle of "from the land, giving back to land", "who protects, who benefits" with the method of town-driving-country, nurturing agriculture by industry. The basic work of cultivated land compensation should be accelerated with the establishment of a compensation model of economic compensation, ecological compensation, and social compensation.

On the basis of strengthening management and control protection for cultivated lands, the economic compensation mechanism for cultivated land protection should be established and improved. By regulating and balancing the interest allocation among relevant economic entities, a complete incentive-restrictive mechanism on cultivated land protection will be formed. Land consolidation should be introduced to the domain of cultivated land compensation mechanism. A constructive cultivated land compensation mechanism should be formed through the method of improving protection by construction, and thus a long-term protection and compensation mechanism, which is to manage, protect, and construct the cultivated land in our country properly, will be formed. What is planned is to effectively increase farmers' income, improve their living standard and eventually, speed up the process of urban and rural integration.

REFERENCES

Alterman. R. (2004). Farmland Preservation. Nell. J.S, Paul. B, International Encyclopedia of the Social and Behavioral Sciences. *London: Elsevier Science Publishers.* 5406–5411.

Chen, X.X. (2011). The Construction of Compensation Mechanism for Cultivated Land Protection. *J. China Land.* (9), 49–50.

Li, J.Y. (2007). Discussion on the legal mechanism of ecological compensation in China: A Case Study of agricultural ecological compensation in the United States. *J. Green vision,* (10), 33–36.

Li, S.P. & Li, P. (2005). Reflections on the protection of cultivated land in China. *J. Anhui Agricultural Science.* 33 (7), 1323–1325.

Liao, H.P. et al. (2011). Study on the Compensation Standard for Farmland Preservation in Chongqing City. *J. China Land Science.* 25(4), 42–48.

Ma, A.H. et al. (2007). An Empirical Study of Cultivated Land Ecological Compensation Based on Choice Experiments Method. *J. Journal of Natural Resources.* 27(7), 1154–1161.

Niu, H.P. & Zhang, A.L. (2009). Externality and Its Calculation of Cultivated Land Protection: A Case Study of Jiaozuo City. *J. Resources Science.* 31(8), 1400–1408.

Xing, K.X. & Wang, Q.L. (2007). German Agricultural Ecological Compensation and Its Implications for Chinese Agricultural Environmental Protection. *J. Agricultural Environment and Development.* (1), 1–3.

Xu, X.M. (2010). Practice of the economic compensation mechanism of cultivated land protection. *J. Guangdong Land Science.* 9(2), 4–8.

Yang, Z.H. (2009). Practices and Reflections of establishment of compensation mechanism for cultivated land protection and in Chengdu city. *J. China Population, Resources and Environment.* (19), 22–25.

Yin, H. (2005). Agricultural protection programs in the United States and the European Union. *J. China Environmental Protection Industry.* (3), 42–45.

Yong, X.Q. & Zhang, A.L. (2012). Discussion on the Compensation Standard of the Arable Land Protection Based on Food Security. *J. Resources Science.* 34(4), 749–757.

Zang, J.M. et al. (2008). Study on regional compensation mechanism of cultivated land protection under dynamic balance of cultivated land. *J. Research of Agricultural Modernization.* 29(3), 318–322.

Zhang, X.J. et al. (2006). Assumption of Regional Cultivated Land Protection Compensation Mechanism. *J. Research of Agricultural Modernization.* 27(2), 144–147.

Zhou, J.C. (2007). Study on the property right and value of cultivated land in China. *J. China Land.* 21(1), 4–9.

Zhou, X.P. et al. (2010). Economic Interpretation on the Compensation for Farmland Preservation. *J. China Land Science.* 24(10), 30–35.

Zhou, X.P. et al. (2016) The Inter-Regional Farmland Protection Compensation Distribution Index System and Application—A Case Study in Fuzhou City. *J. Economic Geography.* 5, 152–158.

Advances in Energy Science and Equipment Engineering II – Zhou, Patty & Chen (Eds)
© 2017 Taylor & Francis Group, London, ISBN 978-1-138-71798-5

Application of a high-precision magnetic survey to the investigation of mineral resources in Huotuolewusu area, Inner Mongolia

Q. Liu & R.M. Peng
School of Earth Sciences and Resources, China University of Geosciences, Beijing, China

ABSTRACT: In this paper, a 1:25,000 high-precision magnetic survey has been performed by using a *GSM-19T* proton magnetometer in Huotuolewusu area. The data processing includes gridding, filtering, reduction to the pole, continuation, derivation, and so on. By precise interpretation and inference of the magnetic anomaly, the distribution characteristics of the basic rock mass was found out. According to the qualitative and quantitative analysis of two well-selected magnetic profiles (*P1* and *P2*), the distribution characteristics of the magnetic ore rock mass is preliminarily concluded. We find two good polymetallic mineralization targets, which achieved the effect of the geophysical prospecting.

1 INTRODUCTION

Huotuolewusu area is located in the western part of the northern edge of North China Craton. Mineralization and geological conditions in this area are favorable. But there is only a 1:200,000 regional geological survey work, and the research degree is relatively low. Small iron ores have been found in this research area, and so the high-precision magnetic measurement would be a good method for researching the iron ore's metallogenic potential in this area (Yan et al. 2009; Matthew et al. 2005). It has important theoretical and practical significance to guide the metallogenic predictions in this area.

2 GEOLOGICAL BACKGROUND

Huotuolewusu area, located at the western end of the Langshan mountain rift trough, belongs to the Langshan trough of the northern margin of North China Craton (Peng et al, 2007). The strata mainly consists of metamorphic rocks, including the Archeozic Wulashan Group, the Mesoproterozoic Zhaertaishan Group, Cretaceous, and Quaternary. The main mineralized strata are the Archeozic Wulashan Group. It mainly composes of gray to gray-black Biotiteplagioclase Gneiss containing garnet, amphibolites, and migmatization of gneiss.

The structures are dominated by folds and fractures. The fractures obvious direction is in this area, which extends to the north-east direction and is consistent with the spreading direction of the rock formations. There are widely distributed intrusive rocks, including acidic intrusive rocks and small basic intrusive rocks.

3 GEOPHYSICAL CHARACTERISTICS AND DATA PROCESSING METHODS

3.1 *Geophysical characteristics*

In order to study the magnetic characteristics of the main rock (ore) in this area, 144 samples were collected to determine its magnetic affinity. Various types of rock samples were classified and measured. The magnetic parameters are shown in Table 1.

Table 1. The statistical table of rock magnetic susceptibility.

Lithology	Number	Magnetic susceptibility($\times 10^{-5}SI$)		
		Minimum	Maximum	Average
Granite gneiss	21	1	5167	2584
Serpentinous marble	33	−26	146	60
Plagioclase amphibolite gneiss	6	4	2752	1378
Amphibolite	48	97	28,866	13,254
Plagioclase amphibolite	36	9	18,761	8413

From Table 1, we find that the magnetic susceptibility average values of serpentinous marble is $60 \times 10^{-5} SI$, which shows weak magnetic properties and cannot cause magnetic anomalies. Granite gneiss and plagioclase amphibolite gneiss have higher magnetic susceptibility average values of $2584 \times 10^{-5} SI$ and $1378 \times 10^{-5} SI$. Plagioclase amphibolites and amphibolites have the highest magnetic susceptibility average values of $8413 \times 10^{-5} SI$ and $13,254 \times 10^{-5} SI$. Overall, the susceptibility which is related to their mineral composition vary widely among different rock types. And so, it has the prerequisite to perform a high-precision magnetic survey.

3.2 Data processing methods

A *GSM-19T* proton magnetometer which is producted in Canada is used to gather field data. Steps of the magnetic prospecting work include arranging the survey grid, checking instruments, establishing a diurnal variation station, observing data, and processing data. Before field work we should check the instrument performance, observation accuracy, and consistency to ensure that it can work normally.

We have used a computer program to treat diurnal variation correction. ΔT corrections accuracy is 0.1 nT.

A contour map was generated by Surfer software by using the Kriging method and the search radius is 400 m. The mapping was completed with MapGIS software.

The data processing has been carried out before interpreting the magnetic body including reducing to the pole and continuation (Smith et al, 2005). Reducing to the pole is to eliminate the influence to the asymmetry of the magnetic anomaly caused by inclination and declination. Based on reducing to the pole, the upward continuation aims to suppress the shallow stratigraphic interference and reflect deep anomalies. It used geomagnetic parameters such that the magnetic declination D = –4.3° and geomagnetic inclination I = 59.7°.

Through qualitative and quantitative interpretation of magnetic profiles, the depth of the magnetic body top, thickness, and other occurrence characteristics of the magnetic rock can be roughly determined (Liu, 1992; Guan et al. 1993).

4 INFERENCE AND INTERPRETATION OF MAGNETIC ANOMALIES

4.1 Characteristics of the magnetic anomaly

There are two magnetic anomaly zones in this area, which named C1 and C2 in this area (Fig. 1 and 2).

Magnetic anomalies in C1 are mainly positive anomalies, which shows a banding distribution along the NNE direction. The maximum of magnetic anomaly of C1 is up to 1900 nT. High value anomalies are corresponding with black coarse-grained amphibolites and plagioclase amphibolites, which have high magnetic susceptibility values (Table 1). By upward continuation, we think that there are magnetic geologic bodies in the bottom of the magnetic anomaly zone.

The maximum in the magnetic anomaly of C2 is up to 2200 nT, which shows a subelliptical shape. The main lithology of the C2 magnetic anomaly zone is gneiss, which contains some amphibolites and plagioclase amphibolites which can cause magnetic anomaly. By upward continuation we think that there are magnetic geologic bodies in the bottom of the magnetic anomaly zone. We speculate that there may be a fracture in the bottom according to the coexistence of positive and negative anomalies and the large anomaly gradient.

4.2 Precision measurement anomaly profile and quantitative interpretation

For understanding the depth of the magnetic body, the space form of the stretch and the strike of main vertical anomalous, we had selected two precision measurement profiles P1 and P2. These

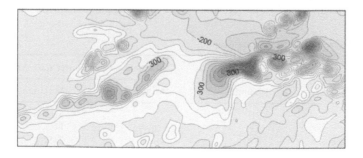

Figure 1. ΔT isoline map of the magnetic anomaly of Huotuolewusu area.

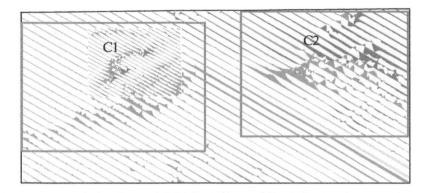

Figure 2.   ΔT section of the magnetic anomaly of Huotuolewusu area.

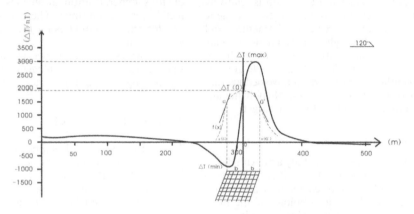

Figure 3.   Curve exploded schematic of the P1 line precision magnetic survey section.

two profiles are arranged in the area where anomalies change greatly, the direction of the two selected profiles are perpendicular to the strike of abnormalies, or crossing the position of positive and negative extreme points.

The maximum magnetic anomaly value of line P1 is 2903 nT and the minimum is -979 nT.

If the curve on both sides of the ΔT anomaly is symmetrical, the anomaly center is below the location of the maximum value. If the curve on both sides of the ΔT anomaly is antisymmetric, the anomaly center is below the location of the zero value between the extreme points. If the ΔT anomaly curve is asymmetrical, the anomaly center biases in favor of the main extremum side between the extreme points (Phillips et al, 2007). According to these principles, we find that the center position of magnetic anomalies of P1 is in the point corresponding to 1924 nT (Fig. 3). The width of the magnetic body is 58 m and depth is 23 m.

The maximum magnetic anomaly value of line P2 is 1576 nT and the minimum is -678 nT. The center position of magnetic anomalies of P2 is in the point corresponding to 898nT (Fig. 4). The width of the magnetic body is 109 m and depth is 94 m.

5   MINERALIZATION TARGET

According to the comprehensive analysis, the geological characteristics, geophysical anomalies, and existing geochemical anomalies in study area, we recognized two mineralization targets C1 and C2.

5.1   C1 mineralization target

It is located in the west of the study area with distribution areas about 600 n × 700 m. The main magnetic anomalies are positive anomalies with a maximum value 1900 nT. There are three distinct positive anomaly bands in ΔT profiles of the magnetic anomaly. The north anomalous zone extends roughly in the EW direction, the distribution of central and southern anomalous zones are nearly along the NNE direction. The central magnetic

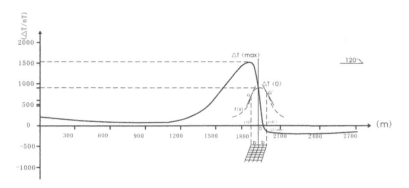

Figure 4. Curve exploded schematic of the P2 line precision magnetic survey section

anomaly zone is small and the south is relatively larger. In general, the stretch of the magnetic anomaly is similar to directions of fractures and formation strike. The high value magnetic anomaly area is well associated with basic rocks. With Cu, Fe, Ni, and Mo multi-element composite anomalies and pentlandite metallization, we deduce that this area is a good Cu, Fe, and Ni polymetallic mineralization area and exploration target.

5.2 *C2 mineralization target*

It is located in the east of the study area with distribution areas about 500 n × 900 m. Positive and negative anomalies have coexisted and have an obvious NNE direction. The main lithology of the magnetic anomaly zone is gneiss, which contains some amphibolites and plagioclase amphibolites which can cause magnetic anomaly. By upward continuation we think that it has a magnetic geologic body in the bottom of the magnetic anomaly zone. With qualitative and quantitative inference, we obtained the width (109 m) and depth (94 m) of the magnetic body. By combining with a small Ni and Cu geochemical anomaly, we think that the region is a good Fe, Ni, and Cu mineralization target area.

6 CONCLUSIONS

The study results demonstrate that a 1:25,000 high-precision magnetic survey in the Huotuolewusu area obtained good effect and the characteristics of magnetic anomalies are obvious.

Based on the anomaly characteristics of profiles P1 and P2 and qualitative and quantitative inference, we deduced that the width of the C1 magnetic body is 23 m and the depth is 58 m and the width of the C2 magnetic body is 109 m and the depth is 94 m.

We recognized two prospecting targets (C1 and C2) in the whole area. According to the anomaly characteristics and the field survey results, targets C1 and C2 are both large-scale targets, which have obvious mineralization anomalies and favorable metallogenic conditions. These two target areas are important candidates for the future geological exploration.

It provides important theoretical and practical significance for the metallogenic prediction and prospecting of Huotuolewusu area in future work.

ACKNOWLEDGMENTS

This research was jointly supported by the Geological Survey of China (grant No. 1212011220923). Professor Peng RM is thanked for him giving me a lot of help and so much amendment during writing article. Thanks for Chen Junlin giving me help.

REFERENCES

Guan, Z.N. Zhang, C.D. & Cheng, F.D. 1993. Theoretical analysis and application of magnetic prospecting important issue. Beijing: Geological Publishing House: 98–126.

Liu, T.Y. 1992. Theory and method gravity and magnetic anomaly inversion. Beijing: China University of Geosciences Press.

Matthew, B.J. Purssl, & James, P. 2005. A new iterative method for computing the magnetic field at high magnetic susceptibilities. Geophysics, 70(5):53–62.

Peng, R.M. Zhai, Y.S. Han, & X.F. et al. 2007. Mineralization response to the structural evolution in the Langshan orogenic belt, Inner Mongolia. Acta Petrologica Sinica, 23(3): 679–688.

Phillips, J.D. 2007. The use of curvature in potential-field interpretation. Exploration Geophysis, 38(2):111–119.

Smith, R.S. Salem, A. & Lemieux, J. 2005. An enhanced method for source parameter imaging of magnetic data collected for mineral exploration. Geophysical Prospecting, 53(5):655–665.

Yan, J.Y. Meng, G.X. Lv. & Q.T. et al. 2009. Regional geophysical field characters structural zonation and deep structure of the Beishan area, Inner Mongolia. Progress in Geophys. 24(2):439–447.

Advances in Energy Science and Equipment Engineering II – Zhou, Patty & Chen (Eds)
© 2017 Taylor & Francis Group, London, ISBN 978-1-138-71798-5

A statistical analysis of rainstorm-flood events and disasters in changing environments in China

S.M. Liu & H. Wang
Beijing Forestry University, Beijing, China

D.H. Yan & W.L. Shi
China Institute of Water Resources and Hydropower Research, Beijing, China

ABSTRACT: Investigating changes in extreme precipitation, i.e., maximum precipitation for multiday events is critical for flood management and risk assessment based on the observed daily precipitation from China's Ground Precipitation $0.5° \times 0.5°$ Gridded Dataset (V2.0) and the Statistical Yearbook of China. Based on the statistical analysis of the rainstorm-flood events, mountain torrent disaster, and the new type of flood disaster, we put forward the problem of the flood disasters in our country. Results show that (1) there is a great difference in rainfall in China, the precipitation is decreasing from the northwest to the southeast, the average rainfall of arid areas in the northwest is less than 200 mm, and the average amount of precipitation in most areas of Central Plains is about 200 mm~800 mm, and the average rainfall in the southern and southeastern coastal areas is more than 800 mm. (2) In China, mountain torrent disasters occur frequently, in the changing environment, the flood of the mountain torrent disaster risk is more severe, the mountain flood disaster prevention situation is very grim, and the governance of task is arduous. (3) Climatic and geomorphological features in China determine the basic pattern of a decrease in the north and an increase in the south did not change. Overall, the extreme precipitation intensity will increase, as a comprehensive effect of temperature and the local precipitation pattern changes; Tibetan Plateau glaciers are retreating glaciers and glacial lakes, and these increase the expansion of potential failure leading to enhanced risk. With the beginning of the global warming in 1980s, especially in the process of severe warming in 1990s, the frequency of the occurrence of the melting and ice type flooding in China showed a significant increasing trend.

1 INTRODUCTION

Climate change and variability have already affected global ecosystems, biodiversity, and social economy (Kotir, 2011). In recent decades, significant changes in climatic extreme events have taken place, which can have devastating effects on the human society and the environment (IPCC, 2007; Min, et al., 2011) and the impacts are likely to be more pronounced in the future (Song et al., 2014). Since 1950, the number of heavy precipitation events over the land has increased in more regions than it has decreased (IPCC, 2013). With the temperatures rising in the future, the extreme precipitation events tend to be more intense and more frequent correspondingly (Benistonet et al., 2007; IPCC, 2013). The extreme precipitation has the potential to trigger floods (Zhang et al., 2008). Additionally, China is more vulnerable to the threats of floods because it is heavily relying on agriculture (Zhang et al., 2012). (Fu et al., 2013) and has demonstrated an increasing trend of extreme precipitation events in the Yangtze River Basin. (Liu et al., 2006) The characteristics of the change of extreme temperature and precipitation in China are analyzed, the results indicate that the precipitation days show an increasing trend; the study on extreme precipitation of the Haihe river basin changes shows that the extreme precipitation variability in the region leads to more uneven conditions in the space—time distribution (Liu et al., 2010). Coupled with recent years, affected by El Nino, the intensity of rainstorm events disaster on frequency, the duration showing increasing trend around in China, for instance, the type of Huaihe and Yangtze rainstorm days occurred in the period of El Nino maintenance or attenuation, and the type of southern rainstorm occurred in the period of development of El Nino (Zhou et al., 2014) Thus, in the storm and flood season, "watching the sea city" phenomenon in China is happening and will be more serious, which brought negative effects in the regional ecological environment and social economy.

But at present, the majority of the research is only a single type that in a rainstorm events analysis, and most of them aimed at the single factor on specific storm events (e.g. extreme rainfall and rainfall duration or certain aspects for a specific precipitation process).The above research for the causation of a single case can be understood in detail, but for the large amount of case regularities that are rarely available (Ronald, 2015). Therefore, in this research, by using statistical methods, we revealed the spatial variability features of rainstorm-flood disasters in the background of climate change from 1951 to 2010, in order to act as a reference in understanding the phenomenon of frequent and wide occurrence of rainstorm disasters in recent years, and to provide a scientific basis to the distribution and building of water conservancy engineering projects in China and the construction of "sponge city".

2 STUDY AREA

In the east part of Eurasia, west to the Pacific Ocean, lies China, which is also located in the north hemisphere sub-tropical and mid-latitude area (135° 2′E~73° 40′E, 3° 52′~53° 33′N). In China, pressure centers in different sea levels shift with seasons because of sea–land thermal gradients. Additionally, high-altitude air currents are strongly affected by the rise of the Tibet Plateau. As a result, monsoons in China occur frequently (Miao et al, 2004), and many different climate types are present. The following four climate zones cover China from the northwest to the southeast: arid, semi-arid, semi-humid, and humid climate zones. China can be divided into the following ten regions by basins: Songhuajiang River Basin, Liaohe River Basin, Haihe River Basin, Yellow River Basin, Huaihe River Basin, Yangtze River Basin, Southeast River Basin, Pearl River Basin, Southwest River Basin, and Northwest River Basin (Figure 1). During 1961–2011, the annual precipitation of China was about 589 mm and most of the precipitation fell in the South: the total precipitation in the Yangtze River Basin, Southeast River Basin, Pearl River Basin, and Southwest River Basin is about 67% of the total precipitation in China.

3 DATA

Daily precipitation data are collected from China's Ground Precipitation 0.5° × 0.5° Gridded Dataset (http://cdc.cma.gov. cn/), which is provided by the Meteorological Records Office of the National Meteorological Information Centre of China. This dataset is based on the daily precipitation records

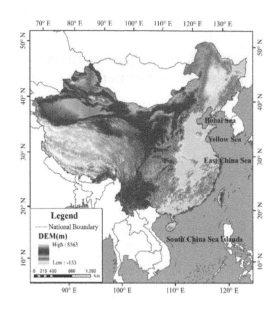

Figure 1. Geomorphological pattern presents a three-layer distribution in China.

of 2474 national meteorological stations and currently, the principal interpolation methods are of three types, Inverse Distance Weighting (IDW), kriging, and spline curve (Tang et al, 2005; Watson et al, 1985; Nalder et al, 1998). Because the meteorological stations are sufficiently dense to capture the surface variations (Garcia et al, 2008), we chose IDW to convert isolated storm events to interpolated events in consideration of elevation. Data are not available for Taiwan and islands in the South China Sea at this time.

Social and economic data were obtained from the Statistical Yearbook of China, Water Resources Bulletin and Emergency disaster database (EM-DAT), whereas related data of the water conservancy project in the basin were obtained from the latest Water Conservancy Survey.

4 RESULTS AND ANALYSIS

4.1 Rainstorm-flood events and disasters in China

4.1.1 Typography for references
In this paper, the statistics of the average annual rainstorm days and the total number of rainstorm days shows that in the past 60 years, the number China's average annual rainstorm days was between 0 and 15 days, the total number of rainstorm days was between 0 and 738 days, the high value areas were in the East and southeast of China, which concentrated in the area of the Huai river basin,

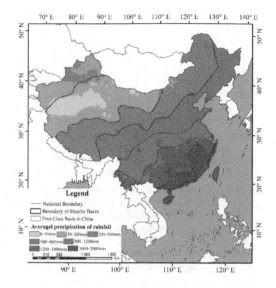

Figure 2. Distribution of mean annual precipitation from the observations in 1961–2011.

the Yangtze River region, the Pearl River Basin, and other regions.

The picture below shows the national average rainfall distribution (Figure 2); it shows great differences in the rainfall of China, the precipitation is decreasing from the northwest to the southeast, the average rainfall of arid areas in the northwest is less than 200 mm, the average amount of precipitation in most areas of the Central Plains from the northwest to the southeast are about 200 mm~500 mm and 500 mm~800 mm, the average rainfall is greater than 800 mm of the area are mostly distributed in the south and southeast coastal areas, the average precipitation from the northwest to the southeast are about 800 mm~1200 mm, 1200~1600 mm, and 1600 mm~3000 mm.

Rainstorm-flood disasters have a direct impact on the crop yield. Situations of the rainstorm events are collected and analyzed in China over the past 60 years. The results show that rainstorm-flood disasters and hazard rates have increases respectively by 1.3% and 0.7% every 10 years. The annual disasters and inundated areas are 9834.68 hectares and 5437.90 hectares, accounting for 8.64% (hit rate) and 4.76% (hazard rate) of the national planting area, respectively. Since 1990, the scope of the impact of rainstorms and floods has significantly expanded. The average hit rate and hazard rate were 12.28% and 6.77% during the period of 1990–2013 are several times higher than the the average 1.90 and 1.89 during the period of 1950–1989. Among them, 1991 and 2003 are most seriously affected by flood disaster years, when the

hit rate reached more than 20% and the disaster rate was above 13% (Figure 3).

There are about two-thirds of the land area that suffer from rainstorm-floods in China, where the downstream areas of the seven major basins (Figure 4) like Yangtze River, Huai River, the Yellow River, Haihe River, Liaohe River, Pearl River, and Songhua River are high-flooding disaster risk areas. These areas are mainly located in eight provinces including Henan, Anhui, Jiangsu, Hubei, Hunan, Jilin, Heilongjiang, and Sichuan of China; there are 738 thousand square kilometers of land area below the flood water level, which concentrated nearly 1/2 of the population, 35% of the arable land, and nearly 3/4 of the GDP.

4.1.2 *Rainstorm and flood disaster*

According to the latest data of Emergency Events Database (EM-DAT) statistics found (Figure 5), in 2006 as the node, the frequency of catastrophic flood disasters rose first and then fell in China. In 1988–2012, a total of 172 major floods occurred in China; among these, 2002, 2005, 2006, and, 2007 were the years in which flood disasters occurred most frequently, more than 10 times per year; since 1990s, a total of nine major floods occurred in China, and in an average of 2.2 years, when compared with the average of 2.7 years since the founding of new China decreased by nearly 20 percentage points.

From 1988 to 2012, China's flood disaster losses relative to the average annual direct economy

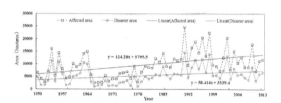

Figure 3. Graph showing the mean annual precipitation from the observations in 1961–2011.

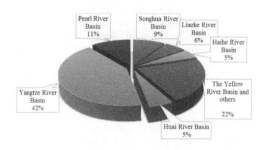

Figure 4. Chart showing the percentage of occurrence frequency of large floods in seven major basins in China.

131

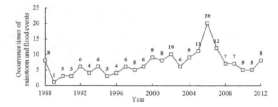

Figure 5. Graph showing the occurrence frequency of large-scale flood disasters in China in the past 25 years.

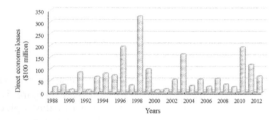

Figure 6. Graph showing the direct economic disaster loss.

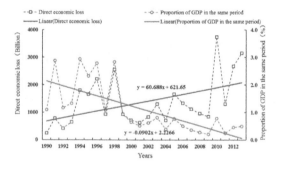

Figure 7. Graph showing the storm flood disaster loss and proportion of the GDP in the same period.

loss was more than about 480 million (Figure 6). According to the Ministry of Civil Affairs announced data, direct economic losses suffered in 1996, 1998, 2003, and 2010 were the largest and were more than 150 million dollars. At the same time, the occurrence of floods and waterlogging has obvious regional and seasonal characteristics; it is closely related to the rainy season of each river basin, precipitation, concentration period, and typhoon activity (Figure 7) in time, mainly concentrated in the summer. Among these, the Yangtze River, Pearl River Basin flood disasters occurred in the time of May to October, and the duration was relatively long; the time of flood occurrence in the Yellow River Basin was from June to August, when more rain occurred in this period; the time of flood disaster occurred in the Song Liao River Basin

from July to August, which was affected by the summer monsoon, and the duration was short.

In 1950–2013, China's average annual flood disaster deaths (including the missing population) was about 4387 people; in addition to the special year, flood disaster tolls in China's death population generally showed a gradual downward trend (Figure 8). In 1950–1960, the average annual death tolls was about 8340; the 1961–1970 average annual death tolls was about 3732 people; the 1971–1980 average annual death tolls was about 5370 people; the 1981–1990 average annual death tolls was about 4337 people; the 1991–2000 average annual death tolls was about 3744 people; the 2001–2013 average annual death tolls was about 1367 people. From this we can see, since 1980s, with the social and economic development level and the level of productivity in our country, the construction of reservoirs, river dredging, dike reinforcement, and river regulation have taken many precautions for flood control, and with the forecast of the flood prevention ability to strengthen gradually, there is a decrease in the number of deaths by flood disasters, and people's lives have been further protected.

From 1840 to 2011, statistics according to the number of townships from 1980 to 1949, as well as statistics according to the areas of disaster from 1950 to 2011, the number of affected counties overall increased from 1980 to 1949, and a number of affected counties obtained peaking values in 1930 (about 600). Since the new China was established, due to an increase in the flood disaster-stricken area year after year, from 1980s, it has increased rapidly, and after 2000s showed a slow decreasing trend, and this shows that with an increase in the number of water conservancy projects, the ability of China to cope with the flood disaster events has improved (Figure 9).

4.2 The situation of mountain flood disaster prevention in China is more serious

The proportion of the national prevented area is up to 4.63 million square kilometers, which covers 29 provinces, autonomous regions, and municipalities,

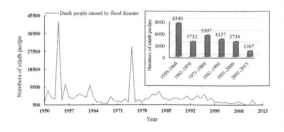

Figure 8. Graph showing the number of deaths due to floods disaster during 1950–2013 in China.

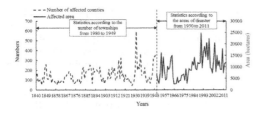

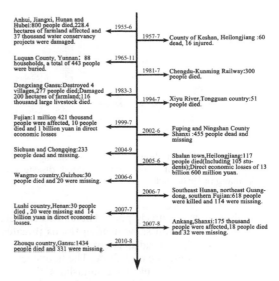

Figure 9. Graph showing the number of affected counties and affected areas.

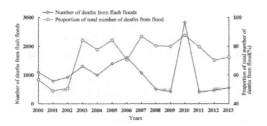

Figure 10. Graph showing the number of flood disaster deaths in 2000–2013.

Figure 11. Timeline showing major flash floods and losses of secondary disasters in the recent 60 years.

274 regional administrative regions, 1836 county-level administrative regions, and 560 million people. Among this, the core area is 970 thousand square kilometers, which influences 130 million people, threatens 74 million people and makes the defensive situation and governing tasks hard. In the period from 2000 to 2013, 1029 people per year died because of torrential floods in China which accounts 73.6% of that was because flood. In the recent five years, 940 people per year died because of torrential floods in China which accounts that 73.6% of that was because of floods. When compared to that from 2000 to 2008, the former one decreased by 12.8% and the latter one increased by 9.5%. Generally speaking, although the number of people who died because of torrential floods decreased, the number of people who died because of floods increased (Figure 10).

In our country, the mountain torrents disaster is also very frequent. According to an uncompleted statistics, from 1950–2000s, there have been 18901 flood ditches, 13409 times of disasters occurrence; 11109 debris-flow disaster ditches and 13409 times of disasters occurrence. Hill town areas are always destroyed and buried by mountain torrents, debris-flow and landslides, which threaten people's lives and property, and seriously restrict the development of the economy and society (Figure 11).

4.3 New problems of flood and waterlogging

At present, the main new flood disasters are mainly displayed as follows: glacial lake outburst, snowmelt/thawing ice storm flood, and "three meet" of typhoon rainstorms, storm surges, and astronomical tides in coastal areas.

4.3.1 Glacier lake outburst

Glacial lake outburst floods or mudslide mainly occurs in the Qinghai Tibet Plateau and the surrounding mountain areas in China. In the background of climate change, Tibetan Plateau glaciers are retreating glaciers, glacial thawing has increased, drift-dam lake experiences rapid expansion, thereby resulting in an increase in the potential outburst risk. In the area of Tibet, by the end of 2006, there were 2483 ice lakes spread over 0.014 km². The average area of a single ice lake is 0.148 km². The largest single ice lake area is 12.87 km². The total area of ice lakes is 366.56 km². Since 1930s, 23 ice lakes have been found to undergo breakage and 27 glacial lake outburst events have been recorded in the literature record or field.

4.3.2 Melting snow/ice melting

The type of snow melting and melting ice flood is formed by the melting of snow or ice, which frequently occurs in middle and high latitudes or mountain areas in the northwest and southwest areas of China. With the global warming in 1980s, particularly the raid raising t process in the 90s, the frequency of our country's snow and ice melting flood was significantly increased and the scope was gradually expanding too. Especially in the Xinjiang region, due to the high and steep mountain and fast river catchment, mountain floods were caused by rapid concentration and water conservancy

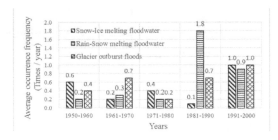

Figure 12. Graph showing the annual average occurrence frequency of snow and ice flood disasters in Xinjiang in 1950–2000.

facilities were relatively poor. Therefore, disasters such as glacier and snow melting floods threaten the safety of local residents as well as the operation of the important line of defense. The frequency of glaciers, debris flow blockage, landslides, flooding block has also obvious increased (Figure 12), and the outburst of glacial floods that happened by an average of 0.7 times per year in the 1980s have increased to 1.0 time per year.

Climatic and geomorphological features in China determine the basic pattern of a decrease in the north and an increase in the south did not change. Overall, the extreme precipitation intensity will increase, in the comprehensive effect of temperature and the local precipitation pattern changes. Tibetan plateau glaciers are retreating glaciers and glacial lakes increase the expansion of potential failure thereby leading to an enhanced risk. With the beginning of the global warming in 1980s, especially in the process of severe warming in 1990s, the frequency of the occurrence of the melting and ice type flooding in China showed a significant increasing trend. In addition, the comprehensive influence on the temperature and the local precipitation pattern changes under the model of drought and flood disasters risk and coastal areas "three chance to meet" will further increase.

5 PROBLEMS AND DISADVANTAGES

The climate and geomorphology determine China's flood control and flood frequency background, to deal with the status quo of China's heavy rain, combined with an international advanced experience of rainstorm and flood disaster response, the four aspects of disaster response in our country mainly has the following problems:

5.1 Shortage of water conservancy projects

China's overall storage capacity is relatively low, and per capita capacity is only 571 m³, only about 20% of the world water power per capita capacity, such as the United States per capita capacity is 2925 m³, Brazil's per capita capacity is 2476 m³. At present, the world average reservoir runoff control capacity is about 40% of the total water resources, developed countries are more than 60%; and the average runoff control ability of reservoirs in China is only 25%, while the control ability in the southwest areas of relative poverty is less than 7%.

The designed irrigation areas in China are about 668 million acres, accounting for only 25.39% of the planting areas in 2010. Among these, the irrigation area ratio is less than 10% of the county which accounted for 42% of the irrigated area while more than 50% counties accounted for only 19%; among these, Tibet, Xinjiang, and Qinghai in the design of the irrigation area account for more than 50%, Shanghai, Beijing, Sichuan, Heilongjiang, Guizhou, Shaanxi, Liaoning, Jilin and Chongqing design of irrigation areas are less than 10%; in the northeast and in other major grain producing areas of the irrigation areas, the ratio is still lower than other areas.

In recent years, frequent drought and flood disasters in the southwest region caused widespread concern in the community. Due to the construction of some water conservancy projects in the past, due to the inadequate number of supporting projects, it is difficult to play the overall efficiency of water conservancy projects. Therefore, to further optimize the construction of supporting facilities of existing water conservancy projects in Southwest China, and to cope with droughts and floods, we also need more deployment of some large-scale water conservancy projects, in order to meet the requirements of the plain areas and water accumulation of regional water security. For the areas with some difficulties in using river water or groundwater, we should pay attention to strengthening the construction of cellars and other decentralized water supply engineering.

5.2 Mismatched water conservancy project layout and areas of flood disaster risk

To calculate the effective volume capacity density and economy, the area ratio can reflect the ability to regulate the water conservancy project in an area of flood disasters to a certain extent; high flood risk density storage areas have higher resistance abilities and the local social and economic development support ability. However, China's overall capacity density is low, and the storage density in many areas and the area of drought and flood disaster risk and the level of economic development does not fully match, for example, there is a high overlap of key areas of China's economic development and flood disasters in high-risk areas and the capacity density in these areas generally possess low ability to regulate the water conservancy project on disaster. It is limited to the social and economic development support, which is insufficient, and this also reflects that water conservancy projects still cannot fully meet the current situation of the social economy development.

5.3 Lower engineering standards and the weak level of management

Lower designed standards are common problems in the construction of water conservancy projects in China. After the "75·8" flood events, the checking job for the calculation of the reservoir's maximum flood protection (PMF) was performed; but in fact, there are still a lot of reservoirs that cannot meet the requirements of small-sized rivers around the country; two-thirds of them cannot meet the requirements of the national flood control standards; more than 300 city constructions have not yet met the requirements of the national flood control standards, accounting for 49% of the total city; 31 key city flood control and 54 important flood control centers are present in the city, which is the standard rate of less than 1/3; the drainage pipe network capacity of more than 70% of the city in less than 1 years (the Ministry of housing standards), more than 90% old urban drainage capacities are even lower than the specification, which is still lower; in the river embankment projects that have been built, most of the flood control standards is only 10 years with 20 years, the Dike Section is thin, crest width and height are not enough, and there are many small rivers in the natural state without regulation of water conservancy projects.

For a long time, China's water conservancy project construction process was affected by the following problems: there are serious heavy and light river tributaries, tendency of heavy construction, light management, governance is not enough for medium and small rivers, especially the river dredging, embankment reinforcement, and pump maintenance engineering seriously affected the governance ability of China's flood control and drought.

5.4 Ageing and out-of-repair and project facilities wear out

The aging degree of China's water conservancy project is serious, most projects are in the final stage of their service period. China's 75% large reservoirs, 67% medium-sized reservoirs and 90% small reservoirs have been built from 1957 to 1977, and the dam engineering design service life is generally 50 years, the sedimentation capacity and metal structure design life is generally 30 years, from the service time, the use of most reservoirs in China has reached 40 to 50 years, and the degree

of aging is very serious. According to statistics, out of China's 500 large pumping stations, there are 350 stations that are seriously aging, and the aging damage rate accounted for 70%. With the passage of time, for a large number of design life close to the flood control, irrigation and water conservancy projects, most of them lack renovation funds, and some lack even maintenance funds, and the project cannot get timely maintenance and renovation.

According to the 2006 National Census dam safety status, the number of existing reservoir dams is 37 thousand, accounting for about 40% of the total reservoirs, and these reservoirs are widely distributed, threats, in addition to Shanghai, the provinces (autonomous regions and municipalities) have dangerous reservoirs; many dangerous reservoir downstream populations are concentrated or located in the upper reaches of the city, serious safety measures should be taken to lower the threat to people's life and property safety and critical infrastructure. As of the end of 2010, 37 thousand reservoirs, a total of 6240 large and medium, and key small reservoirs completed reinforcement, and there are still more than three dangerous reservoirs.

A large number of aging and dilapidated water conservancy projects not only give full play to the engineering benefit of flood control and drought, the project itself has become the weak link of the safety of the flood season, a direct threat to people's life and property safety of the masses.

ACKNOWLEDGMENTS

This work was supported by the Representative Achievements and Cultivation Project of the State Key Laboratory of Simulation and Regulation of Water Cycle in River Basin (No. 2016CG02) and National Key Research and Development Project (No. 2016YFA0601503).

REFERENCES

Beniston M., Stephenson D.B., Christensen O.B., Ferro CAT, Frei C., Goyette S., Halsnaes K., Holt T., Jylhä K., Koffi B., Palutikof J., Schöll R., Semmler T., Woth K. (2007) Future extreme events in European climate: an exploration of regional climate model projections, Clim Change.,81(1):71–95.

Fu G., Yu J., Yu X., Qu Y.R., Zhang Y., Wang P., Liu W., Min L. (2013) Temporal variation of extreme rainfall events in China, 1961–2009. J Hydrol,487:48–59.

IPCC (2007) Contribution of Working Group I to the Fourth Assessment Report of the Intergovernmental Panel on Climate Change. In: Solomon S., Qin D., Manning M., Chen Z., Marquis M., Averyt K.B., Tignor M., Miller H.L. (eds) Climate change 2007: the physical science basis. Cambridge University Press, Cambridge, United Kingdom and New York, NY, USA.

IPCC (2013) Contribution of Working Group I to the Fifth Assessment Report of the Intergovernmental Panel on Climate Change. In:Stocker T.F., Qin D., Plattner G.K., Tignor M., Allen S.K., Boschung J., Nauels A., Xia Y., Bex V., Midgley P.M. (eds) Climate change 2013:the physical science basis. Contribution of Working Group I to the Fifth Assessment Report of the Intergovernmental Panel on Climate Change. Cambridge University Press, Cambridge, United Kingdom and New York, NY, USA.

Kotir J.H. (2011) Climate change and variability in Sub-Saharan Africa: a review of current and future trends and impacts on agriculture and food security. Environ Dev Sustain, 13:587–605.

Liu X.F., Xiang L., Yu C.W. (2010) Characteristics of temporal and spatial variations of the precipitation extremes in the Haihe river basin. Climatic and Environmental Research, 15(4):451–461 (in Chinese).

Liu X.H., Ji Z.J., Wu H.B. (2006) Distributing characteristics and interdecadal difference of daily temperature and precipitation extremes in China for latest 40 years. J Trop Meteorol, 22(6):618–624 (in Chinese).

Miao J, Lin Z (2004) Study on the characteristics of the precipitation of nine regions in China and their physical causes II—the relation between the precipitation and physical causes. J Trop Meteorol 20(1):64–72 (in Chinese).

Min S.K., Zhang X., Zwiers F.W., Hegerl G.C. (2011) Human contribution to more-intense precipitation extremes. Nature, 470 (7334):378–381.

Nalder 1.A., Wein R.W. (1998) Spatial Interpolation of climate norrnal: test of a new method in the Canadian boreal forest. Agric FoL Meteor, 92:211–225.

Ronald I., Dorn (2015) Impact of consecutive extreme rainstorm events on particle transport: Case study in a Sonoran Desert range, western USA. Geomorphology, 250:53–62.

Song F., Qi H., Wei H. et al (2014) Projected climate regime shift under future global warming from multi-model, multi-scenario CMIP5 simulations. Glob Planet Chang, 112: 41–52.

Tang Y., Wang H., Yan D.H. (2005) Research on the Spatial temporal Differentiation of Precipitation in Northeast China in Recent 50 Years. Scientia Geographical Sinica, 25(2):74–76(in Chinese).

Watson D.F., Philip G.A.(1985) Refinement of Inverse Distance Weight Interpolation. Geo-Processing, 2:315–327.

Zhang J.(2012) Research of rainstorm water logging of Zhengzhou city based on GIS and SWMM. A thesis submitted for the degree of Master, Zhengzhou University (in Chinese).

Zhang Q., Gemmer M., Chen J. (2008) Climate changes and flood/drought risk in the Yangtze Delta, China, during the past millennium. Q Int, 176:62–69.

Zhou F., Sun Z.B., Xu X.F., et al (2014) Spatiotemporal characteristics of summer rainstorm days in eastern China and their relationships with the atmospheric circulation and SST. Acta Meteorologica Sinica, 3:447–464 (in Chinese).

Advances in Energy Science and Equipment Engineering II – Zhou, Patty & Chen (Eds)
© 2017 Taylor & Francis Group, London, ISBN 978-1-138-71798-5

Determination of soil parameters and initial values in numerical calculations by using experimental results

Binbin Xu
Tianjin Port Engineering Institute Ltd. of CCCC, Tianjin, China
Key Laboratory of Geotechnical Engineering of Tianjin, Tianjin, China
Key Laboratory of Geotechnical Engineering, Ministry of Communication, Tianjin, China

ABSTRACT: In this paper, the procedure to determine the soil parameters and initial values used in the numerical calculations by using experimental results is discussed in detail. The results of triaxial tests and one-dimensional tests for undisturbed and disturbed samples are utilized step by step to modify the numerical results. It is found that there are several steps to model the whole procedure of sampling from the ground to the experiment equipment. Therefore, the step of obtaining the soil samples from the ground, which is assumed to be an undrained and unloading process until the deviator stress becomes zero, is always ignored in the research.

1 INTRODUCTION

With the development of computational technology, the numerical analyses play an increasingly important role in geotechnical engineering. Usually, in order to obtain reliable results, one of the fundamental factors is the accuracy of the parameters employed in the calculation. Indoor experiments including triaxial compression test, one-dimensional test, permeability test etc. are carried out to help in obtaining detailed soil parameters. However, how to apply these experimental results into the numerical calculation is still debatable and there should be a criterion or bridge to connect two groups of parameters.

The precision of material parameters and initial values in the ground determines the validity of simulation results. In order to determine the values, a series of simulations and reverse analyses should be carried out by taking the disturbance into consideration. In a SYS Cam-clay model, there are totally 12 soil parameters to be determined, which can be classified into two categories. One category consists of five elasto-plastic material constants including compression index $\tilde{\lambda}$, swelling index $\tilde{\kappa}$, critical state constant M, intercept of NCL at $p' = 98.1$ kPa N and Poisson's ratio ν, which are directly succeeding from the original Cam-clay model and can be fixed by conventional triaxial compression tests and oedometer tests. The second one includes seven evolutional parameters containing the degradation parameters of over-consolidation ratio m, degradation parameters of the structure (a, b, c), ratio of plastic shear strain to plastic volume strain c_s, development

of stress-induced anisotropy b_r and limitation of rotation m_b. It should be noted that one of the remarkable properties for the SYS Cam-clay model is that for a given soil sample, only one group of material constants is necessary to represent the soil response under different conditions.

2 DETERMINATION OF PARAMETERS

2.1 Reproduction of the triaxial test

With regard to the five elasto-plastic material constants, they can be acquired by performing undrained triaxial compression tests and oedometer tests on fully remolded soil samples. After obtaining the material constants, the evolutional parameters and initial values are acquired by simulating the response of undisturbed samples with undrained triaxial compression tests and oedometer tests, together with some given state values. The optimum parameters and initial values are predicted by trial-and-error methods. Usually, there is some in situ ground survey datum in advance to evaluate the field characteristics. For clay, the survey datum consists of parameters such as specific volume, unconfined compression strength and so on and for sand, N-value, percentage of fine particles in soil F_c, and unit weight are investigated.

For the computation of undrained triaxial compression tests, the procedure is shown in Fig. 1, which includes four steps. The initial state is marked by a black dot and it can be seen that it is in the K_0 state by lying on the line of K_0 consolidation.

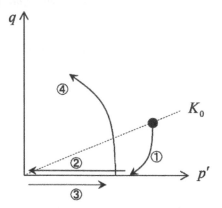

Figure 1. Fitting of undisturbed soil by using the und-rained triaxial test.

During the first step, the soil sample is obtained from its K_0 state to the isotropic consolidation state under undrained conditions, namely during this process, the volume of the sample remains constant. At this moment, the deviator stress q becomes zero and the soil sample is still inside the geotome. During the second step, the soil sample is isotropically unloaded until the mean effective stress is equal to the atmosphere pressure, which implies that the soil sample is taken out of the geotome. And then, it goes into the third step and the soil specimen is isotropically reloaded to a certain mean effective stress, which implies that the soil sample has been installed in the triaxial test equipment and consolidated sufficiently before shearing. The fourth step is well-known as the effective stress path under undrained conditions. The steps of the procedure to obtain the detailed values are given as follows:

1. Determine the initial vertical overburden earth pressure σ_v'. In this work, the soil in the ground is assumed to be totally saturated.

$$\sigma_v'(kPa = kN/m^2) = \gamma' \cdot z = \left(\frac{\rho_s - \rho_w}{1+e}\right) \cdot g \cdot z \quad (1)$$

where γ' is the buoyant unit weight, z is the depth, ρ_s is the soil particle density, and ρ_w is the water density. From the initial state in which the soil is in the K_0 state, the soil specimen is obtained from the ground, namely the "sampling" process, which is simulated as an undrained unloading process with a constant volume until the deviator stress $q = 0$.
2. With regard to specimen preparation, it is regarded as an isotropically unloading process until the mean effective stress $p' = 9.8$ kPa.
3. The specimen is isotropically consolidated to designated consolidation pressure.

4. The undrained triaxial compression test is carried out until the requirement is satisfied.

It should be noted that the above four steps represent the calculation procedure for an undisturbed soil sample. If the specimen is made of disturbed soil, the first step will be omitted and the procedure will be commenced from the second step.

2.2 Reproduction of the oedometer test

The computation procedure of the oedometer test of an undisturbed specimen is shown in Fig. 2. As can be seen, it is composed of three stages and each stage will be explained in detail as follows:

1. The soil is obtained from the ground under constant volume conditions until the deviator stress $q = 0$, which is called "sampling". During this process, the effective vertical stress decreases to some extent.
2. After "sampling", the soil sample is placed into the compression ring as soon as possible and then unloaded until $\sigma_v' = 19.6$ kPa under one-dimension conditions.
3. The one-dimension compression test is carried out.

3 DETERMINATION OF INITIAL VALUES

If there is no experimental result for the deep dense sand sample due to the difficulty in achieving undisturbed sampling, the evolutional parameters and initial values are usually determined from the ground survey datum such as N-value, percentage of fine particles in the soil F_c, and unit weight γ_s.

Firstly, according to N-value and F_c (Isai et al, 1986; Nakase, 1972; Tanaka, 1998) the void ratio e can be determined; and then, the inner friction

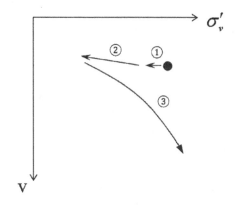

Figure 2. Fitting of an undisturbed soil sample by using the oedometer test.

angle ϕ_d is calculated from the same N-value (Iwasaki, 1990). Now that the inner friction angle ρ_d has been acquired, the maximum deviator stress q_f can be estimated correspondingly. The following steps will be the same as those mentioned in the above undrained triaxial compression tests, such that a series of drained triaxial tests under various confining pressures are simulated by changing the initial values to satisfy the given specific volume $v = 1 + e$ and maximum deviator stress q_f. A brief introduction about the proposed method is given as follows:

3.1 Void ratio

In order to obtain the void ratio e, three variables are involved including the maximum void ratio e_{max}, the minimum void ratio e_{min}, and the relative density D_r. According to the equations, the maximum void ratio e_{max} and the minimum void ratio e_{min} can be acquired from the percentage of fine particles F_c:

$$\begin{cases} e_{max} = 0.020F_c + 1.0 \\ e_{min} = 0.008F_c + 0.6 \end{cases} \tag{2}$$

While for the relative density D_r, the modified equation based on the equation (Meyerhof, 1957) can estimate the relative density D_r and take the percentage of fine particles in soil F_c into consideration.

$$D_r = 21\sqrt{\frac{N}{\sigma'_v + 0.7} + \frac{\Delta N_f}{1.7}} \tag{3}$$

where N is the N-value, σ'_v is the overburden earth pressure expressed by Eq. 1, and ΔN_f is shown, as in Table 1, to represent the influence of F_c and the corresponding figure is shown in Fig. 3. The curve is composed of four linear segments.

After the maximum void ratio e_{max}, the minimum void ratio e_{min} and the relative density D_r are obtained, the following equation can be used to determine the void ratio e:

$$e = e_{max} - D_r(e_{max} - e_{min}) \tag{4}$$

Table 1. Derivation of ΔN_f for fine particle percentage (after Tokimatsu & Yoshimi 1986).

$F_c(\%)$	ΔN_f
0~5	0
5~10	$1.2(F_c - 5)$
10~20	$6 + 0.2(F_c - 10)$
20~	$8 + 0.1(F_c - 20)$

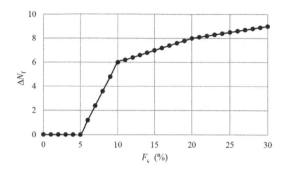

Figure 3. Graph showing the relationship between F_c and ΔN_f.

3.2 Maximum deviator stress

The deviator stress q_f can be determined from the survey datum of N-values (Tanaka et al, 2001), which is related to the inner friction angle ϕ_d:

$$q_f = \sigma'_a - \sigma'_r \tag{5}$$

where the major principal stress σ'_a and minor principal stress σ'_r are given as follows:

$$\sigma'_a = \left(\frac{1 + \sin\phi_d}{1 - \sin\phi_d}\right)\sigma'_r \tag{6}$$

$$\sigma'_r = K_0\sigma'_v \tag{7}$$

where K_0 is the coefficient of earth pressure and its value is taken as 0.6 and ϕ_d is the inner friction angle and determined by the following equation:

$$\phi_d = \begin{cases} \sqrt{20N_1 + 20} & (3.5 \le N_1 \le 20) \\ 40° & (N_1 \ge 20) \end{cases} \tag{8}$$

The relationship between ϕ_d and N_1 is shown in Fig. 4, where $N_1 = N\sqrt{98/\sigma'_v}$. In addition, the slope of the critical state line is calculated by Eq. 9 and the relationship between the deviator stress q_f and N_1 is shown in Fig. 5. As can be seen, there is a similar tendency when compared with Fig. 4. The deviator stress firstly increases slowly while after $N_1 = 20$, there is a sudden increase in the deviator stress.

$$M = 6\sin\phi_d / (3 - \sin\phi_d) \tag{9}$$

Since the given specific volume $v = 1 + e$ and maximum deviator stress q_f have already been determined by the survey datum, the following simulations of drained triaxial tests obey the same procedure to predict the evolutional parameters and initial values by trial-and-error methods.

139

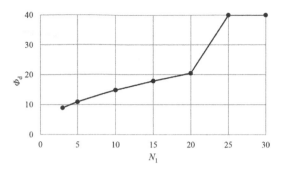

Figure 4. Graph showing the relationship between ϕ_d and N_1.

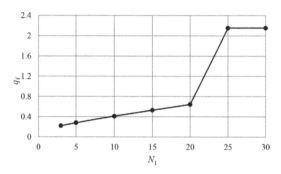

Figure 5. Graph showing the relationship between q_f and N_1.

4 EXAMPLES

In Fig. 6, the triangle represents the displacement constraint in the direction. As can be seen, there are three symmetrical planes consisting of plane *1243*, plane *3487*, and plane *1375*; constant stresses σ_{xx}, σ_{yy}, and σ_{zz} are applied correspondingly. The shear stress is represented by applying a constant velocity on plane *2486*. An undrained boundary is applied to the six surfaces. For the apparent behavior, the equivalent nodal forces at nodes *2, 4, 6* and *8* are measured to quantify the deviator stress; the apparent shear strain is calculated from the ratio of the horizontal displacement of node *2* to the initial height of the element.

The above steps can be calculated by using the model, as shown in Fig. 7 and one of the typical results is shown in Fig. 6. As can be seen, the experiments of the undrained compression test are reproduced with the relationship of deviator stress~mean effective stress, deviator stress~axial strain, specific volume~mean effective stress, and excess pore pressure~axial strain. Following the determination of undrained triaxial test results, the parameters can be obtained step by step.

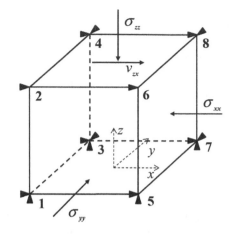

Figure 6. Schematic showing boundary conditions for one 3D element with a uniform deformation field.

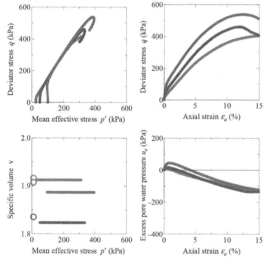

Figure 7. Graphs showing experimental results of reclaimed loose sand under undrained triaxial tests.

5 CONCLUSIONS

In the numerical analysis, the precision of parameters used is the key in spite of whether the calculation results are reliable or not. Many researchers have reproduced the mechanical response of soils numerically in a uniform deformation field. However, the procedure of sampling and its influence on the mechanical properties are rarely considered, especially for undisturbed soil samples. In this paper, the procedures of sampling for disturbed and undisturbed soils under triaxial tests and one-dimensional tests are decomposed step by step and discussed in detail. The conclusions are as follows:

1. For triaxial tests, there are four steps to model the sampling procedure of undisturbed soil samples including the undrained unloading process from the K_0 state to the isotropic state in the geotome, unloading isotropically from the geotome, reloading until reaching the designated pressure, shearing deformation, during which the first step of undrained unloading is mostly forgotten. By using one-element FE analysis, the soil parameters can be obtained by following the above steps.
2. There are three steps to model the sampling procedure of undisturbed soils for oedometer tests, which are similar to the triaxial tests.
3. When there is a lack of experimental results, it is feasible to predict the initial values and soil parameters by using experiential equations. But at the boundary of the value range, we should be cautious while using these equations as there are usually sudden variations, which may yield false results.

REFERENCES

Isai, S. Kowzumi, K. and Tsuchida, H. 1986. Affiliation a new criterion for assessing liquefaction potential using grain size accumulation curve and N-value, *Rept. of PHRI*, 25(3): 125–234.

Iwasaki. 1990. Problems related to standard penetration test, *The Foundation Engineering and Equipment*, 18(3): 40–48.

Meyerhof, G.G. 1957. Discussion on soil properties and their measurement, Discussion 2, *Proc. of the 4th International Conference on Soil Mechanics and Foundation Engineering*, 3: 110.

Nakase, A. Katsuno M. and Kobayashi, M. 1972. Unconfined compression strength of soils of intermediate grading between sand and clay, *Rept. of PHRI*, 11(4): 140–147.

Tanaka, H. Tanaka, M. and Tsuchida, T. 1998. Strengthening characteristics of undisturbed intermediate soil, *Journal of JSCE*, 589(42): 195–204.

The longitudinal evaluation of scale pig breeding carrying capacity

Bi-Bin Leng
School of Economics and Management, Jiangxi Science and Technology Normal University, Nanchang, Jiangxi, China
Center for Central China Economic Development Research, Nanchang University, Nanchang, Jiangxi, China

Qiao Hu
School of Economics and Management, Jiangxi Science and Technology Normal University, Nanchang, Jiangxi, China

Yuan-Yuan Duan
School of Foreign Languages, Jiangxi Science and Technology Normal University, Nanchang, Jiangxi, China

ABSTRACT: With the rapid development of scale pig breeding in our country, the environmental pollution caused by the breeding has drawn widespread attention. In this article, we established a system for the comprehensive carrying capacity evaluation indexes of scale pig breeding, in which factor analysis and hierarchical analysis were combined effectively, and the comprehensive carrying capacity index from 2004 to 2013 was evaluated. The investigation showed that the index was declining. However, with the rise of the resource supply index and social economic development index, there was an effective feedback on the environmental pollution, and the comprehensive carrying capacity evaluation index. In this way, we put forward some suggestions.

1 THE COMPREHENSIVE CARRYING CAPACITY EVALUATION INDEX OF SCALE PIG BREEDING

Scale pig breeding is a complicated system project, and so its comprehensive carrying capacity evaluation should be comprehensively evaluated by using multiple indexes, instead of a single index. Therefore, the establishment of the evaluation index system should obey some basic principles. Firstly, the system must serve the national strategic plan. Secondly, it should guide management departments to conduct related work about the pig breeding industry and environmental protection, and to set up a management target. Thirdly, it is the systematic principle. The next is the representative principle; it means that these indicators should be representative. Finally, it is the principle of measurability. On the basis of the above-mentioned five principles established by the index system—expert consultation, literature review, field investigation and theoretical analysis, and the connotation of the comprehensive carrying capacity of scale pig breeding, we combined the natural resources, social

Target layer	Criterion layer	Index layer
The comprehensive carrying capacity evaluation index A	Resource supply index B_1	The per capita arable area C_1 The per capita water resources C_2 The per capita grain possession C_3
	The social economic development index B_2	The per capita GDP C_4 The promotion of research and experimental development expenditure amount to the gross domestic product C_5 Urban and rural residents' consumption of pork C_6
	Environment carrying capacity index B_3	Scale pig breeding density C_7 (reverse indicator) Fertilizer use per unit area C_8 (reverse indicator) Industrial waste water volume (prefecture-level city) C_9 (reverse indicator)

economy and environment together to establish the comprehensive carrying capacity evaluation index system, which can be divided into three layers, including target layer, criterion layer, and index layer. The evaluation index system is as shown in the following table.

2 CHINA'S SCALE PIG BREEDING COMPREHENSIVE BEARING CAPACITY EVALUATION

Next, according to the construction of the evaluation index system, we had performed a measurement evaluation on China's scale pig breeding comprehensive carrying capacity from 2004 to 2013. The two main evaluation methods adopted were factor analysis and Analytic Hierarchy Process (AHP); we combined them effectively to carry out the evaluation research. First, we used the SPSS statistical analysis to carry out the factor analysis of the index system of the comprehensive carrying capacity index system, followed by the supply index B_1, social economic development index B_2, and environment carrying index B_3. And then, by using the AHP, we subdivided every criterion layer, and worked out the comprehensive carrying capacity index A (all data are obtained from the China Statistical Yearbook). Since each dimension of the index was different, and there were also some inverse indicators, we standardized the original data of every index layer and recorded these as C_{11}, C_{22}, C_{33}, C_{44}, C_{55}, C_{66}, C_{77}, C_{88}, and C_{99}. After processing these index layers positively, we performed a factor analysis on the three indexes of supply resources, the three indexes of social economic development and the three indexes of environmental carrying index, respectively. As a result, we obtained the factor expression of resource supply B_1, social economic development index B_2, and environment carrying index B_3. And then, by normalization processing the variance contribution rate of those factors, we obtained the expressions of B_1, B_2, and B_3, which are as follows. FB_{11} and FB_{12} were the two main factors of the three resource supply indicators after factor analysis. Similarly, FB_{21} and FB_{22} were the two main factors of the three indicators for the social economic development. FB_{31} and FB_{32} were the two main factors of the environment carrying capacity.

Supply index factors and expressions are given as follows:

$$FB_{11} = -0.511\,C_{11} - 0.127C_{22} + 0.528C_{33}$$

$$FB_{12} = 0.031C_{11} + 1.022C_{22} - 0.097C_{33}$$

$$B_1 = 69.73\%*\,FB_{11} + 30.27\%*\,FB_{12}$$

Social economic development index factors and expressions are given as follows:

$$F\,B_{21} = 0.910C_{44} + 0.618C_{55} - 0.781C_{66}$$

$$F\,B_{22} = -0.563C_{44} - 0.189C_{55} + 1.492C_{66}$$

$$B_2 = 91.71\%*\,FB_{21} + 8.29\%*\,FB_{22}$$

Environmental carrying index factors and expressions are given as follows:

$$F\,B_{31} = 0.514C_{77} + 0.569C_{88} + 0.179C_{99}$$

$$F\,B_{32} = 0.011C_{77} + 0.166C_{88} + 1.034C_{99}$$

$$B_3 = 71.65\%*\,FB_{31} + 28.35\%*\,FB_{32}$$

According to the supply resources index B_1, the environmental carrying index B_3, and the social economic development index B_2, we estimated the composite bearing capacity index of scale pig breeding and considered the AHP as a modeling technology of the unstructured decision problem and the AHP is systematic, flexible, and practical. After all these comprehensive considerations of three indexes, we found that it is better to use the AHP to analyze the composite bearing capacity index of scale pig breeding. Therefore, more than 50 experts were invited to analyze B_1, B_2, and B_3. And then, we consulted them over and over again, and amended continually, and finally we obtained the following table.

As can be seen in the table, we can work out the judgment matrix eigenvector W and the maximum characteristic root λ max, CI, CR: λ max = 3, W = (0.4000, 0.2000, 0.4000), CI = 0, CR = 0 < 0.1. Therefore, we knew that the judgment matrix had satisfied consistency, and then we obtained the scale pig breeding composite bearing capacity index expression, which is given as follows:

$$A = 0.4*B_1 + 0.2*B_2 + 0.4*B_3$$

The scale pig breeding comprehensive carrying capacity indexes are indicated in Fig. 1.

On the basis of the original data and SPSS statistic description, we found that the per capita arable area of our country had declined over the last decade. What had happened was that the per

Table 1. (A–B) Judgment matrix table.

A	B_1	B_2	B_3
B_1	1	2	1
B_2	1/2	1	1/2
B_3	1	2	1

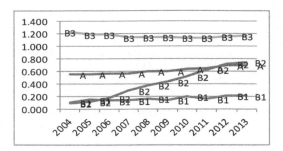

Figure 1. Graph showing the scale pig breeding comprehensive carrying capacity indexes.

capita arable area was reduced from 1.41 Mu in 2004 to 1.34 Mu in 2013, which was less than half of the world's per capita level. Per capita water resources had exhibited a steady decline from 2005 to 2008, but it had been increasing after 2011. However, the per capita grain production had been increasing in the recent decade. Therefore, we may see that the supply index was improving steadily. It showed that with the development of science and technology, the ability of the resource supply and resource development was improving constantly as well. On the other hand, the cultivated land protection and development efforts needed to be further strengthened.

According to the original data and SPSS statistic description, the per capita Gross Domestic Product (GDP) of our country had been on a rise in the recent decade because of the fast development of the economy. For example, the GDP in 2013 was more than three times of that in 2004. The ratio of research and experimental development expenditure to the Gross Domestic Product (GDP) continued to increase, which indicated the increase in the science and technology input and the development of science and technology level. Pork consumption of urban and rural residents exhibited an increasing trend, which made a great difference to promote the living standards of residents, which showed that with the rapid development of social economy, people's consumption ability, and level improved. At the same time, our country supported the scale pig breeding industry by increasing the investment for relative science and technology.

According to the original data and SPSS statistic description, scale pig breeding showed an increasing trend as a whole, which indicated that the number of pigs in our country increased year by year and fertilizer usage of per unit area increase as well. On the other side, industrial waste water emissions had been increasing from 2004 to 2007 due to the rapid development of the industry,

which had brought the increase of water pollution. However, the waste water had decreased to some extent after 2008. On one hand, our country had taken some effective measures of prevention and cure to liquid water and rejected material. On the other hand, some companies had adopted a more advanced technology to control the waste emissions caused by industrial development. The index B3 showed that social environmental awareness had been enhanced. Although we had mitigated pollution such as industrial waste water, the environmental pollution bearing capacity recovered modestly and the declining trend of the environment bearing capacity was still an anxious problem. The environmental situation was still severe. In the meantime, the scale pig breeding pollution prevention efforts needed to be further intensified.

Judging from the viewpoint of the scale pig breeding composite bearing capacity index, it kept rising obviously, which implied that, although the environmental pollution of scale pig breeding was an increasingly serious problem, there was still a positive feedback from the viewpoint of improvement of supply ability because of the progress in the level of science and technology and the development of social economy. Therefore, it presented an increasing tendency of the scale pig breeding comprehensive bearing capacity index.

3 CONCLUSION

This research showed that the development of the economic society, especially the improvement of science and technology, promoted the increase of supply resources and the development of the economic society. However, the environmental bearing capacity index has been declining. Thus, more attention should be paid to the pollution prevention of the scale pig breeding. The scale pig breeding composite bearing capacity index continued to rise, which indicated that although the environmental pollution of scale pig breeding was an increasingly serious problem, positive feedback was still obtained from the viewpoint of improvement of supply ability, the progress of science and technology level and the all-around development of the economic society to the environmental pollution governance. And so, it presented an increasing tendency of the scale pig breeding comprehensive bearing capacity index.

According to the original data and SPSS statistic description, we found that the per capita arable area in our country was declining day by day, even less than half of the world's average. Cultivated land protection and exploitation needed to be further strengthened. Scale pig breeding appeared to be an uplifted trend on the whole, which showed that the number of pigs in our country augment year by year. The fertilizer usage increased as well, thereby resulting in the falling of the environmental bearing capability index. Finally, the scale pig breeding environmental bearing capacity was on the decline. The conclusion is that waste disposal and the environmental protection ability need to be further developed.

ACKNOWLEDGMENTS

The authors gratefully acknowledge the grant of the project Scale Pig Breeding Ecological Energy System Stability Feedback Simulation Study (71501085) supported by the National Natural Science Foundation of China and the Project of Humanities and Social Sciences in Colleges and universities by Jiangxi Province (GL1536).

REFERENCES

Lu Yuanqing, Shi jun. The construction of low carbon competitiveness evaluation index system [J]. Statistics and Decision, 2013(1): 63–65.

Wang Haiwen, Lu Fengjun, etc. Healthy scale pig breeding industry chain main symbiotic mode selection research [J]. Rural Economy, 2014(03): 46–51.

Xu Guiping, Zhao Yuanjun etc. New-generation Migrant Workers' City Adaptation Obstacles Evaluation and Analysis Based On AHP [J]. agricultural economy 2016, 4: 84–86.

Yang Xiuping. Based on the analysis of DIAHP tourism environmental capacity and the relevant countermeasures [J]. Journal of Ecology and Rural Environment, 2008, 24 (1): 20–23.

Zhou Li. Industry cluster, environmental regulation and livestock pollution [J]. Chinese Rural Economy, 2011(2): 60–73.

Advances in Energy Science and Equipment Engineering II – Zhou, Patty & Chen (Eds)
© 2017 Taylor & Francis Group, London, ISBN 978-1-138-71798-5

A study on the scale pig breeding trend and technical efficiency

Bi-Bin Leng

School of Economics and Management, Jiangxi Science and Technology Normal University, Nanchang, Jiangxi, China
Center for Central China Economic Development Research, Nanchang University, Nanchang, Jiangxi, China

Yue-Feng Xu & Jia-Ling Liu

School of Economics and Management, Jiangxi Science and Technology Normal University, Nanchang, Jiangxi, China

ABSTRACT: In this paper, the DEA and DEAP software were used to analyze the different scale pig breeding comprehensive efficiencies due to the advancements of pig breeding on a large scale and its intensification. The research results showed that the comprehensive efficiency and the scale degree of the pig breeding exhibited a positive correlation. Besides, the comprehensive efficiency and pure technical efficiency exhibited a positive correlation as well. We concluded that the appropriate scale breeding exhibited optimal scale efficiency. The scale pig breeding waste disposal problem should be a matter of great concern; increasing the strength of standardized scale pig breeding farms to promote standardized scale pig breeding and appropriate scale breeding was necessary.

1 INTRODUCTION

The boom of China's pig industry, the scale pig breeding, and rapid intensive advancements lead to serious pollution problems due to scale pig breeding. In view of the environmental pressure and pollution problems caused by expanding scale pig breeding, a large number of research scholars carried out some studies on the treatment of scale breeding waste, resource utilization of waste materials, and ecological energy cycle of scale pig breeding. The fruitful research results obtained alleviated the waste pollution problem of scale pig breeding partly. The Party Central Committee and the State Council also attached great importance to the environmental pollution problems of scale pig breeding. Since 2007, the central government had allocated 2,500,000,000 yuan annually to support the construction of standardized scale pig breeding farms (districts), and to promote the construction of standardized scale pig breeding farms throughout the country. In this paper, DEA and DEAP were used for evaluating the different scales of pig breeding technology efficiency to reason out the policy recommendations to promote standardized scale pig breeding, to alleviate the environmental pollution pressure caused by the scale breeding and to promote health and rapid development of China scale pig breeding.

2 THE CALCULATION OF THE INPUT AND OUTPUT INDEXES OF THE SCALE PIG BREEDING

Copy Data Envelopment analysis is an effective method to deal with various objective decision-making methods, and we studied the technical efficiency of the pig scale breeding through data envelopment analysis software (DEAP 2.1). First, we selected the input and output indexes. According to the data from "national agricultural products cost income data compilation 2014" and the main quantitative indexes of pig breeding in our country, the cost index set can accurately reflect the cost of pig breeding in China. After studying related literature and data, we set the number of workers (day/head), the cost of the land (yuan/head), material, and service costs (yuan/head) as input indexes of scale pig breeding technology efficiency evaluation and the yield of main products each hog (kg/head) as output index scale pig breeding technology efficiency evaluation. We analyzed the indicators from 2007 to 2013 and the original data table of input and output is given as shown in Table 1.

The original data are from the "National Agricultural Products Cost Income Data Compilation of 2008–2014". We used data envelopment analysis (DEAP 2.1) software to analyze the technical efficiency of scatter feed, small scale, and large scale

Table 1. The original data table of scale pig breeding technology efficiency evaluation.

Project	Year	Output of major products (kg/head)	Number of workers (day/head)	Land cost (yuan/head)	Material and service costs (yuan/head)	Project	Output of major products (kg/head)	Number of workers (day/head)	Land cost (yuan/head)	Material and service costs (yuan/head)
Scatter	2007	108.80	9.44	0.03	882.01	Small scale	105.30	4.36	1.13	912.11
	2008	112.10	8.66	0.36	1128.68		110.90	3.98	1.54	1192.28
	2009	112.98	7.96	0.11	983.44		112.17	3.60	1.02	1019.60
	2010	111.56	7.63	0.07	1011.15		110.91	3.55	1.17	1049.04
	2011	112.69	7.54	0.03	1274.74		113.73	3.79	1.86	1336.19
	2012	114.74	7.22	0.28	1373.61		116.07	3.48	1.75	1423.42
	2013	115.60	7.05	0.25	1373.57		116.58	3.33	1.79	1433.38
Medium scale	2007	106.50	2.59	2.09	943.23	Large scale	101.60	1.64	2.63	948.46
	2008	111.20	2.58	1.82	1202.74		104.30	1.48	2.41	1181.35
	2009	112.21	2.42	1.89	1056.99		105.74	1.47	2.58	1051.59
	2010	112.45	2.34	2.42	1092.09		107.68	1.42	2.95	1097.19
	2011	112.58	2.33	2.73	1356.57		109.23	1.43	3.19	1368.17
	2012	114.97	2.21	2.93	1452.46		111.69	1.34	3.25	1463.58
	2013	115.96	2.18	2.77	1462.87		112.16	1.31	2.98	1469.32

Table 2. The constant price data sheet of land cost and materials and service costs.

Project	Land cost of scatter feed (yuan/head)	Material and service costs of scatter feed (yuan/head)	Land cost of small scale (yuan/head)	Material and service costs of small scale (yuan/head)	Land cost of medium scale (yuan/head)	Material and service costs of medium scale (yuan/head)	Land cost of large scale (yuan/head)	Material and service costs of large scale (yuan/head)
2007	0.03	1015.78	1.3	1050.44	2.41	1086.28	3.03	1092.3
2008	0.42	1317.05	1.8	1391.26	2.12	1403.47	2.81	1378.51
2009	0.12	1051.69	1.09	1090.36	2.02	1130.34	2.76	1124.57
2010	0.07	1011.15	1.17	1049.04	2.42	1092.09	2.95	1097.19
2011	0.03	1178.82	1.72	1235.65	2.52	1254.5	2.95	1265.22
2012	0.25	1240.55	1.58	1285.53	2.65	1311.76	2.94	1321.8
2013	0.22	1213.41	1.58	1266.24	2.45	1292.3	2.63	1297.99

pig breeding. Because land costs (yuan/head) and the material and service cost (yuan/head) use the same price index of that year in order to increase comparability, we compared the two indicators of land costs and materials and service costs. According to the price of 2010, the constant price data sheet of land cost (yuan/head) and material and service costs (yuan/head) were calculated as shown in Table 2.

3 BREEDING TECHNICAL EFFICIENCY BASED ON DEA BETWEEN SCALE PIGS

Based on DEAP 2.1, firstly we set up a data table, by using the main product of each pig as output indicators yield, and the amount of labor, land costs, material, and service costs as input indicators. The set DEAP 2.1 file is given as follows:

```
1234. dta   DATA FILE NAME
1234.out   OUTPUT FILE NAME
28   NUMBER OF FIRMS
1   NUMBER OF TIME PERIODS
1   NUMBER OF OUTPUTS
3   NUMBER OF INPUTS
0   0=INPUT AND 1=OUTPUT ORIENTATED
1   0=CRS AND 1=VRS 0 0=DEA (MULTI-
     STAGE), 1=COST-DEA, 2=MALMQUIST-
     DEA, 3=DEA (1-STAGE), 4=DEA (2-STAGE)
```

Run DEAP 2.1 executive and get the table of EFFICIENCY SUMMARY as shown in Table 3.

Thereinto, firms 1–7 correspond to the technical efficiency of scatter-feed. Firms 8–14 correspond to the technical efficiency of small-scale pig breeding efficiency. Firms 15–21 correspond to the technical efficiency of medium-scale and firms 22–28 correspond to the large-scale technical efficiency.

Table 3. Pig breeding efficiency technology of free-range, small-scale, medium-scale, and large-scale.

Firm	Crste	Vrste	Scale	Firm	Crste	Vrste	Scale
1	1.000	1.000	1.000	15	0.947	0.986	0.960
2	0.779	0.789	0.987	16	0.903	0.911	0.991
3	0.974	1.000	0.974	17	1.000	1.000	1.000
4	1.000	1.000	1.000	18	1.000	1.000	1.000
5	1.000	1.000	1.000	19	0.916	0.925	0.991
6	0.957	0.960	0.997	20	0.916	0.959	0.955
7	0.998	1.000	0.998	21	0.952	1.000	0.952
Mean	0.958	0.964	0.994	Mean	0.948	0.969	0.978
8	0.941	0.992	0.948	22	0.937	1.000	0.937
9	0.786	0.791	0.993	23	0.865	0.928	0.933
10	1.000	1.000	1.000	24	0.984	1.000	0.984
11	1.000	1.000	1.000	25	1.000	1.000	1.000
12	0.877	0.925	0.948	26	0.945	0.952	0.992
13	0.927	0.969	0.957	27	0.976	0.979	0.997
14	0.955	1.000	0.955	28	1.000	1.000	1.000
Mean	0.927	0.954	0.972	Mean	0.958	0.980	0.978

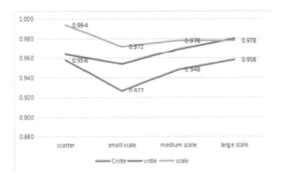

Figure 1. The mean efficiency figure of free-range, small-scale, medium-scale, and large-scale pig farming techniques.

Furthermore, we draw the mean efficiency figure of free-range, small-scale, medium-scale, and large-scale pig farming techniques as shown in Figure 1.

According to the EFFICIENCY SUMMARY results from DEAP 2.1, and the mean efficiency figure of free-range, small-scale, medium-scale, and large-scale pig farming techniques, we find that,

1. The overall efficiency, pure technical efficiency and scale efficiency of the free-range are higher than the scale breeding. This showed the advantages of traditional free-range, mainly due to the effective use of existing backyard gardens of farmers to carry out pig breeding, which did not require additional land costs, but also can effectively use and consume the manure generated from pig breeding.

2. In terms of scale farming, the mean value of overall efficiency and pure technical efficiency of the small-scale, medium-scale and large-scale pig breeding efficiencies showed a significant increasing trend, which indicated that the larger the scale pig farming is, the higher the overall efficiency and technical efficiency are.

3. From the mean value of scale efficiency, the values of small-scale efficiency (0.972), medium-scale efficiency (0.978), and large-scale efficiency (0.976) respectively correspond to the lowest, the highest, and the middle. This indicates that the scale of efficiency in pig farming does not increase with the continuous expansion of the scale. Instead, when the scale farming reaches a certain level, it starts to decrease.

4 CONCLUSION

China, as the largest producer and consumer of pork throughout the world, plays an important role in ensuring food production safety, improving people's living standards, and achieving a well-off society. China's pig industry is rapidly advanced to large-scale intensification and standardization. In view of the environmental pressure and pollution problems caused by expanding scale pig breeding, the different scales of pig breeding comprehensive efficiency were analyzed in this paper. The research results show that the appropriate scale breeding method has the optimal scale efficiency and the strength of the construction of standardized scale pig breeding farms (districts) should be increased to promote standardized and appropriate scale pig breeding and to mitigate the environmental pollution caused by expanding scale pig breeding is necessary.

ACKNOWLEDGMENTS

The authors gratefully acknowledge the grant of the project Scale Pig Breeding Ecological Energy System Stability Feedback Simulation Study (71501085) supported by the National Natural Science Foundation of China and the Project of Humanities and Social Sciences in Colleges and Universities by Jiangxi Province (GL1536).

REFERENCES

Bhattacharya S C, S Abdul P. Low Greenhouse Gas Biomass Options for Cooking in the Developing Countries [J].Biomass and Bio energy, 2002, 22(4): 305–317.

Cao C Z.Hu N.Framework and Influence Factors Analysis of Large—scale Animal Husbandry Breeding Area Change [J]. On Economic Problems, 2014, (1):88–92.

Chen Y, Yang G H, Sweeney S, et al. Household Biogas Use in Rural China: A Study of Opportunities and Constraints [J]. *Renewable and Sustainable Energy Reviews*, 2010, 14(1):545–549.

Jury C, Benetto E, Koster D, et al. Life Cycle Assessment of Biogas Production by Mono Fermentation of Energy Crops and Injection into the Natural Gas Grid [J]. *Biomass and Bio energy*, 2010, 34 (1):54–66.

Leng B B, Tu G P, Jia R A. Dynamic Stability of Scale Pig Breeding and Household Biogas Development System Based on SD Evolutionary Game Model [J]. *Systems Engineering*, 2014, (3):104–111.

Tu G P, Leng B B. Research on the Mode of Development Cooperation Between Scale Pig Breeding and Household Biomass Resources [J]. Rural Economy, 2013, (7):101–104.

Wang C X. Strategy simulation analysis for the scale operation of ecological agricultural using system dynamics [J]. Systems Engineering—Theory & Practice, 2015, (12):3171–3181.

Yang L.Li X F, et al. The mitigation potential of greenhouse gas emissions from pig manure management in Hubei [J]. Resources Science, 2016, (3):557–563.

Advances in Energy Science and Equipment Engineering II – Zhou, Patty & Chen (Eds)
© 2017 Taylor & Francis Group, London, ISBN 978-1-138-71798-5

A study on entropy generation in laminar free convection flow along a vertical surface with uniform heat flux

Li Tian
School of Materials and Metallurgy, Guizhou University, Guiyang, P.R. China

Wenhao Wang
School of Material and Metallurgy, Northeastern University, Shenyang, P.R. China

Fuzhong Wu
The Key Laboratory of Metallurgy and Energy Conservation of Guizhou, Guiyang, P.R. China
State Key Laboratory of Complex Nonferrous Metal Resources Clean Utilization, Kunming, P.R. China

Zengxin Cai
Zunbao Titanium Co. Ltd., P.R. China

ABSTRACT: A numerical analysis of entropy generation, for laminar free convection of air over an infinite vertical plate, due to heat transfer irreversibility has been investigated by solving the continuity, momentum and energy equations in the transient state. Based on the obtained dimensionless stream and temperature function, the variations of total entropy generation number, entropy generation numbers due to heat transfer and fluid friction, and Bejan number as a function of similarity transformation η for Rayleigh number and irreversibility distribution ratio set at $10^2 \leq Ra \leq 10^4$ and $10^{-6} \leq \phi \leq 10^{-4}$ are investigated. The distributions of entropy generation number and Bejan number contour are plotted. It is found that the entropy generation number due to fluid friction and total entropy generation number increase, but the Bejan number decreases with an increase in the Rayleigh number and irreversibility distribution ratio. However, the entropy generation number due to heat transfer increases with an increase in the Rayleigh number, and the irreversibility distribution ratio has no any effect. And entropy generation due to heat transfer is more important than that of fluid friction for low values of Rayleigh number and irreversibility distribution ratio.

1 INTRODUCTION

Free convection heat transfer along a vertical surface with uniform heat flux is an important issue due to its frequent appearance on the surfaces of equipment in engineering. An intensive effort, both theoretical and experimental, has been devoted to solving problems of flat plate, a typical research subject of free convection heat transfer. Ostrach (1952) obtained the numerical solution of flow and temperature distributions for Prandtl number from 0.01 to 1000, and it was shown that the velocities and Nusselt number in forced convection may be obtained from free convection. Kwang-Tzu (1960) established necessary and sufficient conditions for similarity solutions of the unsteady laminar boundary-layer equations for free convection on a vertical plate, and it was found that the available similarity solutions in the literature have essentially covered all steady cases, or even more general conditions. Many interesting effects on velocity, heat and mass trans-

fer, and laminar stability of laminar flows which aroused in fluids due to the interaction between gravity and density caused by the simultaneous diffusion of thermal energy and of chemical species were shown by Gebhart and Pera (1971). According to their results, Fujii & Fujii (1976) established a simple and accurate correlation of Nusselt number with the modified Rayleigh number. And there are also many numerical studies on free convection heat transfer along a vertical surface with isothermal or uniform heat flux (Assunta et al., 2002, Kuehn and Goldstein, 1976, Kuehn and Goldstein, 1980, Wei et al., 2002, Sparrow and Gregg, 1956).

Of course, the current studies have covered a wide range of problems involving the heat and mass transfer of free convection and these are not restricted, from the thermodynamics point of view, to only the analysis of the first law. The contemporary trend in the field of heat transfer and thermal design is the second law of thermodynamics analysis and entropy generation minimization.

Entropy generation due to heat transfer and fluid friction was reported by Bejan (1979) and is closed related with thermodynamics irreversibility, which is encountered in all heat transfer processes. Bejan (1982) analyzed the entropy generation in a circular duct with imposed heat flux at the wall and determined the Reynolds number as a function of the Prandtl number and the other duty parameters. And a dimensionless parameter called Bejan number to indicate the strength of entropy generation due to heat transfer irreversibility was proposed by Bejan (1979, 1982).

Entropy generation has recently been the topic of great interest in fields, such as heat exchangers, electronic cooling, porous media, and combustion. Many studies have been published on entropy generation. Magherbi et al (2003) studied the entropy generation due to heat transfer and friction in the transient state for laminar free convection by solving numerically governing equations, and analyzed the effect on the irreversibility distribution ratio on the maximum entropy generation and the entropy generation in the steady state. Ilis et al (2008) numerically investigated the entropy generation with different aspect ratios. It was found that, for a cavity with a high Rayleigh number, the entropy generation due to fluid friction and entropy generation number increases with an increase in the aspect ratio, attains a maximum and then decreases. Entropy generation due to conjugate natural convection heat transfer and fluid flow has been studied by Varol et al (2008a). They found that the entropy generation due to heat transfer is more significant than that of fluid flow irreversibility for all values of thickness of the solid vertical walls. And Varol et al (2008b) also found that entropy generation due to heat transfer and fluid friction irreversibility was affected by a higher inclination angle of the triangle and length of the heater, which resulted from the study of natural convection heat transfer and fluid flow in isosceles triangular enclosures that were partially heated from below and symmetrically cooled from sloping walls.

To the best of the authors' knowledge, the entropy generation for laminar free convection along an infinite vertical surface with uniform heat flux has not yet been investigated. The objective of this paper is to study the entropy generation for laminar free convection along an infinite vertical surface with uniform heat flux and it is important to energy conservation and environmental sciences.

2 DESCRIPTION OF THE PROBLEM AND THE PHYSICAL MODEL

In a two-dimensional Cartesian coordinate system (x, y), the x-axis is taken in the direction of the

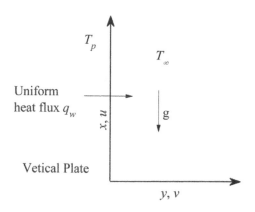

Figure 1.   A schematic view of the considered physical model.

main flow along the plate and the y-axis is normal to the plate with velocity u and v in their directions, a free convection flow caused by the temperature difference between the plate with uniform heat flux from the vertical plate and fluid. The heat transfer occurs by laminar free convection and the effect of radiation is neglected. And the problem is solved for air with $Pr = 0.68$. The considered physical model is schematically illustrated in Figure 1.

3 MATHEMATICAL MODEL

3.1 Governing equations

Under the usual boundary layer approximations, the set of governing equations in the transient state are as follows:

$$\frac{\partial u}{\partial x} + \frac{\partial v}{\partial y} = 0 \qquad (1)$$

$$u\frac{\partial u}{\partial x} + v\frac{\partial u}{\partial y} = -g - \frac{1}{\rho}\frac{dp}{dx} + v\frac{\partial^2 u}{\partial y^2} \qquad (2)$$

$$u\frac{\partial T}{\partial x} + v\frac{\partial T}{\partial y} = a\frac{\partial^2 T}{\partial y^2} \qquad (3)$$

And the relevant boundary conditions are as follows:

$$y = 0, u = v = 0, q = -\lambda\frac{\partial T}{\partial y}\bigg|_{y=0} = h\left(T_p - T_\infty\right) \qquad (4a)$$

$$y \to \infty, u = 0 \qquad (4b)$$

By employing the Boussinesq approximation (Gray and Giorgini, 1976) and the momentum equation, Equation (2) could be rewritten as follows:

$$u\frac{\partial u}{\partial x}+v\frac{\partial u}{\partial y}=g\beta\left(T-T_\infty\right)+v\frac{\partial^2 u}{\partial y^2} \tag{5}$$

3.2 Analytical solution

In order to satisfy the equation of continuity, Equation (1), a non-dimensionless stream function $\Psi(x,y)$ should be chosen like that

$$u=\frac{\partial\psi(x,y)}{\partial y},v=-\frac{\partial\psi(x,y)}{\partial x} \tag{6}$$

Even more, the similarity transformation and dimensionless stream function are introduced as follows:

$$\begin{aligned}\eta&=kyx^{-1/5}\\\Psi(x,y)&=5k\,vx^{4/5}f(\eta)\end{aligned} \tag{7}$$

where k is a undetermined coefficient and

$$k=\left[\frac{qg\beta}{\lambda a\,v}\right]^{1/5}$$

Defining the dimensionless temperature

$$\Theta(\eta)=\frac{T-T_\infty}{T_p-T_\infty} \tag{8}$$

By using Equations (5), (6), (7), and (8), the momentum and energy equations can be written as follows:

$$\begin{aligned}f'''(\eta)+4f(\eta)f''(\eta)-3\left[f'(\eta)\right]^2\\+\frac{Ra^{1/5}}{5NuPr}\Theta(\eta)=0\end{aligned} \tag{9}$$

$$\begin{aligned}\Theta''(\eta)+4Prf(\eta)\Theta'(\eta)\\-Prf'(\eta)\Theta(\eta)=0\end{aligned} \tag{10}$$

which are satisfied by the following boundary conditions:

$$\eta=0, f(0)=0, f'(0)=0, \Theta(0)=1 \tag{11a}$$

$$\eta\to\infty, f'(\infty)\to0, \Theta'(\eta)\to0 \tag{11b}$$

where Ra, Nu, and Pr are respectively the Rayleigh number, the Nuselst number, and the Prandtl number. And the numbers are given by the following relationships:

$$Pr=\frac{v}{a}$$

$$Gr^*=\frac{qg\beta x^4}{\lambda v^2}=\frac{g\beta\left(T_p-T_\infty\right)x^3}{v^2}\frac{hx}{\lambda}=GrNu \tag{12}$$

$$Ra=Gr^*Pr=\frac{qg\beta x^4}{\lambda v^2}\frac{v}{a}=k^5x^4$$

where Gr^* is the modified Grashof number and Gr is the Grashof number.

3.3 Heat transfer relations

The convective heat transfer coefficient h of free convection could be obtained by the Fourier law of heat conduction and Newton's law of cooling:

$$h=\frac{q}{T_p-T_\infty}=\frac{-\lambda\partial T/\partial y\big|_{y=0}}{T_p-T_\infty}=-\lambda kx^{-1/5}\Theta'(0) \tag{13}$$

Heat transfer between the hot wall and cool air fluid would be calculated by the Nusselt number Nu and the mean Nusselt number Nu_m, which are given by the following equation:

$$Nu=\frac{hx}{\lambda}=-Ra^{1/5}\Theta'(0) \tag{14a}$$

$$Nu_m=\frac{1}{l}\int_0^l Nudx \tag{14b}$$

3.4 Solution method and validation

The equations from the free convection of the vertical isothermal wall are programmed in Matlab™ to assess their prediction of the heat and mass transfer rates by using the method of Runge–Kutta.

Validation of the present study is performed by several different studies from Tetsu & Motoo

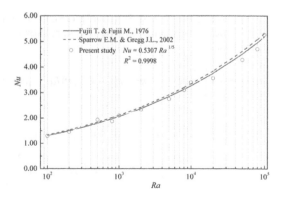

Figure 2. Graph showing the values of the Nusselt number from the literature and the present study.

(1976) and Sparrow & Gregg (1956) with the non-dimensional parameter Nu. The results of comparison are shown in Figure 2. It is decided that present study is valid for further calculations as the obtained results in Figure 2 show good agreement.

4 ENTROPY GENERATION

The local volumetric rate of entropy generation in the current air free convection is given by L.C. Woods (1975)

$$S = S_h + S_f \tag{15a}$$

$$S_h = \frac{\lambda}{T_\infty^2}\left[\left(\frac{\partial T}{\partial x}\right)^2 + \left(\frac{\partial T}{\partial y}\right)^2\right] \tag{15b}$$

$$S_f = \frac{\mu}{T_\infty}\left[2\left(\frac{\partial u}{\partial x}\right)^2 + 2\left(\frac{\partial v}{\partial y}\right)^2 + \left(\frac{\partial u}{\partial y} + \frac{\partial v}{\partial x}\right)^2\right] \tag{15c}$$

It is clearly shown that the two sources of entropy generation, of which the first term S_h on the right-hand side of Equation (15a) is the entropy generation due to heat transfer across the temperature difference, whereas the second term S_f is the local entropy generation caused by fluid friction.

An appropriate expression of L_S called entropy generation number, the excess rate of entropy generation, was defined by a characteristic entropy generation S_0.

$$L_S = \frac{S}{S_0} = \frac{S_h}{S_0} + \frac{S_f}{S_0} = L_{S,h} + L_{S,f} \tag{16}$$

and

$$S_0 = \frac{\lambda \Delta T^2}{l^2 T_\infty^2} \tag{17}$$

By using Equations (5), (6), (7), (15), (16), and (17), the entropy generation numbers are shown as follows:

$$L_{S,h} = \frac{1}{X^2}\left[\frac{1}{25}\left(\Theta(\eta) - \eta\Theta'(\eta)\right)^2 + \Theta'^2(\eta)Ra^{2/5}\right] \tag{18a}$$

$$L_{S,f} = \phi\frac{Ra^{2/5}}{X^4}Pr^2\left\{4Ra^{2/5}\left(3f'(\eta) - \eta f''(\eta)\right)^2 \right.$$
$$\left. + \left[\left(5Ra^{2/5} - \frac{1}{5}\eta^2\right)f''(\eta) + \frac{2}{5}\eta f'(\eta) + \frac{4}{5}f(\eta)\right]^2\right\} \tag{18b}$$

where ϕ is the irreversibility distribution ratio

$$\phi = \frac{\mu T_\infty}{\lambda}\left(\frac{\nu}{l\Delta T}\right)^2 \tag{19}$$

The variation of the energy generation numbers for the Rayleigh number and irreversibility distribution ratio set at $1.00 \times 10^2 \leq Ra \leq 1.00 \times 10^4$ and $1.00 \times 10^{-6} \leq \phi \leq 1.00 \times 10^{-4}$ are investigated.

The irreversibility of heat transfer plays the most important role in free convection problems with low Rayleigh number. The entropy generation number (L_S) is important to profiles or maps in generating entropy, but fails to balance whether heat transfer or fluid friction is more important. The irreversibility distribution ratio ϕ and another parameter called Bejan number (Be), which is the ratio of heat transfer irreversibility to the entropy generation, have been achieving an increasing popularity among the researchers of the Second-Low of Thermodynamics.

Finally, it is noticed that the Bejan number is given as follows:

$$Be = \frac{S_h}{S} = \frac{L_{s,h}}{L_S} \tag{20}$$

The Bejan number ranges from 0 to 1.

5 RESULTS AND DISCUSSION

Entropy generation for laminar free convection along a vertical surface with uniform heat flux has been investigated by solving the continuity, momentum, and energy equations in the transient state. The governing parameters are as follows: Rayleigh number, $1.00 \times 10^2 \leq Ra \leq 1.00 \times 10^4$ and irreversibility distribution ratio, $1.00 \times 10^{-6} \leq \phi \leq 1.00 \times 10^{-4}$. We will first present the results of heat transfer by means of the Nusselt number, and then we will present the distributions of the entropy generation number and Bejan number.

5.1 *Heat transfer*

Heat transfer is calculated by the local and mean Nusselt number as defined in Equations (14a) and (14b), respectively. Variation of the local Nusselt number for different Rayleigh numbers and the mean Nusselt number along the interface of the plate and air for different Rayleigh numbers and positions are shown in Figures 2 and 3, respectively. It is shown that the local Nusselt number increases with an increase in the Rayleigh number due to increasing values of heat flux q in Figure 2. And the mean Nusselt number assumed the same varying tendency in Figure 3. The values of the

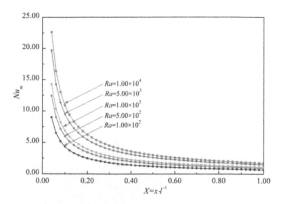

Figure 3. Graph showing the variation of the mean Nusselt number along the interface of the plate and air for different Rayleigh numbers.

mean Nusselt number decrease from bottom to top almost exponentially. But the slope of the mean Nusselt number is the smallest at $Ra = 1.00 \times 10^2$ due to not convection but conduction-dominated heat transfer.

5.2 *Entropy generation and bejan number distributions*

The entropy generation due to heat transfer irreversibility and fluid friction irreversibility are calculated with Equation (18) by using values of stream and temperature functions. Results of entropy generation distributions due to heat transfer irreversibility (on the left) and fluid friction irreversibility (on the right) are shown in Figures 4(a)–(c) for different values of Rayleigh number at $\phi = 1.00 \times 10^{-5}$. The information about the interface of the plate and air and contributions of entropy generation are obtained from the entropy generation distributions. And entropy generation due to heat transfer irreversibility also gives information about the heat transfer through the vertical surface with uniform heat flux to air.

The surfaces on the right in Figure 4 are inclined for low Rayleigh numbers but these are almost parallel with an increase in the Rayleigh number. But the surfaces on the left are more inclined for a higher Rayleigh number. Entropy generation concentrates at the beginning in the corner and it extends further with an increase in the Rayleigh number, as shown in Figure 4, which is due to heat transfer. In another words, more heat is transported from these positions. As it is shown in Figure 4 (on the right), the beginning position is the active sites which generate the entropy. And entropy generation spreads all over the space. Entropy generation due to fluid friction irreversibility (to the right-hand side of Figure 4) shows a very obvious valley.

The main cause for this phenomenon may be due to the decrease in temperature difference between the interface and air. And the velocity along the surface is gradually stabilized.

Iso-Bejan lines are countered in Figure 5 for two different situations as $Ra = 1.00 \times 10^3$, $\phi = 1.00 \times 10^{-5}$ and $Ra = 1.00 \times 10^4$, $\phi = 1.00 \times 10^{-6}$. As we indicated, the above-mentioned Bejan number (Be) is defined as the ratio of heat transfer irreversibility to the entropy generation. It is clear that Be = 0 is the limit at which the irreversibility is dominated by fluid friction effects, and Be = 1 is the opposite limit at which the heat transfer irreversibility dominates. At the top of the red line marked 0.50 in Figure 5, the irreversibility by fluid friction effects is dominated. And on the bottom heat transfer irreversibility does.

5.3 *Bejan number and total entropy generation*

As it is well known, the Bejan number (Be) is the ratio between heat transfer irreversibility and the total irreversibility due to heat transfer and fluid friction. The heat transfer irreversibility and fluid friction irreversibility play the same role when Be = 0.50. In this part, the variation of the Bejan number for different Rayleigh numbers and irreversibility distribution ratios at X = 0.25 are plotted in Figure 6.

This figure shows that the Bejan number highly depends on the Rayleigh number and irreversibility distribution ratio. For a given value of irreversibility distribution ratio, the Bejan number decreases with an increase in the Rayleigh number and therefore fluid friction effects irreversibility becomes significant and begins to dominate heat transfer irreversibility. That is to say that the Bejan number decreases with an increase in the Rayleigh number due to an increase in the domination of the convection model of heat transfer. The irreversibility distribution ratio increases against a decrease in the Bejan number.

And it can be found from Figure 6 that the Bejan number increases rapidly for lower values of the Rayleigh number and higher irreversibility distribution ratio at the very beginning of the similarity transformation η, and reaches a maximum value at practically the same time that entropy generation reaches its minimum at $\eta \approx 1$. This is caused by the convection regime, the air in the hot and cold viscous layers accelerates (rapidly for higher Ra) and the entropy generation due to fluid friction begins to play a significant role.

Finally, the variation of entropy generation with similarity transformation η and X = 0.25 for different Rayleigh numbers and irreversibility distribution ratio is presented in Figure 7. The presentation is preformed for both entropy generations due to

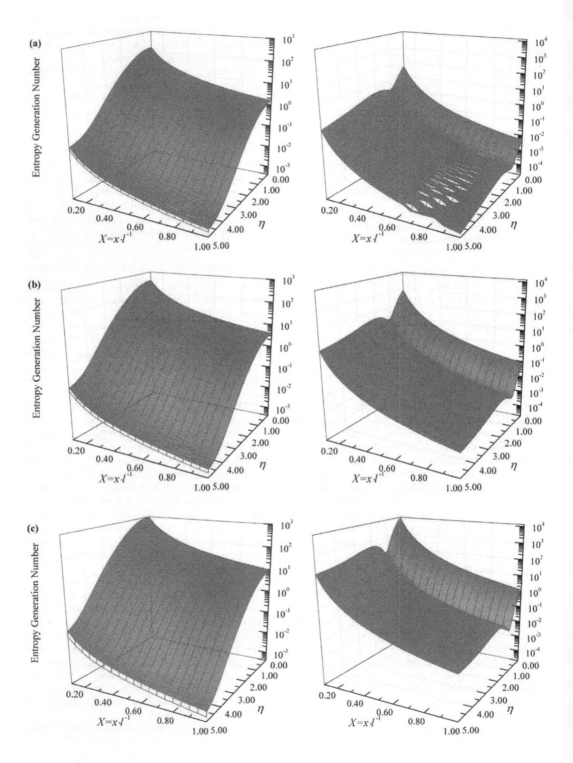

Figure 4. Entropy generation due to heat transfer irreversibility (on the left) and fluid friction irreversibility (on the right) at $\phi = 1.00 \times 10^{-5}$ and (a) $Ra = 1.00 \times 10^2$, (b) $Ra = 1.00 \times 10^3$, and (c) $Ra = 1.00 \times 10^4$.

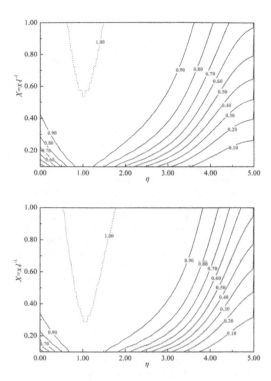

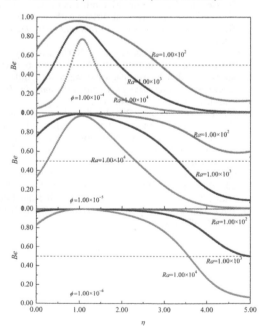

Figure 5. Graphs showing Iso-Bejan number lines at $Ra = 1.00 \times 10^3$, $\phi = 1.00 \times 10^{-5}$ (at the top) and $Ra = 1.00 \times 10^4$, $\phi = 1.00 \times 10^{-6}$ (at the bottom).

Figure 6. Graph showing the variation of the Bejan number for different Rayleigh numbers and irreversibility distribution ratio at X = 0.25.

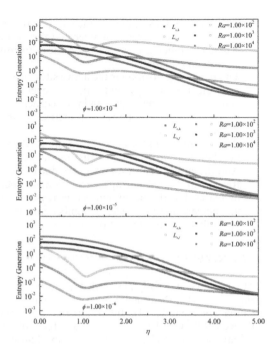

Figure 7. Graph showing the variation of entropy generation due to heat transfer and fluid friction for different Rayleigh numbers and irreversibility distribution ratio at X = 0.25.

heat transfer and fluid friction. The figure shows that the variation of Rayleigh number has a more important role on fluid friction irreversibility than that on heat transfer irreversibility due to changing of the convection model of heat transfer. It is found that the entropy generation number due to fluid friction and total entropy generation number increase with an increase in the Rayleigh number and irreversibility distribution ratio. However, the entropy generation number due to heat transfer increases with an increase in the Rayleigh number and the irreversibility distribution ratio has no any effect.

6 CONCLUSIONS

A numerical analysis of entropy generation, for laminar free convection of air over an infinite vertical plate, due to heat transfer irreversibility has been investigated by solving the continuity, momentum and energy equations in the transient state, by using the method of Runge–Kutta. Results are obtained for different Rayleigh numbers and irreversibility distribution ratios based on dimensionless stream and temperature functions. The main conclusions can be listed as follows:

1. The local Nusselt number increases with an increase in the Rayleigh number and the mean Nusselt number decreases from bottom to top almost exponentially.
2. The beginning position is the active sites, which generate the entropy and entropy generation concentrates at the beginning in the corner and it extends further along with an increase in the Rayleigh number
3. Bejan number highly depends on the Rayleigh number and irreversibility distribution ratio. And the entropy generation number due to fluid friction and total entropy generation number increases, but the Bejan number decreases with an increase in the Rayleigh number and irreversibility distribution ratio.

ACKNOWLEDGMENTS

The authors gratefully acknowledge the financial support of the National Science Foundation of China (No. 51574094), the Industrial Research Project of Guizhou Science & Technology Department in 2013 (No. 3009), and the Laboratory Projects of Guizhou Science & Technology Department (No. 4002 in 2014), and Guizhou Education Department (No. 222 in 2014). The authors also wish to express their very sincere thanks to the reviewers for the valuable comments and suggestions.

REFERENCES

Assunta, A., Oronzio, M. & Biagio, M. 2002. Numerical solution to the natural convection on vertical isoflux plates by full elliptic equations. *Numerical Heat Transfer, Part A: Applications,* 41, 263–283.

Bejan, A. 1979. A study of entropy generation in fundamental convective heat transfer. *J. Heat Transfer,* 101, 718–725.

Bejan, A. 1982. Second-law analysis in heat transfer and thermal design. *Adv. Heat Transfer,* 15, 1–58.

Gebhart, B. & Pera, L. 1971. The nature of vertical natural convection flows resulting from the combined buoyancy effects of thermal and mass diffusion. *International Journal of Heat and Mass Transfer,* 14, 2025–2050.

Gray, D. D. & Giorgini, A. 1976. The validity of the Boussinesq approximation for liquids and gases. *International Journal of Heat and Mass Transfer,* 19, 545–551.

Ilis, G. G., Mobedi, M. & Sunden, B. 2008. Effect of aspect ratio on entropy generation in a rectangular cavity with differentially heated vertical walls. *International Communications in Heat and Mass Transfer,* 35, 696–703.

Kuehn, T. H. & Goldstein, R. J. 1976. An experimental and theoretical study of natural convection in the annulus between horizontal concentric cylinders. *Journal of Fluid Mechanics,* 74, 695–719.

Kuehn, T. H. & Goldstein, R. J. 1980. Numerical solution to the Navier-Stokes equations for laminar natural convection about a horizontal isothermal circular cylinder. *International Journal of Heat and Mass Transfer,* 23, 971–979.

Kwang-Tzu, Y. 1960. Possible similarity solutions for laminar free convection on vertical plates and cylinders *Journal of Applied Mechanics,* 27, 230–236.

L.C., W. 1975. *The thermodynamics of fluid systems,* Oxford, Oxford University Press.

Magherbi, M., Abbassi, H. & Brahim, A. B. 2003. Entropy generation at the onset of natural convection. *International Journal of Heat and Mass Transfer,* 46, 3441–3450.

Ostrach, S. 1952. *An analysis of laminar free-convection flow and heat transfer about a flat plate parallel to the direction of the generating body force,* National Advisory Committee for Aeronautics.

Sparrow, E. M. & Gregg, J. L. 1956. Laminar free convection from a vertical plate with uniform surface heat flux. *Trans. ASME,* 78, 435–440.

Tetsu, F. & Motoo, F. 1976. The dependence of local Nusselt number on Prandtl number in the case of free convection along a vertical surface with uniform heat flux. *International Journal of Heat and Mass Transfer,* 19, 121–122.

Varol, Y., Oztop, H. F. & Koca, A. 2008a. Entropy generation due to conjugate natural convection in enclosures bounded by vertical solid walls with different thicknesses. *International Communications in Heat and Mass Transfer,* 35, 648–656.

Varol, Y., Oztop, H. F. & Pop, I. 2008b. Entropy analysis due to conjugate-buoyant flow in a right-angle trapezoidal enclosure filled with a porous medium bounded by a solid vertical wall. *International Journal of Thermal Sciences,* 48, 1161–1175.

Wei, J. J., Yu, B., Wang, H. S. & Tao, W. Q. 2002. Numerical study of simultaneous natural convection heat transfer from both surfaces of a uniformly heated thin plate with arbitrary inclination. *Heat and Mass Transfer,* 38, 309–317.

Advances in Energy Science and Equipment Engineering II – Zhou, Patty & Chen (Eds)
© 2017 Taylor & Francis Group, London, ISBN 978-1-138-71798-5

Analysis of a Chinese titanium sponge enterprise's energy usage

Huixia Pang
College of Materials and Metallurgy, Guizhou University, Guiyang, P.R. China

Wenhao Wang
School of Material and Metallurgy, Northeastern University, Shenyang, P.R. China

Fuzhong Wu
The Key Laboratory of Metallurgy and Energy Conservation of Guizhou, Guiyang, P.R. China
State Key Laboratory of Complex Nonferrous Metal Resources Clean Utilization, Kunming, P.R. China

Zhongchao Wang
Zunyi Titanium Co. Ltd., Zunyi, P.R. China

ABSTRACT: In this paper, a Gray theory analysis system was used to assess and predict the energy usage and energy intensities of a Chinese titanium sponge enterprise. Analysis of past trends helps modelers to accurately project future changes in energy usage. The results for energy use analysis and prediction are presented and discussed, which include the correlation among comprehensive energy intensity, energy intensity of reduction and distillation, and electrolysis of $MgCl_2$. In the analysis period, the mandatory measures taken by the enterprise resulted in the decrease in the overall energy usage. And the energy intensity exhibited a downward trend from 2012–2015. It was significantly that the energy intensity in 2015 decreased relative to 2011, while the energy intensity of reduction and distillation decreased at a much a lower rate than that of electrolysis. These trends resulted in a 5235.28 tce energy conservation in comprehensive energy use, eventually accomplished the 12th FYP energy intensity reduction task of 4200 tce.

1 INTRODUCTION

The Kroll process, an unique existing industrial process of titanium sponge (Habashi, 1997), matured to an extent that a substantial improvement in the performance has become impossible, which was archaic, cost and energy intensive, bath operated with unfavorable economics (Hartman et al., 1998), owing to technical or economic constraints. But the advancements in the Kroll process have been causing this process become the primary industrial choice in the titanium value chain of China (Marsh, 1996, Van Tonder, 2010, William and Julie, 2013). For this reason, the status and development path of China's titanium industry will greatly affect the market of not only China, but the entire world.

Unlike most industries, the energy consumption pattern of titanium sponge is unique because the main source of energy was electricity power, accounting for about 85% (Lee and Kim, 2003). And the titanium sponge production relies heavily on two energy-intensive steps, named reduction and distillation and electrolysis of $MgCl_2$. An insufficient theoretical understanding of the titanium

system may be an important reason (Nagesh et al., 2004).

Firstly, the Kroll process is operated in a batch fashion in an argon-filled retort with enough liquid magnesium, obtained by electricity-intensive electrolysis, to reduce all the $TiCl_4$ leaving approximately 20% excess magnesium. Secondly, it takes roughly 2 weeks for a batch of ten or more tons, which contributes to the problem that the process is so electricity-intensive (Takeda and Okabe, 2006).

However, for electricity scarcity and electricity security, environmental concerns, and unprecedented market downturn in the global economic crisis in the last decade, China's titanium enterprises would like to see more structural changes away from the energy/electricity-intensive industry to a low-energy intensive industry.

If local technology delivers the energy-efficiency breakthrough, it would acquire a significant increase in momentum, which would stimulate the development of the Chinese titanium industry. A typical titanium sponge enterprise in Guizhou province of China was intensive about its structural change policy in the 11th Five-Year-Plan

(FYP) (2006–2010) and in the current 12th FYP (2011–2015). The enterprise aimed to decrease the comprehensive energy intensity and energy intensity of reduction and distillation from 27000 kWh/t and 8500 kWh/t to 20000 kWh/t and 7800 kWh/t during the 11th FYP, separately. But by the end of 2010, the energy intensity only decreased to 25756 kWh/t and 8174 kWh/t.

A few analyses of the titanium industrial energy balance (Wu et al., 2013) and energy usage trends in China have been conducted, but comprehensive studies including all steps and their role in the historical energy use are scare. This study conducts such analyses.

It was the purpose of this study to provide the decision-making supervisors with energy usage analysis and prediction of the enterprise and steps. The energy intensity studied in this paper is defined as energy usage per unit of product or intermediate good. And a Gray theory analysis system (Hsu and Chen, 2003), based on both qualitative and quantitative techniques, was used to assess, evaluate, and predict the energy usage and energy intensities of a titanium sponge enterprise of China. Analysis of past trends helps modelers to accurately project future changes in energy usage. This also can help them to adjust their policies to meet the 12th FYP energy intensity reduction target in China (Hasanbeigi et al., 2013).

In this paper, the methodology on analysis and prediction of energy usage is first explained. After that, the results for primary energy/electricity usage analysis and prediction are presented and discussed, including the correlation among comprehensive energy intensity, energy intensity of reduction and distillation, and electrolysis of $MgCl_2$.

2 METHODOLOGY

Historical primary energy/electricity usage data of 2012 (the second year of 12th FYP in China and the status was most stable), at equal intervals and in consecutive order without by-passing any data, used in this analysis were obtained from various months of actual production statistical reports. And the month January 2012 was used as the base month for the analysis and forecast.

2.1 Gray relational analysis

The gray relational grade obtained from the Gray Relational Analysis (GRA) was used to solve the Kroll process with the multiple performance characteristics (Tosun, 2006). And the process is shown below.

Denote the raw energy intensity data of comprehensive electricity usage sequence as A_1, electricity usage of reduction and distillation sequence as A_2, and electricity usage of electrolysis sequence as A_3 by:

$$A_i = \{A_i(1) \quad A_i(2) \quad \cdots \quad A_i(n)\} \tag{1}$$

where n is the number of months observed.

And a_i-sequences, namely the gray relational sequence, based on $A_i(1)$ are defined as follows:

$$a_i = \{a_i(1) \quad a_i(2) \quad \cdots \quad a_i(n)\} \tag{2}$$

where

$$a_i(k) = \frac{A_i(k)}{A_i(1)}, k = 1, 2, \cdots, n.$$

In a local gray relation measurement, a_1 is available as the reference sequence, and all other sequences, a_2 and a_3, serve as comparability sequences. Therefore, the gray relational coefficient $\gamma_i(k)$ can be expressed as follows:

$$\gamma_i(k) = \frac{\Delta_{min} + \rho\Delta_{max}}{\Delta_{0i}(k) + \rho\Delta_{max}} \tag{3}$$

where, Δ_{0i} is the deviation sequence of the reference sequence and the comparability sequence.

$$\Delta_{0i} = |a_i(k) - a_1(k)|$$
$$\Delta_{min} = \min_{\forall j \in i} \min_{\forall k} |a_i(k) - a_1(k)| \, \&$$
$$\Delta_{max} = \max_{\forall j \in i} \max_{\forall k} |a_i(k) - a_1(k)|$$

And ρ in Equation 3 is called distinguishing or identification coefficient, the value of ρ may be adjusted based on the actual system requirements, and $\rho = 0.5$ is generally used (Fung, 2003).

Thus, the gray relational grade, taking the average value of the gray relational coefficients, is defined as follows:

$$\gamma(A_i, A_1) = \frac{1}{n}\sum_{k=1}^{n}\gamma_i(k) \tag{4}$$

2.2 Gray analysis and forecast theory

The energy usage and energy intensity for a Chinese titanium sponge enterprise were analyzed and forecasted from 2012 to 2015 in this study. To forecast the energy intensity, we need to have the forecast of energy usage as well as product. Hence, the Gray theory was used to calculate the energy usage and product.

The raw energy use and product sequences are denoted by the following equation:

$$X^0 = \left\{ x_1^0 \quad x_2^0 \quad \dots \quad x_n^0 \right\} \tag{5}$$

According to the Gray theory, by using the one accumulated generation operation (1-AGO), the least square method for fitting, and inverse 1-AGO, the fitted and predicted sequences are obtained (Hsu and Chen, 2003, Tosun, 2006). And the analysis and forecast of energy usage are calculated for each sequence separately. From these, the historical energy intensities in different months were calculated.

In this study, we conduct a retrospective fitted analysis of a typical titanium sponge enterprise by using the historical date from January to December 2012. In addition, we conduct a prospective analysis by using the forecast date calculated based on the method explained above.

3 RESULTS AND DISCUSSION

In this section, we first present the correlation among comprehensive electricity usage, electricity usage of reduction and distillation and electrolysis of $MgCl_2$. And then, analyses of historical as well as forecasted energy usage of a Chinese typical titanium sponge enterprise are presented.

3.1 Gray relational analysis and discussion

The gray relational grade $\gamma_i(A_i, A_1)$ represents the level of correlations between the reference sequence and comparability sequences. And the gray relational grade also indicates the degree of influence that the comparability sequences could exert over the reference sequence.

According to Equation 3 and Equation 4, the GRA results are listed in Table 1.

It is clearly observed from Table 1 that the gray relational grade results are satisfied (more than 0.6), when $\rho = 0.5$.

Since the gray relational grade represents the level of correlation between the reference sequence and the comparability sequences, a greater value of the gray relational grade means that the comparability sequences have a stronger correlation with the reference sequence. In other words, regardless of the category of the energy intensity characteristics, a higher gray relational grade corresponds to better correlation. Therefore, the electricity usage of reduction and distillation is little more important than the electricity usage of electrolysis, even little more significant, mainly due to roughly 30% of the initial magnesium charge, obtained by electricity-intensive electrolysis, would still be unreacted at the completion of the end reduction and a long cycle of reduction and distillation step.

3.2 Analysis of titanium sponge enterprise energy usage

A gray theory analysis was performed for a Chinese titanium sponge enterprise from 2012 to 2015, divided into four time periods by year. These four parts were chosen based on the 12th FYP.

3.2.1 Industry product-add trends

China has become the largest titanium sponge producer since 2007 in the world, which has 45,000 tons production over Japan. However, the share of the enterprise we studied had its total production

Table 1. Calculated gray relational coefficients and gray relational grades for two comparability sequences.

Month	Raw energy intensity			Gray relational coefficient	
	A1	A2	A3	A2	A3
Jan-12	21064.63	8089.60	18176.98	1.0000	1.0000
Feb-12	20581.50	8096.43	17657.41	0.7856	0.9391
Mar-12	18142.95	8086.66	17454.20	0.3865	0.4683
Apr-12	20102.37	8003.31	18104.76	0.7133	0.6763
May-12	19746.39	7990.24	17428.28	0.6340	0.8029
Jun-12	20528.21	8005.46	16815.66	0.8526	0.6381
Jul-12	19099.86	7983.51	16511.30	0.5208	0.9816
Aug-12	21866.28	8100.16	16787.22	0.7033	0.4321
Sep-12	21130.87	7562.83	16068.30	0.5607	0.4224
Oct-12	21684.11	8083.06	17533.28	0.7425	0.5734
Nov-12	23003.95	8298.02	16682.77	0.5679	0.3333
Dec-12	22172.50	8089.65	18771.73	0.6236	0.8143
Gray relational grade				0.6742	0.6735

capacity increase to 24% in 2012. Hence, the status and development path of this titanium industry will greatly affect not only the market, but also the entire titanium chain of China.

The total product-add rates of the enterprise only increased by 6.09% in 2012. This rate of increase is 41.4 times higher than the rate of decrease in energy usage, which decreased by 0.15% over the same period. Overall, the product-add rate decreased during the period, but the products added increased.

Figure 1 shows that the product-add rate in 2013–2015 decreased exponentially. Finally, the product-add will be gradually slower and the product would be controlled in a stable level eventually.

In 2012, there was no major shift in product-add rates. However, even a minor, few percentages change in the product added of the high energy-intensive industry can have a significant impact in the gray theory analysis, especially on the forecast.

Underlying the universal civilian of titanium products, added rate trends would be the demand drivers for Chinese titanium products. During 2006–2010, a strong growth in fixed asset investment (expansion and new production capacity) drove much of the boom in the titanium product add rate, although the growth in national defense strategic industries were also important. As a result, China's titanium sponge industry, before 2012, did not have the distinctive character and import and export balance as a more mature large titanium sponge industrial country.

In the future, less demand pull from an increase in the fixed asset investment is expected when compared to the past. Hence, the slower overall growth in industrial production of the titanium sponge enterprise and a shift in the structure of new demand for titanium products more towards universal civilian consumption are observed. Moreover, the enterprise would occupy a stable market share with the steady growth rate calculated above and the rational development path.

3.2.2 Energy usage trends

By using the gray theory and method presented above for each energy/electricity usage sequence, the energy usage in 2012 of the titanium sponge enterprise mentioned above were calculated and fitted month by month. Since the reduction and distillation and electrolysis of $MgCl_2$ are two key steps of the Kroll process, the energy usage was divided into three parts: reduction and distillation energy usage, electrolysis energy usage, and other steps energy usage. This is one of the unique features of this study since other similar studies typically constructed a forecast for the titanium industry in China.

In December 2012, the total energy usage of the enterprise was 2443.02 ten thousand kWh, which is a 9.61% decrease from the June level (2702.85 ten thousand kWh) and a 12.59% decrease when compared to energy usage in January 2012. The decrease in energy usage during the period of January–December 2012 varied among reduction and distillation, electrolysis, and other steps. Overall, the energy usage of the enterprise in China decreased during this period, due to mandatory measures taken by the enterprise to meet the 12th FYP energy intensity reduction task issued by Guizhou government. Figure 2 shows the trends of energy usage during January–December in 2012. The result in Figure 2 shows that the gray theory method can yield an accurate result and also solve problems resulting from less data.

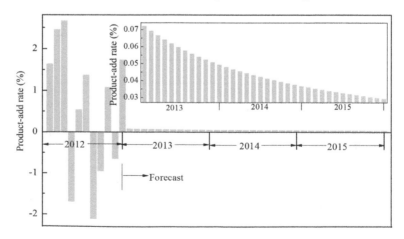

Figure 1. Graph showing the titanium sponge product-add rates of the enterprise that we had chosen during 2012–2015.

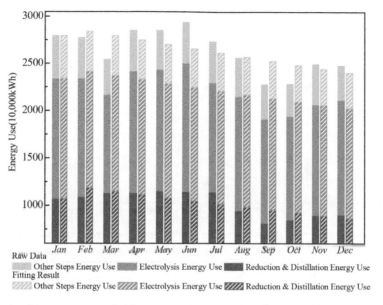

Figure 2.   Chart showing energy usage of different steps during Jan–Dec., 2012.

3.2.3   Energy intensity trends

In the past year, energy usage was calculated by the actual production statistical reports to determine the energy intensity for the titanium sponge enterprise. In future years/months (Jan. 2013–Dec. 2015), the energy intensity will be calculated by using energy usage and product, obtained from the gray theory in Section 2.2. The results of energy intensity calculations are shown in Figures 3 and 4.

Figures 3 and 4 show that during Jan.–Dec. 2012, the electrolysis of $MgCl_2$ method exhibited a 2.25 times higher energy intensity than the reduction and distillation step. Overall, the energy intensity had a downward trend, but the comprehensive energy intensity in several months during that period (e.g. July–December 2012).

Figure 4 shows the energy intensity of the titanium sponge in China during 2012–2015. The 2013–2015 energy intensities are based on the energy usage reduction trends of 2012. Since the steady reduction rates by the end of the 12th FYP (2015) were assumed, the dropping the comprehensive energy intensity, reduction and distillation energy intensity and electrolysis of $MgCl_2$ energy intensity is shown during the period. The reduction rate during the latter 2 years of 12th FYP (2014 and 2015) is lower than that in 2013. The reduction rates assumed for 2013 and 2014 are mostly based on the Guizhou government energy intensity reduction task for the enterprise. The reduction rates for the 12th FYP could mostly be based on expert judgment to meet their energy intensity reduction

task. The 4200 tce (tons coal equivalent) reduction in energy usage of the titanium sponge enterprise in the 12th FYP is in line with the Guizhou government task for energy intensity reduction during this period. It is expected that the reduction and distillation and electrolysis of $MgCl_2$ will contribute the most to achieve this reduction task, maintaining the product does not decrease severely, because they account for approximately 85% of energy usage in not only the enterprise we studied in this article, but also all the titanium sponge enterprises in China and significant energy efficiency potential exists in the step of reduction and distillation.

A gray theory analysis was performed for the titanium sponge enterprise in China for four time periods: 2012, 2013, 2014, and 2015. These four periods were chosen based on the Chinese government's 12th Five Year Plan. A set of energy conservation technology policies that affect the energy intensity were associated every year. And starting in the 12th FYP (2011 and 2012), specific policies, programs, and targets were established with the stated intent of reducing the enterprise's overall energy intensity.

It was significant that the comprehensive energy intensity in 2015 decreased by 7.86% relative to 2011, while the energy intensity of reduction and distillation decreased at a much lower rate than that of electrolysis. These trends resulted in a 5235.28tce energy conservation in comprehensive energy usage over the period of 2012–2015, and thereby eventually accomplished the 12th FYP task, by only relying on the energy-efficiency

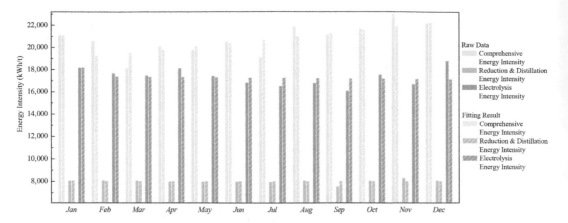

Figure 3.   Chart showing the energy intensity of the titanium sponge enterprise in China during Jan.–Dec. 2012.

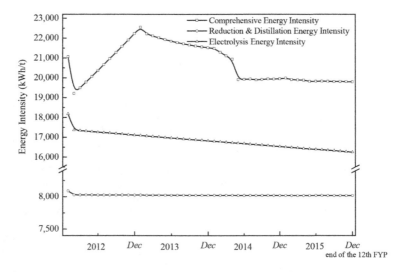

Figure 4.   Graph showing the energy intensity of the titanium sponge enterprise in China during 2012–2015.

measures taken in 2011 and 2012. Figure 5 shows the trends in energy intensity indexes of the titanium sponge enterprise from 2012 to 2015.

3.3   Policy implications

The analysis indicates that if the enterprise wants to shift from an energy-intensive industry to a less energy-intensive industry, the energy intensity reduction rate and strength should be more in line with the above-mentioned answer. The assumed energy intensities reduction is informed by possible growth rates that are foreseen by using the Gray theory. Such a structural change is also a result of shifts in demand for markets in China. The government could influence demand for titanium products indirectly, e. g. universal civilian, but only

to some extent, and generally only temporarily. Hence, the energy-efficiency technology and energy management upgradation would be a more reliable approach to achieve energy conservation and energy efficiency.

The analysis results also show that the intensity effect always reduces energy usage during the analysis and prediction period. The mandatory tasks and fiscal incentives given by the Chinese and Guizhou governments for energy efficiency projects, modernization of the production technologies, phasing out the inefficient and backward technologies, and other aggressive policies and programs to achieve energy conservation and energy efficiency could be important reasons. These reasons along with some influential factors have continued to pressurize the industry or enterprise to improve the

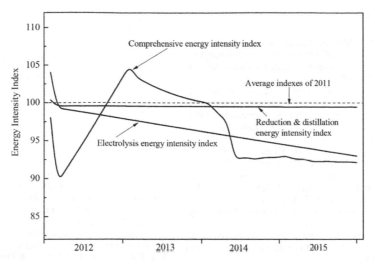

Figure 5. Graph showing the trends in energy intensity indexes of the titanium sponge enterprise (average 2011 value = 100) from 2012 to 2015.

energy efficiency to comply with regulations and to reduce costs. This shows the importance of continued energy efficiency programs and policies in the production process to support continuous reduction in the energy intensity of the titanium sponge enterprise.

The significant changes of the Chinese titanium sponge industry would be seen in the future, delivering the energy-efficiency breakthrough by a local technology, yet such changes would be gradual over the next two decades or longer.

There are a lot of limitations, deviations, and sources of uncertainty in this study, which try to predict future energy usage and energy intensities. Even so, the analysis and prediction in this study can also help to understand how changes in energy intensity can affect the overall energy usage in future.

4 CONCLUSIONS

The Gray theory method can yield an accurate result and also solve problems resulting from less data.

A bottom-up energy usage analysis for a Chinese titanium sponge enterprise is performed in this study using the energy usage data, obtained from various months of actual production statistical reports, and Gray theory. Analysis and prediction are conducted in order to assess, evaluate the energy usage of the enterprise in 2012, and estimate the likely energy usage and energy intensity in the future (2013–2015).

In the analysis period, the mandatory measures taken by the enterprise resulted in the decrease in the overall energy usage. And the energy intensity exhibited a downward trend from 2012–2015. It was significant that the comprehensive energy intensity in 2015 decreased by 7.86% relative to 2011, while the energy intensity of reduction and distillation decreased at a much lower rate than that of electrolysis. These trends resulted in a 5235.28 tce energy conservation in comprehensive energy usage over the period of 2012–2015, and thereby eventually accomplished the 12th FYP energy intensity reduction task of 4200 tce.

The results of this study also show that the energy intensity effect always reduces energy usage during the study period.

ACKNOWLEDGMENTS

The authors gratefully acknowledge the National Science Foundation of China (No. 51574094), the Industrial Research Project of Guizhou Science & Technology Department (No. 3009 in 2013), and the Laboratory Projects of Guizhou Science & Technology Department (No. 4002 in 2014) and Guizhou Education Department (No. 222 in 2014) for financial support. And the authors wish to express their very sincere thanks to the reviewers for the valuable comments and suggestions. And Huixia Pang would also like to acknowledge the support of the Energy & Power Department and R. & D. Center of Zunyi Titanium Corporation Limited.

REFERENCES

Fung, C.-P. 2003. Manufacturing process optimization for wear property of fiber-reinforced polybutylene terephthalate composites with grey relational analysis. *Wear*, 254, 298–306.

Habashi, F. 1997. *Handbook of Extractive Metallurgy*, Heidelberg, Wiley-Vch.

Hartman, A.D., Gerdemann, S.J. & Hansen, J.S. 1998. Producing lower-cost titanium for automotive applications. *JOM*, 50, 16–19.

Hasanbeigi, A., Price, L., Fino-Chen, C., Lu, H. & Ke, J. 2013. Retrospective and prospective decomposition analysis of Chinese manufacturing energy use and policy implications. *Energy Policy*, 63, 562–574.

Hsu, C.-C. & Chen, C.-Y. 2003. Applications of improved grey prediction model for power demand forecasting. *Energy Conversion and Management*, 44, 2241–2249.

Lee, D. & Kim, B. 2003. Synthesis of nano-structured titanium carbide by Mg-thermal reduction. *Scripta materialia*, 48, 1513–1518.

Marsh, E. 1996. *A Technological and Market Study on the Future Prospects for Titanium to the Year 2000*, European Commission.

Nagesh, C.R.V.S., Rao, C.S., Ballal, N. & Rao, P.K. 2004. Mechanism of titanium sponge formation in the kroll reduction reactor. *Metallurgical and Materials Transactions B*, 35, 65–74.

Takeda, O. & Okabe, T.H. 2006. High-speed titanium production by magnesiothermic reduction of titanium trichloride. *Materials Transactions*, 47, 1145–1154.

Tosun, N. 2006. Determination of optimum parameters for multi-performance characteristics in drilling by using grey relational analysis. *The International Journal of Advanced Manufacturing Technology*, 28, 450–455.

Van Tonder, W. 2010. *South African titanium: techno-economic evaluation of the alternatives to the Kroll process*. Master, University of Stellenbosch.

William, M.M. & Julie, C. 2013. ICME—Application of the Revolution to Titanium Structures. *54th AIAA/ASME/ASCE/AHS/ASC Structures, Structural Dynamics, and Materials Conference*. American Institute of Aeronautics and Astronautics.

Wu, F.Z., Gao, C.T., Jin, H.X. & Dou, S.H. 2013. Analysis of energy for deoxidization-distillation process in titanium sponge combination production technology. *Applied Mechanics and Materials*, 368, 697–701.

Risk management of carbon emission rights

Qiu Hong
School of Economy and Management, Tianjin University of Science and Technology, Tianjin, China

ABSTRACT: At present, environmental problems have been paid increasing attention by the whole society. The concept of low carbon economy is also gaining popularity. Energy conservation and emission reduction are not only environmental problems, but these also involve the international, political, and economic aspects. From the point of view of carbon emissions, China is the world's largest carbon emitter. The pressure on the international community to assume the responsibility of emission reduction is increasing day by day. The impact of carbon emissions on China's economic development is one of the key directions of Chinese scholars' research. As China's carbon trading system has just started, there are many problems that lead to the production of risk. China to develop a good carbon emissions trading market must take control of the risk. In this paper, we study the risk management of carbon trading and put forward relevant policy recommendations, in order to improve the carbon trading market.

1 INTRODUCTION

Global warming is one of the most important environmental problems that the world is facing today. There are a variety of approaches to understand global warming. Among these, the feasibility of using a market-oriented approach to global warming has drawn great attention from all walks of life, and has carried out beneficial exploration. Understanding and exploring the issue of carbon emission trading from the perspective of economics is one of the important propositions. Its essence is to solve or weaken the price distortion in the allocation of resources by the market mechanism and achieve the sustainable development of the economy.

With the international community's concern about greenhouse gas emissions, carbon emission rights gradually become a new symbol of the value of activity in international financial markets, the rapid expansion of the carbon trading market. Countries have established the carbon trading market for carbon emission reductions and related derivatives to build the trading platform. China has begun to develop carbon trading in recent years, but it is still in the initial stage of exploration and has not yet established a true sense of the financial trading platform. After the meeting at Copenhagen, China's commitment to carbon emissions per unit of the GDP in 2020 than in 2005 decreased by 40% to 50%, which put forward higher requirements on China's development of low-carbon economy. As an important aspect of the low-carbon economy, the development of the carbon emission trading market is also on the fly.

As China's carbon trading market is not perfect, the relevant policy-making lags; trading brings greater uncertainty and risk to China's enterprises involved in international carbon emissions.

2 THE BASIC DEFINITION OF CARBON EMISSIONS TRADING

Carbon emissions trading refers to the implementation of the national governments under the Kyoto Protocol's commitment to reduce emissions under the premise of their own enterprises to implement carbon dioxide emissions control while allowing them to trade. If a company emits less than the expected amount of carbon dioxide, it can sell the remaining amount and return it. And those companies whose emissions exceed the limit will need to purchase additional permits, so that they can avoid government fines and sanctions, in order to achieve national total control of carbon dioxide emissions.

In developed countries, the cost of carbon dioxide emissions is high; therefore, the Kyoto Protocol encourages developed countries through the provision of financial and technical means, and developing countries are asked to cooperate. In developing countries, emissions are consistent with the requirements of sustainable development policies, but these also produce greenhouse gas emission reduction project investment. Developed countries or enterprises to help developing countries a ton of standard carbon dioxide decomposition of greenhouse gases, you can get a ton of standard greenhouse gas emissions of carbon dioxide.

Emissions trading is one of the environmental economic policies that are concerned by all countries at present. The American economist Dale in the 1960s, and first by the United States environmental protection agency team for air pollution and river pollution management, and then Germany, Australia, Britain and other countries have carried out the practice of the emission trading policy. The emission trading program includes general first determined by the government department of environmental quality standards within a certain region of optimal, and accordingly assessment to determine the environmental capacity, so as to calculate the maximum pollutant emissions in the region can be allowed. It will be divided into a certain number of emission units. And then, the government chose different ways to allocate these property rights, such as public bidding auction, pricing, sale, free distribution, etc.

In the emissions trading market, the sewage companies were allowed from their own interests, to determine the degree of pollution, thereby deciding to buy or sell the number of emission rights. Finally, the government decided in a certain period of time, such as 6 months or 1 year, to check whether the number of sewage companies and the number of sewage permits are consistent with the number of sewage permits, and punish undocumented sewage acts.

The carbon emissions trading system can not only reflect the total control of pollutant control strategies, but also rely on market means to enable enterprises to take the initiative to achieve total control objectives. After the government approved the total amount of carbon dioxide emissions in the region, the right to carbon emissions into the market to trade, reduce carbon dioxide emissions, carbon emissions can be saved in the secondary market trading profit. In this way, carbon dioxide emissions enterprises have the incentive to reduce carbon emissions. It is envisaged that, if the total amount of carbon dioxide emissions in the region was once identified, the carbon emission rights permit to obtain a similar monopoly of resources. Limited carbon emissions will inevitably bring about expensive transactions. Carbon dioxide emissions, in the interests of enterprises, will naturally cherish the limited carbon emissions and reduce carbon dioxide emissions.

Due to the sale of goods, the agreement on carbon emissions trading contract is different from the general goods, services, or investment agreement in that, the process of project operation requires a lot of contract or agreement and the complexity of the transaction will make the whole transaction process full of risks. Risks include contract risk, approval risk, project risk, policy risk, etc. Any one of the potential risks can lead to the failure of carbon emissions trading, which brings huge losses to the enterprise. Carbon emissions trading is so full of uncertainty, since there is no relevant policy support, but also makes a lot of enterprises not to dare to actively participate in carbon emissions trading, and miss the opportunity of the development of carbon emissions trading.

3 THE ESTABLISHMENT OF THE CARBON EMISSION TRADING MARKET

The United Nations Framework Convention on Climate Change, adopted at the United Nations Conference on Environment and Development in Rio de Janeiro, Brazil, in June 1992, is the first in the world to fully control greenhouse gas emissions in response to global warming that affects the human economy and society. At the same time, the Convention is also the representative of the international community in response to global climate change on the issue of international cooperation in an essential framework. The Convention entered into force on 21 March 1994.

The Kyoto Protocol was adopted at the third Conference of the Parties in Kyoto, Japan, in December 1997, for the first time to set greenhouse gas emission reductions target for developed countries during the first commitment period (2008–2012), to be achieved by the end of 2012, where greenhouse gas emissions were aimed to be reduced than the 1990 level of reduction of 5.2%, of which the EU will reduce greenhouse gas emissions by 8%. The Kyoto Protocol also provides for the three trading mechanisms, namely, International Emission Trading (IET), Joint Implementation (JI), and Clean Development Mechanism (CDM). The Kyoto Protocol finally came into effect on February 16, 2005, and three flexible mechanisms were officially launched.

With the encouragement of the above policy, the British Government took the first step and proposed the UK's greenhouse gas Emission Trading Scheme (UK ETS) in the year, which was officially launched in the year, thus opening the global greenhouse gas emission trading market.

4 RISK IDENTIFICATION OF CARBON EMISSIONS TRADING TRANSACTIONS

At the level of commercial activities, enterprises may face risks due to carbon emissions trading and these risks can be divided into two categories: industry risk and business risk. Industry risk is the risk associated with a business in a particular industry. It is critical for companies to choose which industry to run. Some companies are low-carbon types, some companies fall in the high carbon emissions category, and the latter are supervised by the government, the public, the media, and other parties. Operational risks include market risk, political risk, operational risk, legal/compliance risk, project risk, product risk, and reputational risk. The management needs to be fully aware of the risks of carbon reductions that businesses face. One of the biggest problems is the lack of awareness of the potential threats of obstacles. If we are to identify the weak link in the implementation of carbon emission trading in response to climate change regulations and the resulting market changes, we will analyze greenhouse gas emissions. Analysis of emissions and trading conditions from the perspective of the industry's overall value chain is indispensable. Some companies may find that limiting carbon footprints and carbon trading will have only a minor impact, while others may find themselves in a serious threat in this regard.

There are four issues that need to be addressed in the risk identification of carbon emissions trading. These are as follows:

1. Determine the emission list. The development of the joint World Resources Institute, World Business Council for sustainable development of the "greenhouse gas protocol, accounting and reporting standards" is the standard measure of greenhouse gas emissions. The protocol is divided into three categories: direct emissions, indirect emissions from the use of heat, steam, or electricity, as well as other indirect emissions from the industry's downstream emissions sources.
2. Methods and techniques for budgeting emissions. Measuring greenhouse gases is a cumbersome process. Actual carbon emissions can be determined by on-site monitoring and carbon emissions can also be estimated.
3. Management information system. It is the system that implements storage and analysis of information after collecting information. Many companies have developed new information systems to continuously monitor greenhouse gas emissions.
4. Registration of carbon footprint. In order to enable emission reductions to obtain carbon credits (targets), it is necessary to register the enterprises' emission reductions through various mechanisms, so that they can be approved.

5 SUGGESTIONS ON RISK MANAGEMENT OF CARBON EMISSIONS TRADING IN CHINA

5.1 *Play the role of government policy guidance*

First, it should strengthen the carbon emissions reduction, carbon emissions trading, concepts of carbon financial markets such as advocacy, to clarify the potential carbon assets, research-related technology, and market standards. It should play the role of industry-oriented national environmental protection, so that enterprises realize the importance of carbon emissions trading. At the same time, it is necessary to strengthen price regulation and guide enterprises to correct the carbon emissions trading behavior and to avoid a price war. Second, government departments should establish appropriate incentive mechanisms to encourage the introduction of technology and voluntary emission reductions. Actively reduce emissions and actively sell carbon emission rights of enterprises from the capital, taxation, technology, etc. to support and encourage enterprises to implement the carbon emissions trading behavior. Projects involving the introduction and use of new environmental technologies should be encouraged, in particular. Third, we should liberalize the carbon emissions trading restrictions of domestic enterprises to a certain extent, and actively develop the secondary market of carbon emissions trading, in order to grasp the pricing power of carbon emission rights, to avoid the extensive development of carbon emission resources in China.

5.2 *The establishment of the legal system of carbon emissions trading*

First, we need to develop laws and regulations to control carbon emissions, and regulate greenhouse gas emission permits, distribution, fees, transactions, management, and so on. Second, speed up the regulation of carbon emissions trading. Construction management agencies, license issuers, emissions trading agencies, etc, accurately monitor the emissions of units or equipment emissions. Third, clarify the property rights of carbon emission trading rights, so that carbon emission trading can be carried out in a fair, just, and open environment. Fourth, introduce relevant laws and regulations to regulate corporate carbon financing behavior.

5.3 *Further improve the carbon emissions trading market*

First, play a market competition mechanism. The initial allocation of carbon emission rights should be paid by enterprises to purchase the paid way. The government should establish a reasonable allocation system of emission rights. In China's current regulatory situation in the field, the use of auction is more reasonable, not only to avoid the distribution process of black-box operation, but also to allow enterprises to obtain a stable quota of emission rights channels and price guidance.

Second, set up further professional emissions trading intermediaries. The establishment of the relevant information network system is required for the parties to provide information services, improve transaction transparency, and reduce transaction costs. Third, we need to nurture the carbon trading market, by the relevant departments to provide market service information and adjust the unreasonable price trading system, to maintain the stability of carbon emissions trading market.

5.4 Construction of China's carbon financial system

Build a perfect carbon exchange. On one hand, carbon exchange can make carbon trading price fixing open and transparent, and can reduce the transaction risk caused by information asymmetry. On the other hand, since carbon exchange is the main platform for the future of China's carbon financial derivatives trading, investors can make use of hedging methods to avoid investment risks.

5.5 Introduce a risk-sharing mechanism

As the insurance industry plays a certain price-oriented role, determination of the high energy consumption, high pollution, and high carbon emissions industry development can be carried out through the design of insurance products and premiums. This will not only promote the development of low-carbon economy in line with the continuous expansion of the industry, but it is also conducive to the whole society for a low-carbon society to reach a consensus.

Many large-scale clean energy investment projects can sell their unused carbon credits to companies that need more carbon credits. However, since the new energy project itself in the whole process of operation is facing various risks, these risks may affect the smooth delivery of the corporate carbon credit. The establishment of the carbon credit guarantee for the delivery of the project owner or financier to provide security and risk, the risk is transferred to the insurance market to ensure the delivery of carbon emission reductions.

REFERENCES

Cason, T. N, Gangadharan, L. 2003. Transactions costs in tradable permit markets: An experimental study of pollution market designs. *Journal of Regulatory Economics*, 23(2),145–165.

Friedl, B, Getzner M. 2003. Determinants of CO2, emissions in a small open economy. *Ecological Economics*, 45(1),133–148.

Keppler, J. H, Cruciani, M. 2010. Rents in the European power sector due to carbon trading. *Energy Policy*, 38(8),4280–4290.

Lewis, J. 2010. The role of trust in managing uncertainties in the transition to a sustainable energy economy. *Energy Policy*, 38(6),2875–2886.

Malik, A. S. 2002. Further results on permit markets with market power and cheating. *Journal of Environmental Economics & Management*, 44(3),371–390.

Mathews, J. A. 2008. How carbon credits could drive the emergence of renewable energies. *Energy Policy*, 36(10),3633–3639.

Ratnatunga, J, Jones, S. 2012. An inconvenient truth about accounting: the paradigm shift required in carbon emissions reporting and assurance. *Contemporary Issues in Sustainability Accounting, Assurance and Reporting. Bingley: Emerald Group Publishing*, 71–100.

Treasury, H. M. S. 2006. Stern review on the economics of climate change. *London: HM Treasury*, 30.

Advances in Energy Science and Equipment Engineering II – Zhou, Patty & Chen (Eds)
© 2017 Taylor & Francis Group, London, ISBN 978-1-138-71798-5

Study on new energy vehicles application strategies in transportation industry

Luyang Gong
The China Academy of Transportation Science, Beijing, China

ABSTRACT: As a national strategy, development of new energy vehicles is an important measure to reduce fuel consumption, relieve the contradiction of fuel supply and demand, reduce exhaust emission, improve atmosphere environment and promote technical progress and upgrading of the auto industry. The transportation industry is an important field for popularization and application of new energy vehicles. This paper analyzes current situation and problems of the new energy vehicles in the transportation industry, studies and judges popularization and application demand and trend scientifically, and finally puts forward the principle, thinking and specific countermeasures for popularization and application of the new energy vehicles in the transportation industry, providing important decision reference for popularization and application of the new energy vehicles in the transportation industry.

1 STATUS OF NEW ENERGY VEHICLES IN TRANSPORTATION INDUSTRY

1.1 Application scale and feature

1.1.1 Bus

Popularization and application of new energy buses have been accelerated obviously in recent years. By the end of 2015, there were 562,000 buses (633,000, calculated by standardization equivalent) nationwide, including 86,659 new energy vehicles, with year-on-year growth rate of 136.7%, accounting for 15.4% of the total number. Wherein the number of pure electric vehicles is 36,262, and the number of hybrid electric vehicles is 50,397, accounting for 41.8% and 58.2% of the new energy vehicles respectively, as shown in Figure 1.

Seen from the regions, the ratio of the east is more than half, reaching up to 64.4%; the west has least new energy buses, accounting for only 7.7%, as shown in Figure 2.

Seen from the provinces, the top 5 provinces of the new energy buses in number are Guangdong,

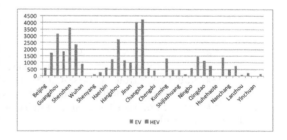

Figure 2. Regional distribution of new energy vehicles and electric vehicles.

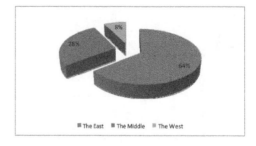

Figure 3. Distribution of new energy vehicles and electric vehicles in 36 central cities.

Shandong, Henan, Jiangsu and Hunan. At present, only Ningxia has no new energy buses.

Seen from the cities, 36 central cities in the country own 2,888 new energy buses in average, wherein Changsha has most operating new energy buses and electric vehicles, 4,270 in total. Hohhot and Yinchuan still have no new energy vehicles, as shown in Figure 3.

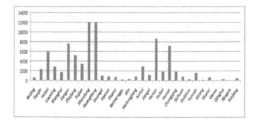

Figure 1. Distribution of new energy vehicles and electric vehicles in provinces.

Table 1. Distribution of fuel types of taxis.

	Gas	ET Gas	Diesel	LPG	CNG	Dual-fuel	EV	Others
Quantity (ten thousand)	49.2	20.1	3.6	0.9	4.4	60.1	0.7	0.3
Ratio(%)	35.3	14.4	2.6	0.6	3.2	43.1	0.5	0.2

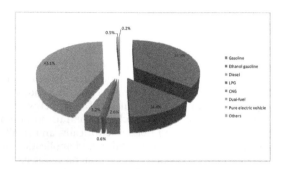

Figure 4. Distribution of fuel types of taxis nationwide.

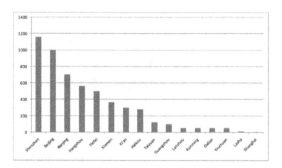

Figure 5. Quantity of pure electric taxis in some central cities.

1.1.2 Taxi

By the end of 2015, there were 492,000 gasoline cars, 201,000 ethanol gasoline cars, 36,000 diesel cars, 9,000 liquefied petroleum cars, 44,000 compressed natural gas cars, 601,000 dual-fuel cars and 6,854 pure electric vehicles, as shown in Table 1 and Figure 4 and 5.

1.2 Application policies

The transportation industry is the most important field for popularization and application of new energy vehicles, where the new energy vehicles are good in popularization and application in various regions. In recent years, Ministry of Transport and related ministries and commissions have formulated some encouragement policies.

Targeted to popularization and application of the new energy vehicles in the public service field

in Beijing-Tianjin-Hebei region, 7 ministries and commissions such as Ministry of Industry and Information Technology of the People's Republic of China (MIIT) and National Development and Reform Commission (NDRC) issued Popularization Proposal of New Energy Vehicles in Public Service Field Like Bus in Beijing-Tianjin-Hebei Region (2014–2015) in 2014 which proposes that the cumulative number of the new energy vehicles popularized and applied in the public service field in Beijing-Tianjin-Hebei region in the year 2014–2015 will reach more than 20,000, by the end of 2015, the ratio of the new energy buses will not be less than 16%, the ratio of new energy taxis in Beijing and Tianjin will not be less than 15%, 94 battery swap stations and 16,200 charging piles will be built.

In 2015, the Ministry of Transport issued Implementation Suggestions on Popularization and Application of New Energy Vehicles in Transportation Industry, proposing the popularization and application targets of new energy vehicles in the transportation industry, i.e. the popularization quantity of new energy vehicles in the fields of bus, taxi and urban distribution will reach 300,000 by 2020, including 200,000 new energy buses and 100,000 new energy taxis and logistic distribution vehicles. The operation efficiency of the pure electric vehicles is not less than 85% of the fuel vehicles with the same length. Meanwhile some relevant technical requirements for use of the new energy vehicles in the transportation industry are identified. The warranty period of the whole vehicles and key components should not be less than 3 years, the vehicles should pass 15,000 km reliability testing; the seat capacity should not be less than 85% of the fuel vehicles of the same length; the ratio of the total mass of the power battery system and the curb weight should not exceed 20%, the battery capacity attenuation in the guarantee period should not exceed 15%, and the cycle life of the battery pack should be more than 1,000. New energy vehicles with long driving mileage and high reliability are preferred. For pure electric buses (exclusive of super-capacitor and lithium titanate fast charging pure electric buses), the vehicle with the driving mileage not less than 200 km should be selected.

In order to solve the problem of cost overhang in operation of new energy buses and common buses, Ministry of Finance, Ministry of Transport and MIIT issued Notice on Improving Urban Bus Petroleum Product Price Subsidy Policy and Quickening Popularization and Application of New Energy Vehicles in May, 2015, which reforms the urban bus petroleum product price subsidy policy. The Notice specifies that the original petroleum product price subsidy is divided into two parts, one part is subsidy for rise in price, subsidized to common buses; the other part is the subsidy for popularization and application of new energy buses. The subsidy for rise in price is reserved as the cardinal of the actual execution number in 2013 and then reduced gradually year by year. 80% of the subsidy for rise in price is paid first, the rest 20% depends on whether the local new energy bus popularization target is completed. The Notice further stipulates that the subsidy condition of the new energy buses must comply with the popularization proportion requirements; the model must be incorporated into "List of Recommended Models for New Energy Vehicle Popularization and Application Project" of MIIT; the annual operating distance should not be less than 30,000 km. The pure electric buses are divided into three classes by length, receiving the subsidies of 40,000 yuan, 60,000 yuan and 80,000 yuan per bus per year respectively; for plug-in hybrid electric buses, the subsidy is cut in halves correspondingly; the super-capacitor and non-plug-in hybrid electric buses receive the subsidy of 60,000 yuan, 20,000 yuan and 20,000 yuan in fixed amount every year. The duration of the subsidy is 2015–2019. The operation subsidy policy will be adjusted after 2020 by considering industry development, cost change condition and preferential electricity price comprehensively.

1.3 Problems

Though the popularization and application work of the new energy vehicles in the transportation industry obtained stage results, there are still some problems to solve and break through.

1.3.1 Lack of electric vehicle charging infrastructure

The existing charging facilities cannot meet the charging demand of the new energy vehicles. Not only insufficiency in total quantity but also large structural contradictions exist in the field of bus. In the existing mode that the charging station is built based on the bus station, construction of the charging station have many difficulties under the condition of lack of land resources and insufficient bus stations. Under the condition that the network of charging facilities to facilitate use is not formed, the bus line network laying, transport capacity increase and operation organization are all restricted greatly.

1.3.2 Poor technology and product quality

The new energy vehicles are in development, and the technologies still need to be improved and developed. The key technology of the new energy vehicles is still not mature completely reflected by the bus enterprise operation status. Compared with that of traditional fuel vehicles, the technical stability, safety and reliability still need to be improved.

1.3.3 Lack of full-life-cycle policy

As for the subsidy range of the financial subsidy policy, currently only the new energy buses and pure electric taxis more than 10 m long have purchase subsidy, the new energy minibuses less than 10 m long are not incorporated into the subsidy scope, and the requirement for developing feeder buses by adopting medium- and small-sized vehicles cannot be met. As for the subsidy link, only the purchase link is incorporated into the subsidy scope, the following operation cost, maintenance cost and personnel training cost are not incorporated. The bus enterprises have large operation cost under the condition that the new energy vehicles have high fault rate and low operation efficiency.

2 DEMAND AND TREND OF NEW ENERGY VEHICLES

Currently, development of new energy vehicles has been identified as national strategy. Energy-saving and New Energy Vehicle Industry Development Program (2012–2020) makes clear that the cumulative number of new energy vehicles will reach 5 million in 2020. Implementation Suggestions of Ministry of Transport on Quickening Popularization and Application of New Energy Vehicles in Transportation Industry also proposes that there will be 200,000 new energy buses in 2020, increased by 126,000 compared with 2015, twice the existing base. Popularization and application of the new energy vehicles in transportation industry have broad prospect.

2.1 Faster application speed

With implementation of new energy vehicle supporting policies, especially urban bus petroleum product price subsidy reform, the operation cost of the new energy buses is greatly reduced; in addition, the new energy charging facility awarding policy also stimulates use of the new energy vehicles and guarantee use of the new energy vehicles in transport enterprise. Therefore, the popularization and application speed of the new energy vehicles will be increased during the "Thirteenth Five Year Plan". Further, with decrease of purchase subsidy of the new energy vehicles, especially in 2016 and 2017, popularization and application of the new energy vehicles will enter the concentrated growth period.

2.2 *More wide application scope*

At first, popularization of the new energy vehicles focuses on the new energy vehicle popularization and application demonstration cities. With full implementation of the supporting policies, more and more medium- and small-sized cities will select the new energy vehicles as the new energy bus popularization and application scale is the important basis for issuing product price subsidy, and the popularization scope will be wider and wider.

2.3 *More application fields*

At present, popularization and application of the new energy vehicles in the transportation industry mainly focus on the urban bus field which is also most easiest to popularize and best in popularization. The cities usually take the urban bus as the sally port to popularize the new energy vehicles. With the new energy vehicle technology becoming mature and infrastructure and supporting policies becoming perfect, application of the new energy vehicles in the fields of taxis, urban logistic distribution vehicles, short trip chartered bus, vehicle leasing and network vehicle booking will also be expanded gradually.

2.4 *More supportive policies for using*

With issuance of Guidance on Quickening Popularization and Application of New Energy Vehicles by General Office of the State Council, policies on popularization and application of the new energy vehicles in terms of infrastructure, product price subsidy, charging electricity price, etc are formulated successively. With gradual decrease of new energy purchase subsidy from 2016 to 2020, the new energy vehicle supporting policies are inclining to the use link from the purchase link. Charging facility land, electricity price discount, battery recycle and technology R&D will be the supporting key points for popularization and application of the new energy vehicles in the future.

3 STRATEGIES AND FRAMEWORK

3.1 *Development principle*

Active and steady steps Give full consideration to technical development of the new energy vehicles, bearing capacity of enterprises, supporting facilities

and policy conditions and use effect of the new energy vehicles, scientifically grasp the popularization and application rhythm, vigorously promote popularization and application of the new energy vehicles to accomplish the national strategic targets avoiding passivity, and scientifically evaluate the popularization and application scale by combining the actual industry avoiding advancing rashly.

Adjustment of measures to local conditions. Perform technology selection demonstration, business mode and popularization and application scale and relevant work of the new energy vehicles under leadership of local people's government by combining actual condition and development demand of local transport organizations, and promote popularization and application work of the new energy vehicles in the transportation industry by adjusting measures to local conditions.

Safety development. Establish the safety development idea, give priority to system safety, facility safety, vehicle safety and operation safety, fulfill safety responsibilities, lay the safety foundation to guarantee operation safety of the new energy vehicles in the transportation industry.

Market oriented. Preserve the leading position of the enterprises, utilize the decisive effect of market allocation of resources, respect the willingness of the enterprises, innovate the popularization and application mode, standardize market operation rules, strive to reduce full-life-cycle cost for purchase, operation, maintenance and battery recycle of the new energy vehicles, and stimulate enterprise's initiative to achieve sustainable application of the new energy vehicles in the transportation industry.

3.2 *Development framework*

Implement the national new energy vehicle development strategy; identify positioning of popularization and application of the new energy vehicles in the industry; accurately grasp the new energy vehicle development direction and popularization and application progress; give full play to the positive role of the government and the enterprises; create a good environment for the development of new energy vehicles by speeding up infrastructure construction, innovating business model, increasing policy support, improving standards and strengthening safety supervision and other measures; eventually build the transportation industry into the main force to promote the popularization and application of new energy vehicles in the public service field; enhance the use effect of new energy vehicles in the transportation industry; provide service for "green traffic" development; promote healthy and sustainable development of the new energy vehicles in the transportation industry.

4 MEASURES

4.1 Improve charging infrastructure

4.1.1 Make charging infrastructure plan

Scientifically prepare charging facility plans for buses, taxis and urban distribution vehicles according to corresponding demands. Strive to incorporate the charging facility plans into urban and rural planning or land use planning. Encourage different bus enterprises, taxi enterprises and urban distribution enterprises in one city to plan charging facility construction as a whole and use the facilities as a whole to improve the utilization efficiency.

4.1.2 Construct inter-city fast-charging network

Quicken construction of quick charging facilities in the expressway service area, build the facilities for "Four horizontal and four vertical expressways" (four vertical expressways: Shenyang-Haikou Expressway, Beijing-Shanghai Expressway, Beijing-Taipei Expressway and Beijing-Hong Kong-Macau Expressway; four horizontal expressways: Qingdao-Yinchuan Expressway, Lianyungang-Khorgas Expressway, Shanghai-Chengdu Expressway and Shanghai-Kunming Expressway), and determine the construction standard of the charging facilities in the expressway service area. Quicken construction of urban quick charging facilities in expressway service areas in urban agglomeration of Beijing-Tianjin-Hebei Region, Yangtze River Delta and Pearl River Delta.

4.1.3 Construct charging infrastructure in hub

Improve construction supporting standard of the bus stations and the taxi stations and confirm the new energy charging facility supporting construction standard. Quicken construction of charging facilities in the urban bus stations and taxi stations. Construct quick charging facilities according to the demand during planning and construction of urban comprehensive passenger hub, public transit hub, taxi operation station, urban logistic distribution center and service area and express logistic park; consider constructing the charging facilities of "slow charging supplemented by quick charging" during planning and construction of urban parking lot, maintenance workshop and taxi stand. Modify the existing bus station, taxi station and urban logistic distribution depot, and quickly construct and improve the charging facilities according to the actual demand if the construction conditions are met.

4.1.4 Unify the interface standards

It's suggested to quicken revision and introduction of charging interfaces and communication protocol, actively promote charging interface operability detection and revision of standards for data exchange between service platforms, modify the existing charging infrastructure, quicken unification of the charging standard, and achieve compatibility and interconnection between charging devices of different manufacturers and electric automobiles of different brands.

4.2 Vehicle technology selection

4.2.1 Identify the standard

By combining industry characteristics of buses, taxis and urban logistic distribution, develop industry standards for new energy buses, taxis and logistic distribution, identify use scope, technical requirements, service guarantee requirements and vehicle configuration requirements of new energy vehicles in the industry, and promote standardization and high efficiency of popularization and application of the new energy vehicles in the industry.

4.2.2 Take evaluation of new energy vehicles usage

Develop dynamic evaluation work of use effect of the new energy vehicles, develop new energy use effect dynamic evaluation index system, evaluate use effects of different technical types under different environments and operation conditions and in different fields, thereby providing policy support for selection of optimal model and optimal solution.

4.2.3 Encourage manufacturers to produce new energy vehicles conforming to industry characteristics

Encourage the transportation industry and the new energy vehicle manufacturing enterprise to research and develop special model for the new energy vehicles applicable to the transportation operation organizations, and intensify popularization of the special model.

4.3 Innovate business model

4.3.1 Utilize PPP model for charging infrastructure construction

Charging facility construction is large in investment, low in rate of return on investment and long in return on investment. Encourage local government to explore to utilize the PPP mode to quicken construction of charging facilities for new energy buses, taxis and urban logistic distribution. Explore to build the win-win mode of cooperation of local government, transportation enterprise and investment enterprise.

4.3.2 Encourage rental outlets and finance lease model

The single purchase cost of the new energy vehicles is high, so many transportation enterprises are lack of sufficient fund. Improve relevant laws and policies, encourage the enterprises and the vehicle producing enterprises to innovate the modes of leasing with option to buy and finance lease and quicken popularization and application of the new energy vehicles. For example, Provisions on Administration of Road Passenger Transportation and Related Stations state that the transportation enterprises developing road passenger transportation line operation must have certain number of self-support passenger transportation vehicles. However, the property right of the vehicle may be reserved by the vehicle manufacture enterprise or the third party leased enterprise in the existing mode of leasing with option to buy, and the property right and the operation right of the vehicle are separated. Therefore, relevant regulations and systems shall be improved to create conditions for the modes of leasing with option to buy and finance lease.

4.3.3 Encourage vehicle and battery separation model

The battery is a most important component of the new energy vehicles and also the component with highest cost of the new energy vehicles. Meanwhile, as the battery technology is not very mature, the distance per charge, the service life and the safety of the battery become the major concern of the transportation enterprise to select the new energy vehicles. Encourage the transportation enterprise and the vehicle manufacture enterprise to innovate the vehicle and electricity separation mode and adopt the transportation enterprise battery leasing mode. The vehicle manufacture enterprise participates in battery full service life management and reduce operation risks of the transportation enterprise.

4.4 Supportive policies

4.4.1 Differentiated traffic management

Strive to perform specialized management on the new energy vehicles by the public security department, better distinguish the new energy vehicles, implement differentiated traffic management policy, and create conditions for formulating new energy vehicle special traffic management policy to better promote development of the new energy vehicles.

4.4.2 Operation priorities

The urban bus and taxi operation right is first awarded to new energy vehicles and inclines to the transportation enterprises high in new energy vehicle popularization and application degree, or special new energy vehicle transportation enterprises are established, and operation right index of the new energy taxis is broadened properly.

4.4.3 Preferential toll

Formulate the new energy vehicle toll preferential policy and promote no toll or roll reduction of new energy vehicles. Formulate the new energy vehicle parking fee preferential policy to reduce or waive the parking fee.

4.4.4 Priority access

Cooperate with the public security department to formulate the new energy vehicle traffic differentiated management policy and the new energy vehicle traffic policy. Promote the cities to cancel or broaden passage time, area and model limit of new energy logistic distribution vehicles and create conditions for popularization and application of the new energy logistic distribution vehicles.

4.4.5 Preferential power price

Implement preferential policy on charge price of new energy buses, taxis and urban logistic distribution vehicles to reduce operation cost of the enterprises. It's suggested to implement "residential electricity" or "general business and others" classification on the charge price of the special charging facilities according to the actual condition. Due to characteristics of urban buses, taxis and urban logistic distribution vehicles that the vehicles need to be charged in daytime, formulate new energy vehicle charge price peak-valley preferential policy, and execute as per the valley price or the fair price for the urban buses, taxis and urban logistic distribution vehicles.

4.4.6 Preferential tax for electric taxi

Consider the pure electric taxis as the supplementary part and the integral part of public transport, and introduce electric taxi purchase tax reduction preferential policy with reference with the bus purchase tax free policy to further reduce purchase cost of the pure electric taxi operation enterprise.

4.5 Technology R&D

4.5.1 Enhance key technology

The technical level of the new energy vehicles is still the key factor of popularization and application of the new energy vehicles in the transportation industry, and distance per charge of the battery, charging time, vehicle performance and stability are still the bottleneck restricting large scale popularization of the new energy vehicles. Suggest MOST and MIIT to accelerate research of new energy power battery and whole vehicle key technology, improve vehicle performance and reduce vehicle cost.

4.5.2 Enhance R&D of new energy operation vehicles

Encourage the transportation enterprise to cooperate with the vehicle manufacture enterprise to research and develop special vehicles suitable for transportation characteristics by combining operation characteristics of urban buses, taxis and urban logistic distribution.

4.5.3 Training for new energy operation vehicles using

Organize transportation management departments, urban bus enterprises, taxi enterprises and logistic distribution enterprises to develop use training of the new energy vehicles to further improve use efficiency and effects.

4.6 Improve standard

4.6.1 Improve operation vehicle evaluation standard

Timely revise road transportation industry vehicle technology rating evaluation standards such as JT/T198–2004 Operation Vehicle Technical Grade Division and Evaluation Requirements and JT/T888–2014 Bus Type Division and Grade Evaluation to create conditions for operation of the new energy vehicles.

4.6.2 Improve vehicle maintenance and detection standard

Improve vehicle technical maintenance detection standards such as GB18565 Operating Vehicle Comprehensive Performance Requirements and Test Methods, GB18832 Items and Methods of Power-Driven Vehicles Safety Inspection and Passenger Transport Station Vehicle Safety Routine Work Technical Specification to guarantee safety operation of the new energy operation vehicles.

4.6.3 Formulate new energy operation vehicle operation guide

Based on use experiences of new energy buses, taxis and urban logistic distribution vehicles in typical regions, formulate operating guide of the new energy vehicle operating vehicles and provide technical guidance for use of the new energy vehicles by enterprises in various places. Identify new energy vehicle operating guide of different fields,

different models and different conditions and improve the use effect of the new energy vehicles.

4.7 Enhance safety and emergency management

4.7.1 Build new energy vehicle monitoring platform

Enhance operation safety monitoring of the new energy vehicles, incorporate the vehicles into the urban traffic intelligent monitoring platform and improve foundation information of the new energy vehicles. Urge the relevant traffic transportation enterprises to add real-time monitoring devices on the new energy buses and taxis to perform real-time monitoring and dynamic management on the vehicle operation technical state, charging state and battery single body, build new energy vehicle operating data collection and statistical analysis system to further provide foundation support for safety operation of the new energy vehicles.

4.7.2 Improve new energy vehicle emergency handling capability

The fault rate of the new energy vehicles is relatively high and has high requirements for use environment. With ceaseless enlargement of popularization and application scale of the new energy vehicles, the proportion of the new energy buses, taxis and urban logistic distribution vehicles is increased continuously, the systematic risks of the new energy vehicles under severe weather and special condition are increased continuously, so we shall enhance new energy vehicle emergency handling capability, improve the ratio of emergency reserved vehicle, build emergency reserve system, and prepare emergency plans to prevent systematic risks.

REFERENCES

General Office of the State Council. Guidance on Quickening Application of New Energy Vehicles. Z. 2014.

MOF P.R. China, MOT P.R. China. Notice on Improving Urban Bus Petroleum Product Price Subsidy Policy and Quickening Popularization and Application of New Energy Vehicles. 2016.

MOT P.R. China. China's Urban Passenger Transportation Development Report (2015). M. China Communications Press. 2016.

MOT P.R. China. Implementation Suggestions on Quickening Application of New Energy Vehicles. 2015

NDRC P.R. China. Electric Vehicle Charging Infrastructure Development Guide (2015–2020). Z. 2015.

Advances in Energy Science and Equipment Engineering II – Zhou, Patty & Chen (Eds)
© 2017 Taylor & Francis Group, London, ISBN 978-1-138-71798-5

Study on the theories and strategies for transportation service in China

Luyang Gong
The China Academy of Transportation Science, Beijing, China

ABSTRACT: As a fundamental, guiding and service-oriented industry, service is the essential attribute of transportation. As an important material foundation and supporting condition for the development of national economy and progress of social civilization, transportation is an important link in the realization of passenger, logistics, capital and information flows. Therefore, the improvement of transportation service level is great significance for building a moderately prosperous society. This text analyzed the theory of transportation services and found out the main problems of transportation services and the reasons for the low level of transportation services. Finally, it put forward the overall idea of improving the quality of transportation service in China and the key tasks on six aspects from the equalization of basic public services, convenience of service, security of service, internet transportation services, standards of service. It provides the important policy suggestions for the improving transportation service in China.

1 TRANSPORTATION SERVICE THEORIES

1.1 *Definition and category of services*

The word "Service" is defined as the work for interest of collective (or others) or a certain career. Service is an invisible thing. From the perspective of economics, services are defined as the exchange of equal value in order to meet the needs of enterprises, public organizations or other social and public service activities.

Services can be divided into three types as public services, business services and quasi public services.

Public services mainly refer to all work engaged by government with public power and public resources. Public services include economic regulation, market supervision, social management and other functional activities.

Business services are the services that can be regulated by market to meet diversified requirements of residents. The governments don't provide this services directly. But governments can encourage and guide social power to provide and operate these services through opening and supervision of market.

Quasi public services are the necessary public services that protect the overall social welfare level. At the same time, quasi public services can be provided or operated by market. But requiring support from various measures taken by government due to no benefit space or insufficient benefit space left owing to government pricing. However, quasi public services should be supported by governments,

because of low or no profit as the government pricing and other reasons.

1.2 *Categories of transportation services*

Transportation services can be divided into three categories:

1.2.1 *Public transportation services*
Public transportation services are the public services that provided by governments. Public transportation services should be suitable for the economic and social development level and stage. There aims are meeting the basic transportation requirements of all citizens. Rail transit, urban public transport, rural road, rural road passenger transport, land-island transport and rural ferry are public transportation services.

1.2.2 *Business transportation services*
Business transportation services are services that meet diversified requirements of publics completely through allocation of market resources. General cargo transportation, car rental, tour passenger transport, road long-distance passenger transport, vehicle maintenance, highway and waterway cargo transportation are business transportation services.

1.2.3 *Quasi public transportation services*
Quasi public transportation services are the public services that there are necessary for the security of social transport welfare level and provided by the

market. Quasi public transportation services need the support by governments because of its low or no profit as the management of governments. Taxi services are the quasi public transportation services.

1.3 Characteristics of transportation services

1.3.1 Fundamentality

Transportation services include the fields of road transport, urban-rural and inter-city road transport, urban transport. There closely related to the people's production and life. The fundamental property of transportation services shows that transportation services should focus on the public fundamental travel demands and freight demands, which means "able to travel" and "able to be transported".

1.3.2 Equalization

Transportation services should make no difference in urban and rural areas, in eastern, central or western regions. There should be possibilities provided for everyone, as the public fundamental services.

1.3.3 Accessibility

Transportation services should be provided for people in a fast and convenient way, so that people could really feel that transportation services are provided "everywhere".

1.3.4 Economy

Transportation services should be provided by minimum resource cost during organization of operation activities acquiring transportation products and services in a certain quantity and quality. Specifically, it is less cost spent on transportation by client during passenger or freight transport process.

1.4.5 Safety

Transportation services should be safely provided, which is the degree transporting people or cargoes to their destination in a complete way. It shows in a more specific way in low accident rate, low damage rate and low potential safety danger.

2 MAIN PROBLEMS

2.1 Insufficient supply

The main contradiction of transportation industry in China is that there is still wide gap between increasing new demands and the relative delayed transportation service. The equalization level of fundamental public services is required to be improved. Transport infrastructure cannot meet the demand of transportation services. Transportation equipment level is still low. Travel information service system still needs to be improved compared with the people's requirements of travel convenience, safety, economy and comfort. For example, problems such as the low development level of public transport, low taxi accessibility and inconvenient driving, parking, loading and unloading in urban distribution process are still serious. The questionnaire survey results show that there are 8%, 6.5%, 10% of the public believe that the road, passenger transportation, taxi service overall level of poor, as shown in Figure 1.

2.2 Inconvenience

The capacity of transportation services to meet people's requirements on convenient transport is still insufficient. Different transportation is not connected with each other in a smooth way. The development of "one-ticket" for passenger transport and of "one order" for freight transport is slow. Integration of urban and rural road passenger transport needs to be further promoted. Travel information services are lagging behind. The public can't use information services conveniently. The content of travel information is neither perfect nor rich. The data of travel information is not timely updated. The way to provide travel information service is relatively single and the release channel is still limited.

The questionnaire survey results show that the convenience of transportation service is still not enough. Especially, in some rural areas and remote mountainous areas, the transportation service level is low, so as "travel inconvenient", "Freight not fast". 35% of the public think that rural passenger transportation is inconvenience. 24% of the public think that bus lines are not enough, transfer is not convenient, travel speed is slow, waiting time is too long. More than 52% of the public believes that it is difficult to take taxi and the taxi service is inconvenience.

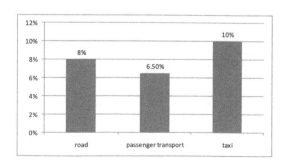

Figure 1.   Survey result of transportation service level.

2.3 Inefficiency

The organization way and means of transportation is backward. The market concentration is relatively low. The operation mode of transportation services is backward and the enterprise competition ability is not strong.

Inland transport ship average tonnage is only 450 tons, the Yangtze River is 800 tons, while the United States is 1200 tons, Germany is 1400 tons, as shown in Figure 2; three inland and above 9894 km waterway, only a total of 7.9% American waterways, three level and above the channel proportion is 61%, 71% in Germany; inland waterway transport enterprises 4600 individual operators, nearly 3 million, the average capacity of only 5 ships, as shown in Figure 3.

2.4 Low quality

People's higher requirements for transportation services are raised from "able to travel" to "able to travel in a safe, fast and comfortable way" and from "able to be transported" to "able to be transported in a fast, safe and efficient way". But with current imperfect devices and equipments in transportation industry and the disordered market behavior, the improvement of service quality is restrained.

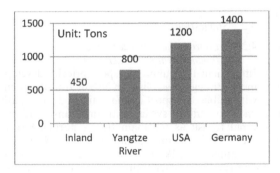

Figure 2. Average tonnage of ships in some areas.

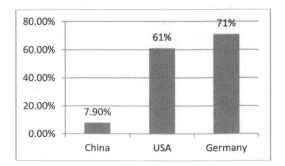

Figure 3. Three level and above the channel rate of China, USA and Germany.

The growing rich masses of the people have put forward higher requirements for transportation services. The requirement of the people are changing from "able to go" to "go safely, comfortably and conveniently" and from "able to transfer" to "transfer fast, safely and efficiently".

According to this survey, 20% of the passengers think that pour off, cheating, discarding behaviour is serious. 55% of respondents think that freight damaged or loss, not on time is serious in logistics service.

2.5 Unsafety

Some key problems restraining the development of transportation safety are not completely solved. Transport market is in disorder. Some transportation enterprises only focus on the benefit, but ignore safety. The effort put in safe production is insufficient. Safety management system is not perfect. Operation procedure cannot be implemented. Especially for those enterprises which compete disorderly, run their vehicles and ships business in affiliating operation or in remote operation. Such problems lead to the absence of enterprise safety management and therefore bear huge accident potential.

The survey results show that nearly 60% of the public think overspeeding, overloading, fatigue driving is serious.

3 CAUSES

3.1 Concept behind

In transportation industry, the concept that focusing on construction but ignoring operation, focusing on management but ignoring service is very strong. Nature of transportation services has not been fully reflected. The service concept has not penetrated into all fields of the transportation industry. The construction of transportation service system lags behind the development of infrastructure construction. Compared with the infrastructure construction, human, material and financial resources put in the improvement of service quality and level are insufficient. Affected by the maximization of benefits, some business entities lack of initiative and enthusiasm for service improvement. The social responsibility awareness of enterprises is not enough. The uneven quality and poor overall quality of workers affect the level of industry service quality.

3.2 Inadequate system and mechanisms

In some areas and regions, systems and mechanisms are not perfect. Sometimes, the coordination between departments, fields and regions cannot be realized to form cooperation. Service supervision mechanism is not perfect. Service supervision function cannot be implemented. Market access and exit mechanism is not perfect with poor implementation. Problems in systems and mechanisms preventing the improvement of service level and quality are not be solved. Therefore, the improvement of transportation services level and efficiency are restrained. There are large space for the transportation management system to improve compared with market demands and requirements of economic society for transportation. The transportation service market mechanism of the survival of the fittest is not perfect. Reward and punishment mechanism, design of regulatory system, market access and exit mechanism, assessment of service quality and credibility awaits to be improved.

3.3 Inadequate regulations, laws and standards

The contents of present laws and regulations on service are insufficient. The basis supporting enterprises with good service quality and high service level is lack. Regulation or laws on the punishment for enterprises with poor service quality and low service level on market access and exit and resource allocation are insufficient. It leads to the lack of enthusiasm and motivation of operators to improve the transport services, so that service level is difficult to be improved. In addition, in some areas there still lack the guidance of the superior law. For example, there is no complete law and regulations in urban public transport field. Besides, the establishment of industry standard is far behind the requirement of industry development. The lack of standards and specifications in some areas. It leads to the lack of effective guidance on services. Service standards and specifications in some areas are old and unable to meet the requirement of transportation service in new period. Service standards and specifications in some areas are difficult to be operated and not applied in a wide range. Therefore, the standards and specifications for transportation services still await to be improved.

3.4 Low informatization level

At present, the overall level of China's transportation information is not high. It can't meet the requirements of the development of modern transportation industry and people's requirements. The integration of information and transportation service is insufficient. The information service level and information resources sharing level are low. Dynamic information acquisition ability is relatively weak. It has not yet been deeply applied in business standardization, process reconstruction and services improvement. The role of information benefit and scale has not yet been fully played. The information service level among ministry, provinces, regions, and departments are unbalanced.

4 IDEAS AND STRATEGIES

4.1 General ideas

The main line is changing the development mode of transportation. The orientation is serving the development of economic society and people's safety and convenience in travel. The key is improving people's satisfaction. The focus is solving the key problems that closely related to people and most urgent to their requirements. Grasp the major process in improving service quality and level. Actively promote the concept of innovation, service innovation and management innovation. Sound system mechanism, strengthen the standard specification, strengthen scientific and technological support, strengthen the moral construction, improve the quality of the team and other comprehensive measures. Finally, according to these measures, strive to build a safe and reliable, convenient and comfortable, economic and efficient transportation service system.

4.2 Suggestions

4.2.1 Improve the equalization level of fundamental public services

Implement the public transport priority development strategy. Establish an urban transit system with public transport as the main body. Establish an urban transit system with rail transit and bus rapid transit as the main frame, normal bus as the main body and community bus, custom bus as supplement. Promote the integration of urban and rural passenger transport. Improve rural and administrative villages accessible rate of passenger car. Improve the service level of urban and rural passenger transport integration. Formulate urban and rural financial support policies. Promote the development of rural logistics to serve the rural economy.

4.2.2 Improve the convencience

Accelerate the construction of comprehensive passenger and freight transport hub and station. Improve the service level of transportation hub and station. Build comprehensive freight hub standard. Accelerate to promote the multi-mode freight transport and passenger connecting transport. Enhance the organization cohesion between different transport modes to improve the service level of comprehensive transportation.

4.2.3 *Ensure the safety*

Enhance the construction of safety production facilities and equipments. Enhance the construction and maintenance of safety production infrastructure. Promote the construction of transport emergency equipment and material reserving center for national and provincial highway. Establish various levels of transport emergency support capacity reserves. Establish large scale transportation infrastructure and management and maintenance safety monitoring system. Establish safety supervision and control system to carry out safety production centralized regulation. Enhance safety management of road passenger transport, hazardous chemicals transport and of port dangerous chemicals to improve emergency handling capacity. Establish product quality supervision system and enhance the quality management of key transportation products. Release supervision categories and improve the quality of key products.

4.2.4 *Promote the deep integration of the Internet and transportation services*

Promote the establishment of ETC network nationwide. Improve the coverage rate of ETC lanes. Expand the user scale. Improve settlement efficiency and promote inter provincial interconnection. Accelerate the formation of a unified national ETC customer service system. Promote the network ticketing system for national passenger transport. Build network ticketing system for national road, waterway transport to provide more convenient way for passengers to book ticket. Promote bus card interoperability. Enhance the popularity and coverage of city bus card application to effectively enhance the public travel and transfer convenience. Promote and regulate the development of online-hailing car. Encourage and support new industry development, such as "Internet + logistics", "Internet + urban and rural passenger transport", "Internet + taxi", "Internet + vehicle maintenance" and "Internet + multi-mode transport", "Internet + passenger connecting transport".

4.2.5 *Improve standardization*

Perfect service standard system. Enhance the revision of major standards in service condition, service process, service quality level, public travel information, logistics facilities and equipments. Establish service standards in quality assessment of highway management services, urban rail transit operation and management service and urban public buses and trams enterprise service. Revise Safety Operation Rules for Highway Maintenance, Standard for Passenger Service of Urban Public Buses. Gradually establish a perfect standard service system of transportation industry. Improve highway traffic sign setting, national highway network naming and numbering system to standardize the management of highway service area. Enhance the organization and management of highway maintenance work. Standardize the organization and management of highway maintenance work. Enhance the supervision and inspection of highway maintenance.

4.2.6 *Improve the efficiency governments*

Promote the transformation of government functions. Cancel and delegate part of reviews and approvals to subordinate departments. For these reviews and approvals, revise relevant laws, regulations and rules to improve the subsequent supervision measures. For the rest reviews and approvals with license approval remained, reduce approval procedures and shorten approval time to improve work efficiency. Optimize the administrative review and approval process and improve quality and efficiency of review and approval to promote online service for administrative license. Enhance the function of hotline 12328, the hotline for traffic and transportation service supervision.

5 CONCLUSIONS

This paper analyzed the current situation and existing problems of the Chinese transportation service, and analyzed the reasons affecting the Chinese transportation service level. Finally it put forward the policy of improving the level of Chinese transportation service.

a. Transportation service can be divided into three types, public transportation service, business transportation service, quasi public transportation service.
b. The Characteristics of transportation service are fundamentality, equalization. Accessibility. Economy, safety.
c. The problems of Chinese transportation service are insufficient supply, inconvenient, inefficient, low quality, unsafe. The causes of the problems are concept behind, inadequate system and mechanisms, inadequate regulations, laws and standards, low informatization level.
d. The measures of improving the Chinese transportation service are improving the equalization level of fundamental public services, improving the convenient, ensuring the safety, promoting the deep integration of the Internet and transportation service, improve standardization, improve the efficiency governments.

REFERENCES

MOT P.R. China. Research on Improving the transportation Service. R. 2013.

MOT P.R. China. Some Suggestions on Improving the Transportation service. Z. 2013.

Tian Yishun. The Development Review on Social Services Capacity of China Transportation. J. 2012.

Wang Yun. Analyse the method of promoting the transportation services. J. 2015.

Xia Zhengnong, Chen Zhili. Word-Ocean Dictionary. M. Shanghai: CISHU, 2010.

Mechanical engineering

Advances in Energy Science and Equipment Engineering II – Zhou, Patty & Chen (Eds)
© 2017 Taylor & Francis Group, London, ISBN 978-1-138-71798-5

A tracked robot system for inner surface scanning of industrial pipelines

H. Zhong, Z. Ling, W. Tan & W. Guo
Zhejiang Provincial Special Equipment Inspection and Research Institute, Hangzhou, China

ABSTRACT: Currently, in-pipe inspection equipment that can run through inside the industrial pipelines is strongly demanded because the industrial pipelines are usually mounted overhead or buried under the ground. The application of robotic systems for in-pipe inspection of industrial pipelines is considered as one of the most attractive solutions available. This work focuses on the application of in-pipe robotic system in industrial pipeline inspection. A tracked robot system for inner surface scanning of industrial pipelines is presented. The construction and the design essential of the tracked robot system are introduced. Inner surface scanning experiments of the industrial pipelines with outer diameter of 159 mm are carried out. The research results show that the presented tracked robot system is effective. It can successfully run through inside the industrial pipelines. The research results also indicate that the presented tracked robot system can implement the inner surface scanning of the industrial pipelines. The photos obtained by the tracked robot system can display the basic conditions of the industrial pipelines.

1 INTRODUCTION

As important transportation facilities, pipelines widely exist in many industrial fields, such as petroleum, natural gas, metallurgy, chemical engineering, etc., They are very useful tools for transporting crude oil, natural gas, tap water, sewage, and other fluids. However, with long-term use of the industrial pipelines, natural deterioration and pipeline failure may occur due to the pipeline defects such as ageing, corrosion, cracks, and mechanical damages (Ahammed 1998, Okamoto et al. 1999, Roh & Choi 2005). Once pipeline failure occurred, unpredictable safety accidents and direct economic losses may arise. Therefore, to eliminate the potential problem and hence to avoid the pipeline failure of the industrial pipelines, defect inspection method or equipment should be studied and periodic inspection of industrial pipelines should be undertaken.

Currently, the in-pipe inspection of industrial pipelines has become an attractive research field. There are a lot of needs for autonomous inspection equipment that can run through inside the industrial pipelines because the industrial pipelines are usually mounted overhead or buried under the ground (Okamoto et al. 1999, Roh & Choi 2005, Park et al. 2011). Up to date, the in-pipe inspection of industrial pipelines is still in difficulty. Besides of the above mentioned installation features, this subject can also be viewed from the following two aspects (Okamoto et al. 1999, Choi & Ryew 2002, Roh & Choi 2005, Tavakoli et al. 2010, Park et al.

2011, Ciszewski et al. 2014): (1) The configuration of industrial pipelines is usually very complicated. Besides straight pipelines and welded joints, there are lots of elbows, branches, and other special components in the industrial pipelines. (2) The inner diameter of the industrial pipelines is usually small which makes it difficult or impossible to carry the inspection equipment inside the pipelines.

Hitherto, many in-pipe inspection methods for industrial pipelines have been reported. Among them, the application of robotic systems for in-pipe inspection of industrial pipelines is considered as one of the most attractive solutions available (Hollingum 1998, Choi & Ryew 2002, Roh & Choi 2005). The development of the in-pipe inspection robotic system can be found in Refs. (Roh & Choi 2005). Although great efforts concerning in-pipe robotic systems have been made, these systems still have certain problems to be solved, such as the dynamic path planning and the industrial applications (Roh & Choi 2005). Therefore, the conventional in-pipe robotic systems cannot completely meet the practical inspection requirements of the industrial pipelines, and further research work should be undertaken.

This work focuses on the application of in-pipe robotic system in industrial pipeline inspection. A tracked robot system for inner surface scanning of industrial pipelines is presented. The construction and the design essential of the tracked robot system are introduced. To test the performance of the tracked robot system, inner surface scanning experiments of the industrial pipelines with outer diameter of 159 mm are also carried out.

2 CONSTRUCTION AND DESIGN ESSENTIAL OF THE TRACKED ROBOT SYSTEM

The tracked robot system mainly includes three parts: a tracked mobile robot, a Charge Coupled Device (CCD) assembly, and a communication-control module, as shown in Figure 1.

The tracked mobile robot, which is the core component of the system, use the spring tensions to press the inside wall of the industrial pipelines and can automatically move through inside the industrial pipelines with three caterpillar bands (driven by six DC motors) as well as the corresponding mechanical structures. The Charge Coupled Device (CCD) assembly is used to photograph the inner surface videos or photos of the industrial pipelines. The communication-control module, which is built up of a micro controller, motor driver and sensor interface, is used to send the control signals to the tracked mobile robot and hence control the motion state of the robot. Meanwhile, the communication-control module can also receive the videos or photos obtained by the CCD camera.

2.1 Construction of the tracked mobile robot

Figure 2 shows the construction of the tracked mobile robot, including the body frame, the corresponding mechanical parts installed in the body frame, and three caterpillar bands spaced at 120°. The three caterpillar bands are driven by six DC motors (each caterpillar band is equipped with two DC motors). The CCD assembly is installed at the front part of the body frame of the tracked mobile robot.

The robot uses the spring tensions (generated by the mechanical structures) to press the inner wall of the industrial pipelines and can automatically move through inside the industrial pipelines. It can also realize the adaption to pipe diameter and the adjustment of wall pressing force.

2.2 The CCD assembly

The CCD assembly mainly includes three parts: a CCD camera for photographing, several lamps for illumination, and the additional mechanism. The additional mechanism can rotate along the circumferential direction and its own axis. Therefore, it can help the tracked mobile robot slide on

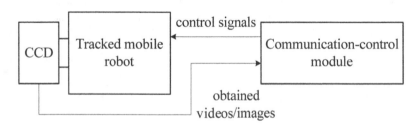

Figure 1.   Scheme of the tracked robot system.

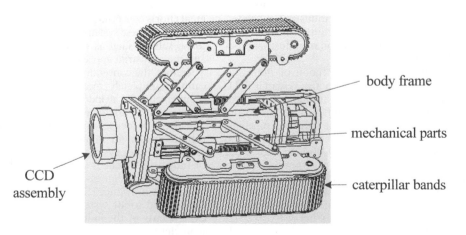

Figure 2.   Construction of the tracked mobile robot.

the wall during steering in the fittings and guides it in the desired direction, and prevents the body of the robot from having direct contact with the wall so that the robot may not be stuck in the industrial pipelines.

2.3 *The communication-control module*

The communication-control module, which is designed on the basis of a micro controller, is used to control the motion state of the robot and receive the videos or photos obtained by the CCD camera. During the movement process of the tracked mobile robot, the communication-control module send the real-time signals to control the six DC motors and hence to the movement state of the three caterpillar bands. This procedure is implemented by the famous Controller Area Network (CAN) bus and the corresponding communication protocol. Meanwhile, the videos or photos are obtained by the CCD camera and then transmits the micro controller.

3 EXPERIMENTAL RESULTS

To test the performance of the tracked robot system, inner surface scanning experiments of the industrial pipelines are carried out. Figure 3 shows the construction of the experimental setup, including a U-type experiment steel pipeline and the steel

fixation apparatus. The U-type experiment steel pipeline is composed of straight pipelines, welded joints, elbows, branches, and a spherical valve. The inner and outer diameters of the steel pipelines are 150 mm and 159 mm, respectively. The fixation apparatus are installed on the ground by using the anchor bolts.

In the inner surface scanning process, the tracked mobile robot automatically run inside the steel experimental pipeline, photograph the inner surface videos or photos of the pipelines and then transmits the videos or photos to the communication-control module. Figure 4(a) to Figure 4(d) show the inner surface scanning photos of a straight pipeline, a welded joint, an elbow and the spherical valve of the steel pipelines, respectively.

As shown in Figure 4(a) to Figure 4(d), the tracked mobile robot can successfully run through inside the industrial pipelines. Inner surface scanning can be implemented by the tracked robot system. The photos obtained by the Charge Coupled Device (CCD) assembly can display the basic conditions, such as straight pipelines, welded joints, elbows and the spherical valve of the industrial pipelines.

4 DISCUSSIONS AND CONCLUSIONS

In this work, a tracked robot system for inner surface scanning of industrial pipelines is presented.

Figure 3.   Construction of the experimental setup.

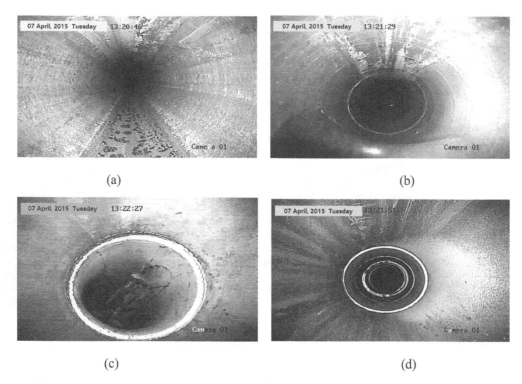

(a) (b)

(c) (d)

Figure 4. Inner surface scanning photos of the steel pipelines. (a) The photo of a straight pipeline. (b) The photo of a welded joint. (c) The photo of an elbow. (d) The photo of the spherical valve.

The construction and the design essential of the tracked robot system are introduced. To test the performance of the tracked robot system, inner surface scanning experiments of the industrial pipelines with outer diameter of 159 mm are carried out.

The research results show that the presented tracked robot system is effective. It uses the spring tensions (generated by the mechanical structures) to press the inner wall of the industrial pipelines and can automatically move through inside the industrial pipelines. It can also realize the adaption to pipe diameter and the adjustment of wall pressing force. The research results also indicate that the presented tracked robot system can implement the inner surface scanning of the industrial pipelines. The photos obtained by the tracked robot system can display the basic conditions, such as straight pipelines, welded joints, elbows and the spherical valve of the industrial pipelines.

ACKNOWLEDGEMENTS

This work was supported by the Science and Technique Plans of Zhejiang Province under Grant No. 2015C33013.

REFERENCES

Ahammed, M. 1998. Probabilistic estimation of remaining life of a pipeline in the presence of active corrosion defects. *International Journal of Pressure Vessels and Piping*, 75(4): 321–329.

Choi, H. & Ryew, S. 2002. Robotic system with active steering capability for internal inspection of urban gas pipelines. *Mechatronics*, 26(1): 105–112.

Ciszewski, M.; Buratowski, T.; Giergiel, M. & Kurc, K. 2014. Virtual prototyping, design and analysis of an in-pipe inspection mobile robot. *Journal of Theoretical and Applied Mechanics*, 52(2): 417–429.

Hollingum, J. 1998. Robots explore underground pipes. *Industrial Robot-An International Journal*, 25(5): 321–325.

Okamoto, J. JR.; Adamowski, J.C.; Tsuzuki, M.S.G.; Buiochi, F. & Camerini, C.S. 1999. Autonomous system for oil pipelines inspection. *Mechatronics*, 9(7): 731–743.

Park, J.; Hyun, D.; Cho, W.H.; Kim, T.H. & Yang, H.S. 2011. Normal-force control for an in-pipe robot according to the inclination of pipelines. *IEEE Transactions on Industrial Electronics*, 58(12): 5304–5310.

Roh, S. & Choi, H.R. 2005. Differential-drive in-pipe robot for moving inside urban gas pipelines. *IEEE Transactions on Robotics*, 21(1): 1–17.

Tavakoli, M.; Marques, L. & de Almeida, A.T. 2010. Development of an industrial pipeline inspection robot. *Industrial Robot-An International Journal*, 37(3): 309–322.

The influence of the fuel supply rate on the power performance of the vehicle diesel engine

Huan Yuan & Jing Jin
School of Mechanical and Electronic Engineering Hubei Polytechnic University, Huangshi, China

ABSTRACT: In order to find out how the fuel supply rate affects the power performance of the vehicle diesel engine, by using the bench tests, in this paper, the influence of the fuel supply rate on the length of torque time and the relationship between the rate and the instantaneous torque, when the torque change is over in the low speed, medium speed, and high speed are studied. The results indicate that engine power performance varies when the fuel supply rate changes. The higher the speed is, the lesser will be the impact of the fuel supply rate on the power performance.

1 INTRODUCTION

The engine cannot be placed under a completely steady condition in the process of vehicle starting, upshift, downshift, brake, acceleration, and deceleration. The engine contributes to the power of the vehicles, so that its performance affects the automotive performance directly. The dynamic condition of the engine is relatively steady, which means that the engine will change while the torque, the speed, or the throttle changes. Parameters such as engine load, speed, etc. will change when a vehicle runs on the road, even while cruising on the highway. Therefore, the dynamic condition accounts for 66%~80% of the whole process (Li, 2001; Zhang, 2000). The performance characteristics of the engine under dynamic conditions are called as the dynamic characteristics of the engine. The fuel supply rate, as one control variable of the engine, has a close relationship with the speed, the torque, the fuel efficiency, and the harmful emission. In this paper, the bench tests have been carried out to know how the fuel supply rate affects the power performance of the vehicle diesel engine. At the same time, it studies the power performance of the vehicle diesel engine in low, medium, and high speeds when the fuel supply is increased and the torque changes (Zhang, 2011).

2 THE REASON OF THE INFLUENCE OF THE FUEL SUPPLY RATE ON THE POWER PERFORMANCE OF THE VEHICLE DIESEL ENGINE

The fuel supply always changes under dynamic conditions. The fuel supply rate has a great impact on the power performance when the vehicles change the speed (i.e., acceleration on the level road) and the torque (i.e., constant speed on the upslope). And it also affects harmful emissions (Yu, 2009).

At present, most cars in our country have the electronic control diesel engines installed in them, which sucks the diesel in three ways: ① oil particles and their steam that directly enter into the airflow from the fuel injector; ② oil vapors are attached to the cylinder wall due to the oil film evaporation; and ③ a small residual amount of the oil film evaporation occurs at the inlet valve.

The diesel engine's characteristic is a regulation of its main performance index and working parameters changed with its working condition, with its proper function as a product and the important basis for people to run the diesel engine correctly. According to the characteristic curve, people can evaluate the diesel engine's motive power and economy under different working conditions and analyze its working feasibility and adaptability in some working circumstance. With this information, people can choose and run the diesel engine effectively, and also this is the important basis for management personnel and maintenance workers, so that people can choose the most reasonable working circumstance and working area and develop its motive power and economy (Peng, 2005).

We analyze the effects and the reasons of the fuel supply rate on the power performance of the vehicle diesel engine according to the data collected from the bench tests by increasing the fuel supply in a constant speed.

3 THE INFLUENCE OF THE FUEL SUPPLY RATE ON THE POWER PERFORMANCE OF THE VEHICLE DIESEL ENGINE

3.1 Mathematical modeling of the experiment

The hysteresis of the fuel supply rate will occur when the fuel supply changes with a change in the pressure of the inlet pipe. As a result, the fuel supply rate will affect the speed and the torque to some extent. Assuming the incremental of fuel supply as k_d during the time from t_0 to t_1 (k_d could be 0 to 1, 1 corresponds to the full throttle), and there is a function relation between k_d and time t. Take a unit time as Δt, the incremental of fuel supply as Δk_d in Δt and therefore, the fuel supply rate can be expressed as follows:

$$\frac{\Delta k_d}{\Delta t} = \frac{d_{k_d}}{d_t} = k \tag{1}$$

where k is the constant in Equation (1). k should be chosen as a more appropriate value in a certain range to study the impact of the fuel supply rate on the power performance of the vehicle diesel engine. In this work, k is set as 0.25, 0.5, 1.0, and 1.5 in the range of 0.25 to 1.5.

3.2 Bench tests and their data

The operating conditions of the varying loads are analogized on the bed. The engine is in the constant speed (such as n = 1500 r/min, n = 1800 r/min, and n = 2000 r/min), and the fuel supply rate is set as k, which is from 25% to 70%. When the speed n is 1500 r/min, the changing situation of the fuel supply rate k can be seen in Table 1. When the speed n is 1500 r/min, k changes with the change in the torque, as is shown Figure 1.

Figure 1 shows that, when the speed n is 1500 r/min, the fuel supply rate k is between 0.25 and 1.5. There are some conclusions drawn which are as follows: ① when we increase the fuel supply with different fuel supply rates k, the final torque is close to a constant value. But when the fuel supply

Table 1. When n = 1500 r/min, the fuel supply rate k changes when torque changes with time.

k	Time(s)									
---	0	2	4	6	8	10	12	14	16	18
0.25	0	15	36	45	63	71	73	74	76	76
0.5	0	13	43	65	73	76	78	78	77	78
1.0	0	11	53	69	77	78	78	78	79	79
1.5	0	13	58	72	78	78	78	78	77	80

rate k varies, the time would be different. The larger the fuel supply rate k, the shorter the time will be; ② the smaller the fuel supply rate k is, the more slowly the torque changes. When the speed n is 1800 r/min, the changing situation of the fuel supply rate k can be seen in Table 2. When the speed n is 1800 r/min, k changes with the change of the torque, as is shown Figure 2.

Figure 2 shows that, when the speed n is 1800 r/min, the fuel supply rate k is between 0.25 and 1.5. There are some conclusions drawn, which

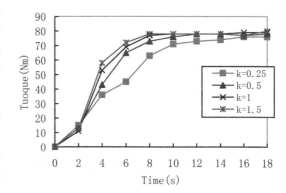

Figure 1. Graph showing that, when n = 1500 r/min, the fuel supply rate k changes when torque changes with time.

Table 2. When n = 1800 r/min, the fuel supply rate k changes when torque changes with the time.

k	Time(s)									
---	0	2	4	6	8	10	12	14	16	18
0.25	0	13	33	48	56	61	65	68	69	68
0.5	0	10	38	50	60	63	68	67	69	68
1.0	0	12	44	55	63	70	70	71	70	70
1.5	0	15	45	56	62	71	69	69	69	71

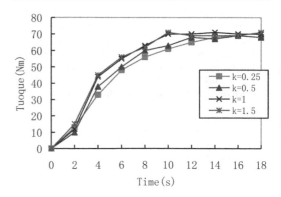

Figure 2. Graph showing that, when n = 1800 r/min, the fuel supply rate k changes when torque changes with time.

are as follows: ① when we increase the fuel supply with different fuel supply rates k, the final torque is close to a constant value. But when the fuel supply rate k varies, the time would be different. The larger the fuel supply rate k, the shorter the time will be. When comparing with the conditions when n is 1500 r/min, the time when the torque is close to be constant is lesser; ② the smaller the fuel supply rate k is, the more slowly the torque changes. When comparing with the conditions when n is 1500 r/min, the torque changes moderately. When the speed n is 2000 r/min, the changing situation of the fuel supply rate k can be seen in Table 3. When the speed n is 2000 r/min, k changes with the change of the torque, as is shown Figure 3.

Figure 3 shows that, when the speed n is 2000 r/min, the fuel supply rate k is between 0.25 and 1.5. There are some conclusions, drawn which are as follows; ① when we increase the fuel supply with different fuel supply rates k, the final torque is close to a constant value. But when the fuel supply rate k varies, the time would be different. The larger the fuel supply rate k, the shorter the time will be. When compared with the conditions when n is 1500 r/min and 1800 r/min, the time is lesser and lesser when the torque is close to the constant, almost inconspicuous; ② the smaller the fuel supply rate k is, the more slowly the torque changes. When comparing with the conditions when n is 1500 r/min and 1800 r/min, the torque changes rapidly.

Table 3. When n = 2000 r/min, the fuel supply rate k *changes* when torque changes with time.

k	Time(s)									
	0	2	4	6	8	10	12	14	16	18
0.25	0	10	30	45	60	62	65	66	65	66
0.5	0	9	31	48	58	63	65	65	65	66
1.0	0	10	30	46	60	65	66	67	66	67
1.5	0	10	35	49	62	63	65	68	67	66

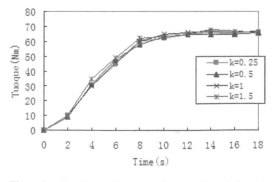

Figure 3. Graph showing that, when n = 2000 r/min, the fuel supply rate k *changes* when torque changes with time.

4 ANALYSIS OF THE DATA

The vehicle engine's dynamic process and control were analyzed, including process of mixtures, combustion process, engine performance index, and so on.

The engine's dynamic performance was analyzed including the influence of the fuel supply rate, speed rate of change influence, engine power performance, and dynamic torque estimate (Miao, 2007).

The data show that we can distinguish the characteristics of the power performance of the vehicle diesel engine when the fuel supply rate k changes according to the time and the torque T which is increased from T_1 (when the fuel supply is 20%) to T_2 (when the fuel supply is 70%). ① At the constant speed, the time of changing torque t and the instantaneous torque T will decrease with an increase in the fuel supply rate k. ② When the fuel supply rate k is maintained as a constant, the higher the speed is, the lesser will the influence be on the dynamic torque; ③ the greater the fuel supply, the lesser the influence of the fuel supply rate k will be on the dynamic torque.

5 CONCLUSIONS

We can draw the following conclusions through the dynamic experiment of the engine: we can lower the engine speed and feather the gas pedal to increase the fuel supply rate, before we change the torque, when the car is moving at constant speed (ie. the car goes up the slope). The torque will increase promptly while the small load changes to the medium load to meet the needs of going up the slope and the power is strong. But the torque will increase slightly and the power becomes insufficient later when the load changes from the medium to the large. By analyzing the data collection of the domestic and international diesel engine bench tests, and joining together the college's reality and development, the plan of the diesel engine bench test data collection and processing system including diesel engine parameter collection, data classification and storage, function analysis, and program control are given in this paper (Miao, 2007).

ACKNOWLEDGMENTS

The research work is supported by the program "*Research on theories and methods of 6-RSPS Platform without Z-axis Rotation Washout for Simulator*". Financial support was obtained from the Natural Science Foundation of Hubei Province (grant number 2014CFB177).

REFERENCES

Chunyu Miao. Dynamic Test Method and the Dynamic Performance of the Vehicle Engines [D]. Wuhan. Wuhan University of Technology, 2007.

Junzhi Zhang, Qingchun Lu. Simulation Research on Hybrid Simulation Method of Engine Dynamic Bench Test [J]. Journal of Internal Combustion Engine. 2000. 18 (1): 1–5.

Qinxue Peng. The Data Collection of the Diesel Engine Bench Tests [D]. Wuhan. Huazhong University of Science and Technology, 2005.

Su Li, Xiaoying Li, et al. Dynamic Process and Control of Automobile Engine [M]. Beijing. People's Communication Press, 2001: 11–20.

Zhipei Zhang. Principle of Automobile Engine [M]. Beijing. People's Communication Press. 2011. 7: 60–75.

Zhisheng Yu. Automobile Theory [M]. Beijing, Mechanical Industry Press, 2009. 3: 1–5.

Advances in Energy Science and Equipment Engineering II – Zhou, Patty & Chen (Eds)
© 2017 Taylor & Francis Group, London, ISBN 978-1-138-71798-5

Multi-phase flow simulation of CO_2 leakage through minor faults

Xialin Liu, Xiaochun Li & Quan Chen
State Key Laboratory of Geomechanics and Geotechnical Engineering, Institute of Rock and Soil Mechanics, Chinese Academy of Sciences, Wuhan, Hubei province, China

ABSTRACT: Carbon dioxide (CO_2) Geological Storage (CGS) is one of the internationally recognized technologies that can effectively reduce the proportion of CO_2 in the atmosphere. CO_2 leakage through the minor faults within the storage reservoir is one of the main risks in the process of CGS, due to the long distribution of the reservoir and the complex geological conditions, combined with the current limited geological survey technology. Therefore, obtaining the distribution of minor faults along the storage reservoir is neither economic nor practical. Therefore, this paper is based on deep underground CO_2 geological storage, taking into account the pre-existing minor undiscovered faults, porosity and permeability of caprocks, porosity and permeability of storage reservoirs, porosity, dip angle, and permeability of minor faults, CO_2 injection rate are assumed to be continuously varied in a certain range. By using TOUGH2-ECO2N to simulate the leakage of CO_2 through the minor faults under 120 different working conditions, through the response surface analysis techniques, the leakage of CO_2 along the minor faults and the relationship between the leakage and the main influence factors are obtained, and then by using the optimization design method, the maximum value of CO_2 leakage in the presence of minor faults is also obtained, which provides a new way for performing site selection and follow-up study of CGS projects and has some reference value for the assessment of the carbon dioxide leakage in the projects.

1 INTRODUCTION

In response to mitigation strategies of climate change, Carbon Capture and Storage (CCS) climate is considered as a key technology to limit greenhouse gas emissions to manageable levels (IEA, 2014). However, the underground stores of CO_2 may escape from geological formations, which is an important issue, not only from the safety perspective, but also from the viewpoint of public acceptance.

In a CO_2 geological storage project, there are three primary sources of leaks (Stephanie et al. 2016): (1) chemical sedimentation caused due to improper maintenance of mechanical failures, (2) the pressure in the CO_2 reservoir exceeds the capillary breakthrough pressure of the caprock, (3) leaks from pre-existing faults or fractures or from the presence of faults or fractures that are not detected. Especially for closed or semi-closure cracks, due to large-scale underground injections, when the pressure rises, it is easy to induce the potential crack to open; however, because of the substantial uncertainty of the original rock stress field, combined with the strength and the structure of the fault are likely to have a high degree of heterogeneity, and the influence of cracks on the seepage ability and water injection after the effect of heterogeneity should not be underestimated (Yuan

Shiyi et al. 2004). Therefore, the influence of faults in the reservoir cannot be ignored.

The structure of the fracture is widespread and exists in the crust; it is proved that the majority of known oil and gas reservoirs in the basin in China at present consists of a large number of fault structures (Fan Jichang et al. 2007). Thereafter, the large-scale geological storage site location is determined, under the action of lithogenesis and tectonic movements thereafter, and a tiny fracture zone can be formed within the deep saline aquifer, may be tiny fault zone is formed within the reservoir, or even form a fault zone that penetrates the entire reservoir.

During geological exploration of large-scale site characterization, the minor faults inside the reservoir may be ignored; however, the presence of these minor faults may cause the change of the reservoir's physical parameters of the local area and the change of hydrogeochemical characteristics, which will influence the spatial distribution of CO_2 in the internal reservoir, and in turn could lead to CO_2 leakage along the fault (Freifeld et al. 2005, Gaus et al. 2005). In addition, in the process of site selection, the undiscovered minor faults within the scope of the reservoir are also likely to induce earthquakes; currently, there is a lot of literature with detailed research on seismic assessment of existing faults (Kanamori 1997, Garagash 2012, NRC 2012). Mazzoldi et al. (2012) have shown how CO_2 injection induced seismicity on undetected faults; Antonio et al. (2014) discussed the influence of different rates of permeability under different injection rates of CO_2 leaks along the fixed angle of minor fault; Jihoon et al. (2015) came up with a method of order implicit equation algorithm by coupling thermal fluid flow, chemical reaction migration, and geological mechanics on fractured reservoir; Jonny Rutqvist et al. (2013) and Frederic Cappa et al. (2011) used TOUGH—FLAC to simulate the influence of minor faults reactivation induced by underground CO_2 injections research. However, these studies are assumed the minor faults with a fixed angle, actually, the minor fault distribution within the reservoir may have significant uncertainty, and the minor faults angle cannot accurately predict, considering a fixed angle of minor faults are thoughtless, and so it is necessary to find a method to study the distribution, migration, and leakage of CO_2 in the reservoir within a certain range of physical property parameters of rock strata that continuously change.

Owing to the target reservoir of CGS projects mostly located in the deep ground, and under complicated geological conditions, the reservoir stretches for several kilometers to tens of kilometers in the horizontal direction; the large fault zones are easy to survey, but the minor faults, however, are not easy to find. At present, there is not a large number of reports on the research of minor faults; in the CGS field, most of the literature is concerned with the large faults; in addition, the seismic reflection surveying method has proved that it is not possible to resolve faults with shear offset $D \leq 10$ m (Gauthier & Lake 1993, Kim & Sanderson 2005); therefore, it is of practical significance for the research on the minor faults that induce earthquake and cause CO_2 leakage, and we cannot ignore the impact of minor faults in the geological storage of CO_2.

In the process of CO_2 injection, the propagation velocity of the pressure disturbance is very fast, and the pressure propagation distance within the reservoir is much larger than that of the point of CO_2 injection (Zhou & Birkholzer 2010), and so, in any location away from the injection point, the minor faults will be affected by the pressure disturbance, the occurrence of a slip even may cause a small earthquake, which in turn could cause the leakage through the faults. On the other hand, due to the mineralization of the fault zone, the low flow characteristics, and the heterogeneity of the temperature and salinity along the fault zone the fault zone will be used as the preferred route for the vertical migration of the fluid (Hooper 1991). The research of the leakage of CO_2 along

the minor faults can be used as a feature of the activation of the faults in the geological storage of CO_2 (Richon et al. 2010, Mazzoldi 2004), which can also be used as a safety index for evaluating the geological storage of CO_2.

A review of Griffith (2011) discussed that the composition, structure, and fluid transport properties of underground salinity aquifers show strong heterogeneity and anisotropy. Although caprocks with low permeability and faults only with caprock properties will be used as the sites of CO_2 storage, however, the fault zone is undoubtedly fragmentation, when the reservoir pressure increased, and these areas have CO_2 leakage risk.

Therefore, before CO_2 injection, it is necessary to evaluate its sealing performance. Based on these issues, in this paper, through a hypothetical model, we carried out a large number of simulations to evaluate the CO_2 leakage along the existing minor faults within the saline aquifer, thereby evaluating the total amount of leakage, which aims to provide a kind of new research approach, and as a reference for the CGS project.

2 RESPONSE SURFACE METHOD

The Response Surface Method (RSM) (David 1982) uses the reasonable experimental design, based on the arrangement of the experiment to obtain the relevant data, and then uses the regression equation to fit the functional relationship between the factors and the response values; and finally, through the analysis of the regression equation, it finds the optimal process parameters. RSM is a statistical method to solve the multi-variable problem.

In the case of n random variables that is, x_1, $x_2, ..., x_n$, the form of the RSM analytic expression can be expressed as (Khuri & Mukhopadhyay 2010) follows:

$$
\begin{aligned}
Y = Y(X) &= Y(x_1, x_2, \cdots, x_n) \\
&= K + \sum_{i=1}^{n} K_1 x_i + \sum_{i=1}^{n} K_{ii} x_i^2 \\
&\quad + \sum_{i=1}^{n-1} \sum_{j=1}^{n} K_{ij} x_i x_j
\end{aligned}
\tag{1}
$$

where K, K_i, K_{ii}, and K_{ij} are determined by the regression coefficient, and ε is the model error.

In statistics, during the analysis of variance of regression coefficient and regression equation, P value is usually chosen as hypothesis testing criteria and marked as the significance level (Nuzzo 2014). In this paper, a threshold value of 0.05 is selected; when the P value is less than 0.05, the regression coefficient and regression equation are significant, and otherwise it is not significant.

In order to preliminarily study the heterogeneity of the sealing layer on the field scale, in this paper, the porosity, permeability, dip angle of the minor fault, the rate of CO_2 injection of the reservoir, and the sealing layer are assumed to be continuously changed in a certain range. In order to reflect the continuous change of these parameters in a given range, the response surface technique is used in this paper. The distribution of the predicted model obtained by the response surface method is continuous, and when compared with the orthogonal test, it is precisely because its prediction model is continuous, it has a lot of advantages; for example, in the process of optimizing the experimental conditions, we can continue to analyze the level of each factor of the experiment, while the orthogonal experiment can only analyze the isolated experimental points.

3 NUMERICAL MODEL

In the process of selecting the site of geological storage of CO_2, the large fault zone that exists in the target area can be studied through the detailed seismic exploration technology, while neglecting the minor faults; one of the reasons is that it has been proved that the seismic reflection surveying method cannot be used to resolve faults with shear offset $D \leq 10$ m; the area that contains minor faults is considered as a homogeneous body; another reason is that, in the CGS project, which stretches from a few kilometers to tens of kilometers, carrying out a detailed geological survey of each location of the fault distribution is neither economic nor realistic.

Therefore, generally only the study of the fault zone of a certain size is focused. However, as a minor fault, its impact on the migration of CO_2 in the reservoir should be paid enough attention.

In this paper, a preliminary discussion is carried out on the heterogeneity in the field scale of the sealing layer, based on the hypothesis that the porosity and permeability of the reservoir and seal layer; porosity, permeability, and dip angle of the minor fault; and the injection rate change in certain ranges continually.

The simulation is basically the one presented by Cappa & Rutqvist (2011b), Antonio et al. (2014), and Mazzoldi et al. (2012), with a minor fault which intersects a 100 m reservoir and is bounded by a 150 m thick low-permeability caprock, as shown in Figure 1.

Figure 1 shows the geometry and initial conditions of the model, which is a two-dimensional model with the size of 2 km × 2 km. Through sensitive analysis, it can be found that the horizontal size does not affect the simulation results in the zone of

interest (Cappa & Rutqvist 2011b). The top level of the model is −500 m, the point P is located in the minor fault section located in the CO_2 reservoir, and details of the strata and their dimensions are shown in Figure 1.

The simulation system has a minor fault, with a thickness of 10 m; this simulation aims to represent a minor fault that cannot be detected by using a seismic survey, since its shear offset is less than 10 m. Since no detailed geological survey data are available, we assume that the range of the minor fault dip angle α is 45°–80° and the elevation of the injection point is −1500 m. At this depth, initial fluid pressure is 4.72 MPa and the temperature is 47.5°C, and the minor fault and the injection point are spaced 500 m apart in the horizontal direction.

The temperature at 500 m depth is 47.5°C and at 2500 m is 72.5°C; with the hydrostatic gradient of 9.81 MPa/km, the initial fluid pressure at 500 m depth is 5 MPa, and the fluid pressure at 2500 m is 24.63 MPa. As for boundary conditions, except for the left boundary, with no flow condition, at boundaries, the pressure, saturation, and the temperature are all constant. It is assumed that the temperature of the whole system remains constant during the simulation.

3.1 Model parameters' decision

The fault is usually composed of the core of the fault and the surrounding zone, and the permeability of the fault core is usually in the range of 1×10^{-17} m²–1×10^{-21} m², and the permeability of the surrounding zone is in the range of 1×10^{-14} m²–1×10^{-16} m² (Cappa & Rutqvist 2011b), as shown in Figure 2.

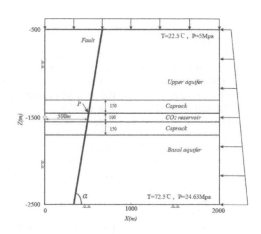

Figure 1. Model geometry for the coupled simulation of CO_2 injection and fault.

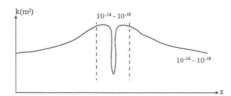

Figure 2. Schematic of the permeability of the fault.

Table 1. List of factors and their corresponding values.

Factors	Lower limit value	Upper limit value
ϕ (caprock)	0.008	0.012
K (caprock), m^2	1E-20	1E-19
ϕ (reservoir)	0.1	0.2
K (reservoir), m^2	1E-14	1E-13
ϕ (fault)	0.08	0.12
K (fault), m^2	1E-16	1E-14
Q, kg/s/m	0.05	0.08
Dip angle (fault),°	45	80

Since the range of the fragmentation zone is much larger than that of the fault core, in this paper, the permeability of the fractured zone is replaced by the permeability of the fault zone.

The calculation parameters are based on the data in Cappa & Rutqvist (2011b). Considering that the distribution of the predicted model obtained by the response surface method is continuous, we assume that the porosity and permeability of the reservoir and sealing layer; porosity, permeability, and dip angle of the minor fault; and the injection rate change in certain ranges continually.

In this paper, in order to study the case of CO_2 leakage under different conditions, we applied the Box–Behnken Design (BBD) test, to consider the range of parameters. In this paper, we consider eight factors that affect CO_2 leakage, which are porosity and permeability of the reservoir and seal layer; porosity, permeability, dip angle of the minor fault; and the injection rate, respectively. By using test design software Expert Design, a total of eight test variables are chosen to carry out 120 sets of simulation experiments. In this work, the TOUGH2 ECO2N simulator is used (Corey 1954).

For other calculation parameters, it is worth noting that there are two critical parameters in this simulation. The first critical parameter is the minor fault permeability, it was varied from 1×10^{-14} m^2 to 1×10^{-16} m^2, although actually the fault zone permeability is much smaller, see Figure 2; but the selected values are representative of the fragmenta-tion damage zone permeability. The second critical parameter is injection rate. In this paper, it was var-ied from 0.05 kg/s/m to 0.08 kg/s/m, and this injec-tion rate is varied to meet the general requirements of the industry (Rutqvist 2002). For a horizontal well that is 1000 m long, this range of injection rate is equivalent to 50 kg/s to 80 kg/s, which is a very high injection rate; however, in this paper, only a method of analysis of the leakage of the minor fault is provided, and the specific value varies from site to site; for the actual project, it should be spe-cific analysis of specific issues, and it should take the actual site data.

A range of the main parameters is shown in Table 1, where φ is the porosity; K is the perme-ability; and Q is the injection rate of CO_2.

In this paper, the BBD design is used; the advan-tage of using BBD is that it can evaluate all the main influencing factors, the interaction between factors, and the relationship between the quadratic effect factors.

3.2 Simulation and results

Apart from Table 1, other calculation parameters are shown in Table 2. This work is carried out by using the TOUGH2 ECO2N simulator (Pruess & Spycher 2007), and the relative permeability is assessed by using the Corey function (Corey 1954) and capillarity pressure is determined by using the van Genuchten function (van Genuchten 1980).

Expert Design is used to design the 120 sets of working conditions, which are simulated by using the TOUGH2-ECO2N module

By using TOUGH2-ECO2N module to simu-late the 120 sets of working conditions designed by Design Expert, we can obtain from simulative calculation of the amount of CO_2 in the reservoir, which has been leaking from the minor faults, that accounts for the total amount of injection after 20 years of continuous injection. In this paper, we define the leakage as the ratio of the CO_2 leak-age along the minor fault to the total amount of the CO_2 injected. And then, through the Expert Design, the obtained 120 sets of working condi-tions are analyzed, and the analysis of variance showed that porosity and permeability of the caprock have no significant effect on the leakage of CO_2 along the minor fault, because their cor-responding P value is greater than 0.05, while the porosity and permeability of the reservoir; poros-ity, permeability, and dip angle of the minor fault; and the injection rate have a great significant effect on the leakage of CO_2 along the minor fault, since their corresponding P values are significantly smaller than 0.05 ($P < 0.0001$).

By using Expert Design to carry out the second order polynomial regression fitting of the data, we can obtain the quadratic polynomial regres-sion equation between the amount of the leakage

Table 2. Values of hydraulic properties.

Parameters	Upper aquifer	Reservoir	Caprock	Basal aquifer	Minor fault
Rock density (kg/m³)	2260	2260	2260	2260	2260
Porosity (–)	0.1	–	–	0.01	–
Permeability (m²)	10–14	–	–	10–14	–
Residual CO_2 saturation (–)	0.05	0.05	0.05	0.05	0.05
Residual liquid saturation (–)	0.3	0.3	0.3	0.3	0.3
van Genuchten, p_0 (kpa)	19.9	19.9	621	621	19.9
van Genuchten, m (–)	0.457	0.457	0.457	0.457	0.457

of CO_2 along the minor faults and the significant factors, as well as the interaction effect among the significant factors, the equation takes the following form:

$$R = 1.36 - 0.18C - 0.21D + 0.18E + 0.43F \\ - 0.10G + 0.12H - 0.092BD + 0.14CD \\ + 0.098CG - 0.13FG - 0.13F^2 \ (R^2 = 0.9230) \tag{2}$$

where R is the leakage of CO_2 along the minor fault that accounted for the total volume injected (mass percent); C is the porosity of the CO_2 reservoir; D is the permeability of the CO_2 reservoir; E is the porosity of the minor fault; F is the permeability of the minor fault; G is the injection rate; and H is the dip angle of the minor fault. The P value of the overall model is less than 0.05, and indicates that the quadratic polynomial regression equation is significant, the correlation coefficient of the regression equation is 0.9230, and adjusted R square is 0.9024.

Through the actual results of the distance correlation analysis, the correlation coefficient between the measured value and the predicted value of the leakage of CO_2 was 0.8452; it is proved that the analytical model is effective, which has certain practical decision-making significance. When the levels of each significant factor are known, then the leakage of CO_2 can be obtained by using the equation to predict the leakage of CO_2 along the minor fault, which can be used as a reference for the evaluation of the CGS project's pre-feasibility study.

By using equation (2) for partial derivatives, it can be concluded that, for the model under the conditions of the simulation of the CO_2 maximum leakage, at this time, the value of each factor is as shown in Figure 3.

It can be seen that under the conditions of this paper, the assumption of the model under the conditions is as shown in Figure 3 and the maximum leakage of CO_2 in the reservoir is 2.655%.

Hepple and Benson (2005) pointed out that, for a project of geological storage of CO_2, from the effectiveness of storage, if the amount of

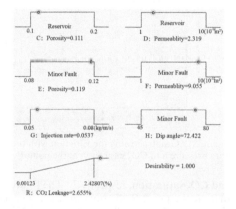

Figure 3. Schematic of the final optimal design for the indicator value.

CO_2 leaked to the shallow layer is less than 0.1% per year, it can be considered effective. There are many stated examples in this paper, in which after continuous injection of 20 years, the theoretical maximum leakage along the minor fault is 2.665%. The value is greater than the recommended value of Hepple and Benson and therefore, relevant technical measures must be carried out to ensure the storage effectiveness, for instance, one should take measures to strengthen the monitoring of leakage in the region of the minor fault.

Under the condition of the maximum value of the leakage of CO_2 along the minor faults, Figure 4 shows the CO_2 migration process at the corresponding working conditions.

From Figure 4, we can see that CO_2 spreads quickly through the minor fault, and after 20 years, the migration distance of the horizontal direction of the gaseous carbon dioxide is about 1100 m; in the vertical direction, carbon dioxide has been leaked into the caprock along the minor fault.

By using TOUGH2, we can obtain the variation with time of the pore pressure and CO_2 saturation at the point P, as shown in Figure 5, where the solid and the dashed lines represent the pore pressure

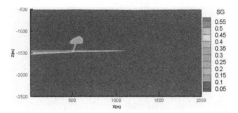

Figure 4. Graph showing CO_2 saturation obtained after 20 years of injection.

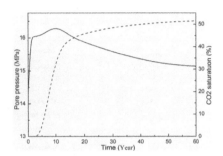

Figure 5. Graph showing the variation with time of the pore pressure and CO_2 saturation at the point P.

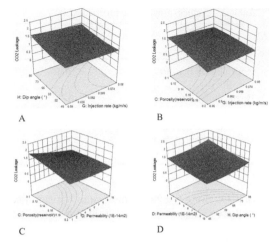

Figure 6. Schematic of the response surfaces and their contour plots.

and CO_2 saturation, respectively. The figure shows that, at the point P, the pore pressure increases at about 1 year and finally reaches 16.384 MPa, and then the pore pressure gradually decreases and reaches about 15.137 MPa at 60 years; this may be due to the presence of the minor fault that has played a role in the relief of pressure. However, at the point P, CO_2 saturation increases quickly at the beginning of the injection, and then increases slowly and finally reaches a value of 51.474%; due to the presence of the minor fault, the pressure in the reservoir is released during the injection, which increases the amount of carbon dioxide storage, and therefore, the presence of minor faults may increase the leakage of carbon dioxide, but it also contributed to the increase in the amount of carbon dioxide storage, although this contribution is not positive correlation.

According to the regression equation of the model, the response surface and contour map can be used to observe the strength of the interaction between factors. Figure 4 lists the factors that are more interactive.

The shape of contour lines can reflect the strength of the interaction effect, the ellipse indicates that the interaction of the two factors is significant, while the circle is the opposite (Muralidhar et al. 2001), Figure 6 shows the response surfaces and contour maps of the CO_2 leakage with the strong interaction factors, from which we can see that the interaction effect of the four factors has a great influence on the response value of the leakage of CO_2 along the minor fault. These are the injection rate and minor fault dip angle (Figure 6A), permeability of the reservoir and minor fault dip angle (Figure 6B), porosity of the reservoir and injection rate (Figure 6C), and injection rate and minor fault dip angle (Figure 6D), respectively. This also shows the actual CGS project according to the actual reservoir physical parameters of a reasonable choice of CO_2 injection speed.

4 CONCLUSIONS

In this paper, a preliminary discussion is carried out on the heterogeneity in the field scale of the sealing layer, based on the hypothesis that the porosity and permeability of the reservoir and sealing layer; porosity, permeability, and dip angle of the minor fault; and the injection rate change in certain ranges continually. In order to reflect the continuous variation of these parameters in a given range, by using Expert Design to design test combination, combined with the numerical simulation program TOUGH2 ECO2N module, with the combination of the different levels of the eight factors, we carried out a large number of simulations (120 simulations in total); the leakage of CO_2 along the minor fault in the reservoir has been calculated under various working conditions, and the results showed that the porosity and permeability of the cap did not significantly affect the penetration of CO_2 along the minor fault, through the response surface analysis, the multiple regression equation of the CO_2 leakage and the other six factors and the cross factor are obtained.

Through the analysis of 120 different conditions, the leakage of CO_2 along the minor fault under the corresponding working conditions were obtained, on the basis of the multiple regression equation, through the optimized analysis, we obtained the maximum leakage along the minor fault and the corresponding value of each factor, this method can be used for the safety evaluation of the pre-feasibility study stage, in the actual operation process, according to different geological conditions, thereby establishing the corresponding model, and then a preliminary determination of the range of relevant parameters can be carried out; finally, this method is used to estimate the amount of leakage of CO_2 along the suspicious minor faults, and to provide a reference for the project's decision-making process.

However, in this paper, it is aimed to present a method for the analysis of this response surface combined with numerical simulation, by using a conceptual model to analyze the problem. Actually there are many factors that affect the leakage of CO_2 along the minor faults, and these are not limited to the factors mentioned in this paper, for instance, the width of the minor fault, fault distance from the injection well, the offset of the minor faults, the layout of the injection/production wells and so on, and the level of each factor also needs to be determined according to the actual project.

Due to the fact that the distribution of the minor faults are not easy to detect in the reservoir, in the concrete application, through the seismic survey technology measures, we can divide the object storage area into several sub-regions which includes the detection of minor faults, and use this method to forecast the analysis of the leakage of CO_2 along the minor fault in the sub-region, and then the extreme value of the CO_2 leakage along the suspicious minor fault can be obtained, which can be saved for pre-feasibility study stage of safety evaluation, and to provide a certain reference for the CGS project.

REFERENCES

Antonio, P. Rinaldi, Jonny Rutqvist & Cappa, F. (2014). Geomechanical effects on CO_2 leakage through zones during large-sacle underground injection. International Journal of Greenhouse Gas Control. 20, 117–131.

Cappa, F. & Rutqvist, J. (2011b). Modeling of coupled deformation and permeability evolution during fault reactivation induced by deep underground injection of CO_2. International Journal of Greenhouse Gas Control. 5(2), 336–346.

Corey, A.T. (1954). The Interrelation Between Oil and GasRelative Permeabilities. Producers Monthly. 38–41.

Dr. David Thompson (1982). Response surface experimentation. Journal of Food Processing and Preservation. 6, 155–188.

Fan Jichang & Liu Mingjun (2007). An approach determining internal structure and physical parameters of fractural zone. Oil Geophysical Prospecting. 42(2), 164–169.

Frederic Cappa & Jonny Rutqvist (2011). Modeling of coupled deformation and permeability ecolution during fault reactivation induced by deep underground injection of CO_2. Inernational Journal of Greenhouse Gas Control. 15, 336–346.

Freifeld B.M. et al (2005). The U-tube: a novel system for acquiring borehole fluid samples from a deep geologic CO2 storage experiment. Journalof Geophysical Research. 110, B10203, 1–10.

Garagash, D.I. & Germanovich, L.N. (2012). Nucleation and arrest of dynamic slip on a pressurized fault. J. Geophys. Res. 117 (B10).

Gaus I, Azaroual M. & Czernichowski-Lauriol I. (2005). Reactive transport modeling of the impact of CO_2 injection on the clayey cap rock at Sleipner (North Sea). Chemical Geology. 217, 319–337.

Gauthier, B.D.M. & Lake, S.D. (1993). Probabilistic modeling of faults below the limit of seismic resolution in Pelican Field, North Sea, off shore United Kingdom. AAPG BULLETIN, 5, 761–777.

Griffith, C., Dzombak, D.A. & Lowry, G.V. (2011). Physicaland chemical characteristics of potential seal strata in regions considered for demonstrating geological saline CO_2 sequestration. Environ. Earth Sci. 64(4), 925–948.

Hepple, R.P. & Benson, S.M. (2005). Geologic storage of CO_2 as a climate change mitigation strategy: performance requirements and the implications of surface seepage. Environ. Geol. 47(4), 576–585.

Hooper, E.C.D. (1991). Fluid migration along growth faults in compacting sediments. Journal of Petroleum Geolo gy. 14(2), 161–180.

International Energy Agency (2014). Energy Technology Perspectives 2014-Harnessing Electricity's Potential. IEA/OECD. Pairs, Frace.

Jihoon Kim, Eric Sonnenthal & Jonny Rutqvist (2015). A sequential implicit algorithm of chemo-thermo—poro-mechanics for fractured geothermal reservoirs. Computer & Geosciences. 76, 59–71.

Jonny Rutqvist, Frederic Cappa, Alberto Mazzoldi & Antonio Rinaldi (2013). Geomechanical modeling of fault response and the potential for notable seismic events during underground of CO_2 injection. Energy Procedia. 37, 4774–4784.

Kanamori, H. (1997). The energy released by great earthquakes. Journal of Geophysical Research. 82(20), 2981–2988.

Khuri, A.I. & Mukhopadhyay, S. (2010). Response surface methodology. Wiley Interdiscip. Rev. Comput. Stat. 2, 128–149.

Kim, S.Y. & Sanderson, D.J. (2005). The relationship between displacement and length of faults: a review. Earth Science Reviwes. 68, 317–334.

Mazzoldi, A. (2004). Hydrogeochemical studies of Val Caffarella, Studies of crustal/magmatic origin of the gas CO_2 found in water of the river Almone, district of the Alban Hills, Rome(I), through the determination of the activity of the radioactive gas Radon–222. M.S. Thesis, University of Roma Tre, unpublished work (in Italian).

Mazzoldi, Albert, Rinaldi, A.P., Borgia, A. & Rutqvist, J. (2012). Induced seismicity within geological carbon sequestration projects: Maximum earthquake magnitude and leakage potential from undetected faults. Int. J. Greenhouse Gas Control. 10, 434–442.

National Research Council (2012). Induced Seismicity Potential in Energy Technologies. National Academies Press, Washington, DC. 300.

Nuzzo, R.G. (2014). Scientific method: statistical errors. Nature. 506, 150–152.

Pruess, K. & Spycher, N. (2007). ECO2N—a fluid property module for the TOUGH2 code for studies of CO_2 storage in saline aquifers. Energy Conv. Manage.

Richon, P., Klinger, Y., Tapponnier, P., Li, C.-X., Van Der Woerd, J. & Perrier, F. (2010). Measuring radon fluxes across active faults: relevance of excavating and possibility of satellite discharges. Radiation Measurements. 45, 211–218.

Rutqvist, J. & Tsang C.-F. (2002). A study of caprock hydromechanical changes associated with CO_2 injection into a brine aquifer. Environ Geol. 42, 296–305.

Stephanie Vialle, Jennifer L. Druhan & Katharine Maher (2016). Multi-phase flow simulation of CO_2 leakage through a fractured caprock in response to mitigation strategies. International Journal of Green house Gas Control. 44, 11–25.

van Genuchten, M.T. (1980). A closed-form equation for predicting the hydraulic conductivity of unsaturated soils. Soil Sci. Soc. Am. J. 44, 892–898.

Yuan Shiyi, Song Xinmin & Ran Qiquan (2004). Fractured reservoir development technology. Beijing: Petroleum industry press.

Zhou, Q. & Birkholzer, J.T. (2010). On scale and magnitude of pressure build-up induced by large-scale geologic storage of CO_2. Greenhouse Gases: Science and Technology. 1(1), 11–20.

Application of staged fracturing technology by using a ball injection sliding sleeve for horizontal wells' open-hole completion in Block Ma 131

Kun Ding, Jianmin Li, Ning Cheng, Haiyan Zhao, Shanzhi Shi & Liring Wang
Engineering Research Institute of Xinjiang Oil Field, Karamay, China

ABSTRACT: Located at the north inclined slope of the central sag and Mahu sag in Junggar Basin, Block Ma 131 Baikouquan Reservoir is the most practical output and reserve increasing region, which belongs to the typical low hole and leakage reservoir in Xinjiang Oilfield at present. Massive production can be gained by fracturing. Practice proves that the research and application of staged fracturing technology for horizontal well open-hole packers can provide better technical support for low cost development, single well output improvement, and stable production target. At present, five wells are dug with staged fracturing technology of the well completion ball sliding sleeve in this block; the accumulated oil output is 5–12 times that of vertical wells and the continuous stable production time is 3–8 times that of vertical wells.

1 INTRODUCTION

Block Ma 131 Baikouquan Reservoir is located at the north inclined slope of the central sag and Mahu sag in Junggar Basin, the reservoir lithology of which mainly includes grey sandy conglomerates and grit stones and can be divided into the following three sections from bottom to top: T_1b_1, T_1b_2, and T_1b_3. The oil reservoir is mainly distributed through T_1b_3 and $T_1b_2^1$, with a mean thickness of each layer of about 9.0 m, the mean porosity being lower than 10.0% and the mean permeability of about 1.0 mD. It is buried at about 3,200 m deep, which belongs to the low hole and leakage reservoir.

To improve the transformation effect of the low leakage oil reservoir in Mabei Oilfield, well completion staged fracturing technology for horizontal well open-holes was used in Block 131 in 2013. The key technologies of which mainly include the following: (1) dividing the horizontal well fracturing length reasonably to ensure that geological fillings can be improved completely; (2) the performance parameters such as thermal endurance and tightness of open-hole packers shall meet requirements for fracturing construction; (3) the dimension of the fracturing ball should be selected which can meet construction displacement requirements and ensure that the sliding sleeve can be opened safely; (4) the well completion string should be selected suitably and satisfy the requirements for string strength by large-scale fracturing.

2 PROCESS DESIGN

2.1 Selection of the sliding sleeve ball seat

To meet demand of construction displacement for reservoir transformation and reduce the risks of opening failure of the sliding sleeve due to ball seat corrosion, the dimension of the ball seat should be selected suitably based on fracturing fluid simulated by ANSYSY-FLUENT software for sliding sleeve ball seat speed (GUO 2009 et al., TANG et al. 2013, Finnie I 1978).

According to data simulated, the fracturing sand carrying liquid can washout the ball sliding sleeve, the most serious part of which is the front cone; the ball seat and the back end are also washed out to a certain extent (see Fig. 1). In case

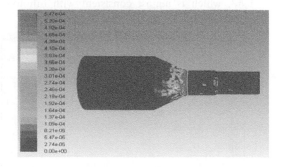

Figure 1. Schematic of the erosion wear of the ball seat.

of large-scale fracturing, the minimum single hole ball seat and back Grade 2~3 sliding ball seat can lose efficacy most probably (Ding et al. 2014).

All in all, the sliding sleeve in Block Ma 131 should be provided with a large nominal diameter ball seat as much as possible to control the ball seat that loses efficiency due to corrosion and ensure the fracturing transformation effect.

2.2 Design for string strength

For the sake of string force conditions in the shaft, the Von Mises yield criterion is used for checking the tri-axial stress strength and analyzing and comparing different types of casings in order to control loss due to "poor safety" and wastes as a result of "over safety" (QU et al. 2009, Ye et al. 2013). Finite element analysis should be used for determining the sliding sleeve string's strength; the throttling differential pressure of the sliding sleeve hole should be simulated and calculated based on FLUENT software and the safety coefficient of the string should be calculated based on the ANSYS Workbench platform.

In consideration of the Φ114.3 mm (4.5") external diameter of the string, 7.37 mm (0.3") wall thickness and 4.5 m^3/min construction displacement, the practical throttling differential pressure of the sliding sleeve is about 0.6 MPa (see Fig. 2) when the total area of the sliding sleeve hole is 1.06 times the sliding sleeve cross-section; for normal construction, the pipes at the sliding sleeve cannot create too large stress concentration and the minimum safety coefficient of pipes at the hole is 2.998 (Fig. 3), and so the string strength can meet requirements for construction.

2.3 Optimization of scale transformation

According to practical growing conditions of reservoirs in Block Ma 131 (see Fig. 4), the mudstone interlayer grows in the oil reservoir in T_1b_3 and $T_1b_2{}^1$, which is almost consistent with lithology

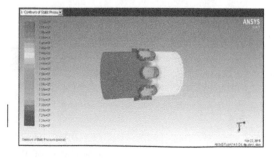

Figure 2. Schematic of the throttling differential pressure of sliding sleeve.

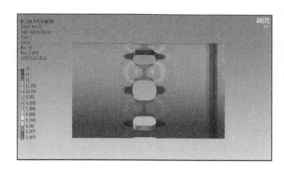

Figure 3. Schematic of the safety coefficient of the sliding sleeve.

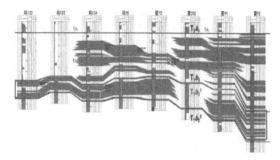

Figure 4. Connected well profile of the oil reservoir across Ma 133–Ma 132–Xia 91–Xia 92.

in $T_1b_2{}^2$. The layer under $T_1b_2{}^1$ and $T_1b_2{}^2$ is more compact than $T_1b_2{}^1$. The whole sand body can be transformed by fracturing, but the vertical effective application of the oil reservoir should be ensured with enough fracturing scale.

The horizontal open-hole section with good physical properties and high gas logging ability should be selected for fracturing and the distance between the sections should be 60~80 m. According to output simulation at different sand scale pressures, the sand for the horizontal section for unit length of the horizontal well designed should be about 1.0 m^3/m; the joint of the efficient support is about 130 m long and 40 m high (see Fig. 5); and the propping agent concentration paved can reach up to 8.5 kg/m^2.

2.4 Optimization of construction parameters

The construction parameters are optimized according to transformation features of the reservoir and staged transformation process features based on analysis and reorganization of the early stage fracturing process. For specific results, see Table 1.

2.5 Optimization of the fracturing process

In light that Block Ma 131 Baikouquan Reservoir is featured by large thickness, the fiber sand-adding

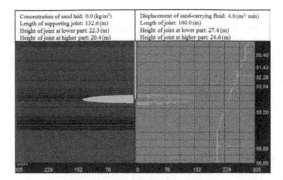

Concentration of sand laid: 0.0 (kg/m³)	Displacement of sand-carrying fluid: 4.0 (m³/min)
Length of supporting joint: 132.6 (m)	Length of joint: 160.0 (m)
Height of joint at lower part: 22.3 (m)	Height of joint at lower part: 27.4 (m)
Height of joint at higher part: 20.4 (m)	Height of joint at higher part: 24.6 (m)

Figure 5. Simulation results of the length of adding 1 m³ sand in the unit length for the horizontal section.

Table 1. List of optimized construction parameters.

Target parameters	Design basis	Recommendation results
Displacement	Meeting reservoir transformation demand and reducing failure risk due to sliding sleeve corrosion.	4.0–4.5 m³/min
Average sand ratio	Improving desertification, increasing oil reservoir support efficiently and improving flow conductivity of the joint. Adding sand at Baikouquan Reservoir and keeping high desertification.	25%
Pad fluid ratio	Increasing pad fluid ratio, supplementing reservoir energy, and improving volume for transformation.	40%

process can be used for improving propping agent payment efficiently and improving the supporting effect and flow conductivity of joints. After 0.2% fiber is added to the fracturing fluid, the gelling process and gel breaking performance are not affected (see Table 2) and the hanging performance of the jelly can be improved in order to prevent the propping agent from sediments (see Figs. 6–8).

Table 2. Comparison between liquid performance before and after addition of fiber.

Target parameters	No fiber	0.2% fiber added
Gelling time	191 seconds	210 seconds
Gel breaking time	75 minutes.	105 minutes

Figure 6. Picture of freshly prepared samples.

Figure 7. Picture of samples after 1 hour.

Figure 8. Picture of samples after 2 hours.

3 APPLICATION FOR BLOCK MA 131

The staged fracturing technology with an open-hole packer is used for five wells in Block Ma 131 Baikouquan Reservoir. The horizontal section is 667.0~1,000.0 m long and the mean section is 55.7~73.5 m long; the volume of the total fracturing liquid is 5,285.3~6,799.5 m³; the total sand capacity is 546.7~1,150.0 m³; the total fiber capacity is 1,496.0~2,751.0 m³; the mean desertification is 21.2~25.7%; and the construction displacement is 4.0~5.0 m³/min.

The oil output of the early stage is 7.52~28.22 t/d while the current oil output is 7.5~11.4 t/d; the flush stage is longer than 1 year. According to general conditions of oil testing and extraction, the fracturing effect is good.

4 CONCLUSIONS

Dramatic output increase is observed at the Block Ma 131 low leakage oil reservoir by applying ball staged fracturing technology for horizontal wells; practice proves that the technique is able to meet the demand of horizontal wells for increasing fracturing output, which can provide a reliable technical guarantee for developing oil reservoirs efficiently.

The open-hole staged fracturing technology of the horizontal well is a high-tech reservoir transformation technology. The plan for the horizontal well can be designed independently and technical and economic restrictions can be improved by co-operating with foreign companies at the aspect of technologies and developing key project subject research in order to develop low leakage oil reserves efficiently.

Since the partial pressure tool is one of the key technologies for horizontal well reservoir improvement, it is helpful to reduce costs for fracturing by improving national tools continuously.

REFERENCES

Ding Kun. et al. 2014. Numerical Simulation of Erosion Wear of Fracturing Fluid on Sleeve's Ball Seat of Horizontal Wells. Finnie I. 1978 Wear.
GUO Jianchun. et al. 2009. Horizontal well ball staged fracturing technique and field application.
QU Zhan. et al. 2009. CAO Feng. Design and implementation of the software for the triaxial stress check of casing string.
TANG Jiapeng. et al. 2014. FLUENT Self-study manual. BeiJing: Post & Telecom Press.
Ye Qinyou. et al. 2013. Mechanical Analysis of Staged Fracturing String with Sliding Packer in Horizontal Wells of Miao 22.

Advances in Energy Science and Equipment Engineering II – Zhou, Patty & Chen (Eds)
© *2017 Taylor & Francis Group, London, ISBN 978-1-138-71798-5*

Numerical analysis of a suitable number of rigid ribs in the hollow torsion test

Binbin Xu
Tianjin Port Engineering Institute Ltd. of CCCC, Tianjin, China
Key Laboratory of Geotechnical Engineering of Tianjin, Tianjin, China
Key Laboratory of Geotechnical Engineering, Ministry of Communication, Tianjin, China

ABSTRACT: In the numerical simulation hollow torsion test, the number of rigid ribs applied on the top surface of the specimen is the key to guarantee the success of the calculation. If the number of rigid ribs is too many, the stiffness difference between the pedestal and the specimen would be too large to get a reliable result. While if the number of ribs is too little, the torque cannot be transferred to the specimen faithfully. A numerical calculation is carried out to determine the suitable number of rigid ribs from the aspects of shear strain, apparent mechanical behavior and torque rate and the results show that 6 ribs are suitable for the numerical hollow torsion test.

1 INTRODUCTION

In practical geotechnical engineering, grounds/soils are generally under a complex stress state and subjected to various loading conditions. In order to grasp strengthening and deformation properties of the ground, many laboratory tests were developed and conducted. Among them, the Hollow Cylinder Torsional shear test (HCT) controlled by four individual external forces is often employed in order to reproduce the actual stress path during in-situ constructions. Via adjusting the loading condition, the soil response can be investigated under the designated stress path by both monotonic and cyclic shear tests (Lade, 1981; Hight et al, 1983; Nataka et al, 1998; Frost & Drnevich, 1994).

For the torsional test, the torque is usually applied on the top surface by several ribs installed at the pedestal, as shown in Fig. 2. The torque control methods can be divided into strain control and stress control. In the numerical simulation, when the specimen is controlled by strain, a constant angular velocity field is applied to each node on the top, which represents that there is no relative slippage between the frictional pedestal and the specimen and all the nodes on the top keep the same pace in the circumferential orientation.

While in practical experiments, relative slippages are still possible even though a frictional porous stone is installed to transfer the torque from the pedestal to the specimen. The common way to enhance the reliability of torque transfer is to install several additional rigid ribs, as shown in Fig. 1. However, how many ribs there should be

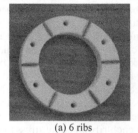 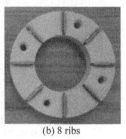

| (a) 6 ribs | (b) 8 ribs |

Figure 1. Porous stone with different rigid ribs.

is always controversial, in this section the stress-control calculation will be carried out to find the appropriate number of rigid ribs.

In this paper, for the torque control a method employing Lagrange constraint conditions (Asaoka et al, 1998) is applied to the nodes on the top surface and the optimum number of rigid ribs is discussed.

2 LAGRANGE CONSTRAINT CONDITIONS

For the simulation under torque control it is impossible to directly assign a constant stress rate to each node on the top because of the unknown distribution of stress rate below the rigid frictional pedestal. (Asaoka et al, 1998) introduced the "no-length change", "no-angle change" and "no-direction change" conditions through the Lagrange multiplier method to represent the friction of the

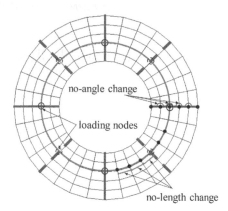

Figure 2. Constraint conditions on the top surface under torque-control method.

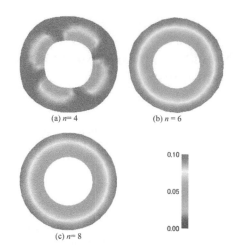

Figure 3. Distribution of shear strain at the top with n ribs at $\gamma_s = 12\%$ ($n = 4, 6, 8$).

pedestal, rigidity of the pedestal and orientation of loading rod in triaxial tests respectively. Here, the "no-length change" condition which means that the distance between any two nodes is kept constant and the "no-angle change" condition which means that the angle between two lines composed by any three nodes is constant are adopted.

The "no-angle change" condition at the top is setting on the nodes along the radius with an equal interval angle θ which is taken as 90 deg, 60 deg or 45 deg, namely 4, 6 or 8 "no-angle change" conditions, as marked by blue lines in Fig. 2. Such conditions can represent the rigid ribs installed on the pedestal. Therefore, there are 4, 6 or 8 ribs used in this calculation and for each rib, an equal stress rate is applied to the node located at the average radius as shown in Fig. 2. For the "no-length change" condition, it is assigned between every two nodes along the circumferential direction from inner diameter to outer diameter, as marked by red lines in Fig. 2. The friction of porous stone which can transfer the torque to the specimen is indicated by the "no-length change" condition. The stress rate will be 1.0 kPa/s for 4 ribs, 0.67 kPa/s for 6 ribs and 0.5 kPa/s for 8 ribs and the other boundary conditions are same with the actual experimental conditions. The following calculation will compare three kinds of conditions with different ribs to determine the proper number in the hollow torsion test.

3 CALCULATION RESULTS

3.1 *Distribution of shear strain*

As the function of the rib in the porous stone is transferring the torque to the top surface of the specimen, one way to check whether the number of ribs is enough or not is to see the distribution of shear strain at the top surface. Fig. 3 shows the shear strain distribution on the top surface at

$\gamma_s = 12\%$ with 4, 6 and 8 ribs respectively. As can be seen, when there is only four ribs there is significant shear strain at the location of ribs compared with other two cases and the shape of circle cannot be maintained when the apparent shear strain is large. While for the cases with 6 and 8 ribs, the circle shape can be kept on the top surface even though the apparent shear strain is 12%. But the circumferential distribution of shear strain is more uniform for the case with 8 ribs compared with case with 6 ribs. Therefore, in the view of uniform shear strain in the circumferential direction, 8 ribs are preferred.

3.2 *Apparent mechanical behavior*

The apparent mechanical behavior behaviors are shown in Figs. 4–6, where "Bottom EPP" and "Top EPP" indicate the excess pore water pressure measured at the bottom and at the top respectively. The apparent mechanical behavior including the deviator stress-axial strain ($q - \varepsilon_a$) relationship, the deviator stress-mean effective stress ($q - p'$) relationship (in other words, the effective stress path), the pore water pressure-axial strain ($u - \varepsilon_a$) relationship, and the specific volume-mean effective stress ($v - p'$) relationship is obtained when the whole specimen is viewed as one mass.

As can be seen, in the case with 4 ribs the deviator stress is smaller than that in "Perfect path", if the friction between the pedestal and the specimen is not strong enough, namely relative slippage occurring between the pedestal and specimen. Also excess pore water pressures measured at the top and the bottom are slight different due to the non-uniform deformation along the circumferential direction on the top. However for the case with 6 ribs, there is a good accordance for the deviator

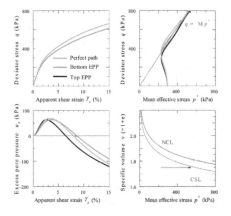

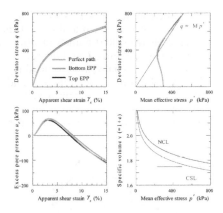

Figure 4. Apparent mechanic behavior of specimen with 4 ribs.

Figure 6. Apparent mechanic behavior of specimen with 8 ribs.

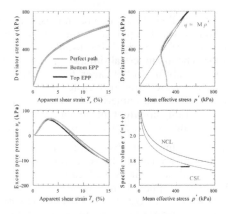

Figure 5. Apparent mechanic behavior of specimen with 6 ribs.

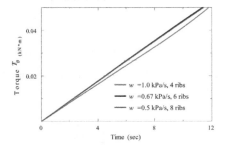

Figure 7. Relation of torque and time for 4, 6 and 8 ribs.

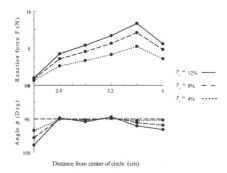

Figure 8. Distribution of magnitude and direction of reaction forces at $\psi = 0°$.

stress and excess pore water pressure. The same result can be seen for the case with 8 ribs. In the view of apparent mechanical behavior, 6 ribs are enough and 8 ribs are better.

In addition, Fig. 7 presents the relation of total torque and time for three cases, in which the torque rates are constant and almost equal due to the different stress rates applied on the loading nodes. There is only slight deviation of the case with 4 ribs from the other two cases. And in view of the torque rate 6 ribs are suitable. Therefore, according to Figs. 3 (b), 5 and 7 it can be reasonable to conclude that 6 ribs are enough for the torque control to transfer the torque reliably under this analysis conditions.

3.3 Reaction force of nodes on the rib

As a particular case with 6 ribs, details about the distribution of reaction forces below the pedestal will be discussed in the following. Fig. 8 shows the magnitude of reaction forces F and direction of reaction forces which is presented by angle ϕ

along the horizontal radius marked by $\psi = 0°$. The lateral axis represents the current radius of the nodes. In ideal situation, the angle ϕ should be 90° which means the reaction force F is always perpendicular to the radius r and tangent to the circle. As can be seen, the distribution of reaction forces is not uniform under the constant torque rate and the maximum locates at $r = 3.6$ cm. Moreover, such non-uniformities gradually grow with the increase of apparent shear strain. From the angle distribution, it seems that the directions of reaction forces at $r = 2.0$ cm and $r = 4.0$ cm deviates remarkably from 90 θ, namely an obtuse angle between the reaction force F and the radius r.

Considering the axial-symmetry loading condition, the distributions of magnitude F and direction of reaction forces at radii marked by $\psi = 15°$, $\psi = 30°$ and $\psi = 45°$ are illustrated in Figs. 9, 10 and 11. Besides the same tendency mentioned above, the distribution of directions in Figs. 9 and 10 should also be noticed. For the radius with $\psi = 15°$, the angle ϕ is much smaller than 90 deg, which means that the direction of reaction force points inward the circle while for the radius with $\psi = 45°$, the angle ϕ is much larger than 90 deg, which means that the direction of reaction force points outward the circle. Considering the anti-clockwise torque direction and the symmetry of ribs in every 60°, the radius with $\psi = 15°$ can be regarded in the forward of the loading rib and the radius with $\psi = 45°$ can be regarded in the backward of the loading rib, which reveals that for the radius close to loading ribs the direction of reaction forces deviates greatly from the ideal 90°.

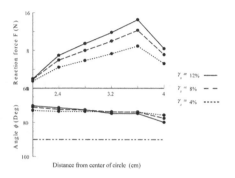

Figure 9. Distribution of magnitude and direction of reaction forces at $\psi = 15°$.

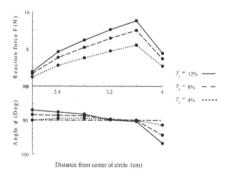

Figure 10. Distribution of magnitude and direction of reaction forces at $\psi = 30°$.

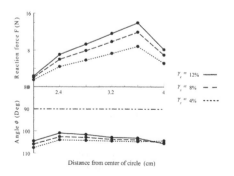

Figure 11. Distribution of magnitude and direction of reaction forces at $\psi = 45°$.

4 CONCLUSIONS

In the hollow torsion experiments, the torque action is usually applied to the soil specimen at the top surface by rigid ribs installed at the porous stone. The in numerical simulation, the number of ribs is doubtful. In the paper, the suitable number of rigid ribs required to transfer the torque faithfully is discussed numerically from the aspects of shear strain, apparent mechanical behavior and torque rate and the conclusions are as follows:

1. If the number of ribs is 4, the circle shape of the top surface cannot be kept and there would be great shear strain at the location of ribs.
2. For the apparent mechanical behavior, there is only slight deviation between the case with 6 ribs and 8 ribs, while the error of case with 4 ribs is greater.
3. The suitable number of rigid ribs is 6 which can satisfy both the mechanical response and deformation requirement.

REFERENCES

Asaoka, A., Noda, T. and Kaneda, K. 1998. Displacement traction boundary conditions represented by constraint conditions on velocity field of soil, *Soils and Foundations*, 38(4): 173–181.

Frost, J. D. and Drnevich, V. P. 1994. Towards standardization of torsional shear testing, *Dynamic Geotechnical Testing II*, ASTM STP 1213, Philadelphia, 276–287.

Hight, D. W., Gen, A. and Symes, M. J. 1983. The development of a new hollow cylinder apparatus for investigating the effects of principal stress rotation in soils, *Géotechnique*, 33(4): 355–383.

Lade, P. V. 1981. Torsion shear apparatus for soil testing, *Symp. On Laboratory Shear Strength of Soil*, ASTM STP 740: 145–163.

Nakata, Y., Hyodo, M., Murata, H. and Yasufuku, N. 1998. Flow deformation of sands subjected to principal stress rotation, *Soils and Foundations*, 38(2): 115–128.

Advances in Energy Science and Equipment Engineering II – Zhou, Patty & Chen (Eds)
© *2017 Taylor & Francis Group, London, ISBN 978-1-138-71798-5*

The mechanism of the micro-EDM process

Shuyang Liu, Zhihong Han & Yonghong Zhu
School of Mechanical and Electronic Engineering, Jingdezhen Ceramic Institute, Jiang Xi, China

ABSTRACT: In this paper, the mechanism of the micro-EDM process was modeled and analyzed based on the electron field emission theory, electron sputtering theory, and the effect of the resistance heat of electron current when the tool electrode is positive. A series of single-hole EDM experiments and the metallographic examinations were designed. This study shows that the concentration ratio of the work energy on the machined surface increases with an increase in the equivalent radius ratio of the real discharging area and the area of the workpiece end surface, and the reduction of the inter-electrode discharging current. It can be observed that the ratio of the resistance heat to the total work energy increases with a decrease in the equivalent radius ratio of the real discharging area and the area of the workpiece end surface, and an increase in the interelectrode discharging current.

1 INTRODUCTION

As a kind of special machining technology, EDM has already been indispensable in processing ultra-hard materials and heat sensitive materials (V. Yadav, 2002; R. Pérez, 2007; M.S. Murali, 2005). However, electrical discharge phenomena in EDM make both observation and theoretical analysis extremely difficult. Therefore, some questions still remain unanswered. First, the material removal mechanisms are not yet well understood. Second, what are the relationships between the energy distribution and working efficiency? Third, how the EDM process results in the energy distribution (B. Lauwers, 2004)? It shows that further research work is necessary for better understanding of the nature of the phenomena involved in the EDM process.

Many scholars' research works and related experiments have already shown that particles in the discharge channel were mainly present on electric streams ejected from the cathode, and the effects of positive ions on the channel were very small (S.Y. Liu, 2004). Modern field-emission theories and thermal field emission theories, especially F–N equation descriptions, demonstrated that electrons escaping from the surface needed to overcome surface barriers to work, which would translate into heat energy, and gave the reactive functions of the material (R.H. Fowler, 1928; E.L. Murphy, 1956; R.H. Hare, 1993). Mao, C et al. stated that with longer pulse-on times, the power available between the anode and cathode becomes greater, and thus strengthens the discharge energy (Y.C. Lin, 2008). Liu et al. (2010) derived electrodes complied with skin effect when flowing through the electrodes, which provided space boundary conditions for energy analysis of the material removal.

In this paper, the energies on the workpiece surface both in the positive polarity EDM process were modeled and its related parameters were analyzed, based on the electron field emission theory and the resistance heat reaction of electron flow when the tool electrode is positive. A set of machining blind-hole experiments with pole electrodes were carried out and investigated, and the related parameters that affect the working efficiencies were discussed. The experimental phenomena and results verify the rationality and effectiveness of the proposed model.

2 EXPERIMENTAL SECTION

In order to analyze the mechanism and parameters of EDM processing, we designed single-hole EDM experiments with pole electrodes as shown in Fig. 1, where Sw_p represents the pulse switch, U_s is the voltage of the discharging tube, R_{pp} and R_{np} are the resistances of the tool electrode and workpiece electrode respectively, R_{pv} is the resistance of the single discharging tube, and a number of pulse tubes is n.

Related parameters of devices applied in the experiments are as follows:

Machine model: *AGIE CHARMILLES SE2;*
Inputting power: $P_{in} = 10kw$;
Dielectric medium: kerosene

The image of the bottom surface of the EDMed blind hole captured by using a Scanning Electron Microscope (SEM) is shown in Fig. 2, where the polarity of the tool electrode is positive.

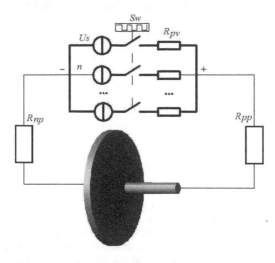

Figure 1. Schematic of single-hole EDM experiments with pole electrode.

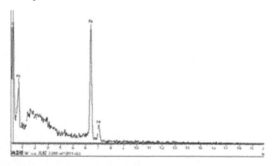

Figure 2. SEM image of the bottom surface of the EDMed blind hole.

It is clearly known by observing the microstructure of the machined surface and analyzing the corresponding metallographic pictures that

1. Fig. 2 shows that the electrode surface machined by the positive polarity EDM process has obviously melted traces and the size distribution of the melted spot is non-uniform, which implies the poorer thermal concentration degree of the positive polarity EDM process during the single discharge cycle.
2. From Fig. 2, it is speculated that the thermal corrosion of positive polarity machining is generated from an inner substrate material to the workpiece surface.

3 THE ANALYSIS OF THE MICRO-EDM PROCESS

According to the skin effect of electric charge, the electron current channel performs shaping into a concentric tube with two diameters $2\pi(R_2-d_a)\times L$ and $\pi(R_2^2-r^2)\times d_a$ respectively, as shown in Fig. 3.

Defining the ratio of the resistance heat to the total work energy on the area of A_1 as $\eta_c^- = \dfrac{Q_s^{(A_1^-)}}{Q^{(A_1^-)}}$ the ratio can be calculated as follows:

$$\eta_c^- = \frac{\dfrac{2\rho_w}{3\pi\cdot d_a}\cdot(\dfrac{R_2}{R_1}-\dfrac{R_1^2}{R_2^2}+\dfrac{3}{4})\cdot\int_o^{T_{on}}J_t^2\cdot dt}{\dfrac{\pi R_1^2\Phi}{e}\cdot\int_o^{T_{on}}J_t\cdot dt+\dfrac{2\rho_w}{3\pi\cdot d_a}\cdot(\dfrac{R_2}{R_1}-\dfrac{R_1^2}{R_2^2}+\dfrac{3}{4})\cdot\int_o^{T_{on}}J_t^2\cdot dt}$$

$$= \frac{C_2\cdot(4-4\lambda^3+3\lambda)\cdot\int_o^{T_{on}}J_t^2\cdot dt}{C_1\cdot\lambda^3\cdot R_2^2\cdot\int_o^{T_{on}}J_t\cdot dt+C_2\cdot(4-4\lambda^3+3\lambda)\cdot\int_o^{T_{on}}J_t^2\cdot dt} \quad (1)$$

Taking the data of the single-hole EDM experiment of positive polarity listed as the example too, the varying tendencies of η_c^- about λ at different steady discharging currents are shown in Fig. 4. Obviously, η_c^- is a decreasing function about λ at certain J_{t3}, and at certain λ, η_c^- gets increases with an increase in J_{t3}.

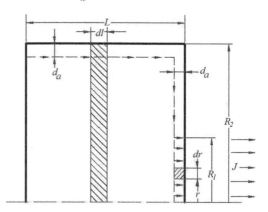

Figure 3. Schematic of the movement of the electron current in the workpiece electrode.

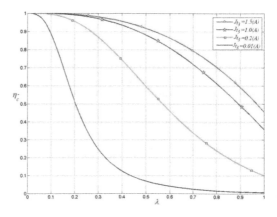

Figure 4. The varying curves of η_c^- about λ at different J_{t3}.

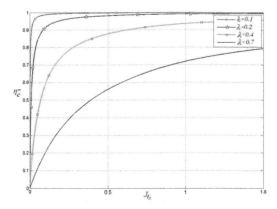

Figure 5. The varying curves of η_c^- about J_{t3} at different values of λ.

Fig. 5 shows the varying tendencies of η_c^- about J_{t3} at different equivalent radius ratios; obviously, η_c^- is an increasing function about J_{t3} at a certain value of λ, and η_Q^- gets smaller decreases with an increase of λ at a certain value of J_{t3}.

4   CONCLUSIONS

This study proposed the calculation equations about the work energy acting on the material removal during positive polarity EDM process and a series of single-hole EDM experiments and the metallographic examinations were designed to verify the model. The following conclusions are drawn from the experimental results:

1. Based on the field emission theory and the effect of resistance heat, the work energy on the machined surface in the positive polarity EDM process was studied, and the relevant equations are obtained. The related parameters that affect the working efficiencies are discussed.
2. The calculation on the distribution of the work energies during the steady discharging stage of a pulse period shows that the concentration ratio of the work energy on the machined surface increases with an increase in the equivalent radius ratio of the real discharging area and the area of the workpiece end surface, and the reduction of the interelectrode discharging current.

ACKNOWLEDGMENTS

This work was financially supported by the National Natural Science Foundation of China (Nos 61563022 and 61164014) and Jiangxi Province Natural Science Foundation of China (No. 20152 ACB20009).

REFERENCES

Fowler, R.H., and L.W. Nordheim. "Electron emission in intense electric fields", .Proc. Roy. Soc. Lund. Ser. A, vol. 119, pp. 173, 1928.
Hare, R.H., R.M. Hill, and C.J. Budd. "Modeling charge injection and motion in solid dielectrics under high electric field", Plays. D: Appl. Phys. vol. 26, pp. 1084, 1993.
Lauwers, B., J.P. Kruth, W. Liu, W. Eeraerts, B. Schacht, and P. Bleys, "Investigation of material removal mechanism in EDM of composite ceramic materials", Mater. Process. Technol. vol. 149, pp. 347–352, 2004.
Lin, Y.C., Y.F. Chen, C.T. Lin, and H.J. Tzeng, "Electrical discharge machining (EDM) characteristics associated with electrical discharge energy on machining of cemented tungsten carbide", Mater. Manuf. Proc. vol. 23, no. 3–4, pp. 391–399, 2008.
Liu Yu, Y., and F.L. Zhao,"Influences of skin-effect on micro-EDM", J. Dalian Univ. Technol.no. 3, pp. 362–367, 2010.
Liu, S.Y., Y.M. Huang, and Y. Li. "A Plate capacitor model of the EDM process based on the field emission theory", Int., J. Mach. Tool. Manuf. vol. 51, no. 7, pp. 653–659, 2004.
Murali, M.S., and S. H. Yeo. "Process simulation and residual stress estimations of micro-electro discharge machining using finite element method", Japanese J. Appl. Phys.vol. 44, no. 7 A, pp. 5254–5263, 2005.
Murphy, E.L., and R.H. Good. "Thermionic emission. field emission and the transition region", Phys. Rev. vol. 102, pp. 1464, 1956.
Pérez, R., J. Carron, M. Rappaz, G. Wälder, B. Revaz, and R. Flükiger. Measurement and metallurgical modeling of the thermal impact of EDM discharg on steel, in Proceedings of the 15th International Symposium on Electro machining, ISEM XV, pp. 17–22, 2007.
Yadav, V., V.K. Jain, and P.M. Dixit, "Thermal stresses due to electrical discharge machining", Int. J. Mach. Tool. Manuf. vol. 42, pp. 877–888, 2002.

Advances in Energy Science and Equipment Engineering II – Zhou, Patty & Chen (Eds)
© *2017 Taylor & Francis Group, London, ISBN 978-1-138-71798-5*

Preparation and performance optimization of phosphorescence materials in white light OLED devices

Huajing Zheng, Quan Jiang, Yadong Jiang & Zheng Ruan
State Key Laboratory of Electronic Thin Films and Integrated Devices, School of Optoelectronics, University of Electronic Science and Technology of China, Chengdu, China

ABSTRACT: In this work, a white light electroluminescence device is constructed by utilizing bis [2-(4-tert-butyl phenyl) benzothiazolato-N, C2'] iridium (acetylacetonate) [(t-bt)2Ir(acac)], a complex of Ir, as the phosphorescence luminescent material and by performing the binding doping process. The device has an onset driving voltage of 3 V, and the maximum luminance is 15,460 cd/m^2 at 16.5 V; the maximum luminous efficiency is 7.5 lm/W at 4 V, the white-light spectrum, sent out by the driven device, has little movement as the voltage changed at a low voltage, but is still in the normal range of white light. When the voltage reached 8 V, the CIE color coordinates, (0.33, 0.32), almost coincide with the best white color coordinates, whose color coordinates are (0.33, 0.33). The spectrum and color coordinates are stable and do not change while the voltage changes.

1 INTRODUCTION

In the past 10 years, White Organic Light-Emitting Devices (WOLEDs) have attracted widespread attention. When compared with mature and cheap luminescence devices, for example incandescent lamps, WOLEDs are becoming a new growth point in the organic electroluminescence field, due to the fact that it can be used as a backlight for flat panel displays and solid-state lighting. Organic electro phosphorescence devices are considered as the point of focus of research in the OLED field[1–3]. According to Spin theory, of all the organic electroluminescent materials, fluorescence materials can exhibit luminescence only by the radiative decay of polarons, and so its maximum internal quantum efficiency is only 25%. However, the phosphorescent material can use singlet and triplet excitons by intersystem crossing while luminescence, and so the maximum internal quantum efficiency in theory could be 100%, which is four times that of the fluorescence material[4]. In preparation of the phosphorescent material, the doping method is the most effective approach of all the methods. This can prevent concentration quenching when the phosphorescent material exhibits luminescence and phosphorescent molecules crystallize, which leads to functional film deterioration[5–8]. It is applied in small molecule materials and has obtained efficient EL devices based on the higher luminous efficiency of the phosphorescent material. According to the previous studies, we focus on the relationship between concentrations of the phosphorescent material and the EL efficiency in different current densities and try to approach the maximum EL efficiency by changing the doping concentration. In this research[9–11], the fluorescent material 4,4'-bis (carbazol-9-yl) biphenyl (CBP) is used as the main luminescent material, a new phosphorescence material bis [2-(4-tert-butylphenyl) benzothiazolato-N, C2'] iridium (acetylacetonate) [(t-bt)$_2$ Ir (acac)] is doped as the yellow light emitting layer, NPB is used as the hole transport layer and ultrathin (1 nm) NPB is used as the blue emitting layer. We prepared an improved and structured white light device with high efficiency and brightness, and indicated its photoelectric properties and discussed the mechanism about the relationship between luminescence spectra, luminous efficiency, and driving voltage.

2 TEST

2.1 Materials and devices

Organic multifunctional film-forming equipment OLED-V was purchased from Shenyang Vacuum Technology Institute, a vertical manifold of clean bench was bought from Nantong Cleaning Equipment Co., Ltd, a spin coater KW-4 A was obtained from Shanghai Chemat Function Ceramics Technology Co., LTD. A Sartorious BS electronic balance was acquired from Beijing Sartorious Mechatronics T&H Co., Ltd, a luminance meter ST-86 LA and spectrometer OPT-2000 were bought from Beijing Normal University, a semiconductor

Table 1. List of experiment medicines.

	Manufacturer	Purity	Energy level(eV) HOMO	LUMO
NPB	Aldrich	99%	5.4	2.4
Alq₃	Aldrich	99.9%	5.7	3.2
CBP	Aldrich	99%	6.0	2.9
PVK	Aldrich	99%	2.0	5.5
Rubrene	Aldrich	99%	5.4	3.2
BCP	Aldrich	98%	6.7	3.2
Mg	Sinopharm Chemical Reagent Co., Ltd	99.9%	3.66	
Ag	Sinopharm Chemical Reagent Co., Ltd	99.5%	4.26	
ITO Glass	Shenzhen Nanpo.,Co., Ltd	11 Ω/Filtered 80%	4.8	

Table 2. List of test reagents.

	Manufacturer	Purity	Content
Acetone	Chengdu Kelong Chemical Reagent Co., Ltd	99.9%	Purity analysis
Ethanol	Chengdu Kelong Chemical Reagent Co., Ltd	99.9%	Purity analysis
1,2 Dichloroethane	Chengdu Kelong Chemical Reagent Co., Ltd	99.9%	Purity analysis

test system Keithley 4200 was obtained from Keithely Company and an ultrasonic cleaner KQ-200DB was purchased from Nantong Ultrasonic Cleaning Equipment Co., Ltd, as shown in Tables 1 and 2.

2.2 Device preparation

Place the cleaned substrate into the preprocessing vacuum chamber of the film-forming equipment Sonicel plus 200 (Sonic, Korea). Prepare the oxygen plasma treatment, by cleaning the surface dirt furthermore, improving the surface oxygen content of ITO to increase functioning. And then, with no damage to the system vacuum, continuously and successively perform vapor plating of organic materials and the electrode. The vacuums of vapor plating organic materials and mental electrodes is 2.1×10^{-4} Pa and 3.8×10^{-3} Pa, the speed is 0.3–1 nm/and 1.1–1.3 nm/s in order. The structures used to prepare the device in the test are as follows: ITO/NPB (40 nm)/CBP: (t-bt)₂ Ir (acac) (3 wt%, 20 nm)/NPB (1 nm)/BCP (3 nm)/Alq (20 nm)/Mg:Ag (200 nm). The thickness in each layer and the evaporation rate for materials are all measured by using an in situ quartz crystal oscillators monitor.

3 RESULTS AND DISCUSSION

Figure 2 depicts the luminance–voltage–current density characteristic diagram of the device. The main CBP material of the emitting layer is mixed

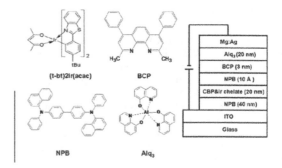

Figure 1. The main material's molecular structure diagram and the device's structure diagram.

with 3% phosphorescence material (t-bt)₂ Ir (acac). From Figure 1, it can be seen that the diagram trend of device luminance and current density are both fitting for the characteristics of the diode. The change trends of the luminance and current density are similar and they are changed in direct proportion within the measured voltage. The turn-on voltage is 3V and the current density is only 0.2 mA/cm². When the luminance reaches up to 300 cd/m², the corresponding driving voltage is 4.5V and the current density is 0.2 mA/cm². Driven by the maximum voltage of 16.5V and the current density of 680 mA/cm², the luminance reaches up to 15.460 cd/m².

For discussing the property of the device further, the luminous efficiency–voltage characteristic diagram of the device is shown in Figure 3. From

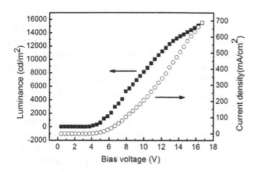

Figure 2. Graph showing the L–V–J characteristic diagram of the device.

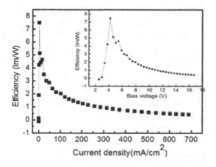

Figure 3. Graph showing the characteristic diagram of the device luminous efficiency–current density. The inset shows the efficiency–voltage characteristic diagram of the device.

Figure 3, it can be observed that the maximum luminous efficiency of the device is 7.5 lm/W and the corresponding voltage is 4V when the luminance reaches up to 170 cd/m². When the voltage is 3V, the luminous efficiency is 2 lm/W, but when the luminance reaches the maximum value, the luminous efficiency is 0.44 lm/W. The relationship of luminous and current density or the relationship of luminous efficiency and voltage are mostly paid attention, but the factors of current density and voltage which lead to the luminous efficiency change are different. For the mechanism of luminous efficiency, we can complete the changing mechanism of the luminous efficiency by considering these two factors. From Figure 3, it is evident that with an increase in the voltage and current density, the luminous efficiency of the device firstly increases sharply and then decreases slowly. However, some previous studies[13] mention that this phenomenon occurs because of the annihilation from the exciton's triplet state to triplet state. The reason for this phenomenon might be studied intensively by combining the energy levels of the spectra and device.

Fig. 4 shows the EL spectrum of the device at different driven voltages, in which there are two obvious peaks that are at 450 and 560 nm respectively, and a shoulder at 600 nm. The peak at 450 nm is blue light which is emitted by NPB, and the peak at 560 nm and shoulder at 600 nm are generated by (t-bt)₂ Ir (acac). The white light spectrum emitted by the device covers the visible spectrum from 400 nm to 700 nm. On the contrary, only difference is observed at 450 nm on the spectrum at different bias voltages. As the bias voltage increases, the relative intensity of blue light increases and the relative intensity of the light emitted by (t-bt)₂ Ir (acac) does not change. When the voltage is raised from 5 V to 8 V, CIE coordinates from (0.35, 0.36) to (0.33, 0.32), the color purity and color coordinates are very close. As the bias voltage keeps rising (>8 V), the emitted spectrum and CIE color coordinates do not vary essentially. Additionally, the emission components of BCP (at 490 nm) and Alq₃ (at 530 nm) are not observed, which proves that the two kinds of organic material in the device do not involve in light emission, but only play the roles of the cavity, excitation blocking layer, and electron transport layer.

Fig. 5 depicts the energy level diagram of the device. Fig. 5(b) is the energy level diagram used by our institute. Fig. 5(a) is the structure of phosphorescent material (t-bt)₂ Ir (acac) based and adulterated for normal white light devices reported in the literature. From Fig. 5 and their structures, it can be inferred that the difference between the two is that an additional ultrathin layer (1 nm) of NPB is used in our device structure. The second NPB layer from the anode is an emission layer, the first NPB layer only plays the role of cavity transport, but the device reported in the literature uses only one NPB layer to play the two roles of cavity transport and light emission. The reflected difference in the spectrum is as follows: there is a shoulder at 480 nm for device (a) and it disappears for device (b). This is a significant reason that affects the color purity for device (a). This is also a constraint in the color

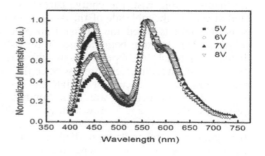

Figure 4. Graph showing the EL spectra under four types of voltages.

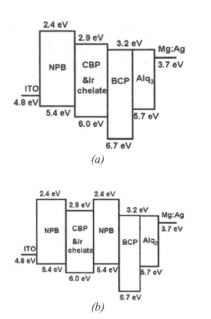

Figure 5. Schematic of the energy level structure diagram of the device.

purity (0.319, 0.342) could not approach any more to achieve the optimal color purity (0.33, 0.33). Furthermore, the electron transport performance of the adulterated layer of the body material CBP and the phosphorescent material is relatively much lower than that of BCP. At low bias voltage, the adulterated layer emits relative intensity light and low intensity blue light. As the ultrathin NPB is added, it helps to increase the intensity of blue light emission, and relatively reduces the probability of electrons entering the adulterated layer, and correspondingly lowers the intensity of yellow light emission, which is critical to enhance the purity of white light.

Generally speaking, in the mixed system, the main luminescent mechanism is the deliveries of carrier trap and energy[14]. In device (b), the two mechanisms still exist, the charge carrier is limited directly in the molecule of the (t-bt)$_2$ Ir (acac) material. This is called the carrier trap process, which leads to excitons being formed in the (t-bt)$_2$ Ir (acac) material. It is reported that[12] the carrier trap is the main luminescent process when the current is low. When the current is high, the carrier trap and energy delivery are both luminescent processes. Figures 2 and 3 show their relationship. At a low voltage, the current density is small and the luminance is relatively high, the luminous efficiency increases and soon reaches the maximum. The maximum efficiency indicates that the device makes maximum use of the injected electric charge.

As the factor is that the carrier trap is the main luminescent process under low current conditions, the carrier trap luminescent process efficiency is higher than that of the energy delivery process. And so, it is shown that the carrier trap luminescent process efficiency is higher than energy delivery process efficiency. With an increase in the voltage, the current density increases along with the luminance. But the efficiency decreases then, by the factor that the energy delivery process is the main mechanism at a high voltage, and it is proved that carrier trap luminescent process efficiency is actually higher than energy delivery process efficiency.

Firstly, under primary voltage conditions, the device luminescence efficiency is high and is mainly attributed to the carrier trap luminescent process. With an increase in the voltage and current density, the main electroluminescent mechanism switches from carrier trap to energy delivery. Secondly, voltage and current density are the two important factors that affect device luminance and the switches of the luminescence mechanism. The increase of the voltage is the primary reason. Although it appears that the current density is the main reason that affects luminance, the fundamental reason is the increase in the injected charge caused by a corresponding voltage increase. With an increase in the voltage and current density, the device efficiency decreases rapidly. Except the factors, the annihilation effect from the exciton triplet state to the triplet state should be considered. Except that the energy delivery process efficiency is low, after the energy of the main material passed on to phosphorescent material, some parts of the exciton triplet state in the phosphorescent material exhibits annihilation, which is another main factor that leads to a decrease in the efficiency.

However, the carrier trap luminescent process efficiency is higher than that of energy delivery. When the carrier trap process is the main factor at the time when the voltage is low, the luminance is not high, and that is because the voltage of the injected carrier is little under the low voltage. The hole and the electron can neither largely get through the barrier to the luminescent zone. The height of the luminescent efficiency is that of carrier usage efficiency. This is compared to the number of the injected carriers but is not in contrast with the main luminescent mechanism efficiency.

In addition, by combining the energy structure of Figure 5, Figure 4 is discussed about the device luminescence spectra. When the voltage and the current density are both low, the NPB luminous intensity is relatively low. That is because the device barrier is high, and the electron and the hole fail in forming excitons to exhibit luminescence. The hole cannot be transported efficiently in the doped yellow layer, by the fact that the NPB layer is the ultrathin film which is only 1 nm and can be used for sending out blue light. The electron can tunnel through easily and reach to the doped emitting layer, which results in the luminous intensity imbalance, and so the color purity is relatively low. With an increase of the voltage and the current density, the NPB luminous intensity is high. The increase in the voltage leads to an inclined energy inside the devices. The higher the voltage is the heavier the lead is, but this is beneficial for the hole in the doped layer to get into the NPB blue luminescent layer. With an increase in the NPB luminous intensity, it stabilizes the best situation where the color purity is approximately located in the white light.

4 CONCLUSIONS

In the mainstream development, the structures used to construct a white light electroluminescence device with yellow phosphorescent material of Ir complex and doped process conditions are as follows: ITO/NPB (40 nm)/CBP: (t-bt)$_2$ Ir (acac) (3 wt%, 20 nm)/NPB (1 nm)/BCP (3 nm)/ Alq (20 nm)/Mg:Ag (200 nm). The device is turned on at around 3V. When the voltage is 16.5V, the device achieves the maximum luminescence which is 15,460cd/m. When the voltage is 4V, the device achieves the maximum luminous efficiency, which is 7.5 lm/W. After turning on the device, the white spectra move a little bit with the voltage change under the low voltage condition, but the change is still within the white light area. When the voltage reaches up to 8V, the CIE coordinates are (0.33, 0.32). The spectra and coordinates are stable and do not change with voltage change and almost coincide with the white light coordinates (0.33, 033).

REFERENCES

[1] Kim, T. H., H. K. Lee, O. O. Park, et al. White-light-emitting diodes based on iridium complexes via efficient energy transfer from a conjugated polymer. Adv. Funct. Mater., 2006, 15: 611–617.

[2] Wang, J., Y. D. Jiang, J. S. Yu, et al. Low operating voltage bright organic light-emitting diode using iridium complex doped in 4,42-bis [N-1-napthyl-N-phenyl-amino] biphenyl. Appl. Phys. Lett., 2007, 91: 131105.

[3] Gong, X., M. R. Robinson, J. C. Ostrowski, et al. High-efficiency polymer-based electrophosphorescencet devices. Adv. Mater., 2002, 14: 581–585.

[4] Tang, C. W., S. A. VanSlyke, C. H. Chen. Electroluminescence of doped organic thin films. J. Appl. Phys., 1989, 65: 3610–3616.

[5] Shaheen, S. E., B. Kippelen, N. Peyghambarian, et al, Energy and charge transfer in organic light-emitting diodes: A soluble quinacridone study. J. Appl. Phys., 1999, 85: 7939–7945.

[6] Chwang, A. B., R. C. Kwong, J. J. Brown, Graded mixed-layer organic light-emitting devices. Appl. Phys. Lett., 2002, 80: 725–727.

[7] Mitsuya, M., T. Suzuki, T. Koyama, et al, Bright red organic light-emitting diodes doped with a fluorescent dye. Appl. Phys. Lett., 2000, 77: 3272–3274.

[8] Chang, W., A. T. Hu, D. K. Rayabarapu, et al. Color tunable phosphorescent light-emitting diodes based on iridium complexes with substituted 2-phenylbenzothiozoles as the cyclometalated ligands. J. Organomet Chem., 2004, 689: 4882–4888.

[9] Fang, J., H. You, J. Guo, et al, Improved efficiency by a fluorescent dye in red organic light-emitting devices based on a europium complex. Chem. Phys. Lett., 2004, 392: 11–16.

[10] Uchida, M., C. Adachi, T. Koyama, et al, Charge carrier trapping effect by luminescent dopant molecules in single-layer organic light emitting diodes. J. Appl. Phys., 1999, 86: 1680–1687.

[11] Adachi, C., M. A. Baldo, S. R. Forrest, et al. High-efficiency organic electrophosphorescent devices with tris (2-phenylpyridine) iridium doped into electron- transporting materials. Appl. Phys. Lett., 2000, 77: 904–906.

Advances in Energy Science and Equipment Engineering II – Zhou, Patty & Chen (Eds)
© 2017 Taylor & Francis Group, London, ISBN 978-1-138-71798-5

A three-dimensional positioning system based on fingerprint positioning algorithm

Yepeng Ni, Jianping Chai, Yaxin Bai & Jianbo Liu
Communication University of China, Beijing, China

ABSTRACT: To improve the positioning accuracy, this paper attempted to propose a parameter adjustable fingerprint algorithm based on the nearest neighbor algorithm in signal space. In addition, the three-dimensional positioning system has been realized through our proposed algorithm. Experimental results show that the positional accuracy of the three-dimensional positioning system based on the parameter adjustable fingerprint algorithm has been obviously improved, which could meet the requirements of most scenes.

1 INTRODUCTION

In recent decades, with the popularity of wireless networks and the development of mobile terminals, the LBS (Location Based Service) in indoor area has become increasingly important. As for indoor localization, only the positioning accuracy approaching a few meters can support relevant applications. For this purpose, researchers have developed a number of systems with different technologies, such as RFID positioning system, infrared ray positioning system, Bluetooth positioning system, ZigBee positioning system, ultrasonic positioning system and fingerprint positioning system, etc. Among them, the fingerprint positioning technology has become a hotspot due to the popularity of WLAN signal in indoor environment.

Nowadays, most fingerprint positioning systems are aimed at two-dimensional space, which could not satisfy the requirements of location services in many indoor scenes. To explore this problem, this paper puts forward a three-dimensional positioning system based on an improved deterministic algorithm.

This paper is organized as follows: the Section 2 introduces the related works. The section 3 proposes a three-dimensional positioning algorithm based on fingerprint location. In section 4, we realize the system and carry out the test. The last section concludes the research in this paper.

2 RELATED WORK

RADAR (Bahl & Padmanabhan 2000, Bahl, Padmanabhan & Balachandran 2000) is the earliest indoor positioning system based on location

fingerprint, which is designed and realized by Bahl and Padmanabhan et al from Microsoft in 2000. RADAR put forward the positioning method based on location fingerprint, which initiated this research field. The Nearest Neighbors in Signal Space (NNSS) positioning method is utilized by RADAR. The estimation coordinates obtained through NNSS algorithm is not the actual coordinate of the measured position, but the coordinate of Reference Point (RP) nearest to measured position.

Horus (Youssef et al. 2003, Youssef et al. 2005) is a classical location fingerprint positioning system developed by Youssef et al in 2003. Different from RADAR, Horus introduced the probability statistical method to construct the fingerprint feature. It deposited the histogram of RSSI of each Access Point (AP) in the location into the location fingerprint database, and the histogram could be translated into Gaussian distribute model.

Zee (Rai et al. 2012) is an earlier location fingerprint positioning system without offline acquisition proposed by A. Rai et al. Zee utilizes the inertial sensors in the smart phone to estimate the users' step size and walking trajectory as well as makes full use of the environment and path information in indoor plan. The above information will be inputted to the particle filter to achieve the positioning and navigation result.

LiFS (Yang et al. 2012) positioning system is based on location fingerprint by Yang Z and Wu C in 2012. LiFS correlates different location fingerprints through the user's moving trajectory, and generates fingerprint space and non-pressure planar graph through the MDS algorithm. It applies the graph theory to construct the location fingerprint database, and realizes the absolute positioning in the room. LiFS achieves the location fingerprint positioning system without offline acquisition through the application of the crowd sourcing theory.

Apart from the novel positioning system, the researchers also improve efficiency and positioning accuracy by adopting other ways. R Dutzler et al put forward an interpolation method and Log-Distance Path Loss Model to optimize the set up procedure of radio map, and reduce the human cost of offline acquisition stage (R Dutzler et al. 2013). Mu Zhou et al proposed that using heuristic Simulated Annealing (SA) algorithm to optimize the AP location and approach the fundamental limit of localization precision, and the positioning accuracy could be improved from the WLAN network layout (Mu Zhou et al. 2015). Mccarthy et al put forward an adaptive localization method through using WLAN and GPS mixing network, and introduced the GPS signal to improve the positioning accuracy (Mccarthy et al. 2014).

3 THREE-DIMENSIONAL POSITIONING ALGORITHM

3.1 *The principles of fingerprint positioning systems*

The basic principle of location fingerprint positioning technology is as follows: the received signal strengths of different physical locations are complex and unique, which can also be distinguished. According to this characteristic, we can construct a mapping relation database, which could correlate the physical location and the RSSI vector. When we obtain the RSSI of the location, the corresponding physical location could be estimated theoretically.

The indoor positioning based on location fingerprint includes two stages, namely, offline acquisition stage and online positioning stage. Offline acquisition and on-line positioning process are shown as Figure 1.

The main purpose of the off-line acquisition phase is to complete the radio map, which correlates the location information and the RSSI of each AP received at this location. In the off-line stage, we will select a number of RPs in the test environment, and record the RSSI received at all RPs and all the RSSI data will be stored in a database in a certain way.

In online stage when the users enter into the positioning area, the RSSI will be collected and the user location will be estimated through some positioning algorithm. Location information can be three-dimensional or two-dimensional space coordinates.

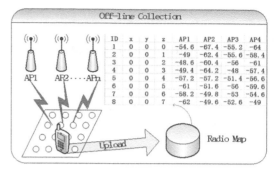

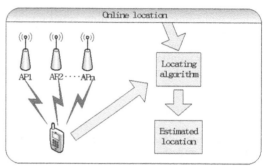

Figure 1. Working process of location fingerprint positioning system.

3.2 *NNSS and NNSS-AVG algorithm*

NNSS is the most classical deterministic positioning algorithm, which was firstly proposed by the RADAR system. NNSS is expressed as follow:

$$\min(D), D = \sqrt{\sum_{i=1}^{n} \left(F^i - S^i\right)^2} \tag{1}$$

The RSSI F is measured in online positioning stage and the RSSI S is stored in radio map constructed in offline stage. The n represents the number of APs, the position of APs which has the minimum Euclidean distance is estimated as the user location.

NNSS-AVG (Nearest Neighbors in Signal Space-Average) is improved on the basis of NNSS algorithm. NNSS-AVG was applied in RADAR-2 system. NNSS-AVG introduces a parameter k, so the user location will be corresponding to the k positions which have the minimum Euclidean distance from user location to RPs. The mean value of the k near positions of the RPs is the estimated user location.

3.3 *NNSS-AP-AVG algorithm*

Aiming at three dimensional space, we put forward a parameter adjustable positioning algorithm

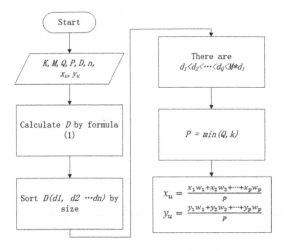

Figure 2. NNSS-AP-AVG algorithm flow diagram.

named as NNSS-AP-AVG (NNSS Adjustable Parameter Average), referring to the improvement of Weyes system (Gu & Lang 2006) on the basis of NNSS-AVG. The NNSS-AP-AVG algorithm flow diagram is shown in Figure 2.

There are two improvements of NNSS-AP-AVG relative to NNSS-AVG. Firstly, the fixed k has been modified to the smaller value P between k and Q. The Q is the number of the RSSI of the Euclidean distance which is less than or equal to M times the minimum Euclidean distance. Secondly, the average estimation is modified to weight average estimation.

The first improvement prevents irrelative location involved in the estimation. Since a larger estimation error will be generated in a positioning estimation if the k has been fixed. Secondly, the comparison between K and Q is used to relieve the positioning accuracy decreasing caused by too much location participating in the estimation. Therefore, we utilize threshold k to limit the number of participating location number.

The second improvement can increase the weight of the similar RSSI, which makes the estimated location closer to the position where the RSSI is more similar. As a result, the position accuracy could be improved.

4 PERFORMANCE EVALUATION

4.1 Experimental testbed

Figure 3 is the plan of experimental test bed, which locates at the west of the four floor of the lab, building in school. The dimension of this area is 17.68 m × 8.27 m. There are eight APs in this area, which have already been marked in Figure 3.

The interval of the RP is 1.2 m, and there are total sixty RPs marked as blue point. The twenty-five red points are the location, requiring to be positioned. We use a Samsung S5 to collect all the RSSI measurements.

4.2 NNSS-AP-AVG algorithm emulation

To test the performance of our algorithm, We input the data into the simulated NNSS-AP-AVG algorithm, and get the estimated location coordinates of each measured point by setting different parameters (k and M). The positioning error will be obtained by calculating the Euclidean distance between the estimated coordinates and the real coordinates. Here, we use the CDF (cumulative distribution function for error) and the RMSE (root-mean-square error) to indicate the positioning accuracy.

At first, when the M is fixed, we simulate the positioning results within different k to observe the variation of positioning accuracy, and the results are shown in Figure 4. When k is 45, the CDF is the highest, namely, the positioning accuracy is the highest. When the k is higher, namely, 90 or 180, the CDF within 1 m is slightly lower than that of 45, and the CDF is also higher in rest range. This is because when k takes a larger value, according to NNSS-AP-AVG algorithm, M plays a determinate role, which will allow all the eligible similar locations to be involved in the estimation.

Then, we simulate the positioning result with different M after determining the threshold k as 45. Simulate the CDF when M is taken as 1, 1.5, 2.25, 3, 3.75 and 4.5. The results are presented in Figure. 5.

As shown in Figure. 5, when M is taken as 3.75, the CDF is the highest, implying that the positioning accuracy is the highest. CDF can reach 96% within 3 m. When M is taken as 1, the NNSS-AP-AVG algorithm will be turned into NNSS algorithm, and its positioning accuracy is significantly reduced. When M value is lower than 3.75, CDF will be increased with the increase of M. This proves that when M value is too small, excess locations will be introduced to participate in positioning estimation and the positioning accuracy will be decreased. When M is larger than 3.75, CDF will be reduced with the increase of M value, proving that when M value is too large, the insufficient estimation locations will reduce the positioning accuracy.

Then, we compare NNSS-AP-AVG with NNSS-AVG. In NNSS-AP-AVG, we set k to 45 and M to 3.75, and in NNSS-AVG, we respectively set k to 1, 27, 45, 90, 180. Through comparing the CDF and the ARMSE of these two algorithms, we obtain the result shown in Figure. 6 and Figure. 7. The

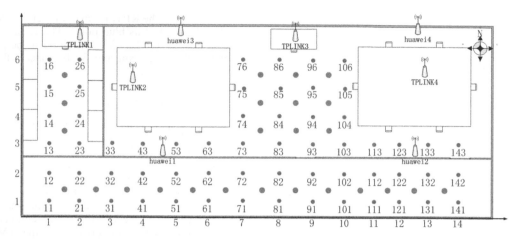

Figure 3.　Experimental testbed.

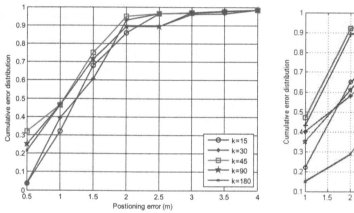

Figure 4.　The influence of threshold k on CDF.

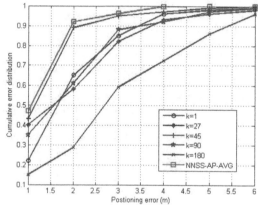

Figure 6.　The CDF comparison of two algorithms.

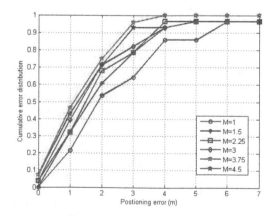

Figure 5.　The influence of M on CDF.

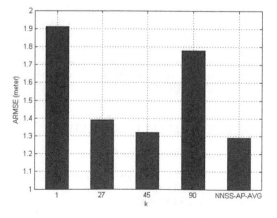

Figure 7.　The ARMSE comparison of two algorithms.

Table 1. The experimental result of 3DLZ system.

Index	Result
ARMSE	1.31 m
CDF	within 3 m 92%
MAX ERROR	3.43 m
DELAY	within 3 second

Table 2. The comparison of positioning systems.

System	RADAR	Horus	3DlZ
Algorithm	NNSS-AVG	MLE	NNSS-AP-AVG
ARMSE	3 m	2 m	1.31 m
3D positioning	No	No	Yes

accuracy of NNSS-AP-AVG algorithm is higher than NNSS-AVG algorithm. When k is taken as 45, ARMSE of NNSS-AVG can reach 1.33 m, and NNSS-AP-AVG can reach 1.29 m. The improved algorithm increases about 3% accuracy.

4.3 *Performance of the three-dimensional localization system*

In order to test our algorithm in the real environment, we implement a three-dimensional localization system named 3DLZ which utilizes NNSS-AP-AVG algorithm to estimate the use position.

The results are shown in Table 1. In our testing environment, CDF could reach 92% within 3 m, 88% within 2 m, and 36% within 1 m. The ARMSE is 1.31 m, and the test accuracy is a little bit lower than MATLAB simulation result.

Then, we compare the positioning system proposed in this paper with the classic location fingerprint positioning system, which is shown in Table 2. Compared with RADAR and Horus, 3DLZ can support three-dimensional positioning after three-dimensional offline sampling. The positioning accuracy is higher than the first two systems. Thus, it can completely meet the requirements of indoor location services.

5 CONCLUSION

In this paper, we proposed a NNSS-AP-AVG algorithm based on the NNSS-AVG algorithm. This algorithm has two improvements: firstly, the positioning error caused by fixed parameter k is mitigated because of the introduction of parameter Q; secondly, the weight of similar RSSI is increased so the positioning accuracy is higher. Through using the NNSS-AP-AVG algorithm, a three-dimensional positioning system is realized in this paper as well. After the simulation and experiments, the results show that our algorithm and system have better performance by means of comparing with the classical fingerprint positioning system.

ACKNOWLEDGMENTS

This work is supported by the National Science-Technology Support Plan Projects of China (No. 2015BAK22B02, No. 2014BAH10F00).

REFERENCES

Bahl, P. & Padmanabhan, V.N. RADAR. 2000. An in-building RF-based User Location and Tracking System. *IEEE Infocom:*775–784.

Bahl, P., Padmanabhan, V.N. & Balachandran, A. 2000. Enhancements to the radar user location and tracking system. *Microsoft Research.*

Dutzler, R., Ebner, M. & Brandner, R. 2013. Indoor Navigation by WLAN Location Fingerprinting Reducing Trainings-Efforts with Interpolated Radio Maps. *Proceedings of the 7th international conference on Mobile Ubiquitons Computer, System, Service and Technologies and networking. IARIA:* 1–6.

Gu, Z.G.C., & Lang, V.L.V. 2006. Multi Space Projection Algorithm for WLAN-based Location Service. International *Conference on Wireless Networks, June:* 290–294. Las Vegas, Nevada, USA.

Ngo, T.F.K., & Mccarthy, T. 2014. Autonomous hybrid WLAN/GPS location self-awareness. US, US8681741.

Rai, A., Chintalapudi, K.K., Padmanabhan, V.N. et al. 2012. Zee: Zero-Effort Crowdsoucing for Indoor Localization. *Proceedings of ACM Mobicom:* 269–280. ACM.

Yang, Z., Wu, C. & Liu, Y. 2012. Locating in fingerprint space: wireless indoor localization with little human intervention. *Proceedings of the 18th annual international conference on Mobile computing and networking.:* 269–280. ACM.

Youssef, M. & Agrawala, A. 2005. The Horus WLAN location determination system. *International Conference on Mobile Systems:*205–218. ACM.

Youssef, M.A., Agrawala, A. & Shankar, A.U. 2003. WLAN location determination via clustering and probability distributions. Pervasive Computing and Communications, 2003. (PerCom 2003). *Proceedings of the First IEEE International Conference on:* 143–150. IEEE.

Zhou, M., Qiu, F., Tian, Z., Wu, H., Zhang, Q. & He, W. 2015. An information-based approach to precision analysis of indoor wlan localization using location fingerprint. *Entropy*(17).

Advances in Energy Science and Equipment Engineering II – Zhou, Patty & Chen (Eds)
© 2017 Taylor & Francis Group, London, ISBN 978-1-138-71798-5

A study on the electrical conductivity of electrospun MWNTs/PMIA nanofiber mats

B. He, T. Liu, C. Liu & D. Tan
Institute of Textile and Fashion, Hunan Institute of Engineering, Xiangtan, China

ABSTRACT: Multi-walled carbon nanotubes/poly (mphenylene isophthalamide) (MWNTs/PMIA) nanocomposites with different MWNTs contents were successfully prepared by using the electrospinning process. Scanning Electron Microscopy (SEM) was used to characterize the nanofibers produced. The results showed that the nanofibers were uniformly distributed with good morphology and with an average diameter of about 140 nm to 180 nm, which decreased a little with an increase in MWNTs contents. From an electrical point of view, it has been observed that the electrical conductivity of nanofiber mats increases by about ten orders of magnitude from lower than 10^{-15} S $\cdot$ cm^{-1} for pristine PMIA to 10^{-5} S $\cdot$ cm^{-1} with the addition of MWNTs. This observed behavior is very interesting in the context of sensor developments.

1 INTRODUCTION

The electrospinning technique is an important and effective method to prepare one-dimensional nanostructure materials (Li 2014). When compared with the conventional fiber, the resultant polymer nanofibers are featured with a high specific surface area, large length–diameter ratio and ease of functionalization for various purposes. The electrospun nanofiber membranes/mats have a unique reticulated structure and high porosity. These unique properties of electrospun nanofibers make a large potential applications in the fields of filtration materials, protection materials, biomedical materials, molecular nanoelectronics, etc (Reneker 2007, Lingaiah 2014).

Carbon Nanotubes (CNTs) have been the focus of numerous investigations, since these were discovered in 1991 (Iijima 1991). The CNTs were divided into Single-Walled Carbon Nanotubes (SWNTs) and Multi-Walled Carbon Nanotubes (MWNTs) according to the different layers of graphene sheets. The CNTs possess many excellent properties, such as exceptional mechanical properties, high thermal stability, very good electrical conductivity, strong chemical resistance, and so on, which have inspired interest in trying CNTs as fillers in polymer composites to improve the properties of polymer materials (Popov 2004, Jiang 2016). Electrospinning was supposed to be a potential method used for isolating and debundling CNTs (Kannan 2008). He et al. (2012) carried out a functional treatment of MWNTs firstly, and then the eletrospun MWNTs/Poly(Mphenylene IsophthAlamide) (PMIA) nanofibers were prepared. The results showed that the MWNTs were well-dispersed and well-aligned along the nanofibers axis, and improved the thermal stability and mechanical properties with lower content of MWNTs. Almuhamed et al. (2012) prepared the eletrospun MWNTs/PolyAcryloNitrile (PAN) nanofibers and it was observed that the electrical volume conductivity increased by about six orders of magnitude from 2×10^{-12} S $\cdot$ m^{-1} for pristine PAN to 4×10^{-6} S $\cdot$ m^{-1} for PAN charged by MWNTs. Wang et al. (2014) prepared the electrospun MWNTs/PolyVinyliDene Fluoride (PVDF) ultra-fine fibers. The electrical conductivity of these fibers reached up to 1×10^{-6} S $\cdot$ cm^{-1} when the loading of MWNTs was 1.2wt%.

Poly (Mphenylene IsophthAlamide) (PMIA) has especially prominent thermal stability, high flame retardancy, self-extinguishing characteristics, electrical insulation ability, chemical resistance, mechanical properties, etc. And so, PMIA is widely applied in special protective clothing, high temperature filter materials, electrical insulating materials, etc. (Machalaba 2002).

In this paper, the MWNTs were grafted with a non-ionic surfactant Triton X-100 to improve their dispersion in the PMIA solution, and then the MWNTs/PMIA nanofiber mats were produced by electrospinning. The effects of the MWNTs on the electrical conductivity of nanofiber mats were investigated in detail.

2 EXPERIMENTAL DETAILS

2.1 *Materials*

Commercial PMIA fibers (tensile strength at breakage was ca. 4.5 cN $\cdot$ dtex^{-1} and density 1.37–1.38 g $\cdot$ cm^{-3}) were provided by Yantai Spandex Co. Ltd. (China).

MWNTs (diameter <10 nm, length 1–2 μm, purity 95%–98%) were purchased from Shenzhen Nanotech Port Co. Ltd. (China). Anhydrous lithium chloride (LiCl) was supplied by Shanghai Jufeng Chemical Scientific Co. Ltd. (China). Triton X-100 ($C_8H_{17}C_6H_4(OCH_2CH_2)n$, n ≈ 10), DMAc, and N,N-DiMethylFormamide (DMF) were purchased from J&K Scientific Co. Ltd. HNO_3 (65%) and H_2SO_4 (95%) were supplied by Shanghai Chemical Reagent Co. Ltd. (China). TetraHydroFuran (THF) and $SOCl_2$ were purchased from Sinopharm Chemical Reagent Co., Ltd. (China). All the chemicals were of analytical reagent grade.

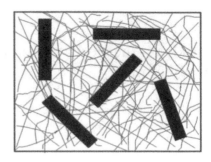

Figure 1. Schematic diagram of the sampling process of the resistance test.

2.2 Functionalization of MWNTs

Firstly, MWNTs were purified with the mixed solution of HNO_3 and H_2SO_4 (v/v = 1:3). Secondly, the MWNTs–COCl were obtained after the treatment of MWNTs in the mixed solution of $SOCl_2$ and DMF. Thirdly, the MWNTs–COCl were placed in the solution of Triton X-100 in DMF under a N_2 atmosphere, and the MWNTs grafted with Triton X-100 were obtained after this process. The detailed process has been reported by our research group.

2.3 Preparation of the spinning solution

LiCl was dried in a vacuum dryer at 120°C for 2 h. Firstly, the dried LiCl was added into DMAc and stirred till it was dissolved completely, and this was followed by placing PMIA into the solution. The solution was stirred at 90°C until a transparent solution was obtained. Meanwhile, MWNTs–Triton X-100 were added into the DMAc solvent and ultrasonicated for 1 h. At last, the two solutions were mixed with different MWNTs mass fractions (0%, 0.6%, 1.0%, 2.0%, 3.0%, and 4.0%) after 1 h of ultrasonic treatment. The mass fractions of LiCl and PMIA in the solution were 3% and 12%, respectively.

2.4 Electrospinning

The spinning apparatus used was the Nanofiber Electrospinning Unit (KATO TECH Co. Ltd., Japan). Spinning parameters were as follows: inner diameter of the spinneret is 0.45 mm, voltage is 23 kV, collecting spinning distance is 11 cm, solution flow rate is 0.26 ml · h^{-1}, spinneret scan speed is 14 cm · min^{-1}, and the collector rotating speed is 6 m · min^{-1}.

2.5 Measurements

The morphologies of the electrospinning MWNTs/ PMIA composite nanofiber mats were observed by performing scanning electron microscopy (Hitachi S-4800, Japan). The mean diameter of fibers was measured by using an image analyzing system (Image-Pro Plus 5.0), and the average value was calculated by analyzing 100 measurement data.

The electrical properties of the nanofiber mats were tested by using an Agilent 4339B high resistance meter. Figure 1 shows the sampling method, in which the sample amount was 5 for each nanofiber mats, and the size of each sample is 10 mm × 2 mm. The electrical conductivity was obtained by using Equation 1.

$$\sigma = \frac{1}{\rho} = \frac{L}{R \times w \times t} \quad (1)$$

where σ and ρ are the electrical conductivity and resistivity of the nanofiber mats, respectively, R is the specimen electrical resistance, L is the specimen length, w is the specimen width, and t is the specimen thickness.

3 RESULTS AND DISCUSSION

3.1 Nanofiber morphology

Figure 2 shows the SEM images of MWNTs/ PMIA composite nanofibers electrospun using the different MWNTs concentrations in spinning solutions of 0 wt%, 0.6 wt%, 1.0 wt%, 2.0 wt%, 3.0 wt%, and 4.0 wt%, respectively. The results indicated that the composite nanofibers were uniformly distributed and exhibited good morphology. The average diameter of the composite nanofibers is about 140 nm to 180 nm which is shown in Figure 3 and was decreased a little with an increase in MWNTs contents. The electrical conductivity of the spinning solution has increased by increasing the MWNTs concentrations, which would cause a decrease in the diameter.

3.2 Electrical properties

Figure 4 shows the electrical conductivity of the electrospun nanofiber mats. The results indicate

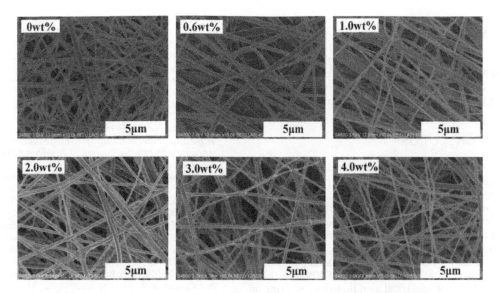

Figure 2. SEM images of the nanofibers with different MWNTs concentrations.

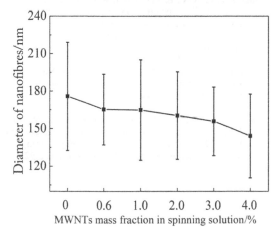

Figure 3. Graph showing the diameters of the nanofibers at different MWNT concentrations.

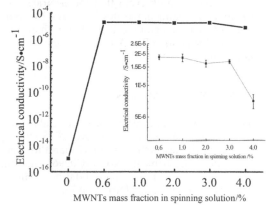

Figure 4. Graph showing the electrical conductivity of nanofiber mats at different MWNTs concentrations.

that the electrical properties of nanofiber mats were greatly improved with the addition of MWNTs, and their threshold value of electrical conductivity occurs at the loading of MWNTs content of 0.6 wt% in the spinning solution when the electrical conductivity is about 1.86×10^{-5} S · cm^{-1}. There are about ten orders of magnitude from lower than 10^{-15} S · cm^{-1} for pristine PMIA to 1.86×10^{-5} S · cm^{-1} for PMIA charged by MWNTs.

The electrical properties of CNTs/polymer composites are mainly determined by the CNTs and the polymer's self-resistance, the tunneling resistance between adjacent CNTs, and the contact resistance. The conductive network is formed inside the nanofiber when the MWNT contents are present up to a certain extent in the composite nanofiber, and then the connected circuit is built among MWNTs. And so, the proportion of PMIA polymer's self-resistance is reduced, which causes the electrical conductivity of the composite nanofiber to be improved greatly. However, the electrical conductivity has been maintained at the same order of magnitude after the threshold appears to continue to increase with an increase in the content of MWNTs (such as 1.0 wt% to 3.0 wt%). According to the threshold theory, the conductive network is formed inside the nanofiber when the threshold occurs, which will not be destroyed when it continues to increase the content of MWNTs in a certain range. Therefore, there is no significant change in the electrical conductivity of the composite nanofibers (Hu 2010). But when the MWNTs content is beyond a certain range (such as 4.0 wt%), the conductive network is destroyed, and the electrical conductivity decreases.

4 CONCLUSIONS

MWNTs/PMIA composite nanofiber mats at different MWNTs concentrations were produced successfully by electrospinning. The results showed that the nanofibers were uniformly distributed with good morphology with an average diameter of about 140 nm to 180 nm, which decreased a little with an increase in the MWNT contents. The electrical properties of nanofiber mats were greatly improved with the addition of MWNTs with the threshold of electrical conductivity occuring at the loading of the MWNT content of 0.6 wt% in the spinning solution. From the electrical point of view, it has been observed that the electrical conductivity of nanocomposites increases by about ten orders of magnitude from lower than 10^{-15} S $\cdot$ cm^{-1} for pristine PMIA to 10^{-5} S $\cdot$ cm^{-1} for the addition of MWNTs.

ACKNOWLEDGMENTS

This work was supported by the Provincial Natural Science Foundation of Hunan [No. 2015 JJ6023], the Scientific Research Fund of Hunan Provincial Education Department [No. 14C0296], and the Doctoral Scientific Research Foundation of Hunan Institute of Engineering [Nos 14082 and 14093].

REFERENCES

Almuhamed S., Khenoussi N., Schacher L., Adolphe D., & Balard H. 2012. Measuring of Electrical Properties of MWNT-Reinforced PAN Nanocomposites. *Journal of Nanomaterials, 2012(6348)*: 3517–3526.

He B., Li J., & Pan Z. 2012. Morphology and mechanical properties of MWNT/PMIA nanofibers by electrospinning. *Textile Research Journal, 82(13)*: 1390–1395.

Hu N., Karube Y., Arai M., Watanabe T., & Yan C. 2009. Investigation on sensitivity of a polymer/carbon nanotube composite strain sensor. *Carbon, 48(3)*: 680–687.

Iijima S. 1991. Helical microtubules of graphitic carbon. *Nature, 354(6348)*: 56–58.

Jiang Q., & Wu L. 2016. Property enhancement of aligned carbon nanotube/polyimide composite by strategic prestraining. *Journal of Reinforced Plastics & Composites, 35(4)*: 287–294.

Kannan P., Young R.J., & Eichhorn S.J. 2008. Debundling, isolation, and identification of carbon nanotubes in electrospun nanofibers. *Small; 4(7)*: 930–933.

Li C., Su D., Su Z., & Chen X. 2014. Fabrication of Multiwalled Carbon Nanotube/Polypropylene Conductive Fibrous Membranes by Melt Electrospinning. *Industrial & Engineering Chemistry Research, 53(6)*: 2308–2317.

Lingaiah S., Shivakumar K., Sadler R. 2014. Electrospun Nanopaper and its Applications to Microsystems. *International Journal for Computational Methods in Engineering Science & Mechanics, 15(15)*: 2–8.

Machalaba N.N., & Perepelkin K.E. 2002. Heterocyclic aramide fibers-production principles, properties and application. J Ind Text, 31(3): 189–204.

Popov V.N. *2004*. Carbon nanotubes: properties and application. *Materials Science and Engineering Reports, 43*: 61–102.

Reneker D.H., Yarin A.L., Zussman E., & Xu H. 2007. Electrospinning of Nanofibers from Polymer Solutions and Melts. *Advances in Applied Mechanics, 41(07)*: 43–195.

Wang S.H., Wan Y., Sun B., Liu L.Z., & Xu W. 2014. Mechanical and electrical properties of electrospun PVDF/MWCNT ultrafine fibers using rotating collector. *Nanoscale Research Letters, 9(1)*: 522–522.

Advances in Energy Science and Equipment Engineering II – Zhou, Patty & Chen (Eds)
© *2017 Taylor & Francis Group, London, ISBN 978-1-138-71798-5*

Research on Changjiang river waterway maintenance vessels' spare parts and materials information coding

Quan Wen
Changjiang Waterway Plan and Design Institute, Wuhan, China

ABSTRACT: In this paper, it is aimed to solve the problem of no unified coding in waterway maintenance vessel spare parts and materials. Firstly, the condition of the previous coding application is analyzed by researchers; meanwhile, this research work concentrates on the basic principle of compile coding and major component. According to methods contrast, this research work adopts the method of line surface classification to compile information coding. At last, on the basis of previous codes' indication and compatibility, the research result provides a part of new codes of waterway maintenance vessel spare parts and materials.

1 INTRODUCTION

The waterway maintenance vessels are the basic important equipment of the Changjiang waterway routine maintenance work. According to those reasons mentioned above, many different types of spare parts have become an important part of the supply security system of waterway maintenance vessels. In order to achieve reasonable management of numerous spare parts, information coding of spare parts establish and apply for efficient management in different stages that include procurement, transportation, inventory management, produce traceability, and so on. Meanwhile, information coding technology should be widely used in "Digital Channel Construction" and as a key security measure to carry out routine work.

2 INFORMATION CODING RULES OF THE WATERWAY MAINTENANCE VESSEL MACHINERY MANAGEMENT SYSTEM AND PROBLEMS

The waterway maintenance vessel machinery management system mainly refers to provide vessel spare parts and materials management, vessel certification management, crew management, audit management, vessel information management, task management, backstage management, and lots of comprehensive vessel transaction management (Xin 2007, Zhang 2007, Gong 2010). For the future development of the waterway maintenance vessel machinery management system, it should concentrate on the future trends of establishing a process-oriented, main body of function, data standardization, data generalization, data modularization, and data sharing's waterway maintenance vessel machinery management system. Therefore, the basic and realizable aim is establish a general standard database which can realize data sharing. At present, the vessel machinery management study mainly focuses on the following aspects: maintenance mode decision, maintenance time prediction mode decision, spare parts procurement, and material storage.

2.1 *The problems of information coding in waterway maintenance vessel machinery management*

In order to ensure the trouble-free operation of waterway dredging maintenance tasks in different periods, the waterway maintenance vessel spare parts needed for various types of material procurement is mainly based on monthly, quarterly, and annual audit declaration quantity. However, in the relative field of waterway maintenance vessel spare parts, information coding has no unified standard, and the current spare parts purchasing and inventory management in the waterway maintenance vessel machinery management system has several function modules that contain material application, loading and unloading management, spare parts procurement, material coding, and so on (Lai 1999, Wang 2009, Jiang 2005). Currently, waterway maintenance vessel spare parts do not have a unified state standard; however, the spare parts information coding systems in available vessel machinery management systems are mainly based on their characteristic to compile. According to the reasons mentioned above, the different information coding rules result in a series of inconveniences in routine work which are displayed as follows:

1. The same type of vessel built by the same manufacturer in different time periods has adopted a lot of different components and parts; meanwhile, with rapid technology development, the same type of vessel selects technical scheme update which results in adoption of different equipment.
2. Data and code in the vessel machinery management system are not based on a unified coding standard; meanwhile, the general spare parts in different manufacturers have different coding rules that easily result in an increase in spare parts species and inventory capital.

2.2 Information coding principles and standards of the vessel machinery management system

At present, information coding research usually uses information classification and code types are mainly divided into three categories. Information classification mainly includes line classification, surface classification, and line surface mixed classification. Code type includes numbers, letters, and alphanumeric hybrids. Base on the state waterway transportation industry's characteristics to compile information codes, the code types of numbers and alphanumeric hybrids are usually adopted by manufacturers and waterway transportation units. In order to compile a unified coding standard for routine work demand, those principles that include system principle, minimization principle, readability principle, compatibility principle, scalability principle, integrity principle, and comprehensive practical principle should be obeyed by the compiling process (Zhou 2007, Liu 2006).

3 CODING SYSTEM AND MODULES MANAGEMENT IN THE VESSEL MACHINERY MANAGEMENT SYSTEM

In 2002, Changjiang waterway bureau issued their owned information classification and code standard for routine work trials. With rapid development of technology and institutional framework change, the old version of information classification and code standard cannot satisfy the current demand.

In the face of the new version of information classification and code standard, this study particularly considers the continuity and convenience of

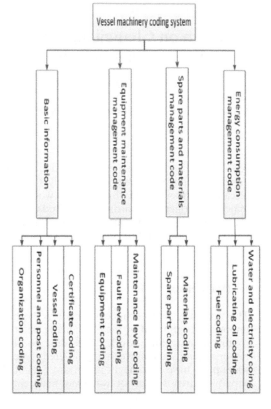

Figure 1. Vessel machinery coding system.

some code field definition in the past; meanwhile, it also strives to obey the coding principles mentioned as follows:

1. The maximum codes retained the original code indicative of important fields.
2. Considering the new code compatibility with existing code.
3. For future new management object demand, the new code should consider its expandability.
4. Utmostly ensure employees' accessibility and familiarity, when they are using those new codes.

According to the research object and goal of this study, the vessel machinery coding system and fundamental coding module should be researched firstly. Obviously, the vessel machinery coding system mainly includes four parts that are divided into basic information, equipment maintenance management code, spare parts and materials management code, energy consumption management coding, and the system function structure, as shown in Figure 1 and Table 1.

232

Table 1. List of fundamental coding modules.

Machinery management coding composition	Coding categories
Basic information management coding	1) Organization coding; 2) Personnel and post coding; 3) Vessel coding; 4) Certificate coding and so on, the part of coding that inherited digital channel public coding, added digital machinery individual coding demand.
Equipment maintenance management coding	1) Equipment coding; 2) Fault coding; 3) Maintenance level coding and so on.
Spare parts and materials management coding	1) Spare parts coding; 2) Materials coding and so on.
Energy consumption management coding	1) Fuel coding; 2) Lubricating oil coding; 3) Water and electricity coding.

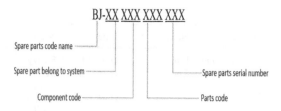

Figure 2. Vessel spare parts coding structure.

Table 2. Vessel spare parts coding table.

Code	Name	Subordinate equipment
BJ-AS 101 150 001	Air filter element	Starter air filter
BJ-AS 101 150 002	CCV filter element	Starter air filter
BJ-BW 102 142 001	Sea water pump overhaul	Main sea water pump
BJ-ES 201 308 001	Electric generator	Variable speed generator
BJ-ES 203 000 001	Belt	Variable speed generator drive mechanism
BJ-FW 101 149 001	Water pressure switch	Fresh water barge valves
BJ-FW 202 154 001	Thermostat	Low temperature fresh water heat exchanger
BJ-FW 202 154 002	Thermostat gasket	Low temperature fresh water heat exchanger
BJ-FW 202 154 003	Pressure cap	Low temperature fresh water heat exchanger

4 VESSEL SPARE PARTS AND FUEL CODING RULES

In terms of compiled rules mentioned above, this study aims to consider the vessel maintenance system code compiling principle (CWBT2009) as a guidance and inherited original coding instruction of the part code. Meanwhile, it also guarantees the expansion and legibility of the new code.

4.1 Vessel spare parts coding

The subsection coding structure of spare parts is adopted by the waterway maintenance vessel. First two letters of "BJ" represent the spare part's code name. In the middle of the eight letters, the CWBT standard code is adopted and it respectively represents the subordinate equipment system, equipment, and component. The last three letters adopt the digital sequence code as the spare parts' serial number (see Figure 2 and Table 2).

4.2 Fuel coding of the waterway maintenance vessel

Fuel as a material of fundamental consumption is widely used in all sorts of fields, and according to different application fields that have several compile code methods. In terms of the vessel spare

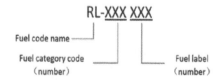

Figure 3. Schematic of the fuel coding structure.

parts code rule mentioned above and its own characteristics, the fuel code structure is divided into three parts. First two letters "RL" represent the fuel code name, and the three numbers in the middle represent the fuel category code, and the last three numbers represent the fuel label (see Figure 3 and Table 3).

Table 3. Fuel coding table.

Code	Name	Code	Name
RL-000 000	Fuel	RL-424 000	Heavy oil
RL-400 000	Coal	RL-424 250	#250 Heavy oil
RL-410 001	Cleaned coal	RL-424 200	#200 Heavy oil
RL-410 002	Clean coal	RL-424 180	#180 Fuel oil
RL-412 003	Coking coal	RL-424 120	#120 Fuel oil
RL-420 000	Petrol	RL-424 007	#7 Fuel oil
RL-421 090	#90 Petrol	RL-424 001	Industrial fuel oil
RL-421 092	#92 Petrol	RL-424 002	Catalytic slurry oil
RL-421 093	#93 Petrol	RL-424 003	Wax oil slurry
RL-421 095	#95 Petrol	RL-424 004	Mixed heavy oil
RL-421 097	#97 Petrol	RL-424 005	Pitch
RL-423 000	Diesel	RL-425 000	Natural gas
RL-423 110	#10 Diesel	RL-425 001	LNG
RL-423 100	#0 Diesel	RL-425 002	CNG
RL-423 010	#-10 Diesel		
RL-423 020	#-20 Diesel		
RL-423 035	#-35 Diesel		
RL-423 050	#-50 Diesel		

5 CONCLUSIONS

In terms of the actual application situation of information code in the Changjiang waterway vessel machinery management, this research firstly studies the structure and modules of vessel information coding and analyses several main compiling methods of coding. Then, the subsection coding structure and alphanumeric hybrids are adopted by compiling a new code. Finally, in this paper, many waterway maintenance vessel spare parts codes and fuel codes are provided that will be widely used by the vessel machinery management system as the main research results.

REFERENCES

Gong Qinhu (2010). Material code-the key technology of the equipment manufacturing industry informationization. Mechanical, 03:37–39.

Jiang Wanjun (2005). Material coding system solutions in the application of the ERP implementation. Metallurgical equipment, 08:54–57.

Lai Ruolan (1999). Standardization of MRPI system material code [J]. Information technology and standardization, 04:37–39.

Liu Chengfeng (2006). Ship assembly and virtual reality system in the research of virtual product coding system. China shipbuilding, 47(Supplement):250–255.

Wang Tailong (2009). Spare parts coding and figure coding solution in special circumstances. Equipment management and maintenance, 03:10–11.

Xin Hong (2007). Research of PCB industry material coding rules. Printed circuit information, 08:51–54.

Zhang Yun (2005). Application of material coding management system in Sinopec. Digital oil and chemical industry, 05:31–32.

Zhou Honggen (2007). The research and application of Marine diesel engine parts coding management system [J]. Marine engineering, 01:80–83.

Advances in Energy Science and Equipment Engineering II – Zhou, Patty & Chen (Eds)
© *2017 Taylor & Francis Group, London, ISBN 978-1-138-71798-5*

Research on the airborne monitoring semi-physical simulation system based on beidou

Yigang Sun
College of Aeronautical Engineering, Civil Aviation University of China, Tianjin, China

Shuangxing Wang
College of Electronic Information and Automation, Civil Aviation University of China, Tianjin, China

ABSTRACT: Airborne monitoring semi-physical simulation system based on beidou (BD) could send the real-time data generated by an airplane to the ground as a secondary system of the ground–air communication system. The purpose of designing this system is to test the function of an aviation electronic communication navigation system. First, the difficulties encountered on the realization of this system are proposed, such as sending a large amount of data via the beidou module within a limited time, predicting the position of the airplane, and displaying the position of the airplane on the ground. By discussing multi-module design methods, algorithms about machine learning and baidu map are adopted to solve the problems. The experimental results about data comparison of the database show that the problems have been solved well. The final semi-physical simulation system has been tested in a laboratory environment and it worked very well in sending messages, predicting positions, and so on. And finally, it has been used in the system of airworthiness verification and evaluation of the aviation electronic communication navigation system.

1 INTRODUCTION

There are about 26 countries involved in the rescue after the Malaysian Airlines MH370 lost contact with the ground. They used satellites, airplanes, ships, and many other methods to search MH370, but up to now we are not able to locate its exact position. Although the ACARS system has been installed in the plane, the ground cannot monitor the plane in real time if the ACARS system is closed by people.

Nowadays, there are three kinds of mature ground–air communication systems applied in the field of civil aviation. They are ACARS, ADS-B, and AFIRS.

With the development of technology, the ACARS will be eliminated gradually. The International Civil Aviation Organization predicts that the ADS-B will be chosen as the main monitoring method in the future. But we also need a secondary system to achieve ground–air communication to strengthen the monitoring of the plane. AFIRS is a reliable secondary ground–air communication system, which is designed on the base of Iridium Communication System. Because the Iridium Communication System is a foreign system, we cannot avoid risk and uncontrollable factors. It will also cost the airline company a lot of money to install these systems and use the service of ground–air communication provided by these systems. Without independent intellectual property rights, we cannot avoid political risks.

The beidou satellite navigation system is a navigation system designed by our country and we have its intellectual property rights. Nowadays, it has been providing some services for the Asia-Pacific region, such as position, navigation, timing, and short message communication. The system introduced in this paper was implemented creatively based on the beidou satellite navigation system. It is a secondary system of ground–air communication which we can control, and we also own its intellectual property rights.

When compared with the three kinds of ground–air communication systems applied in the civil aviation field, this system is totally controlled by us. The user can benefit from the ground–air communication service at a lower cost, which makes it convenient for the ground to monitor airplanes in real time. Meanwhile the rescuers can give assistance referring to the simple prediction of the airplane's position.

The semi-physical simulation technology applied in this system is one of the avionics simulation technologies. It is a kind of simulation technology combining digital simulation with objects.

The semi-physical simulation system can provide physical ports to connect the avionics equipment and it also can simulate the logistical function of avionics equipment via software. It plays a very important role in integrating and testing the avionics system.

The airborne monitoring semi-physical simulation system based on beidou provides an alternative to the avionics equipment for testing the laboratory projects as well as technology and experience for the development of equipment based on the beidou satellite navigation system.

2 RESEARCH ON THE AIRBORNE MONITORING SEMI-PHYSICAL SIMULATION SYSTEM BASED ON BEIDOU

2.1 *The operation principle of system*

The overall structure of the system is shown in Figure 1. The system is divided into two subsystems: airborne system and ground monitoring system.

The function of this system is realized by sending real-time flight data generated by other airborne equipment to the ground. The flight data will be processed, stored, and displayed and then, the position of the airplane will be marked on the map. The hardware structure of the system is shown in Figure 2. The airborne system installed in IPC1will receive the flight data from the aviation bus via an aviation bus board, and then the flight data will be processed, stored in the database in the airborne system, and sent to the ground via the beidou communication module. The ground monitoring system installed in IPC2 will receive the flight data sent by the airborne system via the beidou communication module, and then the flight data will be processed, stored in the database in the ground monitoring system, and displayed on the software interface. The location of the plane will be marked in the map.

2.2 *Detailed description of the subsystem and difficult problems*

2.2.1 *Airborne system*

From Figure 1, it can be observed that the airborne system is composed of the airborne beidou communication module and core software. The beidou communication module will send the flight data to the ground and the core software will receive the flight data, process and store the data, and send the data to the ground by operating the beidou module. When the plane is under good condition, the sample interval of the airborne system is about 15 s, while the fight data will be sent every 60 seconds. When the plane is under emergency conditions, the unsent flight data will be sent quickly. The core software of the airborne system is composed of a data forwarding module, data processing and storage module, emergency judging module, airplane position predicting module, and MySQL database. The functions of all the modules are described as follows:

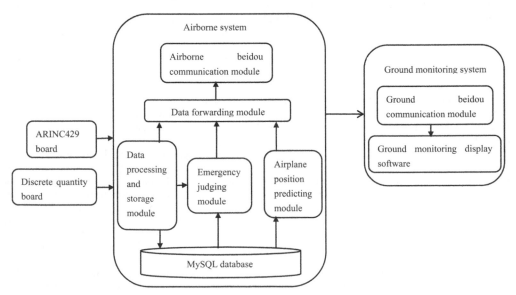

Figure 1.   The overall structure diagram of research on the airborne monitoring semi-physical simulation system based on beidou.

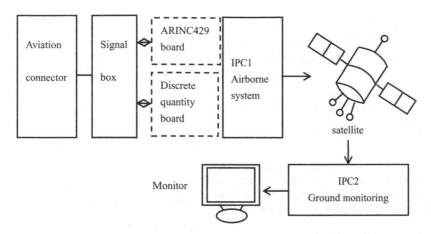

Figure 2. The hardware structure diagram of the research on the airborne monitoring semi-physical simulation system based on beidou.

Content	Length	User address	Information content							Checksum
Communication information $TXXX	16 bit	24 Bit	Information category 8bit	Sender address 24bit	Sending time		Message length 16bit	Message content 1680bit	CRC mark 8 bit	8 bit
					H 8bit	M 8bit				

Figure 3. The protocol format of short messages.

1. Data forwarding module: Forwarding the flight data including longitude, latitude, height, and speed of the airplane by operating the beidou communication module. When the plane is under good conditions, the data will be sent every 60 seconds (functional limitations of the civil beidou module, the time interval between sending two short messages is about 60 seconds). When the plane is under emergency conditions, the large amount of unsent flight data stored in the database will be sent quickly.
2. Data processing and storage module: Processing the flight data gained from boards and inserting the data into MySQL database, while sending them to the emergency judging module.
3. Emergency judging module: Judging whether the plane is in an emergency. When the plane is off course, we can judge that it is in an emergency.
4. Airplane position predicting module: Because of the time interval between two short messages, we may not know the exact position of the airplane in real-time. But we can predict the position of the airplane according to the existing flight data, such as speed, acceleration, time, longitude, or latitude. The rescuers can provide

assistance referring to the simple prediction of the airplane's position.

Airborne beidou communication module: Sending the flight data of the airplane. The time interval between sending two short messages of the civil beidou module is about 60 seconds and the length of each message is about 1680 bit. The protocol format of the short message of the beidou communication module is shown in Figure 3. The flight data should be coded according to the protocol format shown in Figure 3 when they are forwarded by the data forwarding module. And then, the data will be sent in hex via operating the beidou communication module.

The format of information content will be introduced concretely: the information category is appointed to 46H, which indicates that the mode of communication is the form of Chinese characters including keys; sender address is the user address of the communication or message sender; sending time is up-to-the-minute (m). It will be 0 if the communication is not a query communication; Message length is the length of the message content and message content is the flight data; CRC mark 00H indicates that the data are correct and 01H indicates that the data are incorrect.

Because of the functional limitation of the civil beidou module and the requirements of system function, we encountered the following two difficult problems in the realization of the airborne system:

1. How to send large amounts of unsent flight data stored in an airborne database to the ground in a short time?
2. How to predict the position of an airplane via flight data?

We solved these two problems by adopting the multi-module design method, which comprises algorithms about machine learning. The realization methods are provided in the third section.

2.2.2 Ground monitoring system

From Figure 1, it can be observed that the ground monitoring system is composed of the ground beidou communication module and the ground monitoring display software. The main function of the ground beidou communication module is to receive flight data sent by using an airborne system. The main function of ground monitoring display software is to process the captured data, insert data into the database when displaying the longitude, latitude, height, and speed and mark the position of the airplane on the map. The flight data displayed on the software and the position marked in the map will be refreshed in real time.

We encountered the third difficult problem in the realization of an aground monitoring display system:

1. How to mark the position of the airplane in real time on the ground?

We adopted the baidu map to solve this problem. The realization method is provided in the third section.

3 SOLUTIONS TO DIFFICULT PROBLEMS OF THE AIRBORNE MONITORING SEMI-PHYSICAL SIMULATION SYSTEM BASED ON BEIDOU

3.1 How to send large amounts of unsent flight data within a limited time via the beidou module?

The method we adopted is by applying several beidou modules in the system. The core software of the airborne system will create multiple threads to operate multiple beidou modules, when the airplane is in an emergency. The method we adopted might occupy many communication resources; however, it did solve the insufficient communication ability of the beidou module.

3.2 How to predict the position of the airplane?

We solved this problem by adopting linear regression in machine learning. First, we chose speed v, acceleration a, and time t as input features, and then standardized them. Because the speed or acceleration of the airplane will not be changed frequently when the airplane is in the cruise phase, the problem of multiple, mutual, and linear variables will appear. To solve this problem, we adopted the method of ridge regression to estimate the regression coefficients.

We decided to approximate output y (longitude or latitude) as a linear function of x (speed v, acceleration a, and time t):

$$h_\theta(x) = \theta_0 + \theta_1 x_1 + \theta_2 x_2 \tag{1}$$

In this work, the θ_i's are called weights. To simplify our notation, we will remove the θ subscript in $h_\theta(x)$, and write it more simply as $h(x)$ Let $x_0 = 1$, so that

$$h(x) = \sum_{i=0}^{n} \theta_i x_i = \theta^T x \tag{2}$$

θ and x are both viewed as vectors, and n is the number of input variables. Now, we need to choose θ_i to make $h(x)$ close to y. We define the cost function as follows:

$$J(\theta) = \frac{1}{2} \sum_{i=1}^{m} \left(h_\theta(x^{(i)}) - y^{(i)} \right)^2 \tag{3}$$

The problem now is to find θ_i to minimize $J(\theta)$.

The method of least squares was used to minimize $J(\theta)$. Each input x consists of three variables, and define the design matrix x to be the m-by-n matrix that contains the training examples' input values in its rows, which is given as follows:

$$X = \left[(x^{(1)})^T \quad (x^{(2)})^T \quad ... \quad (x^{(m)})^T \right]^T \tag{4}$$

Also, $\vec{y}$ was defined as the m-dimensional vector containing all the target values that form the training set:

$$\vec{y} = \left[y^{(1)} \quad y^{(2)} \quad ... \quad y^{(m)} \right]^T \tag{5}$$

From equation (2), (4) and (5), we can conclude that

$$X\theta - \vec{y} = \left[h_\theta(x^{(1)}) - y^{(1)} \quad ... \quad h_\theta(x^{(m)}) - y^{(m)} \right]^T \tag{6}$$

Simplify (6) to obtain the following equation:

$$\frac{1}{2}(X\theta - \vec{y})^T (X\theta - \vec{y}) = \frac{1}{2} \sum_{i=1}^{m} (h_\theta(x^{(i)}) - y^{(i)})^2 = J(\theta) \tag{7}$$

To minimize $J(\theta)$, we need to use equation (7) to find its derivatives with respect to θ. And then, we set its derivatives to zero:

$$
\begin{aligned}
\nabla_\theta J(\theta) &= \nabla_\theta \frac{1}{2}(X\theta - \bar{y})^T(X\theta - \bar{y}) \\
&= \frac{1}{2}\nabla_\theta(\theta^T X^T X\theta - \theta^T X^T \bar{y} - \bar{y}^T X\theta + \bar{y}^T \bar{y}) \\
&= \frac{1}{2}\nabla_\theta tr(\theta^T X^T X\theta - \theta^T X^T \bar{y} - \bar{y}^T X\theta + \bar{y}^T \bar{y}) \\
&= X^T X\theta - X^T \bar{y} \\
&= 0
\end{aligned}
\tag{8}
$$

From equation (8), we can obtain the normal equations:

$$
X^T X\theta = X^T \vec{y} \tag{9}
$$

Finally, we can find the value of θ to minimize $J(\theta)$:

$$
-\theta = (X^T X)^{-1} X^T \bar{y} \tag{10}
$$

From equation (10), we can know that the existence of θ is under the premise that the matrix X must be non-singular. But we have discussed previously that the chosen variables may be multiple, mutual, and linear which makes matrix X singular. To ensure that there is a solution for equation (10), we adopt ridge regression to estimate the regression coefficients. To ensure that there is a solution for equation (10), we add λI to it:

$$
\hat{\theta}(\lambda) = (X^T X + \lambda I)^{-1} X^T \bar{y} \tag{11}
$$

λ is called the ridge parameter, I is an identity matrix. $\hat{\theta}(\lambda)$ is a function of λ when the ridge parameter λ varies from 0 to ∞. We can determine the range of λ according to the shape of the ridge trace curve and then make it more compatible by minimizing the prediction error. The specific implementation processes are as follows: first, we should write programs according to this machine learning algorithm to calculate a biased estimate of the regression parameters. The data (speed v, acceleration a, time t, longitude, and latitude) required by the training algorithm are stored in the database in an airborne system. The relationship between them is determined by equation (1). And then, we should divide the data into two parts: one is used to the post-test as a test set, the other is used to train regression parameters θ as a training set, so that we can minimize the prediction error.

When the training is completed, we should use the test set to test the performance of equation (1). The data (speed v, acceleration a, and time t) obtained from the test set will be treated as the input, and then we will obtain one corresponding output, such as the longitude or latitude. The comparison between the output and corresponding actual longitude and latitude will be the indicator judging whether the regression parameters can meet demands of the prediction performance or not. λ, which minimizes the prediction error, will be obtained finally by choosing different λ values to repeat the test process mentioned above. When the regression parameters are determined, we can predict the longitude and latitude of the airplane by inputting the corresponding x value (speed v, acceleration a, and time t).

3.3 How to display the position of airplane in real time?

We adopted the baidu map and MFC program to display the flight data and mark the position of the airplane by discussing. We require a baidu developer key before developing. The developer key will be called in the program when we use the baidu map. The baidu map file is named as baidumap.html which is written in JavaScript language. The software flow based on MFC is as follows:

First the software will connect to the beidou module, load the baidu map and start the timer during the initialization phase. The flight data will be read when the time is up. If the flight data are available, it will be processed and stored. Also the flight data will be displayed and the position of the airplane will be marked on the map. If the flight data are unavailable, the software will wait for another chance to read flight data.

The software has been written by VS2010 according to the execution logic above, which has met the demands of the function.

4 SYSTEM IMPLEMENTATION AND PERFORMANCE ANALYSIS

4.1 Implementation of core software in an airborne system

The core software in an airborne system has no display interface whichvruns in the background and works when the machine is switched on. The whole project is developed in the "watcher" folder and each function module is implemented in a different file. The "Makefile" file has been written to make it easy to install the software. In the folder, data processing and storage function are implemented in file "dataProcessSave.c", the emergency judging function is implemented in the file "emergencyJudge.c", operation of MySQL is implemented in the file "useMySQL.c,airplane", the position predicting function is implemented in the file "locationAlgorithm.c", and sending data function via the beidou communication module is implemented in the file "dataSend.c". At last, the main function is implemented in the file "airplaneWatch.c".

The core software flow in an airborne system is as follows according to the file "airplaneWatch.c".

First, global variables should be defined for function calls. And then, the function used to read the data in the configuration file should be written. The data in the configuration file includes the port number of the database, serial number, and so on. The function used to record the important information of software execution should be written. The thread functions used to operate beidou modules should be written. Before the main function calls, the data structure should be initialized. The logic of the main function will be provided with the pseudo code form, which is as follows:

```
int main (int argc, char **argv)
{
    ...

    ...
    if(fail to create thread 1)
    {
        return -1;
    }
    while(1)
    {
        if(airplane is in an emergency)
        {
            for(n=0; n<3; ++n)
            {
                if(fail to create thread n + 2)
                {
                    return -1;
                }
            }
        }
        ...
```

```
    while (airplane is still in an
    emergency)
    {
        ...
        continue to receive flight data,
        process and store;
        if(threads 2,3,4 have not been
        started)
        {
            Start threads 2,3,4;
            Send the unsent data
            stored in database;
        }
        if(m == 4)
```

```
        {
            m = 0;
            Send the newest data by
            main process;
            if(succeeded to predict the    }
            position of airplane)
            {
                Send the forecast data
                by thread 1;
            }
        }
        delay 15 seconds;
        ++m;
    }
    kill threads 2,3,4;
}
receive flight data, process and store;
...
```

```
        if(num == 4)
        {
            num = 0;
            Send the newest data by main
            process;
            if(succeeded to predict the position
            of airplane)
            {
                Send the forecast data by thread
                1;
            }
        }
        delay 15 seconds;
        ++num;
        ...
    }
    return 0;
```

4.2 Test results and analysis in a laboratory environment

The flight data from the flight simulation system, which simulates the flight, which was under emergency conditions, from Shanghai to Beijing in a laboratory was obtained by the system. A part of the flight data obtained in an airborne system database in an hour is shown in Table 1. The corresponding data received in the ground monitoring system database is shown in Table 2.

The data of the time column from Table 1 are the sample times; however, the actual sample intervals are much smaller than the given intervals. To simplify the problem, only parts of the data are listed here. The data of the time column from Table 2 are the times when the ground monitoring system received the corresponding flight data. The comparison between the data of the time column from Tables 1 and 2 shows that there is a delay in the process of receiving data by the ground monitoring system. But it makes no difference to the use of this system.

The data of the longitude column, latitude column, true airspeed column, pressure column, and acceleration column from Table 1 are the data sampled by the airborne system. These were generated by the flight simulation system, which simulated the flight from Shanghai to Beijing. The data of corresponding columns from Table 2 are the received data corresponding to the flight data. We can see that there is no loss of data and it is proved that the system is reliable to some degree.

The data of the predicted longitude column and the predicted latitude column from Table 1 are the data predicted by the airplane position predicting module. The prediction time is 15 seconds later than the corresponding sample time. The data in Table 3 are the data of the actual longitude and latitude corresponding to the predicted times:

The differences between predicted values and actual values are shown in Table 4 (predicted longitude or latitude minus actual longitude or latitude values).

Table 1. Flight data in an airborne system database.

Time	Longitude/°E	Latitude/°N	True airspeed/Knot	Pressure altitude/Ft	Acceleration/Knot/s	Predicted longitude/°E	Predicted latitude/°N
1:00:00	117.6640	35.1793	273.5	30001	0	117.6565	35.1950
1:06:00	117.4553	35.7465	275.3	30000	0	117.4490	35.7636
1:12:00	117.2502	36.2982	273.5	30000	0	117.2428	36.3178
1:18:00	117.0307	36.8481	273.8	30000	0	117.0233	36.8709
1:24:00	116.8468	37.3898	276.3	30001	0	116.8371	37.4152
1:30:00	116.6431	37.9315	266.8	29996	0	116.6322	37.9603
1:36:00	116.6504	38.4892	276	30005	0	116.6526	38.5222
1:42:00	116.7624	39.0614	280.3	23588	0	116.7701	39.1001
1:48:00	116.6939	39.5849	259	13651	1.7	116.6879	39.6136
1:54:00	116.6196	39.9372	207.8	6228	0	116.6147	39.9605
2:00:00	116.5603	40.2182	171.5	4112	4.0	116.5584	40.2269

Table 2. Corresponding data in the ground monitoring system database.

Time	Longitude/°E	Latitude/°N	True airspeed/Knot	Pressure altitude/Ft	Acceleration/Knot/s	Predicted longitude/°E	Predicted latitude/°N
1:00:28	117.6640	35.1793	273.5	30001	0	117.6565	35.1950
1:06:24	117.4553	35.7465	275.3	30000	0	117.4490	35.7636
1:12:27	117.2502	36.2982	273.5	30000	0	117.2428	36.3178
1:18:30	117.0307	36.8481	273.8	30000	0	117.0233	36.8709
1:24:28	116.8468	37.3898	276.3	30001	0	116.8371	37.4152
1:30:42	116.6431	37.9315	266.8	29996	0	116.6322	37.9603
1:36:35	116.6504	38.4892	276	30005	0	116.6526	38.5222
1:42:18	116.7624	39.0614	280.3	23588	0	116.7701	39.1001
1:48:23	116.6939	39.5849	259	13651	1.7	116.6879	39.6136
1:54:27	116.6196	39.9372	207.8	6228	0	116.6147	39.9605
2:00:47	116.5603	40.2182	171.5	4112	4.0	116.5584	40.2269

Table 3. Data of actual longitude and latitude corresponding to the predicted times.

Time	Actual longitude/°E	Actual latitude/°N	Time	Actual longitude/°E	Actual latitude/°N
1:00:15	117.6544	35.2014	1:36:15	116.6547	38.5114
1:06:15	117.4473	35.7682	1:42:15	116.7669	39.0839
1:12:15	117.2423	36.3194	1:48:15	116.6905	39.6012
1:18:15	117.0238	36.8694	1:54:15	116.6171	39.9493
1:24:15	116.8390	37.4109	2:00:15	116.5591	40.2240
1:30:15	116.6351	37.9528			

Table 4. The differences between predicted values and actual values.

Time	Longitude-difference/°E	Latitude-difference/°N	Time	Longitude-difference/°E	Latitude-difference/°N
1:00:15	0.0021	−0.0064	1:36:15	−0.0021	0.0108
1:06:15	0.0017	−0.0046	1:42:15	0.0032	0.0162
1:12:15	0.0005	−0.0016	1:48:15	−0.0026	0.0124
1:18:15	−0.0005	0.0015	1:54:15	−0.0024	0.0112
1:24:15	−0.0019	0.0043	2:00:15	−0.0007	0.0029
1:30:15	−0.0029	0.0075			

From the differences of longitude and latitude in Table 4, we can see that the predicted values fluctuate around the actual values. A pair of predicted values was chosen from the data samples, which were farthest from the corresponding actual values (116.7701°E, 39.1001°N). And then, the actual distance between the two points can be calculated by using the following equations:

Equation (12) is the conversion equation of latitude and distance:

$$S_{la} = 111000m \tag{12}$$

S_{la} denotes the arc length per degree of latitude. When the arc length is short, the conversion equation of longitude and distance is equation (13):

$$S_{lo} = \frac{\cos B \times (R - B \times (R - r)/90) \times 2\pi}{360} \tag{13}$$

B denotes the latitude of the selected point (accurate to degree), R denotes the radius of the equator, and r denotes the polar axis. And in this work, R = 6378137 m and r = 6356752 m. From the literature, it is known that when the latitude is 39 degree, the arc length per degree of longitude is about 90943.728 m.

According to the above-mentioned information, we can calculate that the distance between the two points, which is about 1822 m. And then we can conclude that the predicted position will fall in a circle with a radius of 2000 m, and the center of the circle is the corresponding actual position. From the analysis of data above, we can know that the airborne system has met the demands of function and performance.

5 CONCLUSIONS

With the development of the beidou satellite navigation system, we are gradually breaking the monopoly of GPS in the navigation domain. The application of the navigation system with independent intellectual property rights in China will be in favor of tour security. The short message communication ability of the beidou satellite navigation system has been applied in the design of an airborne monitoring semi-physical simulation system based on beidou. However, there are still some deficiencies in this system, because the beidou satellite navigation system can only provide services to the Asia-Pacific region. But, we believe that as a secondary system of ground–air communication, it will work better with the development of the beidou satellite navigation system.

REFERENCES

An Le (2015). Boeing 777 airplane ACARS and analysis of missing of Malaysia airline MH370. J. Civil Aircraft Design & Research, (116), 62–67.

Chen Ping, Yu Miao, Yan Hong, & Zheng Kaijun (2014). Inspiration of missing Malaysian airlines flight MH370 for air traffic management. J. Command Information System and Technology, (4), 36–40.

Huang Hailan, Niu Ben (2011). Research of determination of ridge parameters. J. Science of Surveying and Mapping, 36(4), 31–32.

Huang Jianqiang, Ju Jianbo (2011). Development introduction of hardware-in-the-loop simulation. J. Ship Electronic Engineering, (7), 5–7.

Li Zhenxi, Li Jiaxun (2013). Quickly calculate the distance between two points and measurement error based on latitude and longitude. J. Geomatics & Spatial Information Technology, 36(11), 235–237.

Lin Lianlei, Jiang Shouda (2007). Design of universal Interface-class simulation platform for bus-style avinics. J. Journal of System Simulation, 19(7), 1485–1488.

Liu Yanyu, Li Jianwei, Liu Xiaogang (2007). Determination of partial ridge parameters. J. Journal of Zhengzhou Institute of Surveying and Mapping, 24(6), 413–418.

Lu Yangwei, Wang Zhenjie (2015). Determination the ridge parameter in ridge estimation using U-curve method. J. Journal of Navigation and Positioning, 3(3), 132–134.

Wang Gaitang, Li Ping, Su Chengli (2011). ELM ridge regression learning algorithm of ridge parameter optimization. J. Information and Control, 40(4), 497–500.

Wang Guofan, Tang Xuefeng, Xue Erjian (2012). Optimization of multivariable linear partial least square regression model and its application in sports measurement. J. Journal of Shanghai University of Sport, 36(2), 37–41.

Zhou Dexin, Peng Nina, Fan Zhiyong, & Zhang Jingyu (2015). Research of flight parameters online automatic monitoring technology. J. Computer Measurement & Control, 23(8), 2689–2695.

Zhu Yongxing, FENG Laiping, JIA Xiaolin, & et al (2015). The PPP precison analysis based on BDS regional navigation system. J. Acta Geodaetica et al Cartographica Sinica, 44(4), 377–383.

Advances in Energy Science and Equipment Engineering II – Zhou, Patty & Chen (Eds)
© *2017 Taylor & Francis Group, London, ISBN 978-1-138-71798-5*

A study on CFD simulation of flow rate, sand concentration, and erosion rate in double cluster hydraulic ejectors on fracturing

Shaokai Tong & Zuwen Wang
CNPC Chuanqing Drilling Engineering Company Limited, Changqing Downhole Technology Company, Xi'an, China

Yihua Dou & Zhiguo Wang
School of Mechanical Engineering, Xi'an Shiyou University, Xi'an, China

ABSTRACT: Hydraulic Jetting Perforation and Fracturing (HJPF) for the horizontal well becomes a highly efficient measure for EOR. The flow rates of fluid and sand concentrations at outlets of nozzles are quite non-uniform. The nozzle which has a larger flow rate and sand concentration may fail due to serious erosion. In this project, the flow distribution of the liquid and solid in a double cluster hydraulic ejector was analyzed by using the Euler–Lagrange multiple phase flow simulation method. The differences of flow rate, sand concentration, and spray gun erosion rate of different nozzles with different diameters and numbers were investigated. The results show that the mass flow and sand concentration through the nozzle are relatively homogeneous as the nozzles number is four and six in the upstream and downstream injectors, respectively. As the diameter of the upstream nozzle and downstream nozzle is 5.5 and 6.3 mm respectively, the sand concentration is more homogeneous between the upstream and downstream nozzle relatively. The simulation results provide a significant reference for the optimization of the structure design of the tool.

1 INTRODUCTION

The hydraulic jetting multiple cluster fracturing technique for horizontal wells is a highly efficient measure for EOR, especially for low permeability reservoirs (Li, 2014). Meanwhile, with an increase in the difficulty of oil and gas reservoir exploitation, large discharge and high sand concentration hydraulic jetting fracturing technology is widely used in the reservoir construction process (Tian, 2008).

A double cluster hydraulic ejector is a typical fracturing tool for horizontal wells; it consists of the upstream and downstream stage guns through two or three tubing connection. The packer is set in the bottom of the downstream spray gun; each spray gun also has four nozzles. In the process of large displacement and high sand ratio hydraulic jet fracturing, the flow rate and sand concentration through the upstream and downstream nozzles are quite non-uniform, which results in an uneven erosion damage on the inside of the gun, on one hand. On the other hand, it causes serious local erosion damage of the gun.

The Discrete Particle Model (DPM) method is widely used in erosion simulation in a multiphase flow (Chen, 2004; Shah, 2004; Bozzini, 2003). In this paper, the liquid and solid two-phase flow field distribution and erosion rate of the inner wall of the ejector is concentrated by using the Euler–Lagrange simulation method for hydraulic jetting conditions. When compared with the flow field distribution and erosion rate of the upstream and the downstream ejectors with different nozzles and diameters, the structural optimization nozzles of the double cluster hydraulic ejector was determined based on the erosion analysis.

2 CFD MODELING AND PROCEDURE

A 3D computer-aided engineering drawing package is shown in Fig. 1. It is reproduced to be identified as the typical computational domain. And then, it is imported into the ANSYS FLUENT CFD software, remodeled into different sections, and refined to generate a finite volume meshing. The total element count is approximately 241,7640, as shown in Fig. 2.

The turbulent flow in the process of hydraulic sand blasting perforation is a complex two-phase flowing process combined with the non-Newtonian fluid and solid particles. For the DPM method, a power law non-Newtonian model is taken as the

Figure 1. Schematic of a single stage spray gun of the hydraulic ejector.

Figure 2. Meshing model of the double cluster hydraulic ejector.

continuous phase and the quartz sand is deemed as the discrete phase. And then, the motion equations of the fluid phase in Euler coordinates are solved, and the particle moving is tracked by solving the particle mobbing equation in the Lagrange coordinate system.

Equation (2) describes the momentum equation of an average two-phase flow. The interaction between the fluid and particle was considered by the drag force. The mass continuity and momentum equation are shown as follows (Atul Bokane, 2013; Huang, 2014):

$$\frac{\partial}{\partial t}(\alpha\rho) + \nabla \cdot (\alpha\rho\vec{V}) = 0 \tag{1}$$

$$\frac{\partial}{\partial t}(\alpha\rho\vec{V}) + \nabla \cdot (\alpha\rho\vec{V}^2) = -\alpha\nabla p + \nabla \cdot (\alpha\vec{\tau}) + \alpha\rho\vec{g} - \vec{F}_d \tag{2}$$

where ρ is the fluid density, kg/m³; $\vec{V}$ is the fluid velocity, m/s; $\vec{g}$ is the gravity acceleration, m/s²; p is the pressure, pa; $\vec{\tau}$ is the fluid stress tensor; α is the fluid volume fraction; and $\vec{F}_d$ is the average fluid drag for particle. The standard k-ε model is used to describe the turbulence flow.

The DPM method is used in the Lagrange coordinate system. The equation of particle motion can be shown as follows (Zhang, 2006):

$$m_p \frac{d\vec{V}_p}{dt} = \vec{F}_d + \vec{F}_m \tag{3}$$

where $\vec{F}_m$ is the mass force considering buoyancy and m_p and $\vec{V}_p$ are the mass and velocity of a single particle. A random walk model (DRW) is used to describe the particle phase turbulence.

The erosion rate is generally related to the particle's impact velocity, impact angle, and particle geometry parameters, and the erosion rate formula is determined by using the following formula (Oka Y, 2005; Oka Y, 2005):

$$ER = \sum_{i=1}^{n} \frac{q_{p,i} C(d_{p,i}) f(\theta) \vec{V}_{p,i}^{b(\vec{V}_{pi})}}{A} \tag{4}$$

where $C(d_p)$ is a function related to particle diameter; θ is the particle impact angle on the wall (invasion angle); $f(\theta)$ is the invasion angle correction function; $\vec{V}_{p,i}$ is the collision speed of the i particles, m/s; $b(\vec{V}_p)_{p,i}$ is a function of particle collision velocity; n is the number of particles; A is the area of the wall calculation unit, m²; $q_{p,i}$ is the mass flow rate of particles, kg/s.

3 MODEL ESTABLISHMENT, MESHING, AND BOUNDARY CONDITION SETTING

The model of the double cluster hydraulic ejector is shown in Fig. 1. Its nozzles are distributed at 120 degree at intervals and the diameter is 5.5 mm. The inner wall of the casing pipe is added to describe. The model is divided into grids according to the domain, an inner ejector uses a tetrahedral mesh and the mesh is encrypted at exit, and the surface layer of the flow field uses the prismatic grid. The meshing model is established, as shown in Fig. 2.

The boundary conditions include using steady turbulent flow, fixed wall, and variable velocity at the inlet and outflow at the outlet. The liquid phase follows non-Power law and is a non-Newtonian fluid and the solid phase is quartz sand, in which the quartz sand density is 2650 kg/m³, while the particle size is 40–60 mesh. The parameters, K and n of the power-law fluid are constant with values such as 0.0277 and 0.82, respectively.

4 RESULTS AND DISCUSSION

4.1 Effect of nozzle number on the flow field and erosion rate of the double cluster hydraulic ejector

Each nozzle flow field distribution of a multistage hydraulic spray gun often exhibits heterogeneity during perforation. The displacement is defined by 2.8 m³/min and the sand content is 160 kg/m³. The difference of liquid and solid two-phase flow field and erosion rate under different nozzle quantities is investigated. The nozzle number uses the following three combination projects.

Scheme A: the upstream and downstream ejectors all use six nozzles;

Scheme B: the upstream ejector uses six nozzles and the downstream ejector uses four nozzles;

Scheme C: the upstream ejector uses four nozzles and the downstream ejector uses six nozzles.

The overall number of nozzles for Scheme B and C is reduced to 10 when compared with Scheme A. It can be seen from Fig. 3 that the mass flow, G, increases when the number of nozzles, N, decreases. The heterogeneity of mass flow is obtained by the difference of the average mass flow of each nozzle, which is shown in Fig. 4. Among the three schemes, the mass flow through the superior nozzle is higher than that of the downstream nozzle.

Particle concentrations of the nozzles with different nozzle numbers are shown in Fig. 5. The sand concentration through the downstream nozzles is larger than that of the upstream nozzle. When the two nozzles of the downstream ejector are blocked (Scheme B), the sand concentration through the upstream nozzles are higher than that of Scheme A slightly. Obviously, the sand concentration of the downstream nozzles increases. When the two nozzles of the upstream ejector are blocked (Scheme C), the sand concentration through the upstream nozzles do not change slightly when compared with A; and the downstream nozzles decrease obviously. The main reason is that the perforation fluid accumulates at the bottom of the ejector, static pressure rises and dynamic pressure reduces, the inlet flow rate of the downstream nozzle is smaller than that of the upstream nozzle, which causes sand to accumulate at the rear end of the gun.

It can be seen from Fig. 6 that when the numbers of the upstream and downstream nozzles are six, the average erosion rate of the wall is uniform

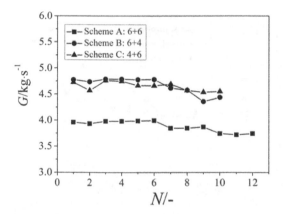

Figure 3. Graph showing the mass flow rates of nozzles with different numbers.

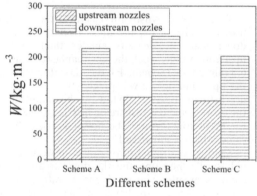

Figure 5. Graph showing particle concentrations of nozzles with different numbers.

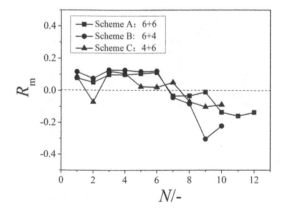

Figure 4. Graph showing the mass flow heterogeneity degrees of the nozzle with different numbers.

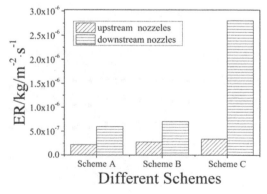

Figure 6. Graph showing the average erosion rates on the inner wall of the ejector with different numbers.

and minimal. Among the three schemes, erosion rates on the inner wall of the downstream ejector are higher than that of the upstream ejector, the average erosion rates of Scheme C are the largest. In summary, Scheme A is better than the other two schemes.

4.2 Effect of the nozzle diameter on the flow field and erosion rate of the double cluster hydraulic ejector

The nozzle diameters of the upstream and downstream spray gun use the following three combination schemes:

Scheme D: The nozzle diameters of upstream and down ejectors are all 5.5 mm;

Scheme E: The diameter of the upstream nozzles is 5.5 mm and the downstream nozzles is 6.3 mm in the ejector;

Scheme F: The upstream ejector uses 6.3 mm nozzles and the downstream ejector uses 5.5 mm nozzles.

When the downstream ejector uses 6.3 mm nozzles (Scheme E), the mass flow through the downstream nozzle was greater than that of upstream nozzle; when the upstream ejector uses a 6.3 mm nozzle (Scheme F), the mass flow through the upstream nozzle was greater than that of the downstream nozzle; obviously, the diameter of the upstream and downstream nozzle are the same, the mass flow through the nozzle is basically the same, which is shown in Fig. 7.

Sand concentration through the upstream nozzle is basically the same and the downstream nozzle changes greatly. The sand concentration of the upstream is much smaller than that of the downstream. When the diameter of the upstream nozzle is 5.5 mm and the downstream nozzle diameter is 6.3 mm (Scheme E), the sand concentration difference is little; when the upstream nozzle diameter is 6.3 mm, and the downstream nozzle diameter is 5.5 mm (Scheme F), there is an obvious difference in the sand concentration, as shown in Fig. 8. Fig. 9 represents sand concentrations on different nozzle diameters. The average value of the sand concentration is shown in Fig. 10.

It can be seen from Fig. 10 that when the diameter of the upstream nozzle is 6.3 mm and the downstream nozzle is 5.5 mm, the erosion rate of the inner wall is uniform and minimal. When the diameter of the upstream nozzle is 5.5 mm and downstream nozzle is 6.3 mm, the erosion rate is heterogeneous and is the largest. Therefore, Scheme F is better than the other two schemes. These results are in agreement with the former simulations (Mclaury B S, 2011; Chong, 2013; Chen, 2004).

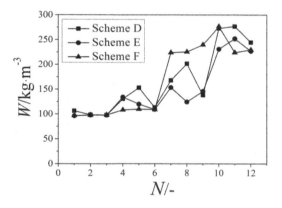

Figure 8. Graph showing sand concentrations of nozzles with different diameters.

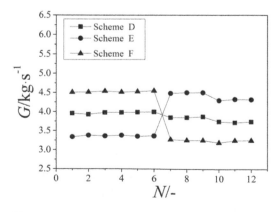

Figure 7. Graph showing the mass flow of nozzles with different diameters.

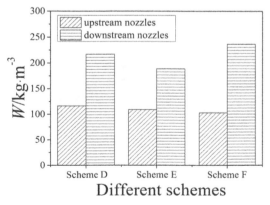

Figure 9. Graph showing particle concentrations of nozzles with different diameters.

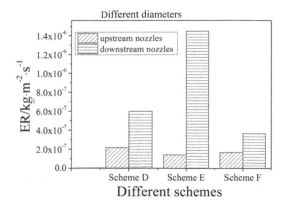

Figure 10.   Graph showing average erosion rates on the inner wall of the ejector with different diameters.

5   CONCLUSIONS

A CFD model and theoretical fluid model are simultaneously used, and the following conclusions can be drawn from the CFD results:

1. Simulation results shows that reducing the nozzle number did not significantly improve the heterogeneity of flow and erosion rates during the nozzles. By comparison, the scheme of four nozzles in the upstream and six nozzles in the downstream has a more uniform flow rate and sand concentration than the other two schemes.
2. It cannot significantly improve the heterogeneity of the flow rate distribution and the erosion rate of the upstream and downstream nozzles by increasing the diameter of nozzles. By comparison, Scheme E with 5.5 mm in upstream and 6.3 mm in downstream nozzles has a more uniform flow rate and sand concentration when compared to the other two schemes.

ACKNOWLEDGEMENT

Project supported by the Natural Science Foundation of Department of shaanxi province, China (2015JM5223).

REFERENCES

Atul Bokane, Siddhath J, Yogesh D, et al. Computational Fluid Dynamics (CFD) Study and Investigation of Proppant Transport and Distribution in Multistage Fractured Horizontal Wells[C]//SPE Reservoir Characterization and Simulation Conference and Exhibition. Society of Petroleum Engineers, 2013.

Bozzini B, Ricotti M E, Boniardi M, et al. Evaluation of erosion–corrosion in multiphase flow via CFD and experimental analysis [J]. Wear, 2003, 255(1–6):237–245.

Chen X, Mclaury B S, Shirazi S A. Application and experimental validation of a computational fluid dynamics (CFD)-based erosion prediction model in elbows and plugged tees [J]. Computers & Fluids, 2004, 33(10):1251–1272.

Chong Y. Wong, Christoppher Solnordal, et al. Experimental and computational modelling of solid particle erosion in a pipe annular cavity [J]. Draft Manuscript for Journal of Wear, 2013:1–6.

Huang Z, Gensheng L I, Tian S, et al. Wear investigation of downhole tools applied to hydra-jet multistage fracturing [J]. Journal of Chongqing University, 2014.

Li Longlong, Yao + Jun, Li Yang, et al. Productivity calculation and distribution of staged multi-cluster fractured horizontal wells [J]. Petroleum Exploration & Development, 2014, 41(4):504–508.

Mclaury B S, Shirazi S A, Viswanathan V, et al. Distribution of Sand Particles in Horizontal and Vertical Annular Multiphase Flow in Pipes and the Effects on Sand Erosion [J]. Journal of Energy Resources Technology, 2011, 133(2):180–190.

Oka Y, Okamura K, Yoshida T. Practical estimation of erosion damage caused by solid particle impact. Part 1: Effect of impact parameters on predictive equation [J]. Wear, 2005, 259:95–101.

Oka Y, Okamura K, Yoshida T. Practical estimation of erosion damage caused by solid particle impact. Part 2: Mechanical properties of materials directly associated with erosion damage [J]. Wear, 2005,259:102–109.

Shah S N, Jain S. Coiled tubing erosion during hydraulic fracturing slurry flow [J]. Wear, 2004, 264(s 3–4): 279–290.

Tian Shouzeng, Gensheng L I, Huang Z, et al. Research on hydrajet fracturing mechanisms and technologies [J]. Oil Drilling & Production Technology, 2008.

Xianghui Chen, Brenton S, et al. Application and experimental validation of a computational fluid dynamics (CFD)-based erosion prediction model in elbows and plugged tees [C]. Computers & Fluid, 2004, 33:1252–1254.

Zhang Yongli. Application and improvement of Computational Fluid Dynamics (CFD) in solid particle erosion modeling [D]. Oklahoma: The University of Tulsa, 2006.

Advances in Energy Science and Equipment Engineering II – Zhou, Patty & Chen (Eds)
© 2017 Taylor & Francis Group, London, ISBN 978-1-138-71798-5

Influence of geometry and operating parameters on the flow field in a cyclone furnace

J.X. Guo, D.Q. Cang, L.L. Zhang, D. Wang & H.F. Yin
School of Metallurgical and Ecological Engineering, University of Science and Technology Beijing, Beijing, China

ABSTRACT: In this paper, the influence of different geometric and operation parameters on the flow field in a cyclone furnace is studied, such as the tuyere positions, the air supply forms, and the air velocity. Under these conditions, the RSM model of the Fluent software was used to simulate the flow field in the cyclone furnace, in order to determine the optimum furnace structure and operation parameters. Simulating results show that the inlet air tuyere is suitably fixed tangentlally on opposite sides, which is conducive to improve the slag captured rate of the cyclone furnace and reduce the slag formation on the wall and air tuyere. Besides, the residence time of fuel particles also can be prolonged, which is beneficial to increase the combustion efficiency of the cyclone furnace.

1 INTRODUCTION

Although the thermal state experiment is necessary to reflect the real performance of fuel burner combustion, it requires lot of manpower, raw materials, and financial support. Therefore, the development of the thermal state experiment is limited, especially for the pulverized coal combustion in a cyclone furnace, in which a complex three-phase flow movement is formed and the flow field distribution is difficult to be obtained. These drawbacks create a lot of difficulties in promoting the development of burners.

However, fluid mechanics, heat transfer, combustion form, chemical thermodynamics, chemical kinetics, and numerical methods provide the basis for the numerical simulation of the combustion process. The numerical simulation has the following advantages such as lower usage of funds, short calculation time, and can reveal the internal structure of a fluid deeply and comprehensively. Besides, it can be limited by the experimental conditions and measurement methods, and also it has strong simulation accuracy. More detailed data can be obtained when compared to the engineering experiment, by which the experimental data can be predicted on time.

Based on the simulated aerodynamic characteristics of the cyclone furnace at different operation parameters, we have found the optimum geometry and air parameters for the cyclone combustion of the pulverized coal, and thus provide a reference for relevant scholars or engineers to the hot stage experiments in the cyclone furnace.

2 MODEL DESCRIPTION

2.1 Geometry and meshing

The geometry of the cyclone furnace used for simulation in this study is presented in Figure 1a, and the parameters investigated are listed in Table 1.

ANSYS Meshing is used for meshing the geometries into hexahedral meshes, with a grid dependency with 300000 cells. In this paper, we chose the discrete QUICK format control equations and the pressure interpolation scheme in PRESTO format. At the initial phase, in this study, no particles are injected. But the influence of particles on the flow field will be detailed and discussed in other studies.

2.2 The choice of the turbulence model

Currently, the numerical simulation model about swirling flow includes the standard k-ε turbulence model, RNG k-ε turbulence model, and RSM (Reynolds Stress Model) turbulence model. The standard k-ε turbulence model is half the empirical model, and is mainly based on the turbulent kinetic energy and turbulent diffusivity equation, and thus the simulation model is only applicable to a fully turbulent flow stream; The RNG k-ε turbulence model is in relation to standard k-ε turbulence model improvements. However, this improvement is still not abandoned on the basis of eddy viscosity hypothesis, and so the forecast result is limited; the RSM turbulence model completely abandoned the eddy-viscosity hypothesis and completely solves the differential transport

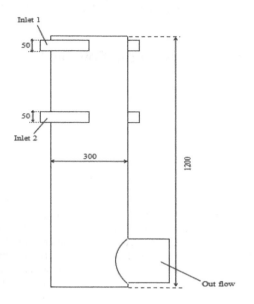

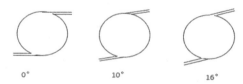

| 0° | 10° | 16° |

Figure 1a. A schematic diagram of the chamber (unit: mm).
Figure 1b. Schematic showing the angle of the inlet 1 and inlet 2.

Table 1. List of parameters investigated in this study.

Structural parameter	Value
Diameter of the cyclone furnace/mm	300
Size of the inlet 1/mm	50 × 8
Size of the inlet 2/mm	50 × 10
The angle of the inlet 1 and inlet 2	0°, 10°, 16°
Blast capacity of inlet 1/ (m³/h)	20
Blast capacity of inlet 2/ (m³/h)	90

equations of Reynolds stress, thereby considering the influence of the wall face of Reynolds stress distribution; when compared with the other two models, the RSM is more compliant with the characteristics of anisotropy in a swirling flow field. The RSM model has obvious advantages with three-dimensional strong swirl and reflux in a cyclone furnace, therefore, the RSM model is used numerical simulation in this paper.

3 RESULTS AND DISCUSSION

3.1 The effect of the flow field with the air supply with different forms of tuyere

In general the air inlet mode of the cyclone furnace comprises unilateral and bilateral modes to receive the air. The internal fluid motion in a cyclone furnace is a complex three-dimensional spiral motion. The velocity of one point can be divided into tangential velocity, axial velocity, and radial velocity; tangential velocity and axial velocity are the main factors, which influence the separation performance of the cyclone furnace particle, and so we mainly study the distribution of these two kinds of velocity. By analyzing the variation of tangential and axial velocities, the difference of the velocity field in the cyclone furnace with different air inlet modes can be obtained when compared with the tangential inlet air, and the best air inlet mode can be determined.

3.1.1 Tangential velocity

Solid particles with the air flow move in circular motion inside the cyclone furnace, and due to the centrifugal force, the solid particles are thrown to the wall. The tangential velocity of the air flow is the supplier of the centrifugal force of the solid particles, and the magnitude of the value represents the ability of the air flow to exert a centrifugal effect on the solid particles; the greater the tangential velocity of the air flow, the greater is the probability that the particles are thrown to the wall, and the better is the particle separation performance of the cyclone furnace.

The tangential velocity distribution of the primary and secondary air inlet is shown in Figure 2. It shows that, as the radius increases, the tangential velocity on the same section increases and the centrifugal force on the particles increases gradually in the forced vortex area of the flow field, which helps the particle to move to the wall; as the radius increases, the tangential velocity on the same section decreases and the drag force of the particles decreases in the free vortex area of the flow field, which helps in collection of the particles.

From Fig. 3, the tangential velocity distributions of the two different inlet modes on the same section can be seen and also it can be observed that the tangential velocity of the bilateral mode in the air is too large at the same coordinate position. Therefore, the particles in the two-side air flow within the cyclone furnace movement can exert a greater centrifugal force and enhanced cyclone furnace particle separation performance.

Although the tangential velocity distributions of the primary and secondary air inlet have a good symmetry in different inlet modes, the position of

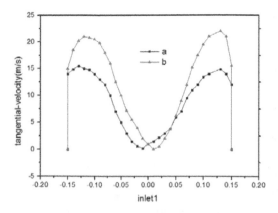

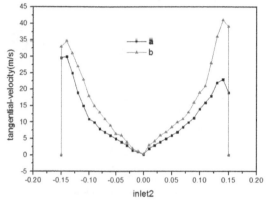

Figure 2. Graphs showing the distributions of tangential velocities in different inlets (a. unilateral into the air and b. bilateral into the air).

the maximum tangential velocity point fluctuates, and the radius of the cutoff point has a certain effect on the particle separation efficiency.

In general, the larger the vortex radius, the smaller is the centrifugal separation zone, and the lower is the separation capacity; The smaller the radius of the vortex core, the greater is the centrifugal separation area, and the higher is the separation capacity; this is because the flow field in a cyclone furnace can be divided into a centrifugal trap zone composed of free vortices and an escape flow area composed of a forced vortex, which is the boundary of the tangential velocity of the largest vortex core radius re. When the radius of the vortex core decreases, it implies that there is an increase of the centrifugal trap area and the reduction of the escape flow area, and an increase in the particle separation efficiency of the cyclone furnace.

From Figure 2, it can be observed that the vortex radius is smaller when the distribution is bilateral into the air, and so the cyclone furnace

centrifugal separation area is greater, and the separation capacity is higher.

3.1.2 Axial velocity

The axial velocity distribution cloud at $Z = 0$ cross-section is shown in Figure 3. When the airflow enters the cyclone furnace, it moves downward, and the particles are separated while moving to the lower part of the cyclone furnace. The downward axial velocity of the air flow plays a role in the downward discharge of particles. In the lower part of the cyclone furnace wall, the downward axial velocity is large, which is advantageous to prevent the powder particles from slagging at the wall surface. Due to the weakening of the powder particles at the wall, slagging can reduce secondary dust flying, thereby improving the efficiency of particle separation.

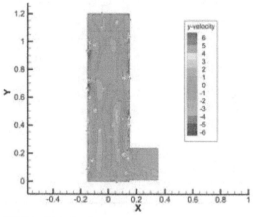

a) Unilateral into the air

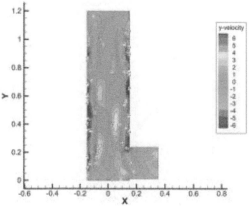

b) Bilateral into the air

Figure 3. Graphs showing the axial velocity distribution cloud at the $Z = 0$ cross-section.

From a comparison of axial velocity changes with a and b in two different ways into the air can be seen that when the tuyere is arranged as a double inlet, the downward velocity area near the wall is larger. Therefore, the bilateral distribution into the air is more conducive to improve particle separation efficiency.

In addition, when the tuyere is arranged on both sides of the inlet air, the area near the center of the lower part of the cyclone furnace appears at the upward flow of the axial velocity. The presence of the recirculation zone prolongs the residence time of the particles in the cyclone furnace, which is beneficial to improve the combustion efficiency of coal particles.

3.1.3 *Summary*

In summary, a comparison of the cyclone furnace tangential velocity and axial velocity changes with different air modes when the air enters into the cyclone furnace with tangential velocity is made.

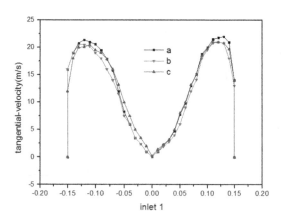

inlet 1

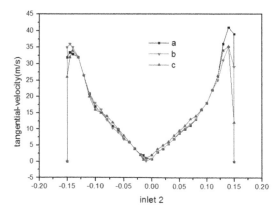

inlet 2

Figure 4.   Graphs showing the distributions of tangential velocities in different cross-sections (a. tangential, b. cutting angle = 10°, and c. cutting angle = 16°).

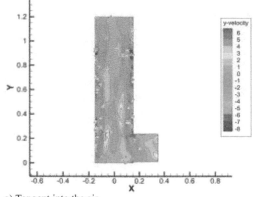

a) Tangent into the air

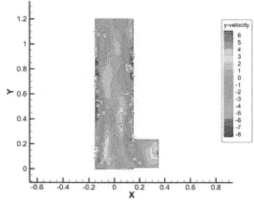

b) Secant into the air (cutting angle =10°)

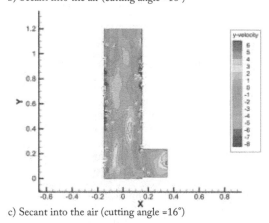

c) Secant into the air (cutting angle =16°)

Figure 5.   Graphs showing the axial velocity distribution cloud in the Z = 0 section.

When the tuyere is arranged on both sides of the inlet, the tangential velocity and the axial downward velocity area are larger, which is helpful to improve the particle separation efficiency of the cyclone furnace. In addition, the reflux zone appears in the cyclone furnace, which is beneficial to prolong the residence time of the coal particles in the furnace and improve the combustion efficiency.

3.2 Influence of different inlet angles on the air flow in the furnace

A cyclone furnace tuyere layout is generally divided into tangent and secant. Because the way of introduction is different, the distance between the center line of the cylinder and the center line of the air outlet is different.

3.2.1 Tangential velocity

The tangential velocity distributions of the primary and secondary air inlet sections of the cyclone furnace are shown in Figure 4 when bilateral air inlet is used with different tuyere layouts. We can see that the tangential velocity distributions of the same cross-section of the cyclone furnace are almost the same, but the maximum tangential velocity value fluctuates. With an increase in the inlet angle, the maximum tangential velocity decreases, and therefore, when the air inlet with tangential inlet, it is helpful to improve the efficiency of particle separation in the cyclone furnace.

3.2.2 Axial velocity

The axial velocity distribution cloud in the $Z = 0$ section is shown in Figure 5. From a comparison of a, b, and c axial velocity changes, it can be seen that the axial velocities change little with an increase in the air inlet angle.

3.2.3 Summary

In summary, with an increase in the inlet angle, the maximum tangential velocity decreases, the axial velocities change little with an increase in the air inlet angle. Therefore, when the air inlet is provided with a tangential inlet, it is helpful to improve the efficiency of particle separation in the cyclone furnace.

4 CONCLUSION

1. The RSM model in the FLUENT software can effectively simulate the flow field of the cyclone furnace inside.

2. Bilateral inlet air is beneficial to improve the particle separation efficiency in the cyclone furnace, reduce the slagging phenomenon of the fuel particles at the wall surface, prolong the residence time of the coal particles in the furnace, and improve the combustion efficiency.

3. When the air inlets are tangential, the maximum tangential velocity of the same section decreases as the inlet angle increases, and there is no obvious change in the axial velocity with an increase in the inlet angle, which is beneficial to improve the particle separation efficiency of the cyclone furnace and reduce the slag formation in the tuyere.

REFERENCES

Cortés, C., & Gil, A. 2007. Modeling the gas and particle flow inside cyclone separators. *Progress in Energy & Combustion Science* 33(5): 409–452.

Cui, B.Y., Wei, D.Z., Gao, S.L., Liu, W.G., & Feng, Y.Q. 2014. Numerical and experimental studies of flow field in hydrocyclone with air core. *Transactions of Nonferrous Metals Society of China* 24(8): 2642–2649.

Delgadillo, J.A., & Rajamani, R.K. 2005. A comparative study of three turbulence-closure models for the hydrocyclone problem. *International Journal of Mineral Processing* 77(4): 217–230.

Hoekstra, A.J., Derksen, J.J., & Akker, H.E.A.V.D. 1999. An experimental and numerical study of turbulent swirling flow in gas cyclones. *Chemical Engineering Science* 54(s 13–14): 2055–2065(11).

Liu, Z., Zheng, Y., Jia, L., & Zhang, Q. 2007. An experimental method of examining three-dimensional swirling flows in gas cyclones by 2d-piv. *Chemical Engineering Journal* 133(1–3): 247–256.

Majumdar, S. 1988. Role of underrelaxation in momentum interpolation for calculation of flow with nonstaggered grids. *Numerical Heat Transfer Fundamentals* 13(1): 125–132.

Mousavian, S.M., & Najafi, A.F. 2009. Influence of geometry on separation efficiency in a hydrocyclone. *Archive of Applied Mechanics* 79(11): 1033–1050.

Shin, T.M., & A.L. Ren. 1984. Primitive-variable formulations using nonstaggered grids. *Numerical Heat Transfer Applications* 7(4): 413–428.

Wang, B., Xu, D.L., Chu, K.W., & Yu, A.B. 2006. Numerical study of gas–solid flow in a cyclone separator. *Applied Mathematical Modeling* 30(11): 1326–1342.

Advances in Energy Science and Equipment Engineering II – Zhou, Patty & Chen (Eds)
© 2017 Taylor & Francis Group, London, ISBN 978-1-138-71798-5

A study on the temperature field simulation of turbocharger turbines

F. Cao, S.R. Wang, G.Q. Wang, P.Q. Guo & X.S. Wang
School of Mechanical Engineering, University of Jinan, Jinan, Shandong Province, China

ABSTRACT: A three-dimensional model of a turbocharger turbine was established. And the temperature change of turbine working in a high temperature environment was simulated by using finite element method. The temperature distribution of the turbine surface and the temperature change within the turbine were obtained. This study provides valuable reference information about thermal stress, strength, and service life of turbines for the automotive industry, and the results can be used to improve the structure and material of the turbine.

1 INTRODUCTION

Turbochargers have been widely used in automobiles as effective devices to improve the engine power; therefore, the auto industry has drawn great attention to the improvement of turbocharger. The turbine, which is the core component of a supercharger, works in a high temperature and high speed environment. The strength of the turbine has decisive influence on the service life of the supercharger. Consequently, it is necessary to analyze the influence of temperature on the reliability of the turbine. It must consider the thermal coupling problem at first when studying the effect of temperature on the reliability of the turbine blades (Librescu, 2005). The thermal stress produced by the alternating change of temperature is the main reason that induces crack initiation and extension, thereby causing failure of the turbine (Jia, 2010).

However, there are few studies that are focused on the simulation of the temperature field and thermal stress of the turbine. Research on the temperature field is a significant technical means in engineering applications. For example, when designing an engine, steam turbine, and turbine, which work under high temperature conditions, the temperature field and thermal stress analysis are used as momentous reference factors.

In this paper, the turbine working at high temperature was simulated and then the surface and internal temperature distribution of the turbine was obtained.

2 THE STEADY STATE THERMAL ANALYSIS BY USING THE FINITE ELEMENT METHOD

The thermal analysis includes temperature, convection, heat flux, and heat generation according to different load types. The steady state thermal analysis is used to analyze the effect of stabilized thermal load on components, and reflect the heat transfer inside the parts. It is defined as, if the heat flowing into the system and the heat generated by the system itself are equal to the heat flowing out of the system, then the system is in a thermally stable state. The steady thermal analysis is independent of the process of temperature change, whereas it is only related to the temperature distribution at last. The energy equilibrium equation of steady thermal analysis is shown as follows (Zhang, 2013):

$$[K]\{T\} = \{Q\} \tag{1}$$

where, $[K]$ is the transmission matrix, including thermal conductivity coefficient, convection coefficient, radiation coefficient, and shape factor, $\{T\}$ is the node temperature, and $\{Q\}$ is the node heat flow rate vector, including heat generated.

3 MODELING AND ANALYSIS

The supercharger turbo used in this study has nine leaves, which have a complex blade structure. The

Figure 1.   The 3D model of the turbine.

size of the turbine was obtained by using a measuring instrument. The diameter of the turbine is 40 mm and the blade thickness is 1 mm. In order to facilitate finite element analysis, the entity structure was simplified appropriately. The 3D model was constructed by Solidworks software, thereby ignoring the small fillet and thread parts, as shown in Figure 1.

The 3D model of the turbine was imported into ANSYS, which is professional finite element analysis software. The thermal module was employed to conduct thermal and stress analysis. The material properties of K418 are listed in Table 1. The model was meshed with tetrahedron elements (see Figure 2).

The turbine is impacted by the high temperature gas produced by the engine. In general, the temperature of the gas is between 600°C and 1000°C, and the heat is transferred from the turbine surface into the interior. Due to the presence of a cooling lubricant on the turbine shaft, heat transfer occurs between the turbine and the shaft. When the gas arrives at the outlet, the gas temperature is just about 70% of the initial temperature. At last, the thermal distribution of the turbine will reach the dynamic steady state. In the present study, the heat convection load was adopted, and five different values of the inlet gas temperature were loaded on the model (Du, 2010). The convective heat transfer coefficient between the gas and turbine is 135 W/(m²·°C), and that between the turbine and the cool-

Table 1. List of material properties of K418.

Density Kg/m³	Elasticity modulus GPa	Poisson's ratio	Thermal conductivity W/m·°C
7800	176	0.3	67

Table 2. List of load parameters.

Number	Entrance gas temperature °C	Exit gas temperature °C
1	600	420
2	700	490
3	800	560
4	900	630
5	1000	700

ing oil is 4448 W/(m²·°C) (Han, 2008; Yang, 2012), as shown in Table 2.

4 RESULTS AND DISCUSSION

After solving, the temperature distribution can be obtained by post-processing, which is shown in Figure 3.

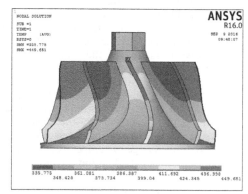

(1) 600°C

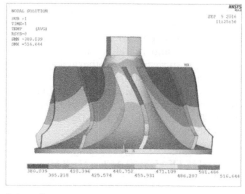

(2) 700°C

Figure 3. (continue)

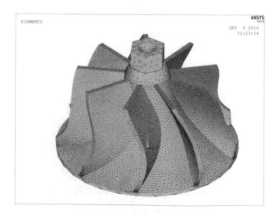

Figure 2. Schematic of meshing of the model.

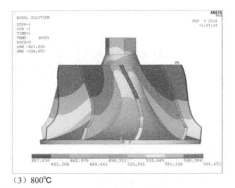

（3）800℃

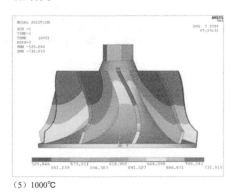

（4）900℃

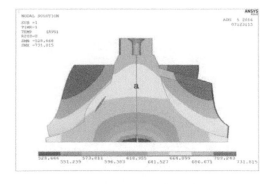

（5）1000℃

Figure 3.   Temperature distributions of the turbine under five load conditions.

It can be seen from Figure 3 that the maximum temperature appeared on the turbine blade tip. This phenomenon is because the double faces of the blade contact the hot gas at the same time and the blade tip is most far from the cooling region. Consequently, the heat was accumulated on the blade tip.

The minimum temperature was recorded on the contact interface of the turbine and shaft. It was because the heat was transferred to the interior of the turbine and shaft, and rare heat was accumulated around the contact interface.

The maximum and minimum temperature changes are shown in Figure 4. K_{max} represents the maximum temperature and K_{min} represents the minimum temperature. As can be seen from Figure 4, K_{min} is about 52–56% of the intake air temperature and K_{max} is about 73–75% of the intake air temperature. The gap between K_{max} and K_{min} increases with an increase in the ambient temperature. The creep temperature of K418 is generally greater than 900°C, and so the effect of the creep was ignored when calculating the service life of the turbine in this research work.

In order to check the temperature distribution within the turbine, the cutaway drawing of the turbine is shown in Figure 5. The variation of temperature along the axial path was extracted, as shown in Figure 6. Obviously, the temperature inside the turbine increased gradually from the bottom to the top, and the lowest temperature was recorded on the contact face. These results are broadly in line with Wang's computational results (Wang, 2012).

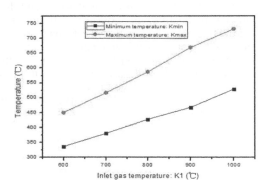

Figure 4.   Graph showing the minimum and maximum temperature curve.

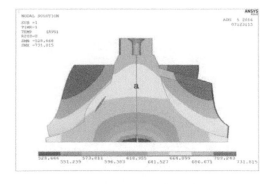

Figure 5.   Cutaway drawing of the turbine.

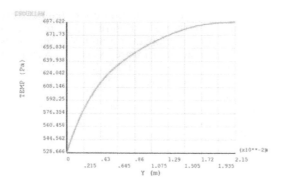

Figure 6.   Graph showing the axial temperature distribution curve.

5   CONCLUSION

In the present study, five different turbine models with different initial gas temperatures were established. The temperature distribution of the turbine surface was obtained by using the finite element analysis method. The maximum temperature is less than the creep temperature of K418, and thus the creep effect was not considered. In addition, the temperature field within the turbine and the axial temperature distribution curve was obtained, which presented the temperature distribution and a changing trend inside the turbine. This study provides a valuable reference to engineers to design and optimize the materials and structure of the turbocharger turbine.

ACKNOWLEDGMENTS

This work is supported by the National Natural Science Foundation of P. R. China (ID: 51372101, 51405195) and Taishan Scholar Engineering Special Funding (2016.01.01–2020.12.30).

REFERENCES

Du Juan. 2010. The analysis of temperature fields of supercharger turbine impeller and test method. *North University of China*: 36–39.
Han, S. 2008. The heat/mass transfer analogy for a simulated turbine blade. *Elsevier Technical Paper Series*51: 5209–5225.
Jia Yanlin, Zhang Yi & Qiao Linhu. 2010. Research on the temperature field of turbocharger turbine. *Diesel Locomotives*10: 16–18.
Librescu, L. Oh, S.Y. & Song, O. 2005. Thin-walled beams made of functionally grade materials and operating in a high temperature environment: vibration and stability. *Thermal Stresses*2: 649–712.
Wang Hao, Song Wenyan & Shi Deyong. 2012. Calculaton Study on the Compressor and Turbine Rotor Three-dimension Temperature Field. *Advances in Aeronautical Science and Engineering*3: 45–49.
Yang Xiaoming, Zhang Li, Jiao Teng & Zhu Huiren. 2012. Three-Dimensional Temperature Field of Turbine Blades Based on Coupling Calculation. *Machinery Design & Manufacture*8: 18–20.
Zhang Jianwei, Bai Haibo & Li Xin. 2013. *ANSYS 14.0 Super learning handbook*. Beijing: Posts & Telecom Press.

Advances in Energy Science and Equipment Engineering II – Zhou, Patty & Chen (Eds)
© 2017 Taylor & Francis Group, London, ISBN 978-1-138-71798-5

An obstacle-avoiding path planning method for robots based on improved artificial potential field

Bin Zhang, Ju Wang, Chen Tang & Qian Wang
College of Metrology and Measurement Engineering, China JiLiang University, Hangzhou, China

ABSTRACT: An obstacle-avoiding path planning method for robots based on improved artificial potential field is presented. Firstly, contours of robotic parts and obstacles are simplified with polygons. Then, potential field force model including attractive forces toward the target position and repulsive forces of the obstacles is constructed according to the traditional artificial potential field method. Besides that, additional control forces are applied to the robot to solve the local extreme position problem. Finally, the velocity control model in the generalized coordinate space is used to generate the obstacle-avoiding path of the robot. The path planning method presented is available not only for mobile robots but also multi-link robots. It takes into account both the robotic geometric contours and the obstacles' geometric contours and avoids solving the complicated robot kinematics and dynamics. The simulation results proved the feasibility of the method.

1 INTRODUCTION

Obstacle avoiding is a key technical problem for path planning of robots in the case of obstacles around. The C-space method and the artificial potential field method are the most common used methods. C-space method is a searching method, which uses Voronoi diagram, regular grid, quaternary tree, octree and so on to represent the motion space of the robot, and then search out the obstacle avoiding path in the space.

Compared with the C-space method, the artificial potential field method is a local planning method suitable for on-line planning. It requires less environmental information, less computation and has a better real-time performance. Besides, it not only considers the obstacle avoidance and trajectory planning of the robot, but also takes the robot's dynamics into account, so that the movement is more natural and flexible.

The artificial potential field method was first proposed by Khatib (1986). The main idea of the method is to apply the virtual potential field force which includes the attraction force of the target position and the repulsion force of the surrounding obstacles to the robot, and then the robot can move from the initial position to the target location under the influence of the forces. Many scholars have proposed some new models of potential field, such as the super quadratic potential field function (Volpe & Khosla 1990), the harmonic potential field function (Kim & Khosla 1992), an analytical potential field function established in the C space (Rimon & Koditschek 1992), Chuang's Newtonian potential field function (1998) and Jwg's S-field potential function (2007) and so on.

The traditional artificial potential field method has the local extreme value problem, that is, when the composition force of the attraction force and the repulsion force is zero, the robot will fall into the equilibrium state at a non-target position. Therefore, some scholars applied additional control forces to the robot (Zhang Jianying et al. 2006) or used orthogonal potential field (Jing Xingjian et al. 2004, Masoud & Masoud 2000) to avoid this problem. Jurgen (1995) combined the artificial potential field method with the sliding mode control algorithm and solved the local extreme problem by forcing the robot to move along the equipotential line of the potential field. All of the methods above can effectively improve the local extreme problem.

At present, the study of path planning based on artificial potential field method is used only for the mobile robots which have simple shapes, such as soccer robot. For multi-link robot, due to the motion coupling of the joints and the interference problem, the path planning is more complicated. Hu Xiaolin (2009) simplified the obstacles as convex polyhedrons, and the robot as a series of line segments. Then, the quadratic programming was used to search the shortest distance between the robot and the obstacle. Sun Shaojie (2011) searched the obstacle-avoiding path of the robot end firstly based on the artificial potential field method, and then calculate the corresponding robot inverse kinematics for each point of the robot end. Wang Junlong (2013) used the artificial potential field method and the genetic algorithm to search out the obstacle-avoiding path in the robot joint space. The researches mentioned above have not taken the contours of the robot into account, and still need to use the searching method, which has a large amount of calculation.

In this paper, an obstacle avoiding path planning method for robots based on improved artificial potential field is presented. Firstly, contours of moving robotic parts and obstacles are simplified with polygons. Then, additional control force is applied besides the traditional artificial potential field force, which makes it more effective to avoid the local extreme problem. Finally, a velocity control model in the generalized coordinate space is used to avoid solving the complicated robot kinematics and dynamics, so that the efficiency of path planning can be improved. The path planning method presented is available not only for mobile robots but also multi-link robots.

2 IMPROVED ARTIFICIAL POTENTIAL FIELD

Whether the robot or the obstacles, their contours are usually complex, which are not conducive to the calculations of potential field. For the planar path planning problem, as it is shown in Figure 1, the polygon bounding box can be used to simplify the contours of the obstacles and the robot. The polygon must completely cover the object. The acquisition of polygon bounding box can be achieved by OBB or FDH method. For the problem in three-dimensional space, the objects can be represented by polyhedrons.

After simplifying, the artificial potential field can be constructed and applied to the robot. The potential field forces include the attraction force of the target position, the repulsion force of the obstacle, and the additional control force to avoid the local extreme problem. In the following, we take the planar problem as an example to introduce the method.

2.1 Attraction force of the target location

Assuming that the dimension of the robot's generalized coordinate space is n, then the attractive force in the dimension i is:

$$F_a^i = k_a(\frac{1}{2} - \frac{1}{1 + e^{-r_a U^i}}) \quad 0 < i \leq n \quad (1)$$

Where k_a is the coefficient determining the size of attraction, r_a is the coefficient determining the varia-

tions of the attraction force near the target position. U^i is the difference of coordinate value between the current position and the target position. The attractive force represented by the Formula (1) is an inverse sigmoid function. When the robot is far from the target position, the magnitude of the attraction force is stable, and when the robot approaches the target position, the force decreases to zero gradually.

2.2 Repulsive force of obstacles

The repulsive force of the obstacle to the robot depends on the distance between them. As shown in Figure 2, there is a point P outside the line segment AB and PH is the perpendicular line. According to the location of point H, it can be divided into two cases: H in AB and H outside AB.

The shortest distance from the point P to the line segment AB is:

$$d(P,AB) = \begin{cases} |PH| & H \in AB \\ \min(|PA|,|PB|) & H \notin AB \end{cases} \quad (2)$$

As shown in Fig. 3, the shortest distance E_1E_2 between two non-intersecting line segments A_1A_2 and B_1B_2 can be calculated according to formula (3):

$$d(A_1A_2,B_1B_2) = \min \begin{Bmatrix} d(A_1,B_1B_2) \\ d(A_2,B_1B_2) \\ d(B_1,A_1A_2) \\ d(B_2,A_1A_2) \end{Bmatrix} \quad (3)$$

Formula (3) shows that the shortest distance between two line segments is the minimum value of the distances of the four end points to the other line. This conclusion can be proved by the apagoge method.

As shown in Figure 4, suppose that A_1A_2 is an edge of the obstacle, B_1B_2 is an edge of the robot, point

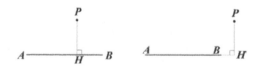

Figure 2. The distance between a point to a line segment.

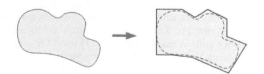

Figure 1. Simplification of planar object.

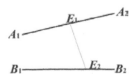

Figure 3. The shortest distance between two line segments.

E_1 and point E_2 are the two nearest points between them, and then according to the FIRAS function, the repulsive force between A_1A_2 and B_1B_2 is:

$$F_r = \begin{cases} -k_r(\dfrac{1}{d} - \dfrac{1}{d_0})\dfrac{1}{d^2} & d \le d_0 \\ 0 & d > d_0 \end{cases} \quad (4)$$

Where d is the distance between E_1 and E_2, d_0 indicates the range of potential field force, and k_r is the gain coefficient of the potential field force.

2.3 Additional control force

In the traditional artificial potential field method, the robot is only subjected to the attraction force of the target position and the repulsive force of the obstacles. However, once the composition of the two forces is zero, the robot may stop moving because of the force balance. In order to solve this problem, an additional control force is added to the robot, and its direction is along the boundary of the obstacle.

As shown in Figure 5, A_1A_2 is the edge of the obstacle, B_1B_2 is an edge of the robot, when B_1B_2 is close to A_1A_2, A_1A_2 will exert the control force on B_1B_2:

$$F_c = k_c F_r \quad (0 < k_c \le 1) \quad (5)$$

Where k_c is the gain coefficient. The direction of F_c has two kinds of possibilities: from A_1 to A_2 or from A_2 to A_1, both of them need to be considered.

The control force is applied when the attraction force and the repulsive force of the robot are in the local equilibrium state. Its direction is orthogonal to the repulsive force, and its magnitude is proportional to the repulsive force, so that the force equilibrium will be broken certainly.

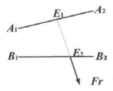

Figure 4. Schematic diagram of repulsive force.

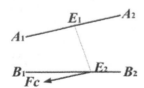

Figure 5. Schematic diagram of additional control force.

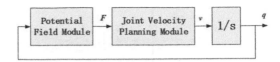

Figure 6. Joint velocity planning model.

3 MOTION CONTROL OF THE ROBOT

In theory, dynamic model should be established based on the artificial potential field forces to generate the robot path. However, most of the robot dynamics is very complex, and it is not conducive to calculation.

Considering that in the artificial potential field method, the force applied to the moving object is essentially virtual force, it is only related to the position, posture and shape of the object. Therefore, we present a generalized force based velocity control strategy for path planning. That is, the moving velocity V_i of each robot joint is the function of the composition force F_i of the potential forces in each generalized coordinate space.

$$V_i = sign(F_i)min(v_{imax}, k_i|F_i|) \quad (6)$$

Where V_{imax} is the maximum motion velocity of the joint and k_i is the gain coefficient.

The joint velocity planning model presented is established in the generalized coordination space, and thus the complex kinematics and dynamics will be avoided, as shown in Figure 6.

4 EXAMPLES

Two path planning examples are shown below using MATLAB Simulink.

4.1 Mobile robot

The path planning model established by MATLAB Simulink is shown as Figure 7. The model is composed of potential field force module, generalized space velocity planning module, integral module and filter module.

This example simulates the process of a triangular mobile robot passing through a rectangular obstacle. The generalized space of the robot includes three degrees of freedom: translation of the X, Y axis and rotation around the Z axis. The result of the planning is shown in Figure 8.

It can be seen from the simulation that when the mobile robot encounters with the obstacle, the attraction force and repulsion force are both vertical, and the robot fall into a local extreme position. When additional control is applied, the robot moves smoothly along the boundary of the obstacle to the target position.

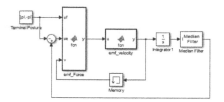

Figure 7. MATLAB Simulink path planning model.

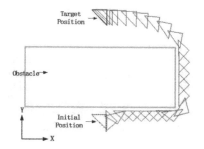

Figure 8. Simulation of the process of a triangular mobile robot passing through a rectangular obstacle.

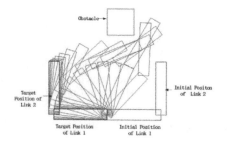

Figure 9. Simulation of the process of a multi-link robot passing through a rectangular obstacle.

4.2 Multi-link robot

Figure 9 shows the simulation of the process of a multi-link robot passing through a rectangular obstacle. The corresponding MATLAB model is similar to the mobile robot planning model in Figure 7. The generalized space here contains the two joint rotational degrees of freedom of the robot.

From the two simulation examples above, it can be seen that the presented obstacle-avoiding path planning method is feasible. It is available not only for mobile robots but also multi-link robots.

5 CONCLUSIONS

In this paper, an obstacle avoiding path planning method for both mobile robots and multi-link robots based on improved artificial potential field is presented. Contours of moving robotic parts and obsta-

cles are simplified with polygons. Additional control forces are applied to the robot to solve the problem of falling into local extreme position. The velocity control model in the generalized coordinate space is used to avoid complicated robot kinematics and dynamics, and thus the efficiency of path planning is improved. Simulation shows the feasibility of the method.

ACKNOWLEDGEMENTS

The project is supported by scientific research fund of Zhejiang provincial education department (Y201328094) and Zhejiang key discipline of instrument science and technology.

REFERENCES

Chuang J.H. (1998). Potential-based modeling of three-dimensional workspace for obstacle avoidance. *J. Robotics and Automation. 14(5)*, 778–785.

Guldner, J. & Utkin V.I. (1995). Sliding mode control for gradient tracking and robot navigation using artificial potential fields. *J. Robotics and Automation. 11(2)*, 247–254.

Hu X., Wang J. & Zhang B. (2009). Motion planning with obstacle avoidance for kinematical redundant manipulators based on two recurrent neural networks. In *Proc. IEEE International Conference on Systems, Man and Cybernetics*, San Antonio, USA, pp. 137–142.

Jing Xingjian, Wang Yuechao & Tan Dalong (2004). Artificial coordinating field based real-time coordinating collision-avoidance planning for multiple mobile robots. *J. Control Theory & Applications. 21(5)*, 757–764.

Khatib O. (1986). Real-time obstacle avoidance for manipulators and mobile robots. *J. The international journal of robotics research. 5(1)*, 90–98.

Kim J.O. & Khosla P.K. (1992). Real-time obstacle avoidance using harmonic potential functions. *J. Robotics and Automation. 8(3)*, 338–349.

Masoud S.A. & Masoud A.A. (2000). Constrained motion control using vector potential fields. *J. Systems, Man and Cybernetics, Part A: Systems and Humans. 30(3)*, 251–272.

Ren J., McIsaac K.A., Patel R.V., et al. (2007). A potential field model using generalized sigmoid functions. *J. Systems, Man, and Cybernetics, Part B: Cybernetics. 37(2)*, 477–484.

Rimon E. & Koditschek D.E. (1992). Exact robot navigation using artificial potential functions. *J. Robotics and Automation. 8(5)*, 501–518.

Sun Shaojie, Qi Xiaohui, Su Lijun, et al. (2011). Obstacle Avoiding Research on the Machine Arm of Robot Based on Artificial Potential Field Method and Genetic Algorithm. *J. Computer Measurement & Control. 19(12)*, 3078–3081.

Volpe R. & Khosla P. (1990). Manipulator control with superquadric artificial potential functions: Theory and experiments. *J. Systems, Man and Cybernetics. 20(6)*, 1423–1436.

Wang Junlong, Zhang Guoliang, Yang Fan, et al. (2013). Improved artificial field method on obstacle avoidance path planning for manipulator. *J. Computer Engineering and Applications. 49(21)*, 266–270.

Zhang Jianying, Zhao Zhiping & Liu Dun (2006). A path planning method for mobile robot based on artificial potential field. *J. Journal of Harbin Institute of Technology. 38(8)*, 1306–1309.

Advances in Energy Science and Equipment Engineering II – Zhou, Patty & Chen (Eds)
© *2017 Taylor & Francis Group, London, ISBN 978-1-138-71798-5*

Target course estimation for a fixed monostatic passive sonar system

Z. Chen, Y.C. Wang & W.G. Dai
Acoustic Center of Navy Submarine Academy, Qingdao, China

ABSTRACT: Aiming at the targets performing uniform linear motion, based on the relative motion model, the target course was calculated by using three azimuths at any different time. Based on the unitary linear regression analysis theory and time-bearing sequence, a highly accurate estimated algorithm of target course was proposed. And finally, the influence factors on estimated results were analyzed. The simulation results show that the algorithm proposed in this paper is highly accurate, which is also practicable when the target bearing has large measured error. With the increase of the data length and the bearing measured accuracy, the estimated result of target course will be more stable.

1 INTRODUCTION

Bearings-Only Target Motion Analysis (BOTMA) is a method to estimate the target motion parameter only by using the target bearing information provided by the observation station. It has good hidden property, long detecting distance and strong anti-interference capacity, which makes it become an important part of tracking and forewarning system in recent years (Northardt et al. 2014, Yu et al. 2016).

For the single sonar observation platform, the target motion parameters can be estimated when the platform moves effectively (Li et al. 2012, Zhang et al. 2010, Huang et al. 2012). If the sonar platform is fixed, the target course, speed and distance can not be calculated at the same time (Jauffret et al. 2010). But when the target performs uniform linear motion, its course could be estimated, which can guide the forewarning system removes interfering targets effectively when there are many observation targets.

At present, most researches on BOTMA for the single platform focus on the observability of system (Yang et al. 2015, Fanf et al. 2013), optimal maneuver trajectory (Ghassemi & Krishnamurthy 2006) and target tracking and positioning algorithms (Fang & Guo 2016, Ho & Chan 2006, Rao 2005). But for the fixed station, its researches are relatively less. A passive location model of fixed single station is built (Liu et al. 2015, Fardad et al. 2014), which can estimate all the target parameters for only one time, but the target's carrier frequency and Doppler frequency rate must be calculated beforehand. Zhang estimates target course by the geometric model, which can be achieved simply, but it requires that the time intervals of different measured bearings must be equal, which has the

certain limitation (Zhang & Luo 2012). Yu assumes that the fixed single station could continuously measure the target bearings with the equal space, and then the target course could be estimated, but the simulation analysis result shows that when the root mean square of measured error is one degree, the estimated accuracy is low (Yu 2014).

In this paper, a kind of target course estimated algorithm with high precision is proposed. It can calculate the target course at any time, which is also practicable when the accuracy of measured bearing is low. Finally, the factors which may affect the estimated results are analyzed under different movement states.

2 MATHEMATIC DESCRIPTION

The geometric state of moving target for the fixed single station is shown in Figure 1. Set the location of observation station as the coordinate origin, which is marked as O.

Assume that the target performs uniform linear motion, F_i is the target bearing measured at the moment t_i, when the target is located in M_i, where $i = 1,2,...n$, C_m is the target course. The mission of

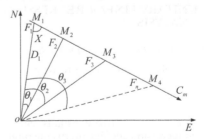

Figure 1. The geometric state of moving target.

this paper is to estimate C_m based on the bearing sequence set $\{F_n\}$.

According to the geometric model above, F_1, F_2 and F_3 are three azimuths at any time, and make $t_{12} = t_2-t_1$, $t_{13} = t_3-t_1$, $\theta_1 = F_2-F_1$, $\theta_2 = F_3-F_1$, we can obtain

$$\frac{V_m \cdot t_{12}}{\sin \theta_1} = \frac{D_1}{\sin(X + \theta_1)} \tag{1}$$

$$\frac{V_m \cdot t_{13}}{\sin \theta_2} = \frac{D_1}{\sin(X + \theta_2)} \tag{2}$$

where V_m is the target speed, D_1 is the distance between observation and the target at the moment t_1, X is the target initial angle with the bound $(0,\pi)$.

Unite these two equations, and eliminate the parameters V_m and D_1

$$\frac{t_{13} \cdot \sin \theta_1}{t_{12} \cdot \sin \theta_2} = \frac{\sin(X + \theta_1)}{\sin(X + \theta_2)} \tag{3}$$

Simplify the equation to

$$\frac{t_{13} \cdot \sin \theta_1}{t_{12} \cdot \sin \theta_2} = \frac{\tan X \cdot \cos \theta_1 + \sin \theta_1}{\tan X \cdot \cos \theta_2 + \sin \theta_2} \tag{4}$$

The cotangent of the target initial angle could be given by

$$\tan X = \frac{(\tan \theta_2 - \tan \theta_1) \cdot (t_{13} - t_{12})}{(t_{12} \cdot \tan \theta_2 - t_{13} \cdot \tan \theta_1)} \tag{5}$$

Here we can see that the target initial angle is only related to the angles of different azimuths at different time, which means the target course could be calculated with three azimuths at any different time by

$$X = a \tan\left[\frac{(\tan \theta_2 - \tan \theta_1) \cdot (t_{13} - t_{12})}{(t_{12} \cdot \tan \theta_2 - t_{13} \cdot \tan \theta_1)}\right] \tag{6}$$

3 UNITARY LINEAR REGRESSION ANALYSIS

3.1 Unitary linear regression model

The model of unitary linear regression is

$$y = a + bx + \varepsilon \tag{7}$$

where a and b are two constants called undetermined regression coefficient, ε is the random error. The observation value can be given by

$$y_i = a + bx_i + \varepsilon_i \ , \quad i = 1,2,\cdots n \tag{8}$$

where $\{(x_i, y_i)\}$ are the observation data, ε_i is the random error.

The purpose is to find the exact line that fits the relation between independent variable and dependent variable based on the samples $\{(x_i, y_i)\}$, and calculate the regression coefficient a and b.

According to the equation (5), the target initial angle is a fixed value, so is cot X. Therefore, combine equation (5) with equation (8), the system functions can be given by

$$\begin{cases} y_i = (\tan \theta_{i+1} - \tan \theta_i) \cdot (t_{i+2} - t_{i+1}) \\ x_i = (t_{i+1} \cdot \tan \theta_{i+1} - t_{i+2} \cdot \tan \theta_i) \\ K = \tan X \end{cases} \tag{9}$$

so we will be able to obtain a linear function in the form of

$$y_i = K \cdot x_i \tag{10}$$

It can be seen that theoretically there is a strict linear relationship between x_i and y_i, so that the target initial angle could be calculated with the regression analysis method.

3.2 Least squares estimation

For formula (7), there are kinds of methods to estimate a and b, and the most common one is the least square theory, which is the optimal linear unbiased estimation of a and b.

By using the formula (10), we can estimate the K based on the samples $\{(x_i, y_i)\}$.

Define

$$\begin{cases} \bar{x} = \frac{1}{n}\sum x_i \ , \bar{y} = \frac{1}{n}\sum y_i \ , \\ l_{xx} = \sum (x_i - \bar{x})^2 = \sum (x_i - \bar{x})x_i \\ l_{yy} = \sum (y_i - \bar{y})^2 = \sum (y_i - \bar{y})y_i \\ l_{xy} = \sum (x_i - \bar{x})(y_i - \bar{y}) = \sum (x_i - \bar{x})y_i \end{cases} \tag{11}$$

the regression coefficient could be given by

$$\hat{K} = \frac{\sum_{i=1}^{n}(x_i - \bar{x})(y_i - \bar{y})}{\sum_{i=1}^{n}(x_i - \bar{x})^2} = \frac{l_{xy}}{l_{xx}} \tag{12}$$

so that the target initial angle is

$$X = a \cot(\hat{K}) \tag{13}$$

Combining the measuring bearing we can calculate the target course.

4 NUMERICAL SIMULATION AND ANALYSIS

4.1 *System model validation*

Based on the system model above, assume that the target performs uniform linear motion, and the target course $C_m = 150°$, the speed $V_m = 12$ kn, the distance between observation station and target $D_1 = 15$ km, the target initial bearing $F_1 = 15°$, the whole duration is 15 min. By these parameters, we can obtain the time-bearing sequence $\{t_n, F_n\}$. Considering the measured errors of target bearing in the practical application, take the random error as 1°, the simulation result is shown in Figure 2.

When $t_1 = 50$ s, $t_2 = 110$ s, $t_3 = 220$ s, the corresponding azimuth are $F_1 = 15.8°$, $F_2 = 16.9°$, $F_3 = 18.9°$. Substituting them in (6), we can calculate the target initial angle $X = 45.8°$ and the target course $C_m = 150°$, which could prove the correction of the system model.

4.2 *Data pretreatment*

If the measured bearing contains errors, we can't get precise result using formula (6), when it needs to make full use of the bearing data $\{F_n\}$, together with the regression analysis model, to calculate the result. In order to reduce the impact of the measured errors, it usually needs to take the filter processing before using the measured data. The common filtering method is polynomial fitting (Ma & Xu 2013). In this paper, we use the 2-order polynomial fitting to process the time-bearing sequence that has errors. The result is shown in Figure 2.

4.3 *Regression analysis*

Unite the formula (5) and formula (9), we can obtain a set of sample point $\{(x_i, y_i)\}$. The two-dimensional figure is shown as Figure 3.

It can be seen that on the whole, the samples are linear. In order to verify the algorithm proposed in this paper, assume that the target moves with different courses, speeds and initial azimuths, we can simulate different motion states and get different measured bearings with errors. And then do the Monte Carlo test 50 times to calculate the target course, the results of mean μ and standard deviation σ are shown in table 1.

It can be seen that the mean μ is very close to the real value and σ is also very small, which means the result estimated by the algorithm is exact.

4.4 *Influence factors*

4.4.1 *The influence of the observation time*

The data length simulated above is 15 min, in order to research the factor of observation time, we suppose the data length are 10 min, 15 min and 25 min, the simulated results based on the states in table 1 are shown in Figure 4. The statistical results are shown in table 2.

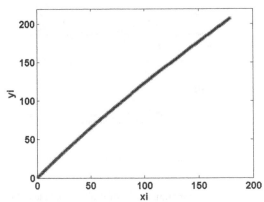

Figure 3. 2-D figure of compositive sample.

Table 1. Estimation results of different states.

State	C_m °	V_m kn	F_1 °	D_1 km	μ °	σ °
1	160	16	25	20	161.2	2.0
2	260	16	25	20	260.2	2.1
3	160	20	25	20	161.3	1.3
4	160	16	90	20	159.0	2.0
5	160	16	25	30	159.9	4.5

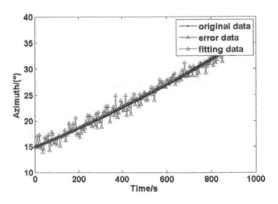

Figure 2. Azimuth observation data.

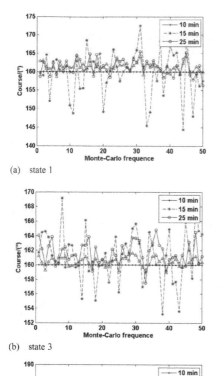

(a)   state 1

(b)   state 3

(c)   state 5

Figure 4.   Simulation results of different lengths.

Table 2.   Estimation results of different data lengths.

State	10 min		15 min		20 min	
	μ	σ	μ	σ	μ	σ
1	159.6	6.2	161.2	2.1	161.4	1.2
3	161.2	3.5	161.5	1.6	160.4	0.6
5	159.9	13.5	161.7	4.7	161.9	2.3

It can be seen from the transverse comparison that the means in different time lengths are close, but the standard deviations are decreasing with the time length increasing, which means that the result is more stable. When the data length is short, the fluctuation is relatively wide, but we can see from the means that it is still accurate to take the average of measured data.

According to the longitudinal comparison, it can be seen that when the target speed is faster and the distance between observation station and target is closer, the estimated result will be more stable.

4.4.2   *The influence of the measuring accuracy*

During the actual application, because of the beam forming and complicated hydrologic environment, the errors of the measured data are even more than 1°, so it needs to analyze the impacts of measured accuracy on the algorithm.

(a), (b), (c) in figure(5) are the simulation results of different measured bearing accuracy when

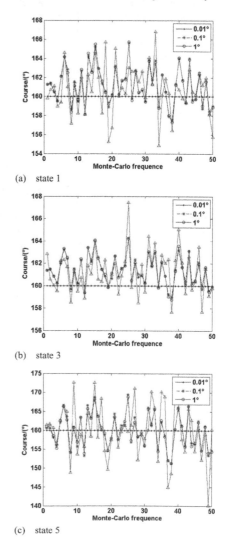

(a)   state 1

(b)   state 3

(c)   state 5

Figure 5.   Simulation results of different precisions.

266

the targets are in state 1, state 3 and state 5. The statistical results are shown in table 3.

It is shown that the algorithm proposed in this paper is relatively accurate. When the precision is low, the standard deviation will increase, which means that the fluctuation of the estimated result is wide and not stable. But we can see from the mean that it is still accurate to take the average of the measured data.

5 ANALYSIS OF THE SEA TRIAL DATA

In order to test the algorithm's performance to process the real data, we use it to analyze the sea trial data. The signal comes from the uniform vector linear array with 16 elements, and the time length is 1200 s. Figure 6 shows the signal's time-bearing curve after beam forming.

Owing to the merchant ships and fishing boats in the sea areas, we can there are some interfering targets on the figure. The course of target ship is 0°, the azimuth range is from 210° to 238°, and the azimuth measured accuracy is 1°.

According to the algorithm proposed in the paper, the estimated results of target course from different periods and time lengths are shown in table 4.

It can be seen that only the result of 0–900 s is relatively poor, while the other periods' are accurate. By the comparison of No.1 with No.3 in the table 4, though the time length of the latter one is short, its estimated result is more precise,

Table 4. Results of trial data.

Number	period	data length	bearing variation	estimated course
	s	s	°	°
1	0–900	900	18	3.6
2	0–1200	1200	31	1.0
3	300–900	600	15	1.6
4	300–1200	900	28	359.8

the reason may be that the target bearings from initial 300 s have no change, which will affect the estimated result of the whole data. To solve this problem, we can use longer data or remove the data which has a small rate of bearing variation. By the comparison of No.1 with No.4, it can be seen that the time lengths of them are equal, but the azimuth range of the latter one is wider, so the result is relatively more accurate, which meet the theory mentioned above.

6 CONCLUSION

This paper is mainly to propose an accurate algorithm to estimate the target course, and analyze the factors that affect the estimated results. Finally, the algorithm is validated by simulation analysis and the sea trial data. The results indicate that the method proposed in this paper can calculate the target course at any time, which is also practicable when the measured accuracy of target azimuth is low. So it could overcome the previous methods' limitations on high measured accuracy.

REFERENCES

Fanf X. et al. 2013. Observability analysis of underwater localization based on fisher information matrix. 2013 32nd Chinese Control Conference. 4866–4869.
Fang F. & Guo H.D. 2016. Simulation analysis of single fixed-station passive location based on TOA and DOA. Shipboard Electronic Countermeasure, 39(2): 10–13.
Fardad M. et al. 2014. A novel maximum likelihood based estimator for bearing-only target localization. 2014 22nd Iranian Conference on Electrical Engineering. 1522–1527.
Ghassemi F. & Krishnamurthy V. 2006. A stochastic search approach for UAV trajectory planning in localization problems. IEEE International Conference on Acoustics, Speech and Signal Processing. 1196–1199.
Ho K.C. & Chan Y.T. 2006. An asymptotically unbiased estimator for bearings-only and doppler-bearing target motion analysis. IEEE Trans. on Signal Processing, 54(3): 809–822.

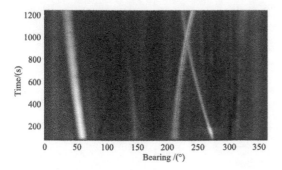

Figure 6. Time-bearing curve of the sea trial data.

Table 3. Results of different measured accuracies

State	0.01		0.1		1	
	μ	σ	μ	σ	μ	σ
1	161.2	2.0	161.2	2.0	160.7	2.6
3	161.3	1.3	161.3	1.3	161.2	1.9
5	160.0	4.5	160.0	4.3	160.0	7.1

Huang Y.Y. et al. 2012. Infrared sensor and passive radar track robustness correlation algorithm. Chinese Journal of Scientific Instrument, 33(11): 2629–2634.

Jauffret C. et al. 2010. Bearings-only maneuvering target motion analysis from a nonmaneuvering platform. IEEE Transactions on Aerospace and Electronic Systems, 46(4): 1934–1949.

Li Z. et al. 2012. Study of automatic continuous tracking and location algorithm for underwater target. *Chinese Journal of Scientific Instrument*, 33(3): 520–527.

Liu D. et al. 2015. Influence and analysis of DDFRC algorithm location error in distance information. Journal of Naval Aeronautical and Astronautical University, 30(4): 326–330.

Ma J.G. & Xu X.N. 2013. An underwater target azimuth preprocessing method based on motion model. Technical Acoustics, 32(3): 243–247.

Northardt, E.T. et al. 2014. Bearings-only constant velocity target maneuver detection via expected likelihood. *IEEE Transactions on Aerospace and Electronic Systems*. 50(4): 2974–2988.

Rao S.K. 2005. Modified gain extended Kalman filter with application to bearing-only passive manoeuvring target tracking. IEEE Proc. of Radar, Sonar and Navigation, 152(4): 239–244.

Yang J. et al. 2015. Overview of observability of bearing-only target motion analysis. Fire Control & Command Control, 40(12): 1–8.

Yu J. et al. 2016. Modified instrumental variable method for bearings-only target tracking. *Journal of Xidian University*. 43(1): 167–172.

Yu T. 2014. Bearing-only estimation for target heading uniform linear motion conditions based on equispaced measurement. Modern Navigation, 6: 446–449.

Zhang W. et al. 2010. Analysis of influencing factors on the optimal observer trajectory in bearings-only tracking. Systems Engineering and Electronics, 32(1): 67–71.

Zhang X.Y. & Luo L.Y. 2012. Target course estimation for single stationary bearing-only observation system. Technical Acoustics, 31(6): 566–569.

Advances in Energy Science and Equipment Engineering II – Zhou, Patty & Chen (Eds)
© *2017 Taylor & Francis Group, London, ISBN 978-1-138-71798-5*

Two-way fluid-structure coupling simulation of oil film characteristics of hollow shaft hydrostatic bearings

W. Sun

School of Mechanical Science and Engineering, Jilin University, Changchun, Jilin, China

X.J. Yu

School of Automatic Control and Mechanical Engineering, Kunming University, Kunming, Yunnan, China

ABSTRACT: To study the oil film characteristics of Hollow Shaft Hydrostatic Bearings (HSHBs) of the large-type ball mill, the fluid-structure coupling simulation model for HSHB is created in this paper. Based on the coupling mechanism and simulation results, the distribution rules of oil film pressure and thickness are analyzed. According to the two-way simulation results, the influences of the bearing capacity, the deflection of the hollow shaft, and the bearing clearance on the stiffness coefficient of the oil film are obtained. The results show that the oil film around the oil cavity is at safe thickness, while the oil film around the sealing edge is close to the minimum oil film thickness. The direct stiffness coefficients are sensitive to parameters such as bearing capacity, hollow shaft deflection, and bearing clearance. The anti-disturbance ability in the vertical direction is strong.

1 INTRODUCTION

Hollow Shaft Hydrostatic Bearings (HSHBs) have the advantages of high bearing capability, low power consumption, and high reliability. Thus, it is widely used in large-type ball mills (Cui, 2009; Zhu, 2009). However, the hollow shaft and the bearing bush are easily deformed during the grinding operation, especially in the start and brake process. The failure of the oil film is caused by the deformation, which may further cause "bush burning".

With the upsizing trends of ball mill, the bearing capability and reliability of the HSHB are the focus in the mining industry (Sun, 2011). Liu, G. (2007) analyzed the fluid simulation of hydrostatic bearing. The relationship between the oil cavity parameters and carrying capacity were obtained. The influence rules of the elastic modulus of materials, lubricating oil viscosity, and oil cavity pressure on the minimum oil film thickness were discussed. Shen, D.P. (2012) carried out the two-way fluid-structure coupling analysis of the HSHB of the large-type ball mill. The pressure and deformation of the hollow shaft and bearing bush were analyzed.

In this paper, a two-way fluid-structure coupling method is applied to the coupling calculation of fluid lubrication and structural deformation (Sun, 2016). Furthermore, both the bearing capacity and the deflection of the hollow shaft have an influence on the pressure and thickness of the oil film, and

the influence is analyzed to compute the stiffness of the oil film.

2 FLUID-STRUCTURE COUPLING MODEL OF THE HSHB

2.1 *Finite element model*

The oil cavity structure of the HSHB is shown in Figure 1. The hollow shaft radius (R) is 1500 mm and bearing width (b) is 733.5 mm. The bearing wrap angle is 120°. The oil cavity width (a) is 400 mm. The oil cavity attitude angle (β) is 11°.

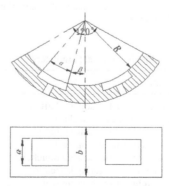

Figure 1. Schematic showing the structure parameters of the HSHB.

The element number of the oil film is 56700, as shown in Figure 2. The finite element model of the HSHB (including bearing bush, bearing bouchon, and bearing seat) is shown in Figure 3.

2.2 *Boundary conditions*

The oil inlet is set as the pressure entrance. The oil outlet is set as the pressure exit. The bottom surface of the bearing seat is fully constrained. The bearing seat and the bearing bouchon are in nonlinear contact. The material parameters are shown in Table 1.

Figure 2. Schematic of the finite element model of the oil film.

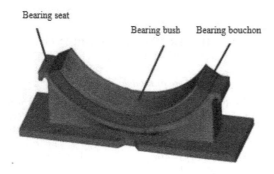

Figure 3. Schematic of the finite element model of the hydrostatic bearing.

Table 1. List of material parameters.

Parameters	Value
Elastic modulus of the hollow shaft	210 GPa
Poisson ratio of the hollow shaft	0.3
Elastic modulus of the bearing bush	1.03 GPa
Poisson ratio of the bearing bush	0.3
Elastic modulus of the bearing bouchon	154 GPa
Poisson ratio of the bearing bouchon	0.27
Elastic modulus of the bearing seat	210 GPa
Poisson ratio of the bearing seat	0.27
Lubricating oil density	895 kg/m³

3 ANALYSIS OF THE OIL FILM CHARACTERISTICS BY USING TWO-WAY FLUID-STRUCTURE COUPLING SIMULATION

3.1 *Analysis of oil film pressure*

The fluid-structure coupling results of oil film pressure is shown in Figure 4. The distribution of oil film pressure at 5900 kN is shown in Figure 4(a). The distribution of oil film pressure at 6600 kN and the deflection angle 0.02° is shown in Figure 4(b). The pressure of the oil cavity is higher than those of other areas. The maximum pressure position moves 90 mm to the direction along which the oil film thickness decreases.

The oil film pressure distribution of the oil film pressure axial symmetry plane and radial symmetry plane under different bearing capacities are shown in Figure 5(a) and (b). The oil film distribution of the oil film pressure axial symmetry plane and radial symmetry plane under different deflection angles of the hollow shaft are shown in Figure 5(c) and (d).

Due to the deformation of the hollow shaft, the distribution of the oil film is not symmetric and causes the self-aligning torque along the film's radial symmetry plane. The upper surface of the bearing seat is cylindrical and the lower surface of the bearing bouchon is a double circle arc.

As shown in Table 2, the oil film bearing capacity is unchanged basically. The maximum pressure position on an axially symmetric plane moves closer to the cylinder side. The moment around the radial symmetry plane can be seen as the oil film self-aligning torque produced by the deformation of the hollow shaft. The deflection angle increases from 0.005° to 0.02°. The self-aligning torque exhibits a 4.25 times increase. It can be seen that the oil film changes the deflection load of the hollow shaft into the self-aligning torque.

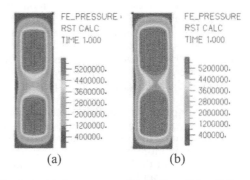

Figure 4. Surface pressure distribution of the oil film.

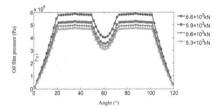

(a) Influence of bearing capacity on oil film pressure in the circumferential direction

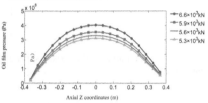

(b) Influence of bearing capacity on oil film pressure in the width direction

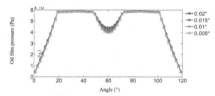

(c) Influence of the hollow shaft deflection angle on oil film pressure in the circumferential direction

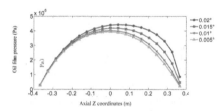

(d) Influence of the hollow shaft deflection angle on oil film pressure in the width direction

Figure 5. Graphs showing the oil film pressure distributions.

3.2 Analysis of the oil film thickness

The basic principle of the Elasto-Hydrodynamic Lubrication (EHL) is defining the minimum film thickness. In engineering practice, the minimum oil film thickness of the HSHB is not less than 0.2 mm.

The vertical displacement distribution of the hollow shaft and the bearing bush at 6600 kN is shown in Figure 6. The oil film thickness distribution of the axial symmetry plane and the radial symmetry plane of the oil film under different bearing capacities are shown in Fig-

Table 2. Influence of the hollow shaft deflection on the performance of the oil film.

Hollow shaft deflection angle (°)	0.005	0.01	0.015	0.02
Bearing capacity (kN)	7174	7225	7293	7330
Z coordinate of maximum axial pressure position (m)	0.01	0.02	0.05	0.09
Self-aligning torque (Nm)	16735	37489	60963	71142

ure 7(a) and (b). Because of the deformation of the hollow shaft and bearing bush, the oil film thickness of the center position increases, and the oil film thickness of both ends reduces. The minimum oil film thickness in the width direction is 0.216 mm and the minimum oil film thickness at 6600 kN in the circumferential direction is 0.201 mm.

When the hollow shaft deflection angle is 0.02°, the displacement distribution of the hollow shaft and bearing bush in the vertical direction is shown in Figure 8. The oil film thickness distribution of the oil film axial symmetry plane and radial symmetry plane under different hollow shaft deflection angles are shown in Figure 9(a) and (b). It can be seen that in the width direction, the oil film thickness of the hollow shaft deflection side obviously decreases. When the hollow shaft deflection angle is 0.02°, the oil film thickness in the width direction of the bearing is 0.099. The minimum oil film thickness in the circumferential direction is 0.134.

3.3 Analysis on the oil film stiffness

As shown in Figure 10, o' is the axis center in the equilibrium state and it changes to o' under the tiny disturbance in the X direction. By using the fluid-structure coupling method, pressure distribution of the oil film is obtained. The components of the oil-film force in the X direction, F_{x1} and F_{y1}, are calculated by integrating the pressure along the

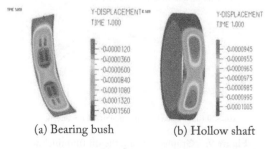

(a) Bearing bush   (b) Hollow shaft

Figure 6. Vertical displacement distribution at 6600 kN.

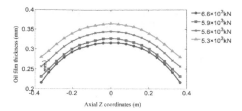

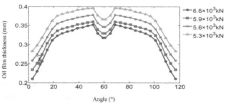

(a) Oil film thickness in the width direction

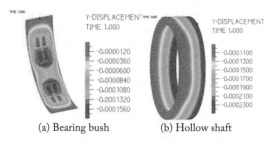

(b) Oil film thickness in the circumferential direction

Figure 7. Graphs showing oil film thickness distributions under different bearing capacities.

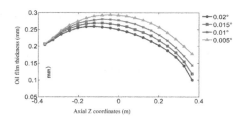

(a) Bearing bush   (b) Hollow shaft

Figure 8. Vertical displacement distributions at a deflection angle 0.02°.

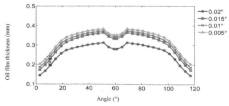

(a) Oil film thickness in the width direction

(b) Oil film thickness in the circumferential direction

Figure 9. Graphs showing the oil film thickness distributions at different hollow shaft deflection angles.

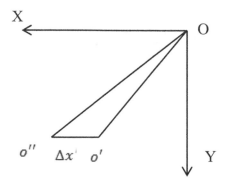

Figure 10. Schematic of the axis displacement of the hollow shaft.

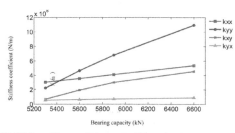

(a) Oil film stiffness coefficient under different bearing capacities

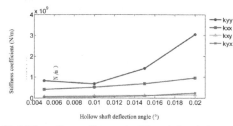

(b) Oil film stiffness coefficient under different hollow shaft deflection angles

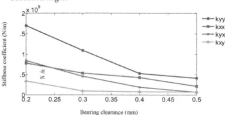

(c) Oil film stiffness coefficient under different bearing clearances

Figure 11. Influence of different variables on the oil film stiffness coefficient.

X and Y direction. Similarly, the components in the Y direction, F_{x2}, and F_{y2} are calculated.

The influence of the displacement disturbance on the oil film bearing capacity can be expressed by the four stiffness coefficients:

$$k_{xx} = \frac{\Delta F_x}{\Delta x} = \frac{F_{x1} - F_{x0}}{\Delta x}$$
$$k_{yx} = \frac{\Delta F_y}{\Delta x} = \frac{F_{y1} - F_{y0}}{\Delta x}$$

(1)

$$k_{xy} = \frac{\Delta F_x}{\Delta y} = \frac{F_{x2} - F_{x0}}{\Delta y}$$
$$k_{yy} = \frac{\Delta F_y}{\Delta y} = \frac{F_{y2} - F_{y0}}{\Delta y}$$

(2)

where, the first subscript of each coefficient indicates the direction of the force and the second subscript indicates the direction of the displacement.

The influences of the bearing capacity, hollow shaft deflection, and bearing clearance on the oil film stiffness coefficient are shown in Figure 11(a)–(c). It can be seen from the calculation results that the direct stiffness coefficients k_{xx} and k_{yy} are greater than the cross stiffness coefficients k_{xy} and k_{yx}. The bearing capacity, deflection angle of the hollow shaft, and bearing clearance have a great influence on the direct stiffness of k_{yy}. When the bearing capacity increases from 5300 kN to 6600 kN, the direct stiffness coefficient increases from 2.3×10^8 N/m to 1.09×10^9 N/m, which exhibits a 4.9 times increase. When the hollow shaft deflection angle increases from 0.005° to 0.02°, the direct stiffness coefficient k_{yy} increases from 7.27×10^8 N/m to 3.14×10^9 N/m, which exhibits a 4.5 times increase. When the bearing clearance increases from 0.2 mm to 0.5 mm, the direct stiffness coefficient k_{yy} reduces from 1.7×10^9 N/m to 0.41×10^9 N/m, which exhibits a 8.9 times decrease.

4 CONCLUSIONS

The simulation analysis method of the two-way fluid-structure coupling of the HSHB is created. The results show that the hollow shaft and the bearing bush deformation have an adverse impact on the film thickness. The oil film around the oil cavity is at safe thickness, while the oil film around the sealing edge is close to the minimum oil film thickness, 0.2 mm, which is hazardous. The influence of the structural deformation on the oil film thickness should be taken into consideration in the design of the HSHB.

Through the calculation and analysis of the oil film stiffness, the influences of the bearing capacity and the deflection of the hollow shaft and the bearing clearance on the oil film stiffness coefficients are obtained. The direct stiffness coefficient is sensitive to the three factors above. The oil film of the HSHB is more undisturbed by the vertical direction. The influence of structural deformation on the oil film stiffness should be taken into consideration in the design of the initial clearance of the bearing.

ACKNOWLEDGMENT

This work was supported by the National Natural Science Foundation of China (No. 51265020).

REFERENCES

Cui, F.K. & Zao, W. & Yang, J.X. (2009). Pressure and streamline field simulation analysis of ball mill hydrostatic bearings. *Bearings,* 01:32–36.
Liu, G. (2007). Study of application of CAE technique on hydrostatic journal bearings of the large-scale ball-grinding mill. Changchun: Jilin University.
Shen, D.P. Guo, B. & Wang, F.F. (2012). Oil recess structure optimization and flow simulation for heavy hydrostatic bearing. *Bearing,* 07:7–10.
Sun, W. (2016). Research on the performance of hydrostatic bearing of large-type ball mill based on fluid-structure interaction. Changchun: Jilin University.
Sun, Y.H. & Ma, X.Z. (2011). Influence of elastic deformation on ball mill bearing. *Lubrication and seal,* 01:65–69.
Zhu, X.L. (2009). Optimum design of hydrostatic bearing bushing structure used in ball-grinding mill. *Lubrication and seal,* 34(7):88–90.

Advances in Energy Science and Equipment Engineering II – Zhou, Patty & Chen (Eds)
© *2017 Taylor & Francis Group, London, ISBN 978-1-138-71798-5*

Design and bench test of the powertrain for series hybrid electric loaders

S.S. Feng
School of Mechanical Science and Engineering, Jilin University, Changchun, China

ABSTRACT: In order to solve the problems of low transmission efficiency and high fuel consumption of traditional wheel loader, the configuration of Series Hybrid Electric Loader (SHEL) is proposed. The front and rear axles of SHEL are driven independently. According to the working characteristics and control strategy of SHEL, the hardware of the SHEL powertrain is selected. Based on the rapid control prototype technology, the test bench of the SHEL is set up and the test program is established. The results show that the hybrid system and control strategy proposed in this paper are reasonable.

1 INTRODUCTION

Due to complex working conditions and frequent violent changes in the load, the wheel loader has the disadvantage of low utilization efficiency, high fuel consumption, and bad emission (Lin, 2010; Sun, 2010; Li, 2013). Hybrid technology is particularly suitable for the energy conservation and emission reduction of the wheel loader. To achieve certain energy-saving effects, recently, companies like Hitachi, Volvo, Caterpillar, Kumatsu, XCMG, Liugong, etc. (Prandi, 2007; Peter, 2008; Sun, 2009) have developed their experimental prototypes or products. Based on the present developing status and tendency of hybrid technology, key components of the series hybrid electric loader reach the level of industrialization. In this paper,

a configuration of SHEL is proposed. The front and rear axles of the SHEL are driven independently. The test bench of the SHEL is set up. The model and control strategy of the SHEL are tested and verified (Feng, 2016).

2 DESIGN OF THE SHEL POWERTRAIN

Due to frequent violent changes in the load and no dynamic torque allocations between the front and rear axles, the wheel loader has a skid phenomenon, which influences the traction performance and tire life. The configuration of the SHEL is shown in Figure 1. In the configuration, the power of the engine can be allocated reasonably and the traction of the wheel loader can be improved.

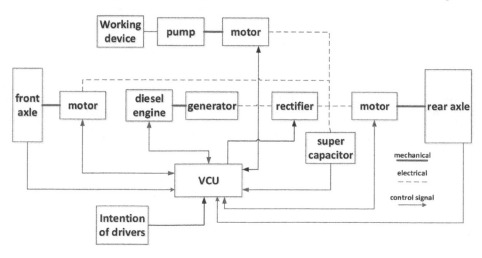

Figure 1. Schematic of the configuration of SHEL.

The direct-current Permanent Magnet Synchronous Motor (PMSM) is selected. The driving motor has two control modes, namely the speed control mode and torque control mode. A super capacitor is selected to regulate the output power of the generator. According to the driving voltage of the motor, the optimum operating voltage of the super capacitor is 290~380 V. Considering the demanded power of the loader and the available power of the super capacitor, the power of generator units is 50 kW and the nominal voltage is 320 VAC. To realize the parallel connection of generator units and super capacitor, the rectifier is selected to convert AC to DC. After research and analysis, a three-phase controlled rectifier is selected. As the rectifier demands automatic regulation according to the demanded voltage, the full bridge thyristor rectifier is selected as the system rectifier.

3 TEST BENCH OF THE SHEL

In the test bench of the SHEL, one motor is used as the driving motor and the other motor is used as the load motor. The test bench is shown in Figure 2. The test program is established as follows:

a. To promote the charging efficiency of the super capacitor and ensure the normal operation of the circuit, a rectifier is set to different modes under different conditions. When the super capacitor is in low voltage, a rectifier is set to constant current mode. When the driving motor is under normal operation conditions, a rectifier is set to constant voltage mode.
b. To realize the accuracy control of the electrify order and working mode, the test bench is only driven by using a super capacitor.

c. To verify the validity of the control strategy of the SHEL, the power is provided by using a rectifier and super capacitor.

4 RESULTS OF THE BENCH TEST

4.1 Super capacitor charged by using a rectifier

When the rectifier is used to charge a super capacitor, the control model is shown in Figure 3. The test results are shown in Figure 4. Figure 4(a) shows the curve of the working mode in the charging process. Figure 4(b) shows the curves of output current "$I_{Control}$" and output voltage "$V_{Control}$" of the rectifier, and voltage "V_{CAP}" of the super capacitor. When the voltage of the super capacitor is low, the super capacitor is charged by using a rectifier in constant current mode. When the voltage reaches a certain value, the super capacitor is charged by using a rectifier in constant voltage mode.

4.2 Electric driving

In a torque control model, the torque is controlled and speed is self-adapting. The target speed and the load torque are controlled by using two accelerator pedals. The torque of the load motor is controlled by using the load acceleration pedal "APS diff". To simulate the target of drivers, the torque difference between two motors is controlled by using the driving acceleration pedal "APS". The torque control model of the driving motor and the load motor is shown in Figure 5.

The test results are shown in Figure 6. The curves of the torque of the driving motor and the load motor and the curve of "APS diff" are

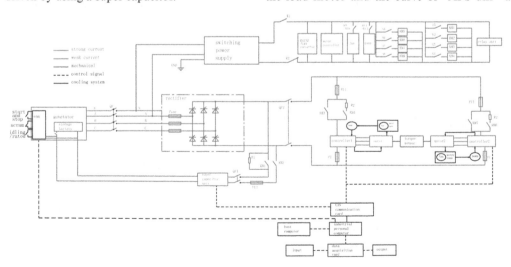

Figure 2. Schematic of the test bench of the SHEL based on the rapid control prototype technology.

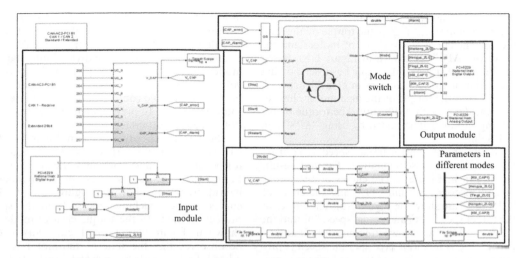

Figure 3. Schematic of the control model of a super capacitor charged by using a rectifier.

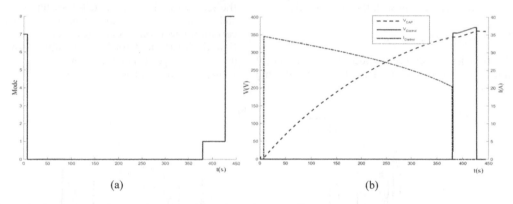

(a)   (b)

Figure 4. Test results of a super capacitor charged by using a rectifier.

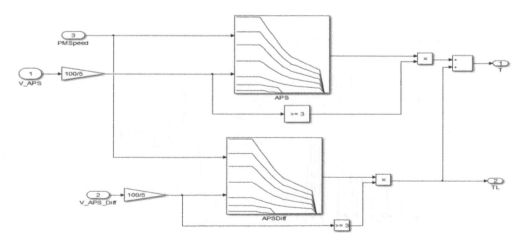

Figure 5. Torque control model of the driving motor and load motor.

shown in Figure 6(a), where "Ta1" is the torque of the driving motor, "Ta2" is the torque of the load motor. The curve of speed "n" of motors and the curve of "APS" are shown in Figure 6(b). Based on Figure 6(a) and (b), it can be observed that the speed of motors changes in coincidence with "APS", and the torque of the load motor changes in coincidence with "APS" diff. The torque of the driving motor is determined by using "APS" and "APS diff".

In the speed control mode, speed is controlled and torque is self-adapting. The test results are shown in Figure 7. The curves of the torque of two motors, "T" and "TL", and the curve of "APS diff" are shown in Figure 7(a). The curves of the target speed "nr" and actual speed "n" and the curve of "APS" are shown in Figure 7(b). Based on Figure 7, it can be observed that the speed is proportional to the "APS" and the load torque is proportional to the "APS diff", and the actual speed is basically in agreement with the target speed.

4.3 Hybrid driving

There is less difference between the speed control mode and torque control mode on the driving motor. The torque control mode is used as an example to test the function of hybrid driving. The flow chart of the control strategy is shown in Figure 8. Control modes are divided into standby mode, constant current charging mode, constant voltage charging mode, precharging mode, enabling mode, torque control mode, fault mode, and shutdown mode. To prevent high motor speed, the over-speed protection mode is added.

The test results are shown in Figure 9. The curves of the motor torque, speed "n", accelerator pedal "APS", and working mode are shown in Figure 9(a). The curves of the state switching in hybrid driving are shown in Figure 9(b)–(d). The curves of the switching process from the speed protection mode to the torque control mode are shown in Figure 9(e). From Figure 9(a), it can be observed that the state of the motor is only related to the accelerator pedal position. From Figure 9(b), it can be observed that the states of the rectifier and super capacitor switch automatically, when the state of super capacitor lies within a certain range. From Figure 9(c), it can be observed that, during the high load current state, if the voltage of the super capacitor is lower than a certain value, the super capacitor will no longer provide

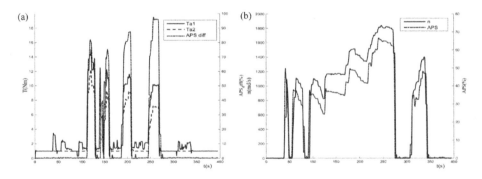

Figure 6. Test results of the torque control mode.

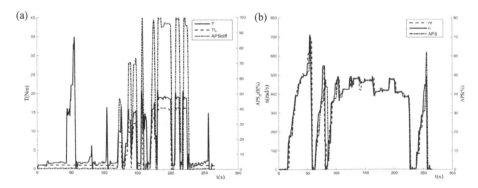

Figure 7. Test results of the speed control mode.

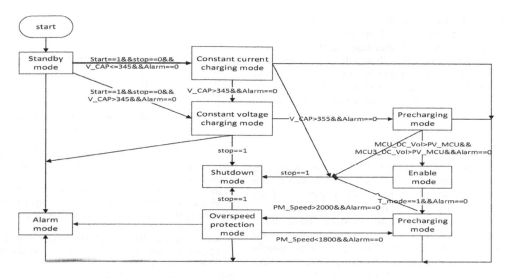

Figure 8. Flow chart of the control strategy of hybrid driving.

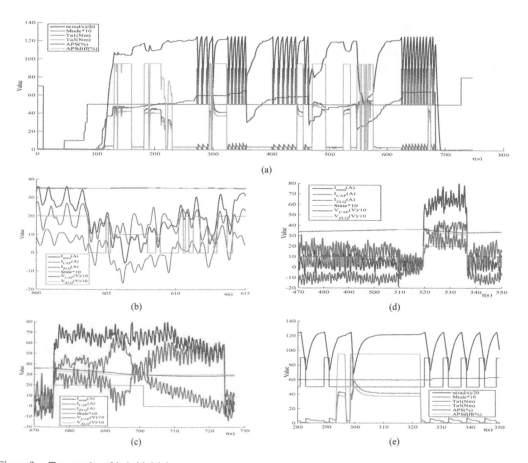

Figure 9. Test results of hybrid driving.

power. From Figure 9(d),), it can be observed that, during the low load current state, if the super capacitor voltage is higher than a certain value, the rectifier will no longer charge the super capacitor. From Figure 9(e), it can be observed that, when the motor is operating in high speed, the working mode switches into over-speed protection mode.

5 SUMMARY

A configuration of the SHEL is proposed. In this configuration, the front and rear axles of the SHEL are driven independently. The power of the engine can be allocated reasonably and the traction of the wheel loader can be improved. Based on the rapid control prototype technology, the test bench of the SHEL is set up. Three conditions, namely super capacitor charged by using a rectifier, electric driving and hybrid driving, are verified. The test results prove that the design and control strategy of the SHEL is reasonable.

ACKNOWLEDGMENT

This work was supported by the National Natural Science Foundation of China (No. 51375202).

REFERENCES

Feng, S. (2016). Research on experimental platform of power system of series hybrid loader. Changchun: Jilin University.

Li, T., Zhao, D. & Kang, H. (2013). Parameter matching of parallel hybrid power loaders. *Journal of Jilin University,* 43(4):916–921.

Lin, T., Wang, Q. & Hu, B. et al (2010). Development of hybrid powered hydraulic construction machinery. *Automation in Construction,* 19(1):11–19.

Peter, A. (2008). A serial hydraulic hybrid drive train for off-road vehicles. *Proceedings of the National Conference on Fluid Power,* (51):515.

Prandi, R. (2007). A true heavy-duty hybrid: Deutz, Heinzmann, Atlas-Weyhausen team up to develop prototype hybrid wheel loader. *Diesel Progress North American Edition,* (10): 34–41.

Sun, H., Jing, J. & Wang, X. (2009). Torque control strategy for a parallel hydraulic hybrid vehicle. *Journal of Terramechanics,* 46(6):259–265.

Sun, H. & Jing, J. (2010). Research on the system configuration and energy control strategy for parallel hydraulic hybrid loader. *Automation in Construction,* 19(2):213–220.

A partial discharge data management platform based on cloud services

Zhixin Pan, Hui Chen, Hui Fu & Yeying Mao
State Grid Jiangsu Electric Power Company, Nanjing Jiangsu, China

Guang Chen
State Grid Jiangsu Electric Power Research Institute, Nanjing Jiangsu, China

ABSTRACT: A novel partial discharge data management platform is introduced in this paper, which is a highly integrated system providing functions of data collection, data query, and data analysis. The platform, which is based on the B/S architecture, can excavate the potential information in the data, and make an accurate analysis. The platform provides technical support to testers who then can quickly and accurately detect partial discharge and determine the type of partial discharge. And also, the platform is combined with cloud services perfectly thereby giving full play to the advantages of cloud services and overcoming shortcomings of earlier systems in stability.

1 INTRODUCTION

With the development of partial discharge detection technology, higher efficiency and accurate location of insulation flaws are highly expected nowadays. To achieve these goals, it is required to accumulate enough data, and find the potential information from these data by using data mining technology, which will provide a strong scientific support for partial discharge detection and judgment.

In this paper, a novel partial discharge data management platform is built to achieve the above-mentioned goals. The platform is designed according to the B/S architecture model. It includes functions such as data collection, data query, and data analysis, which are capable of performing deep data mining on partial discharge data.

The platform is different from the traditional systems on the deployment and implementation. The traditional B/S architecture systems are deployed in the customer's server, which has shortcomings in the stability and performance due to the limitation of server room environment on the server hardware, electricity, network, etc. In this paper, the platform is deployed on the "cloud", giving full play to the advantage of the "cloud". In recent years, cloud technology has become more and more mature, by which cloud service providers can build a stable and powerful cloud server for us. When compared with the traditional server, a cloud server performed more functions in terms of stability and exception handling. This also provides a powerful guarantee for the stable operation of the platform.

2 PARTIAL DISCHARGE DATA MANAGEMENT SYSTEM

The partial discharge data management system is based on the popular B/S architecture. In order to ensure a good user experience of the system, the ExtJs is used as the front-end language. Fully combined with the characteristics of ExtJs, the system is developed into a RIA application. The architecture of the system is shown in Figure 1.

The system consists of the following modules:

1. Data collection module
   The data collection module is used to collect data through partial discharge detection. It provides two accesses to acquire detecting data. One is the data access module and the other is the intelligent tour inspection module. The data access module is used to import the detection data with no plan. The intelligent tour inspection module is used to schedule detection plans and take back detection data

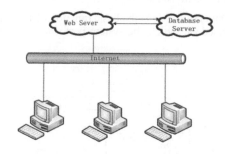

Figure 1. System architecture diagram.

of the plans from the partial discharge detecting equipment. Once the data have been collected, we can query and analyze these data in the data query and analysis module.

2. Data query and analysis module.

The data query and analysis module is used to query and analyze the collected data of partial discharge. It consists of monitoring the data analysis module and fingerprint libraries module. The former analyzes the collected data and sends the data and analysis results to relevant experts. The latter stores typical partial discharge data, by which one can judge the discharge type when similar data are detected. After the analysis of the data by the system, one can send the relevant data and analysis results to the relevant experts to carry out deeper analysis.

3. Expert push module.

The expert push module is used to send the partial discharge data analysis results to experts who will analyze the data and make a conclusion. Some data, in which it is difficult to determine the type of partial discharge after the analysis module, will be sent to experts by using the expert push module. Experts push module also includes the expert diagnosis function, which automatically reminds relevant experts to receive the data to diagnose. After experts analyze and diagnose the data, the data will be fed back into the platform.

The modular structure of the system is shown in Figure 2.

The system is designed to present the following main features.

- RIA applications for a good user experience.
- Flexible data acquisition methods, which can input partial discharge detection data from the detecting devices into the system directly.
- Vivid display interface about query and analysis.
- Timely expert push function, which plays the advantage of expert's professional.
- Combining the advantages of cloud services, thereby ensuring stable, safe, and efficient operation of the system.

2.1 Data collection

Data collection is the basic function of the system, based on which the system performs an in-depth

analysis. It is also the module which users come in contact with firstly. In order to give users a good experience and a deep impression, the system provides a flexible approach of data collection.

Users can upload the local data, which are already in the computer, into the system by using the interface shown in Figure 3.

We can use the intelligent tour inspection module with handheld partial discharge detection equipment, which is flexible and convenient. Through the system, inspection tasks can be downloaded to the handheld testing equipment by using the interface shown in Figure 4.

After the handheld device has completed inspection tasks, the inspection data can be uploaded to the system directly by using the interface shown in Figure 5.

2.2 Data query and data analysis

After the completion of data collection, it is necessary to query and analyze the internal rules of the basic data and find the potential information of the data. Also, the typical data need to be sorted and stored which will provide references for later detection.

Because the module is powerful and complex, simple block diagrams are used to display the function of this module. The operation of the query

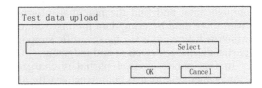

Figure 3. Screenshot of interface used to upload data into the PC.

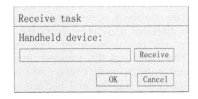

Figure 4. Screenshot of interface used to download the task on the handheld device.

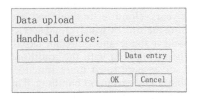

Figure 5. Screenshot of interface used to upload data on the handheld device.

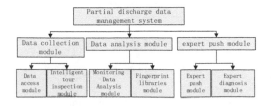

Figure 2. Module structure diagram.

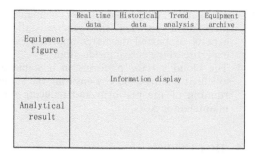

	Real time data	Historical data	Trend analysis	Equipment archive
Equipment figure				
		Information display		
Analytical result				

Figure 6. Test query and analysis.

and analysis module is shown in Figure 6. Users can obtain the relevant details in the "Information display" area by clicking on the "Real time data", "Historical data", "Trend analysis", and "Equipment archive".

In addition to the single data analysis, the system also can compare and analyze multiple data. Some deeper information contained in the data can be mined by the contrast.

After the analysis of huge amounts of data, some representative test data are found, which represent a particular type of discharge. When similar data are detected, the type of partial discharge can be judged quickly. This typical data then are stored as fingerprints in the database.

The system provides flexible supports to collect partial discharge fingerprint data. Some users, who require a rich experience, can input the data into the fingerprint directly in the module or input the analyzed data in the data query and analysis module into the fingerprint.

2.3 Expert's knowledge push

Due to the variability of partial discharge and limitations of the system itself, the system is unable to analyze all the discharge types of some partial discharge data accurately. In order to solve this problem, an expert analysis module is added to the system, which can send uncertain data to the experts and inform the experts in time.

After experts receive the pushed information, they can enter the system, analyze the data, and then feed the results back into the system in time. In this way, the system can learn the experts' professional knowledge and grow smarter.

3 CLOUD SERVICES

3.1 Shortcomings of traditional systems

The traditional systems are usually deployed on the user's own severs. However, the systems stability

could not be fully guaranteed due to the limits of their own servers, which include the following:

1. Server performance is limited. Since the users' server configuration is limited, it is difficult for the disk capacity to meet the demand of the growing volume of data. And the performance of the servers cannot meet the demands when the number of the visitors and concurrent accesses both increase.
2. The computer room environment is variable. The user's server room is vulnerable to disruption of power and network and server downtime can easily occur. In addition, there will be system crash led by some staff misuse.

3.2 The advantages of cloud servers

Cloud servers are a group of multiple virtual servers on a group of cluster servers, in which every server has a mirror of cloud servers. In this way, the security and stability of the virtual server is greatly increased. Cloud servers can be accessible anytime unless all servers in the cluster have problems. Advantages of cloud servers include the following:

1. The cloud server is a WEB-based service, providing an adjustable elastic cloud hosting configuration, integration of the *Iaas* services of the computer storage and network resources. It also has an on-demand use capability and on-demand instant payment capability of cloud hosting services. When compared with traditional servers, it has great improvement in flexibility, controllability, scalability, and reusability of resources.
2. Host service configuration and business scale can be configured according to the user's requirements and can be flexibly adjusted.
3. Host services user applied can be supplied and deployed rapidly, thereby achieving elastic characteristics in the clusters.
4. Each cluster node of the host platform is deployed in one of the Internet's backbone computer room, which can provide Internet infrastructure services independently such as computing, storage, online backup, and hosting.

Based on the above advantages of cloud servers, the partial discharge data management system is deployed on the cloud servers.

4 THE IMPLEMENTATION OF OPERATION

We chose Huawei cloud servers finally after trying different cloud servers in order to ensure the running effects of the system on cloud severs. The configuration of the cloud servers is given as follows.

1. Region: Eastern China
2. CPU: 8 nuclear
3. Memory: 16 G
4. Bandwidth: 10 Mbps
5. System: Windows Server Enterprise 2008 R2, 64 bit
6. System disc: 100 GB
7. Data disc: 500 GB

4.1 *System deployment*

When the cloud servers are determined, system deployment is carried out by the basic steps, which are as follows:

1. Configure cloud servers including configuring security group configuration of cloud servers, thereby setting the server inbound and outbound rules to ensure the safety of cloud servers.
2. Install a Web server which uses IIS7.
3. Install a database server. The database server software is Microsoft SQL Server 2012.
4. Distribute the program with Microsoft Visual Web Developer 2008 and release the system to IIS.
5. Input basic data. Basic data such as the administrator account and the basic business data need to be fed as input into the database before the formal operation.
6. Run the system program. The system can be accessed and the platform can run normally after the completion of the above items.

After the system deployment is completed, the system can be accessed via a browser. If users want to access the platform by using domain names, the domain names should be registered and recorded. Otherwise, users can only access the platform by using IP address. However, if the server is not in China, there is no need to register.

4.2 *Operating effects*

The platform is proved through actual use for some time that it can run normally and meet the requirement of the design. The platform has achieved the desired effects in the following manner:

1. The platform collected a large number of partial discharge data efficiently and accurately which made a good foundation for later data mining.
2. The platform analyzed the partial discharge data accurately by supporting users to take measures about the equipment, which had potential hazards, and saved huge economic losses.

3. The platform established a timely communication mechanism between the general staff and experts of partial discharge applying experts' expertise to the actual production.
4. The cloud service performs in a stable and excellent manner, thereby ensuring the smooth running of the platform and reducing server maintenance costs.

5 DISCUSSION

Data collection function of the platform is easy to use, by which users can input data directly or via the data importing process or via the handheld device according to their needs. The platform can accurately analyze partial discharge data and obtain partial discharge type and give the diagnosis result, thereby providing users with decision support. The platform is safe, reliable, and stable which intercepted several malicious attacks successfully, thus saving the maintenance cost of servers.

REFERENCES

Darabad, VP., M Vakilian, BT Phung, TR Blackburn, An Efficient Diagnosis Method for Data Mining on Single PD Pulses of Transformer Insulation Defect Models, 2013, 20(6):2061–2072.

Ibrahim, K., RM Sharkawy, MMA Salama, R Bartnikas, Realization of Partial Discharge Signals in Transformer Oils Utilizing Advanced Computational Techniques, 2012, 19(6): 1971–1981.

Kantarci, B., L Foschini, A Corradi, HT Mouftah, Inter-and-Intra Data Center VM-Placement for Energy-Efficient Large-Scale Cloud Systems, Globecom Workshops, 2012, 48(11):708–713.

Murakami, K. A New Steganography Protocol for Improving Security of Cloud Storage Services.

Ou, G., Y Liu, X Ma, C Wang, DM-Midware: A Middleware to Enable High Performance Data Mining in Heterogeneous Cloud, 2013, 3:70–73.

Salimijazi, HR., L Pershin, TW Coyle, J Mostaghimi, S Chandra, Practical Applications of Data Mining in Plant Monitoring and Diagnostics, IEEE Power Engineering Society General Meeting, 2007, 16(4):1–7.

Wang, K., J Li, S Zhang, Y Qiu, Time-Frequency Features Extraction and Classification of Partial Discharge UHF Signals, International Conference on Information Science, 2014, 2:1231–1235.

Advances in Energy Science and Equipment Engineering II – Zhou, Patty & Chen (Eds)
© 2017 Taylor & Francis Group, London, ISBN 978-1-138-71798-5

Flow simulation of transmission conductors considering the surface characteristics

P. Li, B. Liu, J.L. Yang, X.Z. Fei, X.P. Zhan, L.C. Zhang, K.P. Ji & B. Zhao
China Electric Power Research Institute, Beijing, China

Y.X. Dong
North China Electric Power University, Beijing, China

ABSTRACT: The section features of three kinds of transmission conductors are summarized and a geometric simplification method and mesh scheme to describe the surface roughness of the conductor are proposed. CFD flow simulation of the conductors is realized based on the DES method. The velocity and pressure distributions of different conductors are obtained. Results show that the refined models can effectively simulate the Carmen vortex characteristics of conductors with different cross sections; the vortex of the smooth round conductor presents most regular and minimum scale, the vortex of the round stranded conductor is relatively larger and easier to spread, and the Carmen vortex's scale and clarity of the molded conductors is in between the former two conductors.

1 INTRODUCTION

Different forms of wind-induced vibration appear in transmission conductors under wind load, and the wind-induced vibration is one of the main disasters that endanger the safety and stability of transmission conductors (Rawlins, 1979). With the rapid expansion of the scale of China's power grid construction and the changing climate and environment, wind-induced disasters (see Figure 1) in conductors are becoming more serious since the beginning of the new century, and seriously threaten the operation of transmission conductors almost every year. Typical wind-induced disasters include aeolian vibration, windage, galloping, sub-span oscillation, wind shift, and so on.

At present, the design value of the wind load on conductors is mainly aimed at the limited results of wind tunnel tests (Lou et al., 2015); most of them are obtained on the basis of flow simulation around the smooth cylindrical cross-section and there exists a certain difference from the real type of conductors (Iwan 1975, Brika 1997, Laneville 1999). Therefore, the precise calculation of the aerodynamic load of the conductors, with consideration of the characteristics of the conductor cross section, is of great significance for enhancing the anti-wind design ability of the transmission conductors. A fine meshing scheme for describing the surface roughness of different types of conductors is proposed based on the ADINA platform in this paper, the CFD simulation models are established

Figure 1. Pictures of wind-induced disasters in conductors.

for three kinds of typical conductors and the flow calculation is carried out, and the flow field distribution and Carmen vortex of all kinds of conductors are obtained.

2 SIMULATION AND SOLUTION METHOD FOR AERODYNAMIC CHARACTERISTICS OF CONDUCTORS

2.1 *Simplified cross-sections of typical conductors*

The transmission conductor, which is commonly used in the present project, includes a smooth round conductor, round stranded conductor, molded conductor, and so on (see Figure 2). The cross-section deformation and loose degree of the conductor can be neglected when the bluff body flow calculation is conducted with fluid mechanics, and consider as a rigid section for modeling. That is only the boundary shape difference of the conductor is considered, the surface roughness is

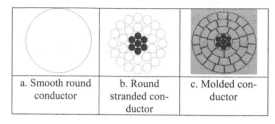

| a. Smooth round conductor | b. Round stranded conductor | c. Molded conductor |

Figure 2. Schematic of commonly used forms of conductors.

described by using different quantities of semi-arc in the numerical model for round stranded conductors, and expressed by the circular cross-section with a V-type small gap for molded conductors.

2.2 Scheme of flow field mesh generation

Two-dimensional flow around the cylinder model of three types of conductors is built with an ADINA platform. In order to accurately calculate the aerodynamic characteristics of the conductor under wind load, reasonable models are necessary to describe the surface roughness of the conductor. The grid near the edge of the conductor should be refined and should be sufficient enough to simulate a small gap at the boundary in the CFD model. In addition, the regularized grid should be mainly distributed in the entire model in order to ensure the efficiency and accuracy of the calculation. For eliminating the boundary effect, the upstream dimensions of the calculation area of the flow field should not be less than 5D, downstream dimensions should not be less than 15D, and height should not be less than 10D. The parameters of the flow field are set as follows: the air density is $\rho = 1.18$ kg/m³, the dynamic viscosity is $\mu = 1.82 \times 10^{-5}$ N•s/m², and the kinematic viscosity is $v = 1.542 \times 10^{-5}$ m²/s. The slip boundary condition is set in the upper and lower walls of the flow field, and a non-slip boundary condition in the outer surface of the conductor, and velocity loads are applied at the inlet boundary.

Grid partitions of the round conductor, round stranded conductor, and molded conductor in the CFD model can be seen in Figures 3–5, respectively.

2.3 The resolution methods of CFD

The simulation of fluid around conductors in this paper is a kind of a high Reynolds number problem, and involves large-scale separation of vortices and fluid-structure coupling. With the LES method, not only the calculation cost is very high, but also the operability is poor and the accuracy

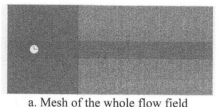

a. Mesh of the whole flow field

b. Refined mesh for the partial flow field

Figure 3. Mesh in the CFD model of a smooth round conductor.

a. Mesh of the whole flow field

b. Refined mesh for the partial flow field

Figure 4. Mesh in the CFD model of the round stranded conductor.

b. Refined mesh for the partial flow field

Figure 5. Mesh in the CFD model of a molded conductor.

is hard to guarantee under the existing platform. The RANS method is limited in calculation precision and the transient solutions for time history analysis are not available. In order to obtain more accurate transient turbulence simulation results and to achieve fluid solid coupling analysis of aerodynamic characteristics of the conductors with a relatively low calculation cost, the Detached Eddy Simulation (DES) is applied to solve the CFD model in this paper, the RANS model is based on the Spalart–Allmaras model of an equation. Detached Eddy Simulation (DES) combines the advantages of LES and RANS method, the RANS turbulence model is applied to the boundary layer near the wall, and taking advantage of the RANS method in the simulation of the wall shear layer and the efficiency of calculation. Besides, large eddy simulations are carried out by using the sub-grid scale (SGS) model in the region away from the wall, so that the separation of large scale vortices can be simulated more precisely and accurately.

The eddy viscosity v_t in the Spalart–Allmaras model is defined as follows:

$$v_t = \tilde{v}f_{v1} \tag{1}$$

where, $\tilde{v}$ is the modified eddy viscosity coefficient and f_{v1} is the dimensionless function of $\tilde{v}$.

The transport equations (SA equation) of $\tilde{v}$ can be deduced based on equation (1) as follows:

$$\frac{\partial \tilde{v}}{\partial t} + u_j \frac{\partial \tilde{v}}{\partial x_i} = C_{b1}[1 - f_{t2}]\tilde{S}\tilde{v}$$
$$+ \frac{1}{\sigma}\{\nabla \cdot [v + \tilde{v}]\nabla\tilde{v} + C_{b2}|\nabla v|^2\} \tag{2}$$
$$- \left[C_{w1}f_w - \frac{C_{b1}}{k^2}f_{t2}\right]\left(\frac{\tilde{v}}{d}\right) + f_{t1}\Delta U^2$$

where

$$f_{v1} = \frac{\chi^3}{\chi^3 + C_{v1}^3}, \chi = \frac{\tilde{v}}{v}, \tilde{S} = S + \frac{\tilde{v}}{\kappa^2 d^2}f_{v2}$$

$$\Omega_{ij} = \frac{1}{2}\left(\frac{\partial u_i}{\partial x_j} - \frac{\partial u_j}{\partial x_i}\right), r = \frac{\tilde{v}}{\tilde{S}\kappa^2 d^2}, f_{v2} = 1 - \frac{\chi}{1 + \chi f_{v1}}$$

$$f_w = g\left[\frac{1 + C_{w3}^6}{g^6 + C_{w3}^6}\right]1\backslash 6, g = r + C_{w2}(r^6 - r)$$

$$S = \sqrt{2\Omega_{ij}\Omega_{ij}}$$

$$f_{t1} = C_{t1}g_t \exp\left(-C_{t2}\frac{w_t^2}{\Delta U^2}[d^2 + g_t^2 d_t^2]\right)$$

$$f_{t2} = C_{t3}g_t \exp(-C_{t4}\chi^2), gt = \min(0.1, \Delta U / w_t \Delta x)$$

The values of the constants in the above equation are as follows:

$$\sigma = 2/3, C_{b1} = 0.1355, C_{b2} = 0.622, \kappa = 0.41,$$
$$C_{w1} = C_{b1}/\kappa^2 + (1 + C_{b2})/\sigma = 3.24, C_{w2} = 0.3, C_{w1} = 7.1,$$
$$C_{t1} = 1, C_{t2} = 2, C_{t3} = 1.1, C_{t4} = 2$$

The d in equation (2) represents the distance from a point to the nearest wall in the flow field. In terms of the DES method, based on the Spalart–Allmaras model, $\tilde{d} = \min[d, C_{DES}\Delta]$ is substituted for d, C_{DES} is the constant coefficient and its value ranges from 0.61 to 0.78 (usually 0.65), Δ is the unit scale, the cube root of unit volume and the square root of area are respectively taken for three-dimensional and two-dimensional meshes. The $\tilde{d} = \min[d, C_{DES}\Delta]$ is used as an automatic switch to control the transformation from a turbulence model to a SGS model in the DES method, the advantages of RANS and LES simulation are combined to not only significantly reduce the grid and computing time, but also to achieve more ideal simulation results of turbulence.

The transient analysis module and the pressure-velocity coupling algorithm SIMPLE are applied in this paper, the time step is taken as 0.02 s, the number of calculation steps is 100, and the scheme of time integration is the composite integration.

3 IMPLEMENTATION OF FLOW CALCULATION

From the view of the fluid, the vortex detaching from the surface of the structure can be produced alternately on both sides of any non-stream conductor object under a condition of certain constant flow velocity. For general cylindrical section structures in engineering, this alternate purging vortex can lead to produce periodic fluctuating pressure on the cylinder with stream-wise and transverse direction. If the cylinder is on elastic support or the elastic deformation of a flexible pipe body is allowed, the pulsating fluid force will cause the periodic vibration of the cylinder (tube). This kind of regular vibration of the cylinder will also change the vortex-shedding pattern of its wake in turn, namely the vortex-induced vibration, which is also the highest frequency of vibration in transmission conductors. In this section, the fine flow simulations of smooth round conductors, round stranded conductors and pattern conductors are fulfilled in accordance with the second section, and the Carmen Vortex of three kinds of conductors is obtained with a condition that the diameter of conductor D is 50 mm and the wind speed V is 5 m/s (see Figures 6–8).

Figure 6a and b show that the flow field passed around the conductor with circular cross-section and produced clear and regular Carmen vortexes, and the scale of shedding vortexes increased gradually and turned round; those vortexes which were in the leeward side of the conductor swung up and down and then fell off. The maximum speed appeared on both upper and lower sides of the

a. Velocity distribution in the whole flow field (m/s)

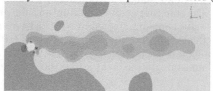

b. Velocity distribution in the partial flow field (m/s)

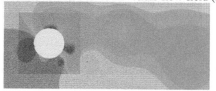

c. Pressure distribution in the whole flow field (Pa)

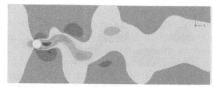

a. Velocity distribution in the whole flow field (m/s)

b. Velocity distribution in the partial flow field (m/s)

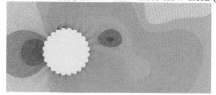

c. Pressure distribution in the whole flow field (Pa)

d. Pressure distribution in the partial flow field (Pa)

Figure 6. Calculation results of the wind flow around the conductor with circular cross-section (t = 2.0 s).

d. Pressure distribution in the partial flow field (Pa)

Figure 7. Calculation results of wind flow around stranded conductors with circular cross-section (t = 2.0 s).

separation zone and the minimum value appeared in the center of vortexes. And what is more, Figure 6c and d show that the maximum stress value of the wind pressure appeared in the center of vortex and in both the windward side and leeward side, the minimum value appeared in the unilateral separation zone.

Figure 7a and b show that the flow field passed around stranded conductors with circular cross-section-produced regular Carmen vortexes. The scale of vortexes was elongated, and the vortex in the downstream showed faster divergence; the maximum value of speed also appeared on both sides of the separation zone, as well as minimum value of speed was also located in the center of the vortex. Figure 7c and d show that it is obvious that the maximum point of the pressure distribution appeared on the windward, as well as the minimum point on the leeward side; in addition, there was a pressure change in a small scale in the edge gear of the conductor, which proved that the model could describe the influence of roughness of the stranded conductor with circular cross-section.

Figure 8a and b show that regular Carmen vortexes also appeared when the flow field passed around a shaped conductor with circular cross-section. The scale of vortexes here was between smooth conductors and stranded conductors. Vortexes' shape was clear and regular. The maximum value of speed was in the unilateral separation zone, and the minimum value of speed was in the center of vortex. The small scale of vortexes in the edge of the conductor was not obvious, but because of the V-type small gap, there were differences between vortexes distributions of shaped conductors and smooth conductors. Figure 8c and d show that there was one extreme point of pressure on both the windward and leeward side, the characteristic of wind pressure distribution was observed between the smooth conductor and stranded conductor.

In summary, Carmen vortexes for smooth conductors with circular cross-section was more clear, regular, and had a minimum scale; Carmen vortex

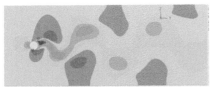

a. Velocity distribution in the whole flow field (m/s)

b. Velocity distribution in the partial flow field (m/s)

c. Pressure distribution in the whole flow field (Pa)

d. Pressure distribution in the partial flow field (Pa)

Figure 8. Calculation results of wind flow around shaped conductors with circular cross-section (t = 2.0 s).

for stranded conductors was easier to diverge, and scale of vortexes in the downstream was larger; the Carmen vortex for shaped conductors had a medium scale and clarity in comparison with the other two vortexes. In this paper, the grid partition scheme of the CFD model as well as the solving method was reasonable, so that they could reasonably describe and reflect the influence of different sections' form and roughness.

4 CONCLUSION

In this paper, flow simulations of different transmission conductors are carried out considering the section features. Mesh schemes to describe the surface roughness of conductors are proposed and the Carmen vortex and flow field distribution for three kinds of conductors are obtained. Results show that the model could effectively simulate the flow field around the conductors with different sections; the vortex of smooth round conductors presents the most regular and minimum scale, the vortex of round stranded conductors is relatively larger and easier to spread, the Carmen vortex's scale and clarity of the molded conductors is in between the former two conductors.

ACKNOWLEDGMENTS

The work described in this paper was supported by the State Grid Corporation Project (GYB17201500154).

REFERENCES

Brika, D. & Laneville, A. (1997). Vortex-induced oscillations of two flexible circular cylinders coupled mechanically. *Journal of Wind Engineering and Industrial Aerodynamics*, 69–71:293–302.

Iwan, W.D. (1975). The vortex induced oscillation of elastic structure elements. ASME, *Journal of Engineering for Industry*, 97: 1378–1382.

Laneville, A. & Brika, D. (1999). The fluid and mechanical coupling between two circular cylinders in tandem arrangement. *Journal of Fluids and Structures*, 13(7–8): 967–986.

Lou, W.J., Li, T.H., Lv, Z.B. & Lu, M. (2015). Wind tunnel test of aerodynamic coefficient of multi split sub conductors [J]. *Journal of air dynamics*, 33 (6): 787–792.

Rawlins, C.B. (1979). Transmission line reference book: *wind-induced conductor motion* [M]. EPRI: Palo Alto, CA.

Research and implementation of vehicle control systems

Zhan Li

Beijing China-Power Information Technology Co. Ltd., Beijing, China

ABSTRACT: The rapid development and wide application of an intelligent transportation system, not only effectively solve traffic congestion, but also have an enormous impact on the process of speeding up the accident and rescue, thereby improving transport management and other aspects of capacity. This article starts with the vehicle control system structure, and proceeds further with research on the control of speed and induction control.

1 INTRODUCTION

The rapid development and application of an intelligent transportation system not only effectively solves traffic congestion but also speeds up the process, and it has a significant impact on improving transportation management and other aspects. The overall analysis of various factors that integrate and take advantage of high-tech thinking in solving traffic problems came into being, which is called as the Intelligent Transportation System.

An intelligent transportation system consists of the following two parts, one is a series of advanced technology for transportation systems, and the other is a variety of services offered by this technology (Lu, 2004). Its essence is using high and new technology, especially utilizing information technology to transform traditional transportation systems, and form "a new type of modern transportation system based on information, by means of modern communication and computer, in a safe, efficient, service as the goal" (Zhang, 2013).

As the core part of an intelligent transportation system, an intelligent vehicle should have independent, autonomous navigation, positioning, and car-driving as well as wireless communications and other functions. A vehicle control system is the basis and guarantees to realize the function.

2 VEHICLE CONTROL SYSTEMS

A vehicle control system feeds driving decision-making as the input and obtains the physical exercise as the output of vehicles. Its executive body is a DC motor, and it obtains a sensor module as the feedback subject, a controller as the main control module of the vehicle motion control system.

An intelligent vehicle control system structure is shown in Figure 1.

The execution parts of a vehicle control system are two DC motors in the front and back. The DC

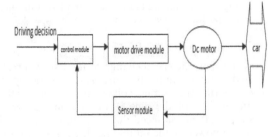

Figure 1. Schematic of a vehicle control system structure.

motor in the front is used to control the smart cars' direction, and the DC motor at the end is used to control the intelligent vehicles' driving state and driving speed.

The vehicle control system aims to achieve the vehicle speed control and direction control, which is divided into the speed control module and the induction control module in this article.

3 SPEED CONTROL MODULE

The speed control module makes the decision of speed, which is fed by the system control module as input, and obtains the actual speed as output. The executive main body is the DC motor, and the feedback main body is the speed sensor, and the controller is a single chip microcomputer control chip.

3.1 *Speed control strategy research*

In the vehicle speed control system, the DC motor's dynamic characteristics or structure will change with the change of load or disturbance factors (Cakir, 2011). Therefore, it is necessary to study the control algorithm.

The control algorithm mainly includes the PID control algorithm and fuzzy control. Fuzzy control is based on rules, and so the controlled object's precise math model need not be set up in the design, and it is mainly based on the control experience of the field personnel or relevant experts', which makes the control mechanism and strategy easy to accept and understand (Zeng, 2013). Therefore, in this paper, the fuzzy control method is used as the speed control system. The speed control system's structure is shown in Figure 2.

From Figure 2, we can see that the control module of the vehicle system makes the driving behaviour decision, after providing the set speed, the vehicle motion control system module compares the set speed with the speed of feedback, and feeds the error and differential value as two input values of the fuzzy control. According to the preliminary fuzzy control rule table, which is the designed output for the corresponding velocity increment, the motor speed is changed by changing the single chip microcomputer to realize the vehicles according to the set speed.

3.2 Research on the DC motor drive module

At present, the speed change of the PWM is realized by adjusting the pressure. The basic idea is by comparing the average voltage and current of the motor, 50% of the time by turning on the power supply of motor speed than 40% time connected to the power of the motor speed; when the motor is not receiving power, it does not consume energy. The vast majority of the DC motor uses switch mode, so that the semiconductor power components work in the state of the switch, through pulse width modulation to control the DC motor armature voltage to achieve the desired speed.

The average value U_a of the voltage at both ends of the armature winding of the DC motor is given by the following equation:

$$U_a = \frac{t_1 U_s + 0}{t_1 + t_2} = \frac{t_1}{T} U_s = \alpha U_s \qquad (1)$$

Type: α is the duty cycle, and the ratio of the length of time and the period of the switch tube conduction is expressed in a cycle T. The change range of α was $0<\alpha<1$. Basic principle of PWM speed control: When the supply voltage U_s is constant, the average value U_a of the voltage at both

ends of the armature winding depends on the size of the duty cycle α; changing the value of α can change the average value of the terminal voltage, so as to achieve the purpose of speed.

α is an important parameter in the PWM speed control. At present, the control of the DC motor mainly uses the method of constant frequency width modulation.

The circuit of this paper is shown in Figure 3.

As shown in Figure 3, direction is the motor's steering control input signal and PWM is the motor's speed control input signal. 3C and 3D are input signals in the motor drive circuit. When the direction feeds high electricity as input, the collector of the triode Q4 output to a low level, through the 3D output level is low, and the 3C output is the PWM signal. Thus, the rotation speed of the motor according to the duty ratio of the PWM is realized. Conversely, when the direction feeds low electricity as input, the triode collector output to a high level, through the door after the output PWM signal, and 3C output is low level, so as to realize the motor according to the PWM's corresponding speed reverse rotation.

3.3 Speed detection module

The accuracy of speed detection will affect the control effect of the vehicle motion control system, whether it is the kind of speed control system, the high accuracy control systems only from the high accuracy detection state.

According to the principle in which the photoelectric sensor output pulse frequency is proportional to the speed, the speed can be measured by using pulse frequency, which is obtained as the output by using the photoelectric sensor, namely using the photoelectric encoder connected with the motor (Yu, 2005). The speed measuring sensor detection circuit is shown in Figure 4.

As shown in Figure 4, the photoelectric unit feeds input signals into the microcontroller sensor through a hysteresis comparator circuit. The photoelectric sensor's location is static; when the direct photoelectric sensor to the encoder gaps between tooth slice, the voltage comparator output is low level, when the photoelectric sensor's light is reflected by

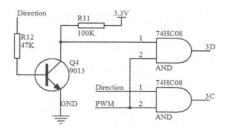

Figure 3. Speed and steering synchronous logic circuit diagram.

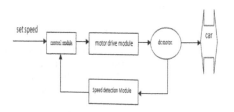

Figure 2. Schematic of the speed control system structure.

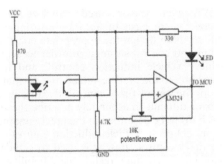

Figure 4. Speed sensor circuit diagram.

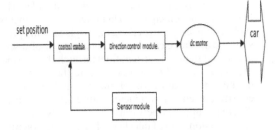

Figure 5. Schematic of the induction control module structure.

the encoder tooth slice, voltage comparator output is high level. According to code rotating, the MCU receives the voltage pulse signal continuously, and so the pulse frequency is also changed.

In a speed measuring system consisting of photoelectric code, cycle frequency synthesis is chosen, namely the M/T method (Xu, 2014). Time T1 is determined in the M/T method, and then on the basis of the T1 extend a changing time ΔT, so that the actual measurement time $T = T1 + \Delta T$. By recording a high frequency pulse number within T, the precise numerical value is obtained, and then the actual speed is launched.

4 INDUCTION CONTROL MODULE

The control object of the induction control system is the vehicle's driving direction, the executive main body is the direction control module, the feedback main body is the induction sensor module, and the controller is the microcomputer control chip of a single chip. The vehicle induction control module structure is shown in Figure 5.

4.1 Induced mode selection

Vehicle induction control technology is the foundation of the car and path's integration. In this work, magnetic induction technology is mainly introduced, and it has high accurate measurement and good repeatability, it is less-affected by the outside world, and what is more, the location of the measurement and control can be used in vehicles. When vehicles pass through the magnetic steel, the automotive magnetic sensor detects the magnetic signal and generates induced voltage, an on-board computer determines the relative position in the target lane of the vehicle according to the size of the induced voltage, in order to control automatic vehicle steering in real-time (Hung, 2002).

4.2 Direction control module

The direction control module is the executive main body which inducts a control subsystem, and it

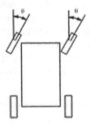

Figure 6. Schematic of a vehicle corner map.

is made of a drive circuit, DC motor, and corner detection feedback device.

The vehicle corner diagram is shown in Figure 6. When the vehicle turns to the right, θ is positive; when it turns to the left, θ is negative. The scope of θ is between -30 and 30. The direction control system assembly is completed, and the angle to the potentiometer output value and the actual measurement are measured and contrast, and it can satisfy the requirement of the detection system for the car steering angle.

Due to fact that the motor's speed itself is not high, the query method is adopted, namely set rotation angle and direction; the corresponding output value is fed as the input value of the direction control module, and the output value is the feedback value of the module through sampling. The system decides motor steering according to the error, and when the error is less than a certain range, the motor is shut down. So the front wheel can rotate to the point set by the system.

4.3 Induction sensor module

An induction sensor module is designed for a magnetic sensing module. The magnetic field intensity can be measured through a variety of technology; most of the magnetic sensors convert magnetic signals into electrical signals, so as to realize the magnet's magnetic field strength measurement.

Different magnetic sensors have different application occasions and measuring ranges, thereby selecting the hall sensor to form a vehicle induction control sensor module in there.

A Hall sensor is the use of Hall element and will be based on the principle of Hall effect measurements such as current, magnetic field, displacement, pressure, etc into an electromotive force output of a sensor. When there is a current flow through the outside, the magnetic field is perpendicular to the conductor; in the vertical direction of the magnetic field and electric current in conductors, an electric potential difference appears between two end faces, and this phenomenon is the Hall effect. The electric potential difference is called Hall electric potential difference.

A HW300B Hall effect sensor detection circuit is shown in Figure 7.

The Hall electric potential sends the difference of the sensor $V_h = V_+ - V_-$ as output in Figure 7, by using a differential amplifier circuit; an enlarged Hall electric potential difference can be received as the output, by using single-chip microcomputer sampling, and is advantageous for the system testing (Jia, 2015).

4.4 Positioning and induction control algorithm research

Major positioning methods use the magnetic field matching method and the sequence algorithm (Xu, 2007). Magnetic field matching algorithms respond significantly by interference and vehicle vibration, and so we use the sequence algorithm for magnetic localization and induction control.

A sequence algorithm is sorting for each size of the sensor signal, according to the sorting results to determine the lateral migration of a kind of algorithm. When receiving the signal of magnetic steel, the nearer the magnetic feet distance is to the magnetic location, the greater is the voltage of the sensor that is outputted. And so, one can detect the magnetic steel with the sensor that is the closest and can calculate the lateral displacement of the vehicle according to the type D.

$$D = 7N - 21 \tag{2}$$

Type D is the distance between vehicle lateral displacements and N is the sensor serial number of the closest sensor to the magnetic steel.

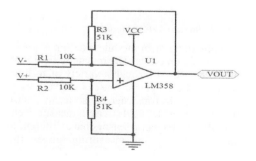

Figure 7.   HW300B detection circuitry.

A magnetic sensor is used to analyze the vehicle lateral, on the basis of accurate positioning, and still needs to design the appropriate control algorithm and control strategy, in order to accurately achieve vehicle induction control, and make it according to the predetermined line, so that it can adopt the way of fuzzy control. Because it is complex or difficult to accurately describe the system, it is hard to be described in the precise mathematical model of the vehicle induction control system, the fuzzy control strategy is used to simulate the pilot lane keep state, realize the reliability of the test system, and improve the security of practice.

5   CONCLUSION

This study, based on the vehicle motion control systems, is based on the two aspects, such as speed control subsystem and the induced control subsystem. Under the precondition of research on the vehicle speed control strategy, research on the DC motor driving module and speed detection module led to better realization of the accurate control of vehicle speed. According to the magnetic localization algorithm, induction control strategy, and magnetic induction sensor module, carrying out the direction control module is to realize the vehicle's precise induction control on the basis of vehicle-induced control mode selection. A vehicle control system which is the foundation of the intelligent control system has a certain practical significance in the actual operation.

REFERENCES

Cakir B, Yildiz A.B, Abut N, Inanc N., DC motor control by using computer based fuzzy technique, Conference Proceedings—IEEE Applied Power Electronics Conference and Exposition, v1, 2011:391–395.

Hung Pham. Integrated Maneuvering Control for Automated Highway Systems Based on a Magnetic Reference/sensing System, California PATH Research Report, 2002:49–50.

Jia bonian, Yu Pu. Sensor Technology 2nd edition, Nanjing, Southeast University Press, 2015.

Lu Huapu, Li Ruimin, Zhu Yin. Intelligent transportation Systems Introduction[M]. Beijing: China Railway Publishing House, 2004.

Xu Haigui, Wang Chunxiang, Yang Ruqing. Magnetic sensing system in the outdoor mobile robot navigation Research, robotics, 2007, 1: 28–32.

Xu Jing, Ruan Yi, based on TMS320F240 M/T speed of implementation and application of law, the inverter world, 2014, 4:41–43.

Yu Pingliang. motor speed measurement method, Shandong Science, 2005, 5:41–51.

Zeng Guangqi. Fuzzy control theory and engineering applications, Wuhan, Huazhong University Press, 2013: 32–34.

Zhang Guowu, Peng Hongqin. Intelligent Transportation Systems Engineering Introduction [M]. Beijing: Electronic Industry Press, 2013.

Advances in Energy Science and Equipment Engineering II – Zhou, Patty & Chen (Eds)
© *2017 Taylor & Francis Group, London, ISBN 978-1-138-71798-5*

Research on high-frequency isolated interconnection converters and their PWM strategy in AC/DC hybrid microgrids

Hujie Xue & Qichuan Tian
Beijing University of Civil Engineering and Architecture, Beijing, China

ABSTRACT: An AC/DC hybrid microgrid can improve the reliability and reduce the cost of hybrid microgrids. How to interconnect the AC and DC sub-grids efficiently is a key problem to be solved. In order to overcome the shortcomings of conventional interconnection technologies, in this paper, a novel AC/DC hybrid interconnection technology based on high-frequency isolated bidirectional converters is proposed. The operation principle and soft-switching mechanism of the converter are analyzed in detail. A novel pulse width modulation strategy is proposed. The proposed PWM strategy utilizes the phase current information to determine the interval of the leg commutation and eliminates the circuit that detects the voltage across the switch, which keeps the merit of soft-switching and simplifies the PWM strategy greatly. In order to verify the proposed strategy, a simulation platform is established and the experiments are performed on the platform. The results show the validity and feasibility of the proposed converter and its PWM strategy.

1 INTRODUCTION

Nowadays, Renewable Energy Sources (RESs) are under rapid development around the world because of their sustainability and cleanness. But the intermittent characteristic of the RESs will adversely affect the stability of the electric power system if the penetration of RESs is too high. A microgrid is an effective solution for the utilization of RES in large scale while maintaining the stability of the electric power system. Many microgrid projects have been built in the world. AC, DC, or both bus voltages can be adopted when a microgrid is constructed. If an AC bus voltage is adopted, the microgrid will be compatible with the power grid easily. However, most RESs, for example, PV, fuel cell, etc., output electric power in DC voltage. In order to interface the AC microgrid, the electric energy from the DC RES has to be converted into AC energy, which needs several stages of conversion with power electronic converters. On the other hand, more and more equipment, for example, the computer, home appliance, and other electronic devices require DC voltage to function normally. And so, an AC bus cannot power the DC load directly and can cause low efficiency, low reliability, and high cost because of necessary DC to AC and AC to DC power conversions. If a DC bus voltage is adopted, the power conversion mentioned above can be eliminated and high efficiency, high reliability, and low cost will be achieved. But the DC voltage is hard to be connected in the power

system and cannot meet the need of AC equipment. Therefore, an AC/DC hybrid microgrid can be built to power both AC and DC equipment meanwhile to take advantage of the DC microgrid (Nejabatkhah, 2015). In a hybrid microgrid, the AC and DC sub-grids can be interconnected through a power electronic converter. After interconnection, the power can flow between the two sub-grids. The AC sub-grid can be a backup for the DC sub-grid and vice versa, which improves the reliability of the whole microgrid. The typical topology of an AC/DC hybrid microgrid is shown in Figure 1.

In this paper, the interconnection converter and its PWM strategy in AC/DC hybrid microgrids are studied and organized as follows: Section II explains conventional interconnection converters and proposes a high-frequency isolated interconnection converter. Section III addresses the PWM strategy of the proposed interconnection converter.

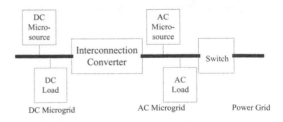

Figure 1. Typical topology of an AC/DC hybrid microgrid.

Simulation results of proposed converter and its PWM strategy in AC/DC hybrid microgrids are provided in Section IV. The conclusion is presented in the end of this paper.

2 INTERCONNECTION OF HYBRID MICROGRIDS

2.1 Conventional interconnection converters

The interconnection converter in Figure 1 can be realized with different topologies (Dong, 2013; Luo, 2013). Some converters are based on non-isolated topology, such as shown in Figure 2.

The converter in Figure 2 consists of two stages, i.e. DC/DC and DC/AC stages. If the power flows from the DC microgrid to the AC microgrid, the DC/DC stage works in the boost mode and the DC/AC stage works in the inverter mode. On the contrary, if the power flows from the AC microgrid to the DC microgrid, the DC/AC stages work as a PWM rectifier and the DC/DC stage works in the buck mode. And so, the interconnection of the hybrid microgrid can be achieved by adjusting the operation mode of the converter. Because there is no galvanic isolation in the converter in Figure 2, high frequency capacitive current from the DC sub-grid to the ground is unavoidable and grounding of the DC sub-grid is impossible. A simple method of providing galvanic isolation is to insert a line frequency transformer between the AC side of the converter and the AC sub-grid. But the transformer is undesirable in many situations because of its bulky volume and inductive characteristic. Another method of galvanic isolation is to adopt a two-stage converter, which consists of a non-isolated AC/DC stage and isolated DC/DC

stage. An example of such a solution is shown in Figure 3, where the CLLC is adopted as the DC/DC stage (Kim, 2013).

The two-stage connection converter still has the shortcomings of low efficiency and density because of multiple power conversion. Moreover, the electrolytic capacitor at the DC bus reduces the lifetime of the converter.

2.2 Proposed interconnection converters

2.1.1 Topology of proposed converters

In order to overcome the shortcomings of the solutions mentioned above, a single stage bidirectional high-frequency isolated converter shown in Figure 4 can be adopted as the interconnection converter. The converter was ever used in many different situations (Norrga, 2002). It has the ability of bidirectional power flow because active devices are used at both sides. It needs only one stage to achieve power conversion between AC and DC microgrids. Moreover, soft switching can be realized with an appropriate PWM strategy, which enables high efficiency and low EMI. And so, it is suitable for application in interconnecting the hybrid microgrid.

The converter works as follows: the difference between the bus voltage of the AC microgrid and the bridge voltage generated by switching the three bridge legs drops on the filter inductor and causes the power flow between the two sides as expected. If the power flows from the AC to DC microgrid, the converter works in isolated boost mode. If the power flows in the opposite direction, the converter works in isolated buck mode. And so, a bidirectional operation can be achieved to interconnect the hybrid microgrid.

2.2.2 Commutation process

ZCS turns-off at the AC side and ZVS turns-on at the DC side of the IGBTs in the converter and can be achieved with an appropriate PWM strategy. Meanwhile, turning off the trajectory can be improved with the capacitors paralleled with the IGBTs at the DC side.

The commutation of the IGBTs at the DC side can be illustrated in Figure 5.

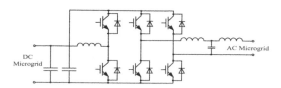

Figure 2. Non-isolated interconnection converter circuit.

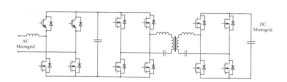

Figure 3. Two-stage high-frequency isolated interconnection converters.

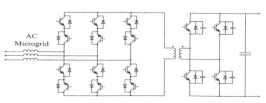

Figure 4. Proposed interconnection converter circuit.

In Figure 5(a), Q2 and Q3 are being switched on and the voltage across the transformer winding is negative. To start a commutation, the drive signals of Q2 and Q3 are removed, as shown in Figure 5(b). Then Q2 and Q3 will be turned off. Because of the clamping effect of the capacitors paralleled with them respectively, the voltages of the two IGBTs will rise slowly. And so, the turn-off loss is reduced greatly. Meanwhile, the capacitors paralleled with Q1 and Q4 begin to charge up. When the voltages across Q1 and Q4 are charged to Udc, the voltages across Q2 and Q3 are discharged to zero simultaneously. Equation 1 shows the time of charging and discharging, where N is the ratio between the turns of the AC and DC windings.

$$t_c = \frac{4CU_{dc}}{N \sum_{k=a,b,c} i_l} \tag{1}$$

And then, the diodes paralleled with Q1 and Q4 begin to conduct and keep zero voltage across corresponding IGBTs. The drive signals are applied at this moment to achieve ZVS turn-on of the IGBTs. Because the direction of the current is not changed, the current keeps flowing through the diodes rather than the IGBTs till commutation begins at the AC side. The leakage L_σ of the transformer plays an important role in commutation of the IGBTs at the AC side and cannot be omitted in the analysis.

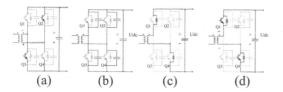

(a)   (b)   (c)   (d)

Figure 5. Circuit showing commutation at the DC side.

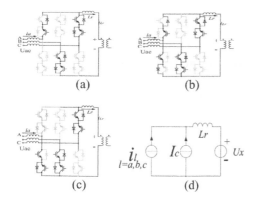

(a)   (b)

(c)   (d)

Figure 6. Circuit diagram showing commutation at the AC side.

Figure 6 shows the process of commutation at the AC side. In order to achieve the ZCS turn-off of the IGBTs at the AC side, the current at the AC side of the transformer should be injected into the positive node of the winding as shown in Figure 6. The commutations of the three legs are the same, and so only leg A is described here.

In Figure 6(a), the upper IGBT and diode are conducting the current from the filter inductor to the transformer and output voltage at the leg is positive. In order to finish the commutation, the lower IGBT and diode should be turned on firstly, which is shown in Figure 6(b), because the current flowing through the filter inductor cannot change abruptly. In Figure 6(b), both upper and lower IGBTs and diodes are switched on. The output voltage of the transformer drops on the leakage completely. And so, the current through leg A decreases linearly as indicated in Equation 2. In the equation, i_g, I_a, U_x, and t represent the current passing through Leg A, the current flowing through the filter inductor A, output voltage of the transformer, and time respectively. Because the commutation time is very short when compared with the line frequency period, the current flowing through the filter inductor A during commutation can be regarded as constant.

$$i_g = I_a - \frac{1}{L_\sigma} \cdot U_x t \tag{2}$$

After time t_c is indicated by Equation (3), i_g reduces to zero, which means ZCS of the upper IGBT and diode. All the current flowing through the filter inductor is transferred to the lower IGBT and diode.

$$t_c = \frac{I_a L_\sigma}{U_x} \tag{3}$$

The process described above can be represented by an equivalent circuit shown in Figure 6(d), in which i_l is the current flowing through the leg which is commutating and I_c is the sum of currents of the other two legs. Obviously, t_c is not constant and varies with the phase of the AC current.

3 PWM STRATEGY OF THE CONVERTER

3.1 Soft-switching and PWM strategy

According to the analysis performed above, the current should be fed as input to the positive pole of the transformer before the commutation to achieve soft-switching. Otherwise, the current in the leakage will increase linearly other than decreasing to

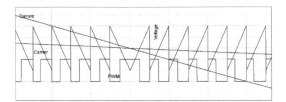

Figure 7. PWM of IGBTs at the AC side.

zero during the commutation, which will cause commutation failure. And so, the PWM strategy depends not only on the leg voltage, but also on the direction of the current at the AC side. There are different methods for achieving the PWM strategy. One method is to use different edge modulations according to the direction of the current. If the current is sent as output to the converter, the front edge of the pulse is modulated and the back edge of the pulse is modulated in the opposite case. This method is shown in Figure 7.

3.2 Realization of the PWM strategy

At the DC side, there exists dead time between the two IGBTs in the same leg. The dead time can be determined by calculating the charging and discharging time of the capacitors paralleled with the IGBTs. At the AC side, there exists an overlap of the conduction of the IGBTs in the same leg, which complicates the realization of the PWM strategy proposed above. Moreover, the overlap time varies constantly with the current through the filter inductor as shown in Equation 2. In one previous study (Norrga, 2008), the voltage is measured across every IGBT to determine the end of commutation to apply the drive signal to or remove it from the IGBT. Although the method is accurate, it needs to sense many channels of isolated voltages and is too complex. In this paper, a novel method is proposed to realize the PWM strategy, which eliminates the sensing and isolation circuit. In the method, a variable time delay is inserted into turning-on and turning-off of the IGBTs in the same bridge leg. Time delay is determined by Equation 4, in which I_m is the peak of the inductor current; t_s is the switching time; and Δt is the error allowing for the worst case.

$$t_o = \frac{I_m L_\sigma}{U_x} \sin(\omega t) + t_s + \Delta t \qquad (4)$$

4 SIMULATION RESULTS

In order to verify proposed topology and its PWM strategy, the simulation platform is constructed on the PSIM. The schematics of main circuit and PWM circuit used in simulation are shown in Figure 8 and 9, respectively.

In Figure 9, there are three parts, i.e. PWM generation circuit, which is at the left upper area of the figure; drive overlap time generation time, which is at the left lower area of the figure; and PWM to the driver conversion circuit, which is at right lower area of the figure. The switching frequency in the simulation is 20 kHz. Figure 10 shows the output current at the AC side, which indicates successful power conversion between the DC and AC microgrids.

Figure 11 shows the voltage across the AC side winding of the transformer. The polarity of the

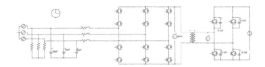

Figure 8. Diagram of the main circuit in simulation.

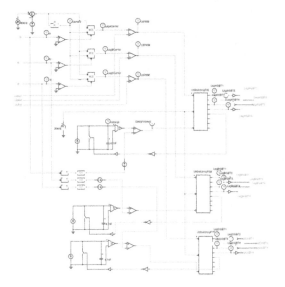

Figure 9. PWM circuit in simulation.

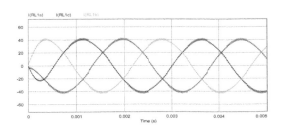

Figure 10. Output current at the AC side.

298

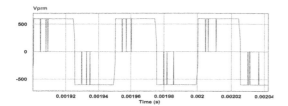

Figure 11. Voltage across the AC side winding of the transformer.

voltage is determined by switching of the DC side IGBTs. There are glitches on the wave. They are caused by an overlap conduction of the bridge legs. Each leg switches once at every half cycle, which meets the analysis above very well.

5 CONCLUSION

An interconnected hybrid microgrid can improve the efficiency of renewable energy sources while integrating them into the power system conveniently. Interconnection of the hybrid microgrid can be implemented with different methods. The proposed single-stage high-frequency isolated converter has the virtues of high efficiency, low profile, and low cost. Soft-switching can be achieved with an appropriate PWM strategy. Simple time delay can be adopted to avoid complex voltage sensing in realization of the PWM strategy. Simulation results verify the feasibility of the proposed interconnection converter and its PWM strategy in hybrid microgrids.

REFERENCES

Dong, D., Cvetkovic, I., Boroyevich, D., Zhang, W., Wang, R., "Grid-Interface Bidirectional Converter for Residential DC Distribution Systems—Part One: High-Density Two-Stage Topology," IEEE Trans. Power Electron., vol.28, no.4, pp. 1655–1666, (2013).

Dong, D., Luo, F., Zhang, X., Boroyevich, D., "Grid-Interface Bidirectional Converter for Residential DC Distribution Systems Part 2: AC and DC Interface Design With Passive Components Minimization" IEEE Trans. Power Electron., vol.28, no.4, pp. 1667–1679, (2013).

Kim, H.S., Ryu, M.H., Baek, J.W, "High-efficiency isolated bidirectional AC/DC converter for a DC distribution System" IEEE Trans. Power Electron., vol.28, no.4, pp. 1642–1654, (2013).

Nejabatkhah F. and Y.W. Li, "Overview of Power Management Strategies of Hybrid AC/DC Microgrid" IEEE Trans. Power Electron., vol.30, no.12, pp. 7072–7089, (2015).

Norrga, S., "A soft-switched bi-directional isolated AC/DC converter for AC-fed railway propulsion applications", Proc., IEE PEMD Conf. pp. 433–438, (2002).

Norrga, S., S. Meier and S. Ostlund, "A three-phase soft-switched isolated AC/DC converter without auxiliary circuit", IEEE Trans. Ind. Appl. vol.44, no.3, (2008).

U.S. Department of Energy, "High Penetration of Photovoltaic Systems into the Distribution Grid, Solar Energy Technology Program", [OL], http://www1.eere.energy.gov/ (2016).

Advances in Energy Science and Equipment Engineering II – Zhou, Patty & Chen (Eds)
© *2017 Taylor & Francis Group, London, ISBN 978-1-138-71798-5*

Research on the detecting technology of DC crosstalk with impedance

Hao Pang, Qiang Yin, Yuan Ding, Hai-bin Yu, Bai-chao Wang & Guo-qiang Pei
XJ Power Co. Ltd., State Grid Corporation of China, Xuchang, Henan, China

ABSTRACT: In view of the problem of DC crosstalk with impedance in substations, a new detecting method is proposed based on DC crosstalk with impedance. The insulation monitoring devices for the DC power system is designed. This kind of grounding fault is prevented. A prototype test shows that it is able to monitor the DC crosstalk with impedance in real time, send out the alarm information and improve the reliability of the DC system, which has the advantages of simple operation, accurate measurement, and good stability.

1 INTRODUCTION

A DC system has a very important position in substations and power plants. A DC system is power equipment for the signal equipment, the protection automation equipments, the emergency lighting, the emergency power, and the operating power supply. A DC system should provide service to the users of the hydraulic power plant, the thermal power plant, the substation, and the other users using DC power equipment.

The two independent storage batteries are chosen in the 110 kV substation and 220 kV substation in view of the importance and the demand of relay protection and circuit breaker tripping mechanism based on the reliability of the DC power supply in the technical code for designing the DC system of power projects.

In case of DC crosstalk appearing in substations, many adverse effects and harm will be introduced. It will bring the double misoperation, cause related equipment refuse operation, affect the precision positioning of the grounding fault point, result in the feed burning or fire in the DC system, shorten the service life of the battery, and affect the safety of production (REN, 2011; LUO, 2008; XU, 2008). Therefore, the DC crosstalk is an urgent problem that needs to be resolved.

It is pointed out that the feed network should use the radial power supply mode instead of the annular power supply mode in the DC system of substations. When DC crosstalk appears in a DC system, the devices should be able to send out the DC crosstalk alarm information and can choose the fault slip road.

Therefore, in this paper, the detecting technology of DC crosstalk and designs of the insulation monitoring devices for the DC power system are proposed, which can monitor the situation of DC crosstalk in real time and can send out the alarm information of DC crosstalk. A prototype tests whether the validity and feasibility of the devices are verified.

2 THE DETECTING PRINCIPLE OF DC CROSSTALK

The equivalent circuit diagram of the DC system is shown in Fig. 1, where, R1 and R2 are the resistors in the equalization bridge, R3 is the resistor in the inspection bridge, S1 is the switch in the inspection bridge, R+ and R− are the resistors of positive and negative electrodes to the grounding, C+ and C− are the capacitances of positive and negative electrodes to the grounding, and the imaginary frame is the insulation monitoring device for the DC power system.

$$R_{\Sigma+}=R1// R+// C+ \tag{1}$$

$$R_{\Sigma-}=R1// R-// C- \tag{2}$$

$$E=(V+)-(V-) \tag{3}$$

where, $R_{\Sigma+}$ is the insulation resistor of the positive electrode to the grounding, $R_{\Sigma-}$ is the insulation

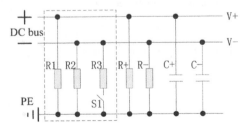

Figure 1. The equivalent circuit diagram of the DC system.

resistor of the negative electrode to the grounding, and E is the voltage between the positive bus and the negative bus.

2.1 The homopolarity interconnection with impedance

The homopolarity interconnection with impedance is the connection mode in such operation between the positive electrode and the positive electrode or between the negative electrode and the negative electrode, even with impedance between the positive electrode and the positive electrode or between the negative electrode and the negative electrode in the two-stage DC system. The mathematical model of the homopolarity interconnection with impedance is shown in Fig. 2. The homopolarity interconnection mode with impedance between the positive electrode and the positive electrode is named as A mode and shown in Fig. 2(a), and the homopolarity interconnection mode with impedance between the negative electrode and the negative electrode is named as B mode and shown in Fig. 2(b).

2.2 The heteropolarity interconnection with impedance

The heteropolarity interconnection with impedance is the connection mode with impedance in such operation between the positive electrode of I stage DC and the negative electrode of II stage DC or between the negative electrode of I stage DC and the positive electrode of II stage DC. The mathematical model of the heteropolarity interconnection with impedance is shown in Fig. 3.

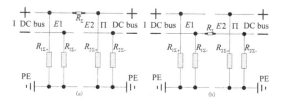

Figure 2. The mathematical model of the homopolarity interconnection with impedance.

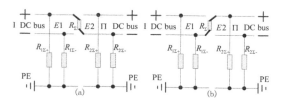

Figure 3. The mathematical model of the heteropolarity interconnection with impedance.

The heteropolarity interconnection mode with impedance between the positive electrode and the negative electrode is named as C mode and shown in Fig. 3(a), and the heteropolarity interconnection mode with impedance between the negative electrode and the positive electrode is named as D mode and shown in Fig. 3(b).

2.3 The mathematical model analysis

While $R_{1\Sigma} \neq R_{2\Sigma}$ and $E1 = E2$, or $R_{1\Sigma} \neq R_{2\Sigma}$ and $E1 \neq E2$, it is no obvious change law with the voltage of the two-stage DC power supply and the voltage to the grounding shown in Fig. 3. The equation based on the loop current is shown in equations (4) and (5).

$$\begin{cases} E1 = R_{1\Sigma+}i1 + R_{1\Sigma-}(i1 - i3) \\ E2 = R_{2\Sigma-}i2 + R_{2\Sigma+}(i2 - i3) \\ (i3 - i1)R_{1\Sigma-} + (i3 - i2)R_{2\Sigma+} = (E1-) - (E2+) \end{cases} \quad (4)$$

$$R_{\Sigma} = \frac{(E2+) - (E1-)}{i3} \quad (5)$$

In view of the DC crosstalk with impedance in the two stage DC power system, it is supposed that it is the kind of types, the voltage of I stage DC between the positive electrode and the negative electrode is set to A, and the voltage of II stage DC between the positive electrode and the negative electrode is set to B. It has a voltage value between A and B, which is named as UAB, and then, UAB is equal to the voltage of the insulation resistance between the positive and negative electrode and the grounding, where UAB = URC = (E1+) + E2− or (E1−) + E2+ or UAB = URC = (E1+) + E2+ or (E1−) + E2−. The grounding point of the equalization bridge is off or the detecting bridge is on in the insulation monitoring devices for the DC power system, where UAB is equal to the voltage of the insulation resistance between the positive and negative electrode and the grounding, which can be used to judge the existence of interconnection systems, but this is not the only criterion. When it is also possible for the voltage relationship on UAB above the paper to appear in the system without DC crosstalk, but the probability is very small, and so it is not the only criterion for the voltage relationship on UAB.

Although the voltage has no obvious change law between the positive and negative electrode and the grounding occurs in a two-stage DC power system, when the grounding point of the equalization bridge is off or the detecting bridge is on in the insulation monitoring devices for the DC power system, the four parameters of the voltage whether or not there are significant changes between the positive and negative electrode and grounding in the two insulation monitoring devices for the DC power system, which can be used to judge the existence of interconnection systems.

In view of the system without the DC crosstalk, the grounding point of the equalization bridge is off or the detecting bridge is on in the insulation monitoring devices for the DC power system; it is an obvious change with the parameter of E1+ and E1−, and the resistor of the insulation resistance from the one device in which the grounding point of the equalization bridge is off. At the same time, there is no obvious change with the parameter of E2+ and E2−, and the resistor of the insulation resistance from another device in which the grounding point of the equalization bridge is on. At this point, the system does not have the DC crosstalk fault.

In view of the system with DC crosstalk, the grounding point of the equalization bridge is off or the detecting bridge is on in the insulation monitoring devices for the DC power system, the voltage between I stage and II stage is equal to the voltage of the insulation resistance between the positive and negative electrode and the grounding, which can be used to judge the existence of interconnection systems. With the grounding point of the equalization bridge is off or the detecting bridge is on in the insulation monitoring devices for the DC power system, there is no obvious change with the parameter of E1+ and E1− from the one device in which the grounding point of the equalization bridge is off and with the parameter of E2+ and E2− from another device in which the grounding point of the equalization bridge is on. And so, it can be seen that the two-stage DC systems are connected with the grounding and are interconnected with each other. At this point, the system has the DC crosstalk fault (Zhao, 2013; Huang, 2014).

3 THE DESIGN OF DC CROSSTALK

The DC crosstalk fault has a common feature such that the voltage varied trend of the electrode to the grounding with interconnection was the same. Therefore, it can be used to judge the existence of interconnection systems by detecting the voltage to the grounding and judging the correlation of the voltage varied trend.

The voltage of the DC system, the voltage of the positive electrode to the grounding and the voltage of the negative electrode to the grounding were sampled, divided, and converted through the signal acquisition unit, and then the signal is sent to the single chip microcomputer for analyzing and processing, and then the processing signal is matched with the DC crosstalk fault models of the devices with built-in fault models. And so, it can be

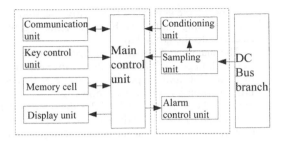

Figure 4. The hardware block diagram.

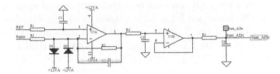

Figure 5. The hardware detecting circuit of DC crosstalk.

used to accurately judge the existence of interconnection systems and the DC crosstalk fault models, and then send out the alarm information of the DC crosstalk. It can significantly reduce the DC crosstalk fault operation time and reduce the DC crosstalk fault found difficulty with the two-stage DC power system in substations or power plants.

The hardware design is mainly divided into two parts. The first part is composed of the main control unit and the external device of the main control unit, the other part is composed of the sampling unit, the conditioning unit, and the alarm control unit based on the DC bus and branch. The hardware block diagram is shown in Fig. 4.

The hardware detecting circuit of DC crosstalk with impedance is shown in Fig. 5, where network labelled Samx can represent the sampling points V+, the sampling points V−, and the sampling points grounding. They form the Sam_ADx signals into the AD pin of the controller through the conditioning circuit, and then, the sampling value of the real-time detection is realized through the software operation.

4 EXPERIMENTAL SECTION

In order to verify the validity and reliability of the detecting technology, the insulation-monitoring devices are designed for the DC power system. Test instruments have the MX30-3PI-400-LF-SNK of the programmable power supply and the FLUKE15B of the multimeter, where, (a) is the page of interconnection type, (b) is the page of the alarm information of interconnection, (c) is the page of the voltage of the positive electrode to the grounding, and (d) is the page of the negative electrode to the grounding.

The diagram of the homopolarity interconnection with impedance between the positive electrode of I stage DC and the positive electrode of II stage DC based on E1 = E2 is shown in Fig. 6. The diagram of the A mode on E1 ≠ E2 is the same as the A mode on E1 = E2. And then, it can be seen that the display status is given in the device displaying unit, and the conclusion is verified above in this paper.

The diagram of the homopolarity interconnection with impedance between the negative electrode of I stage DC and the negative electrode of II stage DC based on E1 ≠ E2 is shown in Fig. 7. The diagram of the B mode on E1 = E2 is the same as the B mode on E1 ≠ E2. And then, it can be seen that

the display status is given in the device displaying unit, and the conclusion is verified above in this paper.

The diagram of the heteropolarity interconnection with impedance between the positive electrode of I stage DC and the negative electrode of II stage DC based on E1 = E2 is shown in Fig. 8. The diagram of the C mode on E1 ≠ E2 is the same as the C mode on E1 = E2. And then, it can be seen that the display status is given in the device displaying unit, and the conclusion is verified above in this paper.

The diagram of the heteropolarity interconnection with impedance between the negative electrode of I stage DC and the positive electrode of II stage DC based on E1 ≠ E2 is shown in Fig. 9. The diagram of the D mode on E1 = E2 is the same as the D mode on E1 ≠ E2. And then, it can be seen that

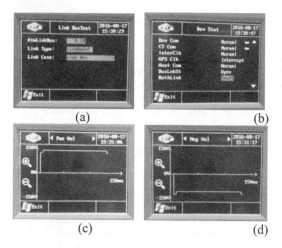

Figure 6. Pictures showing the A mode on E1 = E2.

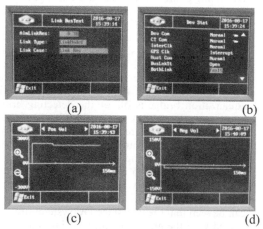

Figure 8. Pictures of the C mode on E1 = E2.

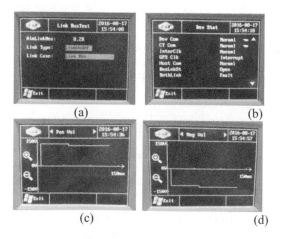

Figure 7. Pictures of the B mode on E1 ≠ E2.

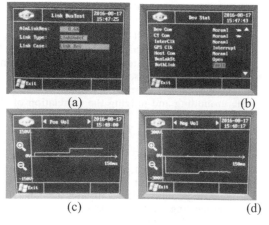

Figure 9. Pictures of the D mode on E1 ≠ E2.

the display status is given in the device displaying unit, and the conclusion is verified above in this paper.

The processing method of the DC crosstalk fault has several suggestions, which are as follows:

1. The switch is off, which can be resolved by establishing the interconnection for the same load in the two-stage DC bus power system.
2. The DC crosstalk fault point is that the interconnection signal enters into the DC bus of another stage by finding the direction point from the interconnection branches to the load. The load that belongs to the bus will be back to the bus.
3. The parasitic circuit can be removed, which has been found out.

5 CONCLUSION

In this paper, a detecting method of DC crosstalk is presented. The insulation monitoring devices for the DC power system is designed. A prototype test that it is able to monitor DC crosstalk in real time and send out the alarm information of DC crosstalk is carried out. This device has been running for more than half a year in the field of substations, with the advantages of reliability, accuracy, stability and innovation, and at the same time, with strong market competitiveness.

ACKNOWLEDGMENTS

This work was financially supported by the Science and Technology Project of State Grid Corporation of China (research on the intelligent on-line monitoring and maintaining technology for AC–DC integration power in substations, and then, research on the DC electrical source supervisor and the decision support system of operation and maintenance in substations).

REFERENCES

DL/T 1392–2014. Technical specification of insulation monitoring devices for DC Power system [S].

DL/T 5044–2004. Technical code for designing DC system of power projects [S].

Huang Dong-shan, Zhou Wei, Yang Li-cai, etal. Study on insulation failure and loop fault in the DC system [J]. Electronic Test, 2014.24:130–132.

Luo Zhi-ping, Xiong Di, Xie Zhi-hao, et al. Looped network problems and their solutions in DC system in substations [J]. Relay, 2008.36(3):71–74.

Ren Dong-hong, FAN Shu-gen. Analysis and measures for protection misoperation caused by grounding of DC power supply system[J]. Distribution & Utilization, 2011.28(3): 44–48.

The 18 items of grid major anti accident measures of SGCC [Z]. Beijing: China Electric Power Press. 2011.

Xu Yu-feng, Yun Chang-an, Yin Xing-guang. Hazards and Handling of Ring Faults in DC Systems [J]. GUANGDONG ELECTRIC POWER, 2008.36(3):71–74.

Zhao Ying-chun, Jiang Xiao-hong, Liu Qing, et al. Design of DC ring alarm and loop find device [J]. Automation Application, 2013.10:65–66.

Research on hybrid electric vehicle power switching speeds based on ADVISOR

Jiatian Guo & Ning Wang
Department of Automotive Engineering, Shandong Vocational College of Science and Technology, Shandong Weifang, China

Zhuming Cao
School of Mechanical Engineering, Beijing Polytechnic, Beijing, China

ABSTRACT: Now, most of the hybrid electric vehicles use speed as the power switch to control parameters. Research on the hybrid electric vehicle's power performance, fuel economy, and emission performance at different power switching speeds has great significance in improving the comprehensive performance of the hybrid electric vehicle. Good power switching speeds can promote the motor and engine power cooperates better. And it can also improve the vehicle's fuel consumption and emissions under the condition of meeting power performance.

1 VEHICLE SIMULATION MODEL

To consider the dynamic fuel economy and emissions of the vehicle, the influence of power switching speed on the vehicle's power performance, economy and the emissions should be studied. ADVISOR simulation software is used in this article to perform simulation and analysis of the vehicle's power performance, economy, and the emissions at different power switching speeds.

The ADVISOR vehicle simulation model is set up based on the basic structure of a parallel hybrid electric vehicle, as shown in Figure 1.

The data of the simulation model can be fed as input by modifying the parameters in the M file.

The vehicle parameters in the M file are as shown below:

veh_gravity = 9.81;

veh_air_density = 1.2;

veh_mass = 865;

vehicle_height = 1.380;

vehicle_width = 1.590;

veh_CD = 0.38;

veh_FA = 1.7;

veh_wheelbase = 2.365;

Some parameters can be set in the parameter input window, as shown in Figure 2.

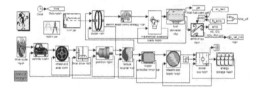

Figure 1. The vehicle simulation model.

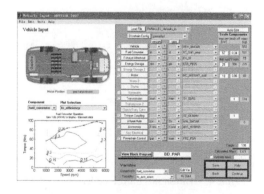

Figure 2. Motor parameters input window.

The effect of a power switch can be achieved by controlling the starting of the engine, as shown in Figure 3.

Choose the Urban Dynamometer Driving Schedule (UDDS), select the speed of the power switch respectively as 25 km/h, 20 km/h, 15 km/h, and 10 km/h. The control model is shown in Figure 4.

3 THE SIMULATION RESULTS

The simulation results are shown in Figure 5.

The economy and emission performance of the vehicle at different power switching speeds is shown in Table 1

Simulation results show that the proportion of the work of the motor is bigger at the average speed under low cycle conditions. But when the required power switching speed is much beyond the actual power switching speed, there will be a big deviation between the actual speed and the target speed.

4 FUZZY LOGIC CONTROL POWER SWITCH

Fuzzy logic controller is added to the engine start control, the control model is shown in Figure 9.

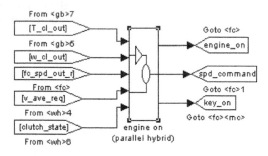

Figure 3. Engine starting control module.

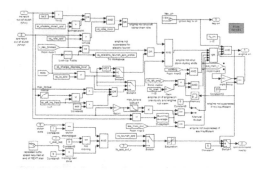

Figure 4. Threshold control model.

The simulation results are shown in Figure 10.

The simulation results that the motor still represent a significant proportion of the work, battery SOC values will be better able to maintain in high efficient areas. When the required speed beyond the power switch speed is large, the

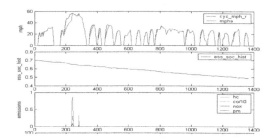

Figure 5. Simulation results of the hybrid electric vehicle when the speed of the power switch is 25 km/h.

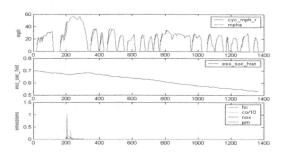

Figure 6. Simulation results of the hybrid electric vehicle when the speed of the power switch is 20 km/h.

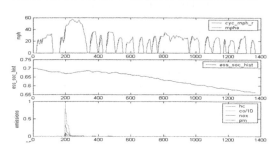

Figure 7. Simulation results of the hybrid electric vehicle when the speed of the power switch is 15 km/h.

Table 1. The economy and emission performance at different power switching speeds.

Power switching speed (km/h)	Operating conditions	Economy performance (L/100 km)	Emission performance (gram/mile)		
			HC	CO	NO$_x$
25	UDDS	0.47	0.422	10.82	0.113
20	UDDS	2.37	0.66	17.968	0.213
15	UDDS	3.59	0.691	18.347	0.241
10	UDDS	9.70	0.777	18	0.38

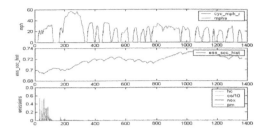

Figure 8. Simulation results of the hybrid electric vehicle when the speed of the power switch is 10 km/h.

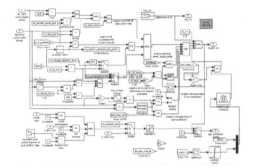

Figure 9. The fuzzy logic control model.

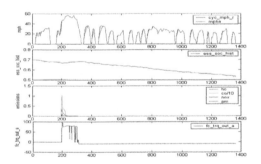

Figure 10. The simulation results at UDDS driving cycles.

actual speed and the target velocity have a little deviation.

Threshold control and fuzzy control of the economy and emissions comparison shown in the table below:

Operating conditions	Control strategy	Economy (L/100 km)	Emission (gram/mile)		
			HC	CO	NOx
UDDS	Threshold control	0.47	0.422	10.82	0.113
	Fuzzy control	2.49	0.666	18.005	0.223

5 CONCLUSIONS

The vehicle's economy is good when the speed of the power switching is at 25 km/h, but the actual speed and the target speed has a big deviation, the driver will feel powerless. With the power switching speed lower, the vehicle's dynamic performance is getting better and better. When the speed of the power switching is 20 km/h, the automobile operating mode is basically consistent with the target conditions. With the power switching speed is lower, the automotive power performance increases, but fuel consumption and emissions are increased.

The logic threshold control strategy can set the power switching speed at 25 km/h under the economy condition and at 15 km/h or 10 km/h under the dynamic conditions. The simulation results state that the economy and emission performance have fallen sharply when the power switching speed is between 20 km/h and 10 km/h. The fuzzy control strategy can select the appropriate power switching speed between 20 km/h and 10 km/h. The appropriate power switching speed can meet the demand of power, and can avoid the power switching speed from dropping directly down to 10 km/h

Therefore, the hybrid electric vehicle power switch control strategy can use the logic threshold control strategy under the economic priority conditions, and use the fuzzy logic control strategy under the dynamic priority conditions.

REFERENCES

Chan-Chiao Lin, Zoran Filipi, Yongsheng Wang, et al. Integrated, Feed-Forward Hybrid Electric Vehicle Simulation in SiMULINK and its Use for Power Management Studies. SAE 2001-01-1334.

Chu Liang, Wang Qingnian. Energy Management Strategy and Parametric Design for Hybrid Electric Family Sedan. SAE paper: 2001-01.

He Ren, Shu Chi. Overview of power-switch coordinated control of hybrid electric car [J]. Journal of Jiangsu University, 2014.7. Vol.35, No.4:373–379.

Hwang H.S, Yang D.H, Choi H.K, et al. Torque control of engine clutch to improve the driving quality of hybrid electric vehicles [J]. International Journal of Automotive Technology, 2011, 12(5):763–768.

Schouten N.J, Salman MA, Kheir N.A. Fuzzy Logic Control forParallel Hybrid Vehicles. IEEE Transactions on Control Systems Technology, 2002, 10(3).

Wipke K.B, Cuddy M.R, Burch S.D. ADVISOR user-friendly advanced power train simulation using a combined backward/forward approach [J]. IEEE Transactions on Vehicular Technology-Special Issue on Hybrid and Electric Vehicles, 1999(5):1–10.

Zeng X, Wang Q, Li J. The Development of HEV Control Strategy Module Based on ADVISOR2002 Software [J]. Automotive Engineering, 2004(26): 394–396.

Advances in Energy Science and Equipment Engineering II – Zhou, Patty & Chen (Eds)
© 2017 Taylor & Francis Group, London, ISBN 978-1-138-71798-5

The design and implementation of PCI express—SPI bus based on FPGA

Tianzhi Lv, Changbo Xiang & Xiaojun Li
The 41st Institute of CETC, Qingdao, China
National Key Laboratory of Science and Technology on Electronic Test and Measurement, Qingdao, China

Feng Wang
The 41st Institute of CETC, Qingdao, China

ABSTRACT: In order to achieve high-speed transmission between CPU and the external devices in the electronic equipment, this paper introduces a design and implementation method of adaptive PCI Express—SPI bus based on FPGA depending on the external devices. The high speed PCI Express bus and adaptive SPI bus were designed and implemented with VHDL in FPGA, realizing the control of the CPU to the different external devices. Transmission function of the SPI bus is validated in the Modelsim simulation. And the correctness and reliability of the PCI Express—SPI bus is verified online by Signal Tap.

1 INTRODUCTION

The architecture of CPU + external devices is commonly used in electronic equipment. CPU is the primary controller and FPGA completes logic control. The high-speed data bus like PCI (Peripheral Component Interconnect), PCI Express and USB (Universal Serial Bus) is mostly used for communication between CPU and the external devices to ensure the system real-time and efficient. The simple communication bus such as SPI (Serial Peripheral Interface), I2C (Inter Integrated Circuit) is commonly used in the external devices. The SPI bus is supported by more and more chips, for it is a high-speed, full-duplex, synchronous communication bus. FPGA is a bridge between the controller and external devices. The high-speed PCI Express bus is used for communication between the FPGA and the CPU. And the SPI bus is used for communication between the FPGA and the devices. Thus communication between the CPU and the external devices is realized. Based on these, this paper designs the PCI Express—SPI bus in the FPGA, to achieve high-speed communication and efficient management of CPU and the external devices.

2 THE OVERALL SCHEME DESIGN

The block diagram of the electronic equipment based on PCI Express—SPI bus is shown in Figure 1. It mainly includes the CPU, FPGA and the external devices. CPU completes control of the

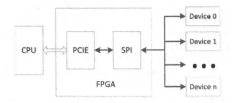

Figure 1.   The block diagram of the electronic equipment based on PCI Express -SPI bus.

entire system, data processing and other functions. Endpoint PCI Express bus and SPI bus are implemented in the FPGA. The external devices may contain many integrated circuit chips like DAC, ADC and amplifier.

Since the SPI bus data format and baud rate for each device varies, adaptive SPI bus module selects different external devices according to the address. SPI data in a specific format in with specific clock rate is output to achieve control of the integrated circuit chip. The specific process of CPU control the external devices is: CPU outputs SPI control information, the external device address and data which are written to the FPGA register via PCI Express bus; FPGA sets SPI bus according to the received SPI control information, including clock rate, data bits, etc.; The SPI module selects the external device according to the received address and transmits the control information and data to each device via SPI bus in accordance with the specific SPI format of the device; If it is a read command, the data is read back to the CPU.

3 DESIGN OF PCI EXPRESS BUS

PCI Express is the next generation bus and can achieve high-speed, point to point, dual-channel high-bandwidth transmission. For PCI Express 1.0, the theoretical transfer rate per channel in each direction is 2.5 Gbit/s, the latest 3.0 up to 10 Gbit/s, up to X32 transmission channels to meet the high-speed data transmission. PCI Express bus for high-speed communication between CPU and the external devices is widely used in the computer control system. Not only with good support in the X86 architecture CPU, the next generation of ARM also supports PCI Express bus. PCI Express bus protocol may be implemented in the PCI Express chip. This approach is simple, but increases hardware complexity. It requires a lot of address and data bus for connection between the PCI Express chip and FPGA. And the PCI Express parameters and data ports cannot be configured flexibly. In this paper, physical layer, data link layer and the transaction layer protocol of PCI Express are implemented by the FPGA IP core. PCI Express parameters, address bus and data bus can be set to suit the application requirements of different occasions. The FPGA used in this design is EP4CGX150F23C7 which is in Altera Cyclone IV GX series. The FPGA has characteristics of low cost, low power, high performance. It is flexible and efficient with integrated PCI Express hard IP core.

The block diagram of PCI Express bus is shown in Figure 2, mainly including three parts: PCI Express IP core, receiving channel and transmitting channel.

The physical layer, data link layer and transaction layer are implemented in the PCI Express IP core. They are automatically generated by Altera's IP core tool in Quartus and you can set the reference clock, the width of the link, the base address register BAR, maximum load and other parameters. The data interface of IP core is Avalon_ST format which is a serial bus interface defined by Altera Corporation.

Receive and transmit user data is implemented in the application layer. In this paper, the PCI Express application layer mainly finishes receiving SPI control information, the external device address and data from the CPU. And completes transmitting the external device's SPI data to the CPU. The application layer includes receiving channel and transmitting channel.

The receiving channel completes for receiving data and analyzing data. That is, to convert the PCI Express IP core's Avalon ST signal to the address and data bus output to the FPGA register. It mainly includes the receiving data conversion module and the analyzing data module. The module of receiving data conversion converts the received Avalon_ST data into the rx_desc (receive descriptor) and the rx_data (receive data). The rx_desc contains read and write commands, the current operation of BAR address space, data length, data address and other information. The rx_data contains transmitting data from CPU to FPGA. The analyzing data module completes the analysis of rx_desc to control the corresponding section and writting the received data to the specified register.

The transmitting channel converts the SPI read back data into Avalon_ST format and outputs it to the PCI Express IP core. It mainly includes the FIFO, the filling data module and the transmitting data conversion module. For the SPI data is serial and the PCI Express data is parallel, the FIFO completes serial to parallel conversion and caching data. In the filling data module, PCI Express control word and address are filled in tx_desc (transmit descriptor) and the SPI read back data is filled in tx_data (transmit data). The transmitting data conversion module completes converting the tx_desc and the tx_data to Avalon_ST format, then outputs them to the PCI Express IP core.

4 DESIGN OF SPI BUS

4.1 Standard SPI bus

SPI bus is a high-speed, full-duplex, synchronous communication bus. And it occupies only four pins on the chip, saving chip pins while saving space on the PCB layout. For these simple-to-use features, now more and more chips support this communication protocol. SPI bus consists of SCK (serial clock), MOSI (master output, slave input), MISO (master input, slave output) and CS (slave select). The input and output of SPI bus are synchronized with SCK produced by the host. The SCK form is determined by the CPOL (clock polarity) and CPHA (clock phase). If CPOL = '0', SCK is low when idle; If CPOL = '1', SCK is high when idle. If CPHA = '0', data is valid in the first SCK clock edge (rising or falling); If CPHA = '1', the data is valid in the second SCK clock edge (rising or falling).

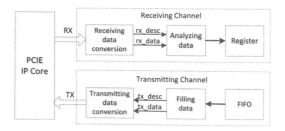

Figure 2.   The block diagram of PCI Express bus.

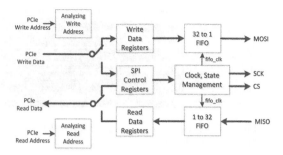

Figure 3. The block diagram of SPI bus.

The block diagram of SPI bus is shown in Figure 3, including SPI transmitting channel, SPI receiving channel and clock, state management of three parts.

SPI transmitting channel realizes the master output and slave input. CPU sends the write address and write data to FPGA via PCI Express bus. FPGA distinguishes the PCI Express write address to determine control address or data address. Here we set the SPI control address as 0X04, the data address 0X08. If it is the control address, then the write data is output to the SPI control registers to set the SPI control parameters, including clock divider ratio, clock polarity and clock phase. If it is the data address, the write data is written to the FIFO for parallel-to-serial conversion. PCI Express application layer clock rate is 62.5 MHz, but SPI clock rate varies with the device, generally less than 50 MHz. After the data is written to FIFO, SPI state management module generates CS and SCK meanwhile at the rising edge of FIFO read clock the data is read to MOSI. SPI receiving channel completes the function of the master input and the slave output. FPGA receives and analyzes the PCI Express read address sent by the host, determining to read data from the SPI control register or the device. If it is from SPI control registers, then the data of SPI control register is written to the PCI Express read data register. If from the device, then the MISO data is read into the FIFO, completing serial-to-parallel conversion and output data to PCI Express read data register.

The control of SPI transmission state machine is shown in Figure 4. SPI data transmission is completed a by controlling the SPI clock and the FIFO read and write.

The PCI Express application layer clock is divided to obtain the SPI master clock SCK_INT. SPI clock SCK is obtained from the phase shift of SCK_INT. After system reset it enters the state IDLE. After SPI start signal is valid it enters the state ASSERT_N1. At the rising edge of SCK_INT it enters the state ASSERT_N2. And

the falling edge enters the state UNMASK_SCK, SPI clock SCK valid. Next SCK_INT enters the state XFER_BIT and output FIFO read enable is valid to output data to MOSI. If the output FIFO almost_empty = 1, it enters the state ASSERT_DONE, and then completes the penultimate bit data transmission. The next clock cycle enters the state CHK_START, and then completes the last bit transmission and sets SCK idle state in accordance with CPOL. The next clock cycle it enters the MASK_SCK, SCK and FIFO read invalid. Finally, after the state HOLD_N1, HOLD_N2 to delay two cycles it enters the reset state IDLE, completing a SPI data transmission.

SPI data transmission is implemented with VHDL, set CPOL and CPHA 0, the output data 10100101. The simulation result is shown in Figure 5.

4.2 Adaptive SPI bus depending on the external devices

SPI communication protocol supports readable, writable, two-way communication. Read and write operations are initiated by the host. When there are multiple devices need to control and read or write, address needs to be defined to distinguish between each device and to indicate read and write operations. This design uses 32-bit SPI bus, including the high 8-bit of slave address and the lower 24 bits of the slave data. The even address is defined as write address and the odd address is

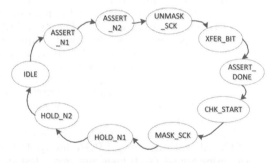

Figure 4. The control of SPI transmission state machine.

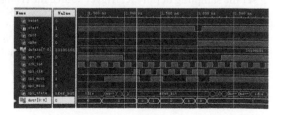

Figure 5. The simulation waveform of SPI data transmission.

Figure 6. The block diagram of adaptive SPI depending on the external devices.

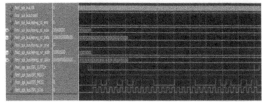

Figure 7. The simulation waveform of PCI Express—SPI bus in Modelsim.

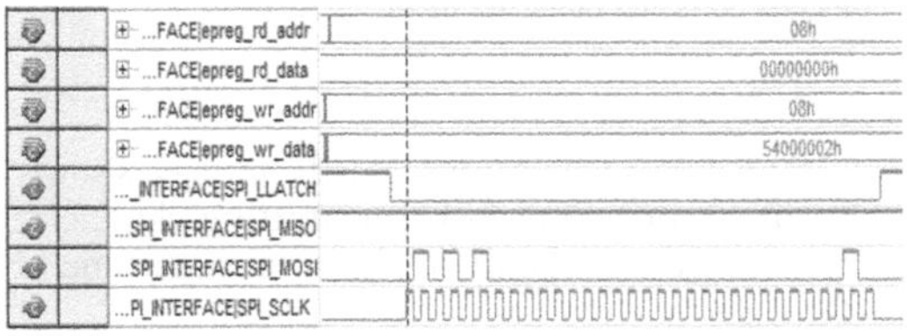

defined as read address in all of the address space. That is the definition of an even address +1 corresponding to the odd address is the address of read back.

The block diagram of adaptive SPI depending on the external devices is shown in Figure 6. SPI receiving data module receives the 32-bit data. The high 8 bits is stored in the address register and the lower 24-bit is stored in the data register. After receiving 8-bit address, selecting module select the appropriate controlled device according to the address. If the controlled device supports SPI bus, then CS, SCK, MOSI is output to the controlled device. If the controlled device is the switch, each bit of data register indicates on or off.

5 SIMULATION AND VERIFICATION

Based on the above design ideas, PCI Express bus module and SPI bus module are implemented with VHDL. The data transmission process of SPI bus is simulated by Modelsim. Here PCI Express write address is 8 bits and data is 32 bits. The input clock is PCI Express application clock with frequency 62.5 MHz. First, write SPI control address (address 0X04), then write data 0x00000000. That is to set the clock divider ratio of 8, the clock polarity and phase 0. Then write SPI data address (0X08), write data 0x5 AF0F0A5. The simulation waveform is shown in Figure 7.

After the program is compiled, the generated. sof file is downloaded to the FPGA through JTAG. And using Signal Tap II logic analyzer to test and analysis PCI Express—SPI bus. In Signal Tap II we set the sampling depth 2K, trigger source the rising edge of spi_bus: SPI_INTERFACE | SPI_SCLK. The host writes data addresses 0X08 and writes data 0x54000002 via PCI Express. The test result is shown in Figure 8. The online waveform of Signal Tap II shows that the host can send data to the slave device via PCI Express—SPI bus.

		...FACE|epreg_rd_addr	08h
		...FACE|epreg_rd_data	00000000h
		...FACE|epreg_wr_addr	08h
		...FACE|epreg_wr_data	54000002h
		...INTERFACE|SPI_LLATCH	
		SPI_INTERFACE|SPI_MISO	
		SPI_INTERFACE|SPI_MOSI	
		PI_INTERFACE|SPI_SCLK	

Figure 8. The online waveform of PCI Express—SPI bus by Signal Tap II.

6 CONCLUSION

In this paper, the design and implementation of adaptive PCI Express—SPI bus depending on the external devices are completed in FPGA. This paper describes the design and workflow of PCI Express bus module and adaptive SPI bus module detailedly. And the communication between the CPU and the external devices is completed in FPGA. This design PCI Express—SPI bus with characteristics of high efficiency, versatility and adaptive can be widely used in electronic devices with architecture of CPU + external devices.

REFERENCES

Chengfu, Y., Zhijun, X. Journal of Military Communications Technology, **25**, 72–76 (2004).

IP Compiler for PCI Express User Guide. http:// www. altera.com

Lin, G., Wengjie, L. China Integrated Circuit, **155**, 34–37 (2012).

Mingguo, X., Mingli, D. Computer Measurement & Control, **24**, 252–254 (2016).

Ning, J., Jianchun, C. Electronic Sci. & Tech., **27**, 188–191 (2014).

Xiaoning, X., Wenqiang, S. Electronic Design Engineering, **20**, 153–156 (2012).

Advances in Energy Science and Equipment Engineering II – Zhou, Patty & Chen (Eds)
© *2017 Taylor & Francis Group, London, ISBN 978-1-138-71798-5*

A study on the emulsifier fault diagnosis of the BP neural network based on EMD and KPCA

Yue Sheng Wang, Liang Wang & Hua Li
School of Hangzhou Dianzi University, Hangzhou, China

ABSTRACT: For the strong noise in the measured signal, complex structure, and long training time of the BP neural network and other problems during the emulsifier fault diagnosis, in this paper, a fault diagnosis method of the BP neural network for emulsifiers is proposed based on Empirical Mode Decomposition (EMD) and Kernel Principal Component Analysis (KPCA). By using EMD to decompose the collected vibration signal of the emulsifier, the original feature parameter set is obtained, and then the KPCA is used to reduce the dimensions of the set parameter. This new parameter set is used as the input of the BP neural network to train the emulsifier fault diagnosis model, and finally in this paper, this model is used to diagnose the samples. The experimental results show that this proposed method can not only effectively improve the signal-to-noise ratio and reduce the complexity, training time, and frequency of the network, but also can improve the accuracy of fault diagnosis.

1 INTRODUCTION

As the key equipment to determine the quality of emulsion explosives in the production line, the safety of the emulsifier is particularly important in the production of emulsion explosives. But an emulsifier is also one of the equipment that fail frequently in the production line; if the fault cannot be handled in time, it may cause major accidents by explosion due to the high temperature, extrusion, and collision caused by the friction between the machinery and equipment. Therefore, it is very important to find an accurate and effective method for emulsifier fault diagnosis for the whole explosive production line.

Vibration signals can well reflect the working state of the emulsifier, and so the current diagnostic methods generally use it to carry out the fault diagnosis (Tong, 2015). For the measured vibration signals also contain a large amount of background noise except the information reflecting the working state of the emulsifier; therefore, it is necessary to pre-process the vibration signal by EMD before extracting the characteristics vectors, to highlight the fault information and improve the signal-to-noise ratio. And then, the KPCA method is used to reduce the dimension of the parameter set, and on this basis, the fault diagnosis model of the BP neural network is established.

The result shows that this method can not only preserve most useful information of vibration signals of the emulsifier, effectively eliminate the influence of the noise, reduce the feature dimension and remove the non-linear characteristics, but also can simplify the neural network structure and obtain the diagnostic results quickly and accurately.

2 BASIC PRINCIPLE OF EMD AND KPCA

2.1 Basic principle of EMD

The EMD method assumes that any signal is composed of different Intrinsic Mode Functions (IMFs), and each IMF can be linear or non-linear. The IMF must meet the following two conditions: (1) in the whole data segment, the number of extreme points and the number of zero points must be equal or mostly differ by one; (2) at any time, the average of the upper envelope curve formed by the local maximum points and the lower envelope curve formed by the local minimum points is zero.

The EMD method is used to obtain the IMF through the screening process; the specific decomposition process is given as follows (Li, 2011):

1. Determine all local extreme points of the signal $x(t)$, and respectively connect all the local maximum and minimum points to form the upper and lower envelope curve by the three spline interpolation, the envelope curve includes all signal data.

2. Calculate the average of the upper and lower envelope, which is denoted as m_1, and then the following equation is computed:

$$h_1 = x(t) - m_1 \qquad (1)$$

If h_1 is an IMF, then it is the first IMF of the signal $x(t)$.

3. If h_1 is not an IMF, use it as the raw data and repeat steps (1) and (2) to obtain the average of the upper and lower envelope curve, which is denoted as m_{11}, and then calculate $h_{11} = h_1 - m_{11}$ and judge whether it meets the conditions of an IMF; if not, then repeat the cycle to calculate $h_{1k} = h_{1(k-1)} - m_{11}$ until h_{1k} meets the conditions of an IMF. Denote $c = h_{1k}$, and c_1 is the first IMF of the signal $x(t)$.

4. Separate c_1 from $x(t)$, and the following equation could be obtained:

$$r_1 = x(t) - c_1 \qquad (2)$$

Replace the original signal by margin r_1 to repeat the steps (1) to (3), and then the second function c_2 of $x(t)$ that meets the two conditions of IMF can be obtained, through cycling n times, and we could obtain n functions of $x(t)$ that meet the conditions of an IMF.

$$\left.\begin{array}{c} r_1 - c_2 = r_2 \\ \cdots\cdots\cdots \\ r_{n-1} - c_n = r_n \end{array}\right\} \qquad (3)$$

When the signal margin r_n becomes a monotonic function and cannot extract IMF from it after looping n times, the loop ends. And so, the original signal can be expressed as follows:

$$x(t) = \sum_{i=1}^{n} c_i + r_n \qquad (4)$$

2.2 Basic principle of KPCA

KPCA is a non-linear extension of the principal component analysis (PCA). Denote the vibration signal sample set $X = [x_1, x_2, \ldots, x_n]$, $x_k \in R^m$, and n is the number of samples. KPCA uses nuclear techniques to transform the input space r_m to a high dimensional space (often called as the feature space) F by non-linear transformation φ, where the PCA is conducted. The covariance matrix of the feature space F is represented as follows (Zhu, 2007; Wang, 2010):

$$C^F = \frac{1}{n}\sum_{i=1}^{n} \varphi(x_i)\varphi(x_i)^T \qquad (5)$$

Feature decomposition of F is given as follows:

$$\lambda v = C^F v. \qquad (6)$$

In the equation, v is the characteristic vector and λ is the characteristic value. And v can be expressed as follows:

The principal component is the projection of the mapping sample on the feature vector.

$$t_k = <v_k \bullet \varphi(x)> = \sum \alpha_{k,j} K(x_i, x). \qquad (7)$$

In the equation, k is the kernel matrix, $K_{i,j} = <\varphi(x_j) \bullet \varphi(x_i)>$, $k = 1, 2, \ldots, l$, $\alpha_{k,i}$ is the i_{th} coefficient of the k_{th} characteristic vector of the matrix K, and t_k is the k_{th} non-linear principal component of non-linear transformation a.

3 FAULT DIAGNOSIS PRINCIPLE OF THE BPNN BASED ON EMD AND KPCA

3.1 Feature extraction and processing

Vibration signals can well reflect the working state of the emulsifier, and so the current diagnostic methods generally use it to carry out the fault diagnosis of emulsifiers. When the emulsifier fails, other parts will be intermittently striking the fault location and forming a series of shocks and vibrations, which turns the original stationary vibration signal into a non-stationary vibration signal. In addition, the measured vibration signals also contain a large amount of background noise of other moving parts and structures except the information reflecting the working state of the emulsifier.

If the collected vibration signals are directly sent to the neural network for training, it will always require a complex network structure and long learning time to get the accurate results (Zhu, 2007). Even so, the generalization ability of the model is still poor, and the diagnostic error rate of the test samples is high. Therefore, feature extraction of the original signal is necessary and the flow chart is shown in Figure 1.

The steps of feature extraction based on EMD and KPCA are as follows:

1. Use a vibration sensor on the emulsifier to measure the original vibration signal $x(t)$;
2. Through the EMD decomposition of the measured vibration signal, the non-stationary vibration signal $x(t)$ is decomposed into a series of IMFs with different characteristic scales $c_1, c_2, \ldots c_n$;
3. Denote eigen vectors $C = [c_1, c_2, \ldots, c_n]$. By using the KPCA method to extract the principal component of n IMF, the eigen vectors after dimensionality reduction are obtained, which are denoted as $C_i = [c_1, c_2, \ldots, c_l]$, which preserve most part features of the vibration signal

and remove the non-linear characteristics of the signal.

4. Take the processed eigen vectors as the input to train the BP neural network model, and finally test the samples.

This method can effectively improve the signal-to-noise ratio, reduce the complexity, training time, and frequency of the network, and improve the accuracy of fault diagnosis.

3.2 The BP neural network model based on EMD and KPCA

Regardless of whether the emulsifier is normal or at fault, the vibration information has a high degree of non-linearity and is non-stationary. In this paper, the feature of the emulsifier vibration signal is extracted by using EMD and KPCA, and the BP neural network model is established.

The fault diagnosis model of the BP neural network based on EMD is shown in Figure 2.

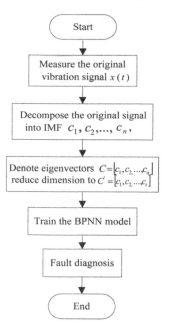

Figure 1. Flow chart of feature extraction based on EMD and KPCA.

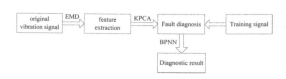

Figure 2. BP neural network model based on EMD and KPCA.

4 EXPERIMENTAL ANALYSIS

4.1 Experimental object

The SRF-2WS horizontal emulsifier provided by Shanghai Chemical Equipment Research Institute is taken as the research object, whose technical parameters are shown in Table 1.

Three acceleration sensors are respectively installed on the horizontal radial, vertical radial, and axial locations of the emulsifier to measure the vibration. The preliminary studies reveal that the vibration sensor on the vertical radial part of the emulsifier has greater vibration intensity and can obtain comparatively good signal analysis results. While the vibration signal collected by the vibration sensor on the horizontal radial part of the emulsifier is subjected from serious noise pollution, and the vibration intensity of the axial vibration sensor is small, the result of the signal analysis of the two is both not good. And so, the vibration sensor on the vertical radial part of the emulsifier is taken as the experimental object (Pan, 2015).

4.2 Data preprocessing and feature extraction

In the experiment, there are three kinds of fault conditions, which are pedestal loosening, the crack of the bearing's outer ring and rotor crack, and one normal working condition. These are carried out through loosening the connection screws on the emulsifier gasket to simulate pedestal loosening, through processing a 2 mm width and 0.5 mm depth groove on the bearing outer ring to simulate the crack of the bearing's outer ring, and through processing a 2.5 mm width and 1.0 mm depth groove on the emulsifier rotor to simulate the rotor crack.

According to the above four situations, 50 sets of data were collected by acceleration sensors for each situation. The raw data contain a large amount of noise signals, and the fault signal is too weak to extract, and so the EMD is used to highlight the fault information in the signal and improve the signal-to-noise ratio. Through decomposition, 12 intrinsic mode functions are obtained, which compose 200 sets of 12- dimensions original feature data.

The original data samples show a high degree of non-linearity; all kinds of signals are mixed together and it is difficult to isolate the useful information. Directly using the 12 intrinsic mode functions as the input of the BP neural network will result in the problem of excessively long training times and frequencies.

Actually, the fault information is mainly concentrated in the high frequency band, and so we only need to select the previous IMF component.

Table 1. Technical parameters of the SRF-2WS emulsifier.

Motor power	18.5 [kW]	Rotating speed	960–1470 [r/min]
Linear velocity	9.3–15 [m/s]	Through put	2–7 [t/h]
Effective volume	$<= 5$ [L]	Jacket working pressure	$<= 0.3$ [MPa]
Axial clearance	3 [mm]	Unilateral radial clearance	6.5 [mm]

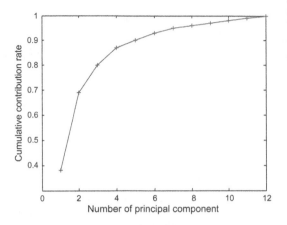

Figure 3. Graph showing the principal component cumulative contribution rate.

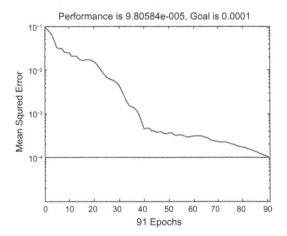

Figure 4. Graph showing the training error curve of the original signal.

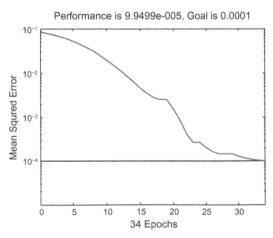

Figure 5. Graph showing the training error curve of IMF by EMD.

In this paper, KPCA is used for feature extraction to reduce the dimension of the feature vector. The principal component cumulative contribution rate of the feature extraction of the emulsifier vibration signal by using KPCA is shown in Figure 3.

As seen in Figure 3, the KPCA method has strong ability in feature extraction, and the first four principal components could have 86.57% information of the original data of the vibration signal, which is consistent with the fact that fault information is mainly concentrated in the high frequency band. And so, the first four IMFs are used as the inputs of the BP neural network instead of the original 12 IMFs, which can further reduce the complexity of the network, and reduce the network training time and frequency.

4.3 Experimental results

In order to verify the validity of the fault diagnosis model of the BP neural network based on EMD and KPCA, in this paper, this model is compared with the fault diagnosis model of the BP neural network without data preprocessing by EMD and KPCA, and the results of these two models are compared.

Taking 25 sets samples of each situation as the training samples and the remaining 25 sets samples as the test samples, the two models use the same training samples and training methods to train the network.

Wherein, the network learning rate is 0.1, the maximum number of trainings is 10,000, the minimum mean square error is 1×10^{-4}, the minimum gradient is 1×10^{-10}, the transfer function of the hidden layer neuron uses the logarithmic transfer function logsig, the transfer function of the output layer neurons use the linear activation function pureliu, and the training function uses traiulm.

To verify the effect of EMD and KPCA on fault signal analysis, in this paper, a comparative experiment is performed in the training phase. As seen in Figure 4 and 5, when the predetermined

318

target accuracy is 1×10^{-4}, the original signal data without preprocessing takes about 90 epochs to reach the target, while the signal data processed by EMD and KPCA only takes about 30 epochs to reach the target. This is because the feature of the original signal is not obvious, and so the neural network needs a long time to grasp the features of the signal, while EMD can decompose the signal into IMFs one by one according to different scales, and separate different local characteristics from the original signal. In addition, KPCA extracts the principal component to reduce the complexity of the structure of the neural network. Therefore, the effectiveness of training is much better than the former. The specific training effect is shown in Table 2.

In the identification process, when comparing with the simple BP neural network, the BP neural network based on EMD is more accurate in determining the type of fault. The specific result is shown in Tables 3 and 4.

As shown in Table 3, the fault diagnosis model of the BP neural network based on EMD and KPCA is more accurate than the simple BP neural

network model. The former can accurately separate the samples of each condition, which means that the EMD method has a very strong ability of the non-linear feature extraction. Therefore, the fault diagnosis model of the BP neural network based on EMD and KPCA is more suitable for the non-linear and non-stationary system, such as emulsifier fault diagnosis.

5 SUMMARY

In this paper, a fault diagnosis model of the BP neural network is proposed for the emulsifier based on EMD and KPCA, which can highlight the fault information and improve the signal-to-noise ratio. Through applying this model into the study of the SRF-2WS horizontal emulsifier, good diagnostic results are obtained. The main conclusions include that that EMD method has a very strong ability of the non-linear feature extraction and is suitable for the non-linear and non-stationary system. When compared with the simple BP neural network, the model of the BP neural network based on EMD and KPCA can effectively improve the signal-to-noise ratio, reduce the complexity, training time, and frequency of the network, and improve the accuracy of fault diagnosis.

Table 2. Results of the training contrast test.

Feature extraction method	Training accuracy (%)	Training times
Without EMD and KPCA	100	91
Within EMD and KPCA	100	34

Table 3. Diagnosis results of the BPNN.

| Model category | BPNN | |
	Misjudgment number	Accuracy (%)
Normal	1	96
Pedestal loosening	6	76
Crack of the outer ring	3	88
Rotor crack	5	80

Table 4. Diagnosis results of EMD_KPCA BPNN.

| Model category | EMD_KPCA BPNN | |
	Misjudgment number	Accuracy (%)
Normal	0	100
Pedestal loosening	2	92
Crack of the outer ring	0	100
Rotor crack	1	96

REFERENCES

Cai yangping, Li aihua, Shi linsuo, Roller bearing fault detection using improved envelope spectrum analysis based on F1VID and spectrun kurtosis, journal of vibration and shock (2011).

Li ningning, Research of Indution Motors Fault Detection based on EMD and ICA, Tianjin science university (2011).

Pan long, Pan hongxia, Chen yuqing, Application of EEMD and Generalized Dimension Optimization Approach in Automaton Fault Diagnosis, Modular Machine Tool & Automatic Manufacturing Technique (2015).

Tong qi, Hu shuangyan, Li zhao, Gear Pump Fault Diagnosis Method Based on KPCA and BP Neural Network, journal of Radio Engineering (2015).

Wang yuesheng, Liujianfan, Lv deyan, fault diagnosis system of Emulsifier based on vibration signal analysis, Mechanical and electrical engineering (2010).

Zhu zhihui, Sun yunlian, Li hong, Power fault detection using empirical mode decomposition, Engineering Journal of Wuhan University (2007).

Advances in Energy Science and Equipment Engineering II – Zhou, Patty & Chen (Eds)
© 2017 Taylor & Francis Group, London, ISBN 978-1-138-71798-5

Design and experiment of the horizontal banana stalks crushing machine

Chao Wang, Enyu Zhang, XiRui Zhang, Yue Li & Dong Liang
School of Mechanics and Electrics Engineering, Hainan University, Haikou, China

ABSTRACT: Banana is one of the most important fruits in tropical and sub-tropical areas. Banana stalks are the by-products of the planting process. Owing to the thick and high level moisture of banana stalks, few mature mechanical types of equipment for banana stalk treatment are presented. Returning stalks to the soil in time can increase the soil's organic content effectively, thereby improving the biological environment's banana yield. To solve the above-mentioned problems, combined with the features of banana stalks such as bulkiness, high moisture content, etc. a horizontal banana stalks disintegrator with double shafts was designed in this work, and the overall design of the machine is described. Field experiments showed that the average efficiency of the machine was 0.4 hm²/h, the shattered ratio of banana stalks ran up to 96.98%, and the field coverage reached 87.87%. According to the processes of the experiments, with a series of advantages including excellent performance, remarkable ecological benefits, and economic benefits, the machine was easy to operate. It can be inferred that this machine had a promising prospect in the banana growing areas, south of China. Having solved the problem of large-scale banana cultivation by facing massive stalks processing, its successful implementation will greatly improve the production efficiency.

1 INTRODUCTION

Bananas have been widely grown in tropical and sub-tropical regions, because of the increasing demand for banana production (Shongweb, 2008; Oliveira, 2009; Chen, 2012). The banana industry occupies an important position in the economic development. In order to ensure bananas' quality, the renewal cycle of banana plantations usually lasts for one or 2 years. Banana stalks are the by-product of the planting process. Therefore, banana plantations piled up a lot of banana stems and leaf wastes every year. According to the statistics data in the year of 2014, 28.30 million tons of banana stalks have been received as output in China (Liu, 2012; Alarcón, 2015).

Banana stalks are rich in nitrogen, phosphorus, potassium, etc. and organic matter. Mechanical crushing directly can increase the soil's organic content, and improve the soil structure of banana plantations to raise banana's quality and yield (Elanthikka, 2010). But limited by mechanization processing technology, the thick and high level moisture of banana stalks piled up in the field after chopped off (Elanthikkal, 2010; Liu, 2009; Zhu, 2012). The traditional method may suffer the potential threat from banana pests and diseases, thereby seriously affecting updates of banana plantations and the development of the banana industry (Gan, 2014; Zhang, 2015).

At present, the studies focus on straw shattering and returning of wheat straw, corn stover, rice, cotton, and other crops' by-products. Zhu Deyun et al. (Liu, 2012) designed a new kind of banana straw-returned machine, and the machine is characterized by smashing the banana stems, leaves, and roots returning to the field, reducing the times tractor should be used across the field and the cost, preventing soil compaction, but there are problems such as the large consumption of machine power, ease of damage of the cutting tool due to the stem fiber winding etc. which demands further improvement to the experiment.

To solve the above-mentioned problems, combined with the features of banana stalks such as bulkiness, high moisture content, etc. (Chen, 2011), a horizontal banana stalks crushing machine with double shafts is designed in this paper. Furthermore, to determine the key structure parameters and working parameters of the feeding device, crushing device, and rolling device, etc. the overall design of the machine was described in order to solve the problem of large-scale banana cultivation facing massive stalks processing, and its successful implementation could not only greatly improve the production efficiency, but also prevents the environment from pollution thereby having important economic and ecological benefits.

2 STRUCTURE AND WORKING PRINCIPLE

2.1 The whole structure

A horizontal banana stalks disintegrator with double shafts consists of the feeding device, crushing device, three-point linkage, and repression device, as shown in Figure 1. The machine's working height can be adjusted to adapt different fields via a three-point linkage (Li, 2011). Banana stalks have been affected by using the feeding device, crushing device, and repression device in sequence.

The feeding device and crushing device are driven by the two output shafts of the gearbox; the roller of the rolling device rolls by friction between the roller and stalk debris on the ground.

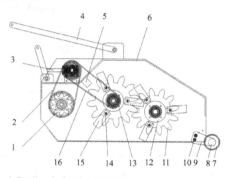

1. Feeding device, 2. belt-driving mechanism, 3. gearbox 4. three-point hitch, 5. stationary knife, 6. housing case, 7. sleeve, 8. roll axis, 9. connecting otic placode, 10. bolt, 11. gear driving mechanism, 12. driving knife roll, 13. driven knife roll, 14. knife holder, 15. crushing knife, and 16. belt-driving mechanism

Figure 1. Schematic diagram of the horizontal type with double shafts machine for banana stalks crushing and returning.

Table 1. Main technical parameters of the horizontal type with double shafts' extracting machine for banana stalks crushing and returning.

Technical parameters	Design values
Power/kW	≥58.8
Working width/mm	1600
Driving speed/(m·s⁻¹)	0.7~1.0
Speed of feeding device /r·min⁻¹	76.5
Horizontal space between two Crushing blade rollers/mm	350
Vertical space between two crushing blade rollers/mm	80
Driving shaft speed/r·min⁻¹	1400
Transmission shaft speed/r·min⁻¹	1600
Efficiency/hm²·h⁻¹	0.36~0.43
Total weight/kg	600

2.2 Working principle and technical parameters

The transmission system consists of a gearbox and belt drive. The power of the output via the gearbox to the drive shaft, and then a drive shaft via the belt drive to the gullied-tooth feed roller, and then to the driven feeding roller completes automatic feeding of the banana stalks. The other drive shaft operates via the gearbox, the drive shaft, belt drive to the crushing device, which can drive the crushing knife rotating at a high speed. The machine's main specifications and technical parameters are shown in Table 1.

3 DESIGN OF THE KEY COMPONENTS

3.1 Feeding device

The feeding device mainly consists of the sulcus gear roller, self-aligning roller bearing, and belt wheel, as shown in Figure 2. The banana stalks are fed between the ground and feeding roller which sends it into the crushing device.

Main technical parameters of the feeding device are as follows: the sulcus gear roller L, diameter of the feed roller D, and the speed of the rotation rate of the sulcus gear roller n_1.

3.2 Crushing device

Crushing devices are key components of the banana crushing machine. The crushing device consists of crushing axes, knife holder, crushing knife, and bolt. Knife holders are distributed along the crushing axes with an equal angle-120°, and crushing knife hinged knife holder by using bolts. While the machine is working, the anti-clockwise front-roll crushed banana stalks, and the clockwise rear-roll crushed the remaining part of the stalks and buried stalks debris simply.

The crushing knife, as main job parts, and straight blades, which are sharp and have a good crushing effect, have been adopted to lower processing cost and reduce winding (Li, 2013; Zhang,

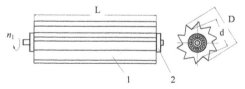

1. Sulcus gear roller and 2. self-aligning roller bearing
Notes: L is the length of the sulcus gear roller, mm; D is the external diameter of the gullied-tooth feed roller, mm; d is the internal diameter of the sulcus gear roller, mm; and n_1 is the speed of rotation rate of the sulcus gear roller, r/min.

Figure 2. Schematic diagram of the feeding device.

2012). Due to the high moisture content of the banana stalks and vile environment, and considering the erosion resistance, the cemented carbide after hardening has been welded on blade parts (Wang, 2011). 65Mn steel with sufficient strength and high wearing resistance has been applied in cutter manufacture.

The driving roller and driven roller rotate in opposite directions. The layout of the ladder-type can effectively perform to guarantee the crushing effect, as shown in Figure 3. This form can not only increase the crushing rate, but also allow the even spreading of the debris on the ground.

3.3 Rolling device

The rolling device that installed behind the machine, and the main function is to compact the banana

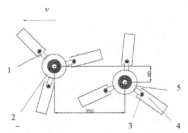

1. Knife holder, 2. driving knife roll, 3. bolt, 4. crushing knife, and 5. driven knife roll

Figure 3.   Schematic diagram of the crushing device.

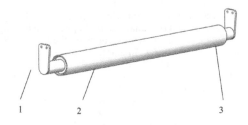

1. Rolling axis, 2. sleeve and 3. connecting otic placode

Figure 4.   Schematic diagram of the roller device.

debris. The device mainly consists of the rolling axis, sleeve, connecting otic placode, etc. as shown in Figure 4. The rolling shaft can turn freely through the bearing and connecting sleeve. The connecting otic placode fixed on both sides of the casing by using the screw. The height of the rolling roller can be adjusted by connecting the otic placode. The switch port with different heights in the case, connecting otic placode can be connected to different switch ports to change the height of the rolling roller.

The roller working length is 1 600 mm. To ensure the rolling strength and reduce the weight of the rolling device, the sleeve adopted a Q235 steel tube in which the diameter is 10 mm and thickness is 200 mm. The rolling device structure is simple and steady.

4   FIELD EXPERIMENTS

4.1   Field experiment conditions

The performance test was carried out in the field of Baodao Island village, Danzhou city, Hainan Province, on October 16, 2015. The soil type is latosol with the features of depth, thickness, and stickiness. The whole landscape is flat with which the slope not more than 5 degrees. Banana cultivars Brazil Williams is the locally grown banana variety.

The banana stalks' average height is 1500 mm and row spacing is 2 m. A Red-LX804 acts as the auxiliary power source.

4.2   Test results and analysis

Performance indicators of the banana stalks disintegrator with double shafts are shown in Table 2.

Field experiments showed that the average efficiency of the machine was 0.43 hm^2/h and the field coverage reached 87.87%. The proportion of chopped stalks length that is less than 100 mm ran up to 96.98%, which could satisfy the requirements of the returning straw crushing agronomy and meet the performance requirements. According to a series of the experiments, with many advantages including excellent performance, good safety, remarkable ecological and economic benefits and great reliability, the machine was easy to operate.

Table 2.   The test results of the main performance indicators.

Test parameters		Efficiency /hm^2·h^{-1}	Rate of the straw coverage /%	Rate of qualified smashing /%
No.	1	0.42	89.24	97.17
	2	0.46	85.92	95.72
	3	0.41	90.23	98.12
	4	0.45	86.54	97.37
	5	0.43	87.43	96.51
Average value		0.434	87.87	96.98

5 CONCLUSION

A horizontal banana stalks crushing machine with double shafts was designed in this paper. The overall design of the machine was described. Furthermore, the methods used to determine the key structure parameters and working parameters of feeding device, crushing device, rolling device, etc. are provided. The main structure and working principle of the crushing device were analyzed. Likewise, the speed of the crushed drive shaft and the driven shaft, and structural parameters of the knife roller were determined. The ladder-type double roll is firstly adapted to the banana crushing machine, which can return the whole banana straw to the field directly and effectually. Stalks debris can be directly dried on the surface. Its humus fertilizes the surrounding areas, increases the supply of soil nutrients and promotes the banana yield greatly.

1. A gullied-tooth feed roller was adapted to feed and hold stalks. Squeeze the feed roller blade and punctured stalks can effectively shorten the length of the fiber. Stalks were fed steadily, which were squeezed by using the blade of the feed roller and the fiber of the stalks was shortened effectively.
2. A ladder-type double crushing knife roll rotates in opposite directions. The anti-clockwise front-roll crushed banana stalks and the clockwise rear-roll crushed the remaining part of the stalks and buried stalks debris simply. The working mode improved the banana crushing efficiency and avoided the common winding on horizontal-axis machines.
3. Field experiments showed that the average efficiency of the machine was 0.4 hm²/h, and the shattered ratio of banana stalks ran up to 96.98%, and the field coverage reached 87.87%. The average chopped stalks length was 61 mm, which could satisfy the requirements of returning the straw crushing agronomy. According to the processes of the experiments, with a series of advantages including excellent performance, good safety, remarkable ecological and economic benefits, and great reliability, the machine was easy to operate. It can be inferred that this machine had a promising prospect in the banana growing areas, such as south of China. Having solved the problem of large-scale banana cultivation facing massive stalks processing, its successful implementation could not only greatly improve the production efficiency, but also prevent the environment from pollution, which has important economic and ecological benefits.

The machine has the advantages of excellent performance, simple operation, safe and reliable work, remarkable ecological and economic benefits, with a broad promotion prospect in the banana growing areas, such as the south of China.

ACKNOWLEDGMENTS

This work was financed by the National Natural Science Foundation of China (Grant No. 51565010), the National Natural Science Foundation of Hainan Province (Grant No. 20163038), Special Fund for Agro-scientific Research in the Public Interest (Grant No. 201503136), and Patent Implementation and Guidance of Haikou (No. 2015-03).

REFERENCES

Alarcón LC, Marzocchi VA. Evaluation for Paper Ability to Pseudo Stem of Banana Tree [J]. Procedia Materials Science, 2015, 8: 814–823.

Chen Liqing, Wang Li, Zhang Jiaqi, et al. Design of 1JHSX-34 straw crusher for whole-feeding combine harvesters [J]. Transactions of the CSAE, 2011, 27(9): 28–32.

Chen Shi, Zhou Hongling Chen Lina, et al. Research advances in comprehensive utilization of banana by-products Research advances in comprehensive utilization of banana by-products [J]. Fujian Agricultural Science and Technology, 2012, 2: 80–82.

Elanthikkal S, Gopalakrishnapanicker N. Cellulose microfibres produced from banana plant wastes: Isolation and characterization [J]. Carbohydrate Polymers 2010, 80: 852–859.

Gan Shengbao, Li Yue, Zhang Xirui, et al. Design and experiment on banana stalk chopper with feeding type spindle flail [J]. Transactions of the Chinese Society of Agricultural Engineering (Transactions of the CSAE), 2014, 30(13): 10–19.

Li Ling, Zhang Jin, Ou Zhongqing, et al. Technology and birotor equipment design of returning pineapple leaves [J]. Hubei Agricultural Sciences, 2011, 30(34): 273–274 2014,19: 4705–4707+4711.

Li Yonglei, Song Jiannong, Kang Xiaojun, et al. Experiment on twin-roller cultivator for straw returning [J]. Transactions of the Chinese Society for Agricultural Machinery, 2013, 06: 45–49.

Liu Gang, Zhao Man quan, Kang Wei dong. Design and Research of 9R-40 Rubber's Feeding Devic [J]. Journal of Agricultural Mechanization Research, 2009, 01: 94–96.

Liu Guohuan, Kuang Jiyun, Li Chao, et al. The research progress about utilization of banana stalks [J]. Renewable Energy, 2012, 30(5): 64–69.

Liu Xiaoliang. Design and Experiment of the new Smashed Straw Machine Blade [D]. Changchun: Jilin University, 2012.

Oliveira L. Evtuguin D. Structural characterization of stalk lignin from banana plant [J]. Industrial Crops and Products 2009, 29: 852–859.

Shongweb V. Tumber R. Soil water requirements of tissue-cultured dwarf cavendish banana (Musa spp. L) [J]. Physics and Chemistry of the Earth, Parts A/B/C 2008, 33: 768–774.

Wang Jikui, Fu Wei, Wang Weibing, et al. Design of SMS-1500 type straw chopping and plastic film residue collecting machine [J]. Transactions of the CSAE, 2011, 07: 168–172.

Zhang Junchang, Yan Xiaoli, Xue Shaoping, et al. Design of no tillage maize planter with straw smashing and fertilizin [J]. Transactions of the Chinese Society for Agricultural Machinery, 2012, 12: 51–55.

Zhang Xirui, Gan Shengbao, Zheng Kan, et al. Design and experiment on cut roll feeding type horizontal shaft flail machine for banana stalks crushing and returning [J]. Transactions of the Chinese Society of Agricultural Engineering (Transactions of the CSAE), 2015, 31(4): 33–41.

Zhu Derong, Chang Yunpeng. Technical Research and Equipment Design of Banana Straw's Returning [J]. Chinese. Agricultural Mechanization, 2012, 01: 140–143.

Design and implementation of the multi-granularity hierarchical control system of the safe based on the mobile intelligent terminal

Yuhui Cao, Dawei Zhang, Weihong Wang & Fuxiang Zhou
Institute of Information and Technology, Hebei University of Economics and Business, Shijiazhuang, China

ABSTRACT: Aiming at the problem that the use of the safe is not convenient and its security level is not high, in this paper, a novel multi-granularity hierarchical control system of the safe based on the mobile terminal, namely MGHS-MT, is proposed. Firstly, based on the research about the safe and the demands on convenience and security, the strategy of MGHS-MT is deeply discussed. Subsequently, MGHS-MT is established according to the strategy and it fully considers the practical needs of the users. Finally, the test results show that the MGHS-MT, when compared with the traditional method in the same field, has good convenience and security.

1 INTRODUCTION

With the development of science and technology, the safe industry has also changed. At present, the traditional mechanical safe is transforming into an intelligent safe. People's lives are more convenient and safe.

But there are still some problems in the current safe. The electronic system of the safe based on fingerprint identification is proposed by Wu (Wu, 2015). From this article, we can understand that the basic principle of fingerprint safe is fingerprint matching based on pattern recognition. Although the security is improved, the accuracy is yet to be improved. At the same time, there is a higher requirement for the equipment. Wang (Wang, 2002) proposed a non-key safe system based on face recognition. The face recognition method and fingerprint recognition method are similar. But, the selection of the classifier is different. Wang combines the traditional classification method with the artificial neural network method to construct a hybrid classifier, which greatly improves the recognition rate. But it still needs a higher device support. Lin (Lin, 2011) developed the remote intelligent safe system. In this article, the problems and solutions of the remote control of the safe are introduced, such as multimedia technology, video surveillance technology, remote monitoring technology, and other technologies. However, due to the rapid development of the network, the security problem still exists in the remote control.

Zhao (Zhao, 2012) put forward the password lock based on FPGA. In this article, we can know that the cipher device has the function of password modification, voice prompt, preventing multiple tests, warning, and so on. Although the stability is high, the password can be easily seen by others. And so, the security needs to be improved. Shi (Shi, 2014) came up with a new concept of smart home safe. From his article, we can see that this safe is improved on the basis of Wu's concept. He has added a mobile phone password to the fingerprint password. Although this safe can enhance the safety of the cabinet, the requirements for the equipment are a bit high. Wu (Wu, 2012) achieved the realization of the user interface and remote database interface operation. All user information is stored in a remote database. Although it is convenient to manage the safe, the user's data are lost easily.

Aiming at some problems existing in the above-mentioned safe, and at the same time, by combining with the current actual situation and consumer demand, in this paper, a novel multi-granularity hierarchical control system of the safe based on the mobile terminal is proposed. The whole system ensures convenience and security, and requirements of the equipment are not high.

2 THE STRATEGY OF MGHS-MT

The popularity of mobile terminals provides a very good platform for this system, because the control of the safe can be realized on the mobile terminal. At the same time, the mobile phone is essential to people's lives, and so it will not increase the burden on consumers. And the mobile terminal can complete authentication to ensure the safety of the safe. At the end, it not only ensures convenience, but also improves the security.

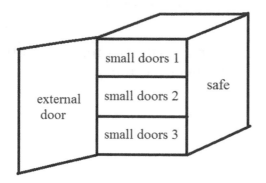

Figure 1. The schematic diagram of the safe.

2.1 *The basic ideas of MGHS-MT*

The basic idea of this paper is to use the mobile terminal to authenticate the user's identity information, and then carry out the next step of the safe operation. In order to guarantee the safety of the user's identity, a number of authentications have been performed in this system. At the same time, different users have different rights.

2.2 *The concrete implementation means of MGHS-MT*

The safe that is described in this article and used is shown below. We can know that the safe has an external door. After opening the external door, we can see three small doors. Each door is relatively independent and has its own lock, and so each safe has four locks. The model of the safe model is shown in Figure 1.

The system has performed a complete job on the user's authentication. First, the reliability of the mobile terminal is verified. Secondly, the user's account and password information are verified. In the end, the control of the safe is realized. User permissions are not the same to open different small doors. And different users can only use their own drawers.

At the same time, the biggest difference with other schemes is that all of the data in this system are stored in the hardware store in the safe. The user's data are not stored in the network or in the phone. This ensures that the user's data information is not easy to be maliciously accessible. The role of mobile phones is just a tool for personal computer interaction.

3 THE OVERALL DESIGN OF MGHS-MT

The MGHS-MT is composed of the mobile intelligent terminal, SCM control module, Bluetooth communication module, motor control module, and infrared detection module. The mobile intelligent terminal achieves the matching and connection of Bluetooth while providing the input function of the account and password, but the most important function is to establish the communication with the Bluetooth module; the SCM control module mainly completes the data processing and the control function. The data include the information of the Bluetooth transmission and the infrared detection information, and main achievement of the control is the motor module status management; the function of the Bluetooth communication module is to transmit information, and to complete the information interaction between the mobile intelligent terminal and the MCU control module; the motor control module is the control of the motor that is used to achieve the management of the safe. The infrared detection module is used to detect the open and close state of the safe and then send to the microcontroller for further processing. The system's structure is shown in Figure 2.

3.1 *The hardware design of the MGHS-MT*

The hardware includes the microprocessor control module, Bluetooth communication module, motor control module, and infrared detection module. The parameters of each part are given as follows:

The SCM control module mainly carries out the data processing and the electric machine control. In this design, the STC89C52RC chip is used as the processing chip, with super interference, high speed, low power consumption, etc. The programmable serial port (UART), two timer/counter, watchdog, and communication with Bluetooth can be achieved by setting the corresponding special register. In addition, the SCM is built in 4 KB EEPROM with a convenient account and

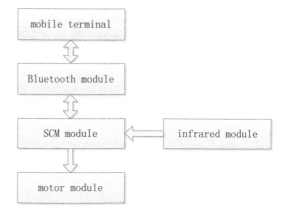

Figure 2. System structure.

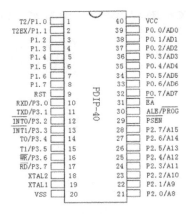

Figure 3. The STC89C52RC pin diagram.

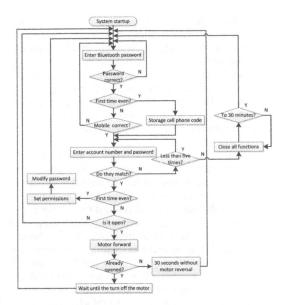

Figure 4. System control flow chart.

password storage, and so the cost of equipment is reduced. The working temperature is minus 40 to 85 degrees Celsius which can be applied to most of the environment. The STC89C52RC pin diagram is shown in Figure 3.

Supporting the AT command, the user may need to change the baud rate, equipment name, pairing password, and other parameters. The module supports the UART interface, which has the advantages of low cost, small size, low power consumption, high sensitivity, and low power consumption. It will be able to achieve its powerful function equipped with only a few peripheral components.

The infrared detection module uses a pair of tubes, such as the infrared transmitting tube and receiving tube. The infrared transmitting tube emits infrared rays. When the direction of detection encounters obstacles, the infrared-receiving tube receives the reflected infrared rays. The comparator circuit processes a digital signal output to the SCM. The measuring distance can be adjusted by using the knob of a potentiometer and the effective range is 2~60 cm, when the detection angle is 35 degrees, and the operating voltage is 3.3~5V. The module has the advantages of convenient assembly, small interference, ease of use, adjustable distance, and meeting the requirements of the system.

The motor control module uses the YL-86 motor module. This module contains two L9110S motor driver ICs. The supply voltage is 2.5~12V, the maximum operating current is 0.8 A, and the working temperature is minus 30~105 degrees Celsius.

3.2 The software design of MGHS-MT

The software design of the system is composed of two parts, the mobile intelligent terminal and SCM. The mobile intelligent terminal software system uses an eclipse open platform and JAVA language.

The single chip computer software system is developed by using the C51 Keil software platform and C language. The following is the system control flow chart, as shown in Figure 4.

With the system's Bluetooth communication module being used to set the detecting state and connection password, you need to enter the correct password to connect to Bluetooth, or you cannot connect to the normal Bluetooth; the mobile intelligent terminal connected to the SCM sends this phone code. If this intelligent terminal first links with Bluetooth, this phone code will be stored in the microcontroller. After entering the password on the account password input interface, it will prompt whether to open the appropriate safe or input error. If wrong information is fed as input five consecutive times,, the entire system will go into sleep mode and re-enter the normal mode after half an hour; After clicking the open button, the corresponding motor will forward rotate for three seconds and open the lock, and then you can open a safe door. If you close the door, the infrared detecting module will detect that the door has been closed, and then it reversely rotates three seconds and the door is locked. If the lock is open, there are special reasons and the infrared detection module does not detect that the door is opened safely; the motor reversely rotates three seconds to lock the door in 30 seconds to ensure security of the system.

3.3 The safety design of the MGHS-MT

The principle of this design is safe and convenient, and contains multiple encryptions to ensure security.

The first layer of security is the Bluetooth pairing password. Only the mobile intelligent terminals that match successfully can be connected to the Bluetooth.

The second layer of security is the verification of the mobile phone code. After connecting to Bluetooth, it will verify that the phone code is correct. Since the mobile phone code for each mobile intelligent terminal is unique in the world, thereby it ensures security.

The third layer of security is the verification of the account and password. Only the correct account and password can go through; otherwise, the account and password should be re-entered. If wrong information is fed as input for five times consecutively, the entire system will go into sleep mode and re-enter the normal mode after half an hour.

In addition, in order to evaluate the security of the password, the system uses the MCU internal encryption storage. The account password and other important information are stored in the STC89C52RC built-in EEPROM after encryption. It can prevent the account password from being extracted if there is virus in the mobile intelligent terminal.

The security can be achieved through the above insurance measures. At the same time, the timer is set to limit the timing of the lock. After connecting the Bluetooth, if there is no operation within the set time, it will automatically disconnect the connection.

4 THE OVERALL IMPLEMENTATION OF THE MGHS-MT

According to the above design, the multi granularity hierarchical control system of the safe is realized based on mobile terminals. The MGHS-MT can completely meet the practical requirements. In this paper, we used an Android system and the control interface is shown in Figure 5.

In addition, according to the actual situation in the real life, in this work, a lot of practical verifications have been carried out, including the normal lock, forgetting the password, password input error more than five times, the super user to lock, and so on. Test results are shown in Table 1.

The results show that the normal rate of the lock is 98% and is fully able to meet the normal needs. One of the two anomalies is the problem of the mobile intelligent terminal connecting with Bluetooth. This situation is of small probability and difficult to avoid, but it does not affect the normal use, because re-connection can solve this problem. Another exception is that the test time is too long and the motor driver chip is not working properly; but in real life, users cannot unlock for a

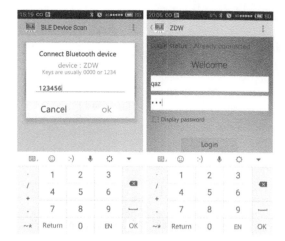

Figure 5. Mobile terminal application platform.

Table 1. List of test results.

Test case	Normal times	Abnormal times	Total times	Accuracy rate
Normal lock	98	2	100	0.98
Forget the password	20	0	20	1.00
Input error 5 times	9	1	10	0.90
The super user to lock	10	0	10	1.00

long time and so this situation is almost impossible in reality. The accuracy rate is 100% when the password is forgotten. The accuracy rate is 90% when the wrong password is fed as input 5 times, which is a little bit low. There is a relationship between the number of tests and the accuracy rate. The only exception is that the power supply problem occurs during the dormant state and causes the entire system to restart. This situation will only occur in the test, because in reality there will be a standby power to ensure the power supply. The accuracy rate of the super user unlock is 100%, which does not appear abnormal.

Overall, the test results are satisfactory and the security is guaranteed. At the same time, the mobile intelligent terminal is used to achieve portability, reduce the input of the hardware, and reduce the cost, so that it can be widely used.

5 SUMMARY

Through the combination of the mobile intelligent terminal and SCM, this system successfully realizes the function of security password control. Bluetooth pairing password, mobile code verification, and account password matching, and other encryption measures are conducted to ensure the overall security. Through the test, the system can be successfully used with the use of mobile intelligent terminals for switching operations to achieve the desired results.

ACKNOWLEDGMENTS

This work is supported by research projects C201400313, C2015003042, and 2014 KYY09.

REFERENCES

Hai-long Zhao, Sun Shaolin, Hu Ming, et al. Design and Achievement of the Password Lock Based on FPGA [J]. value engineering, 2012, 31 (26): 199–200.

Na Lin. Design and implementation of remote intelligent safe system [D]. Zhengzhou University, 2011.

Shu-da Wang, Xuedong Han. Research on key free safe system based on face recognition [J]. Journal of Harbin Institute of Technology, 2002, 34 (2): 270–274.

Xiao-ming Shi. Innovative design and research of intelligent home safe [D]. Southwest Jiao Tong University, 2014.

Yao-wei Wu. Design and implementation of electronic system of fingerprint safe cabinet [D]. Chinese Academy of Sciences (Engineering Management and Information Technology Institute), 2015.

Yin Wu. Study on the function of intelligent measurement and control system of the safe implementation of [D]. Beijing University of Posts and Telecommunications, 2012.

Advances in Energy Science and Equipment Engineering II – Zhou, Patty & Chen (Eds)
© 2017 Taylor & Francis Group, London, ISBN 978-1-138-71798-5

A financial products' innovation method based on modularity technology

Zhijian Wang, Hua Yin, Dan Li & Jun Zhang
Information Science School, Guangdong University of Finance and Economics, Guangzhou, China

ABSTRACT: Product innovation is a key issue to provide high quality financial services. The goal of this paper is to discuss the idea of financial products' innovation based on modularity technology. Principles of production and system reconstruction through module decomposition and reorganization methods are summarized. Common features of financial products are analyzed, and the feasibility of financial products' innovation based on modularity is investigated. By redefining financial products and services according to customer demand and recombining existing elements in new service, financial product and service features' innovation is achieved.

1 INTRODUCTION

Modular technology rose in the 1960s. Modularity refers to the use of common units to create product variants (Huang, 1998). In the past few decades, modular technology and ideas have been further developed and widely used, and modularization has become the way of product design. "Module" is a kind of independent functional unit, which can be linked together according to certain rules to form a more complete and more complex system or product (Haruhiko, 2003; Cetin, 2004; Cetin, 2003). The complex product or system is divided into different modules according to certain rules and methods, and the dynamic integration process of information communication between modules is called modularization. From another point of view, it can be understood that a module is the component of the system or product decomposition, and modularization is the process of product and system decomposition.

Modularity is divided into the narrow and broad sense. In the narrow sense, modularization only refers to the application of a modular theory in products' design and manufacturing, but generalized modularization not only includes the modular theory in the production and design process, but also refers to the dynamic integration process of a system's modular decomposition and module centralization. The system includes the product, production organization, and process. All productions and operations based on the modular theory can be called as generalized modularity, including product modularization, process modularization, and organization modularization.

2 MODULARITY-BASED INNOVATION

The scope of application modules expands gradually over time. The module application ranges from the various components of the product, and gradually to the direction of the system. Module application is becoming more and more common. Modular product design and production methods furthermore affect the industry's modular process, which leads to the modular management. Modularity not only helps in the design of products and organizational structures, but also applies to innovative ways.

A module design consists of the following three parts: (1) determine which modules are the key elements of a system and how they play a role; (2) how to interact with each other, how to arrange and contact each other, how to exchange information, and so on; (3) decide whether the tested module is in compliance with design rules and test module performance with respect to other modules. Each designed module should contain information of these three parts, which can be reflected in the modularization process.

A complex system can be decomposed into several sub-systems to maintain a certain hierarchical relationship between the various subsystems, and these subsystems can also be continued to be decomposed into smaller subsystems, and then these subsystems constitute the module of the whole system. These subsystems are namely the modules used to constitute the whole system. The research of Henderson & Clark shows that there are two ways to innovate by using the modular theory: one is based on the internal innovation of each module; and the second is based on framework innovation (Henderson, 1990).

In the innovation process, we first decompose a system into modules, and then combine these modules according to certain rules, and we can innovate the complex system in this way.

Shenhar put forward a method to study the complex product system (CoPs) (Hobday, 1999) from two dimensions, the first is the product range and the second is technical uncertainty. He divided the product range into three levels and technical uncertainty into four types (Shenhar, 1994). Other researchers put forward a similar view (Chen, 2006). Based on their research, we believe that characteristics of a complex product system can be depicted, as shown in Figure 1.

Modularity innovation is the reorganization of an original system based on the modules, by restructuring the existing elements in the new service, the product and service are redefined according to customer demand, so as to innovate product function and service (Wang, 2014).

Innovation can be achieved by several forms of modular operation: (1) replace the old module design with the updated module design; (2) remove a module; (3) add modules that did not existed so far; (4) sum up elements from several modules, and organize them to form a new one; and (5) change the use of existing modules.

In the modern modular industrial structure, each module is subject only to a fixed standard, and each module's own financing, design, production, and so on can be made inside the module. A sub-module can be designed independently; this is the biggest difference between traditional division and modern modular production, which allows innovation to take place at the level of modules. In addition, the independent sub-module design feature promotes sub-system modules to be innovated continuously; otherwise, they can be replaced easily with modules outside the systems, while modules outside of the system may also try to enter the system through efforts on development. Thus, each module has the pressure of being eliminated, so that the entire chain of production is at a higher level.

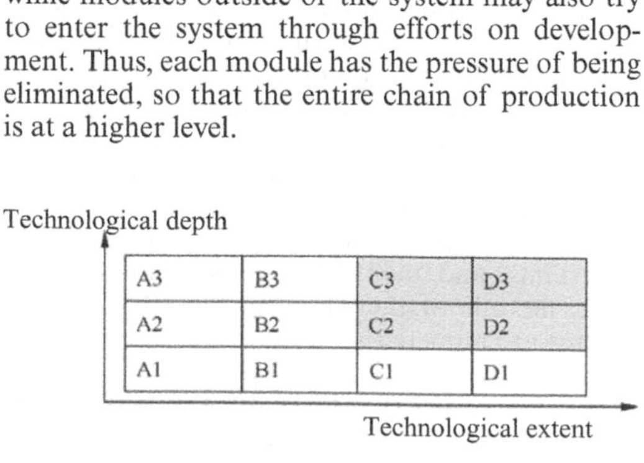

Technological depth

A3	B3	C3	D3
A2	B2	C2	D2
A1	B1	C1	D1

Technological extent

Figure 1. CoPS characteristics.

The module itself has a strong ability to adapt uncertainty from a subordinate system. By adding, split, consolidation, conversion, and other modular operations, it can cope with temporary changes in the various sub-modules. For example, in the case of a module that is out of market value or the emergence of newer technology could replace it, and it can be transformed, updated, or replaced, so as to restore the system to its best effectively and quickly. To respond to market changes, we can improve the entire production chain by a merging module or splitting a module in line with market requirements.

3 COMBINATION IN FINANCIAL INNOVATION

Financial products have four common characteristics: period, liquidity, profitability, and risk. Period is the time limit for the debtor to repay the principal and interest, liquidity refers that holders who can convert the financial products into cash quickly on necessity, profitability refers to the value of the financial products that can be added over time so as to bring additional benefits for holders, and risk refers to the financial products that may make holders lose money; in practice such risk is often proportional to benefits. In addition to above-mentioned common characteristics, innovated financial products also have a number of different characteristics.

1. Value dependency: innovative financial products are based on some current products, no matter how they change the value of a new product following a function of the value of the underlying financial products; this provides a reference for the pricing of the innovative product.
2. Contract nature: financial derivative products are a kind of contract.
3. Virtual: innovative financial products are intangible products, and so they can be easily imitated and mass-produced, but these are difficult to be protected, and so high-quality innovation is important.
4. Out of table: financial derivative products may have the potential of profit or loss, but risk scale cannot be shown in the financial table.
5. Uncertainty and high risk: in addition to the traditional risk factors, when a new financial product is designed, the large number of transactions per day may lead to potential risks.
6. Combination and innovation: the essence of financial derivative products is to break up the elements of the basic assets or basic products, and to carry out a new combination. As a result, derivative financial products can have

numerous combinations, which can be innovative in many different forms. Traders can take different times, different basic products, and different cash flows of a variety of products into different products, and they can also design a certain product according to specific requirements of customers.

Financial product innovation is to change the existing products, the popularity, safety, and profitability. The method includes product decomposition and product synthesis. The basic approach is to modify the terms of the contract for financial assets, derive gradually, and produce new products. Product synthesis innovation is defined as the process that combines two or more than two kinds of financial products to form a new financial product; under ideal conditions, almost any income distribution can be synthesized. You can also break down risk—return characteristics of financial products, change their liquidity, profitability, and security, so as to produce productions with a combination of new liquidity, profitability, and safety.

Internet financial innovation has its principles and laws, and its essence is to simplify complex issues. Simplify complex information to different combinations of 0 and 1, simplify hardware to different combinations of standard parts, and simplify software to different code combinations or middleware; they were then assembled into different systems, trialled, tested, and iterated in practice. Online sales of various financial products also need to be standardized in order to achieve economies of scale.

A simplification principle is the standardization of a simple issue, decomposing elements of things into different units. The product manufacturing process is also broken down into small stages, thereby simplifying it into standard units, and then is combined into a whole, so as to improve efficiency.

Computable financial means establish a financial decision-making model; financial analysis and decision-making are based on this model. Finance equals to algorithm in computable financial times. As finance becomes increasingly digitalized and is more and more computable, it is more and more algorithmic. A good financial theory or a good investment pattern can be described by using algorithms and should be calculable and verifiable.

4 MODULARITY IN FINANCIAL PRODUCTS' DEVELOPMENT

Because of above-mentioned characteristics of innovative financial products, it is difficult to find an innovative way to meet the individual needs of

customers, and to reduce the risk as much as possible at the same time. Especially for the supply chain of financial products, with numerous participants and many complex factors, participants only pay attention to their own interests. The target to achieve interest equilibrium undoubtedly increases the difficulty of supply chain financial product innovation.

According to modularity theory, innovative products can be created by product decomposition and combination. This will not only be able to create innovative products quickly, but also reduce the cost of innovation, and reduce risk effectively. Therefore, modularity theory-based financial product innovation, on one hand, satisfies the market demand and, on the other hand, provides an effective way to enhance banks' competitiveness.

Modularity does not mean all complex products or systems must be modularized; a system or product can be broken down effectively is the prerequisite of modularity. Only in the case that after decomposition it can meet a variety of needs through a combination of modules, modularity is necessary; otherwise, the modularization of the product or system will lose its significance. It is also one of the reasons for the success of enterprise product innovation to reduce the dependence on other modules when designing a product.

Generally, relations between the modular decomposition level and enterprise systems cost was "U" shaped (Yan, 2015) such as Figure 2.

Suppose C is the manufacturing cost of CoPS and y is the number of decomposable modules of the CoPS, then

$$C = ay^2 + by + c \left(a, c > 0, b < 0\right) \quad (1)$$

Let $a=\beta 2$, $b=\beta 1$, $c=\beta 0+\mu$, after variable substitution for the above model, then

$$C = \beta_0 + \beta_1 y + \beta_2 y^2 + \mu \left(\beta_0, \beta_2 > 0, \beta_1 < 0 \right) \quad (2)$$

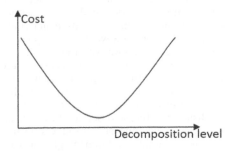

Figure 2. Graph showing relations between the decomposition level and cost.

This is a standard quadratic regression model, where, β_0, $\beta1$, and β_2 are regression coefficients, and these coefficients can be obtained by using the least square method. On the basis of modular decomposition meeting preconditions, designers can decompose a CoPS using this model. The number of modules from modular decomposition should be an optimal quantity corresponding to the best corporate performance.

The current financial product development cycle in our country is too long to meet the quick market demand; the innovation process is isolated and rigid. There is a lack of necessary communication and communication between the departments; the participating departments attend to maintain their own interests; there is no overall awareness and common design standards; product innovation information and knowledge are exclusive. This series of deficiencies weaken the effect of product innovation in different degrees.

Financial products' innovation by using the modularity theory can overcome these shortcomings effectively. Firstly, financial products can be decomposed and the combination of modules can meet diverse needs of the market; this is the prerequisite for designing innovative products by using the modularity theory. Secondly, the decomposition and combination of financial products based on the modularity theory can shorten the product development cycle, reduce the cost of overall products' development, and meet the diverse needs of customers. Finally, product development by using the modularity theory needs to establish a team of professionals from all departments, and communicate in real time. There is no personal interest in the team, only the overall interests. Everyone needs to develop their comprehensive ability and innovative enthusiasm. This overcomes the shortcomings of the traditional product development model, which is isolated and rigid with little information sharing.

The connotation of financial services modularity innovation is that a financial enterprise uses modular ideas and methods in designing of financial services' products and financial services' system, improve service quality, and service efficiency of the existing innovation activities. The creation of an innovative IT management platform supporting innovative products designing is very important. Commercial banks should form a unified product attributes parameter library, combine product attributes and parameters classified according to the market and customer needs, and establish a product and service function module library, assemble innovative products flexibly, and construct a rapid development platform thereby supporting combination innovation. Banks should build product innovation information

systems, grasp details about innovative product development by using market and customer dynamics information timely and comprehensively. Establish scientific management methods, innovative product classification standards, and product catalog maintenance mechanism; standardize the definition of innovative products, thereby forming a unified sound commercial banking product catalog and maintenance management system.

Commercial banks should draw on existing scientific methods, speed up the assessment system about innovative products, and give full scope to its important role in promoting product innovation and restructuring. Product innovation and product management information integration should be strengthened; innovation monitoring and evaluating indicators are set dynamically according to various financial and non-financial factors. From the points of customer satisfaction, market share, degree of risk, inputs, and outputs profitability, product innovation input–output analysis, post-market evaluation and performance measurement can be performed. Commercial banks should establish a suitable product innovation evaluation model, improve customer evaluation of products, and perform market competitiveness evaluation, financial benefit assessment, risk assessment, and a series of product evaluation systems. Commercial banks should also manage product monitoring information centrally, improve conversion, and promotion mechanism for innovative products, and enhance product innovation capability maturity.

5 CONCLUSIONS

The development of the modularity theory is closely related to the globalization and informationization of economic development. With the continuous deepening of the modularization idea, the practice of modularization is constantly innovating. The development of modular practice requires the strengthening of the theoretical research on modularity. This research work shows that the modularity theory has the applicability and feasibility in the finance field.

Previous studies are mainly from the macro-level facing product module partition; there are few studies on the operation process analysis and the module partition of financial products, and even less research on modeling with certain modeling tools, and this can be strengthened in the future.

ACKNOWLEDGMENTS

This work was supported by the Foundation for Technology Innovation in Higher Education of Guangdong Province, China (No. 2013 KJCX0085), Degree and Postgraduate Education Reform Project of GDUFE (2014ZD01), Foundation for Distinguished Young Talents in Higher Education of Guangdong Province, China (No. 2013LYM0032), Foundation for Technology Innovation in Higher Education of Guangdong Province, China (No. 2013KJCX0084), and Guangdong Natural Science Foundation (2014A030313609).

REFERENCES

Aaron Shenhar, A Conceptual Framework for Modern Project Management, 4th Int. Conf. on Management of Technology, Florida (1994).

Ando Haruhiko and Aoki Masahiko, Modulization, China: Shanghai Far East Publisher, (2003).

Chen Jin, Gui Binwan, Modular Complex Product Systems Innovation Process and Management Strategy. R & D management, 18(3) (2006).

Chun-Che Huang, A Kusiak, Modularity in Design of Products and Systems. IEEE Transactions on Systems Man and Cybernetics—Part A Systems and Human, Vol.28(1) (1998).

Hobday, M. Rush, H. Technology management in complex product systems (CoPS), International Journal of Technology Management, 17(6) (1999).

Cetin, OL. Saitou, K. Decomposition-Based Assembly Synthesis for Structural Modularity, Journal of Mechanical Design, 126(2) (2004).

Onur, L. Cetin, Kazuhiro Saitou, Decomposition-Based Assembly Synthesis of Multiple Structures for Minimum Production Cost, Asme International Mechanical Engineering Congress & Exposition, 127(4) (2003).

Henderson, R.M. and Clark, K.B. Architectural Innovation: The Reconfiguration of Existing Product Technologies and the Failure of Established Firms, Administrative Science Quarterly, Vol. 35(1) (1990).

Wang Yu and Ren Ha, Value innovation of modular organization: content and nature, Studies in Science of Science, Vol.2 (2014).

Yan, H F, Zhong W J. Research on the modular decomposition model of the complex product system.Journal of Beijing University of Aeronautics and Astronautics, 41(1) (2015).

Advances in Energy Science and Equipment Engineering II – Zhou, Patty & Chen (Eds)
© 2017 Taylor & Francis Group, London, ISBN 978-1-138-71798-5

Research on hybrid electric vehicle power switching control strategy

Zhuming Cao
School of Mechanical Engineering, Beijing Polytechnic, Beijing, China

Jiatian Guo & Lei Wang
Department of Automotive Engineering, Shandong Vocational College of Science and Technology, Shandong Weifang, China

ABSTRACT: Hybrid Electric Vehicle has the advantages of motor and engine. This article has built two control strategies to control the motor and engine switching. The logic threshold control strategy is built based on the target of the engine's best fuel economy and the motor's best efficiency. The fuzzy logic control strategy is built based on the accelerator pedal opening, battery State of Charge (SOC) value and vehicle speed. The Simulink software is used to analyze the influence of the two types of control strategies for the vehicle economy and power performance. At last, the article put reasonable suggestions forward for improving the cooperation between the engine and the motor.

1 THE RESEARCH STATUS OF HYBRID ELECTRIC VEHICLE CONTROL STRATEGY

The vehicle power switching control strategies is becoming more and more complex with the rapid development of the Hybrid Electric Vehicle (HEV). The strategies can be classified into four categories: the logic threshold control strategy, the fuzzy logic control strategy, the adaptive control strategy and the neural network control strategy.

1.1 *The logic threshold control strategy*

The logic threshold control strategy is based on Boolean logic. The control rules are described by precise value. The logic threshold control strategy is simple. It is mainly utilised in the hybrid vehicle control strategy. The strategy control rules make the engine working in its high efficiency range. When the engine works in the low speed range, it would use the motor instead of engine working. The battery can supplement energy by generators and it also can recycle part of the braking energy to be recharged.

1.2 *The fuzzy logic control strategy*

The fuzzy logic control strategy is a type of control strategy based on fuzzy logic. It works with the logic threshold control strategy in the same control target, and the control strategy is based on rules as well. The main difference between the logic threshold control strategy and the fuzzy logic control strategy is the way to design the limit value. The limits value and the rules are fuzzy under the fuzzy logic control strategy. In recent years, people pay more attention to the fuzzy control method in the field of HEV because of its good control quality. Therefore, this control strategy has a good development prospect.

1.3 *The adaptive control strategy*

Adaptive control is a control method that contains certain knowledge and the ability to adapt to changeable environmental conditions. It is not only able to change automatic correction control action according to the condition but able to make the control effect to achieve or near the optimal level. The objective function of the Dynamic adaptive control should depend on the characteristic of engine fuel economy and emissions. The control strategy minimises the objective function value to improve the purpose of the engine fuel economy and to reduce its emissions. But the dynamic adaptive control needs a large amount of data and calculation, and the optimized procedure is complex.

1.4 *The neural network control strategy*

The HEV control has nonlinear characteristics. Neural network control is able to approximate nonlinear function with any degree of accuracy. It can adapt and learn complex uncertain problems using its powerful information comprehensive ability. But neural network control is uncertain, and it has not been applied on the HEV.

2 POWER SWITCHING LOGIC THRESHOLD CONTROL STRATEGY

The battery SOC value should not exceed the limit when setting the upper and lower limit for the battery SOC. The engine working point should be fluctuated around its highest efficiency curve.

The accelerator pedal connects to the trigger and the motor power voltage is regulated by the pedal opening value. Therefore, the output speed of the motor should adjust depending on the motor efficiency curve to keep the rated output torque.

In order to achieve the goal of energy conservation and emissions reduction, the engine should be accelerated by electric motors to the idle speed, and then the engine can be started and works along the curve of the best economic.

The vehicle illustrated in this article is driven by motor when the accelerator pedal opening under 20%. When the accelerator pedal opening is big, the vehicle should make full use of the advantages of motor torque, and the engine should be operated to work as soon as possible. The 10 km/h speed is selected as the power switch control speed to ensure the rapid acceleration of the vehicle, and this speed can make the engine to achieve the steady state working point as soon as possible.

When the accelerator pedal opening is small, selected the 25 km/h speed as the power switch control speed to energy-saving and emission-reduction, prolong the working time of the motor, avoid engine working under the high fuel consumption.

3 POWER SWITCHING FUZZY LOGIC CONTROL STRATEGY

The two power switching points of logic threshold control strategy are economic and have power performance. If the driver wants to start the car faster, the accelerator pedal opening at 40% can meet the demand of the driver on the power performance. However, the driver has to step the accelerator pedal for more than 60%, otherwise the power switching point will be efficiency as the goal, still with the throttle opening at 20% as power switching control standards. At this point, the vehicle's performance are not up to the driver's demand. If the accelerator pedal is stepped more than 60%, the vehicle power performance will beyond the needs of the driver, and thus the vehicle's fuel economy will be reduced.

Threshold blur can reflect the transition zone between the two threshold values. The method of fuzzy control is much closer to people's way of thinking than the method of traditional threshold control is. Thus, it can better describe the control rules. The fuzzy control strategy illustrated in this article is built based on the threshold control strategy. The vehicle speed of the engine and electric motor is determined by utilising the opening of the accelerator pedal and battery SOC value.

On the one hand, the adjustment target of the power switching speed will achieve economic performance when the accelerator pedal opening is under 20%. At this point, the engine will be start after the vehicle speed up to 25 km/h. On the other hand, the adjustment target of the power switching speed will achieve power performance when the accelerator pedal opening is upper 60%. At this point, the engine will be started when the vehicle speed up to 10 km/h. In addition, when the accelerator pedal opening is between 20% and 60%, with the increase of the pedal opening, the adjustment target of power switching control speed will fluctuate from the economic performance to power performance.

4 SIMULATION ANALYSIS

The engine throttle opening signal in the simulation model get from the accelerator pedal position signal.

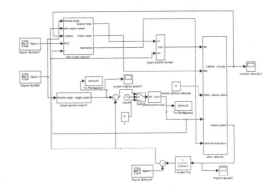

Figure 1.   The vehicle simulation model.

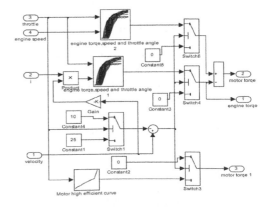

Figure 2.   Threshold power switch control module.

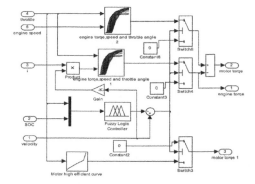

Figure 3. Fuzzy power switch control module.

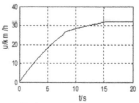

a. The threshold control simulation results under

accelerator pedal opening at 30%

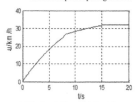

b. The threshold control simulation results under

accelerator pedal opening at 50%

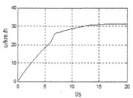

c. The fuzzy control simulation results under accelerator

pedal opening at 30% and 50%

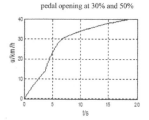

d. The fuzzy control simulation results under accelerator

pedal opening at 30% and 50%

Figure 4. The simulation results under different accelerator pedal opening.

The power switch of logic threshold control strategy and fuzzy logic control strategy respectively established the threshold power switch control module and the fuzzy power switch control module in order to determine the working range of the engine and the motor.

Battery SOC value is 0.6, and the opening of the accelerator pedal is 30% and 50%. The simulation results are shown in figure 4.

From the simulation results of the speed change, the work time of the motor was found relatively fixed under the threshold control strategy. The car's dynamic was enhanced under the fuzzy logic control strategy, and this pattern was more pronounced with the increase of the throttle opening.

Fig a illustrated that although the accelerator pedal opening changed from 30% to 50%, there was no change of the power switching speed under the threshold control strategy. Fig b showed that from the accelerator pedal opening changing from 30% to 50%, the power switching speed under the fuzzy control strategy was obviously decreasing, thereby the car's power performance could meet the demands of the driver.

In conclusion, the hybrid electric vehicle power switch control strategy can use the logic threshold control strategy under the economic priority condition, while the fuzzy logic control strategy could be utilised under the dynamic priority condition.

REFERENCES

Chu Liang, Wang Qingnian. Energy Management Strategy and Parametric Design for Hybrid Electric Family Sedan. SAE paper:2001–01.

Dai Yifan, Luo Yugong, Bian Mingyuan et al. The control strategy for a new full hybrid power train structure [J]. Automotive Engineering, 2009, 31(10):919–923.

He Ren, Shu Chi. Overview of power-switch coordinated control of hybrid electric car [J]. Journal of Jiangsu University, July, 2014, 4(35):373–379.

Hwang H S, Yang D H, Choi H K, et al. Torque control of engine clutch to improve the driving quality of hybrid electric vehicles [J]. International Journal of Automotive Technology, 2011, 12(5):763–768.

Rosen C L. Hybrid Power Train U.S. Patent 3791473, Filed September, 1972.

Schouten N J, Salman M A, Kheir N A. Fuzzy Logic Control for Parallel Hybrid Vehicles.IEEE Transactions on Control Systems Technology, 2002, 10(3).

Sun Dongye, Qin Datong. Control Strategy on Power Conversion of a Parallel Hybrid Vehicle [J]. Transactions of The Chinese Society of Agricultural, 2003, 34(1):5–7.

Advances in Energy Science and Equipment Engineering II – Zhou, Patty & Chen (Eds)
© *2017 Taylor & Francis Group, London, ISBN 978-1-138-71798-5*

Design of a brightness wireless control system for machine vision LED light sources

Xiaojin Yan & Xiaodong Yang
School of Electrical and Control Engineering, North China Institute of Aerospace Engineering, Langfang, China

Chunqing Xu
Shanghai Huilin Image Technology Co, Ltd., KunShan, China

Qian Liu & Xiuliang Ju
School of Electrical and Control Engineering, North China Institute of Aerospace Engineering, Langfang, China

ABSTRACT: In order to further improve the image quality and intelligent control machine vision LED light source, infrared remote control technology is used to control the brightness of the LED light source. The system MCU receives infrared remote control signals, generates a PWM to control the output current of the LED drive circuit, and then realizes the wireless control of brightness. It is shown that the system can accurately adjust to the brightness of the LED light source, and improve the quality of the machine vision's image.

1 INTRODUCTION

Machine vision technology, which has been used in modern manufacturing industry, is becoming more and more mature. The acquisition and processing of the image is the core of machine vision technology, and so image quality is very important to the whole system. The target feature and the background information can be integrally separated from the image with a good light source and lighting solution (T, 2013). And this can reduce the difficulty of image processing and improve the stability and reliability of the system. Currently, the LED light source and linear analog dimming is the most commonly used light source and lighting solution (W, 2015). The linear analog dimming method is simple and direct, but it can reduce the luminous efficiency and cause color cast. And the output current precision is not high. This is only applicable to the occasions when the dimming control requirement is not high. PWM dimming will not cause color cast and its dimming accuracy is higher. In a larger range of dimming, the LED light source will not flicker (C, 2014; X, 2014). In this paper, a brightness wireless control method of the machine vision LED light source is proposed. The machine vision light source is composed of a series and parallelly connected LED with good consistency. An infrared remote-controlled MCU generates a PWM to control the duty cycle of LED current to adjust the brightness.

2 SYSTEM COMPOSITION AND THEORY

In order to realize the wireless control for the LED light source of the machine vision, the infrared remote control technology is used to control the PWM duty cycle to change the current of the LED light source in this system. The system mainly consists of a 5 V regulator module, direct output 5 V power supplied for controllers, and other ancillary devices. An infrared remote control module is used to launch the 38 kHz carrier signal, a HX1838 integrated infrared receiver is used to complete the 38 kHz signal shaping and output reverse signal to the MCU, and MCU internal settings for infrared remote control decoding. A LED brightness controller combines infrared command and regulation brightness together and sends the PWM signal to the LED constant current driver, which drives the LED light source to control the brightness change according to the instruction. The system's block diagram is shown in Figure 1.

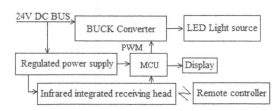

Figure 1. System block diagram.

3 SYSTEM HARDWARE DESIGN

3.1 MCU control circuit

STC12C5410 AD is used as core controller in the control system, which has advantages such as high speed, lower power consumption, and strong anti-interference ability. Its instruction code is fully compatible with the traditional 51 single chip microcomputers. Otherwise, four high-speed 10-bit A/D conversion channels, and four PWM output channels meet the needs of the LED lighting source PWM control. The minimum system includes an external clock circuit, RS232 circuit, and reset circuit schematic, as shown in Figure 2.

3.2 5 V power supply design

Since the system input voltage from the 24 V DC bus and the power supply voltage of each module is 5 V, therefore MC34063 is used to constitute a Buck converter output 5 V voltage. The specific circuit is shown in Figure 3.

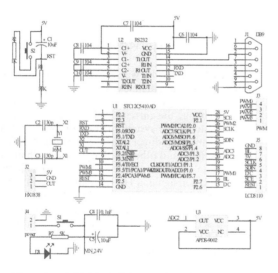

Figure 2. A STC12C5410 AD minimum system.

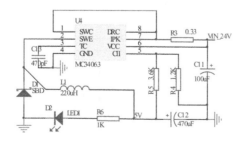

Figure 3. 5 V power supply circuit diagram.

3.3 Infrared remote control and display

Infrared remote control has two components: transmitter and receiver. The transmitter transforms infrared coding modulated binary code corresponding to the button into the pulse string signal, and is emitted by using an infrared transmitting tube. The receiver receives the infra red signal, and then processes the signal automatically, including signal amplification, detection, shaping, and finally demodulating the remote encoder pulse (Z, 2003). In order to improve system reliability, the design used an integrated infrared receiver HX1838, which can receive a 38 kHz infrared signal, and the signal would be changed into the corresponding key value by using an MCU.

An infrared remote control uses an ordinary remote control, which is encoded by using the NEC protocol. Infrared remote control and hex encoding are shown in Figure 4. The buttons EQ, PLAY/PAUSE, VOL+, VOL—are chosen as adjust function keys of the LED light source. The specific function keys table is shown in Table 1.

The display portion uses the Nokia 5110 LCD, which consists of the LPH7366 LCD module and the PCD8544 LCD controller chip. Nokia 5110 could display four lines of Chinese characters, which is widely used in various portable devices' display systems.

3.4 LED constant current drive circuit

The specific design parameters of the machine vision LED light source drive circuit are as follows: input voltage DC is 24 V, output voltage is 17.6 V, output rated current is 350 mA, and the switching frequency is 350 kHz. The driving circuit uses the Buck converter, as shown in Figure 5. In this paper, RT8452 is selected as the drive control chip, whose real-time the PWM dimming function and complete protection can meet the design requirements (Y, 2013).

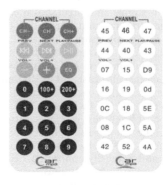

Figure 4. Picture showing the infrared remote control and hex encoding.

Table 1. List of the function keys.

Keys	Name	Function
EQ	ON	Control LED light
PLAY/PAUSE	OFF	Control LED off
VOL+	To increase brightness	Control LED to increase brightness
VOL−	To reduce brightness	Control the LED to reduce brightness

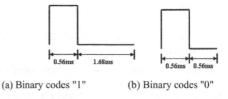

(a) Binary codes "1"   (b) Binary codes "0"

Figure 6. Schematic of binary codes.

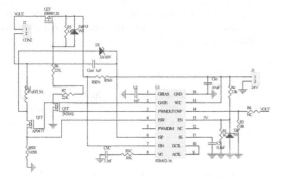

Figure 5. Buck circuit diagram.

4 SYSTEM SOFTWARE DESIGN

4.1 Infrared control program design

An infrared remote control transmitter chip uses Pulse Position Modulation (PPM) encoding; the coded pulse consists of a preamble with 9 ms high level and 4.5 ms low level, 8-bit address code, 8-bit address counter code, 8-bit opcode, and 8-bit opcode anti-code. A pulse represents a binary "1" as shown in Figure 6(a), whose cycle is 2.24 ms and high level width is 0.56 ms. A cycle of 1.12 ms and high level of 0.56 ms pulse represents a binary "0", as shown in Figure 6(b).

The decoding process is carried out to achieve recognition coded pulse and extract user code function procedures. Correctly decoded is effectively identified as "0" and "1". The decoding algorithm was designed based on the coded pulse sequence diagram, as shown in Figure 7.

4.2 Brightness control program design

When the correct infrared signal is detected, the programs determine whether to control the brightness. If the instruction is to control brightness, the duty cycle of the PWM is controlled according to the brightness increase or decrease. The brightness control program flow chart is shown in Figure 8.

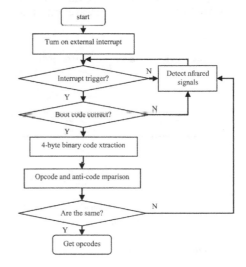

Figure 7. Decoding algorithm flowchart.

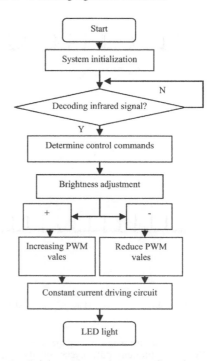

Figure 8. Brightness control program flow chart.

345

5 TEST RESULTS AND ANALYSIS

5.1 *Constant current driving circuit for LED light sources*

The DC bus voltage is 24 V, and the rated current of the LED source is 0.35 A. Figure 9(a) shows the duty cycle of 0.84, Q1T MOSFET drain-source waveform, and at this time the LED current is 0.29 A. Figure 9(b) shows the duty cycle of 0.49, Q1T MOSFET drain-source waveform, and at this time the LED current is 0.17 A. Figure 9(c) shows the duty cycle of 0.3, Q1T MOSFET drain-source waveform, and at this time, the LED current is 0.1 A.

5.2 *Remote control decoding and LCD display test*

When combined with the liquid crystal display module to debug infrared remote control decoding, when the decoding is successful, it shows that IR decoding OK. As shown in Figure 10, the decoding is successful. An LCD can show the power status, duty ratio, and other information, as shown in Figure 11.

5.3 *Brightness control test*

Test condition: to the LED light source, add a lamp shade and a dark cardboard. We shot the test images as shown in Figure 12, and the picture clearly reflects the variation of brightness.

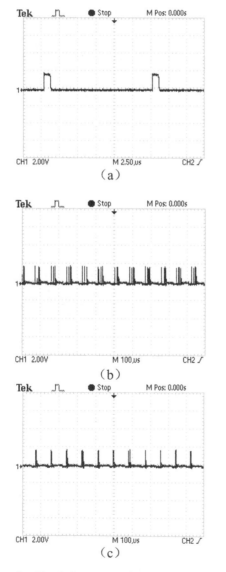

Figure 9.   The drain-source voltage waveform of Q1T MOS under different duty cycles of the buck drive circuit.

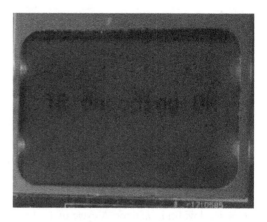

Figure 10.   Picture showing infrared decoding success.

Figure 11.   Picture of the LCD display test.

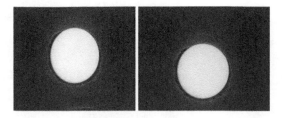

Figure 12. Pictures showing the brightness change test pattern.

6 CONCLUSIONS

The design uses infrared remote control technology to realize the brightness wireless control of the machine vision LED light source, which makes the machine vision LED light source control more intelligent, and also lays a foundation for obtaining high quality visual images. By debugging the hardware circuit and software program, the system can meet the requirements of the machine vision LED light source, and the performance can fully meet the practical engineering applications. In the future, it can be combined with wireless sensor network to realize one machine multi-control, networked control, and other applications.

ACKNOWLEDGMENTS

This work was financially supported by the 2015 Hebei Province Department of Education Science and Technology Research Project (No.: QN2015103). At the same time, the authors gratefully acknowledge the Hebei Provincial Key Disciplines "detection technology and automation" project for its support.

REFERENCES

Aiquan, Z. Shanxi Electronic Technology, 6, 40–41 (2003).
Jimin, T. 2013 machine vision technology and Application Conference, (2013).
Jin-ru, C. Sen-quan, Y. Journal of Shaoguan University, 35, 36 (2014).
Shuaigui, W. Gongyan, L. Yi, Y. CN104797052 A[P]. (2015).
Xiaojin, Y. J. Journal of North China Institute of Aerospace Engineering, 23, 23 (2013).
Xiaoming, X. Heping, X. En-xu, L. Industrial Control Computer, 27, 151–152 (2014).

Advances in Energy Science and Equipment Engineering II – Zhou, Patty & Chen (Eds)
© 2017 Taylor & Francis Group, London, ISBN 978-1-138-71798-5

Path planning for the multi-robot system in intelligent warehouse

DanLu Zhang, Bin Zheng & Shun Fu
Chongqing Institute of Green and Intelligent Technology, University of Chinese Academy of Sciences, China

Xiaoyong Sun
Chongqing DELING Technology Co., Ltd., China

ABSTRACT: Automatic intelligent warehouse is highly desirable because of the developing demand for logistics. Thus, in this paper, we will build a flexible stable and reconfigurable intelligent warehouse using multi robots, which will reduce the human resource and improve the efficiency in warehouse system. Cooperation and obstacle avoidance in multi robots path planning is the main issues to achieve a stable and efficient intelligent warehouse. So, firstly, this paper proposes a reconfigurable intelligent warehouse system with traffic rules. Then, based on the warehouse model, we systematically put forward an improved A* algorithm combined with reservation table to coordinate the multi robots. And finally, we use simulation experiments to prove that the path planning approach proposed in this paper will increase the efficiency of the multi-robot system in terms of delivery time.

1 INTRODUCTION

With the increasing demands for efficiency automatic warehouse, researches on the coordination and obstacle avoidance of the multi robots have become the main issues in intelligent automatic warehouse system (Ma, 2015). So, in this paper, we will propose a method that can solve the path planning problem and improve the efficiency of the whole system.

A robot is sharing a structured environment with other robots and humans, and they are regarded as moving obstacles. In this way, the robots are delivering goods in a highly dynamic environment, thus, considering the safety, traffic rules are used to avoid some collisions in the structured environment. Generally, there are two approaches to control the multi-robot system: centralized approaches and decentralized approaches (see e.g. (Olmi R, 2011; Digani, 2016; Zheng, 2013; Alonso-Mora, 2016) for detail). Centralized approach uses a central control system to coordinate a fleet of robots, which can find optimal solution for robots and reduce the number of collision. But with the growing number of robots, the efficiency of central control system will reduced quickly. On the other hand, decentralized approach lets robots plan their routes which will reduce the burden of central control system. But will also lead more collisions. So we will use a method that combines the two approaches that can make full use of their advantages and avoid their disadvantages. Firstly, we use central control system to deal with the goods and allocate them into group, and collect information of robots, such as their

locations and status. Then, robots can plan their path by themselves in a decentralized way, using the information they get from central control system, which will save communication time between them in the warehouse. A robot can find global optimal solution with A* algorithm (see e.g. (Hart, 1968; A R L, 1990) for detail), so A* algorithm was widely used in path planning. But in intelligent warehouse, robots are moving in dynamic environment. So we use reservation, combining with priority to avoid collision. Using the method above, we proposed an improved path planning algorithm to realize an efficiency intelligent warehouse.

2 INTELLIGENT WAREHOUSE MODEL AND TASK ANALYSIS

Figure 1 shows the intelligent warehouse model (adopted from (Williams, 2014)) and traffic rules. And we use grid-based map (see (Burgard, 1997)

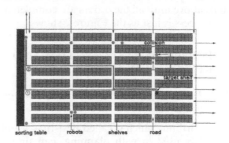

Figure 1. Intelligent warehouse model and some rules.

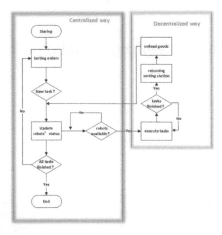

Figure 2. The process of intelligent system.

for detail) to represent the warehouse because of the structured environment. In the figure, green areas represent the goods shelves. Gray areas represent the sorting table that allow staffs to pack the goods. White areas represent one-way roads. And red dots represent a fleet of robots $R = \{r_1, \cdots, r_n\}$, moving around in the warehouse to deliver goods. The direction of each road is just as the arrow shows in the figure. In this way, robots can avoid the situation of head-on collision.

Firstly, the central control system processes the goods in order continuously and sorting them in group according to their information such as time and location (see e.g. (Rodríguez, 2012; Hernández, 2014) for more details). Then central control system sends the information of orders to robots that are unoccupied. After getting information of orders, robots' status become occupied, and they go to pick up goods using the path planning approach we proposed in this paper. Finally, after the orders has been completed, the robots go back to sorting table with goods. And staffs in the sorting table pack the goods in robots. After unloading the goods, the status of robots become unoccupied, and they will execute new tasks assigned by central control system again. The robots will repeat the steps above until all the orders are completed. And the process is just as the Figure 2 shows.

3 IMPROVED A* ALGORITHM AND RESERVATION TABLE

3.1 Improved A* algorithm with improved grid-based map

A* algorithm, which is called best-first search and is widely used in path planning, can find global optimal path according to the estimated function.

$$f^*(n) = g(n) + h^*(n) \tag{1}$$

Where $g(n)$ represents the real cost that moving from the start node to the current node, $h^*(n)$ represents the estimated cost that moving from the current node to the goal node. We can use time distance or weight to calculate $g(n)$ and $h^*(n)$. In this paper, we use distance to estimate $f^*(n)$, and use Manhattan distance to estimate $h^*(n)$ just as function 2 shows:

$$d_{(n,goal)} = abs(n.x - goal.x) + abs(n.y - goal.y) \tag{2}$$

Where n represents the current node, $goal$ represents goal node and xy represent their coordinate.

Staring from start node, a robot calculates the estimated cost $f^*(n)$ of the nodes that are around its' current node, and put their estimated cost into a list. Then the robot chooses a node in the list with smallest estimated cost as the next node it will step into. Finally the robot repeats the steps above until it find the goal node.

Every time a robot stepped into a new node, the robot has to calculate the estimated value $f^*(n)$ of every nodes around it. And because the roads between each two crossroads are one-way street, it is a waste of time under the situation of traffic rules in sometimes. So we proposed an improved A* algorithm. As the Figure 3 shows, we can simplify the grid-based map and named it as improved grid-based map. In the improved grid-based map, the moving state of robots can be divided into three situations:

1. Robots moving from crossroad to another crossroad.
2. Robots moving from the start node to a crossroad.
3. Robots moving from a crossroad to the goal node.

When robots at situation 1, there is no need to calculate the estimated value $f^*(n)$ of each nodes that belong to the one-way street. And robots can just calculate the estimated value of each crossroad node using the improved grid-based map and

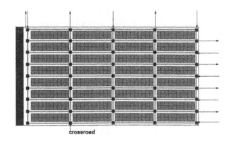

Figure 3. Improved grid-based map.

move a fixed number of nodes to the next crossroad. And when robots at situation 2 and situation 3, they have to calculate every nodes around them to choose a next node to extend using the grid-based map.

3.2 Improved A* algorithm with turning time

As Figure 1 shows, when robots moving from start node to goal node, there are more than one route that have the same length. But a robot need to turn a corner once with route 1 just as the red line shows, and has to turn twice as the blue line shows. Because a robot will spend more time when turning a corner, it will take different time when a robot move along different routes that have the same length. Thus a robot must take the factor of turning into account when plans route, instead of just choosing a route randomly with traditional A* algorithm.

In summary, the process of the improved algorithm is as follow: Firstly, starting from start node, a robot has to calculate the estimated cost $f^*(n)$ of every node around it until the robot find the goal shelf node or a crossroad. Then, when the robot arrived at a crossroad, it will check whether the goal shelf node in adjacent shelves areas. When the goal node is in the adjacent areas, in other words, robot moves to the goal shelf from a crossroad, the robot also has to calculate the estimated value $f^*(n)$ of every node around it using traditional A* algorithm. But when the goal node is not in the adjacent shelves areas, in other words, a robot has to move to a crossroad from another crossroad, the robot can then use improved A* algorithm based on the improved grid-based map. So, the robot calculates the estimated cost $f^*(n)$ of the crossroads next to it, and move a fixed distance to the crossroad that has the smallest estimated cost. Finally, when the robot plans route at crossroad, it has to take the factor of turning into account. Robot will use equation 3 to calculate the estimated value $f^*(n)$ when moving to a crossroad from another crossroad:

$$f^*(n) = g(n) + h^*(n) + \sum_{i=1}^{n} t \qquad (3)$$

Where t represents the time that a robot need to take a turn, and n represents the number of times for turning. The robot repeat the steps above until it find its goal shelf node.

3.3 Reservation table

Although robots can avoid some head-on collisions using traffic rules, there are still some collisions at crossroad as Figure 1 shows. So we put forward to a three-dimensional reservation table to record the

status of every node in the grid-based map during a period of time. At a certain time t_i, a node in the grid-based map can be only occupied by a robot, and the status of the node is unavailable. We will use robots' id to mark the unavailable node until the node is available, just as the Figure 4 shows. And we will save the record during a period of time $t_{(i,j)}$, which forms a three-dimensional reservation table.

If a robot wants to step into a node at time t_i, it has to refer to the three-dimensional reservation table at time t_i. If the node is unavailable at time t_i, the robot has to wait until the status of the node become available. And if several robots want to step into a node that is available, the robot with high priority (priority can be determined by departure time or something else) can step into the node firstly. But if several robots have the same priority, we can choose a robot randomly.

We can also refer to the location of robots during a period of time according to the three-dimensional reservation table. And according to their location, we can know the situation of traffic congestion. For example, at time t_i, we can know the location of a robot is $node_n$ according to the three-dimensional reservation table. And at time t_j, the location of the robot is $node_m$. According to the two location, we can calculate the velocity v_{robot_k} of the robot during time t_i and time t_j. Robots will move around in the warehouse at velocity v when there is no traffic congestion in the warehouse. Comparing the velocity v_{robot_k} to the robot's normal velocity v, we can know the situation of traffic congestion. And we use weight w to represent the situation of traffic congestion. The function 1 and function 3 are modified as function 4 and function 5:

$$f^*(n) = g(n) + w \cdot h^*(n) \qquad (4)$$

$$f^*(n) = g(n) + w \cdot h^*(n) + \sum_{i=1}^{n} t \qquad (5)$$

Using the function 4 and function 5, routes with traffic congestion will have greater estimated value $f^*(n)$. Robots will choose routes with smaller estimated value, so they can avoid traffic congestion.

col row	col₁	col₂	······	colₙ₋₁	colₙ
row ₁	RobotID₁		······	RobotID₃	
row 2			······		RobotIDₖ₋₁
⋮	⋮	⋮	⋮	⋮	⋮
row m-1	RobotID₂		······		
rowm			······		RobotIDₖ

Figure 4.   Reservation table at time t_i.

351

4 SIMULATION AND CONCLUSION

We use MATLAB to simulate the situation that a fleet of robots moving in intelligent warehouse system. Comparing with traditional A* algorithm (see e.g. [6] [7]), we want to prove that the improved A* algorithm with traffic rules and three-dimensional reservation table can save delivery time for multi-robot system. We assumed that the simulation condition are as follows:

1. We compare improved A* algorithm with traditional A* algorithm
2. The order tasks are generate randomly. Each order task has 50 orders and each order has six kinds of goods for robot to carry. A fleet of robots will excute 5 group of randomly generated tasks, and calculate the average time and distance.
3. The number of robots that excute the tasks are: 10 20 30 40.

As the figure 5 shows, when performing the same task, different multi-robot system with different number of robots will move different length of total distance. And using the two kinds of different path planning algorithm, a group of robots with a certain number of robots will move the same length of total distance to finish the same task.

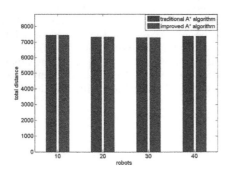

Figure 5.   Total distance.

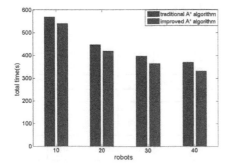

Figure 6.   Total arrival time.

As the figure 6 shows, with the growing number of robots, the total times for multi-robot to perform the same task are reduced no matter what kinds of path planning method they use. And multi-robot system with improved A* algorithm will use less time than the robots using traditional A* algorithm.

Therefore, we can draw the conclusion that, using improved path planning algorithm and three-dimensional reservation table, multi-robot system will use less time to finish the task and move the same distance at the same time, which means improved path planning algorithm can save time for carrying goods without extra cost, comparing with traditional A* algorithm. Finally, it was proved that we can improve the efficiency of intelligent warehouse using the path planning approach we proposed in this paper.

REFERENCES

Alonso-Mora J, Montijano E, Schwager M, et al. Distributed Multi-Robot Formation Control among Obstacles: A Geometric and Optimization Approach with Consensus [J]. 2016.

A R L, Prout K. Automated conformational analysis: Directed conformational search using the A* algorithm [J]. Journal of Computational Chemistry, 1990, 11(10):1193–1205.

Burgard W, Fox D, Hennig D. Fast grid-based position tracking for mobile robots[C]// German Conference on Artificial Intelligence: Advances in Artificial Intelligence. Springer-Verlag, 1997:289–300.

Digani V, Sabattini L, Secchi C. A Probabilistic Eulerian Traffic Model for the Coordination of Multiple AGVs in Automatic Warehouses [J]. Robotics & Automation Letters IEEE, 2016, 1(1):26–32.

Hart P E, Nilsson N J, Raphael B. A formal basis for the heuristic determination of minimum cost paths [J]. Systems Science and Cybernetics, IEEE Transactions on, 1968, 4(2): 100–107.

Hernández-Pérez H, Salazar-González J J. The multicommodity pickup-and-delivery traveling salesman problem [J]. Networks, 2014, 63(1):46–59.

Ma H, Wu X, Gong Y, et al. A Task-Grouped Approach for the Multi-robot Task Allocation of Warehouse System[C]// International Conference on Computer Science and Mechanical Automation. IEEE, 2015: 277–280.

Olmi R, Secchi C, Fantuzzi C. Coordination of industrial AGVs [J]. International Journal of Vehicle Autonomous Systems, 2011, 9(1–2): 5–25.

Rodríguez-Martín I, Salazar-González J J. A hybrid heuristic approach for the multi-commodity one-to-one pickup-and-delivery traveling salesman problem [J]. Journal of Heuristics, 2012, 18(6):849–867.

Williams K. A Fulfilling Position: Villeneuve revolutionizing e-commerce [Amperes: Current Affairs from Around the World] [J]. IEEE Women in Engineering Magazine, 2014, 8(2):12–16.

Zheng K, Tang D, Gu W, et al. Distributed control of multi-AGV system based on regional control model [J]. Production Engineering, 2013, 7(4): 433–441.

Advances in Energy Science and Equipment Engineering II – Zhou, Patty & Chen (Eds)
© 2017 Taylor & Francis Group, London, ISBN 978-1-138-71798-5

Research on non-destructive testing technology for bolt anchoring quality by using the stress wave method

Tongqing Wu

School of Civil Engineering and Architecture, Chongqing University of Science and Technology, Chongqing, China
National Engineering Research Center for Inland Waterway Regulation and Key Laboratory of Hydraulic and Waterway Engineering of the Ministry of Education, Chongqing Jiao Tong University, Chongqing, China

Nianchun Xu & Baoyun Zhao

School of Civil Engineering and Architecture, Chongqing University of Science and Technology, Chongqing, China

Kui Chen & Chengfang Li

Professorate Senior Engineer, Chongqing Construction Science Research Institute, Chongqing, China

ABSTRACT: Based on the one-dimensional wave theory for an anchoring system, according to the energy intensity change and phase shift for stress wave's reflection signals in different wave impedances, research on the non-destructive detection test has been carried out for an anchor system with different locations of defect, defect length, and grouting compactness by using the bolt detector. The results show that due to the influence of the stress wave's reflection signal at the anchor rod's defect section, the deformity is generated on the waveform curve. The characteristic of the tested stress wave curve changes with the defection length and position and the tested defect position is relatively close to the real position. This research work illustrates the practicability of the use of stress wave to perform the anchor quality's non-destructive test.

1 INTRODUCTION

Anchor bolt, a vital component of the supporting system, is widely applied in projects as slopes, underground roadways, and tunnels. As anchoring work is treated as a concealed work, the quality of anchor bolt plays a critical role in meeting the requirements of an engineering design, thereby anchoring the rock mass and stabilizing the anchored rock mass. Consequently, it appears to be particularly significant to test the quality of anchor bolts. At present, there are two test methods which are available—traditional loading method and Non-Destructive Testing (NDT) method. The traditional loading method is listed as a destructive test, which is focused on the pull out test and whose objective is testing anti-pull force of the anchor bolt. It only measures and determines whether the anchoring force is tallying with the requirements of the engineering design, but it fails to give an assessment on the length and quality of the anchor bolt. The procedure of the NDT method is excitation of the stress wave and using a wave detector to gather the signal of the stress wave at the end of the

anchor bolt. It performs time domain or frequency domain analysis, so that the involved parameters will be obtained, including the speed of wave propagation, the length of the anchor rod as well as the position of the defect section of anchoring rock. In this way, a comprehensive assessment will be given for the quality of an anchor bolt.

Currently numerous scholars at home and abroad have studied the quality of anchor bolts. Li Yi (2000), Wang Meng (2013), and Li Zhihui (2009) et al. have validated that the stress-reflected wave method could test the quality of the anchor bolt based on experiments and engineering practice, and have analyzed the feasibility of comprehensive evaluation in this regard. Zhang Shiping et al. (2011) have studied the guided wave propagation law in the anchor bolt and the free anchor through theoretical analysis and experiments. Pei Baolin et al. (2014) have researched the wave propagation law in the anchor bolt through theoretical analysis and experiments. Yin Jian min (2011) and Deng Dongping et al. (2015) have explored the law of the stress wave curve changing along the length of the anchor rod, and the relationship of the consolidation wave and energy attenuation of the reflected wave with the anchor length through experiments and numerical simulation.

In this paper, it is aimed to confirm the validation of NDT of the anchor bolt through the stress wave, and to provide criteria for the application of the anchor bolt NDT in practical engineering applications, based on how the stress wave reflects in different positions and diverse mediums, and also based on the analysis of parameters of the reflection wave like frequency, value of amplitude and consolidation speed, together with analysis of NDT on the anchor bolt quality model under such different conditions when holes are found in grouting the model, no holes exist in the model, and gap stands between the anchoring medium and surrounding rock.

2 THE PRINCIPLE OF STRESS WAVE TESTING THE ANCHORING QUALITY

2.1 One-dimensional longitudinal vibration wave equation of the anchor rod model

Generally, anchor rods simultaneously bear multiple effects of the anchoring medium and the surrounding rock. Upon imposing an impact load or emitting a vibration source at the end of the anchor rod, transmitting and radiating energy all around will occur along the anchor rod, and also reflection and transmission will happen on the interface of the anchor rod, the anchoring medium (mortar), and the surrounding rock. The research of anchor testing intends to make the structure system composed of anchor rods, anchoring medium, and surrounding rocks equivalent to slender rods of uniform section, to presume the anchor rod and its peripheral medium uniform as well as the shearing stress of such medium irrelevant to the depth, which is as shown in Figure 1.

Analysis is performed on the micro-unit of the anchor rod when it suffers an external load. The shearing stress is $dF = kudx + c\partial x/\partial t$, where k and c are the equivalent stiffness coefficient and equivalent damping coefficient, respectively acting on the unit depth medium at the side of the anchor rod. In the light of the equation of motion balances on the micro-unit and excluding the effect of volume

force, the one-dimensional wave equation obtained is as follows:

$$\frac{\partial^2 u}{\partial x^2} = \frac{1}{v_c^2}\frac{\partial^2 u}{\partial t^2} = \frac{ku}{AE} + \frac{c}{AE}\frac{\partial u}{\partial t} \qquad (1)$$

$v_2 = \sqrt{E/\rho}$ will be obtained, where ρ, E, and A are the density, elasticity modules, and acreage of anchoring body, respectively. This equation introduces the stress wave propagating in the anchor rod.

When the stress wave is transmitted from one medium to another medium, then reflection and transmission will happen at the interface of the anchoring section, which is as shown in Figure 2. In this figure, I stands for incident stress wave, R stands for reflected stress wave, T stands for transmitted stress wave, ρ_1, v_1, and A_1 stand for the density, transmitting speed, and cross-sectional area of the anchor rod respectively, and ρ_2, v_2, and A_2 for the density, transmitting speed, and cross-sectional area of the anchored segment.

According to the stress wave's displacement, speed, and continuous conditions at the varying interface in the anchor bolt system together with equation (1), namely the wave equation, the coefficient R of reflected wave and coefficient T of the transmitted wave can be obtained as denoted below:

$$R = \frac{\rho_2 v_2 A_2 - \rho_1 v_1 A_1}{\rho_1 v_1 A_1 + \rho_2 v_2 A_2}, T = \frac{2\rho_2 v_2 A_2}{\rho_1 v_1 A_1 + \rho_2 v_2 A_2} \qquad (2)$$

Hence, the stress and wave velocity acting at the varying section in the anchor bolt system can be expressed as follows:

$$\begin{cases} \sigma_2 = R\sigma_1, v_2 = Rv' \\ \sigma_3 = T\sigma_1, v_3 = \dfrac{\rho_2 v_2 A_2}{\rho_1 v_1 A_1}Tv' \end{cases} \qquad (3)$$

$v' = \dfrac{\sigma_1}{\rho_1 v_1 A_1}$ in equation (3) indicates the velocity increment gained by the mass point when the incident wave reaches the varying section.

Varying Interface MN

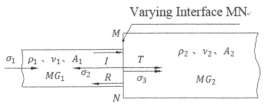

Figure 1. Micro-unit mechanical model of the anchor rod.

Figure 2. Reflection and refraction of the stress wave at the section's varying interface.

When the stress wave is transmitted in the anchoring system, the velocity, stress, and phase value of stress wave at the varying sections can be obtained based on the initial conditions and boundary conditions of the stress wave, the aforementioned equation (1) and equation (3) describing the relations of varying sections in the anchoring system. As a result, the measured wave velocity, stress, phase value, and other indexes can be used to analyze the time domain and frequency domain of the stress wave, and hence a comprehensive assessment for the anchoring length and anchoring quality will be achieved.

3 EXPERIMENTAL INVESTIGATION OF NDT OF ANCHORING QUALITY

3.1 Anchor specimen machining

It is determined to design the anchoring system of the indoor model as different types—holes in grouting (see Table 1), incompact grouting (see Table 2), gap between the anchoring medium and surrounding rock (see Table 3), and completely dense grouting. The technology of "insert anchor rod first, secondary grouting later" is adopted to produce a defective anchoring system. The anchoring body is made of $30 \times 30 \times 50$ cm and $20 \times 20 \times 50$ cm (length $\times$ width $\times$ height) concrete block and the defective position is filled by foam. The anchoring system diagram is shown in Figure 3, and the produced specimen is shown in Figure 4.

3.2 Experimental instruments and equipments

This indoor experiment makes use of the JL-MG(C) anchor quality detector of Wuhan Conourish Engineering Company as illustrated in

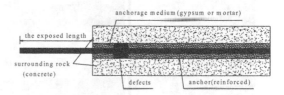

Figure 3. The anchor test schematic diagram.

Table 1. Hole–defect specimen.

Number	MG-2	MG-3	MG-4	MG-6	MG-7	MG-8	MG-10	MG-12
Specification/cm	$30 \times 30 \times 50$ (length × width × height)							
Simulated defect length/mm	20	20	20	50	50	50	80	100
Defect position (distance away from the bottom)/mm	100	150	250	100	150	250	150	150

Table 2. Incompact grouting specimen.

Number	MG-13	MG-14	MG-15	MG-16	MG-17	MG-18	MG-19	MG-20
Specification/cm	$20 \times 20 \times 50$							
Remark	Different values of compactness are listed two groups for each: Compactness of grouting is more than 90%; Compactness of grouting stands between 75% and 90%. Compactness of grouting stands between 50% and 75% With the compactness less than 50%.							

Table 3. Gap between the anchoring medium and surrounding rock.

Number	MG-21	MG-22	MG-23	MG-24	MG-25	MG-26
Specification/ cm	$30 \times 30 \times 50$					
Position of the gap and fracture	The gap and fracture are set at a half, a quarter, and one-eighth of the anchoring medium and surrounding rock.					
Remark	Requirement: the fissure between the anchoring medium and surrounding rock is devoid of the existing law. Wherever the fissure is, the width of the fissure can be varied at diverse heights to produce different regular surfaces. Later, it is necessary to mark the size of fissure in the specimen.					

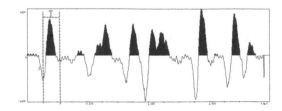

Figure 6. MG-4 time-domain features of the specimen.

(a) 30 × 30 × 50 cm (b) 20 × 20 × 50 cm

Figure 4. Bolt sample preparation.

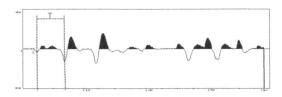

Figure 7. MG-6 time-domain features of the specimen stress waves stress waves.

Figure 5. JL-MG (C) quality bolt detector.

Figure 5. The JL-MG (C) anchor quality detector is composed of a signal acquisition instrument, a stress-wave stimulating seismic source, an exclusive bolting detector, and analytical treatment software. Elastic stress-waves from the seismic source stimulation spread and radiate energy around along the direction of bolting. The exclusive bolting wave detector could detect reflected waves, and its signals could be analyzed and stored by using the signal acquisition instrument.

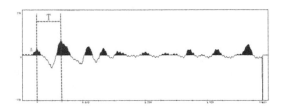

Figure 8. MG-10 time-domain features of the specimen.

4 RESULTS ANALYSIS OF THE INDOOR EXPERIMENT

4.1 Results analysis of different defects on bolting testing

The time-domain analysis on waveform curves of different defective locations in the bolt and anchorage system as shown in Figures 6–8 indicates that Point A is the location where the triggering signal meets variable impedance section. The stress-waves' velocity at the free end of bolting tested by using the bolt integrity testing survey meter is 5120 m/s and the wave velocity of anchorage is 4700 m/s, taking the MG-4 specimen as an example (Figure 6). Intercepting a period in the waveform curve, based on the starting time of stress-waves stimulation and arrival time of reflected waves after spreading

along bolting, the spreading time of stress-waves in bolting will come out: $T = 267$ us. Intercepting (MG-6) is shown in Figure 7, a period of the stress-waves' waveform curve, on the basis of stress-waves' stimulation starting time and arrival time of reflected waves after spreading along bolting, can be calculated as the spreading time of the stress-waves in bolting: $T = 315$ us. Intercepting (MG-10) as shown in Figure 8, a period of the stress-waves' waveform curve, as per the stress waves stimulation starting time and the arrival time of reflected waves after spreading along bolting, can be obtained as the spreading time of stress-waves in the bolting: $T = 310$ us. Obtaining operating parameters of bolting and anchorage' situation on the stress-wave feature curve depends on the time-domain analysis on stress-waves' waveform curves of all bolting models. Consequently, the quality of bolting and anchorage can be evaluated.

While testing the defects of the bolting and anchorage system by using the stress-wave detector, as shown in Figures 6–8, the stress-wave curve distorts at the defecting sites and the response curve displays irregular changes; all these are owing to the reflection and transmission on the defecting fracture surface.

It can be seen from Table 4 that, the defective testing position is close to the actual position, which means adopting the stress wave to non-destructively testify the quality of the rock bolt and anchoring can find the defects and defective position in the anchoring system precisely. Meanwhile, from Table 4, it also can be concluded that all defective positions tested are less than actual positions because these have transformed partially under the self-weight action of their anchoring dielectric materials pouring when using foams to simulate the defect.

4.2 *Result analysis on different grouting density testing*

According to relevant technical regulations, grouting density is the main indicator of the anchor bolt's quality control in the process of construction. Grouting density can be divided into four grades (Grade I: grouting density > 90%, excellent; grade II: 90% > grouting density > 75%, qualified; grade III: 75% > grouting density > 50%, unqualified; grade IV: grouting density < 50%, defective). Table 5 shows the comparison between the real value and testing value of grouting density.

From the results of Table 5, only the estimated value of specimen MG-20 is greater than the real value; because of that, it is hard to control the simulated real compactness of grouting, which

leads to the estimated value that is larger than the real compactness. Test results of the rest specimens are consistent with the real results.

It can be seen from Figure 9 that there are no rules of waveform curve decay; the rate of decay is slow; the stress wave reflects at the defective sites in the anchoring system; waveform curves are distorted; the reflection at the bottom of the anchor bolt almost cannot be identified.

4.3 *Testing analysis of the gap between anchoring medium and surrounding rock*

There are six specimens with different locations existing between anchoring medium and surrounding gap, as shown in Table 6. It can be seen that the detection rate of the gap existing in the void bolts is 100% and the center location of the gap is defined basically accurately.

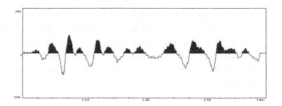

Figure 9. Graph showing the MG-17 stress wave form curve of specimens' stress waves.

Table 4. Defective position vs real value of the anchor bolt.

Number	MG-2	MG-3	MG-4	MG-6	MG-7	MG-8	MG-10	MG-12
defective Position/mm	100	150	250	100	150	250	150	150
Testing value/mm	99.65	148.93	249.86	99.37	148.98	249.76	149.69	149.87
Discrepancy /%	0.35	0.71	0.56	0.63	0.68	0.96	0.21	0.87

Table 5. The estimated value vs experimental testing value of grouting density.

Number	MG-13	MG-14	MG-15	MG-16	MG-17	MG-18	MG-19	MG-20
Real compactness/%	>90	>90	90 ≥ ρ>75	90 ≥ ρ>75	75 ≥ ρ>50	75 ≥ ρ>50	≥50	≥50
Estimated value/%	92.30	94.10	79.50	86.80	72.60	66.50	43.80	52.20

Table 6. Comparison of the gap location and estimated value.

Number	MG-21	MG-22	MG-23	MG-24	MG-25	MG-26
Real defect location/cm	25	25	12.5	12.5	6.25	6.25
Defect estimated value/cm	25.01	25.13	12.52	12.56	6.3	6.25
Discrepancy /%	0.04	0.52	0.16	0.48	0.8	0

5 CONCLUSIONS

By indoor experimental research and analysis on stress wave non-defective testing to different existing defective locations, different grouting compactness values and specimens of anchor bolt and anchoring existing gaps between anchoring medium and surrounding rocks, the following conclusions can be drawn:

By using the time-domain analysis on the stress wave curve of different defective anchor bolts, the defective fracture section displays various features due to different influences of the stress-wave's reflected signals on waveform curves.

The detective testing location is close to the actual location, which means adopting stress-wave to non-destructively testify the quality of the anchor bolt and anchoring can find the defects and defective location in the anchoring system precisely.

Although the stress-wave detection of the quality of anchor bolt and anchoring is a fast, non-destructive, and high-efficient method, the quality of the anchor bolt and anchoring are influenced by several elements such as the status of surrounding rocks, geological conditions, construction craft and so on.

ACKNOWLEDGMENTS

This project is supported by the Fundamental and Advanced Research Projects of Chongqing (cstc2010BB7334, cstc2015jcyjA30003), Research Foundation of Chongqing University of Science & Technology (CK2015Z29), and National Engineering Research Center for Inland Waterway Regulation & Key Laboratory of Hydraulic and Waterway Engineering of the Ministry of Education of Chongqing Jiaotong University (SLK2014B02).

REFERENCES

Deng Dong ping, Li Liang, Zhao Lianheng. Numerical study on dynamic testing of anchorage quality of rock bolt (cable) by stress wave method. Journal of Yangtze River Scientific Research Institute, 32(01) 62–69 (2015).

Li Yi, Wang Cheng. Experimental study on bolt bonding integrity with stress reflected wave method. Journal of China Coal Society, 25(2)160–164 (2000).

Li Zhihui, Li Liang, Li Jiansheng. Research on stress wave method for bolt nondestructive testing technique. Science of Surveying and Mapping, 34(01) 205–207 (2009).

Pei Baolin, Yang Dong. Study on nondestructive testing of bolt anchoring quality. Mining Research and Development, 34(01)81–84 (2014).

Wang Meng, Li Yi, Dong Jia. Stress wave non-destructive testing method of rock bolt and field experimental study. Coal Technology, 32(1)203–204 (2013).

Yin Jian min, Qin Qiang, Xiao Guoqiang, (2011) "Study on sonic propagation laws in anchored body of anchorage bolts with variable lengths," Yangtze River, 42(03)95–98 (2011).

Zhang Shiping, Zhang Changsuo, Bai Yun long, et al. Research on method for detecting integrity of grouted rock bolts," Rock and Soil Mechanics, 32(11) 3368–3372 (2011).

Advances in Energy Science and Equipment Engineering II – Zhou, Patty & Chen (Eds)
© *2017 Taylor & Francis Group, London, ISBN 978-1-138-71798-5*

A study on seat state recognition based on the seating grid

Zaiyu Pang & Mengfei Wu
College of Computer Science and Technology, Jilin University, Changchun, China

Weifeng Yang
College of Software Engineering, Jilin University, Changchun, China

Haotian Guan & Hongwei Zhao
College of Computer Science and Technology, Jilin University, Changchun, China

ABSTRACT: The recognition of seat states in the classroom is beneficial to the teaching supervision work of universities and students' self-study lives. The ordinary recognition method based on the seating grid has many deficiencies, such as being not able to work on the classroom images with short and blurred longitudinal edges. An improved recognition method based on the seating grid overcame the disadvantage and improved the accuracy rate through other methods such as minimum error thresholding and dynamic area thresholding.

1 INTRODUCTION

As the economy develops, the living conditions of people have improved accordingly. An increasing number of densely populated places enter people's lives, such as large offices, university classrooms, theatres, and so on. How to acquire the seat states (whether these are occupied or not) in specific places has become a hot and important spot in the fields of image processing and pattern recognition. The university classroom is a typical place with a substantial number of people; due to the particularity of the educational environment, students' locations and poses are relatively changeless while they are attending classes and studying by themselves. The acquirement of classroom seat states' information becomes easier because there are few students' movements in the classroom. Meanwhile, the recognition results are fairly accurate and have great practical value.

The seat state recognition methods towards the images obtained by vertical shooting can be mainly divided into two categories. One is on the basis of people identification including face detection by using Haar Cascade Classifiers (Savaş, 2016), face detection based on Haar-like features (Zhu, 2015) and face detection with neural networks (Li, 2015). The other one is on the basis of the image subtraction including the method based on the extraction of people's contours and the method based on the seating grid (Zhang, 2011). There will be only introductions of methods in the second category, for the approaches based on people detection

cannot detect other objects such as students' bags whereas those objects are quite important for recognition in a self-study room.

With regard to the approach based on the extraction of people's contours, first, the subtraction between a classroom picture with students and another one without students is performed and then threshold the result image. Through the binary image produced by thresholding, contours can be extracted, which will be used to judge whether there is a person at last. Regarding the method on the basis of the seating grid, first generate the seat grid through the classroom picture without students. And then, acquire the same binary image produced by using thresholding as the above method. Finally, seat states can be obtained by processing the pixels of the binary image in each grid.

The identification of both students and their objects on seats is useful and necessary in a self-study room, because a seat with students' properties usually means it is occupied. Therefore, in this paper, the method based on the seating grid is chosen for its better recognition of both humans and objects. However, the existing method using the seating grid has drawbacks which are as follows:

1. Most classroom pictures are taken by using surveillance cameras, and thus these pictures are mostly of poor quality and have image noises. The existing algorithm does not preprocess the picture, which will have a negative effect on the later recognition.

Figure 1. Vertical edge map obtained by using the sobel operator.

2. In many classroom pictures, the horizontal edges of seats are clear and easy to detect, but the vertical edges are so short and blurred that they are quite difficult to detect (Fig. 1). Even though they are detected, most of them appear in the form of dots that cannot be used later.

3. Because of the variety of students' clothes and objects, classroom images are fairly complex. The complexity implies that thresholding algorithms with general application are required rather than fixed value thresholding. Similarly, it will also lead to inaccurate results to decide whether the seat is occupied with a fixed area threshold value.

To deal with above-mentioned disadvantages, in this paper, an improved recognition method is proposed based on the seating grid.

2 IMPROVED SEAT STATE RECOGNITION METHOD BASED ON THE SEATING GRID

The overall process of the algorithm is shown in Fig. 2.

2.1 Classroom images pre-processing

The process of generating a picture is often interfered by different noises due to external or internal factors, which reduces the quality of a picture (Zhong, 2015). While classroom cameras are shooting at night, pictures are easily polluted by salt-and-pepper noise. Besides, classroom pictures are captured indoors, which means that the light condition is not perfect. Imbalanced light conditions might cause color deviation. To solve these two problems, filtering and light compensation are introduced in the pre-processing stage.

A median filter is a typical non-linear filtering method. It can not only suppress various noises efficiently, but also can preserve the edge details well, and so it is widely spread and utilized

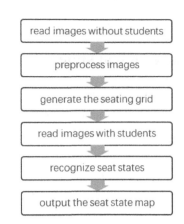

Figure 2. The overall process of the algorithm.

(Ma, 2016). A median filtering algorithm performs very well in eliminating the salt-and-pepper noise in an image, and thus in the recognition method of this paper, we process pictures with a median filter at first. With regard to the light deficiency, we use a positive light compensation algorithm to overcome it in this work. Finally, in order to adjust the images that are overall too bright or too dark owing to the lack of contrast, images are processed with histogram equalization.

The result of pre-processing is shown in Fig. 3 and 4.

Another situation that merits attention is that the seats in some classroom pictures may be not horizontal due to different installation angles of the camera. The situation will bring unnecessary trouble to later recognition. In this case, the seats need to be made horizontal by using affine transformation.

2.2 The seating grid generation

The horizontal grid lines are generated through Canny edge detection and line detection by using Hough transformation. The Canny edge detector is one of the most widely used edge detection algorithms due to its superior performance (Xu, 2014). Hough transformation detects objects (lines in this paper) with specific shapes by using a voting algorithm (Wang, 2016), which has a fairly accurate detection result. Therefore, first, we detect edges in the images after pre-processing by using the Canny operator and then detect lines in the images by using Hough transformation. The horizontal grid lines can be selected from those lines detected by using Hough transformation.

The process of vertical grid lines detection is as follows:

1. Extract the basic seat area by using the color histogram of the image.

Figure 3.   The original image.

Figure 4.   The image after pre-processing.

Figure 5.   Picture of the seat area.

Figure 6.   Picture of the seating grid.

2. Obtain each concrete seat area by using contour detection.
3. Reduce some noise and narrow down the seat area appropriately by using an erosion operation to acquire the final seat area (Fig. 5).
4. Scan horizontal pixels in the center of every row of seats to obtain the outermost vertical grid lines of each row.
5. Acquire all vertical grid lines by calculating with the help of a known column number of seats (Fig. 6).

2.3 Seat state recognition

A fixed threshold value cannot serve the purpose to convert a large number of various classroom pictures into binary images properly. For this reason, other thresholding methods need to be implemented. There are many kinds of automatic threshold methods. According to practice, result binary images of minimum error thresholding (Kittler and Illingworth, 1986) are best suitable for later recognition. The main idea of minimum error thresholding is to optimize the error rate of average pixel classification by using exhaustive searches or iterative algorithms.

To solve the problem of determining whether the seat is occupied through the ratio between the white pixel number and the whole pixel number in a grid, we adopted a dynamic area thresholding approach in this work. In this approach, a mapping relationship between the grid area (the total number of pixels in a grid) and the white point number is established in advance and the ratio threshold is chosen according to the grid area. n_w represents the white point number in a grid. n_G represents the area of the same grid. T_{nG} expresses the dynamic threshold value corresponding to the grid area. Y indicates that the seat is occupied and N indicates that it is not occupied. And then, the method can be expressed as follows:

$$\frac{n_W}{n_G} > T_{nG} ? \, Y : N \tag{1}$$

For example, when the grid area is no more than 300 pixels, we set the threshold value as 0.5; when grid area is more than 300 pixels, we set the value as 0.3. Now assuming that there is a gird whose area is 250 pixels, we should use 0.5 as the threshold value. To some extent, this method has a certain degree of universality owing to getting rid of the specific seating arrangement.

3 EXPERIMENTAL RESULT AND ANALYSIS

3.1 Recognition accuracy test

For the classroom shown in Fig. 7, we take 66 pictures with different distributions of people and

Figure 7. An example of classroom images with students.

361

objects or different poses of people by using an ordinary mobile phone (Lenovo ZUK Z1). All of these photos are scaled to the size of 800*600 pixels. We use these sized images to test our system. In each photo, we only use the second row (including) to the sixth row (including) of the central complete seat area. The sum of seats in an image is 20 (5 rows*4 columns). While testing, all the seats with people as well as objects are regarded as occupied. If a bag is placed in the middle of two seats, we choose one seat randomly as occupied.

The result of the experiment is shown in Table 1.

In Table 1, the test method "Kittler+DAT" means using both minimum error thresholding and dynamic area thresholding; "Kittler" means using only minimum error thresholding; "DAT" means using only dynamic area thresholding. n_R represents the number of the seats recognized correctly. n_W represents the number of the seats recognized wrongly. N expresses the number of seats in all images (1320 in this experiment). n_{WP} expresses the number of the unoccupied seats that are recognized as occupied. n_{WN} expresses the number of the occupied seats that are recognized as unoccupied. And then, the recognition rate, positive error rate, and negative error rate are calculated as follows:

$$recognition\ rate = \frac{n_R}{N} \qquad (2)$$

$$positive\ error\ rate = \frac{n_{WP}}{n_W} \qquad (3)$$

$$negative\ error\ rate = \frac{n_{WN}}{n_W} \qquad (4)$$

As shown in Table 1, the recognition method given in this paper is able to identify the occupancy of the seats in the classroom fairly accurately (91.52%) and both the positive and negative error rates are around 50%, which means that this method does not tend towards certain error recognition. In comparison with the results obtained by not using minimum error thresholding or dynamic area thresholding, the introduction of these two methods significantly increases the recognition rate and retains the positive and negative error rates close to 50%.

Table 1. The test result of 66 images.

Test method	Recognition rate	Positive error rate	Negative error rate
Kittler + DAT	0.9152	0.4821	0.5179
Kittler	0.8447	0.9317	0.0683
DAT	0.8644	0.8827	0.1173

3.2 Algorithm running time

For the pictures (size: 800*600 pixels) in Section 3.1, the time of generating the seating grid is 0.585 seconds. After obtaining the grid, the total time of processing and recognizing 66 images is 0.932 seconds, and thus the average time of processing and recognizing one photo is 0.014 seconds. These data suggest that generating the seating grid requires a long time (this part of the algorithm is the most complicated) but afterwards only recognizing images needs quite a short time. Therefore, the program after debugging and generating the seating grid can recognize classroom pictures swiftly and has great application value.

4 CONCLUSIONS

The improved recognition method based on the seating grid is able to work well on the classroom pictures without clear longitudinal edges, while the ordinary method based on the seating grid cannot do so. And the introduction of minimum error thresholding and dynamic area thresholding augments the recognition accuracy, which further enhances the practical value.

ACKNOWLEDGMENTS

The authors are grateful to the anonymous reviewers for their insightful comments, which have certainly improved this paper. This work is supported by the Undergraduate Training Program for Innovation and Entrepreneurship of Jilin University and Special Funds for Industrial Innovation in Jilin Province (2016C035).

REFERENCES

Li, H., Z. Lin, X. Shen, et al. CVPR, IEEE. 5325–5334 (2015).

Ma, Y.Q., L.S. Wei, P.G. Zhang, T. Ji, J. Anhui Engrg. Univ. 04, 63–67 (2016).

Savaş, B.K., S. İlkin, Y. Becerikli, SIU, IEEE. 2217–2220 (2016).

Wang, Y., J.M. Wu, H.X. Zheng, Comput. Eng. Appl. 17, 203–207+242 (2016).

Xu, Q., S. Varadarajan, C. Chakrabarti, et al. IEEE Trans. Image Process. 23, 2944–2960 (2014).

Zhang, X.Z., N. Liu, G. Wang, Autom. Inf. Eng. 05, 19–21+35 (2011).

Zhong, T., J.G. Zhang, J.Y. Zuo, J. Yunnan Univ. Nat. Sci. 37, 505–510 (2015).

Zhu, G.M., Z.L. Ying, L.W. Huang, AIIE, Atlantis Press (2015).

Advances in Energy Science and Equipment Engineering II – Zhou, Patty & Chen (Eds)
© 2017 Taylor & Francis Group, London, ISBN 978-1-138-71798-5

The influence of suspension parameters on the performance of vehicles

Yanhui Zhao, Dapeng Shi & Mingming Dong
School of Mechanical Engineering, Beijing Institute of Technology, Beijing, China

ABSTRACT: At present, the electric car becomes the trend of new energy vehicles. And the most widely-used driving form is the hub motor. However, when using the hub motor, the unsprung mass increases greatly. The article is based on building a vertical vibration model of single wheel, then analyzing the influence of sprung-unsprung mass ratio, stiffness and damping ratio on riding comfort and handling stability. Then, the article gives some suggestions on how to adjust suspension parameters when unsprung mass is changed. Finally, in the article, it will build a single-wheel platform to test the results of the simulation.

1 INTRODUCTION

At present, the electric car becomes the trend of new energy vehicles. And the most widely-used driving form is the hub motor. But when using the hub motor, the unsprung mass increases greatly, so the ride comfort and handling stability becomes worse. Comparing with internal combustion engine vehicles, vehicles driven by hub motors don't have engine, gearbox, transmission mechanism and fuel tank, but add batteries, vehicle controller and hub motors. So the change of unsprung mass must cause the change of sprung mass. This article analyzes the influence of changes of unsprung mass on performance of vehicle. When the unsprung mass is changed, at the same time, this article analyzes the influence of stiffness and damping ratio, and how to match stiffness and damping.

2 THE INFLUENCE OF MASS RATIO, STIFFNESS AND DAMPING RATIO

2.1 *Building a vertical vibration model of single-wheel*

Assume that the electric car is symmetrical on its longitudinal axis, left and right side have the same rut roughness function, the quality distribution coefficient is equal to 1, and we can ignore the tire damping. So the 2 DOF linear vibration model can be set up, as shown in Figure 1. Parameters of the model are listed in Table 1.

Take Z_b, Z_t as generalized coordinates to describe the system's movement, so the differential equations of motion can be written as:

$$\begin{cases} m_b\ddot{Z}_b + c(\dot{Z}_b - \dot{Z}_t) + k(Z_b - Z_t) = 0 \\ m_t\ddot{Z}_t + c(\dot{Z}_t - \dot{Z}_b) + k(Z_t - Z_b) + k_t(Z_t - q) = 0 \end{cases} \quad (1)$$

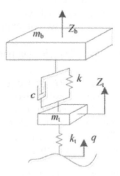

Figure 1. The vertical model of single-wheel.

Where, m_b is sprung mass, m_t is unsprung mass, c is suspension damping, k is suspension stiffness, k_t is tire stiffness, Z_b is the displacement of sprung mass, Z_t is the displacement of unsprung mass, q is road pavement input.

2.2 *Building road model*

According to the request of *GB 7031 Vehicle vibration—describing method for road surface irregularity*, road surface power spectrum density $G_q(n)$ can be expressed by the flowing equation:

$$G_q(n) = G_q(n_0)\left(\frac{n}{n_0}\right)^{-W} \quad (2)$$

where, n is spatial frequency (m^{-1}), n_0 is reference space frequency, $n_0 = 0.1$m^{-1}, $G_q(n_0)$ is the road power spectral density of the reference space frequency n_0, W is the frequency index.

Choose A-class road, of 8 road roughness classification criteria, as a reference. The simulation speed

Table 1. Main parameters of the model.

Parameter symbol	Meaning of parameters	Value	Unit
m_b	Sprung mass	360	kg
m_t	Unsprung mass	40	kg
k	Suspension stiffness	19600	Nm^{-1}
k_t	Tire stiffness	200000	Nm^{-1}
c	Suspension damping	876.5	Nsm^{-1}
	mass ratio	9:1	

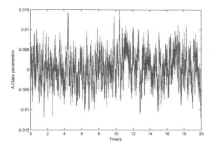

Figure 2. The time-domain model of A-class road.

is 120 Km/h. The spatial frequency power spectral density function and the corresponding time-domain expressions are used to describe the level of random road input of automobile vibration system. According to the random vibration theory, into the road horizontal displacement and the speed of the differential relation, the spatial frequency function for time and frequency conversion function, can get road roughness displacement time domain expression as follows:

$$\dot{q}(t) = -2\pi n_l v q(t) + 2\pi n_0 \sqrt{G_q(n_0)} v w(t) \qquad (3)$$

On A level road, $G_q(n_0)$ equals to 16×10^{-6}, $n_l = 0.01$ (m^{-1}) is the lower cut-off spatial frequency of the road; v is speed, $w(t)$ is white noise series, the random road input model is set up in Matlab/Simulink. The road model that generated in Matlab/Simulink is shown as Figure 2.

2.3 The influence of mass ratio

In this section, the sprung-unsprung mass ratio is changed only, and the stiffness and damping ration stay the same. Analyzing the influence of the changed mass ratio on ride comfort and handling stability. When the mass ratio is 9:1, $\xi_{cl} = 0.165$ $k = 19600$ N/m, and when the mass ratio changes, compare the root mean square value of sprung mass acceleration, dynamic deflection and relative dynamic load.

1. The amplitude-frequency characteristics of sprung mass acceleration $\ddot{Z}_b$ to $\dot{q}$

$$\left| H(j\omega) \right|_{\ddot{Z}_b \sim \dot{q}} = \omega \gamma \left[\frac{1 + 4\zeta^2 \lambda^2}{\Delta} \right]^{\frac{1}{2}} \qquad (4)$$

Where, μ is mass ratio, $\mu = m_b/m_t$, γ is stiffness ratio, $\gamma = k_t/k$.

$$\Delta = \left[\left(1 - \left(\frac{\omega}{\omega_0} \right)^2 \right) \left(1 + \gamma - \frac{1}{\mu} \left(\frac{\omega}{\omega_0} \right)^2 \right) - 1 \right]^2$$
$$+ 4\zeta^2 \left(\frac{\omega}{\omega_0} \right)^2 \left[\gamma - \left(\frac{1}{\mu} + 1 \right) \left(\frac{\omega}{\omega_0} \right)^2 \right]^2$$

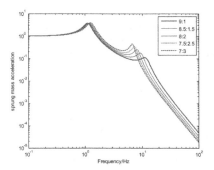

Figure 3. The amplitude-frequency characteristics of sprung mass acceleration.

The root mean square value of sprung mass acceleration is given below:

$$\sigma_{\ddot{Z}_b} = \sqrt{4\pi G_q(n_0) n_0^2 v \int_0^\infty \left| H(j\omega) \right|_{\ddot{Z}_b \sim \dot{q}}^2 df} \qquad (5)$$

2. The amplitude-frequency characteristics of suspension dynamic deflection f_d to $\dot{q}$

$$\left| H(j\omega) \right|_{f_d \sim \dot{q}} = \frac{\gamma}{\omega} \lambda^2 \left[\frac{1}{\Delta} \right]^{\frac{1}{2}} \qquad (6)$$

The root mean square value of suspension dynamic deflection is as below:

$$\sigma_{f_d} = \sqrt{4\pi G_q(n_0) n_0^2 v \int_0^\infty \left| H(j\omega) \right|_{f_d \sim \dot{q}}^2 df} \qquad (7)$$

3. The amplitude-frequency characteristics of relative dynamic load F_d/G to $\dot{q}$

$$\left| H(j\omega) \right|_{F_d/G \sim \dot{q}} = \frac{\gamma}{\omega g} \left[\frac{\left(\frac{\lambda^2}{1+\mu} - 1 \right)^2 + 4\zeta^2 \lambda^2}{\Delta} \right]^{\frac{1}{2}} \qquad (8)$$

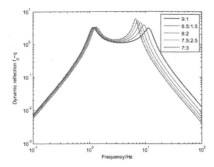

Figure 4. The amplitude-frequency characteristics of suspension dynamic deflection.

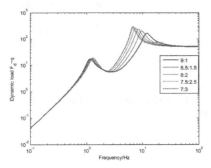

Figure 5. The amplitude-frequency characteristics of relative dynamic load.

The root mean square value of relative dynamic load is as below:

$$\sigma_{f_d/G} = \sqrt{4\pi G_q\left(n_0\right)n_0^2 v\int_0^\infty \left|H\left(j\omega\right)\right|^2_{f_d/G\sim q} df} \qquad (9)$$

For easily analyze the impact of mass ratio on system performance, the RMS of sprung mass acceleration, dynamic deflection and dynamic load will divided by the benchmark time-domain RMS. And we can get the system performance index ratio under different mass ratio, as shown in Figure 6.

Taking sprung-unsprung mass ratio 9:1 as standard, the suspension stiffness and damping ratio stay the same, as the mass ratio decrease, it can be seen form Figure 3~Figure 6: ① the natural frequency of unsprung mass reduces, and peak enhances; the natural frequency of sprung mass and peak changes little; ② the impact of mass ratio on ride comfort and tire dynamic load is more apparent.

2.4 *The influence of damping ratio*

In this section, it will analyze the influence of damping ratio on ride comfort and handling

stability. In different mass ratio, stiffness takes the same value, and change damping ratio. Compare the RMS value of sprung mass acceleration, dynamic deflection and relative dynamic load.

From figure 7 ~ figure 10 shows: (1) in any mass ratio, body acceleration root-mean-square values are in the damping ratio of 0.19 minimum value; (2) the relative damping ratio on vehicle dynamic load and dynamic deflection of the influence of

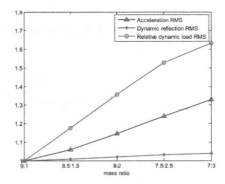

Figure 6. The trend of performance parameter caused by different mass ratio.

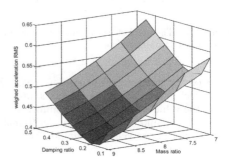

Figure 7. Sprung mass acceleration RMS of different damping ratio.

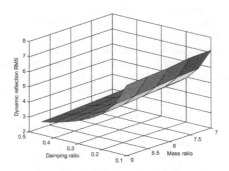

Figure 8. Dynamic deflection RMS of different damping ratio.

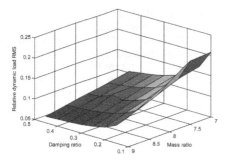

Figure 9. Dyanmic load RMS of different damping ratio.

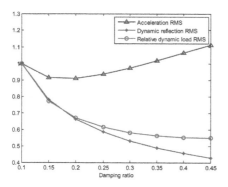

Figure 10. The trend of performance parameter caused by different damping ratio.

ratio on the acceleration of sprung mass; (3) for the wheel hub motor driven vehicles, in order to make the vehicle has good ride comfort, damping ratio 0.19, should be selected in order to make the vehicle has better stability, should choose larger damping ratio.

2.5 The influence of suspension stiffness

In the analysis of section 2.3 and 2.4, the suspension stiffness is kept constant. But the natural frequency of sprung mass is bound to change when the mass ratio has been changed. In order to eliminate the influence of the change in sprung mass' natural frequency, the suspension stiffness is renewed to keep the natural frequency unchanged in this section. The suspension stiffness value is listed in Table 2.

The damping ratio $\xi_c = 0.19$.

In order to facilitate the analysis of the effect of stiffness change on the system performance, the RMS of sprung mass acceleration, dynamic deflection and dynamic load are divided by the reference RMS. The radio of system performance indexes under different stiffness are shown in Figure 14.

Table 2. The stiffness value at different mass ratio.

Mass ratio	9:1	8.5:1.5	8:2	7.5:2.5	7:3
Suspension stiffness k(N/m)	19600	18450	17300	16150	15000

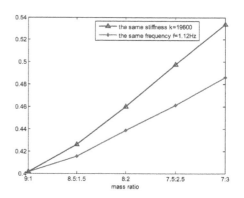

Figure 11. The comparison of acceleration RMS of the sprung mass.

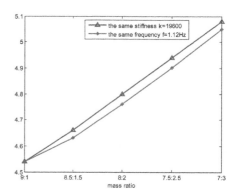

Figure 12. The comparison of dynamic deflection RMS.

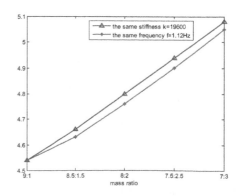

Figure 13. The comparison of dynamic load RMS.

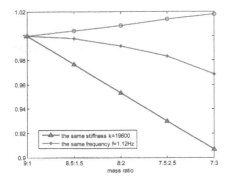

Figure 14. The trends of each performance parameter in different stiffness.

In Fig. 10~Fig. 13, the suspension stiffness is adjusted to keep the natural frequency of sprung mass unchanged when the mass radio has been changed. It can be seen from the pic that:

① The comfort is improved greatly after adjusting. And the deterioration trend is slower than before;

② The change has little influence on the dynamic deflection and dynamic load.

It can be seen from Fig. 14 that with the same mass radio, the RMS of the sprung mass acceleration and dynamic load are reduced and the RMS of the dynamic load is increased when reducing the suspension stiffness.

3 EXPERIMENTAL VERIFICATION

3.1 System composition

In this section, the single-wheel vehicle suspension test platform is designed and built to verify the correctness of the simulation results. The test platform consists of three main parts: input-output device, upper control system and mechanical actuator. The relations between the three parts are shown in Figure 15.

The relationship of the three parts is as follows: The input-output equipment is mainly used to realize the interaction between the operator and the upper control system by which the operator can carry on the artificial intervention to the upper control system in time. The upper control system consists of two parts which are connected by cables: the host computer and the PXI-1336 control box. In order to realize the real-time measurement and control of mechanical actuator, the LabVIEW program in the host computer and the PXI-1336 control box shares the same variable. The PXI-6224 and PXI-6722 board card in the PXI-1336 control box are used for the sensor signal collection and the output control of adjustable damping shock absorber. The

mechanical actuator consists of six sensors which can be divided into three categories. They use the signal and electric current to realize the state information upload and control signal receiving.

3.2 Sensor placement

In this test and control platform, a total of three displacement sensors are used to measure the displacement of the vibration table, the displacement of the sprung mass and the deformation of the tire. The sensors are all Linear Variable Differential Transformer (LVDT). The INV9823 ICP acceleration sensor is used to measure the sprung mass and the vertical acceleration response of the unsprung mass of the vehicle, which is produced by china orient institute of noise & vibration. The sensor placement is shown in Figure 16.

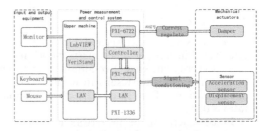

Figure 15. The structure chart of the single-wheel vehicle suspension test platform.

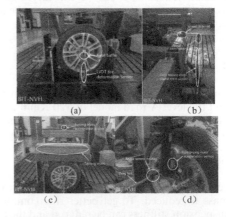

Figure 16. Sensor placement layout of the test platform.

Figure 17. The counterweights layout of the sprung mass and unsprung mass.

367

Table 3. Results of the experiment.

Mass ratio	Sprung mass acceleration RMS m/s^2		Dynamic deflection RMS mm		Dynamic load RMS	
	Simulation	Test	Simulation	Test	Simulation	Test
9:1	0.4025	0.4312	4.41	3.90	0.0812	0.0704
8.5:1.5	0.4268	0.4553	4.53	4.00	0.0966	0.0822
8:2	0.4607	0.4875	4.67	4.13	0.1126	0.0943
7.5:2.5	0.4976	0.5230	4.80	4.25	0.1285	0.1062
7:3	0.5344	0.5641	4.94	4.38	0.1408	0.1178

Some counterweights of different weight are made to change the radio of sprung mass and unsprung mass. The layout of the counter weights are shown in Figure 17.

3.3 Results

Since the single wheel vertical vibration model established in this paper is a linear model, it only needs to verify one kind of condition to determine whether the simulation results are reliable. The data in this section is from section 2.4 whose damping ratio is 0.19.

In consideration of the errors in every step of the experiment, although there are some difference between the experiment results and simulation results, it can be considered that the simulation result is credible according to the results in Table 3.

4 SOLUTION

Firstly, the single wheel vertical vibration model and the level A road model of 120 Km/h are built in this paper to analyze the influence of the radio of sprung mass and unsprung mass, suspension damping ratio, suspension stiffness on the vehicle ride comfort and handling stability.

Since the hub motor has been used in electric vehicles, the ratio of the sprung mass and unsprung mass is reduced. To get better ride comfort, the suspension stiffness can be reduced and the damping radio is taken as 0.19. To get better handling stability, the suspension stiffness can be increased and the damping radio should take a larger value.

In order to verify the correctness of the simulation results, the experiment platform of the single wheel vehicle suspension is set up. The sprung and unsprung mass radio can be changed by adding counterweight. And the road simulation test bench is used to play the road spectrum and collect data. By comparing the experimental data with the simulation data, it can been seen that the simulation data is credible.

REFERENCES

Dong, Y.J. Simulation of Vehicle Ride Comfort and Optimization of suspension Parameters, Chongqing Jiaotong University, 57–59 (2008).

Fang, Y. Influence of In-wheel Motors on the comfort of Electric Vehicles, Jilin University, 90 (2012).

Fu, Y. The Ride Comfort and Handling Stability Simulation if Electric Vehicles. Jilin University, 81 (2014).

Jin, L., Yu Y. & Fu, Y. Study on the ride comfort of vehicles driven by in-wheel motors. Advances in Mechanical Engineering, 8(3) (2016).

Liu, M.C., Zhang, C.N. & Wang, Z.F. A study on the in-wheel motor control strategy for four-wheel independent drive electric vehicles, Journal of Beijing Institute of Technology (English Edition), v23, 177–181 (2014).

Ma, Y. Configuration Analysis and Structural Research of In-wheel motor, Chongqing University, 17–22 (2013).

Tong, W. & Hou, Z.C. Analyses on the Vertical Characteristics and Motor Vibration of an Electric Vehicle with Motor-in-wheel Drive, Automotive Engineering, 398–403 (2014).

Wan, S. & Chen, X.B. Influence of Mass Ratio of In-wheel Motor Electric Vehicle on Vehicle Performance, CCAMMS, 4 (2011).

Wang, Z.Q. Study on the Effect of Hub Motor on Vehicle Handling Stability and Ride Comfort, Jilin University, 53–53 (2014).

Xu, G.H. Analyze and Control of the Ride Comfort and Maneuver Stability for the In-wheel Motor Electric Vehicle, Chongqing University, 43–59 (2014).

Yang, W.H. & Fang, Z.F. Analysis of Dynamic Characteristics of Electric Vehicle with 4 Independently Driven In-wheel Motors, J of China Three Gorges Univ. 96–100 (2015).

Yu, Z.S. Vehicle Dynamic, China Machine Press, 222–231 (2009).

Zhang, C.N., Liu, M.C. & Wang, Z.F. Effects of the unsprung mass on vehicles, Journal of Beijing Institute of Technology (English Edition), 49–53 (2011).

Zhang, Z.H. & Dong, M.M. Analysis on the Optimal Damping Ratio on a 3 DOF Linear Model of Vehicle Suspension, Transaction of Beijing Institute of Technology, 1057–1059 (2008).

Zhang, Z.N. Research on Ride and Safety Performance of In-wheel Motor Drive EV, Nanjing Forestry University, 81 (2014).

Zuo, S.G., Duan, X.L. & Wu, X.D. Vibration test analysis of an electric wheel-suspension system on a test bed, Journal of Vibration and Shock, 165–170 (2014).

Advances in Energy Science and Equipment Engineering II – Zhou, Patty & Chen (Eds)
© *2017 Taylor & Francis Group, London, ISBN 978-1-138-71798-5*

On the graph-spectrum of spatial direction relationships between object groups

Xiaomin Lu, Haowen Yan & Zhonghui Wang
Faculty of Geomatics, Lanzhou Jiaotong University, Lanzhou, China
Gansu Provincial Engineering Laboratory for National Geographic State Monitoring, Lanzhou, China

ABSTRACT: Spatial direction relation between object groups plays an important role in automated map generalization and spatial analysis. In order to express the direction relation between object groups visually and quantitatively, a visible method based on graph-spectrum for determining spatial direction relation between object groups is proposed. Its basic idea is as follows: Firstly, the MBR (Minimum Bounding Rectangle) of the reference object group is computed; Then a morphological transform to the reference object group is made at every specified angle interval starting from and stopping at the due north; After this, the spectrum density is computed by calculating the intersection of the morphological transformed reference object group and the target object group; Last, the corresponding spectrum is drawn based on analyzing the characteristics of the spectrum. The experiments show that the method can describe the spatial direction relations between object groups intuitively and quantitatively and it takes kinds of factors which can influence direction relations between object groups into account.

1 INTRODUCTION

In geographical space, spatial objects always exist in forms of object groups such as elevation points, roads, rivers, habitations, islands and so on (Yan, 2009). The spatial direction relation between object groups plays important roles in process of automatic map generalization, spatial analysis and so on. At present, related researches (Wang, 2013; Zhang, 2009) only realized qualitative and rough description of spatial direction relation between object groups.

Geo-information graph-spectrum, considered as a digital map, a chart, a curve or an image according to certain index periodic rules or classification rules, can reflect spatial information rules of the earth science (Qi, 2004). Graph-spectrum has the capability of visual expression which can express complex phenomenon in brief way, and multi-dimensional spatial information can be shown on two-dimensional map by the graph-spectrum, which can decrease complexity of model simulation (Ji, 2011; HU, 2006; Chen, 2000).

Therefore, graph-spectrum is introduced into the calculation of spatial direction relation between object groups, which aim at describing spatial direction relation between object groups with quantitative way by using angle, density and so forth, as well as expressing the results visually.

2 GRAPH-SPECTRUM DESCRIPTION OF SPATIAL DIRECTION RELATION BETWEEN OBJECT GROUPS

The basic idea of the algorithm is: first, constructing the MBR of the reference object group (habitation group *A* shown in Figure 1), then an unlimited expansion that means boundary translating outward infinitely (Ren, 2004), (shaded part shown in Figure 1) to the MBR is made according to the specified angle α starting from and stopping at the direction of north, and calculating the intersection MBn of the morphological transformed

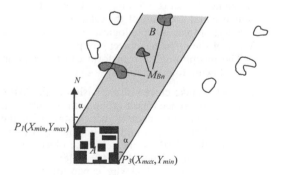

Figure 1. Basic principle of the algorithm.

reference object group and the target object group (lakes group B shown in Figure 1); after this, the spectrum density is computed, the character of spectrum is extracted and the corresponding spectrum is drawn. The detail steps can be described as following.

2.1 Constructing the MBR of the reference object group

Computing the minimum abscissa X_{min}, the maximum abscissa X_{max} and the minimum ordinate Y_{min}, the maximum ordinate Y_{max} of the feature points of the elements constituting the reference object; Then taking the point (X_{min}, Y_{max}) as the upper-left vertex and the point (X_{max}, Y_{min}) as the bottom-right vertex to construct the MBR of the reference object, as habitation group A shown in Figure 1.

2.2 Morphological transforming of MBR of the reference object group

Morphological transform mainly includes expansion, corrosion, opening and closing. The morphological transform used in the paper is expansion which means boundary expanding outward.

According to the basic idea of Gestalt psychology, people tend to connect and organize the adjacent sides of reference object and target object together as an important basis to determine the direction relations between two objects (Wang, 2013). Therefore, with the basis of the above mentioned MBR of the reference object group, the intersection of the specified direction region of the reference object group and the target object group is calculated after transforming the boundaries of the MBR of the reference object group which are adjacent to the target object group. The concrete steps are as following:

Step1: Determine the unit angle of rotation.

Define α as rotation unit angle according to the actual needs; Set $\alpha = 5°$ after considering the convenience and accuracy of the calculation.

Step2: Make morphological transform to the MBR of the reference object groups.

The morphological transform can be made in the following four situations:

1. $\alpha = 0°$, make clockwise expansion to the MBR of the reference object group with the direction of α every five degrees, which start with the upper-left point P_1 and the bottom-right point P_3 of the MBR of the reference object group (Figure 2(a)), denoted as $A \oplus I_\infty^{\alpha_n}$ [8].
2. $\alpha = 90°$, make clockwise expansion with the direction of α every five degrees, which start with the upper-right point P_2 and the bottom-right point P_4 of the MBR of the reference object group (Figure 2(b)).

3. $\alpha = 180°$, make clockwise expansion with the direction of α every five degrees, which start with the upper-left point P_1 and the bottom-right point P_3 of the MBR of the reference object group (Figure 2(c)).
4. $\alpha = 270°$, make clockwise expansion until α reaches 360 degrees with the direction of α every five degrees, which start with the upper-right point P_2 and the bottom-right point P_4 of the MBR of the reference object group (Figure 2(d)).

2.3 Calculating the intersection of morphological transformed reference object group and target object group

Calculate the intersection of morphological transformed reference object group A and target object group B, denoted as M_{Bn} (Wei, 2012), that is:

$$A \oplus I_\infty^{\alpha_n} \cap B = M_{B_n} \left(\alpha = 5°; n = 1,2,\ldots\ldots,[360° / \alpha] \right)_{(1)} \\ -1; \text{set } M_{B_n} \text{ as pixel value of areas}$$

There are 71 transforms as making a circle clockwise morphological transform, that is:

$$A \oplus I_\infty^5 \cap B = M_{B_1}, A \oplus I_\infty^{10} \cap B \\ = M_{B_2}, A \oplus I_\infty^{355} \cap B = M_{B_{71}}$$

2.4 Calculating the characteristics of the spectrum and drawing the corresponding direction relation spectrum

Step1: Calculate the spectrum density distribution of direction relation

The corresponding spectrum density can be gotten by using the following formula based on the above operations.

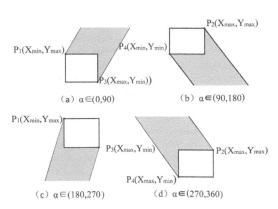

(a) $\alpha \in (0,90)$ (b) $\alpha \in (90,180)$

(c) $\alpha \in (180,270)$ (d) $\alpha \in (270,360)$

Figure 2. Selection of vertices of morphological.

$$\rho = \frac{M_{B_1}}{M_B}, \rho_1 = \frac{M_{B_1}}{M_B}, \rho_2 = \frac{M_{B_2}}{M_B}, \ldots \rho_n = \frac{M_{B_n}}{M_B} (0 < \rho_n \leq 1) \quad (2)$$

In the above formula, $\rho_1, \rho_2, \ldots, \rho_n$ represent the spectrum density of a certain direction in the process of morphological transform (Wei, 2012), that is the ratio of the intersection region in a certain direction of the reference object group and target object group to the whole area of target object group.

According to the above algorithms, the spectrum density of lakes group B relative to habitations group A can be obtained as shown in Table 1.

Step2: Extract the corresponding characteristics of the spectrum

The corresponding characteristics of the spectrum are extracted based on spectrum density analysis and its distribution.

The mean value of spectrum density is calculated as the following formula:

$$\bar{\rho} = \frac{1}{n} \sum_{i=1}^{n} \rho_i \quad (3)$$

The variance value of spectrum density is calculated as the following formula:

$$\sigma = \sqrt{\frac{1}{n} \sum_{i=1}^{n} (\rho_i - \bar{\rho})^2} \quad (4)$$

The spectrum characteristic analysis results are shown in Table 2 by introducing spectrum density into formula 3 and formula 4.

Step3: The spectral vector map of spatial direction relation between two object groups is drawn as Figure 3 shows.

3 EXPERIMENTS AND DISCUSSIONS

In geographical space, the spatial direction relation between object groups will be affected by their spatial form, distribution scope, distribution density and distance relation (GoyalRK, 2000; Peuquet, 1987; Wang, 2013). Therefore, experiments which take the above four factors into account have been done as shown in Figure 4, Figure 6, Figure 8 and Figure 10, the corresponding quantitative calculations of spatial direction relation are shown in Table 3 to Table 6, and the spectrum density distribution maps are shown in

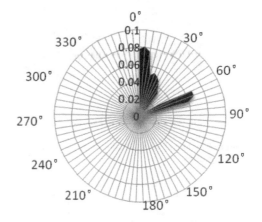

Figure 3. Density distributions of object group A and B.

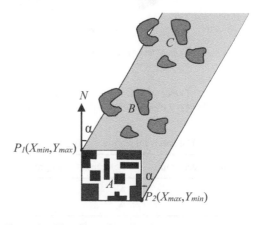

Figure 4. The effect of spatial distance to direction relation between object groups.

Table 1. The spectrum density table.

Serial number	A	ρ_n	Serial number	α	ρ_n
1	0	0.0783	10	45	0.0149
2	5	0.0812	11	50	0
3	10	0.0757	12	55	0.00537
4	15	0.0492	13	60	0.0249
5	20	0.0532	14	65	0.068
6	25	0.0503	15	70	0.068
7	30	0.0462	16	75	0.059
8	35	0.0379	17	80	0.0132
9	40	0.0334	18	85	0
...	...	...			

Table 2. The direction relation spectrum characteristics of object groups A and B.

Target object group	Reference object group	Direction distribution	Mean value	Variance value
B	A	Dir(B, A) = [0°,50°)∪[55°,80°]	0.0446	0.02526

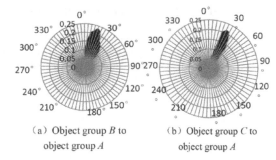

(a) Object group B to   (b) Object group C to
object group A object group A

Figure 5.   The density distributions of spatial direction relation between object groups with different distance.

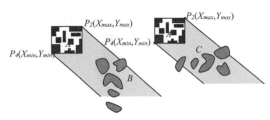

Figure 6.   The effect of spatial form to relation between object groups.

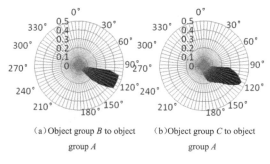

(a) Object group B to object   (b) Object group C to object
group A group A

Figure 7.   The density distributions of spatial direction relation between object groups with different shapes.

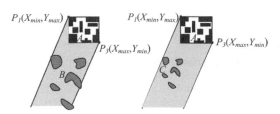

Figure 8.   The effect of shape spatial scope to relationships between object groups.

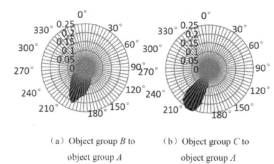

(a) Object group B to   (b) Object group C to
object group A object group A

Figure 9.   The density distributions of spatial direction relation between object groups with different scope.

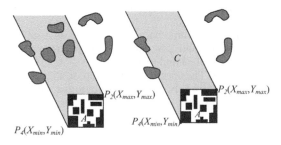

Figure 10.   The impact of density factors to relation between object groups.

Table 3.   The spectrum feature of object groups A and B, C (under the impact of distance).

Target object group	Reference object group	Direction distribution	Mean value	Variance value
B	A	Dir(B, A) = [0,65]	0.1359	0.0717
C	A	Dir(C, A) = (0,50]	0.1064	0.0751

Figure 5, Figure 7, Figure 9 and Figure 11. On this basis, the spatial direction relation between an industrial area and its nearby railway group is calculated, and its quantitative direction relation is shown in Table 7 and Figure 13 shows its spectrum density distribution.

By analyzing the above experiments, the following characteristics of the method proposed in the paper can be concluded:

1. The method can describe the spatial direction relations between object groups intuitively and visually, and the spectrum density distribution

Table 4. The spectrum feature of object groups A and B, C (under the impact of spatial form).

Target object group	Reference object group	Direction distribution	Mean value	Variance value
B	A	Dir(A, B) = [100,155]	0.311	0.1579
C	A	Dir(A, C) = [90,205]	0.276	0.1247

Table 5. The spectrum feature of object groups A and B, C (under the impact of scope).

Target object group	Reference object group	Direction distribution	Mean value	Variance value
B	A	Dir(B, A) = [125,255]	0.0955	0.0611
C	A	Dir(C, A) = [180,240]	0.2069	0.0572

Table 6. The spectrum feature of object groups A and B, C (under the impact of density).

Target object group	Reference object group	Direction distribution	Mean value	Variance value
B	A	Dir(A, B) = [0,35]∪[280,360]	0.0671	0.0371
C	A	Dir(A, C) = [5,35]∪[280,340]	0.0435	0.0249

Table 7. The spectrum feature of industrial district A and railway group B.

Target object group	Reference object group	Direction distribution	Mean value	Variance value
B	A	Dir(A, B) = [0°,40°]∪[275°,360°]	0.3929	0.2027

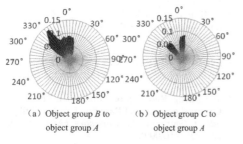

(a) Object group B to object group A (b) Object group C to object group A

Figure 11. The density distributions of spatial direction relation between object groups with different density.

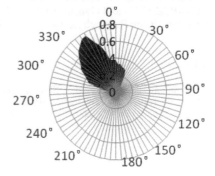

Figure 13. The density distributions of spatial direction relation between industrial district A and railway group B.

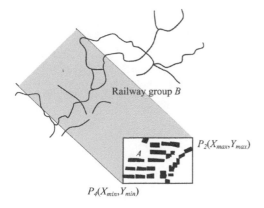

Figure 12. The spatial direction relation between industrial district A and nearby railway group B.

map is consistent with human's spatial cognitive;

2. The method can describe the spatial direction relation between object groups by accurate angles, spectrum density, mean value of spectrum density and variance value of spectrum density;

3. The method can consider the effects of the factors which can affect the spatial direction relations between object groups, and the factors such as spatial distance, distribution scope, distribution shape and distribution density can be expressed in intuitive and accurate way.

4 CONCLUSIONS

The paper uses the morphological transform and spectrum to calculate quantitatively and visualize the spatial direction relation between object groups. Experiments shows that the method takes various influence factors into account well, and it can express the factors intuitively through angles, densities and spectrum.

Compared with existed methods, the method mentioned in the paper calculates and expresses the spatial direction relation quantitatively and visually. Therefore, the method can be used to calculate and express spatial direction relation between object groups intuitively and accurately in the process of automatic generalization to improve the automatic generalization level of object groups, which is also the key point in further research.

ACKNOWLEDGMENTS

The work described in this paper was partially supported by the National Natural Science Foundation of China (Project No.41561090), the National Natural Science Foundation of China (41371435), the National Science Foundation of Gansu Province (148RJZA041) and Gansu Provincial Finance Department Basic Research Service Fee (214146).

REFERENCES

Chen Shupeng, YueTianxiang. Studies on Geo-Informatic Tupu and its application [J]. Geographical Research, **19**(4):337–342(2000).

GoyalRK. Similarity Assessment for Cardinal Directions between Extend Spatial Objeets: Ph.D. Dissertation. Orono, United States: University of Marine:48–55(2000).

HU Shengwu, WANG Hongtao. Research on Geo-Informatics Tupu [J]. Geomatics & spatial information technology, **29**(5):2–4(2006).

Ji Yuhong. The application of Geo-information TuPu in urban ecological information expression [D]. Nanjing University of Information Science & Technology: 6–10(2011).

Peuquet D and Zhan C X. An Algorithm to determine the directional relationship between arbitrary-shape polygons in the plane. Pattern Recognition, **20**(1):65–74(1987).

Qi Qingwen. The latest development on Geo-Info TUPU [J]. Science of surveying and mapping, **29**(6):15–23 (2004).

Ren Huorong. *Mathematical Morphology and its application* [D]. Xi'an: Ph.D. Dissertation of Xian University: 4–6(2004).

Wang Zhonghui, YAN Haowen. Computer of Direction Relations Between Object Groups Based on Direction Voronoi Diagram Model [J]. *Geomatics and Information Science of Wuhan University*, **38**(5): 584–588(2013).

Wang Zhonghui. *A Computational Model for Spatial Direction Relations between Object Groups in Environmentally Geographical Spaces* [D]. Lanzhou: Ph.D. Dissertation of Lanzhou Jiaotong University: 52–62(2013).

Wei Kongpeng, Chen Xiaoyong. Graph-Spectrum analysis on directional relations of spatial targets [J]. Geography and Geo-information science **28**(004): 29–32(2012).

Yan Haowen, WANG Jiayao. *Description and Automatic Synthesis of Object Groups* [M]. Beijing: Science Press, 5–8(2009).

Zhang Lifeng. Research on Directional Relationships of Grouped point object Based on Convex Hull [J]. *Journal of Gansu Lianhe University (Natural Sciences)*, **23**(5):39–41(2009).

Advances in Energy Science and Equipment Engineering II – Zhou, Patty & Chen (Eds)
© *2017 Taylor & Francis Group, London, ISBN 978-1-138-71798-5*

Numerical studies on air-decking charge in the fragmentation of concrete

Xin-jian Li
State Key Laboratory of Explosion Science and Technology, Beijing Institute of Technology, Beijing, China
School of Mechanics and Engineering Science, Zhengzhou University, Zhengzhou, China

Jun Yang & Bing-qiang Yan
State Key Laboratory of Explosion Science and Technology, Beijing Institute of Technology, Beijing, China

ABSTRACT: A numerical method is proposed to study the fragmentation of concrete subjected to detonation load by explosives in borehole under air-decking charge. Four comparisons have been performed with AUTODYN-2D to investigate the effects of several conditions on the fragmentation of concrete, including Air-decking locations, air-decking ratios, initiation locations, and explosives types. The numerical results show that compared to the fully coupling borehole, air-decking charge can reduce the initial borehole pressure and redistribute the explosives energy. More detonation energy can accumulate in air-decking rather than crush rock and soil medium close to the explosive in the blast hole. Therefore, more energy may contribute to the consequent crack propagation.

1 INTRODUCTION

When explosives are initiated in a bore hole more than 20 parameters are involved simultaneously, i.e. charge type, rock characteristics, initiation locations etc. Crush and fragmentation are caused by the impulsive loading generated by the sudden release of the explosive's potential energy, which tremendously exceeds the strength of the rock to be blasted. During this process, a large portion of the explosives detonation energy is devoted to pulverized region immediately around the bore hole, which should be avoided in blasting. This problem can be solved by introducing the air-decking column in explosives, so that initial bore hole pressure will be reduced. The idea of air-decking charge originated in 80 s by Melnikov et al. (1971). Fourney et al. (1981) verified this theory through experiment with thick Plexiglas models. The advantages over the traditional fully coupled charged have been verified by engineering application.

The main purpose in present paper is dealing with the air-decking charge on the fragmentation of concrete by AUTODYN-2D. At last the corresponding results are discussed.

2 GEOMETRY DESCRIPTION AND BOUNDARY CONDITION

The model uses 2000 × 2000 mm block with a pre-drilled hole of 100 mm diameter and 1000 mm

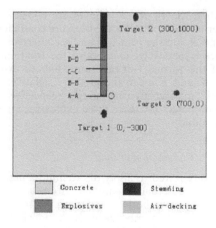

Figure 1. Geometry of the Simulation Model.

depth. The transmitting boundary is applied to the left, right and bottom sides to simulate the infinite case. The transmitting boundary may reduce reflection of shock wave from the boundaries. Only outward traveling waves are allowed without reflected energy back into the computational grid. The number of total grids is 40,000. The length of explosives, air column and stemming material are 400 mm, 200 mm and 400 mm respectively. The initiation location is at the top of explosives column. This geometry is applied in this paper unless otherwise stated. The geometry is illustrated in Figure 1.

3 MATERIALS

3.1 Model of concrete

Rock is heterogeneous due to the presence of pores, micro cracks, grains, minerals, discontinuity as well as joints. Therefore it is difficult to study the failure mechanism and the fragmentation process of rock even though simplified model is applied. As a homogenous material, concrete is ideal for numerical simulation. In this paper, RHT constitutive damage model is used to describe the performances of concrete.

3.2 JWL of explosives

The Jones-Wilkens-Lee (JWL) equation of state is used for explosives, which models the pressure generated by chemical energy of explosives. It describes the relationship between the pressure p and the relative specific volume V. It can be written as follow:

$$p = A\left(1 - \frac{\omega}{R_1 V}\right)e^{-R_1 V} + B\left(1 - \frac{\omega}{R_2 V}\right)e^{-R_2 V} + \frac{\omega E}{V} \quad (1)$$

where A, B, R_1, R_2 and ω are constants. The terms A and B relate the pressure coefficients; R_1 and R_2 are the principal and secondary eigenvalues respectively; ω refers to the fractional part of the normal Tait equation adiabatic exponent; p, V and E mean the pressure, relative volume and specific internal energy, respectively. JWL parameters for these three explosives are shown in Table 1.

3.3 Compaction of sand

It is required to apply the stemming material so that the actual blasting condition can be modeled. Non-cohesive sand with loose compactness is commonly taken as the ideal stemming material (2010). The density of sand used is 2641 kg·m^{-3}.

Different processors are available to describe the performances of the various materials. In present study, the explosives, ideal gas as well as sand are simulated by Euler processor, and Lagrange processor is adopted for concrete.

4 NUMERICAL RESULTS AND DISCUSSION

4.1 Damage evolution of concrete

It is known that the feature of explosives detonation is the super-acoustic reaction rate within an order of 10^{-3} s, resulting in lots of gas products with high temperature and high pressure and irreversible damage to the medium. It is widely accepted that fragmentation relates the stress wave load followed by the immediate detonation gas products load with long duration. Both play a key role in the fragmentation of the medium. Damage evolution of concrete is presented in Figure 2. The resulting concrete fragmentation is apparently characterized by two distinct parts, crushed zone and crazing zone. Crushed zone is created in the immediate vicinity of the bore hole, because stress wave exceeds the dynamic strength of the concrete.

Detonation starts from C-C section, and then propagates towards the bottom and finally arrives at A-A section. Pressure-time curves of the selected section are shown in Figure 3. It can be observed that pressure at C-C section is lower. However, when the stable detonation is reached, the detonation pressure maintains unchanged at a certain level. It can be seen that a smaller oscillation from the pressure-time curve at A-A section. When the detonation propagates downwards, shock wave will also propagate upwards. Pressure at D-D section is apparently lower than that of B-B and C-C sections, only taking up 1/5 of that at bottom. Both pressures at E-E and A-A sections are several times larger than that at the intermediate section D-D. These results are in agreement with the experimental results by W. L. Fourney, et al. (2006). There are two distinct pressure peaks at D-D and E-E section. The second peak of D-D section comes later than that of E-E section which indicates the presence of reflected shock waves or aftershocks from the stemming. The second peak value is a little higher than the first at the E-E section. It means that reflected shock wave exhibits a higher magnitude of pressure.

From the above analysis, it can be inferred that the average pressure in the bore hole with airdecking is lower than that of fully coupling scenario. As it is anticipated, over crushing of the

Table 1. JWL parameters for different explosives used in present paper.

Explosives	A/GPa	B/GPa	R_1/GPa	R_2/GPa	ω	ρ/kg·m^{-3}	D/m·s^{-1}	P_{cJ}/GPa	E/GJ·m^{-3}
EMXs	326.42	5.8089	5.8	1.56	0.57	1145	5165	7.62	2.6738
TNT	373.77	3.7471	4.15	0.90	0.35	1630	6930	21	6
ANFO	49.46	1.891	3.907	1.118	0.333	931	4160	5.15	2.484

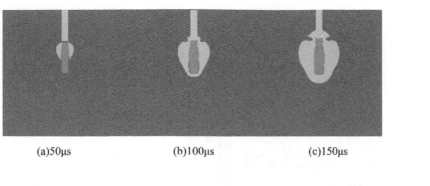
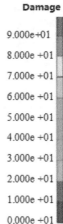

Damage

9.000e +01
8.000e +01
7.000e +01
6.000e +01
5.000e +01
4.000e +01
3.000e +01
2.000e +01
1.000e +01
0.000e +01

(a)50μs   (b)100μs   (c)150μs

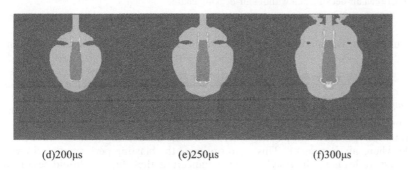

(d)200μs   (e)250μs   (f)300μs

Figure 2.   Damage evolution of concrete.

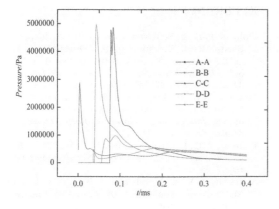

Figure 3.   Pressure-time curve of five sections.

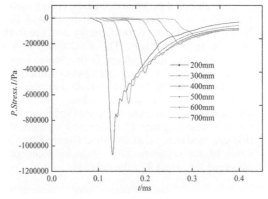

Figure 4.   Stress-time histories.

concrete adjacent to the charge is reduced in case of air-decking because the initial borehole pressure decreases. And the consequent crack is enhanced due to the repeated loading by a series of after-shocks. So use of air-decking technology can pro-vide better control over the resulting excavation than traditional drill and blasting methods.

In order to study the local stress distribution, seven locations are selected at 200 mm, 300 mm, 400 mm, 500 mm, 600 mm and 700 mm from the center of the explosives at the bottom of the hole. As plotted in Figure 4, stress is negative for all the selected point and it dramatically drops off with the distance. It seems to exhibit exponent attenuation.

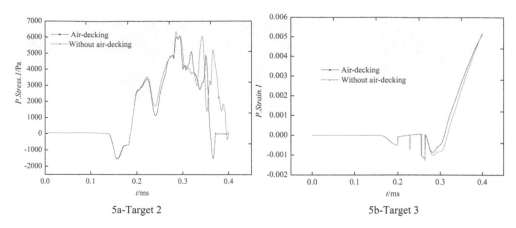

5a-Target 2   5b-Target 3

Figure 5. Comparison between air-decking and without air-decking charge.

However, the action duration is increased with the distance.

4.2 Comparison between air-decking charge and without air-decking charge

The initial borehole pressure with air-decking is lower, since the detonation products propagate through the borehole where air is present. This decreasing borehole pressure leads to the smaller stress load for target 2 in the near field in Figure 5a compared to the fully coupling charge. It has been verified by previous research.

Though initial pressure decreases, energy wasted for undersize medium immediately around bore hole is reduced. This part of energy is stored in the air-decking temporarily. The air-decking now is served as strain accumulator. It first stores energy and then releases it in separate pulses rather than instantly. Therefore, there still stores a greater fraction of the energy within air-decking. This greater fraction of explosion energy is reserved to contribute to useful fragmentation coupled with detonation gas products. So as shown in Figure 5b, larger stress of the concrete with air-decking charge is observed for target 3 in the far field.

5 CONCLUSIONS

This paper mainly deals with the numerical simulation on fragmentation of concrete with air-decking

charge. Initial borehole pressures decreases and the energy wasted for the undersize medium in adjunct of the bore hole is reduced when the air-decking is used. More energy is saved in the air-decking and contributes to the consequent crack propagation. Air-decking locations and initiation locations can result in the specific serious fragmentation of concrete. Suitable air-decking ratio should be selected to optimize the blasting performance. So in blasting engineering, these factors should be considered carefully.

REFERENCES

Dohyun Park & Seokwon Jeon (2010). Reduction of Blast-induced Vibration in the Direction of Tunneling Using an Air-deck at the Bottom of a Blasthole. *International Journal of Rock Mechan-ics & Mining Sciences, 47,* 752–759.

Fourney, W.L., Barker, D.B. & Holloway, D.C. (1981). Model Studies of Explosive Well Stimulation Techniques. *International Journal of Rock Mechanics & Mining Sciences, 18,* 113–116.

Fourney, W.L., Fourney, S. Bihr & U. Leiste (2006). *Borehole Pressures in an Air Decked Situation.* Fragblast, *10,* 47–56.

Melnikov, N.V. & Marchenko, L.N (1971). Effective Methods of Application of Explosive Energy in Mining and Construction. *12th Symp. Dynamic Rock Mechanics.* AIME, New York.

Environmental and architectural engineering

Advances in Energy Science and Equipment Engineering II – Zhou, Patty & Chen (Eds)
© 2017 Taylor & Francis Group, London, ISBN 978-1-138-71798-5

Multi-objective vs. single objective automatic calibration of a conceptual hydrological model

Na Sun, Jianzhong Zhou, Hairong Zhang & Jiang Wu
School of Hydropower and Information Engineering, Huazhong University of Science and Technology, Wuhan, Hubei, P.R. China

ABSTRACT: Hydrological models' calibration is developed to assist hydrologists and related researchers to understand the hydrological system thoroughly. They can be helpful and instructive tools for water resource management. In this paper, a comparative study on the effectiveness of Multiple and Single Objectives Automatic Calibration (MOAC and SOAC) of a conceptual watershed model (Xinanjiang) to simulate streamflow of a Chinese humid basin is presented. A multi-objective automatic calibration routine is developed by using the Multi-Objective Shuffled Complex Differential Evolution (MOSCDE) and a classic single objective technique, SCE-UA, is employed for comparison. The results of MOAC based on the arbitrary combination of three objectives demonstrate that the MOAC is effective in improving the hydrological model's performance than SOAC, could provide more comprehensive evidence for further decision-making in hydrological forecasting, and has good prospects in practical engineering applications.

1 INTRODUCTION

Hydrological models' calibration is a helpful and instructive tool to assist hydrologists and related researchers to understand the hydrological system in depth. However, the watershed models usually have a great deal of parameters, which cannot be measured directly but can be estimated by model calibration. The fundamental work of the calibration is tuning the parameters of a hydrological model in order to obtain higher accuracy simulations which matched observations well. This task is fulfilled traditionally by using the manual trial-and-error method, which is substituted by the automatic calibration with a specific objective function to measure the goodness-of-fit of the calibrated model.

The single objective calibration methods include simplex method, SCE-UA (Duan et al. 1994), Differential Evolution (DE) algorithm (Xu et al., 2008), Genetic Algorithm (GA) (Reshma et al., 2015), Particle Swarm Optimization (PSO) algorithm (Liu et al. 2010), etc. With the development of computer technology, the automatic calibration in hydrologic models has become more and more popular. And a large number of applications have shown that single-objective calibration methods do not usually provide the parameter set that sufficiently satisfies the requirement of modelers. Due to this, many researchers have attempted to increase the objective functions to improve the performance

of hydrologic models to provide most appropriate parameter estimates for modelers. (Bekele and Nicklow, 2007; Guo et al., 2014; Huang, 2014; Rajib et al., 2016; Reshma et al., 2015).

The objective of this study is to calibrate the Xinanjiang model, which was set up to calibrate the flood events streamflow in the Chinese humid basin, by using a novel multi-objective algorithm (MOSCDE) and a classic single objective algorithm (SCE-UA) as a comparison standard to understand in-depth the effectiveness of multi-objective calibration and the influence of different objective function combination.

The remainder of the paper is organized as follows: brief elaboration of the hydrologic model is provided in Section 2. And then, multi-objective and single objective algorithms for the calibration process are stated in Section 3. Section 4 presents a case study and discusses the multi-objective and single objective simulation with the utility of the designed objective combination scheme and the comparative ones. Eventually, conclusions are summarized in Section 5.

2 THE HYDROLOGICAL MODEL

The hydrological model used in this study is the Xinanjiang conceptual hydrological model, which recently became the most widely used model in humid and semi-humid regions in China, to compare the effectiveness of multi-objective and single objective automatic calibrations. The model consists of two main steps; one is the generation of runoff and the other is runoff routing. The study area is often divided into a set of sub-basins according to the distribution of rain gauges and the other characters of the study basin. And first, the runoff transformed into discharge by using a linear system calculated from the water balance component. And then, the outflow from each sub-basin is routed down the channels to the main basin outlet by using the Muskingum method. Variable ranges of the parameters in the Xinanjiang model are listed in Table 1. Readers may refer to Zhao et al. (1992) for detailed description of the Xinanjiang model and its parameters.

3 ALGORITHM

3.1 *The single objective algorithm*

The SCE-UA algorithm is a classic single objective algorithm and developed for parameter optimization of conceptual rainfall-runoff models (Duan et al., 1994). This method was developed based on the combination of deterministic and probabilistic approaches, complex shuffling, and competitive evolution. The synthesis of these three concepts makes the SCE-UA method effective and robust, and the method avoids falling into the local optimal. A more detailed presentation of SCE-UA has been given in the literature (Duan et al., 1994). The steps of SCE-UA are described in Figure 1 briefly.

Table 1. Variable ranges of the parameters in Xinanjiang model.

Model parameters	Range	Model parameters	Range
WUM	5–30	EX	0.5–2.0
WLM	60–90	KG	0.35–0.45
WDM	15–60	KI	0.25–0.6
B	0.1–0.4	CG	0.99–0.998
IM	0–0.03	CI	0.5–0.9
K	0.5–1.1	CS	0.01–0.5
C	0.08–0.18	KE	0–time step
SM	10–50	XE	0–0.5

3.2 *The multi-objective algorithm*

The MOSCDE algorithm is developed by Guo et al. (2010) for the parameter optimization of hydrological models. It is a multi-objective extension of the SCE-UA algorithm and it thoroughly uses the information of the individuals in the evolutionary population and also improves the searching ability of the algorithm by replacing the simplex search in SCE-UA with the Differential Evolution (DE) algorithm. Furthermore, the premature convergence problem of the differential evolution is solved by introducing Cauchy Mutation (CM) operator. For detailed information about the MOSDCE algorithm, refer Guo et al. (2010). A brief description of MOSCDE's steps is illustrated in Figure 2.

4 CASE STUDIES

4.1 *Study area and data*

To demonstrate the effect of the objective function chosen in this study, these were applied to the parameter calibration problem of the Xinanjiang model in the Zhexi reservoir basin (21833.23 km²) of China, which is located in a typical humid climate. The whole data set consists of 12-year precipitation, evapotranspiration, and flow. The first 6 years (2004~2009) were used for model calibration and the rest (2010~2015) were used for model evaluation, and the number of flood events in the calibration and validation period are 29 and 23, respectively.

4.2 *Parameter setting of the algorithm*

The present study offers the results of single, two-objective, and three-objective simulations, and therefore the results of SCE-UA and two-objective functions calibration using MOSCDE are chosen for comparison. The parameter settings of these two algorithms are constructed as Table 2, where dx is the number of parameters, dy is the dimension of objective functions, q is the number of complexes, s is the population size, ss is the number of evolution steps for each complexes before shuf-

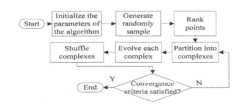

Figure 1. Flowchart of the SCE-UA algorithm.

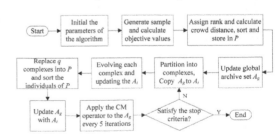

Figure 2. Flowchart of the MOSCDE algorithm.

Table 2. Parameters settings of the two algorithms.

Parameters	dx	dy	q	s	ss	max	CR	F	S_g	S_c	ε	η
SCE-UA	32	1	2	100	11	–	–	–	–	–	–	–
MOSCDE	32	2/3	2	100	5	100000	0.2	0.5	100	100	0.1	0.5

Table 3. The objective functions for the two algorithms.

Schemes	Objectives	Algorithms	RVE	NS	MREP
(1)	RVE	SCE-UA	√	×	×
(2)	NS		×	√	×
(3)	MREP		×	×	√
(4)	RVE vs. NS	MOSCDE	√	√	×
(5)	RVE vs. MREP		√	×	√
(6)	NS vs. MREP		×	√	√
(7)	RVE vs. NS vs. MREP		√	√	√

fling, max is the maximum number of evolutionary generations, S_g is the size of the global archive set and S_c is the size of the complex archive set. Readers may refer to Duan *et al.* (1994) and Guo *et al.* (2010) for detailed settings of the SCE-UA and MOSCDE algorithms.

4.3 Selection of objective functions

Generally, the objective functions should reflect the main aspects of the flood hydrograph, and overall runoff Relative Volume Error (RVE) (Vos *et al.*, 2008; Liu *et al.*, 2010), Nash–Sutcliffe coefficient (NS) (Nash and Sutcliffe, 1970), Root Mean Square Error (RMSE) (Vrugt *et al.*, 2003; Gill *et al.*, 2006; Guo *et al.*, 2010) are adopted frequently. It is generally accepted that the peak flow is vital in a watershed flood event simulation. Therefore, three-objective functions: RVE, NS, and Mean Relative Error of Peak (MREP) (Cheng *et al.*, 2006) are chosen in this study, according to the national criteria for flood forecasting in China, to obtain the Pareto Frontier. The definitions of RVE, NS, and MREP are defined as Equation. (1)–(3):

$$maxF_1 = \left| \sum\nolimits_{i=1}^{n} (Q_{obs,i} - Q_{sim,i}) \right| / \left(\sum\nolimits_{i=1}^{n} Q_{obs,i} \right) \quad (1)$$

$$maxF_2 = 1 - \sum\nolimits_{i=1}^{n} (Q_{obs,i} - Q_{sim,i})^2 / \sum\nolimits_{i=1}^{n} (Q_{obs,i} - \bar{Q}_{obs})^2 \quad (2)$$

$$minF_3 = \left| \sum\nolimits_{j=1}^{N} \left[(P_{obs,j} - P_{sim,j}) / P_{obs,j} \right] \right| / N \quad (3)$$

where n denotes the total number of time series for all flood events, N is the total number of flood events, $Q_{obs,i}$ and $Q_{sim,I}$ represent the observed and simulated runoff at i time, respectively. $\bar{Q}_{obs}$ is the average observed runoff. $P_{obs,j}$ and $P_{sim,j}$ are the observed and calibrated peak flows of the jth flood event.

Table 4. Results of the different objective combinations.

Schemes	Period	RVE	NS	MREP
(1)	Calibration	5.88*	0.85	12.48
(2)		5.96	0.92*	12.15
(3)		6.36	0.88	9*
(4)		5.64	0.93	12.45
(5)		4.92*	0.89	11.65
(6)		6.05	0.92*	9.83*
(7)		6.12	0.93	12.26
(1)	Validation	7.15*	0.84	13.3
(2)		7.45	0.9*	20.46
(3)		7.51	0.85	10.74*
(4)		7.4b	0.89*	16.22
(5)		9.01	0.85	12.58*
(6)		7.68	0.88	12.63
(7)		8.91	0.9	10.59

*Extreme objective function values in different objective combinations.

In this study, the single objective calibration used these three-objective functions separately as a comparison for multi-objective calibration at first. And then, all the combinations of objective functions are employed to calibrate the Xinanjiang model. The objectives for the two algorithms are listed in Table 3, and '√' represents that the objective function is selected and ' × ' shows that this objective is not chosen for calibration.

4.4 Results and discussion

In the study, firstly, the results of three single objective calibrations generated by SCE-UA are provided; later, the results of two-objective and three-objective calibrations found by MOSCDE are provided. These results are presented and located in the corresponding region in Table 4, and the best value of three type calibrations in each column are denoted by using superscript *. The performances were evaluated by using RVE, NS, and MREP.

It can be observed that when one index has the lowest objective function values, the objective function values for other indices are high in the three SOAC schemes, which shows the conflicts of objective functions and implies that single-objective methods could not provide an appropriate solution, which thoroughly reflects the vital aspects of the hydrological system. This conclusion can be seen in Figure 3 in which one flood event hydrograph of the three SOACs are presented. It indicated that the calibrations underestimate the observed streamflow and different schemes have an advantage in the corresponding objective.

There is a distinct character that one scheme possesses a massive feasible solution when its objective functions are more than two. The most appropriate solutions for two-objective calibrations and three-objective calibrations listed in Table 4 show that one-objective function values is improved and it will inevitably lead to deprava-

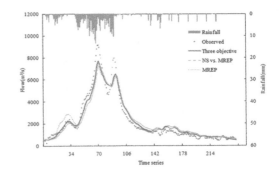

Figure 5. Flood hydrograph for the three-objective scheme and the best scheme of one and two-objectives calibration.

tion of some others. Figure 4 shows a considerable advantage of the two-objective calibration scheme than the single objective scheme. The best scheme of the single objective is the MREP scheme and of two-objective schemes, it is NS vs. MREP scheme. The best two schemes for one and two-objective calibrations and the three-objective calibration scheme, RVE vs. NS vs. MREP are presented in Figure 5. It can be concluded that the RVE vs. NS vs. MREP scheme possesses a distinct superiority which is the same as the NS vs. MREP scheme than the MREP scheme in the distribution of the flood hydrograph and a slight advantage in the flood peak value than the other two schemes.

5 CONCLUSION

In this paper, we compare the effective and efficient model of single and multiple-objectives calibration of the Xinanjiang model in a Chinese humid basin through the design of different number objective functions calibration schemes. The results indicated that the multi-objective automatic routine is capable of incorporating multiple objectives into the calibration procedure to comprehensively reflect characteristics of the hydrology system than single objective schemes. The application of multiple objectives during the calibration process resulted in an improved model performance effectively and efficiently, in particular, than single objective calibration, and the concomitant problem of the multi-objective calibration is to shorten the time for deriving the most appropriate parameter set.

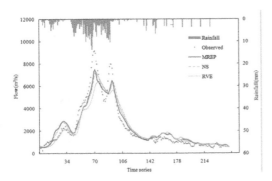

Figure 3. Flood hydrograph for one-objective calibration.

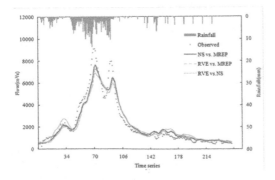

Figure 4. Flood hydrograph for two-objective calibration.

ACKNOWLEDGMENTS

This work was supported by the Key National Natural Science Foundation of China (No. 51239004).

REFERENCES

Bekele EG, Nicklow JW. 2007. Multi-objective automatic calibration of SWAT using NSGA-II. J Hydrol, 341 (3–4), 165–176.

Cheng C T, Zhao M Y, Chau K W, et al. 2006. Using genetic algorithm and TOPSIS for Xinanjiang model calibration with a single procedure [J]. J Hydrol, 316 (1–4), 129–140.

Duan Q, Sorooshian S, and Gupta V. K. 1994. Optimal use of the SCE-UA global optimization method for calibrating watershed models. J Hydrol, 158 (3–4), 265–284.

Gill M K, Kaheil Y H, Khalil A, McKee M, Bastidas L. 2006. Multiobjective particle swarm optimization for parameter estimation in hydrology [J] Water Resour Res, 42: W07417.

Guo J, Zhou J, Lu J, Zou Q, Zhang H, Bi S. 2014. Multi-objective optimization of empirical hydrological model for streamflow prediction. J Hydrol, 511, 242–253.

Guo J, Zhou J, Zou Q, et al. 2013. A novel multi-objective shuffled complex differential evolution algorithm with application to hydrological model parameter optimization [J]. Water Resour Manag, 27(8), 2923–2946.

Huang Y. 2014. Multi-objective calibration of a reservoir water quality model in aggregation and non-dominated sorting approaches. J Hydrol, 510, 280–292.

Liu Suning, Gan Hong, Wei Guoxiao. Application of PSO algorithm to calibrate the Xin'anjiang Hydrological Model [J]. J Hydraulic Engineering, 41 (2010) 5, 537–544.

Rajib M A, Merwade V, Yu Z. 2016. Multi-objective calibration of a hydrologic model using spatially distributed remotely sensed/in-situ soil moisture. J Hydrol, 536, 192–207.

Reshma T, Reddy K V, Pratap D, Ahmedi M, Agilan V. 2015. Optimization of Calibration Parameters for an Event Based Watershed Model Using Genetic Algorithm. Water Resour Manag, 29 (13), 4589–4606.

Vos N J, Rientjes T H M. 2008. Multiobjective training of artificial neural networks for rainfall-runoff modeling [J]. Water Resour Resh, 44, W08434.

Vrugt J A, Gupta H V, Bastidas L A, Bouten W, Sorooshian S. 2003. Effective and efficient algorithm for multiobjective optimization of hydrologic models [J]. Water Resour Res, 39(8), 1214.

Zhao R J. 1992. The Xinanjiang model applied in China [J]. J Hydrol, 135 1–4, 371–381.

Advances in Energy Science and Equipment Engineering II – Zhou, Patty & Chen (Eds)
© 2017 Taylor & Francis Group, London, ISBN 978-1-138-71798-5

A study on measuring methods about pulverized coal concentrations in conveying pulverized coal pipelines

Jialin Wang, Jun Zhao, Ruiping Zhang & Shougen Hu
School of Energy and Power Engineering, University of Shanghai for Science and Technology, Shang Hai, China

ABSTRACT: The precise measurement of pulverized coal concentrations in conveying pulverized coal pipelines is very important to steady combustion and safe operation in the power industry. In this paper, a discussion on several common measuring methods about pulverized coal concentrations, and recommends that the pressure drop method along the pipeline is feasible, which is of practical value in engineering. Simultaneously, in the case of three kinds of resistance components, circular orifice, Venturi pipeline, and airfoil measuring device, in this paper, measurement requirements and the relationship between pressure drop and the factors such as pulverized coal concentration, the throttle ratio of the resistance components, pulverized coal granularity, and velocity of flow are studied.

1 INTRODUCTION

The arrangement of conveying pulverized coal pipelines is complicated, and there are various kinds of flowing patterns in the pipeline, which bring great instability to the conveying pulverized coal. For the complexity and instability of the flowing, there are many difficulties in measuring the parameter of the conveying pulverized coal (L. Cong, 2013; L.D. Chen, 1998).

In the power industry, in spite of the corner tangential firing boiler, W-type flame boiler, and front and rear wall-fired boiler, the parameter of the pulverized coal concentration is significant. For the corner tangential firing boiler, when the pulverized coal concentration, entering the same floor burner, is of large deviation, the combustion conditions of the boiler will be unstable, which may bring about the problems of the oblique tangential firing, fire pressing close to the wall, uneven heat loads, large deviation in the temperature, increase in the emission of NO_x, and the partial water chilling of wall over temperature and burnt. The pulverized coal concentration, symbolizing the running station of the pipeline, is an important parameter. Studying the measurement technology of the pulverized coal concentration and looking for the suitable measurement way of pulverized coal concentration with the feature of detecting on line bring great meaning to the safety and the economy of the running of the large boiler (L.Gao, 2013).

The process of pulverized coal transportation is extremely complex. It is impossible to resolve all the problems of pulverized coal transportation only depending on the theory. Currently, the technology of the concentration of the pulverized coal transportation stays in the stage of half practice and half theory yet. The problems related to the pulverized coal transportation mainly rely on trying various kinds of experiments widely to solve. Among the studied experiments, the techniques of the pulverized coal concentration are extremely important for the pulverized coal transportation.

A coal pulverizing system include direct-fired and storage pulverizing systems. There is only one type, hot air blowing pulverization, for the direct-fired system, while there are two types, hot air blowing pulverization and exhaust air used as primary air, for the storage pulverizing system. The air entering the furnace with the pulverized coal is called primary air. A system that uses high temperature air as primary air is called the hot air blowing pulverization system. The air and coal are separated from the finely-pulverized coal, and this part of air, with the slightly high temperature and considerable water vapor, is called exhaust air. Exhaust air contains little pulverized coal. A system that uses exhaust air as primary air to convey pulverized coal is called an exhaust air used primary air system. When it is difficult for coal, such as the anthracite, meager coal, and inferior coal, to burn, requiring primary air with high temperature to make the combustion stable, the storage pulverizing coal system should choose hot air to blow pulverized coal at this time (L.X. Ding, 2008). To date, the measurement methods as for the pulverized coal concentration in conveying pulverized coal pipelines online are significant of quantity from domestic and foreign studies, which summed up can be divided into two categories, direct measurement and indirect measurement (W.G.Pan, 1999). The direct measurement method is based on inducing the solid in conveying pulverized coal pipelines straightly to determine the concentration of the solid, such as ultrasonic method, microwave method, light fluctuation method, electrostatic method, capacitance method, etc. Indirect measurements, including the heat balance method, energy method, pressure drawdown method, etc. mainly need to establish the corresponding math model, and predigest the model based on reasonable assumptions, basing on the conservation of mass, conservation of energy, and conservation of momentum, etc. (Q.S.Duan, 2009). Next, I would like to introduce several domestic and foreign common methods about the pulverized coal concentration measurement.

2 DIRECT MEASUREMENT ANALYSIS

2.1 Ultrasonic method

The transmission of the ultrasonic waves relies on the medium for it belonging to the mechanical wave. The ultrasonic method is that the attenuation of the ultrasonic changes according to the concentration of the solid. Install an ultrasonic sensor on the two corresponding surfaces of the conveying pulverized coal pipeline. The first pair is used to measure the transmission time of the ultrasonic pulse along two directions, 45° from the direction of the calculating running speed. The other paperback pair of the sensors is used to measure the attenuation of the ultrasonic waves, which is always perpendicular to the direction of the flow (E.G.Shi, 2010). The attenuation of the ultrasonic waves can be affected by the air turbulence, and so we employed β-ray instruments to regulate the air turbulence in practical applications. The ultrasonic measuring device contains several advantages, such as non-contact detection, online measurement, relative lower price, and no influence from the coal type. However, the attenuation of the ultrasonic waves with low frequency is likely to be affected by the nearby factors and the flow state of the conveying pulverized coal pipeline including the factors of the flow pattern and the distribution uniformity requests much higher installation accuracy, and so the ultrasonic method hasn't been used widely (D.G.Liu, 2003;
Q.Liu, 2011).

2.2 Microwave method

The microwave method demands to install a period of measuring pipeline through the flange among the conveying pulverized coal pipeline, and microwave generator and microwave receiver down the pulverized coal flowing direction on the two corresponding surfaces angularly, greater than 90°. Owing to the existence of the gas–solid flow, microwaves, propagating in the waveguide pipeline, will be attenuated while these collide with the pulverized coal in the measuring pipeline. The relation between the attenuation degree and the pulverized coal concentration is monotone increasing. The pulverized coal concentration can be determined uniquely through the microwave attenuation, and then the surefire pulverized coal concentration appears (X. G. Cui, 2003).

The microwave method is a non-contact detection method with the advantage of non-interfering flow, compact structure, and quick response. The disadvantages are that the existence of the place is not measured, requesting much higher installation accuracy, much higher investment, and the probe being contaminated easily, and so the microwave method hasn't been used widely.

2.3 Light fluctuation method

The pulverized coal in the motion passes through a tiny cross section with measuring beam. With passage of time, the number and size of the pulverized coal in the measuring beam keeps changing constantly, which results in the arbitrary fluctuation of the transmitted light intensity. This is the basic principle of the light fluctuation method. For the arbitrary fluctuation of the transmitted light intensity related to the number and size of the pulverized coal pipeline in the measuring beam instantaneously, measuring the arbitrary fluctuation of the transmitted light intensity will provide the average number and size of the pulverized coal pipelines through light fluctuation method analysis (X.S.Cai, 1999). Generally speaking, the diameter of the measuring beam is far bigger than the size of the pulverized coal pipeline. Although the number of the pulverized coal pipelines passing through the measuring area changes constantly, we can still regard the number of the pulverized coal pipelines as constant. If the transmitted light intensity fluctuates lightly for the little transformation of the number and size of the pulverized coal pipelines in the measuring beam, generally measurement repeatedly to get the average value is likely to reduce the impact (X.S.Cai, 2002). However, the nearby factors may affect the measurement result of the light fluctuation method. Much higher investment, and the probe, being in touch with the pulverized coal, being contaminated and wearing easily lead to narrow application in the industrial field.

2.4 *Electrostatic method (JT. Y.iang, 2005; J. Ma, 2000; Y. Yan, 1994)*

As the pulverized coal flows with the air in the pipeline, the pulverized coal comes into collision with pulverized coal pipeline's surface and air, which brings about charge accumulation on the surface of the pulverized coal. The amount of the charge accumulation is related to the velocity, the concentration, and the size of the pulverized coal closely. Therefore, an electrostatic sensor converts the actuated charge signal measured through the probe into voltage signal, which represents the diversification of the amount of the charge. Processing the voltage signal realizes the measurement of the flow parameters of the pulverized coal. While there are many factors infecting the amount of the charge of the pulverized coal, it is difficult to build a model between the pulverized coal concentration and the voltage signal. And so, the variety of the charge is just regarded as a relative instruction, and the electrostatic method tends to be used in qualitative analysis.

2.5 *Capacitance method (Y.P.Zhang, 2000; X.G.Chen, 2014)*

The capacitance method works on the following basic principle. The output capacitance from the electronic survey sensor varies with the corresponding dielectric constant of the medium in the pulverized coal pipeline, and the dielectric constant is related to the concentration of the medium in the pipeline closely, and so the output capacitance can be regarded as the measurement of the pulverized coal concentration. In theory, the capacitance method is secure and reliable, and possesses inexpensive cost, oversimplified installation, and quick response. However, there are some problems found in practical applications. Firstly, the flow pattern is complicated in the pulverized coal pipeline, which keeps changing constantly, and there is a non-linear correlation between the output capacitance and the pulverized coal concentration. Secondly, when the units are operating normally, the output capacitance varies with the fluctuation of the pulverized coal concentration marginally, which results in a relatively higher requirement for the sensitivity of the capacitance sensor. But the higher sensitivity the capacitance sensor possesses, the easier it is being interfered from the nearby factors, which leads to quite a low accuracy rate. Therefore, the capacitance method is still in the stage of the experiment, and there is no mature product widely used in the industrial field.

3 INDIRECT MEASUREMENT ANALYSIS

The heat balance method is based on the energy conservation law. The pulverized coal (T_{coal}) is heated by the hot air (T_{air}). During this process, air is cooled down constantly and the pulverized coal is heated continuously. After a period of time, the air/pulverized coal mixed temperature reaches a balanced temperature (T_{mix}). The sum of the hot air heat and the pulverized coal heat is equal to the total air/pulverized coal mixed heat. The energy conservation law can be written as follows:

$$\mu = \frac{C_{air}T_{air} - C'_{air}T_{mix}}{C'_{coal}T_{mix} - C_{coal}T_{coal}} \tag{1}$$

3.1 *Heat balance method*

Actually, it takes a period of time to make the temperature of the pulverized coal reach up to a certain temperature and the hot air reaching a consensus according to heat transfer, for convective heat transfer between the surface of the pulverized coal and the hot air, and heat conduction inside the pulverized coal. The application of the heat balance method relies on the pulverized coal and the hot air striking a balanced, which brings about the hysteresis effect for the air/pulverized coal mixed temperature. In addition, the coal quality is likely to make changes in the power industry production, which will affect the specific heat of the pulverized coal. In practice, using the empirical value of the pulverized coal has a great influence on the measurement accuracy. Additionally, regarding exhaust air being used as the primary air for the storage pulverizing system, exhaust air used as primary air blows the pulverized coal into the furnace to burn. There is no comparatively large temperature difference before and after mixing. However, using the heat balance method based on the heat balance equation requires a comparatively large temperature difference before and after mixing. Only large temperature difference before and after mixing could ensure the accurate measurement. And so, the heat balance method can't be used in exhaust air used as the primary air system (W.D.Fan, 2002). The storage pulverizing coal system mostly uses the heat balance method at present.

3.2 *Energy method*

In the conveying pulverized coal pipeline, the total energy of the coal air mixture can be equal to the sum of kinetic energy and static pressure energy. Since there are local resistance loss and linear frictional loss in the conveying pulverized coal pipeline, the energy difference is produced before and after mixing. In the case of velocity and temperature determination, the different mixing ratios of the air and pulverized coal lead to the different static pressure energy losses. Therefore, as long as measuring the velocity and static pressure of the air before the mixing and measuring the temperature and static pressure of the mixture after the mixing can be performed, we can obtain the corresponding pulverized coal concentration by using simultaneous thermodynamics partial differential equations.

3.3 Pressure drop method (W.D. Fan, 2003; W.G. Pan, 1999; Z.Q. Yang, 2002)

The pulverized coal falling into the pipeline is accelerated gradually by the air, and the horizontal velocity of the pulverized coal will be the same with the horizontal velocity of the air in theory. That is to say, there is no relative velocity and the heat exchange is completed between pulverized coal and air, and the mixture enters the relatively stable status. In this process, the static pressure and temperature of the system is constantly reducing, and attains the air/pulverized coal mixed temperature eventually. It takes a certain amount of energy during conveying of pulverized coal, which is shown in the pressure drop including speed-up loss, friction loss, suspension loss, local resistance loss, entrance-trailing loss, air updraft loss, etc. The pulverized coal concentration is defined by the ratio of the mass flow of pulverized coal and the mass flow of air. The pulverized coal concentration ranges from 0 to 1. Classically, it is close to 0.5. The pressure drop is measured by calculating local resistance loss $\Delta p\xi$ and linear frictional loss Δp_f and pulverized coal concentration is obtained based on the application of mass conservation, momentum conservation, Bernoulli equation, and ideal gas equation. Although the pressure drop method hasn't been widely used, the application prospect is fine from the writer's analysis.

$$\Delta p_m = (1+k\mu)\lambda\frac{L}{D}\frac{1}{2}\rho_a v_1^2 = (1+k\mu)\Delta p_a$$

The pressure drop of the pipeline transportation is intently related to pipeline features, transportation conditions, and solid particle feature. In the condition of the certain velocity, the pressure drop decreases with an increase in the pipeline diameter, and increases with an increase in the solid concentration, but it is hardly affected by the particle size. The influence between positive inclination and negative inclination is diverse. In the range of small angles, the pressure drop increases with an increase in the inclination angle as the inclination is positive (up-welling), and which is bigger than the pressure drop of the horizontal pipe all the time, and the pressure drop decreases with an increase in the inclination as the inclination is negative (down-welling), which is smaller than the

pressure drop of the horizontal pipe all the while (S.G.Hu, 2002; Lin Jinxian, 2014). To ensure the relation between the pressure drop and pulverized coal concentration, the commonly used resistance components are orifice, Venturi pipeline, and airfoil measuring device. Measuring the pressure of the air/pulverized coal is under the steady flowing condition of air/pulverized coal to reduce the measuring error for flow volatility. The distance between the former and latter cross sections is 8D at least for circular orifice and Venturi pipelines, 4D for the crescent orifice, and 0.6D for the airfoil measuring device. In the meantime, the application of the grid before the resistance component makes the flow field more homogeneous (W.G.Ai, 2000).

4 CONCLUSIONS

Several measurements have been discussed. The direct measurement is mainly influenced by the anti-jamming ability, equipment investment, and signal stability. But for the practical application, the indirect measurement possesses the higher applicability. The indirect measurement uses the physical parameters and flow parameters related to fluid for the calculation, and the calculation result can be compared with the practical concentration to satisfy the requirement of the guiding operation. The pressure drop method is based on the certain relation between the pressure drop and pulverized coal concentration, which is simple, feasible, and precise. Simultaneously, the pressure drop measurement can be successive and real-time without the applicability problem when compared with heat balance method. Therefore, it is feasible to measure the pulverized coal concentration through the pressure drop method. The problem of pulverized coal concentration measurement in the conveying pulverized coal pipeline is in relation to the stability, security, and economy of the boiler operation, which is the complicated problem of the gas–solid two-phase flow originally. Further blameless theory and practice is required to provide reliable dependence to application and improvement of the pulverized coal combustion technology, for the measurement and control level of the boiler parameters to be improved.

REFERENCES

Ai, W.G., Z.Y. Yu, T.M. Xu, et al. Application of fuzzy petri net knowledge representation in fault diagnosis erpert system of gas turbine [J]. *Power Engineering, 2000, 20(5)*:892–895.

Cai, X.S., J. Yu, et al. A study of on-line measurement technology for size, concentration and velocity of pulverized coal [J]. *Power Engineering, 1999, 19(6)*: 466–470.

Cai, X.S., K.Q. Wang, et al. Online measurement of pulverized coal at power station [J]. *Journal of Engineering Thermophysics, 2002, 23(6)*:753–756.

Chen, L.D., Y.S. Shen, D.Q. Chang. Distinguishing flow pattern and monitoring conveying stability of high density pneumatic conveying [J]. *Engineering Chemistry & Metallurgy. 1998, 19(1)*:44–49.

Chen, X.G. Study on online pulverized-coal concentration for power plant by soft-sensing theory [D]. *North China Electric Power University, 2014.*

Cong, X.L. Study on relationship between flow patterns and pipeline pressure signals in dense-phase pneumatic conveying of pulverized coal [D]. *East China University of Science and Technology, 2013.*

Cui, X.G., H.W. Chen, Y.H.Li. Study on the method of coal particle concentration measuring [J]. *Chinese Journal of Scientific Instrument, 2003, 24(4)*:525–529.

Ding, L.X. The principle of power plant boiler [M]. China Electric Power Press, 2008.

Duan, Q.S. Study on pulverized coal volume concentration and distribution measurement in the pneumatic conveyor [D]. *North China Electric Power University (Bei Jing), 2009.*

Fan, W.D., M.C. Zhang, Y.G. Zhou. An investigation of the measurement of solid concentration and air velocity in pneumatic pipes using the pressure drop method [J]. *Fluid Machinery, 2002, 30(6)*:20–25.

Fan, W.D., M.C. Zhang. Experimental study on measurement method of solid concentration and air velocity in pneumatic conveying pipes [J]. *Chinese Journal of Scientific Instrument, 2003, 24(1)*:13–18.

Gao, L. Real-time measurement device of pulverized coal deposition condition in pneumatic pipeline [D]. *North China Electric Power University, 2013.*

Hu, S.G., H.B. Qin, X.N. Bai, et al. Resistance characteristics of particulate materials in pipeline hydro-transport [J]. *Chinese Journal of Mechanical Engineering, 2002, 38(10)*:12–16.

Lin Jinxian, Lin Qi, Lou Chen, et al. Numerical simulation of flow drag characteristics of gas-solid two-phase flow in pipeline [J]. *Oil & Gas Storage and Transportation, 2014(1)*:32–41.

Liu, D.G. Measuring techniques of primary air pulverized coal concentration in thermal power plants [J]. *Guang Dong ELECTRIC Power, 2003, 16(3)*:4–7.

Liu, Q., P. Cui. The application of focused probe in wireline formation test [J]. *Inner Mongolia Petrochemical Industry, 2011 (2).*

Ma, J. and Y. Yan. Design and evaluation of electrostatic sensors for the measurement of velocity of pneumatically conveyed solids [J]. *Special Issue of Flow Measurement and Instrumentation, 2000, 11(3)*:195–204.

Pan, W.G., et al. Study on measurement of the coal powder concentration in pneumatic pipes of boiler with velocity-pressure difference method [J]. *Chinese Journal of Scientific Instrument, 1999, 20(5)*:461–463.

Pan, W.G., et al. Study on measurement of the coal powder concentration in pneumatic pipes of boiler with velocity-pressure difference method [J]. *Chinese Journal of Scientific Instrument, 1999, 20(5)*:461–463.

Shi, E.G. Research on the measurement of volume concentration of pulverized coal in pneumatic pipeline [D]. *North China Electric Power University (Bei Jing), 2010.*

Yan, Y., B. Byrne and J. Coulthard. Radiometric determination of dilute inhomogeneous solids loading in pneumatic conveying systems[J]. *Measurement Science and Technology. 1994, 34(5)*:110–119.

Yang, Z.Q. Experiment research on online detection of solid concentration of two-phase flow by double pressure drop method [D]. *North China Electric Power University (He Bei), 2005.*

Yiang, J.T., Y.H. Xiong. Research in velocity and mass flow rate measurement of gas-solid flow using electrostatic method [J]. *Journal of Huazhong University of Science and Technology (Natural Science Edition), 2005, 33(1)*:93–95.

Zhang, Y.P., F. Jin, Y. Zhang, B.F. Zhang. Phase concentration measurement for two—phase flows [J]. *Journal of Bei jing Institute of Technology, 2002,22(3)*:383–386.

Advances in Energy Science and Equipment Engineering II – Zhou, Patty & Chen (Eds)
© 2017 Taylor & Francis Group, London, ISBN 978-1-138-71798-5

Optimization design of residential buildings in Guangzhou based on energy consumption simulation

Xia Li
Institute of Architecture and Engineering, Guangzhou City Construction College, Guangzhou, China

Xian Guo Wu
School of Civil Engineering and Mechanics, Huazhong University of Science and Technology, Wuhan, China

ABSTRACT: The extensive development has led to excessive growth of energy demand in China. Energy shortages, lack of sustainable supply capacity of conventional energy and other issues have brought tremendous environmental and economic pressure. Energy consumption of buildings is on the rise, and the proportion is 24.5% in 2013 (Building energy research center of Tsinghua University 2016). Building energy consumption has become one of the three main energy consumption sources. High performance buildings are broadly implemented recently to relieve the environmental and economic pressure. This paper presents the work of the thermal performance modeling and analysis for a residential building located in Guangzhou, Guangdong Province, China. The building modeling is developed in DesignBuilder, a software developed for building performance analysis, and conforms to the "Design standard for energy effciency of residential buildings in hot summer and warm winter zone" (JGJ75-2012). The goal of this study is to predict energy consumption of the building for refrigeration, lighting and indoor equipment. Moreover, the study aims to identify sustainable measures to reduce energy consumption and operation cost for the new building. As a result, it is found that in the hot summer and warm winter area as Guangzhou, the design and selection of exterior windows should be paid more attention to than exterior walls and roofs. Better performance windows can significantly reduce the energy consumption of the building. It is not desirable to improve the thermal performance of the exterior wall and roof blindly, or it may increase the building energy consumption, resulting in higher construction costs and waste of building materials.

1 INTRODUCTION

1.1 Background

According to statistics, by the end of 2015, the total volume of existing buildings in Guangzhou has broken 300 million square meters, the city's building energy consumption accounts for nearly 30% of the total energy consumption, and continues to grow. It has seriously affected the sustainable development of the megacity Guangzhou, the energy self-sufficiency rate of which is less than 2%. In order to alleviate environmental and economic pressure, high performance buildings are broadly implemented by the governments.

The purpose of this study is to identify sustainable measures for residential buildings to realize the potential of energy and economic savings and to optimize the energy consumption and cost of residential buildings in hot summer and warm winter areas. Hui Guo et al. (2008) have studied that improvement of thermal performance and the efficiency of air conditioning and heating equipment are the main forms to save energy in "Design standard for energy effciency of resi-

dential buildings in hot summer and warm winter zone" (JGJ75-2012) and related provisions. And the work of Xiaoliang Zhang et al. (2005) have indicated that energy saving optimization design of residential buildings mainly refers to the optimization design of retaining structure. The purpose of this paper is optimization design of energy-saving envelope. The building studied is located in Guangzhou, belonging to the south of the hot summer and warm winter area. In the energy-saving design, the summer air-conditioning should be taken into account, without considering the winter heating.

1.2 Introduction of research object

The object to be studied is a residential building in Guangzhou, North and South orientation. The building has 6 floors, each of which has a height of 3 meters. Each floor has two units. The construction area is 1494 m^2, external wall area 1360.8 m^2, exterior window 408 m^2, and roof area 249 m^2. The relevant parameters are shown in Table 1. Floor plan is shown in Figure 1.

Table 1. Thermal parameters of the basic scheme envelope structure.

Parts to be compared	Scheme*	Performance Criteria
External Window	Monolithic transparent glass, thickness: 6 mm	Heat Transfer Coefficient (W/(m²·°C)):5.70 Shading Coefficient: 0.93
External Wall	Extruded polystyrene foam board, thickness: 35 mm	Conductivity (W/m·°C):0.03 Specific Heat (J / kg·°C):1500
Roof	Extruded polystyrene foam board, thickness: 75 mm	

*Only the materials that need to be improved are listed.

Figure 1. Floor plan.

2 METHOD

Based on the model build by DesignBuilder, the research compares the parameters to analyze energy efficiency and comfort of the building, and to predict building cooling, lighting and indoor equipment energy consumption.

A basic scheme is developed first to compare the energy and economic savings of each sustainable approach. It bases upon the schematic design provided by the design team, plus all the materials used for building components in accordance with "Hot summer and warm winter residential building energy efficiency design standards" (JGJ75-2012) and some engineering assumptions. And then change one or more of the external walls, exterior windows, and roofs to formulate improvement programs while maintaining other measures unchanged. The energy consumption of the modified model is compared with that of the basic scheme, and then the economy of each scheme is analyzed. The best sustainable measures are found by comparatively analyzing energy consumption and economy.

2.1 *Energy simulation software—DesignBuilder*

At present, the softwares used in the simulation of energy consumption are DesignBuilder, DOE-2, EnergyPlus, BLAST and others. EnergyPlus is relatively new. Based on the development of BLAST and DOE-2, It is more reasonable, and more convenient to maintain. DesignBuilder is developed from the user graphical interface based on EnergyPlus, including all the EnergyPlus building construction and lighting system data, and also transplants all the material database. It is more powerful, more convenient to operate (Andy Tindale 2004). Therefore, this paper uses DesignBuilder to build the model for energy consumption simulation.

The model developed is able to provide the details of the energy consumption of this building including heating, cooling, interior lighting, indoor equipment, and hot water supply. It also estimates the heat flux through each surface of the building so that the heat loss through each surface could be studied and the sustainable measures could be made and analyzed (Zhongdi Chen & Ming Qu (2016)).

2.2 *Basic case modeling foundation*

The thermal parameters of the basic scheme envelope structure are in Table 1.

Figure 2 shows the analysis model established by the energy consumption analysis software DesignBuilder.

2.3 *Improved schemes*

Changes are made primarily from the external walls, the exterior windows and the roofing, as compared to the base scheme. Table 2 shows the specific content of the improved schemes.

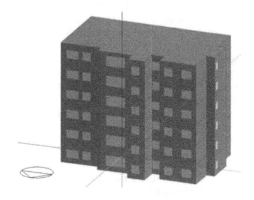

Figure 2. The analysis model established by DesignBuilder.

Table 2. The specific content of the improved schemes.

Number	Schemes	Performance Criteria
1	External Window: Heat-reflective membrane glass, 6 mm	Heat Transfer Coefficient (W/(m²·°C)): 5.40 Shading Coefficient: 0.49
2	External Window: Low-Emissivity insulating glass, 6 mm	Heat Transfer Coefficient (W/(m²·°C)): 1.80 Shading Coefficient: 0.50
3	External Wall: Extruded polystyrene foam board, 15 mm	Conductivity (W/m·°C): 0.03 Specific Heat (J / kg·°C): 1500
4	External Wall: No extruded polystyrene foam board	–
5	Roof: Extruded polystyrene foam board, 35 mm	Conductivity (W/m·°C): 0.03 Specific Heat (J / kg·°C): 1500
6	Roof: No extruded polystyrene foam board	–

3 SIMULATION RESULTS AND ANALYSIS

3.1 Energy consumption analysis

Using DesignBuilder, we can get the annual cooling power consumption, room power consumption, lighting power consumption, domestic hot water and other power consumption of the residential building. The sum of the above four is the annual total power consumption. In the case, without considering winter heating in Guangzhou, annual total power consumption is the total annual energy consumption. Figure 3 shows the process of DesignBuilder energy consumption simulation. Figure 4 shows the total annual energy consumption of the basic scheme.

From Figure 4, the annual power consumption of the basic scheme: cooling 190,445.9 kW·h, Room 68,601.21 kW·h, Lighting 99,143.63 kW·h, Life with hot water and others 6,207.278 kW·h. So the total energy consumption of the basic scheme is 364,398.018 kW·h per year.

The calculation of the six improved schemes of the total annual energy consumption is the same as the basic scheme. The comparison is shown in Figure 5.

As indicated in Figure 5: (1) energy consumption has been reduced by changing the materials of the external wall, window and roof, and the energy-saving effect is most significant by the change of the external window; (2) annual cooling power consumption is decreased by 6.24% in program one, while 2.51% in program two; (3) with the decrease of the insulation layer thickness, the total energy consumption of the exterior wall and roof is decreasing year by year, and without insulation layer it is the lowest.

3.2 Economic analysis

The civilian electricity price is 0.61 yuan / kWh in Guangzhou at present, and the price of the material is based on the investigation of the market price, as is shown in Table 3. Table 4 shows the economic performance of all the schemes.

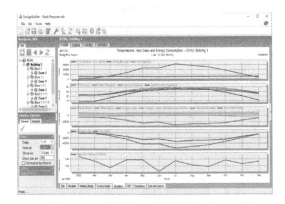

Figure 3. The process of DesignBuilder energy consumption simulation.

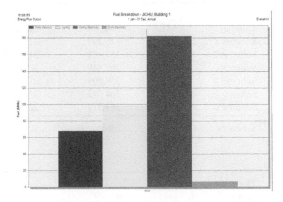

Figure 4. The annual power consumption of the basic scheme.

As Table 4 indicates: (1) compared with the basic scheme: heat—reflective membrane glass and low-emissivity insulating glass can save 4.45% of electricity and 1.90% per year, but due to the higher unit price of these two materials, the cost increases greatly. (2) The energy consumption and

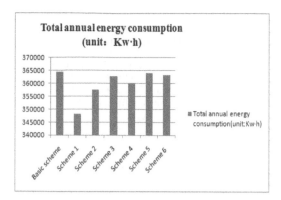

Figure 5. Total annual energy consumption.

Table 3. Price of the materials.

Materials	Unit price $¥/m^2$
Monolithic transparent glass, 6 mm	30
Heat—reflective membrane glass, 6 mm	100
Low-Emissivity insulating glass, 6 mm	210
Extruded polystyrene foam board, 75 mm	85
Extruded polystyrene foam board, 35 mm	45
Extruded polystyrene foam board, 15 mm	35

Table 4. Economic performance of all schemes.

Schemes	Cost ¥	Annual energy consumption kW·h	Electric charge ¥
The basic scheme	94,641	364,398	222,283
Scheme 1	123,201	348,248	212,432
Scheme 2	168,081	357,495	218,073
Scheme 3	81,033	362,850	221,339
Scheme 4	33,405	359,809	219,484
Scheme 5	84,681	364,042	222,066
Scheme 6	73,476	363,379	221,661

Schemes	Cost Saving* %	Electricity Saving %
The basic scheme	0	0
Scheme 1	−30.18	4.45
Scheme 2	−77.60	1.90
Scheme 3	+14.38	0.45
Scheme 4	+64.70	1.26
Scheme 5	+10.52	0.09
Scheme 6	+22.36	0.27

*"+" means the reduction of the cost, "–" means the increase of the cost.

cost of the external wall and roof are decreased with the decrease of the thickness of the insulation layer, and reach the lowest when there is no insulation layer.

4 CONCLUSION

It is found that the building in accordance with "Design standard for energy effciency of residential buildings in hot summer and warm winter zone" (JGJ75-2012) could achieve a great potential for energy and economic savings by using different sustainable measures. For the studied building, the most sustainable measure is to change the thermal performance of the external window. It is not wise to improve the thermal insulation performance of building envelope blindly in the hot summer and warm winter area. It will increase the building energy consumption, and also increase the initial investment, resulting in a waste of building materials.

This study provides a useful reference for the energy-saving design of residential building envelope in hot summer and warm winter area. Although the study is limited by the availability of the detailed design, the method used in this study is sufficient for analyzing any existing or planning building to assist a lot in its design.

REFERENCES

Andy Tindale. 2004. The Building Energy Simulation User News. *DesignBuilder and EnergyPlus*. 25(1).
Building energy research center of Tsinghua University. 2016. *2016 Annual Report on China Building Energy Efficiency*. Beijing: China Architecture & Building Press.
Hui Guo et al. 2008. Energy saving analysis of residential exterior wall in hot summer and warm winter areas of building energy saving under the condition of 65%. *Building Science*. 24(12): 55.
Xiaoliang Zhang et al. 2005. Building environment design simulation analysis software DeST Thirteenth residential simulation optimization examples. *Heating Ventilating & Air Conditioning*. 35(8): 65.
Zhongdi Chen & Ming Qu. 2016. Model-based Building Performance Evaluation and Analysis for a New Athletic Training Facility. *Procedia Engineering*. 145:884–891.

Advances in Energy Science and Equipment Engineering II – Zhou, Patty & Chen (Eds)
© 2017 Taylor & Francis Group, London, ISBN 978-1-138-71798-5

Design and construction of the key technology of the Hengqin subsea tunnel with geomembrane bags

Ming-yu Li, Ping-yuan Jiang & Song Wang
School of Civil Engineering, Zhengzhou University, Zhengzhou Henan, China

ABSTRACT: Taking the Hengqin subsea tunnel project as the research background, the structural design and construction method of the cofferdam with geomembrane bags are studied in saturated deep silt layers. Filling sand is graded and the static-load time is greater than 30 days after multi-stage loading. Hence, the sum of geomembrane bags compression deformation and silt layer consolidation settlement under the cofferdam should be more than 70% of the total settlement. Considering the larger post-construction settlement, more number of geomembrane bags with clay shall be added in time to the sub-cofferdam top, so that the sub-cofferdam elevation is kept above the high-tide level. And geomembrane bags with clay could prevent the tide penetration effectively. Differential settlement of the cofferdam is reduced effectively as geogrids and geotextiles shall be laid on the original silt surface. Anti-seepage geomembranes shall be adopted at the upstream face. In addition, single or multiple rows of plastic anti-seepage curtain walls shall be laid vertically in the middle of the cofferdam.

1 INTRODUCTION

In a large number of ports, wharfs, and tunnels, a geomembrane-bag-cofferdam method is used widely because the geomembrane bag has higher tensile strength and filled sand has higher compressive strength (Tian, 2002; Fu, 2006; Wei, 2013).

As far as the Hengqin undersea open-excavated tunnel engineering of the island-cofferdam method is concerned, greater cofferdam leakage causes water to rise and reduces the soil strength both inside and outside the foundation pit while affecting the construction operation, and then reduces the anti-sliping and heave-resistant stability of tunnel excavation, and even leads to foundation pit collapse and seawater intrusion into the foundation pit. Besides, the saturated low-strength and low-permeability silt makes the cofferdam differential settlement and slip. Therefore, for more similar projects in the future, it will be an important reference value and guiding significance to study the structural design methods and the key construction technologies, such as laying geotextile material, filling sandbags, laying plastic drainage plates, anti-seepage curtain, etc. (Liu, 2012).

2 PROJECT OVERVIEW

Figure 1 shows that the Hengqin undersea tunnel stretches from the planning road around the Hengqin Island in the west, going through the Cross Gate waterway, to the Macau Avenue lotus coastal road in the east, which is constructed by the open-cut method of island cofferdam. It is

Figure 1. Route map of the Hengqin tunnel.

Figure 2. A three-dimensional rendering of the tunnel.

Table 1. Soil parameters in typical strata.

Name of the soil layer	Depth [m]	Natural gravity γ [kNm³]	Cohesion c [kPa]	Internal friction angle φ [°]
Plain soil①₁	2.76	–	5	18
Fine sand①₂	4.23	19	0	20
Silt②	10.63	16	3.7	3.4
Clay③₁	6.54	19.4	20.3	12
Silt soil③₂	5.38	17.8	12.2	9.6
Fine-medium sand③₃	2.35	19.5	0	25
Clay③₄	4.26	19.3	19.8	11.8
Silt soil③₅	6	16.4	12	9.9
Medium-coarse sand③₆	5.16	20.5	0	31
Clay③₇	4.77	19.2	29.1	14.2
Silt soil③₈	8.08	17.7	9.1	7.2
Gravel sand③₉	12.98	26.5	0	36

divided into the Hengqin section, Sea section, and Macau section. The Hengqin semi-section and the Macau section are divided into the open-cut open section and the open-cut buried section (Mileage: K0 + 680 ~ k1 + 210). The whole length of the "Z" shaped tunnel is about 1570 m. The length of the Sea section is about 530 m. The roof elevation is –16.26 m–5.57 m. The floor elevation is –26.36 m–2.01 m.

A three-dimensional rendering of the tunnel is shown in Figure 2. Firstly, two cofferdams are built parallel to the tunnel, which seperated the tunnel construction area and the seawater in the Cross Gate waterway. Secondly, steel trestles are set up for excavation after draining water between two cofferdams. Finally, the cofferdam shall be demolished and the waterway shall be recovered after completing the main structure.

In the tunnel site, the quaternary overburden layer includes plain fill, fine sand, silt of quaternary paralic deposition, quaternary alluvial clay, silt clay, fine-medium sand, coarse sand, gravel sand, quaternary residual sandy clay, late Yanshanian granite in the sub-terrane, fully weathered granite, strongly weathered granite, moderately weathered granite, slightly weathered granite, etc. Soil parameters of typical strata are shown in Table 1.

3 STRUCTURAL DESIGN

3.1 Cofferdam height

The grade of the Cofferdam is 4. Considering tide change and construction settlement, elevation is about 4 m on top of the sub-cofferdam. In addition, in order to meet flood-control requirements, bags filled with clay are stacked in time on top of the sub-cofferdam, thereby maintaining the elevation at 4.33 m.

3.2 Cofferdam section type

The sloping cofferdam structure is divided into five parts, which are toe berm, cofferdam core, casing, anti-seepage curtain, and ground treatment. The slope ratio of the upstream slope and downstream slope are respectively 1:5 and 1:2. In the upstream slope, the geomembrane-bag platform elevation is 0.87 m–2.13 m, the platform width is 7 m–12.65 m. In the downstream slope, the geomembrane-bag platform elevation is 0.87 m–2.13 m, the platform width is 10 m–16 m. Both inside and outside toes are pressed by using an arris body that consists of geomembrane bags. A layer of geotextiles and geogrids is laid at the bottom of the cofferdam. Rocks protecting foundation with the thickness of 1 m are used at the upstream-slope toe. The top width of the bottom protection is 5 m.

3.3 Cofferdam filling material

The cofferdam's filling material shall be filling sand bags. Above the construction water level, from the outside to the inside, the face protection structure in turn is the dry-block-stone revetment (250 mm), gravel cushion (300 mm), a layer of composite anti-seepage geomembrane; below the construction water level, the face protection structure in turn is the filled-stone revetment (800 mm), two rubble (300 mm), gravel cushion (300 mm), a layer of composite anti-seepage geomembrane. A sketch of the typical structural section is shown in Figure 3.

3.4 Cofferdam seepage control

Firstly, a layer of composite anti-seepage geomembrane is laid in the upstream slope. Secondly, a single-row vertical curtain of plastic anti-seepage is

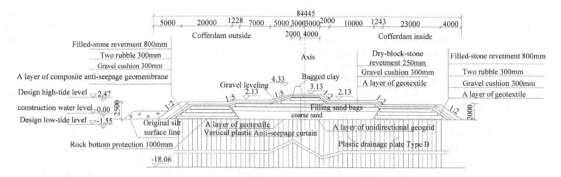

Figure 3. Sketch of the typical structural section.

inserted in the dike body. The design permeability coefficient of the curtain is K = n × 10⁻¹² cm/s (n = 1~9). Finally, a single-row high-pressure jet-grouting pile anti-seepage curtain is embedded in the cofferdam bending position and double-row high-pressure jet-grouting pile anti-seepage curtain is embedded in the junctions of the cofferdam and the dams of Hengqin and Macau. The pile diameter is 0.1 m; the pile spacing is 0.08 m; and the design permeability coefficient of the pile is K = n × 10⁻⁶ cm/s (n = 1~9).

3.5 Cofferdam foundation treatment

First of all, a layer of geogrid and geotextile is laid on the original silt surface under the cofferdam bottom in the consideration of the thick mud layer at the bottom of cofferdam. Secondly, two layers of sand are filled hydraulically, the thickness of which is about 1m. Thirdly, the plastic drainage plate (Type B) with the spacing of 1.0 m is embedded; its plane layout is in the shape of a square. Finally, the dike body is built by using filling sandbags.

3.6 River closure and restoration

The cofferdam is closed by filling sandbags in the low-tide period. The closure point is near the center of the sea. After the completion of the main project of the tunnel, it is required to demolish the cofferdam and restore the navigation.

4 KEY CONSTRUCTION TECHNOLOGIES

The key technologies of the cofferdam construction include laying geotextile material, filling sandbags, inserting plastic drainage plate and vertical-plastic anti-seepage curtain, and embedding high-pressure jet-grouting pile.

4.1 Laying geogrid and geotextile

For reducing the uneven settlement at the bottom of cofferdam, it is required to lay a geotextile and a geogrid at the bottom of cofferdam. The geotextile is laid under the geogrid. The geogrid construction sideline is 2 m wider than the geotextile construction sideline on the inside of the cofferdam; the geogrid construction sideline is 7 m wider than the geotextile construction sideline on the outside of the cofferdam. After filling the dike-body sandbags, a layer of geotextile is laid on the inside of the cofferdam and two layers of geogrids are laid sandwiching a layer of the anti-seepage geomembrane on the outside of the cofferdam. In some shallow water areas, the original silt surface is almost completely bare in low tide areas. For these areas, geogrids and geotextiles are laid artificially. Laying ships are used in the deep water areas.

4.2 Filling sandbags

Throughout the whole process of cofferdam construction, filling sandbags is graded in different sections. Two hundred and ten days after loading, the consolidation degree of the silt layer settlement should be above 70%. It shall be constructed in two

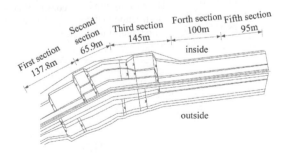

Figure 4. Sketch of the south-cofferdam construction area.

operations in the practical construction, such as laying sandbags and filling sandbags

Layering sandbags in the sea. First of all, processed bags are wound onto the laying-ship roller. Secondly, the sand carrier is docked next to the laying ship positioned by GPS. Thirdly, the sand pump pipe is connected with bags. And then, sand is pumped into the bag. During pumping of sand, sandbags are unfolded slowly and entered the sea.

Artificial paving sandbags on land. First of all, processed bags of upward mouths are wound along the longitudinal direction, and then placed on the ship to the laying site. Secondly, the sandbag is unfolded along the cross-section direction of the cofferdam. It ensures that the unfolding direction is perpendicular to the cofferdam axis by correction at any time. Thirdly, it fixes the steel pipe through the sand ring into the ground, and then the sand pump pipe is connected with bags to fill sand. To ensure that filled sandbags are flat, uniform, and dense, it needs to adjust the sandbag-mouth position at any moment and construction workers moved back and forth on the sandbags to ensure the smoothness, compactness, and evenness of the sandbag. The cofferdam is divided into five construction sections as shown in Figure 4.

Filling-sand presses of the first, fourth, and fifth sections are in shown Figure 5. The first load fills two layers of the sandbag cushion. The second load fills sandbags from the cushion to the first construction platform. The third load fills sandbags from the first construction platform to the cofferdam top.

Filling-sand presses of the second and third sections are shown in Figure 6. Four loadings are used for the cofferdam design height that is higher than 5.83 m; a three-level loading is applied to the cofferdam at a height lower than 5.83 m.

Four-level loading: the first load is laying geogrids and geotextiles, filling sand-bag cushions (including filling inside and outside the cofferdam), hydraulically filling coarse sand, embedding plastic drainage plates, and filling remaining sandbags. The load time is 30 days; the static-load time is 60 days. The second and third load fills sandbags from the cushion to the first construction platform. The load time is 30 days; the static-load time is 60 days. The curtain construction of the vertical plastic antiseepage shall be constructed immediately 10 days after completion of the third filling construction. The fourth load fills sandbags from the first construction platform to the second construction platform. The fifth load fills sandbags from the second construction platform to the cofferdam top. The process and the photograph of filling-sandbag construction are shown respectively in Figure 7 and 8.

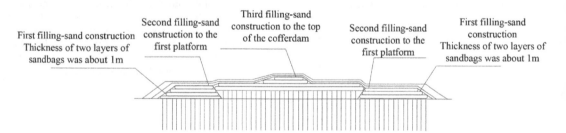

Figure 5.   Construction process of filling sandbags in the first, fourth, and fifth sections.

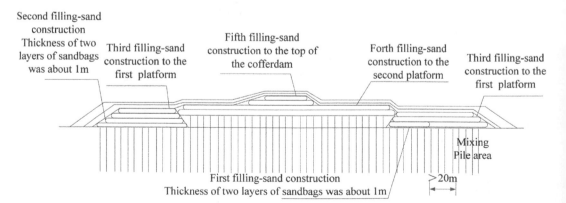

Figure 6.   Construction process of filling sandbags in the second and third sections.

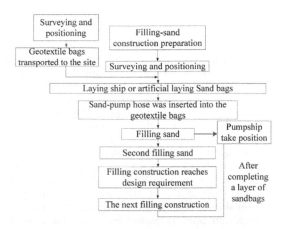

Figure 7. Construction process of filling sandbags.

Figure 8. A construction photograph of filling sand-bag construction.

4.3 *Plastic drainage plate construction*

The plastic—drainage—plate construction shall begin after completion of filling sand cushions and filling sandbags to the dike core. The plastic-drainage-plate construction is divided into land construction and water construction. Its construction process includes machinery in place → plastic drainage plate booting → inserted casing → pulling out casing → cutting off the plastic drainage plate → machinery moved to the next position. Construction photographs are shown in Figure 9.

Land construction: land construction of the hydraulic—crawler -spile machine is applied to some areas of the first platform above the water at low tide areas. All drainage plates need to penetrate through the silt layer into the clay layer. The insertion depth of the clay layer is more than 1.0 m and meets the design requirements.

Water construction: spile machines are installed on both sides of a barge. The modified barge is used in the water region of 1.5m~2m. The pile frame height of the barge is 15 m. GPS is used in water-spile construction. A pre-cut method is used to control the inserting depth of the drainage

(a) Hydraulic crawler spile machine

(b) Water spile barge

Figure 9. Construction photographs of the hydraulic crawler spile machine and water spile boat.

plate, as shown in Figure 9(b). A construction photograph of the water spile barge is shown in Figure 9(b).

4.4 *High-pressure jet-grouting-pile construction*

According to the tide condition and site construction situation, land construction is applied to inserted anti-seepage curtains and drive piles on the junction of the cofferdam and the dams of Hengqin and Macau. Under low tide conditions, an uplifted pile machine is used for the layered sandbag; Pile machines and barges are used for the high-pressure jet-grouting pile in the deep-water area. The high pressure jet-grouting pile is distributed in the separated pile construction method and processed with the jet-grouting construction process of a new two-pipe method. During piling, the cement mortar jet flow with the pressure higher than 36 MPa erupts from the nozzle with a diameter of 2.4 mm. The jet flow is protected by compressed air. With the rotating and lifting of the nozzle, the high pressure cement mortar jet flow cuts off the damaged soil. At the same time, with an increase in the jet flow, the cement mortar and clay are mixed into a pile body; the compressed air plays the role of lifting and displacement, so that the cut soil can be dropped from partial soil and formed into the required loaded body after consolidation of cement mortar and clay, as shown in Figure 10.

Figure 10. A construction photograph of the high-pressure jet-grouting pile on land.

4.5 *Anti-seepage curtain construction*

Multiple anti-seepage measures shall be adopted. Anti-seepage geomembranes shall be adopted at the upstream face. In addition, single or multiple rows of plastic anti-seepage curtain walls are laid vertically in the middle of the cofferdam. The vertical-plastic anti-seepage curtain is laid out along the longitudinal axis of the cofferdam. The anti-seepage curtain of the high-pressure jet-grouting -pile is adopted near the turning points of the cofferdam in the range of 40 m and cofferdam ends in the range of 20 m. The vertical anti-seepage curtain shall be begin 10 days after completion of the third construction static loading.

Vertical-plastic curtain-wall construction is carried out using special machinery with hydraulic-mechanical cutting to dig a continuous groove of 18 mm~30 mm width, in which plastic films are laid to prevent the penetration of sea water. The two key steps in the whole process of construction are laying and overlapping. After the groove depth to meet design requirement, a rolled bundle of plastic films is hung in the groove. And then, it turns and lays the bundle of plastic films at the upstream wall of the groove as it drags the wire rope at both ends of the plastic-film steel pole. In order to prevent the plastic film from sliding into the groove, on top of which is about the 1 m wide plastic film is reserved to be pressed with clay on the ground. It adopts nature lapping between two bundles of plastic films. The lapping width is not less than 2 m.

5 CONCLUSIONS

Cofferdam stability is very important for the foundation-pit safety of the island-cofferdam open-cut tunnel in sea. Taking the Hengqin subsea tunnel project as the research background, the structural design and construction method of cofferdam with geomembrane bags is studied in the saturated deep silt layer. The main conclusions are as follows:

The cofferdam should be closed under low tide conditions. The classification-segmentation construction method should be adopted in the process of hydraulic filling of sand and loading. For reducing post-construction settlement, the static-load time should be greater than 30 days after multistage loading so that the sum of compression deformation of geomembrane bags and consolidation settlement of the silt layer under the cofferdam can be more than 70% of the total settlement. In addition, to ensure that filled sandbags are smooth, compacted and even, the sandbag-mouth's position should be adjusted at any moment and construction workers moved back and forth on the sandbags.

Based on the monitoring data, constantly filling clay sandbags to the sub-cofferdam top ensures that the cofferdam elevation is above the design of the high-tide level in the saturated deep silt layer.

Under the bottom of the cofferdam, layers of geogrids and geotextiles are laid for reducing the uneven settlement of the silt layer caused by classification—segmentation load. In addition, soft foundation can be reinforced by inserting a plastic drainage plate. The drainage plate needs to avoid bending and penetrate through the silt layer into the clay layer. The insertion depth of the clay layer should be more than 1.0 m.

Multiple anti-seepage measures should be adopted. In addition to laying anti-seepage geomembranes at the upstream face, single or multiple rows of plastic anti-seepage curtain walls should be laid vertically in the middle of the cofferdam.

This research work provides a valuable reference and guidance for the future design and construction of the similar projects.

REFERENCES

Fu, H.F. Research on the construction method of the marine cofferdams with geotextile bags filled with solidified soil. (MS., Tianjin University, China 2006).

Liu, T., Y.B. Chen, H. Liu, etc. Case study of ultra-deep foundation pit by island and cofferdam construction in soft soils in coastal areas. *Chinese Journal of Geotechnical Engineering*, **34**, 773–778 (2012).

Tian, Q.L., B.H. Li and Y.H. Zhu. Research on large geotextile tubes containing hardened cement-soil mixture and its use in cofferdams in Nanjiang Area of Tianjin port. *China Harbour Engineering*, **4**, 53–55 (2002).

Wei, X.J., J.Y. Wang, Z. Ding, etc. Displacement analysis of geomembrane bag with sand soil cofferdam in Zhoushan undersea immersed tube tunnel. *Chinese Journal of Rock Mechanics and Engineering*, **32**, 1836–1842 (2013).

Advances in Energy Science and Equipment Engineering II – Zhou, Patty & Chen (Eds)
© *2017 Taylor & Francis Group, London, ISBN 978-1-138-71798-5*

The resolution of lifeline bridges based on the faults position in the perspective of probability: A case study of the Tangshan area

Peiwei Cao & Jiancheng Yu
School of Transport, Southeast University, Nanjing, China

ABSTRACT: Earthquakes are related to faults (Ghassemi, 2016). When earthquakes beyond the design intensity occur, rescue may be delayed because the bridges are vulnerable. In this paper, a model that involves points on faults as the epicenter, ellipses as attenuation law, and considering population distribution, is discussed to complete the statistics of which the bridge is emergent to be reinforced in case of the earthquakes beyond design. The probability model is based on the fault map in the bitmap format of the Tangshan area. Computers would process the lines of bitmap into one-pixel width, so as to resolve the cubic spline interpolation and derivation to determine the direction of ellipses, which represents the influenced area by earthquakes. One element means a district, which will accumulate the weights based on the number of ellipses surrounded. At last, the population data are applied to the weight matrix to resolve the importance value of bridges.

1 INTRODUCTION

Earthquake disaster could damage the safety of lives and properties, which is the main limitation of social and economic developments. The loss due to earthquakes could be classified into three perspectives: the loss of lives; the loss of infrastructure; and the damage of the environment of living.

Tangshan is a heavy industrial area with dense population and infrastructure. In the earthquake of 1976, the destroyed bridges in this earthquake were not designed with a seismatic technique. While facing an earthquake of intensity 11, nearly 50% of bridges are broken, and all highways are cut off. The weakness of bridges is a significant cause of damage from seism. It showed that the design of bridges should retain some extra safety degree, which is called as the secondary seismic design (DING, 2006).

In the Tangshan area, the design intensity is 8 (MCPRC, 2010), but there was an earthquake of even intensity 11 in history (Yang, 1981). Many studies have shown that the transport may be the limitation of rescue (Zhang, 2012; Kaspi, 2015).

In Tangshan Earthquake of 1976, land transport slipped into a state of paralysis, which had a bad influence on the import of medicine and rescue equipment, and the export of the wounded. Local governments organized residents to recover the highway bridges immediately (Li, 2000).

Historical records have shown that the lifeline infrastructure should be the major project in an urban program, especially the key points such as bridges and tunnels.

2 THE INTERPRETER OF THE MODEL

2.1 The theory of geology

The epicenter may focus on the line of faults, which is the gathering of geologic activity points (1), as shown in Figure 1.

The intensity will decay from the epicenter to the surrounding area, and the isoseismal map is used to demonstrate the areas which share the same intensity.

Figure 1. The faults map of the Tangshan area.

Table 1. Attenuation formulation of the model.

No	Type	Axis	Formulation
1	Fixed model	Major	$I_1 = 2.033 + 2.01M - (0.923 + 0.151\text{M})\ln(R_1 + 27.035)$
		Minor	$I_2 = 0.064 + 1.949M - (0.433 + 0.151\text{M})\ln(R_2 + 13.073)$
2	None fixed model	Major	$I_1 = -2.473 + 2.547M - (0.111 + 0.272\text{M})\ln(R_1 + 27.035)$
		Minor	$I_2 = -1.130 + 2.026M - (0.233 + 0.173\text{M})\ln(R_2 + 13.073)$
3	Unified regression model	Major	$I_1 = 3.727 + 1.429M - 1.538\ln(R_1 + 12)$
		Minor	$I_2 = 1.483 + 1.429M - 1.138\ln(R_2 + 4)$

(SHA, 2004).

Table 2. The summary of the data of strong earthquakes.

No	Magnitude	Epicentral intensity	Intensity	Major axis/km	Minor axis/km
1	6.8	10	10	3.8	1.3
			9	13.9	6.0
			8	20.6	12.0
2	7.2	10	10	8.0	5.3
			9	24.1	17.4
			8	50.2	36.4
3	7.8	11	11	5.8	3.0
			10	17.6	7.2
			9	31.0	17.0
			8	58.5	37.6

(SHA, 2004).

Table 3. Data of earthquakes in China of intensity stronger than 8 from 1800 to 2015.

Epicentral intensity	Frequency	Rate
8	74	0.578
9	31	0.242
10	17	0.133
11	5	0.039
12	1	0.008
ALL	128	1

(From the datacenter of national geophysical data center of America).

The isoseismal lines of lower intensity will expand to the external, thereby obeying the law of an epicenter-like model (Gao, 2000; LU, 2013; Szeliga, 2010; SHA, 2004; SHI, 2011). The epicenter attenuation model is more close to the isoseismal map of moderate and strong magnitude, but has more difference in the strong earthquakes, which occur in the large-length faults. The isoseismal lines of strong intensity are close to the shape of the playground while the isoseismals of the moderate and far field intensities are more like epicenters (LU, 2013).

Scholars have regressed the coefficients and given the attenuation formulation, as shown in Table 1.

From the statistics data, the major axis semidiameter is 1.6 times as long as minor axis's, commonly (Gao, 2000), as shown in Tables 2 and 3.

2.2 The theory of probability

Given the construction standards, the degree of destruction could be correlated to intensity. And each pixel element will be valued as a weight, which is accumulated by each ellipse from fault points. The scope of different intensities would contribute various weights. And so, the variables are valued as a bit inflexible.

The final weights will couple with the density of the population, to give the final solution of the lifeline.

3 SIMPLIFICATION AND WEAKNESS

3.1 The process of the bitmap

The original data of faults are in the format of bitmap, and so the information is a bit inaccurate. And it needs to deal with the potential mistakes if the map should be fed as input to calculate.

First, the lines in the bitmap are in different widths, but the width could not demonstrate the severity of faults. If the program takes each point as one epicenter, the result would be disturbed. In practice, the algorithm will simplify lines into one-pixel width, which will reduce some of the errors. And the rate between the bitmap and the real map is nearly 5.88:1, which means that 5.88 pixels represent one kilometer.

3.2 The inaccuracy of the earthquake model

The limited data could not process the effective regression to induce a formulation of the axis radius. Scholars have researched the relationship between intensity and other parameters, but the factors are too complex to use one or two parameters to demonstrate (Szeliga, 2010; Gutenberg, 1956; Wells, 1994). And so, it is hard to take intensity into the probability model, as shown in Figure 2.

 fixed

Figure 2. Example of the line process rule (each point means a pixel point).

Table 4. The population density of the Tangshan area.

Name	Total area	Population density
Qian an city	1208	606
Zunhua city	1509	488
Luan county	999	556
Luan nan county	1270	449
Lao ting county	1308	377
Qian xi county	1439	269
Yu tian county	1165	584
Caofeidian district	700	286
Feng nan district	1568	336
Feng run district	1334	697
Lu nan district	67	3464
Lubei district	112	5237
Gu ye district	253	1415
Kai ping district	252	961

3.3 *The coupling of weights and density of population*

The value of lives could be explained as the amount the members of society are willing to pay to save one (Thaler, 1976), and so it is more of a social issue and we would not discuss the topic in this paper. The model assumes that each life has the same value, and population density would be the direct parameter. Further, the final results are obtained by the weight multiplied by the population density, as shown in Table 4.

4 RESULTS AND CONCLUSION

The resulting map shows what is important, by coupling risks of earthquake and density of population. The red area is the most significant part that should be connected to the lifeline. On combining the road system map in the Tangshan area, the South Lake Avenue goes across this area. And so, the bridges on this avenue may be reinforced, in case of the great earthquakes, as shown in Figure 3.

The interrelationship between the length of the major axis and minor axis, the probability of intensity, and its harm to the society should be further researched.

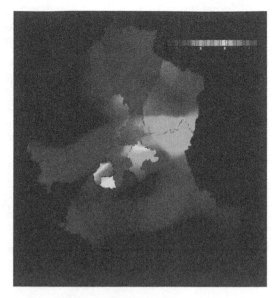

Figure 3. The resulting plot.

REFERENCES

Ding J, Jiang S, Bao F. Review of seismic damage to bridges in Tangshan earthquake [J]. World Earthquake Engineering. 2006;22(1):68–71.

Gao Y-f, Xie K-h, Zeng G-x. The seismic attenuation regularities in the moderate-strong earthquake atea. Journal-zhejiang university engineering science. 2000; 34(4):404–8.

Ghassemi MR. Surface ruptures of the Iranian earthquakes 1900–2014: Insights for earthquake fault rupture hazards and empirical relationships. Earth-Science Reviews. 2016;156:1–13.

Gutenberg B, Richter CF. Earthquake magnitude, intensity, energy, and acceleration (second paper). Bulletin of the seismological society of America. 1956;46(2):105–45.

Kaspi M, Raviv T, Tzur M, Galili H. Regulating vehicle sharing systems through parking reservation policies: Analysis and performance bounds. European Journal of Operational Research. 2015.

Li F. The research of mechanism of rescue in Tangshan Earthquake. Jinan: Shandong University; 2011.

Lu J, Li S, Li W, Song J, Cheng Y. Isoseismal area and long axis radius based intensity attenuation relationship. Journal of Earthquake Engineering and Engineering Vibration. 2013;2(33):11.

MCPRC. GB50011-2010 Code for seismic design of buildings [S]. 2010.

Sha H-j, Lu Y-j, Peng Y-j, Tang R-y, Zhang X-b. A Constrained Model of Elliptical Attenuation of Earthquake Intensity and Its Application in North China. Inland Earthquake. 2004;1(24):9.

Shi J-l, Yan Q-m, Ge q-y. An algorithm for arbitrary engineering site earthquake intensity or motion parameter using ellipsoid attenuation model. Inland Earthquake. 2011;1(25):8.

Szeliga W, Hough S, Martin S, Bilham R. Intensity, magnitude, location, and attenuation in India for felt earthquakes since 1762. Bulletin of the Seismological Society of America. 2010;100(2):570–84.

Thaler R, Rosen S. The value of saving a life: evidence from the labor market. Household production and consumption: NBER; 1976. p. 265–302.

Wells DL, Coppersmith KJ. New empirical relationships among magnitude, rupture length, rupture width, rupture area, and surface displacement. Bulletin of the seismological Society of America. 1994;84(4):974–1002.

Yang L, Chen G. Intensity distribution of the Tangshan Earthquake. Earthquake Engineering and Engineering Vibration. 1981;1(1):9.

Zhang L, Liu X, Li Y, Liu Y, Liu Z, Lin J, et al. Emergency medical rescue efforts after a major earthquake: lessons from the 2008 Wenchuan earthquake. The Lancet. 2012;379(9818):853–61.

Construction techniques of the bridge deck pavement of hot-mix and warm-compact asphaltic concrete under low temperature conditions

Ziqiao Cheng
Power China Road Bridge Group Co. Ltd., Beijing, China

Zhendong Zhao & Xue Song
Zhengzhou Sinohydro Investment and Development Co. Ltd., Henan, China

Yao Liu
School of Civil Engineering, Southwest Jiaotong University, Chengdu, Sichuan, China

ABSTRACT: Based on experience from the bridge deck pavement project of Longhai Expressway in Zhengzhou, in this article, the bridge deck construction applications of hot-mix and warm-compact techniques of the LY warm-mix agent under low temperature conditions are introduced, and the construction process and quality control points for the hot—mix and warm-compact techniques are elaborated. Practice showed that hot-mix and warm-compact techniques can be adapted to the constricted conditions with low temperature, thereby prolonging the effective time of construction, and improving the quality of road and construction conditions dramatically.

1   INTRODUCTION

A hot-mix asphalt mixture is widely used in road engineering because of its good performance. However, the traditional hot-mix asphalt mixture is a kind of hot-mix and hot-paving material, and so it needs high temperature in the process of production and construction, which not only consumes large amounts of energy, but also emits large amounts of waste gas and dust, thereby affecting the environment quality and construction personnel's physical health (Ji, 2006). Hot-mix and warm-compact techniques emerge in the background of vigorously promoting energy conservation and emission reduction. On one hand, hot-mix and warm-compact techniques can reduce the minimum shaping temperatures of the asphalt mixture, solving the problem that construction cannot be carried out at low temperature in winter. On the other hand, the lower paving temperature reduces emissions of toxic gases such as greenhouse gases (carbon dioxide, etc.) and asphalt fume and so on, reducing the pollution to the environment and the damage to the constructors' health (Lü, 2012). For construction purposes, the LY warm-mix agent is an organic complex compounded by a variety of waste materials (Ma, 2012). It is directly added to the asphalt mixture, and needs only simple equipment without a big overhaul upon equipment modification, and so the cost is relatively low.

2   PROJECT BACKGROUND

Longhai Expressway is an east–west main road in Zhengzhou. It is located in an urban core area across the railway and about 29km, west of the West Fourth Ring Road, east of Beijing, Hong Kong, and Macao high-way. The express road includes 38 ramps and four interchanges. As an important part of the express road system of Zhengzhou City, Longhai Expressway plays an important role in balancing urban traffic and connecting eastern and western towns. This project of an elevated bridge deck pavement is part of the Longhai Road Expressway. The elevated bridge is 940 m long and 25.5 m wide, and is two-way six-lane. Its stake ranges from XK4 + 327.000 to XK5 + 267.00. Hot-mix and warm-compact techniques are used in winter under low temperature conditions, and the pavement design is shown in Figure 1.

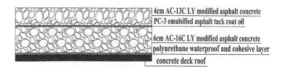

Figure 1.   Bridge deck pavement diagram.

3 CONSTRUCTION PROCESS

A hot-mix and warm-compact asphalt concrete pavement construction process is shown in Figure 2. The main process includes construction of the waterproof and cohesive layer, the mixing and transportation of asphalt mixtures, paving and rolling lower layer of the LY-modified asphalt concrete, and paving and compaction of the LY-modified asphalt concrete of the upper layer.

4 QUALITY CONTROL

4.1 The application of waterproof and cohesive layers

Acceptance of the application shall be carried out in accordance with the following requirements:

1. The bridge deck was first dealt with shot blasting, letting shot impact the work surface, by using flow inside a blasting grinding machine to clean the bridge deck before brushing, coating, and recycling remaining impurities and shot shall meet the following requirements: the base remains fully intact and the wound surface achieves 100%; surface roughness is even and

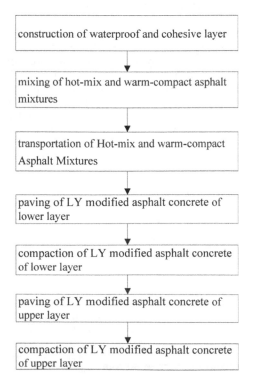

Figure 2. Hot-mix and warm-compact asphalt concrete pavement construction process.

BNP > 85; the finished surface shall meet the appropriate rate of exposed aggregate; surface laitance and sugaring layer shall be completely and evenly removed.

2. The thickness of the deck polyurethane waterproof layer shall be 1.5 mm. The solid content of the waterproof layer shall be no less than 98% and tensile strength shall be no less than 10 MPa. The deck shall be painted 3 times meeting the requirements that the thickness of waterproof coating shall be no less than 1.5 mm. The curing time shall be more than 5 hours and coating shall be painted again after the previous one cures. When using the coating method, each time the direction of brushing shall be consistent with the previous one. The end of the waterproof layer shall be brushed repeatedly a few times.

3. After the waterproof layer is accepted as qualified, a spiral drainage pipe shall be installed. A spring steel pipe of $\phi 15 \times 2.0$ made of Nickel-chromium alloy steel wire is chosen as the drainage pipe. A piece of pipe is 10 meters long, pitch is 1.8 mm, and the compressive strength shall be no less than 350 kN. The pipe shall be buried close to the inner line of the bump wall and the lowest shall be in the same height of the deck. The end of the pipe shall be 50 mm long into the water-collecting well. The hole of the water-collecting well shall be opened by electric welding or gas cutting according to the design requirements, and its diameter shall be no less than 2.5 cm.

4. Before the lower layer of asphalt concrete construction, the bottom of the bump wall shall be brushed with tack coat oil to 10 cm, the design height of asphalt pavement, ensure it is impervious between the bump wall and concrete deck.

4.2 Making of hot-mix and warm-compact asphalt mixtures

Acceptance of the making shall be carried out in accordance with the following requirements:

1. A batch mixer may be used for mixing asphalt mixtures. The proportion of aggregate and asphalt materials shall be in accordance with the mixture ratio report of the construction site. The mixer shall be equipped with device for dusting.

2. During the progress of mixing asphalt materials, fuel hoisting equipment of the mixing system should be wrapped by quits to be thermally insulated. The mixing time shall be based on the trial mixing and it should last as long as the time when mixtures mix evenly and asphalt is all wrapped with mineral aggregate. Mixed asphalt materials shall meet requirements of being uniform without blockage and appearence of aggregate or shall be adjusted in time.

3. Before leaving the factory, mixed asphalt mixtures shall be in accordance with corresponding requirements of the approved mixture ratio report, and its performance indicators shall be within the tolerances of the target value. The finished product can be stored in the storehouse if the mixture is not paved immediately.

4. A series of temperatures of mixing LY-modified asphalt mixtures is shown in Table 1 and the mixing process is shown in Figure 3. The mixtures shall not be overheated during mixing and the site temperature of mixtures shall not be less than 160°C.

4.3 Transportation of hot-mix and warm-compact asphalt mixtures

Acceptance of the transportation shall be carried out in accordance with the following requirements:

1. Asphalt mixtures should be transported by using large-tonnage dump trucks, whose carriage should be clean and applied with a thin layer of oil–water liquid (1:3) before transporting, thereby preventing the mixtures from bonding in a carriage board.

2. In order to reduce the phenomenon of segregation of coarse and fine aggregates, the dumper should move a bit as the mixer unloads a pipeful of mixtures.

3. The dumper should be covered with tarpaulins and wrapped with a fireproof quilt to prevent rain and contamination and maintain warmth.

4. According to mixing ability or paving speed, a suitable dumper load should be chosen to ensure that there is dumper waiting in front of the paver.

5. After receiving asphalt mixtures at the paving position with a list of direct discharge components, the asphalt mixtures shall be inspected, thereby prohibiting using those mixtures that do not meet the temperature requirement of specifications or that hold together.

4.4 Paving of hot-mix and warm-compact asphalt mixtures

Acceptance of the paving shall be carried out in accordance with the following requirements:

1. Paving temperature shall not be less than 130°C, and there should be 2 pavers working at the same time in the form of echelon, between which the longitudinal distance is 10–20 m and the overlapped width is about 5–10 cm. Loose paving coefficients of the material is finally determined by performing a test. The two pavers are equipped with an automatic leveling system of the electronic balance beam and multi-probe ultrasonic sensor for detecting pavement elevation to continuously adjust the paving thickness

Table 1. Mixing temperature of hot-mix and warm-compact asphalt concrete.

Serial number	Project	The range of temperature (°C)
1	Heating temperature of asphalt mixture	190–220
2	Heating temperature of asphalt	160–165
3	Mixing temperature of the asphalt mixture	170–185
4	Temperature dispatched from the factory	160–175

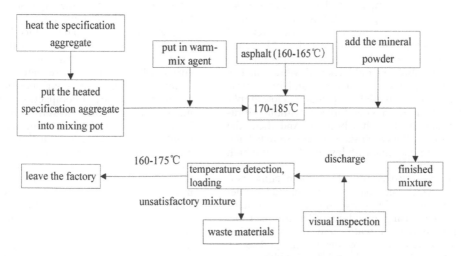

Figure 3. The mixing flow chart of hot-mix and warm-compact asphalt mixtures.

of the surface layer. The elevation and roughness of pavement shall meet requirements.

2. Before paving, the screed of the paver shall be preheated before 0.5 h–1 h at the temperature more than 100°C. During the progress of paving, the vibration frequency and amplitude of the hammer shall be adjusted to make the degree of compaction of the mixtures maximum without affecting the roughness of the pavement.
3. The transport capacity of the dumper shall be a little surplus, and there shall be more than 5 dumpers waiting in front of the paver when paving. In the process of paving, the idled dumper shall stop 10–30 cm in front of the paver and be pushed by the paver to unload slowly thereby avoiding hitting the paver.
4. Before the compaction of hot-mix and warm-compact asphalt mixtures, no one shall be allowed to tread. Asphalt mixtures can be made up or replaced when the pavement has local segregation, but they shall be shoveled away and the paver construction technology is adjusted if the segregation is serious (Geng, 2014).

4.5 The compaction of hot-mix and warm-compact asphalt mixtures

The compaction of asphalt mixtures can be divided into three stages: initial compaction, second compaction, and final compaction. Initial compaction and final compaction are carried out by using steel rollers, while the second compaction is carried out by using tire rollers. The specific requirements of the compaction of the LY hot-mix and warm-compact asphalt mixtures are as follows:

1. Initial compaction: the general temperature shall not be less than 120°C and the two steel rollers shall compact asphalt mixtures side-by-side after the paving distance reaches up to 20 m. The outer-roller shall compact back and forth one time at 2–3 km/h with high frequency and low amplitude, thereby moving forward without vibration and back with vibration. The inner-roller shall compact back and forth 1 time at 2–3 km/h with high frequency and low amplitude, thereby moving forward without vibration and back with vibration. And then, the two steel rollers each shall vibrate strongly back and forth one time with low amplitude and high frequency.
2. Second compaction: after compacting two wheel-width, compact asphalt mixtures side-by-side at 2–3 km/h by using two tyre rollers, and then using one tyre roller compact the whole pavement width and the compaction time is determined by the summary data of the test. There shall be a 1/3–1/4 width of the compaction width overlap between two rollers during compaction.
3. Final compaction: the temperature shall not be less than 70°C and by using steel rollers compact asphalt mixtures at 2–3 km/h to make the road sleek and compact. The road can be open to traffic when the temperature is below 50°C.

4.6 Treatment of road surface joints

Acceptance of the road surface joint shall be carried out in accordance with the following requirements:

1. Two pavers work in the form of echelon, between which the longitudinal distance is 10–20 m and the screed shall be set at the same level. Width of the paved asphalt mixtures is set at 10–20 cm, as the elevation base level datum of later mixtures shall not be compacted and the overlapped paving width is about 5–10 cm. Joint traces shall be eliminated in the form of heat joint construction.
2. When the half range of the road cannot be constructed in the form of heat joint construction, longitudinal joints can be cut tidily by using the parting tool. Before paving the other half road, the edge of the longitudinal joint shall be cleaned and a small amount of asphalt tack cloth is spread. There shall be 5–10 cm width overlap with the paved layer and the asphalt mixtures on the former half road shall be shoveled away after the pavement. When compaction begins, the roller firstly moves on the compacted road, thereby rolling the new paving layer 10–15 cm, and gradually moves across the longitudinal joint, thereby fully compacting the joint.
3. The transverse joint of the bridge deck shall be set in the position of the expansion joint and the upper and lower longitudinal joints shall be staggered over 15cm. The surface longitudinal joint shall be set in the position of the lane line.
4. Transverse joint construction shall be constructed before the mixtures cool in the form of abutment joint. The joint's position shall be set in the final part of compaction in which the maximum gap of the road surface is larger than 3 mm when using a 3 m ruler to measure longitudinally. The transverse joint shall be constructed after the mixtures cool down, and the mixtures shall be shoveled away from the end of the compacted road to the transverse joint. Before the next pavement, the edge of the transverse joint shall be cleaned and some asphalt tack cloth is spread, and then the transverse joint shall be compacted by using a roller at an angle of 45 degrees or transversely first and then longitudinally to become smooth and tight.

5 CONCLUDING REMARKS

The successful use of hot-mix and warm-compact technology of asphalt concrete in Longhai Expressway construction effectively solves the difficulty in the construction of asphalt concrete at low temperature, shortens the construction period of the pavement and ensures the construction quality of the pavement. At the same time, it is also remarkable to save energy and protect the environment and results in good social and economic benefits. The construction method can provide references to other similar engineering cases and it serves as a good local example for the further spreading of hot-mix and warm-compact technology of asphalt concrete.

REFERENCES

Geng Ke, Discussion on Construction Quality Control for Asphalt Road Surface of Expressway in Gansu Province [J], Highway, **9**, 63–69 (2014).

Ji Jie, Wang Ruiying, Xu Shifa, Yan Bin. A comparative study on the performance of high energy saving low emission warm mix asphalt mixture and hot mix asphalt mixture [J]. Journal of Highway and Transportation Research and Development, In Chinese, **11**, 90–92 (2006).

Lü Yi, Analysis of Influence of Warm-Mixed Additives on Asphalt Performance [J], Highway Engineering, **18**, 45–47 (2012).

Ma Guangwen, The influence of LY Warm-mix Agent on Asphalt Mixture [J] Shanghai Highways, **3**, 57–59 (2012).

Ministry of Transport of the People's Republic of China. *Technical Specifications for Construction of Highway Asphalt Pavements* (JTG F40-2004), Communications Press, Beijing, China(in Chinese), 38–42 (2004).

Advances in Energy Science and Equipment Engineering II – Zhou, Patty & Chen (Eds)
© 2017 Taylor & Francis Group, London, ISBN 978-1-138-71798-5

A study on the evolution law of the tunnel seepage field in the vicinity of the compressive torsional fault zone

Saizhou Yang & Zheng Li
MOE Key Laboratory of Transportation Tunnel Engineering, Southwest Jiaotong University, Chengdu, China

ABSTRACT: Based on the connecting line of the Shenzhen eastern border crossing expressway project, which crosses multiple fault zones, the visualization software is applied to establish a 3D numerical model to study the evolution law of the tunnel seepage field in the vicinity of the compressive torsional fault zone and the effect of grouting reinforcement. The results reveal that under certain conditions, the fault presents a significant water barrier effect. The groundwater flows along the fault and does not penetrate the fault zone. Because of the waterproof nature of the tunnel structure and surrounding rocks, the seepage field (after excavation) lies close to the initial seepage field.

1 INTRODUCTION

The tunnel seepage evolution is mainly affected by the permeability of surrounding rocks, the method of tunnel construction, the type of waterproof and drainage system, and so on (He, 2013; Streltsova, 1973; Bouvard, 1969). The theoretical analysis, numerical simulation, and physical experiments are the commonly used research methods. When the geological structures are relatively stable in the tunnel site area, the groundwater seepage could achieve a dynamic balance after the tunnel excavation (Streltsova, 1976; Sembenelli, 1999). The seepage field distribution could be simulated by performing the physical experiment (Chapuis, 2012; Post, 2007; Chiu, 2007). However, sometimes there are special geological sections in the tunnel site, such as the compressive torsional fault zone. The complex rock mass structural plane leads to the discontinuity of rock strata and the different thicknesses of aquifers. The hydrogeological parameters differ greatly in space and time. The flow of groundwater is mostly discontinuous (Shante, 1971). It is hard for the traditional physical test to simulate this case effectively. But the numerical model can be used to simulate it reasonably and to restore the tunnel seepage field in the vicinity of the compressive torsional fault zone as far as possible.

There are three types of the seepage model of fractured rock mass in the fault zone: equivalent continuum model, discrete fracture network model, and dual medium model. The existing studies of the tunnel seepage field are mainly focused on the prediction of water inflow, the relationship between water pressure and displacement, and so on (Bear J., 2012; No A W G, 1989; P. Renard,

2015). The study of the seepage field near the fault zone is limited to a single fractured rock mass. In this paper, based on the connecting line of the Shenzhen eastern border crossing expressway project, which crosses a multiple fault zone, the 3D visualization software is applied to study the evolution law of the tunnel seepage field in the vicinity of the compressive torsional fault zone and the effect of grouting reinforcement. The results could be valuable as reference when dealing with similar projects.

2 SURVEY OF TUNNEL ENGINEERING NEAR THE FAULT ZONE

Influenced by the large fault zone in the southeast coast, the geological structure of the Shenzhen city is complex and mostly is the fault structure. The connecting line of the Shenzhen eastern border crossing expressway project is adjacent to the Shenzhen reservoir. The project is composed of two parallel main tunnels (the south and north lines) and ramps. The total length is about 8.7 km. The hydrogeological conditions of the tunnel site are complex, and the majority of surrounding rocks are of grade IV~VI. The groundwater is rich in resources, and is active in the movements. The groundwater mainly accepts the infiltration of atmospheric precipitation and the Shenzhen reservoir water lateral recharges, and drainages from the hill and slope to the ditch and other low-lying places. The project is located in the Luohu fault zone and its influenced zone, as shown in Fig. 1. The problems such as large-scale water gushing may occur near the fault zones.

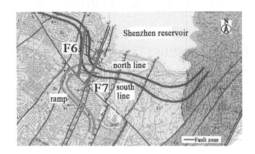

Figure 1.   Project overview.

3   NUMERICAL SIMULATION OF THE TUNNEL SEEPAGE FIELD

3.1   Calculation parameters and working conditions

Visual-modflow software is used to simulate the complex hydrological conditions, and it is usually achieved by changing the hydraulic parameters (Karanth, 1987; Molinero, 2002). The rock mass in the compressive torsional fault zone has been subjected to a strong compression in the long-term geological evolution (Hwang, 2007; Bobet, 2001). Besides making the rock broken, the tectonic compression also makes the rock compact. Therefore, the compressive torsional fault zone presents a characteristic of water barrier. According to the hydrogeological investigation, the main calculation parameters of the model are shown in Table 1.

The 3D calculation model is established by using the Visual-modflow software, as shown in Fig. 2. The calculation cases are shown in Table 2.

3.2   Result analysis

1. The seepage field evolution of the tunnel cross-section in case 1.

Fig. 3 indicates that after tunnel excavation, the underground water shows a trend of influx into the tunnel, within a large range of strata around the tunnel. It is because the tunnel does not have any support measures or grouting reinforcement. However, the hydraulic connection between tunnels and underground water on both sides is cut off because of the existence of compressive torsional fault zones F7 and F6, as shown in Fig. 3(a). The faults F6 and F7 present a strong water resistance. The groundwater flows along the fault and does not penetrate the fault zone. The underground water level above the tunnel structure gradually decreases over time after the excavation. The biggest drop just above the tunnel is about 8 m, which forms an irregular precipitation funnel. No obvious changes in the direction of groundwater flow and hydraulic head are observed in the area from the F7 fault zone to the lower reaches of the Shenzhen reservoir. It further reflects the local water barrier effect of the compressive torsional fault.

2. The seepage field evolution of the tunnel cross-section in case 2 is shown in Fig. 4.

As shown in Fig. 4, the cross-section seepage variation of the tunnel with initial lining and grouting circles is clearly different from the unlined tunnel. Only the groundwater, which is located in the upper strata of the tunnel, presents a trend of flowing into the tunnel. It is clear that the effective waterproof system (primary lining and grouting circle) plays a significant role in maintaining the stability of groundwater and reducing the influence of tunnel excavation on the seepage

Table 1.   Main calculation parameters.

Items	K_x (m/d)	K_y (m/d)	K_z (m/d)	Effective porosity (%)	Specific yield (%)
Artificial fill	0.2	0.2	0.1	3	2.4
Strong weathering metamorphic sandstone	0.5	0.5	0.25	15	11
Medium weathering metamorphic sandstone	0.5	0.5	0.3	12	9
Weak weathering metamorphic sandstone	0.15	0.15	0.075	10	7
Strong weathering rock fragmentation mixed granite	0.5	0.5	0.25	5	3.3
Weak weathering rock fragmentation mixed granite	0.2	0.2	0.1	3	2.1
Strong weathered granite gneiss	0.4	0.4	0.2	4	3
Weak weathered granite gneiss	0.08	0.08	0.04	2	1.5
Fault zone	0.01	0.01	0.01	10	6.8
Dense fracture zone	1.0	1.0	0.5	30	15
Initial lining	8.64E-03	8.64E-03	8.64E-03	—	—
Grouting circles	0.02	0.02	0.02	2	1.5

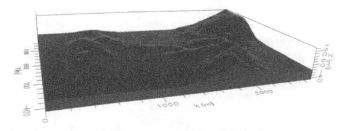

Figure 2. Sketch map of the 3D model of the tunnel.

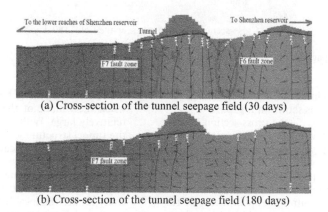

(a) Cross-section of the tunnel seepage field (30 days)

(b) Cross-section of the tunnel seepage field (180 days)

Figure 3. The seepage field evolution of the tunnel cross-section in case 1.

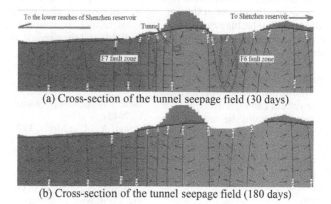

(a) Cross-section of the tunnel seepage field (30 days)

(b) Cross-section of the tunnel seepage field (180 days)

Figure 4. The seepage field evolution of the tunnel cross-section in case 2.

Table 2. List of calculation cases.

Calculation cases	The state of the tunnel
Case 1	Unlined
Case 2	Initial lining + grouting zone
Case 3	Initial lining + secondary lining + grouting zone

Note: The construction of the tunnel in all of the above-mentioned cases is completed.

field. Because the hydraulic relation between the hilly area and the lower reaches of the reservoir is relatively weak, the barrier effect of F7 is not as obvious as F6. The initial lining and the grouting circle further reduce the influence on the seepage field. Therefore, the change range is limited near the fault zone. Due to the smaller affected area, the underground water level drops nearly 4 m right above the tunnel structure, which forms a relatively regular precipitation funnel.

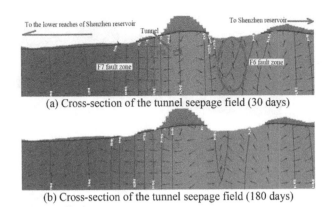

(a) Cross-section of the tunnel seepage field (30 days)

(b) Cross-section of the tunnel seepage field (180 days)

Figure 5.   The seepage field evolution of the tunnel cross-section in case 3.

3. The seepage field evolution of the tunnel cross-section in case 3 is shown in Fig. 5.

As shown in Fig. 5, the cross-section seepage field of the tunnel with a complete waterproof system (initial lining, secondary lining, and grouting circles) is similar with the initial seepage state. The effective waterproof system has an important effect in maintaining the stability of groundwater, and it is helpful in controlling the influence of tunnel excavation on the seepage field to the minimum degree. And the affected area has not increased over time, which means the grouting reinforcement has already taken effect on filling the cracks in the surrounding rock of the tunnel. The biggest drop of the groundwater level is about 2 m. Both the direction of groundwater flow and hydraulic head near F7 and F6 faults are similar to the initial seepage state. The above-mentioned results show that for tunnel projects near the fault, the well-performed tunnel waterproofing system can effectively reduce the local hydraulic contact around the tunnel. The physical and mechanical properties of fractured rock masses can be improved by adopting reasonable grouting reinforcement measures. It also could reduce the disturbance of tunnel excavation to the surrounding seepage field, and then ensure the safety of construction.

4   CONCLUSIONS

1. The fault zone F6 is located in the area where lots of hydraulic connection between the Shenzhen reservoir and the tunnel existed. Regardless of the tunnel structure and the waterproof measures, the fault has shown a significant water barrier effect. The groundwater flows along the fault and does not penetrate the fault zone.
2. The fault zone F7 is located between the tunnel and the lower reaches of the Shenzhen reservoir.

And it shows a stronger water barrier effect only when the volume of the tunnel water inflow is relatively large. With the waterproof nature of the tunnel structure and surrounding rocks, the seepage field (after excavation) lies close to the initial seepage field.

3. For tunnel projects near the fault, the well-performed tunnel waterproofing system can effectively reduce the local hydraulic contact around the tunnel. The physical and mechanical properties of fractured rock masses can be improved by adopting reasonable grouting reinforcement measures. It also could reduce the disturbance of tunnel excavation to the surrounding seepage field, and then ensure the safety of construction.

REFERENCES

Bear, J. *Hydraulics of groundwater* (Courier Corporation, 2012).
Bobet, A. J Eng Mech., **127**(2001).
Bouvard, M., N. Pinto. La Houille Blanche, (1969).
Chapuis, R.P. Bulletin of engineering geology and the environment, **71**(2012).
Chiu, P.Y., H.D. Yeh, S.Y. Yang. Int. J. Numer. and Analytical Methods in Geomech., **31**(2007).
He, C., B. Wang, J. Modern Transp., **21**(2013).
Hwang, J.H., C.C. Lu. Tunn Undergr Sp Tech., **22**(2007).
Karanth, K.R. *Ground water assessment: development and management* (Tata McGraw-Hill Education, 1987).
Molinero, J., J. Samper, R. Juanes. Eng Geol, **64**(2002).
No A W G, de Travail No A G. Tunn Undergr Sp Tech., **4**(1989).
Post, V., H. Kooi, C. Simmons. Ground Water, **45**(2007).
Renard, P. Ground water, **43**(2015).
Sembenelli, P.G. G. Sembenelli. ASCE, **125**(1999).
Shante, V.K.S., S. Kirkpatrick. Advan. Phys., **20**(1971).
Streltsova, T.D. J. Hydrol., **20**(1973).
Streltsova, T.D. Water Resour. Res., **12**(1976).

Advances in Energy Science and Equipment Engineering II – Zhou, Patty & Chen (Eds)
© 2017 Taylor & Francis Group, London, ISBN 978-1-138-71798-5

A study on treating nitrilon wastewater with Expanded Granule Sludge Bed (EGSB)

Huili Li, Ning Dang & Huimin Wang
School of Civil Engineering, Lanzhou University of Technology, Gansu, Lanzhou, China

ABSTRACT: Treating nitrilon wastewater with biological technology is difficult because of CN-toxicity and presence of synthetic oligomers. In this study, nitrilon wastewater was treated with EGSB reactor for improving wastewater quality and enhancing removal efficiency. The key parameters were measured, which are as follows: nitrogen concentration, COD, pH value, and B/C ratio and analyzed the change of molecular weight. The results showed that the average removal ratio of the pollutants reached 30%; and the highest removal was over 45%. The biochemical performance of nitrilon wastewater was improved effectively, and the B/C ratio was enhanced to 0.4. This treating method may reduce the molecular weight of organic matter and change PAN to a small molecular material.

1 INTRODUCTION

With the development of economy in China, the chemical industry has been growing quickly. The situation of wastewater released from a chemical-industry-polluted environment is worse than before. The main components of chemical wastewater are complicated structures, poisonous and harmful nature, and biologically hard-to-degrade organic pollutants (D. Chuncheng, 2005; D. Xiaojun, 2008). It is difficult to treat this kind of wastewater. The nitrilon wastewater is the typical chemical wastewater, whose main components are PAN (PolyAcryloNitrile), SCN^-, and CN^-. PAN is hard to degrade by biological digestion, and CN^- has biological toxicity (G. Dong, 2009; G. Guiyue, 2009; W. Yuan, 2004; W. Gang, 2006; Y. Jianghong, 2006; Y. Tianming, 2008; Z. Chaocheng, 2004). Because of the particularity of nitrilon wastewater, the biological treatment process of nitrilon wastewater is difficult to meet the national requirements of the emission standards in China. And so, nitrilon wastewater caused serious environmental pollution and great environmental stress to the area discharged.

EGSBs (Expanded Granular Sludge Beds) are the third generation of anaerobic reactors (R. Nanqi, 2004). In the 1990s, it was used in the development of lead by Lettinga of Dutch Wageingen agricultural university. The main characteristic is to have a bigger ratio of height to diameter, higher liquid speed of upflow, COD load rate, and resistance to impact load ability (D. Wjbm, 1994; H. T C, 1997; H. Pol L W, 1998; H. Pol L W, 1983; M. T Kato, 1994). An EGSB reactor can be used to deal with high content of SS, microbial toxicity of

wastewater, and high concentration organic waste-water treatment (G. Collins, 2003).

The nitrilon wastewater treatment process of the petrochemical factory is hydrolysis acidification after iron–carbon electrolysis, and then in-depth biological treatment. The treatment effect of hydrolytic acidification is poorer; the pollutant removal rate is lower than 20%. The effluent B/C (BOD_5/COD) ratio of hydrolytic acidification is from 0.1 to 0.2, which had a serious impact on the treatment efficiency of the subsequent biological process. Constantly, the total system of the biological treating process is useless to nitrilon wastewater. According to wastewater quality conditions, the nitrilon wastewater is treated with an EGSB reactor for increasing the pollutant removal rate and improving wastewater quality for making wastewater to fit the subsequent biological treatment.

2 EXPERIMENTAL SECTION

2.1 Equipent

The reactor is made of polymethyl methacrylate. It is separated into two parts: cylindrical main reactor and three-phase separator on the top. The dimension of the EGSB reactor is according to Table 1.

Table 1. The dimension of the EGSB reactor.

Inner diameter (mm)	External diameter (mm)	Total height (m)	Height–diameter ratio	Active volume (l)	Total volume (l)
90	100	1.7	19	10	16.5

2.2 Substrate chemical characterization

The nitrilon wastewater is from production wastewater of the petrochemical factory. The acrylic factory contributes 90% of the wastewater and the rest of the wastewater is from acetonitrile wastewater and acrylonitrile wastewater. The quality of nitrilon wastewater is different every day. The wastewater quality index is as shown in Table 2.

2.3 Operational conditions

Inoculums sludge is obtained from the granule sludge of treating beer wastewater. The concentration of the granule sludge is 19.3 g/l; the ratio of the sludge VSS/SS is 0.71. Because of CN^- toxicity, the domestication time of the granule sludge is nearly 3 months. When the domestication time is over, the concentration of the granule sludge is down to 12.5 g/l and the ratio of sludge VSS/SS decreases to 0.57. When the VFA concentration is less than 4 mmol/l, the EGSB reactor can operate

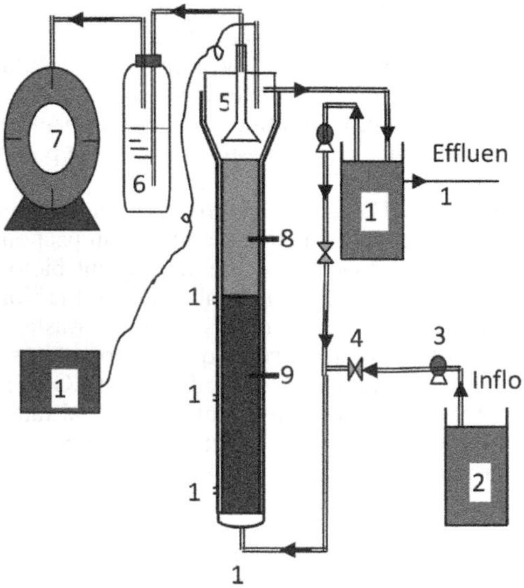

Figure 1. Scheme of the EGSB reactor.
1. Reflux tank; 2. inflow tank; 3. pump; 4. valve; 5. three-phase separator; 6. alkali absorption bottle; 7. gas flow meter; 8. small granule sludge; 9. big granule sludge; 10. discharge outlet; 11, 12, and 13. sampling mouth (1), (2), and (3), 14. EGSB reactor; 15. temperature meter.

normally. The temperature of the reactor is controlled at $20\pm2°C$.

The quantity of inflow is 0.8 l/h. The average volume loading of the EGSB reactor is 12 kg COD/ $(m^3 \cdot d)$, average sludge loading is 0.96 kg COD/ (kg SS $\cdot$ d), HRT is 12 hours, and recycle ratio is 8 to treat nitrilon wastewater, as shown in Figure 1.

2.4 Analysis and measurement

COD values of the inflow and effluent were measured with potassium dichromate method; ammonia nitrogen concentrations (NH_4^+-N) of the inflow and effluent were measured with sodium reagent spectrophotometry; NO_3^--N levels were measured with phenol disulfonic acid spectrophotometry; NO_2^--N levels were measured with N-(1-naphthyl)-ethylenediamine spectrophotometry; the BOD_5 value was measured for 5 days by using the biochemistry method; the pH value was measured with pHS-3C pH meter; the molecular weight was measured with gel chromatography (J. Ahn, 2002). And other related parameters were measured according to the water and waste water detection analysis method, to determine the test methods.

3 RESULTS AND DISCUSSION

Because the component of nitrilon wastewater was polyacrylonitrile (PAN), acrylonitrile (AN), and ammonia nitrogen, the nitrogen source was not included in the experiment. Only the phosphorus source (KH_2PO_4) was added. Because the pH value of nitrilon wastewater was between 7.5 and 8.0, it does not require the addition of an alkaline material for maintaining the pH value stability of the EGSB reactor. In the process of the operation, trace nutrient elements were added for keeping nutrition balance every day one time ($FeCl_2 \cdot 4H_2O$, $CoCl_2 \cdot 6H_2O$, $NiCl_2 \cdot 6H_2O$).

3.1 COD concentration and removal rate

The EGSB reactor was operating more than 2 months under the operational conditions. The result is as shown in Figure 2.

According to Fig. 2, the COD concentration of the inflow varied obviously; but the effluent COD concentration was steady relatively and the average

Table 2. Water quality index of nitrilon wastewater.

COD concentration (mg/l)	NH_4^+-N concentration (mg/l)	CN^- concentration (mg/l)	BOD_5/COD	pH
500~1000	60~100	2.7~3.5	0.10~0.27	7.5~8.0

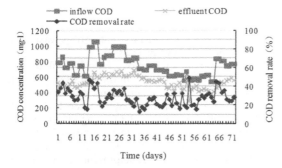

Figure 2. COD concentration and removal rate.

value was 500 mg/l. This meant that the capability of the resisting shock load of the EGSB was good. The influent COD concentration affected the effluent COD concentration of the EGSB reactor by comparing the variation tendency. The variation tendency of influent and effluent COD concentrations was similar in EGSB reactors. But the variation trend was not completely consistent. From 59 days to 62 days, the influent COD concentration was higher but the effluent COD concentration was lower by comparing with data from 56 days to 57 days. From 46 days to 51 days, the influent COD concentration was steady but the effluent COD concentration was wave. Because the tendency of the influent and effluent COD concentration was not in accordance, this meant that the pollutant concentration of the biodegradable substance in wastewater was not in a certain proportional relation with the influent COD concentration.

The average COD removal of the EGSB reactor was about 30%; and the highest removal efficiency was over 45%. The COD removal was not relative with influent COD concentration. The result showed that the part of the pollution substance was a difficult degradation substance because the influent COD concentration was low; the COD removal rate was not high, such as 12 days, 13 days, 37 days, 53 days, and 57 days.

3.2 Nitrogen concentration

The results of the measured nitrogen concentration showed that: the nitrate nitrogen concentration and nitrite nitrogen concentration were very low. Therefore, the nitrogen concentration approximated the ammonia nitrogen concentration. During the experimental process, any form of nitrogen source was not added into the reactor. Under the anaerobic conditions, ammonia nitrogen cannot be converted into nitrate nitrogen or nitrite nitrogen. Under normal operational conditions, the effluent nitrogen concentration of the EGSB is lower than the influent nitrogen concentration because

a large number of anaerobic microbes exist in the EGSB, which need to consume the nitrogen source for metabolism. CN^- cannot turn into ammonia nitrogen under anaerobic conditions. This means that the ammonia nitrogen concentration should drop. In fact, the ammonia nitrogen concentration of the effluent was higher than that of the influent, according to Fig. 3. And so, the increase in the concentration of ammonia nitrogen in the EGSB was perhaps from anaerobic microbes digestion of PAN and AN, because the nitrogen element exists in PAN and AN.

3.3 pH value

From Fig. 4, it can be found that the effluent pH value of the EGSB reactor was higher than the influent pH value in the most of operational times. The pH value of the EGSB reactor should be reduced under normal operational conditions, due to producing acid and consuming alkalinity. The tendency of the pH value proved that the alkaline substance was formed in the process of treating nitrilon wastewater.

3.4 MWD (Molecular Weight Distribution)

The raw wastewater and effluent of the EGSB reactor was treated for measuring the molecular weight. Molecular weight changes of nitrilon

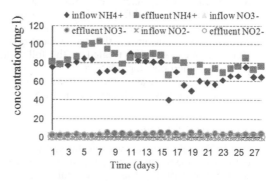

Figure 3. Nitrogen concentration of inflow and effluent.

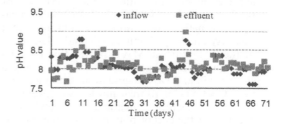

Figure 4. pH value of inflow and effluent.

419

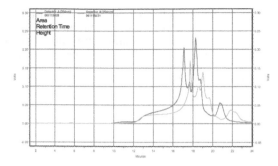

Figure 5.　Graph showing the MWD.

wastewater were analyzed qualitatively. The results are shown in Fig. 5.

The results of gel chromatography show that the molecular weight distribution of nitrilon wastewater yielded three peaks on behalf of three different materials (red lines). The time appearing peak is shorter; the molecular weight of the material is bigger. The first peak of nitrilon wastewater appeared in the molecular weight from 1.5×10^5 to 1×10^5. The second peak of nitrilon wastewater appeared in the molecular weight from 1×10^5 to 2×10^4. The molecular weight of third peak was less than 2×10^4. After anaerobic treatment, the peak wave height of nitrilon wastewater dropped and moved downward; meanwhile, the three peaks of the EGSB effluent were produced (green line). The molecular weight of PAN in nitrilon wastewater decreased with anaerobic treatment, and the concentration of the difficult digestion substance diminished, according to comparing the red line and green line. This meant that the effluent of the EGSB was easier to be treated in the follow-up process of microbial degradation.

The red line indicates raw wastewater; and the green line indicates the effluent of the EGSB.

3.5　BOD_5/COD ratio

The BOD_5 values of raw wastewater and the effluent of the EGSB reactor were measured (we measured 5 times during the operational time). The results show that the original B/C (BOD_5/COD) ratio increased from 0.1~0.27 to 0.33~0.43. An average effluent B/C ratio of the EGSB reactor reached up to 0.4. The change of wastewater quality was beneficial for treatment; this meant that it was easy for the pollutants to degrade. The result of the B/C ratio and MWD proved that the long chain and difficult digestion material were changed short chain and easy degradation substance with the EGSB reactor to treat nitrilon wastewater by anaerobic microbial degradation metabolism. Meanwhile, the part of pollutants was removed by using microbes.

4　CONCLUSIONS

The EGSB reactor is fit to treat nitrilon wastewater as the pretreatment process.

The average COD removal ratio of nitrilon wastewater reached up to 30% in the EGSB reactor. The highest COD removal percentage was over 45%.

Anaerobic microbes may degrade PAN in the EGSB reactor. They cut the long chain material into the short chain material, and make the big molecules turn into small molecules. The change of wastewater quality is helpful for advanced treatment of nitrilon wastewater with sequential biological treating processes.

The quality of nitrilon wastewater became better by using the EGSB reactor. The biochemical performance of nitrilon wastewater was improved effectively, and the B/C ratio of nitrilon wastewater was enhanced from 0.1~0.27 to 0.33~0.43. An average B/C ratio of the effluents of the EGSB reactor reached up to 0.4.

The nitrogen source is not necessary while treating nitrilon wastewater by using the EGSB reactor. An alkaline substance need not be added to the EGSB reactor for control and adjustment of the pH value.

REFERENCES

Ahn, J., T. Daidou, S. Tsuneda, A. Hirata. Modern Scientific Tools in Bioprocessing, Water Research. *Characterization of denitrifying phosphate-accumulating organisms cultivated under different electron acceptor conditions using polymerase chain reaction-denaturing gradient gel electrophoresis assay*. **36**, 403–412(2002).

Beijing: China Environmental Science Publishing House. *Detecting and analyzing method of water and wastewater (third edition)*. 233–237(1997).

Chaocheng, Z., L. Xiaohua, L. Haihong. Industrial Water Treatment. *Study on the biochemical treatment of acrylic fiber wastewater*. **24**, 42–45(2004).

Chuncheng, D., L. Daqian. Journal of Zhejiang University of Technology. *Processes and developments on chemical wastewater treatment*. **33**, 647–651(2005).

Collins, G., A. Woods, S. Mchugh, M. W. Carton, V. O'fiaherty. FEMS Microbiology Ecology. *Microbial community struture and methanogenic activity during start-up of psychrophilic anaerobic digesters treating synthetic industrial wastewaters*. **46**, 159–170(2003).

Dong, G., C. Hong. Industrial Water Treatment. *The treatment of dry-spun acrylic fiber wastewaterr by coagulation-potential three-dimensional electrolysis-facultative aerobic-aerobic process*. **29**, 35–37(2009).

Gang, W., Z. Dongkai, J. Linshi. Contemporary Chemical Industry. *Exploitation and treatment of study on composite flocculants applied in acrylic fiber craft wastewater*. **35**, 29–33(2006).

Guiyue, G., L. Zhongyue, R. Lili. Industrial Water Treatment. *Study on the effect of nano-material on the biodegradability of acrylic fiber wastewater*. **29**, 35–37(2009).

Jianghong, Y., Z. Shihui. Yunnan Environmental Science. *Feasibility analysis of treating wastewater from acrylic fibers industry by two-phase anaerobic process.* **25**,38–39(2006).

Kato, M.T., J.A Field, R. Kleerebezeml, G. Lettinga. Journal of Fermentation and Bioengineering. *Treatment of low strength soluble wastewater in UASB reactors.* **77**,670–686(1994).

Nanqi, R., W. Aijie. Chemical Industrial Press. *Anaerobic biological technology principle and application.* (2004).

Pol H. L W. *The phenomenon of granulation of anaerobic sludge.* PhD thesis, Wageningen Agricultural University, Wageningen, The Netherlands, (1998).

Pol H., L W, L. G, V. C T M, D. Zeeuw W. J. Water Sci. Technol. *Granulation in UASB reactors.* **15**,291–304 (1983).

T C H., T. A, P. R. Water Sci. Technol. *Anaerobic fluidized beds: ten years of industrial experience.* **36**,415–422(1997).

Tianming, Y. W. Changzhen, X. Zhengmiao. Bulletin of Science and Technology. *Treatment of wastewater from acrylic fiber plant, using bentonite and lime.* **24**,424–426(2008).

Wjbm, D., T. Rd Mh, V. Tlfm. Water Sci. Technol. *Experience on anaerobic treatment of distillery effluent with the UASB Process.* 30,193–201(1994).

Xiaojun, D., S. Shumiao, Y. Shuangchun. Liaoning Chemical Industry. *The status of acrylic fiber wastewater treatment.* **37**,673–676(2008).

Yuan, W., X. Zhibing, P. Fangming.Journal of Anqing Teschers College (Natural Science Edition). *Experimental study on the treatment of wastewater mixed with acrylic fiber by SBR process.* **10**, 88–89(2004).

Advances in Energy Science and Equipment Engineering II – Zhou, Patty & Chen (Eds)
© *2017 Taylor & Francis Group, London, ISBN 978-1-138-71798-5*

Experimental study and theoretical analysis on mansard steel beam

J.J. Kouadjo Tchekwagep, Junchao Huang, Zhixin Zhang & Wenyi Cui
Department of Structural Engineering, Yanbian University, Yanji, China

ABSTRACT: The bending test of 5 different slopes of mansard steel beams with both end hinges under a symmetrical concentrated load was carried out. The failure patterns of the specimens, mid-span deflection and the ultimate bearing capacity were observed. Meanwhile, the influence of bearing relative displacement to the reaction force and deflection were studied. Using mechanics, we analyzed the horizontal support anti-force and mid-span deflection of mansard steel beam. The theory and the related calculation formula were put forward, which can provide a theoretical basis for engineering design.

1 INTRODUCTION

Steel construction has a pivotal position in the industry field in China with its larger span, good ductility, toughness, easy to standardize production, easy installation as well as short construction period and could soon become an important investment benefit (Andrews, 1912; Richard, 1997; Newmark, 1951; Zhou, 2013). But steel building also has its own weaknesses: poor fire resistance and corrosion resistance. In contrast, these weaknesses are precisely the advantages of reinforced concrete structures. The steel price is not stable and the amount of steel used in the steel column is greater than the amount of steel in the steel beam. Moreover, fire coating are very expensive. So reinforced concrete columns and H-shaped steel beam structure are produced in order to replace the portal frame light house structure (Nie, 2004; Wang, 2006; Huang, 2008). Reinforced concrete columns and steel beam bearing structure system is a hybrid architecture which is widely used in large span single layer industrial factory building. It also has the advantages of reinforced concrete structure and steel (Leng, 2013).

The concrete column steel beam structure is developed on the basis of portal frame. At present, there is no consensus on the design basis, construction points, application scope and mechanical performance of this kind of structure in the academic and engineering circles (Wan, 2007). The relevant standards, codes and regulations are not clearly defined in China. Due to the needs of engineering construction, concrete columns and steel beam structures gradually obtain the affirmation of the owners and are applied more and more widely. However, the design application shows many problems such as too large steel beam deflection, loose concrete stigma destruction, or even factory collapse accident. Therefore, such systems structure requires further study (Yukio, 1996).

Concrete columns and beam structures are composed of two completely different materials. Steel beam is an elastic material whereas concrete column is an elastic plastic material. If the beam-column joints are designed as rigid joints, the stress is very complex and very difficult to achieve. So the beam-column joints are generally designed to be hinged. In this paper, five mansard steel beams with different slopes and both ends hinged beam were studied. The failure mode, mid-span deflection, ultimate bearing capacity and the effect of relative bearing displacement to force and deflection were studied. Meanwhile, the theoretical analysis of the horizontal support and the deflection of the two hinged chevron steel beam is carried out, and the related calculating formulas are put forward, which can provide the theoretical basis for the engineering design.

2 EXPERIMENT

2.1 Specimen design

In total, we have five different steel beam specimens with 1/5, 1/6, 1/8, 1/10, 1/12 slope. The span of the beam is 2 m, and both ends of the beam are fixed hinge. The cross section of the specimen is HM150 × 100 × 6 × 9 mm, the steel beam is Q235 steel. The break point is welded. Table 1 shows the main parameters of the test specimen sample. All the specimens are manufactured by Yanji Zhongcheng steel color plate Co. Ltd.

The beam end welding is 20 mm thick. The curved end plate is mounted on the rotating shaft of the test frame with the completion of the hinge bearing.

In order to avoid the local damage of the steel beam, a 6 mm thickness stiffener was set at the loading point. In concentrated loading, it is straightened outside by another load plate transfer to the flange welding. In order to prevent the plane instability, the end of the beam is made with two arc plate ends with a 50 mm interval (Figure 1b). All steels used are Q 235.

Table 1. Main parameters of tests.

Specimen	Slope	Girder Dimension	span	Rise/mm	Attended mode
L1	1/5	HM150 ×	2000 mm	20	soldering
L2	1/6	100 ×6 ×		16.7	
L3	1/8	9 mm		12.5	
L4	1/10			10	
L5	1/12			8.3	

2.2 Test stand

In order to reach the constraint form of fixed hinges on both ends of the beam, the specimen was placed on a test stand as shown in Figure 2a. The test frame length is 2.3 m, hot rolled H steel section of size HM150 × 100 × 6 × 9 mm were used and the steel material for is Q 345 steel. The two beam end bracket plates are welded on the surface of the 20 mm thick plate. The end plate of the two ends of the test specimen beam can be installed on the φ 45 mm rotating shaft to realize the restraint of the fixed hinge support. To prevent the end of the test frame web buckling deformation and failure, the lower part of a web plate is provided with three stiffeners. The bearing design is detailed in Figure 2b.

2.3 Material performance test

The test material is cut from the same batch as the steel beam specimens. The interception of

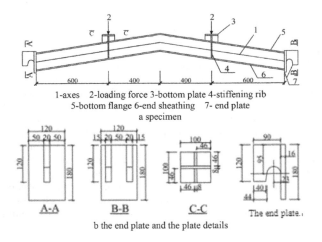

1-axes 2-loading force 3-bottom plate 4-stiffening rib
5-bottom flange 6-end sheathing 7- end plate
a specimen

b the end plate and the plate details

Figure 1. Specimen beams and end-plate detail.

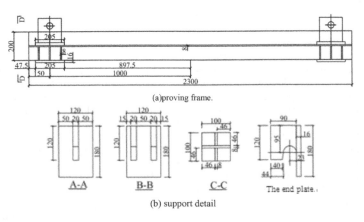

(a)proving frame.

(b) support detail

Figure 2. Test frame and support details.

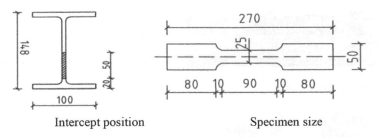

Intercept position   Specimen size

Figure 3. Material test specimen.

Table 2. Test results.

Thickness/ mm	Yield strength f^y/Mpa	Ultimate stress f^u/Mpa	Yield ratio	Modulus of elasticity E/10^5 MPa	elongation/ %
6	295	435.8	0.68	2.04	26.7
6	294	432.7	0.68	2.06	27.8
6	307	440.8	0.7	2.05	27.8
Average value	299	436.4	0.69	2.05	27.4

straightening materials, the body shape of the specimen material and all dimension requirements are all based on the test methods (standard test method for tensile of metallic materials at room temperature, GB/T228–2010). The specimen dimensions and intercept straighteners are shown in Figure 3. The affected area zone of the flame was removed from the web by cutting the specimen from the web. The affected area of the test specimen cut from the web through the planning edge was eliminated from the flame heat. All the material specimens and the thickness were 6 mm. Using the tensile machine, the evaluation of the material mechanical properties was done with a force rate of 10 N/mm². S (Andrews, 1912). The test results are presented in Table 2.

2.4 Adding bearing scheme and arrangement of measuring points

All the tests were completed in the Laboratory of Structural Engineering in Yanbian University. During the test, a 500t electro-hydraulic servo pressure testing machine was used to carry out one direction symmetrical static loading until the piece was damaged. The formal loading prior to the per-loading and per-loading value is 20 KN. At that time, the working state of the inspection equipment includes the pull shift reading meter strain-gauge. After the completion of the per-loading unloading to zero, all the readings are cleared and restarted for recording data. The main content of the liquid is: (1) liquid fixed test piece at all levels

of load point of the vertical pull, (2) determination of the strain under the external load at the point and the beam end support, (3) determination of ultimate bearing capacity of specimens at the same time considering the influence of fabrication error test, the specimens are installed at both ends of the two pull displacement meter measured in support of horizontal tension shift.

3 EXPERIMENTAL RESULTS

3.1 Experimental phenomena

For the 1/12 slope the test load reaches 110 kN. When the flange is crossed there is the sound of crackling extrusion and oxidation skin cracking. The apparent mode of cracking is the crack and break at the point direction bending (Figure 5). A loading to 130 KN results in a drawing on the flange break point. An uplift buckling slightly upturned the ends. After adding 162 KN, the compression buckling deformation of super flange is rapidly developed and it is obvious that there is a loud noise. The buckling failure of the beam on the top of the beam is lost and the buckling failure mode of the supper flange is shown in Figure 5b. The slope 1/10 and 1/8 of the steel beam damage is processed and the 1/12 slope is similar. The difference of the bearing capacity is minimal.

When the 1/5 slope, a load of 140 KN is applied on the shape of the cross and the concentrated force is loaded on the outer edge of the upper flange. When the loading reaches 156 KN, the span flange ends first. The buckling is slightly upturned with the increase of the load on the top of the flange. As shown in Figure 5c, when the buckling load reaches 203 kN the bending deformation of the flange is too large. For 1/6 and 1/5 slopes, the beam test similarly reaches the ultimate bearing capacity. But, the carrying capacity of 188 kN of the uplift height of the middle flange is significantly larger than that of the lower slope. In the test process, the quality of the weld join is good and no damage is found. Figure 5d shows that at the end of the test beam plate, we can observe a plate rotation and

Figure 4.   Specimen failure mode.

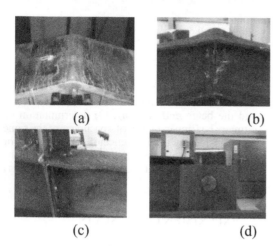

(a)   (b)

(c)   (d)

Figure 5.   Local damage form.

also we can see that the test fixed hinge bearing design is feasible.

3.2   Ultimate bearing capacity

The ultimate bearing capacity of the specimen is shown in table 3. The 1/5 slope has the most bearing capacity with a maximum bearing capacity of 203 KN compared to the 1/8, 1/10, 1/12. The carrying capacity with the decreasing slope of and the particularity of the two hinged chevron steel beam is related to two characteristics. The failure of the steel beams is caused by the interaction of variable pitch and the axial force which causes the cross section to yield to stress. With the increasing slope of 1/5 and 1/8 the axial force of girder cause the axial force of section to increases the stress. As the limit load decrease from 1/8 to 1/12 slope, the axial force of the beam caused by stress decreased and the limit load changes less through experiments.

3.3   Vertical displacement (deflection)

The relationship curve of load and the test piece including the effect of the displacement of the bearing is obtained as shown in Figure 6. The deflection curve of the specimen is shown as two stages along with the increase of the load. First the elastic and second the plastic stage. When the load is smaller, the deflection increases linearly with the increase of the load (Huang, 2015). When the load

Table 3.   Ultimate load.

Specimen	1/5	1/6	1/8	1/10	1/12
Ultimate load/kN	203	188	156	158	162

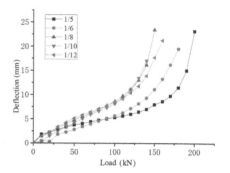

Figure 6.   Load deflection curve.

Table 4.   Beam end displacement (Concentrated force P = 50 kN).

Specimen	1/5	1/6	1/8	1/10	1/12
Finite element/mm	0	0	0	0	0
Test/mm	0.9	0.8	1	0.8	0.5
percentage/%	0.045	0.04	0.05	0.04	0.025

does not exceed 120 kN the deflection changes linearly and the specimen goes into elastic stage along with the further increase of the load. The increase of the vertical deflection is accelerated, the curve is nonlinear and the stage transitions from elastic to plastic stage.

3.4   Beam end pull

During the pull test, the fixed hinge supporting the bearing has moved. But due to the manufacturing errors at both ends of the outer support, there are different degrees of level pull shift. The horizontal beam end pull shift value is shown in Table 4. The finite element analysis of the beam end horizontal pull shift is zero (ideal gauge fixed hinge support). The control error of the beam length direction is ±3 mm and the error of the beam end in this experiment is within the allowable error 1/6 slope of the test piece. The end of the beam pull is 0.8 mm for grader span 0.04% and the deflection increased by 2.86 mm (3.6 times the end of the beam pull) and by 4.31 mm (4.2 times the end of the beam pull). There is a large increase in the level of the beam end to the deflection of the steel beam and the increase degree with the decrease of the slope. The smaller the slope is the more sensitive to the increase of the beam to the end, as shown in Table 5.

Table 5. Vertical deflection (Concentrated force P = 50 kN).

Specimen	1/5	1/6	1/8	1/10	1/12
Finite element/mm	1.64	2.14	3.09	3.92	4.59
test/mm	4	5.0	8	7.3	8.2

4 THEORETICAL PLATE (MENG, 2008)

Articulated steel beam is statically indeterminate in its structure (Bao, 2010). There are two kinds of characteristics this beam. Under the vertical load in addition to the moment, there will be a strictly greater level of support reaction force. The large axial pressure on the horizontal support reaction force, existing to sever reverse bending deflection, may suppress the development by mechanics (Wei, 2015). The two hinged chevron steel beam is divided into plates. The bending deformation and axial deformation are mainly considered and the shear deformation is ignored. The chevron steel beams in the symmetrical concentrated loads elastic range supports the horizontal thrust. The mid span deflection and bearing relative displacement of structure influence are discussed below. Some calculation formulas and other related problem are put forward in the paper for reference.

4.1 Deformation without considering the relative displacement of bearing

1. Horizontal thrust of support

Figure 7 shows the curve of β, n, i, 1 variable relationship with a base loading of 50 KN. The smaller the values of n are, the closer to the concentration of the point of the cross β is. Under the same load, the maximum level thrust bearing (when n = 2), reaches a maximum of 6.64 at the horizontal force of the support for the 332 kN. The law changes the thrust of the two hinged straight arch foot with the change of the slope. The β value for the foot thrust increases with the decrease of the slope to a critical gradient. The β value for the foot thrust with the reduced slope decreases the size of the critical gradient of the beams. For a given concentrated force and sectional characteristic under the same load and point of action, the larger the span is the greater the value of β and the level of the foot thrust are.

2. Mid span deflection

Figure 8 shows the deflection curve of deformation caused by the deformation of the cross section. It is mainly caused by the bending deformation (the two curves are basically coincident).

The deflection caused by the axial deformation is negligible and can be ignored. The bending deformation can only be considered when calculating

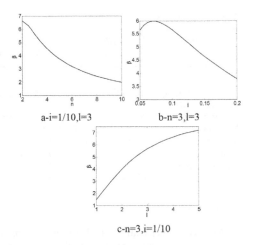

a-i=1/10,l=3   b-n=3,l=3

c-n=3,i=1/10

Figure 7.   β, n, i, l variable curves.

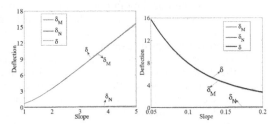

Figure 8.   Deflection curve.

the deflection under the same load. The deflection action position decreases as the slope increases as the span increases.

4.2 General analysis

Formulas (1) and (2) are used to calculate the reaction coefficient of the horizontal bearing of the test case, as shown in Table 6.

With 1/6 slope (p = 50 kN) as an example, in the horizontal displacement of 1 mm (l/2000) in the case of HΔ = −79.7 kN, HA = 168,5 kN, and H = 68,8 kN, the horizontal support displacement is 2 mm (l/1000) (HΔ = −159 kN and H = 9.1 KN). When the horizontal displacement of the bearing is 2.11 mm, the support reaction is 0. The condition that the shape of the steel beam is lost to the arch is changed into a simple supported beam, as shown in Table 7.

For the consideration of the horizontal displacement, the bearings under the cross deflection values are given by the formulas (3) and (4). They can be found through the ratio. The calculated results of the theoretical calculation are the smallest, the value is the largest and the finite element is in the center at design time. The value of theoretical calculation is small which should be taken into account.

Table 6. β values.

Test specimen	1/5	1/6	1/8	1/10	1/12
β	3.09	3.37	3.67	3.70	3.60

Table 7. Mid-span deflection by considering horizontal displacement.

Specimen	1/5	1/6	1/8	1/10	1/12	1/5
Value/mm	4.5	4.9	6.8	6.5	7.0	4.5
Finite element/mm	4.6	5.1	7.1	6.9	7.6	4.6
Experiment/mm	4	5.0	8	7.3	8.2	4

5 CONCLUSION

1. Mansard steel beam with both end hinges is a new type of structure form. It also has the beam and arch characteristics and its failure mode is the bending buckling failure on the top flange of apex point. It is recommended that this kind of welding break points be strengthened to improve the bearing capacity.
2. The horizontal thrust coefficient β is related to the slope, the span, the position of the load and the characteristics of the cross section. In the same case, the thrust of the arch foot increases as the slope decreases. The β value and the arch foot thrust decreases as the slope decrease after the critical slope of 1/13. The deflection decreases as the slope increases. Therefore, considering the support reaction and deflection deformation of the new structure formation, the design of the application should be selected according to the critical slope.
3. The horizontal displacement of the support can reduce the huge support reaction in the mid span deflection. The release of the horizontal displacement of the bearing is more sensitive. A small displacement can produce a Δ/2i deflection increment. The increment increases with the decrease of the slope. The design of the fixed hinge supports is feasible to ensure adequate outer support and moving in the other direction displacement limit test. But to achieve higher accuracy and quality of production, the test needs further improvement.

REFERENCES

Andrews E. S. ElemenmO, Principles of reinforced concrete construction. Scott, Greenwood and Sons, 1912.

Bao Shihua. Structural mechanics. Wuhan: wuhan University of Technology press, 2010.

Cai Yiyan, Chen Youquan. The problems of design and construction of steel inclined arch supported on concrete columns. [J] Steel Structure, 2009,24(10):32–35

Huang Binsheng, Jiang Meng, Huang Guzhong. Experimental Research on Folded Web I—shaped Girder. Journal of Highway and Transportation Research and Development, 2008,8:87–91.

Huang Junchao, Cui Wenyi, Zhang Zhixin. FEA of rotating performance for break point of mansard steel beam. [J] Steel Structure, 2015,30(6):42–45.

Leng Jiabing. Design of Steel Roof and Concrete Substructure world [J], 2013,34(1):65–67.

Meng Huanling. Experimental research on X-shaped joints of concrete-filled steel tube column. Journal of building structure. 2008(z):62–67.

Newmark N. M., Siess, C. P., Viest, I. M, Test and analysis of composite beams with incomplete interaction. Experimental Stress Analysis, 1951,9(6):896–901.

Nie Jianguo, Wang Hongguan, Tan Ying, Chen Ge, Experimental study on composite Steel-HSC beams. Journal of Building Structures, 2004,2:58–62.

Richard Yen J. Y Composite beams subjected to static and fatigue loads. Journal of structure engineering 1997,123(6):795–771.

Wan Chenyao. Approach to structural system of reinforced concrete column and steel beam for single story industrial mill building [J]. Industrial Construction, 2007,37(z):387–390.

Wang Wenda, Han Linhai, Tao Zhong, Experimental research on seismic behavior of concrete filled CHS and SHS columns and steel beam planar frames. Journal of Building Structures, 2006,6:48–58.

Wei Chaowen, Chen Youquan. Design of Straight Steel Arch with Hinge-supported on Concrete Columns [J]. Building Structure, 2015,35(2):31–33.

Yukio Maeda, Research and Development of Steel·Concrete Composite Construction in Japan form 1950 to 1986, Composite Construction in Steel and Concrete III: Proceedings of an Engineering Foundation Conference, 1996, 20–40.

Zhou Xue-jun, XU Yuan, LJ Guo qiang, Yang Rongqian, Hou HPtao. Theoretical Analysis and Experimental Study on Bending Capacity of I-Shaped Steel—Concrete Composite Beams with Corrugated Webs, Progress in Steel Building Structures, 2013,12:36–40.

Advances in Energy Science and Equipment Engineering II – Zhou, Patty & Chen (Eds)
© 2017 Taylor & Francis Group, London, ISBN 978-1-138-71798-5

Simulation and visualization of salty soil water and salt migration

Li Chai, Jiancang Xie, Jichao Liang, Rengui Jiang & Hao Han
State Key Laboratory Base of Eco-hydrologic Engineering in Arid Area, Xi'an University of Technology,
Xi'an, Shaanxi, China

ABSTRACT: Agricultural soil salinization has become more and more serious with environment change, which restricts the healthy development of agriculture and plant growth. With the rapid development of industry and urbanization in Shaanxi, situated in the northwestern part of China, it has become more and more difficult to achieve balance in the requisition of cultivated land. The task of controlling saline alkali soil is imminent. In this paper, we designed and developed a three-dimensional Visualization Simulation System (3D-VSS) for the simulation of agricultural salty soil water and salt migration in saline alkali soil. The results show that the 3D-VSS has strong ability of information integration and expression, user-friendly experience, interactive performance, and strong ability of multisource data analysis. For example, in Lubotan, Shaanxi, the 3D-VSS has been analyzing and simulating the observed data for the last 10 years. It has achieved high-efficiency integration of multisource data and visual demonstration of multiyear data, multiple soil depths, and multiple indices, satisfying the requirements of visualization, credibility, and availability, which provide good guarantee and experience for the improvement and management of saline alkali soil. Therefore, it has good application and promotional value.

1 INTRODUCTION

The rapid economic development has greatly promoted the social progress worldwide. With economic development in the catchment, resource is consumed very quickly. Climate and environment are getting more and more worse, and the discharge of pollutant is continuously increasing. Global environment problems have become a great threat. Soil salinization is an important issue in the process of human development. Saline alkali soil has pH > 9 and contains too much salt. It can suppress the growth of crops and increase the complexity of soil formation. Firstly, the soil can easily become a source of salt accumulation, thereby leading to high mineral composition, which is influenced by natural conditions. Secondly, if the water table in a region is high and it is in a long-term drought, the water level would rise by capillary action and get evaporated in air while the salt would remain in the soil because of less supply of water. This is the procedure of formation of saline alkali soil.

Saline alkali soil is found worldwide. China ranks third in the total area of saline alkali soil after Australia and the Soviet Union. It is mainly distributed in the northwest, northeast, and coastal areas of China in 17 provinces, affecting more than 240 million acres of cultivated land, which is more than 10% of the total area of cultivated land in the country. Shaanxi is one of the six regions in China with wide distribution of saline alkali soil

($\sim$153200 hm^2), which is 7% of the total land area of Shaanxi. With the development of industry and urbanization in Shaanxi, it is difficult to achieve a balance between cultivated lands occupied and cultivated land developed and reclaimed.

Many previous studies have investigated the reuse and reclamation of saline and alkaline soil to improve the productivity. Changhe Chen proposed a technique for the desulfurization of gypsum, which is widely used in northeast China. The development of information technology has facilitated the simulation of saline alkali soil (Ma, 2011). Zenghui Ma developed a simulation software, Hydurs-3D, which can realize the system simulation and numerical simulation of soil water and salt transport. Furthermore, it can achieve real-time simulation, dynamic control, curtailing test cycles, and cost-saving (Zhao, 2006). Jian Zhao adopted a new method, which combined water quality model and Geographic Information System (GIS) technology to evaluate the nitrogen loss in farmland drainage in the irrigation district of Yinnan, Ningxia. Haibo Liu developed a three-dimensional movement model of groundwater based on Gravity Measurement System (GMS). The model is propitious to predict the changing of water-level fluctuation in irrigation (Liu, 2011). This paper focuses on three-dimensional visualization and simulation technology for salty soil water and soil migration. The data of different time and spatial position are fitted, and the contour line of water and salt transport is drawn to demonstrate the simulation process using pictures and animation. The three-dimensional Visualization Simulation System (3D-VSS) is designed and developed to support the salt movement in saline alkali soil.

2 DESIGN OF THE 3D-VSS

2.1 *System framework*

Visual simulation has a strong ability of information integration, performance capability, user-friendly expression, interactive performance, and strong secondary data analysis capability. This paper used the visual simulation method to design the transport simulation system of saline water and salt transport in saline alkali soil. The main requirements are: (1) to establish the mathematical model suitable for saline water and salt transport; (2) to establish the simulation model to characterize saline water and salt transport; (3) to develop the saline water and salt transport simulation program; and (4) to verify the feasibility of the program and the accuracy of the established model and experimental results. The system aims to achieve data visualization performance for water and salt transport, which generally follows the Client/Server (C/S) structure, which has a high computing power and interactive capabilities. The system uses a common interface to interact with other data and to display the evaluated data. Data sources include saline water and salt measured data. The visual representation is divided into two types: pictures and animations. In the case of compatibility with a variety of data sources, the system is designed with strong visual representation for the main objective, using contour drawing, filling, data fitting techniques, and developing on a general purpose personal computer based on J2EE platform. It has good data reliability, ease of operation, bug compatibility, scalability, human–computer interaction, and so on, which can provide users with intuitive visualization services.

The framework of the 3D-VSS is shown in Figure 1. A modular method for connecting, reading and maintaining the data source was used in the system to ensure the effective integration of system and external data. Different parts of the 3D-VSS are loosely coupled to achieve data reading and saving, complete data integrity, and consistency checks, to ensure system data availability and system scalability. The mainly requirements include: (1) to import the monitoring data and external analog data into the appropriate database and store in a Microsoft Excel file; (2) to analyze the data density and fit data from the two dimensions of space and time to enhance the visual

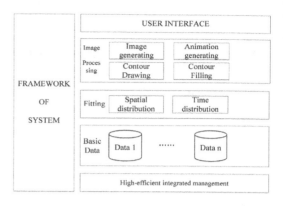

Figure 1. Framework of the 3D-VSS.

effect; (3) to plot the data after fitting, drawing, and filling contours, which increase the credibility of the visual system; and (4) to enable the user to demonstrate the processes of water and salt transport in the display area with locally generated images vividly. Water and salt transport animation display can be used for parallel comparison, such as the same year with different depths or the same depth in different years (Chen, 2012).

Firstly, the 3D-VSS fits the monitoring data with external data and then uses the fitting data to draw and fill contour by using pictures or animations. To achieve visual representation of water and salt transport processes and to establish the database required for storage based on the data source, adopt data fitting, performance data, and the approach users pay separate from each other to build the structural framework of the system. The data source layer provides the raw data for the system, and the database layer can store user setting parameters, fitting intermediate data and results of data. The fitting data layer provides database layer with a standard interface to fit through the calculation of mathematical function. The data presentation layer may receive instructions from the user interaction layer according to user requirements and follow the instructions to organize data in the database and to draw and fill contour lines. The user interaction layer is responsible for encoding and transmitting instructions, mainly including browsing and generating animations and pictures, as well as setting various parameters of the simulation.

2.2 *Data reading and data-storage*

Firstly, the 3D-VSS establishes a connection with the Excel file, with sheets containing simulated and measured raw data. Secondly, stripe to read data in accordance with the rules of reading data and then store in the corresponding database until all the data have been read. Before the Excel spreadsheet recorded the latitude and longitude of soil samples, soil information at different depths represented by each latitude and longitude is recorded at the right-hand side of the table. The borrow depth is measured in centimeters. The combination of borrow depth and latitude and longitude can uniquely identify the spatial position of a soil sample in the study area, which facilitates data classification. In this paper, visual simulation includes three indicators: water content, total salt content, and pH.

The system can deal with Excel files and meet the above specifications through a data source interface. By combining the norm and the above functions in the Excel file, we can analyze and read the file. Because the system uses a modular layered architecture, different modules interact with each other by passing data. To maximize the maintenance of system stability, the data source interface should read the Excel file into memory at first and then pass it to the system database layer for its unified storage. This architecture has a high demand for data storage mode, which must be easy to parse the database layer (Xie, 2012).

2.3 Data fitting

The basis of the fitting method includes spatial fitting and time fitting. Spatial fitting is the hypothesis based on "the first law of geography": for the spots that are closer in terms of spatial position, the possibility for the similarity of characteristic values is bigger; on the contrary, for the spots that are farer in terms of spatial position, the possibility for the similarity of characteristic values is smaller. Spatial fitting is also a kind of data fitting for discrete points and it will be fitting the numerical value in the other time according to the numerical value in the time of its own. Time fitting is a kind of fitting for the data of discrete points within the time while fitting beyond the time is a kind of fitting for the data of discrete points beyond the time. The purpose of time fitting is mainly to increase the time density or to optimize the time distribution. Since the distributions of the measuring points are affected by various factors, sometimes they are not evenly distributed in the study area. The measured data are not good representations. In addition, the uneven distribution of the data, such as dense or deletions, may cause non-smooth jagged and other phenomena when drawing contour or reading data, which will seriously affect the visual expression, greatly affect contour accuracy, and even reduce application value. The visualization effect will be affected when there is no clear schedule for the data acquisition time, such as large acquisition time intervals or different acquisition time span. Fitting the data at different

time and spatial location can significantly improve contouring accuracy. For example, in measuring the total salt in saline soil, the higher a total salt content of the sample, the greater the likelihood that collected soil samples are near the total salt content, which has certain relevance. The total salt content of the soil samples in other places is far away from the sampling points that may also be high or low. The samples are independent. Use the inverse distance interpolation method to achieve fitting space of linear discrete points, given as follows:

$$c(x_0) = \sum_{i=1}^{n} \tau_i c(x_i) \sum_{i=1}^{n} \tau_i = 1 \qquad (1)$$

where the weighting factor is determined by the function $\phi(d(x, x_i))$, which is the weighting function of the inverse distance.

Time fitting refers to fit the value of the other time according to the numerical value of the existing time. This paper aims to study the change of distribution of saline water and salt over time. It is particularly important to obtain a more intensive and equally spaced time series. For the data series with time characteristics, since time itself has a linear characteristic and saline water and salt transport is slow, the time set is supposed to have a good set of linear features. Fitting time is calculated by linear interpolation between two adjacent linear areas as follows:

$$\bar{c}(x_i) = \frac{1}{2}(c(x_{i-1} + x_{i+1})) \qquad (2)$$

For the data of different indicators, the above formula is used to time fitting. The data density is increased and time distribution is optimized to facilitate visual presentation and analysis of water and salt transport processes (Wang, 2009).

2.4 Contours and animation generation

Contour maps are used to show time and spatial distribution of different indices during the transport of saline water and salt. In this paper, the inner grid contours propagation algorithm is used. Drawing contour mapping with different colors between two adjacent contour regions can be more intuitive and significantly show the data change trends, which facilitate the analysis of the data. First, attribute information on the each contours of contour maps should be obtained, to determine the color and correspondence of the contour, structure contour area, and all the contours and attribute step values, and different colors must be used to fill contour zone in order to obtain a contour plot to meet the requirements (Jiang, 2011; GA, 1979).

3 TYPICAL APPLICATION

3.1 General situation of Lubotan

Lubotan is located in Puchengand and Fuping counties, Shaanxi, with a total area of 12.24 hectares, with widths of about 30 km in the east–west direction and 1.5–7 km in the north-south direction. The test aims to study the changes in soil moisture, salinity, pH value, and ion content in different plots, different depths, and at different times so as to summarize the changes in the laws of key factors. Early experiments include: long-term regular or irregular observation, water and salt observation experiments in fields, and water–soil interface indoor simulation experiments. Experimental works include: measurement of total salt content using conductivity meter that measures 10 g +50 ml conductivity of the soil extracts, according to the relationship between the total salt content and electrical conductivity converted into total salt content; determination of soil pH value, by recording pH value of 25 g + 25 ml of the soil suspension measured by pH meter to directly obtain the soil pH; and determination of organic matter content by COD_{cr} method. Cl^- was determined by silver nitrate volumetric method, and SO_4^{2-}, Ca^{2+}, and Mg^{2+} were determined by volumetric method. The HCO^{3-} indicator was determined by double titration. Through a 9-year transformation, soil salinity, pH value, and ionic concentration have significantly changed. Thus, the above-mentioned saline water and solute transport simulation system 3D-VSS was applied to the analysis and simulation of preliminary experimental observation and monitoring data of Brine State Beach saline land to verify the feasibility and practicality of the system (Chen, 2011; Han, 2009).

3.2 Effect assessment

3D-VSS realizes the visualization of transformation and movement processes of water and salt in the way of importing external data and fitting and by adopting the method of drawing and filling of contour line. Therefore, the system is designed as a form in which the data source, system database, data fitting, data representation, and user interaction are separated. The layer of data source is divided into simulated data source and measured data source, which will provide original data for the system. The database of the system is used to store the Excel original data, the intermediate data in data fitting process, and the results data

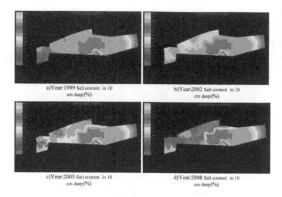

a)Year:1999 Salt content in 10 cm deep(%)   b)Year:2002 Salt content in 10 cm deep(%)

c)Year:2005 Salt content in 10 cm deep(%)   d)Year:2008 Salt content in 10 cm deep(%)

Figure 2. Visualization of total salinity changes at a soil depth of 10 cm in different years.

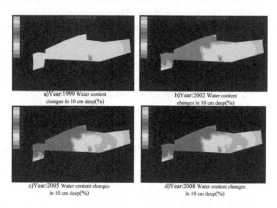

a)Year:1999 Water content changes in 10 cm deep(%)   b)Year:2002 Water content changes in 10 cm deep(%)

c)Year:2005 Water content changes in 10 cm deep(%)   d)Year:2008 Water content changes in 10 cm deep(%)

Figure 3. Visualization of moisture content changes at a soil depth of 10 cm in different years.

as well as some parameters set by the users. The data fitting is composed of some math functions, which provide a standardized interface for database layer so that the fitting of original data can be carried out. The data presentation layer receives the instruction from the interaction layer of users and organizes the data in database according to the instruction as well as draws and fills the contour line. Its main functions include browsing and generating animation and pictures and setting of a variety of simulated parameters. Figures 2 and 3 show the changes in the total salt and water content, respectively, in the saline water and solute transport using the 3D-VSS at a soil depth of 10cm in different years, to provide users with the data of water and salt transport of a particular index in the same depth and in different years. Visualization of index migration is achieved using contour maps so as to meet user demand for a flexible analysis system of contour maps providing locally stored function. When the user visits again, we can read data directly from the local cache to increase the system response speed. To indicate the migration process more vividly at different depths or in different years, the multiple images are compressed and made into animation as a gif file. Showing the process of saline water and salt transport in animation is more intuitive and vivid.

4 SUMMARY AND CONCLUSIONS

In this paper, we designed and developed the saline water and salt transport simulation system 3D-VSS, which can show directly the changes in water and salt transport at different times. By entering the simulation and monitoring data into the system and by using data fitting method, we can improve the accuracy by means of three-dimensional visualization technology. The results show that the simulation system achieved high-efficiency integration of multisource data and visual demonstration of multiyear data, multiple soil depth, and multiple indices, to meet the requirements of visualization, credibility, and availability which provide good guarantee and experience for the improvement and management of saline alkali soil. Therefore, it has good application and promotional value.

ACKNOWLEDGMENT

The study was partly supported by the National Natural Science Foundation of China (51509201, 51479160, and 41471451), Project funded by China Postdoctoral Science Foundation (2016M590964), Scientific Research Program Funded by Shaanxi Provincial Education Department (15JK1503), Young Talent fund of University Association for Science and Technology in Shaanxi, China (20160217).

REFERENCES

Chen Feng, Jiancang Xie, Bo Sun, Ni Wang. Numerical Simulation of Soil Water and Salt Transport Based on Hydrus Model in the Storage Condition [C]. Proceedings of 2011 International Symposium on Water Resource and Environmental Protection (ISWREP 2011), 2011, 2: 830–833.

Haibo Liu. Groundwater simulation and groundwater vulnerability assessment in LuoHui irrigation scheme [D]. Xi'an: Xi'an University of Technology, 2011.

Jian Zhao. Modeling study of nitrogen losses based on GIS in YinNan irrigation district [D]. Xi'an. Xi'an University of Technology, 2006.

Jichang Han, Jiancang Xie, Tao Wang, et al. experimental observation of saline alkali of soil in saline land after changing drainage to impoundment in Lubotan of Shaanxi [J]. Transactions of the Chinese Society of Agricultural Engineering, 2009,06: 59–65.

Rengui Jiang, Jiancang Xie, Jianxun Li, et al. Analysis and simulation of flood inundation based on digital earth [J]. Computer Engineering and Applications, 2011, 47(13): 219–222.

Tianqing Chen. Research on saline-alkali soil integrated service platform and applications in Lubotan Shaanxi [D]. Xi'an University of Technology, 2012.

Wei Wang, Jiancang Xie, Junming Huang, et al. Study on the New Improvement Mode for the Saline Land [J]. Journal Of Water Resources And Water Engineering, 2009,05: 117–119.

Wood GA, Jennings LS. On the use of spline functions for data smoothing [J]. Journal of biomechanics, 1979, 12(6): 477–479.

Yufeng Xie. Water and salt movement simulation and servicecomposition of saline land [D]. Xi'an: Xi'an University of Technology, 2012.

Zenghui Ma, Jichang Han, Jiancang Xie, et al. The modeling method for the transport of water and salt based on Hydrus 3D in Lubotan Shaanxi [J]. Shaanxi Journal Of Agricultural Sciences, 2011,01: 62–65.

Advances in Energy Science and Equipment Engineering II – Zhou, Patty & Chen (Eds)
© 2017 Taylor & Francis Group, London, ISBN 978-1-138-71798-5

Multisource data integration and three-dimensional visualization of the interbasin water transfer project

Xiang Yu, Jiancang Xie, Rengui Jiang, Dongfei Yan & Jichao Liang
State Key Laboratory Base of Eco-hydrologic Engineering in Arid Area, Xi'an University of Technology, Xi'an, Shaanxi, China

ABSTRACT: The interbasin water transfer project is a very complicated system, which involves many stakeholders. Thus, it is of great importance to use effective information technologies to facilitate decision making in the early period of project planning. Therefore, the Digital Earth Platform (DEP), a three-dimensional visualization virtual reality interactive environment, was proposed. The multisource data integration and the fusion scheme of the project area were investigated. These multisource data include spatial geographic information and remote sensing images. Taking the project of water transfer from Jia to Han in Shaanxi as an example, both the application of multisource data integration and the loading and rendering of multisource data on the DEP were realized, which should assist the decision-making process. The results show that the multisource data fusion based on DEP can provide a new method for the design and planning of water transfer project, which is more efficient, convenient, and scientific. Thus, it is valuable for popularized application and has great application prospect.

1 INTRODUCTION

Interbasin water transfer project is crucial to solve the water crisis of water-deficient area by transferring water resources from water-affluent areas, which directly stimulates the economic development of water scarce areas. Many areas in northern China are water-deficit, and water resource is unevenly distributed in both spatial and temporal regions. The south-to-north water diversion project has been implemented to overcome water scarcity in the northern regions, enhance their water resource carrying capacity, increase resource allocation efficiency, and improve the local drinking water quality. It is a promising approach to alleviate the bottleneck restriction of water resource in the development of urbanization in the northern areas of China (BEN, 2000; CHEN, 1986).

Interbasin water transfer project will redistribute the relevant river basin water resources and may change the lives or ecological environment of the people living in the surrounding areas. Therefore, it is significant to make an overall analysis and comprehensively coordinate the contingent contradiction between adjoining areas. In other words, in the early period of the water transfer, project planning is essential, and further optimization and comparative selection of both water diversion line and the construction arrangement are needed, provided the area of the project covers the scope and impact on the environment. How to use effective information technologies to assist in the decision-making process? This has important realistic meanings to increase the efficiency of the project, save investment, and reduce the running cost of the construction (LIU, 2006).

In this study, the integration and fusion of multisource data in the water transfer project was investigated using the three-dimensional visualization technology. Integration technology of multisource data was used to develop the integrated application of the interbasin water transfer project. It is helpful to provide effective multisource data organization and management and decision support.

2 KEY TECHNOLOGIES

2.1 DEP framework

DEP is composed of a series of model frameworks in progressive layers with the support of computing power provided by a high-performance computing environment. On the basis of a topographic and geomorphic model within the space constructed by remote sensing image data and digital elevation data, it establishes a basic three-dimensional simulation and virtual reality environment of water conservancy elements. Then, by seamlessly connecting GIS and remote sensing images, it establishes a 3S (GIS, RS, and GPS) integration platform to realize the connection between GIS and remote sensing images and enhance the service efficiency of the

GIS system. After that, it integrates a variety of spatial information standards and water conservancy standards and effectively incorporates spatial information resources and river management data by means of title pyramid and data middleware so as to improve data access. Finally, supported by a comprehensive service interoperability environment and oriented toward business application, it provides an interface for the virtual reality environment and then facilitates the implementation of business application services including basic data query, emergency response, and decision making. In the process of implementing the DEP, a parallel algorithm is first established for projection transformation, and then, remote sensing images are sliced as required by the pyramid model to set up remote sensing image tile description and a WMS service environment. A three-dimensional GIS service environment oriented toward digital earth is set up to index space image tiles and build a neighborhood search mechanism and hierarchical model so as to realize the transformation algorithm between pixels within the visual field and the latitude and longitude, and implement a variety of application interfaces, including digital earth interoperability service interface (e.g., amplification, minifying, and translation), spatial information upgrading and maintenance interface, dynamic spatial information creation interface (e.g., text production and pixel drawing), digital earth extension and control interface (e.g., flight and positioning), application service data interface for integrating business resources, non-latitudinal and non-longitudinal project transformation interface, GIS connection interface, and high-definition video interface. In addition, in the framework of DEP, by effectively integrating data communication and exchange protocol, a caching mechanism is set up to guarantee smooth information display (Boschetti L, 2008; Bell, 2007; JIN, 2007). The framework of the DEP is shown in Figure 1.

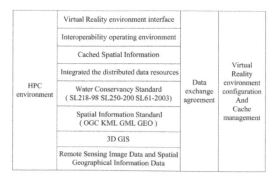

Figure 1.   Framework of the DEP.

2.2   *Multisource data integration*

1. GIS data integration

GIS data are acquired from aerospace, aviation, satellite, and remote sensing as well as the geological map, topographic map, and other professional image data obtained by scanning digitizing map image. Then, these data structure, format, and projection coordinates transform processing and extract the data information, finally generating 4D data (digital raster graphic and digital line graphic, digital orthophoto map, digital elevation model). The data must follow the code and standard of national unity. Then, it is integrated and applied.

The technologies of image pyramid and data integration were used to create the earth model and to realize the integration of GIS data of water conservancy and digital earth information with map server through network map service following OGC specifications. Compatible with the characteristics of open source WMS (Web Map Service) and WFS (Web Feature Service) following OGC standards, it can support PostGIS, Shapefile, MapInfo, and other data formats and can output network map into OpenLayers, KML, PDF, SVG, GeoRSS, and other formats. Map server is compiled using pure Java, based on J2EE framework and Selvlet framework that can be deployed in the Web application server. It itself contains complete configuration management components and can realize operation of management, configuration, and sample these three components after security login of users through fixed URL access, in which the configuration component can realize settings and generation of network map service.

Map Styled Layer Descriptor (SLD) is the key technology of map server to realize network map service. It is a text file based on XML formulated by OGC and is used to describe the representation of characteristic data. It can be used in cross-platforms and can accurately control representation of each characteristic element in the layer and display space elements' flexibly in different conditions. Through the fusion of above information, we can conveniently display rivers, notes, roads, and many other layers of GIS in the mode of transparent overlaying on the digital earth (Garnett, 2002).

2. Standardized interoperability environment

Digital earth information service platform has a large amount of data, many of which are from heterogeneous platform. A standardized data interoperability environment is needed for the effective integration of data in different formats. Combining the characteristics of the World Wind and application requirements, this paper adopts open geographical data interoperability specification as the data interoperability standards. Open geographical data interoperability specification (hereinafter referred to as OpenGIS) was proposed by the American "Open Geospatial Consortium" (OpenGIS Consortium, OGC for short.), whose goal is to provide a set of universal components with open interface specifications. Developers develop interactive components according to these specifications, which can realize transparent access among different types of geographical data and geographical processing methods. Data manipulation based on the OpenGIS specifications has a better scalability, portability, openness, interoperability, and easy manipulation. OpenGIS enhances interoperability of space information and location technology in the network environment. It describes space interface specifications, and developers develop interactive components according to these specifications and realize transparent access among different kinds of geographical data and geographical processing methods.

Standardized interoperability, which will be implemented in this paper, mainly includes WMS, WFS, WPS, GML, and other specifications. WMS mainly provides image releasing, superimposing, and rendering of map data. WFS mainly specifies data editing of OpenGIS simple elements so that the server and client can communicate at the element level. WPS is a kind of service providing space algorithm and processing; it also realizes the business logic of GIS using web service and extends space algorithm on this basis. GML is a kind of specification for geographic information encoding based on XML and mainly used for modeling, transmission, and storage of geographic information (Jiang, 2011).

3. Remote sensing image integration

Remote sensing image data integration is mainly composed of image pyramid roof tiles. Image pyramid refers to the reference in the same space, where the user can store according to the needs and display in different resolutions, from coarse to fine, and data amount from small to large pyramid structure. Image pyramid hierarchy contains multiple data layers, where the bottom layer stores the original of the highest resolution data. With the increase in the number of pyramid from bottom to top, the resolution of the data reduces in turn. If the image can be seen as a pyramid abstract iterative transformation that is associated with filtering and sampling process, the iteration process to the image of the original data is decomposed into different resolution image tiles, as shown in Figure 2, thereby becoming suitable for the organization of raster data, image data, and multiresolution DEM data. Under the environment of digital earth, tile image is composed of all kinds of spatial information, image information, and rendering to the surface of a sphere earth model is the smallest

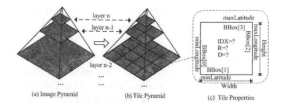

Figure 2. Image pyramid model.

Figure 3. High-resolution remote sensing image of the project of water transfer from Jia to Han.

unit, which has a fixed size, raster images, and six groups {independent IDX, D, R, W, H, and BBOX}. Through the tile image, we can construct a multiresolution hierarchical model (described in XML). Under the uniform spatial reference and according to the level of resolution, this model can establish a set of remote sensing images and elevation data. The image of the entire picture or deposited DEM data is divided into blocks, and according to the latitude and longitude records, it establishes a spatial index, whose position changes in response to different resolution data access and storage requirements, thereby exchanging time cost through space, to improve the efficiency of access of DEP. In addition, the image pyramid hierarchical data provided by the management technology, can conduct massive geographic data organization and management easily and can be implemented easily, which has nothing to do with the data content and display area of multiresolution smooth display. Figure 3 shows the process of water transfer from Jia to Han areas with high-definition image data after the tiles (XML description as shown below), established in accordance with latitude and longitude record each tile spatial index values, the implementation of the different resolution of data storage and access requirements, and through the way of image pyramid provides project area topography (Xie, 2011).

The XML description of the high-definition image multiresolution hierarchical model is presented as follows:

```xml
<?xml version="1.0" encoding="UTF-8"?>
<LayerSet Name="yjrh" ShowOnlyOneLayer=
"false" ShowAtStartup="true" xmlns:xsi="http://
www.w3.org/2001/XMLSchema-instance" xsi:no
NamespaceSchemaLocation="LayerSet.xsd">
<QuadTileSet ShowAtStartup="true">
    <Name>yjrh</Name>
<DistanceAboveSurface>0</Distance-
AboveSurface>
<BoundingBox>
    <North><Value>33.472690192667</Value>
</North>
    <South><Value>33.109948297894</Value>
</South>
    <West><Value>105.985107421875</Value>
</West>
    <East><Value>106.534423828125</Value>
</East>
</BoundingBox>
<Opacity>255</Opacity>
<TerrainMapped>true</TerrainMapped>
<RenderStruts>false</RenderStruts>
<ImageAccessor>
    <LevelZeroTileSizeDegrees>0.08789062 </Level-
ZeroTileSizeDegrees>
    <NumberLevels>5</NumberLevels>
    <TextureSizePixels>512</TextureSizePixels>
    <ImageFileExtension>png</ImageFileExtension>
    <ImageTileService>
    <ServerUrl>http://127.0.0.1:8080/tiledImages/
senderPNG.jsp</ServerUrl>
    <DataSetName>yjrh</DataSetName>
    </ImageTileService>
</ImageAccessor>
</QuadTileSet>
</LayerSet>
```

3 TYPICAL APPLICATION

3.1 Study area

The water transfer project from Jia to Han was designed to transfer water from JiaLing River to HanJiang River by drawing water near the LueYang county, located upstream of JiaLing River, along about 30 km then downstream into JuHe (Baihe) River, a branch of HanJiang River. This is the complement and extension of the project of water transfer from south to north in China and the project of water transfer from Han to Wei in Shaanxi province. It also plays a significant role in ensuring the volume of water transfer and mitigating the impact in swap area. The multiyear average water diversion of this project is 1.35 billion cubic meters, compared to the volume of the project of water transfer from Han to Wei, which is 1.55 billion cubic meters, accounting for 88%. Therefore, it can eliminate the influence of the project of water transfer from Han to Wei and the middle line of water transfer project from south to north. The project of water transfer from Jia to Han will reduce the operating cost and increase the assurance of cascade hydroelectric station on HanJiang River. In addition, it will also complement the volume of water diversion for the national middle line of transfer water from south to north, decrease the effect for the downstream ecological environment, and create conditions to optimize water allocation in the central area of Shaanxi province.

3.2 Multisource data visualization services

First, setting up the DEP of the water transfer project from Jia to Han provides the basic visualization service environment. Through the acquisition and processing, the project area of remote sensing image data and spatial geographic information data release server map data by using GeoServer, allowing the user to update, delete, and insert the feature data. Furthermore, GeoServer can facilitate rapid sharing of spatial geospatial information between users.

Interbasin water transfer project planning and design needs remote sensing image data and spatial geographic information data support, mainly related to the elevation, geology, rivers, pipelines, roads, buildings, and other important information. All of these data represent a layer. When we need information of some rivers, we can add the corresponding river layer. Therefore, according to the experience of the designer in the DEP, choose relevant data layer and provide auxiliary decision making for planning.

Finally, the DEP calls the GeoServer service for multisource data loading and rendering, to pro-

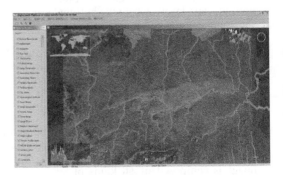

Figure 4. Remote sensing image layer and spatial geographic information layer loading DEP.

vide users with image and geographic information data visualization applications, geographic service, and map operation. Remote sensing image data and spatial geographic information data are loaded into the DEP, as shown in Figure 4.

4 SUMMARY AND CONCLUSIONS

Interbasin water transfer project involves complicated problems in a very complicated system. In this paper, we developed a digital earth platform, taking the project of water transfer from Jia to Han as an example to build visualization service environment. Virtual reconstruction of the whole project and the surrounding terrain is carried out on the computer. These multisource data include remote sensing image data, DEM data and reservoir, river, roads, and other geographic information data. Through the study of the project of multisource data integration and fusion, we implement multisource data integration in the project area to provide a 3D visualization platform for the planning and design of the project of water transfer from Jia to Han. The results show that the DEP can be used for the multisource data integration for efficient organization, management, and to assist in decision making as well as to provide more efficient, convenient, and scientific new ways for water transfer project design and planning. However, a variety of steps in the application of this technique, such as measurement, data conversion, and data output, are likely to cause a number of errors, which need to be calibrated. Furthermore, the function of this technique tends to be confined to the assistance to the planning and design, and thus the results mainly depend on the planner rather on the technique.

ACKNOWLEDGMENTS

The study was partly supported by the National Natural Science Foundation of China (51509201, 51479160, 41471451), Project funded by China Postdoctoral Science Foundation (2016M590964), and Young Talent fund of University Association for Science and Technology in Shaanxi, China (20160217).

REFERENCES

Ben Ke-ping. Experience and prospect of large scale water diversion from neighbor basin overseas. Hunan Hydro and Power, 2000(6):26–29,34.

Bell, D.G., Kuehnel, F., Maxwell, C., Kim, R., Kasraie, K., Gaskins, T., Hogan, P., Coughlan, J., "NASA World Wind: Opensource GIS for Mission Operations", Aerospace Conference, 2007 IEEE on 3–10 March 2007 Page(s):1–9.

Boschetti L, Roy D P, Justice C O. Using NASA's World Wind virtual globe for interactive internet visualization of the global MODIS burned area product[J]. International Journal of Remote Sensing, 2008, 29(11):3067–3072.

Chen Chun-huai. Problems about inter-basin water transfer. Water Resources and Hydropower Engineering, 1986(6):6–8.

Jiancang Xie, Rengui Jiang, Jianxun Li, et al. Three-dimensional visualization system facing sudden water pollution incidents[J]. Journal of Natural Disasters, 2011, 47(13):219–222.

Jin Hong-chang, Wang Tie-qiang, Zhu Qing-li. Study on Electronic Sand Table System of the East Route of the South-to-North Water Diversion[J]. South-to-North Water Transfers and Water Science&Technology, 2007, 5(02):31–35.

Jody Garnett, "Welcome GeoServer [EB/OL]", http://geoserver.org/display/GEOS/Welcome.

Liu Qiang, Huang Wei, Sang Lian-hai. Discussion on inter-basin water transfer management of China. Journal of Yangtze River Scientific Research Institute, 2006, 23(6):39–43.

Rengui Jiang, Jiancang Xie, Jianxun Li, et al. Analysis and simulation of flood inundation based on digital earth [J]. Computer Engineering and Applications, 2011, 47(13):219–222.

Advances in Energy Science and Equipment Engineering II – Zhou, Patty & Chen (Eds)
© 2017 Taylor & Francis Group, London, ISBN 978-1-138-71798-5

Numerical simulation for the propagation laws of residual shock wave after protective door destruction in tunnel

Zhen Liao, Degao Tang, Zhizhong Li & Ruliang Zheng
School of Defense Engineering, PLA University of Science and Technology, Nanjing, China

ABSTRACT: Because of the development of modern high-damage weapons, it becomes much more likely for protective doors in tunnel to be destructed. Moreover, residual shock waves formed may cause severe damages to personnel and facility instrument in the tunnel. In order to study propagation laws of residual shock waves after the protective door is destructed, LS-DYNA dynamic analysis software is adopted to establish a 3D finite element model for tunnel and reinforced concrete protective door. In addition, the numerical simulation results are also compared to experimental results. It is also showed that during the propagation, peak pressure attenuation of residual shock waves is gentle, while obvious changes occur to wave forms. At the initial stage of propagation, a platform pressure phase of millisecond magnitude exists before the shock wave pressure rises to the peak pressure. However, while the platform pressure basically remains unchanged with the increase in propagation distance, its duration gradually drops and completely disappears ultimately. It is also indicated that destruction of protective doors and residual shock waves under the action of shock waves in tunnel can be rather accurately predicted by the dynamic finite element software LS-DYNA, which provides an effective method for further study of residual shock waves.

1 INTRODUCTION

As the chief means of implementing surgical strike in modern war, the emergence and application of precision guided weapon may attack the mouth of national defense tunnel engineering (L, 2001). Being limited by tunnel wall, shock wave propagating inside the tunnel has a high peak pressure and long duration, which causes much more serious damage effects than the shock waves propagating in the free field. Over the years, many scholars have conducted extensive research on the blast wave propagation and attenuation law in tunnel engineering worldwide. Kong et al. (Lin, 2012) studied the explosion shock wave propagation of different explosives inside tunnels and found that when the initial blast in tunnel is propagating at a far distance, a steady plane wave is formed, and the impulse is getting higher in some distance. Li et al. (L, 2007) conducted tunnel proportion model explosion experiments 20 times and analyzed the propagation law of waveform and impulse of shock wave and presented the empirical formula for shock wave impulse calculation. Miao et al. (M, 2015) established a three-dimensional numerical calculation model for explosive shock wave in tunnel based on the finite element software ANSYS/LS-DYNA and studied the effects of factors such as grid size, artificial viscosity coefficient, and tunnel length on

numerical calculation results. Yu et al. (Y, 2013) comparatively analyzed the parameter calculation methods worldwide for tunnel shock waves produced by tunnel internal explosion, and adopted the finite element software AUTODYN to numerically simulate the scale model test of a tunnel.

However, the constantly developing weapons technology brings a growing threat to tunnel projects. As the important protective equipment of the passageway of protective engineering, tunnel protective doors, especially the first protective door, face the threat of direct damage under the effect of shock wave in the tunnel. After the failure of protective doors, shock wave propagation continues to the internal tunnel in the form of blast leakage through protective doors, and may cause damage to personnel, facilities, and equipment in the internal tunnel. The propagation law of blast leakage may change to some extent due to the resistance of protective door. The propagation and attenuation law of shock wave in the tunnel has been studied more thoroughly based on numerical calculation methods, and many computational formulas of peak pressure and impulse for shock wave in the tunnel have been fitted out based on it (Welch, 1997; Scheklinski-Glück G, 1993; Y, 2003). However, there is little research on the property and propagation law of blast leakage after failure occurs in protective doors.

The objective of this paper is to study the interaction between protective doors and shock waves in the tunnel and the formation and propagation law of residual shock waves when damage occurs in protective doors using the LS-DYNA dynamic analysis software based on the experimental findings in Z (2007). A three-dimensional finite element model of the tunnel and reinforced concrete protective door has been established, which can provide a reference of residual shock wave for the protection of staff and facilities and the set of multiple protective doors in the tunnel.

2 NUMERICAL SIMULATION

2.1 The FE model

The cross section of the model tunnel was 067 m², and the length was 12 m, which was opened at both ends. A cubic charge with 4.6 kg mass TNT was placed in the center of the entrance to the tunnel, as shown in Fig. 1. The reinforced concrete door was set at a distance of 2 m from the entrance. The thickness of the protective layer was 20 mm. The strength grade of concrete was C30, and the reinforcement ratio was 0.2%. The configurations of reinforcements were 6@250 with a yield strength of 335 MPa. Structure dimension and reinforcements are shown in Fig. 2. Considering the requirements of computational efficiency and accuracy comprehensively, a 1/2 model has been established with symmetry. The total length of the model is 12.3 m, including 12 m tunnels and 0.3 m mouth air layer.

It is often assumed that there is no relative slip between steel and concrete because the duration time of blast load is very short (W, 2012). Thus, the separate finite element model of reinforced concrete protective door is established in this paper using the common node method. Lagrange grid is adopted for concrete while ALE grid for explosive and air. The grid size of concrete and air is 1 cm while that of air is 2 cm. The fluid structure interaction algorithm is adopted to simulate the interaction between the blast wave in the tunnel and the

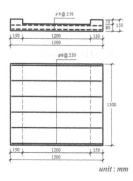

Figure 2. Structural dimension and reinforcement of the protective door.

reinforced concrete protective door. According to boundary restriction conditions of experiments conducted in Z (2007), the vertical wall and baseboard of the tunnel is fully constrained while vault surface is radially constrained so as to simulate the restriction effect of tunnel wall surface on blast wave. The corresponding constraints of the protective door outside the tunnel are imposed and the nonreflecting boundary condition is applied on the outer surface of air layer at the tunnel entrance and tunnel end to simulate infinite air field. The finite element model is shown in Fig. 3.

2.2 Material model

The *MAT_PLASTIC_KINEMATIC material model is used to simulate rebar (P. O. LS-DYNA, 2007; C, 2015). This model adopts Cowper–Symonds strain rate formula to consider the strain rate effect of rebar. The relationship between yield stress and strain rate can be expressed as follows:

$$\sigma_y = \left[1 + \left(\frac{\dot{\varepsilon}}{C}\right)^{\frac{1}{P}}\right]\left(\sigma_0 + \beta E_p \varepsilon_p^{eff}\right) \quad (1)$$

where σ_0 is the initial yield stress, $\dot{\varepsilon}$ is the strain rate, C and P are strain rate parameters, β is the hardening parameter, ε_p^{eff} is the effective plastic strain, E_p is the plastic hardening modulus determined by $E_p = E_{tan}E/\left(E - E_{tan}\right)$. As the improved version of K&C material model, the *MAT_CONCRETE_DAMAGE_REL3 material model is used to simulate concrete in LS-DYNA (P. O. LS-DYNA, 2007). Several concrete parameters, such as density, Poisson's ratio, the uniaxial compressive strength, and strain rate effect curve, should be inputted by the users. Other material parameters can be automatically generated. The calculation formula for strain rate effect curve can be found in C (2015). Because this model can get

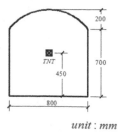

unit : mm

Figure 1. Cross section of the model tunnel.

pretty good dynamic response of concrete materials under high strain rate and has the characteristics of convenient parameter input, it is widely applied to simulate the damage of concrete under explosion shock wave in recent years.

The TNT explosive is modeled by *MAT_HIGH_EX-

PLOSIVE_BURN material model and the *EOS_JWL state equation in LS-DYNA (P. O. LS-DYNA, 2007; Y, 2008):

$$p = A\left(1 - \frac{\omega}{R_1 V}\right)e^{-R_1 V} + B\left(1 - \frac{\omega}{R_2 V}\right)e^{-R_2 V} + \frac{\omega E_0}{V} \quad (2)$$

where V is the relative volume, E_0 is the internal energy per unit volume, and A, B, R1, R2, and ω are the parameters of state equation.

The air is modeled by the *MAT_NULL material model and *EOS_LINEAR_POLYNOMIAL to define its state equation (P. O. LS-DYNA, 2007; Y, 2008):

$$p = C_0 + C_1 \mu + C_2 \mu^2 + C_3 \mu^3 + (C_4 + C_5 \mu + C_6 \mu^2)E \quad (3)$$

where E is the initial energy, ρ / ρ_0 is the ratio of current density to reference density, $C_0 = -0.1$ MPa, $C_1 = C_2 = C_3 = C_6 = 0$, $C_4 = C_5 = 0.4$, and air density is 1.3 kg/m³. The values of the remaining material parameters are shown in Table 1.

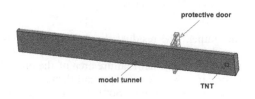

Figure 3. Finite element model.

2.3 Erosion algorithms

In order to stimulate failure phenomena such as cracking, stripping, and crushing of concrete materials subjected to blast wave and avoid numerical computation problem caused by excess distortion of grid, the software provides *MAT_ADD_ERO-SION keyword, which can be used to define an erosion algorithm to delete those elements distorted seriously. The maximum principal strain is usually adopted as the failure criterion for concrete subjected to blast load. The expression of failure strain is as follows:

$$\varepsilon_f = K_1 K_2 \varepsilon_s \quad (4)$$

where ε_s is the static tensile peak strain of concrete, which is usually set at 0.0002, and K_1, K_2 is amplification coefficient, which considers concrete softening effect and strain rate effect, respectively. In this paper, these values are taken as 5 and 10, respectively, according to Xu K (2006). When the maximum principal strain of concrete element equals 0.01, it will be deleted immediately.

3 CALCULATION RESULTS ANALYSIS

3.1 Analysis of interaction process between tunnel shock waves and protective door

Fluid structure coupling algorithms are adopted to simulate the whole process from the charge initiation at the entrance of the tunnel to the coupling effect between shock waves and protective door. The algorithm can also be used to describe dynamic propagation process of residual shock wave after the door is completely damaged. The overpressure nephograms of the shock wave in the tunnel at different times are shown in Fig. 4. These nephograms displayed the dynamic interaction process between shock waves and the protective door as well as the propagation of the residual shock wave.

Table 1. Material properties.

TNT	A (GPa)	B (GPa)	R1	R2	ω
	371.2	3.231	4.15	0.95	0.3
	E_0 (GPa)	ρ (kg/m³)	Detonation velocity (m/s)	C-J pressure (GPa)	
	7	1630	6930	21	
Rebar	ρ (kg/m3)	E (GPa)	υ	f_y (MPa)	
	7800	210	0.3	335	
	C	P	fs		
	40	5	0.12		
Concrete	ρ (kg/m3)	υ	f'_c (MPa)		
	2500	0.2	30		

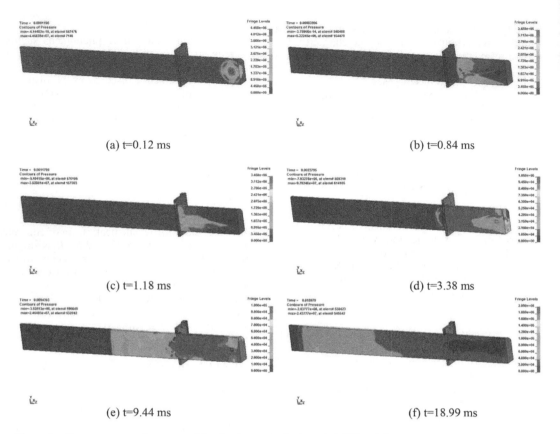

(a) t=0.12 ms

(b) t=0.84 ms

(c) t=1.18 ms

(d) t=3.38 ms

(e) t=9.44 ms

(f) t=18.99 ms

Figure 4. Overpressure nephograms of the shock wave in the tunnel at different times.

At t = 0.12 ms, detonation products spread from the explosion center to all around.

With the increase of propagation distance, a wave front with approximately spherical symmetry was formed. At t = 0.84 ms, reflection occurs when the shock waves encounter the tunnel wall. Because of the straight wall and arch-shaped tunnel, the reflected wave is relatively complex and pressure field near the entrance of tunnel is in a state of disorder. At t = 1.18 ms, the front wave propagating in the direction of internal tunnel begins to interact with the protective door, forming the reflection wave, which propagates in a direction opposite to that of the initial shock wave. At the same time, pressure of the protective door increases from zero to maximum value at a very short time. However, the shock wave has not entered the interior of the tunnels through the protective door. At t = 2.38 ms, by the continuous effect of shock wave, the elements of protective door turn into failure due to the deformation exceeding the threshold value of concrete material. A small amount of shock wave runs into inside tunnel through the hole of the protective door and continues to spread inside

the tunnel. The residual shock wave front has an approximately spherical symmetry. At t = 9.44 ms, with the enlargement failure area of the protective door, larger shock waves spread through the door to the internal tunnel. Some fragments generated by the shock wave can also be thrown at a certain speed to the interior tunnel, which may threaten the safety of the personnel and equipment in the tunnel projects. The protective door is seriously damaged, thus loosing the further ability to resist the explosion shock wave. After t = 18.99 ms, the residual shock wave propagates to a certain distance and gradually forms a stable plane wave front. The propagation law of residual shock wave is similar to tunnel shock wave without a protective door when the propagation distance is far enough.

3.2 Analysis of overpressure time history curve of residual shock wave

In order to conduct quantitative study of attenuation law of the residual blast wave, elements within the range of 1.75–6.75 m away from the protective door are selected for analysis based on the above

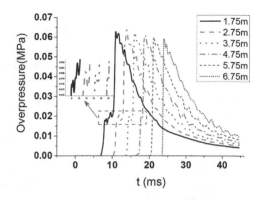

Figure 5. Overpressure time history of residual shock wave.

analysis. The center of these elements are 0.45 m above the ground and as high as the blast center. The interval of these elements along the tunnel length is 1 m. The overpressure time history curves are shown in Fig. 5.

From Figure 5, we can find that, with the increase of propagation distance, the form of residual overpressure time history curve changes significantly. Within a distance of 4.75 m away from the protective door, the residual shock waves have an approximation of overpressure platform stage before rising to the peak pressure, which value is about 30% of the peak overpressure. The platform stage has a duration of 2.3 ms in 1.75 m, accompanied by a slight concussion. However, as the residual shock wave continues to spread in the internal tunnel, the duration of the platform stage and the oscillation amplitude gradually decrease. When the residual shock wave reached a distance of 5.75 m away from the protective door, the platform stage completely disappears. After that, the residual shock waveform is similar to the typical tunnel shock wave without a protective door. The rise time of the pressure almost equals zero. After reaching the peak pressure value, the overpressure decays like saw tooth. The plateau pressure of the residual shock wave and the peak overpressure of the corresponding point in Z (2007) were measured to be 0.01 and 0.05–0.06 MPa, respectively. Plateau pressure by numerical calculation is about 0.02 MPa, while the peak overpressure is about 0.059–0.064 MPa, which quite agrees with the experimental values.

From Figures 4 and 5, after being reflected by the tunnel surface for many times, the blast wave generated by blast at the tunnel formed a complicated and disordered pressure field. After the front wave acted on the blast side of protective door, reflections occurred and the protective door began to deform. The propagation of the reflected blast wave to the tunnel reduced its pressure. As acting time increased, some parts of the protective door turned out to be failure due to excess plastic strain, which led to an increasingly expanding penetrated hole. At this moment, the pressure was transmitted to the inner side through the hole. However, as the initial hole was smaller than the cross section of the tunnel, the pressure of the residual blast wave was reduced, which made the value of pressure in the platform phase lower than the peak overpressure value. When the door is completely broken, the shock wave in the front of protective door was transmitted to the inside tunnel, and constantly pursued the shock wave in front of it. As a result, the duration of the pressure platform stage gradually decreased and finally disappeared.

4 SUMMARY

In this paper, numerical simulation was conducted to study the generation, propagation, and attenuation laws of residual shock waves. Some conclusions are as follows:

1. LS-DYNA finite element dynamic analysis software can well predict the characteristics of residual shock wave, providing an effective method for the further study of residual shock wave.
2. When the protective door is subjected to the tunnel blast wave that is much greater than its resistance, it might be heavily damaged. Some energy, in the form of residual blast wave, will continue to propagate to the interior of the tunnel. The restriction effect of protective door and tunnel would lead to a disordered pressure field of residual blast wave at the initial stage. However, after reaching a distance far away from the protective door, the residual blast wave would form a stable plane wave.
3. The peak overpressure of the residual blast wave attenuates slowly in the propagation process, and the waveform will be changed significantly. At the initial stage of propagation, a platform pressure phase of millisecond magnitude exists before the shock wave pressure rises to the peak pressure. However, with the increase in propagation distance, the platform pressure value remains unchanged, while its duration gradually drops and completely disappears ultimately.

REFERENCES

Chaoyang M, Xiudi L, Wei S. Three-dimensional Numerical Simulation and Experimental Verification of Blast Wave in Tunnel. Journal of Logistical Engineering University 31, 5(2015).

Journal of PLA University of Science and Technology **8**, 5(2007).

Kezhi Y, Xiumin Y. Shock waves propagation inside tunnels. Expl Shock Wave **23**, 1(2003).

Lin K, Jianjun S, Zhirong L. Test Study on Explosion Shock Wave Propagation of Different Explosives inside Tunnels. Initiators & Pyrotechnics, 7(2012).

P.O. *LS-DYNA Keyword User's Manual Volume II: Material Models* (Livermore Software Technology Corporation, California, 2007).

Scheklinski-Glück G. Blast in tunnels and rooms cylindral HE-charges outside the tunnel entrance (Proceeding of the 6th Symposium on the Interaction of Nonnuclear Munitions with Structure, Floride, 1993).

Wei W. *Study on Damage Effects and Assessments Method of Reinforced Concrete Structural Members under Blast loading* (National University of Defense Technology, Chang Sha, 2012).

Welch C R. *In-tunnel air blast engineering model for internal and external detonations* (Proceeding of the 8th International Symposium on Interaction of the Effects of Munitions with Structures, Virginia, 1997).

Wenhua Y, Yadong Z. Study on Propagation Laws of Explosion Shock Wave in Tunnels. Explosive Mater **42**, 3(2013).

Xiaojun L. *Conventional Weapons Destruction Effect and Engineering Protection Technology* (The Third Engineer Scientific Research Institude of the Headquarters of the General Staff, Luo Yang, 2001).

Xiaowei C, Yi L, Xinzheng L. Numerical investigation on dynamic response of reinforced concrete to impact loading. Eng Mech **32**, 2(2015).

Xin Y, Shaoqing S, Pengfei C. Forecast and Simulation of Peak Overpressure of TNT Explosion Shock Wave in the Air. Blasting **25**, 1(2008).

Xiudi L, Yingren Z, Yunmu Z. Scale model tests to determine blast impulse from HE-charges in-tunnel. Explosive Mater **36**, 3(2007).

Xu K, Lu Y. Numerical simulation study of spallation in reinforced concrete plates subjected to blast loading. Comput Struct 84(2006).

Yingren Z, Xiudi L. Model tests to determine In—tunnel blast leakage with protective doors destruction.

Advances in Energy Science and Equipment Engineering II – Zhou, Patty & Chen (Eds)
© *2017 Taylor & Francis Group, London, ISBN 978-1-138-71798-5*

Design and treatment of landslide engineering in the Three Gorges Reservoir area

Xu Zhang & Zhuo-Ying Tan
School of Civil and Environment Engineering, University of Science and Technology Beijing, Beijing, China

Bing-Yong Gao
Chinaroads Communication Science and Technology Group Co. Ltd., Hubei Branch, Wuhan, China

ABSTRACT: Landslides frequently occur in the Three Gorges Reservoir area. The water level in this area has been in long-term changes and fluctuations, while the stability of the landslide mass is in dynamic changes. According to the on-site engineering investigation and indoor experiments, the basic rock and soil mechanic parameters of the landslide mass are obtained, and the stability analysis of the regional landslide mass engineering under the condition of the stable water level and dropping water level is carried out. In the natural condition, it is in a stable condition. In the working condition of dropping water level, the stability coefficient of the landslide is low. Both the main landslide mass and the front secondary landslide mass are likely to slide. Targeting at the actual characteristics of engineering, different preventing and treating measures are analyzed. By combining the antisliding supporting with draining and protecting slope, relevant processing and designing methods are proposed. The treatment plan is economical and reasonable. After being treated, the stability is improved and enhanced.

1 LANDSLIDE SITUATION

Majiagou landslide occurred at the foot of Woniushan Mountain in the east coast of Yangtze River. The east coast of Zhaxi River, the branch stream of Yangtze River, is 2.1km away from the Yangtze River's estuary and is the low valley between medium-high gorges mainly formed by erosion and partially by corrosion. Its geographical coordinates are N31°01′08″–31°01′17″, E110°41′48″–110°42′10″. In terms of its administrative division, it belongs to 8th Team, Pengjiapo Village, Guizhou Town, Zhigui County, Hubei Province. Guizhou Town is located in the rainstorm area of the Three Gorges. The large and intensive precipitations are the important factors that cause landslide in this area.

Majiagou landslide mass consisted of two adjacent landslide masses: (1) landslide's front part faces Zhaxi River Valley, and is 135 m under the water level of the river. It is a project involving hydraulic system below 156 m. According to the results of investigation, the first landslide is currently in a stable state. When the reservoir's water level rises to 156 m, it is under stable conditions. The front of the second landslide mass faces Majiagou Valley. The bottom of the valley is narrow and the front of the sliding mass is located 156 m above the water level. This project does not involve hydraulic

systems below 156 m. When the water level of the reservoir reaches 175 m, only part of the landslide mass of the lower reach of Majiagou is affected by water. According to the results of investigations, the second Majiagou Valley landslide mass is in a stable state overall at present and when the reserved water is over 156 m. This paper mainly investigates the first landslide mass.

2 STABILITY EVALUATION

According to the results of the investigation of engineering [1], both the natural and saturated unit weights of the landslide are obtained in the experiment of digging holes and injecting water on the site. During calculations, the numeric averages of both the values are respectively obtained.

Natural density of slip mass is $\gamma = 21.1$ kN/m^3 and the saturated density is $\gamma_{sat} = 21.9$ kN/m^3.

Combining comprehensive inverse computations with the experiment's results and taking reference from the indoor test of mudstone, the stability of the landslide and the strength parameters of the thrust calculation are as follows:

Natural condition: $c = 18$kPa, $\phi = 14.5°$; saturation condition: $c = 16$kPa, $\phi = 12.5°$.

The stability calculation results of the first landslide's 2–2′ cross section in different working

Table 1. Stability calculation results of the first landslide's 2–2' cross section.

Water level of reservoir operation	Working condition	Load combination	The minimum safety factor	Stability coefficient
Static water level	1	Weight + surface load + static water level of 139 m + 5 days of heavy rain once 20 years	1.15	1.12
	2	Weight + surface load + static water level of 156 m + 5 days of heavy rain once 20 years	1.15	1.07
Dropping water level	3	Weight + surface load + water level drops from 162 m to 145 m + 5 days of heavy rain once 20 years	1.10	1.06
	4	Weight + surface load + water level drops from 175 m to 145 m + 5 days of heavy rain once 20 years	1.10	1.04

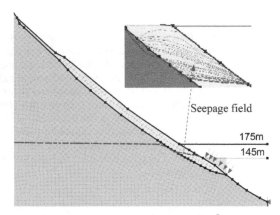

Seepage field

175m
145m

Figure 1. Seepage calculation result.

conditions are shown in Table 1, and the seepage field of the landslide is shown in Figure 1.

The calculation results show that the mass of the first Majiagou landslide is currently in a stable condition, and when the reservoir's retained water is over 156 m or the water level changes, this landslide is under a stable state and even a critical state. Therefore, the landslide may be under-stable when the retained water is over 156 m.

3 DESIGN AND TREATMENT

For different landslide mass projects, the treating means are also different. For example, the case presented in Zhang (2016) adopted antisliding poles to support and block the middle part of the slope mass, which achieves significant outcomes; the case presented in Zhang (2016) adopted bolt engineering to support and protect the part where the shearing force of sliding is large. According to the characteristics of sliding engineering, draining can effectively prevent water on the surface from penetrating the landslide. The presence of less amount of water in the landslide mass and on the sliding surface blocks the surface runoffs in the upper stream of the landslide area and removes the surface water within the sliding area, which drained over 80% of surface runoffs from the sliding area out during rainy periods. Its engineering measures are relatively easy and less expensive; however, they achieve significant outcomes. The draining engineering includes the peripheral retaining ditch, slant retaining ditch, vertical torrential tank, and absorption basin. Moreover, it includes treating measures, such as retaining wall, anchor cable frame, antisliding pole, boring, and grouting.

According to the technical requirements of designing, the following measures are used to treat the landslide.

1. Draining system—intercepting ditches and torrential gutters

It consists of two lines of horizontal intercepting ditches and four lines of vertical draining gutters. Position setting: the first line of horizontal intercepting ditches are located in the inner side of the first village-level road on the Guishui Road and runs from east to west; the second line of horizontal intercepting ditches are constructed along the second country-level road's inner side. Vertical draining gutters are arranged between two lines of horizontal intercepting ditches. Vertical intercepting ditches: rectangle with dimension 0.4 m × 0.4 m (Fig. 2). Horizontal draining ditches: trapezoids with top length, bottom length, and height of 1.8, 1.0, and 0.8 m, respectively (Fig. 3). Torrential gutters: rectangle with width and height of 0.8 and 0.5 m, respectively (Fig. 3).

2. Slope-protecting and—retaining wall

The retaining wall is set on the inner side of Guishui Road and is used to support and protect the slope toe on the inner side of the road.

3. Antislide piles

Plan 1: Set into two lines

The first line consists of anchoring anti-sliding poles, which are set on the inner side of Guishui Road and is used to block the remaining sliding force over this part. The anchoring section is 10 m.

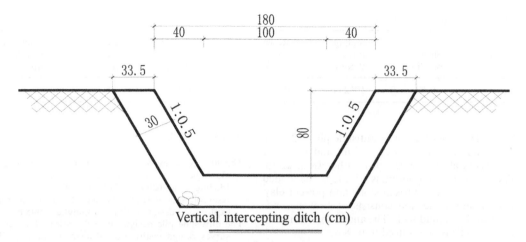

Figure 2. Design of vertical intercepting ditch.

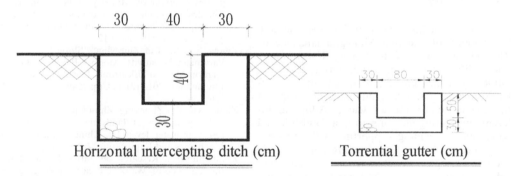

Figure 3. Design of horizontal intercepting ditch and torrential gutter.

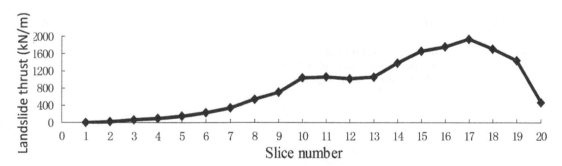

Figure 4. Landslide thrust line of the first landslide's 2–2' cross section.

The second line consists of common antisliding poles and is set on the platform that is 183 m above the bottom side of Guishui Road. It is used to block the remaining sliding force between the upper line of poles and this part. After filling the pole and leveling it, the land can be utilized. The position of this line of poles is between the No. 11 and No. 12 sections. This part is still in the sliding phase. Without considering the antisliding force of the earth before the pole, the remaining sliding force is 1062.66 kN/m and the design load is 1063 kN/m, as shown in Figure 4 and Table 2.

Plan 2: Set into one line

Design one line of common antisliding pole on the first landslide mass and set it on the platform that is 200 m above the bottom side of Guishui

Table 2. Design load of pile body.

Cross section	Slice position	Sliding force (kN/m)	Resistance force (kN/m)	Residual thrust (kN/m)	Design load (kN/m)	Rock-socket length (m)
2–2′	11–12	1062.66	0	1062.66	1063	8

Road. The cross section of antislide pile is 2×3 m; the distance between piles is 7 m, and there are 17 piles in all, whose length is either 18 or 22 m. The C30 concrete pouring is adopted. The loaded section consists of broken stone and powder clay. The anchoring section consists of broken sandstone and soft mud rock. The anchoring section of antislide pile is guaranteed to be 8 m.

4 SUMMARY

By adopting the limited equilibrium slice method, the stability of the first Majiagou landslide of Zhigui County in the Three Gorges Reservoir area under different working conditions is analyzed. Targeting at measures for treatment and the different treatment and designing plans are proposed to carry out comparative research. Adopting different engineering measures to treat landslide and combining antisliding supporting with measures for protecting the slope & draining can effectively prevent geological disasters from happening and safeguard people's lives and assets, which have important practical significance.

ACKNOWLEDGMENT

This study was financially supported by the National Natural Science Foundation of China (Grant No. 51174013 and Grant No. 51574015).

REFERENCES

The 906 project exploration design institute of Qinghai. Geologic survey report of the control engineering of MaJiagou landslide in ZiGui county in three gorges reservoir area of Hubei Province [R]. The 906 project exploration design institute of Qinghai, 2005.6

The anti-slide pile design and calculation. The second survey design institute of Ministry of railways [M]. China railway publishing house, 1983.3

The geological hazard control headquarters in three gorges reservoir of Ministry of land and resources. The design technical requirements of the third period of geological hazard control engineering in three gorges reservoir [S], 2004.12

Zhang Xu and Tan Zhuoying: "Slope Excavation and Parameter Sensitivity Analysis Based on Grey Correlation Method" Electronic Journal of Geotechnical Engineering, 2016 (21.12), pp 4549–4558. Available at ejge.com

Zhang Xu, Tan Zhuoying, Zhou Chunmei. Seepage and stability analysis of landslide under the change of reservoir water levels [J]. Chinese Journal of Rock Mechanics and Engineering, 2016, 35(4): 713–723.

ZhouChun-mei. Research on anti-slide pile design of the landslide of WanZou area in three gorges reservoir area [D]. PH.D Thesis. China university of geosciences, 2007.

Advances in Energy Science and Equipment Engineering II – Zhou, Patty & Chen (Eds)
© 2017 Taylor & Francis Group, London, ISBN 978-1-138-71798-5

Infiltration analysis to evaluate the stability of two-layered slopes

Jia-dong Huang
Guangzhou University, Guangzhou, Guangdong, China
Guangxi Transportation Research Institute, Nanning, Guangxi, China

Hua Tian
Guangxi Transportation Research Institute, Nanning, Guangxi, China
Guangxi University, Nanning, Guangxi, China

Zhong-zhi Zhu
Guangxi University, Nanning, Guangxi, China

ABSTRACT: Shallow slope failures are common in China. Using a one-dimensional infiltration model and an infinite slope analysis, an approximate method to determining the effect of infiltration on the stability of two-layered slopes is proposed. The slope stability factor–depth of the wetting front curve was calculated. The results revealed the failure mechanism of two-layered slopes.

1 INTRODUCTION

Rainfall leads to the development of a perched water table, a rise in the main groundwater level, and surface erosion. Furthermore, the unit weight increases with the increase in moisture content. Moreover, the instability of the unsaturated residual soil slopes during wet periods is common worldwide (Vargas Jr, 1986). There are many shallow failures, and the failure surfaces are usually parallel to the slope surface. Therefore, a two-layered slope was studied to estimate the effect of infiltration on the stability of the two-layered slopes.

In the analysis of rainfall-induced two-layered slope failure, the soil hydraulic properties should describe the physical phenomenon of water behavior characteristic in unsaturated soil (Mukhlisin, 2014). Furthermore, soil moisture distribution has a great influence on the property of soil strength. Therefore, it is necessary to select the compatible soil hydraulic properties model.

Many domestic and foreign scholars have conducted a large number of studies on the rainfall infiltration process of the slope. Huang conducted an experimental investigation of rainfall criteria for shallow slope failures. Rahimi studied the effect of hydraulic properties of soil on rainfall-induced slope failure (Rahimi, 2010).

However, these studies assume that the slope is composed of homogeneous soil, and there is little research on rainfall infiltration of stratified slope. Therefore, in this paper, the stability of the rainfall in a two-layered slope under infiltration process is analyzed.

2 THEORETICAL CONSIDERATION

The factor of safety can be calculated by the limit equilibrium method as shown in Figure 1.

For two-layered soil slopes with a surface layer of thickness H1, the shear resistance associated with the net normal stress and the matric suction within the slice mass can be characterized by the modified Mohr–Coulomb failure criterion for unsaturated soil as follows (Cho, 2009):

$$F_s = \frac{\tau_f}{\tau_m} \tag{1}$$

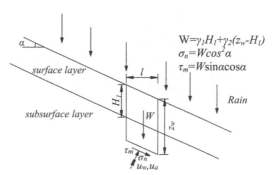

Figure 1. Infinite slope of a two-layered soil profile.

$$\tau_f = c' + (\sigma_n - u_a)\tan\phi' + (u_a - u_w)\tan\phi^b \qquad (2)$$

$$F_s = \cfrac{2c'}{[\gamma_1 H_1 + \gamma_2(z_w - H_1)]\sin 2\alpha} \\ + \left[\cfrac{\tan\phi'}{\tan\alpha} - \cfrac{2u_a\tan\phi'}{[\gamma_1 H_1 + \gamma_2(z_w - H_1)]\sin 2\alpha}\right] \\ + \cfrac{2(u_a - u_w)\tan\phi^b}{[\gamma_1 H_1 + \gamma_2(z_w - H_1)]\sin 2\alpha} \qquad (3)$$

where F_s = factor of safety, τ_f = shear strength at the corresponding point, τ_m = shear stress at any point along the slip surface, γ_1 = unit weight of top-soil, γ_2 = unit weight of deep soil, W = weight of a slice with unit width, α = slope angle, c' = cohesion intercept, u_a = pore air pressure, u_w = pore water pressure, σ_n = total normal stress, $(\sigma_n - u_a)$ = net normal stress, $(u_a - u_w)$ = matric suction, ϕ' = effective angle of friction, and ϕ^b = angle that defines how the shear strength increases with the increase in matric suction.

The rise in the groundwater table induces the condition of a flow parallel to the slope and hydrostatic state, resulting in the following expression for the pore pressure at a depth of z_w from the water table:

$$u_w = \gamma_w z_w \cos^2\alpha \qquad (4)$$

where γ_w is the unit weight of water. Then, Eq. (1) can be reduced to the classical solution for infinite slope in the saturated slope.

Figure 2 shows the conceptual two-layered soil profile used in the derivation of Moore's model. H_1 is the thickness of surface layer and $\Delta\theta_1$ is the initial moisture deficit. The soil is characterized by hydraulic conductivity (k_1). Below this layer, the soil is homogeneous and semi-infinite with an initial moisture deficit ($\Delta\theta_2$). The soil is also

characterized by hydraulic conductivity in the wetted zone of k_2 (Lee, 2009).

3 SLOPE SELECTION AND OVERVIEW

In order to facilitate research and analysis, slope selection is indeed necessary. When slope is chosen, the following basic principles should be followed: (1) Typical spatial and temporal distribution of two-layered structure slope should be reflected when landslide is triggered by rainfall; (2) the landslide would have a greater impact on the lives of local people; (3) the place should have many slope failures; and (4) there is much research information available for reference.

Target regions of the slope located in tectonic denudation belonged to eroded hilly topographical area. Furthermore, the meteorological data are detailed enough. They contend monthly precipitation from 1961 to 2010, each month's maximum daily precipitation in every year, and the number of days each month's daily precipitation levels ≥ 0.1 mm, ≥ 10 mm, ≥ 25 mm, ≥ 50 mm in every year. The maximum daily rainfall in this area is 180.8 mm while the maximum monthly rainfall is 300 mm. Drilling results show that the average thickness of the surface layer of the slope is 1 m and the subsurface is bedrock. Therefore, in this study, two types of soil, namely weathered soil and bedrock, were considered. Figure 3 shows finite element model of slope used in this study. The geometry of the slope of the soil is assumed to be in two layers: surface layer and subsurface layer. The soil slope has an inclination of 38° and the surface layer has a thickness of 1 m. The geometry was discretized by triangular mesh with 1296 elements. Extreme rainfall is necessary to compute the intensity of the rainfall that should be applied to a slope corresponding to a specific rainfall duration and geographical location. The ratio of rainfall intensity is 2.78×10^{-6} m/s, so it would take 13h for wetting front to reach the subsurface layer. The

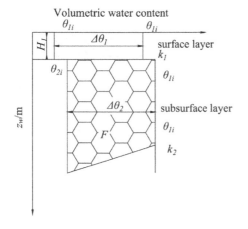

Figure 2. Conceptual water content profile for two-layered soil.

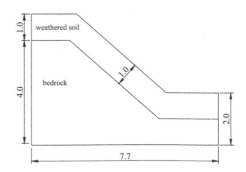

Figure 3. Finite element model of slope (unit: m).

mechanical properties of the soils to calculate the factor of safety are presented in Table 1.

It was assumed that the initial matric suction was 30 kPa through the depth and that the thickness of the surface layer was 1 m. Figure 4 shows the Soil–Water Characteristic Curve (SWCC) and the relative hydraulic conductivity function for weathered soil. The properties have been selected as a representative for weathered soil since two-layer soil systems that consist of weathered soil and bedrock have been studied due to the contrasting hydraulic characteristics of the soils. Soil–water characteristic curve for weathered soil may be relatively steep such that as soon as the soil de-saturates, the hydraulic conductivity drops dramatically.

Table 1. Physical and mechanical parameters of the soil at various stratigraphies.

Layers	E_s (MPa)	k (10^{-6} m/s)	γ (kN/m³)	c (kPa)	φ'	φ^b
Weathered soil	11	2.5	20	22	28	23
Bedrock	52	0.96	23	28	30	26

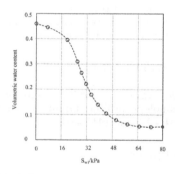

(a) Soil–water characteristic of weathered soil

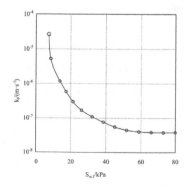

(b) Hydraulic conductivity curve of weathered soil

Figure 4. Hydraulic properties for analysis.

4 RESULTS AND DISCUSSION

As shown in Fig 5, when $t_2 = 13$ h, the wetting front reaches the subsurface. However, the surface of the bedrock did not reach saturation. With the infiltration of rainwater into the bedrock, the subsurface gradually becomes saturated. Because of the low penetration of the bedrock, the time of water production is very short. Time has a linear relationship with wetting front depth. When the wetting front depth reaches the contact surface of weathered soil and bedrock, the curve becomes very smooth, which indicates that the infiltration rate of precipitation becomes low.

Figure 6 shows the typical results from a one-dimensional finite element analysis using SEEP/W under the condition of k1 < I (ratio of rainfall intensity). The results show that the rainfall infiltrates the weathered soil and that surface ponding occurs eventually (Figure 6(a) and (b)). If the rainfall continues after the ponding, positive pore pressure (a perched water table) develops in the upper layer (Figure 6(a)). Figure 6(c) shows that the hydraulic gradient increases sharply at the wetting front and that the largest gradient occurs when the wetting front passes the interface. The gradient in the lower layer is not so high because the hydraulic conductivity of the silt changes more gradually.

Figure 7 shows the slope stability factor–depth of wetting front curve. From the numerical analysis results, it was found that before the wetting front to reach the surface of the second soil layer, the factor of safety would decline sharply to 1 with the increase of the wetting front. The factor of safety has a mutation when the wetting front reaches the contact surface of the second soil layer. Strength parameters change when the sliding surface by the transition to the grass roots soil topsoil. Fig 8 shows the security factor versus time curve of the second surface. The arrival of the wetting front to the contact surface is the most dangerous moment.

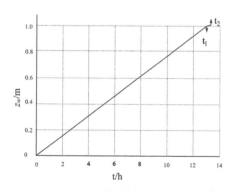

Figure 5. Depth of wetting front–time curve.

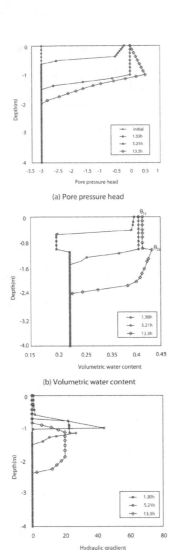

(a) Pore pressure head

(b) Volumetric water content

(c) Hydraulic gradient

Figure 6. Results of numerical analyses for coarse-over-fine stratification.

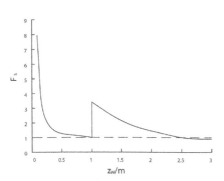

Figure 7. Slope stability factor–depth of wetting front curve.

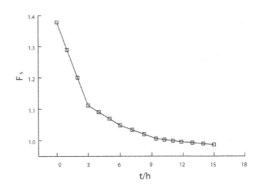

Figure 8. Safety factor–time curve at subsurface.

5 CONCLUSIONS

In the factor of safety analysis of two-layered slope, surface failures occur because of the positive pore water pressure or the reduced suction when the pore pressure is still negative.

The gradient in the lower layer is not so high because the hydraulic conductivity of the silt changes more gradually.

To assess how the rainfall infiltration affects the slope stability in two-layered soil, one-dimensional infiltration model was reviewed.

The factor of safety would be minimal when wetting front reaches the contact surface.

ACKNOWLEDGMENT

This study was supported by the Guangxi science and technology project (Grant Nos. 1355008-2, 14124004-4-12, and 14124004-4-5).

REFERENCES

Cho, S. E. (2009). Infiltration analysis to evaluate the surficial stability of two-layered slopes considering rainfall characteristics. Engineering Geology, 105(1), 32–43.

Lee, L. M., Gofar, N., & Rahardjo, H. (2009). A simple model for preliminary evaluation of rainfall-induced slope instability. Engineering Geology, 108(3), 272–285.

Mukhlisin, M., Baidillah, M. R., Ibrahim, A., & Taha, M. R. (2014). Effect of soil hydraulic properties model on slope stability analysis based on strength reduction method. Journal of the Geological Society of India, 83(5), 586–594.

Rahimi, A., Rahardjo, H., & Leong, E. C. (2010). Effect of hydraulic properties of soil on rainfall-induced slope failure. Engineering Geology, 114(3), 135–143.

Vargas Jr, E., Oliveira, A. R. B., Costa Filho, L. M., & Campos, T. P. (1986). A study of the relationship between the stability of slopes in residual soils and rain intensity. In International Symposium on Environmental Geotechnology. Envo Publishing, Leigh, USA (pp. 491–500).

Advances in Energy Science and Equipment Engineering II – Zhou, Patty & Chen (Eds)
© 2017 Taylor & Francis Group, London, ISBN 978-1-138-71798-5

Bridge damage pattern identification based on perturbation theory

Duo Wu & Lai-jun Liu
School of Highway, Chang'An University, Xi'an City, China

ABSTRACT: At present, the mainstream of bridge damage identification methods is the inversion calculation based on the change rate of the stiffness matrix and flexibility matrix. In this paper, we introduced perturbation theory into the damage identification of bridges, simulated the damage of the components according to the small changes of the structural parameters, employed MATLAB software to program, and verified the effectiveness of the perturbation theory on damage identification through practical examples. The result shows that in the case of small damage degree, perturbation theory can effectively identify bridge damage and satisfy engineering requirements. This can provide numerical guidance for real-time monitoring of large bridges and provide security for the safety operation of the bridge.

1 INTRODUCTION

At present, various types of bridges are suffering different degrees of damage after years of wearing and tearing, which seriously affect their durability and stability. In order to ensure their normal operation, the use of health monitoring means to monitor the real-time condition of the bridge has attracted increasing attention.

2 DAMAGE IDENTIFICATION BASED ON STIFFNESS MATRIX AND FLEXIBILITY MATRIX

At present, a relatively mature damage identification method is necessary to calculate the damage degree according to the change of dynamic characteristics in bridge structures worldwide. Many scholars have carried out substantial researches on bridge damage, and some of them have made in-depth research on the change of the stiffness matrix. For example, LIN Xian-kun analyzed the stiffness matrix changes before and after the structural damage and found that the specific location and extent of the damage can be effectively identified based on the changes and change degrees of the stiffness matrix at the specific local unit. Huang Yu-kun improved the traditional stiffness method, introduced the concept of damage degree coefficient, and further studied the functional relationship between the change of the physical parameters and the damage degree. By testing the dynamic parameters of a curved continuous rigid frame bridge during operation and comparing them with those of the theoretical model, he inversely worked out the actual performance of the bridge structure and identified the damage degree directly.

As the structure frequency is proportional to the stiffness, to obtain more accurate results, it is necessary to provide the higher-order frequency of the structure, which increases the difficulty of detection to a certain extent. However, the structure frequency is inversely proportional to the flexibility, and thus only lower frequency is needed to identify the exact damage location and degree. This has drawn the attention of some scholars. For example, YANG Hua identified the damage location according to the change rate of the diagonal elements of the flexibility matrix before and after damage and determined the damage degree based on the variation of the stiffness matrix. SUN Guo proposed a refined flexibility matrix method, took the change rate of the diagonal elements of the flexibility matrix as the damage indicator function, mapped the stiffness changes to the stiffness matrix, and used the localization feature of the damage indicator function for local measurements. He further derived the calculation formulas, carried out numerical simulation for different damage conditions, and compared the results of different methods. Both the aforementioned scholars took advantage of the characteristics of flexibility matrix and contributed to damage identification of bridge structures.

3 PERTURBATION THEORY OF DYNAMIC CHARACTERISTICS OF BRIDGES

Rayleigh pioneered to introduce the modification and analysis of structural parameters into structural dynamics. Later, R.H. Rogers, L.C Plaut, and others also conducted in-depth research on the perturbation theory. In China, Hu Hai-chang and Chen Su-huan also conducted systematic study on the perturbation theory.

It is a common phenomenon that small changes occur in the structural parameters such as cross section, materials, and stiffness during the construction of a bridge. Small changes in structural parameters can lead to changes in structural vibration characteristics, which affect the performance and durability of the structure. Therefore, it is important for optimizing the design and identifying damage of engineering structures to study the small variation of structural parameters. It is a new direction of bridge damage research that how to make use of the real-time variation of the small and medium parameters of the large bridge structure to determine the real-time damage of the bridge structure.

3.1 Principle of perturbation method

At present, two main factors are attributed to the bar system damage like bridges: one is the decrease of elastic modulus caused by the failure of the material and the other is the decrease of structural rigidity caused by the dead load of the bridge and the vehicle load. However, usually there is little influence on the mass of the bridge. Therefore, it is assumed that the damage of bridge structure only leads to the change of stiffness matrix, but not the mass matrix. Thus, the change of stiffness matrix is taken into consideration.

Let ε be a random parameter that causes bridge damage. The damage of the component itself surely is accompanied by the change of the dynamic characteristic and is reflected by the change of mass matrix and stiffness matrix:

$$[M] = [M_0] + \varepsilon[M_1] \tag{1}$$

$$[K] = [K_0] + \varepsilon[K_1] \tag{2}$$

In the above two equations, $[M_0]$ and $[K_0]$ represent the initial mass matrix and stiffness matrix before damage, respectively, $[M]$ and $[K]$ represent the initial mass matrix and stiffness matrix after damage, respectively, $[M_1]$ and $[K_1]$ represent mass matrix and stiffness matrix of the damaged local units, respectively, and ε is the perturbation value of the bridge.

Here, only the stiffness of the bridge is considered, and let n damages occur in the bridge and then the eq. (2) can be rewritten as:

$$[K] = [K_0] + \sum_{m=1}^{n} \varepsilon_m[K_1] \tag{3}$$

For the structures of bar system like a bridge, $\varepsilon[K_1]$ belongs to small parameter variations. Therefore, the eigenvalue and eigenvector of the dynamic characteristics also belong to small parameter variations. According to the perturbation theory, ε can be expanded in the following ways based on the provisions of the small parameter method:

$$\lambda_i = \lambda_{io} + \varepsilon\lambda_{i1} + \varepsilon^2\lambda_{i2} + o(\varepsilon^3) \tag{4}$$

$$\chi_i = \chi_{io} + \varepsilon\chi_{i1} + \varepsilon^2\chi_{i2} + o(\varepsilon^3) \tag{5}$$

On the basis of the eigenvalue of the structural components in its broad sense, we obtain the following equation:

$$K_0\chi_{i0} = \lambda_{i0}M_0\chi_{i0} \ (i = 1 \sim n) \tag{6}$$

Combining eq. (6) with eq. (2), we obtain:

$$(K_0 + \varepsilon K_1)\chi_{i0} = \lambda_{i0}(M_0 + \varepsilon M_1)\chi_{i0} \tag{7}$$

Combine eq. (4) and eq. (5) with eq. (7) and extract ε with the same order at both sides of the equation as follows:

$$\varepsilon^0 : K_0\chi_{i0} = \lambda_{i0}M_0\chi_{i0} \tag{8}$$

$$\varepsilon^1 : K_0\chi_{i1} + K_1\chi_{i0} = \lambda_{i0}M_0\chi_{i1} + \lambda_{i0}M_1\chi_{i0} + \lambda_{i1}M_0\chi_{i0} \tag{9}$$

$$\varepsilon^2 : K_0\chi_{i0} + K_1\chi_{i1} = \lambda_{i0}M_0\chi_{i2} + \lambda_{i0}M_1\chi_{i1} + \lambda_{i1}M_0\chi_{i1} + \lambda_{i1}M_1\chi_{i0} + \lambda_{i2}M_0\chi_{i0} \tag{10}$$

Equations (8)–(10) are the whole basic equations used to analyze bridge damage under the framework of perturbation theory. Using eq. (8), we can work out the initial values of eigenvalue λ_{i0} and eigenvector χ_{i0} of the bridge structure.

3.2 First-order perturbation formula

To obtain the accurate value, we can expand χ_{i1} as $\chi_{10} \sim \chi_{n0}$:

$$\chi_{i1} = \sum_{k=1}^{n} \alpha_{k1}\chi_{ko} \tag{11}$$

In eq. (11), α_{k1} represents n undetermined coefficients. Combining eq. (11) with eq. (9) and multiplying the left-hand side of the obtained equation by x_{k0}^T and simplifying, we obtain:

$$\alpha_{k1}\lambda_{k0} + x_{k0}^T K_1 x_{i0} = \alpha_{k1}\lambda_{i0} + \lambda_{i0}x_{k0}^T M_1 x_{i0} + \lambda_{i1}x_{k0}^T M_0 x_{i0} \tag{12}$$

where $i = k$, $\lambda_{k0} = \lambda_{i0}$, and $\lambda_{k0}^T M_0\chi_{i0} = 1$. From eq. (12), we can obtain the eigenvalue's first-order perturbation formula:

$$\lambda_{i1} = \chi_{i0}^{T}\left(K_1 - \lambda_{i0}M_1\right)\chi_{i0} \tag{13}$$

3.3 Second-order perturbation formula

To obtain more accurate value, we may resort to second-order perturbation and expand χ_{i2} as $\chi_{10} \sim \chi_{n0}$:

$$\chi_{i2} = \sum_{k=1}^{n} \alpha_{k2}\chi_{k0} \tag{14}$$

Combining eq. (14) with eq. (10), multiplying the left-hand side of the obtained equation by χ_{k0}^{T}, and simplifying, we obtain:

$$\begin{aligned}
&\alpha_{k2}\lambda_{k0} + \chi_{k0}^{T}K_1\chi_{i1} = \alpha_{k2}\lambda_{i0} \\
&+ \chi_{k0}^{T}\left(\lambda_{i0}M_1\chi_{i1} + \lambda_{i1}M_0\chi_{i1} + \lambda_{i1}M_1\chi_{i0}\right) \\
&+ \lambda_{i2}\chi_{k0}^{T}M_0\chi_{i0}
\end{aligned} \tag{15}$$

where $i = k$, $\lambda_{k0} = \lambda_{i0}$, and $\chi_{k0}^{T}M_0\chi_{i0} = 1$. From eq. (15), we can obtain the eigenvalue's second-order perturbation formula:

$$\lambda_{i2} = \chi_{k0}^{T}\left(K_1\chi_{i1} - \lambda_{i0}M_1\chi_{i1} - \lambda_{i1}M_0\chi_{i1} - \lambda_{i1}M_1\chi_{i0}\right) \tag{16}$$

4 DAMAGE IDENTIFICATION OF BRIDGE BASED ON PERTURBATION THEORY

Let the initial frequency of the bridge opened to traffic be λ_{0i} and the eigenvalue after the bridge being damaged be λ_i^0, which equals to λ_i in equation (4), namely:

$$\lambda_i^0 = \lambda_{0i} + \varepsilon\lambda_{1i} + \varepsilon^2\lambda_{2i} + o\left(\varepsilon^3\right) \tag{17}$$

Not considering high-order accuracy, we can obtain the following equation:

$$\lambda_{2i}\varepsilon^2 + \lambda_{1i}\varepsilon - \left(\lambda_i^0 - \lambda_{0i}\right) = 0 \tag{18}$$

where ε is a small parameter, which can represent the damage condition, whose value can be obtained as follows:

$$\varepsilon_{1,2} = \frac{-\lambda_{1i} \pm \sqrt{\lambda_{1i}^2 + 4\lambda_{2i}\cdot\left(\lambda_i^0 - \lambda_{0i}\right)}}{2\lambda_{2i}} \tag{19}$$

When the bridge structure bears no damage, $\varepsilon = 0$ and when the local structure is totally damaged, $\varepsilon = -1$; therefore, $\varepsilon \in [-1,0]$.

On the basis of the above assumption, we can abnegate the redundant root and determine the value of ε:

$$\varepsilon = \frac{-\lambda_{1i} + \sqrt{\lambda_{1i}^2 + 4\lambda_{2i}\cdot\left(\lambda_i^0 - \lambda_{0i}\right)}}{2\lambda_{2i}} \tag{20}$$

Considering the accuracy of first-order perturbation, the equation $\lambda_{1i}\varepsilon - \left(\lambda_i^0 - \lambda_{0i}\right) = 0$ is established, and the following equation is available:

$$\varepsilon = \frac{\left(\lambda_i^0 - \lambda_{0i}\right)}{\lambda_{1i}}, \varepsilon \in [-1,0] \tag{21}$$

Therefore, we have the following calculation formula regarding the accuracy of first-order and second-order perturbation:

$$\varepsilon = \begin{cases} \dfrac{-\lambda_{1i} + \sqrt{\lambda_{1i}^2 + 4\lambda_{2i}\cdot\left(\lambda_i^0 - \lambda_{0i}\right)}}{2\lambda_{2i}} & \text{(second-order)} \\[2mm] \varepsilon \in [-1,0] \\[2mm] \varepsilon = \dfrac{\left(\lambda_i^0 - \lambda_{0i}\right)}{\lambda_{1i}} & \text{(first-order)} \end{cases} \tag{22}$$

According to the above equation, the value of ε can be calculated, and its accuracy can be equivalent to the damage degree of the bridge structure.

5 EXAMPLES

Figure 1 shows the structure of a truss bridge and Table 1 shows the material properties of its bars. To simplify the calculation, this paper only selected semi-structure to investigate the following working conditions: single-unit damage, multiunit damage, and single unit of different damage degrees.

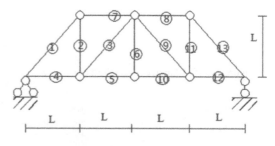

Figure 1. Truss bridge structure.

Table 1. Material properties.

Cross section A	Elastic modulus E	Mass density ρ
1.0×10^{-4} m^2	2.1×10^{11} Pa	7300 kg/m^3

1. Let the third bar be damaged at the single cell and its stiffness decreases by 15%, as shown in Fig. 2.
2. Let the bars be damaged at multiple cells: the third and fifth bars are damaged simultaneously and their stiffness decreases by 20%, as shown in Fig. 3.

The first and second working conditions indicated that the perturbation method was able to identify damages satisfactorily. Moreover, this method could identify both single-cell damage and multicell damage fair enough.

3. Let the third bar be suffered from different degrees of damage: its stiffness decreases by 5%, 10%, 15%, 20%, 25%, and 30%. Table 2 shows the comparison of the perturbation values under different degrees of damage of the third bar. Figure 4 shows the different degrees of damage of the third bar (the horizontal coordinates representing the designed damage degrees, the longitudinal coordinates the actual damage degrees).

As Table 2 shows, within 3–30% damage degree, the perturbation methods could perform good identification of the structural damage. When the damage degree was about 5%, the first-order perturbation theory performed the best identification. When the damage degree was about 10%, the second-order perturbation theory had the highest

Table 2. Comparison of different damage degrees of the third unit.

Actual damage degrees ε	First-order perturbation values	Second-order perturbation values
3%	2.95%	2.86%
5%	5.07%	5%
7%	7.32%	7%
10%	11%	10%
12%	14%	12%
15%	18.05%	15.60%
20%	26.50%	21.77%
25%	36.80%	28.62%
30%	49.57%	36.36%

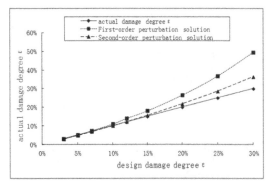

Figure 4. Different damage degrees of the third unit.

identification rate. When the damage degree was 30%, the identification error of first-order perturbation theory was up to 19.57% and that of the second-order perturbation theory was only 6.36%, which could satisfy the engineering precision requirement.

Figure 4 indicates that in the case of small damage, both first-order and second-order perturbations were able to identify damage. With the increase of the damage degree, the recognition results had certain "drift," and in the case of medium damage, the second-order perturbation performed identification job well. Therefore, it is concluded that perturbation method was useful for the small damage identification.

On the basis of the perturbation theory, this paper applied the modification of the stiffness parameters to simulate the bridge damage and verified its effectiveness by practical examples. The results show that:

1. The first-order perturbation frequency of the structure was able to quickly identify the local unit damage of the bridge structure.

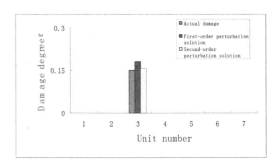

Figure 2. 15% damage of the third unit.

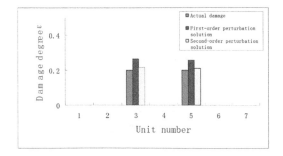

Figure 3. 20% damage of third and fifth units.

2. In the case of small damage (3–10%), the first-order perturbation method was of good accuracy in identifying damage. In the case of medium damage (10–20%), the second-order perturbation method was of good accuracy. In the case of large damage, neither first-order nor second-order perturbation method was applicable, and we may resort to a higher-order perturbation method.

6 CONCLUSIONS

The main idea of perturbation theory is to use the inherent dynamic characteristics before damage to approximately represent the eigenvalue after damage. Compared to the conventional stiffness matrix identification method, this theory is clearer and simpler, as it only needs the first-order frequency to acquire the damage condition. In this paper, we employed practical examples to verify that this method has high accuracy in damage identification. Therefore, perturbation theory can be widely used in identifying and calculating damage of bridge structure.

REFERENCES

Cheng Yong-chun, Tan Guo-jin, Liu Han-bing, et al. Damage identification of bridge structure based on statistical properties of eigen-solution [J]. Journal of Jilin University (Engineering and Technology Edition), 2008, 38(4):812–816.

Chen Su-huan. 1999. Matrix Perturbation Theory Structural Dynamic Design. Beijing: Science Press.

Hu Hai-chang, 1987. Multi-DOF structure inherent vibration theory. Beijing: Science Press.

Huang Yu-kun, Xu Lin, Liu Lai-jun. Quantitative evaluation of bridge damage based on improved direct stiffness method [J]; Journal of Highway and Transportation Research and Development (Application Technology Edition), 2013,4:243–244.

John Wm. Strutt (Baron Rayleigh), The Theory of Sound, 2nded. (London:Macmillan, 1896), vol. 2, pages 231–234.

Lin Xian-kun, Qin Bo-ying, Liang Xing-hua. Researches on Bridge Damage Identification Based on Stiffness Matrices [J]. Journal of Guangxi University of Technology, 2010, 21(3):13–18.

Plaut R. H. and Huseyin K. 1973. Derivatives of Eigenvalues and Eigenvectors in Non-Self-Adjoint Systems. AIAA journal. 11(1),11.

Ren Wen-feng. Wang Xing-hua, TU Peng. Simulation and monitoring of high-speed railway tied-arch bridge construction with arch first and beam late method. [J]. Journal of Traffic and Transportation Engineering, 2012, 12(5):28–36.

Rogers L.C. 1977, Derivatives of Eigenvalues and Eigenvectors. AIAA J. 8(5), 943–944.

Sun Guo. Improved Flexibility Matrix Method for Damage Identification of Multi-span Beams [J]. Engineering mechanics, 2003, 20(4):50–54.

Wang Wen-liang, Hu Hai-chang. 1993. Small parameter method multiple eigenvalues. Journal of Fu dan University (Natural Science), 32(2):168–175.

Yang Hua. Structural Damage Identification Based on Flexibility Matrix Method. Journal of Jilin University:Science Edition, 2008, 46(2):242–244.

Yu Yan-lei, Gao Wei-cheng. Liu Wei. 2012. Simplified perturbation method for analyzing the mode jumping of close mode structure. Engineering Mechanics, 29 (3): 33–40.

Advances in Energy Science and Equipment Engineering II – Zhou, Patty & Chen (Eds)
© 2017 Taylor & Francis Group, London, ISBN 978-1-138-71798-5

Characteristics of debris flow in Jian Shan-bao ravine

An Chen & Shu-yan Xi
Faculty of Land Resource Engineering, KunMing University of Science and Technology, KunMing, China

Ran Tao
Swan College, Central South University of Forestry and Technology, ChangSha, China

Ke-wen Liu
The 14ᵗʰ Metallurgical Construction Yunnan Invstigation and Design Co. Ltd., KunMing, China

Rong-jun Ding
Faculty of Land Resource Engineering, KunMing University of Science and Technology, KunMing, China

ABSTRACT: To study the characteristics of debris flow in Jian Shan-bao ravine, site investigation is needed to study topography and geomorphology, geological structure, stratum lithology, solid material source, and ravine characteristics based on comprehensive collection of geological and meteorological data. The results show that the extreme temperature difference can be up to 40.6°C and the extreme ground temperature can be up to 77.7°C in Jian Shan-bao ravine area. The stratum lithology is soft rock and very soft rock, and the total amount of loose solid matter is about 2.0×10^7 m³ in the Jian Shan-bao ravine drainage basin. The relative height difference is up to 1328 m, and the average gradient of ditch bed is 132.21‰. The average annual precipitation is 1058 mm and the 24-h rainfall intensity can be up to 111.5 mm in the area. Rich solids source, abundant rainfall, storm rainfall intensity, and steep terrain are very conducive to the outbreak of debris flow. The range of debris flow volume at different rainstorm frequencies is 573.53–260.54 m³/s and that of the total process volume is 2605.38–103235.65 m³. The velocity of debris flow is 4.77 m/s and its overall pressforce is 3.71 kPa. The maximum height of debris flow is 1.16 m, which can reach up to 1.85 m when blocked.

1 INTRODUCTION

Jian Shan-bao ravine is located on the left bank of the reservoir area of Wudongde Hydropower Station at a distance of about 10.3 km from the dam. There is a diluvial fan in the Mizoguchi area of Jian Shan-bao ravine, which imports Jinsha River, whose length and width are 710–760 m and 65–120 m, respectively. Its longitudinal gradient and cross gradient are 5–8° and 4–6°, respectively. The deposits can be divided into three layers: the bottom is composed of reddish-brown gravel with angular to sub-rounded shape, with grain diameter of 2–200 mm, and the gravel is sandstone mixed with a small amount of pink sandstone and limestone. The middle layer is composed of grayish-brown gravel with angular shape, with grain diameter of 5–200 mm mainly, but contains chunk, and the gravel is sandstone and mudstone. The upper layer is composed of boulder and chunk with angular shape, with grain diameter greater than 200 mm mainly, and the boulder and chunk are sandstone and limestone. The diluvial fan led to the shift of the streamline of Jinsha River to the right bank, which indicates the occurrence of debris flow in the past and its possibility in the future (Fig. 1).

In fact, as early as 1928, Blackwelder (an American scholar) discussed debris flow (mudflow as a geologic agent in semi-arid mountains) (Blackwelder, 1928). Then, Fryxell (1943), Sharp (1953), and

Figure 1. Drainage basin and its tributaries distribution of Jian Shan-bao ravine (picture from Google earth).

Curry (1966) recorded and described the debris flow phenomenon. In 1969, Anderson (an American scholar) specially reported expansive soil debris flow in Alaska alpine areas. In 1974, Beaty (Hunt, 1974) (a Canadian scholar) published "Newtonian fluid mechanics treatment of debris flows and avalanche". In 1983, Innes (1983) (a United Kingdom debris flow expert) published "Debris Flows," which is a review article on debris flow. In 1997, Costa (1984) (an American geologist) published "Physical geomorphology of debris flows," which proposed the research direction of debris flow. In 1997, Iverson (1997) (an American debris flow expert) published "The physics of debris flows."

At present, debris flow research is focused on refinement (Yu, 2010; Gan, 2012) and in-depth study (Guo, 2012; Zhang, 2010). For example, Gou Yinxiang (2012) studied the dynamic characteristics of debris flow, Liming Xu (2013) studied the risk assessment and prevention of debris flows, Niu Cen-cen studied the index selection and rating of debris flow hazard, Li Zhi (2014) studied the initiation and motion characteristics of debris flow, and He Xiaoying (2014) studied the shock characteristics of debris flow.

China is one of the countries in the world worst affected by debris flow hazards. The annual direct loss of economic due to debris flow is up to 20 billion RMB. Because each debris flow ravine has unique characteristics, their harmfulness are also different. It is necessary to study the causes and hazard characteristics of specific debris flow.

2 CAUSES CONDITIONS

2.1 Climate

Topographic relief of the study area is huge. The maximum relative elevation is up to 2510 m. The extreme temperatures difference is up to 40.6°. The extreme ground temperature difference is up to 77.7°C. The enormous temperature and ground temperature differences lead to the strong weathering of rockmass and the formation and accumulation of a large number of loose solids, which provide sufficient solid material for debris flows. The average annual precipitation is 1058 mm. The rainfall concentrated from June to October constituted about 85% of the annual rainfall. The rainfall intensity during 24 h is up to 111.5 mm (Table 1).

Great rainfall is conducive to the saturation of loose deposits, which reduces their stability. Great rainfall intensity easily converges to torrent, which provides dynamic conditions for the flow of loose material and lead to debris flows.

2.2 Topography and geomorphology

Jian Shan-bao ravine is a primary tributary of Jinsha River, which is located on its left bank. The drainage basin area is 16.8 km². The main ravine length is 10.2 km. The maximum relative elevation is up to 1328 m. There are five tributaries in the drainage basin, which is dendritic distribution and cutting deeply. The tributary development density is 0.67/km (Fig. 1). There are water flows in tributary annually.

The bed average gradient of Jian Shan-bao ravine is 132.21‰. The bed average gradients of the tributaries are 157.9‰ and 432.8‰. The ravines in drainage basin are of asymmetrical "V" type. The slope gradients distributed on both sides of the ditches are 25° and 35°. From the gully head to Mizoguchi, the bed gradients and the slope gradients are increased (Table 2). The above features are conducive to water collection and provide advantageous terrain conditions for debris flow formation and flow.

2.3 Geological conditions

The strata in Jian Shan-bao ravine are mudstone, silty mudstone, marlite, siltstone, shale, and coalbed. The thickness of strongly weathered rockmass ranges from 11 to 26 m. The superiority attitude of the strata is 305°∠14°, and there are dip slopes in partial area. The broken soft rock and extremely

Table 2. Characteristics of Jian Shan-bao ravine and tributaries.

Name	Length (m)	Elevation difference (m)	Bed gradient (‰)	Slope gradients
Main ravine	1020	1328	132.2	19°–22°
Tributary No.1	1664	789	432.8	20°–35°
Tributary No.2	1487	621	417.1	20°–25°
Tributary No.3	1502	350	233.0	10°–20°
Tributary No.4	991	178	179.1	8°–15°
Tributary No.5	1900	300	157.9	8°–15°

Table 1. Characteristics of average annual precipitation (mm).

Month	Jan.	Feb.	Mar.	Apr.	May	Jun.	Jul.	Aug.	Sep.	Oct.	Nov.	Dec.
Monthly averages	8.2	7.5	11.5	17.7	77.5	229.7	221.8	174.9	173.2	104.9	23.4	7.7
Day maximum	49.0	23.4	29.5	27.6	46.2	111.5	105.5	100.1	106.3	50.0	53.2	23.3

soft rock distributed in Jian Shan-bao drainage basin provide sufficient solid material for debris flows. In 1992, some small landslides took place due to heavy rain in Jian Shan-bao drainage basin, which led to debris flow that lasted about 30 min.

2.4 *Solids source*

The loose solid materials in the Jian Shan-bao drainage basin are landslide deposits, ravine deposits, and slope deposits.

1. Landslide deposits

A landslide is located at a distance of 760–800m to Mizoguchi in main ravine. Its volume is about 7.0×10^6 m³. There are two landslides located near tributaries No.3 and No.4 in main ravine, whose volumes are 5.0×10^6 and 4.0×10^5 m³, respectively. A landslide is located near tributary No.5 in main ravine with volume of about 5.5×10^5 m³.

2. Ravine deposits

The volume of early debris flow deposit in ravine is about 6.0×10^4 m³. The volume of flash floods alluvial deposit is about 2.0×10^6 m³. The volume of the deposit in ravine is about 3.0×10^5 m³ and that of other loose solid materials is about 3.0×10^6 m³.

3. Slope deposits

The volume of slope deposits is about 3.6×10^6 m³.

3 ACTIVITIES CHARACTERISTICS

The debris flow activity can divided into two districts according to bed average gradient and loose solid distribution. The section from Mizoguchi to elevation + 890m is the deposit area. The section from elevation + 890m to Fandango Reservoir is forming region to moving region. The section above Fandango Reservoir is water confluence area (Fig. 2).

1. Deposit area

The ditch bed average gradient is 61.27‰. Its extending direction is 135°–145°. The maximum width of diluvial fan is 450–500m and the minimum width is 20–25m. The gullies in diluvial fan form water erosion. The trench wall is steep.

2. Forming region–moving region

The ditch bed average gradient is 206.32‰. Its extending direction is 125°–135°. The ravine width is 10–25m. There are four landslides in this section. The volume of loose solid is 1.3×10^7 m³, which is the solid material source of debris flow.

3. Water confluence area

The ditch bed average gradient is 101.13‰. Its extending direction is 65°–75°. The slope gradients distributed on both sides of the ditch are 20°–30°. The vegetation is developed, whose coverage rate is 45–55%. The solid is less in this section, which is the hydrodynamic region of debris flow.

4 DYNAMIC CHARACTERISTICS

4.1 *Flow characteristics*

The drainage area is 16.8 km². The ravine length is 10.2 km. The ditch bed average gradient is 132.21‰. The drainage basin characteristic coefficient is 6.526. The confluence parameter is 0.324. The tributary parameter is 2.106. The ravine maximum flood flows at different frequencies as shown in Table 3.

1. Debris flow volume

The rain flood method is used to compute debris flow volume:

$$Q_C = (1 + \phi) \times Q_P \times D_C \qquad (1)$$

where Q_P is maximum flood flow, D_C is obstructive coefficient of debris flow, which is determined by the relations of main ravine and tributary, ravine

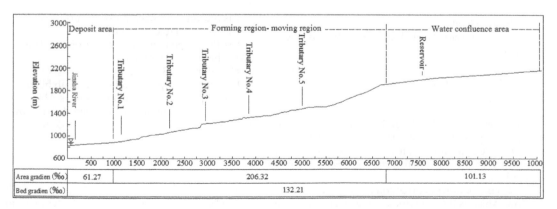

Figure 2. Ditch bed gradient distribution of Jian Shan-bao ravine.

463

Table 3. Calculation of maximum flood flow.

Frequency p (%)	0.2	0.5	1	2	5	10
F.P.C Ψ	0.931	0.935	0.932	0.916	0.911	0.886
Confluence time τ(h)	3.134	3.237	3.326	3.433	3.616	3.787
Max. flow Q_p (m³/s)	171.3	150.4	136.0	117.6	95.9	77.8

Note: F.P.C. is flood peak runoff coefficient.

Table 4. Flow calculation table of debris flow.

Frequency p (%)	10.0	5.0	2.0	1.0	0.5	0.2
Volume Q_C (m³/s)	260.54	321.06	393.60	455.30	503.67	573.53

width, ravine curvatures, and the viscosity of debris flow. Here, it is 2.1.

Φ is debris flow correction coefficient, which is calculated by the following formula:

$$\Phi = (\gamma_C - \gamma_W)/(\gamma_H - \gamma_C) \tag{2}$$

where γ_C is the debris flow density (t/m³), γ_W is the water density (t/m³), γ_H is the solid proportion of debris flow (t/m³).

Here, γ_C is 1.60 and $\gamma_H = 2.61$ according to the composition and particle composition of the debris flow. Therefore, $\Phi = 0.594$. The calculation results of the debris flow volume at various frequencies are shown in Table 4.

2. Process total volume

Debris flow has sudden and sharp rising and nosedive characteristics. The formula of total discharge is $Q = KTQ_C$ according to the duration (T, s) of debris flow and the maximum debris flow discharge (Q_C, m³/s). K = 0.0378 when the drainage area is greater than 10 km² but less than 100 km².

The calculation results of the total volume at various frequencies are shown in Table 5.

4.2 Dynamic characteristics

1. Flow velocity

The flow velocity of dilute debris flow is calculated as:

$$V_c = \frac{1}{n} \frac{1}{\sqrt{r_H \varphi + 1}} R^{2/3} I^{1/2} \tag{3}$$

where R is the hydraulic radius, which is equal to the average water depth (H). Here, it is 2.8m.

I is the hydraulic gradient of debris flow, which is equal to the ditch bed vertical gradient. Here, it is 132.21‰.

φ is the sediment correction coefficient of debris flow, which is 0.600 by look-up table according to γ_H and γ_C.

Table 5. Total process volumes calculation of debris flow.

Duration (min) Volume (10⁴ m³) Frequency	10	30	60	90	120	150	180
10.0	0.59	1.77	3.55	5.33	7.09	8.86	10.64
5.0	0.73	2.18	4.37	6.57	8.74	10.92	13.11
2.0	0.89	2.68	5.36	8.06	10.71	13.39	16.07
1.0	1.03	3.10	6.20	9.32	12.39	15.49	18.59
0.5	1.14	3.43	6.85	10.31	13.71	17.13	20.56
0.2	1.30	3.90	7.80	11.74	15.61	19.51	23.41

$1/n$ is the riverbed roughness coefficient, which is related to sediment characteristics, riverbed flatness, and riverbed vertical gradient. Here, it is 10.57.

Therefore, the debris flow velocity of Jian Shan-bao ravine is 4.77 m/s.

2. Total impact force

The total impact force of debris flow is:

$$F = \lambda(\gamma_c/g) V_c^2 \sin\alpha. \tag{4}$$

where g is the acceleration due to gravity (9.8m/s²).

α is the angle between pressure surface and pressure direction. Here, it is 90°.

λ is the shape factor of impact object. It is 1.00 when the shape is circular.

Therefore F = 3.71 (kPa).

3. Debris flow impact height and climb height

The impact height of debris flow is $\Delta H = V_c^2/2g = 1.16$ (m). The climb height when debris flow is blocked is given by $H = 0.8 V_c^2/g = 1.85$ (m).

5 CONCLUSIONS

1. The extreme temperature difference is up to 40.6°C. The extreme ground temperature difference is up to 77.7°C. The broken soft

rock–extremely soft rock is distributed in the Jian Shan-bao drainage area. The volume of loose solid in the Jian Shan-bao drainage area is 1.3×10^7 m³, which provides sufficient solid material for debris flows.

2. The drainage basin of Jian Shan-bao ravine is 16.8 km². The ravine length is 10.2 km. The maximum relative elevation is up to 1328m. The ditch bed average gradient is 132.21‰. The above features are conducive to water collection, which provides favorable terrain conditions for the formation and flow of debris.

3. The average annual precipitation is 1058 mm. The rainfall concentrated during June to October is about 85% of the annual rainfall. The rainfall intensity during 24 h is up to 111.5 mm. Great rainfall intensity easily converges to torrent, which provides dynamic conditions for the flow of loose material and lead to debris flows.

4. The debris flow volume is related to rainstorm frequency. The debris flow volume is 260.54 m³/s when rainstorm frequency is 10. The process total volume is 3.55×10^4 m³ when rainstorm time is 60 min. The debris flow volume is related to rainstorm frequency. The debris flow volume is 573.53 m³/s when rainstorm frequency is 0.2. The process total volume is 7.80×10^4 m³ when rainstorm time is 60 min.

5. The flow velocity is 4.77 m/s. The total impact force of debris flow is 3.71 kPa. The impact height of debris flow is 1.16 m. The climb height when debris flow is blocked is 1.85 m.

REFERENCES

Blackwelder E. Mudflow as a geologic agent in semi-arid mountains [J]. Geological Society of America Bulletin, 1928, 39: 465–487.

Costa J E. Physical geomorphology of debris flows [A]. In: J. E. Costa and P. J. Fleisher et al. Developments and Applications of Geomorphology [C]. Berlin: Springer-Verlag, 1984, 268–317.

Curry R. R. Observations of alpine mudflows in the Ten mile range, central Colorado [J]. Geological Society of America Bulletin, 1966, 77: 771–776.

Fryxell F. M. and Horberg L. Alpine mudflows in grandTeton national Park, Wyoming [J]. Geological Society of America Bulletin, 1943, 54: 457–472.

Gan Jianjun, Sun Haiyan, Huang Runqiu, Tan Yong, Fang Chongrong, Li Qianyin and Xu Xiangning. Study on Mechanism of Formation and River Blocking of Hongchuangou Giant Debris Flow at Yingxiu of Wenchuan County [J]. Journal of Catastrophology, 2012, 27(1): 5–9.

Guo Xiaojun, Cui Peng, Xiang Lingzhi, Zhou Xiaojun and Yang Wei. Research on the Debris Flow Hazards in Gaojia Gully and Shenxi Gully in 2011 [J]. Journal of Catastrophology, 2012, 27(3): 81–85.

Hunt B. Newtonian fluid mechanics treatment of debris flows and avalanche [J]. Journal of Hydraulic Engineering, 1974, 120(12): 125–129.

Innes J L. Debris flows [J]. Progress in Physical Geography, 1983, 7(1): 469–501.

Iverson R M. The physics of debris flows [J]. Reviews of Geophysics, 1997, 35(3): 245–296.

Ministry of Land and Resources of the People's Republic of China. Specification of geological investigation for debris flow stabilization (DZ/T 0220–2006) [s], BeiJing: Standards Press of China.

Sharp R. P. and Nobles L. H. Mudflow of 1941 at Wrightwood, southern California [J]. Geological Society of America Bulletin, 1953, 64: 547–560.

Yu Bin, Yang Yonghong, Su Yongchao, Huang Wenjie and Wang Gaofeng. Researchon the Giant Debris Flow Hazards in ZouQu County, GanSu Province on August 7, 2010 [J]. Journal of Engineering Geology, 2010, 18(4): 437–444.

Zhang Guoping, Xu fengwen, Zhao linna. Review of the study of rainfall triggered Debris Flows [J]. Meteorological Monthly, 2010, 36(2): 81–86.

Advances in Energy Science and Equipment Engineering II – Zhou, Patty & Chen (Eds)
© 2017 Taylor & Francis Group, London, ISBN 978-1-138-71798-5

Influence of loess property on ancient circumvallation deformation laws induced by metro shield tunnelling

Junling Qiu
School of Highway, Chang'an University, Xi'an, Shaanxi, China

Hao Sun
School of Civil Engineering, Xi'an University of Architecture and Technology, Xi'an, Shaanxi, China

Jinxing Lai
School of Highway, Chang'an University, Xi'an, Shaanxi, China

ABSTRACT: Shield construction in loess area can induce the ground surface subsidence and the impact on the surrounding environment with many monuments. Based on the metro 2# line tunnel through the circumvallation in Xi'an, FLAC3D simulation was conducted to study the influence of loess properties on the stratum settlement. Research results are shown that: as the loess cohesion and internal friction angle increased, the city wall foundation settlement and ground subsidence are slowed. But when the friction angle is over 21°, it has little effect on the city wall foundation formulation and the ground subsidence.

1 INTRODUCTION

Loess, which is derived from unconsolidated materials, is sensitive to sliding upon wetting and being broken down due to its metastable structure. Loess has a significant character of randomly open and loose particle with high porosity. When a tunnel or underground space is constructed in loess ground, it inevitably disturbs the in situ stress field, and then causes large ground displacements. Generally, loess is a multi-phase porous medium and develops complex stress and strain variation while executing a tunnelling project.

The research of stratum subsidence and deformation law in tunnel excavation has been performed deeply, and a lot of research has been carried out through analytical and numerical simulation method (Lee, 1991; Liu, 2012). For the theoretical analysis: According to the analytical solution of the circular hole unloading under axisymmetric condition in the linear elasticity—perfect plasticity media, Clough and Schmidt (Clough, 1981) presented strata deformation induced by tunneling. According to the analytical solution of the spherical pore unloading in the linear elasticity—perfect plasticity media, Mair and Taylor (Mair, 1993) proposed the description method of ground subsidence induced by working face unloading. Shi et al (Shi, 2002) established the semi analytical function of soil and liner by using the semi analytical method with axial and transverse analysis,

vertical dispersion. For the numerical analysis: Liu and Sun (Liu, 2001; Sun, 2002) calculated the influence of various factors on the formation of the size of the settlement by using the three-dimensional finite element method, and summed up the law of strata subsidence. Based on Guangzhou metro line 2 project, Liu et al (Liu, 2002) analyzed the influence of the strata subsidence induced by shield tunnelling on the shopping mall foundation employing FLAC software. Chen et al (Chen, 1999) studied the horizontal and vertical response of pile foundation in the construction of the adjacent tunnel using the analytical solution of the stratum displacement caused by tunnelling and the simplified boundary element method. Ng et al (C.W.W. Ng, 2014) studied the influence of shield tunneling on adjacent pile foundation by using three dimensional elastoplastic finite element method. Surface subsidence, stratum displacement and ground building damage induced by tunnel construction are the complicated problems, especially tunnelling in the loess area (Lambrughi, 2012).

Due to the complexity of the surface environment and the peculiar features of the strata, construction conditions, the settlement caused by Xi'an metro 2# line shield tunnelling and the impact on the surrounding environment are particularly prominent. Xi'an metro 2# line under-crosses the ancient circumvallation, which is located in the downtown area with rectangular shape, 13.91 kilometers in length. Existing circumvallation has more than 600 years of history (LEI, 2010). Therefore, it is very important to study the law of surface movement and deformation caused by shield tunnelling through the circumvallation. In this study, based on the metro 2# line tunnel through the circumvallation in Xi'an, the influence of loess properties on the stratum settlement was studied.

2 PROJECT OVERVIEW

The starting mileage of this range is YCK13+264, and the ending mileage is YCK14+496, with the overall length of 1232 m. The distance between the left and right tunnel axis is 17.0 m. When under-crossing the circumvallation, these two tunnels were excavated through the bypass way with the big spacing of 70–150 m. The design rail top elevation of the tunnel is 383.52–391.12 m, and the corresponding buried depth is 16–25 m. The strata distribution of the study area is shown in Figure 1.

3 NUMERICAL INVESTIGATION ON THE LOESS PROPERTY

3.1 Numerical model

The ancient circumvallation with height of 12 m, base width of 15–18 m and top width of 12–14 m is filled with the loess as the main material. The circumvallation foundation is rammed, and its depth is about 3–5 m. The relative relation of tunnel and circumvallation is shown in Figure 2. In the numerical calculation range, the tunnel was taken as the straight line. The average depth of tunnel is 16 m, and the diameter is 6 m. The segment with the length of 1.5 m and the thickness of 0.3 m composed of five conventional block segment and a wedge-shaped segments. Additionally, the layout of the measuring points on circumvallation foundation and ground is shown in Figure 3.

3.1.1 Modeling and unit division

As is well-known, the circumvallation and the tunnel structure are the plane strain problems. According to the Saint Venant's principle, the model size is taken as 3–5 times the diameter of the subway tunnel, which can not only meet the accuracy requirements, but also meet the requirements of calculation speed. Therefore, the FLAC^{3D} calculation model with size of 60 m × 36 m × 36 m (Width × Height × Length) was established with 19288 elements and 22035 nodes.

3.1.2 Calculation condition

In the numerical investigation of the shield tunnelling, segment support and back-filled grouting were performed. Moreover, the chemical grouting reinforce-

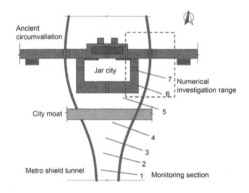

Figure 2. Circumvallation and shield crossing.

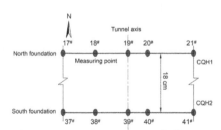

(a) Subsidence measuring point for I - I section

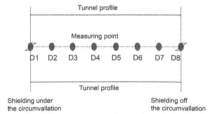

(b) Ground settlement measuring point along tunnel axis

Figure 3. Layout of the measuring points on circumvallation foundation and ground.

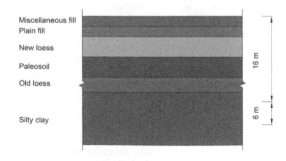

Figure 1. Strata distribution.

468

ment was carried out in the soil near the three gate holes. Considering the effect of the moat, conjunct bored piles with the specification of φ1000@1300 were installed on the south foundation of the circumvallation, which were 8 m away from both sides of the tunnel, 5 m away from the south foundation of the circumvallation (c.f. Figure 4). And the parameter of the grouting body are shown in Table 1.

3.1.3 Calculation parameter

Based on the in-situ investigation report of this project, the basic physical and mechanical

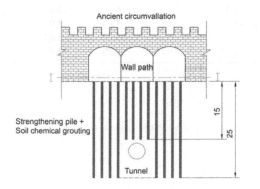

Figure 4. Stratum strengthening under circumvallation.

Table 1. Parameter of the grouting body.

Volume weight (kN/m³)	Compressive modulus (MPa)	Poisson ratio	Cohesive force (kPa)	Internal friction angle (°)
20	20	0.2	50	35

parameters of the surrounding soil were determined. The parameters of each soil layer are shown in Table 2. Bored pile was simulated by the specific pile element in the FLAC³ᴰ software, and C25 concrete was used with the linearly elastic hypothesis. Specifically, the elastic modulus was 28 GPa, the poisson ratio was 0.17, and the volume weight was 25 kN/m³.

3.2 Influence of loess properties on the stratum settlement

The physical properties of loess have significant influence on the stratum settlement during shield tunnelling. The natural moisture content of loess is related to the collapsibility, and the lower the water content, the stronger the loess collapsibility. With the increase of water content, the collapsibility is gradually weakened. Generally, when the water content is more than 23%, the collapsibility has been largely disappeared, and the corresponding compressibility increases. When the natural water content of loess is lower than the plastic limit, the water change is the most influential to the strength. When the water content increases, the internal friction angle and cohesive strength of the soil are reduced.

3.2.1 Influence of the cohesive force

To study the stratum settlement induced bu shield tunnelling under the special loess soil condition, series of simulation calculation were performed with different cohesive force of the loess, and the calculation results are shown in Tables 3 and 4.

From the comparison curves of the ground settlement with the different cohesion force of the loess (Figures 5–7), with the increase of the loess cohesion force, the trend of the ground settlement

Table 2. Numerical calculation parameter.

Name	Density (kg/m³)	Bulk modulus (MPa)	Shear modulus (MPa)	Cohesive force (kPa)	Internal friction angle (°)	Poisson ratio	Layer thickness (m)
Brick wall	1900	2875	2156	900	26.6	0.2	
Ramming soil	1800	6.83	3.15	26	18	0.3	
Miscellaneous fill	1875	5.44	3.27	11	12.5	0.25	2
Plain fill	1780	7.8	3.6	14.5	19.5	0.28	2
New loess	1870	4.9	2.3	24	16	0.3	4
Paleosoil	2010	7.8	3.6	26	17	0.3	3
Old loess	2010	8.8	4.1	28	17	0.3	4
Silty clay	2050	10.7	5	35	18	0.3	21
Pile body	Elastic modulus 2.80 × 10⁴MPa					Poisson ratio 0.17	
Tunnel segment	Elastic modulus 3.45 × 10⁴MPa					Poisson ratio 0.17	

Table 3. Vertical displacement of measuring point (unit: mm).

Measuring point	17#	18#	19#	20#	21#	38#	39#	40#	41#
C = 25kPa	0.17	−0.37	−1.88	0.18	0.93	−0.23	−2.2	−1.89	−0.23
C = 28kPa	0.18	−0.34	−1.65	0.16	1	−0.24	−1.9	−1.82	−0.2
C = 31kPa	0.18	−0.31	−1.47	0.16	0.9	−0.22	−1.6	−1.41	0
C = 34kPa	0.14	−0.23	−1.24	0.11	0.81	−0.2	−1.45	−1.1	−0.1

Table 4. Ground vertical displacement on tunnel axis (unit: mm).

Measuring point	D1	D2	D3	D4	D5	D6	D7	D8
C = 25kPa	1.57	0.6	−2.17	−3.4	−4.1	−4	−3.92	−3.7
C = 28kPa	1.38	0.51	−2.1	−3.2	−3.9	−3.88	−3.9	−3.6
C = 31kPa	1.2	0.39	−1.8	−3	−3.7	−3.59	−3.89	−3.3
C = 34kPa	1.04	0.22	−1.3	−2.61	−3.48	−3.26	−3.29	−3.27

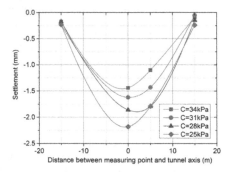

Figure 5. South foundation settlement.

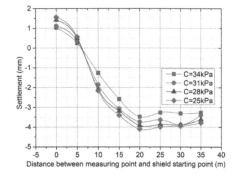

Figure 7. Ground settlement on tunnel axis.

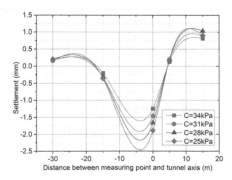

Figure 6. North foundation settlement.

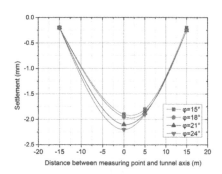

Figure 8. South foundation settlement.

curve is basically the same. The ground surface settlement in the area far away from the tunnel axis is not very obvious. But the ground settlement at the top of the tunnel axis (the maximum value) significantly reduces (Figure 8), which is relative to the allowable settlement safety value. It can be seen that in the reasonable range of mechanical properties, the greater the cohesive value of the loess

layer, the smaller the maximum surface subsidence induced by the shield tunnelling, that is, the better the formation stability.

3.2.2 *Influence of the internal friction angle*

From the comparison curves of the ground settlement with the different internal friction angle of the loess (Tables 5 and 6, Figures 8–10), it can be seen that the ground settlement curve is basically

Table 5. Vertical displacement of measuring point (unit: mm).

Measuring point	17#	18#	19#	20#	21#	38#	39#	40#	41#
$\varphi = 15°$	0.18	−0.37	−1.89	0.20	0.95	−0.24	−2.2	−1.91	−0.25
$\varphi = 18°$	0.17	−0.35	−1.78	0.19	1	−0.22	−2.1	−1.89	−0.23
$\varphi = 21°$	0.16	−0.30	−1.72	0.19	1	−0.20	−1.96	−1.83	−0.2
$\varphi = 24°$	0.16	−0.29	−1.71	0.193	1.05	−0.19	−1.93	−1.8	−0.17

Table 6. Ground vertical displacement on tunnel axis (unit: mm).

Measuring point	D1	D2	D3	D4	D5	D6	D7	D8
$\varphi = 15°$	1.59	0.62	−2.3	−3.45	−4.23	−4.11	−3.95	−3.96
$\varphi = 18°$	1.5	0.54	−2.2	−3.36	−4.15	−4.01	−3.91	−3.88
$\varphi = 21°$	1.32	0.39	−1.92	−3.07	−3.84	−3.71	−3.86	−3.82
$\varphi = 24°$	1.3	0.35	−1.9	−3	−3.8	−3.66	−3.77	−3.73

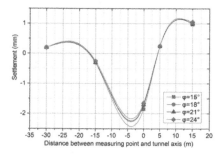

Figure 9. North foundation settlement.

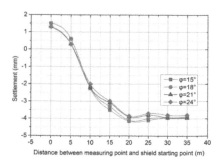

Figure 10. Ground settlement on tunnel axis.

the same along with the increase of the friction angle. The ground settlement is less obvious in the area far away from the tunnel axis. But the ground settlement at the top of the tunnel (the maximum value) reduces obviously. Therefore, the increase of internal friction angle of the loess has a certain influence on the ground surface settlement induced by shield tunnelling, but it is not obvious. When the internal friction angle of the loess is more than 21°, the influence of the control settlement has been gradually reduced.

4 CONCLUSIONS

1. In the reasonable range of mechanical properties, the greater the cohesive value of the loess layer, the smaller the maximum surface subsidence induced by the shield tunnelling, that is, the better the formation stability. The increase of internal friction angle of the loess has a certain influence on the ground surface settlement induced by shield tunnelling, but it is not obvious. When the internal friction angle of the loess is more than 21°, the influence of the control settlement has been gradually reduced.

2. The ground surface settlement induced by shield tunnelling is a process with strong temporal and spatial effects. The ground and circumvallation foundation subsidence during shield construction was simulated through the numerical investigation. However, the time-space effect on the calculation properties was not considered in this study. How to consider the whole process of ground settlement should be further studied.

ACKNOWLEDGMENTS

This work is financially supported by the Special Fund for Basic Scientific Research of Central Colleges of Chang'an University (No. 310821165011).

REFERENCES

C.W.W. Ng, M.A. Soomro, Y. Hong. Tunnelling and Underground Space Technology **43**, 350–361, (2014).
G.W. Clough, B. Schmidt. In Soft Clay Engineering, Elsevier, 569–634, (1981).

J. Sun, H.Z. Liu. Journal of Tongji University, 2002, **30**, 379–385.

J.F. Liu, T.Y. Qi, Z. Wu. Tunnelling and Underground Space Technology **28**, 287–296, (2012).

J.Y. Shi, J. Zhang, C.G. She, Y.W. Fan. Journal of Hohai University (Natural Sciences), 2002, **30**, 48–51.

K.M. Lee, R.K. Rowe. Canadian Geotechnical Journal, **28**, 25–41, (1991).

L.T. Chen, H.G. Poulos, N. Loganathan. Journal of Geotechnical and Geo-environmental Engineering, ASCE, 1999, **125**, 207–215.

Lambrughi, A., Medina, L. Comput. Geotech. **40**, 97–113, (2012).

Lei Yong-sheng. Rock and Soil Mechanics, **31**, 223–228, (2010).

Liu B., Ye S.G., Tao L.G., Tang M.X. Coal Science and Technology, 2002, **30**, 75–80.

Liu H.Z., Sun J. Modern Tunnelling Technology, 2001, **38**, 24–28.

R.J. Mair, R.N. Taylor, J.B. Burland. Proceedings of the International symposium. On Geotechnical Aspects of Underground Construction in Soft Ground. London Balkema, 713–718, (1996).

R.J. Mair, R.N. Taylor. Bracegirdle A. Geotechnique, **43**, 315–320, (1993).

Advances in Energy Science and Equipment Engineering II – Zhou, Patty & Chen (Eds)
© 2017 Taylor & Francis Group, London, ISBN 978-1-138-71798-5

Effect of axial direction on the stress state of a circular roadway in highly vertical in situ stress fields

Xun Xi, Qifeng Guo & Peng Lv
School of Civil and Environmental Engineering, University of Science and Technology Beijing, Beijing, China

ABSTRACT: On the basis of the theories of elastic–plastic mechanics, a three-dimensional mechanical model of circular roadway was built and the principal stresses of surrounding rock were obtained. According to the in situ stress fields of Chinese mines, on the basis of three strength theories, the effect of axial direction on the maximum principal stress, difference between principal stresses, and the maximum principal deviator stress of a circular roadway surrounding rock were analyzed. The results show that when the axial angle between roadway and the maximum horizon in situ stress increases from 0° to 90°, the maximum principal stress, difference between principal stresses, and the maximum principal deviator stress of surrounding rock in roadway roof and floor obviously increase. In an in situ stress field, the vertical in situ stress is the maximum stress, the maximum principal stress and the difference between principal stresses and the maximum principal deviator stress of surrounding rock in the sides of roadways are all higher than these of roadway roof and floor, and it is better to make the axial angle close to 0°.

1 INTRODUCTION

Most of the roadways in underground mines are designed to be circular. Circular roadways have advantages in stress state and bearing capacity (Ma, 2015; Miao, 2009). In situ stress is the natural stress in the formation of earth, which is a main force that leads to the deformation and destruction in underground mines. When analyzing the stress state of circular roadway, scholars often treat a circular roadway as a circular hole in a thin plate, that is, a plane strain problem. This method does not consider the axial angle between roadway and maximum horizon in situ stress. In fact, the stress state of circular roadway is a three-dimensional problem. Therefore, it is necessary to probe the effect of axial direction on the stress state of circular roadway.

Australian scholars proposed the maximum principal stress theory to design and support roadways, which put forward that in situ stress has a minimum influence on roadway stability when the angle between roadway and maximum horizon in situ stress is 0 (Gale, 1993). However, they did not calculate the value of stress of rock mass surrounding roadway, and different in situ stress fields were not considered. As the depth of developing underground mines is increasing, the vertical in situ stress becomes higher. And in some underground mines, it has become the maximum principal stress (Lv, P., 2015). Several scientists studied the stress state in high vertical in situ stress field. Dong pointed out that when the vertical in situ stress is the maximum principal stress, the maximum principal stress theory is not suitable (Dong, 2000). Sun, Zheng, and Zhao completed numerical simulations and pointed out that the reasonable angle between roadway and maximum horizon in situ stress varies from the in situ stress state (Sun, 2010; Zheng, 2010; Zhao, 2015). Here, we obtained the analytical solution of principal stress of circular roadway by a three-dimension model. On the basis of the theories of elastic–plastic mechanics, the stress state of the circular roadway in different directions was analyzed.

2 ANALYTICAL SOLUTION OF PRINCIPAL STRESS

In reference to the in situ stress measurement of hydraulic fracturing, the in situ stress state of roadway is described in Figure 1 (Cai, 2006).

In Fig. 1, σ_1, σ_2, and σ_3 are maximum principal stress, intermediate principal stress, and minimum principal stress, respectively; α is the angle between roadway and maximum horizon in situ stress whose value is 0–$\pi/2$; and R is the radius of roadway.

On the basis of stress translation formula in elastic theory, six components of stress resemble those in Equation 1: $\sigma_{x'}$, $\sigma_{y'}$, $\sigma_{z'}$, $\tau_{x'y'}$, $\tau_{y'z'}$, and $\tau_{x'z'}$.

$$\begin{cases} \sigma_{x'} = \sigma_1 \sin^2 \alpha + \sigma_3 \cos^2 \alpha \\ \sigma_{y'} = \sigma_2 \\ \sigma_{z'} = \sigma_1 \cos^2 \alpha + \sigma_3 \sin^2 \alpha \\ \tau_{x'y'} = 0 \\ \tau_{y'z'} = 0 \\ \tau_{x'z'} = 0.5 \sin 2\alpha (\sigma_1 - \sigma_3) \end{cases} \quad (1)$$

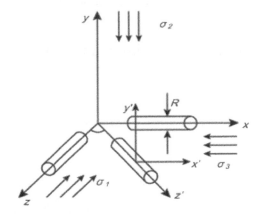

Figure 1. In situ stress state of roadway.

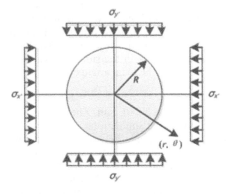

Figure 2. Mechanical model of circular roadway.

The stress state of roadway does not vary from locations in direction. Therefore, we take a section of roadway as object. Assuming that the surrounding rock of roadway is homogeneous, isotropic, and elastic, the stress solution in polar coordinates was obtained. In Equation 2, σ_r, σ_θ, and $\tau_{r\theta}$ are radial stress, circumferential stress, and shear stress, respectively, as shown in Figure 2.

$$
\begin{cases}
\sigma_r = \dfrac{1}{2}(\sigma_1 \sin^2 \alpha + \sigma_3 \cos^2 \alpha + \sigma_2)\left(1 - \dfrac{R^2}{r^2}\right) \\
\quad + \dfrac{1}{2}(\sigma_1 \sin^2 \alpha + \sigma_3 \cos^2 \alpha - \sigma_2) \\
\quad \cos 2\theta\left(1 - \dfrac{R^2}{r^2}\right)\left(1 - 3\dfrac{R^2}{r^2}\right) \\
\sigma_\theta = \dfrac{1}{2}(\sigma_1 \sin^2 \alpha + \sigma_3 \cos^2 \alpha + \sigma_2)\left(1 + \dfrac{R^2}{r^2}\right) \\
\quad + \dfrac{1}{2}(\sigma_1 \sin^2 \alpha + \sigma_3 \cos^2 \alpha - \sigma_2)\cos 2\theta\left(1 + 3\dfrac{R^4}{r^4}\right) \\
\tau_{r\theta} = \dfrac{1}{2}(\sigma_1 \sin^2 \alpha + \sigma_3 \cos^2 \alpha - \sigma_2) \\
\quad \sin 2\theta\left(1 - \dfrac{R^2}{r^2}\right)\left(1 + 3\dfrac{R^2}{r^2}\right)
\end{cases}
\tag{2}
$$

We can translate the stress solution in polar coordinates to rectangular coordinates as follows:

$$
\begin{cases}
\sigma_{x'} = \sigma_r \cos^2 \theta + \sigma_\theta \sin^2 \theta - \tau_{r\theta} \sin 2\theta \\
\sigma_{y'} = \sigma_r \sin^2 \theta + \sigma_\theta \cos^2 \theta + \tau_{r\theta} \sin 2\theta \\
\tau_{x'y'} = (\sigma_r - \sigma_\theta)\sin \theta \cos \theta + \tau_{r\theta} \sin 2\theta
\end{cases}
\tag{3}
$$

All six components of stress could be obtained by the above three equations. On the basis of simple cubic equations theories, we can obtain utility equations of principal stress as follows (Wang, 2014):

$$
\begin{cases}
\sigma_1 = \dfrac{I_1}{3} + 2\sqrt{-\dfrac{p}{3}}\cos \dfrac{\beta}{3} \\
\sigma_2 = \dfrac{I_1}{3} - \sqrt{-\dfrac{p}{3}}\left(\cos \dfrac{\beta}{3} - \sqrt{3}\sin \dfrac{\beta}{3}\right) \\
\sigma_3 = \dfrac{I_1}{3} - \sqrt{-\dfrac{p}{3}}\left(\cos \dfrac{\beta}{3} + \sqrt{3}\sin \dfrac{\beta}{3}\right)
\end{cases}
\tag{4}
$$

$$
I_1 = \sigma_{x'} + \sigma_{y'} + \sigma_{z'}
$$
$$
I_2 = \sigma_{x'}\sigma_{y'} + \sigma_{y'}\sigma_{z'} + \sigma_{z'}\sigma_{x'} - \tau_{x'y'}^2 - \tau_{y'z'}^2 - \tau_{x'z'}^2
$$
$$
I_3 = \sigma_{x'}\sigma_{y'}\sigma_{z'} + 2\tau_{x'y'}\tau_{y'z'}\tau_{x'z'} - \sigma_{x'}\tau_{y'z'}^2
$$
$$
\quad - \sigma_{y'}\tau_{x'z'}^2 - \sigma_{z'}\tau_{x'y'}^2
$$
$$
p = \dfrac{3I_2 - I_1^2}{3}; \quad q = \dfrac{9I_1I_2 - 2I_1^2 - 27I_3}{27};
$$
$$
\beta = \arccos\left[\dfrac{q}{2}\left(-\dfrac{p^3}{27}\right)^{-\frac{1}{2}}\right] \quad (0 \leq \theta \leq \pi)
$$

Combining equations 1, 2, 3, and 4, the analytical solution of principal stress of circular roadway could be obtained.

3 ANALYTICAL METHODS TO STRESS STATE

3.1 *In situ stress field case of underground mines*

With the increase in the demand of resources and energy, more and more underground mines in China have increasing depths. In deep mines, vertical in situ stress gradually increases to maximum in situ stress[4]. We assume a case that in situ stress field of the location of roadway is described in Table 1, which could be used to calculate the

Table 1. In situ stress fields of the location of roadway.

Vertical stress	Maximum horizon stress	Minimum horizon stress
20	18	8

stress value. In fact, the value does not reflect a true stress state, but it can reflect the effect of axial direction.

3.2 Elastic–plastic theories and MATLAB tool

The maximum stress theory holds that the maximum tensile stress or maximum compressive stress determines destruction of materials. Single shear strength theory holds that the difference between principal stresses determines the destruction of materials. Furthermore, principal deviator stress theory holds that plastic deformation of materials is determined by the maximum principal deviator stress, which can be calculated as follows (Yu, 2004):

$$\sigma_1' = \sigma_1 - \frac{1}{3}(\sigma_1 + \sigma_2 + \sigma_3) \qquad (5)$$

On the basis of the above equations and elastic–plastic theories, the maximum principal stress, difference between principal stresses, and the maximum principal deviator stress of roadway in designed in situ stress field are analyzed. Using MATLAB tool, the stress value of rock in different locations ($r=R$, $r=1.5R$ and $2R$) are calculated, which varies with the angle (α) between roadway direction and the maximum horizon in situ stress and position angle (θ).

4 ANALYSIS RESULTS OF STRESS STATE

Figures 3, 4, and 5 show change of stress with the angle between roadway direction and maximum horizon in situ stress.

From these figures, we can find that: When the vertical in situ stress is the maximum in situ principal stress, with the angle (α) between roadway direction and maximum horizon in situ stress increasing from 0 to $\pi/2$, the maximum principal stress, difference between principal stresses, and the maximum principal deviator stress of rock in roof and floor of circular roadway are gradually increasing, which is obviously lower than those of rock in two sides of roadway. Furthermore, the maximum principal stress, difference between principal stresses, and the maximum principal deviator stress of rock in two sides of roadway have the same tendency when the rock is far away from the roadway periphery. Therefore, the suitable angle (α) between roadway direction and maximum horizon in situ stress is nearly 0. The stability of rock on the two sides of the roadway should be considered.

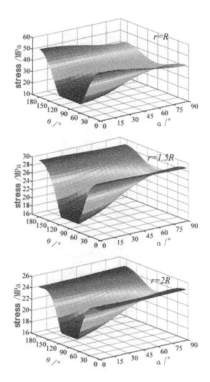

Figure 3.   Maximum principal stress in the in situ stress field.

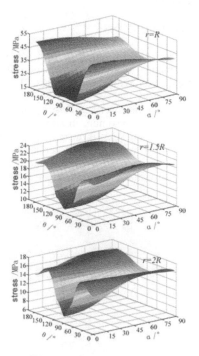

Figure 4.   Difference between principal stresses in the in situ stress field.

475

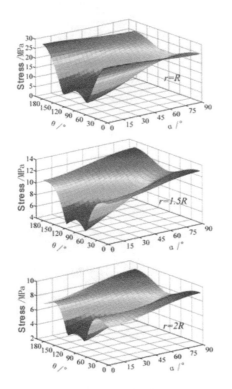

Figure 5. Maximum principal deviator stress in the in situ stress field.

5 CONCLUSIONS

1. The analytical solution of principal stress was obtained, which could be used to analyze the effect of axial direction on stress state of circular roadway.
2. In deep mines, vertical in situ stress gradually increases to maximum in situ stress. When the vertical in situ stress is the maximum in situ principal stress, rocks on the either sides of roadway have a higher stress concentration, and more attention should be paid to ensure their stability.

3. When the vertical in situ stress is the maximum in situ principal stress, the suitable angle (α) between roadway direction and the maximum horizon in situ stress is nearly 0.

REFERENCES

Cai, M.F., Chen, CH.Z., Peng, H., et al., "In-situ stress measurement by hydraulic fracturing technique in deep position of Wanfu coal mine," Chin J Rock Mech Eng **25**(5), 1069–1074(2006).

Dong, F.T., "The support theory of maximum horizon stress and its application," Bolting Support **3**:1–4 (2000).

Gale, W.J., Fabjanczyk, M.W., "Design approach to assess coal mine roadway stability and support requirements". *Proceedings of the VIII Australian Tunnelling Conference "Finding Common Ground"*, 153 (1993).

Lv, P., Xi, X., Shao L.L., "Study of roadway layout in the σ_{HV} type in-situ stress field," *4th International Conference on Green Building, Materials and Civil Engineering* 513–516 (2015).

Ma, N.J., LI, J., Zhao, Z.Q., "Distribution of the deviatoric stress field and plastic zone in circular roadway surrounding rock," J China u Min Techno **44**(2), 206–213 (2015).

Miao, X.X., Qian M.G., "Research and prospect of China green mining of coal resources," Mining & Safety Eng **26**(1), 1–14 (2009).

Sun, Y.F., "Effects of in-situ horizontal stress on stability of surrounding rock roadway," J China Coal Soc **35**(6), 891–895 (2010).

Wang, K., "Calculation formulas for principal stresses," Mechanics in Engineering **36**(6), 783–785 (2014).

Yu, M.H., "Advances in strength theories for materials under complex stress state in the 20th century," Advances in Mechanics **34**(4), 529–560 (2004).

Zhao, W.S., Han, L.J., Zhang, Y.D., et al., "Study on the influence of principal stress on the stability of surrounding rock in deep soft rock roadway," Mining & Safety Eng **32**(3), 504–510 (2015).

Zheng, S.B., "3D geostress field distribution and roadway layout optimization in Sihe Mine," J China Coal Soc **35**(5), 717–722 (2010).

Advances in Energy Science and Equipment Engineering II – Zhou, Patty & Chen (Eds)
© *2017 Taylor & Francis Group, London, ISBN 978-1-138-71798-5*

Modeling of hydraulic fracture for concrete gravity dams considering hydromechanical coupling

Sha Sha & Guoxin Zhang
Department of Hydraulic Engineering, Tsinghua University, Beijing, China
China Institute of Water Resources and Hydropower Research, Beijing, China
State Key Laboratory of Simulation and Regulation of Water Cycle in River Basin, Beijing, China

ABSTRACT: Hydraulic fracturing with high pressure in the crack is an important part of the safety assessment of the concrete dams. The study of the hydromechanical coupling in the process of cracking is the key to accurately predict the crack propagation path and the corresponding bearing capacity in all cases. In this study, an isotropic continuum damage model is used to simulate the static fracture behaviour of the concrete gravity dams. The tensile damage variable is to describe the fracture mechanism of the concrete. Regarding the dam concrete as a saturated porous medium, based on Biot's classic coupling equations, the effect of the damage on the permeability coefficients and the pore pressure influence coefficients is considered. In this way, the hydromechanical coupling effect during the cracking process is considered. Using this model, the hydraulic fracturing of Koyna concrete gravity dam is simulated. The results show that: considering the hydraulic coupling effects, the deflection angle becomes smaller; the total propagation distance becomes longer and deeper. Both the critical load from linear to nonlinear and the ultimate load are reduced significantly.

1 INTRODUCTION

From the late 20th century, China has entered a golden age of water conservancy and hydropower construction. A large number of dams under bad hydrological and geological conditions such as high water head, large depth have started construction. Under the influence of factors such as design, construction and maintenance, the cracks occurring on the surface of the dam during construction was almost inevitable. These cracks affect the safety, practicality and durability of the structures. Concrete dam is a kind of concrete structure contacting with high pressure water, a s the dam age grows, water going into the cracks, these cracks may further develop under the high pressure and temperature difference, the hydraulic fracturing due to cracks on the surface is particularly prominent.

The interaction between the water in the cracks and the concrete gravity dam is a typical fluid-structure interaction problem. On one hand, the water pressure load on the crack surface, causing the crack deformation or further expansion, on the other hand the deformation and the change of the crack length result in fluid boundary changes, which affect the water pressure distribution. Therefore, consideration of fluid-structure coupling during the crack propagation is the key to accurately predict the crack path and the degree of risk in all cases.

The water pressure distribution during the crack propagation is studied by many scholars. Li Zongli et al. (2005) derived the internal water pressure distribution in rock and concrete fractures formed by hydraulic fracturing based on the law of mass conservation and momentum principle by the finite control volume approach. Fang Xiujun et al. (2007) used an extended finite element method to simulate the cracking process of concrete structures under water pressure. The fluid flow is simulated based on the cubic law and the cracking behavior of the concrete is described by the cohesive crack model. Based on the Navier-Stokes equations describing the flow, Yang Xiufu et al. (2002) simplified the equations by the boundary conditions and got the pressure distribution satisfying the Laplace equation in the crack. It is still more difficult to solve with the boundary conditions. Barani et al. (2016) proposed a partially saturated finite-element algorithm to numerically model the water pressure distribution within a propagating cohesive crack.

Currently there are a few of numerical methods for hydraulic fracture analysis, such as the finite element method, the boundary element method and the meshless method. Among them, the most widely used is the finite element method. There are four typical crack models for the finite element analysis: the discrete crack model, the smeared crack model, the special model embedding the crack through the transformation of the shape function and the continuum damage model. Many scholars have analyzed the static fracture response of the concrete gravity dams. They don't consider the water pressure in the crack or just regard it as a uniform or triangular distributed pressure (Gioia, 1992; Barpi, 2000; Shi, 2003; Shi, 2013; Vargas-Loli, 1989; Bhattacharjee, 1994; Wang, 2000; Calayir, 2005; Horii, 2003; Zhang, 2013; Wang, 2015; Ghrib, 1995; Cervera, 1995; Lee, 1998; Calayir, 2005; Omidi, 2013; Oudni, 2015). While the studies considering the hydromechanical coupling effect in the crack propagation of the concrete gravity dams are still few (Bhattacharjee, 1995; Plizzari, 1998; Bary, 2000; Alfano, 2006; Barpi, 2008; Wang, 2015).

In this study, the static fracture behavior of the concrete gravity dams is simulated by an isotropic damage model. A tensile damage variable is introduced to describe the fracture mechanism of the concrete. Based on the fracture energy conservation principles, the damage model is combined with the fracture mechanics. So the fracture energy dissipation is not affected by the finite element mesh size. Considering the concrete as a saturated porous medium, by the principle of the effective stress of the porous media, the hydraulic coupling effect during the cracking process of the concrete gravity dams is considered. The effect of the damage on the hydraulic parameters (the permeability coefficients and the pore pressure influence coefficients are reflected through the damage variable.

2 THE DAMAGE MODEL

In this study, the following nonlinear behavior are defined in the damage model for the saturated porous media: (1) the stress-strain relationship of the damaged elements; (2) change of the pore pressure influence coefficients b_i with the evolution of damage; (3) change of the permeability coefficients k with the evolution of damage. In the calculation, the compressive stress-strain relationship is assumed to be linear and elastic, because the compressive stress of the gravity dams normally does not exceed the compressive strength of concrete. The characteristics of this model are presented in the below.

2.1 The stress-strain relationship

About the expressions of the stress-strain relationship of the concrete under uniaxial tension, most scholars claim that the ascending part is linear, the main difference is the descending part, including linear descending, segment declining, curve descending. No matter what form is used, the fracture energy of the uniaxial tensile curve should be same. In this study, the exponential descending expression proposed by Jiang Jianjing (Jiang, 2013)

is adopted. Therefore under uniaxial tension, the stress-strain relation is expressed as follows:

$$\sigma(\varepsilon_t) = \begin{cases} E_0 \varepsilon_t & , \quad \varepsilon_t \leq \varepsilon_0 \\ f_{t0} e^{-\alpha(\varepsilon_t - \varepsilon_0)} & , \quad \varepsilon_t > \varepsilon_0 \end{cases}$$

where f_{t0} is the tensile strength; ε_0 is the strain corresponding to cracking, $\varepsilon_0 = f_{t0}/E_0$; α is the softening coefficient controlling the descending part and it is related to the fracture energy G_f of the materials. In order to avoid the grid size effecting the uniqueness of the analysis results, the softening coefficient α should be corrected according to the characteristic length l_t. l_t is the characteristic length of the crack band, for the plane element taking the square root of the related area for each integration point, for the solid element taking the cube root.

2.2 The damage evolution equation

According to the principle of the equivalent strain put forward by Lemaitre (Lemaitre, 1984), so there is

$$\sigma = (1 - D) E_0 \varepsilon \tag{2}$$

where E_0 is the elastic modulus of the material when it is undamaged; D is the damage variable. $D = 0$ corresponding to no damage, $D = 1$ corresponding to the complete damage, $0 < D < 1$ corresponding to different degrees of damage.

From the perspective of the damage mechanics, the nonlinearity of the stress-strain relationship of the rock and concrete is caused by the initiation and propagation of micro cracks due to the damage under the load. The brittleness is more obvious under the tension, so it is proper to describe the mechanical characteristics using the elastic constitutive equation of damage mechanics. Some researches and experiments show that there is no difference between the results analyzed by the elastic-plastic damage constitutive model and that of the elastic constitutive model. Therefore, the elastic damage model is used in this study.

By formula (1), (2), under uniaxial tension, the damage evolution equation of concrete can be expressed as:

$$D = \begin{cases} 0 & , \quad \varepsilon_t \leq \varepsilon_0 \\ 1 - \dfrac{\varepsilon_0}{\varepsilon_t} \cdot e^{-\alpha(\varepsilon_t - \varepsilon_0)} & , \quad \varepsilon_t > \varepsilon_0 \end{cases} \tag{2}$$

2.3 Hydromechanical coupling model

The pore water pressure p and porosity variation Δn are listed as state variables in Biot's (Biot, 1941; Biot, 1955) seepage-stress coupling equations, constituting the unknowns of the materials' constitutive equations with the stress tensor σ_{ij} and the strain tensor ε_{ij}, the basic equations for three dimensional problems are as follows:

$$\sigma_{ij,j} + X_i = 0 \tag{3}$$

$$\varepsilon_{ij} = \frac{1}{2}\left(u_{i,j} + u_{j,i}\right), \quad \varepsilon_v = \varepsilon_{11} + \varepsilon_{22} + \varepsilon_{33} \tag{4}$$

$$\sigma'_{ij} = \sigma_{ij} - bp\delta_{ij} = \lambda\delta_{ij}\varepsilon_v + 2G\varepsilon_{ij} \tag{5}$$

$$\Delta n = \frac{p}{Q} - \alpha\varepsilon_v = \frac{p}{Q} - \frac{\sigma_{ii}}{3H} \tag{6}$$

$$k_{ij}\nabla^2 p = \frac{1}{Q}\frac{\partial p}{\partial t} - \alpha\frac{\partial\varepsilon_v}{\partial t} \tag{7}$$

where ρ is the density; ε_v and ε_{ij} are the volumetric strain and the strain respectively; δ_{ij} is Kronecker constant; Q is Biot constant; G and λ are the shear modulus and Lame coefficient; ∇^2 is the Laplace operator.

In Biot's classic coupling equations, the item representing the effect of the stress on seepage is added in equation (7). It is the characteristic item of Biot's consolidation theory, reflecting the impact of stress on the fluid mass conservation. In the calculation of steady flow, the right item of the equation is zero, ignoring the interaction between the total stresses and the pore water pressure. According to the principle of the effective stress, the item of the pore water pressure is added in the medium deformation, reflecting that the deformation characteristics of the medium is under the influence of the pore water pressure. Moreover the stress causing pore deformation and the pore water pressure are discussed separately. However, the permeability changes induced by the stress are not taken into account in this theory, thus it cannot meet the momentum conservation. In this study, on the basis of Biot's theory, the changes of the permeability coefficients and the pore pressure influence coefficients caused by the damage are considered. The relationship between the pore pressure influence coefficients and the damage variable is as follows

$$b = b_0 + D(1 - b_0) \tag{8}$$

where b_0 is the initial isotropic Biot's coefficient.

Because of the uncertainty of the parameters in the coupling expression of seepage-stress and seepage-damage, usually the test is needed. As the rock mass is regarded as homogeneous medium composed by the skeleton grains and pores in the porous continuum model of the seepage in the rock mass, and in the micro level the concrete is also formed by the skeleton grains and pores, so in this study, the seepage-stress coupling model of the rock mass is adopted. In the fully coupling analysis of rock mass, usually the permeability coefficients are

regarded as variables, expressed as the function of the stress, strain, porosity ratio and damage. The change of porosity ratio usually means the deformation that is the change of the strain, so the permeability coefficients of the rock mass can be expressed as the function of the strain. Assuming that the change of the volume of the rock mass is equal to that of the pores, by the definition of the volumetric strain and the relationship between the porosity ratio and the porosity, the relationship between the volumetric strain and the porosity is as follows:

$$n = (\varepsilon_v + n_0)/(1 + \varepsilon_v) \tag{9}$$

where $\varepsilon_v = \varepsilon_x + \varepsilon_y + \varepsilon_z$, ε_v is the volumetric strain; n is the porosity, representing the ratio of the pore volume and the total volume, $n = e/(1+e)$; e is the porosity ratio representing the ratio of pore volume and grains volume. n_0 is the original porosity.

Based on Carman-Kozeny formula (Taylor, 1948), the relationship between the permeability coefficients and the volumetric strain is as follows:

$$k = k_0(1 + \varepsilon_v / n_0)^3 /(1 + \varepsilon_v) \tag{10}$$

Formla (10) not only includes the change of the permeability coefficients due to the variation of pore volume caused by the structural deformation when there is no damage, but also contains the change of the permeability coefficients when the damage and the large structural deformation occur. Therefore, equation (10) representing the relationship between the volumetric strain ε_v and permeability coefficient k reflects the relations of seepage-stress and seepage-damage. In this study, formula (8) and formula (10) reflect the effect of damage on the pore pressure influence coefficients and permeability coefficients, thereby taking the hydromechanical coupling into account during the cracking process.

3 HYDRAULIC FRACTURE ANALYSIS OF KOYNA CONCRETE GRAVITY DAM

In the past decades, Koyna gravity dam has been a typical engineering case for the static and dynamic analysis. A lot of researchers have studied it [5,8,10,15,20]. Koyna concrete dam was attacked by a strong earthquake and the neck of the dam was seriously damaged. The geometry and the finite element model of the dam are shown in Figure 1 and Figure 2. The height of the dam is 103 m. The width at the bottom is 70 m. The width at the top is 14.8 m. The width d at the neck is 19.3 m. In this study, a horizontal crack with the width $0.1d$ is set on the upstream side at the elevation of the downstream slope change. Crack at the location has been reported to be most critical to the

479

ultimate structural resistance. In the calculation, a plane stress state is assumed. The loads include the self-weight of the dam, the hydrostatic pressure corresponding to a full reservoir and an increment of water pressure due to reservoir overflow until the ultimate failure of the dam occurs. The dam is fixed at the bottom. The mechanical parameters of the dam concrete are shown in Table 1. In this study, two conditions that are without considering the hydromechanical coupling and considering the hydromechanical coupling are analyzed.

The relationship between the horizontal displacement at the top and the overflow is shown in Figure 3. From Figure 3, it is clear that without considering the coupling, the critical overflow from linear to nonlinear is 8.0 m and the ultimate overflow is 10.2 m. While considering the coupling, the critical overflow from linear to nonlinear is 5.0 m and the ultimate overflow is 8.2 m. Considering the coupling, the critical load decreases by 37.5% and the ultimate load decreases by 19.6%. Both the critical load from linear to nonlinear and the ultimate load are reduced significantly. Figure 4 and Figure 5 show the damage distribution of the two conditions at the failure. From Figure 4 and Figure 5, it can be seen that considering the coupling, the horizontal propagation distance gets shorter, the deflection angle gets smaller and the total propagation distance gets longer. The results agree well with that in literature (Wang, 2015).

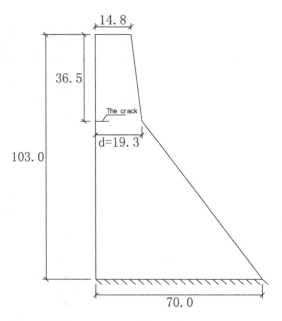

Figure 1.   The geometry of Koyna dam (Unit: m).

Table 1.   The mechanical parameters of the dam concrete.

E (GPa)	ν	ρ (kg/m³)	f_t (MPa)	G_f (N/m)
25	0.2	2450	1.0	100

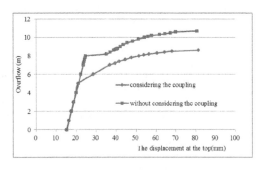

Figure 3.   The relationship between the horizontal displacement at the top and the overflow.

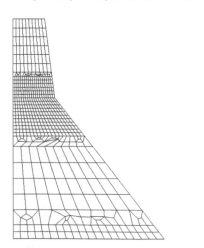

Figure 2.   The finite element model of Koyna dam.

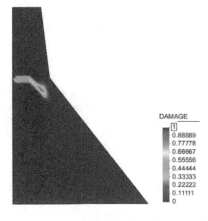

Figure 4.   The damage distribution without considering the coupling at the ultimate overflow 10.2 m.

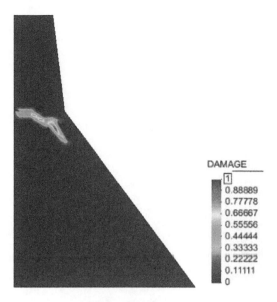

DAMAGE

1
0.88889
0.77778
0.66667
0.55556
0.44444
0.33333
0.22222
0.11111
0

Figure 5.   The damage distribution considering the coupling at the ultimate overflow 8.2 m.

4   CONCLUSION

In this study, a continuum damage model is adopted to simulate the static fracture behavior of concrete gravity dams. Based on the fracture energy conservation principle, the model is combined with the fracture mechanics, so the fracture energy dissipation is not influenced by the mesh size. Regarding the dam concrete as a saturated porous medium, the hydromechanical coupling effect is considered during the crack propagation process. Taking Koyna concrete gravity dam as an example for the hydraulic fracture analysis, the results show that the hydromechanical coupling effect has influence on the crack propagation paths, the propagation distance and the ultimate bearing capacity of the dam. Considering the coupling effect, the deflection angle becomes smaller; the total propagation distance becomes longer and deeper. Both the critical load from linear to nonlinear and the ultimate load are reduced significantly.

ACKNOWLEDGEMENT

This paper is sponsored by National Foundation of China (GN:51579252)

REFERENCES

Alfano, G., S. Marfia, E. Sacco, Comput. Method. Appl. M. **196**, 192–209, (2006).
Barani, O.R., S. Majidaie, M. Mosallanejad, J. Eng. Mech. **142**, 04016011, (2016).
Barpi, F., S. Valente, J. Eng. Mech. **126**, 611–619, (2000).
Barpi, F., S. Valente, Eng. Fract. Mech. **75**, 629–642, (2008).
Bary, B., J.P. Bournazel, E. Bourdarot, J. Eng. Mech. **126**, 937–943, (2000).
Bhattacharjee, S.S., P. Léger, J. Struct. Eng. **120**, 1255–1271,(1994) 3.
Bhattacharjee, S.S., P. Léger, J. Struct. Eng. **121**, 1298–1305, (1995).
Biot, M.A., J. Appl. Phs. **12**, 155–164, (1941).
Biot, M.A., J. Appl. Phs. **26**, 182–185, (1955).
Calayir, Y., M. Karaton, Comput. Struct. **83**, 1595–1606, (2005).
Calayir, Y., M. Karaton, Soil. Dyn. Earthq. Eng. **25**, 857–869, (2005).
Cervera, M., J. Oliver, R. Faria, Earthq. Eng. Struct. Dyn. **24**, 1225–45, (1995).
Fang, X.J., F. Jin, J. Hydraul. Eng, **38**, 1466–1474, (2007).
Ghrib, F., R. Tinawi, Earthq. Eng. Struct. D. **24**, 157–173, (1995).
Gioia, G., Z.P. Bazant, B.P. Pohl, Dam. Engrg. **3**, 23–34, (1992). 1.
Horii, H., S.C. Chen, Eng. Fract. Mech. **70**, 1029–1045, (2003).
Jiang, J.J., X.Z. Lu, *Finite element analysis of concrete structures* (Tsinghua Press, Beijing, 2013).
Lee, J., G.L. Fenves, Earthq. Eng. Struct. Dyn. **27**, 937–956, (1998).
Lemaitre, J., Nucl. Eng. Des. **80**, 233–245, (1984).
Li, Z.L., Q.W. Ren, Y.H. Wang, J. Hydraul. Eng. **36**, 656–661, (2005).
Omidi, O., S. Valliappan, V. Lotfi, Finite. Elem. Anal. Des. **63**, 80–97, (2013) 5.
Oudni, N., Y. Bouafia, Eng. Fail. Anal. **58**, 417–428, (2015).
Plizzari, G.A., Eng. Fract. Mech. **59**, 253–267, (1998).
Shi, M.G., H. Zhong, E.T. Ooi, C.H. Zhang, C.M. Song, Int. J. Fracture. **183**, 1–20, (2013) 2.
Shi, Z.H., M. Suzuki, M. Nakano, J. Struct. Eng. **129**, 324–336, (2003).
Taylor, D.W. *Fundamentals of Soil Mechanics* (John Wiley & Sons, Inc., New York, 1948).
Vargas-Loli, L.M., G.L. Fenves, Earthq. Eng. Struct. D. **18**, 575–592, (1989).
Wang, G.H., W.B. Lu, C.B. Zhou, J. Earthq. Eng. **19**, 991–1011, (2015) 4.
Wang, G.L., O. Pekau, C.H. Zhang, S.M. Wang. Eng. Fract. Mech. **65**, 67–87, (2000).
Wang, K.F., Q. Zhang, X.Z. Xia, Appl. Math. Mech. **36**, 970–980, (2015).
Wang, K.F., Q. Zhang, X.Z. Xia, Eng. Fail. Anal. **57**, 399–412, (2015).
Yang, X.F., M. Chen, X.S. Liu, J. South Petro Ins, **24**, 87–90, (2002).
Zhang, S.R., G.H. Wang, X.R. Yu, Eng. Struct. **56**, 528–543, (2013).

Advances in Energy Science and Equipment Engineering II – Zhou, Patty & Chen (Eds)
© 2017 Taylor & Francis Group, London, ISBN 978-1-138-71798-5

Overview of experimental studies on the seismic performance of rectangular hollow reinforced concrete bridge piers

Huige Wu, Xiuling Cao & Jinsheng Fan
College of Engineering, Hebei GEO University, Shijiazhuang, China

Yan Zhao
Institute of Disaster Prevention, Sanhe Hebei, China

ABSTRACT: Testing the seismic performance of bridge piers is the basis of the seismic analysis of a bridge. Taking into account the advantages of hollow sections and the multidimensional nature of earthquake ground motion, through the quasi-static test, the research of rectangular hollow reinforced concrete bridge piers under biaxial loading is of great significance. From the seismic performance of bridge piers under one-way loading, the seismic performance of bridge piers under biaxial loading, and the comparison of seismic performances of bridge piers under one-way loading and under biaxial loading, this paper makes a comprehensive analysis.

1 INTRODUCTION

Bridge, included in the transportation hub project, is an important part of lifeline engineering (Xiang, 2000; Fan, 1997; Han, 1999). The hollow section of the piers occupies a considerable proportion in large bridge projects, and these are located in high-seismic-intensity region. The destruction of reinforced concrete bridge piers leads to serious damage or even collapse of bridges, and it has become the main characteristic of bridge seismic damage during earthquakes. Once a bridge is destroyed, it will cause indirect economic losses such as traffic confusion, connection interruption of regional economy, delay in relief supplies, transport wound, and so on. Therefore, the research on seismic behavior of the bridge engineering has very important theoretical significance and engineering application value. During earthquake, ground motion has multidimensional nature.

Analysis of the seismic performance of bridge pier is the basis of the research on seismic performance of a bridge. From the seismic performance of rectangular, hollow section bridge piers under one-way loading and biaxial loading and the comparison of seismic performance of bridge piers under one-way loading and under biaxial loading, this paper makes a comprehensive generalization and summary. Taking into account the multidimensional nature of earthquake ground motion and the advantages of the hollow cross-sectional bridge piers, it is close to the engineering practice and thus plays an important role as engineering reference to

study the seismic performance of widely used reinforced concrete piers with different axial compression ratio subjected to combined uniaxial load and biaxial bending.

2 RESEARCH STATUS OF DOMESTIC AND INTERNATIONAL STUDIES

As a common form of bridge pier, the antiseismic capability of reinforced concrete bridge piers depends on three factors: ductility capability, bending resistance ability, and shear ability. Many factors influence the antiseismic capability of reinforced concrete bridge piers, including section type and size; axial compression ratio concrete strength; strength and ratio of longitudinal reinforcing steel; and strength, ratio, and forms of transverse reinforcing steel.

2.1 *Experimental research on the seismic performance of reinforced concrete bridge piers under horizontal uniaxial loading*

The damage forms of pier under horizontal seismic action include bending failure, bending shear failure, and shear failure. Bending failure is better than shear failure.

Hysteresis curve types include arch, fusiform, and S-shaped. Because the influence shear failure is larger than the bending failure to some extent such as wide extension, speed, and the effects on the structure of the safety, shear failures should be avoided in the seismic design (Sun, 2003). Scholars worldwide have conducted more research on the seismic performance of reinforced concrete bridge piers under the action of one-way horizontal load of solid.

2.1.1 Energy dissipation ability

The hollow, rectangular piers have a stable, effective way to dissipate energy through the plastic hinges formed at the bottom of the piers. The ability of energy consumption in elastic–plastic deformation process is one of the important indicators to measure its seismic performance. The higher index of consumed energy is related to the high energy consumption during earthquake. This contributes to the structure of the earthquake (Sun, 2003).

1. Effect of stirrup ratio

With the increase in the number of steel bars, the energy dissipation ability of the bridge generally increases. High stirrup ratio makes the structure still retain the higher ability of energy consumption when structure is in the near-failure stage. Considering the failure phenomenon, stirrup densification should be appropriately considered in plastic hinge area (Sun, 2003).

2. Effect of shear span ratio

Shear span ratio markedly affected energy consumption. When the reinforcement ratio is higher and shear span is small, in order to ensure energy consumption, a large number of stirrups and transverse reinforcement are necessary. When reinforcement ratio is small, requirements may be less (Sun, 2003).

3. Effect of reinforcement ratio

When stirrup or transverse reinforcement is enough, the increase of reinforcement ratio can increase the energy dissipation. Instead, the accumulated energy will decrease with the increase of reinforcement ratio.

2.1.2 Skeleton curve

1. Effect of reinforcement ratio

The descent dot of framework curve represents the beginning of the resistance and stiffness degradation. It indicates that the structure will soon damage after entering the stage in which the curve falls fast. Instead, even if the structure entered the stage of degradation it still maintains seismic performance (Fan, 1997). Stirrup ratio does not greatly affect the ultimate strength of the structure, but it has a significant effect on the decline of the skeleton curve. With the decrease in the stirrup ratio, skeleton curve falling speed significantly increases and there is obviously a turning point.

2. Effect of stirrup arrangement

The effect of stirrup form is little on skeleton curve. However, under the same conditions, it can improve the seismic performance of bridge piers to some extent that horizontal stirrups are changed into spiral stirrups.

3. Effect of shear span ratio

There is only little effect of the shear span ratio on the ultimate strength. However, when the shear span ratio is very small, the influence is evident on the skeleton curve. It is more apparent that the shear failure should be avoided.

2.2 Experimental research on seismic performance of reinforced concrete bridge piers under horizontal biaxial loading

Analysis of seismic performance of bridge pier is the basis of the research on seismic performance of bridge. Most of these studies only consider the one-way horizontal earthquake action. The force and deformation hysteresis model under one-way relationship under the horizontal action can only be used for nonlinear analysis of plane structure. When it is used to simulate the space problem, there will be a great discrepancy with the actual situation. This is because it cannot consider the effect of multidirectional coupling. Research results show that the component performance is significantly different under biaxial load and one-way load (Lau, 2001; Skjrbk, 1997; Park, 1985; Low, 1987; Du, 1999).

1. Seismic load-carrying capacity of the main shaft direction will decrease with the increase of deformation in the orthogonal direction, even if the corresponding displacement remains unchanged.

2. Damage accumulation of components in one direction may adversely affect the performance in another orthogonal direction. This kind of adverse effects shows that the degraded degree of strength and stiffness on two-axis direction is much more serious than the uniaxial loading condition.

Low et al. (1987) found that different combinations of various loadings have a great influence on the resistance, stiffness, and damage of columns.

Du (1999) carried out tests on reinforced concrete cantilever columns loaded at a certain angle and analyzed the effect of axial compression ratio, stirrup ratio, and reinforcement ratio of the column as well as loading angle on its seismic performance under inclined horizontal loads. He also simulated the characteristics of the restoring force of frame column in the two-way horizontal earthquake; however, there is a certain difference between the simulation and the actual two-way seismic action.

Qiu et al. (2001) studied the interaction in two principle directions and analyzed the cumulative hysteretic energy and damage under different loading paths by conducting two-way quasi-static tests following six different loading paths.

Plastic hinges were formed and rotated fully at the bottom of the model piers, which were subjected to combined axial load and biaxial bending. The longitudinal reinforcement bars at the bottom of the piers yielded and finally flexural failure occurred.

The seismic performance in the two horizontal directions is quite different under combined axial loading and biaxial bending. The ductility in the X-direction is superior to that in the Y-direction; however, its hysteretic energy and stiffness are smaller than those in the Y-direction.

An increase in the axial compression ratio can increase the overall bearing capacity of the specimen.

High axial compression ratio can effectively improve the energy dissipation capacity of the pier columns. However, the final total cumulative hysteretic energy is not magnificent because the specimen failed in the ninth loading cycles.

In conclusion, the response of components under multidirectional load is complicated. Despite some macroscopic phenomena and the preliminary model, many problems remain to be solved.

2.3 Comparison of seismic performance between uniaxial and biaxial compression-bending actions

Results of seismic damage analysis and test show that the damage of horizontal biaxial seismic action is much greater than that of the uniaxial seismic action to the structure. This is due to the fact that damage from one direction has direct influence on the seismic performance of another direction. In addition, the torsional vibration of the structure caused by the asymmetry of geometrical characteristics and physical characteristics also exacerbates the seismic response of structures.

2.3.1 Accumulated energy dissipation
Under the same conditions, the total accumulated energy consumption of the action of horizontal biaxial earthquake is greater than that of the horizontal uniaxial earthquake (Park, 1985).

2.3.2 Accumulated energy dissipation and accumulated damage (Park, 1985)
1. Under the same conditions, the accumulated damage of the action of horizontal biaxial earthquake in any direction is less than that of the horizontal uniaxial earthquake.
2. Under the same conditions, the total accumulated damage of the action of horizontal biaxial earthquake in two directions is greater than that of the horizontal uniaxial earthquake.

2.3.3 Degree of strength and stiffness degradation
Under the same conditions, the degree of strength and stiffness degradation of the action of horizontal biaxial earthquake is greater than that of the horizontal uniaxial earthquake (Park, 1985).

3 INTRODUCTION OF SEISMIC ANALYSIS METHOD OF BRIDGE PIERS

The problem of seismic analysis method of bridge piers has attracted great importance by bridge engineering worldwide. At present, the following five types of methods are commonly used in the analysis of the seismic performance of bridge pier.

3.1 Static analysis method

This method is more used for the calculation of bridge abutment and soil retaining structure. The concept and calculation formula are simple in the static analysis method. The structure is regarded as a rigid body in approximate calculation. The earthquake force is equal to the product of the earthquake coefficient and structure of the weight.

3.2 Nonlinear analysis method

This approach is a static analysis method considering elastic–plastic deformation ability under seismic action. In recent years, this method has more application and development. Using this method, we can evaluate the inelastic deformation process from yield to failure state of the structure, especially the nonlinear seismic response structures. It mainly includes two parts. The first is the analysis of the structure of the load displacement curve. The second is the evaluation of the seismic performance with the increase in monotonic lateral force. Compared with static method, it can estimate the deformation capacity of the whole or part structure.

3.3 Response spectrum method

Response spectrum analysis method is widely applied in the seismic design of structures. It can calculate the maximum seismic response of structures using the concept earth seismic load. Because of the consideration of the dynamic characteristic of the structure and the characteristics of ground motion, it has a great progress compared with the static analysis method. Response spectrum is the outer envelope curve in a certain natural frequency of the maximum displacement, velocity, and acceleration response under the condition of different ground motion. Its object is SDOF of different inherent frequencies.

3.4 Power spectrum method

Because it is plagued by the calculation method, the power spectrum method could not become popular, thereby limiting its application for the analysis of complicated bridge structure model. In recent years, pseudo-excitation method has appeared as a new method for the analysis of random response. This method can accurately and efficiently solve the heterogeneous incentive problem. Therefore, it is expected to be used in seismic design.

3.5 Dynamic history analysis method

Dynamic history analysis method is also called direct integral method. When the ground motion records or artificial wave act on the structure, it can obtain the seismic response values at any time through direct integration of motion equation. It can consider all kinds of problem such as complicated nonlinear factors, pile–soil interaction, and damping block. Dynamic history analysis method can calculate and analyze the elastic–plastic state structure. Compared with other methods, it is the most reliable method for structural seismic analysis. At the same time, it has also obvious advantages although there are more troubles of modeling, ground motion input, and postprocessing. Because of the development of computer technology, the processor speed becomes higher and the computing time of dynamic history method becomes much shorter. At present, dynamic history analysis method is recommended for the analysis of long-span or complex important bridge in most countries.

3.6 Finite element method

The calculation process of both linear and nonlinear dynamic history analyses is quite complicated. A great amount of calculation must be done through large-scale finite element analysis software. This is known as the finite element analysis method.

Currently, large analysis software used for the study of seismic performance of reinforced concrete bridge piers include Ansys, Abaqus, and OpenSees. A complex problem model can be established by software, and the efficiency and accuracy of the model will be high.

4 CONCLUSION AND DEVELOPMENT TRENDS

4.1 Conclusion

At present, scholars worldwide have done a great deal of fruitful research on the seismic performance of bridge. However, there are still some problems as follows:

1. Studies on seismic performance are less under the action of multidimensional earthquake about bridge piers. There is a lack of basic experimental data and theoretical research methods.
2. On the basis of experimental study, restoring force models are generally brought in a bending state or geometric features of a specimen. Thus, they are applied on a certain geometry and stress state of specimens.
3. Biaxial loading is difficult when the seismic performance of rectangular, hollow reinforced concrete bridge piers is tested through pseudo-static test of horizontal biaxial earthquake. Coupling effect is difficult to control from two-way force and displacement. Designing small load equipment and making detailed loading plan are necessary.

4.2 Development trends

Guidelines for Seismic Design of Highway Bridges (JTG/TB 02-01-2008) and Code for Seismic design of railway engineering (GB501111-2006) have not provided seismic design methods of hollow bridge, but they provided general principles of solid pier seismic design.

Some important research directions in the future include anticracking measures of hollow bridge pier under the water, residual displacement control of hollow bridge pier, improving the seismic behavior of hollow bridge pier by news structures and materials, and researching these misbehaviors of hollow bridge pier by modern experimental techniques.

A calculation method with high precision, short time, and numerical consideration will be studied and established.

ACKNOWLEDGMENTS

This study was financially supported by Scientific Research Plan Projects of Hebei Education Department (QN2014020); Scientific Research Plan Projects of Hebei Science and Technology Department (15275430); China and Australia cooperation research projects (16394507D); and Hebei overseas training project for excellent experts.

REFERENCES

Du Hongbiao. Seismic Behavior of reinforced concrete columns subjected to biaxial bending and compression [J]. Journal of Harbin University of Civil Engineering and Architecture, 1999, 32(4): 47–52.

Fan Lichu. Seismic Design of Buliding [M]. Shanghai: Tongji University Press, 1997.

Han Dajian, Ma Wentian. Prospect and Innovation of the Modern Suspension Bridge [J]. Journal of South China University of Technology 1999, 7(11): 57–67.

Lau Kintak, Zhou Limin. The mechanical behaviour of composite-wrapped concrete cylinders subjected to uniaxial compression load [J]. Composite Structures, 2001, 52(2): 189–198.

Low S S, Moehle J P. Experimental Study of Reinforced Concrete Column Subjected to Multiaxial Loading [R]. Report No. UCB/PEERC-87/14, California: University of California, Berkeley, 1987.

Park Y J A, Ang H S, Wen Y K. Seismic damage analysis of reinforced concrete buildings [J]. ASCE Journal of Structural Engineering, 1985, 111(4): 740–750.

Priestley M J N, Seible F, Xiao Y, Verma R. Steel jacket retrofitting of reinforced concrete bridge columns for enhanced shear strenth-part1: Theoretical considerations and test design [J]. ACI Structural Journal, 1994, 91(4): 394–405.

Priestley M J N, Verma R, Xiao Y. Seismic shear strength of reinforced concrete columns [J]. Journal of Structural Engineering, ASCE.

Qiu Fawei, Li Wenfeng, Pan peng, et al. Quasi-static Test Research of Reinforced Concrete Column Under Biaxial Loading [J]. Journal of Building Structures, 2001, 22(5): 26–31.

Shen Jumin, Liu Zhuqing, Weng Yijun. The experimental investigation of the seismic resistance behavior of reinforced concrete hollow core columns [J]. Journal of Buliding Structures, 1982, 3(5): 21–31. (in Chinese)

Skjrbk P S, Taskin B, Kirkegaard P H,et al. An experimental study of a midbroken 2-bay, 6-storey reinforced concrete frame subject to earthquakes [J]. Soil Dynamics and Earthquake Engineering, 1997, 16(6): 373–384.

Sun Zhuo, Yan Guiping, Zhong Tieyi, et al. Experiment Study on Anti-seismic Performance of Reinforced Concrete Bridge Piers Part I:Brief Introduction and Result of the Experiment [J]. China Safety Science Journal, 2003(1): 59–62.

Sun Zhuo, Yan Guiping, Zhong Tieyi, et al. Experiment Study on Anti-seismic Performance of Reinforced Concrete Bridge Piers Part II: Experimental Result Analysis and Conclusion, 2003(3): 46–49.

Xiang Haifan. Prospect of Worlds Bridge Projects in 21st Century [J]. China Civil Engneering Journal, 2000, 33(3): 1–6.

Advances in Energy Science and Equipment Engineering II – Zhou, Patty & Chen (Eds)
© 2017 Taylor & Francis Group, London, ISBN 978-1-138-71798-5

Deformation analysis and waterproof safety estimation of GINA gaskets in immersed tunnels

Lunlei Chai, Yongdong Wang, Xingdong Chen, Yongxu Xia & Xingbo Han
Key Laboratory for Bridge and Tunnel of Shaanxi Province, School of Highway, Chang'an University, Xi'an, China

ABSTRACT: GINA is a significant water proof barrier of immersed tube tunnel joint. The method of calculating the minimum allowing compression during operation was put forward by considering the GINA relaxation in condition of long-term tight compression. Two methods based on the deformation of Ω steel plate and the subsidence of immersed tube were employed to calculate the current compression. Then, the joint water proof performance was evaluated based on the GINA current compression, and the immersed tunnel safety classification was determined according to GINA residual compression. It is indicated that the working states of GINA gaskets in Yongjiang tunnel are stable currently and can meet the waterproof demand. The research method presented in this paper can be applied for similar engineering purposes.

1 INTRODUCTION

Immersed tunnels usually work under water for several decades. However, because of water flow, foundation of uneven subsidence, overburden thickness changes, and other factors, it is easy to produce floatation and sinking or translational motion, which would result in joint disengagement and leakage phenomenon, and even cause severe joint damages, thereby leading to scrapping of the entire tunnel (HU, 2014; Grantz, 2001; Grantz, 2001). Therefore, a key issue of the immersed tunnel is to ensure the water tightness of closure joints (LIU, 2011; LIU, 2009; LIU, 2008; LU, 2004). Since the early 1960s, VREDESTEIN ICOPRO company in the Netherlands has set the first watertight part between two sinking pipes, called GINA gaskets in a subway tunnel project in Rotterdam. This flexible waterproof material was widely used in traffic tunnels, cooling pipes, and other similar immersed channels worldwide (Peng, 2002; Zhao, 2007). In this paper, we take an immersed tunnel in China as an example to analyze its settlement, which had been in operation for 21 years. Deformation of GINA gaskets was calculated, according to the demand for water tightness of GINA gaskets and the effect of several errors, to assess its waterproof security. The methods presented in this paper are simple and easy to implement and can provide theoretical calculation for immersed tunnel construction and operation.

2 RELYING ENGINEERING

Yongjiang immersed tunnel was constructed in soft soil, which began its operation in September 1995. The length of the tunnel is 1019.7 m, with a 360.44 m approaching road in the northern part supported by a diaphragm wall, a 224.53 m approaching road in the southern part supported by a masonry retaining wall, and a 420 m immersed tube. The schematic diagram of the vertical section is shown in Figure 1. The immersed tube consists of five rectangular reinforced concrete sinking pipes (85 m +80 m +3 × 85 m). Flexible joints were used between tubes. The profile structure is shown in Figure 2. The outboard GINA rubber gaskets and W-type rubber-sealing belt behind GINA were equipped as two waterproof barriers of joints.

3 CALCULATION OF THE MINIMUM ALLOWING COMPRESSION

The compression value of GINA gaskets must ensure water tightness under the actual water pressure and can overcome pressure relaxation (Huang, 2010; Guan, 2004; Wang, 2007; Akitomo, 2002), uneven error, and butting error. Therefore, the calculation formula of allowance of minimal compression value of waterproof belt is (Liu, 2009; Shen)[5, 14]:

$$\delta = \delta_1 + \delta_2 + \delta_3 + \delta_4 \tag{1}$$

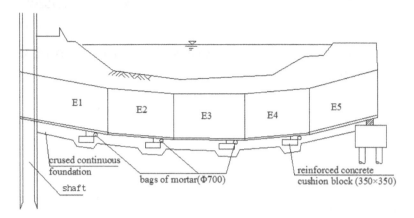

Figure 1. Vertical section of Yongjiang immersed tunnel.

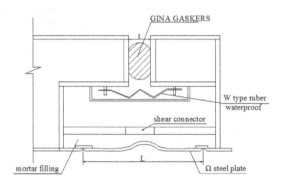

Figure 2. Joint structure of immersed tube.

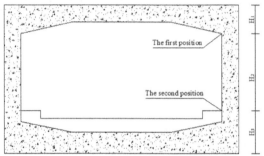

Figure 3. Measuring position of immersed tunnel.

where δ is the allowance of the minimal compression value of waterproof belt; δ_1 is the required compression value of water tightness under certain depth, which increases with the average water pressure, and generally, it will be given in product description; δ_2 is the relaxation of waterproof belt under long-term compression, which accounts for about 15% of total compression value; δ_3 is uneven error because of the unevenness of the immersed tube contact surface, and generally, it is 5 mm for each section; δ_4 is the construction error because of the nonparallel sides when sinking tube butting, and it can be acquired by calculating specific displacement error when sinking tube butting.

4 GINA GASKETS DEFORMATION ANALYSIS

4.1 Calculation based on the deformation of Ω steel plate

Figure 2 shows the joint structure of an immersed tube. The thickness and stiffness of the concrete of flank wall is much greater than that of GINA gaskets; therefore, the joint section can be treated as a rigid plate. In this way, the variation of GINA gaskets is equal to the variation of the distance between the two sections. That means:

$$\Delta t = t - t_d = L - L_d$$
$$\text{Or} \quad t = \Delta t + t_d = L - L_d + t_d \quad (2)$$

where Δt is the variation of GINA gaskets; t and L are the current thickness of GINA gaskets and width of Ω steel plate, respectively; and t_d and L_d are the thickness of GINA gaskets and width of Ω steel plate after construction, respectively.

On the basis of the geometric similarity principle, the thickness of GINA gaskets at top and bottom can be calculated from two data at the middle position in different heights, as shown in Figures 3 and 4 and formula (3)–(8).

$$x = \frac{\left(H_1 + : H_2\right)a - H_1 b}{H_2} \quad (3)$$

$$y = \frac{\left(H_3 + H_2\right)a - H_3 b}{H_2} \quad (4)$$

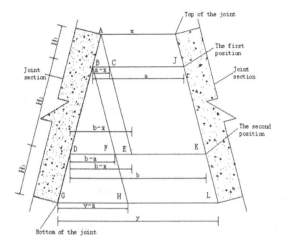

Figure 4. Deformation of GINA joints and calculation principle.

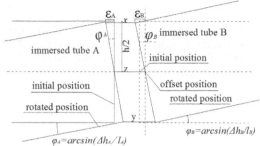

Figure 5. Joint sections rotation of immersed tube.

$$z = \frac{x+y}{2} \qquad (5)$$
$$\Delta t_t = x - t_d \qquad (6)$$
$$\Delta t_b = y - t_d \qquad (7)$$
$$\Delta t_m = z - t_d \qquad (8)$$

where x and y are the current thickness of GINA at the top and bottom, respectively; z is the average thickness of GINA in vertical direction, which is equal to the thickness of GINA at the middle, which can be regarded as the sum of initial thickness of GINA and the translation of joint sections caused by temperature variation, dry shrinkage of concrete, and so on. The difference between the top and bottom is caused by the rotation of the joint section, which can be calculated by the subsidence of the immersed tube.

4.2 Calculation based on the subsidence of immersed tube

As shown in Figure 5, when the joint sections are rotated, GINA at the top and bottom rebounded or compressed and the thickness of GINA gaskets is changed.

When the lengths of tubes are assumed as l_A and l_B, the differential settlements are Δl_A and Δl_B, respectively; and the variation of GINA, ε_A, and ε_B can be calculated by formulas (9) and (10):

$$\varepsilon_A = \frac{h}{2}\sin\varphi_A \qquad (9)$$

$$\varepsilon_B = -\frac{h}{2}\sin\varphi_B \qquad (10)$$

Considering the translation of tubes, the thickness of GINA at top and bottom can be calculated by formula (11) or (12):

$$t_t = z + \varepsilon_A + \varepsilon_B \qquad (11)$$

$$t_b = z - \varepsilon_A - \varepsilon_B \qquad (12)$$

where t_t and t_b are the thickness of GINA at top and bottom, respectively. Then, the maximum variation of the thickness of GINA (Δt) can be calculated by:

$$\Delta t = \max(t_t, t_b) - t_d \qquad (13)$$

Compared with the minimum allowing compression calculated by formula (1), the residual compression can be obtained as:

$$t_r = t_i - \Delta t - \delta \qquad (14)$$

where t_i and t_r are the initial compression and residual compression, respectively.

4.3 Waterproof safety analysis of joints

On the basis of the ratio of the residual compression to the initial compression, the waterproof safety can be classified into four levels, as shown in Table 1.

The first level is the condition of normal operation, and we only need to monitor the operation of immersed tunnels. However, in others, some measures must be taken. Specifically, the frequency of monitoring should be increased and the forecast of displacement is needed in the second level. And

Table 1. Classification of the waterproof safety about GINA.

Levels	I	II	III	IV
t_r/t_i	>0.5	≤0.5	≤0.3	≤0.1

prewarnings should be issued, measures must be taken to control variations in the third level. The forth level is alerting, and measures must be taken to ensure the safety of the tunnel and the vehicles, and, if necessary, the tunnel may be closed until the measures can ensure safety.

5 CASE STUDY OF YONGJIANG IMMERSED TUNNEL

5.1 Calculation of the minimum allowing compression

The minimum allowing compression value δ can be calculated based on formula (1) and the results are listed in Table 2. Specifically, the minimum compression value (δ_1) satisfying water tightness can be obtained according to the product manual of the GINA gaskets and the water depth of each immersed tube. On the basis of the flabby value of the joint after working 100 years is 15% of its initial compression, The flabby value (δ_2) after working 21 years can be obtained by the method of interpolation. The error of δ_3 caused by the uneven contact surfaces between tube segments is assumed equal to 5 mm × 2. The connection error (δ_4) between two segments can be calculated according to the displacement in the axial direction of tube segments.

5.2 Calculation of variations in GINA

According to the data measured in April 2015, the variations in the thickness of GINA gaskets were calculated based on the deformation of Ω steel plate (method 1) and the subsidence of immersed tube (method 2). The results are shown in Tables 3 and 4. In order to reduce error, we take the average of the results of the two methods as the maximum variations of GINA gaskets, as shown in Table 5.

According to Tables 3 and 4, the maximum variation of GINA is between E4 and E5, which reaches 12.56 mm (11.63 mm). It means that the unsubstantial position maybe here. Therefore, the measurement of E4~E5 should be enhanced. And the variation of E2~E3 was different from others,

namely the compression was increased, which was beneficial for the waterproof characteristic of GINA.

5.3 Waterproof safety estimate

According to the original compression after the construction, the compressions of GINA at present and after residual compression can be calculated, as show in Table 6. Compared with the minimal compression, the levels of GINA gaskets are given in Table 7.

According to Tables 6 and 7, waterproof safety levels of GINA in Yongjiang tunnel are in the

Table 3. Thicknesses variations of GINA gaskets based on method 1 (mm).

Joints position	E1~E2	E2~E3	E3~E4	E4~E5
t_d	111	110	115	119
x	117.05	108.44	109.35	118.10
y	99.74	104.60	126.66	131.56
z	108.39	106.52	118.01	124.68
Δt_t	6.05	−1.56	−5.65	−0.90
Δt_b	−11.26	−5.40	11.66	12.56
Δt_m	−2.61	−3.48	3.01	5.68

Table 4. Thicknesses variations of GINA gaskets based on method 2 (mm).

Joints position	E1~E2	E2~E3	E3~E4	E4~E5
z	108.39	106.52	118.01	124.68
$\varepsilon_A + \varepsilon_B$	8.53	−1.5	−7.98	−5.95
Δt_t	5.92	−4.98	4.97	−0.27
Δt_b	−11.13	−1.98	10.99	11.63

Table 5. Maximum variations of GINA gaskets (mm).

Joints position	E1~E2	E2~E3	E3~E4	E4~E5
Method 1	6.05	−1.56	11.66	12.56
Method 2	5.92	−1.98	10.99	11.63
Average	5.99	−1.77	11.33	12.10

Table 2. Calculation for allowing minimal compression (mm).

Joint position	E1~E2	E2~E3	E3~E4	E4~E5
δ_1	18	13	10	10
δ_2	1.86	1.86	1.71	1.53
δ_3	10.0	10.0	10.0	10.0
δ_4	1.4	4.9	0.42	2.8
δ	31.26	29.76	22.13	24.33

Table 6. Compressions of GINA at present and after residual compression calculation (mm).

Joints position	E1~E2	E2~E3	E3~E4	E4~E5
Original	62	62	57	51
Variations	5.99	−1.77	11.33	12.10
Currently	56.01	63.77	45.67	38.90
Residual	24.75	34.01	23.54	14.57

Table 7. Critical value of Yongjiang tunnel (mm).

Joints position levels	E1~E2	E2~E3	E3~E4	E4~E5
I	15.37	16.12	17.435	13.34
II	15.37	16.12	17.435	13.34
III	9.22	9.67	10.46	8.00
IV	3.08	2.22	3.49	2.67

normal operation state. The residual compression of GINA gaskets are enough for the joints to keep the waterproof safety of Yongjiang tunnel. Monitoring of joints is still important.

6 CONCLUSION

1. GINA gaskets are the significant waterproof barrier of the immersed tube tunnel joint. The methods of calculating the variation of GINA gaskets presented in this paper provide a set of effective research ideas. The specific calculation method can be applied for similar project.
2. The methods proposed in this paper are simple and easy to implement. The waterproof safety can be easily monitored in real time through simple procedure and measuring device for the deformation of Ω plate and settlement of tube.
3. The residual compression of GINA gaskets is enough for joints to keep the waterproof safety of Yongjiang tunnel.
4. The differential settlement between tube segments is a basic factor, which affects waterproof safety of joint. We can control and even adjust tube settlement by adopting some measures such as top dredging and bottom grouting, to reduce the open content of GINA joint to ensure its safety.

REFERENCES

Akitomo, K., Y. Hashidate, H. kitayama, Proceedings of the 2002. International Symposium on Underwater Technology. Onohama: IEEE New York, 81–86 (2002).
Guan, M.X. Modern Tunnelling Technology, 6, 57–59 (2004).
Grantz, W.C. Tunneling and Underground Space Technology, 16, 195–201, 2001.
Grantz, W.C. Tunneling and Underground Space Technology, 16, 203–210, 2001.
Hu, Z.N., P. Yang, C. Shan, Y.Z. Ren, D. Dang., Chinese Journal of Underground Space and Engineering. 34, 937–943 (2014).
Huang, F. Structural Engineers, 1, 96–102 (2010).
Liu, Z.G., H.W. Huang, Chinese Journal of Underground Space and Engineering, 2, 347–353 (2009).
Liu, Z.G., H.W. Huang, D.M. Huang, Chinese Journal of Underground Space and Engineering, 7(4), 691–694 (2011).
Liu, Z.G., H.W. Huang, Y.H. Zhao, Chinese Journal of Underground Space and Engineering, 4(6), 1110–1115 (2008).
Lu, M., Z.Y. Lei, Modern Tunneling Technology, 2, 236–242 (2004).
Peng, S.y., D. Guo, Port & Waterway Engineering, 9, 13–16 (2002).
Shen, H.C. Tunneling and Underground Works, 14(1), 31–35 (1993).
Wang, Y.D., Y. Zhao, C.H. Sun, The National Highway Tunnel Academic Conference Proceedings in 2007, Chongqing University Press Club, 129–134 (2007).
Zhch, Zhao, J.Q. Huang, Modern Tunnelling Technology. 44, 5–8 (2007).

Advances in Energy Science and Equipment Engineering II – Zhou, Patty & Chen (Eds)
© 2017 Taylor & Francis Group, London, ISBN 978-1-138-71798-5

In situ observation of loess landslide affected by rainfall

Guangbo Du & Wankui Ni

College of Geological Engineering and Surveying, Chang'an University, Xi'an, Shaanxi, China

ABSTRACT: Monitoring stations are established on a loess slope in Xiji County, Ningxia Hui Autonomous Region, and the temperature, rainfall, evaporation, and soil moisture rate are observed. The results show that the variation of soil moisture content within the depth of 1 m is influenced significantly by evaporation and rainfall, wherein greater depth leads to more gentle change in the curve of soil moisture content. The rainfall over 19 mm/day intensity causes a surge in soil water content, and a surge in soil water rate is generally observed 48 h after heavy rainfall. The pull of the slope surface is the main channel for the rainwater to enter the slope body. Rainwater is poured into the slope through the ground fissures, which leads to the reduction in the shear strength of the soil. The saturated seepage flow in some areas of the slope generates dynamic water pressure, which results in an increase in the sliding force of the slope. When the sliding force is larger than the antisliding force, slope slips down.

1 INTRODUCTION

Rainfall is one of the most important and active factors to induce landslide, and the rainy season is the peak period for landslide disaster. Several scholars worldwide have devoted extensive research effort to the investigation of the mechanism of rainfall-induced landslides. Jian (2013) put forward the rainfall infiltration model of landslide in the Three Gorges Reservoir area. Liu (2008) analyzed the rainfall infiltration mechanism of heavy rainfall-type landslide, and it was considered that the existence of cracks in the soil slope resulted in landslide over a large area after heavy rainfall. Li (2012) proposed a calculation model for the stability of shallow landslide induced by rainfall. Wang (2014) proposed the slope shape condition and formation condition of the landslide in loess area under the condition of continuous rainfall.

In terms of the infiltration of rainfall in the loess slope, Zhang (2011) studied the field survey and found that under continuous rainfall conditions, rainwater infiltration was in the loess depth of about 1 m. Tu (2009) conducted artificial rainfall experiment, and the result indicated that the depth of rainfall infiltration was not more than 3 m. Liu (2008) proposed by artificial rainfall experiment that the depth of the impact of rainfall was not more than 2.7 m. Wu (2006) conducted the water seepage test and reported that the maximum influence depth of water seepage was 6 m. Li (2014) conducted field monitoring and reported that soil moisture within 1.2 m below the surface was affected by rainfall and evaporation markedly.

In terms of the critical rainfall-induced landslide, Brand (1984) proposed that the rainfall intensity threshold of Hong Kong landslide was 70 mm/h. Wang (1982) put forward the statistics of Sichuan Basin rainfall-induced landslide data, wherein the landslide rainfall threshold was 200 mm/day. Zhang (1993) collected the records of the Yunyang area rainfall-induced landslide and reported that the landslide was induced by critical rainfall intensity of 140–150 mm/day.

Most of the studies have adopted statistical methods to obtain the information about the rainfall-induced landslide. In this study, observation equipment was installed on a loess slope in Xiji County, Ningxia, and the meteorological information and the soil moisture information were recorded under natural rainfall conditions to analyze the law of rainwater infiltration in loess slope and the mechanism of rainfall-induced landslide.

2 GENERAL SITUATIONS IN MONITORING AREA

The monitoring site is located at Baowei Village, Group 2, Shizi town, Xiji County, Ningxia; the geographical position is E105° 55' 30", N35° 43' 20"; the land form is hilly; and there is a gully region in the Loess Plateau. Xiji County has a typical continental semi-arid climate. Average annual rainfall is 570.2 mm and the evaporation is 1480.9 mm. The rainfall is more concentrated from July to September every year. The annual maximum temperature can reach 42°C, the minimum temperature can be 21.8°C, and the annual average temperature is around 12.7°C.

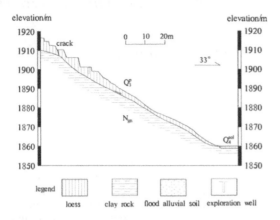

Figure 1. The slope profile.

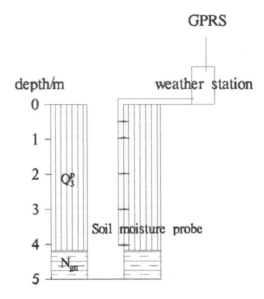

Figure 2. The equipment layout.

The monitoring target is Baowei village slope, which is 82 m long, 154 m wide, and 51 m high. The upper Pleistocene loess appears brown-yellow and the exposed thickness is about 3–5 m. The vertical joint is developed, the structure is loose, and the strength is low. The underlying strata of the Neogene mudstone appear orange-red with water disintegration and poor permeability. The edge of the slope is fractured, which leads to the formation of sinkholes, thus resulting in the development of a gully slope on the right-hand side from the top extending to the foot of the slope. It is 0.5 m wide and 0.4 m deep. There are grassland and crop on the surface of the slope. The slope profile is shown in Figure 1.

3 MONITORING SITES AND EQUIPMENT LAYOUT

The exploratory well was dug in a relatively flat part of the upper slope, and when the depth reached 4.2 m, the underlying new tertiary mud stone was exposed. Soil moisture probes were equipped on the walls of wells at depths of 0.5, 1, 2, 3, and 4 m, respectively, followed by back-filling of the wells. Soil moisture meter with measuring range of 0–100% and accuracy of 0.1% is provided by Changsha Yituo Sensing Technology Co., Ltd. Soil moisture information was recorded every hour. The equipment layout is shown in Figure 2.

A small weather station was established near the exploratory wells to monitor and collect the real-time weather information. In this research, TRM-ZS1-type meteorological instrument, manufactured by Jinzhou Yangguang Meteorological Science and Technology Co., Ltd., was adopted The collected meteorological information included temperature and rainfall. Weather information was recorded every hour.

The data collected by the soil moisture meter and the meteorological science were transmitted to the server in real time through the general packet radio service network to realize the remote reading.

4 ANALYSIS OF THE MONITORED RESULTS

In this paper, we analyzed the annual data from 26 August 2012 to 25 August 2013. Monthly evaporation data were collected from the local meteorological department during the observation period. The observed data are shown in Figure 3.

Analysis of the observed data produces the following results:

1. The amount of evaporation in the research area is much larger than that of rainfall; thus, evaporation is the main cause of water loss. Temperature, rainfall, and evaporation in the study area vary annually. Temperature reaches the lowest level in December and then increases gradually. Temperature acquires the highest value from May to August and then decreases gradually. The annual change of evaporation follows the trend similar to that of air temperature. The maximum evaporation occurs from May to June. Rainfall is mainly concentrated from July to September. The precipitation is little from November to March in the following year.

2. The rate of change of the moisture content of the soil below the earth's 1 m surface is obvious and is significantly influenced by evaporation

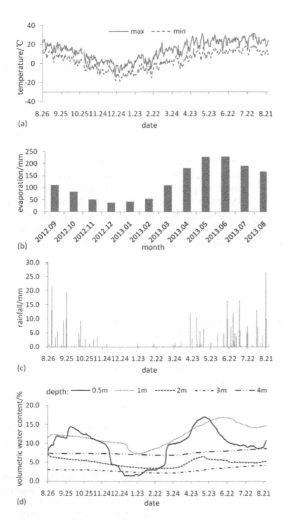

Figure 3. The observed data.

moisture content at 4 m was greater than that at 2 and 3 m in this area.

4. The soil moisture rate curve, exhibiting obvious changes at 0.5 m depth, indicates that the soil moisture rate on 4 September 2012, 27 September 2012, and 22 August 2013 increases dramatically. The amounts of rainfall on 1 September 2012, 25 September 2012, and 20 August 2013 are 21.4, 19.2, and 26.4 mm/day, respectively. Therefore, clearly, the rainfall over 19 mm/day can cause soil moisture surge, and surge in soil moisture rate is generally observed 48 h after heavy rainfall. The water from rainfall less than 19 mm/day quickly evaporates, which does not cause soil moisture surge.

5 ANALYSIS OF THE MECHANISM OF RAINFALL-INDUCED LOESS LANDSLIDE

The actual survey made us conclude that most landslides occurred in the research area in the rainy season. Therefore, rainfall has an important impact on the occurrence of landslides. According to the previous analysis, the depth of infiltration of atmospheric precipitation in soil is about 1 m. Smaller change in soil moisture content is observed at a greater depth. If we consider the infiltration of only rainwater through the surface, the influence of slope stability is limited. Most landslides occurred within a few days after heavy rainfall. Therefore, it was speculated that the role of rainfall in causing landslides must be in other forms.

According to field investigation, prior to the occurrence of the landslide, there must be obvious deformation damage phenomena, such as tensile cracks on the slope surface and sinkholes. If these channels extend to the soil rock contact surface, they become rain circulation channels during a rainy day, which are sometimes extended to the foot of the slope. These flow channels make rainwater enter the slope, which is the main factor of rainfall-induced landslide. The process and mechanism of rainfall-induced loess landslide is analyzed as follows:

1. The research object is the geotechnical slope structure; soil moisture migration is blocked in soil rock contact surface and gathered in the area. Weak zone in the rock and soil contact surface is formed due to water disintegration characteristics of Neogene mudstone.
2. The studied area has the characteristic "upper thick and lower thin," being the thinnest at the foot of the loess. The soil moisture at the contact zone of the slope rock can be influenced by gravity and moves down, thus forming a range of saturated zones at the foot of the slope. Moreover, it reduces the strength of the loess.

and rainfall, which shows the features of annual changes. The water content of soil was the lowest in December and then increased gradually. It reached the maximum from May to June and then decreased gradually. Affected by evaporation, the soil water content at 0.5 m depth is usually lower than that at 1 m depth. Only when heavy rainfall occurs, it is higher than that at 1 m depth for a short time.

3. Soil moisture content at 2–4 m depth also exhibited a changing trend. The water content reached the lowest level in March and then increased gradually. It reached the peak level in August and then decreased gradually. With the increase in the depth, more gentle change in the soil moisture content was observed. Water infiltration was blocked at the depth of 4.2 m due to the impermeable mudstone. Therefore, the soil

Furthermore, during heavy rainfall day, the slope gully erosion of the water moves down to airport surface, which results in the shear stress concentration at the foot of the slope. When the shear stress at the foot is higher than the effective shear strength of the weak zone, partial damage occurs at the foot of the slope. The stress concentration caused by the initial damage along the slope in the weak zone gradually moves to the uphill direction so that local failure propagation moves to the uphill direction (Urciuoli, 2007). The progressive destruction leads to the occurrence of the slope soil relative displacement from the foot to uphill direction, thus forming the slope surface tension cranny.

3. When heavy rainfall occurs, which is beyond the infiltration capacity of the surface layer soil, the slope surface generates runoff and moisture falls from slope surface cracks and the water hole into the contact surface between the slope and rock. If water is present in the cracks, the pressure penetration further increases the infiltration of rainwater. The rain forms a saturated zone in the soil, and the shear strength of the soil can be reduced. At the same time, the saturated seepage flow in some areas of the slope generates dynamic water pressure, which results in an increase in the slope sliding force. When the slope of the sliding force is greater than that of the antisliding force, the slope faces stability failure.

According to the previous analysis, when the rainfall intensity is more than 19 mm/day, the rain infiltration leads to soil moisture surge. For the surface-fractured slope, rainfall with intensity greater than the critical rainfall intensity forms several moisture surge regions on the inner slope, suddenly decreasing the antislide force of the overall slope, which readily causes landslide. Therefore, the rainfall intensity of 19 mm/day can be adopted as an early warning rainfall intensity of rainfall-induced landslide in the research area.

6 CONCLUSIONS

1. The amount of evaporation in the research area is much larger than that of rainfall, thereby becoming the main cause of water loss. The variation of soil moisture content within the depth of 1 m is significantly influenced by evaporation and rainfall, with the characteristics of the annual changes. With the increase in the depth, the change in soil moisture content becomes moderate.

2. Rainfall intensity more than 19 mm/day can cause soil moisture surge, and surge in soil moisture rate is generally observed 48 h after heavy rainfall. Water from rainfall with intensity less than 19 mm/day quickly evaporates and does not cause soil moisture surge.

3. Rainfall is an important factor inducing loess landslides. The effect of rainwater infiltration from the surface is limited on slope stability. The free face at the slope toe causes tensile cracks, which are the main flow paths of rainwater infiltration.

4. When rainwater pours into the slope through the ground fissure, it forms a certain range of saturated area in the slope body; therefore, the shear strength of the soil is reduced. The saturated seepage flow in some area of the slope generates dynamic water pressure, which results in an increase in the sliding force on the slope. When the sliding force is larger than the antisliding force, the slope slips down.

REFERENCES

Brand, E.W., J. Premchitt, H.B. Phillipson. Relationship between rainfall and landslides in HongKong. In *Proc. Proceeding of 4th International Symposium Landslides*, Toronto, 377–384 (1984).

Jian, W.X., J. Xu, L.Y. Tong. Rainfall infiltration model of Huangtupo landslide in Three Gorges Reservoir area. *Rock and Soil Mechanics* (12): 3527–3533 (2013).

Li, N., J.C. Xu, Y.Z. Qin. Research on calculation model for stability evaluation of rainfall-induced shallow landslides. *Rock and Soil Mechanics* (05): 1485–1490 (2012).

Li, P., T.L. Li, Y.K. Fu. In-situ observation on regularities of rainfall infiltration in loess. *Journal of Central South University* (10): 3551–3560 (2014).

Liu, H.S., W.K. Ni, H.Q. Yang. Site Test on Infiltration of Loess Subgrade under Rainfall Circumstance. *Journal of Earth Sciences and Environment* (01): 60–63 (2008).

Liu, L.L., K.L. Yan. Analysis of rain fall infiltration mechanism of rainstorm landslide. *Rock and Soil Mechanics* (04): 1061–1066 (2008).

Tu, X.B., A. Kwong, F.C. Dai, H. Min. field monitoring of rainfall infiltration in a loess slope and analysis of failure mechanism of rainfall-induced landslides. *Engineering Geology* 105: 134–150 (2009).

Urciuoli, G., L. Picarelli, S. Leroueil. Local soil failure before general slope failure. *Geotechnical & Geological Engineering* (25): 103–122 (2007).

Wang, L.S., Y.G. Li, Z. Zhan. Development characteristics of the rock landslide in the Sichuan basin during the heavy rain in 1981. *Discovery of Nature* (01): 44–51 (1982).

Wang, N.Q., W.J. Wu. Landslides in Gansu province. (2006): Lanzhou University Press.

Wang, T., X.L. Zhao, P.R. Qi, Effects of continuous heavy rainfall on landslide geological hazards in the Loess Area. *Geological Survey and Research* (03): 224–229 (2014).

Zhang, M.S., T.L. Li. Triggering factors and forming mechanism of loess land-slides. *Journal of Engineering Geology* (04): 530–540 (2011).

Zhang, N.X., Z.P. Sheng, A.Z. Sun. Study on the bedding bank slope in the Yangtze river three gorges reservoir area. (1993): Earthquake Press.

Advances in Energy Science and Equipment Engineering II – Zhou, Patty & Chen (Eds)
© *2017 Taylor & Francis Group, London, ISBN 978-1-138-71798-5*

Experimental study on parallel strand bamboo lumber beams

Yulin Li
College of Civil Engineering, Nanjing Forestry University, Nanjing, China

Xiaohong Xiong
Jiangxi Precious Bamboo Company Ltd., Ganzhou, China

Haitao Li
College of Civil Engineering, Nanjing Forestry University, Nanjing, China

Kangwen Wu
Jiangxi Precious Bamboo Company Ltd., Ganzhou, China

Xiaochen Mei, Zhixia Zang, Wei Qin, Chenwei Wang & Yuan Yuan
College of Civil Engineering, Nanjing Forestry University, Nanjing, China

Yeqing Xie & Ziwen Yan
Jiangxi Precious Bamboo Company Ltd., Ganzhou, China

ABSTRACT: Parallel strand bamboo lumber beam tests have been performed for investigating the bending properties of structural parallel strand bamboo lumber. The PSBL beams are characterized by the brittle tensile damage because the tensile strength of the bamboo is sensitive to the stress concentration emerged in the area with defects. The average strain for the cross-section of the parallel strand bamboo lumber beam is linear distribution during the test process. Thus the plane cross-section assumption can be used for calculating the glued bamboo beam when designing. Compressive strength for PSBL hasn't been brought into full play. All the deflections for the ultimate load point are more than 65 mm and far bigger than the value 7.2 mm (L/250) which is the maximum allowable deflection required by the Chinese wood structure design specification (GB50005–2003). The test results proved that the critical design criteria for parallel strand bamboo lumber structures should be deflection rather than strength.

1 INTRODUCTION

Bamboo is a green construction material which has the advantages of sustainability, environmental friendliness, and the capability of being reused or recycled. Usable bamboo can be harvested in 3–6 years from the time of planting, as opposed to traditional timbers which need decades between planting and harvesting (Mahdavi, 2011; Gottron, 2014). Compared with other common building materials, bamboo is stronger than timber, and its strength-to-weight ratio is greater than that of common wood, cast iron, aluminum alloy, and structural steel (Correal, 2010). Owing to remarkable mechanical response along with ease of access and environmental friendliness, bamboo has attracted numerous attentions, so far, to be used in construction industries as a sustainable construction material. In this regard, lots of attempts have been made so far to explore its mechanical

behavior in different ways (Verma, 2012; Sinha, 2014; Chen, 2015; Li, 2015; Li, 2013; Li, 2016; Li, 2016; Su, 2015; Li, 2016; Li, 2015; Li, 2015; Su, 2015; Li, 2016; Li, 2015; Huang, 2009; Cheng, 2009; Nugroho, 2000; Nugroho, 2001; Cui, 2012; Malanit, 2011; Ahmad, 2011; Wei, 2012).

Although bamboo is a promising wood substitute, structural forms in which it can be used are limited by the diameter of the bamboo culm and the low rigidity of the bamboo. To solve the limitation of member size and to enhance dimensional consistency, strength, and uniformity, the bamboo culm can be disassembled into bamboo filament bundles by passing through a roller press crusher and then glued together with adhesive to form certifiable structural members. The composite material is called Parallel Strand Bamboo Lumber (PSBL) (Su, 2015).

Su et al. (Su, 2015) investigated the mechanical performance of PSBL columns under axial compression. Li et al. (Li, 2016; Li, 2015) carried out a preliminary investigation of PBSL under axial compression for different directions based on large-scale specimens and also discussed the experimental study and analysis on parallel strand bamboo lumber column under eccentric compression preliminarily. Huang (Huang, 2009) has studied the aging resistant performance of parallel bamboo strand lumber from three aspects, including the basic material performance of PSBL. Cheng (Cheng, 2009) investigated the bamboo bundle preparation, glue immersion, forming and hot pressing process as well as the its performance, based on the foreign manufacturing processes of reconstituted lumber, to utilize the entire bamboo through improving equipment and produce technology.

Nugroho et al. (Nugroho, 2000; Nugroho, 2001) have conducted a study to determine the suitability of zephyr strand from moso bamboo (Pyllostachys pubescens Mazel) for structural composite board manufacture. Thirty-two $1.8 \times 40 \times 40$ cm Bamboo Zephyr Boards (BZB) were produced using four diameters of zephyr strand (9.5, 4.7, 2.8, and 1.5 ram) and four target densities (0.6, 0.7, 0.8, and 0.9 g/cm³). Results indicate that BZB exhibits superior strength properties compared to the commercial products. The size of the zephyr strand and the level of target density had a significant effect on the moduli of elasticity and rupture, internal bond strength, water absorption, and thickness swelling, but they did not have a significant effect on linear expansion. With regard to the physical properties, BZB exhibited less thickness swelling and exhibited good dimensional stability under dry-wet conditioning cycles.

Cui et al (Cui, 2012) has investigated the flexural characteristics of Parallel Strand Lumber (PSL), in which bamboo slivers with viscoelastic deformation were mixed with poplar veneer strands at a weight ratio of 1:4 as raw materials. The presence of prestress in bamboo slivers reinforced PSL significantly enhanced its flexural stiffness and toughness, while it had no obvious effect on its flexural strength.

Malanit et al (Malanit, 2011) produced thirty-six lab boards from bamboo strands with two manufacturing parameters varying, i.e., four resin types (MF, MUPF, PF, and pMDI) and three levels of resin content (7, 10, and 13%). The results indicate that Oriented Strand Lumber (OSL) made from bamboo strands exhibits superior strength properties compared to the commercial products made from wood for the building sector. The resin type has a significant effect on board properties. M. Ahmad et al (Ahmad, 2011) have analyzed the physical and mechanical properties of Parallel Strand Lumber (PSL) made from Calcutta

bamboo, which proved the suitability of Calcutta bamboo as raw material for structural composite products. A recent study by Wei et al (Wei, 2012) examined the failure of bamboo scrimber beams in detail, and concluded that the cross-sectional stiffness was the control condition for design load.

The majority of studies into the properties of PSBL have been concerned with boarding rather than structural members. As one new building material, a lot of basic theories haven't been solved. Thus, the further study is needed in order to prompt the use in civil engineering area. This study aims to examine the bending behaviour of PSBL structural members. Details of the systematic experimental investigation will be provided in the subsequent section. The test setup particulars and procedures used in studying the behavior of PSBL beams will be explained.

2 EXPERIMENTAL

The source Moso bamboo (Phyllostachys pubescens) was harvested at the age of 3–4 years. Bamboo strips from the upper growth heights, of a 2000 mm tall culm were selected. The culms cut from assigned growth portions were then split into 20 mm wide strips, aeamnd the outer skin (epidermal) and inner cavity layer (pith peripheral) were removed using a planer. All culm strips were split into bamboo filament bundles by passing through a roller press crusher. Then these bamboo filament bundles were dried and charred under the temperature of 165 degree centigrade and air pressure of 0.3 MPa. Finally the bundles (Fig. 1) were made into parallel bamboo strand lumber lumbers.

Phenol glue was used to manufacture the PSBL specimens. All bundles were put into many molds (Fig. 2), and these were then were pressed together to

Figure 1. Bamboo strand bundles.

form the blocks. A transverse compression of 20 MPa was applied for the blocks under the normal pressing temperature. The final moisture content was 8.22% and the density was 1018 kg/m³ for the laminate sourced from the upper portion. According to the compression tests for the specimens with the dimension of 96 mm × 96 mm × 285 mm, the compression strength for the PSBL is 65.3 MPa, with the modulus of elasticity of 11328 MPa, ultimate compression strain of 0.02 and the poison's ratio of 0.35.

PSBL beam specimens were constructed with the same design cross section of 50 mm x 107 mm. The span for all beams is 1800 mm.

The test arrangement is illustrated in Fig. 3. The displacement of the two supports, the loading points, and the mid-span deflection were measured by five laser displacement sensors (LDS type: Keyence IL-300, Japan). The beams were strain gauged longitudinally at the middle cross section, with five strain gauges pasted on one side face at even spacing through the depth, and one strain gauge pasted

on each of the bottom face and the top face, as shown in Fig. 3. The test was performed using a microcomputer-controlled electro-hydraulic servo testing machine (popwil instrument, Hangzhou, Zhejiang, China) with a capacity of 300 kN and a TDS data acquisition system. Tests were performed according to GB/T 50329–2012 (2012). Fig. 3 plots the experiment for beam specimen.

3 TEST RESULTS AND ANALYSIS

3.1 *Failure modes*

All the test specimens showed similar behaviour during the whole loading process and the final failure phenomenon are similar (Fig. 5). Each specimen behaved elastically at the beginning of loading. The deflection value becomes bigger and bigger with the increase of loading, and then the specimens showed a small amount of plastic deformation. Cracks (accompanied by a slight noise) appeared on the bottom tensile surface as the deflection became more and more obvious, where there were defects such as bamboo joints. More and more cracks appeared with the increasing of the loading. Finally one or two main cracks appeared in the middle bottom area of the beam specimens (accompanied by loading decreasing suddenly).

The final failure photos for four surfaces of the PSBL beam specimens can be seen from Fig. 5. The compressive strength for PSBL hasn't been given full play and that is why none failure phenomenon can be seen from the top face. Cracks can be seen clearly for all other three surfaces in Fig. 5, and the crack distribution area lies in the middle one third part of the beam.

3.2 *Load against strain*

Fig. 6 plots the typical load against the strain for the mid-span cross-section of parallel strand bamboo

Figure 2.   Molds for PSBL.

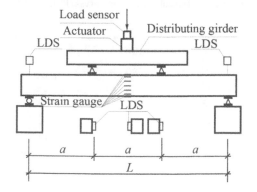

Figure 3.   Test scheme.

Figure 4.   Experiment for beam specimen.

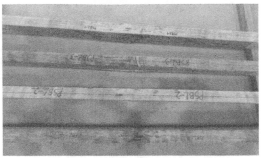

(a) Top face

(b) Side face I

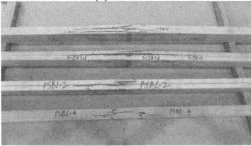

(c) Bottom face

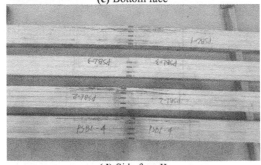

(d) Side face II

Figure 5.   Failure photos for four surfaces.

lumber beam specimens. Seven strain gauge values from the top surface to the bottom surface of the beam can be seen from left to the right in Fig. 6. Two main stages can be seen clearly for

the beam before the loading decreasing which are elastic stage and elastic-plastic stage. The ultimate compression strain value is 0.013 which is smaller than the ultimate compression strain of 0.02 from compression test results. It means that compressive strength for PSBL hasn't been brought into full play which is conformed to the failure photos for the top face in Fig. 5. This conclusion is similar as that for laminated bamboo lumber beams studied by Li et al. (Li, 2016; Li, 2015).

3.3   *Cross-section strain*

Fig. 7 plots the typical strain profile for the mid-span cross-section. Similar as laminated bamboo lumber beams studied by Li et al. (Li, 2016; Li, 2015), each test shows that the average strain for the cross-section of the parallel strand bamboo lumber beam is linear distribution during the test process. Thus the plane cross-section assumption can be used for calculating the parallel strand bamboo lumber beam when designing.

3.4   *Load-displacement response*

Figures 8 plot the load against displacement for all the beam specimens. It can be seen learly

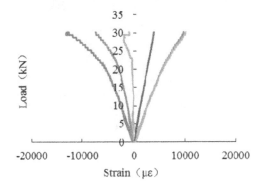

Figure 6.   Typical load-strain curves (PSBL-4).

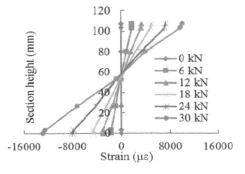

Figure 7.   Typical strain profile (PSBL-4).

that there was an excellent consistency in the load-displacement response for each of the four specimens. However, there was some scatter in the ultimate failure loads, which ranged from 27.9 to 32.5 kN. These specimens showed clear elastic behaviour up to a load of approximately 8 kN, followed by non-linear softening behaviour until the specimens failed suddenly and shed most of the load. The load decreased suddenly after the ultimate load. The specimens do not show any purely plastic behaviour prior to failure, so the failure is relatively brittle in nature.

4 TEST RESULTS ANALYSIS

The test results for all specimens are presented in Table 1. The average deflection at ultimate load was 85 mm (standard deviation 12 mm), which was far bigger than the maximum allowable design value of 7.2 mm. The serviceability limit state design specified in the Chinese wood structure design specification (GB 50005–2003) states that the middle deflection should be less than L/250 (where L is the span of the beam). The smallest percentile value for the test deflections was 69.8 mm, which is still 9.69 times the design value of 7.2 mm; however, the safety factors for timber structures are less than 9.69. These results show that both critical design

criteria (deflection and strength) can be used as the design criterion for shallow PSBL beams. However, the critical design criterion for PSBL structures is normally deflection rather than strength. In addition, the advantage of deflection over strength is that it can be measured directly on the beam. Consequently, the bending of modulus of elasticity is of interest. From another point of view, the stiffness of the beam needs to be improved to reduce the deflection and to make full use of the strength. These conclusion are similar as that for laminated bamboo lumber beams studied by Li et al. (Li, 2016; Li, 2015).

5 CONCLUSIONS

Parallel strand bamboo lumber beam tests have been performed for investigating the bending properties of structural parallel strand bamboo lumber. Similar as laminated bamboo lumber beams studied by Li et al. (Li, 2016; Li, 2015), the PSBL beams are characterized by the brittle tensile damage because the tensile strength of the bamboo is sensitive to the stress concentration emerged in the area with defects. The average strain for the cross-section of the parallel strand bamboo lumber beam is linear distribution during the test process. Thus the plane cross-section assumption can be used for calculating the glued bamboo beam when designing. Compressive strength for PSBL hasn't been brought into full play. All the deflections for the ultimate load point are more than 65 mm and far bigger than the value 7.2 mm (L/250) which is the maximum allowable deflection required by the Chinese wood structure design specification (GB50005–2003). The test results proved that the critical design criteria for parallel strand bamboo lumber structures should be deflection rather than strength.

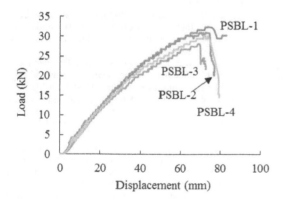

Figure 8. Load-displacement response.

Table 1. Test results.

Specimen	Ultimate load /F_{max} (kN)	Bending strength (MPa)	Ultimate deflection w /mm
PSBL-1	32.5	99.7	82.4
PSBL-2	30.9	95.5	70.3
PSBL-3	27.9	86.1	69.8
PSBL-4	30.1	92.9	71.3

ACKNOWLEDGMENTS

The material presented in this paper is based upon work supported the Project of Ministry of Housing and Urban-Rural Development of the People's Republic of China (No. 2014K4023), Project of the Housing and Urban-Rural Development Bureau of Jiang-su Province (No. JS2012ZD34), Doctoral Program Foundation of the Ministry of Education under Grant No. 20123204120012, A Project Funded by the Priority Academic Program Development of Jiangsu Higher Education Institutions, Students practical National University students practical and innovation training project (201610298003Z), and innovation training project of Nanjing Forestry University (2015sjcx095).

Any opinions, findings, and conclusions or recommendations expressed in this material are those of the writer(s) and do not necessarily reflect the views of the foundations. The writers gratefully acknowledge Nian-qiang ZHOU, Hui-juan ZHU, Jing MAO, Rong Liu, Yuan-long PENG, Sai-bo WANG, An PAN and others from the Nanjing Forestry University for helping with the tests.

REFERENCES

Ahmad, M., F. A. Kamke. Properties of parallel strand lumber from Calcutta bamboo (Dendrocalamus strictus). Wood Sci Technol, **45**(1): 63–72 (2011).

Chen, G., H. Li, T. Zhou, C. Li. Experimental Evaluation on Mechanical Performance of OSB Webbed Parallel strand bamboo I-Joist with holes in the web. Construction and Building Materials, **101**: 91–98 (2015).

Cheng, L. Manufacturing Technology of Reconstituted Bamboo Lumber. Huhehaote, China, Master degree thesis of Inner Mongolia Agricultural University, (2009).

China Building Industry Press. Chinese wood structure design specification (GB 50005–2003), Beijing. (2003).

China Building Industry Press. Standard for test methods of timber structures (GB/T 50329–2012), Beijing, (2012).

Correal, J.F., L.F. Ramire. Adhesive bond performance in glue line shear and bending for glued laminated guadua bamboo. Journal of Tropical Forest Science, **22**(4): 433–439 (2010).

Cui, H., M. Guan, Y. Zhu. The Flexural Characteristics of Prestressed Bamboo Slivers Reinforced Parallel Strand Lumber (PSL). Key Engineering Materials, **517**: 96–100 (2012).

Gottron, J., K.A. Harries, Q. Xu. Creep Behavior of Bamboo. Construction and building materials, Vol. **66**: 79–88 (2014).

Huang, X., The study on accelerated aging method and aging resistant performance of parallel bamboo strand lumber. Nanjing China, Master degree thesis of Nanjing Forestry University, (2009).

Li, H., A.J. Deeks, Q. Zhang, G. Wu. Flexural performance of laminated bamboo lumber Beam. BioResources, **11**(1): 929–943 (2016).

Li, H., G. Chen, Q. Zhang, M. Ashraf, B. Xu, Y. Li. Mechanical properties of laminated bamboo lumber column under radial eccentric compression. Construction and Building Materials, **121**: 644–652 (2016).

Li, H., G. Wu, Q. Zhang, G. Chen. Experimental study on side pressure LBL under tangential eccentric compression. Journal of Hunan University (Natural Science), **43**(5): 90–96 (2016).

Li, H., J. Su, A.J. Deeks, Q. Zhang, etc. Eccentric compression performance of parallel bamboo strand lumber column. BioResources, **10**(4): 7065–7080 (2015).

Li, H., J. Su, D. Wei, Q. Zhang, G. Chen. Comparison study on parallel bamboo strand lumber under axial compression for different directions based on the large scale. Journal of Zhengzhou University (Engineering Science), **37**(2): 67–72 (2016).

Li, H., J. Su, Q. Zhang, A.J. Deeks, D. Hui. Mechanical performance of laminated bamboo column under axial compression. Composites Part B: Engineering, **79**: 374–382 (2015).

Li, H., J. Su, Q. Zhang, G. Chen. Experimental study on mechanical performance of side pressure laminated bamboo beam. Journal of Building Structures, **36**(3): 121–126 (2015).

Li, H., Q. Zhang, D. Huang, A.J. Deeks. Compressive performance of laminated bamboo. Composites Part B: Engineering, **54**: 319–328 (2013).

Li, H., Q. Zhang, G. Wu. Stress-strain model under compression for side pressure laminated bamboo. Journal of Southeast University (Natural Science), **45**(6): 1130–1134 (2015).

Mahdavi, M., P. L. Clouston, S. R. Arwade. Development of laminated bamboo lumber: Review of processing, performance, and economical considerations. Journal of Materials in Civil Engineering, **23**(7): 1036–1042 (2011).

Malanit, P., C.B. Marius, A. Frühwald. Physical and mechanical properties of oriented strand lumber made from an Asian bamboo (Dendrocalamus asper Backer). European Journal of Wood and Wood Products, **69**(1): 27–36 (2011).

Nugroho, N., N. Ando. Development of structural composite products made from bamboo I: fundamental properties of bamboo zephyr board. J wood Sci, **46**: 68–74. (2000).

Nugroho, N., N. Ando. Development of structural composite products made from bamboo II: fundamental properties of laminated bamboo lumber. J wood Sci, **47**(3): 237–242 (2001).

Sinha, A., D. Way, S. Mlasko. Structural Performance of Glued Laminated Bamboo Beams. Journal of Structural Engineering, **140**(1): 04013021-1-8 (2014).

Su, J., F. Wu, H. Li, P. Yang. Experimental research on parallel bamboo strand lumber column under axial compression. China Sciencepaper, **10**(1): 39–41 (2015).

Su, J., H. Li, P. Yang, Q. Zhang, G. Chen. Mechanical Performance Study on laminated bamboo lumber column pier under axial compression. China Forestry Science and Technology, **29**(5): 89–93 (2015).

Verma, C.S., V.M. Chariar. Development of layered laminate bamboo composite and their mechanical properties. Composites part B: Engineering, **43**(3): 1063–1069 (2012).

Wei, Y., G. Wu, Q. Zhang, S. Jiang. Theoretical analysis and experimental test of full-scale bamboo scrimber flexural components. Journal of Civil, Architectural & Environmental Engineering, **34**: 140–145 (2012).

Advances in Energy Science and Equipment Engineering II – Zhou, Patty & Chen (Eds)
© 2017 Taylor & Francis Group, London, ISBN 978-1-138-71798-5

Study of the mechanical characteristics of the heavy-haul railway tunnel structure

Ziqiang Li, Mingnian Wang & Li Yu
School of Civil Engineering, Southwest Jiaotong University, Chengdu, China
Key Laboratory of Transportation Tunnel Engineering, Ministry of Education, Southwest Jiaotong University, Chengdu, China

ABSTRACT: In this paper, lining and base structure of the heavy-haul railway tunnel are studied as the research objects, according to the definite expression of heavy-load train, using the finite element software to analyze and study the stress characteristics of the single-line heavy-haul railway tunnel structure. The calculation results show that the dynamic response of structural internal forces of the tunnel lining and basement structure under the axle 27t passes in different surrounding rocks. Finally, the mechanical safety of the structure and surrounding rock is determined, providing theoretical basis for the design and construction of heavy-haul railway tunnel.

1 INTRODUCTION

Heavy-haul railway has some characteristics of axle load, major total transport, high-density traffic, and large volume. In general, the heavy-rail tunnel structure will have a greater impact than the ordinary rail tunnel (Hong, 2000; Yadong H, 2015). Heavy-load train axle is significant and the dynamic stress amplitude is large, thereby increasing the structural internal force and depth of the vertical transfer. Traffic density and large volume increase the load on the heavy-load railway tunnel structure, the static and dynamic responses play a significant role in the performance, and fatigue damage of the structure is obvious in train fatigue. Maintenance of heavy-haul railway tunnel is difficult, which will in turn affect the transport efficiency and safety. Especially when the surrounding rock is broken, the severity of the problem increases, which may affect the safety of the tunnel structure (L, 2015; L, 2016; Cdtr, 2005). Therefore, it is necessary to study the dynamic and static distribution of heavy-load railway tunnel under heavy-load conditions. In this paper, we simulate and analyze the single heavy-haul railway tunnel structural internal force of the static response and dynamic response under 27t train axle in different surrounding rocks using the finite element software. From this analysis, the stress characteristics of the structure of the heavy-haul railway tunnel are obtained, which provides reference and basis for the research of the heavy-haul railway tunnel.

2 THEORETICAL BASIS

2.1 Engineering situation

Taihangshan Mountain tunnel is located at the boundary of Shanxi Province and Henan Province. The starting and end points of the tunnel are marked by the stone town of Changzhi City and Anyang City, respectively. The scheme of the tunnel is double holes and single line and the line spacing is 30 m. The lengths of the left and right lines of the tunnel are 18.125 and 18.108 km, respectively. The tunnel adopts ballast-less track concrete layer, the curve is set in the inlet and outlet sections, and the other parts are straight lines. The lining section is shown in Figure 1.

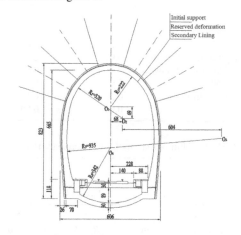

Figure 1. Lining section (Unit: cm).

2.2 Theoretical analysis of dynamic load

The results of the great deal of theoretical research and experimental work in the UK Railway Technology Center for many years show that the rail force of the train lies mainly in three frequency ranges: (1) Low-frequency range (0.5–5 Hz), which is almost entirely due to the relative motion of the car body and the suspension section; (2) intermediate-frequency range (30–69 Hz), which is created by the rebound effect of the spring group quality and rail; and (3) high-frequency range (200–400 Hz), which is due to the resistance of the movement of rail by wheel rail. Therefore, it can use an exciting force function to simulate the train load, which includes static and dynamic loads formulated with a series of sine function superposition:

$$F_t = A_0 + A_1 \sin\omega_1 t + A_2 \sin\omega_2 t + A_3 \sin\omega_3 t \quad (1)$$

where A_0 is the wheel static load; A_1, A_{12}, A_3 are peaks of vibration load corresponding to the circular flat rate of rail vibrations ω_1, ω_2, and ω_3, respectively.

If the unsprung mass of train is m, the corresponding amplitude of vibration load is:

$$A_i = m\, \alpha_i\, \omega_i^2 \ (i = 1, 2, 3) \quad (2)$$

where α_i is the typical vector height corresponding to the three conditions in Table 1 and ω_i is the circular frequency of vibration under the condition of irregularity control, $\omega_i = 2\pi v/L_i$. The wavelength is selected according to Table 1.

According to numerical simulation, the train axle is 27 tons. According to the above exciting force function, the dynamic load of the train can be obtained as shown in Figure 2.

2.3 Stress control standard

2.3.1 Structural stress safety control standard
The ultimate strength of the concrete and reinforced concrete structures of the *Railway tunnel design code* (TB10003-2005) are shown in Table 2.

Therefore, the axial compressive ultimate strength, flexural compressive ultimate strength, and ultimate tensile strength distribution of C20 jet concrete are 15, 18, and 1.3 MPa, respectively, and the elastic modulus is 21GPa.

2.3.2 Standard for safety control of the surrounding rock stress
According to the *tunnel engineering design points set*, the limit tensile stress of the surrounding rock is about 0.2 times of the cohesion of the surrounding rock, and the ultimate tensile stress of the surrounding rock at all levels is shown in Table 3.

2.3.3 Determination of section strength safety
When the earthquake force is not considered, the safety criterion of section strength is selected according to the railway tunnel design code (TB10003-2005). Criteria for determining the safety factor are listed in Tables 4 and 5.

3 FINITE ELEMENT MODEL

The distance between the left and right boundaries of the model to the structural section of the tunnel

Table 1. Irregular management value of track geometry.

Control conditions	Wavelength (m)	Versine (mm)
According to the ride comfort	50.0	16.0
	20.0	9.0
	10.0	5.0
Dynamic additional load on the railway line	5.0	2.5
	2.0	0.6
	1.0	0.3
Corrugation	0.5	0.1
	0.05	0.005

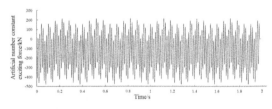

Figure 2. Exciting force of train vehicles.

Table 2. Ultimate strength of concrete (MPa).

| Type of strength | Symbol | Ultimate strength of concrete | | | | | |
		C15	C20	C25	C30	C40	C50
Compression	R_a	12.0	15.5	19.0	22.5	29.5	36.5
Bending compression	R_w	15.0	19.4	24.2	28.1	36.9	45.6
Tensile	R_l	1.4	1.7	2.0	2.2	2.7	3.1

Table 3. Ultimate tensile stress of surrounding rock.

Surrounding rock level	Cohesion (Mpa)	Ultimate tensile stress of surrounding rock (Mpa)
II	1.5–2.1	0.3–0.42
III	0.7–1.5	0.14–0.3
IV	0.2–0.7	0.04–0.14
V	0.05–0.2	0.001–0.04

Table 4. Strength safety factor of concrete structure.

Load combination	Main load	Main load + Additional load
The ultimate compressive strength of concrete	2.4	2.0
The ultimate tensile strength of concrete	3.6	3.0

Table 5. Strength safety factor of reinforced concrete structure.

Load combination	Main load	Main load + Additional load
Reinforced calculation strength or the concrete compressive or shear ultimate strength	2.0	1.7
The ultimate tensile strength of concrete	2.4	2.0

is three times the diameter of the tunnel. The lateral length of the model is about 44 m. The distance from the model to the top of the tunnel is three times the height of the tunnel and the distance between the model and the base of the tunnel is three times the height of the tunnel. The height of the model is about 63 m. The longitudinal length of the model is taken as the fixed distance of two heavy-haul train vehicles, so the length is 16 m. The model size is 16 m × 63 m × 44 m, as shown in Fig 3.

The buried depth of the tunnel is simulated by the equivalent gravity field. Monitoring points of the arch wall structure of single-line ballast-less track tunnel and surrounding rock monitoring points are shown in Fig 4. The monitoring points of the base structure (including the bottom plate and the invert) are shown in Fig 5.

4 FINITE ELEMENT ANALYSIS

Since the distribution of the monitoring points of the finite element model is symmetrical, the calculation results for the left-hand side image are presented.

4.1 Composite lining of III

Using FLAC3D separately on the tunnel excavation process and the train passes through Type III composite lining, the safety of surrounding rock and structure under 27t axle dynamic loads of heavy-haul train is analyzed.

4.1.1 Structural safety analysis
When the heavy-haul train passing Type III composite lining produces the maximum principal stress and minimum principal stress of early supporting and secondary lining arch wall structure as shown in Table 6, the maximum and minimum

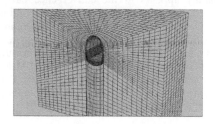

Figure 3. Exciting force of train vehicles.

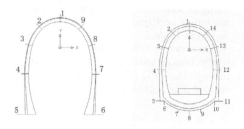

Figure 4. Structure monitoring point distribution of arch wall and surrounding rock.

Figure 5. Distribution of filling of inverted arch.

Table 6. The main stress of the inverted arch structure when the train passes through (MPa).

Serial number	Initial support		Secondary lining	
	Max.	Min.	Max.	Min.
1	0.00521	−0.08415	0.09031	−0.04233
2	0.04871	−0.08617	0.00965	−0.06046
3	0.24280	−0.15998	0.01926	−0.02391
4	−0.01855	−0.12581	0.03440	−0.05550
5	−0.01795	−0.05975	0.04142	−0.11142

principal stresses generated on the basement structure are shown in Tables 7–8.

Table 6 shows that when the train passes, the maximum tensile stress of the initial supporting concrete is mainly concentrated in the arch and the maximum tensile stress is 0.29 MPa, which is less than the ultimate tensile stress of the initial supporting concrete. Secondary lining produces maximum tensile force of the parts in the vault (0.09 MPa), which is less than the limit of secondary lining concrete tensile stress. Therefore, when the train passes the Type III composite lining, the arch wall structure is safe.

As shown in Tables 7–8, the tension stress mainly appears in the convergence part of the invert, the corner of the wall, and the middle position of the inverted arch. The maximum tensile stress is 0.05 MPa, inverted arch structures on the train after process are in alternating pressure state, and the maximum tensile stress is 0.04 MPa, which is less than concrete limit tensile stress. Therefore, when the train passes the Type III composite lining, the basement structure is safe.

4.1.2 Safety analysis of surrounding rock

The heavy-haul train passing Type III composite lining produces the maximum principal stress and minimum principal stress of surrounding rock, as shown in Table 8.

As shown in Table 9, all surrounding rock experience compressive stress when the train passes through the Type III composite lining. Hence, the surrounding rock is in safe state.

Table 7. The main stress of the invert filling structure when the train passes through (MPa).

	1	2	3	4	5
Max.	0.0383	0.0402	−0.0213	−0.0261	0.0200
Min.	−0.0684	−0.0990	−0.1790	−0.1600	−0.0794

Table 8. The main stress of the invert structure when the train passes through (MPa).

	1	2	3	4	5
Max.	0.0078	0.0413	0.0314	0.0230	0.0057
Min.	−0.0493	−0.0928	−0.1388	−0.0920	−0.0347

4.2 Composite lining of IV

Using FLAC3D separately on the tunnel excavation process and passing the train through Type IV composite lining, the safety of surrounding rock and structure under 27t axle dynamic loads of heavy-haul train is analyzed.

4.2.1 Structural safety analysis

When the heavy-haul train passes through Type IV composite lining, the produced maximum and minimum principal stresses of early supporting and secondary lining arch wall structure are shown in table 10, and the maximum and minimum principal stresses generated on the basement structure are shown in Tables 11–12.

From Table 10, when the train passes, the maximum tensile stress of the initial supporting concrete is mainly concentrated in the arch, and the maximum tensile stress is 0.19 MPa, which is less than the ultimate tensile stress of the initial supporting concrete. Secondary lining maximum tensile force of the parts in the vault is 0.12 MPa, which is less than the limit of secondary lining concrete tensile stress. Therefore, when the train passes through Type IV composite lining, the arch wall structure is safe.

As shown in Tables 11–12, the tension stress mainly appears in the convergence part of the invert, the corner of the wall, and the middle position of the inverted arch. The maximum tensile stress is 0.29 MPa, inverted arch structures on the train after process are in alternating pressure state, and the maximum tensile stress is 0.16 MPa, which is less than the concrete limit tensile stress. Therefore, when the train passes through Type IV composite lining, the basement structure is safe.

4.2.2 Safety analysis of surrounding rock

The heavy-haul train passing through Type IV composite lining produces the maximum principal stress and minimum principal stress of surrounding rock, as shown in Table 13.

As shown in Table 8, all the surrounding rock experience compressive stress when the train passes through Type IV composite lining. Hence, the surrounding rock is in safe state.

4.3 Composite lining of V

Using FLAC3D separately on the tunnel excavation process and the train passing through Type V com-

Table 9. The main stress of the surrounding rock when the train passes through (MPa).

	1	2	3	4	5	6	7
Max.	−0.03723	−0.02028	−0.01430	−0.02422	−0.01994	−0.00702	−0.01401
Min.	−0.59896	−0.48715	−0.42162	−0.38273	−0.49606	−0.78557	−0.65233

posite lining, we analyze the safety of surrounding rock and structure under 27t axle dynamic loads of heavy-haul train.

4.3.1 Structural safety analysis

When the heavy-haul train passes through Type V composite lining, it produces maximum and minimum principal stresses of early supporting and secondary lining arch wall structure, as shown in Table 14. The maximum and minimum principal stresses generated on the basement structure are shown in Tables 15–16.

From Table 14, when the train passes, the maximum tensile stress of initial supporting concrete is mainly concentrated in the arch, and the maximum tensile stress is 0.09 MPa, which is less than the ultimate tensile stress of the initial supporting concrete. The secondary lining maximum tensile force of the parts in the vault is 0.14 MPa, which is less than the limit of secondary lining concrete tensile stress. Therefore, when the train passes through

Table 10. The main stress of the inverted arch structure when the train passes through (MPa).

Serial number	Initial support		Secondary lining	
	Max.	Min.	Max.	Min.
1	0.07982	−0.30324	−0.01673	−0.07352
2	0.19426	−0.08289	0.07984	−0.03088
3	0.07250	−0.16981	0.12367	−0.13109
4	0.10373	−0.36954	0.07270	−0.11346
5	−0.03187	−0.21899	0.05448	−0.29259

Table 11. The main stress of the invert filling structure when the train passes through (MPa).

	1	2	3	4	5
Max.	0.0841	0.0695	0.0096	−0.2247	−0.0175
Min.	−0.0739	−0.0475	−0.1298	−0.3444	−0.1752

Table 12. The main stress of the invert structure when the train passes through (MPa).

	1	2	3	4	5
Max.	0.0078	0.0413	0.0314	0.0230	0.0057
Min.	−0.0493	−0.0928	−0.1388	−0.0920	−0.0347

Table 14. The main stress of the inverted arch structure when the train passes through (MPa).

Serial number	Initial support		Secondary lining	
	Max.	Min.	Max.	Min.
1	0.02622	−0.17544	0.14405	−0.08318
2	0.05559	−0.09154	0.08753	−0.07981
3	0.01591	−0.21109	0.03641	−0.11171
4	0.05717	−0.28603	0.07121	−0.13445
5	0.01970	−0.23699	0.05295	−0.09999

Table 15. The main stress of the invert filling structure when the train passes through (MPa).

	1	2	3	4	5
Max.	0.0352	0.0777	−0.1380	−0.0368	−0.1543
Min.	−0.2186	−0.1210	−0.2439	−0.2543	−0.4063

Table 16. The main stress of the invert structure when the train passes through (MPa).

	1	2	3	4	5
Max.	0.04386	0.04716	0.02839	0.13074	0.03844
Min.	−0.1359	−0.1000	−0.1437	−0.1539	−0.1431

Type V composite lining, the arch wall structure is safe.

As shown in Tables 15–16, the tension stress mainly appears in the convergence part of the invert, the corner of the wall, and the middle position of the inverted arch. The maximum tensile stress is 0.11 MPa, the inverted arch structures on the train after process are in alternating pressure state, and the maximum tensile stress is 0.13Mpa, which is less than concrete limit tensile force. Therefore, when the train passes through Type V composite lining, the basement structure is safe.

4.3.2 Safety analysis of surrounding rock

When the heavy-haul train passes through Type V composite lining, it produces maximum and minimum principal stresses of surrounding rock, as shown in Table 17.

As shown in Table 17, all surrounding rock experience compressive stress when the train passes

Table 13. The main stress of the surrounding rock when the train passes through (MPa).

	1	2	3	4	5	6	7
Max.	−0.05084	−0.02618	−0.02445	−0.02545	−0.01575	−0.05577	−0.09006
Min.	−0.92917	−0.63079	−0.50771	−0.63536	−0.32789	−1.2688	−1.2625

Table 17. The main stress of the surrounding rock when the train passes through (MPa).

	1	2	3	4	5	6	7
Max.	−0.05142	−0.02574	−0.00982	−0.00515	−0.05325	−0.05809	−0.06775
Min.	−0.33031	−0.23851	−0.22429	−0.22283	−0.27413	−0.21525	−0.30319

through Type V composite lining. Therefore, the surrounding rock is in safe state.

5 CONCLUSION

In this paper, we analyzed the Taihangshan Mountains heavy-haul railway tunnel and simulated and analyzed the dynamic response of each structural part of the single heavy-haul railway tunnel under 27t train axle load. The main conclusions are as follows:

1. When the heavy-haul train passes through the tunnel section, for the upper structure, the maximum dynamic stress is mainly distributed in the arch waist to vault position and its value is 0.09–0.29 MPa. For the base structure, the maximum increment of the dynamic stress is mainly distributed in the middle of the line and the maximum value is 0.25 MPa.
2. When the heavy-haul train passes through the tunnel section, the maximum tensile stress of each section arch wall structure and basement structure is less than the ultimate tensile stress. Therefore, the various types of lining section are safe.
3. The surrounding rock is in the compression state, and the compressive stress value is far less than the ultimate compressive strength value. Therefore, when the train passes, the surrounding rock is in a safe state.

REFERENCES

Hong L. Cdohhra. J. R. E. **4** (2000).
TB10003-2005 Cdtr. C. R. P. H. 2005.
Wuming L, Wenjie L, Chunyan, Z. Erdfrccgsfhhrs. R. S. M. **3**(2015).
Yadong H. Csdtsrhhtc. C. R. S. **2** (2015).
Ziqiang L, Mingnian W, Li Y, et al. Dprfbshhratunnel. C. R. S. **1** (2016).

Advances in Energy Science and Equipment Engineering II – Zhou, Patty & Chen (Eds)
© 2017 Taylor & Francis Group, London, ISBN 978-1-138-71798-5

Characteristics of climate change in Aksu River Basin in the last 50 years

Qingqing Chen
College of Hydrology and Water Resources, Hohai University, Jiangsu, Nanjing, China
State Key Laboratory of Simulation and Regulation of Water Cycle in River Basin, China Institute of Water Resources and Hydropower Research, Beijing, China

Zhe Yuan
State Key Laboratory of Simulation and Regulation of Water Cycle in River Basin, China Institute of Water Resources and Hydropower Research, Beijing, China

Keke Sun
College of Hydrology and Water Resources, Hohai University, Jiangsu, Nanjing, China
State Key Laboratory of Simulation and Regulation of Water Cycle in River Basin, China Institute of Water Resources and Hydropower Research, Beijing, China

Jie Feng
State Key Laboratory of Simulation and Regulation of Water Cycle in River Basin, China Institute of Water Resources and Hydropower Research, Beijing, China

ABSTRACT: On the basis of the meteorological data in the Aksu River Basin from 1961 to 2011, in this paper, unary linear regression and Mann–Kendall (M-K) mutation test were used to analyze the temporal and spatial evolution of the two main meteorological elements in the Aksu River Basin: temperature and precipitation. The results show that: (1) the climate of the Aksu River Basin showed a growing trend of temperature and humidity in the last 50 years, especially in the 2000s. The mutation points of annual average temperature and precipitation took place in 2001 and 1998, respectively. (2) The temperature evolution had a good consistency in the spatial distribution, average temperature increase rate reached 0.21°C/10a, and it also increased from south to north. (3) The spatial difference of precipitation evolution was significant and the average precipitation growth rate was 15.5 mm/10a. Only the east of Kumalake Mountainous area had a slight decreasing in precipitation, and all other regions showed an increasing trend, that is, an increase by 20 mm/10a, which is the most significant increase in the precipitation of the river basin.

1 INTRODUCTION

Global climate change is one of the major issues of social concern in recent years. Climate change has a direct or indirect impact on the natural environment, agriculture, and even the living environment of human beings. Natural ecosystems are susceptible to serious or even irreparable damage due to the limited adaptation capacity. The domestic and international research on climate change mainly uses meteorological data to statistically analyze the temperature and precipitation in recent years through a variety of mathematical analysis methods, and combining the climatology model, using DEM as the spatial interpolation analysis data. However, in China, 85% arid and semi-arid areas distribute in the northwestern region and hence the spatial distribution of temperature and precipitation of these areas varies greatly. Recently, scholars worldwide have conducted a lot of research on the hydrological and meteorological factors in the arid area of Northwest China and made a series of progress. Li Qihu proposed that the average temperature in the northwest arid region shows a significant upward trend in the last 50 years and found that the warming rate of 0.33°C/10a is far higher than the rates of warming of global temperature (0.13°C/10a) and national air temperature (0.22°C/10a). Through the study of climate change in Xinjiang, Hu Ruji et al. revealed that Xinjiang has been experiencing a wet and warm climate since 1987, particularly more severe in the west Tianshan region.

In recent years, the climate of Xinjiang has changed significantly, which has a profound impact on the ecological environment, economic development, flood and drought disasters, and livelihood. However, the vast territory of Xinjiang and a variety of complex natural conditions make the climate evolution obvious in regional characteristics. Located in the inland arid and semi-arid area, Aksu River is one of the largest rivers of runoff in the southern slope of Tianshan as well as the main origin of Tarim River. Through the analysis of the variation patterns of climate factors in the Aksu River Basin, this paper provides a theoretical basis in further revealing the drought evolution regulation in Aksu River under the changing environment and acts as a reference to the recovery of local eco-environment and sustainable development of future ecology.

Temperature and precipitation, two of the most important climate factors, directly affect the ecosystem from the aspects of structure and function. In this paper, from the analysis of changing characteristics of climate factors in the Aksu River Basin, collecting climate lattice data from 1961 to 2011, and using the unary linear regression and Mann–Kendall test method, we comprehensively analyzed the evolution of main climate factors for the last 50 years.

2 STUDY AREA AND METHODS

2.1 Study area

The Aksu River Basin is located in the Xinjiang Uygur Autonomous Region of China, the south of the Tianshan Mountains, and the northern margin of Tarim Basin, covering an area of 71000 km². Its geographic coordinates are 75°35'~80°59'E and 40°17'~42°27'N. It is characterized by temperate continental arid climate, and the average annual temperature is between 9.2 and 11.5°C. The law of vertical zone in the Aksu River Basin is extremely obvious. Near Mount Tuomuer and Khan Tengri, the annual precipitation is above 900 mm. The regional annual rainfall at an altitude of about 1000 m is only around 50 mm, the annual precipitation with the elevation of the increasing rate is about 16.8 mm/100 m, which is relatively scarce.

In order to better reflect the climate change in the Aksu River Basin, its topographic and geomorphic features are combined using DEM data extracts of the topography prominence and river network distribution information. According to the mountain export control hydrological stations Sharik Fank and Xehela, as shown in Figure 1, the basin was divided into three parts: Tuoshigan Mountain (19500 km²), Kumalake Mountain (13000 km²), and Plain Area (38700 km²).

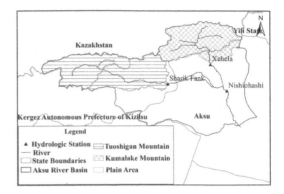

Figure 1.   Location and zoning of the Aksu river basin.

2.2 Methods

Daily temperature and precipitation data are collected from China's Ground meteorology 0.5° × 0.5° Gridded Dataset (V2.0) (http://cdc.cma.gov.cn/), which is provided by the Meteorological Records Office of the National Meteorological Information Centre of China. On the basis of the grid data analysis, a series of continuous meteorological observation data (January 1961 to December 2011) was selected for the study as the basic data. With GIS technology as the key data-processing platform, through the sub-decompression, format conversion, resampling, and so on, we generated daily 0.5° × 0.5° grid point of ground meteorological data.

The values of average temperature and precipitation are obtained by a series of time analysis method, with time (x) as the independent variable, elements (y) as the dependent variable, and $y(x) = ax + b$ for the establishment of a meta-regression equation. Here, $10 \times a$ is called the rate of change tendency, $a > 0$ shows that the time period is on the upward trend, $a < 0$ shows that the calculation period is on the downward trend, the magnitude of a can be used to measure different degrees of the rising or falling evolution tendency. In order to better reflect the climatic factors in spatial variations, we analyze the distribution of average annual climatic factors change tendency rate in the space on the Aksu River Basin, which is based on the raster image of the annual change in climate elements by unary linear regression method:

$$Slope = \frac{n \times \sum_{i=1}^{n}(i \times K_i) - \sum_{i=1}^{n}i \sum_{i=1}^{n}K_i}{n \times \sum_{i=1}^{n}i^2 - \left(\sum_{i=1}^{n}i\right)^2} \qquad (1)$$

where *Slope* is the tendency rate of time series variation; *Slope*>0 shows that the trend of time series is increasing, whereas it is reduced; i is the year number; and K_i is the value on Year i.

Mutation analysis of climate change is a phenomenon in which climate changes from a steady-state (steady trend) fast conversion to another steady-state stability (or a sustained trend). It is realized as a rapid climate change in temporal and spatial scales from one statistical characteristic to another. In this study, using the Mann–Kendall test method to reflect the dramatic changes in climatic factors from a mean state to another, we analyzed the spatiotemporal variation trend of temperature and precipitation in the Aksu River Basin from 1961 to 2011, and compared the mutation change before and after the point of change. This method has the characteristics of wide test range, high degree of quantification, small human impact factors, and so on. It is one of the most widely used methods and is full of theoretical significances in the mutation test.

3 RESULTS AND DISCUSSION

3.1 Annual average temperature analysis on the time series

The trend of average annual temperature change in Tuoshigan Mountain is shown in Figure 2, as well as the Kumalake Mountain, Plain Area, and

the whole Aksu River Basin from 1961 to 2011. On the basis of the temperature data in different areas, there is a regional difference in the change of temperature at both high and low altitudes. In the last 50 years, the annual variation of temperature in the three areas of the Aksu River Basin and the Whole Basin showed a certain increasing trend, but the increasing amplitude varies between subdivisions. The amplitude of the annual temperature change in the Tuoshigan Mountain area was 0.23°C/10a, the annual minimum temperature was −4.10°C (1972), and the annual maximum temperature was −1.80°C (1998); in the Kumalake Mountain area the average temperature increased at minimum speed, inclination rate was 0.21°C/10a, annual minimum temperature was −5.70°C (1974), and the annual maximum temperature was −3.50°C (2007). The tendency rate of average annual temperature in Plain Area was 0.23°C/10a, the annual minimum temperature was 6.50°C (1974), and the maximum temperature was 8.70°C (2007).

A significant increase trend of annual average temperature on the three zones of the Aksu River Basin and the Whole Basin was observed from 1961 to 2011, and the maximum value appeared in the 2000s (Table 1). It is evident from the data that it is the largest increase in nearly 10 years, and the increases in the amplitudes were 17.2%, 12.7%, and 6.6% in the Tuoshigan Mountain, Kumalake Mountain, and the Plain Area.

3.2 Annual average precipitation analysis on the time series

Summarizing the daily precipitation to get the annual precipitation, the trend of precipitation in the last 50 years is obtained as shown in Figure 3. The average annual precipitation values of the Tuoshigan Mountain, Kumalake Mountain, Plain Area, and the Whole Aksu River Basin were 350.2, 528.3, and 164.6 mm, respectively. The annual variation of precipitation in the three subdivisions and the Whole Basin has a certain increasing trend during 1961–2011. Tuoshigan Mountain had the largest increase in annual precipitation, whose extent of the rise was about 19.2 mm/10a. The lowest annual precipitation was 234.0 mm in 1976. The highest annual precipitation was 567.5 mm in 1972. The annual precipitation tendency rate of Kumalake

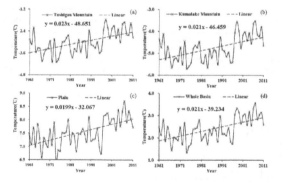

Figure 2. Changing trend of annual average temperature in the Aksu river basin during 1961–2011. (a) Tuoshigan Mountain; (b) Kumalake mountain; (c) Plain; and (d) Whole basin.

Table 1. Time series analysis of temperature (°C).

Zone	1960s	1970s	1980s	1990s	2000s
Tuoshigan mountain	−3.3	−3.3	−3.3	−2.9	−2.4
Kumalake mountain	−5.0	−5.0	−5.1	−4.7	−4.1
Plain	7.2	7.3	7.4	7.6	8.1
Whole basin	2.1	2.2	2.2	2.5	3.0

River was 13.1 mm/10a. The annual minimum precipitation values in 1979 and 1972 were 355.6 and 877.0 mm, respectively; the average annual change tendency in the precipitation rate of the Plain Area was 14.4 mm/10a, the annual minimum precipitation was 92.9 mm in 1975, and the maximum precipitation was 304.0 mm in 2010.

The annual changes of precipitation in the three basins and the Whole River Basin also showed a significant increasing trend in 1961–2011 and reached the maximum value in the 2000s (Table 2).

The largest increase was in the 1990s. In addition to the Kumalake Mountain, in which the highest increase was in the 2000s, the increase in the amplitudes of Tuoshigan Mountain, Kumalake Mountain, and Plain Area were 15.8%, 18.4%, and 14.0%, respectively.

3.3 *Spatial variation analysis*

The spatial variation of climate change in the Aksu River Basin can be seen in Figure 4. The spatial distribution pattern can be seen from the variation tendency rate. The temperature and precipitation in the study area are mainly increasing. Furthermore, the precipitation in the east of Kumalake Mountain has a slightly decreasing trend. The annual average temperature change in space can be seen in Figure 4 (a). The fastest growing region distribution in the west of Tuoshigan Mountain and Kumalake River Mountain Area is greater than 0.20°C/10a, increasing at a higher rate. The region's least growth rate (0.13°C/10a) was also reached, but the region was relatively small, only in the middle of the plain. The spatial variation of the precipitation tendency rate can be seen in Figure 4 (b). Precipitation increase in the upper reaches of the river basin was extremely significant, as high as 20 mm/10a. The precipitation in the southeast of the Plain Area in the lower reaches of the river

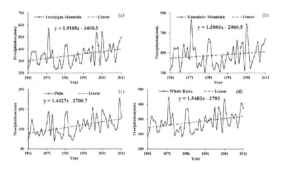

Figure 3. Changing trend of annual average precipitation in the Aksu river basin during 1961–2011. (a) Tuoshigan mountain; (b) Kumalake mountain; (c) Plain; and (d) Whole basin.

Table 2. Time series analysis of precipitation (mm).

Zone	1960s	1970s	1980s	1990s	2000s
Tuoshigan mountain	327.2	315.7	323.7	374.8	404.2
Kumalake mountain	515.7	520.1	479.8	532.4	587.8
Plain	134.9	148.8	153.4	181.7	200.7
Whole basin	257.0	262.3	259.6	298.5	327.1

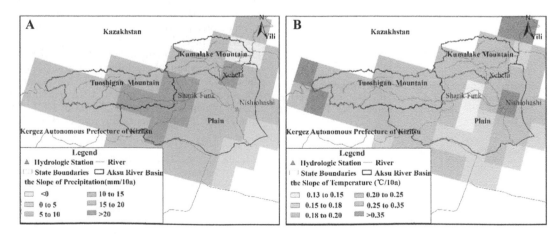

Figure 4. Spatial variation of temperature and precipitation.

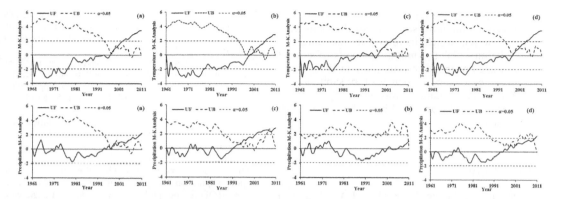

Figure 5. Mutation analysis of climate change in the Aksu river basin during 1961–2011. (a) Tuoshigan mountain; (b) Kumalake mountain; (c) Plain; and (d) Whole basin.

Table 3. Mutagenicity test of climate change in different regions of the Aksu River Basin during 1961–2011.

Zone	Temperature (°C)		Precipitation (mm)	
	Mutation year	Increment	Mutation year	Increment
Tuoshigan mountain	1999	0.78	1997	75.88
Kumalake mountain	2001	0.83	–	–
Plain	1999	0.69	1995	47.03
Whole basin	2001	0.73	1998	49.82

basin also increased in varying degrees, increasing in the range of 0–10 mm/10a.

3.4 M-K mutation analysis

The UF_k and UB_k curves are drawn by the M-K method, value of UF_k or UB_k greater than 0 indicates that the sequence is on the upward trend, while less than 0 indicates a downward trend. A significant intersection between UF_k and UB_k in the critical value of ±1.96 (a = 0.05), exceeding the critical line, indicates that the rise or fall trend is significant, while the moment of the intersection point is the mutation of time.

Abrupt changes in temperature and precipitation can be seen in the Aksu River Basin during 1961–2011 (Figure 5). UF and UB have a point of intersection around 2000, and then UF gradually increases over the border, which showed the annual average temperature and precipitation in the Aksu River Basin changed from low to high before and after 2000. The M-K mutation curve and data analysis can be seen through the three partitions and the Whole Basin in Table 3. Tuoshigan Mountain and Plain Area passed the M-K test for temperature and precipitation; however, Kumalake Mountain passed the temperature test, but the trend of precipitation was not obvious.

4 CONCLUSION

On the basis of the daily meteorological data from 1961 to 2011 in the Aksu River Basin, taking the GIS technology as the key data-processing platform, using the grid data, the unary linear regression method, and the inspection of mutation analysis method, this paper analyzed the change of main climatic factors in the last 50 years. The following conclusions are drawn:

1. During 1961–2011, the tendency of annual average temperature and precipitation in the whole Aksu River Basin were 0.21°C/10a and 15.48 mm/10a, respectively. The three partitions of the Aksu River Basin of the annual average temperature and precipitation were significantly increased, and Tuoshigan Mountain increased most obviously.

2. The average temperature in Tuoshigan Mountain and the southwest of Kumalake Mountain increased rapidly (>0.20°C/10a). However, the lowest growth rate of the whole study area was also reached (0.13°C/10a); the temperature in the Aksu River Basin was changed from low to high abruptly in 2001.

3. From the point of the whole study area, the precipitation in Plain Area increased greatly, especially in the upper reaches of the river

basin, where the rate reached 20 mm/10a. The precipitation in the southeast of the Plain Area was lower, which increased in the range of 0–10 mm/10a; the mutation points of annual average precipitation took place in 1998. However, the change tendency of precipitation in Kumalake Mountain was not very obvious. The rest of the area had been changed in 2000 basically.

REFERENCES

Groisman P.Y., Legates D., Rankova E.Y., Changes in the probability of heavy precipitation:Important indicators of climatic change. *Climatic change*, **42**(3): 243–283. (1999).

Huang S.C., Krysanova V., Zhai J.Q., Su B.D., Impact of Intensive Irrigation Activities on River Discharge under Agricultural Scenarios in the Semi-Arid Aksu River Basin, Northwest China. *Water Resources Management*, **29**(3):945–959. (2015).

Luo Q.H., Guo S.L., Li T.Y. et al. 2011. Precipitation and temperature change trend analysis for a long time in Jianghan plain area. *Journal of Yangtze River Scientific Research Institute*, 03:10–14. (2011).

Miao J., Lin Z., Study on the characteristics of the precipitation of nine regions in China and their physical causes II-The relation between the precipitation and physical cause. *J Trop Meteorol*, **20**(1):64–72. (2004).

Tong J.L., Wu H., Hou W.et al, Early warning signals of abrupt temperature change in different regions of China over the past 50 years, *Chinese Physics B*, **23**(4):49201–49209. (2014).

Vasilis D., Marten S., Egbert H., Victor B. et al. Slowing down as an early warning signal for abrupt climate change. *Proceedings of the National Academy of Sciences of the United States of America*, **105**(38):14308–12. (2008).

Yi X.S., Li G.S., Yin Y.Y., Temperature variation and abrupt change analysis in the Three-River Headwaters Region during 1961–2010, *Journal of Geographical Sciences*, **22**(3):451–469. (2012).

Yuan W., Liu S., Yu G. et al. Global estimates of evapotranspiration and gross primary production based on MODIS and global meteorology data. *Remote Sensing of Environment*, **114**(7): 1416–1431. (2010).

Zagat A., Sadiqi R., Khan F.I., Uncertainty-driven characterization of climate change effects on drought frequency using enhanced SPI. *Water Resources Management*, **28**(1):15–40. (2014).

Zhang Q., Sun P., Singh VP., Chen X., Spatial-temporal precipitation changes (1956–2000) and their implications for agriculture in China. *Global Planet Change*, 82:86–95. (2012).

Zhang Q., Xu C.Y., Tao H., Variability and stability of water resource in the arid regions of China: a case study of the Tarim River basin. *Frontiers of Earth Science in China*, **3**(4):381–388. (2009).

Zhao F.F., Xu Z.X., Long-term trend and jump change for major climate processes over the upper yellow river basin. *Acta Meteorological Sonica*, **64**(02): 52–56. (2006).

Advances in Energy Science and Equipment Engineering II – Zhou, Patty & Chen (Eds)
© 2017 Taylor & Francis Group, London, ISBN 978-1-138-71798-5

Study on the influence of tunnel structure on the upper drum deformation of the tunnel bottom structure

Yong Luo & Li Yu
Key Laboratory of Transportation Tunnel Engineering, Ministry of Education, School of Civil Engineering, Southwest Jiaotong University, Chengdu, China

Yu Yu & Zhu Yuan
China Railway Eryuan Engineering Group Co. Ltd., Chengdu, China

Tianyuan Xu & Ziqiang Li
Key Laboratory of Transportation Tunnel Engineering, Ministry of Education, School of Civil Engineering, Southwest Jiaotong University, Chengdu, China

ABSTRACT: The upper drum deformation will seriously affect the safety of train operation. In order to ascertain the influence of tunnel structure on the upper drum deformation, effective control measures are taken against the upper drum deformation of the tunnel bottom structure. On-site investigation, survey, and analysis show that the cross tunnel and the shape of the tunnel bottom structure are the main structural factors that influence the upper drum deformation of the tunnel bottom structure. After construction of the cross tunnel, the contact pressure between the tunnel bottom structure and the surrounding rock will increase, thereby increasing the upper drum deformation of the tunnel bottom structure, but the increase is not very high. The shape change of the tunnel bottom structure will greatly reduce the carrying capacity of the tunnel bottom structure; thus, under the same conditions of surrounding rock pressure, the shape change of the tunnel bottom structure will increase the upper drum deformation. The research results provide theoretical support for the analysis of the cause of the upper drum deformation and can provide reference for similar projects.

1 INTRODUCTION

In recent years, because of the upper drum deformation of the tunnel bottom structure, safety cases of train operation have occurred. The upper drum deformation of the tunnel bottom structure is a complicated physical and mechanical process. Exploring the influence factors is beneficial to take effective measures to control the deformation of the bottom structure. Because the tunnel bottom structure of the mine tunnel is mainly used for the bottom plate, there is no inverted arch and the drum deformation can easily occur. In the tunnel, the tunnel bottom structure adopts invert structure, and there is less drum deformation. Therefore, at present, the research of the upper drum deformation in tunnel is mainly focused on controlling the drum deformation of the floor in the mine roadway (C, 1994; B, 2011; C, 2011; H, 1995; C, 2011; Y, 2009). Wang Yang (2010), Zhong Zuliang (2012), and Deng Tao (2014) focused on the stress analysis of the upper drum deformation on the tunnel bottom structure and the estimation of the upper

drum deformation, but the influence factors of the upper drum deformation are less studied.

By site reconnaissance and measurement of the relying on tunnel engineering, effect of analysis of tunnel structure on the upper drum deformation of the tunnel bottom structure, and through numerical simulation, we ascertain the influence law of cross tunnel and the tunnel bottom structure shape on the upper deformation, provide theoretical support for putting forward the effective control measures of the drum deformation, and provide reference for similar engineering.

2 ENGINEERING SURVEY

The total length of the tunnel is 7858 m. It is used for both passenger and cargo transport. The speed of the passenger train is 200 km/h, and the tunnel is a double-track one. The length of the exit section is 17.81 m and the curve radius is 8000 m. The rest were situated in a straight line. The range of tunnel slope is +4‰ (4646 m) to –4‰ (3212 m).

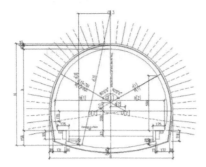

Figure 1. Lining section of tunnel design (cm).

a. The reticular crack b. Boiling and mud

c. The large deformation d. Lateral groove cracking

Figure 2. Field reconnaissance of the uplifting deformation of the bottom structure.

The tunnel adopts composite lining, initial support was provided by C20 corrosion-resistant airtight concrete, the second lining uses C25 corrosion-resistant airtight concrete, and the tunnel bottom structure filling material uses C20 concrete. The tunnel bottom thickness is 45 cm, and the bottom structure pouring is a 5 cm M10 cement mortar leveling layer. The construction of the tunnel was completed in June 2009 and put into use. The design of the tunnel is shown in Figure 1.

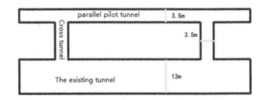

Figure 3. Relative position of the existing tunnel and cross tunnel.

3 SITE SURVEY OF UPPER DRUM DEFORMATION

On-site investigation shows that there were 5225 transverse cracks in the tunnel monolithic roadbed. The filling layer and the joint seam within the overall bed take the mud phenomenon and side wall fracture. The side of the tunnel wall and overall bed-filling layer had cracks with side ditch wall fracture, and crack width is 3 mm. The train, to some extent of the tunnel, often experienced different degrees of the car shaking phenomenon. Preliminary survey data showed that there is a filling layer, with the tunnel bottom structure uplift phenomenon. Reconnaissance scene diagram is shown in Figure 2.

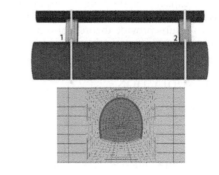

Figure 4. Numerical calculation model and monitoring section.

4 INFLUENCE OF CROSS TUNNEL ON THE UPPER DRUM DEFORMATION OF THE TUNNEL

Drilling and investigation data showed that the construction of cross passage has a certain effect on the deformation of the tunnel floor. In order to prove that the presence of cross tunnel can affect the force and displacement of the tunnel bottom structure, a finite element software numerical calculation model is established, according to the construction process selected, two monitoring sections, and spacing of 50 m (Section 1, 2). The

contact element is used to simulate the contact pressure between the tunnel bottom structure and the surrounding rock, and the influence of construct cross tunnel on the upper drum deformation of the tunnel bottom structure is analyzed.

4.1 Calculation model and parameters

According to the design data, the relative position between the existing tunnel and the cross tunnel is shown in Figure 3. The three-dimensional model is shown in Figure 4. The related physical and mechanical parameters are presented in Tables 1 and 2.

Table 1. Physical and mechanical parameters of lining.

Parameter	Value
Thickness of the lining (cm)	40
Thickness of the bottom structure	measure
Elastic modulus (GPa)	29.5
Density (kg/m³)	2500
Poisson ratio	0.2

Table 2. Physical and mechanical parameters of surrounding rock.

Parameter	Value
Surrounding rock level	III
Specific weight (KN/m3)	25
Elastic reaction coefficient (MPa/m)	500
Poisson ratio	0.25

Table 3. Vertical displacement of tunnel bottom structure.

Position	Displacement (mm)	Added value (mm)	Increase percentage
No cross channel	14.04	/	/
Section 1	15.01	0.97	6.91%
Section 2	15.01	0.97	6.91%

Note: Positive value indicates that the displacement is upward.

Table 4. Contact pressure of tunnel bottom structure and surrounding rock.

Position	Contact pressure (kPa)	Added value (kPa)	Increase percentage
No cross channel	48.21	/	/
Section 1	54.82	6.61	13.71%
Section 2	54.82	6.61	13.71%

Note: Positive value indicates pressure.

4.2 Calculation results and analysis

Through simulation, we can obtain the contact pressure between the tunnel bottom structure and surrounding rock after the construction of cross tunnel. By comparing the result of the analysis with the data before the construction of cross tunnel, we can obtain the effect of constructing the cross tunnel on the upper drum deformation. The specific calculation results are shown in Tables 3 and 4

Tables 3 and 4 show that, after cross tunnel construction, deformation of the tunnel bottom structure is still the deformation of the upper drum, which increases to 0.97 mm, with an increasing percentage of 6.91%, and the contact pressure increases to 6.16 kPa between the tunnel bottom structure and surrounding rock, with an increasing percentage of 13.71%. It is indicated that the contact pressure between the tunnel bottom structure and the surrounding rock will be increased so that the tunnel structure can be generated on the upper drum deformation, but the deformation is not very large.

5 INFLUENCE OF THE SHAPE OF THE TUNNEL BOTTOM STRUCTURE ON THE DRUM DEFORMATION OF TUNNEL

Through the field survey and borehole data, it is known that some differences exist in the bottom structure of the tunnel due to some reasons (see Figure 5). Therefore, the carrying capacity of the tunnel bottom structure is influenced. In order to prove the effect of bottom structure shape change on the upper drum deformation of the tunnel bottom structure, an elastic and plastic model of load structure was established using the finite element software, and numerical simulation of three different structures is carried out. It can be concluded that the shape change of the tunnel bottom structure can affect the bearing capacity of the tunnel bottom structure.

5.1 Calculation models and parameters

The load structure model is used to calculate the load, and the elastic reaction of the surrounding

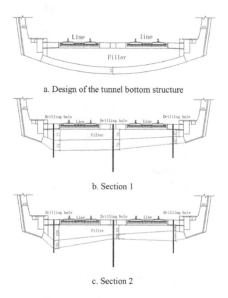

a. Design of the tunnel bottom structure

b. Section 1

c. Section 2

Figure 5. The bottom structure diagram (cm).

rock is applied to the lining structure. The first strength theory is used to analyze the structure of the lining and the bottom plate, and the first principal stress is used as the criterion to determine whether the structure is damaged. The calculation parameters are presented in Table 5, and the numerical calculation model is shown in Figure 6.

5.2 Calculation results and analysis

Through numerical simulation, we obtained the standard section and construction section at the bottom of the tunnel structure, and by comparison analysis, the bottom of the tunnel structure and the shape change of the tunnel bottom structure bearing capacity were studied. The calculated results are shown in Table 5.

It is evident from the above analysis that the shape change of the tunnel bottom structure can reduce the bearing capacity of the tunnel bottom structure, and the structure with special shape is more easily damaged than straight structure of the tunnel bottom structure, and the load bearing by lower edge inner concave structure is very small. The failure load of section 1 is 53 kPa, which reduces 32 kPa than the damage load of the design section, and the ratio is 37.65%. The failure load of section 2 is 47 kPa, which reduces 38 kPa than the damage load of the design section, and the ratio is

44.71%. The failure load of section 1 reduces 6 kPa than that of section 2.

6 CONCLUSIONS

On the basis of the field investigation and survey data, the following conclusions are obtained by numerical analysis:

1. After construction of the cross tunnel, the contact pressure between the tunnel bottom structure and the surrounding rock will increase; thus, the upper drum deformation of the tunnel bottom structure will increase, but the increase is not very high.
2. The shape change of the tunnel bottom structure will greatly reduce the carrying capacity of the tunnel bottom structure; thus, under the same conditions of surrounding rock pressure, the shape change of the tunnel bottom structure will increase the upper drum deformation.
3. Because of construction and other reasons, the bearing capacity of tunnel structure is greatly reduced, and the structure with special shape is more easily damaged than straight structure of the tunnel bottom structure. Under the same conditions, the upper drum deformation is greater. Therefore, construction quality is the life of the project. It is necessary to take corresponding measures to improve the construction quality.

Table 5. Failure load of tunnel bottom structure in each section.

Calculation section	Failure load (kPa)
Design section	85
Section 1	53
Section 2	47

REFERENCES

Chao jiong H, Yanan H, Xiao L, et al. Rfscgcfh. J.C.C.S. 3 (1995).
Jianbiao B, Wenfeng L, Xiangyu W, et al. Mfhctrm. J.M & S.E.1 (2011).
Jucai C, Guang-xiang X, Fhmoegbtrrdm. J.M & S.E. 3 (2011).
Mingxiang C, Qingbiao W, Junmin X. Sfmfhgsrpt. R.S.M.S 2 (2011).
Tao D, Ming H, Jinwu Z, et al. Sfmfhdslt. J.E.G. 3 (2014).
Xifeng Y. Mctfhrdg. (2009).
Yang W, Xiongjun T, Xiankun T, et al. Mafhyl T.R.S.M. 8 (2010).
Yanguang C, Shiliang L. Csrcrc (1994).
Zuliang Z, Xinrong L, Daoliang W, et al. Mafhtsytpt. C.J.G.E. 3 (2012).

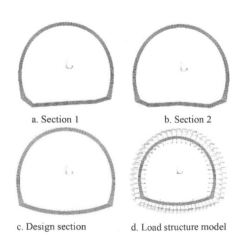

a. Section 1   b. Section 2

c. Design section   d. Load structure model

Figure 6. Numerical model diagram.

Advances in Energy Science and Equipment Engineering II – Zhou, Patty & Chen (Eds)
© 2017 Taylor & Francis Group, London, ISBN 978-1-138-71798-5

Study on reservoir-induced seismicity in Three Gorges Reservoir

Xinxiang Zeng, Xiao Hu, Kezhong Su & Tinggai Chang
Earthquake Engineering Research Centre, China Institute of Water Resources and Hydropower Research, Beijing, China

Hongbing Zhu
China Three Gorges Corporation, Beijing, China

ABSTRACT: Considering that the focal depth of Reservoir-Induced Seismicity (RIS) is very low (usually less than 5 km), the damage of RIS may probably be much bigger than that of a natural earthquake. Along with the development of Chinese hydropower in recent years, a large number of huge reservoirs have been constructed and new events of RIS occurred especially in southwestern China, such as Three Gorges, Longtan and Xiaowan reservoirs. In this paper, we have collected and studied the earthquake catalog (Jan. 1st, 2001–Sep. 9th, 2015) in Three Gorges Reservoir (TGR) as well as introduced the seismic network of Three Gorges Reservoir. Through the comparison of spatial distribution of hypocenters before and after impoundment (135 m, 1 June 2003), we found that the hypocenters tend to move toward the reservoir region. In fact, the biggest earthquake ever occurred in the Three Gorges region is ML5.0 (16 December 2013), which is in line with the forecast (ML ≤ 5.5). Therefore, we are quite confident that the works for constructing hydropower plants in China are solid, and the risks of RIS in the Three Gorges Reservoir region are manageable.

1 INTRODUCTION

Reservoir-Induced Seismicity (RIS) means typically minor earthquake sequence resulting from reservoir impoundment or variation of water level. Considering the focal depth of RIS is very low (usually less than 5km), the epicenter intensity is much higher than that of natural earthquake, which rated exactly the same magnitude. Therefore, the damage caused by RIS may probably be much severe than that of natural earthquake, particularly near the epicenter.

By the end of the twentieth century, there were about 100 cases of earthquakes that were identified as RIS (Jiang, 2014; Gupta, 2002; Yang, 1996) worldwide, of which 19 (Yang, 1996; Chen, 1998; Yang, 2012) took place in China. Along with the recent development of hydropower in China, large quantities of huge reservoirs were constructed and also new events of RIS occurred especially in southwestern China, such as Three Gorges, Longtan, and Xiaowan reservoirs.

The Three Gorges Dam/Reservoir and its associated infrastructure is the largest artificial integrated hydro project (Liu, 2009; Gleick, 2008; Yi, 2012) in terms of installed capacity. Also, in the generation of electricity, the Three Gorges Dam is intended to reduce the flood risk downstream the Yangtze River and will also improve the waterborne transport capability upstream the dam. Detailed

Table 1. Specifications of Three Gorges Dam/Reservoir.

Parameter	Value
Length	2309.5 m
Height	181 m
Installed capacity	22,500 MW
Maximum storage capacity	3.93×10^{10} m^3
Maximum length (reservoir)	600 km
Surface area	1084 km^2

specifications of the dam and the Three Gorges Reservoir are listed in Table 1.

2 SEISMIC NETWORK

It is known that some human activities such as artificial lakes (reservoir), mining, geothermal energy, and hydraulic fracturing will induce earthquakes. In China, considering the recent extensive development of hydropower, Reservoir-Induced Seismic (RIS) has attracted much attention, especially the Three Gorges Reservoir has attracted the most attention because it is not only the world's largest hydropower station in terms of installed capacity, but it also has incredibly large volume of water.

Considering the importance of the Three Gorges project, seismic observation in the reservoir region

Figure 1.　Seismic network in Three Gorges Reservoir region (Yellow concentric circle: dam site, Red line: Gaoqiao fault, Yellow line: Jiuwanxi fault, Blue line: Xiannvshan fault).

was started in 1958. A brand new digital telemetered seismic network containing 24 fixed seismic stations officially started construction at the end of the 1990s and started operation in 2001. Locations of these 24 seismic stations are shown in Figure 1.

In general, the Three Gorges Reservoir region is divided into three sections. Section I: from dam site to Miaohe (16km upstream the dam), where crystalline rock plays a dominant role and no regional fault is observed. Section II: from Miaohe to Baidicheng with a length of 141km, where regional faults (such as Gaoqiao, Jiuwanxi, and Xiannvshan, see Figure 1) are well developed, and the biggest earthquake in history (ML = 5.1 in 1979) occurred in Gaoqiao fault. The karsts along the river are developing intensively and deeply embedded. Section III: from Baidicheng to the tail of TGR with a length of 492km. Geological structures are very simple such that the seismic activity maintains at low level in this area.

3 ANALYSIS AND DISCUSSION

In this research, a seismic catalogue containing 19489 records is used to study the seismic activity of the Three Gorges Reservoir region, which was recorded from 1 January 2001 to 9 September 2015. The details and statistical features are shown in Table 2 and Figures 2 and 3.

3.1 Time variation

As shown in Table 2, before the impoundment, the annual events (2001 and 2002) recorded by TGR seismic network remind in a lower level (not more than 200/year). After the first impoundment period, in June 2003, the water level has been raising up to 135m, and the number of recorded seismic events increased about five times (nearly 1000/year). The secondary impoundment period was from September 2006 to October 2006. Accompanied with increase of water level up to 156m, the seismic events increased about two times. In the third impoundment, the water level raised up to 170+m (from September 2009 to October 2010, and reached 175m on 26 October 2010), and there is relatively severe fluctuation of seismic activities. These features are revealed much more clearly in Figure 2. Furthermore, the seismic events also changed associated with the changes in water level, even measured in months. In general, the seismic frequency reaches zenith after the water level raised up to the top of a year's impoundment cycle.

As shown in Table 2 and Figure 3, the seismic activities in the Three Gorges Reservoir region are mainly micro earthquakes or infinitesimal earthquakes. In fact, earthquakes with intensity smaller than 2 constitute about 94% of the total seismic events occurred after the impoundment.

In addition, the biggest seismic event occurred after the impoundment is $M_L = 5.0$ (16 December 2013), which is still smaller than the maximum historic earthquake occurred in the research area[8].

3.2 Spatial distribution

The spatial distribution of epicenter in the Three Gorges Reservoir region is shown in Figure 4.

As shown in Figure 4-(i), the spatial distribution of the epicenter before the first impoundment

Table 2. Seismic catalogue in the Three Gorges Reservoir region.

Year	Frequency	Frequency in different magnitude						Maximum magnitude
		$M_L<1$	$1<=M_L<2$	$2<=M_L<3$	$3<=M_L<4$	$4<=M_L<5$	$5<=M_L$	
2001	175	15	95	59	5	1	0	4.0
2002	153	5	72	67	9	0	0	3.8
2003	402	235	133	26	7	0	1	5.3
2004	934	630	279	24	1	0	0	3.0
2005	831	491	314	25	1	0	0	3.3
2006	789	419	326	42	2	0	0	3.7
2007	1607	770	740	93	4	0	0	3.3
2008	2586	1286	1138	148	12	2	0	4.5
2009	2462	1361	984	106	11	0	0	3.3
2010	1441	807	574	54	5	1	0	4.2
2011	640	414	203	22	1	0	0	3.0
2012	1618	941	586	80	11	0	0	3.8
2013	1932	1107	739	76	8	1	1	5.0
2014	3222	1649	1374	188	7	4	0	4.9
2015	697	400	258	39	0	0	0	2.9
Total	19489	10530	7815	1049	84	9	2	/

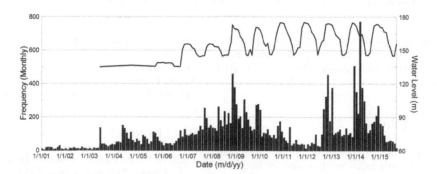

Figure 2. Monthly variations of seismic frequency (bottom bar) and water level (top curve).

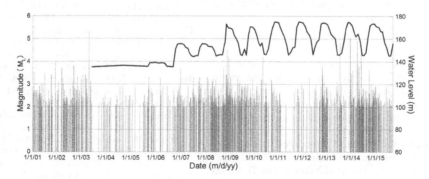

Figure 3. Time variations of magnitude (bottom vertical line) and water level (top curve).

is relatively homogeneous. And after the first impoundment in June 2003, there is a certain trend in the distribution of epicenter, which is moving toward the reservoir area. (see Figure 4-(ii)). Furthermore, the earthquakes are mainly grouped under Section II of the Three Gorges Reservoir. Further enlarging Section II of the reservoir (see Figure 4(iii)), it is very clear that three clusters of earthquakes are formed, of which the larger two clusters are corresponding to the locations

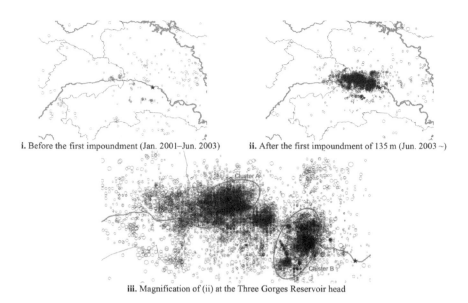

i. Before the first impoundment (Jan. 2001–Jun. 2003)   ii. After the first impoundment of 135 m (Jun. 2003 ~)

iii. Magnification of (ii) at the Three Gorges Reservoir head

Figure 4.   Spatial distribution of epicenter in the Three Gorges Reservoir region.

of the Gaoqiao fault segment (cluster A) and the Xiannvshan–Jiuwanxi fault segment (cluster B). The maximum earthquake (M_L = 5.0) after impoundment also occurred in Section II.

4   CONCLUSION

Seismic activities in the Three Gorges Reservoir region are mainly micro earthquakes or infinitesimal earthquakes.

Features about time variations of magnitude and space distribution of epicenter after impoundment can be summarized as follows.

1. The frequency of earthquake increases after impoundment.
2. The biggest seismic event occurred after the impoundment has not exceeded the maximum historic earthquake of the reservoir region.
3. There is a certain trend in the distribution of epicenter, which is moving toward the reservoir area.
4. After impoundment, the recorded earthquakes are mainly grouped in Section II of TGR, especially in the Gaoqiao and Xiannvshan–Jiuwanxi fault segments.

On the basis of the measured data, both the magnitude and location of epicenter are consistent with the prediction of TGR region. Furthermore, we are quite confident that the works for constructing hydropower plants in China are solid, and the risks of RIS in the Three Gorges Reservoir region are manageable.

ACKNOWLEDGMENT

We would like to express our gratitude to China Three Gorges Corporation for providing us the earthquake catalogue and related data.

REFERENCES

Chen, L.Y., P. Talwani. Reservoir-Induced Seismicity in China, *Pure and Applied Geophysics*, Vol. **153**, pp. 133–149, (1998).

Gleick, P.H. Three Gorges Dam Project, Yangtze River, China, The World's Water 2008–2009, *Island Press*, Washington, D.C., pp.139–150, (2008).

Gupta, H.K. A review of recent studies of triggered earthquakes by artificial water reservoirs with special emphasis on earthquakes in Koyna, India, *Earth-Science Reviews*, Vol. **58(3–4)**, pp. 279–310, (2002).

Jiang, H.K., X.D. Zhang, X.J. Shan. The reservoir-induced seismic of China continent, *Seismological Press*, Beijing, China, (2014).

Liu, W.Q., F. Huang. Analysis of seismic activity of Fengjie section in Three Gorges Reservoir area, *Journal of Geodesy and Geodynamics*, Vol. **29(6)**, pp. 49–51, (2009).

Yang, M.L., D.Q. Chen. Statistical Research of Time and Maximum Magnitude in the Reservoir-induced Earthquake of China Continent, *South China Journal of Seismology*, Vol. **32(4)**, pp. 1–9, (2012).

Yang, Q.Y., Y.L. Hu, X.C. Chen, L.Y. Chen. Catalogue of reservoir induced seismic events in the world, *Seismology and Geology (China)*, Vol. **18(4)**, pp. 453–461, (1996).

Yi, L.X., D. Zhao, C.L. Liu. Preliminary study of Reservoir-Induced Seismicity in the Three Gorges Reservoir, China, *Seismological Research Letters*, Vol. **83(5)**, pp. 806–814, (2012).

Advances in Energy Science and Equipment Engineering II – Zhou, Patty & Chen (Eds)
© 2017 Taylor & Francis Group, London, ISBN 978-1-138-71798-5

Water inrush and mud bursting scale prediction of tunnel in water-rich fault

Yinghua Tan, Shucai Li, Qingsong Zhang & Zongqing Zhou
Geotechnical and Structural Engineering Research Center, Shandong University, Ji'nan, Shandong, China

ABSTRACT: Water inrush and mud bursting disaster is a common geological hazard in the construction process of tunnel in water-rich fault. The scale-predictive expert system of water inrush and mud bursting in water-rich fault tunnel is established to predict the scale of this disaster. According to existing related research and hydrogeological factors that affect the severity of water inrush and mud bursting disaster of tunnel in water-rich fault, the evaluation items of the scale-predictive expert system are identified as the infilling materials, width of the fault fracture zone, water source of water inrush and mud bursting disaster, vertical and horizontal hydrodynamic zone of tunnel location. The scale of water inrush and mud bursting in Yonglian Tunnel on the line of Jilian Expressway is predicted by the scale-predictive expert system. The results show that the prediction results of the scale-predictive expert system agrees well with the excavation conditions. Therefore, the scale-predictive expert system is verified effective for the prediction of water inrush and mud bursting in water-rich fault.

1 INTRODUCTION

Tunnel engineering is the critical control project for national infrastructure construction such as traffic, water conservancy, and hydroelectricity engineering. It is related to several engineering fields such as roads, railways, hydropower, and mine resources. Efficient and safe construction of tunnels is crucial. Because of the influence of geological factors and engineering factors, all kinds of bad geological disasters will often happen in the process of tunnel construction. Water inrush and mud bursting is one of the geological disasters that produces much harm to the society. Furthermore, it is a serious threat to the safety of construction crew and national property.

Li and Zhou presented a veracious and feasible method to systematically evaluate the risk of water inrush in karst tunnels 38 (2013) 50058. Rao built a three-grade fuzzy comprehensive evaluation model for highway tunnel structure safety in the karst area 20 (2016) 124201249. Li and Lei established a software system for risk assessment of water inrush 8 (2015) 184301854.

From the above, the existing research is mainly about the karst tunnel, but the research about tunnel in water-rich fault tunnel is rare. In this paper, the scale-predictive expert system of water inrush and mud bursting in water-rich fault tunnel is established to predict the scale of this disaster. It provides effective means to predict the scale of water inrush and mud bursting. Early warning of disaster is of great significance.

2 THE SCALE-PREDICTIVE EXPERT SYSTEM OF WATER INRUSH AND MUD BURSTING IN WATER-RICH FAULT TUNNEL

On the basis of the existing expert judging system 23 (2004) 2130218, and combined with the hydrological geological factors, the scale-predictive expert system is established to predict the scale of water inrush and mud bursting in water-rich fault tunnel. The predictive expert system predicts the scale of water inrush and mud bursting through five evaluation items: infilling materials, width of the fault fracture zone, water source of water inrush and mud bursting disaster, and vertical and horizontal hydrodynamic zones of tunnel location.

2.1 *The infilling materials of the fault fracture zone*

The infilling materials of the fault fracture zone are the main source of mud flow when water inrush and mud bursting disaster occurs. The nature of the infilling materials directly affects the scale of the water inrush and mud bursting disaster. The nature of the infilling materials is expressed by the surrounding rock classification. According to government standard "Standard for engineering classification of rock mass" (1994), from good to bad, the surrounding rock is divided into five levels: grade I, grade II, grade III, grade IV, and grade V. The surrounding rock grade of fault fracture zone is generally lower. For fault fracture zone, it is thought that there is no grade I surrounding rock, and grade II surrounding rock is extremely rare.

According to the surrounding rock grade of fault fracture zone, the fault is divided into three types: type A, type B, and type C. When surrounding rock is of grade V, the fault is defined as type A; when surrounding rock is of grade IV, the fault is defined as type B; and when surrounding rock is of grade III, II, the fault is defined as type C. Type A fault is one of the necessary conditions causing large-scale water inrush and mud bursting disaster. Type B fault can lead to medium and small-scale water inrush and mud bursting disaster. Type C fault will only cause small-scale water inrush and mud bursting.

2.2 The width of the fault fracture zone

The muddy mixture is formed by the softening effect and the role of mud on the fault fracture zone in the process of water saturation. The gravel present in the fault fracture zone is the main source of the materials pouring out during water inrush and mud bursting disaster. The fault is also the main channel of water inrush and mud bursting in tunnel. Therefore, the scale of the fault is one of the determining factors influencing the scale of water inrush and mud bursting disaster.

According to the width of the fault fracture zone, the scale of fault can be divided into three grades: grade A, grade B, and grade C. When the width of the fault fracture zone $b > 10$ m, the scale of fault is grade A; when $5 \text{ m} \leq b \leq 10$ m, the scale is grade B; and when $b < 5$ m, the scale is grade C. Grade A fault is another necessary condition causing large-scale water inrush and mud bursting disaster. Grade B fault can lead to medium- and small-scale water inrush and mud bursting disaster. Grade C fault will only cause small-scale water inrush and mud bursting disaster.

2.3 The water source of water inrush and mud bursting disaster

Water source is another determining factor affecting the scale of water inrush and mud bursting. The water-bearing structure of fault fracture zone and the type of make-up water system in fault are identified as the two evaluation indices, where the latter is expressed by the area of water supply S.

According to the water-bearing structure and the type of make-up water system, water source of water inrush and mud bursting is divided into three types: type A, type B, and type C. When a large underground river system exists within fault and $S \geq 10 \text{ km}^2$, water source is defined as of type A; when the water of fault is from a smaller tubular water source and the range of S is $5-10 \text{ km}^2$, the water source is defined as of type B; when the fault is a small-scale water-bearing structure and $S \leq 5 \text{ km}^2$, the water source is defined as of type C. Type A water source is the third necessary condition causing large-scale water inrush and mud bursting disaster. Type B and type C water sources will only cause medium- and small-scale water inrush and mud bursting disaster.

2.4 Vertical and horizontal hydrodynamic zones in the tunnel location

Water pressure is one of the main powers of water inrush and mud bursting. The presence of hydrodynamic zones in the tunnel location affects the scale and the risk of water inrush and mud bursting seriously. The division standard of hydrodynamic zone and its harm to the tunnel can be found in Han 23 (2004) 2130218; however, the description could not be found. Vertical and horizontal hydrodynamic zones in the tunnel location are identified as the two evaluation items to predict the scale of water inrush and mud bursting.

The division standards of vertical hydrodynamic zone in the tunnel location are as follows: the shallow saturated water zone and the pressure-saturated water zone are type A vertical hydrodynamic zone; the seasonal variation zone is the type B vertical hydrodynamic zone; and the epikarst zone and the aeration zone are the type C vertical hydrodynamic zone. Location of the type A vertical hydrodynamic zone is the fourth necessary condition causing large-scale or super-large-scale water inrush and mud bursting disaster in tunnel construction. Locating in the type B and type C vertical hydrodynamic zones will only cause medium—and small-scale water inrush and mud bursting disaster in tunnel construction.

The division standard of horizontal hydrodynamic zones in the tunnel location are as follows: discharging area is the type A horizontal hydrodynamic zone; recharge and runoff area is the type B horizontal hydrodynamic zone; and recharge area is the type C horizontal hydrodynamic zone. Location of the type A and type B horizontal hydrodynamic zones is the fifth necessary condition causing large-scale or super-large-scale water inrush and mud bursting disaster in tunnel construction. Locating in the type C horizontal hydrodynamic zones will only cause medium—and small-scale water inrush and mud bursting disaster in tunnel construction, as shown in Table 1.

According to Table 2, the more the evaluation items that satisfying the condition of causing a certain scale water inrush and mud bursting disaster, the more likely this scale of water inrush and mud bursting disaster will happen in tunnel construction.

Table 1. Evaluation items and criterion for the scale-predictive system of water and mud bursting in water-rich fault tunnel.

Grade	Surrounding rock grade of fault fracture zone	Width of the fault	Water source	Vertical hydrodynamic zone	Horizontal hydrodynamic zone
A	Grade V	>10 m	Large underground river system $S > 10$ km^2	Shallow saturated water zone the pressure saturated water zone	Discharging area
B	Grade IV	$5 \leq b \leq 10$ m	Smaller tubular water source 5 km$^2 \leq S \leq 7.5$ km^2	Seasonal variation zone	Recharge and runoff area
C	Grade II, III	$b < 5$ m	Small-scale water-bearing structure $S < 5$ km^2	Epikarst zone the aeration zone	Recharge area

Table 2. Criterion for the scale-predictive system of water and mud bursting in water-rich fault tunnel.

Evaluation items	Large scale or super-large scale	Medium and small scale	Small scale
Surrounding rock grade of fault	Type A	Type B	Type C
Width of fault	Grade A	Grade B	Grade C
Water source	Type A	Type B, C	–
Vertical hydrodynamic zone	Type A	Type B, C	–
Horizontal hydrodynamic zone	Type A, B	Type C	–

3 ENGINEERING APPLICATIONS

3.1 Engineering background

Yonglian tunnel (formerly called Zhongjiashan tunnel) is located in the western segment of Jian-Lianhua expressway. It is a separate long tunnel, whose left and right lines are 2486 and 2494 m in length. Hydrogeological conditions of Yonglian tunnel are complicated, and engineering geological properties of surrounding rock is poor. The faults such as F1, F2, F3, F4, and F5 are developed in tunnel-setting area. The first four faults, F1, F2, F3, and F4, are connected by the fault F5, and the network structure is formed. A weathered fracture zone is developed between F2 and F3 faults. The lithology of the weathered fracture zone is shale with intercalations of sandstone. Therefore, large-scale water inrush and mud bursting disasters may easily happen. In this paper, risk evaluation of water inrush and mud bursting in ZK91 + 310~ZK91 + 380 section of Yonglian tunnel is evaluated by attribute recognition model established in this paper. The geological profile of Yonglian tunnel is shown in Figure 1. It is evident from the figure that the tunnel pierce through the Devonian system Shetianqiao

Formation, Carboniferous system Datangian water measurement section, Carboniferous system Datangian Zimenqiao section, and the quaternary system in turn from import to export. The surrounding rock of the evaluated section is mainly exposed as Shetianqiao Formation fracture aquifers. The fault F2 is located within the evaluated section.

3.2 Scale prediction of water inrush and mud bursting in Yonglian tunnel

The established scale-predictive expert system is used to predict the scale of water inrush and mud bursting in Yonglian tunnel on the line of Jilian Expressway.

3.2.1 The infilling materials of the fault fracture zone

From the hydrogeological exploration report and Figure 1, the fault F2 is located within the evaluated section. The fault fracture zone is mainly composed of sandy shale. The rock mass is very fragmented, and mixed with rock soil mass with extremely low strength. According to the lithology, hardness, and the degree of softening by water surrounding rock, the surrounding rock of the fracture zone is identified as of grade V. Therefore, the fault within the evaluated section is of type A.

3.2.2 The width of the fault fracture zone

According to the drilling information, the width of the fault F2 located within the evaluated section is 15–35 m. If the width is larger than 10m, the scale of the fault is of grade A.

3.2.3 The water source of water inrush and mud bursting disaster

The high-density resistivity method is adopted to detect the geological structure characteristics of the section under construction. The detection line is set on the left-hand side of the tunnel, paralleling to the tunnel axis, whose total length is 900m.

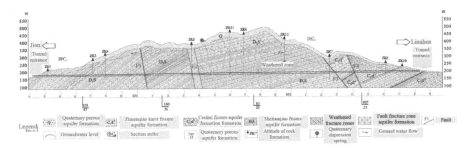

Figure 1. Geological profile of Yonglian tunnel.

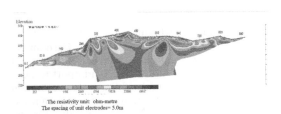

Figure 2. The resistivity section map.

Figure 3. The scene of the excavation.

The detection range is ZK91 + 930~ZK91 + 30. The resistivity section map obtained through the inversion of software is shown in Figure 2. From the figure, a low-resistance zone is developed at 580m (namely ZK91 + 350). This zone is just the location of F2 fault fracture zone.

There are two categories of water source for water inrush and mud bursting in Yonglian tunnel: surface water and groundwater. Multiple long-term rivulets are developed in the tunnel area. Rivulets of some rivers will supply water to the tectonic crevice water and become the water source of the water inrush and mud bursting disaster. Because of the existence of two water conduction zones (weathered zone, F2 fault), weathered fissure water in local area will be linked up and become the water source of the water inrush and mud bursting disaster. Because water source of tunnel site area is rich, the value of S is set to 12 m². The water source is defined as of type A.

3.2.4 *Vertical and horizontal hydrodynamic zone in tunnel location*

From the hydrogeological exploration report and Figure 1, Yonglian tunnel is located under groundwater level. The vertical hydrodynamic zone and horizontal hydrodynamic zone located in the tunnel are saturated water zone discharging areas. Furthermore, both of them are of type A vertical hydrodynamic zone.

The prediction results show that the tunnel in the evaluated section is of type 5A. Therefore, super-large-scale water inrush and mud bursting disaster will easily occur in the evaluated section.

3.3 *Engineering verification*

When mud was dug from the ZK91 + 316 in the entrance of left tunnel, an irregular round hole of mud bursting appeared on the upper left area of the tunnel. The diameter of the hole was about 1m. Large-scale water inrush and mud bursting disaster occurred eight times since 2 July 2012. The quantity of water inrush and mud bursting is about 14000 m³. Thereinto, the outflow of water inrush and mud bursting for the first three times is mainly muddy water. Then, it is mainly silt and only a small amount of muddy water. The condition of tunnel after water inrush and mud bursting is shown in Figure 3.

The results of the scale-predictive expert system are conformed to the excavation conditions of the construction site. Therefore, the scale-predictive expert system is verified effective to predict water inrush and mud bursting disaster in water-rich fault tunnel.

4 CONCLUSIONS

1. The evaluation items of the scale-predictive expert system are determined, and a system of water inrush and mud bursting in water-rich fault tunnel is established. It provides an effective way to predict the scale of water inrush and mud bursting in water-rich fault tunnel.
2. The scale of water inrush and mud bursting in Yonglian tunnel is predicted by the established

scale-predictive expert system. The conditions of the construction site show that the prediction results agree well with excavation situation. The results show that the scale-predictive expert system is effective. It is of great significance to the early warning of water inrush and mud bursting in the tunnel of water-rich fault.

REFERENCES

Han Xingrui. Karst water bursting in tunnel and expert judging system. Carsologica Sinica, **23**, 3 (2004): 213–218.

Liping Li, Ting Lei. Risk assessment of water inrush in karst tunnels and software development. Arabian journal of geosciences, **8** (2015): 1843–1854.

Rao Jun-ying Fuzzy Evaluation Model for In-service Karst Highway Tunnel Structural Safety. KSCE Journal of Civil Engineering, **20**, 4 (2016): 1242–1249.

Shu-cai Li, Zong-qing Zhou. Risk assessment of water inrush in karst tunnels based on attribute synthetic evaluation system. Tunnelling and Underground Space Technology, **38** (2013): 50–58.

Standard for engineering classification of rock mass (Ministry of Construction of the People's Republc of China, 1994).

Incorporation of SiC nanoparticles into coatings formed on magnesium by plasma electrolytic oxidation

Yan Liu, Fuwei Yang, Dongcheng He, JianBing Zhang, BingFeng Bai & Feng Guo
Department of Chemistry, Tianshui Normal University, Tianshui, Gansu, China

ABSTRACT: In this paper, a kind of environmentally friendly anodizing routine for AZ91D magnesium alloy based on SiC nanoparticles was studied. The effect of SiC nanoparticles on the properties of PEO coatings was investigated by Scanning Electron Microscopy (SEM) and X-Ray Diffraction (XRD). The morphology of PEO coatings is similar with and without the addition of SiC nanoparticles in the electrolyte. The temperature of the PEO process also increases to 2873K, so SiC nanoparticles were not observed due to melting.

1 INTRODUCTION

Plasma Electrolytic Oxidation (PEO) is well known and widely employed to produce anticorrosion film on the surface of magnesium alloy. PEO can produce effective protective film on the surface of magnesium alloy and is therefore considered one of the most promising technologies. The temperature of plasma electrolytic oxidation is a very important parameter. Because of the melting behavior of plasma, the mechanism of oxide formation is quite different from that of conventional anodization. In particular, the temperature of the arc during the PEO process is a critical parameter.

The nature of discharge phenomena and their effect on the mechanisms of the formation of coatings by PEO were investigated by Krysmann (2001), S. Ikonopisov (1997), and Albella (1984). For example, the phase transformation from Mg (OH)$_2$ to MgO is thermodynamically induced by the high temperatures produced during the PEO process.

The possibility of contributions from PEO processes has considerable uncertainty about the precise mechanism of the formation of coatings. Studies of coatings formed in the electrolytes with the nanoparticles offer another route to understand the mechanisms of the formation of coatings. Suspensions of SiO$_2$ (V. N. Malyshev, 2007), Fe (Fan, 2009), graphite (H. L. Guo, 2009), and ZrO$_2$ (R. Arrabal, 2011) have been used to study the mechanism of oxide formation. Krysmann (2001), S. Ikonopisov (1997), and Albella (1984) put forward the mechanism of anodic oxidation technology, but there are still many unknowns.

The actual temperature was indirectly detected in the process of PEO. Therefore, the nanoparticles as the tracer atom were added into the electrolytic solution. The temperature was estimated by means of the change in the composition of PEO coatings.

In this study, we incorporated SiC nanoparticles into the magnesium oxide layer as the tracer atom and investigated the variation of the resulting microstructures. On the basis of the results of those investigations, the actual temperature of molten oxide during the PEO process was indirectly estimated.

2 EXPERIMENT

2.1 Materials and specimen preparation

The chemical composition of AZ91D magnesium alloy is as follows (in wt.%): Al (8.77), Zn (0.74), Mn (0.18), Ni (0.001), Cu (0.001), Ca (<0.01), Si (<0.01), K (<0.01), Fe (<0.001), Mg (remaining). Prior to the PEO treatment, samples with a size of 0.5 cm^2 were polished up to 2000 grit, degreased with acetone, washed with distilled water successively, and dried in air. The PEO equipment consisted of an AC power supply and a stirring and cooling system to control the solution temperature at 30°C by a miniature refrigerant equipment (model YT-8A, China). The duration of the PEO process was 3 min. The basic anodic electrolyte containing 50 g/L of NaOH, 35 g/L of Na$_2$B$_4$O$_7$, 20 g/L of H$_3$BO$_3$, and 2.0g/L of SiC nanoparticles (particle diameter: 20 nm) was used as additive. All solutions were made of analytical grade reagents and distilled water.

2.2 Specimen examination

Surface morphology of the coatings was characterized by scanning electron microscopy (SEM, FEI SIRION-100). EDS attached to SEM was used to detect the elemental composition of the PEO coating on the magnesium substrate. The phase structure of the coatings was determined by X-ray diffractometry (XRD, AXS D8 ADVANCE) using Cu Kα (1.5418Å) radiation source.

3 RESULTS AND DISCUSSION

3.1 *Effects of density transient*

Fig. 1 shows the current density change of the PEO process. In general, the current density change can be divided into two periods:

In the first period, the current density increases and reaches the maximum value rapidly. In the absence of SiC nanoparticles, the current density changes with time. Violent sparking and gas release can be observed on the surface of the magnesium alloy specimen. In the presence of SiC nanoparticles, the change in the current density is mitigated and the intensity of the sparking and gas release is slightly reduced.

In the second period, the current density decreases gradually until a stabilized value is reached. During this period, vigorous sparking and gas evolution can still be observed in the absence of SiC nanoparticles. In the presence of SiC nanoparticles, sparking and gas evolvement can still be observed.

Suitable anodizing current density is important for the PEO process. Vigorous sparking and gas evolvement at excessive high current density often result in poor PEO coatings (Liu, 2011), while PEO coatings are difficult to form at too low anodizing current density (Zhang, 2002; Yao, 2009). In general, the addition of SiC nanoparticles to the silicate and phosphate electrolytes had small influences on the current density response during PEO treatment.

In this study, the variation trend of current density in the PEO process is similar with and without the addition of SiC nanoparticles in the electrolyte.

However, with the addition of SiC nanoparticles into the electrolyte, the largest current density was reduced, but the degree of reduction is less.

This result indicates that the addition of particles did not significantly affect the variation of oxidation behavior during the PEO process.

3.2 *XRD*

The XRD results (Fig. 2 (a), (b)) show that the PEO coatings consist mainly of MgO and Mg, with the former resulting from the PEO process. The diffraction peaks of Mg are obtained from the magnesium alloy substrate. X-ray can penetrate the film and detect the metal magnesium in the substrate. Otherwise, the diffraction peaks of SiC were not detected.

3.3 *Morphology of the PEO coatings*

Fig. 3 shows the surface morphology of the PEO coatings with and without SiC nanoparticles obtained by SEM. Crater-like holes of various sizes can be observed in the coatings deposited in the solution with and without SiC nanoparticles. It could be seen that all the PEO coatings have porous microstructure, and the micropores are often caused by spark discharge. The surface is completely covered by PEO coatings. The temperature rises to at least 2873K (Qiu, 2003), which results in transformation from SiC to Si at the PEO coatings. During the PEO process, the temperature also increases to 2873K, so SiC nanoparticles were not observed due to melting.

Fig. 4 shows EDS maps of PEO coatings formed in the solution with 2.0 g/L of SiC. The composition on the sample surface is shown in Fig. 4. The elements Al, Mg, and O are derived from the magnesium substrate. C element cannot be detected by DES, and Si is derived from the SiC nanoparticles. The distribution of Si in the coatings suggests that

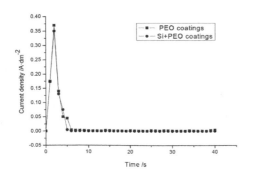

Figure 1. The current density change of the PEO process in the electrolyte.

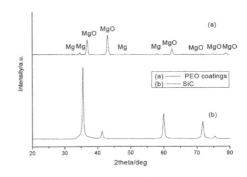

Figure 2. XRD of PEO coatings formed with basic electrolyte with 2.0 g/L SiC nanoparticles and (b) XRD of SiC nanoparticles.

particles can be incorporated at the surface of the coatings.

3.4 *Formation mechanism of the coatings*

The temperature of the arc generated during the PEO process is a critical parameter. The actual temperature of molten oxide during the PEO process was indirectly estimated. The incorporation behaviors of SiC nanoparticles were also systematically

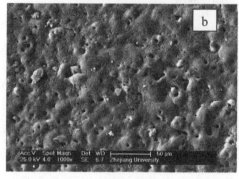

Figure 3. SEM images of the PEO coatings formed in the electrolyte (a) without 2.0 g/L SiC nanoparticles and (b) with 2.0 g/L SiC nanoparticles.

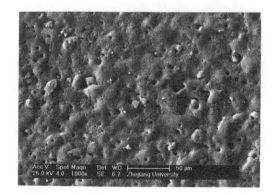

Figure 4. (*Continued*)

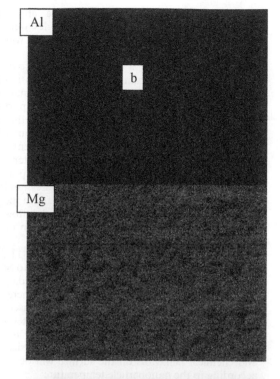

Figure 4. EDS maps of PEO coatings formed in the solution with 2.0 g/L of SiC nanoparticles.

investigated. From the analyses made above, the diffraction peaks of SiC nanoparticles were not detected in XRD. Otherwise, Si element was detected in the EDX.

The distributions of Si within the present coatings suggest that the main route of incorporation of SiC nanoparticles into the coatings is through short-circuit paths in the outer layer. The transport paths are probably breakdown channels, together with cracks and pores in the coatings. The outer and inner coating materials are heated sufficiently by the high local current density. Under high temperature, diffusion of coatings species, transport of coatings species along cracks and pores, and the formation of new phases may take place in the outer layer. The coatings material may also flow under the stresses generated by the formation of coatings to fill pores and cracks.

The temperature rises to at least 2873 K [11], which results in the transformation from SiC to Si at the PEO coatings (eq. 1). CO_2 was synthesized by the reaction of carbon and oxygen (eq. 2). Therefore, the porosity of the PEO coatings increased, and the surface is slightly rough. In fact, the temperature of the nanoparticles might be more relevant, because structural changes or other physical and chemical behaviors on the surface will occur according to the nanoparticle temperature:

$$SiC = Si + C \qquad (1)$$

$$C + O_2 = CO_2 \uparrow \qquad (2)$$

4 CONCLUSION

Si nanoparticles are incorporated into the coatings by AC PEO treatments of magnesium alloy in the basic electrolyte at the surface of the coatings and within the inner layer of the coatings. The main coatings comprise an outer compact layer and inner finely porous layer, and a relatively thin barrier layer is present at the magnesium–coatings interface. The morphology of PEO coatings is similar with and without the addition of SiC nanoparticles in the electrolyte. During the PEO process, the temperature also increases to 2873K, so SiC nanoparticles were not obtained by melting.

ACKNOWLEDGMENTS

Financial support was provided from the project of Gansu Scientific Higher Education Research (2013A-105), and the Laboratory Project of Tianshui Normal University is gratefully acknowledged.

REFERENCES

Albella M., J. I. Montero, D. J. M. Martínez. J. Electrochem. Soc, 131, 1101(1984).

Arrabal, R., E. Matykina, F. Viejo, P. Skeldon, G. E. Thompson, M. C. Merino, Appl. Surf. Sci., 254, 6937 (2008).

Fan, Y. J., K. C. Paul, H. T. Hong, Appl. Surf. Sci., 253, 863 (2006).

Guo, H. L., C. Huan, C. G. Wei, R F. Wen, L. Li, W. N. Er, H. Z. Xian, Z.Y. Si, Curr. Appl. Phys., 9, 324 (2009).

Khaselev, O., Weiss, D. Yahalom. Corrosion Sci. 43, 295 (2001).

Konopisov, S. Electrochim. Acta. 22, 1077 (1977).

Liu, Y., Z. L. Wei, F. W. Yang, Z. Zhang, J. Alloys Compd., 509, 6440 (2011).

Malyshev, V. N., K. M Zorin, Appl. Surf. Sci. 254, 1511 (2007).

Qiu, H. P., H. Z. Song, L. Liu. Chin. J. Mat. Res., 17, 186 (2003).

Yao, Z. P., Y. J Xu, Z. H. Jiang, F. P. Wang, J. Alloy Compd., 488, 273 (2009).

Zhang, Y. J., C. W. Yan, F. H. Wang, H. Y. Lou, C. N. Cao, Surf. Coat. Technol., 161, 36 (2002).

Advances in Energy Science and Equipment Engineering II – Zhou, Patty & Chen (Eds)
© *2017 Taylor & Francis Group, London, ISBN 978-1-138-71798-5*

Study of the FSI characteristics of horizontal axis wind turbine under yaw

Xiaoming Chen
China HuaDian Engineering Co. Ltd., Beijing, China
School of Energy, Power and Mechanical Engineering, North China Electric Power University, Beijing, China

Xiaodong Wang
School of Energy, Power and Mechanical Engineering, North China Electric Power University, Beijing, China

Xiaoqing Feng, Lei Liu & Xiaochun Hu
China HuaDian Engineering Co. Ltd., Beijing, China

ABSTRACT: In this paper, a two-way Fluid–Structure Interaction (FSI) analysis has been performed on the NREL Phase VI horizontal axis wind turbine, based on three-dimensional unsteady CFD simulation and CSD analysis. The FSI analysis results are compared with the experimental results to validate the simulation reliability. Under the conditions of 7 m/s wind speed and 30° yaw angle, the blade aerodynamic load and flapwise and edgewise structural deformation are investigated in detail as well as the influence mechanism of FSI characteristics under yaw and blade deformation is contrastingly analyzed.

1 INTRODUCTION

Wind energy is one kind of clean renewable energy with the fastest development and most large-scale and commercial utilization potentials. Recently, it has attracted increasing attention worldwide (Global Wind Energy Council, 2014). Wind turbines often encounter a variety of complex wind conditions during operation in real wind farm. Yaw is one of the complex conditions, which is defined as the angle between flow velocity and axial angle. In yaw, the horizontal axis wind turbine rotor blades have different inflow angles while rotating to different circumferential positions, leading to wind speed fluctuations at the wind rotor rotation plane. This load fluctuation gives rise to mechanical vibration of the wind turbine blade. Yaw will also affect the quality of output power severely (Hansena M O L, 2006).

At the same time, with the increase of wind turbine scale, blades become more and more slim and flexible. The safety and stability of wind turbines is of utmost importance for the future development of wind power industry (Moeller, 1997; Hu, 2011). The coupling between variable load and flexible structure will lead to blade deformation and structure vibration. Blade vibration exceeding the maximum allowable value leads to vibration instability, thereby resulting in fatigue damage. Therefore,

large-scale wind turbine has higher requirements of efficiency, security, and reliability.

Blade deformation in wind turbine is inevitable. How to take advantage of it to improve the characteristics of the wind turbine should be taken into consideration. Except the traditional variable-pitch control, passive control such as bend-twist coupling and aero-elastic cut have become the focus of research (Larwood, 2006; Veers P, 1998; Karaolis, N. M., Mussgrove, 1988). A. Beyene and T. Ireland put forward the concept of soft blades, which can auto-tune blade shape against the wind, and then improved the aerodynamic performance (Beyene A, 2006).

Therefore, deeper insights into the Fluid Structure Interaction (FSI) analysis of wind turbine are necessary. The FSI can be used to analyze the interaction between blade aerodynamic loading and deformation, which has more practical meaning for blade design and performance analysis from engineering perspective.

2 NUMERICAL METHODS

2.1 Physical model

A widely studied case, NREL Phase VI wind turbine rotor, is used in this paper, which has two blades. The rotor diameter (R) is 10.058 m.

Taking advantage of the variable-cross-section twisted-blade and stall control strategy, the rated power can reach 19.8 kW with 71.6 rpm speed. The full-scale Unsteady Aerodynamics Experiment (UAE) was performed in the wind tunnel of NASA Ames Research Center in 2000, whose width and height are 24.4 and 36.6 m, respectively (Hand, 2001; Simms, 2001).

2.2 *Computational domain and mesh*

The computational domain is illustrated in Fig. 1(a), with height and width of 30.2 (6R) and 30.2 m, respectively, and length of 90 m (18R) in the streamwise direction. The domain consists of two parts: rotational part, including the blades and a round domain, and stationary part.

Two grids with different grid densities are generated for grid-independent studies. The fine grid with 5.6 million nodes in total was generated in fluid domain generated by the ICEM CFD software, including 1.2 million nodes for stationary domain. For the rotational domain, a grid with 4.4 million grid nodes was generated using the grid mesh generator AutoGrid5 of the NUMECA software package, as shown in Figure 1 (b). The grid distribution around the rotor blade is shown in Figure 1 (c). The rectangle surrounding the two blades was composed of O4H grid topology meshes with 20 inflation layers on the blade surface, with a spacing ratio of 1:1 in the normal direction. The height of the first-layer cell is set to 0.03 mm to capture the boundary layer region accurately. Y^+ value was set to ~1 at the blade tip and decreased toward the blade root, to meet the needs of the turbulence model. The coarse mesh has 2.35 million nodes with the same grid topology. For the solid domain, a grid with 120,000 elements is generated.

2.3 *Boundary conditions*

The wind speed and the air temperature are given at calculation domain inlet. The outlet is set as the pressure far-field boundary. The blade surfaces are set to no-slip boundary conditions. The sliding mesh method is adopted for rotating and stationary interface data transfer.

2.4 *Numerical methods*

Commercial software Fluent is employed to solve the Reynolds-Averaged Navier–Stokes equations (RANS) and Unsteady Reynolds-Averaged Navier–Stokes equations (URANS). The one-equation turbulence model, Spalart–Allmaras (SA) model, is used for turbulence modeling. For the unsteady-state calculation, the software adopts the dual time-step approach to solve the URANS. The physical time step adopted corresponds to the time for a rotational angle of 9°. The finite element method is used for structural analysis. On the basis of Ansys CFX physical field method, the two-way coupling iteration is performed between the fluid field and solid field.

2.5 *Convergence criteria*

All calculation residuals decline by more than four orders of magnitude, and the overall performance parameters are stable. Unsteady-state calculation performance parameters showed the periodic changes.

3 FLUID STRUCTURE COUPLING CHARACTERISTIC WITHOUT YAW

The aerodynamics and structural characteristics of the wind turbine rotor without yaw are shown and analyzed firstly to validate the computation model and methods.

3.1 *Aerodynamic characteristic*

In order to study the grid independence, first, the axis uniform inflow condition with 7 m/s wind speed is calculated using both the fine grid (5.6 million) and the coarse grid (2.35 million). Figure 2 shows the pressure coefficient distributions of four spanwise positions, 30%, 47%, 63%, and 95%, comparing with the experimental data. It is seen from the comparison that both the coarse grid and fine grid in the five spanwise exhibitions

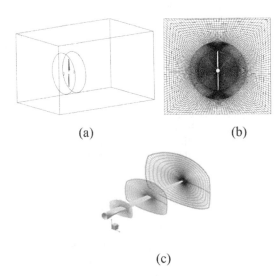

(a)   (b)

(c)

Figure 1.   Computational domain and grids.

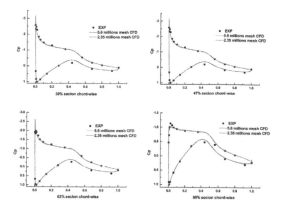

Figure 2. Pressure coefficient distribution at different spans.

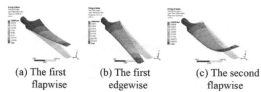

(a) The first flapwise   (b) The first edgewise   (c) The second flapwise

Figure 3. Modal shapes of FE model.

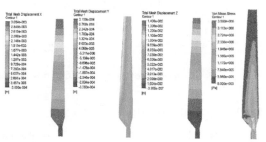

Figure 4. Displacement and stress distribution.

Table 1. Natural frequency of FE model.

Frequency	EXP	Shell	Entity
The first flapwise	7.313	7.2891 ($\Delta = 0.33\%$)	7.2524 ($\Delta = 0.83\%$)
The first edgewise	9.062	13.111 ($\Delta = 44.68\%$)	13.114 ($\Delta = 44.7\%$)
The second flapwise	30.06	31.926 ($\Delta = 6.2\%$)	32.503 ($\Delta = 8.1\%$)

have good agreement with the experimental data. In general, fine mesh result is more close to the experimental value. However, the difference is negligible. Therefore, the coarse grid will be used for rest computations as it requires much less computational labor.

3.2 Structural characteristics

Two different models for solid field are compared in this section: shell model and entity model. According to the material properties and stiffness distribution, natural frequency of FE model is used to validate the structural characteristics (Li, 2013). Table 1 describes the natural frequency of FE model, comparing with the experimental data denoted by EXP. As the results show, shell structure model results are closer to the experimental values than those of the entity structure model; Especially, in flapwise direction, the result is much closer to the experimental value. Figure 2 shows the equivalent structure model using shell structure model of the first wave, shimmy, and second-order wave model.

3.3 Fluid structure coupling characteristic

Figure 4 shows the distributions of the flapwise, edgewise, and spanwise structural displacements and the stress of the blade. It can be seen from this figure that the blade deformation along the flapwise, edgewise, and spanwise displacements all show the first-order vibration form. The displacements increase from blade root to tip, whereas stress is mainly located at the root and lower part of the blade.

4 YAWED WIND TURBINE ROTOR FLUID STRUCTURE COUPLING CHARACTERISTIC

4.1 Convergence analysis

On the basis of the validation shown in the last section, analysis of the FSI characteristics with a yaw angle of 30° is performed in this section. The inflow wind speed is still 7 m/s. The unsteady numerical simulation started with converged steady simulation as the initial field. Then, the fluid structure numerical simulation started with the unsteady initial field. Under yaw condition, the blade azimuth angle at the 0 clock position is defined as 0°. The 6 clock position is defined as 180° according to the counterclockwise rotation. Thus, 9 clock position and 3 clock position are corresponding to 90° and 270°, respectively.

Figure 5 shows the convergence history of the load under yaw condition. Figure 6 shows the convergence history of the blade deformation along the flapwise and edgewise deformation, which illustrates the periodic variation of blade deformation from large vibration to a stable state.

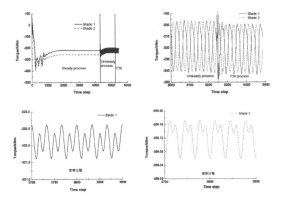

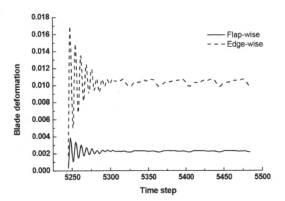

Figure 5.   Convergence history of load (7 m/s, 30° yaw angle).

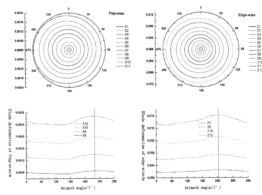

Figure 7.   Flapwise and edgewise deformation change of a cycle (7 m/s, 30° yaw angle).

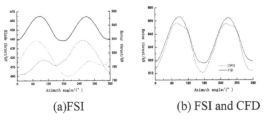

(a)FSI   (b) FSI and CFD

Figure 8.   Blades and rotor thrust curve of a cycle (7 m/s, 30° yaw angle).

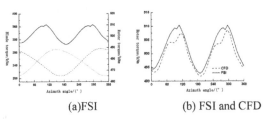

(a)FSI   (b) FSI and CFD

Figure 9.   Blades and rotor torque curve of a cycle (7 m/s, 30° yaw angle).

Figure 6.   Deformation convergence history (7 m/s, 30° yaw angle).

4.2   Fluid structure coupling characteristic

Figure 7 shows the variation of the deformation along the flapwise and edgewise deformation in one cycle. The wind rose diagram shows the 12-section deformation diagram along different azimuth angles. The sections are distributed with 0.4 m segment along the spanwise deformation. The variations at the 6th, 8th, 10th, and12th sections in one cycle are analyzed in detail. The results show that the deformation trend at the different sections have the same characteristic, which increase form the blade root to blade tip. The deformation of the flapwise deformation is larger than that of the edgewise for one order of magnitude. The deformation trends of both the flapwise and edgewise deformation are the same, and the thrust development trend is shown in Figure 8(a). The azimuth angle of the maximum thrust lags behind the azimuth angle corresponding to the maximum deformation. The position of azimuth angle related to the maximum deformation shows obvious difference between the flapwise and edgewise deformation.

Figures 8 and 9 show the variation of the thrust and torque, respectively, along one blade and rotor in one cycle. It is seen from these two figures that the thrust and torque of one blade and rotor show obvious periodicity. The change trend of FSI results is reasonable.

5   CONCLUSIONS

This paper investigated the fluid–structure coupling characteristics of a wind turbine blade under yaw condition. On the basis of the comparison with experimental data and the FSI characteristics without yaw, the following conclusions can be drawn:

1. Utilizing the equivalent structure modeling, the computational result of shell structure model is better than entity structure model.
2. With the axial uniform flow, the blade deformation along the flapwise, edgewise, and spanwise deformation all show first-order vibration forms. The maximum stress along the chordwise deformation is located on the position of maximum thickness. The wind turbine rotor leading edge load increases slightly, closer to the experimental value.
3. Under the circumstance of yaw, wind turbine aerodynamic loads and blade structure deformation show obvious periodicity. The deformation trends in different sections are similar, which increases from blade root to blade tip. The azimuth angle of the maximum thrust falls behind the angle at the maximum flapwise deformation angle. The maximum deformation along the flapwise and edgewise directions has different azimuth angles.

ACKNOWLEDGMENTS

This study was supported by the National Natural Science Foundations of China (No. 51576065) and the Fundamental Research Funds for the Central Universities (No. JB2015RCY05).

REFERENCES

Beyene A, Mangalekar D, Miller R, Alexander S, Villegas S. A new turbine concept, the oceanic engineering society of the institute of electrical and electronic engineers. In: OCEANS 2006 ASIA conference/exhibition, May 16–19, Singapore.

Global Wind Energy Council. Global Wind Report-Annual market update 2014 [R].

Hand M, Simms D, Fingersh L, etc. Unsteady Aerodynamics Experiment Phase VI: Wind Tunnel Test Configurations and Available Data Campaigns [R]. NREL/TP-500-29955, NREL, 2001.

Hansena M O L, Sensena J N, Voutsinasb S, et al. State of the Art in Wind Turbine Aerodynamics and Aeroelasticity [C]. Progress in Aerospace Sciences, 2006, 42: 285–330.

Hu Ying, Retrospect and prospect of our country wind power industry development at 2010. Energy equipment 2011.02, 17–20.

Karaolis, N.M., Mussgrove, P.J., and Jeronimidis, G. (1988). Active and Passive Aeroelastic Power Control using Asymmetric Fibre Reinforced Laminates for Wind Turbine Blades. Proc. 10th British Wind Energy Conf., D.J. Milbrow Ed., London, March 22–24, 1988.

Larwood, S., Zuteck M. Swept Wind Turbine Blade Aeroelastic Modeling for Loads and Dynamic Behavior. Wind power. 2006.

Li Yuan. Numerical Simulation of Fluid Structure Coupling of Wind Turbine Blades [D]. Bei Jing: North China Electric Power University Dissertation for the Doctoral Degree in Engineering, 2013, 108–111.

Moeller, T., 1997. Blade cracks signal new stress problem. Wind Power Monthly 25.

Simms, D.A., Schreck, S., Hand, M.M., and Fingersh, L. J, 2001, "NRELUnsteady Aerodynamics Experiment in the NASA-Ames Wind Tunnel: A Comparison of Predictions to Measurements," NICH Report No. TP-500-29494.

Veers P, Bir G, Lobitz D. Aeroelastic Tailoring in Wind-Turbine Blade Applications. Wind power. 1998.

Advances in Energy Science and Equipment Engineering II – Zhou, Patty & Chen (Eds)
© 2017 Taylor & Francis Group, London, ISBN 978-1-138-71798-5

Brief discussion on rural landscape design in Chengdu Plain based on the demand investigation of farmers

Tao Jiang & Shu Li
Landscape Architecture, College of Landscape Architecture, Chengdu, China

Chunnong Li
Institut Entwerfen von Stadt und Landschaft, College of Karlsruher Institut für Technologie, Karlsruhe, Germany

Junzhuo Li, Luyuan Fan & Qibing Chen
Landscape Architecture, College of Landscape Architecture, Chengdu, China

ABSTRACT: National urbanization construction has presented modern-day challenges to the rural landscape in China. Unprecedented challenges are typified by environmental pollution, farming cultural loss and changing landscape characteristics. The typical rural settlement landscape in Chengdu Plain, Linpan, which has long benefited from Dujiangyan irrigation system founded 2000 years ago, is under the above mentioned threat. Farmers are the heart of Linpan and it makes sense to understand their demand in Linpan protection and development. Therefore, this paper focuses on the existing difficulties of Linpan's rural community based on the demand of farmers through questionnaire surveys and field investigations, which were conducted in Dujiangyan city, Wenjiang District, Longquanyi District, Xindu District, Shuangliu County, Jintang County, Dayi County, Pengzhou County, Pi County, and Qionglai city as the representative samples in Chengdu city, Sichuan Province. Rural residents were asked to rate personal satisfaction as related to requirements for their rural areas. Results followed indicating priorities on monetary income, living conditions, production conditions (traditional or modern), and environmental quality. Discovery of these priorities may help develop some design strategies in establishing a balance between conservation and development of the rural landscape.

1 INTRODUCTION

Sichuan is located upstream of the Yangtze River and is home to the Minjiang River, a main river in Sichuan and a tributary to the Yangtze River. The culture of western Sichuan originates from the Minjiang River, naturally descending from the Yangtze River civilization. The origin of the Yangtze River Hydraulic civilization is 1000–2000 years later than the Nile Valley civilizations and Mesopotamia. In 221BC, Ying Zheng, the first empire in Qin dynasty, ended a period of fighting country, which lasted over 250 years and set up the Qin, the first uniform and centralized feudal countries comprising multiple nations. Some large-scale irrigation projects were started in Qin dynasty, such as Zhengguo Canal, ZhangShui Twelve Canal, and Dujiangyan Irrigation System. Designed and built by Li Bing and his son, the Dujiangyan Irrigation System enhanced the connection of Minjiang River and Chengdu Plain to the Yangtze River and changed the natural and social forms of Chengdu

Plain, making Chengdu a political, economic, and cultural center in southwest China (Chen, 2011).

Chengdu Plain is located in the Dujiangyan irrigation core area with the Minjiang River flowing into the farmland of the plain, forming a highly networked irrigation system. Located here is Linpan in Chengdu Plain, a unique farming civilization to China and a model of ecological and livable habitation (Duan, 2004). Dating from the ancient Shu period, Linpan typifies early settlements that form a unique rural landscape in Sichuan.

Development through national urbanization has produced changes in the traditional Linpan, both in the number of Linpan quantity and Linpan architecture form. Quantity has been gradually reduced, especially in the second circle layer referring to Wenjiang District, Longquanyi District, Xindu District, Shuangliu County, Pi County, and Qingbaijiang District. Rural residents moving into the town or the rural communities have converted the traditional Linpin into an empty nest. Therefore, it is important to consider the growing demand of farmers of Linpan in western Sichuan for the protection and development of the rural landscape.

2 MATERIALS AND METHODS

2.1 *Description of the study area*

Chengdu city with a total area of about 12,390 km² is located in the vicinity of the largest plains in the southwest region of Chengdu Plain. Chengdu is in a subtropical humid monsoon climate zone, with climate characteristics of early spring, hot summer, cool autumn, and warm winter. The annual average temperature is 16°C, and the annual rainfall is about 1000 mm. The city's permanent population was 14.298 million at the end of 2013. Communities are divided into city, county, township, and village. Township is the smallest administrative unit. Beginning from the Chengdu city center and going out, there are three circular zones. The first circle zone is the central area of the city, the second circle zone includes Longquan District, Pi, Wenjiang District, Shuangliu County, Xindu District, and Qingbaijiang District, and the third circle zone includes Dayi County, Xinjin County, Jintang County, Pujiang County, Dujiangyan, Pengzhou, Chongzhou, and Qionglai.

Considering the urbanization rate of different districts and counties in Chengdu released by Chengdu Bureau of Statistics in 2012 and combining the existing Linpan distribution statistics according to Chengdu Urban and Rural Construction Committee, as well as the distribution of Dujiangyan irrigated area, we selected 11 of the surrounding counties and cities of Chengdu, randomly selecting a village from each city and county to conduct the research. Among them, Qinbaijie Village, Xinfan Town, Xindu District; Shuanghuai Village, Luodai Town, Longquan District; Yufu Village, Wanchun Town, Wenjiang District; Long'an Village, Ande Town, Pixian County; and Jiahe Village, Huanglongxi Town, Shuangliu County, are in the second circle zone. Dujiangyan Cuiyuehu Town, Pengzhou Xinxin Town, Shishan Village, Qionglai Sangyuan Town, Xiangyang Village, Xinjin County, Xingyi Town, Sanhe Village, Dayi County, Wangsi Town, Xi'an Village, Jintang County, Guangxing Town, Baota Village are in the third circle zone. A relatively high rate of urbanization was focused on when the research points were selected. As for the status quo, every research point has good and bad Linpan landscape. When the village was selected, the method of random sampling was utilized to reduce the number of variables for study efficiency.

2.2 *Research of farmers' satisfaction and needs in their production and life*

Early studies of the rural landscape illustrated an important point that farmers play a vital role in the whole rural landscape transformation process (Kristensen S, 2003). As mentioned above, because of the development and urbanization in Chengdu in recent years, farmers who formed the core of Linpan abandoned the way of life they have had for thousands of years and flocked to cities, converting Linpan into an "empty nest." For example, in Pixian County, 11,000 Linpan were present several years ago, but only around 8,000 are currently left, with 10 or more households retaining less than 900. The present conservation and development efforts, according to the remaining few farmers, must be people-oriented (Yang, 2011).

This survey, which took family as the unit, was conducted, combining the questionnaires, interviews, the field study, image acquisition, and so on. A total of 550 survey questionnaires involving 11 districts and counties in Chengdu (79% of the second and third circle zones covering the villages and towns that preserve Linpan in all standards) were issued and all were recovered. Questionnaires were divided into two components: production and living. The production component included water conservancy infrastructure, agricultural science and technology training, agricultural production conditions (traditional or modern), and economy. The living component includes infrastructure for sectors such as transportation, communication, education, health, culture, and expenses. The interview can make up questionnaire insufficiency and act as a reconnaissance to improve the comprehensiveness and accuracy of the survey. The overall insight and analysis toward comprehending the farmers' viewpoint of the Linpan environment will guide further research.

3 RESULTS

3.1 *Satisfaction evaluation*

As mentioned above, with the rapid development of urbanization and new rural construction in China, the traditional Linpan landscape has been declining.

The questionnaire survey results indicate that modern rural residents currently confirm many problems such as the loss of local rural residents, the loss of the agriculture and farming culture, the ecological environment destruction, and the loss of the landscape character.

The research effort issued 550 questionnaires with 533 valid returns. Following interviews with rural residents and analysis of the results for the

533 questionnaires, this paper further analyzes the discontent of the rural residents. Leading factors contributing to discontent include the external environment such as rapid urbanization and new rural construction of China, as well as the Linpan internal environment including poor building quality, poor courtyard environment, and infrastructure differences. For example, some Linpan lack trash collection and libraries while some lack community park and network infrastructure.

3.2 Demand analysis

Results of future planning viewpoints of farmers as related to expectation for new rural construction indicate an urgent need to solve the problem (Yang, 2012). It follows from the analysis of the farmers' viewpoints with Linpan in Chengdu Plain that there are many issues leaving the land unable to meet the modern needs of residents. On the basis of the summarized 533 questionnaires and interviews, the following conclusions were considered major problems in the forefront of the planning and development of villages: low economic income, difficulties in employment, burden in children's education, and environmental pollution. Secondary issues for planning and development are economic income, production conditions, and living conditions.

4 DISCUSSION AND CONCLUSION

As an important farming area in southwest China, Chengdu Plain is one of the places of origin of farming civilization in China. The agricultural landscape of Chengdu Plain has obvious regional characteristics with the Linpan settlement network and the drainage system. Construction of modern rural cities worldwide was a concept actually introduced by the government of Chengdu, which provided an opportunity for the protection and development of Chengdu Plain landscape. Follow through of this modern rural city planning has theoretical and practical meaning for intensive land use planning.

The goal of landscape planning is not only to provide a healthy urban environment but also to provide a protected rural environment (Lewis, 1998). The analysis and summary of farmers' demand is of utmost importance to get some design strategies to realize the sustainable development of rural landscape.

4.1 Settlement perspective

Conservation includes the ecological environment of the upper reaches of Minjiang River and the ecological environment of Chengdu Plain (natural ecology aspect), the traditional lifestyle and customs of the western Sichuan people (social culture aspect), and the traditional agriculture (industry aspect). Developments included the construction of ecological barrier of the Longmen Mountains and Longquan Mountains (natural ecology aspect), traditional farming and folk culture (social culture aspect), and modernized agriculture and characteristic industry tourism (industry aspect).

The practical design methods of Linpan community and the landscape infrastructure constitute municipal infrastructure and public service infrastructure. As for the former, it reflects in water supply (from the town within 1 km): Using the same water supply system as the town; Far away from the town: Building the water tower, extracting groundwater, and centralizing water supply pipe network after the treated compliance; Without the above conditions: pipe pressure wells with different household, extracting groundwater, supplying raw water after the treated compliance drainage (from the town within 1 km); Using the same water supply system as the town. According to the above conditions, three to five users build one biogas digester for a single family; build integrated sewage treatment pool with the permitting conditions; use agricultural irrigation after the treatment compliance, power, and signal communication; use the same system as the town; improve and standardize the power lines and communication lines layout; change leaving chaos phenomenon; and eliminate fire hazards. Power load of 400 W per capita is used to plan communication line in every household energy; combine natural gas with biogas digester gas; replace fuel gas cans, wood, coal, straw, and sanitation; prohibit littering; establish garbage collection rooms; transport landfill treatment at regular time interval; and set up public toilets. As for the latter, it contains (1) Administration (establish housing management of the residents, with construction area no less than 90 m^2, possessing security, sanitation, and other functions); (2) Entertainment (keep the traditional tea, add the square, the square land area no less than 100 m^2; (3) Education (construct the kindergarten at least and configuring a kindergarten with more than 800 persons, with an area of about 10 m^2 per capita); (4) Medical facilities (construct a health station, with construction area no less than 40 m^2); (5) Culture (construct the activity room, according to the science and technology service point, the construction area is 50–200 m^2, take the lower limit value under 1000 person and combine the management room of the comprehensive configuration; (6) Library (the construction area is 50–100 m^2, take the lower limit value under 1000 person, combine the activity room of the comprehensive configuration); (7) Physical fitness infrastructure (combine the small square green to put up, with the land area no less than 100 m²); and (8) Municipal (public toilets, with construction area of about 30 m²; garbage collection room, with service radius not over 70 m and construction area pf about 4 m²; switch board room, with area of 50 m²; water pump house, in the area of noncentralized water supply; biogas pool, according to the circumstances).

The green construction includes landscaping, green spot, green ribbon, and green sheet. The green spot involves the following strategies: planning, reformation with the use of the existing rivers, ponds, nurseries, orchards, and small forest; emphasis on rural feeling and natural features rather than rules; using native plants, such as Platycladus orientalis (Linn) Franco, Phoebe zhennan S. Lee, C. lanceolata (Lamb.) Hook., Pterocarya stenoptera C. DC, Eucalyptus robusta Smith, Camptotheca acuminate, Pinus massoniana Lamb, Ginkgo biloba L., Chimonanthus praecox (Linn.) Link, Magnolia denudata Desr., Prunus Cerasifera Ehrhar f. atropurpurea (Jacq.) Rehd., Metasequoia glyptostroboides (Hu & W. C. Cheng), etc. The green ribbon strategy involves the following methods: tree species selection and mixed types. Trees with high stability, long life, and high resistance are selected; the most advisable are excellent native tree species (The configuration of Protection forest). According to the purpose of the protection and geomorphic type to build the configuration shelterbelt, water, and soil conservation, forest should be configured into a sheet, strip, or block Water and soil conservation forest system should remain a complete system (The Management of protection, prohibited deforestation and update). The green sheet involves the construction of ecological forest, with afforestation density not less than 100 strains (35 strains of each mu of bamboo). The water system planning contains plane morphological (To solve the local bottleneck phenomenon, to expand some bends in the river; to retain the natural form of river for main canal and branch canal and the ditch with agricultural functions can be linear), section form (set different shades of waters); Under different water levels in the form of steps, according to the water plant configuration, such as Phragmites australis (Cav.) Trin. ex Steud, Cyperus glomeratus L., Acorus calamus L., Metasequoia glyptostroboides (Hu & W. C. Cheng), Salix babylonica L, etc.), bulkhad form (rigidity revetment including stone revetment, plate stone revetment, and pebble revetment; Flexible revetment including pile revetment, natural revetment, and wetland revetment) and ditch greening (give priority to the local plant; selection of aquatic plants should be considered beneficial to the improvement of the water self-purification ability; choice should still consider the seasonal and ornamental plants). Road landscape contains road network (combine the existing roads and add necessary road; clear the road level; road pavement width of the community of foreign road: 8–10 m; the main township road pavement width: 6–8 m; branch road width: 2–4 m), greening the wayside (keeping farmers spontaneously to grow crops which can form the productive landscape with local characteristics; In no planting districts, most roads choose native tree species, some parts of the road section need tall trees) and combining with drainage (The footpath of waterfront district; Combined with water plant configuration).

The space of public activity contains square and dam (according to the residential scale, it should not be too large; proper arrangement of featured landscape and leisure infrastructure; square, it should avoid too neat geometric designs as well as apparent axis of symmetry; public sun dam should be combined with the yard layout, can be combined with square set, not planting trees), hydrophilic space (transform the pond losing production function; combined with the ditch and the stream to set up the waterscape), seniors' activity space (increase the elderly fitness infrastructure; add the function such as chess room (combined with the public service infrastructure) and children activity space (equipped with a children's game play infrastructure such as slide, swing, seesaw, and bunkers).

4.2 Unit perspective

Conservation: Lin (the historical trees and traditional plant landscape), Shui (the source of water and the quality of water), Zhai (the house & courtyard layout architectural characteristics), Tian (farmland form, farmland texture and boundary & Corner). Development: We need to increase plant community diversity and landscape appreciation, broaden the edge of farmland, increase the farmland comer planting, and so on.

Practical design methods of Linpan community. Lin: Retain the traditional species such as historical trees and bamboo; Strengthen the breeding of new varieties of bamboo such as hybrid bamboo; Introduce a small amount of ornamental plants, including small trees, shrubs, and herbaceous wild flowers; Keep a small vegetable field and renovate it; Plant design between the buildings should be in small scale.

Shui: Select aquatic plants in landscape configuration to purify the water. Zhai: including house & courtyard (reduce demolition; keep the natural geological features and should not cut the mountain; reduce the damage to the existing plant community and adjust measures to local conditions; retain the traditional "—," "L," and three-courtyard Sanheyuan), quadrangles sihey-uan form, Supplemented by "工," zigzag, and other derivative forms, and architectural characteristics (The storey is not higher than three layers and eaves height is not more than 10 m; keep the traditional style of the western Sichuan folk houses; use brick wall to insulate heat and prevent moisture; Roof: use unified form of slope roof and small Chinese style tiles; Doors & Windows: keep traditional forms of the western Sichuan folk house doors and windows, and use symbols reflecting the regional characteristics.

Consider the tools in and out and the diversified processing window lintel and casement; Handrail: Use native material such as the bamboo and wood, with simple color and hollow form; Wall: Metope material can choose stone, brick, brick veneer, paint, and so on, decorate hanging fish—xuanyu, window cover, door cover, tracery, railings, and so on. Metope color should be a cool one), structure (Use brick or frame construction with high-safety performance) and materials (stone, air bricks, and concrete). Tian: Plant dwarf trees and hybridization at farmland corners and connect the corners through the thickets; Broaden the farmland edge with grassland; Broaden the farmland edge with forest.

ACKNOWLEDGMENTS

This paper is a partial product of the research project funded by the National Natural Science Foundation of China subsidization project. The project is studied using the coupling mechanism of the human physiological and psychological response to the ornamental bamboo and its health function (project code 31570700).

REFERENCES

Chen, Q.B., Research on landscape resources protection and development mode of Linpan in Chengdu Plain. China Forestry Press, 2011.

Duan, P., Liu, T.H., Ecological homland of Shu culture. Sichuan Science and Technology Press, 2004.

Kristensen S, "Multivariate analysis of landscape changes and farm characteristics in a study area in central Jutland, "Ecological Modelling, vol.168, pp. 303–318, October 2003.

Lewis, P.H., Tomorrow by design—a regional design process for sustainability. John Wiley & sons, Inc, 1988.

Yang, X.Y., "Innovation mode of Linpan protection and development", Journal of Chengdu University social science edition, vol.186, pp. 50–53, October 2011.

Advances in Energy Science and Equipment Engineering II – Zhou, Patty & Chen (Eds)
© 2017 Taylor & Francis Group, London, ISBN 978-1-138-71798-5

Study of mechanical characteristics of prestressed concrete polygonal line cable-stayed bridge

Yan-feng Li
Traffic Engineering College, Shenyang Jianzhu University, Shenyang, China

Li Liang
School of Resources and Civil Engineering, Northeastern University, Shenyang, China

Hao Zhang
Traffic Engineering College, Shenyang Jianzhu University, Shenyang, China

Shuang Sun
School of Resources and Civil Engineering, Northeastern University, Shenyang, China

ABSTRACT: On the basis of a fold line tower cable-stayed bridge project in Shenyang and the mechanical characteristics of concrete fold line tower, the mechanical characteristics of the key parts involved were studied by numerical analysis on the fold area. The results show that, in this area, there is serious stress concentration, although high-stress domain is little. After setting stiffening rib, the stress concentration coefficient decreases, and the out-of-plane bending of the middle-span-side tower wall is eliminated. At the same time, the vertical tensile stress appears in the stiffening rib of the fold area. Therefore, the stiffening rib size should be increased and the principle tensile steel and anticracking mesh reinforcement should be appropriately strengthened in the design.

1 INTRODUCTION

With the development of economy and the improving requirement of urban landscape construction, many cable-stayed bridges with diagonal tower have been built worldwide (Miyamoto, 2001; Brozzetti, 2000; Chrimes, 1996). Wang Bohui (Ma, 2008) introduced the leaning tower cable-stayed bridge project and summarized the design experience. Liu Yongjian et al. (Jiang, 2009) used the method of combining plate, beam, and three-dimensional solid element on the tower beam root region of Changsha Liuyang River Hongshan Temple bridge and then compared the results of the finite element analysis with the experimental results. Zhang Xue et al. (Tian, 2013) analyzed the structure's integral equilibrium and local equilibrium in dead load based on a leaning tower asymmetric cable-stayed bridge. Man Niujing et al. (Guo, 2012) studied the mechanical characteristics and leaning angle on the Banfu No. 2 leaning tower cable-stayed bridge. Jiang Chenqiang et al. (Tang, 2014; Zhang, 2010; Wei, 2012) used the least bending energy method, the minimum sum of moment square, and the rigid support continuous beam method to study the

cable forces in the completed stage of a project in Xiamen. The above studies were carried out around the straight-line tower cable-stayed bridges; however the mechanical properties of fold tower cable-stayed bridges have not been studied.

2 RESEARCH BACKGROUND

Fumin Bridge in Shenyang is a single-plane prestressed concrete cable-stayed bridge, with length of 420 m and span arrangement of 89 + 242 + 89 m (shown in Figure 1). The 4# bridge pier adopted

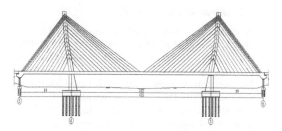

Figure 1. General layout of the bridge.

the tower-girder-pier rigid frame. The 5# bridge pier adopted the fixed-tower-girder, separated pier-girder system. The tower is 67.5m high with box section using C50 concrete. The fold angle is located at 33.9 m above the bridge deck, the angle between the lower part of the tower and the horizontal plane is 75°, and the angle between the upper and lower parts of the tower is 7.5°. The tower's root section is 8.148 m × 3.5 m, the top section is 6.994 m × 3.5 m, the thickness of the front and back wall is 1.4 m, and that of the side wall is 0.7 m. Cable spacings at the tower are 4, 3, 4, 4.5 m. In the section of the tower wall, prestressing steel strands are arranged longitudinally and transversely at the tower wall section, using a φ32 mm rolling rebar. The external force produced by prestressing steel strands is used to balance the internal force acted on the tower wall caused by the horizontal force of cables. The tower at the 4# bridge pier is analyzed afterward.

3 FINITE ELEMENT MODEL

3.1 Finite element model calculation hypothesis

The following assumptions were made in the calculation: (1) the tower is considered as a homogeneous elastic body, whose material properties are represented by the elastic modulus of 34.5 GPa and Poisson ratio of 0.1667; (2) the main loads are cable tension, wind load, and the horizontal pressure of the prestressed steel, which are applied in stages. The wind load applied on the tower is 1.26kN/m². There are three ways of imposing prestressed force in finite element calculation: (1) add directly on the elements; (2) add directly on the key points; and (3) be equivalent by temperature change.

As the model analysis focuses mainly on the whole characteristics, to simplify the analysis process, the prestressed force is added directly on the key points, and the values of the prestressing force are 80% of the design force on account of the prestressing loss.

3.2 Finite element model

According to the actual situation of the project above, using Madis/Civil finite element software, the tower is discretized into a set of 3D solid elements; the model is divided by an eight-node hexahedron element, with each node having three directions of the linear displacement freedom. The simplified tower model includes 49963 nodes and 36124 elements, and the horizontal direction and the vertical and angular displacements of the tower root are restrained.

4 RESULTS AND ANALYSIS

4.1 Analysis result if there is no stiffening plate at the fold area

The normal compressive stress at the fold area (there is no stiffening plate) is shown in Figure 2. The stress concentration at the region can be clearly seen in the figure, and the maximum compressive stress is −19.80 MPa. Afer the calculation, the high-stress area is small and the maximum compressive stress is −17.50 MPa at 0.6 m above the fold section, which has been obviously decreased.

Stress contour of fold section is shown in Figure 3, and stress contour of 0.6m above the fold section is shown in Figure 4. After contrasting the two charts, for there's no stiffening plate at the fold section, the axial force of tower wall at middle span and side span will produce a transverse force, which will result in an out-of-plane bending, with much of the bending influence on the tower of middle span, which also increase the stress gradient.

4.2 Analysis result of stiffening plate at the fold area

In order to improve the stress distribution at the fold area, the stiffening plate is set here, whose structure is shown in Figures 5 and 6.

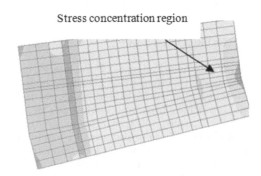

a) —Elevation view;

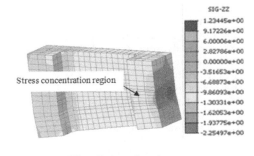

(b) —Perspective view.

Figure 2.   Normal compressive stress of fold section.

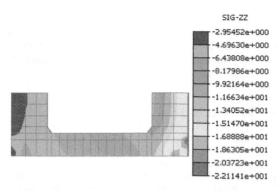

Figure 3.   Normal stress contour of fold section.

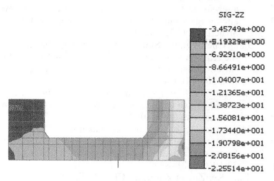

Figure 4.   Normal stress contour of 0.6 m above fold section.

Figure 5.   Elevation of stiffener of bevel section of tower.

Figure 7 shows the normal stress contour of fold area after setting the stiffening plate; the maximum compressive stress of the fold corner is −18.97 MPa and the maximum compressive stress 0.6 m above the corner is −17.00 MPa, which are both lower than those without stiffening plate. It can be seen that the effect of reducing the compressive stress of the fold section is limited, although in spite of setting the stiffening plate, but from the stress contour, the use of the stiffening plate can eliminate the out-of-plane bending of the middle-span-side tower wall, to ensure that the stress dis-

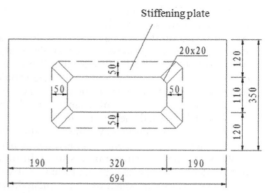

Figure 6.   Plan view of stiffening plate of tower's fold area.

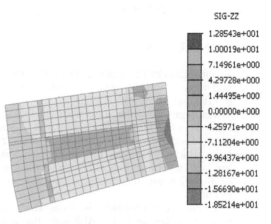

(a) —Elevation drawing;

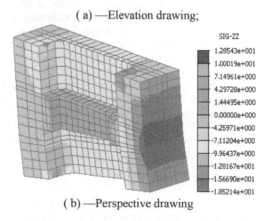

(b) —Perspective drawing

Figure 7.   Normal stress contour of fold area after setting the stiffening plate.

tribution of the tower accords with the hypothesis of plane cross. Vertical tensile stress appears at the inner surface of the stiffening plate. The maximum value is 1.35 MPa, which is born by tower side wall

549

in mechanical interpretation. The concrete triaxial medium transverse expanding deformation happens because of the vertical extrusion. Principal stress flow produces stress vortex at the mutation position, and the local tension will produce cracks. Therefore, the stiffening plate size should be increased, and the inner principle tensile steel and anticracking mesh reinforcement should be appropriately strengthened in design.

The maximum compressive stress of the concrete is a local phenomenon from the finite element analysis and the high-stress range is small. In addition, under high-stress conditions, the concrete will creep to some extent, which results in the redistribution of the stress redistribution and decline of the high stress. It can be seen that the compressive stress of the tower fold area can satisfy the design specification requirements.

5 CONCLUSIONS

1. The fold area of tower is the key part in design, and there is serious stress concentration here, but the high stress area is small.
2. When there's no stiffening plate at the fold section, the axial force of the tower wall at middle and side spans will produce a transverse force, which will result in an out-of-plane bending, with much bending influence on the tower of middle span, thereby increasing the stress gradient.
3. The stress concentration is lower than that without stiffening plate. Although the effect of reducing the compressive stress of the fold section is limited, the use of stiffening plate can eliminate the out-of-plane bending of the middle-span-side tower wall to ensure that the stress distribution of the tower accords with the hypothesis of plane cross and vertical tensile stress appears at the inner surface of the stiffening plate. Therefore, the stiffening plate size should be increased, and the inner principle tensile steel and anticracking mesh reinforcement should be appropriately strengthened in design.

ACKNOWLEDGMENTS

This work was supported by the National Natural Science Foundation of China (Grant Nos. 51308356 and 51474048), the General Project of Education Department in Liaoning Province (Grant No. L2015442), and the General Project of Shenyang Jianzhu University (Grant No. 2015050).

REFERENCES

Brozzetti J. Design development of steel-concrete composite bridges in France [J]. Journal of Constructional Steel Research, 2000, 55(1/2/3):229–243.

Chrimes M M. The development of concrete bridges in the British Isles prior to 1940 [J]. Structures and Buildings, 1996, 116(3/4):404–431.

Guo Fan, Yang Yongqing, Zhou Houbin,et al. Test and analysis on full scale segmental model of cable tower anchorage zone in cable stayed bridge [J]. Building Structure, 2012, 42(10):71–75.

Jiang Cheng-qiang, Sun Xue-xian, Yuan Qi. Research on the reasonable cable force of the leaning tower cable-stayed bridge [J] Urban Roads Bridges & Flood Control, 2009(4):41–44.

Ma Niu-jing, Li Ping-jie. Analysis of mechanics for cable stayed bridge with diagonal tower [J]. Northern Communications, 2008(10):24–27.

Miyamoto A, Kawamura K, Nakamura H. Development of a bridge management system for existing bridges [J]. Advances in Engineering Software, 2001, 32(10/11):821–833.

Tang Ke, Zheng Zhoujun, Chen Kaili, et al. Test Study of Force Mechanism of Pylon Anchorage Zone of Actual Bridge of Jingyue Changjiang [J]. Bridge Construction, 2014, 44(6):52–56.

Tian Zhonghui, Chen Yaozhang. Full-scale Model Experiment on Cable Pylon Anchorage Zone of 3-pylon Cable-stayed Bridge [J]. Journal of Highway and Transportation Research and Development, 2013, 30(7):89–96.

Wei Jianhua, Liao Hailiang. On parameter design for anchor block of concrete beam segment stayed cable on Jiujiang Changjiang Bridge [J]. Shanxi Architecture, 2012, 38(9):179–181.

Zhang Hui, Desrochesb R. Experimental and Analytical Studies on a Streamlined Steel Box Girder [J]. Journal of Constructional Steel Research, 2010, 66(7):906–914.

Advances in Energy Science and Equipment Engineering II – Zhou, Patty & Chen (Eds)
© *2017 Taylor & Francis Group, London, ISBN 978-1-138-71798-5*

Study on the reliability of dynamic evaluation of highway slope stability

Yanfeng Li
School of Civil Engineering and Transportation, Hebei University of Technology, Tianjin, China

Zeng Guo
Department of Civil Engineering, Zhangjiakou Vocational and Technical College, Zhangjiakou, China

Zhaowei Liu
China Road and Bridge Corporation, Beijing, China

Shuwei Wang
Beijing Transportation Engineering Key Laboratory, Beijing University of Technology, Beijing, China

ABSTRACT: In order to solve the problem of early warning to highway slope real-time stability analysis, the evaluation grading standards were built based on the limit equilibrium theory. The displacement value of highway slope is calculated first, and then, the evaluation grading standards are obtained. The experiment conducted on the sample data of highway slope in Beijing demonstrates that the grading standards can be able to meet the need of real-time monitoring engineering.

1 INTRODUCTION

Highway slope instability is one of the main disasters of highway construction, which is related to people's lives and property safety, and has been the focus of research in geotechnical engineering.

In the engineering application and monitoring of dangerous points in highway slope, the deformation value is usually adopted to send early warning of slope instability. There are several advanced real-time monitoring systems for slope stability: Permanent scatterer interferometric synthetic aperture radar of slope deformation monitoring system from Italy's Tele—Rilevamento Europa company, which was successfully used in Assisi landslide real-time monitoring (Lv Leting, 2009); Fiber Bragg grating; displacement monitoring system from America's Geokon Inc. has been successfully applied in highway bridge monitoring, hydropower dam monitoring, structure monitoring projects, and so on (Xu, 2007). Photogrammetry and rock mass structure analysis system through continuous photography of highway slope for a certain period and the comparative analysis and deformation from Australia's Sirovision are used to implement slope stability evaluation in real time (Li, 2008).

The highway slope stability evaluation methods are divided into two categories: static evaluation and dynamic evaluation. Static studies include: Yu Xiaoma (Yu, 2011) analyzed the Zhejiang Province

on the high liquid limit of an expressway cutting slope disease caused in 2011; Xie Gang (Xie, 2011) introduced anisotropy of shear strength parameters to limit equilibrium analysis method; Wei Jun (Wei, 2004) adopted the Bishop method in limit equilibrium analysis in 2004 to get expansive soil slope stability analysis of Nan Yang; Jin Hai Yuan (Jin, 2011) built the slope monitoring and early warning by using analytic hierarchy process (ahp) and entropy method; Huang Run Qiu (Huang, 2008) put forward the engineering principle of rock high slope deformation; and Wang Guang Jun (Wang, 2007) applied a high rock slope safety monitoring to 318 national road and obtained the stability of the slope state judgment results.

Dynamic real-time monitoring mainly studies the current highway slope stability monitoring dangerous points with the change of highway slope displacement. When the displacement monitoring value is higher than a defined value, warning occurs. The relevant personnel on the scene determines the stability of highway slope after the exploration. However, the research on the definition of early warning threshold is limited. Because of the complexity of geological conditions and factors influencing the stability of highway slope, a definite value of real-time monitoring is not given in the current various standard specifications and relevant literature.

In view of the above problems, this paper adopts the reliability theory, and the dynamic evaluation of highway slope stability is obtained by the displacement value classification criteria.

2 APPLICATION OF LIMIT EQUILIBRIUM THEORY

The stability of highway slope is closely related to the geological structure, rock mass structure, mechanical properties of rock and soil, hydrological and geological conditions, human activities, vibration, weather conditions, vegetation conditions, and other factors. The failure can be divided into three types: landslide, collapse, and peeling. Regardless of the form of failure, the slip occurred in the front and the slope instability results due to local displacement. Therefore, the process of dynamic monitoring of the highway slope cannot take into account the evolution of its internal instability, the standard of judging whether the highway slope is unstable or not and whether the slope has the displacement quantity and quality.

2.1 The mechanical graphics and assumptions

The mechanical analysis diagram of the plane strain condition analysis is shown in Fig. 1.

Meanwhile, the following assumptions are made:

1. Impervious surface and the initial groundwater level depths (d and dw, respectively) are parallel to the slope.
2. Quasi-slippery soil uniform and isotropic elastic–ideal plastic material yield obedience Mohr–Coulomb criterion, and follow the flow rule to adapt.
3. The groundwater recharges only by rain. Without evaporation and other losses, the soil above groundwater levels is completely saturated. Therefore, the quasi-slippery soil is severely saturated.
4. Excluding the fact that the initial plastic strain before rain may exist due to the lower part of the soil with a plastic zone, which is initially assumed parallel to the slope of the initial

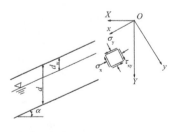

Figure 1. Mechanical analysis diagram.

effective stress σ'_{xo} perpendicular to the initial slope is proportional to the effective stress, σ'_{yo}, namely as Eq. 1:

$$\sigma'_{x0} = k_0 \sigma'_{y0} \tag{1}$$

2.2 Calculation of the safety factor of stability

For simplicity, let rainfall intensity I over the entire period T is constant. According to Iverson (2000) at time t, depth Y of the pore pressure head follows Eq.2:

$$\phi(Y,t) = (Y - d_w)\cos^2\alpha$$
$$+ 2\frac{I}{K}\sqrt{D_1 t}$$
$$\sum_{m=1}^{\infty}\left\{ \begin{array}{l} ierfc\left[\dfrac{(2m-1)d-(d-Y)}{2\sqrt{D_1 t}}\right] \\ +ierfc\left[\dfrac{(2m-1)d+(d-Y)}{2\sqrt{D_1 t}}\right] \end{array}\right\}$$
$$- 2\frac{I}{K}H(t-T)\sqrt{D_1(t-T)}$$
$$\sum_{m=1}^{\infty}\left\{ \begin{array}{l} ierfc\left[\dfrac{(2m-1)d-(d-Y)}{2\sqrt{D_1(t-T)}}\right] \\ +ierfc\left[\dfrac{(2m-1)d+(d-Y)}{2\sqrt{D_1(t-T)}}\right] \end{array}\right\} \tag{2}$$

where K is the saturated hydraulic conductivity; H(η) is the heavyside step function; α is the slope toe; and D0 is the saturated hydraulic diffusivity. D1 is calculated as:

$$D_1 = D_0 \cos 2\alpha \tag{3}$$

The function f (x) is defined as follows:

$$ierfc(\eta) = \frac{1}{\sqrt{\pi}}\exp(-\eta^2) - \eta erfc(\eta) \tag{4}$$

where $erfc(\eta)$ is the complementary error function.

2.3 Calculation of displacement value

Referring to Fig. 1, at depth Y, the plane parallel to the slope of the total normal stress and shear stress are calculated as:

$$\begin{cases} \sigma_y = \gamma_{sat} Y \cos^2\alpha \\ \tau_{xy} = \gamma_{sat} Y \cos\alpha\sin\alpha \end{cases} \tag{5}$$

At time t and depth Y, pore water pressure u(Y, t) is calculated as:

$$u(Y,t) = \phi(Y,t)\gamma_w \qquad (6)$$

Therefore, the planes parallel and perpendicular to the slope of the effective normal stress are:

$$\begin{cases} \sigma'_x(Y,t) = k_0\sigma'_{y0} - \phi(Y,t)\gamma_w \\ \sigma'_y(Y,t) = \sigma_y - \phi(Y,t)\gamma_w \end{cases} \qquad (7)$$

3 THE APPLICATION OF RELIABILITY ANALYSIS

3.1 Generation of random numbers

The distribution function F (s) with highway slope displacement values of discrete distribution, a displacement value of random variables S for the probability of si PI (I = 1, 2,...) is:

$$F(s) = P(S \le s) = \sum_{s_i \le s} P_i \qquad (8)$$

The inverse function method is used to generate a random number, that is, when the (0, 1) random variable of uniform distribution on the stochastic simulation resulting from u and u_i contents Eq. (9. a) is required Eq. (9. b), the desirable s_i as a discrete random variable S and a random number of calculation method, is shown in Eq. (9. c):

$$F(s_{i-1}) < u_i \le F(s_i) \qquad (9.a)$$

$$P\{F(s_{i-1}) < U \le F(s_i)\} = F(s_i) - F(s_{i-1}) = P_i \qquad (9.b)$$

$$s_i = F_s^{-1}(u_i) \qquad (9.c)$$

3.2 Convergence error estimation and judgment

Probability convergence and error estimates are restricted by simulation times N and the central limit theorem. The experiment simulation times when N is large, and the failure probability P_f content are given in Eq. 10:

$$(P_f - P'_f)/\sigma_{pf} \sim N(0,1) \qquad (10)$$

where P'_f is the expected value of the failure probability P_f and σ_{pf} is the standard deviation of the failure probability P_f.

3.3 Determination of displacement threshold

Combined with the relevant specification documents, in this paper, highway slope stability can be

Table 1. Highway slope stability safety coefficient of the scale.

Classification indices	
Slope condition	Stability factor of safety (Fs)
Attention (level I)	>1.20
Warning (level II)	1.20~1.10
Alert (level III)	1.10~1.05
Alarm (level IV)	≤1.05

Table 2. Displacement value of highway slope stability.

Classification indexes		Displacement
Slope condition	Stability factor of safety (Fs)	values (× H‰, mm)
Attention (level I)	>1.20	<0.30
Warning (level II)	1.20~1.10	0.30~0.60
Alert (level III)	1.10~1.05	0.60~1.00
Alarm (level IV)	≤1.05	≥1.00

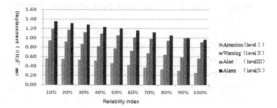

Figure 2. Displacement value distribution map of different reliability indices.

divided into four levels: note, warning, warning, and alarm, denoted by I, II, III, and IV, respectively, They correspond to the slope stability from good to bad, namely at level I, the stability of the slope is higher than that of level IV. The corresponding grading standards of stability safety factor F_S are shown in Table 1.

By using limit equilibrium theory and reliability method, we can calculate the covering layer of highway slope reliability coefficient under the condition of displacement values as shown in Fig. 2.

From the aspects of safety and economy, 90% of reliability indices correspond to the displacement value for covering highway slope stability and dynamic evaluation of early warning indicators, as shown in Table 2.

4 CASE ANALYSIS

Situation of city road through a suburban countryside area in Beijing, K1 + 280~K1 + 320 section of the road on the west slope by the geological exploration evaluation of soil quality/medium slope, the top of building, and concrete are shown in Fig. 3.

Figure 3. Site condition of highway slope.

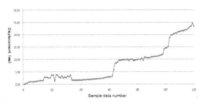

Figure 4. Displacement sample data.

In order to ensure the safety of plant and construction site during the construction stage, the displacement of slope monitoring is carried out. A total of six slope deformation monitoring stations were set up. For a certain period, slope displacement value was subject to a maximum of six monitoring points from 22 February 2010 to 31 August 2010. The monitoring period was 1 day initially, which was changed to 2 days later. Rescue and prestressed anchor cable in slope safety were tested during the construction twice a day. Under special circumstances, such as heavy rain, it is appropriate to increase the frequency of monitoring, in accordance with the order of the observation time from front to back to the serial number of slope displacement value, as shown in Fig. 4.

During the evaluation of the highway slope stability monitoring, in the whole process, six of the measuring point displacement values appeared three times in the early warning. The first mutation occurred on 26 May, the second on 16 June, and the third mutation occurred on 15 July, as shown in Table 3.

In accordance with the most unfavorable principle, the stability of highway slope is predicted by comparing the displacement value index with the result presented in Table 2.

The first mutation occurred on 26 May, then highway slope was in a state of alert (level III). Through the scene reconnaissance, it was found that the highway slope is in unsafe condition, new cracks emerge on the top of the slope, original crack develops, locally sliding failure occurs, construction supports body with an outer drum, that is, cracking phenomenon.

The second mutation occurred on 16 June, then highway slope was in a state of alert (level II). After exploration, the presence of highway slope and building certain deformation was found.

The third mutation occurred on 15 July, then highway slope was in a state of alert (level II). Site

Table 3. Displacement value of the measuring points statistics ($\times$ H‰, mm).

Date Measuring point number	May 26th	June 16th	July 15th
1	0.52	0.26	0.15
2	0.45	0.21	0.10
3	0.28	0.07	0.11
4	0.41	0.12	0.16
5	0.71	0.44	0.32
6	0.87	0.57	0.50

survey found that highway slope toe shows signs of loosening.

5 CONCLUSIONS

A calculation method of the displacement value of highway slope is established using the limit equilibrium theory. Combined with reliability theory, the displacement threshold criterion for the stability of highway slope is obtained. It is in accordance with the characteristics of the slope rock and soil. At the same time, it made up and improved the real-time evaluation of the current highway slope stability problem, providing the basis for real-time stability evaluation.

REFERENCES

Huang Runqiu. The dynamic process of rock high slope development and its stability control [J].

Jin Haiyuan, a comprehensive evaluation method for early warning and early warning of rock high slope [J], Journal of Yangtze River Scientific Research Institute, 2011, 28 (1): 29–33.

Journal of rock mechanics and engineering, 2008, 27 (8): 1525–1544.

Li Huanqiang. Typhoon storms and highway flood characteristics and mechanism of slope damages in Hangzhou [D]. Zhejiang University, 2008.

Lv Leting. D-InSAR detection of surface deformation in Chongqing area [J]. Jilin University, 2009.

Wang Guangjun, Yao Ling Kan, Yang Ming. The high rock slope deformation monitoring technology of [J]. subgrade engineering, 2007,(3): 108–110.

Wei Jun, Xie Haiyang, Li Xiaodui. Stability analysis of expansive soil slopes [J]. Chinese Journal of rock mechanics and engineering, 2004,23 (17): 2865–2869.

Xie Gang, Fu Hongyuan, Jiang Zhongming. Research on the stability of the layered rock slope of the highway [J]. Chinese and foreign highway, 2011,31 (4): 15–17.

Xu Wenjie, Yue Zhongqi, Hu rui-lin. Based on digital image of soil, rock and concrete quantitative analysis of internal structures and mechanical numerical research progress [J]. Engineering Journal of geology, 2007,(3): 289–313.

Yu Xiaoma, Xu Guofeng, Wu Hao. Analysis of highway soil cutting slope stability of high liquid limit highway [J]. Highway, 2011,8: 138–140.

Advances in Energy Science and Equipment Engineering II – Zhou, Patty & Chen (Eds)
© 2017 Taylor & Francis Group, London, ISBN 978-1-138-71798-5

Architectural form and original research of Keji Hall

Lin Chen

School of Architecture, South China University of Technology, Guang Zhou, Guang Dong, China
School of Arts, Sanya University, San Ya, Hai Nan, China

Lingyun Lang

College of Engineering and Technology, Zhongzhou University, Zheng Zhou, He Nan, China

Jianjun Cheng

School of Architecture, South China University of Technology, Guang Zhou, Guang Dong, China

ABSTRACT: Located in Haikou City, Qiongshan District, Qiu Jun's former residence is an important architectural element constructed by Yuan and Ming Dynasties in northern Hainan. The architectural heritage and shape of its construction and the main framework are the characteristics of Yuan and Ming Dynasties, which have a high research value. This paper first studies the history of the Qiujun residence, architectural and cultural characteristics explained on the whole, and then the main building Keji Tang as the main research object, for its architectural shapes and structural characteristics. In comparison with the characteristics of construction during the origin of Qiu Jun, seeking immigrant culture under the influence of the evolution of construction practice and the relationship between the origin, the local architectural features and adaptability in northern Hainan are explored.

1 INTRODUCTION

Qiu Jun's former residence is located in Jinhua Town Village, Qiongshan District, in the national historical and cultural city of Haikou City. It was the Ming Dynasty "Neo Confucianism ministers" Qiu Jun's birthplace, originally founded in late Yuan Dynasty and early Ming Dynasty. As the local extant earliest wooden dwelling in Hainan, the architectural layout, structure features, and artistic characteristics have transitional significance. As a proof of the architectural history and evolution of the Yuan and Ming Dynasties, it has a high historical, scientific, and artistic value. For this reason, it was declared in 1996 as a national relic protection unit. This paper intends to analyze the main building Keji Tang in three steps: (1) to analyze the history of the architecture, construction situation, and structural characteristics; (2) to research the construction practice of evolution and the relationship between the origin under the influence of the immigrant culture; and (3) to make preliminary discussion of the local architectural features and adaptability in northern Hainan.

2 ARCHITECTURAL FORM OF THE KEJI HALL

2.1 Historical evolution

After research for about 5 years, the Qiu family ancestral home was established in 1369, according to Qiu's family genealogy, which was rebuilt by Tingpei Qiu representing the 18th generation of Qiu Jia. Qiu's residence construction group located in Xia Tian Cun Qiongshan County (now Jin Hua Cun Haikou City Town) consists of two parts: (1) Keji Hall and the accessory building, which was founded by Qiu's ancestor Junlu Qiu in late Yuan and the antechamber, following the hall to be expanded in Qiu Jun's period, such as Yuan Feng Xuan and Po Chi building; (2) Qiu's Ancestral Shrine and Qiu Jun's elder brother, Qiu Yuan's former residence. Qiu's buildings are usually connected by galleries and form a series of rectangular courtyards, in addition to the increasing number of ethnic groups in Qiu Jun capital during his tenure, reaching eighteen of such on an unprecedented scale. Among the surviving Qiu family ancestral home, only the antechamber and Keji Tang's structure is still "the original structure," whereas other buildings were damaged or reconstructed.

2.2 Building profiles

Keji Hall is the construction of the main building. The name Keji Hall is extracted from the poem written by Qiu Jun's grandfather. —"Sigh no son to the pension, but fortunately there are two grandchildren can inherit the will of the ancestors". Because the Keji Hall had collapsed by the pressure, Qiu Jun rebuilt the church, but there is no fundamental change to the subject. Keji Hall is a three-bayed structure with width of 12.96 m, with the typical layout of the common residence, that is, the so-called "one bright, two dark and three bay." The central bay's width is 4.7 m, which is divided into two parts. The antechamber occupied more than two-thirds of the area for meetings with people and worship, after the screen for the back hall. The width of the side bay is 4.1 m, which is divided into two bedrooms (Fig. 1). The Keji Hall's height from the ground to the main ridge is 5.8 m, and the eaves height is 2.6 m. The eaves is just in the midline at the building facade, and the renovation the main ridge used fish Chi-wei, which reflects the Hainan unique architectural style of the late Yuan and the early Ming Dynasties.

2.3 Structure features

The beam structure of Keji Hall belongs to the timber frame hall in the Ying-tsao-fa-shih. Because of economic reason and the appearance position, front eaves used the crescent-moon beams and Tou-kung, while the after eaves used the straight beam and short column. Interior column supports the main beam at intermediate point. Furthermore,

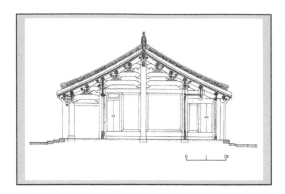

Figure 2. Section of Keji Hall.
Source: Rui Wu, Yi-ping Wang, Pei-ping Huang, Hainan Qiujun home repair Engineering Report, **201**(2003).

the cave purlin, the next purlin, the ridge purlin. And the architraves strengthen the horizontal connection (Fig. 2). The profile of the roof plane is determined by means of a chu, or "raising" of the ridge purlin, and a che, or "depression" of the rafter line. The distance from front to the rear purlin is 9.18 m, the straight height from the ridge to the cave purlin is 2.43 m, and the height of the raise is called the chu-kao, 1:3.78, close to the chu-che rules for a big hall specified in the Ying-tsao-fa-shih.

3 STUDY OF THE ORIGINAL SOURCE

3.1 Immigrant culture in Fujian province

Qiujun ancestral home is located in Gushi of Henan Province, as the Tang Dynasty migrated to Jinjiang County, Quanzhou Prefecture, Fujian. Later, Qiu Jun's great grandfather Qiu Junlu, served Qiongshan government in the Yuan Dynasty, because he could not return home from the war and settled down in Xia Tian Cun Fu Cheng Zhen Qiongshan xian, becoming Qiu's Joan ancestor of Hainan. As one of the earliest immigrants in Hainan Island, the main immigrant population is in the northern region from Putian and other regions of Fujian. The new way of living is affected by the way of living in the original place of residence, such as the most basic units are used, "one bright and two dark" model, combined with a combination of courtyard style. Qiu Jun's former residence in the tradition of the Southern Fujian residential architectural style is gradually local assimilation, evolving to adapt to the local climate characteristics. The flushed gable with flat roof of Keji Hall is the result of accumulated experience in the practice after the typhoon has been hit by the typhoon for a long time. In spite of the political

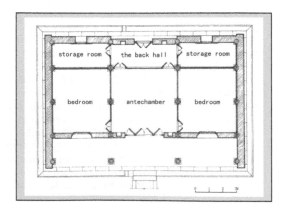

Figure 1. Plan of Keji Hall.
Source: Rui Wu, Yi-ping Wang, Pei-ping Huang, Hainan Qiujun home repair Engineering Report, **199**(2003).

and geographical distance, Hainan Island is separated from the Strait of Joan. The land transport inconvenience rarely involved in the vortex of the north. The lag of migrant culture reflected in the building is to retain the early prototype of the Central Plains culture. The treatment of the structural framing adopted the crescent-moon beams, camel-bumps, brackets, top-chords, and so on, characterized in the architecture of the Sung and Yuan Dynasties. It is a specimen of the transition, embodying the virtues of both periods.

3.2 The influence of coastal culture

The ornaments (fish-shaped) on the ridge ends of the main hall and the antechamber are similar to the shape of the turtle. Fish ornaments are inseparable with the coastal culture. People in coastal regions of the ancient Bai Yue who "live on fishing" were far away from the center of ancient Chinese civilization. The fish and Ao Yu, which bring them more enormous benefits, and the threat naturally become the object of worship to them, and then transformed into a totem. It is comparatively common in the map of Wushanshisha of the Southern Song Dynasty retaining some structural method of earlier periods. In Jiangsu and Zhejiang provinces, enormous fish-Chiwei are found, such as the main hall of Jinshan in Jiangsu, and the typical style of Japanese Zen architecture. Therefore, this practice may be a local characteristic of the coastal culture circle (Figs. 3–4)

3.3 Origin and evolution of structural practices

3.3.1 Inserted Gong and Timu style portrait framework joints

The horizontal timber skeleton members support the main hall and the antechamber, which are consisted of columns, beams, lintels, ties, brackets, purlins, and so on. The vertical lintel component of its tail tenon is inserted into the hypostyle column or center column, and the upper camel-bump brackets supports the main Lin-chuan, with the use of inserted Gong and Timu under the purlin to increase the longitudinal stability of support system. This inserted gong of column in arch longitudinal supporting for Timu and Purlin is called inserted Gong and Timu style, and influenced by the through-jointed frame of Fujian dwellings method (Figs. 5–6). This is also a long-term origin communication in building construction of Min Joan area.

3.3.2 Full tenon made the circle crescent-moon beam

The crescent-moon beam is also called rainbow beam because the structure is more complex, and

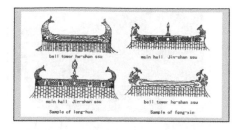

Figure 3. Fish-shaped ornaments on the ridge.
Source: Shi-qing Zhang, Wushan ten map with the Southern Song Dynasty Jiangnan temple, **66**(2000).

Figure 4. Fish-Chiwei of Keji Hall.
Source: This photo was taken by the author.

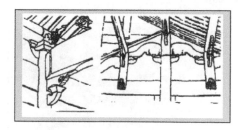

Figure 5. Inserted Gong and Timu style portrait framework joints.
Source: Yu-yu Zhang, The stable supporting system of timber frame and the region distributing of vernacular dwellings in Fujian, Architectural history collection, **26** (2003).

more difficult to process and manufacture, thus becomes a more advanced beam. In the central plains before Song Dynasty, for the higher level of construction, there is no ceiling, and all the structural members supporting the roof are exposed to the crescent-moon beam, but in the official construction of the Ming and Qing Dynasties, it is rarely used, only in the south of the Yangtze River and the Lingnan architecture still continues.

Crescent-moon beam of Keji Hall is vivid and full. As a typical feature in the hall, bright beam using

Figure 6. Beam framing of antechamber.
Source: This photo was taken by the author.

Figure 7. Circle crescent-moon beam.
Source: Yu-yu Zhang, The stable supporting system of timber frame and the region distributing of vernacular dwellings in Fujian, Architectural history collection, **32** (2003).

Figure 8. Crescent-moon beam front interior gallery of Keji Hall.
Source: This photo was taken by the author.

rectangular and piano surface obviously on beams and shoulder-side of the beam is greatly "arched." The practice of bottom side of beam is similar to drawing an illustration of Ying-tsao-fa-shih, but the standard form is different. Firstly, the ratio of depth to width is different. The ratio of the depth to width of Ying-tsao-fa-shih is 42/28 = 1.5, but Keji Hall ru-fu beam's ratio is 2.5, and that for the length beam ratio actually reached 3.0. Secondly, the tail beam is different, Ying-tsao-fa-shih specifies that the shoulder beam with entasis and the tail height is

only half of original beams, but the tail beam height of Keji Hall into the tenon does not decrease the thick part of the top-side tenon as the Shua-tou, while the bottom side is the curve neck. Construction method can find the prototype in Fuding area houses of Fujian Mindong, with the influence of long-term population migration, cultural and technological exchanges coincide in Min Qiong area.

4 CONCLUSION

In summary, Qiu Jun's former residence, as an early representative of the northern Hainan timber houses, reflects the Min Qiong area migration, social development, and cultural exchanges on the side of the layout of buildings, structures, and construction methods. This just shows the interaction effect between different areas on the culture and construction fields. According to its special geographical location and climatic conditions in Hainan Island, building materials, moisture, wind, and other technical aspects of the treatment maintained its own characteristics. Reflecting in the construction form of housing is simple and low, the height of the raise is not high, with the short extent of eaves projection and the east–west direction design. It not only reflects the special practice of the local traditional architecture, but also reflects the heritage and change of the building in the process of immigration. It could be argued that Qiu Jun's former residence carries architectural culture mark of central plains, and adapts to the impact of the local natural climatic conditions.

ACKNOWLEDGMENTS

This study was supported by the National Natural Science Fund (51278196) and the National Key Laboratory of Subtropical Building Science (2014ZA05).

REFERENCES

Ding-hai Yang, Study on Spatial Morphology of Traditional Settlement and Architecture in Hainan Island, A Dissertation Submitted for the Degree of Doctor of Philosophy, (2013) (In Chinese).
Jun Qiu, Qiu wen-zhuang gong set, Siku quanshu cunmu set 406th copies, (1997) (In Chinese).
Rui Wu, Yi-ping Wang, Pei-ping Huang, Hainan Qiujun home repair Engineering Report, (2003) (In Chinese).
Yang-can Ou, Wanli Qiongzhou records, (1990) (In Chinese).
Yu-yu Zhang, The stable supporting system of timber frame and the region distributing of vernacular dwellings in Fujian, Architectural history collection, **26~36** (2003) (In Chinese).

Advances in Energy Science and Equipment Engineering II – Zhou, Patty & Chen (Eds)
© 2017 Taylor & Francis Group, London, ISBN 978-1-138-71798-5

Protection and utilization of Guangzhou's industrial building heritage in the context of urban renewal

Chao Jia & Li-peng Zheng
School of Architecture, South China University of Technology, Guangzhou, China
State Key Laboratory of Subtropical Building Science, Guangzhou, China

Meng-han Wang
Department of Landscape Architecture, Private Hualian College, Guangzhou, China

ABSTRACT: The paper collates the basic process of Guangzhou's industrialization and its industrial heritage resources through literature review and field research; analyzes the spatial distribution and development rules of Guangzhou's industrial heritages and illustrates the significance of the industrial heritage to the studies on urban development. Moreover, it proposes protection strategies according to the distribution characteristics of Guangzhou's industrial heritage spaces.

1 INTRODUCTION

From the policy of "Permitting foreign trade only in Guangdong" to reform and opening up, Guangzhou industry has experienced hundreds of years of development and left a lot of industrial heritages which have witnessed its urban development. In 2009, Guangzhou issued "Opinions on Accelerating Transformation of Old Towns, Old Factories, and Old Villages", which initiated the process of urban renewal. The protection and utilization of industrial heritages has become an important content of Guangzhou's urban renewal.

2 DISTRIBUTION OF GUANGZHOU'S INDUSTRIAL HERITAGE

Guangzhou's industrial heritages are mainly located in four districts including the Xicun, White Swan Pond.

2.1 *Xicun district*

Located in the northwest side of Guangzhou, Xicun was the first industrial zone in Guangzhou. There're seven high-value industrial heritages in Xicun, wherein Zengbu Waterworks, Xicun Power Plants are still in use; industrial restructuring of Guangzhou Beverage Plant, Huaqiao Sugar Factory, and Guangzhou Lime Plant has been completed. They've been transformed into creation gardens and parks where industrial architectural heritages are partially preserved.

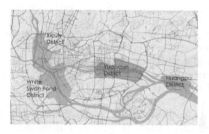

Figure 1.   The main districts of industrial heritages in Guangzhou Huangpu and Yunacun. (Figure 1).

The historic sites of Zengbu Gunpowder Factory and Xicun Cement Factory are preserved.

2.2 *White Swan Pond district*

White Swan Pond is an area with the highest concentration of industrial heritage in Guangzhou. (Figure 1) There're about 27 industrial heritages with certain values, of which 9 heritages are cultural relics protection units and 6 are included into the list of recommended historic buildings. With the industrial transformation and implementation of "withdrawing secondary industry and introducing tertiary industry" policy in Guangzhou, protection and use of industrial heritage has become an important content in the urban renewal of White Swan Pond area (Figure 2). The improvement projects of industrial heritages like Hongxin 922, Lutheran Hall, and 1850 creativity garden have been successively completed. During the transformation of Guangzhou Steel Plant which covers a large area,

Figure 2. Bird's-eye view of White Swan Pond district.

Figure 3. Design rendering of Guangzhou North Shore dock creative industrial park.

the core workshops and part of the equipment are retained, the central area was transformed into a theme park, which maintained the industrial specialties and contributed to public's participation.

2.3 Yuancun district

Since 2008, the industrial remains in the area began to undergo industrial restructuring and the functions of many Soviet-style factories built in 1950s and 1960s were displaced.With proximity to the Zhujiang New Town and the International Convention Centre, Yuancun combined cultural industry advantages and successively developed a series of new cultural industry parks like Redtory, North Bank Creativity Cultural Industry Park, T.I.T Creativity Park, Guangzhou Textile Union Creativity Park, and North Shore dock Creative Industrial Park. (Figure 3) The business contents covers multiple fields like design, clothing, art, film and television, which achieves great economic benefits, smoothly promotes the urban renewal, and provides references in China.

2.4 Huangpu district

Huangpu's industrial heritages mainly include military buildings of late Qing dynasty, modern shipbuilding sites, modern and contemporary ports and wharfs, wherein the Couper Shipyard invested and constructed by Englishmen in 1845 is China's first mechanized modern factory. Military buildings relics including Yuzhu Barbette, Xieshan Barbette, Baihegang Barbette, Dapodi Barbette are witnesses of the opium war. With urban development of Huangpu industrial district, problems like serious environmental pollution, shortage of matching facilities gradually emerges, which makes industrial transformation a pressing issue. The government combined barbette sites with surrounding tourism resources and transformed them into green parks to improve the regional environment; while Huangpu Port and Huangpu Power Plant will be developed into industrial site parks

in the future planning to increase the green spaces and facilitate public's participation.

3 DISTRIBUTION RULES OF GUANGZHOU'S INDUSTRIAL HERITAGE

3.1 Urban planning and distribution of heritage

The distribution of industrial heritages in Guangdong is closely related to modern and contemporary city planning. In 1919, the "southern port" plan in "saving the nation by engaging in industry" by Sun Yat-sen proposed the strategy of locating the industrial districts in the area between Huadi and Foshan and building the docks in the region between Back of Pearl River and Huangpu. In 1922, Cheng Tiangu submitted "Provisional Guangzhou Regional Map" to the municipal government. The plan took the southwest area of the main city as the center of industrial development after analyzing Guangzhou's environmental characteristics; in 1932, the Guangzhou municipal government published the first urban planning administrative document "Draft of Guangzhou Urban Design Brief", which considered the needs of citizen life and the economic development, divided the city into four functional areas—industrial, residential, commercial, and mixed area. Wherein the industrial zone was mainly located in Baihedong, Fangcun, and Huadi areas, which successively formed Xicun and Henan province-owned industrial zones. After 1949, the central government put forward the construction policy of transforming Guangzhou into "a light industry-based production city with certain heavy industry foundation"and intensively built three industrial zones—Yuancun, Huangpu and Miaotou. In addition, Fangcun and Donglong were developed into steel bases. (Figure 4).

The distribution of industrial heritage can demonstrate the characteristics of Guangzhou's modern and contemporary city planning, as well as reflect the trend of urban spatial morphology's expansion. (Figure 5) The city space takes the ancient city as the

Figure 4. The scope of Guangzhou in different stages and industrial heritage distribution.

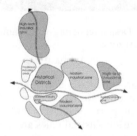

Figure 5. Guangzhou industrial division.

core, first forms commercial and industrial clusters in the western suburbs, then forms industrial and commercial districts to the northwest, southwest of the city. Next, it gradually expands from the southwest to east and south, forming industrial zones with specific functions; Therefore, early industrial heritages are mostly located in the western suburbs and modern industrial heritages mainly lie in the northwest and southwest; while the contemporary industrial heritages concentrate in the eastern and southwest suburbs.

3.2 Transportation systems and distribution of heritage

Transport is the main driving force of industrial development, therefore railways and rivers are the prerequisite of industrial site selection in modern times. Guangzhou is located in the heart of Pearl River Delta, so the Pearl River was the main transport channel before the construction of railways, which is why Couper Shipyard, Xicun Power Plant and Taigucang were all built along the Pearl River. Between 1901 and 1936, Canton–Sanshui Railway, Canton-Hankow Railway and Kowloon-Canton Railway were successively completed, they connect the supply of materials needed for industrial development, trade ports and inland markets; Shi Weitang Station, Huangsha Station and Dashatou station are all located in the industrial clusters by the Pearl River. After the founding of the PRC,

several railway lines were specially constructed for the Yuancun, Fangcun, and Henan area, which reinforced the links between industrial districts.

Consequently, the railways and Pearl River formed a complete transportation system in which the Pearl River is the main channel for the transport of goods and export of finished products, while the rail transport was used to develop the inland market and ensures supply of raw materials. Guangzhou's industrial heritages mainly are concentrated along the bank of Pearl River and expand from west to east. The west section focuses on modern warehousing, transport; the middle section primarily accommodates new-type light industry, chemical industry; while the east section mainly accommodates large ports and manufacturing industries. The railways were constructed along the Pearl River and expand to both shores where the industrial districts center on the railway network. Therefore, Guangzhou's industrial heritages take the Pearl River as the clue and present the characteristic of zonal distribution and local diffusion around the stations, which reflects the characteristics of industrial development and urban planning in Guangzhou as a city which focuses on port trade.

4 OPPORTUNITY TO UTILIZE GUANGZHOU'S INDUSTRIAL HERITAGE

4.1 Implementation of "transforming old town, factory and village" strategy and the government's policy support

In 2008, Guangzhou initiated the "withdrawing secondary industry and introducing tertiary industry"

Table 1. Guangzhou "old factory" reconstructed area summary.

| Name | "old factory" reconstruction | | Urban renewal | |
	Proportion	km2	Total quantity	km2
Yuexiu	3.2%	0.30	125	9.16
Haizhu	29.4%	7.94	728	27.00
Liwan	28.1%	6.94	451	24.72
Tianhe	30.8%	8.04	829	26.14
Baiyun	32.4%	43.29	4537	133.74
Huangpu	29.2%	6.60	273	22.57
Huadu	54.8%	27.84	1487	50.78
Panyu	43.7%	47.12	2617	107.78
Luogang	14.2%	3.42	2107	24.02
Nansha	34.8%	6.56	635	18.86
Conghua	26.9%	11.09	2354	41.16
Zengchen	35.1%	34.38	1554	97.81
Total	34.9%	203.53	17697	583.74

policy to adjust industrial structure of industrial districts; encourage industries with high energy consumption and pollution to exit the downtown; and develop the tertiary industries which mainly include cultural industries. In 2009, the "Transforming Old Town, Factory and Village" strategy clarified the ownership of factory after transformation and carried out industrial restructuring by encouraging enterprises to participate in the transformation of plants. By the end of 2014, the number of "Old Factory" transformation projects in Guangzhou had reached 251 and the projects had covered an area of about 12.9 square kilometers. Wherein 116 government purchasing and storage development projects involve 5.99 square kilometers of lands, 105 owner-led reconstruction projects involve 5.97 square kilometers of lands. Government's encouragement and policy support safeguarded the industrial heritages, preserved a large number of industrial structures, and achieved efficient utilization of industrial spaces. (Table 1).

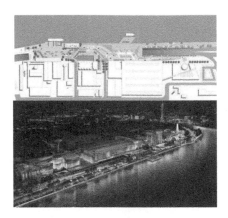

Figure 6.   Design rendering of Zhujiang party pier beer culture & art zone.

4.2   *Distribution along the pearl river is conducive to the transformation of riverside landscapes*

The Guangzhou municipal government has been focusing on creating an urban landscape belt—Pearl River riverside landscape, a clues that connects old and new north-south axis in Guangzhou. Industrial heritages are the main targets of transformation in the Pearl River landscapes. Since 1993, the government has conducted a number of overall planning for the Pearl River waterfront spaces. These plans focused on transforming industry-based reaches into an urban landscape belt which mainly features green space, plazas and supporting facilities. In 2012, Guangzhou Municipal People's Congress adopted the resolution of "Building the Pearl River Golden Coastline" to build a public entertainment-oriented urban landscape belt and preserve memories of the city through transformation of industrial heritages. Currently, Taigucang wharf, Zhujiang Party Pier Beer Culture & Art Zone, (Figure 6) Guangdong Cement Factory have become landmarks along the Pearl River.

In addition, silos and Changgang oil deposit of South Flour Plant have also been included in the transformation plan. Industrial heritages and many outstanding modern architectures jointly form the Pearl River cultural landscapes and constitute the intercity lines in Guangzhou, which reflects important landscape values.

4.3   *The tertiary industry has obvious superiority and high utility value*

The third industry in Guangzhou is enjoying a rapid development. In 2014, the output of tertiary industry grew by 9.4% and 1086.294 billion yuan. Firstly, the foundation of tertiary industry provides financial support for reuse of industrial heritages. Guangzhou's industrial relics are mostly located in the downtown; the surrounding areas feature dense population and frequent commercial activities, which is conducive to the development of creativity cultural industries. Introduction of tertiary industry is the major method to utilize industrial heritage. High-value industrial heritages are often developed by governmental organizations by focusing on its historical value and cultural connotation; building commemorative cultural landscapes, tourist attractions, and museums with industrial characteristics. Secondly, enterprise can adapt to the laws of the market and rationally use resources by attracting investment to develop projects. For example, T.I.T Creativity Park is jointly developed by Guangzhou Textile Machinery Factory and an investment company in Shenzhen. It has become a center for fashion design, R&D, publication, display and successfully attracted well-known domestic and foreign garment companies. In addition, the independently developed industrial plants are able to make full use of the cultural industry's advantages; Redtory is exactly a cultural creative park whose construction was mainly led by artists, and it has been developed into the arts center of Guangzhou.

5   PROTECTION STRATEGY OF GUANGZHOU INDUSTRIAL HERITAGE

5.1.   *Strengthening field research on industrial heritages and improving the system of research on industrial heritages in Guangzhou*

As the national census of historical relics and transformation of industrial heritages proceeds,

Guangzhou's industrial heritages have begun to receive constant attention from the government and society. The archaeological surveys concerning industrial heritage have also commenced. However, damages to industrial heritage due to the lack of protection are also commonplace. Value research is a prerequisite for the protection of industrial heritage. Therefore, the government should strengthen cooperation with related agencies and establish a sound value judgement system; form a ladder-like protection system by establishing value evaluation criteria for industrial heritages in Guangzhou to provide a theoretical basis for the rational protection and utilization.

5.2. *Establishing standards for the conservation of industrial heritage and extending methods to utilize Guangzhou's industrial heritage*

Protection of the industrial heritages should be distinguished from simple architectural design and protection of ancient architecture. We need to combine both methods and generalize a protection system which can adapt to the characteristics of industrial heritage through practice and research. Protection of the industrial heritage should not be independent from the surrounding environment. Thereby we should fully consider the development advantages and constraints; maximize the utility value of cultural heritage after transformation on the basis of ensuring the authenticity of them; draw on the experiences in conservation and utilization at home and abroad on the basis of practices to establish standard protection criteria of industrial heritage and modification specification.

5.3. *Strengthening social supervision and protecting endangered industrial heritages*

In the protection of Guangzhou's industrial heritages, there's a lack of complete protection policy, scientific governmental protective development, public's awareness of participation and sufficient supervision. In the transformation of the industrial heritage, we should respect the views of all social stratum; establish a perfect social supervision system; consider the needs and suggestions of grassroots; increase public's awareness of protecting industrial heritages; recognize the difficulties facing industrial heritage protection, so as to urge relevant departments to carry out salvageable preservation of the endangered industrial heritages, fundamentally reducing the loss of industrial heritage resources. (Figure 7).

6 CONCLUSION

As an important component of Guangzhou's modern and contemporary architectural heritage, industrial heritage witnessed the historical process of the urban development, carrying profound cultural connotations. The spatial distribution of industrial heritages in Guangzhou has close relationships with urban planning, industrial development, economic characteristics, and regional cultures, and it is of great significance to the study of Guangzhou's urban history, industrial history, economic history, and cultural history. Therefore, in the protection and utilization of Guangzhou's industrial heritages, the overall planning should be carried out from the perspective of spatial distribution to protect urban context and continue urban characteristics.

REFERENCES

Anthony Coulls, Railways As World Heritage Site (Occasional Papers for the World Heritage Convention, Paris, 1999).

Chao Jia, Li-peng Zheng, Lingnan Charateristics of Industrial Architecture Heritage in Guangzhou, South Architecture. **172**. 6–11(2016).

Jian-guo Wang, The Protection and Renovation of Industrial Architectural Heritage in Post-modern Industrial Ages, Beijing: China Building Industry Press, (2008).

Kong-jian Yu, Wan-li Fang, The Preliminary Exploration of China's Industrial Heritage. Architectural Journal, **8**.12–15(2006).

Qiang Zhu, Industrial Heritage Corridor Construction of the part of the Great Canal. Beijing University. (2007).

Figure 7. Wuxianmen power station after fire; present situation of Guangzhou glassworks & Guangzhou first rubber factory.

Advances in Energy Science and Equipment Engineering II – Zhou, Patty & Chen (Eds)
© 2017 Taylor & Francis Group, London, ISBN 978-1-138-71798-5

The calculation of the anti-floating ability for uplift piles

Mengxiong Tang
Guangzhou Institute of Building Science, Guangzhou, China

Hang Chen
Guangzhou Institute of Building Science, Guangzhou, China
School of Civil Engineering, Guangzhou University, Guangzhou, China

ABSTRACT: During the design of uplift piles, inconsistent partial coefficients for load determination are given by different technical codes due to the confusion of load classification. Therefore, how to classify the floating of the underground water and choose the corresponding partial coefficients is one of the key factors to the anti-floating design of underground structures. In this paper, a clear criterion is proposed that the floating of water should be considered as permanent load when the water pressure fluctuation over time by less than 20% during the whole life period of the building and otherwise as variable load. It is found that scientifically determining the critical water level for anti-floating fortification related to the safety of the structures is very important to the calculation of floating. During checking on anti-floating ability, the permanent loads acted on the basement as well as its gravity should be represented by standard value, and characteristic value for the pullout resistance of uplift piles. Instead of the basic combination of load effects and the corresponding partial coefficients, the limit state design method with reasonable and economic single safety factor 1.05, should be employed for checking of anti-floating stability.

1 INTRODUCTION

Uplift piles are commonly used tools for anti-floating in underground engineering. For the design of anti-floating, the fortification of water level should be determined first (as shown in Figure 1), which is combined with the structure depth to guide the calculation of the floating. The classification of the floating is not clearly indicated in the existing Chinese corresponding national standard GB50009 (GB50009-2001, 2001) so that confusion and inconsistent methods are found in other related technical codes.

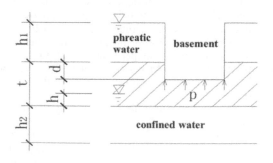

Figure 1. Schematic diagram of the fortification of water level during the design of anti-floating.

In China, the structural design codes for special structures (GB50069) (GB50069-2002, 2002) and pipelines (GB50332) (GB50332-2002, 2002) of water supply and waste water engineering classfy the loads induced by underground water as variable loads. But, the code for design of metro (GB50157) (GB50157-2013, 2013) defines the floating as permenent load. Accroding to the national technical measures for design of civil construction (National technical measures for design of civil construction (Structure), 2003), the underground water pressure should be treated as permenent loads for stable water level, while as variable load for dramatically changing case, which is resonable but difficult to apply to the practice engineering due to the characterization for the fluctuation of water level.

In the standard system in Europe, the code for basis of structural design (EN1990: 2002) (EN1990: 2002, 2001) and actions on structures (EN1991-1-6: 2005) (EN1991-1-6: 2005, 2004) also consider the floating as permenent or variable load dependent on the water level fluctuation with time and the environmental factors respectively. For some centain circumstances, the extrem floating also can be treated as accidental load by the code for geotechnical design (EN1997-1: 2004) (EN1990: 2002, 2003).

It is found that the inconsistence for theunder-grond floating exists between the technical codes in China and Europe, and no clear defination present in the load code for the design of building structures GB50009 (GB50009-2001, 2001). Therefore, how to classify the floating of the underground water and choose the corresponding partial coefficient is one of the key factors to the anti-floating design of underground structures.

During the design of underground structures and foundations, the design value for standard combination effects S_k under the limit state of normal usage is

$$S_k = S_{Gk} + \Psi_{c1} S_{Q1k} + \Psi_{c2} S_{Q2k} + \ldots\ldots + \Psi_{cn} S_{Qnk} \quad (1)$$

where S_{Gk} is the effect of the standard value of permanent forces G_k, S_{Qik} is the effect of the standard value of the ith variable forces Q_{ik}, and Ψ_{ci} is the coefficient for combination of the ith variable forces Q_i, which can be referred in [1]. For the ultimate limit state, the design value for basic combination effects controlled by variable forces S_d is calculated as

$$S_d = \gamma_G S_{Gk} + \gamma_{Q1} \Psi_{c1} S_{Q1k} + \gamma_{Q2} \Psi_{c2} S_{Q2k} + \ldots\ldots$$
$$+ \gamma_{Qn} \Psi_{cn} S_{Qnk} \quad (2)$$

Where γ_G is the partial coefficient of the permanent force, and γ_{Qi} is the partial coefficient of the ith variable forces Q_i, which are also referred in (GB50009-2001, 2001). At the same time, the design value for basic combination effects controlled by permanent forces can be simplified as

$$S_d = 1.35 S_k \quad (3)$$

It is often confusing to apply the combination from the above formulae to the calculation of underground floating.

At the same time, the uplift bearing capacity of the piles is not only decided by the resistance from soil and rocks, but also the stretching strength of the pile material, which belong to two different design systems. Partial safety factors are employed to calculate the resistance, while the strength determined value is calculated by partial coefficient method. The roles of different factors to calculate the reinforcement ratio for the uplift piles should be investigated. Furthermore, the theory of crack controlling in the technical code for building pile foundations (JGJ94) (JGJ94-2008, 2008) is complex. If it is replaced by a simplified method, how to guarantee enough safety for preventing cracks is another important question.

In this paper, a clear criterion is established to calculate the anti-floating ability for uplift piles as

well as the discussion of the questions proposed in this section.

2 THE CLASSIFICATION OF FLOATING

According to the Clause 3.2.5 in Ref. (GB50009-2001, 2001), it is reasonable and uncontested to consider the weight of the structure, which is to resist the floating as a permanent load and take the corresponding partial coefficient as 1.0. However, the applicable circumstances for the floating to be treated as permanent or variable load are not clearly explained in Ref. (GB50009-2001, 2001). The Clauses 5.2.2 and 5.2.3 of Ref. (GB50069-2002, 2002) indicate that, for the design of underground structure, the forces induced by underground water should be the first variable load with partial coefficient 1.27 (i.e. the design value of basic combination for floating is 1.27 times as large as the standard value during the structural strength calculation). According to the commentary of Ref. (GB50009-2001, 2001), the underground water pressure should be considered as permenent load if the water level is stable, otherwise as variable load, which is provided by the unified standard for reliability design of engineering structures (GB50153) (GB50153-2008, 2008). In other words, during the calculation of anti-floating, the partial coffecient is taken as 1.2 when the highest flood level below the outdoor terrace, which is also considered as the fortification of water level, and as 1.4 when the water pressure may increases as the anti-floating water level is below the outdoor terrace or the highest flood water level.

The life period of the architecture is also reasonable to be considered into the classification of floating. If the normal life period of the architecture is 50 years or 100 years for important sites, the investigated fluctuation of the water pressure should cover the whole life period (i.e. 50 or 100 years). The relative difference between the partial coffecients for permenent loads (1.2) and variable loads (1.4) is about 16.67%. Therefore, if the relative water pressure fluctuation is less than 20% over the whole life period of the architecture, the floating can be treated as permenent load. Otherwise, the floating should better considered as variable loads for large enough water pressure fluctuation (i.e. ≥ 20%).

The underground floating can play the role of advantageous load as well as disadvantageous load for different circumastances, and can converse from one to another in different stages. For the low-rise podium, multi-floor deep basement, undergrond parking lot, sunken square etc., the underground floating will unfavorably lead to the inadaqute safety factor for structural anti-floating stability. However, for the running stage of high-rise tower or building, the vertical pressure acted on the bottle plate of the basement is far larger than the floating, which is in turn helpful to balance some upper loads. But if the underground water level is not well controlled during the construction of basement, the floating deleteriously lead to the upward displament overall the underground structures.

The undergound water level is stable in the area with rich water supply. Although the negligible fluctuation benifit the design for foundations of high-rise buildings, almost all of the calculations do not contain this positive effects. Only the the upper structural loads are involved in the reaction forces acted on the bottle of foundation and in the loads on the pile head, instead of the combination with floating, which therefore is considered as reserve for safety.

Both the natural factors and human activities can account for the fluctuation of the underground water level. The dramatically dropping of undergrond water has been found in London and Beijing (as shown in Figure 2) due to overexploitation (Sheng, 2013). Although scholars are denoted to monitoring the dynamics of local underground water, it is difficult to predict the situation of the construction sites in decades. If the benifits of the underground water are considered, the safety factor will decrease once the dropping of water level is out of prediction.

Therefore, strict limitations should be put forward first to consider the floating as advantagous load during the design of high-rise architectures: 1) enough long-term hydrological data are monitored to predict the maximum and minimum water level during the whole life period; 2) the coefficient for combination value of benifical floating should better to be taken as 0.9 during the calculation of loads combination; 3) the design of pile and composite foundtions for high-rise building should not involve the benefits of floating.

Besides, during the anti-floating design of underground structures, the weights of upper structure and soil layers are benificial forces instead of resistance (such as the uplift bearing capacity of the uplift piles).

Figure 2. The underground water level monitored in Beijing in the last decades. (Sheng, 2013).

3 ANTI-FLOATING DESIGN OF UNDERGROUND STRUCTURES

3.1 Counterexample for load combination

Reference (GB50009-2001, 2001) indicates that the partial coefficient of permanent load should be taken as 1.2 and 1.35 for the loads combination controlled by the effects of variable and permanent loads respectively, while as 1.0 when its effect is advantageous to the structure. Many engineers wrongly applied this combination to the calculation of anti-floating as (Pei, 2010).

$$1.20F - 1.00G \leq nR_t \qquad (4)$$

$$1.35F - 1.00G \leq nR_t \qquad (5)$$

where F is the standard value of floating corresponding to the design of maximum water level, G is the standard value of the gravity of foundation and upper structure, R_t is uplift bearing capacity, which is selected as the design value ($R_t = R_u/1.6$), characteristic value ($R_t = R_u/2$) or directly the ultimate value R_u and n is number of uplift piles involved in the calculated area of the foundation.

According to the Ref. (JGJ94-2008, 2008), the uplift bearing capacity for local failure (i.e. not globle failure) presents as

$$N_k \leq T_{uk}/2 + G_p \qquad (6)$$

where N_k is calculated by standard combination controlled by load effects as $N_k = (F - G)/n$ and T_{uk} is the ultimate uplift bearing capacity determined by static test, which have included the weight of the pile. Therefore, Eq. (6) can be rewritten as

$$N_k = (F - G)/n \leq T_{uk}/2 \qquad (7)$$

Example: a two-floor basement with three-floor commercial upper structure is buried 10 m depth from the ground. An open room of 8.0 m × 8.5 m is to be calculated. If the fortification of water level is parallel to the ground, the floating is $F = 10$ kN/m³ × 10 m × 8.0 m × 8.5 m = 6800 kN. The standard value of the weight of the column in the open room and the foundation plate is $G = 6270$ kN. If pre-stressed pipe piles with diameter 400 mm are chosen as the uplift piles, thier uplift bearing capacity is $R_u = 800$ kN for each. Accoring to Eqs. (4), (5) and (7), if R_t is takens as the design value and characteristic value respetivly in Eqs. (4) and (5), five different calculation approches give the number of piles needed for each open room as 3.8, 4.7, 5.8, 7.3 and 1.3 repectively, which have significant discreteness.

In the example above, the ratio of the maximum (7.3) to minimum (1.3) number of piles needed for each open room is 5.6, because Eqs. (4)~(7) employ the partial coefficients for the load combination of other structural calculation to the design of anti-floating so that the fixed safety factor turns to be a undetermined variable. Sometimes, the load combination lead to the design of overmuch piles, while, if the ultimate uplift bearing capacity is used, the anti-floating safety factor of whole foundation will be less than 1.0.

The first and fourth items of the clause 3.0.5 in the technical code for the design of building foundations (GB50007) (GB50007-2011,2011) give clear explainations that during the calculation to determined the height of foundation or pile cap, internal stresses in the foundation structure and reinforcement ratio and checking of the material strengh, the effects from upper structure and corresponding reaction force should comply with the basic combination under ultimate bearing state and corresponding partial coefficients (i.e. Eqs. (2) and (3)), while standrad combination (i.e. Eq. (1)) for the calculation of the piles number based on the bearing capacity of single pile, which also should be presented as characteristic value.

3.2 The checking of pile strength and random detection

The uplift bearing capacity of piles are determined only by the resistance from the soil but also by the material strength of the pile. The former element is represented by the characteristic value of the uplift bearing capacity, which is calculated by safety factor method and the corresponding safety factor is 2.0. But the latter is characterized by standard or design value of the uplift bearing capacity, which is decided by the cross-section reinforced ratio and solved by material partial coefficients method. Reference(JGJ94-2008,2008) suggests the normal cross-section stretching bearing capacity of a reinforced concrete axial uplift pile is written as

$$N_k \leq f_y A_s + f_{py} A_{py} \tag{8}$$

where N_k is basic combination of load effects for the standard value of the axial streching force acted on the pile head, which should be calculated by Eqs. (2) and (3), f_y and f_{py} are the design value of strething strengh of normal and pre-stressed steels respectively, and A_y and A_{py} is the cross-section area of normal and pre-stressed steels respectively.

Usually, the uplift piles are just reinforced with single kind of steels (i.e. normal steels or pre-stressed steels). Therefore, the cross-section area of the steels can be calculated by (JGJ94-2008, 2008; GB50007-2011, 2011)

$$A_s = N/f_y \text{ or } A_s = N/f_{py} \tag{9}$$

where A_s is the area of steels, N is the design value of bearing capacity, which is calcualted by the simpified combination controlled by permenent loads (i.e. $N = 1.35N_k$) for the convenience to transter Eq. (9) into the form with safety factors (Gu, 2004) and further compare the safety factors between two kind of bearing capacity described above. Accoring to the code for design of concrete sutructures (GB50010) (GB50010-2010, 2010), the partial coefficients for hot-rolled steels and pre-stressed steel twisted cable are 1.1 and 1.2 respectively, i.e. $f_y = f_{yk}/1.1$ and $f_{py} = f_{pyk}/1.2$, where f_{yk} and f_{pyk} are standard values of strething strengh of normal and pre-stressed steels respectively. Combining them with Eq. (9), one can get

$$A_s f_{yk} /N_k = 1.485 \text{ or } A_s f_{pyk} /N_k = 1.62 \tag{10}$$

It is found from Eq. (10) that the safety factor for the uplift bearing capacity based on the material strengh ranges from 1.485 to 1.62, which is lower than that based on soil resistance (i.e. 2.0). Therefore, the safety seems to be decided by the material strengh. But in reality, the bearing capacity provided by the resistence from the soil and rocks around the pile side is complex and with large uncertainty.

Due to the difference of safety factors between the bearing capacity based on material strengh and soil resistance, two issues are presented for the design and testing of uplift piles.

1. For the basic test to obatin the ultimate frictional resistance of the soil before the design of uplift piles, to make uniform safety factors for the two kinds of bearing capacity (i.e. 2.0), the reinforced ratio of the testing piles should be increased.
2. During the acceptance testing of the uplift piles, the maximum testing load should be 1.5~1.6 times the characteristic value of the bearing capacity controlled by the material strengh, which means that the ultimate uplift bearing capacity can not be guaranteed to be more than 2 times of the characteristic value. But, if the testing load is 2 times of the characteristic value, the risk of cracking is so large that the reinforced ratio of the selected piles for acceptance testing should be increased, which is unfavorable for random detection.

To establish well solutions for the two issues above, we suggest the modification of corresponding codes and standards as bellow:

1. The maximum testing load during acceptance tesing is obtained as 1.5 and 1.6 times of the estimated charcteristic value of the uplift bearing capacity for piles reinforced with normal

568

and pre-stressed steels respectively, which is an effctive way to avoid the risks of inadquate ultimate bearing capacity.

2. The reinforced ratio of all the uplift piles are increased so that the safety factor for the material strength decided bearing capacity reaches 2.0. Then, the maximum testing load during acceptance tesing is obtained as 2.0 times of the estimated charcteristic value of the uplift bearing capacity. The increased reinforeced area presents as

$$A_s = 2N_k / f_y \text{ or } A_s = 2N_k / f_{py} \tag{11}$$

which are of 19%~26% larger than the results of Eq. (10).

3.3 The checking of cracks and replacement method

The checking of the cracks width should employ the standard combination under the limit state of normal usage. When the uplift piles are used in anti-floating design, the uplift bearing capacity should satisfy the controlling condition of displacement. The design of uplift piles according to the characteristic bearing capacity calculated by the method given in Ref. (GB50007-2011, 2011) can satisfy the requirement of most engineering. However, the calculation of the deformation should be carried out for the engineering with strictly requirement for deformation.

Reference (JGJ94-2008, 2008) suggest the standard combination of load effects should be applied to the cracking control of uplift piles.

1. For the pre-stressed concrete piles of first grade cracking control (i.e. strictly no crack), no stretching stresses exists in the concrete

$$\sigma_{ck} - \sigma_{pc} \le 0 \tag{12}$$

2. For the pre-stressed concrete piles of second grade cracking control (i.e. generally no crack), the stretching stresses in the concrete is no more than the standard value of the concrete stregth

$$\sigma_{ck} - \sigma_{pc} \le f_{tk} \tag{13}$$

3. For the pre-stressed concrete piles of third grade cracking control (i.e. cracks are permitted), the calculated maximum crack width should not exceed the limition

$$w_{max} \le w_{lim} \tag{14}$$

where σ_{ck} is the normal stress of the normal cross-section calculated by the standard combination of load effects, σ_{pc} is the effective pre-stress of the pile concrete, f_{tk} is the standard value of the concrete axial stregth, w_{max} is the maximum crack width calculated by the standard combination of load effects, and w_{lim} is the limition for maximum crack width.

The checking of cracks and steel corrosion are the important problems in the design of anti-floating for underground structure. Currently, two methods are commonly used to solve the above problems, i.e. 1) calculation of cracks width and 2) reserving allowance for steel corrosion. According to the modification of the district code for design of foundation in Shanghai DGJ08-11-2010 (DGJ08-11-2010, 2010), enlarging the diameter of the reinforced steel is relatively economical reinforced method. The latest edition of the technical code for tall building box foundations and raft foundations JGJ6 (JGJ6-2011, 2011) also take the same measure to enlarge steel diameter by 3 mm, which avoid complex calculation of the cracks width. Therefore, revised standards or technical codes prefer to the simplified method of enlarging steel diameter.

For the corrosion of the steel of the piles, Shanghai Baogang project imported pipe pile in 1970s, where the design of the reserving allowance (i.e. 2 mm at each side) for steel corrosion is referred to the Japanese experience. The engineering is still safe and works well after 40 years.

In summary, if the underground water accounts little for the corrosion of the reinforced steel, based on the strength calculation, it is safe to appropriately increase the level of steel, instead of the checking of cracks width. If the underground water makes strongly effects on the corrosion of the reinforced steel, special anti-corrosion design should be carried out.

4 THE CHECKING OF ANTI-FLOATING STABILITY OF UNDERGROUND STRUCTURES

4.1 The comparison of domestic codes

Underground structures should have enough anti-floating safety factor under the influence of water buoyancy. The corresponding clauses in national and local codes are different. It does not reach a consensus that whether to assign partial coefficient for the floating and the value of anti-floating stability safety factor, which is the main reason for the wrongly using load combination formula by designer recently.

Generally, the limit state design method of single safety factor is used in the checking of anti-floating. The clauses 3.0.2 and 3.0.5 of Ref. (GB50007-2011, 2011) respectively indicate that the checking

569

of anti-floating stability should be carried out for the basement and underground structures if the underground water is buried in shallow soil layer and the load effects should refer to the basic combination under ultimate state with the partial coefficient 1.0. The clause 9.0.4 of the technical code for waterproofing of underground works GB50108 (GB50108-2008, 2008) suggests the anti-floating stability safe factor should be more than 1.05~1.1 when the weight of the underground structure constructed by cut and cover method is larger than the floating of static water pressure. In Ref. (GB50069-2002, 2002), the resistance, instead of variable forces and frictions acted on the structure sides, should be classified to permanent load with its standard value during the checking of anti-floating stability and the anti-floating stability safe factor should be more than 1.05, but whether to use the standard value for the floating is not clearly explained. However, the clause 4.2.10 gives a clear provision that all the effects during the checking of anti-floating stability should represented by the standard values, and the anti-floating stability safe factor should be more than 1.10.

If the the requirement of anti-floating stability is dissatisfied, measures should be taken to increase the upper weights or install anti-floating piles/bolting. Increasing the structural stiffness is also an effective method for local enhancement of anti-floating stability.

For the installation of uplift piles, the anti-floating stability of the foundation should satisfy the equation as

$$(G_k + nR_t)/N_{w,k} \geq K_w \tag{15}$$

where G_k is the weight of the architecture and additional pressure, $N_{w,k}$ is the floating, which is calculated by Archimedes principle, K_w is the anti-floating stability safe factor generally as 1.05, and R_t is the characteristic value of the uplift bearing capacity for single pile.

Then one can determine the number of uplift piles as

$$n \geq (K_w N_{w,k} - G_k)/R_t \tag{16}$$

which gives 2.10 for the example in Section 3.1.

In summary, for the anti-floating stability, only the formula with safety factor should be employed and it is safe to assign 1.05 for the global anti-floating factor. The basic combination of load effects and the corresponding partial coffecients should be avoided in the formula of the structural anti-floating checking.

For the example in Section 3.1, if the three-floor basement is buried to 15 m depth, the numbers of uplift piles required by the basement at fortification

Table 1. The number of uplift piles required by the 3-floor and 15 m depth basement at different fortification of water level.

Fortification of water level (m)	0	1	2	3	4	5	6	7
Number of uplift piles required	11	9.3	7.5	5.7	4.0	2.2	0.39	0

of water level as 0 m ~ 7 m are listed in Table 1 respectively.

From Table 1, it is found that the fortification of water level has significant effects on the number of uplift piles required and no uplift pile is need when the fortification of water level is larger than 7 m. Therefore, accurately predicting the water level fluctuation during the architectural life period and scientifically determining the fortification of water level are most important to the anti-floating safety of underground structure and engineering costs.

Besides, if the basement is buried deep so that it passes through the composite soil layers of permeable stratum and aquitard, the reduction factor to consider the water pressure attenuation in aquitard should have large effects on the number of required uplift piles.

4.2 The comparison between Chinese and European codes

The ultimate state design method with partial coefficients is employed to the calculation of anti-floating in Ref. [8] (i.e. EN1997-1) as

$$V_{dst,d} = G_{dst,d} + Q_{dst,d} \leq G_{stb,d} + R_d \tag{17}$$

where $V_{dst,d}$ is the design value of the basic combination of disadvantageous permanent load and variable load, $G_{dst,d}$ is the disadvantageous permanent loads, $Q_{dst,d}$ is the variable load, $G_{stb,d}$ is the design value for the sum of the weight of upper structure and soil and R_d is design value for other anti-floating forces (such as bolt resistance).

In order to compare with Chinese codes, one can rewrite Eq. (17) in the form with safety factor as

$$(G_{stb,d} + R_d)/(G_{dst,d} + Q_{dst,d}) \geq K_w \tag{18}$$

Reference (EN1990: 2002, 2003) advises corresponding partial coefficients as 0.9, 1.0 and 1.5 for advantageous permanent load, disadvantageous permanent load and variable load respectively. During the calculation of anti-floating, the partial coefficient for resistance (such as uplift piles and bolting resistance) is 1.4. The effects of the weight of upper structure and soil are negligible when compared to uplift bearing capacity of piles and bolting resistance. If ignoring the disadvantageous permanent load, the partial coefficients for resistance 1.4 and floating (as variable load) 1.5 lead to the corresponding safety factor as $1.5/1.4 = 1.07$, which is consistent with that in Chinese codes (i.e. 1.05~1.10). If the other resistance is not included, floating is considered as disadvantageous permanent load with partial coefficient 1.0 and that for advantageous permanent load is 0.9, the corresponding safety factor is obtained as $1.0/0.9 = 1.11$, which also fits fairly with that in Chinese codes. Only if the other resistance is not included, floating is considered as variable load with partial coefficient 1.5 and that for advantageous permanent load is 0.9, the corresponding safety factor is $1.5/0.9 = 1.67$, which is obviously higher than that in Chinese codes.

In summary, the checking of anti-floating stability is different between China and foreign countries. The Chinese codes prefer to ultimate state design method with single safety factor, and that of Europe employs partial coefficients. After the comparison and analysis, it is reasonable to follow the method given in Ref. (GB50007-2011, 2011) with anti-floating safety factor 1.05.

5 CONCLUSIONS

In this paper, the Chinese national and local technical codes and standards are compared with that in Europe and recent research progresses in the field of the classification of floating, the checking of anti-floating stability and the design of uplift piles. The reasons for wrongly assignment for the partial coefficient of floating are summarized. The main conclusions are listed below:

1. It is easy to make mistakes to classify the loads of underground structure under the effects of underground water. In this paper, if the relative water pressure fluctuation is less than 20% over the whole life period of the architecture, the floating can be treated as permenent load. Otherwise, the floating should better considered as variable loads for larger water pressure fluctuation.
2. It is found that the fortification of water level has significant effects on the calculated number of uplift piles required. For the the composite soil layers of permeable stratum and aquitard, the reduction factor of the water pressure attenuation in aquitard also makes large effects on the calculated number of required uplift piles.

3. The using of different loads combination is confusing during the calculation for the bearing capacity, deformation, strength and stability checking. In order to calculate the number of required uplift piles, the load effects acted on the foundation or the bottle of the cap should utilize standrad combination under the limit state of normal usage. During the checking of anti-floating stability, the permanent loads acted on the basement and the gravity of itself should be taken as standard value, and characteristic value for the pullout resistance of uplift piles, instead of basic combination of load effects and the corresponding partial coefficients.
4. If the underground water accounts little for the corrosion of the reinforced steel, based on the strength calculation, it is feasible to appropriately increase the level of steel, instead of the checking of cracks width.

ACKNOWLEDGEMENTS

The authors acknowledge the supports from Guangdong Provincial Science and Technology Foundation (Grant Nos. 2004B36001028 and 2015B020238014) and China Postdoctoral Science Foundation (Grant No. 2016M592471).

REFERENCES

DGJ08-11-2010, *Foundation design code*, China Architecture & Building Press (2010).
EN1990: 2002, *Eurocode 7: Geotechnical design*, CEN (2003).
EN1990: 2002, *Eurocode: Basis of structural design*, CEN (2001).
EN1991-1-6: 2005, *Eurocode 1: Actions on structures*, CEN (2004).
GB50007-2011, *Code for design of building foundation*, China Architecture & Building Press (2011).
GB50009-2001, *Load code for the design of building structure*, China Architecture & Building Press (2001).
GB50010-2010, *Code for design of concrete structures*, China Architecture & Building Press (2010).
GB50069-2002, *Structural design code for special structures of water supply and waste water*, China Architecture & Building Press (2002).
GB50108-2008, *Technical code for waterproofing of underground*, China Architecture & Building Press (2008).
GB50153-2008, Unified standard for reliability design of engineering structures, China Architecture & Building Press (2008).
GB50157-2013, *Code for design of metro*, China Architecture & Building Press (2013).
GB50332-2002, *Structural design code for pipelines of water supply and waste water*, China Architecture & Building Press (2002).
Gu, J., J. Hou, Build. Struct. **44**, 133 (2004).

JGJ6-2011, *Technical code for tall building box foundations and raft foundations*, China Architecture & Building Press (2011).

JGJ94-2008, *Technical code for building pile foundations*, China Architecture & Building Press (2008).

National technical measures for design of civil construction (Structure), China Planning Press (2003).

Pei, J., Chin. J. Geotech. Eng. **32**, 290 (2010).

Sheng, X., H. Zhou, J. Wang, X. Han, *Groundwater and Anti-buoyancy of Engineering Structures*, China Architecture & Building Press (2013).

Advances in Energy Science and Equipment Engineering II – Zhou, Patty & Chen (Eds)
© 2017 Taylor & Francis Group, London, ISBN 978-1-138-71798-5

Polysemous real concepts and practices of traditional commercial blocks

Xiana Hou & Lipeng Zheng
School of Architecture, South China University of Technology (SCUT), Guangzhou, China

ABSTRACT: Protection of historical buildings has to put stress on their authenticity. It should be noted that the real reducibility of historical senses differs from the practical production of historical buildings, matching behavior, and experiences with the national culture. Traditional commercial blocks should seek further extension and expansion of their reality between modern commercial behavior and historical experiences, putting great stress on the recovery and protection patterns for the producibility of historical perception. According to not only the principle of protecting the reality of a legacy but also the clue for street-lane, morphology, original-shape styles, left segments, and continuing function, the author carried out multivision analyses of subjectively perceptional factors in the reality of historical legacies, thus putting forward a proposal for reforming traditional commercial blocks by using historical information values as key experiences, which will make up a foundation for designs and studies of the same buildings.

1 INTRODUCTION

The faster the development of cities, the closer attention should be paid to their historical contrast. The commercial functions of the traditional commercial blocks should be left in the commercialized process so that their sited sensation and cultural connotation will make ordinary residents moved in the aspects of special memories and native feelings, which will arouse the memories for their lovely native land. However, a common urban phenomenon is that modern commercial street and traditional commercial blocks are difficult to advance or retreat, that is, whether to maintain the original styles of blocks, which are lacking in modern commercial vigor or to excessively expand the block styles so as to lose the traditional remembrance for them. In fact, the continuous commercial vitality of traditional streets depends on their historical background, and it is necessary to put the historical perception of commercial blocks and their continuous commercial activities together. The protective transformation and the recurrent historical perception of traditional blocks widen the theoretical range of the real occurrence of historical buildings, including space, place, contextuality, symbols, and fold customs, thus resulting in the thought into an essential issue: protection and usage; technique and concept; and culture and aesthetics.

On the basis of real recurrence in the practice of protecting legacies, this paper deals with the effective utilization of the historically left information on the block itself, which may be tangible or intangible, complete or incomplete, by using the psychological mechanism of the real personal experiences as a starting point, thus forming a scene of the traditional block, which is of historical perception and authenticity. Taking Dongjiao Village, Liwan District, Guangzhou City, as an example, its whole area is planned again in coordination with the urbanized transformation, constructing the resident high-rise buildings and exercising the residential management. As it is difficult for buildings to be preserved or moved to their original places, in order to continue the historical authenticity and protect them, the buildings in the traditional blocks were individually pulled down, moved back, and rearranged according to their original styles. The above mentioned engineering project is representative and typical of the social practice so that it is beneficial to and is used as a sample for our own research.

2 MULTIPLE INTERPRETATIONS AND CHOICES OF AUTHENTICITY

The word authenticity has multiple interpretations: reality, trustworthiness, and original, and by the former half of the 19th century, it was drawn in the field of the legacy protection. In the protection of cultural of legacies, the concept of authenticity goes through different historical periods, thus carrying out the concept transforming procedure from both the sense of principle and multiple sciences and reconstruction to integration and gradually from theory to practice (Cao, 2005). The conception to tolerate and protect modern legacy buildings puts emphasis on the relativism of the authenticity standards, which behaves mainly like the multiple interests on the formation of authenticity about the protection of traditional buildings. This authenticity includes objective authenticity, which emphasizes the primitive inherent styles (Wang, 1999), constructive authenticity, which pays close attention to the original impressionable experiences (Cohen, 1988), and postmodernism authenticity, which seeks the hyper reality with the high technical (Liu, 2008) and existential authenticity with prominence aseity (Wang, 2000), and even includes customized authenticity, which simultaneously takes into account three gradations with object, subject, and home-related feelings (Zhang, 2008). The author thinks that specifying the limits to the diversity in historical legacy authenticity is related to the recognition of its true origins. It must be emphasized that, although varied understanding of authenticity makes differences due to different philosophical thinking, there is no good and bad judgment in the aesthetical evaluation, that is, the need to choose a standard of authenticity according to a special case.

"Not to change the original styles" is the basic cultural relics-protecting principle in China. Nevertheless, the authenticity is allowed to understand and reconstitute flexibly and subjectively in the protection or repair of traditional commercial blocks field. As for a city with special traditional commercial activities, the style of traditional block should be properly readjusted in order to meet the requirement of modern commercial activities. The true site and time sensation of a block should be laid down, not just limiting to seek objective authenticity. At the choice of authenticity standards, the transmission of time and case characteristics depend on the collection and composition.

3 TRANSMISSION AND EXPRESSION OF AUTHENTICITY

The true perception is of some uncertainty. The quantity of historical information sources—styles and designs; materials and quality; functions and usages; traditions, locations, and environments; spirit and perception; and other internal and external factors—will influence the weight and determination for authenticity perception factors. The conditions before reconstruction are characterized by the fact that the traditional commercial blocks are gradually losing their commercial vigor and traditional styles, thus their incomplete and scattered information is mixed up with used and reconstructed marks in different periods. This paper aims at the proper use of the left information, recurrence, and transmission of authenticity.

3.1 Authenticity of original native-land experiences of street-lane space

The connection of street-lane spaces forms longitudinal and transverse open-type traffic; the former is a crowd route for the commerce world and the latter is a material–circulate route for daily life, production, storage, and logistics. At early stage, the formation and development of traditional commercial streets depended on convenient traffic, densely populated growth point, and closely distributed street-side shops, which are operated by separate families and horizontally extended, thus forming linearly distributed commercial streets. Commercial activities can expand or extend a long streets and lanes, forming fishbone or lattices in the geographical layout, as shown in Fig. 1, which displays the urban texture matching the tree-shaped main street contours with freely grown small streets and lanes (Shi, 2007). All kinds of pitchmen or neighbor commercial and living scenes there exist in fishbone or twisty streets, and the lanes become the public-recognized traditional market-place life. Customized authenticity means that recognition tends to seek a native land on a strange land or to seek a familiar place in a strange place so that oneself-thought true perception can be obtained through immigrated native land. In fact, true experiences are a sense of nostalgia returning to a native land.

On the basis of public experiences, street-lane styles can be custom-made introducing into some legacy factors and cultural components. Taking the removal and reconstruction of a traditional block in Liwan District, Guangzhou, as an example, several traditional street-lane buildings were arranged in a new red-line range, using comb-shaped plane layout suitable for Guangdong Prefecture fold

Figure 1.   The layout of traditional blocks in the case.

houses, which have zigzag and extended styles of streets and lanes, slight space changes arousing from the closed or turned streets, and lanes that further strengthen a strong impression on native lands. In the determination of street-lane sizes, there is a need to put into account the traditional and modern commercial space sizes, with proper lane widths less than 3 m and the heights of the buildings along streets less the two stories for the purpose of having a proper scale and ratio of height to width (Lu, 2006), thus leading to the sense of close relationship, commerce, and native land (Fig. 2).

3.2 *Authenticity of recognizing original symbols*

The symbolic authenticity is gradually formed depending on experienced people's impressions, slices of a specialized area history, and word translation. A long-term utilization of building styles in specialized areas arouses symbols for people to recognize and form historical pictures, which become a medium to establish historical authenticity and decorate ancient buildings. In the process of recognition, stress is always put on style senses, building materials, parts, and decorative patterns. Used in this paper is a three-room and two-corridor style, shown in Fig. 3, which is equipped with factors of fold house traditions,

Figure 2. The appropriate scale and the interface of traditional streets in the case.

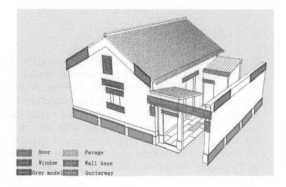

Figure 3. Characteristic component of traditional building in the case.

such as herringbone- and earflap-shaped gable walls, cornice boarding, window lattices, and bars. Joint usages of traditional symbols benefit the identification of areas and strengthen the sense of reality.

However, a large part of architectural parts have lost their values of utilization at present, thus producing changes in the recognition of symbols. Therefore, the use of original words does not stick to the repeat of nostalgia rather than recreate their real connotations, thus the building parts with traditional symbols are a new adaptability. For example, the outer walls built up of gray and red bricks and decorated with hollowed out horizontal Chinese inscription not only beautify the environments but also improve natural lighting and ventilation; the wall holes with earthen cover outdoor air conditions to keep the integrity of environments. All of these construct hazy scenes between objective reality and history (Fig. 4).

3.3 *Authenticity of subjective recognition of original incomplete relics*

Relic buildings are the historical important true origins. Because the values of historical styles of traditional blocks are higher than historical data themselves, the expression of objective authenticity with reconstruct historical buildings are allowed to have a compromise and flexibility via people's true experiences of the new and old buildings. On the one hand, it is necessary to possibly reserve the old buildings with present styles, namely not to excessively restore or renew their old styles in those days. On the other hand, a portion of buildings which influences the usage function can be renewed through the modern architectural techniques so as to sufficiently guarantee continuous utilization of the building. The portion or integration of all buildings presently left show sense of their existence by means of aesthetics ways, event impressions, segment, or composition of historical information. The present reinforced steel structural parts, branches, and leaves growing on walls as well as the revolution-time slogans whitewashed on walls are really the historical footnotes of old buildings, which bring the past to today and provide the past that can still be experienced at present (Steven, 2006) (Fig. 5).

Part of reconstructed buildings and their originals make some differences in their styles, which does not mean the only utilization of present-day materials but means the repeated utilization of the past materials. In the practical cases, the original structural materials and parts were unchanged and moved to new address using traditional techniques. The partial or complete movement of historical buildings to new address and reconstruction reproduce the original atmosphere with black brick-stone walls, block-stone yards, red-sand stone wall footing, uncommon lanes built up of chiseled stones, and so on. In this way, the reconstruction can also keep the same dimensionality of the block texture, harmonious atmosphere, and the past styles of the blocks.

3.4 *Functional continuation of original authenticity*

Existence means reality. As they are the joints between traditional blocks and urban areas, their functional features should be markedly shown in order to make the blocks complete urban joints in the residents' mind. The traditional blocks will last until the future via the last true experiences and present-day urban functions. The practical samples taken in the resident areas of the newly built cities are the types that are traditional blocks put into modern cities. The new blocks have two traditional parts: (1) public buildings, namely flat-grouped centers and (2) public buildings, namely organization offices and commercial centers. Besides, the newly built blocks are jointed so as to form what is

Figure 4. The new use of traditional symbols in the modern context.

Figure 5. Historical marks (shoot in the case of new field of lingnan).

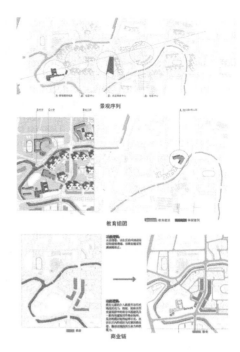

Figure 6. The integration of traditional street blocks and realistic environment.

called "unreal" and "real" spaces by using the longitudinal scenery axes and to show the public functions of blocks. On the contrary, newly built blocks along streams or channels are of educational significance and scenic areas too. Commerce is the original function of traditional blocks, which are accompanied by memorial temple, block squares, and water-side streets, thus forming typical traditional living scenes (Fig. 6). Real traditional commercial blocks must not be located in an urban greenhouse. Their urban and social functions should be changed into commercial function, thus showing their true historical sensations.

4 CONCLUSIONS

Strictly speaking, the traditional commercial blocks differ from general cultural or historical blocks, for their true significance does not stick to original legacies, either or that objective true standard but to the experienced environmental and urban cultural connotations in accordance with the needs for the modern styles and commercial activities of traditional blocks. In accordance with tolerant and opening evaluations of values, it is important that reconstruction must be of authenticity and the living scenes of blocks can make the residents experience the authenticity of legacies.

The protection of historical relics at present is to understand how for them to reasonably exist for the purpose of not just preserving them but sharply adjusting their changes. The above compromise discussions about authenticity aim at not only the organic, gradual, complex and careful protection as well as renovation of historical relics but also flexible expression of their changing processes and possible existence, all of which are on the premise of devotion and reverence toward true history.

REFERENCES

5 Wang. Ning. Tourism and Modernity: A Sociological Analysis. Published in Oxford: Pergamon, 2000.

Cao Juan. Original True Conception and Protections of Chinese Cultural Legacies] (in Chinese), thesis for master's degree, Graduate Shool of Chinese Academy of Social Science, 2005.

Chang Qing. Authentiity inHistoic Preservation and Restoration (in Chinese), Journal of the Time Archrtecture. 2009(3):118–121.

Cohen, E. Authenticity and Commoditization in Tourism. Journal of Annals of Tourism Research, 1988, 15(3):371–386.

(England) Steven. Deyastle, Dymshtis (Turkey) Tarnel Akey, Recovery of Urban Historical Blocks, translated by Zhang Meiying and Dong Wei, published in Publishing House of China Architectural Industry, April, 2006.

Liang Jiang; Sun Hui. Space Style and Pattern Analyses of Traditional Chinese Feudal Commercial Blocks (in Chinese), Journal of Mid-China Architecture, February, 2006.

Liu Zhengjiang. Cultural Authenticity and Infuential Factors of Culturals (in Chinese), Journal of Cultral Architcture, April. 2008, P65–67.

Luyan Xingyi, Steet Aesthetics, published in publishing house of Baihua literature and art, 2006, pages, 46, 57–59, 70–83, 207.

Meng Chunxiao. Study of Perception on Heritage Authenticity of Historical Block. thesis for master's degree, Beijing Forestry University, June, 2012.

Pan Yanling, Research on the Toenism Expericnce of Historic District based on the Symbols Coition, thesis for master's degree, Shanghai Normal University, April, 2012.

Shi Kunlong, etc., Continuation and Reproducitility of Contextuality of Traditional Commercial Blocks at Visual Angle to Urban Space Styles (in Chinese), published at the Forum for Continuous Development (3), 2007, P155–164.

Wang, Ning. Rethinking the authenticity in tourism experience, Journal of Annals of Tourism Research, 1999, 26(2): 349–370.

Xu Song-ling. Heritage Authenticity,Preference of Tourists' Values and Authenticity of Heritage Tourism (in Chinese), Journal of Tourism Tribune, 2008, 23(4):35–42.

Zhang Chaozhi. Understanding for Original Authenticity: Changes and Differences in Visions of Travel and Culturals Legacies (in Chinese), Journal of the Tourism Science, Vol. 22, No. 1, February, 2008.

Advances in Energy Science and Equipment Engineering II – Zhou, Patty & Chen (Eds)
© *2017 Taylor & Francis Group, London, ISBN 978-1-138-71798-5*

Development of early warning criteria for landslide based on internal force calculation method of rigid piles

HuanHuan Li, WanKui Ni & YanLei Zhang
College of Geological Engineering and Geomatics, Chang'an University, Xi'an, China

YanHui Wang
The Third Railway Survey and Design Institute Group Corporation, Tianjin, China

ABSTRACT: Anti-slide pile is the main method to control the landslides of highway, and the supporting effect is an important part of the landslide warning. Based on the method of internal force calculation of rigid piles, the early warning criterion of the anti-slide piles along the highway in Shanxi province has been studied. Taking the effects of the soil resistance and the pile's spacing into account, for a given displacement of a pile head, the loading on the pile has been calculated and then the bending moment of the pile has been calculated. After a series of tentative calculations, the relationship among the top displacement of the anti-slide pile, the maximum bending moment, the foundation coefficient, the proportional coefficient of the foundation coefficient increases with increasing depth, anchorage length, cantilever length and pile diameter has been developed. The formula is suitable for the grade C30 concrete anti-slide piles whose sizes are 3 m × 4 m, 2 m × 3 m and 1.5 m × 2 m. By applying the formula to an example from an earlier study, its reliability has been proven. The formula comprises elementary functions. It is easy to use in an early warning system. In the future, according to the design scheme of a concrete anti-slide pile, based on the monitored data of landslide and by inputting the relevant parameters, the maximum displacement criterion of pile head can be determined which can be used for early warning and forecasting.

1 INTRODUCTION

There have been many landslides along the highways in Shaanxi province, China. Even though almost all the slopes have been protected during the construction of the highway, in recent years, some of them have been found to be unstable, which have caused loss of human lives and properties and threatened the safety of the highway operation. Therefore, it is important to study the supporting effect evaluation of landslide supports and protections structure. Some scholars, both in China and other countries, have done a lot of research on the landslide supports and protections effect.

Gimzburg carried out field tests on the resistance of the anti-slide pile, the pile bending moment and the displacement of the pile top. He also analyzed the effect of the reinforcement (Gimzburg, 1996). By conducting a model test, Liu (Liu, 2012) studied the cantilever anti-slide pile, and found out the stress characteristic and destruction model of the cantilever anti-slide pile. By analysing the anti-slide pile monitoring data, Shen (Qiang, 2015)protected the second sliding surface from sliding, and has assured the stability of the side slope. Using the method of fuzzy mathematics to describe quantitatively or semi-quantitatively, Zheng (Zheng, 2006; Zheng, 2006) selected 11 post-evaluation factors of landslide control effect. The fuzzy mathematics comprehensive evaluation model was established for the first time, and the gray prediction theory was used to predict the pile top displacement. By conducting the micro-pile reinforcement landslide model test, Yan (Yan, 2011) studied the bearing mechanism, stress condition and failure mode of the micro-pile reinforced landslide. Using a numerical calculation method, Zheng (Zheng, 2007) evaluated the supporting effect of the anti-sliding pile with pre-stressed anchor cable. Using the Brillouin Optical Time Domain Reflectometry (BOTDR) monitoring technology, Liu (Liu, 2012) monitored the anti-slide pile strain.

Combining with the available research results, a warning criterion for landslide stability has been determined, which is aimed at the specific supporting structure and can be used widely. In practice, the anti-slide pile displacement is easy to monitor. Based on the rigid pile internal force calculation method, the maximum displacement criterion of the pile head has been developed, which can be used for early warning of landslide along the highway in Shaanxi Province.

2 THEORETICAL DERIVATION

There are many methods to calculate the internal forces of a rigid anti-slide pile. At present, the common method used to calculate the lateral stress is to assume the medium around the pile and below the sliding surface is an elastic body. Thereafter, the internal forces of the pile are determined. When the pile is placed in a homogeneous soil or rock weathering layer, the foundation coefficient is small. For an anti-slide pile which is designed according to the calculation method of a rigid pile, there are three types of foundation coefficient distribution with foundation depth (Zhao, 2008), as shown in Figure 1. The third type is the most commonly used calculation method for rigid piles.

1. The foundation coefficient of the anchorage section increases with increasing foundation depth.
2. The foundation coefficient is constant with foundation depth;
3. The resistance coefficient of the elastic sliding surface is constant, and the foundation coefficient of the surface increases linearly with increasing foundation depth.

In this study, based on the third case, the loading on an anti-slide pile has been calculated. The calculation is for a given pile top displacement, with rectangle, triangle and trapezoid distribution forms. As shown in Figure 2, the effects of the soil resistance and the pile spacing are considered. Therefore, it is suitable for a cantilever pile and an embedded pile. The calculation procedure is as follows.

Calculation Process:

M_0 is the bending moment at the apex of an anchorage segment of an anti-slide pile. Q_0 is the shear force at the apex of an anchorage segment of an anti-slide pile. The calculation formulas for M_0, Q_0 under three types of loading are as follows:

$$\text{Rectangular loading}(q): M_0 = \frac{1}{2}qh_1^2; Q_0 = qh_1 \quad (1)$$

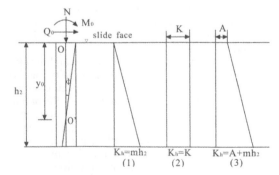

Figure 1. Calculation diagram of rigid pile.

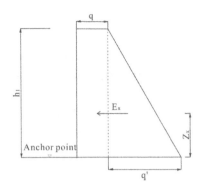

Figure 2. Distribution of landslide thrust.

$$\text{Triangle loading}(q'): M_0 = \frac{1}{6}q'h_1^2; Q_0 = \frac{q'h_1}{2} \quad (2)$$

$$\text{Trapezoidal loading}: M_0 = E_x \times Z_x; Q_0 = E_x \quad (3)$$

The expressions for q and q' are as follows:

$$q = \frac{6M_0 - 2E_x \times h_1}{h_1^2} \quad (4)$$

$$q' = \frac{6E_x \times h_1 - 12M_0}{h_1^2} \quad (5)$$

Z_x is the distance between the action point of the external force on the pile and the anchor point. Based on the results of earlier studies (Dai, 2002), it is

$$Z_x = \frac{7}{20}h_1; \quad (6)$$

h_1 is the length of the pile above the sliding surface, (m).

The horizontal displacement of the pile top can be calculated by the following formula.

$$x = (h_1 + y_0) \times \varphi \quad (7)$$

where y_0 is the distance between the position of the center of rotation of an anti-slide pile and its anchor point, (m), and φ is the rotation angle of an anti-slide pile, (rad). The specific formulas are as follows (Zhao, 2008).

$$y_0 = \frac{h_2[2A(3M_0 + 2Q_0h_2) + mh_2(4M_0 + 3Q_0h_2)]}{2[3A(2M_0 + Q_0h_2) + mh_2(3M_0 + 2Q_0h_2)]} \quad (8)$$

$$\varphi = \frac{12[3A(2M_0 + Q_0h_2) + mh_2(3M_0 + 2Q_0h_2)]}{B_ph_2^3[6A(A + mh_2) + m^2h_2^2]} \quad (9)$$

where **A** is the sliding surface coefficient of the foundation, (kPa/m); **m** is the proportional

580

coefficient under sliding surface which increases with increasing foundation depth (kPa/m^2); and h_2 is the anchorage length (m).

Combining Equations (1), (7), (8), and (9), the relationship between the rectangular loading and the displacement of the pile top is as follows:

$$q = \frac{xB_p h_2^3[6A^2 + 6Amh_2 + m^2h_2^2]}{6Ah_1(6h_1^2 + 4h_2^2 + 9h_1h_2) + 18mh_1h_2(h_1 + h_2)^2} \quad (10)$$

Combining Equations (2), (7), (8), and (9) the relationship between the triangle loading and the displacement of the pile top is as follows:

$$q' = \frac{xB_p h_2^3[6A^2 + 6Amh_2 + m^2h_2^2]}{12h_1 A(h_1 + h_2)^2 + mh_1h_2(6h_1^2 + 9h_2^2 + 16h_1h_2)} \quad (11)$$

Combining Equations (3), (4), (5), (6), (7), (8), and (9) the relationship between the trapezoidal loading and the displacement of the pile top is as follows:

$$E_x = \frac{5xB_p h_2^3[6A^2 + 6Amh_2 + m^2h_2^2]}{126Ah_1^2 + 243Ah_1h_2 + 120Ah_2^2 + 63mh_2h_1^2 + 162mh_1h_2^2 + 90mh_2^3} \quad (12)$$

Then, M_0, and Q_0 can be determined by Equations (1), (2), and (3).

The bending moment (M_y), and the shear force (Q_y) at any point in the pile body can be determined by the following formulas:

If $y < y_0$,

$$Q_y = Q_0 - \frac{1}{2}AB_p \varphi y(2y_0 - y) - \frac{1}{6}B_p m\varphi y^2(3y_0 - 2y) \quad (13)$$

$$M_y = M_0 + Q_0 y - \frac{1}{6}AB_p \varphi y^2(3y_0 - y) - \frac{1}{12}B_p m\varphi y^3(2y_0 - y) \quad (14)$$

If $y \geq y_0$,

$$Q_y = Q_0 - \frac{1}{6}B_p m\varphi y^2(3y_0 - 2y) - \frac{1}{2}AB_p \varphi y_0^2 + \frac{1}{2}AB_p \varphi (y - y_0)^2 \quad (15)$$

$$M_y = M_0 + Q_0 y - \frac{1}{6}AB_p \varphi y_0^2(3y - y_0) + \frac{1}{6}AB_p \varphi (y - y_0)^3 - \frac{1}{12}B_p m\varphi y^3(2y_0 - y) \quad (16)$$

For the landslide along the highway in Shaanxi Province, the concrete of the anti-slide pile is mainly grade C30, the commonly used pile sizes are 3 m × 4 m, 2 m × 3 m and 1.5 m × 2 m. So, in this study, the concrete for the anti-slide pile is grade C30. After many displacement trials, and using a software to fit the relationship, the relationship among the maximum bending moment, the top displacement of the anti-slide pile, anchorage length, cantilever length, pile diameter, foundation coefficient, and the proportional coefficient of the foundation coefficient which increases with the depth of soil has been developed, as shown in the Equation (17),(18),(19), which is suitable for an anti-slide pile of the size 3 m × 4 m, 2 m × 3 m or 1.5 m × 2 m. The values of the coefficients in Equation (18),(19) are shown in Table 1.

$$y_{max} = \frac{y_1}{y_2} \quad (17)$$

$$y_1 = p_1 + \begin{pmatrix} 1 \\ 1 \\ 1 \\ 1 \\ 1 \\ 1 \\ 1 \\ 1 \\ 1 \\ 1 \end{pmatrix}^T \begin{pmatrix} p_3 & p_5 & p_7 & p_9 & p_{11} & p_{13} & p_{59} & 0 & 0 & 0 & 0 & 0 & 0 & 0 \\ 0 & 0 & 0 & 0 & 0 & 0 & 0 & p_{15} & p_{17} & p_{19} & p_{21} & p_{23} & p_{25} & p_{61} \\ D_1 & 0 & 0 & 0 & 0 & 0 & 0 & 0 & 0 & 0 & 0 & 0 & 0 & 0 \\ 0 & D_2 & 0 & 0 & 0 & 0 & 0 & 0 & 0 & 0 & 0 & 0 & 0 & 0 \\ 0 & 0 & D_3 & 0 & 0 & 0 & 0 & 0 & 0 & 0 & 0 & 0 & 0 & 0 \\ 0 & 0 & 0 & D_4 & 0 & 0 & 0 & 0 & 0 & 0 & 0 & 0 & 0 & 0 \\ 0 & 0 & 0 & 0 & D_5 & 0 & 0 & 0 & 0 & 0 & 0 & 0 & 0 & 0 \\ 0 & 0 & 0 & 0 & 0 & D_6 & 0 & 0 & 0 & 0 & 0 & 0 & 0 & 0 \\ 0 & 0 & 0 & 0 & 0 & 0 & D_7 & 0 & 0 & 0 & 0 & 0 & 0 & 0 \end{pmatrix} \begin{pmatrix} \ln(x_1) \\ \ln(x_2) \\ x_3 \\ x_4 \\ x_5 \\ x_6 \\ \ln(x_7) \\ (\ln(x_1))^2 \\ (\ln(x_2))^2 \\ x_3^2 \\ x_4^2 \\ x_5^2 \\ x_6^2 \\ (\ln(x_7))^2 \end{pmatrix} \quad (18)$$

$$y_2 = 1 + \begin{pmatrix} 1 \\ 1 \\ 1 \\ 1 \\ 1 \\ 1 \\ 1 \\ 1 \\ 1 \\ 1 \end{pmatrix}^T \left\{ \begin{pmatrix} p_2 & p_4 & p_6 & p_8 & p_{10} & p_{12} & p_{58} & 0 & 0 & 0 & 0 & 0 & 0 & 0 \\ 0 & 0 & 0 & 0 & 0 & 0 & 0 & p_{14} & p_{16} & p_{18} & p_{20} & p_{22} & p_{24} & p_{60} \\ D_8 & 0 & 0 & 0 & 0 & 0 & 0 & 0 & 0 & 0 & 0 & 0 & 0 & 0 \\ 0 & D_9 & 0 & 0 & 0 & 0 & 0 & 0 & 0 & 0 & 0 & 0 & 0 & 0 \\ 0 & 0 & D_{10} & 0 & 0 & 0 & 0 & 0 & 0 & 0 & 0 & 0 & 0 & 0 \\ 0 & 0 & 0 & D_{11} & 0 & 0 & 0 & 0 & 0 & 0 & 0 & 0 & 0 & 0 \\ 0 & 0 & 0 & 0 & D_{12} & 0 & 0 & 0 & 0 & 0 & 0 & 0 & 0 & 0 \\ 0 & 0 & 0 & 0 & 0 & D_{13} & 0 & 0 & 0 & 0 & 0 & 0 & 0 & 0 \\ 0 & 0 & 0 & 0 & 0 & 0 & D_{14} & 0 & 0 & 0 & 0 & 0 & 0 & 0 \end{pmatrix} \begin{pmatrix} \ln(x_1) \\ \ln(x_2) \\ x_3 \\ x_4 \\ x_5 \\ x_6 \\ \ln(x_7) \\ (\ln(x_1))^2 \\ (\ln(x_2))^2 \\ x_3^2 \\ x_4^2 \\ x_5^2 \\ x_6^2 \\ (\ln(x_7))^2 \end{pmatrix} \right\} \quad (19)$$

$D_1 = p_{27}\ln(x_2) + p_{29}x_3 + p_{31}x_4 + p_{33}x_5 + p_{35}x_6 + p_{63}\ln(x_7);$

$D_2 = p_{37}x_3 + p_{39}x_4 + p_{41}x_5 + p_{43}x_6 + p_{65}\ln(x_7)$

$D_3 = p_{45}x_4 + p_{47}x_5 + p_{49}x_6 + p_{67}\ln(x_7)$

$D_4 = p_{51}x_5 + p_{53}x_6 + p_{69}\ln(x_7)$

$D_5 = p_{55}x_6 + p_{71}\ln(x_7)$

$D_6 = p_{73}\ln(x_7)$

$D_7 = p_{57}\ln(x_1)\ln(x_2)x_3x_4x_5x_6$

$D_8 = p_{26}\ln(x_2) + p_{28}x_3 + p_{30}x_4 + p_{32}x_5 + p_{34}x_6 + p_{62}\ln(x_7)$

$D_9 = p_{36}x_3 + p_{38}x_4 + p_{40}x_5 + p_{42}x_6 + p_{64}\ln(x_7)$

$D_{10} = p_{44}x_4 + p_{46}x_5 + p_{48}x_6 + p_{66}\ln(x_7)$

$D_{11} = p_{50}x_5 + p_{52}x_6 + p_{68}\ln(x_7)$

$D_{12} = p_{54}x_6 + p_{70}\ln(x_7)$

$D_{13} = p_{72}\ln(x_7)$

$D_{14} = p_{56}\ln(x_1)\ln(x_2)x_3x_4x_5x_6$

where y_{max} is the displacement of a pile top, (m); x_1 is the design bending moment of a pile (M_{max}), (kN.m); x_2 is the foundation coefficient of the sliding surface (K), (kPa/m); x_3 is the anchorage segment length of an anti-slide pile (h_2), (m); x_4 is the cantilever segment length of an anti-slide pile (h_1), (m); x_5 is the width of an anti-slide pile section (b), (m); x_6 is the height of an anti-slide pile section (h), (m); x_7 is the proportional coefficient of the foundation coefficient increases with the depth below the slip surface (m), (kPa/m²);

In practical engineering applications, based on the design drawing of an anti-slide pile, the bending moment of the pile can be calculated as follows:

$$M_{max} = \frac{2\alpha_1 f_c b h_0 A_s f_y - A_s^2 f_y^2}{2\alpha_1 f_c b} \quad (20)$$

where M_{max} is the design bending moment of a pile, (kN.m); α_1 is the equivalent rectangular stress coefficient of a concrete compression zone (for concrete strength $\leq$ C50, $\alpha_1 = 1.0$); As is the cross-sectional area of tensile reinforcement, (m²); f_c is the design concrete axial compressive strength, (kN/m²); f_y is the design steel tensile strength, (kN/m²); b is the width of the anti-slide pile section, (m); h is the height of the anti-slide pile section, (m); h_0 is the effective height of the pile section (h-α_s), (m); α_s is the distance from the joint point of all the steel bars to the tension zone of the cross-section, (m).

3 ENGINEERING EXAMPLE

In this study, to verify the reliability of Equation (17), the formula has been applied to the design

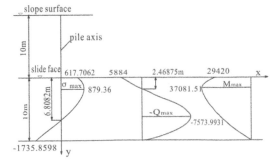

Figure 3. The calculated internal forces of anti-slide piles.

582

Table 1. The values of the coefficients for the fitted formula (equation 18, 19).

Coefficient	Rectangular loading	Triangle loading	Trapezoidal loading	Coefficient	Rectangular loading	Triangle loading	Trapezoidal loading
p_1	0.897722871	2293.976634	16.36039627	p_{38}	2.83953E-10	72.46794655	0.008426758
p_2	−0.26793889	0.0256937	−0.106534156	p_{39}	−5.23899E-05	−6.472170206	−0.001088459
p_3	0.148423918	0.280200506	0.217130605	p_{40}	0.029901727	947.5964473	0.123927889
p_4	−3.41085E-06	−1268.679136	0.397831757	p_{41}	−0.001353544	0.796010261	−0.000159137
p_5	−0.071744285	−0.093095898	−0.053786916	p_{42}	1.15705E-09	771.8049598	−0.004007793
p_6	0.394419926	1410.854546	0.517573662	p_{43}	0.006466596	−43.47225841	0.002786693
p_7	−0.045859662	−119.3185004	−0.062677301	p_{44}	0.002301104	42.87992296	0.0052504
p_8	−0.115251921	−976.8755644	−0.181996021	p_{45}	−0.000141311	−2.307838	−0.000426644
p_9	−2.13229E-07	0.00890562	0.006294991	p_{46}	0.072527918	340.412236	0.104056423
p_{10}	−1.07977895	0.001366667	−1.085873853	p_{47}	−0.001731376	3.079830974	1.07253E-06
p_{11}	3.266196233	−848428.7882	−12.31640386	p_{48}	−0.05191507	543.3801572	−0.026788217
p_{12}	−3.65718E-14	−4872.728108	−1.595607376	p_{49}	0.003751002	−26.37244983	0.000666971
p_{13}	−2.83838304	635517.0302	−2.765984788	p_{50}	0.011365028	85.61891699	0.011024948
p_{14}	−0.007249925	114.6441935	−0.037057813	p_{51}	0.000796315	6.545929523	0.000661258
p_{15}	0.019982575	86.92433717	0.031159098	p_{52}	−0.004533398	−56.8592743	−0.006665279
p_{16}	0.046718046	70.78391197	−0.012877541	p_{53}	−0.000601876	0.034067126	−0.000691281
p_{17}	−0.009094498	−58.07401268	−0.01804497	p_{54}	0.111278879	1531.033439	0.447819833
p_{18}	−0.010551668	−82.98614973	−0.017124719	p_{55}	−0.245644727	−320944.3586	3.057657589
p_{19}	0.001289226	8.443954665	0.002110766	p_{56}	−1.37361E-07	−0.001587935	−1.65481E-07
p_{20}	0.000314256	5.360820345	0.001590202	p_{57}	−4.3304E-09	−7.82308E-05	1.7788E-10
p_{21}	−1.12171E-11	0.091924054	−1.83573E-05	p_{58}	0.111706322	993.878422	−0.145837155
p_{22}	−0.131718203	−2172.685125	−0.689981143	p_{59}	−0.074428048	−5.80609E-05	−0.134145503
p_{23}	−0.653737189	425853.7107	−1.156457019	p_{60}	0.01728954	−62.94201365	−0.014956549
p_{24}	0.117078285	823.5446808	0.454816077	p_{61}	−0.014031103	−114.6185879	−0.024493144
p_{25}	0.611212504	1367.544788	0.35766967	p_{62}	−0.026798534	−37.92984626	−1.4015E-06
p_{26}	−0.00549307	118.7624591	0.031131926	p_{63}	−0.0159689	−8.431278073	−0.023081463
p_{27}	−0.021493826	−112.7251164	−0.036010404	p_{64}	−0.026270066	−198.3633878	0.018503543
p_{28}	−0.015902184	−125.7792655	−0.02070629	p_{65}	0.040643038	212.11729	0.069657436
p_{29}	−0.008208686	−36.84255508	−0.012970744	p_{66}	0.052853548	278.1326849	0.063422893
p_{30}	0.008504049	−45.43386194	0.003084336	p_{67}	0.002447878	−5.799337504	0.002473464
p_{31}	0.000648115	10.18159433	0.001607834	p_{68}	−0.005193189	23.30381722	−0.006744952
p_{32}	−0.007924893	−1526.00944	−0.14495686	p_{69}	−0.00031039	0.012892206	−0.000283125
p_{33}	0.003861986	84.60465728	0.009816585	p_{70}	0.027762729	1306.449142	0.137055797
p_{34}	−0.009324881	−1451.533518	−0.055535765	p_{71}	−0.000109341	−37.40690528	−0.004786772
p_{35}	−0.012623079	24.939547	−0.014529047	p_{72}	−3.63464E-06	221.5416097	0.000171008
p_{36}	−0.019788408	−162.0225904	−0.025841588	p_{73}	0.00387423	−26.35449799	0.006617792
p_{37}	0.005465921	34.19456114	0.00984044				

example on p. 95 of literature (Chen, 2011). The form of the loading distribution is rectangular. The calculated internal forces in the anti-slide piles are shown in Figure 3. The example shows that A = 78543 kN/m³, m = 39227 kN/m⁴, $b = 2$ m, $h = 3$ m, $h_1 = 10$ m, $h_2 = 10$ m. Using Equation (7) to calculate the pile top displacement, $y = 0.01941616$ m.

Substituting the parameters into Equation (17), the calculated $y_{max} = 0.017627411$ m.

The absolute error $\delta = (y - y_{max}) = (0.01941616 - 0.017627411) = 1.789 \times 10^{-3}$ m = 1.789 mm. The error is reasonable.

4 CONCLUSIONS

Based on the method of internal force calculation of rigid piles, for a given pile top displacement, after calculating the loading on a pile body, the pile bending moment can be determined. After many displacement trials, and using a software set to fit the relationship, the relationship among the maximum bending moment, the top displacement of the anti-slide pile, anchorage length, cantilever length, pile diameter, foundation coefficient, and the proportional coefficient of the foundation coefficient which increases with the depth of soil has been developed. The maximum displacement of the pile can be determined by inputting the parameters of the pile and the soil layer. Equation (17) is mainly applicable to the geological conditions of the landslide along the highway in Shaanxi province, and it is suitable to the grade C30 concrete anti-slide piles with a diameter of 3 m × 4 m, 2 m × 3 m or 1.5 m × 2 m. By applying Equation (17) to a design example, the formula's reliability has been proven. Combining with relevant monitoring data, Equation ?? can be used in an early warning system to predict the stability of the slope along the highway in Shaanxi province.

ACKNOWLEDGMENTS

This research has been supported by the special funds of scientific and technological innovation for the provincial state-owned capital operating budget (2013 gykc-018) and the transportational research projects of traffic and transportation department of Shaanxi province in 2013 (13–28 K).

REFERENCES

Chen, H.K., Theory and control of geological disasters (Science Press, 2011).

Dai, Z.H., Chin J. Rock Mech. Eng, 21, 517(2002).

Gimzburg, L.K., Landslides, 1699(1996).

Liu, H.J., Y.M. Men, X.C. Li, T. Zhang, Rock. Soil Mech, 33, 2960(2012).

Liu, Y.L., Y.H.Sun, Y. Yu, W. Yan, Y.Q. Shang, J. Zhejiang Univ (Eng Sci), 46, 243(2012).

Qiang, Q., C.X. Chen, R. Wang, X.G. Liu, Chin. J. Rock Mech. Eng, 24, 934(2015).

Yan, J.K., Y.P. Yin, Y.M. Men, J. Liang, Chin. Civ. Eng. J, 44, 120(2011).

Zhao, Q.H., S.P. Qin, Rock and earth retaining and anchoring engineering (Sichuan University Press, 2008).

Zheng, M.X., X.L. Jiang, Z.Z. Yin, J.M. Wu, Rock. Soil Mech, 28, 1381(2007).

Zheng, M.X., Z.Z. Yin, J.M. Wu, Chin J. Rock Mech Eng, 25, 2150(2006).

Zheng, M.X., Z.Z. Yin, J.M. Wu, Y.F. Du,Y. Luan, Chin J. Geotech Eng, 28, 1224(2006).

Advances in Energy Science and Equipment Engineering II – Zhou, Patty & Chen (Eds)
© 2017 Taylor & Francis Group, London, ISBN 978-1-138-71798-5

Three-dimensional stability analysis of highway embankment in the karst region

Zhongming He
Key Laboratory of Special Environment Road Engineering of Hunan Province, Changsha University of Science and Technology, Changsha, China
School of Communication and Transportation Engineering, Changsha University of Science and Technology, Changsha, China

Senzhi Liu & Qingfang Liu
School of Communication and Transportation Engineering, Changsha University of Science and Technology, Changsha, China

ABSTRACT: In order to analyse the stability of embankment in the karst region of Hunan Province Yi-Lou highway, the FLAC3D software was used to build the three dimensional numerical model of actual geological, then the stability of embankment before karst cave treatment and the treatment effect of karst cave filled with gravel were analyzed. The results show that: When the cave is not treated, the stress concentration took place at the top, the uneven settlement of top and the bottom surfaces of the embankment present funnel shapes, shear failure occurred in the roof of karst cave, tensile failure occurred in the embankment. After karst cave filled with gravel, the uneven settlement of embankment is greatly reduced, the settlements of the top and bottom surface of the embankment are changed from "funnel shapes" to "semicylinder shapes", and the maximum settlement was reduced by 33.7% and 42.5% respectively, which guarantee the stability of the embankment and karst cave.

1 INTRODUCTION

Karst area has become the main occurrence area of highway diseases in our country. Karst cave and soil hole along the line often cause insufficient bearing capacity, Unreasonable design and disposal is very easy to cause uneven settlement of roadbed, sliding and collapse deformation and other issues, during construction or operation, which seriously affects the safety of traffic operations (Kang, 2008; Fu, 2010). Therefore, the stability analysis, and evaluation and the treatment of the foundation in karst area, have become the key to the success or failure of the highway construction, and are also one of the major technical problems of the highway construction in karst area at home and abroad (Liu, 2008).

At the present stage, many scholars have made a lot of achievements in the research of the stability of the roadbed in karst area from the theoretical analysis, laboratory tests and numerical calculation. For example, Y J Zhang (Zhang, 2011) set up the interval fuzzy evaluation method for the stability of the highway roadbed in karst area, and comprehensively analyzed the unfavorable factors of the stability of the roadbed in karst area; X Z Jiang (Jiang, 2005) used the model test in the laboratory to analyze the effect of the karst water on the stability of the embankment.

In the numerical simulation, most scholars used the numerical software to establish a simplified model of karst cave of two-dimensional circular (Chen, 2011; Zheng, 2015; Wang, 2015) or three-dimensional cylindrical (Xiao, 2009). However, in practical engineering, the hidden karsts tend to approach to three-dimensional cavity ellipsoid (Liao, 2010). Therefore, the three dimensional model of actual geological was built which has important significance for the actual stability analysis of embankment and the reasonable evaluation of cave treatment.

In this paper, a construction area of karst region in Hunan Province Yi-Lou highway as the research background, the FLAC[3D] software was used to build the three dimensional numerical model of actual geological, and the stability of embankment before treated and the treatment effect of karst cave filled with gravel were analyzed, which provide a reference for engineering practice.

2 THE ESTABLISHMENT OF A NUMERICAL MODEL

2.1 General situation of the study area

The geological condition of K26+692~K39+00 in the fourth contract section of Hunan province is relatively complicated, Karst development, some areas formed the dissolution channel, karst collapse, karst collapse, etc. According to the engineering geological investigation and the results of the geophysical exploration, surveying the area covering layer is mainly silty clay caused by residual slope, gravel and crushed stone soil and boulder caused by flood. The underlying bedrock is mainly limestone, dolomitic limestone and marl of Permian, Carboniferous limestone, dolomitic limestone, etc. Bedrock cave is mainly distributed in stratum contact zone, fault fracture zone or dense joints zone. The majority of ellipsoidal and line to a large intersection, local road along the direction of distribution, is in the karst collapse area and the surrounding soil to backfill rock collapse pit after pretreatment with plate treatment. A typical geological section (Figure 1) is selected as the research background.

2.2 Calculation model and analysis method

In order to analyze the stability of embankment construction and the effect of "advance treatment" of karst cave, the stability of embankment and karst cave are calculated by numerical method. According to the survey report, the shape of the underlying karst cave is close to the ellipsoid. In order to simplify the calculation, the karst cave is simplified as a regular ellipse, and it is assumed to be in the bottom of the embankment, which can be obtained for the embankment of the most unfavorable situation. Based on the finite difference software FALC3D comes with the fish language (Chen, 2016) to establish the calculation model as shown in Figure 2.

According to the actual engineering conditions, the calculation model is divided into 4 layers, and the bottom rock is slightly weathered limestone. The middle rock is silty clay and soil covering, and the top layer is embankment. According to the principle of symmetry, the model length is 50 m, the width is 15 m, and the height is 26 m. The ellipsoidal cavity is located between the bedrock and the silty clay, embankment cross section width size

is 14 m, embankment longitudinal section direction length size is 7 m, and vertical height size is 7 m. The thickness of the silty clay and planting soil above the bedrock is 8.5 m and 0.5 m. The width of the embankment is 26 m, the height is 6 m, and the slope rate is 1:1.5. The material properties of each layer of the model are consistent with the Mohr-Coulomb yield criterion, and its parameters are shown in Table 1.

In the process of numerical calculation, the initial stress balance is firstly carried out, and the stress condition before the construction of embankment is obtained, then the embankment is filled again (for the simplified analysis the vehicle load is not considered in this paper). Finally, the deformation of embankment and the destruction mechanism of karst cave are analyzed. Based on the existing literature, this paper sets up a method for the initial condition of the foundation of the underlying karst cave: Assuming that the cave does not exist, the stress condition of the model under the action of gravity is calculated, the null command is used to set the cave to the empty model, the stress state of the karst cave is set in the form of the excavated foundation, and the velocity field and displacement field are cleared. In the whole calculation process,

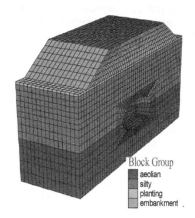

Figure 2. Finite difference calculation model.

Table 1. Physical and mechanical parameters of rock and soil layer.

Type of rock and soil layer	E/MPa	μ	C/kPa	φ/(°)	γ/(kN·m⁻³)	h/m
Embankment fill layer	15.60	0.30	14.9	12	18.9	*
Planting soil layer	16.39	0.30	24.0	13	19.2	0.5
Silty clay layer	15.90	0.30	22.0	13	18.1	8.5
Aeolian limestone	1.9×10^4	0.25	100.0	40	26.4	–

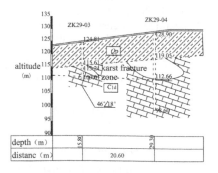

Figure 1. Geological profile of karst cave.

the boundary condition is that both sides of the X direction displacement constraints, front and back sides of the Y direction displacement constraints, at the bottom of the Z direction displacement constraints, embankment slope and the toe of the embankment on both sides are free boundary.

3 CALCULATION AND ANALYSIS

3.1 The stress state analysis of the cave before the treatment

Figure 3 and Figure 4 are the minimum and maximum principal stress distribution of model before the treatment. Figure 3 and Figure 4 shows that under the action of embankment filling, the state of stress distribution of embankment and karst cave is: tensile stress concentration at the top and bottom of karst cave, of which maximum tensile stress is 15 KPa. The stress concentration at the boundary between the two sides of the karst cave is the silty clay and the underlying bedrock, of which the maximum compressive stress is 150 KPa.

3.2 The displacement trend of embankment before the treatment

In order to analyze the displacement trend of embankment before the treatment, the settlement and displacement of the monitoring points of the top surface and the bottom surface of the embankment are obtained in this paper. The displacement trend of two monitoring surface are shown in Figure 5 and Figure 6. Figure 5 and Figure 6 shows that: The settlement trend of the top and bottom surface of embankment generally presents "funnel shape", of which surface settlement amount

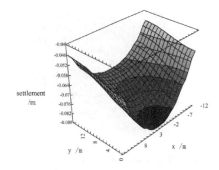

Figure 5. The settlement of the top surface of the embankment before treatment.

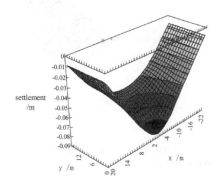

Figure 6. The settlement of the bottom surface of the embankment before treatment.

increased with decreasing cave distance, and the maximum settlement amount were 87.5 mm and 86.9 mm. Settlement difference reached 21 mm to 35 mm and 30 mm to 79 mm, and the uneven settlement of the embankment is large, which has great harm to the operation of vehicle.

3.3 The failure modes before the treatment

Under the load of embankment, the silty clay in the upper part of karst cave is destroyed, and the plastic zone extends upward along the top. Figure 7 shows that the plastic zone extends and penetrates through roof of cave which it occurs shear failure, tensile failure at the top of the embankment. The embankment and cave are unstable, which are necessary to treat cave.

3.4 The reinforcement effect analysis

The above analysis shows that: in the case that the karst cave is not treated, the embankment and karst cave were destroyed, and can not meet the requirements of bearing capacity and deformation. In this paper, the karst cave filled with gravel was used to treat, the Physical and mechanical parameters of gravel are that: shear modulus (K) is 9.9 MPa, bulk

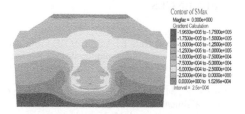

Figure 3. The distribution of maximum principal stress.

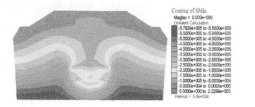

Figure 4. The distribution of minimum principal stress.

Figure 7. The distribution of plastic zone.

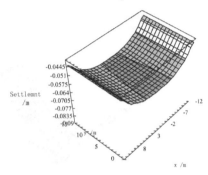

Figure 8. The settlement of the top surface of the embankment after treatment.

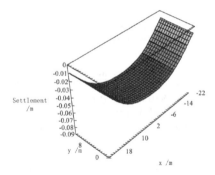

Figure 9. The settlement of the bottom surface of the embankment after treatment.

modulus is 21.45 MPa, internal friction angle (ϕ) is 35, cohesion (C) is 260 KPa, and density (γ) is 20.5 N·m^{-3}. The settlement of the top and bottom surface of the embankment after treatment was got. They are shown in figure 8 and figure 9.

Figure 8 and Figure 9 shows that: the uneven settlement of the top and bottom surface of embankment decreases obviously after the cave filled with gravel. The settlement of the top and bottom surface of the embankment is changed from "funnel shape" to "semicylinder shape", and the maximum settlement amount were 58 mm and 50 mm, which they were reduced by 33.7% and 42.5% respectively. So the treatment effect is good, which can guarantee the stability of embankment and cave.

4 CONCLUSION

1. A construction area of karst region in hunan province Yi-Lou highway as the research background, the FLAC3D software was used to build the three dimensional numerical model of actual geological, to quantitative analysis on the stability of highway embankment in karst area.
2. When the cave is not treated, the stress concentration took place at the top, the uneven settlement of top and the bottom surface of the embankment presents a funnel shape. The maximum settlement amount is 87.5 mm and 86.9 mm. Shear failure occurred in the roof of karst cave, and tensile failure occurred in the embankment.
3. After karst cave filled with gravel, the uneven settlement of embankment is greatly reduced, the settlement of the top and bottom surface of the embankment is changed from "funnel shape" to "semicylinder shape", and the maximum settlement was reduced by 33.7% and 42.5% respectively, which guarantee the stability of the embankment and karst cave.

ACKNOWLEDGEMENTS

This work was financially supported by National Natural Science Foundation of China (No. 51508042), Hunan Provincial Transportation Department Scientific & Technological Progress and Innovation Plan (No. 201417) and Open Fund of Key Laboratory of Special Environment Road Engineering of Hunan Province, Changsha University of Science & Technology (No. kfj140501).

REFERENCES

Chen, J. B. High. Eng, **36**, 124 (2011).
Chen, Y. M., D. P. Xun. FLAC/FLAC3D foundation and engineering examples (second edition) (China Water Resources and Hydropower Publishing House, 2103.
Fu, C. J. Rock. Soil. Mech, **31**, 288 (2010).
Jiang, X. Z., M. T. Lei, Y. Li, J. L. Dai, Y. Meng, Carsologica Sinica, **24**, 96 (2005).
Kang, H. R., Q. Luo, J. M. Ling, *Theory and practice of highway construction in krasy region* (China Communications Press, 2008).
Liao, L. P., W. K. Yang, Q. Z. Wang, Rock. Soil. Mech, **31**, 138 (2010).
Liu, Y. S., G. B. Wang, J. Mount. Sci, **26**, 453 (2008).
Wang, L. C. Min. Metall. Eng, **35**, 17 (2015).
Xiao, J. Q., S. F. Qiao, J. Rai. Sci. Eng, **6**, 33 (2009).
Zhang, Y. J., W. G. Cao, M. H. Zhao, H. Zhao, Chin. J. Geotech. Eng, **33**, 38 (2011).
Zheng, M. X., L. Ouyang, W. J. Huang, J. Henan. Univ. Sci. Tech (Nat. Sci), **36**, 65 (2015).

Advances in Energy Science and Equipment Engineering II – Zhou, Patty & Chen (Eds)
© *2017 Taylor & Francis Group, London, ISBN 978-1-138-71798-5*

Characteristics of satellite-based vegetation health indices in Shaanxi, China: 1982–2015

Rengui Jiang & Jiancang Xie
State Key Laboratory Base of Eco-hydraulic Engineering in Arid Area, Xi'an University of Technology, Xi'an, China

Fawen Li
State Key Laboratory of Hydraulic Engineering Simulation and Safety, Tianjin University, Tianjin, China

Honghong Liu
Xi'an University of Finance and Economics, Xi'an, China

Bin Li
State Key Laboratory Base of Eco-hydraulic Engineering in Arid Area, Xi'an University of Technology, Xi'an, China

ABSTRACT: Vegetation Health Indices (VHIs) have been widely used to estimate vegetation health, moisture and thermal conditions, soil saturation and drought, etc. This paper used the VHIs to monitor terrestrial vegetation productivity and plant growth. Using five types of VHIs including smoothed Normalized Difference Vegetation Index (sNDVI, eliminated noise with respect to NDVI), smoothed Brightness Temperature (sBT, eliminated noise with respect to BT), Vegetation Condition Index (VCI), Temperature Condition Index (TCI) and Vegetation Health Index (VHI) derived from the radiance observed by the NOAA'S Advanced Very High Resolution Radiometer (AVHRR), this paper investigated the spatial and temporal variations of VHIs using non-parametric Mann-Kendall monotonic test and Thiel-Sen's slope, and the spatial distribution of typical VHIs were investigated using Kriging interpolation for the period of 1982–2015 in Shaanxi province, located in the northwestern part of China. The results showed that the sNDVI, sBT and VHI had increased over the past thirty-four years. Five (two) apparent periods were detected for severe-to-exceptional (moderate-to-exceptional) drought. The reasons of drought might be caused by the high temperature or the lack of soil moisture. The results give insights of better understanding impacts of climate change on vegetation productivity and plant growth in Shaanxi province and provide reference for policy makers and stakeholders.

1 INTRODUCTION

Climate change and its impacts on vegetation production and socio-economic systems are important issues, which need to cope with in the 21st century worldwide (Piao, 2010). The Intergovernmental Panel on Climate Change Fourth Assessment Report (IPCC AR4) indicated that the global average surface temperature has increased by $0.74 \pm 0.18°C$ over the last 100 year (1906–2005), and the warming trend has nearly doubled ($0.65 \pm 0.15°C$) over the last 50 years (1956–2005), than that of 1906–2005. More temperature increase occurs at higher northern latitudes, especially the Arctic, where average temperature has raised almost twice the rate of global average over the past 100 years. Drought is a climate phenomenon occurred naturally that affects human, agricultural and environmental activity across the world. It is one of the most severe

natural disasters that cause serious consequences to economic, agricultural, ecological and environmental. Many subjective indices have been proposed to quantify, monitor and analyze drought for the past twentieth century, because it is hard to quantify drought characteristics in accordance with intensity, duration, and spatiotemporal patterns objectively. In general, drought mainly originates from precipitation deficiency for an extended period and the influences of drought usually accumulated slowly. The definition of drought has been one of the most important issues for drought detection, monitoring and analysis. However, it is difficult to determine the onset, end and severity of the drought, which make it difficult to identify and quantify the effects on economy, society and environment (Vicente-Serrano SM, 2010).

Drought indices, including Palmer Drought Severity Index (PDSI), Crop Moisture Index (CMI),

Drought Area Index (DAI), Standardized Precipitation Index (SPI), Standardized Precipitation Evapotranspiration Index (SPEI), have proposed to quantify drought. However, those drought indices probably lead to some differences in values and change patterns, especially at relatively small scales (Dai, 2013). A new vegetation health method of satellite-based Vegetation Health Indices was proposed to estimate the thermal, moisture and vegetation health conditions using the AVHRR data. Those indices including Vegetation Condition Index (VCI), Temperature Condition Index (TCI), and Vegetation Health Index (VHI) were developed to character the cumulative temperature, moisture and vegetation health conditions (Kogan, 1997). For example, the droughts in 2000 affected the agricultural areas of up to 40 million hectares of China, and significant increasing area of drought was detected in most of northern China, especially since the later 1990s, when some of the regions experienced unprecedented severe droughts (Zou, 2005). Droughts, such as the long-term events in western United States and northeast China in the decade of 2000, the severe drought in Yunan province, southwestern of China, and across Russia and Ukraine in 2010, have caused devastating effects on regional agriculture, environment and social-economic, which have attached wide attention (Qiu, 2010; Sheffield J, 2010).

On the other hand, the photosynthetic activity has increased either by an earlier beginning or by a lengthening of the growing season due to the global warming subject to climate change, therefore improves the vegetation growth. Normalized Difference Vegetation Index (NDVI) is one of the most popular vegetation indices for monitoring short and long term variations of terrestrial vegetation productivity, land cover and climate because it reflects Leaf Area Index (LAI) and vegetation biomass. Previous studies have found that the NDVI is strongly subject to climate, and therefore NDVI can be used to evaluate the dynamic response of vegetation to hydroclimatic variations and fluctuations at both global and regional scales. For example, using the Pathfinder Advance Very High Resolution Radiometer (AVHRR) NDVI data set, the significant correlations were found between interannual NDVI and temperature variability in the growing season in the northern mid- and high latitude regions, between NDVI and precipitation (temperature) in northern/southern semiarid regions (Ichii K, 2002). For Asia, Fang et al (2001) found significant relationships between annual NDVI Coefficient of Variation (CV) and precipitation CV for five biome groups from 1982 to 1999 in China, using CV to represent the magnitudes of inter-annual variability of NDVI and precipitation (Fang, 2001).

The objectives of this study are to investigate the variations of monthly VHIs including sNDVI, sBT, and VHI, the spatial distribution of the typical VHIs including VHI and TCI. The paper is organized as follows: material and methodology are given in Section 2, results and discussions are described in Section 3, summary and conclusions in Section 4.

2 MATERIALS AND METHODOLOGY

2.1 Study area

Shaanxi province, situated in the northwestern of China, locates between 31.42°N–39.35°N latitude and 105.29°W–111.15°W longitude. It has a land area of about 205, 800 km², and nearly 40% is loess plateau, mostly of which lie in the north. The landscapes vary from windy desert region in the north to Qinling mountain region in the central and Dabashan mountain region in the south.

The hydrological conditions of Shaanxi province are multifarious due to the diversity of its physiographic features and climate, which vary seasonally and regionally. The mean annual temperature is about 13°C, and the mean annual precipitation is about 580 m, mostly of which concentrated in the monsoon. Both precipitation and temperature decrease from southern to northern Shaanxi province.

2.2 VHIs data

Five types of VHIs including smoothed Normalized Difference Vegetation Index (sNDVI, eliminated noise with respect to NDVI), smoothed Brightness Temperature (sBT, eliminated noise with respect to BT), Vegetation Condition Index (VCI), Temperature Condition Index (TCI) and Vegetation Health Index (VHI) were used in the paper. The VHIs were derived from the AVHRR of seven polar-orbiting satellites: NOAA-7 (launched on Jun. 23, 1981), NOAA-9 (Dec. 12, 1984), NOAA-11 (Sep. 24, 1988), NOAA-14 (Dec. 30, 1994), NOAA-16 (Sep. 21, 2000), NOAA-18 (May 20, 2005) and NOAA-19 (Jun. 2, 2009), produced from the NOAA's Satellite and Information Service (NESDIS) Global Area Coverage dataset, which can be obtained from website http://www.star.nesdis.noaa.gov/. The VHIs, computed for each pixel at 16 km spatial resolution and 7-day composite temporal resolution over a period of 34-year (1982–2015), were appropriate for trends and variation analysis of vegetation at both global and regional scales due to their global coverage and long-term availability (Jong, 2011).

NDVI was usually used as a proxy of vegetation activity, which represented the property of green vegetation to emit and reflect solar radiation, and it thus provided a measurement to monitor vegetation surface because of the chlorophyll, leaf interior tissues and water content in green vegetation. It is defined as the reflectance difference between visible (VIS) and Near Infrared (NIR) sections of solar spectrum.

$$NDVI = (NIR - VIS) / (NIR + VIS) \qquad (1)$$

The values of NDVI range from −1.0 to 1.0, and increasing positive values represent increasing vegetation greenness, while negative and near zero values represent non-vegetated conditions (e.g., water, snow, barren surfaces). The bigger value of NDVI represents that the increasing difference between the NIR and VIS, which indicates that vegetation becomes greener, denser and more vigorous. The short and long term noises caused by atmosphere constituents (e.g., cloud, aerosol, water vapor), unusual event (e.g., volcanic), view geometry, pre—and post-launch calibration, satellite orbital drift and sensor degradation, were removed for a smoothed NDVI (sNDVI) data set.

The VCI, TCI and VHI were developed to character the cumulative temperature, moisture and vegetation health conditions, which can be calculated as follows:

$$VCI = 100 \times (sNDVI - sNDVI_{min}) / (sNDVI_{max} - sNDVI_{min}) \qquad (2)$$

$$TCI = 100 \times (sBT_{max} - sBT) / (sBT_{max} - sBT_{min}) \qquad (3)$$

$$VHI = a \times VCI + (1-a) \times TCI \qquad (4)$$

where the sNDVI and sBT are the smoothed weekly NDVI and BT. The max(min) represent the multi-year absolute maximum(minimum) values of NDVI and BT. The parameter a is the coefficient of VCI and TCI contribution in vegetation health. These indices range from zero to 100, and higher value indicates better vegetation condition (Kogan, 2012).

2.3 Methodology

The non-parametric Mann-Kendall monotonic test at 5% significant level was used for trend analysis of sNDVI, sBT and VHI. The trend magnitude is determined using Sen's slope estimator β, defined as the median value of all possible combinations of pairs for the whole data set.

$$\beta_k = median[(X_{jk} - X_{ik})/(j-i)] \qquad (5)$$

where $X = \{x_1, \cdots, x_i \cdots x_n\}$, n is the length of X, and $i < j$, $i, j \in [1,n]$. A positive (negative) value of β indicates an increasing (decreasing) trend [13].

The spatial distributions of the VHIs were investigated using contour by Kriging gridding method.

3 RESULTS AND DISCUSSIONS

3.1 Monthly variations of typical VHIs

Figure 1 shows the 34-year (1982–2015) monthly mean values and trend magnitudes of sNDVI. The biggest value of mean sNDVI was found in August (0.2930), and sNDVI in winter were smaller (between 0.0989 and 0.1136) than other seasons for the period of 1982–2015. The trend magnitudes of monthly sNDVI were positive around the year, indicating that the sNDVI had increased over the past 34-year period. The biggest slope of trend magnitude was found in May, and surprisingly, the slope decreased after May, and reached the smallest in August, which might caused by the decreased temperature or other climate variables.

Monthly sBT and its pattern were similar to that of sNDVI, as shown in Figure 2. The largest value of mean sBT was detected in June (25.4715°C).

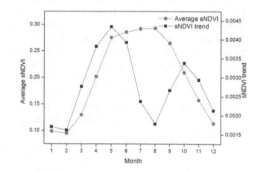

Figure 1. Mean values and trend magnitudes of monthly smoothed NDVI (sNDVI) for the period of 1982–2015.

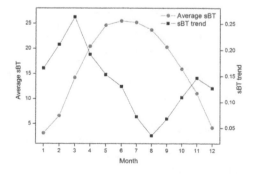

Figure 2. Mean values and trend magnitudes of monthly smoothed BT (sBT) for the period of 1982–2015.

However, the biggest slope of monthly sBT was detected in March with an annual rate of 0.2630°C, which indicated that mean sBT in March had increased up to 8.9°C over the past thirty-four years. Similar with that of sNDVI, the smallest slope of trend magnitude was detected in August (0.0356). The reason for the smallest slopes of trend magnitude for both sNDVI and sBT occurred in August needs to be further investigated.

Figure 3 shows the 34-year monthly mean values and trend magnitudes of VHI. Being different from that of sNDVI (Figure 1) and sBT (Figure 2), the mean values of monthly VHI fluctuated for the period of 1982–2015. The biggest value of mean VHI was detected in August with value of 52.7397, followed by November with value of 52.1415. Relative smaller values of VHI were found during winter, which was expected, because the vegetation does not grow as well as that during spring or summer. Again, the trend magnitudes of monthly VHI were positive for all month, indicating that the VHI had increased for the period of 1982–2015.

3.2 Drought detection using VHIs

The satellite-based VHIs are used to investigate the intensity, duration, frequency and area of drought in Shaanxi province from 1982 to 2015. Two levels of drought intensities, including Severe-to-Exceptional (SE) and Moderate-to-Exceptional (ME), were determined based on the VHI criteria of 15% and 35%, respectively. The percentage of drought area was computed with respect to the total number of pixels of Shaanxi province.

Figure 4 shows evolution of severe-to-exceptional and moderate-to-exception drought area from 1982 to 2015 in Shaanxi province. On a whole, the mainly drought areas were found at the later 1986 and early 2008.For severe-to-exceptional drought (VHI < 15%), five periods were detected including winter of 1986 and 2007, summer of 1995, 1997 and 2001, indicating the drought was mainly occurred

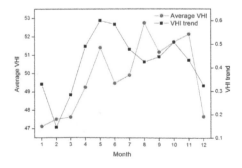

Figure 3. Mean values and trend magnitudes of Vegetable Health Index (VHI) for the period of 1982–2015.

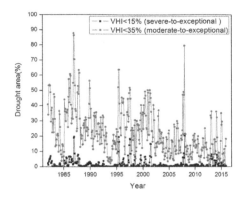

Figure 4. Evolution of severe-to-exceptional (black square line) and moderate-to-exceptional (red dot line) drought areafor the period of 1982–2015.

during summer and winter for the period of 1982–2015 in Shaanxi. The drought occurred during summer might be caused by the high temperature, while the drought occurred during winter might caused by the lack of soil moisture. For moderate-to-exceptional (VHI < 35%) drought, two apparent periods with the percentage of drought area exceed 70% were found during 1987–1988 and winter of 2007.

4 SUMMARY AND CONCLUSIONS

Using Mann-Kendall test and Sen's slope, this study investigated the variations of VHIs for the period of 1982–2015 in Shaanxi province, the drought conditions were also analyzed based on the VHIs. Our findings are summarized as follows:

1. In general, the sNDVI, sBT and VHI had increased over the past thirty-four years (Figure 1~Figure 3). The biggest values of both monthly sNDVI and sVHI were detected in August, while in June for sBT. The smallest slopes of trend magnitude were found in August for both sNDVI and sBT, but the mean values and trend magnitudes of monthly VHI fluctuated for the period of 1982–2015.
2. Two levels of drought intensities, including severe-to-exceptional and moderate-to-exceptional, derived based on the VHI criteria of drought area less than 15% and 35%, respectively, were used to investigate the drought conditions for the period of 1982–2015 in Shaanxi province. The results show that five apparent periods were detected for severe-to-exceptional drought, and two apparent periods were found for moderate-to-exceptional drought. The reasons of drought might caused by the high temperature or the lack of soil moisture.

ACKNOWLEDGEMENTS

The study was partly supported by the National Natural Science Foundation of China (51509201, 51479160, and 41471451), Scientific Research Program Funded by Shaanxi Provincial Education Department (15JK1503), Project funded by China Postdoctoral Science Foundation (2016M590964), Young Talent fund of University Association for Science and Technology in Shaanxi, China (20160217). The VHIs is obtained from the website at http://www.star.nesdis.noaa.gov/.

REFERENCES

Dai AG. Increasing drought under global warming in observations and models. Nature Climate Change, 2013. 3(1): 52–58.

de Jong R, et al. Analysis of monotonic greening and browning trends from global NDVI time-series. Remote Sensing of Environment, 2011. 115(2): 692–702.

Fang JY, Piao S, Tang Z, et al. Interannual variability in net primary production and precipitation. Science, 2001. 293(5536): U1–U2.

Gan TY. Hydroclimatic trends and possible climatic warming in the Canadian Prairies. Water Resources Research, 1998. 34(11): 3009–3015.

Ichii K, Kawabata A, Yamaguchi Y. Global correlation analysis for NDVI and climatic variables and NDVI trends: 1982–1990. International Journal of Remote Sensing, 2002. 23(18): 3873–3878.

Kogan F, Salazar L, Roytman L. Forecasting crop production using satellite-based vegetation health indices in Kansas, USA. International Journal of Remote Sensing, 2012. 33(9): 2798–2814.

Kogan F, Adamenko T, Guo W. Global and regional drought dynamics in the climate warming era. Remote Sensing Letters, 2013. 4(4): 364–372.

Kogan FN. Global drought watch from space. Bulletin of the American Meteorological Society, 1997. 78(4): 621–636.

Piao SL, Ciais P, Huang Y, et al. The impacts of climate change on water resources and agriculture in China. Nature, 2010. 467(7311): 43–51.

Qiu J. China drought highlights future climate threats. Nature, 2010. 465(7295): 142–143.

Sheffield J, Wood EF, Roderick ML. Little change in global drought over the past 60 years. Nature, 2012. 491(7424): 435–438.

Vicente-Serrano SM, Begueria S, Lopez-Moreno JI. A Multiscalar Drought Index Sensitive to Global Warming: The Standardized Precipitation Evapotranspiration Index. Journal of Climate, 2010. 23(7): 1696–1718.

Zou X, Zhai P, Zhang Q. Variations in droughts over China: 1951–2003. Geophysical Research Letters, 2005. 32(4), doi: 10.1029/2004GL021853.

Advances in Energy Science and Equipment Engineering II – Zhou, Patty & Chen (Eds)
© 2017 Taylor & Francis Group, London, ISBN 978-1-138-71798-5

Compression capacity of bolt and screw-connected built-up cold-formed steel stud column section

Boon-Kai Tan, Jin-Sheng Lim, Poi-Ngian Shek & Kueh-Beng-Hong Ahmad
UTM Construction Research Centre, Universiti Teknologi Malaysia, Malaysia

ABSTRACT: Cold-formed steel is formed by cold-rolling or brake-pressing of flat steel sheet at ambient temperature. This type of steel section is mainly used for roof truss design and lightweight construction. This paper focuses on experimental investigation of compression strength capacity of a back-to-back cold-formed steel column section using bolts and screws. Two identical channel sections of depth 250 mm and length 0.75 m were built back-to-back to form a double-channel I shape steel section by bolts or screws. A 150 mm deep and 0.6 m long channel was connected to the aforementioned specimens. A screw and high-strength bolt of M12 Grade 8.8 served as fasteners used in the research. A total of 12 test samples were prepared and subjected to axial compression load. The compression capacity of double C-section stub column-connected back-to-back is in the range of 331.67–378.50 kN. An increase in the section depth will increase the compression capacity. The behavior of cold-formed steel section is independent of the type of connection used.

1 INTRODUCTION

Light steel framing is referred to as cold-formed steel section or lightweight section to form steel frame structure. Cold-formed steel section is formed below recrystallization temperature or room temperature. The properties of hot-rolled and cold-formed steel significantly differ in parameters like strength and moment capacity, durability, failure mode, and yield strength. Cold-formed steel section is manufactured from steel sheet by press-breaking, bending brake operation, and cold-rolling method. Cold-formed steel sections are used as main structure members like frames and second structural members like roof purlin. In construction, cold-formed steel can provide several advantages compare with hot-rolled steel, for example, lightweight, ease of prefabrication, and time and cost efficiency. The thickness of cold-formed steel section is less, typically ranging from 0.9 to 3.2 mm (Lee, 2014).

The major structural advantage of cold-formed steel members is their low thickness, which facilitates their use in extremely lightweight construction. Cold-formed steel members can combine with higher-strength steel, which is devoid of failure modes (Macdonald, 2008). Thin-wall behavior of cold-formed steel section, especially premature buckling and instability affect its structural ability (Lee, 2014). Currently, cold-formed steel section is applied only on secondary structural members like roof purlin and is galvanized to avoid corrosion.

Hancock (2003) stated that cold-formed steel sections were flat sheets of steel that can be bent into certain shape of cold-formed steel structure at ambient temperature (Hancock, 2003). The manufacturing process was press-breaking, bending brake operation, and cold-rolling. The use of cold-formed steel members originated in Great Britain for building construction purpose in 1850. Recently, cold-formed steel members have become common in the design of roof truss.

Compare with other materials such as concrete and timber, the advantages of cold-formed steel were higher stiffness, durability, and strength. It is also lighter than hot-rolled steel. It can mitigate mass production and prefabrication process. In industrial field, installation of cold-formed steel section is fast and easy. Therefore, it saves cost in transportation and handling. Most importantly, cold-formed steel has higher weather resistance and requires less maintenance.

The properties of open sections like back-to-back I-shaped section have been studied by many researchers such as Stone et al. (2005) and Li et al. (2014). Compared with closed-section cold-formed steel, open-section one had less torsional rigidity (Young, 2009). This caused closed-section steel to likely to fail in twisting by other failure modes.

Stone's study (2005) connected two C-section members back-to-back with a single self-driving screw, which assessed the provision of North America Design Specification (Stone, 2005). The analysis data of this study showed that slenderness ratio was suitable for the calculation of strength and design standard was verified.

Li et al. (2014) compared the strength of built-up I section with the design strength of American Iron and Steel Institute (AISI) (Li, 2014). The experimental investigation was conducted to obtain strength capacity of built-up I section and box section. The failure mode of built-up section was observed and compared to that of single-channel section. It is concluded that numerical simulation and design method was verified as accurate, whereas ANSYS analysis overestimated the ultimate compression capacity of built-up cold-formed steel section.

Young's study (2007) conducted compression tests on cold-formed simple lipped channel and double-lipped channel with intermediate stiffener (Young, 2009). These channels were fabricated from high-strength cold-formed steel plate with thicknesses of 0.6 and 0.8 mm. A total of 28 columns were tested under axial compressive load to the failure. The theoretical ultimate strength was calculated by direct strength method. The modified strength formula for the direct strength method showed that the ultimate strength of column sections with the interaction between local buckling and distortional buckling was predicted.

Nguyen et al. (2012) presented compression tests of cold-formed plain and dimpled steel columns (Nguyen, 2012). They conducted compression tests of cold-formed plain and dimpled steel column. In addition, cold-formed plain and dimpled steel column samples were specified with cold-rolling and press-braking. The theoretical buckling load and ultimate compression load values were obtained from finite strip analysis and von Karman's theoretical formula. These samples conducted axial compression load until failure. The compression test results showed that the buckling and ultimate loads of the dimpled column specimens were higher than those of the plain column specimens. The theoretical buckling load from finite strip method showed that it was overestimated. The calculation of theoretical ultimate compression load from von Karman's formula showed that it differed insignificantly with the experimental value.

Some research studies have been already conducted to increase the application of cold-formed steel (Tan, 2011; Lee, 2013; Lee, 2014). Many research works are in progress to increase its safety and stability because of its increasing application over the years. Built-up cold-formed steel section was commonly used in low-rise buildings. This section was formed by two or more cold-formed steel sections. Design standard AISI (America), AS (Australia), NZS (New Zealand), and Eurocode (Europe) provide design guidelines and requirement for cold-formed steel section subject to axial compression force. However, because of the slenderness of cold-formed section and premature buckling behavior, the code of practice may have underestimated or overpredicted the capacity of the section. Therefore, these studies need to present the experimental investigation of the compression capacity of bolt and screw-connected back-to-back cold-formed steel stud column under axial load.

2 EXPERIMENTAL INVESTIGATIONS

The laboratory works of this study consist of compression test for built-up cold-formed steel section.

2.1 Specimen preparation

All cold-formed steel column sections were manufactured in the laboratory. In the study, two identical channel sections with depth of 250 mm and length of 0.75 m were built back-to-back to form double-channel I-shaped steel section by bolt or screw. A 150 mm deep and 0.6 m long channel section was connected to the aforementioned specimens. The nominal bolt spacing of the I-shaped section was 200 mm. The bolt near the ends of the steel section was 100 mm thick.

Screw and high-strength bolt of M12 Grade 8.8 served as fasteners used in the research. The bolt holes were fixed at 12 mm to prevent oversized hole-spaces between the steel members caused by sudden deformation (Tan et al., 2011). The specimens with bolted connection and screw connection were named as SB and SS, respectively, as shown in Table 1 and Figure 1.

2.2 Compression test

Compression test in a column experiencing axial compression load is conducted to determine the behavior and capacity of materials. The test sample was fixed between two plates of equipment and the loading experiment was conducted in the samples. DARTEC universal testing machine was used to conduct compression tests. The built-up section was fixed by top support and bottom support of the load

Table 1. Dimension of the specimens.

Section name	Section depth (mm)	Top flange width (mm)	Bottom flange width (mm)	Stiffener depth (mm)
SB150	150	63	67	16
SS150	150	63	67	16
SB250	250	76	80	20
SS250	250	76	80	20

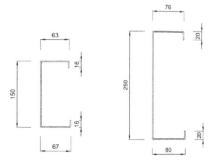

Figure 1.   Dimension of cold-formed steel.

Figure 2.   Set-up of compression test equipment.

frame of universal testing machine. Two steel end plates were used to transfer axial load and compress the built-up column section. The load was increased by hydraulic jack with a rate of 0.2 kN/s until the sample met failure mode. A computer recorded data from load cell by data logger, as shown in Figure 2.

3   RESULTS AND DISCUSSIONS

This section lists the results of this study and discussion. Table 2 shows the experimental results of each set of specimen. Figures 3 and 4 show the failure mode of bolt and screw specimens. Figure 4.3 shows the graph of load versus stroke of each specimen. In this section, every specimen gets its name. For example, SB150-1 stands for the specimen of depth 150 mm and bolt connection.

Table 2 shows the experimental results of each set of specimen. There are four sets of specimens: SB150, SS150, SB250, and SS250. SB stands for

Table 2.   Experimental results of each set of specimen.

Specimens	Maximum loading (kN)	Average (kN)	Deflection (mm)	Average (mm)
SB150-1	299	331.67	16	15.33
SB150-2	353		13	
SB150-3	343		17	
SS150-1	323	337.67	18	18.67
SS150-2	356		20	
SS150-3	334		18	
SB250-1	370	378.50	18	16.5
SB250-2	387		15	
SB250-3	NA		NA	
SS250-1	395	362.00	15	15.33
SS250-2	350		15	
SS250-3	341		16	

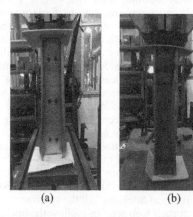

Figure 3.   Failure mode of 150 mm deep. (a) bolted section and (b) screw section.

Figure 4.   Failure mode of 250 mm deep. (a) bolted section and (b) screw section.

bolt connection and SS stands for screw connection. Testing for each set of specimen is conducted thrice to obtain the average compression strength value in order to increase the accuracy of the test result.

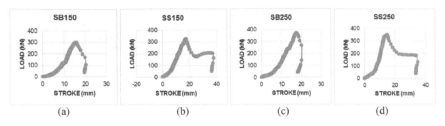

Figure 5.   Load vs stroke. (a) SB150, (b) SS150, (c) SB250, and (d) SS250.

The third specimen for SB250-3 is invalid because of the imperfection at the cutting of the edge of specimen SB250-2.

The average loadings of SB150, SS150, SB250, and SS250 are 331.67, 337.67, 378.50, and 362 kN, respectively. By comparing these results, it is shown that bolt and screw connection of specimens does not have significant difference in compression strength. The result shows that specimens SB250 and SS250 have higher compressive strength than that of SB150 and SS150. This is because increase of the section depth leads to the increase of compression capacity. The average deflections of SB150, SS150, SB250, and SS250 are 15.33, 18.67, 16.50, and 15.33 mm, respectively. By comparing these results, the deflection is not significantly different. It means that the deflection of cold-formed steel section is independent of the type of connection used.

Figures 3 and 4 show the failure modes of each specimen. Specimens SB150, SS150, SB250, and SS250 are local buckling. All the specimens tend to buckle at their web area and follow the buckling at the flange zone. In this paper, it is shown that the built-up cold-formed steel sections still retain their local buckling behavior.

Figure 5 shows the graph of load vs deflection of the tested specimens. All specimens show similar curves. It is evident from the figure that the curve is proportional until reaching maximum loading. This proportional line at elastic range is because the specimen undergoes elastic compression stress. After the maximum loading, plastic deformations of specimens occur at its yielding region.

4   CONCLUSIONS

On the basis of the above findings of this paper, several conclusions can be made:

a. The compression capacity of double C-section stub column connected back-to-back is in the range of 331.67–378.50 kN.
b. The increase of the section depth will increase the compression capacity.
c. The behavior of cold-formed steel section is independent of the type of connection used.

The experimental data show that all tested columns failed at local buckling. The deflection increased gradually before local buckling occurred.

It is suggested that more tests be conducted for further investigation. Various column depths and lengths can be used to increase the reliability of compression capacity for cold-formed steel column design.

REFERENCES

Hancock, G.J. Cold-formed steel structures, *Journal of Constructional Steel Research*, Volume 59, Issue 4, (2003).
Lee Y.H., Tan C.S. Lee Y.L. Tahir M.Md. Shahrin M. Shek P.N. *Numerical modelling of stiffness and strength behaviour of top-seat flange-cleat connection for cold-formed double channel section.* Applied Mechanics and Materials. 284–287, (2013).
Lee Y.H., Tan C.S. Tahir M.Md. Mohammad S. Shek P.N. Lee Y.L. *Influence of angle thickness towards stiffness and strength prediction for cold-formed steel top-seat flange cleat connection*, Applied Mechanics and Materials. 479–480, (2014).
Lee, Y.H., C.S. Tan, Shahrin Mohammad, Mahmood Md Tahir, P.N. Shek, *Review on Cold-Formed Steel Connections*, The Scientific World Journal, (2014).
Li, Y.Q., Y.L. Li, S.K. Wang, Z.Y. Shen, *Ultimate load-carrying capacity of cold-formed thin-walled columns with built-up box and I section under axial compression*, Thin-Walled Structures, Volume 79, (2014).
Macdonald, M., M.A. Heiyantuduwa, J. Rhodes, *Recent developments in the design of cold-formed steel members and structures, Thin-Walled Structure* 46, (2008).
Nguyen, V.B., C.J. Wang, D.J. Mynors, M.A. English, M.A. Castellucci, *Compression tests of cold-formed plain and dimpled steel columns*, Journal of Constructional Steel Research Volume 69, Issue 1, (2012).
Stone, T.A., R.A. Laboube, *Behavior of cold-formed steel built-up I-sections*, Thin-Walled Structures, 43 (12) (2005).
Tan, C.S., M.M. Tahir, P.N. Shek, Ahmad Kueh, *Experimental Investigation on Slip-in Connection for Cold-formed Steel Double Channel Sections*, Advanced Material Research Vols 250–253, (2011).
Young, B.K., S.K. Bong, G.J. Hancock, *Compression tests of high strength cold-formed steel channels with buckling interaction*, Journal of Constructional Steel Research Volume 65, Issue 2, (2009).

Advances in Energy Science and Equipment Engineering II – Zhou, Patty & Chen (Eds)
© 2017 Taylor & Francis Group, London, ISBN 978-1-138-71798-5

Analysis of multi window spraying merged flame height of buildings based on FDS

Hong Hai
School of Civil Engineering, Shenyang Jianzhu University, Shenyang, China

Jun-Wei Zhang
School of Civil Engineering, China Aerospace Construction Engineering Company Limited, Beijing, China

ABSTRACT: In the vertical fire prevention design of buildings, window spraying flame height of buildings is one of the key factors. The study of the merged height of multiwindow spraying flame and external wall thermal insulation material combustion flame is more significant during a multiwindow three-dimensional fire. FDS fire simulation software is used in this paper to build a seven-storey building model of foot size and conduct a simulation study on the merged flame of multiwindow in two cases of commercial and civil buildings, according to the distribution of temperature field to determine the change rule of flame height. The research results can provide strong support for the prevention and control measures of the longitudinal spread of building fire.

1 INTRODUCTION

With the increase of high-rise buildings, the threat of high-rise building fire is also increased. According to the characteristics of high-rise building fire, analysis of multiwindow merged flame height can provide theoretical basis for building fire protection design.

In northern China, the external wall thermal insulation layer is the main way of building insulation measures, and almost all heat insulating material would burn. When a room fire occurs, window spraying flame ignites the external wall thermal insulation material, thereby making the fire spread longitudinally. When the fire is raging, it will ignite the superstructure, and the merged effect of multiwindow spraying flame and external wall thermal insulation material combustion flame will result. Recently, scholars have conducted in-depth research on the mechanism of interaction and integration of flame systems (Naian, 2013; Koyu, 2011; Qiong, 2010; Wenguo, 2004). In the field of building fire, most research has focused on the single-window spraying flame and heat preservation materials. In line with this, Su Lang (Su, 2010) analyzed the influence of the balcony on window flame inhibition and carried out a comparative analysis of different balcony depths and balcony forms of window fire suppression to obtain the best balcony mode. Li Long (Li, 2013) analyzed the fireproof eave and windowsill wall as measures to control the vertical spread of fire window. Chen Aiping

(Ai Ping, 1998) analyzed the characteristics of the outer wall of the flame and achieved a strong heat transfer, especially the radiation heat transfer. The interactions between multiwindow flame, window flame, and heat preservation material burning flame should be studied deeply. Related results could not be found elsewhere.

FDS fire simulation software is used in this paper to conduct simulation research on the merged flame of multiwindow in two cases of commercial and civil buildings and assign the external contour and the height of the flame through the temperature field of slice to get the change rule of flame height. The main purpose is to provide theoretical support for building vertical fire spread prevention and control.

2 ESTABLISH THE MODEL

In this paper, a seven-storey residential building is selected as the research object. As worst conditions are selected, the external area of the building does not have any longitudinal fire protection measures. Room size is 3.7 m × 4.2 m, window size is 1.5 m × 1.8 m, and window sill height is 0.9 m. Natural ventilation is adopted, and outdoor temperature is set to 20°C. The walls and floors of the building are all reinforced concrete materials, and a polyurethane foam board insulation layer is furnished on the exterior wall of the building. According to the condition of FDS application software, the size of mesh is defined as 0.25 m × 0.25 m × 0.25 m, as shown in Figure 1

The development of fire is taken as unsteady t^2 fire. According to the Shanghai municipal engineering construction standard DGJ08-88-2008 and the civil antismoke technical rules (DGJ08-88-2000[S], 2000), the size of the fire source is set to 3MW and 6MW, which represent commercial buildings with sprinkler facilities and civil buildings without spraying facilities, respectively. As there will be flammable items in the bedroom, hotel, and office internal such as wooden tables, chairs, and other furniture, it is considered as the worst principle; therefore, in this paper, $\alpha = 0.0468$ KW/s^2 is assumed. The times of fire reaching maximum heat release rate are 80 and 358 s. Fire location is set at the center of the room.

The external profile of the flame can be assigned by the temperature field of the flame, and it is suggested that the flame shape be assigned by the isothermal surface temperature of 540°C (Yu, 2009), so as to determine the height of the flame. Therefore, each layer of the research object in the grid is arranged with a thermocouple in the center, which has the main role of creating a window of the activation settings, as shown in Figure 2. At the horizontal and vertical positions of the window center, the highest temperature and average temperatures

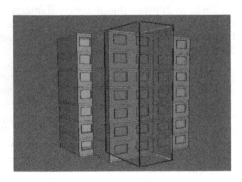

Figure 1. Mesh sketch map.

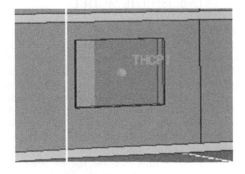

Figure 2. Layout of detecting equipment.

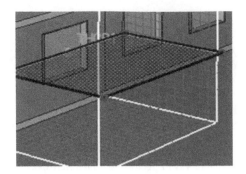

Figure 3. Layout of exploration area.

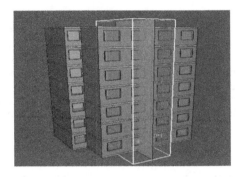

Figure 4. Layout of slice.

of the horizontal plane are set up, and gas-related parameters are recorded in the area, as shown in Figure 3. The slice files are arranged in the center of the grid X-axis along the Y-axis, and the temperature field and the velocity field of the merged flame are obtained mainly to decompose the data, as shown in Figure 4.

3 ANALYSIS OF SIMULATION RESULTS

3.1 Conditions setting

This paper sets two sizes of fire. Fire max HRR of 3MW represents commercial buildings with spraying facilities, and the fire max HRR of 6MW represents civil buildings without spraying facilities. At the beginning, we do not know how many windows to simulate until through comparative analysis the limits of multiwindow spouts merged flame height and other parameters are obtained, and the specific conditions are shown in Tables 1 and 2.

3.2 Analysis of temperature field

When the fire max HRR is 3MW in case 1 as an example. Figure 5 shows the variation of temperature of a layer of window center above the layer

Table 1. Fire max HRR of 3MW conditions.

Case number	Max HRR (MW)	Fire protection measures	Number of the window
Case 1	3	No	1
Case 2	3	No	2
Case 3	3	No	3
......	......	......	......

Table 2. Fire max HRR of 6MW conditions.

Case number	Max HRR (MW)	Fire protection measures	Number of the window
Case 1	6	No	1
Case 2	6	No	2
Case 3	6	No	3
......	......	......	......

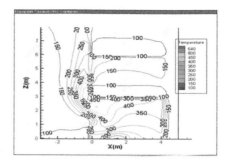

Figure 6. Temperature field distribution when one window is on fire.

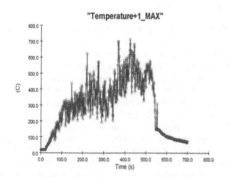

Figure 5. Maximum temperature tendency chart of F+1 when one window is on fire.

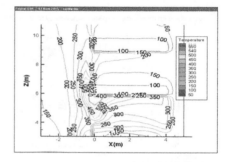

Figure 7. Temperature field distribution when two windows are on fire.

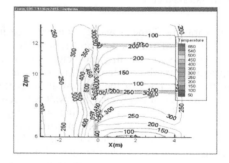

Figure 8. Temperature field distribution when three windows are on fire.

on fire, and we can obtain the highest temperature in 400–450 s. Because of space limitations, period of time chosen should not be enumerated one by one, and the highest temperature appears in 220–270 s and 200–250 s for conditions 2 and 3. Therefore, temperature field distribution corresponding to time section can be obtained as shown in Figures 6, 7, and 8.

Through analysis, we can know that because the flame shape is determined by the isothermal surface temperature of 540°C, when one window is on fire, there is no spraying flame, but when two and three windows are on fire, a 0.5 m high merged flame will be formed. Thus, when fire max HRR is 3MW, the highest merged flame is 0.5 m.

When fire max HRR is 6MW, the highest temperature appeared in the three conditions in 400–450 s, 330–380 s, and 300–350 s, respectively (Figs. 9, 10, and 11). When one widow is on fire, no effective flame will be formed along the wall and

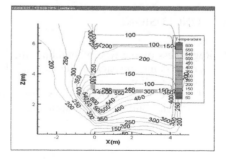

Figure 9. Temperature field distribution when one window is on fire.

601

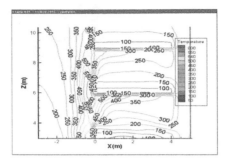

Figure 10. Temperature field distribution when two windows are on fire.

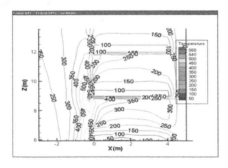

Figure 11. Temperature field distribution when three windows are on fire.

the spraying flame is mainly concentrated in the upper part of the window on fire. However, when two widows are on fire, the height of the flame spraying from the window is 1.7 m, which is spread to the upper layer of the window at the center. At this time, it can easily cross the wall between windows and directly set the building on fire layer-by-layer. When three widows are on fire, no effective flame will be formed along the wall. In this way, it could be concluded that the merged flame of two windows spraying at the same time reaches the height limit, that is, 1.7 m.

4 CONCLUSIONS

1. Regardless of the type of building, the highest merged flame appears in the case of the two windows on fire.
2. When fire max HRR is 3MW (representing commercial buildings with spraying facilities),

in the case of two and three windows on fire, a 0.5 m high merged flame will be formed. The overall level of temperature field increase is 50°C, and the temperature field in the case of three windows on fire has not been improved significantly.
3. When fire max HRR is 6MW (representing civil buildings without spraying facilities), in the case of two windows on fire, a 1.7 m high merged flame will be formed and the temperature field will expands significantly.

ACKNOWLEDGMENTS

This study was financially supported by Shenyang Science and Technology Planning Project (F14-186-1-00) and the Planning Project of State Administration of Safety Supervision (2012-105).

REFERENCES

Ai Ping, C. Research into the Problem of Extrusive Flame through Window in Exterior Wall [J]. Fire Safety Science, 7(3), 55–59 (1998).
Koyu, S., L. Naian, X. Xiaodong, et al. (2011). CFD study of huge oil depot fires-generation of fire merging and fire whirl in arrayed oil tanks [C]. Proceedings of the Tenth International Symposium of Fire Safety Science, 693–706 (2011).
Li, L. The control measure of vertical fire spread through high-rise building window plume [D]. Sheng yang, Shenyang Aerospace University, 31–6 (2013).
Naian, L., L. Qiong, et al. Multiple fire interactions: a further investigation by burning rate data of square fire arrays [J]. Proceedings of the Combustion Institute, 34 (2), 2555–2564 (2013).
Qiong, L. Dynamical Mechanism and Behaviors of Multiple Fires Burning [D]. Anhui: University Of Science And Technology Of China, 4–13 (2010).
Su, L. Study on the simulation of the vertical spreading of the balcony blocked fire [J]. Fire science and technology, 29 (7), 565–568 (2010).
Wenguo, W., D. Kamikawa, Y Fukuda, et al. Study on flame height of merged flame from multiple fire sources [J]. Combustion Science and Technology, 176 (12), 2015–2123 (2004).
Yu, J., xi, R. Yu xin, et al. CFD Simulation of the External Burning in the Compartment Fire [J]. Fire Safety Science, 15 (2), 93–95 (2006) DGJ08-88-2000[S], 11–12 (2000).

Advances in Energy Science and Equipment Engineering II – Zhou, Patty & Chen (Eds)
© 2017 Taylor & Francis Group, London, ISBN 978-1-138-71798-5

Bearing capacity and inversion analysis of long screw drilling uplift pile

Ke-wen Liu, Wei Ling & Jia-ren Shen
The 14th Metallurgical Construction Yunnan Invstigation and Design Co. Ltd., KunMing, China

An Chen
Faculty of Land Resource Engineering, KunMing University of Science and Technology, KunMing, China

ABSTRACT: The bearing capacity of long screw drilling uplift pile is studied using low maintained load test and inversion analysis of five piles. The results of test and inversion analysis show that the bearing capacity of long screw drilling uplift pile can reach 1800 kN, which is located in clay strata with a length of 33.0 m and diameter of 500 mm. At the same time, the maximum displacement of pile top is 27.25 mm, and the maximum springback is 5.84 mm. The load transfer function of uplift pile is $S = 0.111 f/(1 - 28.8 f)$. The ultimate shaft resistance is 21.24 kPa.

1 INTRODUCTION

Kunming Dianchi International Convention and Exhibition Center is a high-end office building and hotel. It is designed with 68 storeys, with a height of 330 m. It has a three-layered basement with a depth of 17 m. The elevation of the basement floor is 1879.3 m. The antifloating water level is 1891.0 m.

The surrounding strata of long screw drilling uplift pile from top to bottom are as shown in Table 1.

The uplift pile is constructed after building foundation pit excavating to the design elevation. Great depth and large-scale excavation of foundation pit can lead to foundation pit bottom unloading and rebounding. In fact, in 1977, Parry conducted an uplift pile model test (Luo, 2013). In 1979, Kulhawy performed an uplift test of bored uplift pile in sand (Luo, 2013; Li, 2012). Poulos and Davis (1980) believed that the bearing capacity of uplift pile is two-thirds of that of the compression pile (Luo, 2013; Huang, 2008). Chandler (1982) found that the failure surface between pile and soil is located at approximately 0.5 mm in soil to pile surface (Luo, 2013; Chen, 2008). In 1985, Chattopadhyay proposed the bearing capacity prediction theory. Anderson (1988) found that the different construction process and pile type affect the fracture plane location when the uplift pile is destroyed (Li, 2012; Wan, 2007). Turner (1990) found that the bearing capacity of uplift pile gradually reduced under cyclic loads. In 1993, Nicola and Randolph found that Poisson's ratio of soil affect the bearing capacity of uplift pile. In 2000, Du Guang-yin deduced the formula of pile side resistance. In 2002, Zhang Shang-gen

Table 1. The strata surrounding uplift pile.

Strata	Consistency	A T/m	C_{uu}/kPa
Clay	Low plastic to nonplasticity	3.74	9.3
Silty soil	Moderately dense to dense	1.37	49.0
Peaty soil	Low plastic to nonplasticity	1.04	7.1
Clay	Nonplasticity to low plastic	2.46	19.4
Silty soil	Dense to moderately dense	2.32	58.3
Peaty soil	Low plastic to nonplasticity	2.05	7.4
Silty sand	Dense	2.31	58.3
Clay	Nonplasticity to hard	2.66	21.4
Peaty soil	Low plastic to nonplasticity	1.05	9.1
Clay	Nonplasticity to hard	9.42	19.2
Silty sand	Dense	2.13	58.3
Clay	Hard	17.06	26.6

Note: C_{uu} is the unconsolidated and undrained shear cohesion of triaxial shear test and AT is the average thickness.

deduced the load displacement theoretical solution of uplift pile based on shearing displacement method. In 2005, Zhang Jie found that the uplift position affects the pile friction distribution. In 2007, Sun xiao-li analyzed the deformation uplift pile and piles using modified variational approach based on the minimum potential energy principle.

In 2007, Sun xiao-li deduced the axial force and deformation elastoplastic analytical expressions of uplift pile. In 2011, Zhang Qian-qing thought that the side resistance of uplift pile is 0.7 times of that of the compression pile. In 2012, Li Hai-xia studied the bearing capacity property of socketed rock uplift pile, and Li Lan-yong studied the load transfer mechanism of uplift pile.

The above studies focused on the properties of conventional uplift pile. It is necessary to study and conduct inversion analysis of the bearing capacity of long screw drilling uplift pile and to simulate the actual stress state of pile.

Five long screw drilling uplift piles are subjected to uplift static load test, with length and diameter of 33.0 m and 500 mm, respectively.

2 TEST PROGRAM

2.1 Test load

The slow maintained load test is conducted. Each stage test load is one-tenth of the maximum test load. The estimated uplift ultimate bearing capacity of single pile is 1800 kN, so the maximum test load is 1800 kN and the stage test load is 180 kN. However, the first-stage test load is two times of the test load, that is, 360 kN.

Each loading test stage can be regarded as stable when the settlement of pile top is not more than 0.1 mm of two successive readings within 1 h.

The interval of readings test data are: the first reading at 5 min after loading, then reading once in 15 min. After 60 min, readings were taken once in 30 min.

2.2 Test unload

Unload test is carried out to test the springback of the pile when test load reaches a predetermined value. After unloading, the amount of each discharge load is twice the amount of loading, that is, 360 kN. Each stage of loading is maintained at 60 min and are readings taken at 15, 30, and 60 min after unloading. After unloading to zero, maintain the test for 180 min and continue measuring and reading test data. The readings were taken at 5, 10, 15, and 30 min and then once in 30 min.

2.3 Termination load

One of the following occurs after load termination:

1. The displacement of pile top under certain load level is more than five times of the displacement of pile top under previous load level.
2. The accumulated displacement of pile top exceeds 100 mm.

3. The tensile strength achieved the yield strength standard value of reinforced system.
4. The displacement of pile top reached the maximum uplift of the designer's requirements.

3 TEST RESULTS

Five long screw drilling uplift piles are subjected to uplift static load test, and the results (Table 2) are as follows:

Test pile No. 1: The maximum displacement of pile top is 20.52 mm and the maximum springback is 2.73 mm. The rebound rate is 13.3%. The unrecoverable displacement of pile is 17.79 mm, which is 86.7% of the total amount of displacement.

Test pile No. 2: The maximum displacement of pile top is 25.86 mm and the maximum springback is 3.28 mm. The rebound rate is 12.7%. The unrecoverable displacement of pile is 22.58 mm, which is 87.3% of the total amount of displacement.

Test pile No. 3: The maximum displacement of pile top is 18.92 mm and the maximum springback is 3.20 mm. The rebound rate is 16.9%. The unrecoverable displacement of pile is 15.72 mm, which is 83.1% of the total amount of displacement.

Test pile No. 4: The maximum displacement of pile top is 27.25 mm and the maximum springback is 5.84 mm. The rebound rate is 21.4%. The unrecoverable displacement of pile is 21.41 mm, which is 78.6% of the total amount of displacement.

Test pile No. 5: The maximum displacement of pile top is 23.91 mm and the maximum springback is 3.10 mm. The rebound rate is 13.0%. The unrecoverable displacement of pile is 20.81 mm, which is 87.0% of the total amount of displacement.

The Q-s curves of long screw drilling uplift piles are parabolic overall (Figs.1–5), which is similar to compression friction pile. Table 2 and Figures 1–5 show that the Q-s curves are approximately linear under the previous three load grades, which mean that the deformation of uplift pile is of elastic type. From the fourth load grade, the curvatures of Q-s curves are larger significantly and their increasing rates also increase, which means that the deformation of uplift pile is mainly of plastic type. The bearing capacity of long screw drilling uplift pile can reach 1800 kN.

The five pile Q-s curves of unloading resilience are very close to linear, which indicates that the deformations of unloading stage are elastic deformation of concrete.

Table 2. Results of uplift pile static load test.

No.	Load Q/kN	Test duration/ min	Pile top displacements/mm									
			Test pile No. 1		Test pile No. 2		Test pile No. 3		Test pile No. 4		Test pile No. 5	
			Single	Overall	Single	Overall	Single	Overall	Single	Overall	Single	Overall
0	0	0	0.00	0.00	0.00	0.00	0.00	0.00	0.00	0.00	0.00	0.00
1	360	120	0.72	0.72	0.61	0.61	0.47	0.47	0.97	0.97	0.48	0.48
2	540	240	0.82	1.54	0.99	1.60	0.55	1.02	1.46	2.43	0.92	1.40
3	720	360	1.06	2.60	1.72	3.32	0.94	1.96	2.19	4.62	1.47	2.87
4	900	480	1.60	4.20	2.45	5.77	1.54	3.50	2.98	7.60	2.17	5.04
5	1080	600	1.95	6.15	3.00	8.77	1.99	5.49	3.32	10.92	2.70	7.74
6	1260	720	2.58	8.73	3.59	12.36	2.46	7.95	3.68	14.60	3.32	11.06
7	1440	840	3.21	11.94	4.19	16.55	3.03	10.98	3.94	18.54	3.76	14.82
8	1620	960	3.68	15.62	4.50	21.05	3.69	14.67	4.23	22.77	4.30	19.12
9	1800	1080	4.90	20.52	4.81	25.86	4.25	18.92	4.48	27.25	4.79	23.91
10	1440	1140	−0.38	20.14	−0.48	25.38	−0.28	18.64	−0.80	26.45	−0.45	23.46
11	1080	1200	−0.47	19.67	−0.56	24.82	−0.37	18.27	−0.88	25.57	−0.50	22.96
12	720	1260	−0.58	19.09	−0.63	24.19	−0.46	17.81	−1.12	24.45	−0.59	22.37
13	360	1320	−0.64	18.45	−0.76	23.43	−1.51	16.30	−1.40	23.05	−0.74	21.63
14	0	1560	−0.66	17.79	−0.85	22.58	−0.58	15.72	−1.64	21.41	−0.82	20.81

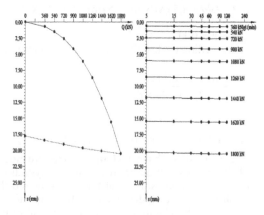

Figure 1. Test pile No. 1 load–displacement curve.

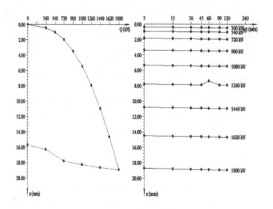

Figure 3. Test pile No. 3 load–displacement curve.

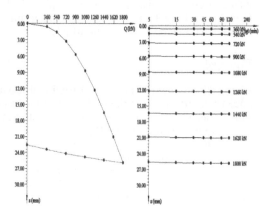

Figure 2. Test pile No. 2 load–displacement curve.

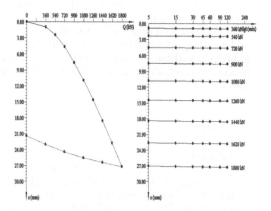

Figure 4. Test pile No. 4 load–displacement curve.

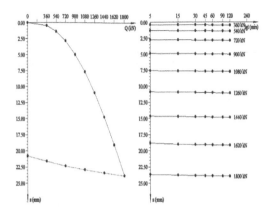

Figure 5. Test pile No. 5 load–displacement curve.

4 INVERSION ANALYSIS

4.1 Load transfer function

The loading deformation of uplift pile is of nonlinear type, which includes elastic and inelastic deformation stages. The elastic deformation stage is very short generally, which has no relative sliding stage between the pile and the soil. When the deformation of pile top is more than soil yield displacement, it is of inelastic deformation type. At this time, there appeared plastic region in soil around the pile.

The bearing capacity of uplift pile during elastic deformation stage (Luo, 2013; Li, 2012; Ding, 2008; Chen, 2008) is given by:

$$p_u = \pi dL\tau \tag{1}$$

According to the results of uplift pile test, the side friction of uplift pile during elastic deformation stage is $\tau = 720/(3.14 \times 0.5 \times 33) = 13.9$ (kPa).

The average displacement of uplift pile top is 3.07 mm at the third-grade load. The shear modulus between uplift pile and soil around pile is: G = $\tau/s = 13.9$ kPa/3.07 mm = 4.5 (MN/m).

According to Gardner theory, in the case of uplift pile without destruction, the load transfer function is (Luo, 2013; Li, 2012; Ding, 2008; Chen, 2008):

$$S = \frac{af}{1 - bf} \tag{2}$$

where a and b are hyperbolic parameters, f is the side friction of uplift pile, S is the relative displacement between uplift pile and soil around uplift pile.

Taking the limit of different variations of equation (2) (Luo, 2013; Li, 2012; Ding, 2008; Chen, 2008), we have:

$$\lim_{s \to \infty} f = \lim_{s \to \infty} \frac{S}{a + bS} = \frac{1}{b} = f_{max} \tag{3}$$

$$\lim_{s \to 0} \frac{f}{S} = \lim_{s \to \infty} \frac{1}{a + bS} = \frac{1}{a} = K_{st} \tag{4}$$

where f_{max} is the ultimate frictional resistance of the uplift pile, N/mm².

For this test, $f_{max} = Q_{max}/\pi dL = 1800000/(3.14 \times 500 \times 33000) = 0.035$(N /mm²), so b = 28.8.

K_{st} is the initial shear stiffness between uplift pile and soil around uplift pile. According to Randolph (Li, 2012; HE, 2014):

$$K_{st} = \frac{G}{d \ln(\frac{d_m}{d})} \tag{5}$$

When the deformation between uplift pile and soil around uplift pile is of elastic type, $d_m \approx d$, then:

$$Kst = G/d = 4.5 \text{ (MN/m)}/0.5 \text{ (m)} = 9.0 \text{ (MN)}$$

So, a = 0.111.
According to this, the load transfer function is:

$$S = \frac{af}{1 - bf} = \frac{0.111f}{1 - 28.8f} \tag{6}$$

4.2 Ultimate lateral frictional resistance

According to Kulhawy F.H, the main failure modes of same cross section uplift pile are the cylindrical destruction along surface of the pile body and the small range of cylindrical soil destruction outside the pile body. The ultimate lateral frictional resistance of uplift pile is different from the ultimate lateral frictional resistance of bearing pile. Xu He (1994) and Ling Hui (2004) solved the ultimate lateral frictional resistance of uplift pile:

$$\tau^u_{max} = \alpha C_u \tag{7}$$

where τ^u_{max} is the ultimate lateral frictional resistance of uplift pile, α is the sticking coefficient of uplift pile, and C_u is the undrained shear strength of saturated clay around uplift pile.

α is an empirical coefficient, which changes with the changes in basement categories, soil quality, and construction technology. It is related to the length-to-diameter ratio of pile and the plasticity index of soil. For drilling uplift pile:

$$\alpha = 1.0 - 0.0075C_u \quad C_u \leq 80 \text{ kPa.}$$

Table 3. Triaxial undrained shear strength of soil around uplift pile.

Strata	Clay	Silty soil	Peaty soil	Clay	Silty soil	Peaty soil	Silty sand	Clay	Peaty soil	Clay	Silty sand	Clay
Thickness/m	3.74	1.37	1.04	2.46	2.32	2.05	2.13	2.66	1.05	9.42	2.13	2.7
C_{uu}/kPa	9.3	49.0	7.1	19.4	58.3	7.4	58.3	21.4	9.1	19.2	58.3	26.6

According to the statistical results in Table 3, the weighted average of undrained triaxial shear strength is 26.5 Kpa. Then:

$$\alpha = 1.0 - 0.0075\,Cu = 1.0 - 0.0075 \times 26.5 = 0.8$$

$$\tau^u_{max} = \alpha C_u = 0.8 \times 26.5 = 21.24 \ (kPa)$$

5 CONCLUSIONS

1. In clay strata, the bearing capacity of long screw drilling uplift pile is 1800 kN, whose length and diameter are 33.0 m and 500 mm, respectively.
2. The maximum displacement of five uplift piles is 27.25 mm and the maximum springback is 5.84 mm. The rebound rate is 21.4%.
3. The load transfer function of long screw drilling uplift pile is $S = 0.111 f/(1 - 28.8 f)$.
4. The ultimate shaft resistance of long screw drilling uplift pile is 21.24 kPa.

REFERENCES

Ding Ming. Study on the Characteristic of the Uplift Capacity and Deformation of Tension Piles [D]. Huangzhou, China:Zhejiang University, 2008. (in Chinese).

He Hong-nan, Dai Guo-liang, Gong Wei-ming. A Review of Computation of Ultimate Bearing-capacity of Uplift Piles with Uniform Cross-section [J]. Journal of Highway and Transportation research and Development, 2014, 31(6):63–68. (in Chinese).

He Si-ming, Wu Yong, Li Xin-po. Research on mechanism of uplift rock-socketed piles [J]. Rock and Soil Mechanics, 2009, 30(2):333–337. (in Chinese).

Huang Mao-song, Li Jian-jun, Wang Wei-dong, Chen Zheng. Loss ratio of bearing capacity of uplift piles under deep excavation [J]. Chinese Journal of Geotechnical Engineering, 2008, 30(9):1291–1297. (in Chinese).

Li Hai-xia. Research of uplift bearing capacity behavior of rock socketed plies [D]. Hefei, China:Hefei University of Technology, 2012. (in Chinese).

Li Lanyong. Study on load transfer mechanism of pile-sail for uplift pile [D]. Guangzhou, China:South China University of Technology, 2012. (in Chinese).

Liang Fayun, Song Zhu, GhenHaibing. Analysis for Load Transfer Behaviors of Uplift Single Pile with an Integral Equation Method and Parameter Studies [J]. Journal of Tongji University (Natural Science), 2013, 41(7):977–983. (in Chinese).

Ling Hui. Pile bearing mechanism of soft soil uplift in Shanghai [D]. Shanghai, China:Tongji University, 2004. (in Chinese).

Luo Ning. Analysis of Uplift Pile Bearing Capacity and Deformation Behavior [D]. Guangzhou, China:South China University of Technology, 2013. (in Chinese).

Shang-Rong Chen. Study of the characteristics of capacity and deformation properties pile bearing of uplift [D]. Shanghai, China:Tongji University, 2008.

Wan Dongli, Luo Yuping, Wang Baoliang. Experimental Analysis of Force on Uplift Piles in Rock Mass [J]. Journal of Shijiazhuang Railway Institute, 2007, 20(2): 65–68. (in Chinese).

Xu He; Liu Yunyun. Bearing capacity forecast of uplift piles [J]. Journal of Tongji University (Natural Science), 1994, 22(3):385–389. (in Chinese).

Zhu Bitang, YangMin. Calculation of displacement and ultimate uplift capacity of tension piles [J]. Journal of Building Structures, 2006, 27(3):120–129. (in Chinese).

Advances in Energy Science and Equipment Engineering II – Zhou, Patty & Chen (Eds)
© 2017 Taylor & Francis Group, London, ISBN 978-1-138-71798-5

A simplified method of obtaining the seismic damage index of single steel-concrete composite building

Shuang Wang & Yangbing Liu
School of Civil Engineering, Nanyang Institute of Technology, Nanyang, China

ABSTRACT: Owing to the lack of the seismic damage data of the steel-concrete composite structure, it is difficult to use the avaible empirical methods to get the seismic damage index of the composite structures to predict the seismic damage. According to the definition of mean seismic damage index and the seismic fragility method of composite structures, a simplified method of calculating seismic damage index of single composite building was proposed. In this method the uncertainty due to variability in ground motion characteristics and structures was considered. Based on this method, seismic fragility analysis was carried out for two composite frames, then the vulnerability matrix was obtained, and finally the seismic damage index was calculated to evaluate and compare the damage sate. The simplified method may be a very useful tool to estimate the seismic safety and predict the seismic loss.

1 INTRODUCTION

Earthquake disaster has a great influence on the safety of our country and even the world city. Chile earthquake (2010), Wenchuan earthquake (2008), and Northridge earthquake (1994) etc. all have caused huge losses. From the previous earthquake damage, economic losses and casualties due to earthquake were always correlated with the damage condition of buildings. Therefore, a reasonable method should be proposed to predict the aseismic behavior and earthquake damage.

The steel-concrete composite structure consists of the steel structure and the concrete structure with different material property. And its aseismic behavior is more complicated than that of the steel structure or the concrete structure. Although the composite structure has been widely used in midrise and high rise buildings, the study on the aseismic behavior of the whole composite structures wasn't much (Tao, 2013; Nie, 2011; Liu, 2010) and research about the damage predication of the composite structures was not available (Liu, 2010). With the application of the composite structures in the high-intensity earthquake region, how to make them safe has been an urgent problem.

Up to now, many methods have been proposed to evaluate the seismic damage of buildings. The seismic damage index proposed by Chinese scientists is used to quantitatize the extent of damage. The index and its evaluation methods are adopted in Chinese Seismic Intensity Sale and have been used in the multiple destructive seismic damage investigation. Many researchers have studied the seismic damage index of reinforced concrete structures and masonry structure. Bo Jingshan et al. (Bo, 2012) summarized the problems about investigation of seismic damage index and pointed out the considerable problems in the evaluation methods; Zhou Guangquan (Zhou, 2011) studied the seismic damage Index of simple-buildings in Yunnan; Xia Shan et al. (Xia, 2009) proposed the seismic intensity scale of the mean damage index for RC frame structure and masonry structure with different earthquake design level according to the relationship between seismic damage matrix and mean damage index; Based on the available building damage data, Xu Jinghai et al. (Xu, 2002) put forward a fuzzy seismic damage index prediction method for building seismic damage.

Although fragility curves could give the probability of different structural damage state caused by different level of ground motion, people would prefer to get the damage state under different level earthquake intensity in the field of seismic damage predication. Therefore, a simplified method of obtaining the seismic damage index of the single steel-concrete composite building was proposed based on the available fragility analysis method proposed in literature (Liu, 2010).

2 SEISMIC DAMAGE INDEX

The seismic damage index is a dimensionless quantity to evaluate the extent damage of the members or the whole building. And it also provides the important theoretical basis for handling decisions

Table 1. Defination of seismic damage index.

Damage level	Basically good	Slight damage	Middle damage	Severe damage	Collapse
Middle value	0	0.2	0.4	0.7	1.0
Range	[0, 0.1]	[0.1, 0.3]	[0.3, 0.55]	[0.55, 0.85]	[0.85, 1.0]

of damage buildings after earthquake. The damage states of buildings are divided into five levels: basically good, slight damage, middle damage, severe damage and collapse. Seismic damage index is the quantification for the five damage levels: 0 denotes basically good of buildings, 1 denotes collapse of buildings, and the value between 0 and 1 denotes the others damage levels. The middle values and the range of seismic damage index were shown in Table 1 (Sun, 2008).

3 METHOD OF OBTAINING THE SEISMIC DAMAGE INDEX

There were many methods of calculating the seismic damage index of the single building and many calculation models for seismic damage index were proposed. But the models hadn't a unified standard, and the index range for different damage level of different models was also different. Furthermore many models were only applied to sections and members, and the seismic damage index of the structure was difficult to be inferred from the index of the member. Besides, the deterministic analysis method was used to calculate the seismic damage index.

Due to the strong random of the earthquake ground motion, it is difficult to evaluate the seismic damage by the deterministic method. And the uncertainty of the structure and the ground motion needs to be considered during the calculation of the seismic damage index. If sufficient date of seismic damage were available for the structures, the method (Sun, 2008) had been proposed to calculate the seismic damage index. As a new type of structures, the steel-concrete composite structure almost didn't experience earthquakes, seismic damage date was insufficient, and the vulnerability matrix wasn't available. So based on the uniform standard of seismic damage index defined in Table 1 and referred to the definition of mean seismic damage index, the method of obtaining the seismic damage index was proposed as followed.

1. Giving the structural information including the type of structures, site classification, ground motion etc.
2. Calculating the vulnerability matrix of the structure. According to the structural information

of (1) and considering the uncertainty of the structure and ground motion, seismic performance levels based on members were determined, probabilistic seismic demand analysis was performed, fragility curves was formed (Liu, 2010), and then the vulnerability matrix of the composite structure was gained.
3. Based on the definition of mean seismic damage index, use the following equation (1) to calculate the seismic damage index of the single structure.

$$d_j = \sum_P D_P \cdot P(D_P / J) \qquad (1)$$

Where $D_P \in [0, 1]$ denotes the middle value of the seismic damage index shown in Table 1. $P(D_P/J)$ denotes the probability of structural failure of level P (divided into five level: basically good, slight damage, middle damage, severe damage, and collapse),when the PGA is equal to J. It can be gained from the vulnerability matrix in the step (2).
4. According to the relationship between PGA and the fortification intensity I shown in equation (2).

$$PGA = 10^{(I \lg 2 - 0.1047575)} \qquad (2)$$

Then the seismic damage index underdifferent fortification intensity was gained.

After using the above method to calculate the seismic damage index, the level of structural failure was evaluated according to the range of the seismic damage index in Table 1.

4 EXAMPLE ANALYSIS

The proposed method in this paper was used to get the seismic damage index of the two 15-storey composite frames. And one (CB-CFST) consisted of composite beams and concrete filled square steel tube columns and the other (SB-CFST) consisted of steel beam and concrete filled square steel tube columns. The composite action between the steel beams and the floor slabs was considered in CB-CFST frame, and the composite action wasn't considered in SB-CFST frame.

4.1 Calculation model

The bottom storey height of the two composite frames was 4.5 m, and the others were all 3.6 m, and the total height was 54.9 m. The plane and vertical section were shown in Figure 1 and Figure 2. The steel beams (Q235 steel) were all welded I-beams; the transverse section was $750 \times 300 \times 13 \times 24$, and the longitudinal sections were $700 \times 300 \times 13 \times 24$.

The columns are concrete-filled square steel tube columns, and the side length was 600 mm. The material of columns was C40 concrete and Q345 steel. The floor and roof slabs were all 140 mm-thick C30 concrete slab. The composite beam was designed by full shear connection. Floor dead load and roof dead load were both 4.5 kN/m², while live load was 2.0 kN/m².

4.2 Fragility curves and vulnerability matrix

The method of fragility analysis proposed by the author in literature(Liu, 2010) was applied to get the fragility curves of the two frames. The fragility curves based on the quantitative index of the story-drift-angle for the two frames were shown in Figure 3. The common used structural performance levels (Federal Emergency Management Agency (FEMA), 2000)were Normal Occupancy (NO), Immediate Occupancy (IO), Life Safety (LF) and Collapse Prevention (CP). And the whole region was divided into five gradation of earthquake damage by the four structural performance levels in Figure 3.

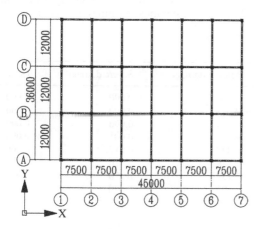

Figure 1. Plane (mm).

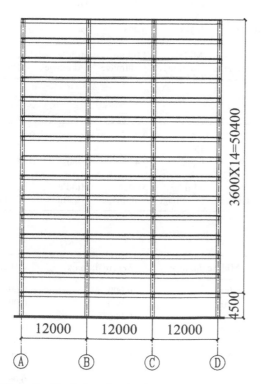

Figure 2. Vertical section (mm).

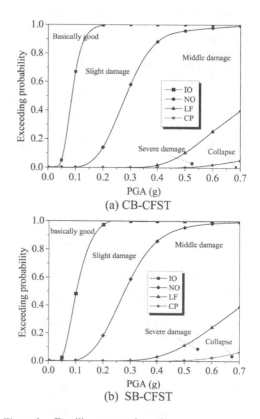

(a) CB-CFST

(b) SB-CFST

Figure 3. Fragility curves of two frame.

From the results of fragility analysis, the vulnerability matrix for the two frames was calculated as shown in Table 2 and Table 3.

4.3 *Calculating the seismic damage index*

Take the according values in Table 2 and 3 into Equation (1), the values of seismic damage index corresponding to PGA were calculated. And then according to Equation (2), the seismic damage index corresponding to the different fortification intensity was gained. The seismic damage index for the two frames was shown in Table 4 and Table 5. Finally compare the calculated values with the range of seismic damage index in Table 1, the gradation of earthquake damage corresponding to the PGA and the fortification intensity was also shown in Table 4 and Table 5.

From the above two tables, the middle damage was found in the two frames under rare earthquakes on the zone of fortification intensity 8 in China. And the two composite frames both had excellent

Table 2. Vulnerability matrix of CB-CFST frame.

PGA (g)	Basically good	Slight damage	Middle damage	Severe damage	Collapse
0.05	0.950	0.050	0.000	0.000	0.000
0.1	0.329	0.669	0.002	0.000	0.000
0.2	0.004	0.813	0.183	0.000	0.000
0.3	0.000	0.349	0.650	0.001	0.000
0.4	0.000	0.085	0.897	0.017	0.000
0.5	0.000	0.037	0.848	0.110	0.004
0.6	0.000	0.014	0.739	0.226	0.020
0.7	0.000	0.005	0.605	0.340	0.051

Table 3. Vulnerability matrix of SB-CFST frame.

PGA (g)	0.05	0.1	0.2	0.3	0.4	0.5	0.6	0.7
Seismic intensity	VI	VII	VIII		IX			
Seismic damage index	0.010	0.135	0.236	0.331	0.388	0.428	0.477	0.531
Damage gradation	Basically good	Slight damage	Slight damage	Middle damage	Middle damage	Middle damage	Middle damage	Middle damage

Table 4. Vulnerability matrix of CB-CFST frame.

PGA (g)	0.05	0.1	0.2	0.3	0.4	0.5	0.6	0.7
Seismic intensity	VI	VII	VIII		IX			
Seismic damage index	0.005	0.097	0.230	0.339	0.392	0.432	0.480	0.537
Damage gradation	Basically good	Slight damage	Slight damage	Middle damage	Middle damage	Middle damage	Middle damage	Middle damage

Table 5. Vulnerability matrix of SB-CFST frame.

PGA (g)	Basically good	Slight damage	Middle damage	Severe damage	Collapse
0.05	0.976	0.024	0.000	0.000	0.000
0.1	0.519	0.477	0.005	0.000	0.000
0.2	0.110	0.631	0.259	0.000	0.000
0.3	0.001	0.310	0.686	0.003	0.000
0.4	0.000	0.089	0.878	0.032	0.001
0.5	0.000	0.025	0.859	0.111	0.005
0.6	0.000	0.008	0.742	0.227	0.023
0.7	0.000	0.003	0.599	0.335	0.062

aseismic behavior. Compared Table 4 with Table 5, the damage index values of CB-CFST frame was greater than those of SB-CFST frame in 8 or less than 8 degree seismic fortification intensity and the damage of CB-CFST frame was severer than that of SB-CFST frame. But when PGA was greater or equal to 0.3 g, the damage index values of SB-CFST frame was greater than those of CB-CFST frame. Therefore, the composite action of floor slabs and steel beams should be considered, otherwise the real damage state of the structure couldn't be really reflected, which would make the design unsafe.

5 CONCLUSION

A simplified method of obtaining the seismic damage index of the single steel-concrete composite building was proposed. And the seismic damage index of two frames was calculated to predict the gradation of damage under different earthquake intensity. This method applies to not only composite structures but also other structures such as RC structures, steel structures etc. And the method is simple and practical; it can be used assess the gradation of earthquake damage under different fortification intensity; it can also be used to not only the urban earthquake prevention and disaster mitigation planning but also disaster losses assessment.

ACKNOWLEDGEMENTS

This work was financially supported by Key scientific research projects of Henan Province (No. 16B560005).

REFERENCES

Bo, J.S., J.Y. Zhang, P.S. Sun, W. Li, P. Li, Discussion on seismic damage index and relevant problems. Journal of Natural Disasters, 21, 6 (2012): pp. 37–42.

Federal Emergency Management Agency (FEMA), *FEMA 356 Commentary on the guidelines for the seismic rehabilitation of buildings*. Prepared by American Society Of Civil Engineers, Washington, D.C. (2000).

Liu, J.B., Y.B. Liu, B. Guo, Seismic behavior analysis of steel-concrete composite frame structure system. Journal of Beijing University of Technology, 36, 7 (2010): pp. 934–941.

Liu, J.B., Y.B. Liu, H. Liu, Seismic fragility analysis of the composite frame structure based on performance. Earthquake Science, 23, 1 (2010): pp. 45–52.

Nie, J.G., Y. Huang, J.S. Fan, Experimental study on load-bearing behavior of rectangular CFST frame considering composite action of floor slab. Journal of Building Structures, 32, 3 (2011): pp. 99–108.

Sun, B.T., D.Z. Sun, New method of seismic damage prediction of single building. Journal of Beijing University of Technology, 34, 7 (2008): pp. 701–707.

Tao, M.X., J.G. Nie, Fiber model of composite frame systems considering slab spatial composite effect. Journal of Building Structures, 34, 11 (2013): pp. 1–9.

Xia, S., A.W. Liu, Assessment of seismic intensity with mean damage index in an earthquake- resistant region. Acta Seismologica Sinica, 31, 1 (2009): pp. 92–99.

Xu, J.H., W.Q. Liu, M.X. Deng, Fuzzy seismic damage index prediction method for building seismic damage. Earthquake Engineering and Engineering Vibration, 22, 6 (2002): pp. 84–88.

Zhou, G.Q. Seismic-damage index of simple-buildings in Yunnan. Journal of Seismological Research, 34, 1 (2011): pp. 88–95.

Advances in Energy Science and Equipment Engineering II – Zhou, Patty & Chen (Eds)
© 2017 Taylor & Francis Group, London, ISBN 978-1-138-71798-5

Using building information modeling for effective planning of construction workspaces

Jongsoo Choi & Seong mok Paik

Architectural Engineering Department, Dongguk University, Seoul, South Korea

ABSTRACT: Proactive construction workspace planning is a challenging task because of project-based specific characteristics and requirements. It is a combination of physical space occupied by building components, site layout, and human labor workspace. Accordingly, developing a proactive site plan for workspace has critical implications from the cost and scheduling perspective. However, many efforts have been made by adopting Building Information Modeling (BIM) to improve the efficiency of project management. Adopting BIM techniques into capital projects can provide diverse benefits, especially for workspace planning. With BIM, design conflicts can be resolved in the early stage of life cycle, such as clash detection, workspace congestion, and work task sequence. This study presents the outcomes obtained from the preliminary study of effective planning of construction workspace. In particular, this research is focused on the identification and categorization of design conflicts and definition of the issues and tasks occurred in design phase. During the 6-month analysis period of eight times coordinating, a substantial set of observations was recorded. A total of 336 validation issues were found for each design segment, and 868 coordination issues occurred with architectural among the other segments: structural, electrical, hydraulic, mechanical, and fire. Also, 458 coordination issues observed or identified in service area, while a 574 issues had been resolved through the coordination during the research period. This study addresses the significance of early involvement to avoid or mitigate clash detection, interference, and deficiency, which will impact the workspace planning.

1 INTRODUCTION

The construction workspace is a combination of physical space occupied by building components, site layout such as material storage, construction site, temporary spaces, site offices, and human labor workspaces. However, complex and dynamic working environment frequently leads to ineffective construction workspace planning. Design errors and conflicts are the most important risks in the design process, which will cause serious problem through the life cycle. Preventing design conflicts and space issues in the early stage of projects can save cost and time by eliminating unnecessary and awkward reworks significantly. Visualizing the design and the construction workflow and sequence through a BIM makes every project participant to communicate and coordinate to detect design risks such as clash detection, work task sequence, and work space congestion. However, to date, BIM has not been effectively utilized to support adequate management of construction workspace planning. This research builds on a case study of office building project using BIM to analyze the validation issues and coordination issues associated with design conflict and clash detections. This study

addresses the significance of early involvement to avoid or mitigate clash detection, interference, and deficiency, which will impact the workspace planning.

2 CONSTRUCTION WORKSPACE

Planning construction workspace can have an influence on the cost, time, quality, safety, and other management fields. Wu et al. (2010) worked on 4D workspace conflict detection and proposed an analysis system, and Moon et al. (2009) proposed an integrated method to allocate workspace using an automatic generation method based on resource requirements. Kuan-Chen et al. (2009) proposed an algorithm which identifies conflicts on static or dynamic construction sites and determines the distance between large dynamic 3D objects in virtual construction sites using different scenarios. Dawood and Mallasi (2006) developed an algorithm to assist project managers in the assignment and detection of workspace conflicts, and Rajiv et al. (2011) developed the construction workspace management as application of nD planning and using tools. The aforementioned studies showed the importance of planning and managing the construction workspace proactively, while few studies have been conducted to automate the process of updating design information in BIM models for site workspace planning. Therefore, consideration and application of various parameters is needed for the follow-up research.

2.1 Designing and planning the workspace

2.1.1 Risks in the design process

The risks in the design process are clash detection, workspace congestion, and work task sequence. BIM coordination is an activity that aims to check the interferences in order to proactively detect and resolve clashes and potential errors (Paik, 2009). The key research for effective construction workspace planning is coordination of consultant information, early clash avoidance, and detection of noncompliance, interference, and deficiency.

2.1.2 Conflict detection and analysis

According to previous studies, there are four major types of conflicts: design conflict, safety hazard conflict, damage conflict, and congestion conflict. Most design conflicts are caused by miscommunication between different designers through building components. Safety hazard conflict is a hazard space generated by an activity conflict with a labor crew space required by another activity. Damage conflict is a labor space, equipment space, or hazard space required by an activity conflict with the protected space required by another activity. Finally, congestion conflict occurs when a labor crew and a piece of equipment or material required by an activity needs the same space at the same time, causing a lack of space or space overlap.[13] This study analyzes the major conflicts in design phase with BIM objects to prevent conflicts during the project life cycle.

3 FINDINGS AND DISCUSSION

BIM coordination mainly focuses on the coordination between different disciplines in terms of design errors and design inconsistencies. With design coordination before the real construction, major clashes and design discrepancies can be discussed and solved. Consequently, early decision making and rehearsal virtually reduce rework and changes during the construction, which have significant impact on construction workspace planning.

3.1 Data description

Data for the identification of validation and coordination issues are collected from a 22-storey building site of 34,450 m². The project includes a commercial tower and hotel tower in Perth CBD, Australia, and will be completed in the early 2018. During the eight coordination meetings, a total of 1,662 items were found and 574 issues had been resolved. Participants in the BIM project, such as architectural, structural, mechanical, electrical, hydraulic and fire designers, and consultants are responsible for performing quality control checks of their designs, data sets, and model properties. After modeling with Autodesk Revit, all files were checked by Solibri Model Checker (SMC) and Navisworks. Figure 1 shows an example of issues checking by SMC.

3.2 Data analysis

During the research period, 336 validation issues were found. These are the number of validation issues of each design segment: architectural, structural, mechanical, electrical, hydraulic, and fire. Above all, 139 structural issues were the highest contributor (41%) of all validation issues, followed by 94 architectural designs (28%). The numbers of each item issues are shown in Figure 2.

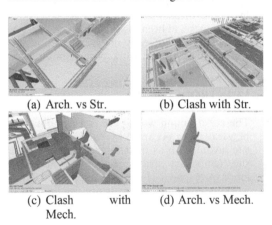

(a) Arch. vs Str.   (b) Clash with Str.

(c) Clash   with   (d) Arch. vs Mech.
Mech.

Figure 1.   Model coordination (SMC).

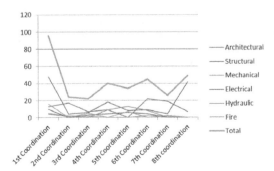

Figure 2.   Number of validation issues.

The coordination issues are between architectural and other disciplines. Figure 3 shows the number of coordination issues and percentages during the period.

There were 868 coordination issues: architectural versus each segment. Issues with hydraulic were dominant, representing 27%, followed by fire, contributing to 25% of all coordination issues. The result shows that services such as hydraulic, fire, and electrics are the significant factors to control with architectural.

Figure 4 shows the number of coordination issues related to services. Issues are categorized to eight sets of issues covering all designs, issues with item itself independently, and cross-item issues.

A total of 458 issues were observed and hydraulic versus mechanical issues was the dominant category with 122 observations (27%). The case data show that the mechanical issue is the most important category to be considered in coordination.

Figure 5 shows the issue resolution progress showing the number of days taken to solve the issues recorded. The duration of start and finish date has been divided to seven categories, starting from less than 5 days and ending with more than 30 days. A total of 574 issues had been solved through the coordination during the analysis period. The workflow for coordination is 1 week. For that reason, 39% of 235 issues had been solved in 6–10 days, 47 issues took less than 5 days, and

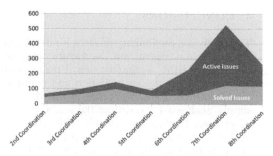

Figure 5. Resolution progress.

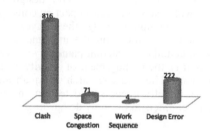

Figure 6. Number of issues analyzed.

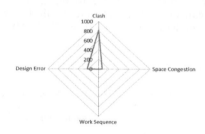

Figure 7. Distribution of issues.

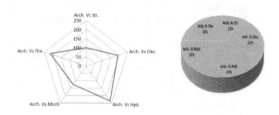

(a) Number of issues   (b) Percentages

Figure 3. Number and percentages of coordination issues.

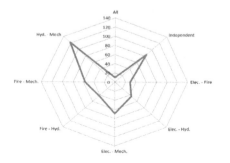

Figure 4. Number of service coordination issues.

113 issues took less than 2 weeks to solve. The data also show that 69% of validation and coordination issues had been solved in 15 days before it happened. Considering the workflow of coordination, more effective improvement of process would be needed for the rest of 31% that took more than 2 weeks to solve.

Consequently, Figure 6 shows the number of issues analyzed in four categories: clash detection, space congestion, work sequence, and simple design errors. It is evident from the figure that most of them (816, 73%) are clash issues, followed by simple design errors (222, 20%). Figure 7 shows the distribution among the four items; however, this project is in the design phase now, so most issues are related to clash and simple design errors, while few issues were related to space congestion and work sequence issues.

4 CONCLUSIONS

This study is a preliminary research for developing an effective construction workspace planning model using BIM. As the first step of the research, we analyzed the design validation and coordination from BIM data. The result of raw data was simply analyzed for counting the issues in terms of changes for future research[9]. This case study explored the management sequences when design conflicts were identified and occurred in project that extensively adopted BIM tools. This research discovered 1,113 issues with clash detection, space congestion, work sequence, and design errors, which will have significant impact on construction workspace planning through the life cycle. Follow-up studies will focus on the implementation of BIM data and construction planning and scheduling, that is, the management of activity execution workspace. The future research will quantitatively analyze the effectiveness of detection and resolution of conflicts in the early design stage.

ACKNOWLEDGMENT

The authors would like to thank the National Research Foundation of Korea for the financial support under Grant NRF-2015R1D1A1A01060168.

REFERENCES

Bansal, V.K., Use of GIS and topology in the identification and resolution of space conflicts, Journal of Computing in Civil Engineering, vol. 25, no. 2, pp. 159–171, 2011.

Benghi, C. and Dawood N., Integration of design and construction through information technology from programme rehearsal to programme development, 8th International Conference on Construction Applications of Virtual Reality, Lisbon, Portugal, 2008.

Chavada, R., N. Dawood, M. Kassem, Construction workspace management: The development and and application of a NOVEL nD planning approach and tool, Journal of information technology in construction, 2012.

Chavada, R.D., M. Kassem, N.N. Dawood, K.K. Naji, A framework for construction workspace management a serious game engine approach.

Chua, D., K. Yeoh, Song Y., Quantification of spatial temporal congestion in four-dimensional computer-aided design, Journal of Construction Engineering and Management, vol. 136, no. 6, pp. 641–649, 2010.

Dawood, N., Z. Mallasi, Construction workspace planning: assignment and analysis utilizing 4D visualization technology, Computer-aided Civil and Infrastructure Engineering, vol. 21, no. 7, pp. 498–513, 2006.

Guo, S., Identification and resolution of work space conflicts in building construction, Journal of Construction Engineering and Management, vol. 128, No. 4, pp. 287–299, 2001.

Kuan-Chen, S.K., Collision detection strategies for virtual construction simulation, Automation in Construction, vol. 18, No. 6, pp. 724–736, 2009.

Lin, Y., S. Chou, I. Wu, Conflict impact assessment between objects in BIM system, National science council Taiwan, 2010.

Moon, Kang, Dawood, Configuration method of health and safety rule for improving productivity in construction space by multi-dimension CAD system, ICCEm/ICCPM, Jeju, korea, 2009.

Paik, S., P. Leviakangas, D. Morison, L. Naas, X. Wang, Building Information Modelling in Change management: A Case Study, (to be published).

Rwamamara, R., H. Norberg, T. Olofsson, O. Lagerqvist, Using visualization technologies for design and planning of a healthy construction workplace, Construction innovation, 2010.

Wu, I., Y. Chiu, 4D Workspace conflict detection and analysis system, 10th International conference on construction applications of virtual reality, 2010.

Zhang, J.P., Z.Z. Hu, BIM and 4D-based integrated solution of analysis and management for conflicts and structural safety problems during construction: 1. Principles and methodologies, Automation in construction, vol. 20, pp. 155–166, 2011.

Advances in Energy Science and Equipment Engineering II – Zhou, Patty & Chen (Eds)
© 2017 Taylor & Francis Group, London, ISBN 978-1-138-71798-5

Adaptive response of occupants to the thermal environment in air-conditioned buildings in Chongqing, China

Hong Liu & Yuxin Wu

Joint International Research Laboratory of Green Building and Built Environment (Ministry of Education), Chongqing University, Chongqing, China
National Centre for International Research of Low-Carbon and Green Buildings, Chongqing, China
Faculty of Urban Construction and Environmental Engineering, Chongqing University, Chongqing, China

Zixuan Wang, Diyi Tan, Yuxin Xiao & Wenshuang Zhang

Faculty of Urban Construction and Environmental Engineering, Chongqing University, Chongqing, China

ABSTRACT: Behavioral response of occupants to different indoor thermal conditions could affect building performance and energy consumption directly. To better understand this, a field study was conducted in office and residential buildings in Chongqing, China, from April to October. Questionnaire survey was conducted in 1304 office staffs and 1171 residents. The data showed a significant difference in the behavioral pattern and subjective comfort between office and residential buildings. Occupants in office buildings preferred to use Air Conditioner (AC) than those in residential buildings; the proportion of open window reduced obviously when AC was used in residential building, but that did not change obviously in office buildings; the prevalence of SBS in residential buildings was obviously much lower than that in office buildings; the mean thermal sensation did not have significant difference between these two types of buildings when AC was used. However, when AC was not used, the mean thermal sensation vote was higher in office buildings than that in residential buildings at the same thermal conditions.

1 INTRODUCTION

International Energy Agency (IEA) notes that energy consumption in buildings accounts for 35% of the global terminal energy consumption. In 2012, China's energy consumption by buildings was around 0.69 billion tons of standard coal. This is continuously increasing (Jiang, 2014), which makes people worried about the fact that the burden of energy consumption aggravated by increasing indoor thermal comfort demands will be unacceptable in China in the immediate future. Therefore, the key issue is how to reduce energy consumption without decreasing occupants' thermal comfort level. The building performance of actual energy consumption is quite different from that of original designs in most cases, which is affected greatly by the management of the available equipment (Peng, 2015) and the different behavioral habits and operating strategies of the occupants (Tanner, 2013). Researchers found that the range of acceptable indoor temperature will be wider due to occupants' adaption, and such adaption also allowed them to set a higher indoor temperature in summer, so as to realize the goal of energy conservation (Nicol, 2002; Kwong, 2014).

The overall percentage of dissatisfaction among a large group of people in the same thermal environment can also be reduced by occupants' adaptive behaviors. In the 1960s, Fanger proposed PMV–PPD model and found that even if the indoor temperature is neutral by average, there are still a small amount of people feel dissatisfied (5%). Besides, studies have recently found that the thermal comfort conditions are also different among different ages (Liu, 2015) and genders (Karjalainen, 2012). These studies indicate that a single indoor temperature standard without considering personal adaption cannot meet everyone's need for thermal comfort. As van Hoof. J (Hoof, 2008) said: "thermal comfort for all can only be achieved when occupants have effective control over their own thermal environment."

When people feel uncomfortable, they will take actions to restore their comfort. In general, thermal comfort and SBS (Sick Building Syndrome) are two important parts of occupants' overall comfort, which are affected by indoor thermal environment in air conditioner space. Besides, work efficiency in office buildings can be also affected by overall comfort (Liu, 2012). Thus, by analyzing data from the field study in different buildings, we will find out the feature of occupants' subjective comfort and adaptive behavioral pattern and give advices for better design of parameters in HVAC to realize comfort for all.

2 RESEARCH METHODS

This study was conducted in Chongqing, which has a typical hot humidity climate in summer. Field studies were conducted in selected offices and residential buildings from April to October, including onsite test of thermal environmental parameters and subjective questionnaire investigation for occupants. Environmental test methods and questionnaire design were in accordance with Chinese code "Evaluation standard for indoor thermal environments in civil buildings GB/T50785-2012" (GB/T50785-2012, 2012). The states of window (open/close) and air conditioner (used/not used) were recorded during the survey.

Because behaviors like opening window or switching air conditioner on are characteristic of bi-point distribution, and their errors do not fit normal distribution. It is inappropriate to analyze it by adopting typical least-squares method. Therefore, this study adopts logistic regression equation similarly to previous studies [11] as:

$$p = \exp(a + bx) / [1 + \exp(a + bx)] \qquad (1)$$

In the questionnaire, the questions about SBS are selected from 32 different types of sick building syndrome recommended by EPA. The question (in Chinese) is as follows: Do you often have the following symptoms in this environment? Including: general symptoms: drowsy/fatigue, nausea/dizziness, upset to hot, hard to concentrate; mucosal symptoms: eye irritation, throat discomfort, stuffy or runny nose; skin symptoms (dry, rashes, or itch). In addition, considering the comprehensive influence of thermal environment on overall discomfort, the research also adds a new choice: "upset or annoying (about thermal environment)."

Thermal sensation adopts ASHERA 7 level index: −3, −2, −1, 0, +1, +2, and +3 to represent cold, cool, slightly cool, neutral, slightly warm, warm, and hot, respectively.

Subjective comfort and thermal adaptive behaviors of occupants in different types of air conditioner can be studied and compared by analyzing the results of investigation.

3 RESULTS

3.1 Thermal environment

The analysis of measured thermal environment is shown in Table 1. The mean outside air temperature reached 31.3°C, both in office and residential buildings when the air conditioners were operated (i.e. AC). Meanwhile, the mean indoor air temperatures were 26.9 and 27.6°C, while mean clothing insulations were 0.27 and 0.21 clo, respectively, in office and residential buildings. When air conditioners were not used (i.e., FR), the mean outside air temperatures were 25.9 and 24.0°C in office and residential buildings, respectively, with the same mean indoor air temperature around 26°C and the same mean clothing insulation of 0.37 clo. The outdoor and indoor relative humidity of these two types of buildings was between 60% and 70%, and lower when air conditioner was used. The mean outdoor air velocity ranged from 0.3 to 0.6 m/s, while the mean indoor air velocity was below 0.3 m/s.

3.2 Behavioral responses

The Proportion of AC used (PAU) is defined as the ratio of the number of air conditioners being used to the total surveyed number at one air temperature bin, which reflects the possibility of a control being used at certain air temperature. The change of proportion of AC used by occupants with outdoor air temperature is shown in Figure 1.

The expressions of the curve by logistic fitting are as follows:

Office buildings:

$$PAU = \exp(-22.53 + 0.83\ t_{out}) / \\ [1 + \exp(-22.53 + 0.83\ t_{out})]\ R^2 = 0.90 \qquad (2)$$

Residential buildings:

$$PAU = \exp(-11.04 + 0.33\ t_{out}) / \\ [1 + \exp(-11.04 + 0.33\ t_{out})]\ R^2 = 0.93 \qquad (3)$$

Occupants in office buildings prefer to use air conditioner. According to equation (2), PAU reached 20% in office buildings only when the outdoor air temperature was above 25.5°C, and reached 50% when the outdoor air temperature was above 27.1°C. According to equation (3), PAU reached 20% in residential buildings when the outdoor air temperature was above 29.3°C, and reached 50% when the outdoor air temperature was above 33.5°C.

When air conditioners were not used, the proportion of window opened in office buildings was slightly higher than that in residential buildings, both of which were ranging from 0.7 to 1.0 and changing with indoor air temperature. When air conditioners were used, the proportion of window opened was dropped, especially in residential buildings.

Table 1. Summary of thermal environment in the field study.

Type (No.) Parameter	Office buildings		Residential buildings	
	AC used (606)	FR (698)	AC used (186)	FR (985)
t_{out} (°C)	31.3 ± 4.1	24.0 ± 3.9	31.3 ± 4.1	25.9 ± 3.8
RH_{out} (%)	63.5 ± 13.5	69.7 ± 13.6	66.9 ± 15.0	71.9 ± 12.9
v_{out} (m/s)	0.55 ± 0.36	0.39 ± 0.27	0.32 ± 0.31	0.48 ± 0.62
t_a (°C)	26.9 ± 1.2	25.9 ± 2.3	27.6 ± 2.5	26.2 ± 3.2
RH_a (%)	60.0 ± 9.6	64.2 ± 11.5	62.8 ± 10.9	70.6 ± 9.7
v_a (m/s)	0.16 ± 0.14	0.12 ± 0.10	0.23 ± 0.22	0.17 ± 0.16
I_{cl} (clo)	0.27 ± 0.11	0.37 ± 0.13	0.21 ± 0.14	0.37 ± 0.20

Notes: Mean values ± standard deviation; FR—free running model (when AC was not used); t_{out}—outdoor air temperature; t_a—indoor air temperature; RH_{out}—outdoor relative humidity; RH_a—indoor relative humidity; v_{out}—outdoor air velocity; v_a—indoor air velocity; I_{cl}—clothing insulation.

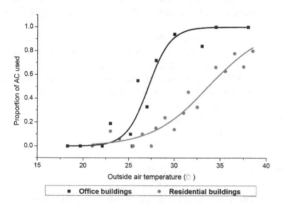

Figure 1. PAU changing with outdoor air temperature.

Table 2. Prevalence of SBS in buildings.

Symptoms	Office —AC	Office —FR	Res. —AC	Res. —FR
Drowsy/fatigue	27.20%	28.40%	4.50%	5.20%
Nausea/dizziness	15.20%	12.20%	0.00%	2.40%
Upset to hot	18.30%	25.60%	9.10%	7.10%
Hard to concentrate	10.70%	12.50%	1.50%	3.10%
Eye irritation	22.10%	26.60%	3.00%	3.60%
Throat discomfort	16.70%	16.60%	3.00%	3.30%
Nose stuffy or runny	13.40%	10.90%	4.70%	3.10%
Skin symptoms	9.40%	10.50%	3.00%	3.50%

3.3 Sick building syndrome

It is evident from Table 2 that the prevalence of SBS in residential buildings was obviously much lower than that in office buildings. In general, the prevalence of each SBS in residential buildings was lower than 10%. However, the prevalence of each SBS in office buildings was between 10% and 30%. The highest prevalence of SBS, over 25%, was drowsy, followed by upset to hot and eye irritation in office buildings with AC un-operated.

3.4 Thermal sensation

By analyzing the questionnaire and environment test data, the linear fit relationship between TSV and indoor air temperature was obtained (Figure 3).

It is noted that under the same indoor temperature, the mean thermal sensation vote in office buildings was higher than that in residential buildings when air conditioner was not used, and was also higher than that when air conditioner was used in office buildings. This indicated

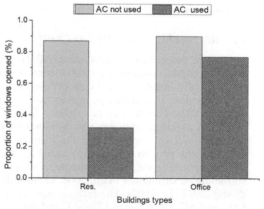

Figure 2. Proportion of window opened.

that occupants in residential buildings had more effective adaptive methods to restore their thermal comfort when AC was not used than those in office buildings.

The equation is obtained as follows:
Office—AC:

$$TSV = 0.17t_a - 4.20; R^2 = 0.55 \qquad (4)$$

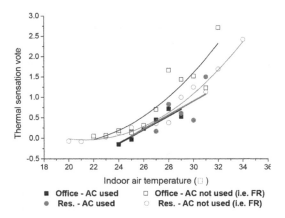

Figure 3. Thermal sensation with indoor air temperature.

Res.—AC:

$$TSV = 0.17\, t_a - 4.14;\ R^2 = 0.87 \qquad (5)$$

Office—FR:

$$TSV = 0.016\, t_a^2 - 0.60\, t_a + 5.67;\ R^2 = 0.84 \qquad (6)$$

Res.—FR:

$$TSV = 0.014\, t_a^2 - 0.60\, t_a + 6.14;\ R^2 = 0.97 \qquad (7)$$

4 DISCUSSION

4.1 Relationship between PAU and APD

In fact, the percentage of dissatisfied (APD) is the rate of occupants who feel thermal discomfort. On the basis of adaptive thermal comfort theory, it is the discomfort that drives people to take action (Nicol, 2002). People take an active role in the operation of air conditioner and windows based on their level of comfort, and the balance was rebuilt in the process of changing variables, such as heat input/removal, fresh air supply, wind speed, and sound transmission (Yun, 2008; Haldi, 2009; Zhang, 2012; Li, 2007). The relationship between PAU and APD is shown in Figure 4, and the equation is as follows:

$$\text{Office: } PAU = 1.58 \times APD + 0.04,\ R^2 = 0.87 \qquad (8)$$

$$\text{Res.: } PAU = 0.90 \times APD - 0.19,\ R^2 = 0.41 \qquad (9)$$

Occupants were more sensitive to thermal discomfort in office buildings than in residential buildings. In office buildings, occupants prefer AC because they need have to pay the electricity bill. However, in residential buildings, more occupants

prefer not to use AC, which implies the effect of psychology (the so-called "alliesthesia") and freedom to choose adaptive behaviors such as changing clothes, seating in different location, and taking a bathe.

4.2 Windows open

Because occupants in office buildings do not pay bills for air conditioning systems as residents do, they lacked the motivation to close window to reduce heat loss, resulting in a high percentage of windows opened when AC is used (Figure 2). Only when thermal discomfort of occupants increases because of poor performance of air conditioner, they tend to close the window in order to maintain indoor temperature by reducing heat loss, and also increased discomfort symptoms caused by lack of fresh air.

Although a high percentage of open windows was maintained in office buildings, which reflected their effort to improve air quality. The prevalence of SBS is still much higher than that in residential buildings. This fact implied the performance of fresh air supply systems and the design of natural ventilation were poor in the office buildings in Chongqing.

5 CONCLUSIONS

The main conclusions drawn from this paper are as follows:

1. Occupants in office buildings prefer to use air conditioner than those in residential buildings. About 20% of the residents did not prefer to use air conditioner even when the outside air temperature reached 35°C. However, in office buildings, almost all occupants choose to use air

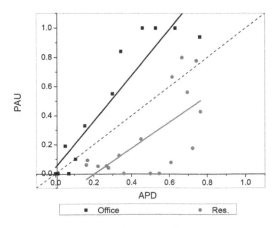

Figure 4. Relationship between PAU and APD.

conditioner when outside air temperature was high.

2. The proportion of opened window reduced obviously when AC was used in residential building, which did not change obviously in office buildings.

3. The mean thermal sensation did not have significant difference between these two types of buildings when AC was used. However, when AC was not used, the mean thermal sensation vote was higher in office buildings than that in residential buildings at the same thermal conditions.

ACKNOWLEDGMENTS

Project supported by Construction Program of S&T Research and Development Base of Chongqing (International S&T Cooperation Project) (Grant No. cstc2013 gjhz90002). The authors also would like to thank the Graduate Scientific Research and Innovation Foundation of Chongqing, China (Grant No. CYS16006) and Students Research Training Program (SRTP) of Chongqing University for the financial support, and the contribution of the 111 Project (Grant No. B13041).

REFERENCES

Fanger, P.O., *Thermal Comfort: Analysis and Applications in Environmental Engineering* (Danish Technology Press, Copenhagen, 1970).

GB/T50785–2012, *Evaluation standard for indoor thermal environments in civil buildings* (China Building Industry Press, Beijing, 2012) [In Chinese].

Haldi, F., D. Robinson, Build. Environ. **44**, 12 (2009): pp. 2378–2395.

Jiang, Y. *Research report on development of building energy efficiency in China* (China Building Industry Press, Beijing, 2014) [In Chinese].

Karjalainen, S., Indoor Air **22**, 2 (2012): pp. 96–109.

Kwong, Q.J., N.M. Adam, B.B. Sahari, Energy Build. **68**(2014) pp. 547–557.

Li, Z.J., Y. Jiang, Q. P. Wei, J. HVAC **37**, 8(2007): pp. 67–71+45 [In Chinese].

Liu, H., Y.X. Wu, H. Zhang, X.Y. Du, J. HVAC **45**, 6(2015): pp. 50–58 [In Chinese].

Liu, J., R.M. Yao, J. Wang, B.Z. Li, Appl. Therm. Eng. **35**(2012): pp. 40–54.

Nicol, J.F., M.A. Humphreys, Energy Build. **34**, 6(2002): pp. 563–572.

Parys, W., D. Saelens, H. Hens, J. Build. Perform. Simu. **4**, 4(2011): pp. 339–358.

Peng C., B. Hao, J. HVAC, **45**, 9(2015): pp. 1–6 [In Chinese].

Tanner, R.A., G.P. Henze, Architectural Engineering Conference 2013, (American Society of Civil Engineers: State College, Pennsylvania, United States, 2013).

Yun, G.Y., K. Steemers, N. Baker, Build. Res. Inf. **36**, 6 (2008): pp. 608–624.

Zhang, Y., P. Barrett, Build. Environ. **50**(2012): pp. 125–134.

van Hoof, J., Indoor Air **18**,3 (2008): pp. 182–201.

Advances in Energy Science and Equipment Engineering II – Zhou, Patty & Chen (Eds)
© 2017 Taylor & Francis Group, London, ISBN 978-1-138-71798-5

Oscillatory region and effect of the impact factor of memory in a happiness model with commensurate fractional-order derivative

Zhongjin Guo

School of Mathematics and Statistics, Taishan University, Taian, P.R. China
College of Mechanical Engineering, Beijing University of Technology, Beijing, China

Wei Zhang

College of Mechanical Engineering, Beijing University of Technology, Beijing, China
Beijing Key Laboratory on Nonlinear Vibrations and Strength of Mechanical Structures, Beijing, P.R. China

ABSTRACT: Fractional-order derivative features the memory effect of the past states. In this paper, the integer-order model of happiness is firstly extended to a commensurate fractional-order pattern which is found to be more suitable to depict the happiness behavior. Then the nonlinear vibration of the happiness model with commensurate fractional-order derivative is studied analytically and ascertained by numerical ones. The boundary between oscillatory and non-oscillatory parametric regions, and asymptotic dynamics of the model in vibration region is predicted and illustrated by residue harmonic balance method. It is found that when one's past experiences have much influence on his/her present and future life, the model response can generate periodic vibration. Finally, we can find that all the obtained stable solution branches match well with numerical ones.

1 INTRODUCTION

An interesting dynamical model of happiness composed by a 3-order differential equation has recently been reported by Sprott (Sprott, 2005). This model describes the time-variation of happiness (or sadness) displayed by individuals under certain external circumstance, the governing equation is given as

$$\frac{d^3R}{dt^3} + \alpha\frac{d^2R}{dt^2} + \beta(1-R^2)\frac{dR}{dt} + R = 0 \qquad (1)$$

Where α and β are the parameters of the model and β denotes the nonlinear damping coefficient. R represents what others presume one's happiness to be based on one's circumstances, and $E = dR/dt$ means one's true feeling of happiness. Depending on different parameter values of the model, a different range to behaviors have been obtained, even with chaos induced.

Fractional-order derivative (Oldham, 1974; Podlubny, 1999), a generalization of the traditional integer-order, has a history of over 300 years old. The earliest research of the fractional calculus was presented by Leibniz and Hospital in 1695. Since then, a lot of investigations, both on general theory and application had been issued by many authors (Rossikhin, 2010; Sun, 2010; Monje, 2010;

Song, 2010; Ahmad, 2007). It has been found that fractional-order derivative is an adequate tool for the study of the so-called "anomalous" social and physical behaviors, in reflecting the non-local, frequency and history dependent properties of these phenomena (Sun, 2010; Monje, 2010). The integral-order derivative of a function is only related to its nearby points, while the fractional derivative has relationship with all of the function history information. As a result, a model described by fractional-order equations possesses memory. In fact, real world processes generally or most likely are fractional-order systems, namely dynamical systems governed by the fractional order derivative equations. As we all known, the emotion of happiness has relationship with all of the past factors, that is, it is influenced by memory. Thus, the characteristic of fractional-order models and the practical emotion of happiness coincide with each other well. So it is more proper describing happiness model with fractional-order equations rather than with integer-order equations. Song et al (Song, 2010) have demonstrated such a dynamical model of happiness described through fractional-order differential equation via numerical simulations which could exhibit various behaviors with and without external circumstance. Moreover, control and synchronization problems of this model have been also discussed. In addition, Ahmad and

Khazali (Ahmad, 2007) examined fractional-order dynamical model of love and obtained the strange chaotic attractors under different fractional orders by using numerical simulations.

The purpose of this work is firstly to predict the parametric function for the boundary between oscillatory and non-oscillatory regions of the happiness model by the zeroth-order approximation using just one Fourier term in the procedure of residue harmonic balance approach (Guo, 2011; Guo, 2016). Then the highly accurate solutions are obtained for steady state of fractional happiness model using residue harmonic balance and the effects of nonlinear damping coefficients are illustrated on the steady state response. The results show that when one's past experiences have much influence on his/her present and future life, the model response can generate periodic vibration.

2 FRACTIONAL-ORDER MODEL OF HAPPINESS

Fractional-order derivative features the memory effect of the past states. In this section, the integer-order model (1) is extended to a commensurate fractional-order pattern which is found to be more suitable to depict the happiness behavior. In Eq. (1), replacing the integer-order derivative with a fractional-order $0 < \lambda \leq 1$ in the sense of Caputo definition, a commensurate fractional-order happiness system is given as below

$$D_t^{3\lambda}R + \alpha D_t^{2\lambda}R + \beta(1 - R^2)D_t^{\lambda}R + R = 0 \qquad (2)$$

Where $D_t^{\lambda}R(t)$

$$= \frac{1}{\Gamma(m-\lambda)}\int_0^t (t-\tau)^{m-\lambda-1}\frac{d^m}{d\tau^m}R(\tau)d\tau, m-1 <$$

$\lambda \leq m$, $m \in N, t > 0$, $R \in C^m{}_{-1}$. From Ref (Song, 2010), R represents one's responses to external events, namely, what others presume one's happiness depending on one's circumstance. The parameter pair (α, β) is to categorize people with different personality. i.e. different (α, β) represents different kind of individuals, and the order λ represents the impact factor of memory of an individual. It is noteworthy that the conception of the impact factor of memory is proposed here to denote a measurement of how influence an individual is by his\her past experiences. When one's IFM is low, his/her past experiences may have little influence on his/her present and future life; while when IFM is high, it might be difficult for him/her to escape from past experiences, in despite of nightmares or sweet memories.

3 ANALYSIS OF OSCILLATORY PARAMETRIC REGIONS

Assume that Eq. (2) exists undamped oscillation and the steady state response of form

$$R(t) = a\cos(\omega t) \qquad (3)$$

Where $|a|$ and ω denote the amplitude and angular frequency of oscillation, respectively.

Substituting Eq. (3) into Eq. (2) and using the Galerkin procedure result in the following harmonic balance equations.

$$1 + \beta\omega^{\lambda}\cos(\frac{\lambda\pi}{2}) + \alpha\omega^{2\lambda}\cos(\lambda\pi) + \omega^{3\lambda}\cos(\frac{3\lambda\pi}{2})$$
$$- \frac{3}{4}\beta\omega^{\lambda}\cos(\frac{\lambda\pi}{2})a^2 = 0 \qquad (4)$$

$$\beta\omega^{\lambda}\sin(\frac{\lambda\pi}{2}) + \alpha\omega^{2\lambda}\sin(\lambda\pi) + \omega^{3\lambda}\sin(\frac{3\lambda\pi}{2})$$
$$- \frac{1}{4}\beta\omega^{\lambda}\sin(\frac{\lambda\pi}{2})a^2 = 0 \qquad (5)$$

Simplifying Eqs. (4) and (5) to get positive real values of amplitude and angular frequency, we can obtain the conditions in which the limit cycle can be generated. Figures 1 and 2 show the boundary between oscillatory and non-oscillatory regions in the (β, λ) plane for the fractional-order system (2) with $\alpha = 1, 2, 5, 10$. In Fig. 1, as the increasing of nonlinear damping coefficients β, the minimum IFM λ increases along with straight line $\lambda = 0.81813 + 0.20338\beta$ and the limit cycle is generated in region I in which the value of commensurate fractional order exceeds straight line. Likewise, Figure 2 shows the effect of parameter α on boundary between oscillatory and non-oscillatory regions. With the increasing of parameter α, the

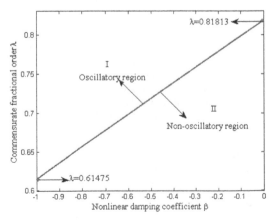

Figure 1. Oscillatory and non-oscillatory regions of the system (11) for $\alpha = 1$.

626

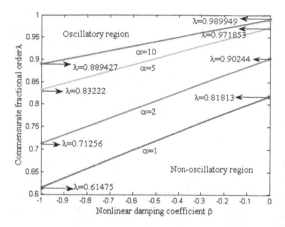

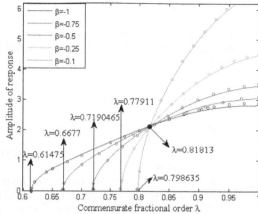

Figure 2. Oscillatory and non-oscillatory regions of the system (11) for $\alpha = 1,2,5,10$.

Figure 3. Comparisons of amplitude and effects of nonlinear damping coefficient.

oscillatory region in which limit cycle generates gradually becomes small and the straight slope of boundary gradually becomes small. The following we give out the corresponding straight line of boundary. For $\alpha = 2,5,10$, the straight lines of boundary are

$$\lambda = 0.90244 + 0.18988\beta,$$

$$\lambda = 0.971853 + 0.13963\beta$$

$$\lambda = 0.89949 + 0.10052\beta$$

respectively.

4 ANALYTICALLY ASYMPTOTIC SOLUTION UNDER DIFFERENT FRACTIONAL ORDER

In this subsection, we take $\alpha = 1$ to illustrate the effect of nonlinear damping coefficients β on the steady state response. In Fig. 3, the solid line denotes the presented residue harmonic approximation, dots denotes the corresponding numerical result. From this figure we can find that

1. with the increasing of the impact factor of memory λ, the periodic solution gradually appears, and for bigger nonlinear damping coefficient β, the IFM λ need become larger to bring periodic oscillation. That is, only when one's past experiences have much influence on his/her present and future life, the system can generate periodic vibration.
2. All the amplitude curves intersect at a common point in which $\lambda \approx 0.81813$. In fact, this point is from the limit where the Hopf bifurcation occurs when nonlinear damping coefficient tends to zero.
3. The present analytical solutions to amplitude and frequency match well with numerical ones.

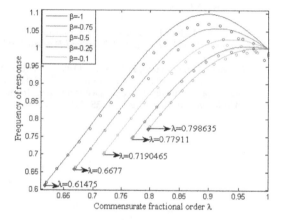

Figure 4. Comparisons of frequency and effects of nonlinear damping coefficient.

4. All the curves of frequency first increases and then decreases in Fig. 4. That is, when the impact factor of memory of an individual increases from low to high, the periodic vibration gradually appears and the vibration frequency first increases and then decreases.

5 CONCLUDING REMARK

The autonomous model of happiness with commensurate fractional-order derivative is explored in detail by using the method of residue harmonic balance. The boundary between oscillatory and non-oscillatory parametric regions in the (β, λ) plane is first predicted, where the effect of parameter α on boundary is shown and formulated. Then asymptotic dynamics of system in vibration region is illustrated and compared with numerical ones. The obtained results show that only when one's past experiences have much influence on his/her present and future life, the model response can generate periodic vibration and the present analytical solutions match well with numerical ones.

ACKNOWLEDGEMENTS

This work is supported by grant Nos.11290152, 11427801 and 11502160 of the National Natural Science Foundation of China, No. ZR2014 JL002 of the Natural Science Foundation of Shandong Province, China, No. 2015zz-18 of the Beijing Postdoctoral Research Foundation, No. J15LI13 of the Higher Educational Science and Technology Program of China.

REFERENCES

Ahmad, W.M., R.E.I. Khazali, Fractional-order dynamical models of love, *Chaos, Solitons and Fractals* 2007; **33**: 1367–1375.

Guo, Z.J., A.Y.T. Leung, H.X. Yang, Oscillatory region and asymptotic solution of fractional van der Pol oscillator via residue harmonic balance technique, *Applied Mathematical Modelling* 2011; **35**: 3918–3925.

Guo, Z.J., W, Zhang. The spreading residue harmonic balance study on the vibration frequencies of tapered beams, *Applied Mathematical Modelling*, 2016, **40**: 7195–7203.

Monje, C.A., Y. Chen, B.M. Vinagre et al., *Fractional-order systems and controls: fundamentals and applications*. London: Springer-Verlag, 2010.

Oldham K.B., J. Spanier, *The fractional calculus*, Academic Press, New York, USA. 1974.

Podlubny, I., *Fractional differential equations: an introduction to fractional derivatives, fractional differential equations, to methods of their solution and some of their applications*, Mathematics in Science and Engineering, Academic Press, New York. 1999.

Rossikhin, Y.A., M.V. Shitikova, Application of fractional calculus for dynamic problems of solid mechanics: novel trends and recent results. *Applied Mechanics Reviews* 2010; **63**: 010801-1-010801-52.

Song, L., S.Y. Xu, J.Y. Yang, Dynamical models of happiness with fractional order, *Commun Nonlinear Sci Numer Simulat* 2010; **15**: 616–628.

Sprott, J.C., Dynamical models of happiness. *Nonlinear Dynamics, Psychology, and Life Sciences* 2005; **9**: 23–26.

Sun, H.G., W. Chen, C.P. Li, Y.Q. Chen, Fractional differential models for anomalous diffusion, *Physica A* 2010; **389**: 2719–2724.

Advances in Energy Science and Equipment Engineering II – Zhou, Patty & Chen (Eds)
© 2017 Taylor & Francis Group, London, ISBN 978-1-138-71798-5

The principles of road-climatic zoning on the example of Western Siberia

Vladimir Efimenko & Sergey Efimenko
Tomsk State University of Architecture and Building, Tomsk, Russia

ABSTRACT: The paper presents the methodical scheme of zoning and execution of activities aiming at specification of calculated values of soil property indicators used in the design of flexible road pavements. The scheme includes a two-stage research. The first stage includes steps to define homogeneous areas (road district). The second stage includes the work on validation of the calculated values of subgrade clay soils properties for the road district defined earlier. The boundary lines of road-climatic zones were specified on the basis of the taxonomic system "zone-subzone-road district". The paper reveals the necessity of taking into account the features of water and thermal regime of roadbeds of the West Siberian auto roads in substantiation of calculated values of the elasticity modulus and strength properties of clay soils. The calculated values of moisture content, strength, and deformability characteristics of clay soils established and differentiated according to road-climatic zones will ensure the required level of transport infrastructure facilities reliability during their life cycle.

1 INTRODUCTION

Research experience in the area of design and construction of auto roads in complex under-studied natural conditions shows that a rational account of regional factors in the specification of road-climatic zoning to ensure operational reliability of roadbeds and road pavements must include the following steps:

– careful study of local environmental conditions of the territory of a particular administrative entity, and identification of zonal factor features as compared to previously studied conditions. The stage of work includes the selection and the analysis of sectorial cartographic zoning materials (climatic, vegetation, hydrogeological, geomorphological, engineering-geotechnical).
– distinguishing of homogeneous road districts and the assignment of numbers to these districts on the basis of the analysis and the generalization of zonal and intrazonal features within the existing road-climatic zones, and the specification, if necessary, of borders of the latter (Sidenko, 1973).
– substantiation (specification) and the assignment of individual road districts to territories, with regard to their zonal, intrazonal, and regional differentiation, of a complex of calculated values of soils and materials characteristics, designs, specifications, etc. in the design and construction of sturdy roadbeds and road pavements.

2 BRIEF INFORMATION ON THE PROBLEM AREA

The solution of problems related to zoning of certain areas, for example, for the purposes to ensure the quality of auto roads design, is carried out by the researchers, as a rule, by applying the experience of a componentwise overlay of cartographic distribution diagrams of zonal, azonal, intra-zonal, and regional geocomplexes. The experience in the development of road zoning principles (Zapata, 2008; Russam, 1961; Yarmolinsky, 2014) shows that a rational consideration of the complex of regional climatic conditions may be based on the use of the element "road district" of the taxonomic system previously shown in (Sidenko, 1973)— "zone—subzone—district—site". It shall be noted that the taxon "site" can be used in the design and construction of unique structures, but in terms of mass road construction it can be neglected. *Road district* is a genetically homogenous territory characterized by typical and peculiar climate, geology, topography, and other physical elements, which has the same type of road constructions (primarily roadbed and road pavements), in similar soil and hydrological conditions should be characterized by similar strength (the ability to withstand loads without breaking the continuity) and resistance (the ability to change its state so that the deformation associated with these changes does not exceed the permissible one). The concept of *subzone* implies a territory with a uniform microrelief, and the subsystem *zone* characterizes the earth's surface with a uniform distribution of heat and moisture, determining the development of specific and interrelated types of soil and vegetation.

For example, in the US the Interagency Guidance for Road Research (Aldrich, 1973) has distinguish between 97 physical-geographic districts according to the bearing capacity of soils and their changes under the influence of moisture, as well as according to the availability of road-building materials. (Efimenko, 2015).

Analyzing the traffic climatic zoning of the territory of China it is possible to conclude that it is based on methodological techniques previously used on the territory of the former Soviet Union, which is why the borders of the two countries have the same road-climatic zones. Depending on moisture degree, depth of soil freezing, and temperature, the territory of China is divided into seven road-climatic zones. There are 33 road districts allocated within each road-climatic zone, taking into account the terrain and climate conditions (The Ministry of Transport of the People's Republic China highway natural zoning standards, 1986). The most extreme road districts in terms of climatic conditions are located in northern China. Therefore, depending on the temperature zoning, three zones are additionally allocated in China according to permafrost conditions: very heavy permafrost zone, heavy permafrost zone, and average permafrost zone (The Ministry of Transport of the People's Republic China, 2006).

Our proposed methodological scheme for rationalizing the territorial location of boundary lines of Road-Climatic Zones (RCZs) in the system "zone–subzone–district" is based on several stages of research (Figure 1) (Efimenko, 2015; Efimenko, 2014). At the first stage, a database for the modeling of geocomplex parameters is established.

The second and third stages of the work on road climatic-zoning of separate territories are aimed at clarifying the geographical distribution of boundary lines of zones, subzones, and road districts, and positioning the boundaries of districts, subzones, and zones for adjacent territories of administrative units. This can be accomplished by either a component wise overlapping of distribution schemes of the geocomplex elements, or with the help of mathematical methods for processing the characteristics included in the database. The fourth stage of the work is aimed at designating a complex of calculated values of moisture, strength, and deformability of clay subgrade soils to be used in the design of flexible road pavements within allocated territories in the process of road-climatic zoning.

An intermediate result of this methodological scheme is a schematic map of the boundary distribution of RCZs in Western Siberia. For example, West Siberian region includes 14 administrative units (the Republic of Altai and Khakassia, Altai and Krasnoyarsk territories, Kemerovo, Kurgan, Novosibirsk, Omsk, Sverdlovsk, Tomsk, Tyumen, Chelyabinsk regions, Khanty-Mansiysk and Yamalo-Nenets autonomous districts). The territories of zones are divided into subzones by the type of the relief and homogeneous by geocomplexes road districts with a detailed characterization of main geographic features. For the territory of Western Siberia four road-climatic zones (I, II, III, and IV), three sub-zones (F—flat, U—undulated, M—mountainous), and 112 traffic areas are recommended (Efimenko, 2014).

Thus, for example, the territory of the administrative entity "Kemerovo region" includes two road-climatic zones (II and III), three subzones (flat, undulated, and mountainous), and 8 road districts (from 1 to 5, depending on the zone and the subzone) (Figure 2).

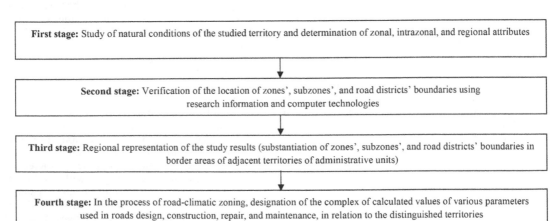

First stage: Study of natural conditions of the studied territory and determination of zonal, intrazonal, and regional attributes

Second stage: Verification of the location of zones', subzones', and road districts' boundaries using research information and computer technologies

Third stage: Regional representation of the study results (substantiation of zones', subzones', and road districts' boundaries in border areas of adjacent territories of administrative units)

Fourth stage: In the process of road-climatic zoning, designation of the complex of calculated values of various parameters used in roads design, construction, repair, and maintenance, in relation to the distinguished territories

Figure 1. The scheme of validating the territorial distribution of boundaries of road-climatic zones, subzones, and road districts.

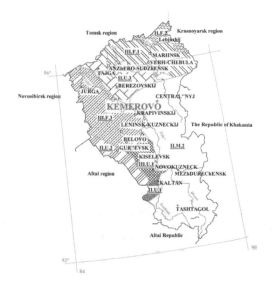

Figure 2. The schematic map of road-climatic zoning of the territory of Kemerovo region. II, III: road-climatic zones based on the results of TSUAB research; F, U, M: the subzone by the type of relief (flat, undulated, or mountainous); 1–5: numbers of the road districts.

Then let us consider the works aimed at substantiation of the calculated values of clay soils used in road surfaces design in the territory of the established road districts.

3 SUBSTANTIATION OF REGIONAL NORMS IN THE DESIGN OF AUTO ROADS

It is necessary to emphasize that the main task of the existing road-climatic zoning specification in relation to individual regions or territories of Russia is distinguishing such areas, within the limits of which the same type of road constructions are characterized by homogenous strength and resistance. The target zoning in the design and construction of auto roads requires the determination of calculated values of soil characteristics for appropriation of roadbeds and road pavements with a predetermined level of reliability (Efimenko, 2016).

The main parameters of mechanical properties of the roadbed soil, used by experts in the calculation of road pavement strength, are strength and deformability characteristics: the calculated relative moisture content W_{rel}; the elasticity modulus E_{sl}; the angle of internal friction φ_{sl}, and the specific cohesion C_{sl}. Characteristics of soil strength and deformability depend on moisture content,

density, structure, genesis and the loafing mode. Therefore, their purpose is carried out in the following sequence. First, the calculated moisture content is determined, and then values E_{sl}; φ_{sl} and C_{sl} at the calculated moisture content are set.

The results of experimental studies of the water and thermal regime of road constructions in the conditions of Western Siberia region were taken into account in the assignment of values of the calculated soil moisture content W_{rel}. The analysis of the results of our studies (Efimenko, 2016) allowed approaching the problem of forecasting the calculated values of soil characteristics a little differently, in contrast to the existing calculation and probabilistic methods (Zolotar, 1971). In particular, it has been established that spring soil moisture of the working layer of the roadbed, as applied to the studied area, to a large extent depends on the freezing mode, which can be indirectly characterized by the character of accumulation of negative air temperatures during October–December.

Considering the fact that a significant portion of the studied area is characterized by excessive moisture content, and the soil-geological and climatic conditions determine the redistribution of moisture in the roadbed in the liquid form, the method of prof. I.A. Zolotar has been accepted for the assignment of the calculated moisture content of roadbed soils in conditions of close groundwater occurrence (Zolotar, 1971; Zolotar, 1972). The scheme for the appropriation of calculation values of subgrade soils includes several sequentially performed stages. In particular, the duration of periods of autumn moisture accumulation (τ_{moist}) and freezing (τ_{fr}) are established. Then, towards the end of autumn, the moisture accumulation and the average moisture content of subgrade soils within the limits of the active zone are calculated (W^{aut}). Then, the characteristics of the freezing rate of the working layer of the subgrade (α) are determined, and the value of the calculated moisture content of soil, in view of moisture migration during freezing of the subgrade, is calculated (W^{wint}). Finally, the values of the unknown spring moisture content of the active zone of the subgrade are determined (W_{rel}).

The evaluation of the reliability of the used methods for the assignment of the calculated moisture content of clay soils for areas with close and deep level of groundwater occurrence was carried out by comparing the results of mathematical modeling with field data for the water and thermal regime of the roadbed on auto roads of Tomsk and Kemerovo regions. It has been found that at the reliability of the forecast P = 0.95 the error in determining the calculated moisture content does not exceed 4%, which is within the accuracy limits of the measurement carried out using the existing methods (Efimenko, 2016).

The actual values and patterns characterizing the course of the water-and-thermal regime of the subgrade and road pavements on the road network of the West Siberian region were taken into account during the specification of the computational algorithm of values of the calculated moisture content of the subgrade.

The complex of field and laboratory works performed by the authors in Western Siberia has allowed establishing the differences in the composition of clay soils of the working layer of the subgrade from the earlier obtained research results in the European part of Russia (Efimenko, 2014). These differences largely explain the reason for the non-compliance of the recommended calculated values of moisture content, strength, and deformability of subgrade soils, obtained in the territory of the West Siberian region.

The statistical material gained by the authors of the paper for more than a 40 year period of observations on moisture content, strength, and deformability characteristics of subgrade soils of auto roads of the West Siberian region has been obtained from more than 100 sites equipped for instrumental observations on the existing road network.

Given that the information base for the construction of mathematical models describing the properties of clay soils common in Western Siberia included the information obtained at different times, the processing of the results of field and laboratory studies was carried out with the introduction of statistical methods of interpretation. The statistical analysis included the evaluation of a set of values obtained during tests for the presence of "pop-up" variants and the verification of the possibility of combining a series of tests carried out for individual administrative units into one statistical number.

As a result of a statistical series processing it has been found that nearly 3% of the experimentally obtained values of moisture content, strength, and deformability of soils must be culled as non-related to the general population.

Due to a combination of statistical series of observation results the functional and graphical dependencies E_{sl}; φ_{sl} and $C_{sl} = f(W_{rel})$ have been established for clay soils that are most common in the investigated area. These dependencies are best approximated by an exponential curve. An example of such a dependence obtained for the territory of Kemerovo region is shown in Figure 3.

In particular, the elasticity modulus of the clay soil (dusty sandy loam) for territories of Kemerovo region can be determined using a functional dependence (1):

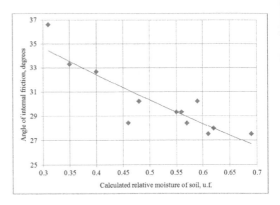

Figure 3. The dependence graph of the angle of internal friction of the clay subgrade soil (dusty sandy loam) on its calculated relative moisture.

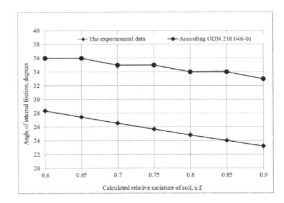

Figure 4. The comparison of the research results with the recommendations reflected in the ODN 218.046-01 carried out on the example (dusty sandy loam).

$$\varphi_{sl} = 42,41 \cdot e^{-0,671W_{rel}} \qquad (1)$$

where φ_{sl} is the elasticity modulus of soil, degrees; W_{rel} is the calculated relative moisture of the working layer of a subgrade, u. f.

The correlation coefficient between the two investigated parameters, in this case, was $R^2 = 0,79$.

The comparison of the research results on the properties of clay soils located in the territory of Kemerovo region with the Recommendations of ODN 218.046–01 (Design of nonrigid pavement, 2001) we shall carry out by means of the overlay of dependencies of the internal friction angle of the subgrade working layer on the calculated relative moisture content on each other. Based on the obvious discrepancy between the data obtained during the experiment and the ones given in standards (see Figure 4) it is possible to argue about their similarity or differences.

4 CONCLUSION

The results of experimental studies on the determination of calculated values of strength and deformability of clay soils peculiar to South-East of Western Siberia show that data ODN 218.046-01 [14] are understated in terms of moisture content by 7…10% (observing the type of terrain by the nature and the degree of moisture content) and overstated in regards to strength and deformability (e.g., the elasticity modulus by 25…30%) (Efimenko, 2015).

ACKNOWLEDGMENTS

The work is supported by the Russian Foundation for Basic Research (project No. 14-07-00673 A), some items are executed within R&D plan of the Federal Road Agency (Rosavtodor).

REFERENCES

Aldrich, H.P. "Filing system" of physiographic units helps to resolve local design criteria, Highway Res. News, **51**, pp. 42–60, (1973).

Design of nonrigid pavement, ODN 218.046-01, (2001) (in Russian).

Efimenko, S.V., M.V. Badina, *Road Zoning of the West Siberia Territory*, (2014) (in Russian).

Efimenko, V.N., S.V. Efimenko, A.V. Sukhorukov Sciences in Cold and Arid Regions, **7**, Issue 4, pp. 307–315, (2015).

Efimenko, V.N., S.V. Efimenko, A.V. Sukhorukov, *IOP Conf. Series: Materials Science and Engineering* **71**, 012049, (2015).

Efimenko, V.N., S.V. Efimenko, A.V. Sukhorukov, Key Engineering Materials, **683**, pp. 250–255, (2016).

Russam, K., J.D. Coleman, The Effect of Climatic Factors on Roadbed Moisture Condition, Geotechnique, **11**, No. 1, pp. 22–28 (1961).

Sidenko, V.M., O.T. Batrakov, M.I. Volkov, Highways (Improvement of design techniques and construction), (1973) (in Russian).

The Ministry of Transport of the People's Republic China highway natural zoning standards, JTG J003-86, China Communications Press, (1986) (in Chinese).

The Ministry of Transport of the People's Republic China, JTG D20-2006 highway route design specifications, China Communications Press, (2006) (in Chinese).

Yarmolinsky, V.A. Vestnik of Tomsk State University of Architecture and Building, **5**, pp. 152–158, (2014) (in Russian).

Zapata, C.E., W.N. Houston, Calibration and Validation of the Enhanced Integrated Climatic Model for Pavement Design (NCHRP report 602), p. 63, (2008).

Zolotar, I.A. Cold Reg. Res. and Eng. Lab. AD0750591, pp. 1–19 (1972).

Zolotar, I.A., N.A. Puzakov, V.M. Sidenko, *Water-and-thermal regime of roadbeds and pavements* (1971) (in Russian).

Advances in Energy Science and Equipment Engineering II – Zhou, Patty & Chen (Eds)
© 2017 Taylor & Francis Group, London, ISBN 978-1-138-71798-5

A study on landscape architecture design based on regional culture—case study of the plan of Wuyishan National Resort Amusement Park

Chang Li
Huaqing College, Xi'an University of Architecture and Technology, Xi'an City, China

ABSTRACT: Under the globalization and multi-nationalization background, the development of Chinese tourist scenic area has been deeply influenced, resulting in serious convergence. Many tourist scenic areas, when making landscaping designing, completely neglect their unique and magnificent regional culture. Based on the phenomenon of serious deficiency in culture at regional cultural landscape, this thesis is taking Wuyishan National Resort Amusement Park as a case, through the study of the idea of the theme, the overall layout, landscape design and plant design, to explore how to plan and construct a tourist scenic area presenting its original, unique, regional culture landscape characteristic.

1 THE CONCEPT OF REGIONALISM OF REGIONAL CULTURE AND LANDSCAPE ARCHITECTURE

As the construction of landscape architecture prospers in China, landscapes of parks and gardens are becoming increasingly identical, resulting in the disappearance of the essential distinctions of a sizable share of gardening works. In those works, local climatic conditions are neglected, popular tree species are planted blindly, and natural features of the sites are destroyed to pursue the artificialization of landscapes. Such practices are reasons behind the absence of natural forms, spatial framework and historical context in urban landscape.

To address the problems of characteristics absence and identification in landscape design, it is imperative for the designers, under the universal principles of culture and landscape design, to underline the particularity of the site, start with regional natural environment and create venues with cultural and affective commitment for people to relax mentally and physically.

1.1 *Regional culture*

Regional culture refers to all of the material and spiritual fruits that people of a certain region create, accumulate, develop and sublime uninterrupted through their manual and mental work in the course of their history. Regional culture, first of all, reflects the regional natural environment and covers different aspects of the society, including religion and faith, social conventions, building technology, economic level and lifestyle.

Different regions, due to the distinctions of their physic-geographical environment and people's different ways in nature remaking, breed different cultures.

1.2 *Regional culture*

The regionalism of landscape architecture indicates the features presented by special associations between landscape architecture and the natural and cultural environment of a designated region and it is a reflection of the features of the territory and the landscape of the site. The essence of landscape design lies in its utilization of natural elements, building parks and gardens with affective commitment that are agreeable to humans. In this regard, designs taking into consideration the regionalism of the site could give expression to the natural and ecological features of the parks and gardens.

2 LANDSCAPE ARCHITECTURE DESIGN BASED ON REGIONAL CULTURE

2.1 *To keep to the environment of the site*

Applying proper landscape elements to constitute diversified landscape space counts as a concrete content of landscape architecture design and the construction of landscape space is based on the overall background of the site environment. French landscape master Michel Corajound once said that the three principles of landscape design are site, site and site. The design must follow the site environment by investigating natural elements including local terrain, climate, hydrology and vegetation and site environment otherness also must be analyzed, hence, connected yet discriminating landscape space could be constructed.

2.2 *Inheritance of regional culture*

Traditions, customs, rites, morality, and habits originating from life experience of a certain region can arouse sense of intimacy and acceptance in people of this place. Nevertheless, identical, patterned and emotion-absent modern designs fill our life space, as a result, cultural memory of the land is lost, emotional connections between people and parks are severed. The inheritance of regional culture by landscape architecture design should make full use of the traces on site, for instance, architectural composition, traditional artisan craftsmanship, and tools of life and production to build a landscape with place spirit.

2.3 *Colour illustrations*

Materials are separated into plant material and hard material. As regards plant material, designers should fully acquaint themselves with seeds of trees and select and match them. Designers must focus on both visual effect and plant communities that are suitable for local natural environment. In terms of hard material, designers should use local materials and give full play to their characteristics like performance, texture, color and shape. This kind of practice is innovative use of local features and inheritance and development of local cultural context. Moreover, adoption of local material could lower construction cost.

2.4 *Application of appropriate technologies*

Culture covers science and technology. It is necessary for the development of technology to take humans into account. Alvar Aalto once said that only by extending the connotation of technological function to psychology scope, can architecture be human's. This is the only approach to humanize architecture. This demand is more evident in gardens and parks because people in different regions have differed adaptive capacity to technologies. New and traditional technologies must be integrated to solve the problem arising in environmental construction. Application of technologies in landscape design must pinpoint the differences of people's adaptive capacity to technologies and technologies conforming to geographical features must be found out to construct parks and gardens.

3 DESIGN PRACTICES OF REGIONAL LANDSCAPE ARCHITECTURE

Hereby, the writer takes Wuyishan National Resort Amusement Park as an instance to explore how construct modern urban parks that could reflect regional characteristics.

3.1 *To keep to the environment of the site*

Wuyishan National Resort Amusement Park, designated to meet the demands of visitors for recreational facilities, is a favorable complement to tourists recreations of Wuyishan scenic spot and holiday resort. To this end, planning targets of the park are as follows.

It is a place for people of Wuyishan to trace cultural origins and it is also a great place to speak for its own features. In the planning, tourism development and natural ecological environment should be integrated; conflicts between new technologies, new materials and traditional cultural structure should be addressed; inheritance and development of traditional culture in new age should be emphasized and new culture should be created by valuing ethnic culture.

In line with natural geographical characteristics of the site, with natural mountains, water bodies and vegetation as established conditions, co-existing, complementary and organically-unified integrity between mankind and nature should be fully demonstrated.

3.2 *Functional structure*

The whole amusement park is approximately divided into 3 parts. In consideration of cultural traditions of Wuyishan, the features that combine Taoism, Buddhism and Confucianism are integrated into the environment.

Taoism-fairyland-bliss: in the southwest of the parks, recreational events such as visitors' center, indoor and outdoor performances and fairy tale world are primary projects, aiming to create relaxing and joyful atmosphere and help visitors take off their seriousness and feel pleasure.

Buddhism-Buddha's Pure Land-openness and quietness: in the north, relatively quiet programs are built. Hushanmiao Park in current environment is a memorial park and the atmosphere is expected to be solemn and respectful. Thereby, plane form with obvious features of manual work is designed. Transformed terrain and vegetation are principal elements employed to build the environment.

Confucianism-Neo-Confucianism-deepness: culture and historical buildings locate in the southeast part. Traditional culture of Wuyishan is diversified, colorful, extensive and profound, and

Figure 1. Architectural layout of the form.

Neo-Confucianism is the most prestigious part. The cultural street and a variety cultural buildings are designed in the same area to cultivate thick cultural atmosphere.

In accordance with natural terrain, the main building is arranged in a curve along the mountain with visitors' center as its head and the center for performance as its closure. The design is interrupted in shape but connected in spirit by taking into consideration the possibility of stage constructions. The curve is extracted from folk houses and structures of Wuyishan and deduced from the stories of the totem of dragon-snake of Wuyishan. Reorganization and reemergence of folk culture and customs serves as a metaphor of Wuyishan folk culture. Parts of the buildings are embedded in the mountain and the exposed area of buildings are lessened and the buildings are made part of the mountain. The relationship between buildings and roads are diversified, as shown in Figure 1.

3.3 Planning content

3.3.1 Landscape arrangement

The arrangement of landscape axis and corridor of sight line is in 3 tiers: the first tier starts from Hushanmiao in the north of the site and ends at the monument at the top of the mountain, also known as memorial axis; the second tier is the one that runs through the center for performance along Route One and the landmark of visitors' center; the third tier starts at visitors' center, passes the stone square and closes at the tea art center.

In the planning of landscape section division, in the west, the natural terrain is moderately altered, the rough mountain transformed into mild and gentle forms, where the fairy tale world and grass skating are set. Meanwhile, the horizon is broadened and building group of visitors' center is highlighted in the amusement park; in the north, the design is combined with Hushanmiao and in regular style, focused on plants and the solemn and

respectful memorial atmosphere is lighted; in the east, a streamline layout is presented combining the natural terrain with cultural buildings placed along the routes to exhibit profound culture and tradition of Wuyishan, as shown in Figure 2.

3.3.2 Plant planning

The original vegetation is mainly retained but in order to underline the characteristics, vegetation in large area is artificially divided. Specifically speaking, strip vegetation zones with a strong flavor of geometry transformed by manual work are integrated into the original natural vegetation. The artificial work is strengthened and at the same time, the original natural conditions are not disturbed, which is both easy to operate and remarkable in effects.

3.3.3 Architectural planning

In the design of single buildings, local traditional styles are primarily adopted with parts of the buildings interconnecting with the mountain. The visitors' center, though with a relatively strong sense of modernity, some designs of top of slope and some features of local traditional buildings successfully make it a harmonious part of the wholeness, as shown in Figure 3.

3.3.4 Water system planning

The amusement park channels water from the north, on one hand, it makes itself independent of Hushanmiao memorial park naturally, on the other hand, kinds of aquatic recreational amenities are built to form the recreational line of the park. It is also a symbol of Jiuqu Brook.

Figure 2. Visitors center building community.

Figure 3. Architectural style.

4 SUMMARY

The planning and design of Wuyishan National Resort Amusement Park, on the basis of in-depth field visits, utilizes the existing landscape features of the site, regards the space of natural elements as the main shape and explores cultural landscapes profoundly. A natural tourist and holiday resort amusement park demonstrating regional landscape features and presenting colorful leisure and entertainment activities is founded.

Contemporary designs are expected to return to nature with space designs conducted in nature and suitable to regional features. Designs, if neglect on-site natural conditions and regional features, go after what is big and foreign, loading landscapes in a vulgar taste that can not be in harmony with surroundings. As a result, garden works will be likened as water without a source, lacking life and vitality.

REFERENCES

Han Bingyue, Shen Shixian. Landscape Architecture Design based on Regional Characteristics [J]. Chinese Landscape Architecture: 2005(7):61–67.

Lin Qing, Wang Xiangrong. Regional Features and Landscape Forms [J]. Chinese Landscape Architecture, 2005(6):16–24.

Meng Gang, Li Lan, Li Ruidong, Wei Shu. Design of Urban Parks [M]. Shanghai: Tongji University Press, 2005.

Shen Sanling, Wang Yizhi. Architectural Innovation and Regional Culture-on Planning and Design of Huanglong Scenic Spot [J]. Architectural Journal, 2003(4):52–54.

Yu Kongjian, Wang Zhifang, Huang Guoping. On Rural Landscape and its Significance to Modern Landscape Design [J]. Huazhong Architecture, 2004(4):123–126.

Zhang Chuan. Place Design Based on Regional Culture [D]. Nanjing: Nanjing Forestry University, 2006.

Zhou Jianming. Develpment Trend and Planning Characteristics of Tourist Holiday Resort [J]. Urban Planning Overseas, 2003,18(1):25–29.

Zhu Jianning, Ma Huiling. Landscape Architecture Art Returning to Natural Culture [J]. Landscape Architecture, 2005(3):25–30.

Advances in Energy Science and Equipment Engineering II – Zhou, Patty & Chen (Eds)
© *2017 Taylor & Francis Group, London, ISBN 978-1-138-71798-5*

Analysis of separation model and current status of the energy consumption of large public buildings in Changchun, China

Junyan Dong
School of Architecture, Harbin Institute of Technology, Harbin, China
School of Architecture and Design, Changchun Institute of Technology, Changchun, China
International Asia Art Academy, Banggkok Thonburi University, Bangkok, Thailand

Wen Cheng
School of Architecture, Harbin Institute of Technology, Harbin, China

Kechao Li, Yuenan Xing & Tenglong Tan
School of Architecture and Design, Changchun Institute of Technology, Changchun, China

ABSTRACT: Large public buildings constitute a large proportion of the total energy consumption of civil buildings in China. On the basis of the investigation in more than 100 large public buildings in Changchun, China, the annual electricity consumption, system equipment consumption, and operation status are obtained. According to the research results and data analysis, we summarized the features of energy consumption in these buildings, proposed the different features of different types, and analyzed the current status of energy consumption of large public buildings by calculating their energy consumption.

1 INTRODUCTION

At present, the energy consumed by buildings in the urban areas of China is 22–26% of the total commodity energy. The building energy consumption we talk about here is the energy that we use to run the building, excluding the energy consumed for material production, transportation, and maintenance or repair cost. Therefore, building energy consumption is the energy consumed for illuminating, heating, and cooling purposes in the building. Large public buildings generally refer to buildings with construction area of more than 20000 m² (Dong, 2011).

It is evident from Table 1 that buildings occupying only about 4% of the total area of China consume 22% of the total energy. The energy consumption per unit area of large public buildings is 10 times more than general public buildings (Dong, 2014), which have enormous energy-saving potential.

2 ANALYSIS OF SEPARATION MODEL OF LARGE PUBLIC BUILDINGS IN CHANGCHUN

2.1 *The constitution of the project model for energy consumption*

The energy consumption of large public buildings depends on the different energy systems equipped.

Table 1. Classification and status of building energy consumption in China.

		Average (million m²)	Status
Rural area	(commodity-energy only)	240	0.4 million tonnes of coal/year, 900 GWh/year
Heating in the northern urban area		65	1.3 million tonnes of coal/year
Urban area (exclude the energy of heating)	Dwelling house	90	2600 GWh/year
	General public buildings	65	1500 GWh/year
	Large public buildings	5	900 GWh/year
	Total	160	5000 GWh/year
Total	1.7 million tonnes of coal, about 12% of total; 5900 GWh, about 29% of total; energy consumption accounts for 20–22% of commercial energy		

Natural gas, coal, and steam can act as supplements to electric power. This energy split model is mentioned only for the consumption of electric power in this paper.

Power consumption of large public buildings is much higher than that of general public buildings; whose energy-saving potential is enormous (Dong, 2011). However, distribution systems of large public buildings are heterogeneous, that is, for single buildings, how could we find out where the energy consumption is too high? Is it the air-conditioning energy consumption, lighting energy consumption, or the power consumption? We do not know. Only by making the project model of the energy consumption of public buildings, high-energy branch can be found and reformed. Considering all factors, the large public building energy consumption model is divided into four major systems: lighting system energy consumption, air-conditioning system energy consumption, power equipment energy consumption, and special region energy consumption (Dong, 2015). The split project model of energy consumption of large public buildings is shown in Figure 1.

The principles of the project are as follows:

① In order to understand the power consumption of the various parts of buildings, the project model should be simple, clear, and easy to analyze.

② Split project model of energy consumption is the basis of monitoring system in the future. Energy consumption data collection should be considered during the research of the split project model of energy consumption.

2.2 Calculation method of split project model of energy consumption

As the distribution system of large public buildings is extremely complex, we cannot directly measure

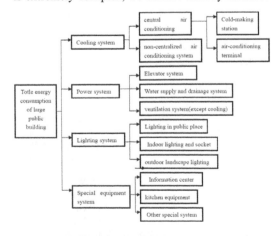

Figure 1. Split project model of energy consumption.

the real-time energy consumption for each terminal unit. In general, only the main power distribution branches are measured. In a single major branch, the devices linked to it are not of the same type. Therefore, energy split is needed (Dong, 2015).

There are two ways to collect energy consumption data: direct acquisition and indirect acquisition.

In direct acquisition, we can measure the data by an ammeter equipped with the branch. In indirect acquisition, no ammeter is present in the branch. The data can be obtained by subtracting the number gained by direct acquisition from the total value. For these two methods, we conclude the following three basic algorithms for energy consumption split model.

1. Subtraction principle

The subtraction principle is the method we used in the indirect acquisition way. In the branches with no electric meter, the energy consumption is the obtained by subtracting the directly measured data from the total value. Subtraction principle is relatively simple and accurate. There are two cases of using the subtraction principle.

① Case one: Only one branch named "j" has no electric meter:

$$E_j = E_t - (E_1 + E_2 + E_3 + ...) = E_t - \sum_{i=1}^{n} E_i \quad (1)$$

where E_j is the power consumption of the j branch. In this case, the use of subtraction principle is equivalent to direct measurement. Data are accurate.

② Case two: n branches are not under measurement:

$$E'_j = E_1 + E_2 + E_3 + ... E_n = E_t - \sum_{i=1}^{n} E_i \quad (2)$$

where E'_j is the total power consumption of these n branches. To get the power consumption value of these branches, we need to split the calculation of its energy consumption. In this case, the results are not as accurate as that of case one. We need the second method here, which is called "the proportion of capacitance."

2. The proportion of capacitance

In the design of electrical system, capacitance of every branch is designed under its equipment load (it usually has a safety factor). If the system runs according to the design, the actual proportion of the energy consumption should be as much as the proportion of designed capacitance. Using this characteristic, we can do the split calculation with the branches in case two.

Following is the expressions of "the proportion of capacitance,"

$$E_{n1} = \frac{Q_{n1}}{\sum\limits_{i=1}^{n} Q_n} E'_j \qquad (3)$$

where E_{n1}—the calculated capacitance of branch "n1," one branch in n branches;

Q_{n1}—designed capacitance of branch "n1";

E_j—the actual power of remaining branches, which we get from "subtraction principle."

Although we get the calculated capacitance of each branches, the data are not absolutely correct. We need to modify the result by using modifying coefficient through the following the expression:

$$k_n = \frac{E'_{n1}}{E_{n1}} = \frac{E'_{n1}}{\dfrac{Q_{n1}}{\sum\limits_{i=1}^{n} Q_n} E'_j} s \qquad (4)$$

where E'_{n1}—the actual energy consumption of branch n1.

Some transmutation to expression 4 gives:

$$E'_{n1} = k_n \frac{Q_{n1}}{\sum\limits_{i=1}^{n} Q_n} E'_j \qquad (5)$$

where E'_{n1} is the modified energy consumption of branch n1. Modifying the calculated result by using the modifying coefficient has two conditions: ① the modifying coefficient can be obtained by testing; ② the percentage of the actual energy consumption in total energy consumption is steady. If we could not reach these two conditions, we can use k_n. Although it will reduce the precision, it is still very important to the contrast of different items of building energy consumption.

3. Optimization splitting calculation

In increase the precision of the calculation, we use the optimization method in mathematics to deal with the energy consumption splitting problems. It mainly focuses a single branch, which has different kinds of terminal equipment attached to it.

Basic principle: ① give the terminal equipment's estimated number of energy consumption; ② introduce the energy consumption of the branch as restriction; ③ use parameters to estimate the precision.

2.3 *Methods of raising precision*

The energy consumption splitting calculation between several complex branches or different kinds of equipment in one single branch will usually have some calculation mistakes. We should also focus on increasing the precision during the calculation.

① The power consumption of one branch is too small, or one kind of power consumption in a single branch is comparatively too small to another, and it can be ignored.

② One branch is comparatively stable; we can split it directly by testing results such as information center of a large public building.

③ For branches in which real-time energy consumption changes are unstable (such as fan coil units and the light-socket mixed slip), we cannot directly measure one instrument's actual power consumption. By measuring running gear and running time, we estimate the power consumption by rated power.

3 CONTRAST OF ELECTRIC POWER USING ACTUALITY OF A LARGE PUBLIC BUILDING IN CHANGCHUN

3.1 *Overview of large public buildings' energy consumption in Changchun*

As the capital of Jilin Province, Changchun has developed enormously in recent years. According to the data provided by Changchun Bureau of Statistics, the city's GDP is 271.91 billion yuan in 2015, which is a 14.5% increase from that of the previous year. According to the city's resident population of 8,324,600, GDP per capital in 2015 is $4737. Area of public buildings is of 8.69 million m^2 in Changchun, with 5.41 m^2 per capital. The amount of electric power used in Changchun in 2011 increased 12.62% from that of the previous year. It is the first time in history, more than 100 million kWh, up to 10.908 billion kWh, energy was consumed, much more than that of other cities in the northwest area.

The investigation of large public buildings' energy consumption in Changchun started in 2013. We have investigated more than 100 public buildings, including shopping malls, hotels, office buildings, and cultural venues. The investigation includes basic information, energy consumption structure, equipment of the building, and energy consumption per month in the last 3 years. In addition, we measured every building for continuous temperature and humidity.

3.2 *Comparative analysis of large public buildings' power consumption in Changchun*

In order to measure a single building's energy consumption level in similar buildings, here we adopt

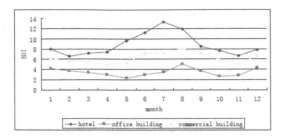

Figure 2. EUI values of three different types of building per month.

the energy consumption of unit floor area (EUI) as a parameter to analyze the energy consumption of buildings:

$$EUI = \frac{E_{sh}}{S} \tag{6}$$

where EUI—the actual energy consumption of building;

S—total area of the building.

In order to analyze the power consumption characters of different types of buildings, we randomly selected 15 typical buildings, including five hotels, five office buildings, and five commercial buildings.

It is evident from Figure 2 that the average electric power consumption of hotels is the highest class of building energy consumption, followed by commercial buildings and office buildings, the lowest. Three types of constructions' peak value appear during the hottest time in summer. The average peak value of hotels is the highest, which appears around July and is unique. The curves of commercial buildings and office buildings are relatively flat.

Reasons for these power consumption characteristics are:

① There are off-season and mid-season in hotels. During the peak season, the room occupancy rate is high.

Power consumption naturally increases with increased room size, which is opposite in the off-season.

② Commercial buildings' business hours affect the power consumption level. The business hours of commercial buildings end at 20:00, that is 10:00 pm, longer than office buildings.

③ Commercial buildings require a higher quality of environment, resulting in waste of energy. In order to provide customers with better shopping environment, the power consumption of lighting and air-conditioning in commercial buildings is much larger than that of office buildings. There is serious waste of energy in existence.

4 CONCLUSION

This paper started form the discussion of the split model of large public buildings' energy consumption and then went on to the analysis of current situation of large public buildings in Changchun. Changchun is the core city of the exploitation of the western area, whose total energy consumption is much less than that of developed cities in the eastern area, such as Beijing and Shanghai. However, the power consumption of large public buildings is almost the same. There are serious risks of unreasonable use of energy whether from the system or human factors. With the deepening of building energy-saving ideas, establishment of a comprehensive open-energy sub-metering system is necessary. It will provide a solid theoretical foundation for energy-efficient monitoring and system alteration of large public buildings in the future.

ACKNOWLEDGMENTS

This work was financially supported by the Science and Technology Project Foundation of Ministry of Housing and Urban-Rural Construction of the People's Republic of China (No. 2016-R2-005), Jilin Provincial Social Science Foundation (No. 2015BS83), Youth Foundation of Changchun Institute of Technology (No. 320140006), and Teaching reform project of Changchun Institute of Technology.

REFERENCES

Junyan Dong, Hong Jin, Jian Kang, Xi Chen. A pilot study of the acoustic environment in residential areas in Harbin, towards the questionnaire design. Journal of Harbin Institute of Technology. 2011,18(2), pp. 319–322.

Junyan Dong, Hong Jin. The design strategy of green rural housing of Tibetan areas in Yunnan, China. Renewable Energy, Vol. 49, pp. 63–67.

Junyan Dong, Wen Cheng. Based on the Characteristics of Respondents and the Voice of the Urban Neighborhood Public Space Business Facilities Noise Environment Evaluation Research. Journal of Harbin Institute of Technology. 2014,20(4), pp. 103–109.

Junyan Dong, Wen Cheng. Research on Optimized Construction of Sustainable Human Living Environment in Regions where People of a Certain Ethnic Group Live in Compact Communities in China. World Renewable Energy Congress XIV. 2015.6

Junyan Dong. The Acoustic Environment Research of Construction Exterior Based on the Ecology Idea. Materials Engineering and Environmental Science. 2015.9

Advances in Energy Science and Equipment Engineering II – Zhou, Patty & Chen (Eds)
© *2017 Taylor & Francis Group, London, ISBN 978-1-138-71798-5*

Study of the conservation technology of Chinese architecture heritage

Xi Chen

Architecture Department, Soochow University, China

ABSTRACT: The values of the Yingxian Wooden Pagoda can be viewed from three aspects: the building technology; the architecture form; and the historical and cultural meaning. All these aspects lie in the authenticity of the Pagoda itself. Through an analysis of the defects of the pagoda and the causes of such defects, we find that the damages to the wooden parts of the Pagoda are caused by its weak wooden structure, particularly its insufficient compressive strength. On the basis of the principles of authenticity, necessity, reversibility, and identity, we propose the addition of a freestanding steel framing between the internal and external colonnades of the Pagoda. Details of structure reinforcement and maintenance are elaborated in this paper.

1 VALUES OF THE YINGXIAN WOODEN PAGODA

Yingxian Sakyamuni Pagoda of Fogong Temple (generally known as the Yingxian Wooden Pagoda), which is the oldest and the most grandiose multi-storied wooden pagoda in China, was constructed in 1056. In 1961, it was listed among the first batches of the Conserved Core Heritage Site at the National Level. The Pagoda has experienced a history of over 900 years and various adverse environments. Being entrusted, China Academy of Urban Planning and Design undertook its site investigation based on the previous studies and submitted a report. Several explorations and studies were made since the start of the reinforcement of this scheme.

The values of the Yingxian Wooden Pagoda can be viewed from the following three aspects:

First, Yingxian Wooden Pagoda, as one of the few examples in Chinese ancient timber-framed buildings and the world's existing tallest timber-framed building, is the masterpiece in the history of world architecture. Consequently, it has an important value as an existing sample for studying the structure and construction. It shows the limit of strength, height, and dimensions that a timber-framed system can provide and the length of time it can sustain. Therefore, it is the most important sample of this kind of buildings that remains at present.

Second, Yingxian Wooden Pagoda is an important representative of the early palatial architecture of Chinese traditional timber-framed building. The feature of this Pagoda is constituted by super-posed layers with a hollow center, that is, the "palatial form," which is recorded in the structural

carpentry model in Volume 31 of *Yingzao Fashi*. Throughout this kind, the Yingxian Pagoda is the most complicated and magnificent example.

Third, Yingxian Wooden Pagoda is also a carrier of the memories of the 900-year-long history. Moreover, this pagoda, a living specimen of this kind, is the only one that has survived around the world. The damage condition after nearly 1000 years, the signs of various incidents over the years, and the traces of repeated restorations in history all bear irreplaceable and important information of historical and scientific value.

The three values of the Pagoda are ensured by the authenticity of the building itself. The relationship between various members and their deformation condition after bearing sustained stress over years provide us with important technical information, and the remaining information on many members provide us with traces of possible changes concerning the architectural form over the years. The Pagoda is also a physical witness of over 900 years of history. Therefore, its authenticity is more important. The historic appearance of the Pagoda itself has become an important component of the information it carries. As only material evidence can provide us with such emotional value, it is also an irreplaceable aspect of architectural cultural relics.

2 THE DAMAGES AND THE CAUSES

According to the "Analytical Report of Defects Survey for Yingxian Wooden Pagoda" submitted by Shanxi Institute of Ancient Architecture Conservation in May, 2000, and the site survey, the main damages lie mainly in the longitudinal and latitudinal compression of the whole structure, splitting off, and leaning of the columns (Fig. 1), breaking of the tenon at column feet, breaking of the head of leaning huagong, and the tearing of the end of pupaifang. Those problems cause the inclining and twisting of the internal and external colonnade, and the horizontal displacement of the beam frame. As a result, the loading system has been damaged to a great extent, thus posing a serious instability to the structure of the Pagoda.

During the 900 years of existence of the Pagoda, six major maintenances were done, including the first maintenance, which was undertaken between 1191 and 1195AD, and also others respectively in 1320, 1508, 1722, 1866, and 1928 through 1929. It has been affirmed that the first maintenance and that done in 1508 benefitted the reinforcement of the Pagoda. The Pagoda had stood immovable from its completion up to the maintenance, although it had experienced several major earthquakes; this was a strong proof of the maintenance result. However, the reinforcement efforts in the past have failed to solve the problems of the strength of the materials.

Comparing the damages of different floors, it was found that the lower parts were much worse than the upper parts, and this evidently proves that the insufficient strength of the wooden structure with insufficient compressive strength was the main cause that brought about the damages of the wooden members. In fact, the wooden pagoda has

Figure 1. Damages of the columns.

already met or even surpassed the limit of the wood stress, and that is the bottleneck, which should be solved in the first place for conservation plan and reinforcement measure.

3 THE CONSERVATION TECHNOLOGY

3.1 Maintenance and reinforcement scheme

The primary objective of the present maintenance and reinforcement is to ensure structural safety by completely solving the fundamental problem of insufficient strength of the wood. On this premise, the whole historical information and the value of historical relics of the Pagoda shall be maintained as much as possible. As a conservation project, in addition to achieving this objective, the principles of integrated conservation, the maximum preservation of authenticity, necessity, reversibility, and identifiability of architectural cultural relics that are internationally recognized shall also be adhered to.

Judging from the features of the structural system and force mechanism of the Pagoda itself, the Yingxian Wooden Pagoda belongs to the early palatial architecture of Chinese traditional timber-framed buildings. The Pagoda can be divided horizontally into 10 superposed structural floors one after the other: five body floors, four platform floors, and one roof. As holistic structure floor of the pagoda, the platform floor plays the leading role of the structure and directly rest on the pupaifang of the colonnade. The colonnades, which support the bucket arch, form the main part using space. The continuous superposition of the two major parts constitutes the basic elements of the structural system. The platform floor is very rigid while the various visible floors formed mainly by colonnades are much lower in strength for this kind of structure. Taking all these features into consideration, we consider that it is possible to erect an intervention of a stand-alone steel frame inside the Pagodato support to every respective platform floor that plays the role as the main load-bearing structures from the bottom. The columns of the steel frame shall replace the original weak colonnades that are badly damaged and help these structural members to release from heavy load.

3.2 Foundation

The intervened steel frame shall have an independent foundation, which is to be located in between the walls of the internal and external colonnades on the ground floor. The superstructure is to be constituted by eight vertical column frames and three beam frames. The beam frames, which are directly under every platform floor and connected with the small beams, are also octagonal in plan. It will bear the vertical load above the four platform floors of the Pagoda and also the deadweight of the floors underneath in order to relieve the load of the original supporting members. Furthermore, it will keep the stability of the entire pagoda by its connection with the bottom beams of the platform floors.

The small beams are directly supported by the beam frames as the supporting flexural members shall have ample stress reserves, and it is evidently most beneficial for the steel frame supports to be located in here. Because of the much less damages to the key structural members over the third floor, the first-phase steel frame shall be erected only up to the fourth platform floor.

The foundation of the steel frame is to be constructed in the tamped silt at the middle of the Pagoda foundation (the octagonal platform) (Fig. 2). Raft foundation with wider underside shall be adopted in between the walls, and upstanding beams shall be adopted at short sides. As there should not be much intervention for the original structure and its foundation during the reinforcement, and the original buried depth of the pagoda plinth is quite shallow (the experience of nearly 1000 years proves that such shallow foundation is quite feasible), it has been finally decided that the underside of the raft foundation of the steel frame shall be flushed with the foundation of the original main columns.

3.3 Column and beam frame

There shall be 16 main columns for the steel frame, and every two columns shall constitute a column frame to be erected onto a foundation upstanding beam and welded with the embedded element of the beam through sprag (Fig. 3). Between various columns, there shall be load-carrying I-beams and collar beams; besides, there shall be sprags made of unequal-angle steel inside the narrow ground floor passage and second- and third-floor platforms to increase the vertical rigidity of the whole Pagoda. The main columns shall be arranged near the corner columns of the internal and external colonnades as near as possible to the beams being supported. The columns are to be arranged in the remaining spaces of each floor, which are left after the superposition of the main structural members. While passing through various floors, the columns shall not interdict any structural entities with the exceptions of the floor slabs and some removable elements (such as staircases). As the whole framework of the Pagoda has been twisted and deformed, the final column arrangement shall be in the shape of two anomalous octagons. The material and top elevation of the collar beams between the columns are the same as that of the main beams. Some sprags of the column frames shall be removed or relocated due to the arrangement of the staircases.

There shall be three different beam frames in vertical directions, and each beam frame is composed of eight horizontally arranged trusses of anomalous octagons as the support of each platform floor. Every horizontally arranged truss is composed of two I-beam girders and some sprags. The top elevation of the I-beam girder of the beam frame should be determined according to the bottom elevation of the small beams joining the

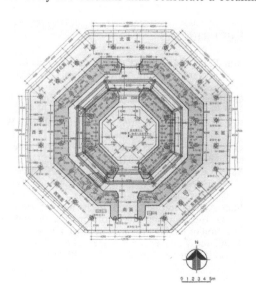

Figure 2. Foundation of the steel frame.

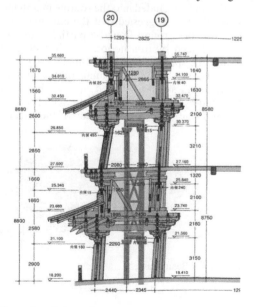

Figure 3. Column frame.

645

bucket arch of internal and external colonnades for each platform floor (Fig. 4). The space between the top of the girder and the bottom of the small beam is to be adjusted by wedge stow wood. If the space exceeds 100 mm, another rectangular hard wood shall be filled up. Filling in wedge stow wood in the space between the top of the girder and the bottom of the small beam at different elevations can relieve the concentrated stress at the supporting point, which can serve as a flexible connection between the steel frame and the original structure, and the adjustment of the tightness of the wedge stow wood is actually the adjustment of the actual load of the steel beam.

3.4 *Staircases*

The original staircases inside the Pagoda are located outside the external colonnade in the clockwise direction from the ground floor in the southwest up to the fifth floor. In general, there is a direction change for each run of the staircase, with only few exceptions. The original staircases are independently structured, whereas after erecting the steel frame, some stairs shall be adjusted, for example, some timber platforms shall be supplemented or the original staircases be relocated. Detailed mapping and camera recording will have to be done for all staircases that are to be relocated, and the temporary removed elements such as floors and footplates have to be properly numbered and stored to facilitate later possible restorations.

After the steel frame is erected, the main body of the Pagoda is reinforced, and the remaining work can be included into the routine maintenance schedule. It is suggested that the lane and pupaifang on the ground floor be partially supported or reinforced by iron hoops. Another suggestion is that the spaces between the small beams, rough small beams, and shunfuchuan of the four platform floors should be filled up with blocks and fastened by iron hoops so that they can become an integral whole. Thus, the function of the integrated

Figure 4.   Beam frame.

beam frame can be brought into full play. It may also be a good suggestion to underlay a long V-iron under the main small beam when it is necessary to increase the carrying capacity at its end.

4   CONCLUSION

The foregoing is a suggested scheme for the maintenance and reinforcement of the Yingxian Wooden Pagoda with a supporting steel frame, which has the advantage of less intervention with the original elements, thorough elimination of the hidden danger of further damage to the Pagoda, and maximum preservation of the historical information and relic value. This is a "reversible" design, which features in lower cost, less construction materials, shorter construction period, and evident identity.

Proposed here is a working scheme, which can be executed right away with the local manufacturing capacity and technical availability. In case high-strength materials and new technologies can be adopted, the cross sections can become smaller, which will result in less visual interference. Necessary treatment can be done to the shape of the steel frame to serve as a foil to the original wooden structure to emphasize the features of the two to gain even better result.

ACKNOWLEDGMENT

This work was supported by the National Natural Science Foundation of China (Grant No. 51508361).

REFERENCES

Brandi, C., 2005. Theory of restoration, trans C. Rockwell, NardiniEditore, Istituto Centrale per ilRestauro/NardiniEditore, Florence.

Chen Mingda. Yingxian Wood Tower. Cultural Relics Publishing House. **19** (1980).

Feiden, B. M, Conservation of historic buildings. World Architecture, 3, **7–10** (1986).

Giovannoni, Gustavo, L'urbanisme face aux villesanciennes. editons du seuil (1998).

Shanxi Institute of Ancient Architecture Conservation, Analytical Report of Defects Survey for Yingxian Wooden Pagoda (2000).

State Administration of cultural heritage of the rule of law. International Charter of conservation and restoration of monuments and sites, **164**. (1993).

Tian Xun, Wang Yourong. State records of Shanxi province. Yingxian County County Office, (1984).

Study on the startup and sliding mechanism of landslide using discontinuous deformation analysis method

Dong-dong Xu

Key Laboratory of Geotechnical Mechanics and Engineering of Ministry of Water Resources, Yangtze River Scientific Research Institute, Wuhan, China

ABSTRACT: Discontinuous Deformation Analysis (DDA) method is a discontinuum-based numerical method, which is developed especially for modelling the movements of discrete rock blocks. Utilizing the functions of DDA to model real time and large deformation, the whole motion process of the landslide including starting to move, accelerating, decelerating and finally forming the deposits was reproduced through the exhibitions of displacement vector diagrams, velocity contour plots and the curve of average velocity versus time. Meanwhile the reasons for causing different motion characteristics were also revealed. The motion of landslide lasted for 23 s and there were two obvious velocity crests. The results suggest that the main reason for causing the landslide is the reduction of mechanical parameter of the controlling structural planes under the effect of rainfall and lack of rock mass strength at the bottom of the landslide.

1 INTRODUCTION

Rock mass is a complex geologic body, which includes various discontinuities such as crack, joint and bedding. Many economic construction activities are relative to rock mass engineering, such as the exploitation of groundwater resource, constructing reservoir or digging tunnels. However, analytical resolutions for rock mass engineering problems are not so easy to obtain. With the development of computer technology, numerical method has become a main tool for solving the rock mass engineering problems. As known, Discrete Element Method (DEM) (Mishra, 1992) and DDA (Shi, 1985) are the two main discontinuum-based numerical methods developed especially for modelling the mechanical behaviours of discontinuum such as the rock mass.

DEM was firstly proposed by Cundall in 1971, which is based on the Newton's second law. The explicit time integration scheme adopted in the classic DEM can ensure the high computation efficiency. Rock mass is viewed as a series of rigid or deformable blocks generated by the intersection of discontinuities in DEM. The calculation model of contact force is based on the relationship between force and displacement and contact force is represented by the penetration between contact blocks. At a given time, the interest block will suffer with the force and moment of force when subjected to some external disturbances, so the acceleration can be calculated by the Newton's second law.

Furthermore, the velocity, displacement and even the deformation can be obtained. When the interest block adjusts to the new position, it will again suffer with the new force and moment of force. The process will never stop until all the blocks have reached a certain equilibrium. DEM has been widely used in the rock mass engineering and the universal two-dimensional and three-dimensional commercial software UDEC and 3DEC have been developed. However, a small enough time step is required to improve computation efficiency and stability in DEM.

DDA is based on the principle of minimum potential energy, which integrates the items of block deformation and contact treatment into the same global stiffness matrix. No tension and no penetration between blocks are strictly satisfied after the open-close iteration. The implicit time integration scheme is adopted in DDA and it needs to integrate and modify the global stiffness matrix at each time step. Abundant research results about DDA were obtained, such as the contact problems (Zheng, 2009; Bao, 2014), crack propagation based on DDA (Tian, 2014), three-dimensional DDA (Jiang, 2004; Wu, 2005; Beyabanaki, 2010; Grayeli, 2008) and engineering applications (Wu, 2006; Wu, 2007).

DDA is very suitable for modelling large deformation and real time. Take a slope in hydraulic engineering in China as an example, the paper focuses on modelling the whole failure process of landslide including starting, accelerating, decelerating and finally forming the deposits using DDA and trying to reveal some important deformation mechanisms and phenomena.

2 THE FOUNDATIONS OF DDA

Different from the continuum-based numerical method such as the Finite Element Method (FEM) limited to small deformation analysis, DDA is developed especially for modelling the large displacement and deformation, which is realized by accumulating the incremental displacement and deformation at each time step.

2.1 The displacement functions of DDA

DDA is a special case of Numerical Manifold Method (NMM) (Zheng, 2014; Zheng, 2015) in essence. A block in DDA can be viewed as a physical patch in NMM whose corresponding weight function is 1. Degrees of freedom defined on the DDA block or says physical patch in NMM should be an incremental form exactly (Jiang, 2009), which can be expressed as

$$\{\Delta D_e\} = \{\Delta u_0 \quad \Delta v_0 \quad \Delta \gamma_0 \quad \Delta \varepsilon_x \quad \Delta \varepsilon_y \quad \Delta \gamma_{xy}\}^T \quad (1)$$

where $(\Delta u_0, \Delta v_0)$ are the incremental translational displacements at the centroid (x_0, y_0) of the block, $\Delta \gamma_0$ is the rotation increment, and $(\Delta \varepsilon_x, \Delta \varepsilon_y, \Delta \gamma_{xy})$ are the strain increments.

Displacement increments at arbitrary point (x, y) of the block can be expressed as

$$\begin{Bmatrix} \Delta u \\ \Delta v \end{Bmatrix} = [T_e]\{\Delta D_e\} \quad (2)$$

where

$$[T_e] = \begin{bmatrix} 1 & 0 & -(y-y_0) & (x-x_0) & 0 & (y-y_0)/2 \\ 0 & 1 & (x-x_0) & 0 & (y-y_0) & (x-x_0)/2 \end{bmatrix} \quad (3)$$

2.2 Equations of motion and its discrete form in DDA

The equations of motion in DDA can be expressed as

$$[M]\{\ddot{D}\} + [C]\{\dot{D}\} + [K(\{D\})]\{D\} = \{F(t, \{D\})\} \quad (4)$$

Where $[M]$, $[C]$ and $[K]$ are the mass matrix, damping matrix and stiffness matrix respectively, $F(t, \{D\})$ is a time-varying load. The detailed expressions can be found in reference (Doolin, 2004).

Denote $\{D_n\}$ and $\{D_{n+1}\}$ as the displacements $\{D(t)\}$ at time t and $\{D(t+h)\}$ at time $t+h$ respectively, where h is the time step. So the discrete form of the equations of motion can be represented as

$$[M]\{\ddot{D}_{n+1}\} + [C]\{\dot{D}_{n+1}\} + [K]\{D_{n+1}\} = \{F_{n+1}\} \quad (5)$$

with the initial conditions as

$$\begin{cases} \{D(0)\} = \{0\} \\ \{\dot{D}(0)\} = \{D_0\} \end{cases} \quad (6)$$

The solution of equation (5) is given by Shi

$$\{\ddot{D}_{n+1}\} \approx \frac{2}{h^2}(\{D_{n+1}\} - h\{\dot{D}_n\}) \quad (7)$$

$$\{\dot{D}_{n+1}\} = \frac{2}{h}\{D_{n+1}\} - \{\dot{D}_n\} \quad (8)$$

Substitute equations (7) and (8) into equation (5), we obtain that

$$[\hat{K}]\{D_{n+1}\} = \{\hat{F}\} \quad (9)$$

where

$$[\hat{K}] = (\frac{2}{h^2}[M] + \frac{2}{h}[C] + [K]) \quad (10)$$

$$\{\hat{F}\} = \{F_{n+1}\} + (\frac{2}{h}[M] + [C])\{\dot{D}_n\} \quad (11)$$

It should be pointed that the item of damping $[C]$ is not included in the implicit DDA, as the result that the time integration in DDA belongs to the constant acceleration method which has introduced numerical damping that is beneficial to the solution stability. Obviously, the above expressions are the same to that of DDA when no damping item $[C]$ is considered.

DDA belongs to the updated Lagrangian formulation, which means that the configuration at the end of each time step should be updated. Therefore it would be specially mentioned that the equation $\{D_n\} = \{0\}$ should be satisfied at the beginning of each time step, which is not pointed out in reference (Qu, 2014). Obviously $\{D_{n+1}\}$ is the incremental displacements.

3 INTRODUCTION TO SLOPE ENGINEERING

Geological section of the slope is shown in Figure 1. The slope is in the region of strong unloading. Lithology of the slope in the modelling region is

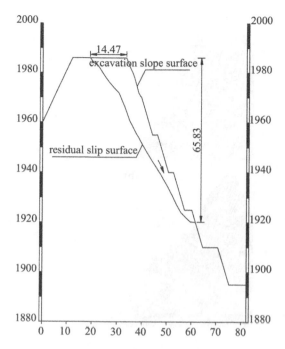

Figure 1. Geological section map of the slope.

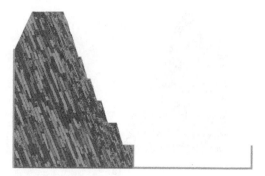

Figure 2. DDA model.

single, which is the thin metamorphic quartz sandstone and splint rock belonging to the cataclastic structure.

Strike of the stratum is 27°, dip direction is SE and dip angle is 67~71°. There are 3 groups of fractured structural planes:

The first group: strike 65~75°, dip direction NW, dip angle 75~80°. The lengths of the cracks are 15~20 m and the spaces are 2~3 m.

The second group: strike 290~300°, dip direction NE, dip angle 68~70°. The lengths of the cracks are 3~6 m and the space are 1.2~2 m. The defects in the thin sand slate are mainly in the form of micro-cracks whose lengths and spaces are generally 0.5~1 m and 0.1~0.3 m respectively.

The third group: strike 330°, dip direction SW, dip angle 28~30°. The lengths of the cracks are generally less than 3 m and the space are 1.2~3 m.

4 MODELLING LANDSLIDE USING DDA

DDA model as shown in Figure 2 is established according to the geologic conditions mentioned above. A U-shaped groove is set in order to observe the piling pattern of the landslide when it stops. The mechanical parameters considering the negative effect of rainfall are adopted.

The displacement vector diagrams in the process of landslide movement are shown in Figure 3.

(a) The landslide wholly presents the trend of downward movement, because the rainfall has weakened the mechanical parameters of the structural planes. The front of landslide is about to slip off along the slip plane under the compression of the upper landslide as the result of the lack of rock mass strength. (b) Most of the landslide is slipping along the slip surface whose dip angle is about 60°. (c) The middle and posterior of the landslide is still sliding along the above slip plane, whereas the front has slipped out of a certain distance and the slip direction is parallel to the plane whose dip angle is about 48°. (d) The landslide slides in the direction parallel to the plane whose dip angle is 52°. (e) Half of the landslide has rushed into the ground and the left is still slipping along the slope surface. (f) The landslide has stopped on the ground and formed the deposits which is accordance with the general knowledge. While little part of it is still on the horse roads.

Velocity contour plots during the landslide movement are shown in Figure 4. (a) The landslide presents nearly the same starting speed on the whole. (b) During the acceleration stage, the sliding velocities are decreasing gradually from the surface of the slope to the slip surface. (c) Some blocks in the front of landslide have fell on the horse roads and the landslide is about to enter the deceleration stage as the result that both the horse roads and blocks locating on it will hinder the motion of landslide. (d) The landslide is going to start again when horse roads are filled with blocks. (e) The landslide accelerates again and rushes to the ground, while some blocks in the front has stopped. (f) The velocity of the landslide is 0, so it has stopped.

The curve of average velocity versus time in the whole process of landslide is shown in Figure 5. It is easy to see that there are two obvious crests. The landslide starts to move gradually under the self-weight and accelerates to the first peak value 16 m/s, then the average velocity rapidly diminishes to 2.35 m/s. At first, the dig angle of the initial

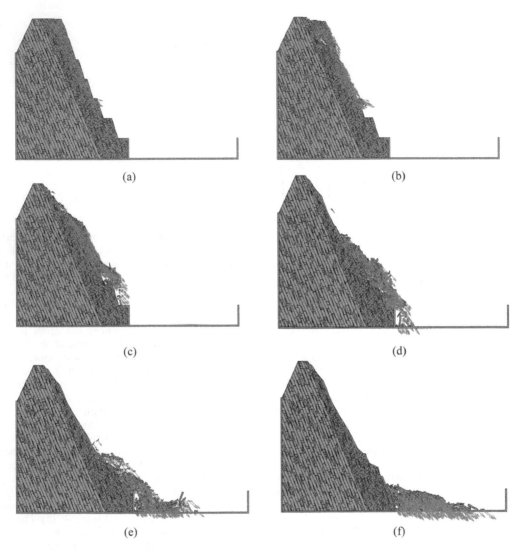

Figure 3.   Displacement vector diagrams in the process of landslide movement.

slip surface is very large and the sliding surface is relatively smooth, so landslide can ascend to the peak value of velocity smoothly after starting to move. But the horse roads below the landslide will hinder the movements of the landslide greatly when the landslide moves off the initial relatively steep sliding surface, so the average velocity will decrease rapidly. A new sliding surface with a gentle slope is formed when the horse roads are filled with the blocks, so the landslide will start, accelerate and decelerate once again. Finally, the landslide rushes to the ground and forms the deposits at the toe of slope. The second peak value about 11 m/s is less than the first peak value, because the gradient of the sliding surface is gentle and the resistance of the sliding surface is large.

Velocities versus time for the measured blocks are shown in Figure 6. There are three measured points, whose coordinates are (32.6, 1984.0), (50.2, 1940.0) and (58.4, 1923.0) respectively. The corresponding blocks where the measured points locate are Block-32, Block-14 and Block-52 respectively. Block-32 is in the upper of the landslide, Block-14 is in the middle and Block-52 is in the lower. There are two obvious crests owning nearly the same value in the movement of Block-32, so it has experienced the motion process from accelerating to decelerating for two times. Only one apparent crest is observed for the Block-14 which is nearly coincident with that of Block-32. For the rest time, the sliding velocity has high fluctuations and no obvious crests.

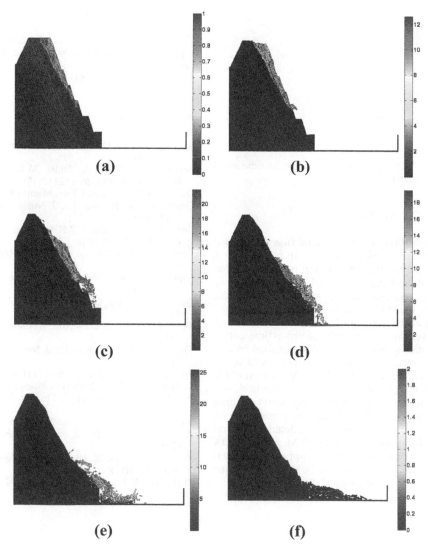

Figure 4.   Velocity contour plots during the landslide movement.

Block-52 slides for 6 s only, which may have fell on the horse roads. The fluctuations of velocities for the three measured blocks can be observed, because blocks will not only translate but also rotate in the sliding process when subjected to the force and moment of force from the surrounding blocks.

Movements of the three measured blocks at three different locations of the landslide individually are different from the motion law observed in the curve of average velocity in Figure 5, but they are complying with it as a whole. They both reflect the real motion process of the landslide from the startup to finally forming the deposits.

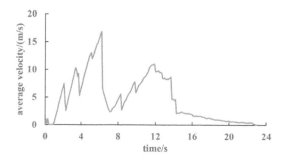

Figure 5.   Average velocity versus time in the whole process of landslide.

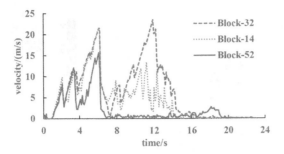

Figure 6. Velocities versus time for the three measured blocks at different locations.

5 CONCLUSIONS

The whole process of the landslide in the real hydraulic engineering from startup to stopping is reproduced by DDA. Numerical results show that startup of landslide is incurred by the reduction of mechanical parameters of controlling sliding plane caused by the rainfall and the insufficient rock mass strength at the bottom of landslide. The curve of average velocity of the landslide versus time, curves of velocities versus time for the measured blocks, velocity contour plots and the displacement vector diagrams are given and analysed. Movement of the landslide lasts for 23 s, which means landslide will cause a huge damage within a very short time once it starts to move. The sliding mechanism during the motion process is analysed in detail, which has beyond the applying range of FEM. Finally, DDA can not only provide technical support for predicting the landslide, but also can be used to assess the loss caused by the geological disasters.

ACKNOWLEDGEMENTS

The research work of this paper was financially supported by National Natural Science Foundation of China under Project grant No. 11502033, 51579016, 51379022, 51539002.

REFERENCES

Bao, H., Z. Zhao, Q. Tian, Int. J. Numer. Anal. Met. **38**, 551 (2014).
Beyabanaki, S.A.R., A. Jafari, M.R. Yeung, Int. J. Numer. Meth. Bio. **26**, 1522 (2010).
Doolin, D.M., N. Sitar, J. Eng. Mech. **130**, 249 (2004).
Grayeli, R., K. Hatami, Int. J. Numer. Anal. Met. **32**, 1883 (2008).
Jiang, Q.H., C.B. Zhou, Z.F. Qi, Chin. J. Rock Mech. Eng. **28**, 2778 (2009) (in Chinese).
Jiang, Q.H., M.R. Yeung, Rock Mech. Rock Eng. **37**, 95 (2004).
Mishra, B.K., R.K. Rajamani, Appl. Math. Model. **16**, 598 (1992).
Qu, X.L., G.Y. Fu, G.W. Ma, Eng. Anal. Bound. Elem. **48**, 53 (2014).
Shi, G.H., R.E. Goodman, Int. J. Numer. Anal. Met. **9**, 541 (1985).
Tian, Q., Z. Zhao, H. Bao, Int. J. Numer. Anal. Met. **38**, 881 (2014).
Wu, A.Q., X.L. Ding, S.H. Chen, G.H. Shi, Chin. J. Rock Mech. Eng. **25**, 1 (2006) (in Chinese).
Wu, J.H., C.H. Juang, H.M. Lin, Int. J. Numer. Meth. Eng. **63**, 876 (2005).
Wu, J.H., Int. J. Numer. Anal. Met. **31**, 649 (2007).
Zheng, H., D. Xu, Int. J. Numer. Meth. Eng. **97**, 986 (2014).
Zheng, H., F. Liu, X. Du, Comput. Method Appl. M. **295**, 150 (2015).
Zheng, H., W. Jiang, Sci. China Ser. E. **52**, 2547 (2009).

Advances in Energy Science and Equipment Engineering II – Zhou, Patty & Chen (Eds)
© 2017 Taylor & Francis Group, London, ISBN 978-1-138-71798-5

Technologies of the building information model analysis

Elena Ignatova

Department of Information Systems, Technology and Automation in Construction, Moscow State University of Civil Engineering, Moscow, Russian Federation

ABSTRACT: This paper gives classification of methods and tools of the Building Information Model (BIM) analysis. The experience of the building information model analysis in a variety of specialized tasks is described. An example of the use of Application Program Interface (API) for creating new functions for BIM analysis is given. The material consumption of monolithic slabs with removable shuttering of metal profiled sheets is analyzed. A method to manage parameters of BIM based on alternative software is described. Program Dynamo is used to analyze different façades of a building model created by program Revit. The problems of the export of BIM data to other analysis programs are discussed. An example of the implementation of IFC data transfer standard for the analysis of used materials and constructions is given. Special program converters are suggested to be used in BIM analysis.

1 INTRODUCTION

Building information modeling is a modern rapidly developing technology to create a digital model of the construction object and to support it during its entire life cycle. Building information model is designed to generate and store a large amount of diverse information about the construction object, for example, its geometrical properties, parametric relations, construction elements, information about materials, and information about costs. Building information model includes data describing the behavior of the object and can be used to analyze work processes, for example, to estimate strength and stability of the object. Autodesk Revit, Nemetschek Allplan, Graphisoft ArchiCAD, and other similar programs are software that allows you to organize BIM data management. The information model is established and analyzed with the help of tools created by the program vendors. A set of tools is quite large and may depend on the software, but it is limited and cannot provide the solution to all emerging problems.

There are two main methods of additional analysis and processing of data (Figure 1): (1) creation of new tools inside BIM application and (2) use of external software applications. The paper describes the experience of using these methods to solve some tasks of civil and architectural engineering.

2 INTERNAL ANALYSIS OF BIM

BIM applications allow performing quantitative analysis of the parameters of the building.

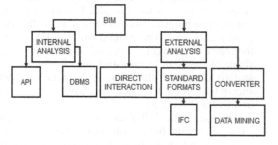

Figure 1. Methods and tools of analysis.

Information about them can be presented in the form of specifications. Specification values can be computed. However, this analysis does not provide a solution to a large number of tasks. Internal analysis is carried out by adding new functions for analysis and processing of the database information model. The results of this analysis also remain inside of the building information model.

2.1 *The use of API*

To program new functions to generate and analyze information model data, we can use the Application Program Interface (API). This approach was applied to the problem of the exact calculation of the amount of metal and concrete in the construction of monolithic slabs with removable shuttering of metal profiled sheets (Figure 2).

Building information model was generated in program Allplan (Mishra, 1992). The standard functions of Allplan were used to select the desired type of profiled sheet, to place the sheets

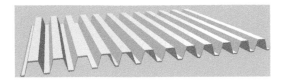

Figure 2. The metal profiled sheet.

at the area of overlap, and to create technological holes. The layout of profile sheets was produced in accordance with the shape of the monolithic slab of a given building. In the process of layout, profile sheets can overlap each other. As the layout of profile sheet for monolithic slabs is labor-consuming and requires significant precision, manual errors may occur. For this case, there is an automatic verification of the correct location of the profile sheets. The user sets values of the overlap of adjacent profiled sheets. Then, the program checks the overlap between all the relevant sheets.

Additional algorithms in the C++ language were developed using API Allplan software. In the main Allplan menu, a button was added to start a special plugin. The plugin allows us to automatically generate a concrete layer in the form of a 3D object, to compute the exact volume of concrete to fit the shape of profiled flooring and holes in the ceiling, and to determine the amount of metal profiled sheets with regard to their overlapping in length and width (Figure 3).

The use of the API allows us to solve problems within the main BIM applications, but it requires programming skills. In addition, not all vendors reveal the internal structure of data storage and provide the ability to use the API.

2.2 Use of DBMS

There are different ways to access BIM data with the help of Database Management System (DBMS). For example, there is a program Dynamo related to Revit, which has access to the format of Revit data storage and allows to manage these data. The management is based on visual programming, which allows us to automate the creation and analysis of information models. Visual programming happens not at the level of code but at the level of algorithms, so it does not require the help of professional programmers (Shi, 1985; Zheng, 2009). Each application created in Dynamo is related to the current Revit project and has access to Revit databases.

Program Dynamo was used to investigate variants of formation and calculation of front wall tiles of the building (Figure 4).

The color options of facade of the building were reviewed. First, the location of the tiles with

Figure 3. A fragment of the ceiling structure (concrete and metal profiled sheet).

Figure 4. Analysis of wall variants.

specific color and size was randomly selected. Then, the options for relief of the facade with tiles of the same color but different thickness were considered. After all, quantitative analysis of these variants was conducted.

3 EXTERNAL ANALYSIS OF BIM

Another possibility for data analysis (e.g., to calculate the strength, value, and cost of resources, project management, etc.) is the export of digital building models in specialized programs.

3.1 Direct export

If the BIM application and the specialized program are preconfigured for mutual data transmission, the direct transmission of data proceeds with maximum retention of information and relationships. Usually, the BIM application has a button function for converting data into a file for the specialized program. The data analysis program reads this file and works in accordance with its algorithms. Often, both programs belong to the same vendor. In this case, the reverse data transfer is also possible. However, if we change the version of any of the programs, the correctness of mutual data transfer may be disrupted.

3.2 Use of standard formats

BIM applications contain built-in functions for detecting intersections that allow architects to test models at the design stage. The lack of filter options and the lack of additional settings of conflict tests do not allow these tools to detect the following problems: checking the size of a room for accessibility for disabled people, checking the gaps between components, and checking the presence/absence of insulation. These problems require the use of additional software of data analysis. If software from different vendors interact, the transmission of data is organized based on the neutral formats of data transfer. Each BIM application has a set of transmission formats, among which there are transmission formats only for graphic information (e.g., dxf, dwg, and SAT) and there are formats that convey complete set of information about the information model (e.g., RVT, PLA, and IFC). The program of information model analysis opens the file and works with it in accordance with its capabilities. Examples of such programs are Naviswork and Solibri, using which many file types can be opened and their collisions be checked. With the help of these programs, we can compile models related to architecture, structures, and engineering equipment of the building in a common coordination model and check the correctness of the result. Models can be checked for spatial–temporal conflict, for visual errors, and other types of errors.

The software Solibri Model Checker was used to implement automated quality verification of the coordination model of a public building (Bao, 2014). As a part of research, compliance of elements of architectural models and models of water supply, sewerage, and heating was verified. Solibri Model Checker detected more than 1000 collisions. All collisions were automatically analyzed and were grouped according to certain parameters.

In addition to the standard problem of collision search, we stated the problem of analyzing evacuation routes in case of fire. We generated new queries to solve this problem with the help of Solibri Model Checker. We created classifications of rooms, doors, lifts, and steps. Then, we checked the length and width of escape routes for compliance with security standards.

We also used program Navisworks Manage to detect conflicts in the coordination model of the same public building. Navisworks Manage has its own advantages but does not allow group conflicts. Hence, we applied a method to automate the sorting and grouping of results. The method was based on the search filters.

3.3 Export in IFC

Most BIM applications have the ability to download data in the IFC data exchange format. This format allows us to see the contents of the file in ASCII codes, but it has object-oriented structure and does not allow us to easily find necessary object parameters. Vision analysis of IFC data is possible based on freely distributed software modules provided by IFC viewers (Tian, 2014). In addition, the data format is not fit for storing information about parameter dependencies between objects of the information model (Jiang, 2004).

The idea is to transmit the IFC information to an object-oriented database and then to obtain the necessary information about the objects by OQL-queries (Wu, 2005; Beyabanaki, 2010; Grayeli, 2008; Wu, 2006). OQL is an extension of SQL for object-oriented databases.

This technology was tested for extracting data about structural elements and materials of the building information model created in ArchiCAD and exported in IFC file (Figure 5). Program ArchiCAD was used because it had convenient tools for data export to IFC file. This software (Beyabanaki, 2010) had toolbox that only allowed working with Java classes. We used ikvm.net (Grayeli, 2008) to enable Java and NET interoperability to use C# classes that were more convenient for our purposes.

The information searching system was created on the basis of a combination of given modules and our own algorithms. First, the system opens, reads, and parses the IFC file. Depending on the queried model objects, the system forms the structure of the object-oriented database. For information about objects on the OQL language, we can create our own queries that are offered to the user.

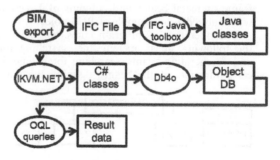

Figure 5.   The technology of extracting data.

Search results of information are presented in tabular form. A simple interface was created for the user. This technology can be used in the formation of specifications of the properties of BIM objects.

3.4 Use of program converter

Data processing can be implemented using a program converter, which reads data in one format and writes in another format, whose structure is exactly specified. Data transfer usually requires additional data processing (sampling, sorting, systematization, adjustment, etc.).

Data processing can be carried out:

– in BIM application before data transmission;
– in the program converter during transmission of the data;
– in the target program after data transfer;
– in combination.

Before the transmission of data within the BIM application, we can select objects that will be passed for further analysis. For example, program ArchiCAD has great settings for the process of partial transmission of information.

If BIM program has not enough functions to manage data transfer, we can use the API and create the necessary plugin. This possibility has been implemented to upload information from Revit model. The developed plugin allocates the necessary objects, creates a specifically structured file, and writes the parameters of these objects in the file.

A program converter was developed to solve the problem of transferring data from BIM applications to the program of Finite Element Method (FEM) strength analysis (Figure 6). A text file generated in Revit was used as the source file. An output file is again a text file, which is readable for the program of the strength analysis. The FEM program was developed for educational purposes. The format of its data was accurately known (Wu, 2007). Text file formats were used for visual checking of the data transformation process. A flat frame of metal beams and columns of I-section was analyzed as an example of BIM object.

The operation of program converter includes the following steps: opening file, searching for and enumerating beams, searching for geometrical parameters of the beams, defining the units

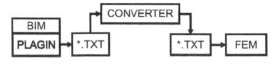

Figure 6. The scheme of converting.

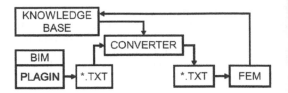

Figure 7. The scheme of data mining system.

of measurement, calculating the scale factor, calculating missing coordinates of the ends of the beams, searching for module of elasticity of the material, searching for the geometric characteristics of the beam crossing, and recording the received information in a new file.

3.5 Data mining

Data mining was successfully applied in engineering (Zheng, 2014). However, nowadays these methods are not widely used.

Program converter can be an element of a data mining system for data analysis (Figure 7). Simultaneously with data transmission, it can control the data and make a decision in conflict situations. The system can make decisions based on expert systems or neural networks. In the first case, the converter must be connected to the knowledge base, which holds options for different situations. In the second case, it is necessary to organize the return of information from the program into which the data is being transferred. If input information is subjected to further correction in this program, the converter needs to know about this and include this situation into the neural network training.

4 CONCLUSIONS

BIM technology is rapidly developing. However, the software still cannot cover the solution of all engineering problems at the stages of design, construction, and operation of the building. The software that allows user to add new features and tools for internal data analysis of information model of construction object has significant advantages.

The disadvantage of external data analysis in specialized programs is usually the lack of postback of the results of the analysis into the information model.

The use of IFC is a promising and universal method of data transfer. The format requires object-oriented data analysis technologies. The quality of information transmission through the IFC can be affected by the settings of the export options in BIM applications. Not all the properties of the model can be transferred via IFC.

Combining the described methods and tools for data analysis can give the best result of the analysis of the building information model in the solution of engineering problems.

The future development of BIM technologies is related to data mining methods.

REFERENCES

Apstex: IFC Tools PROJECT-Viewer. Available at: http://www.ifctoolsproject.com

IFC Toolboxes (buildingSMART), Available at: http://www.buildingsmart-tech.org/implementation/get-started/ifc-toolboxes

Ignatova, E., H. Kirschke, E. Tauscher, K. Smarsly, Parametric geometric modeling in construction planning using industry foundation classes (Proceedings of the 20th International Conference on the Applications of Computer Science and Mathematics in Architecture and Civil Engineering, Weimar, 68–75, 2015.).

Ignatova, E.V., Reshenie zadach na osnove informatsionnoy modeli zdaniya [Problem Solving on the Basis of Information Models of Buildings] (Vestnik MGSU, Proceedings of Moscow State University of Civil Engineering, 9, 241–246, 2012).

Ikvm.net (Open Source), Available at: http://www.ikvm.net/db4o Open Source Object Database, Available at: http://www.odbms.org/category/downloads/object-databases/object-databases-free-software

Solibri Inc.: Solibri Model Viewer v9. Available at: http://www.solibri.com/products/solibri-model-viewer

Tauscher, E., M. Theiler Getting Started—Java Toolbox IFC2 × 3/IFC4. v2.0. 2013. Available at: http://www.ifctoolsproject.com/

The Dynamo primer, Available at: http://dynamoprimer.com/index.html

V.P. Ignatov Modelirovanie snroitelnogo proertirovaniya na osnove intellektualnyh tehnoogiy [Modeling of structural design on the base of intelligent technologies], (Knizhnyi Mir, Moscow, 2012).

Vasshaug, H., Learn Dynamo, Available at: https://vasshaug.net/2015/09/18/learn-dynamo

Zotkin, S.P., N.S. Blokhina, I.A. Zotkina, About development and verification of software for finite element analysis of beam systems (Procedia Engineering 111, 902–906, 2015).

Advances in Energy Science and Equipment Engineering II – Zhou, Patty & Chen (Eds)
© 2017 Taylor & Francis Group, London, ISBN 978-1-138-71798-5

Innovative negative Poisson's ratio cable-sensing system and its slope monitoring warning application

Zhigang Tao, Chun Zhu & Bo Zhang
State Key Laboratory for Geomechanics and Deep Underground Engineering, Beijing, China
School of Mechanics and Civil Engineering, China University of Mining and Technology, Beijing, China

ABSTRACT: A newly developed anchor device is introduced in this paper, that is, Negative Poisson's Ratio cable (NPR) incorporated with four-in-one function: reinforcement, controlling and prevention, monitoring, and forecasting of the side slope of the soft rock masses. On the spot, first, a hole is drilled in the slope, then NPR cable is put across the sliding surface and fixed in the stable bedding by cement. When the deformation damage of rock mass occurred, the deformation energy can be transferred to the constant-resistance body along the cable. There are two possibilities: when the axial force applied on the cable less than the design force with constant resistance, the material deformation of cable will occur to resist the failure deformation of engineering rock mass. When the axial force is greater than the design force, the constant-resistance body will slide friction along the constant resistance casing, and through the structural deformation of NPR to absorb the energy, and prevent the cable from tensile failure. In this paper, NPR cable has been used in landslide monitoring in Nanfen open pit iron mine, and successfully forecast "11-1005" landslide disasters, giving advance warning signal before landslide according to the monitoring curve.

1 INTRODUCTION

Real-time monitoring of the surface displacement or the sliding force for the sliding mass has been one of the hot topics in the geomechanics community. As the difficulties encountered in monitoring large-scale rock masses, monitoring of ground deformation and displacement, underground water flowing, and fault and fracture changes have been playing a major role in landslide monitoring projects over years (Xu Likai et al., 2007; Toshitaka et al., 1998; Lulseged Ayalew et al., 2005). However, researchers and engineers who have conducted many studies in landslide problems using the surface displacement monitoring techniques found that they often failed to capture the evolution features and precursory information during the sliding process solely relying on monitoring the surface displacement of the sliding mass. It was well known that the sliding force is the driven force for the landslide. As the sliding force, however, belongs to the natural mechanics system existing in the sliding surface between the sliding mass and bedrock, it cannot be detected directly. This immeasurable nature of the sliding force in the natural mechanics system has been the obstacle hindering accurate monitoring and thus the predicting and forecasting of the landslide disasters. Monitoring of the sliding force is a

more effective method for prevention and mitigation of the landslide catastrophe based on the following facts: (1) the information from the sliding force monitoring can reveal the evolution features of the acting force in the sliding surface of the slope; and (2) sliding force monitoring provides high flexibility for establishing the mechanical models.

Predicting the probability of landslide by observation and inspection of the ground surface changes and the underground water flowing level using manpower had been widely used worldwide, whose limitations are the early waning of technology and equipment (Xia Bairu et al., 2001; Hu Wei et al., 2002). With the development of the technology in recent decades, the ground surface displacement monitoring has been widely applied in the slope monitoring projects with the aid of varied commercial monitoring instruments such as total station, theodolite, leveling instrument, and drilling tiltmeter as well as GPS mobile found in the market in recent years (Leonardo Zan et al., 2002; B.A. Reevea et al., 2000; Riki Ohbayashi et al., 2008). These methods in the landslide monitoring and predicting projects have achieved numerous meaningful findings and provided many new ideas and techniques. However, because of the inconsistencies of the ground surface displacement and the sliding surface displacement, an accurate prediction or forewarning for the landslide would be impossible using the aforementioned monitoring methods. Although locations of the sliding surface could be identified by deep displacement monitoring method, the tilt displacement monitoring will fail when a relatively larger sliding takes place and the sliding displacement data are unable to be captured in the late sliding process. Thus, this method is limited in its "inherent deficiency" and cannot capture the displacement data in a "faithful way."

According to above shortcomings of the landslide monitoring methods and equipment, a new kind of method and equipment of landslide monitoring is introduced in this paper. At present, the new method and the equipment have been used in Nanfen open pit iron mine, and successfully forecasted several landslide disasters and avoided the casualties and property losses.

2 WORKING PRINCIPLE

The movement of the sliding mass, a natural mechanical system by nature, depends on the changes of the balance state between the sliding force and sliding-resistant force. Landslide is a dynamic process including stable deformation, unstable deformation, and catastrophic failure. Monitoring the evolution features of the acting fore in the sliding surface of the slope during landslide will obtain the database for establishing the sufficient and necessary conditions for the evaluation of slope stability and giving a high accuracy in forecasting the landslide disasters.

However, the sliding forces in the natural mechanics system is difficult to measure, it is very important to study the measurement of this important parameter. It was found that the artificial mechanics system can be measured easily, and we insert a cable device, the so-called constant-resistant and large-deformation cable, into the nature mechanics system as shown in Figure 1. Thus, a measurable artificial system was obtained. On the basis of this system, we can obtain the sliding force calculated by the measured perturbation force.

The landslide monitoring principle shown in this figure is to use "puncture perturbation" technology, and make the NPR cable through the landslide surface, meanwhile loading a small prestressing force on the cable. Then, this force participated in the mechanical system of landslide. Therefore, the functional relationship between the prestressing force and the sliding force can be established.

The mathematical model could be obtained according to the model of the sliding system shown in Figure 1. The sliding force can be written as:

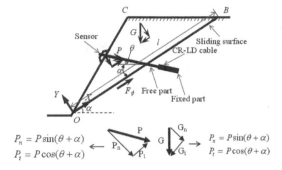

Figure 1. Mechanical model of landslide.

$$F_s = k_1 P + k_2 \tag{1}$$

The parameters k_1 and k_2 are derived as:

$$\begin{cases} k_1 = \cos(\alpha + \theta) + \sin(\alpha + \theta) \cdot \tan \overline{\phi} \\ k_2 = G \cdot \cos \alpha \cdot \tan \overline{\phi} + c \cdot l \end{cases}$$

In Eq. (1), $\overline{\phi}$ is the internal friction angle and c is the cohesion. Geometry influence coefficients are l, α, and θ, with l being the length of the sliding surface, α the angle between the sliding surface and the horizontal axis, and θ being the anchor inclined angle below the horizontal. According to Eq. (1), the immeasurability force G_t is a function of P, which is a measurable quantity. Thus, the force G_t can be calculated according to Eq. (1).

3 MONITORING SYSTEM

3.1 Topology structure of monitoring system

Landslide remote monitoring and warning system works in the way of wireless transmission, and the transmission models belong to the point-to-point mode, that is, each monitoring system operates independently and does not interfere with each other. The monitoring points are distributed in the field within its transmission range. Monitoring information can be remote-transmitted, and multi-host can receive monitoring data at the same time. Figure 2 shows the schematic map of the network of the system.

3.2 System components

Landslide remote monitoring and warning system based on the NPR cable is composed of the data sensing–acquisition–transmission system, which is installed in the field, and the data processing–analysis system, which is installed indoor.

1. Data sensing–acquisition–transmission system

Figure 3 presents a prototype of the intelligent data sensing–acquisition–transmission system, which is composed of mechanical sensing equipment and mechanical signal acquisition–transmission equipment. The mechanical sensing equipment is composed of NPR cable and the mechanical sensor. The NPR must pass through the sliding surface and fixed in the sliding bed. Therefore, it is important to gather detailed geological investigation data (including the geological section and the mechanical parameters). Originally, conventional cable is used for data transmission carrier; however, on the basis of the large number of field experiments, we find that the conventional cable is often destroyed when a landslide occurred and the monitoring system fails. In order to overcome this problem, a new cable is developed, which can resist the large deformation (maximum around 2 m) without being damaged. Thus, the NPR cable solves the monitoring bottleneck of the whole landslide process.

The NPR cable is composed of a constant-resistance device (including the CR casing, the CR body (frictional cone unit), baffle, and fill material), steel strand, face plate, and anchorage device (Figure 4).

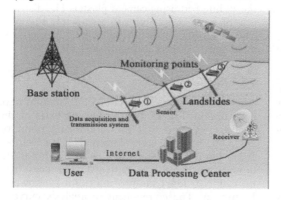

Figure 2. The structure map of network topology.

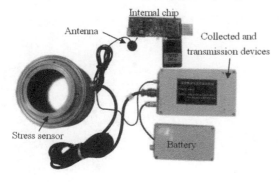

Figure 3. System configuration.

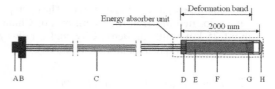

Figure 4. Schematic of the data sensing–acquisition–transmission system.
A—Data collection and transmitting equipment; B—face plate; C—steel tendon, composed of six steel strands with φ15.24 diameter; D—baffle; E—fill material; F—casing tubes; G—frictional cone unit; H—stop element (φ133 mm).

The external diameter of constant-resistance device is 133 mm, installed at the end of conventional cable composed of six anchor cables (the diameter of each cable is about 15.24 mm). The anchorage device and the plate are installed on the other end of conventional cable. Here, we should discriminate the notations for "deformation" and "elongation." The term "deformation" has a generalized meaning, which is used in the case of a narrow scope referring to the elastic or plastic deformation, or in the case of a broad sense referring to the motion of displacement. However, the term "elongation" in this paper specially refers to the "deformation" of the NPR cable, which is actually the displacement of each anchor cable relative to the slide track sleeve in the casing tubes. We want more "deformation" of the NPR cable, whereas small deformations of each anchor cable itself. Hence, rated resistance for NPR cable is designed as much as 80% of the yield strength of anchor cable material to ensure in any case that the deformation of the anchor cable itself within the elastic range under the external load is larger than its yield strength.

2. Data processing and analysis system

This system consists of data receiver, data processor, and data analysis software. Data analysis software is used for communication with the data receiver, and it stored the recovering data into the database. These data will be classified and calculated automatically, and illustrated by monitoring curve and warning level, which can be searched and downloaded by the authorized users.

4 CASE STUDY

4.1 *Project profile*

Nanfen open pit iron mine is a large Anshan-type sedimentary metamorphic iron deposit. Iron deposit occurred in the iron rock of the Anshan group and the Archean group. The ore body presents a monoclinal structure, and is the largest monomer open pit iron mine in Asia. The open pit mine is located in the northeastern part of China, about 25 km south of Benxi City, and a railway connected the mine area and the Benxi city.

4.2 Regional geological characteristics

Landscape of the Nanfen open pit iron mine presents a monoclinal structure, which is composed of metamorphic series strata, belonging to the middle and high mountains geomorphology. The trend of the mountain is from east to west, valleys locate everywhere, and vegetation is scarce. Average elevation of the mountain is about 500–600 m and the relative height is about 300–400 m. Survey data show that Nanfen open pit iron mine is located in the northern Taizirevier sag of north China platform and Liaodong platform anticline Yingkou-Kuandian uplift. Archaeozoic Anshan group develops extensively, followed by Algonkian Liaohe Group, Sinian stratum, and Cainozoic Quaternary stratum.

The main fault in the mine area is F1 fault, which is the SE extension of the NE trend compress-shearing fault, and in the mine area, the fault is of NNE trend or close to SN direction, and toward west, the dip angle is 45°. It had a length of about 10 km and width of 5–20 m. F1 fault threats the stability of open pit footwall slope.

Groundwater in the study area is composed of pore water and fissure water. The pore water is the alluvial soil phreatic water, which has hydraulic connection with bedrock fissure water. The fissure water is supplied from precipitation above 298 m, and it has contact with the surface water below 298 m. Therefore, there is a close relationship between the groundwater regime and the distribution of rainfall.

4.3 History of landslides

The slope in Nanfen open pit iron mine is cut by five to six sets of joint; however, these joints do not have much influence on the stability of the slope. However, there are two sets of joint in the footwall that have much impact on the stability of slope. The altitude of the first joint is 295°/48° and the coefficient of roughness is 2–4. The surface of this joint has a sliding trace and almost penetrates the entire rock mass. Therefore, this surface has constituted the main sliding surface in the footwall. The altitude of the second joint is 291°/13°, which has a small dip angle and water trace in the exposure area. This set of joint constituted the sliding surface at the bottom of sliding mass.

Since 1999, influenced by the special landform and long-term mining, many large-scale sliding masses have been formed in the footwall. By field investigation, we found that the sliding mass in the footwall has a length of about 192 m and width of 250 m, the sliding direction is around 250°, and the volume is around 52×10^4 m^3. In order to increase the resistant force, around 10 million tons of iron ore beneath the sliding mass dare not extracted in a decade, which has brought large economic losses.

4.4 Distribution of monitoring points

In order to extract the ore body beneath the sliding mass successively and safely, we have set up 20 remote real-time monitoring systems for sliding force in the footwall of the slope in the Nanfen open pit iron mine. This system can indirectly measure the sliding force in deep sliding surface by mechanical calculation with the help of special mechanical features of NPR cable. Therefore, we have realized the premeasurement and monitoring of the sliding force (He Manchao, 2009).

During the installation in the field, first, we drill hole in the slope with a certain incidence angle based on the information of detailed reconnaissance. Then, we put the NPR cable into the drill hole and make sure that the end of the NPR cable can penetrate the potential sliding surface (He Manchao et al., 2009). Finally, we fix the cable in the sliding bed using cement plaster. The distribution of monitoring points is as shown in Figure 5. The monitoring points in the Nanfen open pit iron mine began to work normally on 4 June 2010.

4.5 Results

About 10000 precious pieces of information have been obtained in 15 months. On the basis of the analysis, we find that the monitoring point of No. 334-1 began to seem abnormal from 2 October 2011. Until 6 October 2011, after the landslide, monitoring curve was flattened. The cross section of No. 334-1 monitoring point is shown in Figure 6a. The comprehensive analysis curve is shown in Figure 6b, and the curve includes three

Figure 5. Distribution of monitoring points.

parameters: sliding force, accumulative amount of mining, and rainfall curve.

From Figures 6 and 7, it can be seen that under the comprehensive influence of mining and rainfall, the sliding force monitoring curve has shown significant warning characteristics from the occurrence of cracks to the happening of landslide disaster. The process of monitoring and predicting for "11-1005 landslide" in the Nanfen open pit iron mine is described as follows:

1. Before point A, on 1 October 2011. We can see that the sliding force curve on No. 334-4 point is linear, indicating that the slope is stable.
2. After point A, on 1 October 2011. Sliding force monitoring curve increases with time, and the sliding force variance ratio began to increase. However, there is no crack and sliding phenomenon on the ground.
3. When the sliding force monitoring curve arrived at point B, at 12:00 on 2 October 2011. There is a sudden drop in the sliding force monitoring curve (called mutation), and the first warning signal is issued. On the spot, many microfractures began to appear on the 322–334 bench, and three obvious cracks (from the north to the south: numbered ①, ②, and ③) appeared in the surface of unbroken slope rock mass. It has a

length of about 1.5–5 m and width of 15–40 cm; however, the depth is unknown.

4. At point C, at 9:00 on 3 October 2011. The increasing rate of sliding force monitoring curve had been continuing increasing. On the spot, we found that the excavator is located at a distance of 5 m from No. 334-4 monitoring point in 322 bench. Obviously, it over-exploited, and the 334 bench has been damaged. At the same time, we find that the original microfracture on the 322–334 bench began to expand and break. The obvious three cracks become longer and grow wider, with length of about 3–9 m, width of 25–60 cm, and crack depth is unknown. In

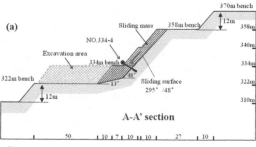

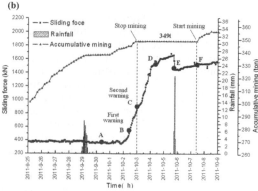

Figure 6. Synthesis of warning curve of landslide. (a) 11-1055 landslide section; (b) synthesis of warning curve.

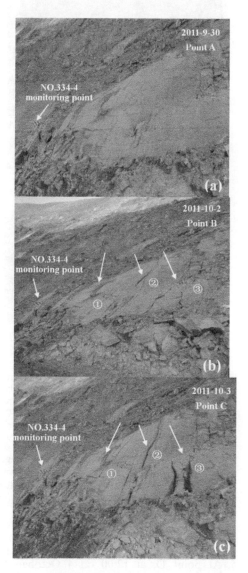

Figure 7. (*Continued*)

Figure 7. Whole process picture of crack-cut through-slide. (a) Rock mass of slope is intact on 334 bench; (b) microfracture development (width is 0.15–0.4 m and length is 1.5–5 m); (c) cut through (width is 0.25–0.6 m, length is 3–9 m); (d) rock extrusion along the original crack; (e) landslide occurred; (f) landslide clean-up.

order to ensure the safety of sliding mining, the decision maker of the open pit mine gave the order to stop mining based on the monitoring curve of sliding force and evacuate personnel and equipment. This information can be demonstrated by the accumulative mining curve shown in Figure 7. From Figure 7, we can see that mining quantity is zero from the point C.

5. At point D, at 12:00 on 4 October 2011. No. 334-4 monitoring curves continue to increase,

and the increasing rate is relatively lower. On the spot, we find that the rock mass in footwall 322–334 bench along the crack has been damaged by extrusion, forming loose rock mass, part of the rolling stones along the slope face impact to 322 m bench. As the personnel and equipment were evacuated on time, no casualties and property losses or accidents has occurred.

6. At point E, at 8:00 on 5 October 2011. According to the data recorded in the rainfall-monitoring system, we find that Nanfen open pit iron mine received rainfall at 19:00 on 5 October 2011. Cumulative rainfall in 3 h is around 31.8 mm. One hour after the rainfall, monitoring curve of No. 334-4 in Nanfen open pit iron mine dropped suddenly and the sliding force suddenly changed from 1700 to 1400 kN. The next day, at 9:00, on the spot, we find that the landslide occurred on 322–358 bench; the vertical height difference of sliding mass is around 36 m, and the length from south to north is around 50 m.

7. At point F, on 7 October 2011. In order to resume mining as soon as possible, decision maker arranged few equipment and person to clean up the mess in part of the sliding body.

CONCLUSION

1. The principle of monitoring the relative motion for the sliding body and sliding bed was proposed. The advantage of this monitoring technique lies in force measurement, that is, the difference between the sliding force and the resistance force, which is a function of many factors such as raining, underground water level, excavation, and blasting. This novel force-monitoring technique, instead of dealing with many factors with complicated interactions, makes the prediction explicit and definite by using a single parameter, that is, the sliding force measured during the sliding processes.

2. NPR cable incorporated with four-in-one function: reinforcement, controlling and prevention, monitoring, and forecasting of the side slope of the soft rock masses. It can provide enough defamation by the structural deformation and the material deformation to absorb the deformation energy triggered by rock mass sliding.

3. NPR cable has been used for landslide monitoring in Nanfen open pit iron mine, and successfully forecast "11-1005" landslide disasters, by giving advance warning signal before landslide according to the monitoring curve so as to avoid casualties and property losses.

4. Because of the use of the NPR cable, landslide remote monitoring warning system has been

applied in open pit iron mine, open pit mine, underground gold mine, slope monitoring along the gas pipeline, and so on, thereby achieving remarkable economic benefits and social benefits.

REFERENCES

He Manchao, Tao Zhigang & Zhang Bin. 2009. Application of remote monitoring technology in landslides in the Luoshan mining area. *Mining Science and Technology* 19(5): 609–614.

He Manchao. 2009. Real-time remote monitoring and forecasting system for geological disasters of landslides and its engineering application. *Chinese journal of Rock Mechanics and Engineering* 28(6):1081–1090.

Hu Wei & Li Shulin. 2002. A study on the acoustic emission technology in the analysis of rock slope stability. *Mining research and Development* 22(3).

Leonardo Zan, Gilberto Latini & Evasio Piscina, et al. 2002. Landslides early warning monitoring system. *Geoscience and remote sensing symposium, In IGARSS' 02. 2002 IEEE International*: 188–190.

Lulseged Ayalew, Hiromitsu Yamagishi & Hideaki Marui, et al. 2005. Landslides in sado island of japan: Part I. Case studies monitoring techniques and environmental considerations. *Engineering Geology* (81):419–431.

Reevea, B.A., G.F. Stickley & D.A. Noon, et al. 2000. Developments in monitoring mine slope stability using radar interferometry. *Geoscience and remote sensing symposium, 2000. Proceedings. IGARSS 2000. IEEE International* 5: 2325–2327.

Riki Ohbayashi, Yasutaka Nakajima & Hideto Nishikado, et al. 2008. Monitoring system for landslide disaster by wireless Sensing node network. *SICE Annual Conference 2008 IEEE International*: 1704–1710.

Toshitaka Kamai. 1998. Monitoring the process of ground failure in repeated landslides and associated stability assessments. *Engineering Geology* (50):71–84.

Xia Bairu, Zhangy an & Yu Lihong. 2001. Monitor and treatment technology of landslide geological disaster in China. *Exploration Engineering*.

Xu Likai, Li Shihai & Liu Xiaoyu. 2007. Application of real-time telemetry technology to landslide in Tianchi Fengjie of Three Gorges Reservoir Region. *Chinese Journal of Rock Mechanics and Engineering* 26 (supp. 2):4477–4483.

Advances in Energy Science and Equipment Engineering II – Zhou, Patty & Chen (Eds)
© *2017 Taylor & Francis Group, London, ISBN 978-1-138-71798-5*

Congestion analysis of labor crew workspace based on 4D-BIM

Zeng Guo, Qiankun Wang & Qianyao Li
School of Civil Engineering and Architecture, Wuhan University of Technology, P.R. China

ABSTRACT: The Architecture, Engineering, and Construction (AEC) industry is typically labor-intensive with much manual work done by labor crew. Because the workspaces of labor crew and building elements are close to each other, workspace congestions are likely to occur and can have bad effects on workers' productivity and even incur hazards. This paper provides a 4D-BIM-based approach to check and visualize workspace congestions of labor crew. In order to realize this approach, a prototype tool based on Autodesk Navisworks and Revit is developed and tested in a case study. The result shows that the proposed approach and the prototype are useful in finding the potential workspace congestion and thus can provide valuable information to support decision making in construction plan optimization.

1 INTRODUCTION

The Architecture, Engineering, and Construction (AEC) industry is typically labor-intensive. During the construction process, much work is done manually by labor crew. Because of the nature of the work and the limitation of site conditions, workspaces of the labor crew are very close to building elements and likely to be congested; if such congestions occur, labor productivity will be affected or even hazards may occur. Thus, it is crucial to carry on a preconstruction analysis of labor crew workspaces so as to minimize or eliminate the congestions caused by other building elements (Thomas, 2006). As the labor workspace is highly related to construction process, Building Information Modeling (BIM), especially 4D-BIM, which integrates both product (building) information and construction process information, is the most suitable solution for workspace analysis.

4D-based workspace analysis has raised much attention. Brucu Akinci et al. designed a tool (4D-spaceGen) to generate workspace as well as to check if clashes among workspaces exist (Akinci, 2002); Xin Su and Hubo Cai proposed a lifecycle approach of modeling workspace. In this method, some certain geometry shapes (cube, column, and sphere) are used to reassemble them based on the feature of the activity linked. Furthermore, they developed a tool for workspace allocation and analysis (Su, 2014). In order to address the issue related to site workspace congestion, M. Kassem et Al. designed an IFC-based tool that integrated schedule and workspace management so as to provide a solution for workspace allocation and congestion detection, and provide heuristic solutions

for construction plan optimization (Kassem, 2015). Hisham Said et al. developed a method that can make full of the indoor workspace for material storage, which is significant for situations with land use limitation (Said, 2013). Sijie Zhang et al. proposed a BIM-based site workspace management method, and they also proposed a method for workspace allocation based on their relative positions to the target elements, namely top, surround, and front and under (Zhang, 2015). Hyounseok Moon et al. described a cell-based workspace overlap detection method to check the existence and severity of clashes among resources (worker, material, and machine) in site. Its principle is casting the objects to both horizontal and vertical planes, which have been divided into several cells. The clashes can be detected by numbering those cells based on whether or not they are included by shadows casted from the objects (Moon, 2015).

The aforementioned studies are of enlightening significance for 4D-BIM-aided workspace analysis and, furthermore, optimization; however, some problems still remain. For instance, the mechanism of spatial-temporal clashes has not been clarified; there has not been enough effort to focus on workspaces for labor crew. To address such issues, this paper will propose a method for labor crew workspace analysis and develop a 4D-BIM-based tool to facilitate automatic analysis of labor crew workspace.

This paper is organized as follows: first, the research background and related research is illustrated and then a framework of 4D-BIM-based workspace analysis is proposed. A prototype is then developed based on the framework. In order to test its validity, a case study is conducted.

2 A FRAMEWORK OF 4D-BIM-BASED WORKSPACE ANALYSIS

A framework of 4D-BIM-based workspace is proposed in Figure 1. It includes five parts: 4D-BIM modeling, workspace definition, workspace generation and allocation, congestion test, and result visualization.

2.1 4D-BIM modeling

4D-BIM modeling refers to the process of linking construction schedule to 3D-BIM model so as to generate a 4D-BIM model. 4D here refers to a model that contains both 3D spatial information and time information. The approaches of converting 3D-BIM to 4D-BIM is as follows: first, prepare a 3D-BIM model; second, arrange a construction schedule; third, link the 3D-BIM model to the construction activities by matching their keys (Benjaoran, 2009). These steps can be carried out easily using 4D modeling software such as Microstation, Digital Project, and Navisworks.

2.2 Workspace definition

In most situations, labor crew is near the building components they are working at (target elements). Therefore, the workspace regions can be determined based on the components. In this paper, the horizontal region of a workspace is a rectangle generated from the bounding box of the target elements with certain expansion (offset) to the allow labor crew to work. The vertical region of the space can be determined by two parameters: workspace elevation and workspace height. The former refers

to the bottom elevation of the workspace and the latter refers to the height needed by labors. It is shown in Figure 2.

2.3 Workspace generation and allocation

Once the parameters described in Section 2.2 are determined, the workspace can be generated and allocated accordingly. In this paper, a cell-based method is utilized. It works as follows: given a workspace and the width of each cell, the workspace will be divided into several cells. In this way, each cell can act as an independent unit to detect the spatial-temporal clash.

2.4 Cell-based congestion test

Congestion occurs when a cell has spatial-temporal clash with building elements. The process of is as follows: for each cell in the current workspace, an extrusion solid will be created based on the profile of the cell. The extrusion's bottom elevation is equal to the workspace elevation, and its height will be equal to the workspace height. If the solid intersects with some elements, then spatial clash exists. If it also has time to overlap with one or more of those elements, then spatial-temporal clash exists because the spatial and temporal clashes occur simultaneously. In this paper, the cells having spatial-temporal clashes are called congested cells, and others are called free cells.

2.5 Result visualization

This step is used to visualize the level of congestions of workspaces by assigning different colors to different types of cells. In this paper, for example, the congested cells will be shown in red, while the free cells will be shown in green. Therefore, a workspace full of red cells means that it is highly congested, and hence some measures should be taken to alleviate the congestion situation. Otherwise, the productivity of labor crew will be affected.

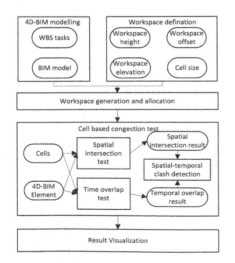

Figure 1. Framework of workspace analysis approach.

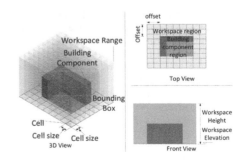

Figure 2. Workspace parameters.

3 DEVELOPMENT OF A LABOR CREW SPACE ANALYSIS TOOL

In order to realize the function of congestion test, a tool consisting of a database and two plug-ins has been developed. The database is used to bridge the information between the two software. The two plug-ins are for Navisworks and Revit, respectively. The former one is called PFN while the latter one is called PFR. The tool works as follows: first, the 3D-BIM model created by Autodesk Revit is imported to Navisworks, where it will link to an MS project file to form a 4D-BIM model. Second, the necessary information is transferred from the 4D-BIM model to the database by PFN. Third, PFR will get the information back to Revit workspace congestion test; finally, the result can be visualized in a plane view of Revit, as shown in Figure 3.

3.1 Database design

In this tool, four data tables, namely tblElement, tblWork, tblElem_Work, and tblElem_Time, are created. Their functions are as follows: *tblElement* is used to store the basic information of elements, *tblWork* stores the information of WBS items, relationships between elements and WBS items are saved in *tblElem_Work*, and *tblElem_Time* is used to store the time information of each element, as shown in Figure 4.

3.2 Plug-in for Navisworks

This plug-in is based on the widely used 4D-BIM platform software Autodesk Navisworks Manage. 3D models in various data formats can be imported in this software and further, by linking schedule plan document created by MS Project, be transferred to 4D-BIM. Furthermore, Navisworks provides Application Programme Interface (API), which supports secondary development by using classes provided by Autodesk.Navisworks.API.dll.Autodesk.Navisworks.Timeliner.

dll, the information of elements, work, and their relationships can be obtained. The process logic of this plug-in is shown in Figure 5.

3.3 Plug-in for Revit

This plug-in is based on Autodesk Revit. As a well-known BIM modeling software, Revit has well-performed visualization effect and well-organized information structure. Its API supports customized geometry intersect test, which is very useful for spatial-temporal clash detection.

When launching the plug-in, it will retrieve data from tblWork in the database, after which the User Interface (UI) will pop up (Figure 6). The user first selects a WBS Item as the activity for workspace analysis. Its start and finish time will be the same as those of the workspace. Then, the spatial information including workspace elevation, workspace height, workspace offset, and cell size are inputted manually by users. After clicking the OK button, it will first get all the elements linked to the activity and then generate a rectangle region based on each element's bounding boxes and workspace offsets. After that, cells will be created within the region.

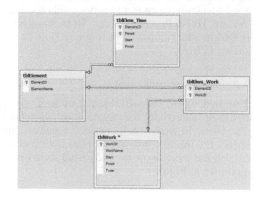

Figure 4.   Data table in SqlServer.

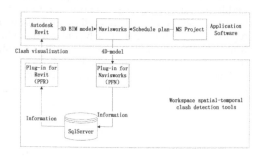

Figure 3.   Structure of the tool.

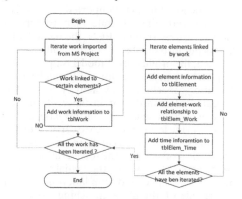

Figure 5.   Process logic of plug-in for Navisworks.

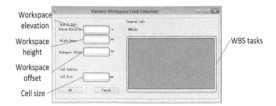

Figure 6. User Interface of plug-in for Revit.

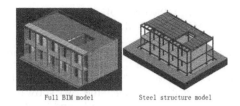

Figure 7. BIM model of the case project.

The congestion test will begin on each cell following the method described in Section 2.4.

If the spatial-temporal clash exists in a cell, it will be shown in red; otherwise, it will be shown in green. In this way, the user can find the availability of workspace so as to make a decision on whether or not to alter the schedule plan.

4 CASE STUDY

4.1 *Case background*

The low-rise modular and ecological steel structure experiment project of China Construction Steel Structure Corp is a two-floor villa located in Wuhan, Hubei Province, China. Its building height is 6.64 m and building area is about 311.44 m², as shown in Figure 7.

4.2 *Initial construction plan*

The first floor, second floor, and roof are all designed based on the module of 9800*3400. Therefore, in order to increase the construction speed, the initial construction plan is to prefabricate all the module offsite and install them on site, as shown in Figure 8.

4.3 *Workspace congestion test*

First, the Revit model is imported into Navisworks together with the schedule plan to form a 4D model, then by running PFN, the related information is put into database automatically.

Then, the original BIM model is opened in terms of plan view in Revit before PFR can be activated. The user can choose any activity in the UI that appeared afterward. For instance, if the selected activity is "bolt screwing", in this case, as workers crouch on the foundation slab to carry on their work, the workspace elevation can be set as −0.54 m, with a line vertical offset from the top elevation of the slab (−0.55 m) so as to exclude the spatial-temporal clash between workspace and foundation slab. According to the rule of thumb, a crouching worker's space requirement is 900 mm*900 mm*1200 mm, so the

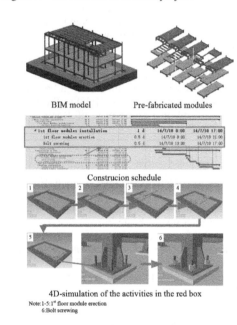

4D-simulation of the activities in the red box
Note:1-5:1ᵗ floor module erection
6:Bolt screwing

Figure 8. Initial construction plan.

offset and height of the workspace can be set as 900 mm and 1200 mm, respectively. The cell size is set to 200 mm*200 mm (Figure 9); when all preliminary conditions are ready, click OK to run the spatial-temporal clash detection.

The result of the spatial-temporal clash detection is shown in Figure 10. According to the result, a great number of red areas exist, which means that the workspaces are heavily congested. In some middle-location areas (e.g., areas around the cross between axes 2 and C), green cell exists in the work region. It is obvious that in such construction plan, workers may be unable to do their work normally, so the plan needs to be modified.

The initial construction plan is modified as follows: In the improved plan, the first floor only has the steel framework-prefabricated offsite and other works are done on-site, so the work *bolt screwing* is between *framework erection* and *profiled sheet installation*, as shown in Figure 11.

In order to run the workspace analysis of the new construction plan, the only manipulation needed

670

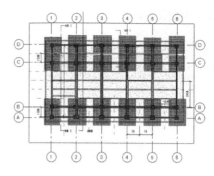

Figure 9. Parameter import interface.

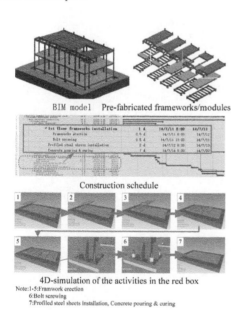

Figure 10. Spatial-temporal clash detection result of initial schedule plan.

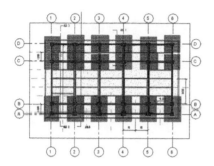

BIM model Pre-fabricated frameworks/modules

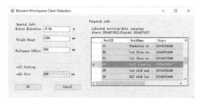

Construction schedule

4D-simulation of the activities in the red box

Note:1-5:Framwork erection
6:Bolt screwing
7:Profiled steel sheets installation, Concrete pouring & curing

Figure 11. Modified construction plan.

is to refresh the schedule plan in Navisworks and follow the aforementioned process again. The new result is shown in Figure 12.

Comparing with the initial construction plan, the congested cells (red cells) are dramatically reduced so the constructability of this plan is much better.

Figure 12. Spatial-temporal clash detection result of the improved schedule plan.

5 CONCLUSION AND DISCUSSION

5.1 Conclusion

During the construction process, workers are often among various building elements, the spatial-temporal clash between them will have a bad effect on the productivity of workers and incur hazard. As traditional clash detection techniques do not allow time factor, and many studies related to workspace clash analysis do not focus on the nature of the workspace, extensive manual input is needed to define a workspace. This paper provided an approach to carry out the workspace analysis so as to detect potential workspace congestions and then developed a prototype to realize it. A case study is also conducted to validate the prototype tool. The results show that the spatial-temporal analysis of workers' workspace is correct and practical, which will be helpful in finding potential classes and eliminate them before they actually occur.

5.2 Limitations

However, as is confined by time and knowledge, there are some limitations to this paper. For instance, the prototype needs to be tested on more projects; the running speed of the tool is directly influenced by the number of WBS items, scale of BIM model, and cell size. Therefore, further optimization on program is needed for a higher speed.

5.3 Further research

The tool provided in this paper realized the visualization of workspace congestion. However, the automation optimization mechanism has not been described here. Besides, although the manual input in spatial-temporal clash has been reduced, it has the potential to be further reduced to obtain information automatically from existing construction code. This will be described in the future.

671

REFERENCES

Akinci, B., M. Fischer, J. Kunz, J. Constr. Eng. Manage. 128. 306 (2002).

Benjaoran, V., S. Bhokha, Engineering Construction & Architectural Management. 16. 392 (2009).

Kassem, M., N. Dawood, R.Chavada,. Automat. Constr. 52. 42 (2015).

Moon, H., H. Kim, H, R. Kamat, L. Kang, J. Comput. Civ. Eng. 29. 1 (2015).

Said, H., K. El-Rayes, Automat. Constr. 31. 293 (2013).

Su, X., H. Cai, J. Constr. Eng. Manage., 140. 615 (2014).

Thomas, H.R., D.R. Riley, S.K, Sinha, Pract. Period. Struct. Des. Constr. 11. 197 (2006).

Zhang, S., J. Teizer, N. Pradhanange, C. Eastman, Automat. Constr. 60. 70 (2015).

Advances in Energy Science and Equipment Engineering II – Zhou, Patty & Chen (Eds)
© *2017 Taylor & Francis Group, London, ISBN 978-1-138-71798-5*

Analysis of moderate–strong earthquake damages of rural residential buildings and seismic inspiration

Ping He & Ting Wang
Earthquake Administration of Guangdong Province, Guangzhou, China
Guangdong Research Center for Destructive Earthquake Emergency, Guangzhou, China

You-jun Cai & Gang Yu
Earthquake Administration of Guangdong Province, Guangzhou, China

ABSTRACT: In this paper, on the basis of the field investigation data analysis of typical rural residential buildings damaged by recent moderate–strong shocks, it is concluded and summarized that the buildings with different structures have different seismic damage characteristics and damage reasons. Finally, combined with the phenomenon of the rural residential damages, we get the preliminary understanding and enlightenment of improvement in the earthquake resistance of rural buildings. The research results may provide preliminary theoretical basis and solution for the antiseismic design of the rural residential buildings in the future.

1 INTRODUCTION

Earthquake is an unavoidable natural phenomenon. With the rapid development and enlargement of the living space of people in earth, casualties and economic loss due to earthquake become more and more serious and enormous. However, the antiseismic performance of domestic rural residential building is still poor and they also have a lot of seismic fortification defects. Therefore, in recent years, the rural housing damages due to moderate–strong earthquakes are particularly serious. In order to realize the damage characteristics of various structural buildings in the disaster area and provide elemental information for the study on the antiseismic buildings in the future, our research team investigated the meizoseismal area of Minxian-Zhangxian Ms6.6, Ludian Ms6.5, and Jinggu Ms6.6 earthquakes in August 2013, October 2014, and July 2016, respectively. Finally, we have accumulated a large amount of building damage data and photographs. The analysis of the collected data led to the conclusion that rural buildings with different structures have different seismic damage characteristics and damage reasons. Finally, combined with the phenomenon of rural residential damages, we get the preliminary understanding and enlightenment of improvement in the earthquake resistance for rural buildings.

2 GENERAL SITUATION OF THREE MODERATE–STRONG EARTHQUAKE DISASTERS

2.1 *The Minxian–Zhangxian Ms6.6 earthquake of Gansu Province (104.2 °E, 34.5 °N)*

M6.6 earthquake occurred in Minxian–Zhangxian neighborhood at 7:45 pm on 22 July 2013. Epicenter was located at Meichuan, Minxian, Gansu. The disaster intensity of the meizoseismal area is VIII, and total area is 706 km^2. In this earthquake, 93 people were killed, more than 1,300 people were injured, and more than 320,000 houses were severely damaged. Because of the local thick loess overburden and complex mountainous terrain, secondary geological disasters are widely distributed in the VIII area. The Adobe Wood structural rural housing accounts for more than 70% of the building ratio there, thereby resulting in the collapse and serious damage of several buildings. Moreover, many buildings were buried by the loess landslide and collapse. This may be the main reason for the serious casualty in this earthquake (Wang, 2013), as shown in Figure 1.

2.2 *The Ludian Ms6.5 earthquake of Yunnan Province (103.3 °E, 27.1 °N)*

At 16:30 pm on 3 August 2014, an Ms6.5 earthquake occurred in Ludian, Zhaotong, Yunnan Province. The epicenter was located at Longmenshan town.

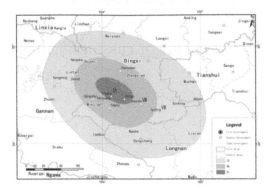

Figure 1. Intensity distribution of the Minxian–Zhangxian Ms6.6 earthquake of Gansu Province.

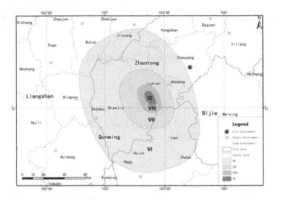

Figure 2. Intensity distribution of the Ludian Ms6.5 earthquake in Yunnan Province.

The disaster intensity of meizoseismal area is IX and the total area is up to 90 km², including Longtoushan, Huodejiang of Ludian County, and Baogunao of Qiaojia County. A total of 616 people were killed, 3,143 people were injured, 80.900 houses were collapsed in this earthquake (report on the assessment of Ms 6.5 earthquake disaster losses in Ludian, 2014), as shown in Figure 2 and Tables 1 and 2.

2.3 The Jinggu Ms6.6 earthquake of Yunnan Province (100.5°E, 23.4°N)

At 21:49 pm on 7 October 2014, an M6.6 earthquake occurred in Jinggu County, Puer City, Yunnan Province, whose focal depth is 5.0 km. Epicentral intensity is VIII, and this shock did not cause serious secondary disasters. There was only one victim caused by house collapse, and 323 people were injured. Compared with the Ludian earthquake, disasters and losses caused by this quake are

Table 1. Ratio of various types of housing structure in the Ludian disaster area (unit%) (Li, 2015).

Frame	Brick-concrete	Brick-wood	Adobe Wood
0.33	17.82	1.41	80.44

Table 2. Damage ratio of various types of housing in the IX area of the Ludian Ms6.5 earthquake (unit%) (Li, 2015).

Housing structure	Building earthquake damage level			
	Destroyed	Serious	Moderate	Slight
Frame	0	9.55	26.38	37.24
Brick-concrete	0	12.30	20.93	40.92
Brick-wood	10.98		73.89	
Adobe Wood	26.18		62.91	

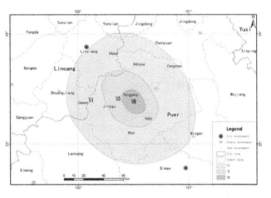

Figure 3. Intensity distribution of the Jinggu Ms6.6 earthquake in Yunnan Province.

Table 3. Ratio of various types of housing structure in Jinggu disaster area (unit%) (Report on the assessment of the Ms 6.5 earthquake disaster losses in Jinggu, 2014).

Frame	Brick-concrete	Brick-wood	Adobe Wood
0.86	18.00	48.54	32.6

Table 4. Damage ratio of various types of housing structure in VIII area of Jinggu Ms6.6 earthquake (unit%) (Report on the assessment of Ms 6.5 earthquake disaster losses in Jinggu, 2014).

Structure type	Building earthquake damage			
	Destroyed	Severe	Moderate	Slight
Frame	0	9.55	26.38	37.24
Brick-concrete	0	12.30	20.93	40.92
Brick-wood	10.98		73.89	
Adobe Wood	26.18		62.91	

much less. The phenomenon had been discussed by experts from different perspectives (Hou, 2015; He, 2015), as shown in Figure 3 and Tables 3 and 4.

3 ANALYSIS OF SEISMIC DAMAGE CHARACTERISTICS OF RURAL BUILDINGS

3.1 Adobe wood structures

Rural Adobe Wood structure houses generally refer to which wooden frames and adobe wall act as load-carrying member. In the previous earthquake cases, the buildings damaged by the quake are more and heavier than others. However, the earthquake resistance of these buildings is still different according to the different bearing ways of wooden truss, construction technology, and materials. Based on the wooden truss bearing way, there are almost three concrete structural types, which have been found in earthquake-stricken areas.

The first one is the through-type timber frame building. The load-carrying member of the house includes wooden columns, beams, purlins, and rafters. There is a connection between each other measures, and the architectural integrity is better.

Adobe walls only act as the partition and filler walls. This building is less ruined by the quake and did not collapse, except filling out the wall. Photos of seismic-damaged buildings are shown in Figure 4.

For the second one, adobe walls act as load-carrying members. For this kind of building, wooden truss is loaded on the adobe walls without any wooden columns, which is frequently found in the rural area of China. The size of the adobe brick differs from one another in various regions. In light of the coldness and the demands for heat preservation in northern China, the adobe brick there is thicker than that in the south of China. It is found that the earthquake brings less damage to this kind of the house in Jinggu area. Photos of the seismic-damaged building are shown in Figure 6. Heavier damage was observed in Gansu and Ludian. Photos of the seismic-damaged building are shown as

Figure 5. Load-bearing walls were slightly damaged (VIII area in Yongping, Jingu).

(a)VIII area in Yongguang, Minxian

(b) VIII area in Yongping, Jingu

Figure 4. Adobe wall damaged in the building with Adobe Wood structures.

Figure 6. Partially collapsed building with Adobe Wood structure (VIII area in Yongguang, Minxian).

in Figures 5–7. Particularly, two-storey buildings are more likely to be built in Ludian area. The higher the building height (almost 6 m) is, more strongly the seismic action effect increases. In such a situation, the shearing-bearing capacity of the loam walls becomes lower, thus resulting in the collapse.

The third is the common Adobe Wood structure, in which adobe walls and wooden frames act as load-carrying members together. This kind of building has simple wooden columns, but they are usually smaller, bearing the weight of the wooden roof. Simple wooden columns or brick columns are set. The wooden columns, which are relatively thin, simply bear part of the load of the roof. Most of the columns refer to "column in wall" with purlin loaded on the wall. This kind of building is the common rural residential housing in northwest China. It is easier to be subjected to moderate damages when suffered an earthquake intensity of VII. It is shown to be severely damaged when suffered an earthquake intensity of VIII. This situation is also found in the Ms6.6 earthquake in Gansu, where housing is severely damaged or collapsed in the meizoseismal area. Photos of seismic-damaged buildings are shown in Figures 8–9. Although the

Figure 7. Load-bearing adobe walls partially collapsed (IX IX area, Babao, Longtoushan, Ludian).

Figure 8. The Adobe Wood structure building collapsed (VIII area in Yongguang, Minxian).

Figure 9. The Adobe Wood structure building partially collapsed (VIII area in Suigu, Minxian).

structure of the housing has enough bearing capacity, it also shows lower tensile strength and is more prone to brittle failure. Under the effect of seismic force, this kind of housing is easier to collapse in the wall at the junction.

3.2 Brick-wood structures

Brick-wood rural houses mainly refer to houses constructed with brick walls or wooden frame bearing, using rafters, purlins, beams, tiles, and so on. In previous earthquake cases, the damage of this kind of building is mainly shown as wall damage (such as different levels of wall cracking, partial or overall collapse of the wall, etc.), roof damage (off-tile, roof deformation, etc.), and other accessory damage. According to the different bearing classifications, we found now that there are two different bearing types of brick-wood house in the disaster areas.

One is "through-type timber frame." The wooden frame of this type acts as load-carrying member and the brick walls mainly play the role of enclosure. This kind of houses in Jinggu accounts for a large proportion. Wood with good mechanical properties, light weight, good plasticity and ductility, good integrity, good plastic deformation capacity, and good energy dissipation & seismic mitigation effect are chosen as bearing components. Furthermore, there are a lot of wooden pillars in the Jinggu disaster of this type of house and the frame has five pillars, which further enhanced the stability. Enclosure of brick masonry is built in the outside of the beam column in a form of outsourcing. During earthquake, the wooden frame withstands overturning, thereby effectively protecting the indoor crowd (Hou, 2015). This is also an important reason for the small casualties in the Jinggu earthquake. Photos of the seismic-damaged building are shown in Figure 10. Only few buildings of this kind are damaged in the Min county earthquake, and the damage is relatively light, with most damages showing that the filling wall penetrates cracks in wall junctions.

The other is brick wall bearing and enclosure house with wooden structure roof. This kind of building brick wall is the main bearing component; wall cracking under earthquake is the most common damage phenomenon, which is generally presented as X-shaped cross cracks and diagonal cracks or other damage forms. Most roof damage performances are as shown as spindle tile phenomenon or tile serious slippage. Photos of the seismic-damaged building are shown in Figures 11–12.

3.3 Brick-concrete structures

With the development of economy, brick-concrete structural houses are becoming increasingly the first choice for newly built houses. However, because of its poor warmth retention property, in the region of the northwest China, Adobe Wood or brick-wood structure houses still occupy a certain proportion. Especially, the rural residential buildings or low-rise commercial houses in the townships usually adopt this kind of structure form. Currently, because of the lack of supervision and guidance of farmers in building, they often only consider cost-saving and spatial applicability to build a house, resulting in the decline of the seismic capacity of the house.

(a) Frontal illumination of the damaged building

(b) Detail photo of the damaged building

Figure 11. Cracks in the connection of load-bearing wall (VIII area in Jiemashan, Meichuan, Minxian).

(a) VIII area in Yongguang, Minxian

(b) VIII area in Mangfei, Yongping, Jingu

Figure 10. Brick-wood structure building with less damaged through-type timber frame.

（a）Appearance of a damaged building

(b) Magnification of the damaged part

Figure 12. Cracks at the junction of load-bearing brick walls (VIII area in Mangfei, Yongping, Jingu).

As a result, in recent years, earthquake damage of this kind of house has been extreme and the loss has been quite high. In addition, the use of heavy materials should be avoided, so as to prevent casualties. Therefore, we should pay more attention to the seismic resistance of such houses. In this paper, we will discuss the damage forms of the three types of buildings in the earthquake area.

The first form is brick-concrete structure house with seismic construction measures. The so-called seismic construction measures refer to:

1. Set reinforced concrete constructional column;
2. Set reinforced concrete ring beam and constructional column connections;
3. Strengthen the connection between walls, floors, and beams with enough length and reliable connection;
4. Strengthen the integrity of the staircases.

Judging from the three earthquake sites, the kind of the house has withstood the test of strong shock. At present, in the epicenter area, we have not yet found a case of damage to the secondary damage. Photos of such buildings are shown in Figures 13–15.

The second is the brick-concrete structure houses without antiseismic structural measures (some of the above seismic structural measures are adopted or are not used at all). It can be seen from the three quakes that damages to single-storey building are not severe, but the destruction of multistorey buildings is extreme. This phenomenon is particularly prominent in Ludian earthquake. In the epicentral area, multistorey buildings without aseismic structure measures generally suffered serious damage. Photos of the seismic-damaged building are shown in Figures 16–17. Mainly because of the failure to take the aseismic measures, the tensile strength, shear strength, bending strength, and deformation ability of multistorey masonry buildings are relatively low. Therefore, under the earth-

Figure 14. Comparison of two kinds of brick-concrete structure buildings in seismic damage (IX area, Longtoushan, Ludian).

(VIII area in Yongping, Jingu)

Figure 15. Report on the assessment of the Ms 6.5 earthquake disaster losses in Ludian, 2014). Intact building with seismic structure.

Figure 16. (Report on the assessment of the Ms 6.5 earthquake disaster losses in Ludian, 2014). Multiple X-shaped cracks in load-bearing walls.

Figure 13. Intact building with seismic structure (VIII area in Dalong, Minxian).

quake force, because of the weight and stiffness of the building, the shortness of self-vibration period, naturally, it is easy to produce brittle failure, which can cause serious damage to the building, as shown in Figure 18.

The third is the use of the current domestic and rural popular bottom-reinforced concrete frame

Figure 17. (Report on the assessment of the Ms 6.5 earthquake disaster losses in Ludian, 2014). Bottom wall cracked, ruined.

Figure 18. Bottom wall penetrating cracks.

（a）Frontal illumination of a damaged building

（b) The bottom structure columns are staggered.

Figure 19. Bottom frame structure of a damaged building.

Figure 20. Bottom frame structure of a partially collapsed building (IX area Longtoushan, Ludian).

structure. The first storey of the building adopted the frame structure. The second floor and above are used in the form of ordinary brick structure. This form of housing structure can create a large space at the bottom of the building, and all kinds of rooms & spaces can be set in the second floor and above, which is particularly popular in the street commercial housing. However, due to the large horizontal shear force generated by the quake, the combination of the bottom frame and masonry became the weak part of the structure. Its damaged forms show column frame damage, such as column bases and capitals crushing of concrete, resulting in the framework of the underlying overall collapse or serious tilt. There are several such damage cases in Ludian earthquake, as shown in Figures 19–20.

3.4 *Frame structure*

In the above mentioned earthquake affected area, there are only few houses with pure frame structure. The frame structure mostly exist in government offices, schools, hospitals, and other long-span and multistorey buildings. Compared with other structures, the ratio of seismic damage of frame structure is relatively small, and it is generally in the area of VIII. We will discuss it in the next paper.

4 INSPIRATION OF THE EARTHQUAKE DAMAGES OF RURAL HOUSES

1. From the above description of earthquake damages of the rural house, it can be seen that the lack of professional guidance of the formal design, building material selection, and

construction of each link during the previous domestic rural construction lead to the poor hysteresis behavior of overall residential buildings. In general, moderate–strong earthquakes bring devastating casualties and huge economic losses. Therefore, it is the most direct and effective way to improve the seismic capacity of rural buildings and reduce the casualties and property losses by extending the scientific seismic fortification, specification design, construction technical service, and management in the rural area. Fortunately, recently, Government of China has paid attention to the question and begun to promote the implementation of rural residential earthquake safety project. We believe that the seismic capacity of rural houses will be greatly improved in the future.

2. At present, the awareness of earthquake disaster prevention and mitigation of people is still poor. The consciousness of the people living in frequent earthquake area can significantly enhance after disasters. Because people have faced less or lighter earthquake disasters, they have poorer sense. It is essential to promote the implementation of rural house earthquake safety project to improve the knowledge of earthquake disaster, thereby increasing safety emergency response knowledge and enhancing the popularization of science aseismic construction.

3. It can be concluded from the disasters mentioned above that the earthquake resistance of the rural residential building is not simply determined by the structure style of the building, regardless of the seismic capacity the building, but depends largely on the following conditions: site selection, reasonable design, proper aseismic measures, correct construction craft, and reliable building materials. Therefore, when designing rural buildings architects should take account the local traditional customs and available building material. They should improve the seismic performance based on the local building type. Only in this way it is easier for local people to accept the earthquake safety project, and the project is more likely to become widely promoted.

4. Attention should be paid to construction skills training and management of the local architectural craftsmen. At present, the houses have been built by familiar local craftsmen in the rural area. The level of architectural sophistication is largely decided by the capacity of the local craftsmen. In promoting the implementation of rural residential earthquake safety project, the long-term training and management of the craftsmen is crucial.

REFERENCES

He Jia-ji, etc. Analysis of Difference of Seismic Capacity of Buildings [J]. *Journal of seismological research.* 38 (2015).

Hou Jian-sheng, et al. Cause analysis of M6. 6 Jinggu & M6. 5 Ludian [J]. *Journal of Catastrophology*, 30 (2015).

Jia Qiang, etc. Reports on the damages of bottom reinforced concrete frame structures [J]. *Journal of Shan-Dong Jianzhu University,* 23 (2008).

Li Zhao-long etc. Characteristic of the Death Age Distribution of Ludian Earthquake [J]. *Journal of seismological research*, 38 (2015).

Report on the assessment of M_s 6.5 earthquake disaster losses in Ludian, Yunnan [R]. *Earthquake Administration of Yunnan Province* (2014).

Report on the assessment of Ms 6.5 earthquake disaster losses in Jinggu, Yunnan [R]. *Earthquake administration of Yunnan province* (2014).

Wang Lan-min, etc. Earthquake Damage Characteristics of the Minxian- Zhangxian [J]. *China Earthquake Engineering Journal.* Vol. 35 (2013).

Wang Lan-ming, etc. Guidelines for Earthquake Resistant Construction of Rural Buildings of Northwest, China [M]. *Seismological Press* (2011).

Advances in Energy Science and Equipment Engineering II – Zhou, Patty & Chen (Eds)
© *2017 Taylor & Francis Group, London, ISBN 978-1-138-71798-5*

Theoretical analysis and experimental study on a natural ventilation window with functions of purification and sound insulation

Junli Zhou, Cheng Ye & Peng Nie
School of Civil Engineering and Architecture, Wuhan University of Technology, Wuhan, China

Hejun Li
School of Energy and Power Engineering Institute, Wuhan University of Technology, Wuhan, China

ABSTRACT: A new type of window with functions of purification and sound insulation was proposed based on natural ventilation. The airflow path of the window was simplified, and the total resistance was analyzed by using the principle of fluid mechanics. The feasibility of natural ventilation for the window was also analyzed. Finally, experiments on the window were carried out and the effects of ventilation, purification, and sound insulation were analyzed.

1 INTRODUCTION

In recent years, outdoor air pollution has become an important threat to human survival, which requires long-term efforts. Outdoor air quality has a direct impact on the quality of indoor environment. For large public buildings, with air-conditioning equipment, people can install mechanical ventilation system and air purification equipment and close the doors and windows to improve indoor air quality.

However, natural ventilation is a main mean of passive architectural technology. For residential building, natural ventilation is always the first choice in view of thermal environment and psychology. According to the survey conducted on residents in Xi'an, the window-opening rate of residents is up to 61% in the morning and 44% at noon (Qiu, 2009).

Meanwhile, traffic noise pollution has become more serious with the increase of urbanization. Noise directly affects the central nervous system, thereby increasing heart rate, blood pressure, and so on. A proportion of 90% of indoor noise passes through the windows, so the installation of sound insulation window is the most direct method to solve the noise problem (Ji, 2015).

Under this background, many researchers have begun to study natural ventilation window, hoping to achieve ventilation, purification, noise removal, and other multieffects (Yu, 2012; Zhai, 2004; Liu, 2015; Liu, 2015; Huang, 2011). On the basis of the existing natural ventilation and sound insulation window, this paper presents a multieffects window, which includes ventilation, sound insulation, and purification.

2 DESIGN

2.1 *Size and structure*

The size and structure of the window are shown in Fig. 1, 2, and 3. Channel width is 200 mm width, height of the inner window is 280 mm, the size of the inlet is 200×200 mm, and that of the outlet is 150×150 mm. The overall width is 1500 mm and height is 1200 mm. The whole window is divided into two parts and connected by the upper and lower plates.

2.2 *The section of sound insulation*

As shown in Fig. 3, the window is composed of a top-suspension window and a bottom-suspension window. A double structure is used inside the top-suspension window. The multilayer transparent

Figure 1. Size of front window.

Figure 2. Size of side window.

Figure 3. Structure of window.

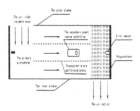

Figure 4. Structure of noise channels.

micro-perforated plate resonance is installed to form a channel for noise elimination, as shown in Fig. 4. The active device for noise reduction is mainly composed of a reference microphone, secondary sound source, a controller, and an error microphone.

2.3 The section of purification

The early effect strainer is installed in the inlet, which is mainly used for filtering dust particles above 5μm. An electrostatic electret layer and an activated carbon adsorption layer are installed in the outlet. When the preliminary filtered air passes the electrostatic electret filter layer, the electret could generate static electricity by itself to attract charged submicron particles and capture them, which also induces polarity generated by neutral submicron particles to capture them, and to filter inhalable particles and bacteria. Finally, the air flows in the room through the activated carbon filter layer, which adopts high adsorption filter and can quickly dispose most harmful gas and clean air.

3 ANALYSIS ON RESISTANCE

3.1 Computational model

When air passes through the window, the window can be regarded as a local resistance member, as shown in Fig. 5.

As shown in Fig. 5, the window can be regarded as two quarter bends: A and B. The local resistance coefficient of A is 0.805 (Lu, 2008) and that of B is 2.5 (HVAC Knowledge Training).

Figure 5. Schematic of airflow.

Supposing that inlet wind speed is V_1, according to the cross-sectional area ratio, outlet speed is 1.78 V_1. The air density is a constant, equal to 1.29 m³/kg, and the total resistance is the sum of two quarter bends (ignoring the lateral frictional loss):

$$\Delta Pa = \frac{1}{2} \times 1.29 \times 0.805 V_1^2 + \frac{1}{2} \times 1.29 \\ \times (1.78 V_1) V^2 = 2.56 V_1^2 \qquad (1)$$

The calculation results show that the relationship between the total resistance and the wind speed meets square-law.

3.2 Analysis of the worst case

Wind pressure and the total resistance are related to wind speed. Therefore, the feasibility of the ventilation window will be discussed under the worst case.

The air change rate of single-sided ventilation is far less than the cross-ventilation, and the wind speed increases with height; therefore, it can be assumed that the worst case is single-sided ventilation and one-layer building.

3.2.1 Theoretical calculation of wind speed

It is supposed that the window is installed in a height of 4 m in a single building on the windward side. The infinity of the wind speed can be described by exponential function (Tang, 2003):

$$U_h = U_{ref} \left(\frac{H}{H_{ref}} \right)^{\alpha} \qquad (2)$$

where U_{ref} and H_{ref} represent wind speed in reference height and reference height, respectively; reference height is 10 m; and α is the surface roughness coefficient. If reference speed (annual average wind speed in Wuhan) is assumed to be 2.8 m/s, the wind speed on building height can be calculated as 2.06 m/s according to the above method.

3.2.2 Theoretical calculation of wind pressure

Wind pressure coefficient of this building can be obtained: the windward side is +0.8; leeward side is −0.5 (Association of China Construction

Standards, 2010). The wind pressure difference of the building is:

$$P_W = \frac{1}{2} \times 1.29 \times 0.8\, U_h^2 = 2.19\, kPa \qquad (3)$$

3.2.3 *Theoretical calculation of inlet velocity*

Considering the problem of single-sided ventilation, the inlet average speed is (Phaff, 1982):

$$V_1 = \sqrt{0.001 U_{ref}^2 + 0.0035 H |\Delta T| + 0.01} \qquad (4)$$

where H is the height of the building and ΔT is the difference between indoor and outdoor temperatures (0°C is adopted here). Assuming the outdoor standard speed to be 2.8 m/s, the average inlet wind speed can be calculated as 0.13 m/s. In fact, because of the effect of building thermal environment, there will be a temperature difference between indoor and outdoor in natural ventilation buildings; then, the actual wind speed will be greater than it.

3.2.4 *Analysis on feasibility*

When the inlet velocity is 0.13 m/s, the total resistance is 0.046 kPa, according to equation (1). The resistance of the cleaning equipment and other local components of the window is 2.14 kPa. If purification equipment and other local resistance are greater than 2.14 kPa, negative pressure will appear in the outlet, which is bad for the flowing.

The local resistance component includes two parts: (1) noise elimination component and (2) purification components. The active noise reduction equipment is small, and the perforated resonance board is placed parallel to the flow direction, so their local resistances can be ignored. For the purification components, the total resistance is composed of three parts: early effect strainer, electrostatic electret layer, and activated carbon adsorption filter. Early effect strainer can select all-metal mesh filter, with the initial resistance of 20–50 Pa and final resistance of 150–250 Pa[13], so its normal working resistance can be supposed as 100 Pa. Polypropylene can be used as the electret filtration material, with air pressure drop of 28.4 Pa and filtration efficiency of 84% (for 0.1 μm NaCl testing). The resistance of other filtration material is generally lower than this value (Xie, 2005). The resistance of activated carbon filter is around 40 Pa (product features of active carbon air filter). Then, the total resistance of the purification equipment is about 168 Pa, which is much lower than 2.14 kPa. It is considered that air can still flow in with installation of purification equipment.

4 ANALYSIS OF RESULTS

After completion of production, an experimental analysis was made on its ventilation performance, purification performance, and sound insulation performance. The first and second functions are carried out in the laboratory of Wuhan University of Technology, and the third in Acoustic Research Laboratory of Nanjing University.

4.1 *Ventilation test*

Outdoor flow was provided by fan, and the inlet and outlet velocities were measured by a universal wind speed recorder. Valid data are shown in Table 1.

Table 1 indicates that the outlet velocity increases with accelerating inlet velocity. Fresh air per capita is 30 m³/h (The State Administration of Quality Supervision, 2002). It is evident from the table that the inlet velocity is 1.88 m/s, the outlet velocity is 0.46 m/s, and ventilation quantity is 37.26 m³/h. It shows that the window can meet the fresh air requirement of one person in this building.

In addition, the inlet velocity calculated in Section 3.2.3 is relatively low. However, the actual wind speed will be greater than it. According to the test data of Guangzhou in summer, when the average outdoor wind speed is 1.7 m/s in Guangzhou, wind speed at a distance of 5 m from the window is 0.8–1.7 m/s (Tang, 2003). It can be judged that the window has a certain effect in ventilation.

4.2 *Purification test*

Gas from burning coal was used to approximately simulate atmospheric pollution, and aerosol monitor was used to test the concentration of pollutants every 10 min in inlet and outlet to determine the cleaning effect of the window. The test results are shown in Figs. 6 and 7.

It is evident from the above figures that the concentration of pollutant decreases gradually with time, and concentrations of PM10 and PM2.5 decrease in both outlet and inlet, whose average purification rates can reach 62.3% and 42.3%, respectively, which indicates that the purification units of the window work well.

Table 1. Ventilation experiment data.

Air inlet velocity (m/s)	Outlet velocity (m/s)	Ventilation volume (m³/h)
1.88	0.46	37.26
1.54	0.36	29.16
0.78	0.24	19.44

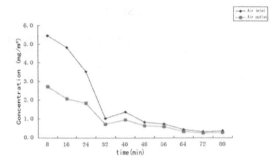

Figure 6. Concentration of PM10.

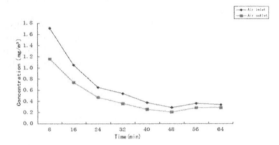

Figure 7. Concentration of PM2.5.

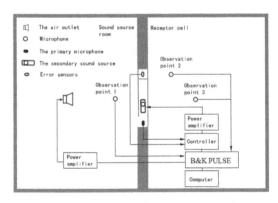

Figure 8. Plan view of soundproof room.

4.3 Sound insulation test

In order to measure the effect of sound insulation, the window was mounted in the middle of the wall between the source room and reception room. A loudspeaker was placed in the sound source room to produce 0–1600 Hz white noise for the simulation of urban traffic noise, meanwhile a microphone was put in the observation point 1, from the outside of the window surface 10 cm for observing sound pressure of sound source room. A microphone was placed in points 2 and 3, from the inside of the window surface 10 cm and

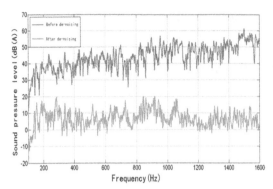

Figure 9. Noise spectrum.

2 m, respectively, for monitoring sound pressure changes of the receiving chamber.

Fig. 9 shows the change of sound pressure level. The average sound pressure level of the sound source room is 80.5 dB(A), that of the receiving chamber is 38.3 dB(A), and the overall sound insulation is 42.2 dB(A). The window reaches grade II of the insulation standard (Science & Technology Department of State Environmental Protection Administration, 1996).

5 SUMMARY

Natural ventilation is an important mean of passive technology. For residential buildings, the paper proposed a new type of natural ventilation window with functions of purification and sound insulation. The resistance of the window was analyzed by using fluid mechanics principle, and the feasibility of natural ventilation was discussed considering the resistance of purification equipment. In addition, the ventilation performance, cleaning performance, and sound insulation performance were also studied in experiments. From the measuring results, it can be seen that the window can meet the requirement of fresh air of the building, and has some ability of purification and sound insulation. Therefore, the application of the window in the civil residential buildings can improve the building thermal environment, acoustic environment, and air quality to a certain extent.

ACKNOWLEDGMENTS

This work was financially supported by ESI Discipline Promotion Foundation (No. 35400664), and the Research on Energy-saving and Simulation Technology of Steel Structure Buildings in Hotsummer and Cold-winter Area (Wuhan Urban And Rural Construction [2012] NO. 308).

REFERENCES

Association of China construction standards. GB50009–2012. Beijing: China Architecture & Building Press Construction Structure Load Standard. (2010) (in Chinese).

HVAC Knowledge Training (air system). http://www.doc88.com/p-9039394084626.html. (in Chinese).

Huang. H., Nanjing University. Active Noise Attenuation in Natural Ventilation Window. 55–66 (2011) (in Chinese).

Ji, Y., D. Liu, Z. Chen, C. Chen, etc. Programmable Controller & Factory Automation. A Natural Ventilation Purification Sound Insulation Window Based on the Passive and Active Noise Reducation Technology. 31–34, **31**(2015) (in Chinese).

Liu, S., H. Zou, X. Qiu. Journal of Nanjing University (Natural Sciences). *An acoustical model for the staggered-structure window of natural ventilation and sound insulation.* 51–59, **51**(2015) (in Chinese).

Liu. X., Taiyuan University of Technology. Study on Noise Reduction of the Optimization Design of Natural Ventilation Acoustic Proof Window. 66–81 (2015) (in Chinese).

Lu. Y., Practical Handbook of heating and air conditioning design Second Edition. 1102 (2008) (in Chinese).

Phaff, H., W. De Gids, Air Infiltration Review. Ventilation Rates and Energy Consumption due to Open Windows: a Brief Overview of Research in the Netherlands. 4–5, **4**(1982).

Product Features of active carbon air filter. http://www.kqglm.com/kqguolvqi/hxtkqglq.html

Qiu, S., A. Li, X. Zhang. Building Energy & Environment. *Investigation on Rate of Window Opening under Natural Ventilation in Xi'an.* 58–61, **28**(2009) (in Chinese).

Science & Technology Department of State Environmental Protection Administration. The standard of sound insulation window HJ/T 17 (1996) (in Chinese).

Tang. Y., South China University of Technology. *The Relationship Between Natural Ventilation and Residential Windows.* 29–71 (2003) (in Chinese).

The State Administration of quality supervision, inspection and quarantine. Interior Air Quality Specification GB/T 18883 (2002) (in Chinese).

The characteristics of metal mesh filter products. http://www.songfengkou.com/36/. (in Chinese).

Xie, X., Xiang Huang, Yuhui Di. Contamination Control & Air-Conditioning Technology. *Disussion of Electret Air Filtration Material Utilizing Electrostatic Electret.* 41–44, **2**(2005) (in Chinese).

Yu, H., K. Chen, H. Hu. et al. Noise and Vibration Control. Experimental Investigation on Influence Factors of Active Noise Attenuation of Sound-insulation Ventilation Window. 191–197, **31**(2012) (in Chinese).

Zhai, G., Noise and Vibration Control.Bangjun Zhang. The Dsign of Ventilation and Sound Insulation Window. 45–46, **1**(2004) (in Chinese).

Advances in Energy Science and Equipment Engineering II – Zhou, Patty & Chen (Eds)
© *2017 Taylor & Francis Group, London, ISBN 978-1-138-71798-5*

Identification of structural natural frequency based on microtremor test

Jianqi Chen, Yong Sun & Xiaomin Zhang
College of Water Conservancy and Civil Engineering, Shandong Agricultural University, Tai'an, China

Nan Yang
Shandong Provincial Research Institute of Coal Geology Planning and Exploration, Ji'nan, China

Hui Wang
College of Water Conservancy and Civil Engineering, Shandong Agricultural University, Tai'an, China

ABSTRACT: In this paper, we obtain the vibration response of a six-floor steel structure under the environmental random excitation by means of microtremor test. Before the test, the sensitivity analysis of each vibration pickup is realized through a pre-experiment and the principle of normalization. This refers to the analysis of the vibration data based on the spectrum analysis, and discusses the applicability of Nakamura's technique to the estimation of the natural vibration frequency of the steel structure. The results show that using the vibration response of steel structure as excitation signal to check sensitivities of the vibration pickups fits better to the real working environment, and it can reflect the performance of instrument more stringently. Nakamura's technique can eliminate noise from the input signals and then improve signal-to-noise ratio. The natural frequency of the steel structure can be accurately estimated with this method.

1 INTRODUCTION

The 941B-type ultralow frequency vibration gauge has been widely used in practical engineering applications since its invention. It is mainly used in the pulsation measurement of the ground and structure, such as the industrial vibration measuring of the general structure and the low-frequency measurement of structural dynamic characteristics like tall and slender structure, which provides technical support for antiquake architectural structure design, site seismic safety assessment, and so on (Yang, 2005; Liu, 2011; Luo, 2010; Huang, 2010).

The 941B-type vibration gauge (Figure 1) is a type of precision instrument that has many components, which may causes many problems if not managed properly, such as poor line contract, low vibration pickup accuracy, and destruction of collecting instrument channel and amplifier caused by line aging. Maintenance works brought great inconveniences to users. The work of returning to the factory for repair will be greatly simplified by timely screening and repairing the failed parts of the equipment.

This paper applies microtremor test and gains vibration response of six-floor steel structure in a random excitation environment. Before the test, through pre-experiments, the sensitivity of vibration pickups were checked by normalization

principle and intuitive judgment method (Wang, 1995). Then, on the basis of the analysis of the test data spectrum, this paper deals with the effectiveness of Nakamura's technique in estimating the natural frequencies of steel structure.

2 MICROTREMOR TEST

Microtremor testing helps researchers to obtain the modal parameters (Xu, 2005). It regards natural random vibration as load inputting to the structure and acquires the structural vibration response by field measurement. It has been widely applied in engineering practice (Huang, 2006; Luo, 2011). Using pulsated method test, Liu Hongbiao obtained the distribution rule of the site predominant period of Xichang, Sichuan Province and Huang Yu measured structure dynamic characteristics of buildings in Shanxi, based on the response of the structure of pulse noise.

2.1 Equipment debugging

The testing location is in the new six-floor steel structure laboratory of Shandong Agricultural University. Test equipment are connected in accordance with Figure 1, and the vibration pickups are fixed on the top surface of the five-layer

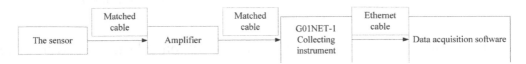

Figure 1. Composition of 941B vibration data acquisition system.

Table 1. Maximal amplitude and the relative ratio of six channels.

Vibration pickup	H10063	H10064	H10065	H10066	V10105	V10106
Channel number	CH 1	CH 2	CH 3	CH 4	CH 5	CH 6
Maximal amplitude ($\mu m/s^2$)	2.723	10.888	11.135	11.016	64.579	58.122
Relative ratio	3.999	1.000	1.023	1.012	1.111	1.000

structure consistent with the measured point movement. The device was started into the test sample and the oscilloscope signal was observed.

2.2 Normalized validation

The vibration pickups used in the test are unified calibration before leaving the factory and it is consistent in principle, but there are still some minor differences in the whole system composed of the vibration pickups, wires, amplifier, and acquisition instrument. In order to reduce the impact of these small differences on the test results, before the formal test, the six vibration pickup is normalized to verify in the scene (Huang, 2006). The results are shown in Table 1.

The corresponding channel readings of H10063 have considerable variation those of the other three channels. Measurement errors of other vibration pickups are within the scope of the permit. Hence, it can be concluded that H10063 is broken down.

Replace it and begin the test. On the basis of the theory of pulse method fortesting dynamic characteristics of the structure, the measuring point should be put in the stiffness center of the underside of the top structure; thus, the translational signal can be merely picked up by the vibration pickups, and the impact of structural torsional vibration can be avoided. The sensors of three different directions of NS, EW, and UD will be placed in the measuring point. Start the instrument and collect the structural vibration data.

3 ESTIMATING STRUCTURAL NATURAL FREQUENCY BY NAKAMURA'S TECHNIQUE

Nakamura, a Japanese scholar, put forward a method to evaluate site characteristics based on the ratio of Fourier amplitude spectrum of ground

pulsation's horizontal component and vertical component in the same surface of the station (H/V) (Nakamura Yutaka, 1989). In recent years, the research domain of Nakamura's technique has been expanded from the site effect study further into the research of architectural structure's vibration response with the environmental noises (Lu, 2006; Luo, 2014).

3.1 Principle of Nakamura's technique

The fundamental assumptions of Nakamura's technique are as follows. First, the spectrum characteristics of ground are the same in earthquakes tests or ground pulsation measurements, and the site amplification effect is mainly related to field dynamic characteristics, and the spectral ratio value of H/V is 1 in the bedrock. Second, the vertical component remains unchanged when the horizontal component gets amplificated, and the vertical transfer function is considered as 1.

Fourier amplitudes of the horizontal component and vertical component corresponded with a frequency on the bedrock surface and free surfaces are S_{BH}, S_{BV}, S_{SH}, and S_{SV}, respectively. According to the method of transfer function, the transfer function of any horizontal component can be expressed as follows:

$$TF_H = \frac{S_{SH}}{S_{BH}} = \frac{S_{SH}}{S_{SV}} \times \frac{S_{SV}}{S_{BV}} \times \frac{S_{BV}}{S_{BH}} = \frac{S_{SH}}{S_{SV}}$$
$$\times \frac{S_{SV}}{S_{BV}} \times \frac{1}{S_{BH}/S_{BV}} \qquad (1)$$

According to the fundamental assumptions above, $S_{BH}/S_{BV} = 1$, $S_{SV}/S_{BV} = 1$, the formula can be changed as:

$$TF_H = \frac{S_{SH}}{S_{SV}} \qquad (2)$$

3.2 Estimating structural natural vibration frequency

Guichun Luo et al. proved the correctness of Nakamura's technique applied to estimate structural natural vibration frequencies (Luo, 2014), but it cannot calculate the structural vibration magnification. The principle of the method to calculate the structural natural vibration frequency is as followed. Structure can be regarded as equal to the soil, the basal surface of the structure can be regarded as the top of bedrock, and different height of floors can be compared to different thicknesses of soil layers. Ground pulsation from the basement as the input signal can be considered as a superposition of different-frequency harmonic component, and the harmonic components

whose frequency are close to the structural natural frequency get magnified in different degrees when they go through the structure. The amplitudes of other harmonic components remain the same or get attenuated. Through the spectrum, the frequency values corresponding to magnification peaks can be pointed out, which is also the structural natural vibration frequency.

The spectrums of structures in three directions of NS, EW, and UD can be acquired through the measurement of the ground pulsation data of the six-floor steel structure by fast Fourier transform, which is shown in Figure 2. It is evident from the figure that the vibrational frequency achieves peaks at 2, 17, 20, and even 40 Hz. However, according to the engineering experience, frequency in the range

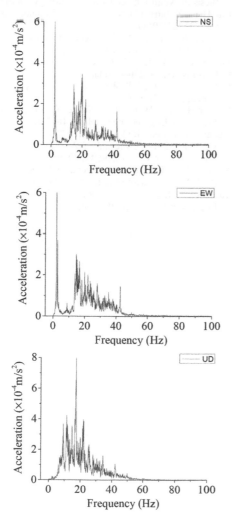

Figure 2. Acceleration amplitude of six-floor steel structure.

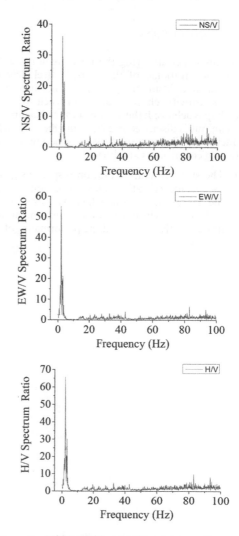

Figure 3. NS/V, EW/V, and H/V spectrum ratio.

of 2–3 Hz is the structural natural frequency, and that in the range of 17–40 Hz is the frequency of the structural vibration that responded the surrounding strong vibration source such as the traffic interference. To verify the conclusion above, the ratio of horizontal and vertical Fourier amplitude is calculated using Nakamura's technique, and the spectrum ratio figures of S_{SNS}/S_V, S_{SEW}/S_V, and S_H/S_V can be calculated, as shown in Figure 3. Vibration signal with a frequency of 2–3 Hz gets remarkably magnified, but signals in other frequency ranges have no significant changes. Therefore, it can be concluded that the structural natural vibration frequency is in the range of 2–3 Hz. Through this method, the interference in the input signal can be eliminated, and the structural natural vibration frequency can be estimated rapidly and accurately.

4 CONCLUSION

In this paper, applying the microtremor test, the vibration response of the six-floor steel structure is measured. Using the principle of normalization, the sensitivity check analysis of each vibration pickup is achieved; the effectiveness of Nakamura's technique is discussed in estimating the natural frequency of vibration of the steel structure. The following conclusions are drawn:

1. The steel structure vibration response signal is used as the excitation signal to accurately check the vibration pickup, which is closer to the real-work environment of the instrument, and more strictly to respond the performance of the instrument.

2. When the direct Fourier method is used to make the spectrum analysis, the existence of the surrounding strong source interference in the structure vibration response signal will increase the difficulty of the spectrum analysis.

3. Nakamura's technique can eliminate the noise in input signals and improve the signal-to-noise ratio. This method can accurately estimate the natural frequency of the steel structure.

REFERENCES

Huang, R.X., A.Q. Li, Z.Q. Zhang, Chinese Journal of Special Structure, 27, 49(2010).
Huang, Y., Analysis of dynamic characteristics and seismic responses of high-rise building (Chinese Master's thesis of Xi'an University of Technology 2006).
Liu, H.B., X. Guo, Chinese Journal of Engineering Geology, 31, 24(2011).
Lu, T., Z.H. Zhou, Y.N. Zhou, et al., Chinese Journal of Engineering Geology, 26, 43(2006).
Luo, G.C., L.B. Liu, C. Qi, et al., Chinese Journal of Geophysics, 54, 2708(2011).
Luo, G.C., X.J. Li, Y.S. Wang, et al., Acta Seismologica Sinica, 36, 491(2014).
Luo, Y.H., Y.S. Wang, F.H. Wang, Q. Deng, Chinese Journal of Engineering Geology, 18, 27(2010).
Nakamura Yutaka, RTRI, 30, 25(1989).
Wang, Y.L., Chinese Journal of Aviation Metrology & Measurement Technology, 15, 20(1995).
Xu, J.C., W.B. Jian, Y.Q. Shang, et al., Chinese Journal of Rock Mechanics and Engineering, 39, 33(2005).
Yang, Q.Y., L.Q. Lou, L.Z. Yang, Chinese Journal of Engineering Geology, 25, 174(2005).

Advances in Energy Science and Equipment Engineering II – Zhou, Patty & Chen (Eds)
© 2017 Taylor & Francis Group, London, ISBN 978-1-138-71798-5

Incentive mechanism analysis of energy-saving renovation for existing residential buildings based on EMC

Siqi Jia
Commercial College, Jiangnan University, Wuxi, China

Xiaoping Feng, Chengyi Bao & Yun Zhu
School of Environment and Civil Engineering, Jiangnan University, Wuxi, China

ABSTRACT: On the basis of the operation mode of EMC and the development of Contract Energy Management mode in terms of domestic existing residential buildings, the game model of incentive mechanism about energy-saving renovation for existing residential buildings is constructed from the perspective of principal-agent theory and a relatively hawkish policy including both punitive and incentive measures is put forward in this paper. By analyzing the relevant relationship between governments, the heating company, EMCO, and householders participating in the EMC model, the results show that energy crisis and subjective energy-saving barriers of householders are the two main factors in drawing up punitive measures. Faced with the increasingly severe energy crisis, the conclusions of the research on punitive measures can be used as a reference for the government intending to implement relatively tough macro-regulation for householders of existing residential buildings.

1 INTRODUCTION

Energy Management Contract in China has been growing quickly since 2003; however, the development of EMC in terms of existing residential buildings is particularly slow. This phenomenon results from several factors such as mutual sense between householders and energy-saving company, low energy-saving consciousness of householders, and technical problems in energy-saving reconstruction implemented in existing residential buildings. Therefore, in order to promote the development in existing residential buildings, it is urgent that a viable economic incentive mechanism for energy-saving reconstruction should be adopted to existing residential buildings. At present, most scholars in China are studying incentive mechanism about urging EMCO and energy-intensive enterprises to participate in energy-saving reconstruction. Chengzhi Li studies mainly the way of incentive stimulation for energy-intensive enterprises and analyzes the types of energy-intensive enterprises that adapt to annual incentive subsidies and to a lump sum during the contract period. Xiaojun Liu and WeiHu formulated EMC modes in the domain of energy-saving reconstruction on existing residential buildings under different technical economic conditions and studied subjects of distribution of benefits under half central heating condition, solving the mode constructed by Shapley value method,

the minimum core method, and MCRS method. Most of the current papers focus incentive mechanism based on promotional measure; nevertheless, punitive incentives are still less studied by Chinese scholars. However, tough punitive measure should be added to domestic incentive mechanism due to increasingly severe energy condition and generally low awareness of saving energy. The author thus focuses on incentive mechanism with incentive and punitive measures as the main part and takes punitive policies into account.

2 OBSTACLES TO THE DEVELOPMENT OF EMC MODEL IN TERMS OF DOMESTIC EXISTING RESIDENTIAL BUILDINGS

Promotion of energy management contract in terms of existing residential buildings in China still poses the following obstacles.

First, there are serious moral hazard problems between householders and energy service companies, such as default and delayed protest. Technical problems and lack of experience in project management are the main problems of EMCO. These problems indirectly lead to frequent moral hazard problems (Li, 2013).

Second, compared with existing buildings of industrial enterprises, existing residential buildings will be more difficult to be transformed. In the process of reconstruction, householders' initiative energy-saving reconstruction is less at present. Simultaneously, daily life and work of householders will be influenced during the construction. Therefore, negative punitive policies and positive incentive policies should be implemented to enhance the awareness of saving energy.

Third, in contrast with developed countries, domestic heating way in existing residential buildings are more concentrated, and residential system with high-rise buildings as the main part is very common in our country. In this situation, the heating company takes charge of heating residential buildings, and heating fees is regularly delivered to the residential property company according to the amount of energy consumed. After energy-saving reconstruction, residents can obtain energy-saving benefits. Heating companies gain benefits from the upgrade of capacity equipment and the conspicuous decline of routine maintenance and management expenses of productive fixed assets (Zhao, 2016). Compared to industrial enterprises, the number of beneficiaries of energy-saving reconstruction on existing residential buildings is significantly larger. EMCO has to consult with householders before the energy-saving contract is signed; however, early great cost of signing contract indicates that EMC operates very slowly in the actual process of the execution and the concentration of residential buildings greatly increased the difficulty of energy-saving transformation (Liu, 2015).

3 MODEL ASSUMPTIONS AND CONSTRUCTING

The interests of the related participants, including government, energy service companies, heating companies, and heads of existing residential buildings, is mainly analyzed in the mode. The reconstruction funds and costs are covered by energy service company together with the heating company, and the provision of technical service and equipment is primarily taken charge of by EMCO, and heating company is responsible for the renovation of funds needed for the old heating equipment. Energy service companies' money-back and profits proceed from the energy-saving effect in monetary form, and heating company's earnings come from the reduction of relative cost and heating benefits resulting from the update of energy-saving equipment.

Faced with domestic residents' low consciousness of saving energy, whether the government adopts incentive policies or not, punitive policies will be imposed, that is, heads not saving energy will be responsible for the negative externalities

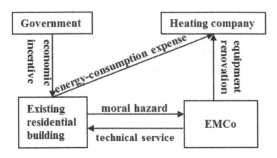

Figure 1. Relationship between four subjects in EMC.

arising therefrom. Government decides whether to carry out incentive policies on householders on the premise that punitive policies have been added to incentive mechanism. In the mode, they already know whether the government has taken bonus incentives before heads decide whether to carry out energy-saving reconstruction, Therefore, dynamic game with incomplete information is formed between governments and householders. Because of the low percentage of energy-saving buildings in the residential area, we suppose that when the government does not take incentive policy, optimal strategy of heads registers as not saving energy. How to design incentive mechanism to ensure that the heads do not take the moral hazard in the process of energy-saving reconstruction is mainly studied in the mode.

Hypothesis 1 heads of existing residential buildings are divided into moral energy-saving individuals and unethical energy-saving individuals, as the proportion of p and 1–p. In the energy-saving benefits-sharing mode, householders do not cover the costs of reconstruction on existing residential buildings, but their awareness of energy-saving is not still so strong. It is a certain amount of revenue made by energy-saving reconstruction or loss raised from punitive measures that can stimulate householders to make energy-saving reconstruction. A specific amount of revenue will be regarded as preservation of subjective energy-saving barriers named "V" in the mode.

Hypothesis 2 during the contract period: duration of the contract is "n" years and the benefit-sharing proportion of EMCO to householders is q: 1–q. When householders show unethical behaviors, energy consumption of existing residential buildings is down to X′ per year after energy-saving renovation, and the additional return obtained from unethical behaviors is represented by "T." When householders do not take moral hazard, energy consumption is down to X per year after the renovation; moreover, the estimated residual value of energy-saving devices is supposed to be 0. When householders do not participate in energy-saving reconstruction, residential energy consumption is supposed to be X'' per year ($X'' > X' > X$), and subjective energy barriers preservation will be acquired. The expense-paying proportion of EMCO to heating company is β: $1-\beta$. The total cost is represented by "C" during the renovation. Heating company's cost reduction and increment of heating benefits is represented by "R_X" altogether, R_X is the related function with annual comprehensive energy-saving benefits as the independent variables.

Hypothesis 3: Incentive and punitive measures are made by setting two demarcation points of energy consumption, X_1 and X_2 ($X_2 > X_1$), to motivate householders. The demarcation point of penalty is set to X_2 and the demarcation point of reward is set to X_1. When householders do not make energy-saving reconstruction, energy consumption will exceed X_2 per year; thus, the government will punish them. Supposing punitive coefficient is set to a_2 ($0 < a_2 \geq 1$), householders will be mulcted in $a_2 (X''-X_2)$ per year by the government. When energy consumption reaches between X_1 and X_2 for householders showing moral hazard behaviors, the government would not reward them. When householders do not show moral hazard behaviors during the transformation, the energy consumption will be less than X_1 per year, thus the government will reward them. Supposing incentive coefficient is a_1, the government will pay $a_1 (X_1-X)$ per year for rewarding.

In the premise of meeting the above three assumptions, without considering the effect of time value of money, incentive mechanism designed by the government for householders is built as follows:

$$\text{Max}[np(X''-X) + n(1-p)(X''-X')-n(1-p) \\ a_1 (X_1-X)] \tag{1}$$

s.t.

$$(IR)n(1-q)(X''-X) + T \geq V-na_2 (X''-X_2) \tag{2}$$

$$(IC)\ n(1-q)(X''-X) + n a_1 (X_1-X) \geq n(1-q) \\ (X''-X') + T \tag{3}$$

$$pqn(X''-X) + (1-p)qn(X''-X')-\beta C \geq 0 \tag{4}$$

$$R_X- (1-\beta) C \geq 0 \tag{5}$$

Formula (1) represents the objective function of the incentive mechanism designed by the

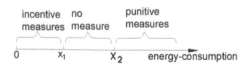

0   X_1 X_2 energy-consumption

incentive measures — no measure — punitive measures

Figure 2. Demarcation points of energy consumption in the mode of EMC.

government for householders. The objective function will be maximized by computing the relationship among X_1, X_2, a_1, and a_2 on the premise of meeting other four equations. Formula (2) shows that the effect of reservations representing the expected utility when householders show moral hazard behaviors should not be less than that when not doing energy-saving renovation. Formula (3) means that effectiveness of reservations when householders do not show moral hazard behaviors should not be less than that when householders show hazard behaviors in order that householders do not show moral hazard behaviors during the energy-saving renovation. Formula (4) represents the effectiveness of EMCO after energy-saving renovation. Formula (5) represents participation constraints of heating company. The incentive mechanism should maximize the objective utility function of the government on the premise of satisfying participation constraints of EMCO, heating company, householders, and incentive compatibility constraints as well as incentive compatibility constraints of householders.

4 MODEL SOLUTION

Participation constraint condition of pledging heating company along with EMCo and householders to participate in energy-saving renovation will be firstly calculated. Then, the relationship among a_1, a_2, X_1, and X_2 will be determined. Ultimately, the maximum expected utility achieved by the government will be calculated according to the mode. The specific process is as follows:

Simplifying formula (5), we obtain: $\beta \geq 1-R_X/C$ (6)

The inequality formula represents that when the cost-sharing ratio of EMCO during the renovation is not less than $(1 - R_X/C)$, heating company's benefits from energy-saving renovation is not less than the cost paid during the contract period, which is a prerequisite for heating company participating in energy-saving renovation, yet incentive policy taking effect.

On the premise that heating company participates in energy-saving renovation, benefits which EMCO gains from participating in energy-saving renovation should not be less than 0, that is, simplifying formula (4), we obtain:

$$q \geq \frac{\beta C}{np (X'' - X)+n(1-p)(X'' - X')}, (\beta \geq 1-R_X/) \tag{7}$$

Formula (7) indicates that only if annul benefits-distributing ratio of EMCO promised in advance cannot be lower than $\dfrac{\beta C}{np (X'' - X)+n(1-p)(X'' - X')}$

693

on the premise that heating company participating in energy-saving renovation, energy-saving services, and equipment for householders will be provided by EMCO. Therefore, formulas (6) and (7) are basic conditions that householders participate in energy-saving renovation.

For the purpose of ensuring that householders can participating on their own initiative in energy-saving renovation, formula (2) must be established. Simplifying formula (2), we obtain:

$$a_2 \geq \frac{V - n(1-q)(X''-X')}{n(X''-X2)} (X_1 < X_2 < X'') \qquad (8)$$

Simplification of formula (8) indicates that there is a positive correlation between penalty coefficients a_2 and the demarcation point of penalty X_2. The demarcation point of penalty depends on the magnitude of energy crisis estimated by the government. When the government tempers the demarcation point of penalty X_2, penalty coefficient a_2 should be reduced to ensure that the punitive policy can proceed smoothly. Conversely, when the demarcation point of penalty X_2 starts to ascend, penalty coefficient a_2 should be increased to ensure that householders will participate actively in energy-saving renovation. In addition, when subjective energy-saving barriers preservation V starts to fall behind, penalty coefficient a_2 should be reduced appropriately. Since the increase of energy-saving awareness of householders, there has been no need that the punitive stimulation with the original strength proceeds to perform.

Formula (3) shows that $\Delta U = n(1-q)(X''-X) + n\, a_1(X_1 - X) - n(1-q)(X''-X') + T$; the government should guarantee that ΔU is not less than zero in order to certify that householders select to become moral energy-saving individuals. Then, simplifying Formula (3), we obtain:

$$a_1 \geq \frac{T - n(1-q)(X'-X)}{n(X1-X)} (X_1 > X) \qquad (9)$$

The formula $n(1-q)(X'-X)$ actually means the balance of heating benefits between immoral and moral householders during the contract period. The additional revenue which householders gain from the unethical behaviors is represented by "T." Formula (9) illustrates that there exists negative correlation between incentive coefficient a_1 and the demarcation point of reward X_1, that is, when demarcation point of reward X_1 is improved by the government, incentive coefficient a_1 should be tempered to maintain the efficiency of incentive policy, otherwise incentive coefficient a_1 should be increased.

On the condition that three formulas,
$$a_1 = \frac{T - n(1-q)(X'-X)}{n(X1-X)}, \quad \beta = 1 - R_x/C, \quad \text{and}$$
$$q = \frac{\beta C}{np(X''-X) + n(1-p)(X''-X')} \quad \text{and formula}$$
(8) are met, calculating the government's maximum expected utility in the incentive mode:

$$MaxU = np(X'-X) + n(X''-X') - T(1-p) \\ + n(1-q)(X'-X)(1-p)$$

From the above solving process, we can deduce that average subjective energy barrier V and the expected set energy-saving effect should be taken into account when the government implements punitive measures to heads of existing residential building. The government's reasonable estimation of the degree of energy crisis will be reflected by the expected set effect of energy-saving physically. If the government holds optimistic attitude to the present energy consumption, the expected set energy-saving effect will be in the relatively low level, so the demarcation point of penalty X_2 can be a higher energy consumption value. Average subjective saving barrier V can be evaluated from the minimum energy-saving benefits entailed to ensure that householders would like to participate in the transformation. Greater V value shows that householders are less willing to take energy-saving transformation; therefore, penalty coefficient a_2 should be increased. In the final analysis, when there is severe energy crisis in society, a lower energy consumption value as the demarcation point of penalty should be selected. Given that householders have a relatively low average subjective energy barrier, the government can select a smaller value of penalty coefficient a_2. Evaluating average subjective energy barrier in advance is essential for enacting penalty coefficient a_2 on different areas.

5 CONCLUSIONS

On the basis of the energy-saving reconstruction on existing residential buildings in EMC mode, economic incentives with reward and punishment measures as the main part is studied deeply in this paper, and the authors propose that subjective energy barriers V and the government's emphasis on energy crisis are two important factors in setting punitive measures; moreover, studies how the relationships between various stakeholder have impact on the mechanism of rewards and punishments through constructing the model, consequently reaching the following conclusions:

The cost-sharing ratio between heating company and EMCO and the percentage of profit distribution between householders and EMCO should be taken into account when the government institutes incentive mechanism including rewards and punishments. These two factors directly affect the relationship between incentive and punitive boundaries and incentive and punitive coefficient.

For domestic areas in which energy reserves are large, energy efficiency is relatively high as well as environmentally sound, the government should choose high energy consumption as the demarcation point of penalty X_2. For severely polluted domestic areas, such as Shijiazhuang, Xingtai, the low utilization ratio of energy by the government should choose a lower energy consumption as the demarcation point of penalty X_2, or perform other relatively tough macro-control, for example, the Japanese government prescribes mandatorily efficiency standards for high-energy-consuming enterprises and commercial buildings and requires them to submit energy consumption state reports, which has opened up the energy market for the EMC model to a certain extent (Zhao, 2016).

Subjective energy barrier V is an important factor in hindering the development of domestic energy-saving reconstruction in existing residential buildings. When average subjective energy barrier V is evaluated to be relatively high, the government should raise penalty coefficient a_2, increasing the opportunity costs of not carrying out energy-saving transformation. In addition, the government should take the lead in energy-saving renovation and call on the public to engage actively in energy-saving modification through the news media in order to reduce the householders' subjective energy barrier V. The development of EMC has always been inseparable from the government's support in policy throughout the United States, and entails enforcement and stability policy of the government (Deng, 2011). With the reduction of social subjective average energy barriers, penalty coefficient decreases, even to 0, and punitive measures will no longer be put into effect.

REFERENCES

Chengzhi Li. Journal of WUT. Governments Economic Incentive to Energy—using Units in Energy Performance Contracting (J). 35, 749–753(2013).

Qianqian Zhao, Handing Guo, Yongxing Chen. Constuction Economy. A Review on Incentive for ESCO under EPC Mode (J). 37, 78–83(2016).

Qianqian Zhao, Handing Guo, YongxingChen. Construction Economy. A Review on Incentive for ESCO under EPC Mode (J). 37, 78–83(2016).

Xiaojun Liu. Construction Economy. Study on Profits Allocation of Existing Residential Buildings Energy Saving Renovation Under Energy Management Contracting: Based on the Analysis of North China (J). 36, 92–96(2015).

Zhijian Deng, Xiao Wang, Wei Wang. Journal of Engineering Management. Incentive Mechanism Analysis of Energy Saving Renovation for Public Buildings Based on EMC (J). 1, 37–40(2011).

Advances in Energy Science and Equipment Engineering II – Zhou, Patty & Chen (Eds)
© *2017 Taylor & Francis Group, London, ISBN 978-1-138-71798-5*

Regional inequality of population in high-earthquake-risk areas and suggestions, based on Theil index

Cong Lang, Xinyan Wu & Mengtan Gao
Institute of Geophysics, China Earthquake Administration, Beijing, China

ABSTRACT: Earthquakes represent a particularly severe risk to China. To reduce earthquake risk, High-Earthquake-Risk Regions (HERRs) and earthquake mitigation policy have been come up through decades of practices and explorations. This paper, using the territory of HERRs in China and National Population Census, measures the regional inequality of population in HERRs by Theil index, gives an index decomposition by subgroups, calculates the contribution of each part, and demonstrates the regional risk variations in China from a population perspective. The result shows that although the total Theil index is almost unchanged, the inner Theil index in eastern China is slightly increased, but dramatically increased relatively. The whole inequality is first contributed by the inner inequality of eastern China and secondly the between-group inequality. The result indicates that it is necessary to pay attention to the earthquake risk in eastern China and apply prospective, corrective, and compensatory risk management, such as scientific planning, guides for building codes, new earthquake-safety technology, and categorized disaster mitigations according to different regions and different risk sources.

1 INTRODUCTION

1.1 *Earthquakes and earthquakes in China*

Earthquakes represent a particularly severe threat to human beings among all types of natural disasters. According to GAR2015, on the basis of 82 sample countries' EM-DAT data set from 1970 to 2013, earthquakes have killed 1,055,888 people, taking 31.8% of all fatalities, the highest source of death toll, and caused 34% of all losses in them, also the highest source, close to the combined losses of the second and third causes (cyclones and floods—41%) (UNISDR, 2015). Earthquake risk contributes US$113 billion to the global average annual loss (UNISDR, 2015).

In China, earthquake-prone areas are large and dispersed. Earthquakes strike the country frequently and strongly (H, 2006; L, 1981). As the largest developing country, China has the largest population and the second largest GDP in the world. Therefore, earthquakes impose a particularly severe damage here. In the top 10 deathful earthquakes from 1900 to 2014, three happened in China (USGS, 2015). The Shaanxi earthquake of 1556, the deadliest earthquake ever recorded, also happened in China, killing approximately 830,000 people (USGS, 2004; Utsu, 2002; NOAA, 2010). After economic reform, the 2008 M7.9 Wenchuan earthquake caused enormous casualty and property loss, killed 88,287 people, injured 3,650,001,

affected over 60 millions of people (EM-DAT, 2011), and led to 692 billion RMB loss (Y, 2008).

Decision making, based on the benefit-pursuit principle, always leads to significantly unacceptable social risk (G, 2014). As China has a rapid growth in both economy and population, the damage brought by earthquakes becoming bigger than ever before. Therefore, it is important to understand the earthquake risk, making effective and problem-oriented disaster mitigation policy.

1.2 *Risk theory and High-Earthquake-Risk Regions (HERRs) in China*

According to the hazard-of-place risk concept, risk is a place-based interaction among vulnerability, exposure, and hazard (Cutter, 1996). It is usually described as below:

$$Risk = Hazard \times Vulnerability \times Exposure \qquad (1)$$

As GAR2009 reported, China is the top country on the Mortality Risk Index for earthquakes, with a value of 8.5. China has the fourth most absolute population and the seventh most absolute GDP exposed to earthquakes per year (UNISDR, 2009). Since economic reform, growing regional inequality, increasing hazard exposure, rapid urbanization, and national planning such as Go-West Campaign, One Belt, and One Road project, all have made

China a rapidly changing economy. Therefore, it is more important to focus on the population in High-Earthquake-Risk Regions (HERRs) and take on prospective measures to planning and preventing development against itself (UNISDR, 2015).

GAR provides a valid approach to identify intensive and extensive risks. Earthquakes are characterized as intensive risk for their very low frequency but high-severity losses. For intensive risk, it is hazard and exposure that dominate the risk equation (UNISDR, 2009). Julio Serje (Serje, 2010) summarized that disaster risk including earthquake risk is decided by three factors: hazard, exposure, and vulnerability. The risk is highly affected by the factors' spatial patterns and temporal trends, especially the dynamics of territorial occupation, land use, urbanization, environmental change, poverty distributions, governance, and development. GAR 2011 (UNISDR, 2011) also noted that in extreme hazards, the seriousness of disaster risk is contributed more by exposure than by vulnerability. This means that in terms of the risk of earthquakes, population and wealth are the domestic factors that affect the consequences. Although we cannot prevent natural phenomena such as earthquakes and cyclones, we can limit their impacts (UNISDR, 2015). That is, we can reduce the risk of earthquakes through a set of development-embedded risk management policies including prospective, corrective, and compensatory risk management.

To reduce the risk of earthquakes in China, on the basis of hazard of place concept (Cutter, 1996), HERRs and earthquake mitigation policy have been come up through decades of practices and explorations (R, 1990; H, 1990; T, 1990; T, 1990; C, 1995). The HERRs are regions affirmed by the State Council that have high risk of earthquakes and need mitigation actions to be taken. Now, the HERRs take 10% of the mainland's territory and control 60% of the national earthquake risk (Z, 2006).

However, the mitigation policy in the HERRs is still not sufficient. Besides the institutional drawbacks (G, 2014; C, 2014; L, 2014; L, 2014; S, 2014), other weaknesses such as lack of correct understanding of earthquake risk; neglecting the seriousness of highly rising earthquake risk in China (S, 2014; Z, 2014); insufficient distinction of the risk caused by hazard, exposure, or vulnerability (L, 2014); and the changing risk due to urbanization, population migration, and economic polarization. In addition, regional inequality of earthquake risk should also be focused on, for the tremendous gap between different parts of the country.

1.3 Literature review

GAR2015 emphasized that at present there is high probability of disaster risk reaching a tipping point beyond which the effort and resources necessary to reduce it will exceed the capacity of future generations (UNISDR, 2015). Population is an essential part of risk analysis. Almost all the studies contained it.

The Global Risk Analysis in Hotspots Project, initiated by the World Bank and Columbia University, System of Indicators for Disaster Risk Management, initiated by Inter-American Development Bank and Columbia University (Cardona, 2005), the risk estimation of ESPON's project, the spatial effects and management of natural and technological hazards in general and in relation to climate change (ESPON, 2003), the Social Vulnerability Index (SoVI) for the United States (S. Cutter, 2003; HVRI,, 2011), and the Assessment of Provincial Social Vulnerability to Natural Disasters in China (Z, 2013) all used population to estimate the disaster risks.

Furthermore, there are also some studies paying attention to earthquake risk, such as EDRI (Earthquake Disaster Risk Index) (R. Davidson, 1997), FEMA's Hazus-MH Earthquake Model (FEMA, 1999), the integrated earthquake vulnerability assessment framework for urban Turkey (S, 2011), the brief analysis and case study of population and GDP (C, 1991; C, 2001; C, 1997; C, 1997), the quantitative assessment of seismic mortality risks in China (X, 2012), GIS-based population earthquake disaster risk analysis in China (L, 2012), macro-assessment of seismic population vulnerability in China (N, 2012), seismic risk assessment in mainland China(L, 2008; L, 2009), and the study of population growth in China's earthquake-prone areas (H, 2016).

According to the existing research on the topic of population in HERRs, four more considerations need to be conducted to fully understand population and rising risk in HERRs. They are (1) investigation of population in HERRs, but not only population in hazardous areas, (2) investigations covering all the HERRs around China, but not only some parts of the HERRs, (3) quantitative analysis about population, but not only its summation, and (4) comparative research about regional variations with four dividing regions, but not three regions previously, in order to represent northeast China, which is also the current division used in China Statistical Yearbook.

To solve the above shortcomings, this paper, using the territory of HERRs in China and National Population Census, measures the regional inequality of population in HERRs by Theil index, gives an index decomposition by subgroups, calculates the contribution of each part, and demonstrates the regional risk variations in China from a population perspective.

2 RESEARCH DATA AND METHODS

2.1 *Reserch data*

2.1.1 *Population*
The national bureau of statistics offers national population census data in 2000 and 2010. The 2000 census contains 2873 counties and that of 2010 contains 2871 counties.

2.1.2 *HERRs*
China Earthquake Administration offers the HERRs data, a county list noting which county is bearing high earthquake risk.

2.2 *Methods*

2.2.1 *How to divide the regions*
Because there are significant hazard variances and vulnerability gaps among regions in China (Z, 2013), it is reasonable to investigate by regions. According to the latest China Statistics Yearbook, four regions are divided as follows: (1) Eastern 10 provinces (municipalities) including Beijing, Tianjin, Hebei, Shanghai, Jiangsu, Zhejiang, Fujian, Shandong, Guangdong, and Hainan; (2) Central six provinces including Shanxi, Anhui, Jiangxi, Henan, Hubei, and Hunan; (3) Western 12 provinces (autonomous regions and municipalities) including Inner Mongolia, Guangxi, Chongqing, Sichuan, Guizhou, Yunnan, Tibet, Shaanxi, Gansu, Qinghai, Ningxia and Xinjiang; and (4) Northeastern three provinces including Liaoning, Jilin, and Heilongjiang.

2.2.2 *Theil index*
The Theil index (H, 1967) is a statistic primarily used to measure economic inequality, named after Henri Theil in 1967 when he used it to calculate the income inequality based on information theory. It was then used to measure the inequality of natural resource (R. Sampath, 1988). Compared with Gini coefficient, Theil index can be decomposed by subgroups without a residual. It can also contribute to each subgroup and between groups as well. As our purpose is to give the regional inequality of population in HERRs and demonstrate the regional contribution, Theil index is more proper here.

The final value of the Theil index (I_{Theil}) is made of two components: Theil index within groups ($I_{withingroup}$) and Theil index across groups ($I_{acrossgroup}$). To decompose Theil index, it can be displayed as:

$$I_{Theil} = I_{across\ group} + I_{within\ group}$$
$$= \sum_{i=1}^{n} \left[\left(\frac{y_i}{y} \right) * log \frac{y_i / y}{P_i / P} \right] \quad (2)$$
$$+ \sum_{i=1}^{n} \left\{ \frac{y_i}{y} * \sum_{j=1}^{m} \left\{ \left[\left(\frac{y_{ij}}{y} \right) * log \frac{y_{ij} / y}{P_{ij} / P} \right] \right\} \right\}$$

where N represents the number of the subgroups, which is 4, indicating the number of the regions here, y_{ij} is the population in HERRs of county j in region i, y_i is the population in HERRs of region i, y is the population in HERRs all over the country, p_{ij} is the population of county j in region i, p_i is the population of region i, and p is the total population of China.

The ratio for each group represents its contribution to the total inequality.

3 RESULTS

To begin with, we first give the Theil index of population inequality in HERRs overall (Table 1).

Table 2 gives the decomposition of Theil index.

As Tables 1 and 2 show, the cross-group Theil index decreased slightly and the inner-group Theil index was almost unchanged from the absolute value. That means the total inequality is slightly reduced. However, from the regional decomposition, the inequality of northeastern China and western China decreased dramatically, dropping by 27.3% and 32.5%, relatively.

Table 3 gives the contribution of each part.

It is evident from the Table 3 that in both 2000 and 2010, the main contribution to the population inequality in HERRs is the cross-group inequality and the inner-group inequality of eastern China.

The contributions of cross-group Theil index are 42.7% and 40.8% in 2000 and 2010, respectively, while those of Theil Index within eastern China are 59.2% and 63.0% in 2000 and 2010, respectively, and the inequality of eastern part is still rising. This means that the regional gap and the inner gap in eastern China are the main source to the whole inequality.

4 CONCLUSION AND DISCUSSION

1. Although the total Theil index is almost unchanged, the inner Theil index in eastern China is slightly increased. The whole inequality is first contributed by the inner inequality of eastern China and secondly by the across-group inequality.

Table 1. Theil index of population inequality in HERRs (2000 and 2010).

Year	I_{Theil}	Growth rate
2000	0.056136	–
2010	0.054463	3.0%

Table 2.   Decomposition of Theil index of population inequality in HERRs for four regions (2000 and 2010).

Year	$I_{acrossgroup}$	$I_{withingroup}$ Northeastern China	Eastern China	Central China	Western China
2000	0.0239653	0.00078154	0.0332494	−0.005036576	0.003176434
2010	0.0221945	0.00056843	0.034304	−0.004748543	0.002145066

Table 3.   Contributions of Theil index of population inequality in HERRs for four regions, (2000 and 2010).

Year	$I_{acrossgroup}$	$I_{withingroup}$ Northeastern China	Eastern China	Central China	Western China	I_{Theil}
2000	42.7%	1.4%	59.2%	−9.0%	5.7%	100.0%
2010	40.8%	1.0%	63.0%	−8.7%	3.9%	100.0%

2. It is necessary to pay attention to the earthquake risk in eastern China. Central and local governments need to apply risk management in consideration of the art of development and come up with integrated risk management policy combining prospective, corrective, and compensatory risk management, such as scientific planning, guides for building codes, research and utilization of new earthquake safety technology, and so on.

3. Categorized disaster mitigations are required for regional inequality because of tremendous gap between regions in China.

ACKNOWLEDGMENTS

This research was supported by the Basic Research Program 2015 of Institute of Geophysics, China Earthquake Administration (DQJB15C09). The authors also would like to thank reviewers who would give valuable suggestion to help improve the quality of this manuscript.

REFERENCES

Baoxia, Z., S. Wenzhuang, M. Ming, *Earthquake Research in China*. 30(3): 373–381. (2014).

Cardona, O., System of Indicators for Disaster Risk Management in the Americas. 250th anniversary of the 1755 Lisbon earthquake. (2005).

Chengjing, N., Y. Linsheng, L. Hairong. *Progress in Geography*. 31(3): 375–382. (2012).

Chunyang, H., H. Qingxu, D. Yinyin, et al., Environ. Res. Lett, 11(7). (2016) http://iopscience.iop.org/article/10.1088/1748-9326/11/7/074028/pdf.

Cong, L., W. Guochun, G. Mengtan, *Earthquake Research in China*. 30(3): 324–329. (2014).

Cutter, S., B. Boruff, W. Shirley, *Social Science quarterly*. 84(2): 242–261. (2003).

Cutter, S., *Progress in human geography*. 20(4): 529–539. (1996)

Davidson, R., HC. Shah, EARTHQ SPECTRA. 13(2): 211–223. (1997).

Duzgun, S., M. Yucemen, H. Kalaycioglu, et al., NAT HAZARDS. 59(2): 917–947. (2011)

EM-DAT, The OFDA/CRED International Disaster Database. http://www.emdat.net.

ESPON, 1st Interim Report of Project 1.3.1: The spatial effects and management of natural and technological hazards in general and in relation to climate change. (2003) http://www.espon.eu/export/sites/default/Documents/Projects/ESPON2006Projects/ThematicProjects/NaturalHazards/1.ir_1.3.1.pdf.

FEMA, HAZUS: User's Manual and Technical Manuals. Vol. 1–4 (1999).

Guomin, Z., F. Zhengxiang, W. Xiaoqing, et al., *Earthquake Research in China*. 22(3): 209–221. (2006) (In Chinese)

HVRI, Social Vulnerability Index for the United States— 32 Variables. (2011). http://webra.cas.sc.edu/hvri/products/sovi_32.aspx.

Hongtai, C., G. Mengtan, L. Bo, et al., *Earthquake Research in China*. 30(3): 306–316. (2014).

Huan, L., X. Zhongchun, W. Shaohong, et al., *Progress in Geography*. 31(3): 368–374. (2012).

Jian, L., X. Xinmin, L. Bicang, et al., *Earthquake Research in China*. 30(3): 330–340. (2014).

Jifu, L., C. Yong, S. Peijun, *Journal of Beijing Normal University (Natural Science)*. 44(5): 520–523. (2008).

Jifu, L., C. Yong, S. Peijun, *Journal of Beijing Normal University* (Natural Science). 45(4): 404–407. (2009).

Mengtan, G., W. Guochun, W. Xinyan, et al., *Earthquake Research in China*. 30(3): 300–305. (2014) (In Chinese).

NOAA, the National Oceanic and Atmospheric Administration (NOAA) Online Catalog. (2010) Http://www.ngdc.noaa.gov/hazard/earthqk.shtml.

National Bureau of Statistics of China, China Statistical Yearbook 2015. (2015). http://www.stats.gov.cn/tjsj/ndsj/2015/indexch.htm.

Qifu, C., C. Ling, *Acta Seismologica Sinica*. 19(6): 640–649. (1997)

Quan, T., H. Weibin, Z. Lingren, et al., *South China Journal of Seismology*. 10(1): 95–99. (1990) (In Chinese).

Quan, T., Recent Developments in World Seismology. (11): 4–6. (1990) (In Chinese).

Sampath, R. *Water International*, 13(1): 25–32. (1988).

Serje, J., Preliminary extensive risk analysis for the Global Assessment Report 2011. (2010). http://www.preventionweb.net/english/hyogo/gar/2011/en/bgdocs/Serje2010a.pdf

Shanbang, L., Earthquake in China. (1981) (In Chinese).

Theil, H., Economics and Information Theory. (1967).

UNISDR, GAR Disaster Loss Data Universe, including a description of disaster datasets, how the extensive/intensive risk threshold was calculated and other related topics. Annex 2 to Global Assessment Report on Disaster Risk Reduction: Risk and Poverty in a Changing Climate. (2009).

UNISDR, Global Assessment Report on Disaster Risk Reduction: Making Development Sustainable, The Future of Disaster Risk Management. Geneva, Switzerland: UNISDR. (2015).

UNISDR, Global Assessment Report on Disaster Risk Reduction: Risk and Poverty in a Changing Climate. (2009).

UNISDR, Loss Data and Extensive Risk Analysis. Annex 2 to Global Assessment Report on Disaster-Risk Reduction: Making Development Sustainable, The Future of Disaster Risk Management. (2015).

UNISDR. Global Assessment Report on Disaster Risk Reduction: Revealing Risk, Redefining Development. (2011).

USGS, Historic World Earthquakes. (2004) http://earthquake.usgs.gov/earthquakes/world/historihis.php.

USGS. Earthquakes with 1,000 or More Deaths 1900–2014. (2015). http://earthquake.usgs.gov/earthquakes/world/world_deaths.php.

Utsu, T., A List of Deadly Earthquakes in The World: 1500–2000, in International Handbook of Earthquake and Engineering Seismology, 691–717. (2002).

Weibin, H., *South China Journal of Seismology.* 10(2): 87–90. (1990) (In Chinese).

Wenzhuang, S., Z. Baoxia, M. Ming, et al., *China Earthquake Engineering Journal.* 36(1): 195–200. (2014).

World Bank Institute, Introduction to Poverty Analysis. (2005) http://siteresources.worldbank.org/PGLP/Resources/PovertyManual.pdf.

World Bank, Natural Disaster Hotspots: A Global Risk Analysis: Synthesis Report. (2005) http://reliefweb.int/sites/reliefweb.int/files/resources/86A116CE3BF467D4C12571730051B49F-wb-gen-mar05.pdf.

Xi, R. *Earthquake Research in Shanxi.* (4): 33. (1990) (In Chinese)

Yang, Z., L. Ning, W. Wenxiang, et al., NAT HAZARDS. 71(3): 2165–2186. (2013).

Yifan, Y., J EARTHQ ENG ENG VIB. 28(5): 10–19. (2008).

Yong, C., C. Qifu, C. Ling, NAT HAZARDS. 23(2): 349–364. (2001).

Yong, C., C. Qifu, C. Ling, World Wide Seismic Risk Analysis Based on Limited Data. First International Earthquake and Megacities Workshop. (1997).

Yong, C., Estimating Earthquake Damage Losses in China: a Decade-Scale Perspective. (1995) (In Chinese).

Yong, C., Z. Hongren, *Recent Developments in World Seismology.* (5): 5 (1991) (In Chinese).

Yuxian, H., Earthquake Enginering. 2nd Edition. Beijing: Earthquake Publish Company. (2006) (In Chinese).

Yuyu, S., C. Mingjin, L. Qiang, *Earthquake Research in China.* 30(3): 341–350. (2014).

Zhongchun, X., W. Shaohong, D. Erfu, Et al., *Journal of Resources and Ecology.* 2(1): 83–91. (2012).

Advances in Energy Science and Equipment Engineering II – Zhou, Patty & Chen (Eds)
© *2017 Taylor & Francis Group, London, ISBN 978-1-138-71798-5*

Analysis of the effect of thermal bridges on the average heat transfer coefficient of the external wall of rural building in cold regions

Guochen Sang & Yan Han
School of Civil Engineering and Architecture, Xi'an University of Technology, China

ABSTRACT: The research of rural construction in Ningxia found that the thermal bridge phenomenon is common in existing buildings, which seriously affects the quality of indoor thermal environment. In this paper, a building model in Zhongwei was taken as an example to reveal the influences of the thermal bridge on the average heat transfer coefficient of exterior wall by the calculation and analysis of the average heat transfer coefficient of the exterior wall. The analysis results show that the proportion of wall area in the total area of the heat bridge (FB/F) and the wall construction type have a significant impact on the exterior wall average heat transfer coefficient. In this paper, the design basis and measures of rural building are proposed to avoid the thermal bridges phenomenon in local building by selecting an appropriate window-to-wall ratio, room bay, and the external thermal insulation composite wall as circumstances permit. In order to improve the quality of indoor thermal environment of local rural buildings, the results obtained from this paper can provide references for the architectural design of similar rural buildings.

1 INTRODUCTION

Thermal bridge is caused by the heat transfer coefficient of the parts with much greater adjacent regions and poor thermal insulation properties, which is a very common phenomenon in the envelope (WU, 2014; SUN, 2010). Thermal bridges will cause much damage to the buildings such as wall condensation, plaster off, bacteria breeding, structure security breach, and increased energy consumption (LIU, 2002; LI, 2007). Therefore, reducing the effects of thermal bridges on heat transfer performance of building envelope is an urgent problem to be solved in the building structure design of rural buildings in cold region. Currently, relevant studies on the effects of thermal bridges on the average heat transfer coefficient of exterior wall are the focus of qualitative analysis, whereas quantitative analysis is rare. This is an important factor restricting construction thermal design of rural energy-saving building. Therefore, in this paper, the quantitative analysis will be focused to analyze the impact of thermal bridges on the average heat transfer coefficient of exterior wall of rural building in cold regions.

2 SITUATION OF LOCAL RURAL BUILDINGS

2.1 Basic characteristics of typical rural buildings

In order to understand the effect of thermal bridges on the indoor thermal environment quality of rural buildings, the research group assesses the basic characteristics of typical rural buildings by conducting actual research test on rural buildings in Zhongwei City, Ningxia Province, China. In this paper, the following living room characteristics are assumed: south window-to-wall ratio, 0.3; exterior wall is clay solid brick; heat transfer coefficient, 1.54 $W/(m^2 \cdot K)$; plastering, 20mm; 370 × 240 reinforced concrete ring beam (instead of lintel); and 370 × 370 reinforced concrete constructional column.

2.2 Wall surface temperature distribution of the research model

The wall surface temperature distribution of the typical rural building by thermal imager is shown in Figure 1.

Figure 1 shows that the inner surface temperature of the building wall angle and the T-shaped wall structure column and ring beam (instead of the lintel) was significantly lower than that in other parts of the wall. Thus, it can be seen that the

(a) Physical picture　　　(b) Temperature distribution

Figure 1.   Temperature distribution of wall surface.

coefficient of thermal conductivity of these parts of Ningxia rural buildings is large, the heat transfer capability is strong, heat flux is dense, and hence the thermal bridge phenomenon is serious. A large amount of indoor heat dissipating outward through the thermal bridge that is a weak link, which will increase building energy consumption and reduce the quality of the indoor thermal environment during improper handling. In order to increase the building energy efficiency, quantitative analysis of the effect of thermal bridges on the heat transfer performance of building envelope is carried out.

3 ANALYSIS OF AVERAGE HEAT TRANSFER COEFFICIENT FOR EXTERIOR WALL

The average heat transfer coefficient of the exterior wall is an average value of heat transfer coefficient of some main parts of the exterior wall, which includes the main wall and its surrounding structural thermal bridge (T-shaped wall structural columns, ring beams, and lintels). In general, main wall and the surrounding thermal bridge are different in terms of building materials and structure. The heat transfer coefficient of thermal bridge is often greater than that of the main wall. Therefore, in building energy-efficient design standards, the average heat transfer coefficient of the exterior wall is used as an index to evaluate the heat transfer performance of exterior wall. In the calculation of the average heat transfer coefficient of the exterior wall, this paper intends to use the one-dimensional heat transfer method (area-weighted average method) to calculate (ZHAO, 2012; HU, 2003). The calculation formula is as follows (LIU, 2013; YU, 2003):

$$K_m = \frac{K_p \cdot F_p + K_{B1} \cdot F_{B1} + K_{B2} \cdot F_{B2} + K_{B3} \cdot F_{B3}}{F_p + F_{B1} + F_{B2} + F_{B3}} \quad (1)$$

In the formula (1), K_m refers to the average heat transfer coefficient of building exterior wall W/(m²·K); K_p, K_{B1}, K_{B2}, and K_{B3} refer to the heat transfer coefficient of the exterior wall body and thermal bridge positions, construction column, ring beam, and the lintel, respectively (W/(m²·K)); F_p, F_{B1}, F_{B2}, and F_{B3} refer to the area of the exterior wall body and thermal bridge positions, construction column, ring beam, and the lintel, respectively (W/(m²·K)).

3.1 *Analysis of the heat transfer coefficient of the main wall*

In this paper, the heat transfer coefficient of the nontransparent exterior wall body is calculated as (LIU, 2013):

$$K_p = \frac{1}{R_0} = \frac{1}{R_i + \sum R + R_e} \quad (2)$$

where R_i and R_e refer to the heat transfer resistance of internal and external surfaces, respectively. In the general engineering practice, $R_i = 0.11$ (m²/K)/W and $R_e = 0.04$ (m²·K)/W, where R refers to the thermal resistance of each layer material (LIU, 2013):

$$\sum R = R_1 + R_2 + L + R_n = \frac{d_1}{\lambda_1} + \frac{d_2}{\lambda_2} + L + \frac{d_n}{\lambda_n} \quad (3)$$

where the material is homogeneous, d is the thickness of the material, and λ is the thermal conductivity coefficient of the material.

For this study model, the main wall heat transfer coefficients for the external and internal thermal insulation are calculated using formulae (4) and (5), respectively:

External thermal insulation:

$$K_p = \frac{1}{R_i + R_1 + R_2 + R_3 + R_e}$$
$$= \frac{1}{R_i + \frac{d_{ip}}{\lambda_{ip}} + \frac{d_w}{\lambda_w} + \frac{d_{il}}{\lambda_{il}} + R_e} \quad (4)$$

where d_{ip} refers to the thickness of the internal plastering layer for exterior wall (m); λ_{ip} refers to the thermal conductivity coefficient of the internal plastering layer for exterior wall (W/(m·K)); d_w refers to the thickness of the exterior wall (m); λ_w refers to the thermal conductivity coefficient of the exterior wall material (W/(m·K)); d_{il} refers to the thickness of the thermal insulation layer for the exterior wall (m); and λ_{il} refers to the thermal conductivity coefficient of the thermal insulation layer for the exterior wall (W/(m·K)).

Internal thermal insulation:

$$K_p = \frac{1}{R_i + R_1 + R_2 + R_3 + R_e}$$
$$= \frac{1}{R_i + \frac{d_{il}}{\lambda_{il}} + \frac{d_w}{\lambda_w} + \frac{d_{ep}}{\lambda_{ep}} + R_e} \quad (5)$$

where d_{ep} refers to the thickness of the external plastering layer for the exterior wall (m) and λ_{ep} refers to the thermal conductivity coefficient of the external plastering layer for the exterior wall (W/(m·K)).

3.2 *Heat transfer coefficient of T-shaped wall structure column parts*

Thermal bridge mainly includes T-shaped wall structural columns, ring beams, and lintels. There is a certain particularity for the calculation of the heat transfer coefficient for T-shaped wall structural columns. The heat transfer coefficients of the structural column part for the external and internal thermal insulation walls are calculated using formulae (6) and (7):

External thermal insulation:

$$K_{B1} = \cfrac{1}{R_i + R_1 + R_2 + R_3 + R_e}$$
$$= \cfrac{1}{R_i + \cfrac{d_{ip}}{\lambda_w} + \cfrac{d_c}{\lambda_c} + \cfrac{d_{il}}{\lambda_{il}} + R_e} \tag{6}$$

where d_c refers to the thickness of the structural column (m) and λ_c refers to the thermal conductivity coefficient of the reinforced concrete column (W/(m·K)),

Internal thermal insulation:

$$K_{B1} = \cfrac{1}{R_i + R_1 + R_2 + R_3 + R_e}$$
$$= \cfrac{1}{R_i + \cfrac{d_{il}}{\lambda_w} + \cfrac{d_c}{\lambda_c} + \cfrac{d_{ep}}{\lambda_{ep}} + R_e} \tag{7}$$

3.3 Heat transfer coefficient of ring beam parts

In this study model, ring beam (instead of the lintel) is set up on the top of the door and window openings, the calculation of the heat transfer coefficient of which is similar to that of the main wall. The formula for the calculation of the heat transfer coefficients of ring beam parts for external thermal insulation and internal thermal insulation wall are shown in formulae (8) and (9), respectively:

External thermal insulation:

$$K_{B2} = K_{B3} \cfrac{1}{R_i + R_1 + R_2 + R_3 + R_e}$$
$$= \cfrac{1}{R_i + \cfrac{d_{ip}}{\lambda_{ip}} + \cfrac{d_{rb}}{\lambda_{rb}} + \cfrac{d_{il}}{\lambda_{il}} + R_e} \tag{8}$$

where d_{rb} refers to the thickness of the ring beam (m) and λ_{rb} refers to the thermal conductivity coefficient of the reinforced concrete ring beam (W/(m·K)).

Internal thermal insulation:

$$K_{B2} = K_{B3} = \cfrac{1}{R_i + R_1 + R_2 + R_3 + R_e}$$
$$= \cfrac{1}{R_i + \cfrac{d_{il}}{\lambda_{il}} + \cfrac{d_{rb}}{\lambda_{rb}} + \cfrac{d_{ep}}{\lambda_{ep}} + R_e} \tag{9}$$

4 THE EFFECT OF BUILDING THERMAL BRIDGE ON THE AVERAGE HEAT TRANSFER COEFFICIENT OF EXTERIOR WALL

Further arrangement of formula (1):
Making:

$$F_B = F_{B1} + F_{B2} + F_{B3} \tag{10}$$

$$K_B = \cfrac{K_{B1} \cdot F_{B1} + K_{B2} \cdot F_{B2} + K_{B3} \cdot F_{B3}}{F_{B1} + F_{B2} + F_{B3}}$$
$$= \cfrac{K_{B1} \cdot F_{B1} + K_{B2} \cdot F_{B2} + K_{B3} \cdot F_{B3}}{F_B} \tag{11}$$

$$F = F_P + F_B \tag{12}$$

Then, there is:

$$K_m = \cfrac{K_P \cdot (F - F_B) + K_B \cdot F_B}{F}$$
$$= \cfrac{K_P \cdot F - K_P \cdot F_B + K_B \cdot F_B}{F}$$
$$= \cfrac{K_P \cdot F + (K_B - K_P) \cdot F_B}{F} \tag{13}$$
$$= K_P + (K_B - K_P) \cfrac{F_B}{F}$$

where F_B and F, respectively, refer to the total area of the exterior wall and thermal bridge (m²) and K_B refers to the average heat transfer coefficient of the thermal bridge (W/(m²·K)).

Formula (13) shows that the effect of thermal bridge on the average heat transfer coefficient of the exterior wall are influenced by the factors of the difference between the average heat transfer coefficient of thermal bridge and the heat transfer coefficient of exterior wall body (ΔK); the total area of thermal bridge accounts for the proportion of the total area of wall (F_B/F).

The average heat transfer coefficient of the exterior wall is proportional to the difference between the average heat transfer coefficient of the thermal bridge and the heat transfer coefficient of the exterior wall body. To reduce the average heat transfer coefficient of the building exterior wall, the building envelope structure provides some insulation. In general, exterior wall structure can be divided into internal insulation, external insulation, and sandwich insulation according to the position set of the insulation material. The thermal bridge of the internal insulation wall cannot get effective heat preservation treatment, of which the heat transfer coefficient is higher, increasing the average heat transfer coefficient of exterior wall significantly. However, the thermal bridge of the external insulation wall can get good insulation, which can reduce the heat transfer between thermal bridges and outside, decrease the heat transfer coefficient, thereby increasing the average heat transfer coefficient increased slightly. Therefore, to select a main material of exterior wall, the heat transfer coefficient has a small difference between the thermal bridge parts, which can effectively reduce the heat transfer coefficient of exterior wall caused by the thermal bridge and ensure the quality of the indoor thermal environment.

In addition, the total area of thermal bridge accounting for the proportion of the total area of wall (F_B/F) is affected by the window-to-wall area ratio, housing bay, and other factors. The following is the analysis of the influence of window-to-wall area ratio and housing bay on the total area of thermal bridge, in order to understand its influence on the heat transfer performance of the exterior wall.

4.1 South windows-to-wall ratio

The size of the south window directly influennces the size of the total area of the wall, which in turn affects the size of F_B/F, making the average heat transfer coefficient of the exterior wall different. For this research model, the change law of F_B/F, the average heat transfer coefficient of the exterior wall, and the increase percentage of the heat transfer coefficient of the exterior wall ($(K_m-K_p)/K_p$) with the south window-to-wall ratio are shown in Figure 2.

Figure 2 shows that, with the increase in window-to-wall area ratio, F_B/F and the average heat transfer coefficient of the exterior wall increased, but the increase was smaller; and the window-to-wall area ratio has a significant effect on the increase in percentage of the heat transfer coefficient caused by thermal bridge. At the same width of a room, window-to-wall area ratio is larger; the increase in the percentage of heat transfer coefficient was

due to the large thermal bridge and increase in the growth rate. $(K_m-K_p)/K_p$ can be as high as 18.9% while the window-to-wall area ratio is 0.5, comparing with the same period of window-to-wall area ratio, 0.25, which is a 67.3% increase. This is mainly due to the larger window-to-wall area ratio. The proportion of the total area of the main wall is smaller. The south window-to-wall area ratio should not be too large, within the scope of the standard, and under the condition of meeting the demands of the indoor thermal environment.

4.2 Housing bays

The survey found that the existing new houses of local rural mostly have a large bay. Housing bay directly affects the size of the total area of the wall, thereby affecting the size of F_B/F, so that the average heat transfer coefficient of the exterior wall is different. Selecting six kinds of housing bay that are often used in rural residential, the change law of the F_B/F, the average heat transfer coefficient of the exterior wall, and $(K_m-K_p)/K_p$ with the housing bay is shown in Figure 3.

Figure 3 shows that, with the increase in housing bay, F_B/F and the average heat transfer coefficient of the exterior wall are reduced. The housing bay has a significant effect on the increase in percentage of the heat transfer coefficient caused by the thermal bridge. With the same window-to-wall area ratio, housing bay is larger. The increase in the percentage of the heat transfer coefficient caused by the thermal bridge is smaller, and the rate of decrease is also low. $(K_m-K_p)/K_p$ can be as low as 13.1% while the housing bay is 7.8. In the same period, housing bay is 4.8, which is a 68.4% decrease. This is mainly because the housing bay is larger. The proportion of the total area of the main wall is larger. The housing bay can be increased appropriately within the scope of the standard and

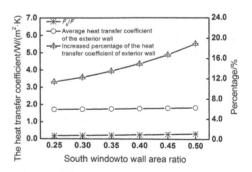

Figure 2. Change regulation of thermal parameters of the exterior wall with the south window-to-wall area ratio.

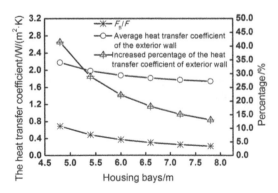

Figure 3. Change regulation of thermal parameters of the exterior wall with the housing bay.

under the condition of meeting the demands of the indoor thermal environment.

5 CONCLUSION

The investigation and testing results show that the thermal bridge phenomenon of rural buildings in Ningxia Autonomous Region is serious, making the reduction of thermal insulation performance of the building envelope, leading to the poor quality of indoor thermal environment. The effect of thermal bridge on the average heat transfer coefficient of an exterior wall in rural buildings is mainly affected by the influence of exterior wall structure, F_B/F, ΔK, and other factors. The total area of thermal bridge is mainly affected by the construction, window-to-wall area ratio, housing bay, and other factors. In order to reduce the influence of thermal bridge on building energy efficiency, it is necessary to choose the main wall materials, the heat transfer coefficient of which has a small difference between the thermal bridge as more as possible, and carry out some external thermal insulation. Meanwhile, the window-to-wall area ratio should not be too large and the housing bay can be increased appropriately.

ACKNOWLEDGMENT

This work was supported by the National Natural Science Foundation of China (51278419).

REFERENCES

GB/T 50824-2013 Design standard for energy efficiency of rural residential buildings. Beijing: China Architecture and Building Press, (2013).

Hu Pingsheng, Hu Qingsheng. Analysis on Effect of Heat Bridge to Heat Transfer of Wall. Journal of Huazhong University of Science And Technology (Urban Science Edition). 4, (2003).

JGJ26-2010 Design standard for energy efficiency of residential buildings in severe cold and cold zones. Beijing: China Architecture & Building Press, (2010).

Li Kuishan, Zhang Xu. Simulation Analysis on Three Dimensional Unsteady Heat Conduction of Building Thermal Bridges. Building Science. 12, (2007).

Liu Jiaping. *Architectural Physics, Fourth ed.* [M]. (Beijing: China Architecture and Building Press, 2013).

Liu Yansong. Analysis about indoor condensation in winter in severe cold regions. Housing OUSING Materials & Applications. 4, (2002).

Sun Daming, Zhou Hazhu, Tian Huifeng. Research Status and Prospect of Building Thermal Bridge. Building Science. 2, (2010).

Wu Shan. Analysis of the effect of Heat Bridge on the heat transfer coefficient of wall in residences. Heilongjiang Science and Technology Information. 8, (2014).

Yu Lihang. Calculation & analysis of average coefficient of heat transfer of outer wall. New Building Materials. 11, (2003).

Zhao Xiaoyan. Effect of Heat Bridge on the heat transfer coefficient of wall in cold zones. zhonghua minju. 1, (2012).

Advances in Energy Science and Equipment Engineering II – Zhou, Patty & Chen (Eds)
© 2017 Taylor & Francis Group, London, ISBN 978-1-138-71798-5

The study of the technology of seismic reinforcements and adding layers of the active-duty brickwork buildings

Xiuling Cao
Hebei GEO University, Shijiazhuang, China

Muci Yue
Suzhou University of Science and Technology, Suzhou, China

Ying Yuan, Huige Wu, Jinsheng Fan & Guangli Sun
Hebei GEO University, Shijiazhuang, China

ABSTRACT: Most of old brickwork buildings, subject to conditions such as financial constraints and construction standards at that time, cannot meet the requirements of the current seismic code. In addition, as the time passes, some buildings need to change the original use of functions, which also needs the reinforcement of the building structure. In this paper, we analyze the multi-storey masonry structure within the framework of a building, focus on additional storeys and renovation of houses seismic strengthening degisn. Then we put forward the methods to solve such problems and provide reference for the similar projects.

1 INTRODUCTION

In recent years, China was hit by several catastrophic earthquakes. A tremendous number of buildings suffered from serious damage, among which masonry structure houses are struck most severely. Experts have researched into the characteristics of the masonry structure building and analyzed the cause of the damage. At present our country has a large number of masonry structure buildings in active service. Caused by a lack of design and construction, aging of building materials, and design specification, the reliability of the structure is greatly reduced. In order to improve the ability of the seismic disaster prevention, reduce the losses after the disaster, this type of building needs to be reinforced to ensure the reliability of the building structures and prolong its service life. Therefore, the reinforcement technique of structures in modern architectural technology becomes more and more important.

2 BUILDING STRUCTURE REINFORCEMENT TECHNOLOGY

Building structure reinforcement technology is through a variety of effective technical measures to make a building with better reliability, security, applicability and durability to meet the requirements of the current specification and the owner(Gui, 2010). Possible measurements include: Improving the structural strength, stiffness, stability and durability, reducing accident hidden trouble, extending the lifespan of the building. The reinforcement of current buildings should be in accordance with the mechanical characteristics of the structure, the new features and surrounding environments. Moreover, seeing from the perspective of safety, applicability and economy, it should comply with the following principles:

1. Building structure reinforcement scheme should be determined according to the reliability of the building and its appraisal result of seismic capability. (2) Building structure reinforcement scheme should combine characteristics of the original structure, technical and economic indicators, new technology and material to avoid unnecessary economic losses. (3) The overall performance of the construction structure must be considered. Newly added components and the original structure should be connected reliably; reinforcement or new components should prevent local strengthening that leads to structural strength or stiffness mutation.

There are many types of structure strengthening methods. According to the mechanical characteristics, they can be divided into direct reinforcement method and indirect reinforcement method (Gui, 2010). Direct reinforcement method is to improve bearing capacity and stiffness of the original structures, which mainly includes: enlarged cross section method, the external sticking steel strengthening method, concrete replacement method, pasting fiber composite materials strengthening method, wiring method, pasting steel plate reinforced method, etc.; Indirect reinforcement method is to change the structure of the mechanical system, adjust its force transmission route or change the internal force distribution of the structure components and reduce the component method of load effect, which mainly includes: adding protection strengthening method and external pre-stressed reinforcement method, etc. In addition, technology that is used ancillary with building structure reinforcement are: technology, planting reinforce bar technique, bar and concrete surface treatment technology, crack repair, resistance to rust and anchoring technology, etc.

2.1 Enlarged cross section method

Enlarged cross section method is to increase the original sectional area or increase the reinforcement structures, improving the bearing capacity and stiffness of structure or changing structure of natural frequency of vibration. According to the mechanical characteristics of the components and reinforcement requirements, we choose unilateral or bilateral, three sides or four sides outsourcing reinforcement measures. The key of this reinforcement technology is the solid combination between the old and new materials, and the effective transfer of shear force.

The strength grade of the original structures determined by the field test should not be lower than C15. Moreover, the reinforced ends must have reliable anchorage measures. The main advantage of this method is simple construction, low cost and wide applicability; the downside is larger workload under wet condition, long time limit for a project and the influence for the clearance and appearance of the building after the increase of cross section.

2.2 Pasting steel plate reinforced method

Paste the steel plate reinforced method is to paste steel plate in the original component surface with special adhesive, making the whole force to work together, improving the bearing capacity of the structure, and enhancing the ductility of reinforced method. Paste the steel plate looks like in vitro do steel, increasing the bearing capacity of component of the cross section. Its advantages are that it has simple rapid construction benefit, does not need wet operation, and does not affect the appearance and the use of space and structure. The key to this technique is to choose special configuration of the modified epoxy resin adhesive, which requires structural adhesive to be non-toxic, high strength, corrosion and aging resistance, strong cohesive force, small string cable expansion coefficient and high elastic modulus, etc.

2.3 The external sticking steel strengthening method

The external sticking steel strengthening method is to wrap flat steel in the original structures, steel welded structure, and then pressure filling adhesive, achieved with the original structure overall stress, mutual constraint original component method. This original component cross section size is not much, and easy to implement "strong node weak component" design principle of "strong column weak beam". This reinforcement method is suitable for the need to greatly improve the structure's seismic performance and carrying capacity of the building structure.

2.4 Fiber embedded reinforcement method

Fiber embedded reinforcement method is to groove in the original component surface in advance, and add the FRP (Fiber Reinforced Polymer) fiber reinforced composite material, and injecting binder to make it as a whole to improve the bearing capacity and seismic behavior of structure strengthening methods. Compared with the method of external sticking reinforcement method, this reinforcement method has a better grip of and the construction is simple, material saving, corrosion resistance, easy to combine with adjacent components anchorage and resistant to high temperature and humidity, when suffers accidental load is not easy to damage, etc. This method of construction should, first of all, according to the design requirements in the original component surface carve groove, width is about 1.5 times the diameter of the FRP tendons, depth is generally not less than 20 mm, and then clean up debris and dust in the concrete, in order to avoid influence on bonding FRP tendons and the original structure; After about 1/2 the slot injection of the binder, put FRP reinforcement in the slot and apply light pressure, making the adhesive fully wrap FRP tendons. Then inject all of the binder. After the solidification of the binder, we can smooth the surface of the components.

3 ENGINEERING EXAMPLE AND ANALYSIS

3.1 Project profile

A commercial street building was built in 1988 with building area of $5200\,m^2$, 2 layer reinforced concrete frame structure, 6.6 m span, 6.6 m column spacing beam span (time), exterior wall thickness 360 mm. First floor has a height of 7.000 m and the second floor has a height of 6.000 m. Basis for column is soil rigid foundation, strip foundation under wall, pre-stressed porous plate cover. Its seismic fortification can't satisfy the current "Code for Seismic Design of Buildings" GB50011-2010 requirements. Because the use of the building is changed, it needs to add layer reconstruction to meet the new needs of using function. To make full use of the original building's indoor space, the constructors decided to build two stories between elevation of 3.500 m and 10.000 m. The load change is bigger after adding the layer. In order to ensure it to be structurally safe, economically reasonable, and meet the requirements of new functions, layer adding and seismic strengthening design are needed.

3.2 Layer adding and seismic strengthening design

House indoor layer adding typically have four basic structures and forms: separate, integral, cantilever and hanging type. Through calculation and analysis, the layer adding scheme of this building adopts the integral indoor floors structure scheme. Integral layer adding is to connect the bearing structure of the newly added layer to that of the existing structure, so that they could all together to share the total load of the building. This solution can make good use of existing building walls, column and basic potential. It also has good entirety and anti-seismic ability. However, the original house needs to be strengthened. Brick-concrete building aseismic reinforcement has internal reinforcement method, external reinforcement method and the internal-external mixing reinforcement method.

External reinforcement method mainly strengthens the external walls, involving little in-door operation; Internal reinforcement method is the inverse of the external reinforcement method.

Internal-external mixing reinforcement method strengthens both internal and external walls at the same time. We will discuss each part of the component reinforcement scheme as follows.

3.2.1 Cross wall reinforcement and vertical wall connection

According to the requirements of building structure seismic code, the adding or reinforcement of cross wall requires reliable connection between seismic cross walls and external vertical walls, which connect the old and the new walls as a whole.

According to the requirements of earthquake-resistant calculation, the original internal cross wall has insufficient ability to stand horizontal seismic force, which means the transverse rigidity

needs to be strengthened by reinforcing part of the wall. This project uses double-sided mesh, cast-in-situ, spraying or pressure of fine stone concrete reinforcement method, which strengthens aseismic cross wall and the stiffness of the structure, improving its resistance to horizontal seismic action. (Particular way as shown in the figure 1 and 2.)

3.2.2 The vertical wall reinforcement measures

Masonry structure building uses mostly external reinforcement method, namely adding buttresses constructional column along external walls vertically and adding ring beams along lateral direction to improve the integrity of the building and comprehensive carrying capacity. This reinforcement method requires less workload and convenient construction. However, if the cross section of column and ring beam is too large, it will cause elevation and affect the beauty of the building. Solution is to reduce double-sided reinforcement of cross section of constructional column and ring beam. The buildings that are on the frontage or have higher demand of the opposite-side can use double-sided reinforcing mesh, spraying or wiping the fine aggregate concrete reinforcement method. Hidden combination of constructional column and ring beam wall stirrups uses borehole drilling machines, so that it will not destroy walls and ensure that the main reinforcement of hidden constructional column and ring beam and wall stirrups can work together, and exterior ring beam, constructional column and wall can be inte-

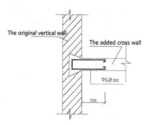

Figure 1. New cross wall connected to the vertical wall of structure diagram.

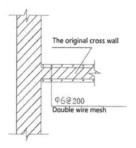

Figure 2. The original cross wall of reinforced.

grated together. The currently use of inside a wall wear more reinforcement construction method is not very solid. Hole increase followed by grouting method can be used to secure the whole connection. Buttress constructional column connected to the base should be put at the foot of the column of the foundation.

3.2.3 *Strengthening methods in relation to the frame beam column*

Layer adding floor has primary and secondary beam structure system. Longitudinal reinforcement in beam and column joint structure is complex, which needs to add ferrule in the bearing place of outer column or make a hole on the column to apply anchoring solid processing. After full consideration and careful comparison, the team decided to change rectangular section to the Π type cross section. Beam reinforcement is shown in figure 3. The cross section of the internal column of the original house is 400 × 500 mm. The column section size is 600 × 700 mm after reinforcement.

4 CONCLUSION

Due to the construction of economy level and the factors such as construction standards, most current houses of masonry structure cannot meet the requirements of the seismic code, which means that they need to be strengthened in accordance with the requirements of the current specification for reinforcement. Because of the changes over time, the building in this project needs to change its original function. Thus this building needs to add new layers and apply seismic strengthening process. Through the analysis of the building reinforcement scheme, the team chose enlarged cross section method to reinforce the house. After reinforcement, the bearing capability of the house is greatly increased and the house itself could meet the force distribution requirement. Meanwhile, after the reinforcement,

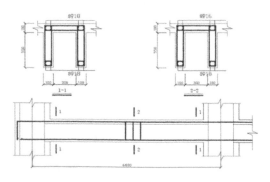

Figure 3. Beam reinforcement figure.

the anti-seismic performance and integrity of the house is also enhanced, which reach the current seismic code. Engineering practice showed that this reinforcement method is economic, safe, reasonable and effective. Similar projects could refer to this method.

ACKNOWLEDGMENTS

This work is financially supported by China and Australia cooperation research projects (16394507D); Hebei overseas training project for excellent experts; Research Plan Projects of Hebei Education Department(QN2014020); Scientific Research Plan Projects of Hebei Science and Technology Department(15275430);

Author: Cao Xiuling, female, Ph.D. professor, Mainly engaged in researching work in structural engineering.

REFERENCES

Cao Xiu-ling, Seismic Reinforcement Method Study on Adding Layer Transformation of Masonry-concrete Structure. Civil Engineering in Current Practice and Research Report 2010 international Conference on Civil Engineering. 2010, 19–20: 1163–1166.

Cao Xiu-ling, et al. Effect of Door Opening Size and Position in Seismic Performance of Autoclaved Aerated Concrete Block comeposite Walls, Advanced Research on Material Engineering, Chemistry, Bioinformatics II. 2012, 531: 539–452.

Cao Xiu-ling, et al. Experimental Study of the Heat Transfer Process of the Building Envelope. China University of Mining and Technology (Beijing) Graduate Education Academic Forum Proceedings (the first prize). 2008, 310~316.

Cao Xiu-ling, et al. The Evaluation Grading of The potential about Hebei Rural Residential Land Reclamation. Transactions of the Chinese Society of Agricultural Engineering. 2009, 25(11): 318–323.

Cao Xiu-ling, et al. The Selection and Procurement for Material of Building Waterproof. Chinese market. 2007, 32: 34–36.

Gui-zhenxi, Building aseismic accrediting standards and strengthening technology handbook [M]. Beijing: China building industry press. 2010.6.78 ~ 90.

People's Republic of China ministry of construction. Building aseismic design code [Z]. GB 50011–50011. Beijing: China building industry press.

People's Republic of China ministry of construction. Building aseismic reinforcement technique regulations [Z]. JGJ 116–116. Beijing: China building industry press.

People's Republic of China ministry of construction. Concrete design specification [Z]. GB 50010–50010. Beijing: China building industry press.

People's Republic of China ministry of construction. Structural reliability design unified standard [Z]. GB 50068–50068. Beijing: China building industry press.

Advances in Energy Science and Equipment Engineering II – Zhou, Patty & Chen (Eds)
© 2017 Taylor & Francis Group, London, ISBN 978-1-138-71798-5

Design and research on sewage source heat pump system of rotary two anti device

Long Xu, Rong-hua Wu, Xiao Wu & Zhong-Xiu Shi
Qingdao University, China

ABSTRACT: Sewage source heat pump system of rotary two anti device is committed to solve the difficult problem of heating blockage and scaling. Firstly, it reviews the sewage source heat pump system, the anti blocking and scaling problem, the application of sewage source heat pump system at present and a detailed explanation of the rotary prevention device structure and operation principle, and gives the examples of engineering application. Through the practical application, we prove that type two anti device can effectively solve the heat transfer process in sewage blocking and scaling problems.

1 INTRODUCTION

Using sewage source heat pump heating and cooling for the building has an important value on energy saving, environmental protection and economy (Sun, 2005; Wu, 2006). China's annual building energy consumption is nearly 10 million tons, of which air conditioning contains more than 80%, and the primary energy efficiency is less than 60%. Thus the research and the development of new energy saving technologies on buildings not only concern to the construction industry, but also will be related to the sustainable development of the whole national economy and society (Sun, 2009). Establishing the use of renewable clean energy air conditioning heating mode has important practical significance in realizing the coordinated development of energy and the environment.

2 PROBLEMS ON CLEANING AND ANTI CLOGGING OF SEWAGE SOURCE HEAT PUMP

The key issues of cooling and heating source of sewage or surface water are the impurities clogging the heat transfer equipment and growth of internal dirt reducing heat transfer efficiency.

2.1 *Problems on anti-clogging*

If sewage heat exchange or evaporator have anti-clogging function, the sewage need to have sufficient water flow cross-section in the heat exchange, such as immersion, spray heat exchange, a clogging problem can be avoided. However, too large flow cross-section cause the flow rate too small or even zero and the immersion means low heat transfer efficiency. Therefore, anti-clogging and improving heat transfer efficiency are an "antagonistic" contradiction (Wu, 2015). How to resolve this conflict, we propose the following two methods:

First, filter and then heat exchange. It means filtering sewage first, filtering out dirt and impurities it contains, to achieve that even a small flow cross-section will not be blocked, in order to ensure the efficiency of heat transfer.

Second, exchange heat by diverting. Based on the large flow cross-section, and then try to improve heat transfer efficiency.

2.2 *Problems on cleaning*

Sewage source heat pump system also has serious dirt problem besides clogging problems. We need to implement decontamination cleaning process on the basis of the two blockades ways.

It is divided into regular cleaning and online cleaning according to continuous of time. Regular cleaning is easy to implement and it is simple and mature, but with a large amount of dirt and short cleaning cycle (Wu, 2007; Xu, 2007), it is difficult to meet the operational requirements of the system, so it needs to develop to online cleaning. In order to timely, effective and easy to implement short-term or on-line cleaning, heat transfer equipment should have the following two characteristics:

First, the small number of sewage flow path, a large cross-sectional area of the flow path and small cleaning workload will help implement manual cleaning and online cleaning.

Second, the sewage flow path is mainly based in series. To maximize the implementation of the series, to maximize flow rate and ensure a larger cross-sectional flow area can improve heat transfer efficiency and anti-blocking ability.

3 THE FORM OF SEWAGE SOURCE HEAT PUMP SYSTEM

Sewage and surface water source heat pump system divide into the filter after the heat transfer system and direct divert heat exchange system according to whether take the filtering measures.

3.1 Direct divert heat exchange system

Sewage or surface water is delivered to the intermediate heat exchange by the pump, then an intermediary heat transfer medium extracts heat or cold of surface water into the unit. The form of the system is shown in Figure 1.

Intermediate heat exchange to divert heat exchange, which has blockades and enhanced heat transfer capabilities. Divert heat exchange is taken King Yu idea that "diverting not trapped," let sewage or dirt and impurities in surface water flowing without clogging in the heat exchange.

3.2 The filter after the heat transfer system

Sewage or surface water is delivered into the filtering and anti-clogging device by the pump, then into the intermediate heat exchange after filtering. Then an intermediary heat transfer medium extracts heat or cold of surface water into the unit. The form of the system is shown in Figure 2.

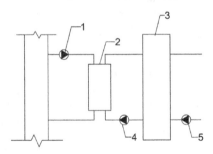

Figure 1. Direct channel heat transfer system.
1 sewage pump; 2 heat exchange; 3 unit; 4 medium pump; 5 end pump.

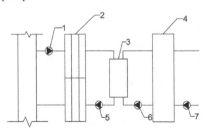

Figure 2. The filtered first heat exchange system.
1 first stage sewage pump; 2 heat exchange; 3 intermediate heat exchange; 4 unit; 5 Two sewage pump; 6 intermediary pump; 7 end pump.

The key components of this system is the filtering device. That is how to achieve "effective filter" to ensure that the intermediate heat exchange is not blocked. Filtering can divide in to static and dynamic filter.

Static filter means filter surface is stationary when the suspended solids and impurities covering in the surface of the filter cause the water flow reducing to a set value or the formation of a pressure differential before and after filter surface, and then repeatedly washing filter surface. Static filter has been applied very early at the sewage source heat pump system, such as automatic screen filter, automatic back-washing.

Dynamic filter is continuous filter and reduction technology, which is different from the traditional filtering programs (Sun, 2005; Wu, 2005). The filter surface is divided into inlet portion and outlet portion, and the area of the inlet portion is larger than that of the outlet portion. Inlet portion guides water decontamination, and outlet portion drains water backwash. Filter surface rotates about an axis, and each mesh conversion can be kept on the inlet portion and outlet portion.

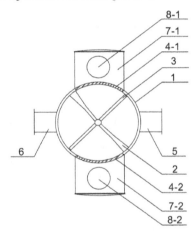

Figure 3. Rotary two defense equipment.
1 valve housing; 2 filter element; 3 scraper; 4-1 filter element I; 4-2 filter element II; 5 inlet; 6 outlet; 7-1 connection box I; 7-2 connection box II; 8-1 connector I; 8-2 connector II; 9 shaft; 10 rotating member.

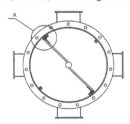

Figure 4. Configuration diagram of the four-way reversing device.

4 STRUCTURE, PRINCIPLE AND CHARACTERISTICS OF ROTARY TWO ANTI DEVICE

Sewage sources two anti rotary device system is first filtration and then heat exchange system. Rotary two anti device is a pre-filtration device belonging to a static filter.

4.1 The composition and operation principle of rotary two anti device

As is shown in Figures 3–5, rotary two anti device consists of a housing, a valve body, scraper, filter elements, inlet, outlet, connection box, shaft and rotating component.

As is shown in Figure 6, rotary two anti device system is made up of the four-way valve, two connecting line defense equipment, heat pump or heat exchange devices, pumps, drainage lines and connections heat pump or heat exchange device.

Operating principle: sewage or surface water entering the inlet, impurities are filtered out through the filter element, out from connection port 1, and then into the heat pump or heat exchange means, out from connection port 2 after the heat exchange, and discharge from the outlet after through filtration 2. When filter element clogged, the rotation member drive body and the blade body movements, the body reversing the two water flows of connection1 and connection 2. Once the scraper cleans a filter element, water flows out from the connection port 2 and reverse into heat exchange or heat pump equipment, then entering connection port1 and making repeatedly washing to filter element1; when the filter element 2 clogged, and rotating member drive body and the blade movement again, and so forth.

When the body movements of rotary two anti device change flow direction or frequent actions affect the sewage or surface water heat exchange, four-way valve and the valve of rotary two anti device can work at the same time in changing the flow direction of filter elements in rotary two anti devices without changing the direction of the heat exchange device.

4.2 Structure and operating parameters

1. Working filter surface area, which is a filter surface area during normal operation, with S_A, in units of m².
2. Contamination density in filtering surface refers to the amount of contamination on filter surface in unit area, with W in units of kg/m².
3. The maximum density of the filter surface, the maximum amount of dirt in the unit area of the filter surface can be covered, that is, the

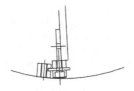

Figure 5. The local amplification figure of A area on the four-way reversing device.

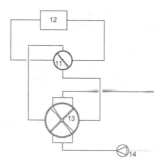

Figure 6. Two anti-rotary device system.
11 Four-way reversing device; 12 unit; 13 rotary two prevention device; 14 pump.

surface pollution density when filter surface is completely blocked filter, with W_M, the unit is kg/m².
4. Clogging coefficient, the rate of the area that the mesh is clogged and the area that the mesh is not clogged, with ψz, in the filtering process, the coefficients are varying by time, $0 \le \psi_z \le 1$
5. A filtering section modulus, the rate of the total mesh area in filter surface and the area of the work surface, with S.

By the above definition, we can get:

$$dV = [1 - \varphi_z(t)]S_A sudt \tag{1}$$

And the average flow on filter surface within 0 ~ t is V (m³/h)

$$V = \frac{S_A S W_M}{Ct}(1 - e^{-\frac{CVu}{W_M}t}) \tag{2}$$

4.3 Characteristics of rotary two anti device

Rotary anti two device system can effectively solve the problem that sewage or surface water impurities clogging the heat transfer equipment, and reducing the growth of the thickness of the heat exchange fouling equipment, to achieve the anti-two, ensuring that surface water or sewage no plug and efficient operation.

Which has the following characteristics:

Solving the clogging problems well, at the time of changing the flow direction of the heat pump or heat exchange device, it can be backwash, reducing dirt, taking into account anti-clogging and cleaning. Through the filter element, filtering while backwash, at the same time cleaning the surface of the heat exchange.

Fixed setting the filter surface, can design the shape and size of perforations according to need, easy to manufacture. This device is a static filter solving the problem of mixing with water between filtration zone and recoiling zone caused by rotation of sewage.

The system can be driven by electric, can be fully automated, and there is substantially no energy consumption.

The four-way valve body is sealed with turntable and type O rings and rubber head.

5 ENGINEERING DESIGN CASE

5.1 Project overview

The project takes the urban sewage source heat pump system to heat for the winter.

Project is located in Qingdao City Mission Island Quad, a building area of 40,000 m². The original coal-fired central heating, due to environmental pollution problems, the implementation of community water source heat pump heating system. The urban district near trunk sewer, the minimum flow 1000t/h or more, the minimum temperature in winter water 10°C, adequate water, away from the main canal buildings within 40 m. End system for the radiator system.

5.2 Design parameters and main equipment

The total construction area of 40,000 m², the area of heat index 50W/m², the total heat load 2000kW. Selected heating capacity 2000kW screw compressor, a high-temperature unit, the refrigerant is R134a, the maximum water temperature up to 63°C (theoretically 65°C), as shown in Table 1.

5.3 Project characteristics and related issues

The project by the untreated urban sewage, winter temperature actual in more than 13°C is because some industrial waste water discharged into the sewer, but consider to the future industrial waste water flow

Table 1. System design conditions.

Water side seasonal temperature	Winter/°C	Summer/°C
Sewage side	10~6	–
Terminal	60~50	–

problems, so also use the usual urban original sewage general temperature as a design basis.

6 CONCLUSION

Sewage source heat pump system rotary two anti device is a pre-filter unit, and has been proven its technical feasibility in engineering practice. It is an ideal method to anti-clog in sewage and surface water source heat pump system in addition to its low initial investment, stable and reliable operation.

ACKNOWLEDGEMENTS

This research was financially supported by the "12th Five Year" science and technology support program", "Waste water use—strengthening heat transfer technology research and equipment research development", 2014BAJ02B03.

REFERENCES

De-Xing Sun, Rong-Hua Wu. Energy saving and environmental protection evaluation method of urban sewage source heat pump system. China Water and Wastewater, 2005, 21(12):103–106.

De-Xing Sun, Rong-Hua Wu. Flow blockage and heat transfer characteristics of urban primary sewage in heat pump cooling and heat source. Heating, Ventilating and Air Conditioning. 2005, 35(2):86–88.

De-Xing Sun, Ying Xu. Key technology and engineering practice of primary sewage source heat pump. Energy Conservation Technology. 2009, 27(1):74–77.

Energy Information Administration. International energy outlook 1999, DOE/EIA0484(99).

Qingdao KeChuang New Energy Technology Limited company. A reversing device for non clean water heat exchanger and system: China, 203009906. U, 2013-06-19.

Qingdao KeChuang New Energy Technology Limited company. Sewage or surface water heat rotary two anti device: China, 101935112. A, 2010-09-02.

Rong-Hua Wu, Yin Xu. The coupling model of canal temperature and soil temperature in urban sewage heat pump system. The 5th international symposium on heating, vertilation and air-conditioning, Tsing Hua, China Beijing, 2007.9.

Rong-Hua Wu, Zhi-Bin Liu, Huang Lei, De-Xing Sun, Nan-Qi Ren. The design of sewage and surface water source heat pump system. Heating, Ventilating and Air Conditioning. 2006, 36(12):63–69.

Rong-Hua Wu. Research and application of urban sewage source heat pump system. [Master's thesis] Harbin: Harbin Institute of Technology, 2005.

Rong-Hua Wu. Technology and system of sewage and surface water heat pump. The Science Publishing Company. 2015:20–21.

Yin Xu, Rong-Hua Wu. Restorability and determination of canal temperature in sewage heat pump system. The 5th international symposium on heating, vertilation and air-conditioning, Tsing Hua, China Beijing, 2007.9.

Advances in Energy Science and Equipment Engineering II – Zhou, Patty & Chen (Eds)
© 2017 Taylor & Francis Group, London, ISBN 978-1-138-71798-5

A study on the way of activity rejuvenation of the historic district—taking Zhili Governgeneral-west street historic district in Baoding as an example

Yan Chen
Agricultural University of Hebei Province, China

Fenxia Fan
Xingtai Planning and Design Research Institute, China

ABSTRACT: In recent years, the historic district faces a problem that the unbalance between functional structure and decay of vitality. Many researches have been made on conservation and renewal of historic district. However, under the influence of New Humanism and New Urbanism, the challenges of continuing the historical context, reshaping historical district humanistic vitality under the conservation are important parts in a research. This paper discusses concrete measures for reconstruction of cultural vitality case study as the historic district in Baoding to promote the value unity about historic and culture, economy and society.

1 INTRODUCTION

The concept of "city cultural crisis" is emerging in recent years. In city's cultural development and urbanization, it is more and more difficult to maintain the independence, diversity and autonomy of the cultural forms and elements. And the historic and cultural district reflects the city culture exactly, which plays an important role in extending and inheriting the historic context. Therefore, the protection of historic and cultural districts is crucial.

Historic and cultural district is the most dynamic and characteristic area in a famous historic and cultural city, it is also a dynamic city heritage, can provide material and spiritual service for residents. Under the influence of the new urbanism, the protection of the historic district pays more attention to the relationship between the people and the environment, culture and society. But in recent years the protection puts more emphasis on "material form" and ignores the problem of the city connotation and cultural vitality, even fades gradually. Therefore, on the premise of protection for historic and cultural heritage, how to continue the historical context and reshape the vitality of historic district is the current research focus.

2 CONCEPTS OF RESHAPING THE CULTURAL VITALITY OF HISTORIC DISTRICTS

2.1 *The organic renewal theory*

Wu Liangyong proposed Organic Renewal Theory based on a long-term study on the old city plan of Beijing, advocating that according to internal rules of urban development, conforming to the urban fabric, to seek the urban renewal and development on the basis of sustainable development (X. M. Zhong, 2011). It is mainly composed of three aspects: organic nature of the whole city, of the urban cellular and urban renewal, and of the renewal process (K. Fang, 2000).

2.2 *The new humanism theory*

The New Humanism planning thought is taken from the human spirit and human care theory in the original humanism planning thought, which combines democratic centralism thought with humanistic planning on the basis of existing national condition (Z. U. Wang, 2009). The thought is characterized by social justice in urban planning. Therefore, adhering to the "people-oriented", improving the quality of people's life are effective ways of realizing the vitality reconstruction.

2.3 *The new urbanism theory*

The New Urbanism is a kind of traditional inheritance, and a revival of traditional development, it also pays more attention to taking advantage of the convenience of modern city. It is considered that the humanistic value of the community should be combined with the convenience of modern transportation organically to revive the urban community (D. F. Jiang, 2007).

3 THE OVERVIEW OF ZHILI GOVERNGENERAL-WEST STREET HISTORIC DISTRICT

3.1 *The development process*

Baoding is the second batch of national historic and cultural cities announced by State Department. More than two thousand years of history has produced its profound cultural background. The West Street was built in Song Dynasty, which was formed with partners and commercial space as the main function after the Song, Jin, Park, Ming and Qing five dynasties and the Republic of China. After liberation, the old city experienced the renovation of roads system construction and large-scale urban renewal in recent years, formed its present size and streets morphology. Nowadays, the West Street is famous with unique style of commercial culture and the diverse architectures, which has an important status in our country, as shown in Fig. 1.

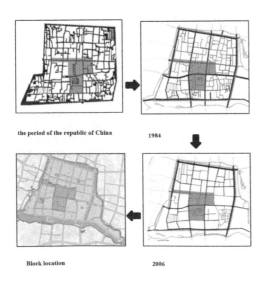

the period of the republic of China   1984

Block location   2006

Figure 1.   The development process of historic district.

3.2 *The situation*

Zhili Governgeneral-west street historic district is located in the center of the old city. The north street, the east street and the west street through building block, lotus fanaw street, Xinghua road and south street, that form a demarcation line of the district. Within the district there are 11 existing bump units, 10 outstanding historic buildings, two traditional and commercial Streets, many old and valuable trees. Features of historic buildings are mainly for the Ming and Qing dynasty palace architecture, landscape architecture and religious buildings, these outstanding historical buildings with different periods have an important value of achievement, and high art value, reflecting the various architectural style and genre, which is the important cultural architecture landscape of Baoding Zhili Governgeneral-west street historic district.

4 REASONS FOR THE DEADLINE OF THE CULTURAL VITALITY OF ZHILI GOVERNGENERAL-WEST STREET HISTORIC DISTRICT

4.1 *The bias of value recognition*

Aldo Rossi has pointed out: "The formation of the urban space is related to the value." With the rapid development of urban economy, large-scale urban renewal in Baoding has damaged the spatial framework and fabric of historic district badly. Due to the bias of economic value (J. D. Zhang, 2016), complex new buildings has covered a large number of historic relics located in Zhili Governgeneral-west street historic district, that makes the historic buildings lost the glory of the past.

4.2 *The ignorance of living problem*

Zhili Governgeneral-west street historic district in the architectural form includes commercial buildings, as well as residential buildings, public buildings and other forms. The renovation for the west street of Baoding at 2010 was focused on recovering the commercial function, but did not improve the living conditions of indigenous people effectively. Parts of the historic district are crowed with houses, rubbish and rough roads. So some residents moved out of the district gradually due to the lack of infrastructure construction and backward of living environment. The migration has a great damage on the social network structure and district space environment.

4.3 *The confusion of management*

As the functional properties of historic buildings are different, the buildings are belonged to different

Figure 2. Messy yard.

Figure 3. Broken virtuous temple.

units. So the management protection is in a state of confusion because of complex property rights and contradiction of interest. Thus parts of important historic buildings need to be repaired urgently, and it is not conducive to the implementation of integrated protection measures such as centralized classification, as shown in Figs. 2 and 3.

5 REASONS FOR THE DEADLINE OF THE CULTURAL VITALITY OF ZHILI GOVERNGENERAL-WEST STREET HISTORIC DISTRICT

Zhili Governgeneral-west street historic district is the typical representation of the historic and cultural characteristic in Baoding, although the vitality in the district in recent years is gradually being declined, it contains many induced dynamic motivations, but also includes the material entity of historic districts, the space of cultural value, and residents' social status living in the neighborhood. Therefore, in the process of protection, we should make full use of the dynamic factors to realize the reshaping of the cultural vitality.

5.1 *The physical form level*

5.1.1 *Protecting the historic building practically*
To implement the control for historic buildings the plan takes kinds of measures such as planning protection units of cultural relics, historic buildings and scope of historic districts, form a complete purple line protection system effectively, and put forward the protection requirements of the targeted. But the protection for historic buildings is "static theory + dynamic practice mode" (Y. Huang, 2015). The protection and renovation for each building is divided into repairment, renovation, demolition and other different ways, coordinating the relationship between the local and global of construction. From

the protection of architectural point to expand to the revival of area; From the behaviour of the historical heritage preservation extended to continuing innovation of urban context, as shown in Fig. 4.

5.1.2 *Protecting the spatial framework and fabric*
Zhili Governgeneral-west street historic district is an important part of the ancient city pattern. The plan determines a spatial pattern, that is "one area three parts (Museum of Zhili Govern-general's, lotus pond and Daci Court as the core of traditional culture), two horizontals (west street and yuhua), a longitudinal (lotus fanaw street).

The plan retains the characteristics of alleys, protects the orderly layout of the primary and secondary in present construction yards, to further protect the spatial framework and fabric of continuation. At the same time, to ensure the coordination of historic style the plan puts forward some specific control requirements on building height, visual corridor, street interface and open space of historic district, as shown in Fig. 5.

5.2 *The social and economic level*

5.2.1 *Developing multivariate mixed formats and creating commercial slow space*
According to the survey found that the west street is mainly composed of bicycle and funeral sup-

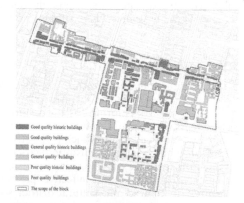

Figure 4.   Construction quality evaluation.

Figure 5.   Renovated Catholic church, Daci court.

plies business, the messy format is extremely unco-ordinated with historic and cultural style. So the business format adjustment of the west street is introduced in specialty restaurants, bars, entertainment and other fashion formats in function, then put it connect with Yuhua Road business district, Museum of Zhili Govern-general's, Lotus Pond together to form a complete slow space with the character of cultural business. This will provides a dynamic guarantee for reshaping the district vitality.

5.2.2 Introducing cultural creativity industry and improving the cultural industry chain

The protection for Zhili Governgeneral-west street historic district takes the reuse of historic buildings and intangible cultural heritage as an opportunity to introduce the creative cultural industry adapted to the new industry of modern life. And fully taking advantage of cultural and commercial resources surrounded areas to form a multifunctional historic district composed of culture and creativity, shopping, tourism, leisure exhibition industry. Through functional replacement and dynamic way of protection, by using historic buildings create a theme museum, a theme culture square, etc., to endow them with new social function, promote the development of tourism industry, improve the cultural industry chain, which makes the historical district structure more reasonable, the function more effective, the culture more vibrant, as shown in Fig. 6.

5.3 The social and livelihood level

5.3.1 Improving the infrastructure, protecting the social network of historic districts

The social network is formed by the relationship of residents, economic structure, life style, social communication and so on, which is the structure and form of the vitality of historic district. Perfecting infrastructure and improving the residents' living environment are the effective protection measures of historic and cultural district social network.

The New Urbanism theory emphasizes the value of urban cultural heritage, at the same time, pursuits the combination with the organic and convenient modern transportation. The plan set up vehicle forbidden area, pedestrian area, bus stops and public parking lots in Zhili Governgeneral—west street historic district to form the organic connection between district and modern traffic. Meanwhile, by improving the water supply, drainage and other facilities to improve the living environment, protect the social network in the historic cultural district, as shown in Fig. 7.

Figure 6. Features street space.

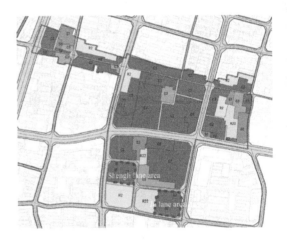

Figure 7. Land use plan.

5.3.2 Arranging for indigenous people properly, proving employment security

The new humanism theory is focus on "people-oriented" to realize the social justice. Ju lane and Shengli lane area are important parts of lane culture in Baoding, but the status quo is not harmonious with the historic surroundings seriously. So the plan reserves specialty streets and historic buildings, adjusts the land nature for commercial land, combined with the functional orientation of the commercial district. In the process of area transformation the government will provide houses for indigenous people to make them get good placement. By converting 2.06 hectares residential land into commercial, green space and other functions, there will bring more diversified and cultural activities, thus enhance the historical and economic value of the land, which can be used for the restoration of historic heritage to realize the long-term protection mechanism of historic district.

6 CONCLUSIONS

The historic district has unique advantages and value on history, economy and social culture, and the value unity is an effective way of reshaping the cultural vitality. It can balance the historic district protection and development, inject new vitality into the historical block to improve the comprehensive value, then becomes a real "live" heritage providing a guarantee for the sustainable development of the district. Reshaping the vitality of historic and cultural districts still need to adhere to the following principles:

1. The whole of the humanism environment. The reshaping of the historic district is a whole project, includes physical form, such as spatial pattern, spatial texture, historic buildings, etc., and also includes the social economy, people's livelihood level to restore the vitality of the ancient city.
2. The reasonable use of sustainable principles. In the process of protection, it should sees the relationship between protection and development block in the sight of the sustainable development, pay more attention to the renaissance, maintain and enhance its vitality and the dynamic and sustainability of reshaping process.

3. The operational principles of the program management. According to the different history, science and art value, the different sound levels of present situation, different types of city space and landscape features of historic buildings, the plan adopts the method of classification protection, formulates the corresponding protection regulations and control measures, to keep the diversity of the historical landscapes and make plan operable.

REFERENCES

Fang, K. The renewal of the old city in Contemporary Beijing:investigation study exploration, 195–196, (2000).

Huang, Y., Shi, Y.L. Review On Historical District Preservation and Renovation Practice, (31):98–104, (2015).

Jiang, D.F. Theory of urban form vitality, 67–69, (2007).

Wang, Z.U. New exploration of urban development based on the new concept of the new humanism, (5):22–24, (2009).

Zhang, J.D. Review and Revelation on the foreign historic district protection and utilization, 70–73, (2016).

Zhong, X.M. Vigorous Renaissance of Historic Precincts, (5):44–48,(2011).

Advances in Energy Science and Equipment Engineering II – Zhou, Patty & Chen (Eds)
© 2017 Taylor & Francis Group, London, ISBN 978-1-138-71798-5

A brief analysis on the application of roof greening to buildings in North China

Yixuan Cai
Urban and Rural Construction College, Agricultural University of Hebei, Baoding, China

ABSTRACT: The pedestrian definition of roof greening is all planting technologies away from earth, with coverage including not only roof planting but also greening for special spaces in various buildings and structures not connected with ground, nature, and soil. In recent years, roof greening has appeared in many places, which gave more vital force to the stereotyped roof. The popularity of a green roof is not only because of esthetic appearance but also its practicality. Exploiting green patches on roof of buildings has become an important method to add spice to buildings, improve urban appearance, insulate heat for buildings, improve greening, and purify air. As a special greening mode, roof greening has much ecological effect, landscape benefits, and social benefits. Hence, we should widely advocate its application. This paper analyzed the actuality of roof greening and put forward some pertinent suggestions for climatic conditions in North China.

1 INTRODUCTION

Roof greening is a form of building green landscape on a buildings' roof and special spaces through certain technology after selecting plants with growth habit matching characteristics of buildings' roof structure, load, and ecological environment on roof. It can also be generally comprehended as a general designation of gardening and planting on roof of ancient and modern buildings, city wall, bridges, terraces, rooftops, balconies, or large artificial rockery. Roof greening is of great significance to increase urban green area, improve deteriorating living environment of humans, change the scene of numerous high buildings and large mansions and hard pavement instead of natural land and plants, improve urban heat island effect caused by excess deforestation and various exhaust pollution, eliminate hazard of sandstorm, improve people's housing conditions and quality of life, beautify urban environment, and improve ecological effect.

2 CHARACTERISTICS OF ROOF GREENING

Roof greening appeared with increase of city density and multiple stratification of buildings, and is a greening beautifying form, in which the urban greening develops toward stereoscopic space, expands green space, and adds multidimensional natural factors in cities. The current roof greening is no longer limited to roof garden, but raises and extends the roof garden to a new industry. Similarly to ground greening, it also belongs to city gardening and is part of city gardening, a new highlight and new stage in the development of gardens nowadays. Meanwhile, its close relation with buildings themselves determines its status of an important link in building design. It aims to create an activity space with more creative and novelty for citizens, increase natural factors and green coverage ratio, beautify environment, protect and improve urban environment, perfect urban ecological system, and promote sustainable development of urban economy, society, and environment. Furthermore, roof garden can edify sentiment and build a good city image.

2.1 Advantages of roof greening

As a special greening form, roof greening stores water and covers buildings' top platform with soil to build garden landscape. With adjustment of current industrial structure, acceleration of urbanization, steep rise, and growing of urban population, the environmental quality of cities is increasingly worsened, and urban green area is in extreme shortage. Roof greening is not only an effective method to extend green area to air, economize land, and exploit urban space, but also is a perfect combination of architectural art and landscape art, and plays a more nonnegligible role in protecting urban environment and improving residence environmental quality.

2.1.1 *Environmental advantages*
a. Purify air and improve climate:
Relative to plants on the ground, the plants in roof are placed high to form stereoscopic and multi-tiered purification system so as to absorb CO_2 and sulfur dioxide and absorb dust in air to effectively reduce the content of solid particles in air. Roof greening can also improve microclimate environment in local region and mitigate urban heat island effect to purify air and improve urban air quality and ecological environment, which is incomparable to plants on the ground. Added to this, roof greening can enhance sound insulation and noise reduction of top level of buildings by reducing the average noise by 3–8 db.

b. Store rainwater and alleviate pressure of municipal drainage system:
Roof greening can store more rainwater through soil horizon and drainage layer to maintain moisture and meet the requirements of irrigation based on the precondition that the supporting capacity of buildings permits so. Meanwhile, roof greening can effectively reduce roof drainage by water retaining effect of plants and water collection to prevent overrunning sewage from entering lake and stream and reduce pressure of drainage system in urban sewer.

c. Low carbon and energy conservation:
The roof makes a traditional pitch against heat rejection and thermal insulation, thereby aggravating urban heat island effect. However, if green plants are planted on roof, soil and vegetation can effectively absorb heat of sun. And the evaporation of leaves can lower air temperature in surroundings and absorb CO_2 through photosynthesis, which can buffer the trend of temperature rise in built-up cities compared with suburbs (Sun, 2009). Roof greening has ecological effect of common greening and serves as the "air conditioner" of buildings. The green roof can absorb and reflect solar radiation through plants to cut air-conditioning cost in summer. In summer, the day temperature of green roof is 30% lower than that of the traditional roof. In winter, the additional heat-insulating layer can cut warming cost. Statistics shows that green roof can reduce the average cost of buildings' energy consumption in winter by 10%.

d. Keep ecological balance:
The plants on green roof can provide habitat for some species, connect ecological island between cities and villages, and provide habitat for various birds and insects, such as the disappearing bee population. In addition to providing service for cities, it offers growth space for animals and plants, thus serving two ends.

2.1.2 *Commercial value*
a. Prolong roof's lifetime:
Roof greening is capable of protecting buildings' waterproofing effect and prolonging its lifetime. The plant protective blanket is more effective than waterproof membrane, so the green roof has a lifetime at least twice that of traditional roof. Roof greening can also shield ultraviolet radiation, alleviate damage to roof by shock heating and water accumulation. From a long-term view, green roof can save money for owners themselves and the whole cities (Shi, 2006).

b. Enrich city landscape and economize land:
Roof greening and ground greening constitute city landscape with hierarchical structure, enrich contour line of building complex, and beatify urban environment. Roof greening makes the most of space and saves urban land. Besides, the seasonal change of plants vests the buildings with aesthetic sense in time and space to conceal the inaesthetic roof and decorate the prim buildings into artworks with high appreciation value.

c. Bring fun to people's life:
Roof garden can be the "hanging garden" open to the public in real sense, which has no bustling street or cars' noise. People can walk, have parties, read, and enjoy serenity in obstreperous cities. Roof greening appears as a special greening form to bring novelty and uniqueness, visual aesthetic fun to users and neighbors, which reduces the distance between people and nature, and alleviates exhaustion and pressure in work. Besides, roof greening covers urban architectures with vivid overdress to substitute prim color of constructional materials and make cities more energetic.

d. Multifunctional roof greening brings diversified economical benefits:
Roof greening has varying types and functions, such as roof garden, roof greenhouse, and roof farm, according to which different economic values can be created. The hotels can plant seasoning herbs, flowers, and vegetables for the use by itself to save procurement cost. A restaurant can have operation space on green roof to provide more comfortable space for diners and give them more serenity and comfort. As a big selling point, roof greening is more competitive in the market and the apartment blocks with green roof will enjoy a popularity boom (He, 2007).

2.2 *Difficulties of roof greening*

The roof, which is called as "the fifth facade of buildings," has always been virgin soil to be reclaimed in cities, and ignored and forgotten for long. On the one hand, urban greening area and water surface area are being encroached by more and more dense buildings. On the other hand, many roofs are still undecorated and unutilized, which is a dead spot in urban construction and management.

2.2.1 *Natural conditions*

a. Temperature and humidity condition:
As roof is at high point where wind velocity is greater than that on the ground and evaporation of water is quick; the higher the roof is from the ground, the worse the greening condition is.

b. Limitation in gardening and plants selection:
As roof is not provided with identical soil with ground, so when selecting plants, one must avoid deep-rooted or fast-growing megaphanerophytes, thereby limiting the option of plant types.

2.2.2 *Technical difficulties*

a. Problem of roof greening's waterproofing work:
The main reasons of housing leak include waterproof layer damage, housing cracks, and long-term soak in the place where there is not waterproof. Housing cracks is the problem of housing quality; the roof floor has the waterproof function, and leak cannot happen when there is water inside houses. As long as the housing quality is ensured, the waterproof issue of the green roof is to prevent the destruction of the waterproof layer. Therefore, the problem can be solved totally in choosing plants and increasing the corresponding protective measures. The plants must have shallow roots and no crocheting ability, or other ways can prevent the puncture and damages of plants' roots to buildings. On the contrary, roof greening avoids the direct exposure of the floor to sunshine and plays the role of heat insulation, which greatly reduces the expansion and contraction caused by the temperature difference between day and night and avoids the leaks of floor cracks caused by the expansion and cold contraction.

b. Construction and maintenance cost of roof-greening project:
The common ways of roof greening include roof landscape, roof garden, roof vegetable farm, and roof grassland. The construction cost of roof landscape is very high, and it is one of the roof green modes that is close to natural ecology. Roof garden is more applied in low-rise buildings or more high-end places. Roof vegetable farm can plant vegetable in addition to roof greening, which let people experience the pleasure of planting in the their spare time. The cheapest way of roof greening is the roof grassland. The maintenance cost should be considered more than the construction cost. For a regular roof garden, it needs to hire a specific person to maintain and manage. Because

Table 1. Weight of each type of roof greening.

Type of roof greening			
Roof landscape	Roof garden	Roof vegetable farm	Roof grassland
Weight ($/m^2$) 300 kg	150–200 kg	60–100 kg	20–50 kg

it is on the roof, the maintenance and management fee are higher than those of land greening. Low cost, less maintenance fee, and high efficiency are the advantages of the large-scale promotion way-out for urban roof greening.

c. Problem of the bearing force of roof greening:
The average bearing capacity of a standard floor is 200 kg/m^2, which is up to 300 kg/m^2 for some business buildings and 100 kg/m^2 for old buildings. The average weight per square meter of roof landscape is over 300 kg, the one of roof garden is between 150 and 200 kg, the one of roof vegetable farm is between 60 and 100 kg, and the one of roof grassland is from 20 to 50 kg (Table 1). Only roof vegetable farm and grassland have no bearing concerns, and the large-sized roof greening must consider the bearing capacity (Zhu, 2013).

3 MEASURES OF ROOF GREENING IN NORTHERN CITIES

In North China, the four seasons are distinct with windy and dry spring and autumn, hot and rainy summer, and frosty and snowy winter. Thus urban ecological environment changes greatly. Roof greening in northern cities cannot only help to save energy and insulate the buildings, but also is conductive to maintaining urban climate and balance of whole ecosystem, which is of great significance to environment's sustainable development. However, the drastic change of climate and vegetables in the four seasons there also lead to great difficulty in the design and implementation of green roof, and requires the designer to adjust measures to local conditions, prudently selecting greening form suitable for North China.

Because of the development level of productive forces, people's lack of understanding on roof greening and other factors, the exploration of roof green in China, especially in North China, developed slowly. From the development of roof greening in North China, the completion of the outdoor roof garden of Great Wall hotel in Beijing in 1983 marked the start of roof greening in the region. With the continuous development and progress of society, people gradually strengthened the understanding of roof greening. However, because of the relatively harsh geographical environment in North China, changeable climatic conditions, and increasing high-rise buildings, more factors must be taken into account in the process of promoting roof greening in the northern cities.

3.1 The plants of roof greening should be in line with specific geographical environment and climate in North China. Difficulties for roof greening

Plants of roof greening are the necessary condition to realize the urban roof greening. The choice of the plants is directly related to the success or failure of roof greening in northern cities, which is due to the special geographical environment and climatic conditions in the north. From the overall environment of the north, the climate is relatively dry and the soil is relatively poor, especially the winter weather is cold, so the choice range of plants is relatively small (Yang, 2000). Especially, in recent years, because of the excessive deforestation of natural forests and more pollution, a strong heat island effect forms, which needs the environment in northern cities to be changed. Therefore, while making roof greening, it is necessary to have knowledge about sunshine; especially, the exposure, big winds, and relative barren soil environmental conditions should be suitable for the growth of plants with perennial roots. In that case, the investment is small and a better landscape effect can be achieved; planting once can last for years. At the same time, the plants should be able to clean the air and enhance urban water drainage and storage, which can beautify the environment and play the role of greening. Roof greening is a green mode of water saving, energy saving, and land saving, but it is designed especially for the roof with specific bearing capacity so as to meet the requirements of lightweight of the floor, easy moving, and simple installation and stability in the process of roof greening. In addition, if the solar energy storage equipment is fixed, the roof greening cannot be adopted, because the growth of climbing plants may affect the operation of the equipment. In that case, some short and erect flowers can be planted in the gaps. Without the installation of solar energy, if the roof of buildings inclines to a certain angle, the climbing plants will cover the roof, which is more practical and safe than the potted flowers and grass.

3.2 Choose different greening modes according to different building styles

Buildings in North China have their own unique styles with simple, honest, and decent layout. The design of roof greening should conform to the characteristics of the buildings and echo with the entire environment. The design should be considered from aesthetics and landscape construction, because it has the strong ecological role and is closely related to the development of the whole city. From the overall environment in the north, because of the influence of environmental factors, irrational urban planning, and economic development in the urban development, the appearance is often neglected. The design of old buildings did not consider the later roof greening, so the harmony with the original architectural style must be considered to suit the overall environment. If roof greening can play its role in urban environment, the urban solid landscape can be enriched and people's psychological and visual experience can be adjusted. Meanwhile, combining the real condition in northern cities, the roof greening can alleviate the heat of summer and the cold of winter and change the living conditions of the public. Promoting the roof greening in the north needs to make land greening, air greening, and land greening echo with each other, which will produce a good effect.

3.3 Choose proper green mode according to the real roof condition

From the practice of roof greening, not all roofs are suitable for roof greening, especially that the building styles in the region and the initial designs did not think of the roof greening to decorate buildings and need scientific guidance in roof green. For example, in the real process of roof greening, professional identification of the force, bearing capacity, and slope is needed. In same time, referring to the scientific investigation data provided by the greening departments, instructions from the professionals in the process are needed, which can make roof greening better (Zhou, 2013).

As for the light roof with low bearing capacity and no green design in advance, it is more likely to plant a small amount of dense planting grass. For the roofs with strong bearing capacity, it is suitable to plant the shrub and garden greening trees, and the method is applicable for the roof with large areas, like high-end hotels and high-rise buildings, which make the roof space changeable and produce multiple levels and colorful effect. The combined methods can be used, placing the plants freely, mainly on the four corners and potted flowers are placed on the side of load-bearing walls, which is more flexible. The choice of method is determined by the load of the roof, location of the load walls, flow of people, the surrounding environment, and use purpose.

4 CONCLUSIONS

Recently, roof greening as a design method of the environment and economic win-win has gradually attracted attention and welcomed by the people. The progress of roof greening in the north is still in the stage of exploration, which needs designers to shift thoughts and concepts, improve their professional knowledge, actively innovate, and explore more ways to solve the problems. Every one improves the ecological environment consciousness so that it can promote the overall improvement of urban ecological environment, creating a better future for urban development.

REFERENCES

Chengtian Sun, Jiong Shi, Zhonghua Zhu, Weidong Li. Environmental Pollution & Control, *Analysis of Environment benefit and Promotion of Urban Roof Greening*, 8 (2009).

Dongqing Zhu, Zhiyuan Zhu. The Ninth International Conference on Green and Energy-Efficient Building, *Overview on system technology of green building's roof*, 4 (2013).

Jian He, Qing-ming Chen. Construction Conserves Energy, *Construction Conserves Energy and Roof Greening*, 8 (2007).

Linyuan Zhou, Yuhui Di. Contamination Control & Air-conditioning Technology. *Present Research Situation and Prospect Analysis of Green Roofs*, 9(2013).

Tao Shi, Hui-sheng Shi, Wilfried Schumacher. Journal of Zhejiang University of Technology, *Research on Roof Planting*, 2 (2006).

Yupei Yang, Min Jin. Chinese Landscape Architecture, *Developing roof greening, increasing urban afforestation*, 6 (2000).

Advances in Energy Science and Equipment Engineering II – Zhou, Patty & Chen (Eds)
© 2017 Taylor & Francis Group, London, ISBN 978-1-138-71798-5

Dependence of tourism development planning of a village on a scenic spot—taking Nan Jiaopo village as an example

Shihao Lin
Urban and Rural Construction College, Agricultural University of Hebei, Baoding, China

Yanbo Zhao
Xingtai Planning and Design Research Institute, Xingtai, China

ABSTRACT: Similarly to the villages around the traditional scenic spots, those located on the way to the scenic spots would benefit from them. With this advantage and combining the characteristics of local area, these villages, which are defined as "tourism village depends on scenic spot," could also get a good opportunity for further development. Nan Jiaopo village is located in Pingshan County, Shijiazhuang, Hebei Province, China. The main purpose of this paper is to introduce the aim and methods of designing and planning village tourism by taking Nan Jiaopo village as an example. The layout of the design would be based on the current situation of the natural conditions of the village. Meantime, through the analysis of tourist sources, cultural mining, ecological protection, facilities construction, industrial transformation, and other aspects of research, the planning and layout of the village would be clear, which is helpful for the further development of this village.

1 INTRODUCTION

1.1 *The concept of a tourism village depends on the scenic spot*

The primary definition of Tourism Village Depends on Scenic Spot is the villages around the scenic area, taking the scenic spot as the core, and the development of villages relies on the tourism resources of the scenic area and rural tourism resources.

In this paper, the definition of Tourism Village Depends on Scenic Spot refers to the village that is on the way to the scenic spot, with agricultural characteristics, cultural characteristics, landscape features, and other characteristics. The main function of these types of villages is to provide leisure facilities and accommodation for the tourists.

1.2 *The current situation of a tourism village depends on the scenic spot*

After several years of operation, some problems such as poor management and low-level expansion have occurred in the Tourism Village Depends on Scenic Spot. However, because of the increasing demanding of getting closer to nature, to return to innocence, and the consumer psychology to get relax and entertainment, the development of some tourism villages is in a lopsided way with a high speed.

2 KEY POINTS OF THE TOURISM VILLAGE'S DESIGN AND PLANNING

2.1 *Definition of tourist groups*

The main tourist source of the Tourism Village Depends on Scenic Spot come from the scenic spot and the rest from the nearby core city, as shown in Figure 1.

2.2 *Rural feature mining*

The attraction of rural to urban residents is mainly based on the local flavor farmhouse characteristics. In order to maintain long-term attraction for visitors from the city, we must combine with market

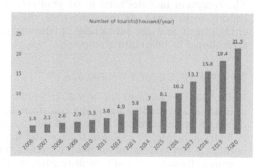

Figure 1. Number of visitors and forecast of Nan Jiaopo village.

demand and resources condition, developing native characteristics, including residential construction, agricultural industry, ruins, folk traditions, food, and folk art. To display and experience the resources, so as to make it have a high cultural taste, artistic style, and popular science significance, the local culture can be protected, developed, and spread. Meanwhile, it can also develop a variety of types, abundant local characteristics of organic agricultural products and tourist arts and crafts, which contains the characteristics of the local agricultural tourism resources and the characteristics of agricultural culture.

2.3 Rural ecological protection

The carrying capacity of resources and environment must be considered and the protection of ecological environment must be strengthened to achieve reasonable development in the capacity of the environment. Specific methods include: before development, evaluate the rural tourism resources, environment carrying capacity, and tourist market scale and estimate the passenger flow, which would be used as the basis for the development scale design and the formulation of environmental protection strategy. In addition, during the development process, make a long-term monitoring and management for the resources and environment. For the natural environment, which has been damaged, emergency measures should be taken for rectification and repairing.

2.4 Planning and construction of tourism facilities

Tourism facilities can provide accommodation, travel, entertainment, shopping, and other services for tourists. The service objects include tourists and local residents. Tourism Village Depends on Scenic Spot has the dependence on the core scenic spot, thus the planning should be based on the combination of tourism development with the core area. The reception facilities need to be gradually and orderly constructed. At the same time, the rural environment and the protection of the original style should be taken into account.

2.5 Rural industrial transformation

With the development of rural characteristic industry as the leading force, the transformation of single agricultural and forestry to the tourism industry is helpful to mobilize the tourism development power and realize the planning effect. The resource characteristics and rural tourism market should be analyzed, and the local characteristics of industrial development must be considered.

3 CASE STUDY—THE DEVELOPMENT PLANNING OF NAN JIAOPO VILLAGE, PINGSHAN COUNTY, SHIJIAZHUANG

3.1 An overview of the current situation of Nan Jiaopo village

Nan Jiaopo village is located in the west of Pingshan County, adjacent to the Xibaipo red spot and Nandian County. It has a good traffic system passing through by 202,241,301 provincial highways.

3.2 Design principles

Ecological principles: On the basis of the natural situation and local conditions, resources are reasonably used to set up facilities.

The principle of human nature: According to the behavior of visitors, the rational organization of the relationship between functional areas straightens the traffic system.

The principles of sustainable development: Reasonable controlling of timing and scale of development. Setting aside flexible development space for the development of scenic spots.

3.3 General layout

3.3.1 Planning and positioning
With mountains and rivers as the background, with rich natural resources and diverse landscape as the carrier, with folk culture, health care, pastoral scenery, and leisure experience as features, the village would gradually develop into the first development characteristic village surrounding Shijiazhuang, as shown in Figure 2.

3.3.2 Planning structure
According to the situation of tourism resources and future tourism development goals in the planning area, the layout of the planning area is defined as

Figure 2. Aerial view of the village.

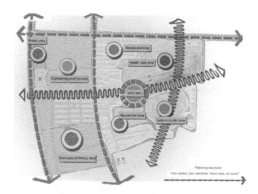

Figure 3. Village structure analysis diagram.

a "one center, two corridors, three axis, six zone" of the dynamic development pattern, as shown in Figure 3.

3.3.3 *Traffic organization*
The village road is divided into traffic road and tourist road. The traffic road is mainly used for village transport links, which would form a convenient transportation system. The tourist road is mainly used for tourism, which is used to create a unique and interesting tour route.

3.3.4 *Green planting*
1. Road plant disposition
   In the village, the road are the links to connect each functional area, leading visitors to different attractions one by one. Therefore, on both sides of the road would plant various plants and flowers with different size and vivid color.
2. Building perimeter plant disposition

Within the village, the main building community is concentrated in the residential area and the supporting facilities area. The vegetation in this area is dominated by arbor and shrub, which have good growth. Some flower beds, flower pot, and flower diameter would be built around the buildings as embellishments.

3.3.5 *Supporting tourism facilities*
In order to improve the function of the planning area facilities and provide humanized service, some shops, signage, bins, public toilets, telephone booths, and rest area would be built in the planning area.

3.4 *Industrial development planning*

3.4.1 *Agriculture industry*
The main agricultural industry in Nan Jiaopo village is the cultivation of fine fruits, vegetables, and ecological farming. With the increase in the

production and sale of green organic products as the main line, to create a characteristic agricultural industry combines with tourism and picking.

3.4.2 *Leisure industry*
The leisure industry of Nan Jiaopo village mainly includes rural life and agricultural sightseeing picking experience area, folk theme park, and catering accommodation area, which, combined with the local cultural and natural landscape construction features of the industrial park, the tourists would have a reality version of happy farm.

3.4.3 *Healthcare industry*
The healthcare industry in the Nan Jiaopo village mainly relies on the status quo of the Happiness Institute for expansion. To make the people in the Happiness Institute from the rural elderly to the trend of integration of rural and urban elderly, improve the living environment of the elderly and increase the activity places and activity projects.

4 CONCLUSIONS

The development of the Tourism Village Depends on Scenic Spot is not only the opportunity of economic development for rural areas, but also the objective needs for the opening and diversified development of the tourist attractions. In recent years, scenic tourist service facilities have been relocated. From one aspect, it reduces disturbance to the ecological environment of the scenic spot. Meanwhile, it also provides an opportunity for the development of tourism service industry in the edge zone of the scenic spot. In addition, one of the key points for the development of rural tourism is that the villagers are required to attend for the protection of local scenery and culture. Therefore, in the construction of the tourism village, the interests between the enterprises and local villagers should be fully coordinated, and the government should play an active role in the coordination, hence that the interests of all parties would achieve a balanced and win-win situation.

REFERENCES

Hejiang Shen, Xuejing Wang, Min Wang. The Development and Research of Rural Tourism Based on "Leisure Experience" [J]. International Journal of Business and Management, 2009, 2(6).

Jingchun Duan, Problems and Countermeasures in the development of rural ecological tourism in China [J]. journal of anhui agricultural sciences. 2008(10).

Lin Hui, Zhong Hua, Countermeasures of Tourism Development on Rural Issues Dabie Mountain Scenic Area—Taking Yingshan Taohuachong Scenic Farm as an example, Journal of Guizhou Commercial College, 2013, 26(3):54–57.

Liu Cailing, The key points and ideas of the development of the Tourism Village Depends on Scenic Spot—Taking Nanzhao County of Henan province stone kiln village as an example, Journal of Zhengzhou Institute of Aeronautical Industry Management, 2011, 29(4):134–137.

Mimi Yang, Maoxing Long, Jianping Liu, Discussion on the development of rural tourism in the area of scenic spot [J]. Ecological economy. 2009(01).

Shi Guoling, Research on The Planning and Design of Tourism Village Depends on Scenic Spot—Taking ShangHu village of BaoShan Scenic Spot for Example, Fujian Architecture & Construction, 2014(6).

Advances in Energy Science and Equipment Engineering II – Zhou, Patty & Chen (Eds)
© *2017 Taylor & Francis Group, London, ISBN 978-1-138-71798-5*

Distance measurement on gnomonic projection charts based on equidistant circles

Gaixiao Li & Yidong Zheng
Department of Hydrography and Cartography, Dalian Naval Academy, Dalian, China
PLA Key laboratory, Dalian, China

Xiaoguang Dong
Navy Press, Tanggu Tianjin, China

Zhiheng Zhang
Department of Hydrography and Cartography, Dalian Naval Academy, Dalian, China
PLA Key Laboratory, Dalian, China

Guimin He
Navy Press, Tanggu Tianjin, China

ABSTRACT: In view of the very serious length deformation in the Arctic region on gnomonic projection charts, which makes it unable for distance measurement to be used on the medium- and small-scale charts, and on the basis of the analysis of the characteristics of the deformation of gnomonic projection, we discover the law that the length deformation is equidistant with the tangent point, which is on the equal-altitude circle, presenting an exact and simple distance measurement method for measuring the distance of great circle route on gnomonic projection charts based on equidistant circle. The results of the experiment indicate that in the range of error allowed, the method can quickly measure the distance of great circle route on Arctic gnomonic projection charts and meet the increasing urgent demand of navigation in the Arctic region.

1 INTRODUCTION

There are abundant natural resources in the Arctic region. The Arctic region abounds in natural resources, and the recession of Arctic ice in recent years has made it possible to explore new economical and convenient sea lane (Zhang, 2009; Global, 2008). These new routes will provide convenient shipping routes for Chinese ships to both sail in the Arctic Ocean and explore Arctic resources, with a prospect of enormous potential economic benefits. Charts are the foundation of safe ship navigation on the sea, and at present, Mercator projection is mainly used in most navigational charts, but due to the high latitude of the Arctic region, fictitious graticule fast converge to the pole, and the complex geographical environment, thereby making the Mercator projection, which has serious length deformation, inapplicable in this area (Global, 2008; Wen, 2015). At present, polar charts published outside China mostly employ two types of chart projection: Mercator Projection and Gnomonic Projection. Generally,

in accordance with China's rules and regulations in choosing projection patterns for sea areas of high latitude, Mercator Projection is used in plotting sea areas ranging from 70° to 80°, and Gnomonic Projection is adopted in sea areas from 80° to 90°. Therefore, before any suitable polar projection is found, in 80°–90° north latitude area, Gnomonic projection is still the main projection for the polar chart (Li, 2012; Wang, 2002).

Because the great circle route is a straight line on the gnomonic projection plane, it is very suitable for the ship on oceangoing voyages. Therefore, the gnomonic projection is very suitable for making small-scale polar chart. However, as a result of the length deformation of gnomonic projection, which sharply increases with the decrease in the latitude, it is difficult to measure the distance of great circle route directly on the gnomonic projection chart. The current distance measurement method is not suitable for middle- and-small-scale gnomonic projection charts, which seriously affects the actual application of the gnomonic projection charts. By analyzing the gnomonic projection deformation and the limitations of the measuring methods of current gnomonic projection charts and according to the laws of the gnomonic projection length deformation, this paper proposes a method of distance measuring on gnomonic projection charts based on equidistant circles to measure the distance of the great circle route on the gnomonic projection chart rapidly.

2 GNOMONIC PROJECTION DEFORMATION ANALYSIS AND THE CURRENT DISTANCE MEASUREMENT METHODS

2.1 Gnomonic projection deformation analysis

Gnomonic projection is one of the projection patterns commonly used in high-latitude navigation. Its distinctive feature is that the great circle route is a straight line on the gnomonic projection plane. At present, the gnomonic projection can plot the entire polar area. Its projection plane has got a tangent point at a certain point on the earth and a point of sight at the earth's spherical center. Its projection formula is as follows (Li, 1993; Hua, 1985):

$$\begin{cases} x = \dfrac{R(\tan u - \tan u_0 \cos \omega)}{\tan u \tan u_0 + \cos \omega} \\ y = \dfrac{R \sec u_0 \sin \omega}{\tan u \tan u_0 + \cos \omega} \end{cases} \quad (1)$$

where x and y are the coordinates on the chart, u and ω represent spherical latitude and parallel, u_0 and ω_0 represent, the latitude and parallel of the tangent point, respectively, and R is the radius of the sphere.

For the shape of parallel and latitude on the gnomonic projection plane, the equation of parallel can be acquired by removing the latitude u, and the equation of latitude can be acquired by removing the parallel ω in formula (1). Therefore, the equation of latitude and parallel is derived as follows:

$$y = \tan \omega (R \cos u_0 - x \sin u_0) \quad (2)$$

$$\sin^2 u(1 + x^2 + y^2) = (\sin u_0 + x \cos u_0)^2 \quad (3)$$

It can be seen from formula (2) that the equation of parallel on the gnomonic projection plane is a linear one. This shows that parallel is a straight line on the gnomonic projection plane and its slope is—$\csc u_0 \tan \omega$.

When $y = 0$, it can be obtained that $x = R \cot u_0$; this point is the intersection of each parallel and it is $R \cot u_0$ away from the tangent point.

It can be seen from formula (3) that the equation of latitude on the gnomonic projection plane is elliptic, hyperbolic, or parabolic. Its shape is determined by the tangent point.

It is known from the map projection that the projection principal direction is consistent with the direction of altitude circles and vertical circles of ellipsoid, the length deformation of the vertical circle is μ_1, the length deformation of the altitude circle is μ_2, and the extremal length deformation of two directions is, respectively, as follows:

$$\begin{cases} \mu_1 = -\dfrac{d\rho}{Rdu} \\ \mu_2 = \dfrac{\rho}{R \cos u} \end{cases} \quad (4)$$

The extremal length deformation of two directions is acquired by integrating formula (4) as follows:

$$\begin{cases} \mu_1 = \csc^2 u \\ \mu_2 = \csc u \end{cases} \quad (5)$$

The area deformation P and maximal angle deformation W are, respectively, as follows:

$$P = \mu_1 \cdot \mu_2 = \csc^3 u \quad (6)$$

$$\sin \frac{W}{2} = \frac{\mu_1 - \mu_2}{\mu_1 + \mu_2} = \tan^2 (\frac{\pi}{4} - \frac{u}{2}) \quad (7)$$

According to the deformation formula of the gnomonic projection, the deformation table is calculated as shown in Table 1.

It is evident from Table 1 that this projection falls into arbitrary projection, whose azimuth, distance, and area all have deformation and the length deformation sharply increases with the decrease in latitude, the length deformation reaches infinity at the equator. Although the length deformation is larger, there is a regularity to follow, parallel to altitude, which has equal distance from the tangent point, the length deformation of each point's length deformation is consistent, and their length can be measured with the length deformation ratio μ_2 of the parallel of altitude.

2.2 The current distance measurement method on the gnomonic projection chart

The current distance measurement method on the gnomonic projection chart is done mainly by putting the kilometers or sea miles drawing on the parallel, which is through the tangent point. In the process of measurement, the distance of the great

Table 1. Length deformation of gnomonic projection.

u	0°	15°	30°	45°	60°	75°
μ_1	∞	14.928	4.000	2.000	1.333	1.072
μ_2	∞	3.864	2.000	1.414	1.155	1.035
P	∞	57.676	8.000	2.828	1.540	1.110
W	180°00′	72°09′	38°57′	19°45′	8°59′	1°59′

circle route on the chart may be taken as radius and the tangent point as the center of a circle, thus making a circle and crossing the parallel to one point, whose value is the actual distance of the great circle route. However, this distance measurement method can only be used on large-scale charts, and the limit value of its scale is calculated as follows.

On the parallel, which is through the tangent point, if putting ω = 0 into the x coordinate of formula (1), it can be obtained as follows:

$$x = R\frac{\tan u - \tan u_0}{\tan u \tan u_0 + 1} = R\tan(u - u_0) = R\tan \Delta u \qquad (8)$$

where $\Delta u = S/R$, S is the complete mileage of each scale on the parallel, R is the radius of the earth measured in kilometers.

Assuming that the tangent point is located in the map border of the gnomonic projection chart and setting the maximum measurement distance at 100 cm on the chart, the radius of the earth is 6371 km. If the distance from the tangent point to any point on the chart is treated as actual distance according to the scale zoom out and the difference between them is less than 0.01 mm, which is the minimum chart drawing error (Zhang, 2015) that can be obtained as follows:

$$R\tan \frac{S}{R} - S \le 0.0001C_0 \qquad (9)$$

Formula (1) is divided by the radius of the earth, R:

$$\tan \frac{S}{R} - \frac{S}{R} \le 0.0001\frac{C_0}{R} \qquad (10)$$

The actual distance of 100 cm on the chart is exactly C_0 m. Putting C_0 as S into formula (10), it can be obtained as follows:

$$\tan \frac{C_0}{R} - \frac{C_0}{R} \le 0.0001\frac{C_0}{R} \qquad (11)$$

$\tan \frac{C_0}{R}$ is expanded into Taylor series, and by taking its first three values, it can be obtained as follows:

$$2(\frac{C_0}{R})^4 + 5(\frac{C_0}{R})^2 - 0.0015 \le 0 \qquad (12)$$

Formula (12) is derived using the MatheMatica software, and the corresponding value can be obtained as follows:

$$\frac{C_0}{R} \le \sqrt{\sqrt{0.00075 + 1.5625} - 1.25} = 0.017319469 \qquad (13)$$

By putting $R = 6371000$ m into formula (13), the following result is obtained:

$$C_0 \le 110342.337 \qquad (14)$$

It is viewed from the result that when the scale of the gnomonic projection chart is 1:11 0 000 or larger with the current chart size of 980 mm × 660 m (General, 1999), the distance from the tangent point to any point on the chart can be treated as the actual distance according to the scale zooming out; however, in practice, when seamen plot the great circle route on the gnomonic projection chart, they adopt the small- and medium-scale chart (1:500,000 to 1:1000,000), which results in existing limitations in practical applications. Therefore, other methods must be explored to measure the distance on the small-scale gnomonic projection charts.

3 DISTANCE MEASUREMENT ON GNOMONIC PROJECTION CHARTS BASED ON EQUIDISTANT CIRCLES

According to the deformation of gnomonic projection, the distance measurement of the great circle route can be treated as that on the parallels bearing the same distance from the tangent point because they are on the altitude circle, which has the same distance from the tangent point, the actual distance of the corresponding distance on the chart is equal. Therefore, the latitude value on the parallel has been subdivided for easy reading of the corresponding actual distance. In order to find the parallel, which has the same distance from the tangent point, the perpendicular line, which is from the tangent point to all parallel, must be drawn so that all the tracks of vertical foot can be determined. With the track determined, assume the tangent point as the center of a circle and the distance from the tangent point to the great circle route as radius. Through drawing an arc cross with the track of vertical foot, the parallel, which has the same distance from the tangent point, will be located, then the distance of the great circle route can be measured on the parallel. The ideas are presented in detail below.

The slope m of the parallel is $-\csc u_0 \tan \omega$ on the gnomonic projection plane, which is calculated by formula (2), the slope of the meridian perpendicular is negative reciprocal of the slope of the meridian and it passes through the origin of coordinates; therefore, the equation of the perpendicular is as follows:

$$x = -\frac{1}{m}y \tag{15}$$

Combining formulas (2) and (15) and canceling the longitude, we obtain:

$$(x - \frac{1}{2}R\cot u_0)^2 + y^2 = (\frac{1}{2}R\cot u_0)^2 \tag{16}$$

It is evident that formula (16) is a circle, whose center is $(\frac{1}{2}R\cot u_0, 0)$ and radius is $\frac{1}{2}R\cot u_0$. As the circle shows the track of vertical foot from the tangent point to all parallel, it can be used to find the parallel, which has the same distance from the tangent point, so it can be called the equidistant circles.

The method of distance measuring on gnomonic projection charts based on equidistant circles is as follows:

As shown in Figure 1, line AB is the great circle route between A and B on the chart; if the actual distance of the great circle route is measured, draw a line perpendicular to line AB from the tangent point O, whose vertical foot is point K. Measure the length of OK with compasses, assume the tangent point O as the center of the circle and the length of OK as the radius, and draw an arc that crosses the equidistant circle at point K_1. On the parallel, which is through the point K_1, make the length of KB equal to the length of K_1B_1 and KA equal to K_1A_1; thus, the latitude difference of A_1B_1 is the actual distance of the great circle route AB with sea mile as its unit.

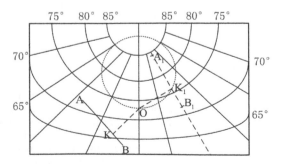

Figure 1. The method of distance measuring on gnomonic projection charts.

4 CASE ANALYSIS

In order to verify the correctness and feasibility of the above methods, this thesis selects the Arctic gnomonic projection chart for experimental analysis.

As shown in Figure 2, the case area is located in north latitudes between 75° and 85°, and the distance of the great circle route AB is reckoned. The scale of the chart is 1:1 000 000, the coordinates of the tangent point is (80°N, 80°E), the radius of the earth (R) is 6,371,116m, the starting point A of the great circle route is (79°N, 70°E), and the end point B is (82°N, 65°E).

1. According to formula (16) and the gnomonic projection chart parameters, the center (x, y) and the radius r of the equidistant circle are determined as follows:

$$\begin{cases} r = x = \dfrac{R}{2C_0}\cot u_0 = 56.170 \text{ cm} \\ y = 0 \end{cases}$$

2. According to the method proposed in Part 2, which measures the great circle route, the results are shown in Table 2.

It is evident from Table 2 that the measurement result error is small and completely meets the requirements of modern chart application. In addition, this method is simple and can be used for

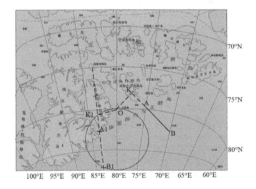

Figure 2. Distance measurement of great circle route on the chart.

Table 2. Error results of the distances.

	Distance (sea mile)
Measured distance on chart	186
Actual distance	186.6
Error value	0.6
Relative error	0.3%

measuring the distance of the great circle route on small- and medium-scale gnomonic projections.

5 CONCLUSIONS

Charts have been deemed as the eyes of sailors, and the support of charts is indispensible to Arctic navigation. The thesis analyzes the gnomonic projection deformation and the limitations of the distance measurement method of current gnomonic projection charts. According to the laws of the gnomonic projection length deformation, this thesis proposes a method of distance measurement on gnomonic projection charts based on equidistant circles to measure the distance of the great circle route on the gnomonic projection charts.

REFERENCES

General Administration of Quality Supervision, Inspection and Quarantine of the People's Republic of China. Chinese Nautical Charts' Complication Specification [S]. Beijing: Chinese Standard Press, 1999.

Global Business Network (GBN). The Future of Arctic Marine Navigation in Mid-Century [R]. Scenario Narratices Produced for the Protection of the Arctic Marine Environment (PAME) Working Group, Singapore, 2008.

Hua Tang. Mathematical Foundation of Navigational Charts [R]. Navigation Guarantee Department of Naval Headquarters, Tianjin, 1985.

Li Guozao, Yang Qihe, Hu Dingquan. Map Projection [M]. Beijing: People's Liberation Army Press, 1993.

Li Shujun, Zhang Zhe, Li Huiwen, et al. Research on Compilation of Nautical Charts of Arctic Regions [J]. Hydrography and Charting, 2012, 32(1):58–60.

Wang Qinghua, E Dongchen, Chen Chunming, et al. Popular Map Projection in Antarctica and Their Application [J]. Chinese Journal of Polar Research, 2002, 14(3):226–233.

Wen Chaojiang, Bian Hongwei, Bian Shaofeng. A Distance Measuring Method on Polar Stereographic Charts Based on Equidistant Circles [J]. Wuhan University Gernal, 2015, 40(11):1504–1508.

Wen Chaojiang, Bian Hongwei, Chen Zhihong, Qian Duo. Study on Availability of Mercator Projection in Polar Nautical Charts' Complication [J]. Hydrography and Charting, 2014, 34(3):56–59.

Zhang Xia, Tu Jingfang, Guo Peiqing, et al. The Economic Estimate of Arctic Sea Routes and Its Strategic Significance for the Devolpment of Chineses Economy [J]. China Soft Science, 2009(S2):87–89.

Zhang Xiaoping, Bian Shaofeng, Li Zhongmei. Comparisons Between Gauss and Gnomonic Projection in Polar Regions [J]. Geometrics and Information Science of Wuhan University, 2015, 40(5):667–672.

Advances in Energy Science and Equipment Engineering II – Zhou, Patty & Chen (Eds)
© 2017 Taylor & Francis Group, London, ISBN 978-1-138-71798-5

A study of college construction engineering project management pattern

Xijin Liu

Infrastructure and Planning Department, Beijing Jiaotong University, Beijing, China

ABSTRACT: This paper summarized the commonly used mode of construction engineering project management in China and around the world and analyzed their advantages and disadvantages in detail. It analyzed the characteristics of three kinds of situation: renovation projects, single project, and new campus construction project. Then, the paper puts forward three management patterns of former situations, which provides the university capital construction management theoretical basis, with a high reference value.

1 INTRODUCTION

The state council issued "overall plan of pushing forward the world first-class university and the first-class discipline construction" on 24 October 2015. The construction of the world first-class university and the first-class discipline is a major strategic decision made by the CPC central committee and the state council, which has a very vital significance to enhance the level of education development in our country, enhance the national core competitiveness, and establish long-term development foundation. The first-class university discipline construction is included in the university engineering construction; therefore, study of university construction engineering project management pattern is particularly important.

2 COMMONLY USED PATTERNS OF CONSTRUCTION ENGINEERING PROJECT MANAGEMENT

2.1 BOT (Build–Operate–Transfer)

It refers to a country's consortium or sponsor for the project, obtains the construction of the infrastructure project concession from the government of a country, forms a project company independently with the other party, and is responsible for project financing, design, construction, operation, and transfer.

2.2 CM (Construction–Management)

It refers to the owner-entrusted CM unit, as a contractor, adopts the conditional "design and construction" and shortens the project cycle, also known as the fast path method.

2.3 DB (Design–Build)

After the project is determined in principle, the owner selects a company that is responsible for the design and construction of the project. This method is based on price contract when bidding and contracting.

2.4 EPC (Engineering–Procurement–Construction)

It refers to project decision-making stage, starting from the design, the tender, entrusting an engineering company general contracting to design–procurement–construction.

2.5 PMC (Project–Management–Consultant)

It refers to the project management contractor taking the full range of project management in the whole process of engineering project on behalf of the owner, including the overall plan for engineering, project definition, project bidding, choosing the EPC contractor, and a comprehensive management of the design, procurement, construction, and commissioning to conduct. The contractor is generally not directly involved in the project design, procurement, construction, and commissioning phase of the specific work.

2.6 DBB (Design–Bid–Build)

It refers to the owners entrusting architect or consulting engineers to do work (such as a chance study and feasibility study) and design after project evaluation.

2.7 PM (Project–Management)

It refers to the engineering project management enterprises participating in the whole or part of a phase of the construction organization and management on behalf of the owner in accordance with the stipulations of the contract.

2.8 TKM (Turn Key–Method)

The contractor provided the owner with a full range of services, including project feasibility study, finance, land purchase, design, and construction until completion that handed over to the owner.

3 ANALYSIS OF THE ADVANTAGES AND DISADVANTAGES OF CONSTRUCTION ENGINEERING PROJECT MANAGEMENT PATTERNS

3.1 BOT (Build–Operate–Transfer)

Advantages: It can solve the problem of inadequate infrastructure, and the shortage of construction funds can lead to preferential policies and broaden the financing channels.

Disadvantages: The prequalification and tender process is complex, and the project sponsor must have high economic strength (consortium).

3.2 CM (Construction–Management)

Advantages: It can shorten the project cycle from planning, design, construction, delivery to the owner, and save construction investment; reduce the investment risk; and make the owner get early benefit.

Disadvantages: Subdivisional bidding results in higher cost of contracting, so people should analyze, compare, conduct a detailed study of engineering, and select the best result.

3.3 DB (Design–Build)

Advantages: Its efficiency and oneness of responsibility solve the problem of overstaffing in organizations, hierarchical overlapping, and imbalance of management personnel.

Disadvantages: Bidding and bid evaluation is much more complicated than traditional pattern. It requires the owner to be well prepared in the early planning stages. The DB pattern requires the owner to have high management level and strong ability of project supervision. It also requires builders to have strong economy and technical ability, strong ability to resist risk, and ability that can provide comprehensive design, construct management, and implement work.

3.4 EPC (Engineering–Procurement–Construction)

Advantages: The owner's management is simple, the contractor's work is coherent because of single general contractor, which can prevent the buck-passing between designers and builders, improve work efficiency, and reduce the workload of coordination. Because the total price is fixed, the cost of claim is zero and there is no additional project cost.

Disadvantages: The guarantee of the quality is based on the contractor's self-consciousness; they can adjust the design scheme to reduce cost and we cannot avoid the actuality. It is very important for the owner to supervise the contractor; however, in the EFC pattern, the owner cannot be involved in the details of the design too much.

3.5 PMC (Project–Management–Consultant)

Advantage: The project management company takes management of the project from early stage, which is conducive to give full play to the project management company's management expertise, experience, and advantage. Project owners can make the necessary change in the design and construction in the project construction process so that they can take control of investment, progress, and quality of the engineering construction project.

Disadvantages: The project owner has no contractual relationship with the construction and procurement contractor, which would reduce owners' influence on the project construction cost, quality, and schedule that mainly depend on the project management contractor control. If the contractor management quality is not in conformity with the requirements, then the project management could fail.

3.6 DBB (Design–Bid–Build)

Advantage: The owner can choose consulting designer freely, the design requirements can be controlled, and the owner is free to choose engineer and standard contract forms, which is advantageous to the contract management, risk management, and reduce the investment.

Disadvantages:

- The project cycle is longer, the owner signs contract with design and the construction separately, and the management fee is higher.
- The constructability is bad, and engineers' ability to control project goal is not high.
- It is not conducive to accident allocation during engineering accident.

3.7 PM (Project–Management)

Advantages:

- It can give full play to the management of contractors in project management skills, unify coordination and management of the project design and construction, and reduce the contradiction.
- It is advantageous to savings of the construction project investment.
- It can optimize the project design and lower the cost in the period of project.
  Disadvantages:
- The owner participation in engineering is less, the right change is limited, and it is difficult to coordinate.
- It is the key for the owners to chose a high-level project management company.

3.8 TKM (Turn Key–Method)

Advantages: Keeping the single contract responsibility in the process of the project, considering the construction factors at the beginning of the project, reducing management costs, and reducing the cost due to design errors and negligence.

Disadvantages: The owner cannot participate in the choice of the architect/engineer, the representative of owner takes the role as a kind of supervision, and the owners' right of monitoring of the project is less.

4 CONSTRUCTION ENGINEERING PROJECT MANAGEMENT PATTERN SUITABLE FOR COLLEGES AND UNIVERSITIES

From 2006 to 2013, infrastructure investment quota of colleges and universities directly under the ministry of education is a maximum of 16.1 billion Yuan, as shown in Figure 1. Infrastructure construction department of colleges and universities

chooses the appropriate project management pattern and makes the full use of money.

4.1 Renovation projects

4.1.1 Characteristics of renovation projects
- Project cost is less.
- Construction cycle is short.
- Repair object is fixed, construction work moves.
- Repairing object is generally existing house.
- Construction work is affected by user department.

4.1.2 Recommended management pattern
Because of the characteristics of repair project, we commend the college take self-management pattern. That means the specialized departments of college (e.g., infrastructure department) entrust department of design, bidding, construction, and management of engineering construction maintenance to manage the project according to the needs of the user departments. The communication with infrastructure department and user department can maximize the limit to meet the requirement of user department, as shown in Figure 2.

4.2 Single project

4.2.1 Characteristics of single project
- Project construction cost is high.
- Project construction cycle is long.
- The object of project is the new house, which is independent.
- Construction site is fixed.
- Less effect by the user department.

4.2.2 Recommended management pattern
On the basis of the characteristics of single project, it is suggested that infrastructure department of college uses the DBB management pattern, the school infrastructure management department entrusts engineers' work such as feasibility study and project evaluation of the re-entry after project design, and then choose contractors and construction supervision company by bidding. They can divide tender scientifically, make full use of competition mechanism to reduce the contract price, and choose professional contractors.

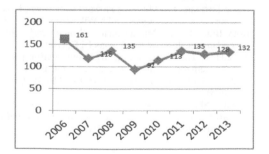

Figure 1. Investment statistics of college infrastructure construction from 2006 to 2013 (unit: 100 million).

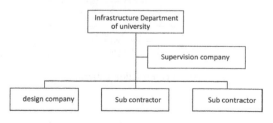

Figure 2. Self-management pattern.

4.3 New campus construction project

4.3.1 Characteristics of new campus construction project
- Project construction cost is high.
- Project construction cycle is long.
- The construction site is fixed, and the utilization rate of construction machinery is high.

4.3.2 Recommended management pattern
Because of its high cost and long construction cycle, the agent construction pattern is suggested through public bidding and choosing the professional agent construction company that is responsible for project investment management and construction organization implementation. Under the authorization of the owner, the agent construction company makes suggestions and solutions for the formulation of project planning, feasibility study, preliminary design, design standards for the project, design parameters, overall layout, structure, investment, and time limit control of the project. Organize engineering design and prepare for project construction on behalf of the owner. Directly organize construction engineering equipment such as bidding, sign a contract with the bid on behalf of the owner. The agent construction company prepares, organizes, and implements annual engineering project investment plan. Project schedule is responsible for project investment, quality, time limit, and safety management and control. By processing the general design change of construction project, the major design change implementation has to be obtained by consent from the owner. Organize project completion acceptance of the contract and completion acceptance certificate issued to the contractor and handover certificate. Hand over all engineering and engineering-related documents to the owner according to the project management contract, as shown in Figure 3.

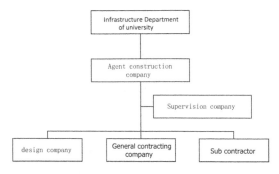

Figure 3. Agent construction pattern.

5 CONCLUSION

In conclusion, there are similarities and differences between the management mode of the university construction project and the one of the construction project in general construction market. For the three common types of construction projects in universities, this paper presents the following management patterns.

Renovation projects. Because the renovation projects are complex, it would be better for the university to adapt the self-management mode. The infrastructure construction department in colleges and universities is responsible for communication with the user department to understand the customers' needs and then carries out commissioned design, construction bidding, supervision bidding, and other following work.

Single project. Because the construction of the single project needs to follow the logic of the construction requirements, it is better for universities to adapt the DBB management mode and follow the project construction order strictly to manage the project. The order is design–bidding–building.

New campus construction project. Because new campus construction project needs high investment, long construction periods, and more managers involved, it is better for the university to adopt the agent construction pattern. The agent company authorized by the university will carry out the management work on behalf of the university.

REFERENCES

Graham Winch and Elisabeth Campagnac, The organization of building project: Anglo/French Comparison, *Construction Management and Economics* 13:3–14 (1995).

Haiyong Yang, Discussion on the management mode of university capital construction, *Construction economy,* 02:33–37 (2013).

Junjie Cao, University infrastructure project management mode analysis and optimization, *Science and technology innovation and application,* 13:292–293 (2013).

Kevin Forsburg, Hal Moozang Howard Cotterman, *Visualizing Project Management,* John Wiley Sons, Ine., (2000).

Rory Buke: Project Management, *Planning and Control Techniques.* John Willey & sons LTD, (2000).

Xu ZHou, The discussion of large construction project management pattern in colleges and universities, *Journal of central south university,* 11(04):464–468 (2005).

Xuyuan Shen, A probe into the Managent Model of the Campus Capital Construction in college, *Journal of jiaxing university,* 7:68–71 (2007).

Advances in Energy Science and Equipment Engineering II – Zhou, Patty & Chen (Eds)
© 2017 Taylor & Francis Group, London, ISBN 978-1-138-71798-5

A study of conservation methodology of architecture heritage—taking Notre Dame Cathedral in Paris as example

Xi Chen

Architecture Department, Soochow University, China

ABSTRACT: This article introduces briefly the features and the construction history of the Notre Dame Cathedral in Paris, which is a masterpiece of Gothic architecture. It reviews the conservation or the so-called 'restoration' history of this cathedral, illustrates the conservation methodology in France that is derived from the work of Viollet-le-Duc, and demonstrates the progress from the creation of the Department of Historic Monuments to the establishment of the integrated system of supervision and research institutions. This article also summarizes the development of conservation technology during the past two centuries.

1 INTRODUCTION

In the early 12th century, Western society underwent a profound change prompted by the emergence of a middle class whose economic role was to be become dominant; by a transformation in religious practices; and by the quest for political balance that was to reach an equilibrium between the power of the crown and that of the church. These changes combined led to persistent urban growth that resulted in the transformation of the city and the passage from archaic, obscure and massive architectural forms of palaces and cathedrals to light-filled, soaring buildings that were a triumphant illustration of society's renewal: Gothic architecture.

2 FEATURES OF NOTRE DAME CATHEDRAL IN PARIS

Gothic architecture did not come from a particular invention, but from the inspired combination of pre-existing architectonic components the properties of which it "optimized" in a new way: the groined arch and the flying buttress. It's fast, economic and magnificent characteristics delivered an inventive answer to emerging aspirations.

It was in Ile-de-France that the new Gothic cathedrals were born. In 1230, 25 construction sites were in progress, and that number rose to nearly 80 by the end of the century.

In this evolution, Notre Dame Cathedral, started in 1160 with the choir, lies at the crossroads between the era of the pioneering primitive Gothic and High Gothic that had reached maturity.

The great work is complete in 1230, and despite all the adaptations, it offers the spectacle of extraordinary homogeneity and architectural unity thanks to the abnegation of five anonymous architects. At 127 metres long and 45 metres wide, 33 meters under the arches, Notre Dame was at the time the largest church in Western Christendom. The chapels between the abutments of the flying buttresses were rebuilt between 1250 and 1270 by the architects Jean de Chelles and Pierre de Montreuil. In 1318, work was finally completed.

In 1708, the Canonry Choir was replaced with a remarkable ensemble of panelling and marble sculptures. The interior finishings of the Cathedral were shortly after covered with a yellow wash; in 1753, the medieval stained glass windows were replaced; in 1787, the statuary, gargoyles, finials, and the pier and lower third of the tympanum of the central portal of the Last Judgement were removed; in 1787, the transept spire was removed. The 1793 revolutionaries were no greater vandals than the clergy who replaced the chapel gables with pediments, or "restored" the sculptures with Molesmes cement.

3 THE GREAT RESTORATION IN HISTORY

The dilapidation was breath-taking in 1831 when Victor Hugo published his novel "The Hunchback of Notre Dame". It was at this same time that François Guizot launched the great inventory of French monuments, created the Inspectorate-General, and then the Historic Monuments Commission.

In 1842, tenders were published for the restoration of Notre Dame, and an initial budget of 2,650 million francs was allocated. Jean Baptiste Lassus and Eugène Viollet-le-Duc won the contract. Work started on 20 April 1844, and was completed on 31 May 1864, after Lassus' death in 1857.

Significant work was undertaken: restoration of the façades, abutments and flying buttresses, the sculptures, the central portal, the roofs, etc; restitution of stained glass windows, the mural polychrome, etc; construction of the Sacristy and the Presbytery… the operation reportedly cost 8 million francs.

The entire operation was directed with extreme rigor by Viollet-le-Duc; the most eloquent illustration of this being the restoration of the spire. By carefully studying the 13th century vestiges still present in the roof, he managed to deduce the original arrangement: historic analysis. He then observed the weaknesses and the probable causes (wind) of ruin, and then devised the restitution project, incorporating into it the necessary corrections and reinforcements: structural analysis; lastly, he assessed the impact on the Cathedral's silhouette, observing that it was necessary to extend it by 13 metres: architectural analysis. The approach that he adopted for himself was that of absolute authority, "The aim is not to make art, but to submit to the art of a time that is no more", in order to "regain and follow the thought that presided over the work's construction". He placed himself with the greatest discipline and conviction in the shoes of the 13th century architect, and set any notion of conservation aside to proceed with restoration.

The building as it was after this work had undeniably gained in homogeneity; and while, as Marcel Aubert says, "some of the principles are open to discussion, one must still bow to the conscience and mastery with which it was carried out…" This is one the most emblematic restorations of the 19th century.

4 CONSERVATION METHODOLOGY

A certain youthfulness regained put the cathedral out of harm's way in the final decades of the 19th century and the first decades of the 20th, which went no further than "cosmetic" measures. But falling stones reported in 1936 were a reminder of the structure's fragile nature, and the building's difficulty with ageing. Work in the 20th century was of three kinds: conservation, restoration and safety.

4.1 Conservation

The first conservation work was undertaken in 1938 by Ernest Herpe on the western towers; interrupted during the Second World War, it was completed in 1955.

Launched in 1968 by André Malraux, Minister of Cultural Affairs, the great campaign to clean Parisian monuments was an opportunity to wash and consolidate the cathedral's facing from August 1968 to November 1970, under the direction of Bernard Vitry, Architect-in-Chief of Historic Monuments.

In 1988, taking advantage of the progress made in the knowledge of the pathologies and mechanisms of deterioration, applied by the Historic Monuments Research Laboratory (LRMH) created in the final years of the 1960s, a systematic health review of the cathedral was performed by the new Architect-in-Chief of Historic Monuments, Bernard Fonquernie. While it did not reveal any structural problems, it did show that the building's conservation was compromised by several factors, the most significant of which was the diversity of types of stone used in the 19th century for the restoration of the facing (up to 18 on the West Façade, eight on the Last Judgment portal alone). We know today, with hindsight not available to Viollet-le-Duc, that they are the source of serious deterioration issues, with aggravating factors due in particular to the use of highly hydraulic mortars. The exposed parts are in particular suffering, affecting both the original and restoration stonework. (Fig. 1).

The search for replacement stone with the same characteristics as the original stone was carried out with the assistance of the LRMH; it was used for the West Façade restoration campaign from 1994 to 2000, followed by that of the North Tower from 2003 to 2006; mortars were developed with very low hydraulic limes, the aim being to work towards new homogeneity across the parts exposed to the weather. During these campaigns, it was possible to measure the considerable progress made since 1968 for cleaning the facing: dry micro-abrasion or nebulization, wet or ammonium carbonate compresses, laser where needed, etc. The conservation of the facing skin is markedly improved as a result, to the extent that rare vestiges have been found of the poly-

Figure 1. The restoration stonework.

chromies that covered the portal sculptures, statues in the gallery of kings, the rose window and the first level windows, providing precious information about colour in the architecture of the Middle Ages.

In tandem with this "de-restoration" work, scrupulous conservation work on the Viollet-le-Duc structures was undertaken, for example, on the lead ridge crests that crown the high roofs revealing their skeletal silhouette against the sky. Present in Gothic architecture, as evidenced in illuminations and goldsmithed reliquaries, there remained no example on Notre Dame. So adopting the rigorous method that he imposed on himself, Viollet-le-Duc reconstituted the most probable configuration, in moulded lead on iron frames. The latters' corrosion led to extensive and dangerous deformations. Their identical replacement required the use of the same skills and expertise of lead roofers, foundrymen and carpenters, providing at the same time an opportunity to conserve these trades (Figs. 2–3). Begun in the 1960s at the western end of the central nave, continued in 1980 on the choir and the southern arm of the transept, the operation was completed in 2009 on the remainder of the building.

4.2 *Restoration*

Today's efforts are focused on conservation, but do not reject opportunities for restoration. Two examples:

Like any church, Notre Dame has a carillon that had up to eight bells and two great bells, housed successively from the 13th to the 15th century in the two western towers. In the French Revolution, the bells, except for the great bell, were melted down. In 1851, the two belfries were reconstructed by Viollet-le-Duc to house a new carillon, never completed: only the great bell and four poorly tuned bells were installed. To mark the Cathedral's eighth centenary, it was planned to complete the carillon (Fig. 4). Using the very extensive studies held in the archives, the characteristic of each bell–size, weight and note–was precisely identified; and the structural effect of bell ringing on the belfries and masonry was carefully checked. After it was consulted, The National Commission for Historic Monuments agreed to allow work to proceed; it should be finalized by Easter 2013.

In the second example, Notre Dame played a truly pioneering role. The windows created between 1235 and 1245, removed in 1753, were recreated by Viollet-le-Duc from contemporaneous models of Gothic stained glass windows (notably in Bourges Cathedral), which he distributed according to a progression starting from pale greys in the main nave through to the sanctuary with its legendary highly colored windows. But in 1937, young glass-making painters wanting to correct the "pale and wan" greys in the main nave suggested replacing them with modern stained glass windows. This generous offer led to the first great discussion about the serving to amplify the vigor and heat of the discussions between the then two opposing camps: those in favor of retaining Viollet-le-Duc's

Figure 2.   The conservation of wooden structure.

Figure 3.   The conservation of lead surface.

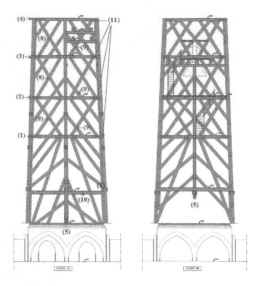

Figure 4.   The mechanical monitoring of the towers.

windows in the name of the Cathedral's homogeneity, and a certain idea of restraint and moderation; and the interventionists arguing for a modern and living church; the press took hold of the subject, and so of public opinion... It took thirty years of dogged arguing, several models, trial panels, temporary installations and six sessions of the High Commission for Historic Monuments, before finally in 1964 the Minister for Cultural Affairs André Malraux approved the final project comprised of irregular greys surrounded by a colored border: a happy compromise respecting the balance and progression of light inside the cathedral. Work was completed on 17 June 1965. While this example opened the doors of historic monuments to modern art, it is not sure, now that the work of Viollet-le-Duc has been recognized, that his windows would have been sacrificed today.

4.3 Safety

Each year, Notre Dame Cathedral, both a place of active worship, and an emblematic monument of French heritage, receives 13 million visitors. This necessarily entails safety obligations for the public and the building, which explains why work is currently being undertaken to provide fire protection (especially for the timber roof frames dating from 1180 and 1220), and against lighting strikes, using the most modern techniques and equipment, while conserving the building's architectural integrity.

The same applies to the public: measures to protect against falling and suicide along the tour circuit, and optimum disabled access.

5   CONSERVATION PLAN FOR NOTRE DAME

The cathedral is today the subject of constant surveillance by the Ministry of Culture, and in particular its architects; the Architect-in-Chief of Historic Monuments (ACMH); the Architect of Buildings of France (ABF). The most modern methods are employed by the State funded out of its own budget, with the assistance of the LRMH, specialist engineering firms, and the Heritage Archives.

Any need for work is identified and then ordered according to the degree of urgency; the preliminary and prior studies are launched under the supervision of the ACMH; the conclusions are reviewed by experts and submitted for approval to the Inspectorate-General of Historic Monuments and, if necessary to the National Commission under the scientific and technical supervision of the Ministry of Culture.

The work is then performed within the context of public calls for tender, with the participation of contactors and tradesmen highly specialized both in the most traditional techniques, and also the most modern, implemented in the light of active research and the results from previous campaigns.

Today, the schedule is drawn up for the coming ten years including the conservation of the high elevations of the flying buttresses of the central nave, transept, and choir (5 years), and the restoration of the high sections of the sacristy (2 years). Above all, we must not lose a year.

ACKNOWLEDGMENTS

Supported by the National Natural Science Foundation of China (grant no. 51508361).

REFERENCES

Camille M, 1992. The gargoyles of Notre-Dame. Chicago: The University of Chicago Press.

De Musset A, 1960. Oeuvres completes: Prose. Gallimard. Paris.

Feiden, B. M, 1986. Conservation of historic buildings. World Architecture.

Glendinning, Miles, 2013. The Conservation Movement: A History of Architectural Preservation—Antiquity to Modernity, Routledge. London.

Jokilehto, Jukka, 2002. A History of Architectural Conservation, Butterworth Heinemann, Oxford.

Violet-Le-Duc E-E. 1866. Dictionaire raisonné de l'architecture française, vol. VIII.

Advances in Energy Science and Equipment Engineering II – Zhou, Patty & Chen (Eds)
© *2017 Taylor & Francis Group, London, ISBN 978-1-138-71798-5*

Technical and economic analysis of new hollow girderless floor structure with thin-walled bellows

Hui Lv & Libo Huang
Jiangxi Institute of Economic Administrators, Nanchang, China

Dae Lv
State Gird Jiangxi Economic Research Institute, Nanchang, China

Feng Yang
School of Civil and Architecture Engineering East China Institute of Technology, Nanchang, China

ABSTRACT: As a new type of construction technology, the hollow girderless floor structure has been widely used in many projects. The paper studied economic efficiency of the new hollow girderless floor structure system (with thin-walled bellows) and the common beam-slab structure system. By comparing and analysing four project cases with difference functions and different loads, that is, underground garage, comprehensive office building, multi-storey plant building, and teaching building,, it is concluded that the new hollow girderless floor structure technology would substantially cut construction material consumption and direct project cost and outperform the common beam-slab structure system. The longer the column span is and the larger the vertical load is, the more obvious the economic benefits are.

1 INTRODUCTION

1.1 *Technical background of application of hollow girderless floor structure*

In the middle of 1960s, Leopold Muller, an engineer from Heidlberg in the former Federal Republic of Germany, first put forward a structure of cast-in-place concrete hollow girderless floor, which was then called "B-Z system" (that is "Belonn Ze Uenplatte" in German, meaning honeycomb type concrete hollow floor structure). Such structure appeared in the United States later and was applied in bridge projects on bridge deck slabs (Muller, 1969).

Professor G Franz made an experimental study on the said honeycomb type concrete hollow floor structure in 1964 and believed that under static load the calculation method for solid girderless floor structure with the same rigidity could be applied in the design of hollow girderless floor structure (Steven, 2016). In 1967, Edgar. Hendler, an engineer in the United States, constructed a hollow girderless floor structure filled with massive foam plastics. He studied the construction methods and structural measures of such structure, and then presented the thickness values of hollow girderless slabs in different spans and loads based on engineering practices. However, the values were conservative without any basis of experimental study (Behairy, 1990). The foam plastics were pollutant and its construction

technology for the hollow girderless floor structure was complicate, so the structure was not spread among projects (Huffington, 1964). From the beginning of 1990s, Chinese engineers improved the filling members of the hollow girderless floor structure by using light and intensive thin-walled tubes, which greatly reduced the difficulty for construction, so the hollow girderless floor structure was once again developed and applied.

1.2 *Significance of application of hollow girderless floor structure*

The hollow girderless floor structure has a small height and thus increases floor clearance with the same floor height, so it not only meets function requirements of the building but also reduces the dead weight of the structure and helps in solving the problem of floor height for buildings with large span and large space (Abdel, 1984). The hollow girderless floor structure greatly reduces the dead weight of the slab and thus reduces the load of slab, column, foundation, and other components to various degrees. Moreover, with such floor structure, the partitions could be arranged optionally to solve the contradiction between free arrangement and fixed stress of the building. The girderless floor structure has no girder in the floor structure. The floor structure and the columns form a plate-column structural floor structure under bidirectional stress. The floor structure is directly on the column (and its edges may be also on the wall). Therefore, the height of such building structure is smaller than that of a ribbed floor structure. The smooth slab bottom improves lighting, ventilation, and hygiene conditions (Oduyemi, 1988). Moreover, the hollow girderless floor structure is of good sound insulation effects because the cavity reduces noise and the structure performs well in thermal insulation and fire prevention.

The cast-in-place reinforced concrete hollow floor structure is formed mainly by its inside filling members, including light fill core materials (rectangle foam plastics), thin-walled boxes (box), round tubes (intensive thin-walled tubes, steel wire mesh cement hollow tube), etc. The application of hollow girderless floor structure in construction engineering reduces the consumption of concrete, rebar, and other materials and thus brings good economic benefits. Therefore, various hollow internal models enter into the market soon and have been applied to buildings in more than 20 provinces and cities across the country (Oduyemi, 1990). In recent years, China started to obtain certain practical experience in engineering application of hollow girderless floor structure.

Figure 1. Site of hollow girderless floor structure—thin-walled bellows.

1.3 Economical efficiency of hollow girderless floor structure

The cast-in-place hollow girderless floor structure is a brand new structure system with high void ratio and it is easy to be constructed. Such floor structure is of good integrity, excellent spanning capacity, and high aseismic capacity. It not only reduces structural layer height to obtain larger space but also reduces the consumption of concrete materials and cost to achieve better economic benefits. Moreover, the hollow concrete girderless floor structure needs no more multi-girder structure design nor girder formwork construction, so the formwork laying and pouring is simpler and easier and it costs less manpower and material resources in girder steel bar tie-in. Therefore, the construction progress speeds up on the whole. The hollow girderless floor structure is being applied widely to large-spaced shopping malls, libraries, teaching building, large-bay residential buildings, and other industrial and commercial buildings.

Based on the existing formwork filler of the girderless hollow floor structure, the project team has independently researched and developed a new type of thin-walled bellow as the filling members of hollow girderless floor structures (refer to Figure 1 and Figure 2). The paper studied 4 different projects: underground garage roof engineering, comprehensive office building, multi-story industrial plant building, and teaching building roof engineering. The authors calculated the quantities and prepared

Figure 2. Flat-die construction of hollow girderless floor structure.

the engineering cost budgets based on construction drawings of the new hollow girderless floor structure and the common beam-slab floor structure. Afterwards, the authors compared the cost budgets and the member quantities of the two different construction technologies in engineering projects. The prices of the materials are based on No. 9 Jiangxi cost information of 2015. The software for price calculation is GYJ of Jiangxi Gongyou Software Co., Ltd.

2 ECONOMIC ANALYSIS

2.1 *Economic analysis of application of new hollow girderless floor structure in underground garage*

One underground garage in Nanchang was designed with the new hollow girderless floor structure system in its construction drawings. The actual drawings (Figure 3 and Figure 4) of the project indicate that the axis net is 8.1 m × 8.4 m, the hollow plate is 450 mm thick, the void ratio is 49.65%, the soil on the slab top is 1.2 m, the floor area is 5618 m², and the load is 22.8 KN/m² for dead load and 5 KN m² for live load, and between the frame columns is a 1 m wide hidden beam. The design drawing (Figure 5) of the corresponding common floor structure indicate the plate is 200 mm thick,

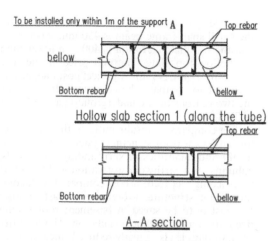

Figure 4. Arrangement and details of thin-walled bellows in hollow girderless floor structure.

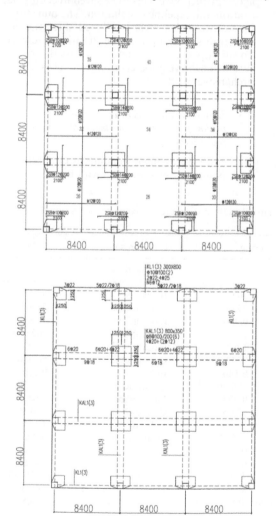

Figure 3. Hollow girderless floor structural drawing.

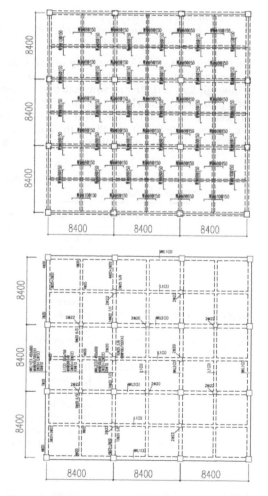

Figure 5. Common beam-slab floor structural drawing.

749

the large span frame beam is 500 mm × 800 mm, the small span frame beam is 450 mm × 600 mm, and the top plate elevation is 400 mm lower than the hollow top plate. The unit price of the core formwork is 40CNY/m. A project cost budget was prepared to compare the economic efficiency of the two options of the underground garage (Refer to Table 1).

The comparison results indicate that such hollow girderless floor structure system reduces floor height and quantities of surrounding shear walls, columns, and earthwork excavation and filling. It saves as high as 15.68% cost compared with common floor structure. Moreover, at least another 5% cost is to be saved in basement roof ceiling decoration and pipeline installation. The quantity of man-days is also greatly reduced since the construction speeds up and the duration decreases. To sum up, it will achieve good economic benefits.

2.2 Economic analysis of application of new hollow girderless floor structure in comprehensive office building

One comprehensive office building in one industrial park Nanchang was designed with the new hollow girderless floor structure system in its construction drawings. The actual drawings of the project indicate that the axis net is 8.1 m × 5.4 m, the hollow plate is 300 mm thick, the hollow rate is 33.35%, the soil on the slab top is 1.2 m, the

floor area is 2800 m², and the load is 12.8 KN/m² for dead load and 4 KN/m² for live load, and between the frame columns are framed girders but no primary beam nor secondary beam between the framed girders. The design drawing of the corresponding common floor structure indicate the plate is 120 mm thick, and the frame beam is 300 mm × 500 mm. The unit price of the core formwork is 15 CNY/m. A project cost budget was prepared to compare the economic efficiency of the two options of the comprehensive building (Refer to Table 2).

The comparison results indicate that such hollow girderless floor structure system reduces floor height and quantities of columns and beam-slab. It saves as high as 7.78% cost compared with common floor structure. Moreover, at least another 5% cost is to be saved in ceiling decoration and pipeline installation. The quantity of man-days is also greatly reduced since the construction speeds up and the duration decreases. To sum up, the new hollow girderless floor structure will achieve good economic benefits in comprehensive buildings.

2.3 Economic analysis of application of new hollow girderless floor structure in multi-story plant building

One multi-storey plant building in one industrial park Nanchang was designed with the new hollow

Table 1. Comparison of economic efficiency of hollow girderless floor structure and common beam-slab floor structure of underground garage.

Economic comparison of underground garage roof engineering			Floor area m²	5618
	Common beam-slab		Hollow floor	
Beam-slab rebar	Index kg/m²	70.07	Index kg/m²	60.18
	Quantities kg	393683	Quantities kg	338085
Beam-slab concrete	Index m³/m²	0.340	Index m³/m²	0.2899
	Quantities m³	1909.44	Quantities m³	1628.64
Beam-slab formwork	Index m²/m²	1.699	Index m²/m²	1.002
	Quantities m²	9547.20	Quantities m²	5630.00
Comprehensive of column rebar/formwork/concrete	Index m³/m²	0.140	Index m³/m²	0.110
	Quantities m³	786.00	Quantities m³	617.70
Comprehensive of shear wall	Index m³/m²	0.095	Index m³/m²	0.082
	Quantities m³	533.52	Quantities m³	460.40
Backfill earthwork on roof	Index m³/m²	2.233	Index m³/m²	1.675
	Quantities m³	12544	Quantities m³	9408
Core formwork	Index m/m²	0	Index m/m²	2.568
	Quantities m	0	Quantities m	14428
Total & comparison	Total cost (CNY)	4706362.62	Total cost (CNY)	3968343.43
	Unit cost (CNY/m²)	837.73	Unit cost (CNY/m²)	706.36
	Cost saving rate of hollow floor structure than common floor structure			15.68%
Quantity of man-days	Man-day	14117	Man-day	9779

Table 2. Comparison of economic efficiency of hollow girderless floor structure and common beam-slab floor structure of comprehensive building.

Economic comparison of comprehensive building			Floor area m²	2800
	Common beam-slab		Hollow floor structure	
Beam-slab rebar	Index kg/m²	44.80	Index kg/m²	37.75
	Quantities kg	125436	Quantities kg	105711
Beam-slab concrete	Index m³/m²	0.230	Index m³/m²	0.198
	Quantities m³	635.66	Quantities m³	554
Beam-slab formwork	Index m²/m²	1.745	Index m²/m²	1.028
	Quantities m²	4886	Quantities m²	2878
Comprehensive of column rebar/ formwork/concrete	Index m³/m²	0.049	Index m³/m²	0.045
	Quantities m³	138	Quantities m³	125
Core formwork	Index m/m²	0	Index m/m²	4.677
	Quantities m	0	Quantities m	13095
Total & comparison	Total cost (CNY)	1475175.49	Total cost (CNY)	1360611.93
	Unit cost (CNY/m²)	526.85	Unit cost (CNY/m²)	485.93
	Cost saving rate of hollow floor structure than common floor structure			7.78%
Floor height	To be reduced by around 200 mm			
Other decoration and installing pipelines	Cost to be reduced by 5%			
Reduction of foundation load	Cost to be reduced by 5%			
Comparison of total cost	To be reduced by 10%			
Quantity of man-days	Man-day	4946	Man-day	3634

girderless floor structure system in its construction drawings. The actual drawings of the project indicate that the axis net is 12 m × 12.6 m, the hollow plate is 400 mm thick, the void ratio is 49.79%, the soil on the slab top is 1.2 m, the floor area is 22927 m² with 5 storeies above the ground, the load is 9.8 KN/m² for dead load and 4 KN/m² for live load, and between the frame columns are 400 mm × 1000 mm framed girdersbut no primary beam nor secondary beam between the framed girders. The design drawings of the corresponding common floor structure indicate the plate is 120 mm thick, the frame beam is 250 mm × 600 mm, and the normal beam is 200 mm × 500 mm. The unit price of the core formwork is 30 CNY/m. A project cost budget was prepared to compare the economic efficiency of the two options of the plant building (Refer to Table 3).

The comparison results indicate that such hollow girderless floor structure system reduces floor height and quantities of columns and beam-slab. It saves as high as 7.37% cost compared with common floor structure. Moreover, at least another 5% cost is to be saved in ceiling decoration and pipelines installation. The quantity of man-days is also greatly reduced since the construction speeds up and the duration decreases. To sum up, the new hollow girderless floor structure will achieve good economic benefits in cast-in-place multi-story industrial plant buildings.

2.4 Economic analysis of application of new hollow girderless floor structure in teaching building roofing system

One teaching building in Nanchang was designed with the new hollow girderless floor structure system in its construction drawings of roofing system. The actual drawings of the project indicate that the axis net is 10.2 m × 8.4 m, the hollow plate is 300 mm thick, the void ratio is 42%, the roof area is 1768 m², the load is 15.8 KN/m² for dead load and 0.5 KN/m² for live load, and the unit price of core formwork is 15 CNY/m. A project cost budget was prepared to compare the economic efficiency of the two options of the teaching building (Refer to Table 4).

The comparison results indicate that such hollow girderless floor structure system reduces floor height and quantities of beam-slab. It saves as high as 8.39% cost compared with common floor structure. Moreover, at least another 5% cost is to be saved inr ceiling decoration and pipeline installation. The quantity of man-days is also greatly reduced since the construction speeds up and the duration decreases. To sum up, the new hollow girderless floor structure will achieve good economic benefits in cast-in-place teaching buildings' roofing system.

Table 3. Comparison of economic efficiency of hollow girderless floor structure and common beam-slab floor structure of multi-story plant building.

Economic comparison of multi-story plant building			Floor area m²	22927
	Common beam-slab		Hollow floor structure	
Beam-slab rebar	Index kg/m²	32.03	Index kg/m²	29.01
	Quantities kg	734412	Quantities kg	665019
Beam-slab concrete	Index m³/m²	0.23	Index m³/m²	0.20
	Quantities m³	5301.45	Quantities m³	4585.42
Beam-slab formwork	Index m²/m²	1.80	Index m²/m²	1.10
	Quantities m²	41311.6	Quantities m²	25245.1
Comprehensive of column rebar/ formwork/concrete	Index m³/m²	0.030	Index m³/m²	0.025
	Quantities m³	688.5	Quantities m³	589.05
Core formwork	Index m/m²	0	Index m/m²	1.6
	Quantities m	0	Quantities m	36683
Total & comparison	Total cost (CNY)	10266217.76	Total cost (CNY)	9509749.03
	Unit cost (CNY/m²)	447.78	Unit cost (CNY/m²)	414.78
	Cost saving rate of hollow floor structure than common floor structure			7.37%
Floor height	200 mm			
Other decoration and installing pipelines	Cost to be reduced by 5%			
Reduction of foundation load	Cost to be reduced by 5%			
Comparison of total cost	To be reduced by 9%			
Quantity of man-days	Man-day	34557	Man-day	24869

Table 4. Comparison of economic efficiency of hollow girderless floor structure and common beam-slab floor structure of teaching building roofing system.

Economic comparison of teaching building roofing system			Floor area m²	1768
	Common beam-slab		Hollow floor structure	
Beam-slab rebar	Index kg/m²	38.09	Index kg/m²	31.82
	Quantities kg	67350	Quantities kg	56265
Beam-slab concrete	Index m³/m²	0.257	Index m³/m²	0.220
	Quantities m³	455	Quantities m³	389.55
Beam-slab formwork	Index m²/m²	1.8	Index m²/m²	1.24
	Quantities m²	3189.9	Quantities m²	2190
Core formwork	Index m/m²	0	Index m/m²	3.96
	Quantities m	0	Quantities m	7000
Total & comparison	Total cost (CNY)	750230.64	Total cost (CNY)	687270.09
	Unit cost (CNY/m²)	424.34	Unit cost (CNY/m²)	388.73
	Cost saving rate of hollow floor structure than common floor structure			8.39%
Floor height	200 mm			
Other decoration and installing pipelines	Cost to be reduced by 5%			
Reduction of foundation load	Cost to be reduced by 1%			
Comparison of total cost	To be reduced by 10%			
Quantity of man-days	Man-day	2446	Man-day	1774

3 CONCLUSION

The analysis of actual projects indicates that with reasonable design, the hollow girderless floor structure is of better building functions and better economic efficiency than the common floor structure. Moreover, the construction of the hollow girderless floor structure is faster than that of the common floor structure, that is, 3~5 days less for wood works and concrete works for each story. The duration is reduced. The turnaround materials' availability is increased. Therefore, the total cost is reduced. The technology of new hollow girderless floor structure is of obvious advantages in comprehensive economic efficiency. With the same functions and axis net arrangement, the integrated project cost of buildings with hollow girderless floor structure technology is 5%~15% lower than that with common floor structure. Compared with common floor structure, its direct project cost is greatly reduced. Compared with the common beam-slab structure system, the longer the column span is and the larger the vertical load is, the more obvious the economic benefits are.

ACKNOWLEDGEMENTS

Fund program: Science and technology research (key) funded project of the Education Department of Jiangxi, No.: (Grant No.GJJ151585).

The Natural Science Foundation of Jiangxi Province, China (Grant No. 20161BAB216142):

The Scientific Research Foundation of the Education Department of Jiangxi Province, China (Grant No. GJJ150575).

REFERENCES

Abdel Rahman H.H, Hinton E.Linear And Nonlinear Finite Element Analysis of Reinforced And Prestressed Concrete Voided Slabs. Proceedings of the International Conference on Computer-Aided Analysis and Design of Concrete Structures. 1984.

El Behairy, S.A, Soliman, M.I., Essawy, A.S., Foud, N.A.Nonlinear Finite Element Analysis Of Voided Reinforced Concrete Slabs. Annual Conference and 1st Biennial Environmental Speciality Conference. 1990.

Muller L. Concrete Ceiling Plate:, US3455071 [P]. 1969.

Huffington, N.J. Bending Athwart A Parallel Stiffened Plate. Martin Mariatta Corporation, Baltimore, Md. 1964.

Oduyemi T.O. S, Clark L.A. Prediction of Crack Widths in Circular Voided Concrete Slabs Subjected to Longitudinal Bending. Proceedings of the Institution of Civil Engineers. 1988.

Oduyemi T.O. S, Clark L.A. Tension Stiffening In Longitudinal Sections Of Circular Voided Concrete Slabs. Proceedings of the Institution of Civil Engineers (London).

Steven Foubert, Karam Mahmoud, Ehab El-Salakawy. Behavior of Prestressed Hollow-Core Slabs Strengthened in Flexure with Near-Surface Mounted Carbon Fiber-Reinforced Polymer Reinforcement [J]. Journal of Composites for Construction, 2016.

Advances in Energy Science and Equipment Engineering II – Zhou, Patty & Chen (Eds)
© 2017 Taylor & Francis Group, London, ISBN 978-1-138-71798-5

Analysis of the phenomenon of affiliated subcontract and illegal subcontracting in the process of construction and their countermeasures

Xijin Liu
Infrastructure and Planning Department, Beijing Jiaotong University, Beijing, China

ABSTRACT: In this paper, we analyzed the reasons for affiliated subcontract and illegal subcontracting activities in the construction market from five aspects and pointed out the danger of these activities. Furthermore, this paper put forward eight suggestions to avoid the former phenomenon from the angle of both government functional departments and the construction management enterprise. This paper has great reference value for government functional departments and the construction management enterprise to reduce the affiliated subcontract and illegal subcontracting in the process of construction.

1 INTRODUCTION

As the pillar industry of the national economy of China, production of construction industry increases every year. Currently, the construction market in China presents the situation of oversupply, low threshold of the construction market, and high profit rate. Although the state law bans the affiliated subcontract, in fact, subcontracting phenomenon exists, as shown in Table 1.

2 THE MEANING OF AFFILIATED SUBCONTRACT AND ILLEGAL SUBCONTRACTING

2.1 *The meaning of affiliated*

Affiliated refers to a construction enterprise that allows others take the name of their enterprise to carry out engineering practice in a project or a certain time period. The affiliated basically has the following forms: the enterprise, individuals with

Table 1. 2006–2015 Construction industry added value accounted for the proportion of GDP.

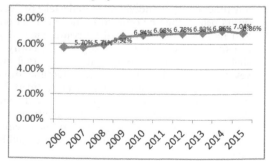

no qualification certificate, or the enterprise with low level of qualification. It undertakes the project under the name of enterprise with qualification certificate or higher-level qualitative, allowing others take the name of the enterprise to undertake projects by leasing or loaning certificate or taking management fee.

2.2 *The meaning of subcontract and illegal subcontracting*

Illegal subcontracting refers to that the contracting enterprise does not perform the responsibility and obligation of the contract, delivers the whole contracted construction project to a third person, or divides the projects in parts, respectively, in the name of the subcontract to a third person. Subcontract has the following forms: Contractor of construction projects does not have the contract responsibility and obligation of the agreement and delivers the project to the third enterprise or individuals. Contractor of construction projects does not have the contract responsibility and obligation of the agreement and divides the construction projects in parts, respectively, in the name of the subcontract to the third enterprise or individuals.

3 CAUSES OF THE AFFILIATED SUBCONTRACT AND ILLEGAL SUBCONTRACTING PHENOMENON IN THE PROCESS OF CONSTRUCTION

3.1 *The lure of economic benefits*

In affiliated phenomenon, the affiliated enterprise has no qualification but has the ability to undertake construct project. They can undertake the project through affiliated to a qualified enterprise. The qualified enterprise could take management fee with low risk. As a result, both sides, winning enterprise delivers all or part of the construction project to any other enterprise or individual by charging management fee and the subcontracting enterprise, can obtain economic benefits.

Construction is a field full of the duty crime and commercial bribery. In 2008, for example, more than 60% of the duty crimes belonged to the field of engineering construction. The proportion of crime is up to 67.4% among all crimes. The number of cases in construction accounted for more than 40% of the total number of cases of commercial bribery, as shown in Table 2.

3.2 Lack of regulation

Part of the relevant functional departments pay sufficient attention to the program, but they pay less attention to the management and neglect the management after the bidding. Especially there is absence of the performance management of the construction contract. After the bid, the tenderer and the department in charge of the management are in shortage of the commitment to the fighting against the bidder and contract tracking supervision. Affiliated subcontract and illegal subcontracting after the bid are in shortage of investigation and punishment.

Table 2. Duty crime proportion of industries under analysis from 2010 to 2012.

Unit nature of crime	2010	2011	2012	Proportion of the total
Department of land and resources, Construction Department	5	7	10	36.7%
Government	9	3	3	25.0%
National Development and Reform Commission, Financial Department	3	4	4	18.3%
Police, traffic, education, health, environmental protection department	3	5	2	16.7%
Other departments	0	1	1	3.3%

3.3 The malign competition in the construction market environment

Because the construction market is a buyer's market, which makes the tenderer has certain control over the process of bidding. The bidder would take various ways (including collusion and together-conspired bidding) for winning in the face of fierce market competition. If the bidder is not actually a real enterprise or individual engaged in construction, it would easily cause the affiliated subcontract and illegal subcontracting phenomenon.

3.4 The owner enterprise considering some relationship or some kind of interests

After the bid, considering some relationships or some kinds of interest, the owner enterprise expresses or suggests that the construction delivers some projects subcontracted to his/her associates. The general contractor has to listen to the owner enterprise considering to the project funds.

3.5 The existing laws are in shortage of penalties to affiliated subcontract and illegal subcontracting

According to the Construction Law of the People's Republic of China Article 66, a construction enterprise which transfers or lends its certificate of qualification to others or which allows others to contract projects in its name shall be ordered to correct itself and imposed fine penalties, with all illegal incomes confiscated. It can also be ordered to stop business operations and to have its level of qualification reduced. An enterprise that is found to have serious violations shall be revoked the certificate of qualification. The construction enterprise, the organization, or the individual that has used the name of the construction enterprise shall assume the associated liabilities for losses incurred from the sub-par quality of the contracted project.

According to the regulation on the Quality Management of Construction Projects Article 61, any survey, design, construction, or project supervisory entity that violates this regulation by allowing any other entity or individual to undertake projects on its behalf shall be ordered to make corrections and its illegal proceeds shall be confiscated. If it is a survey, design, or project supervisory entity, it shall be fined one to two times the contractual survey fee, design fee, or supervision remuneration. If it is a construction entity, it shall be fined 2% up to 4% of the contractual project price. It may be ordered to stop business for rectification, and its qualification shall be degraded. If the circumstance is severe, its qualification certificate shall be revoked.

According to the regulation on the Quality Management of Construction Projects Article 73, where a fine is imposed on an entity under this regulation, a fine of not less than 5% but not more than 10% of the said amount of fine shall be imposed on the directly liable person-in-charge and other directly liable persons of this entity.

According to the Provisions on the Administration of Qualifications of Enterprises in Construction Industry (Order No. 22 of Ministry of Housing and Urban-Rural Development of the People's Republic of China) Article 37, enterprises have violated one of these provisions Article 23 of the act, the construction law of the People's Republic of China, the Regulation on the Quality Management of Construction Projects, and other relevant laws and regulations, concerning way of punishment and punishment, in accordance with the provisions of the laws and regulations. If there are no relevant provisions in the laws and regulations, the department of housing and urban–rural development or other relevant departments in local people's government shall give warning. They shall be ordered to correct, and concurrently be fined from 10000 to 30000 yuan.

The penalties of current laws and regulations for the affiliated subcontract and illegal subcontracting are not serious enough and lack maneuverability. It is difficult in practice accordingly for illegal cognizance and punishment law enforcement.

4 DETRIMENT OF AFFILIATED SUBCONTRACT AND ILLEGAL SUBCONTRACTING IN THE CONSTRUCTION PROJECT

4.1 Causes of corruption

The process of the affiliated subcontract and illegal subcontracting leads to corruption in the construction industry. Some of the enterprise or individuals would take affiliated or illegal subcontracting, subcontract, regardless of the individual ethics and the professional ethics of construction industry, in order to win the trust of the contractor or the owner enterprise.

4.2 Lower the engineering quality

Construction bidding projects require a level of qualification of construction enterprise because the project construction has certain difficulty and requires certain accuracy. Some affiliated construction enterprise that cannot meet the requirement of the construction difficulty and precision of the construction projects would frequently lead to unqualified construction projects because of its low management level and

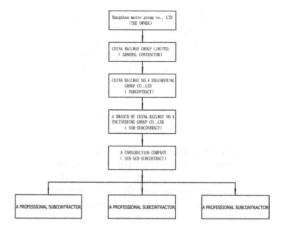

Figure 1. Hangzhou subway project is subcontracted four times.

personnel qualifications. For other projects that have been subcontracted for many times, the real construction enterprise or individual would cheat on workmanship and materials in pursuit of higher profits regardless of the relevant provisions of the laws and regulations in our country, which may leave a hidden danger to the quality of the project.

4.3 Lead to engineering accident

Because of the low management quality of affiliated enterprise and individuals or for the economic benefits, they would invest less manpower and material resources into the production safety, which can easily cause safety accidents. For example, in the Hangzhou subway accident, the project was subcontracted four times, resulting in accidents, as shown in Figure 1.

4.4 Delay the project

The construction enterprise transfers the project to other entity only for charging management fee. They do not involve in the project. The construction enterprise's only purpose is to pursue the economic interest. They will not consider the period of the project, and hence will not consider reputation losses caused by delaying the project. As for the subcontracted projects, especially for the projects that subcontracted for many times, the general contractor is difficult to guarantee the time limit for a project because it involves more enterprise and the personnel is more complex.

4.5 Lead to wage arrears for workers

In general, affiliated enterprise or illegal subcontracting enterprise is a private sector. It is difficult to ensure that the enterprise pay workers on time. Especially, payment of projects that are subcontracted for many times is made to the contractor by the owner, then to the subcontract enterprise, and finally to workers. It is difficult to pay workers on time.

4.6 Effect on the healthy development of construction industry

Project affiliated subcontract and illegal subcontracting is a serious violation of the management order of the construction industry. It harms the interests of others, and there is a negative impact on other enterprises. It affects the healthy development of construction industry.

5 HOW TO PREVENT THE AFFILIATED SUBCONTRACT AND ILLEGAL SUBCONTRACTING PHENOMENON IN THE PROCESS OF BUILDING ENGINEERING CONSTRUCTION

5.1 The government departments

5.1.1 The government departments should supervise the construction enterprises, the affiliated enterprise subcontract and illegal subcontracting to learn the regulations

The government departments should use network, media, newspapers, and other propaganda ways to spread rights and obligations that the tenderer and the bidder shall be followed. Furthermore, they should organize relevant construction enterprises to learn laws and regulations for affiliated subcontract and illegal subcontracting, make the tenderer, the bidder, and the winning bidder fully aware of their rights and obligations that they must follow, and understand the severe consequences of broking the law.

5.1.2 Tracking management after winning the bid
Relevant government functional departments should strengthen the tracking management of the enterprise that wins the bid. The government departments should check the enterprise in the followings two ways: check the project manager, technical engineer, and main operation management of employment and their social insurance relations; and check whether the bidding enterprise contracts project manager, technical director, and the main management personnel indeed engaged in the work.

5.1.3 Increase penalties for the affiliated subcontract and illegal subcontracting
Part of the construction enterprise has the fluky psychology. They think that it is difficult to find

the affiliated subcontract and illegal subcontracting in construction project and the current laws, and regulations of the punishment are not strong enough. The government department should make tougher punishment regulations that stop affiliated subcontract and illegal subcontracting effectively.

5.1.4 The government should set up special organizations to be responsible for affiliated subcontract and illegal subcontracting in the process of construction
Government departments should set up special organizations or establish special association that is responsible for organizing domestic large-scale construction enterprises, mutual communication-related research institutions, and well-known universities to develop more detailed and feasible rules and regulations.

5.1.5 Establish platforms for the report on affiliated subcontract and illegal subcontracting
The government functional departments could not work in the project construction site every day. However, the enterprises workers and management personnel understand the actual situation of construction project. As a result, the government departments should use the network or telephone channels to set up a reporting platform for people to report on affiliated subcontract and illegal subcontracting and give report entity or individual economic rewards.

5.2 Construction management side

5.2.1 The construction enterprise should strengthen their own team, introduce senior management personnel and technical personnel, improve the management ability and the technical level, and enhance the competitiveness of the enterprise, so they can succeed in the fierce competition in construction market.

5.2.2 Supervision company site management personnel should complete the duty; resolutely implement the relevant national laws and regulations. In the construction process, affiliated subcontract and illegal subcontracting should be active to the superior leadership department once found.

5.2.3 Owner enterprise should examine whether the winning entity is the affiliated entity, communicate with project manager and main management personnel to help confirm whether the winning entity is the affiliated entity at the same time. Owner enterprise

should also actively participate in the management of the project construction process to prevent the general contractor subcontracting. Also, the owner enterprise should avoid expressing or implying construction general contractor to transfer the project to its connections or community of interests.

6 CONCLUSION

In conclusion, affiliated subcontract and illegal subcontracting involve with many entities in interests and responsibility. It is difficult to manage. However, the affiliated subcontract and illegal subcontracting severely disrupted the order of construction market and seriously affected the healthy development of construction industry in China. As a result, we should give full play to resolutely stop these illegal activities.

REFERENCES

Bei, Zhang, Analysis of illegal subcontracting cases in the field of Construction Engineering, *Lanzhou University*, 3:43–46(2011).

Jin, Wang, Construction subcontracting and the illegal subcontracting issues countermeasures, *China Housing Facilities*, 1:36–37(2012).

Jinlu, Fu, Illegal subcontracting of construction projects in the illegal subcontracting and coping strategies, *modern enterprises research*, 18:50–53(2015).

Wei Song, The empirical study of engineering construction field characteristics of corruption, *Henan Social Sciences*, 21(05): 9–13(2013).

Yinghong, Zhang Illegal subcontracting, harm and prevention, *building economy*, 11: 24–27(2013).

Zhiyun Wang, Status and Countermeasures of illegal subcontracting in construction market, *China science and technology review*, 5:81–81(2009).

Advances in Energy Science and Equipment Engineering II – Zhou, Patty & Chen (Eds)
© 2017 Taylor & Francis Group, London, ISBN 978-1-138-71798-5

Research on industrial planning of sustainable development new town—a case study of Hengshui Lakefront New Town

Shibo Qi & Hongwei Li
Urban and Rural Construction College, Agricultural University of Hebei, Baoding, China

ABSTRACT: With the proposal of new-type urbanization, a round of new town planning and construction in China is in full swing. New town planning and construction can rapidly improve the level of urbanization in the short term. Industrial location and layout optimization is the basic work of new town construction. It determines the nature of the new town and the future direction of development. It is also closely related to the sustainable development of the new town. Taking Hengshui Lakefront new town industrial planning as an example, this paper analyzes the present situation and existing problems of new town industry and explores the methods of industrial structure location and spatial layout optimization. In addition, the paper also puts forward several strategies to promote the industry sustainable development.

1 ANALYSIS ON INDUSTRY DEVELOPMENT STATUS OF HENGSHUI LAKEFRONT NEW TOWN

Currently, there are two major problems in the industry development of Hengshui Lakefront new town: (1) The population of Hengshui Lakefront new town is scattered and the industry infrastructure is on a low level. Hengshui Lakefront new town is developed from villages and towns. It mainly includes two towns: Weijiatun and Pengduxiang. Because of the dispersion of the villages, the distribution of enterprises and population living is also dispersed. Therefore, the industry infrastructure construction is restrained. (2) The industry structure is unreasonable and the scale is small. At present, the proportion of the new town industrial structure is 16:61:23. The service industry and low-end industry is dominant. The development of the primary industry is extensive. However, the combination of agricultural planting and agricultural products processing industry is not highly developed. Products have low added value. Intensive agriculture is wanting. The second industry is mainly composed of rubber, machinery manufacturing, medical equipment, textile, and other industries. The main mode of operation is the family workshop style, and hence the scale is small. According to the existing data from Hengshui City Province Statistical Yearbooks (2014), the Industrial Enterprises above designated size accounted for only 3% of the annual output of 715 million yuan. The Industrial Enterprises below designated size accounted for 97% of the annual output of

Table 1. Different scale of industrial enterprises statistics.

	Industrial enterprises above designated size	Industrial enterprises below designated size
Ratio of the amount	3%	97%
Output value (million yuan)	715	496
Output value ratio	59%	41%

496 million yuan, as shown in Table 1. The service industry is the main service for high-income groups and is in shortage of mid-market to retain visitors. To solve these problems, this paper explores the methods for industrial structure location and spatial layout optimization.

2 INDUSTRIAL STRUCTURE LOCALIZATION OF NEW TOWN

2.1 Locating new town industrial structure based on regional economy

Industrial Structure localization is a fundamental job of the new town planning. It affects the direction of development and functions of the new town. The reasonable localization of new town industrial structure is important for the development of regional economy. As a result, the industrial structure localization should not only coordinate with the regional industrial layout to generate promotion forces for the regional economic development, but also highlights the advantages and characteristics of its own resources to avoid the similarity of industry.

2.2 Choose leading industry based on comprehensive advantages

After comparing and analyzing each industrial sector of the new town with the same industrial sectors in this region, this paper measures its advantages from the following aspects: natural resources, regional location, transportation costs, labor costs, land use price, adaptability to the market, resource exploitability, and technical feasibility. This paper chooses the industrial sector with the most comparative advantage and the largest development potential as the leading industry of new town.

2.3 Industrial structure localization of Hengshui Lakefront New Town

Hengshui Lakefront new town has the advantage of location and natural resources. It connects the city center district and Jizhou District with close distance to the Hengshui Lake, as shown in Figure 1. In the planning stage, we should strengthen the linkage between the new town and its neighboring area, coordinate all kinds of resources, enhance their function and position, and form the regional economic resultant force. Considering full play to the core role of Hengshui Lake, it determines the tourism industry as leading industry and reasonably develops the middle- and high-end service industry, modern ecological agriculture, and so on.

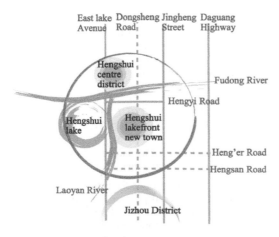

Figure 1. Analysis of location advantages of Hengshui Lakefront New Town.

3 ANALYSIS OF THE SPATIAL LAYOUT OF NEW TOWN INDUSTRY

3.1 Influencing factors for the spatial layout of new town industry

In addition to the traditional theory of land rent, resource advantages, and traffic factor, the influences of ecological environment, technological innovation, and environmental factors also become more and more outstanding.

3.1.1 Differential land rent
Simply speaking, differential land rent refers to the net income of land use (Xu, 2015). In general, the closer the city is to the center, higher the land-intensive utilization rate is. Furthermore, the land net income is higher and land acquisition rate is higher. Therefore, under the influence of land rent theory, the business is located in the center of the city, then industry, office, and agriculture.

3.1.2 Resource advantage
Resource advantage directly influences the spatial layout of the new town industry, and is the material basis for the development of the industry. To develop tourism, it is necessary to have rich cultural resources and landscape resources. The area that is rich in agricultural resources can develop "Farm Fun" and ecological agriculture sightseeing and picking industry. The area that is rich in mineral resources is capable of developing into industry assembling district. Therefore, we should analyze and classify the regional resources, then carry out the industrial spatial layout planning according to the development potential of all types of resources.

3.1.3 Ecological and environmental factors
With the awakening of people's environmental awareness and the implementation of the sustainable development strategy, much attention has been paid to the ecological environment. Industrial spatial layout with the direct relationship to the environment is becoming more and more important. Industrial spatial layout should not only consider the factors such as industrial agglomeration and transportation convenience, but also pays attention to the ecological environment factors. Therefore, giving full attention to the advantages of ecological environment is necessary to achieve the layout of the green industry.

3.1.4 Science and technology innovation ability
Scientific and technological innovation capability is related to the depth and breadth of the utilization of resources. It is also the driving force for the formation and change of industrial

spatial layout. Because of the improvement in the scientific and technological innovation capability, the transportation tools and channels, production tools, and network technology are optimized. Therefore, the effect of a series of natural factors, such as distance, on industrial spatial layout is decreased and the industrial spatial layout changes too.

In addition, transportation infrastructure, preferential policies, human capital, especially the high-quality human resources supply status, local cultural uniqueness, and many other factors also affect the industrial spatial layout. In the industrial spatial layout, we should comprehensively analyze the influence of various factors and realize the optimization of industrial spatial layout.

3.2 *Principle of industrial spatial layout*

Industrial spatial layout is a process that considers all kinds of resources and factors. It is also the process of selecting the best location for the industry and the enterprise in the region. We should follow the following principles to promote the sustainable development of the new town industry.

3.2.1 *Ecological priority*
The core of eco-priority principle is to pursuit a new development mode of green industry, which can unify economic benefits, social benefits, and ecological benefits. Only the industrial spatial layout that adhere to the principle of eco-priority can realize sustainable development of the new town industry.

3.2.2 *Regional overall development*
From the level of regional economy to carry out new town industrial spatial layout planning (Zhao, 2010), it is necessary to establish the concept of regional development and the industrial linkage mechanism between new town and old city. Finding appropriate industrial division of labor in the whole regional industrial chain is also important. Considering only the development of the new town itself, ignoring the impact of the surrounding regional environment and industry will inevitably lead to a series of problems, such as the fragmentation between new town and its neighboring areas, the repetitive construction of industry, and the shortage of characteristics. As a result, only following the principle of regional overall development can realize sustainable development of the new town industry.

3.2.3 *Intensive land use*
According to the principle of intensive use of land, industrial enterprises gradually focus on industrial agglomeration (Liu, 2015). Rural industrial enter-

prises should transform from family workshop style development to industrial agglomeration development in order to form the agglomeration effects, reduce the land waste caused by dispersed layout, and realize regional association industry resource sharing.

3.3 *Industrial spatial layout planning of Hengshui Lakefront New Town*

Hengshui Lakefront new town is the ecological protection zone and ecological economic zone with Hengshui Lake as its core. It is the ecological center of Hengshui City. Therefore, Hengshui Lakefront new town should focus on the development of ecological leisure, cultural tourism, leisure health, sports leisure, and other tourism and leisure industries.

Hengshui Lakefront new town takes Hengshui Lake resources as its core and makes industrial division according to the theory of differential rent theory. Land rent decline curve of different industrial is different. It takes Dongsheng Road and Sulin Line as boundary, as shown in Figure 2.

The tourism industry has the highest yield to the west of Dongsheng Road, but it is more dependent on Hengshui Lake, reducing at the fastest rate. Urban agriculture income and cutting speed are lower than those of tourism. It achieved higher yield than other industries between Dongsheng Road and Sulin Line. For these industries, the Hengshui Lake has little effect on it. As a result, it is dependent on the Hengshui Lake resource advantages, developing the travel enterprise and modern services with characteristics of the lake. It is focused on the development of urban agriculture sightseeing and picking industry between Dongsheng Road and Sulin Line. Industrial and logistics industry will move to the east of Sulin Line with the exploitation and utilization of the lake, as shown in Figure 3.

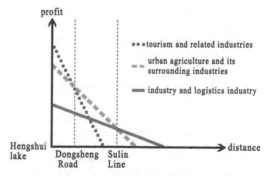

Figure 2. Industrial layout under the influence of land rent.

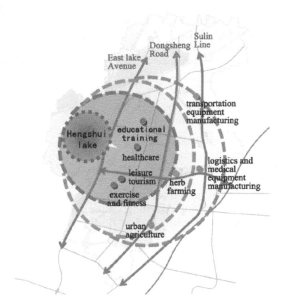

Figure 3. Development trend analysis of Hengshui Lakefront New Town industry.

4 OTHER ISSUES SHOULD BE CONSIDERED FOR INDUSTRIAL SUSTAINABLE DEVELOPMENT

4.1 *Emphasizing the combination of market leading and government guidance*

Industrial development should follow the market rules, fully consider the leading role of the market, and promote the rational allocation of production factors. However, industrial planning for sustainable development is not a by-product of unfettered market forces. It also requires the government to provide guidance in the market system at a very high level. The government should develop a series of compensatory policies to encourage the development of green industries, cultural and creative industries, as well as promote the construction of urban cultural ecology. Only in this way, we can avoid the dangerous consequences caused by the blind pursuit of economic growth.

4.2 *Realizing in situ urbanization via employing idle labor*

One of the purposes of the new town construction is to ease the pressure on the old city population and to solve the problem of employment of landless peasants. On the one hand, the new town should enhance its industrial attractiveness, so that the labor force can return hometown to gain employment. Industrial development should not only create economic benefits, but also pay more attention to the development of labor-intensive industries to create more jobs and provide adequate employment opportunities for surplus labor (Chang, 2004). On the other hand, in order to improve the professional quality and employment skills of the rural labor force and expand employment opportunities for them, related organization should carry out pre-occupation training for farmers according to the market demands and business requirements. Only the industry planning is able to solve the employment problem of landless peasants and is capable of achieving the farmer in-place urbanization to propel a new steady, fast, and healthy development.

4.3 *Stimulate collaborative development of new technique industry and traditional industry*

The traditional industries of the town developed from the township have a considerable foundation and occupy a larger proportion of the regional economy. However, the traditional industry development mode and technology are backward. The comprehensive utilization rate of resources is low. Therefore, we should actively support the development of high-tech industry and use technological innovation and advanced technology to realize the traditional industry transformation and upgrading. As a result, it is necessary to improve the competitiveness of traditional industries and promote coordinated development of high-tech industries and traditional industries.

In addition, it is important to determine construction funding source to ensure industrial planning meet the requirements (Yang, 2014). It is also significant to coordinate the relationship between development and environment, as well as adhere to the low-carbon city construction road.

5 CONCLUSION

City is prospered by industry, and industry is supported by city. Industrial development is based on the development of cities and towns. It is also the foundation of urban development. New town industrial planning should comprehensively analyze the various factors that influence the development of industry and tap the advantages of resources. Industrial spatial layout planning should coincide with the principles of ecological priority, regional overall development, land-intensive use, ecological and regional development, and the principles of intensive land use. Industry planning should look for opportunities from regional transferring industries and emerging industries to form industrial scale advantage look for opportunities from its own resources endowment. It should develop practical competitiveness to form the industrial difference advantage. Building reasonable town industrial structure and layout is conductive. Forming the green industry development mode can unify the economic benefits, social benefits, and ecological benefits, which is beneficial to promote the sustainable development of the new town industry.

REFERENCES

Chang, Y.Q., B. Li, Discussion on the Employment of Rural Labour Force in the Process of City, (01): 84–86, (2004).

Liu, G., M.H. Wang, J.T. Wang, Y.T. Fang, Plans for Country Land Industry against the Backdrop of Urban-rural Integration, (02): 82–88, (2015).

Xu, Q.W. Study on Industrial Layout of Industrial Towns in Hexi Region, 29–32, (2015).

Yang, L., Research on the Industrial Structure of Small Town Planning under the Background of Urbanization, (10): 383–383,(2014).

Zhao, Q., Research on Small Town of Hefei Industry Layout and Optimization of space, 28–35, (2010).

Advances in Energy Science and Equipment Engineering II – Zhou, Patty & Chen (Eds)
© 2017 Taylor & Francis Group, London, ISBN 978-1-138-71798-5

Artificial bee colony algorithm for the Tianjin port tug scheduling problem

Renhai Yu
Dalian Maritime University Dalian, Liaoning, China

Zhichao Yan
Tianjin port Barge company, Tianjin, China

Wei Li, Qingming Wang & Zuojing Su
Dalian Maritime University Dalian, Liaoning, China

ABSTRACT: Aiming at the severe problem that the development of Tianjin port faced regarding the tugboat scheduling, this paper mainly focuses on the Tianjin port tugboat scheduling and the principle of optimal time in the harbor. First, a tugboat scheduling model that accords with the request of Tianjin port is established based on the characteristics of the tugboat scheduling problem. Then, we employ the artificial bee colony to solve the tugboat scheduling problem. Finally, according to the parameters of the tugboat of Tianjin port, an optimized scheme is achieved and the accuracy and feasibility of the scheme are analyzed by simulation as well.

1 INTRODUCTION

Most ships would need one or more tugs to help assist entering and leaving port operation. Modern ships are of larger dimension and deeper draft, which bring about inflexibility of ships. For this reason, tug assistance has become more important than ever before. Therefore, solving the tug scheduling problem with a method that can reach a balance with least time consumption and cost has drawn increasing attention.

With the in-depth study of intelligent algorithm in recent years, a large number of optimization algorithms have emerged in the field of scheduling, such as Artificial Bee Colony (ABC), Ant Colony algorithm (ACO), Particle Swarm Optimization (PSO), Genetic Algorithm (GA), and Neural Network (NN) algorithm (Yan, 2015; Kim, 2003; Pietro, 2007; Kim, 2003; LIU, 2012; MUSTAFA, 2013; CHEN, 2012; GAO, 2013). Artificial Bee Colony (ABC) algorithm was proposed in 2005 by Turkish scholar Karaboga. The original purpose of this algorithm is to settle function optimization problem. The original idea is inspired by the way a swarm exchange information among individuals, and collect honey through mutual cooperation. Compared with classical optimization methods, ABC algorithm does not require constrains or objective function but needs only a small number of parameters; is easy to use; and has high search precision (Karaboga, 2009; Karaboga, 2008). Therefore, ABC algorithm is suitable to solve

tug scheduling problem. This article introduces artificial colony algorithm to solve tug scheduling problem and simulate the algorithm according to relevant parameters of Tianjin port tugs. It then reaches an optimized dispatch scheme and analyzes the feasibility of the accuracy of experimental results and solution.

2 TUG SCHEDULING MODEL

2.1 *Model assumption*

During the modeling process, we transform the unresolved problems into building a single berth base and considering the least time consumption of tug operation model. The main assumptions are as follows:

1. Consider only berthing, unberthing, shifting berth condition, shifting berth could be transformed into re-berthing after unberthing, regardless of salvage or malfunction.
2. The operation of entering and leaving the port or shifting berth would not be affected by other ships, and this operation would be definitely successful.
3. All tugs would be directly turned to execute another task once the former one is accomplished. The time spent on the former task could be estimated by historical data.
4. There would be no malfunction or unforeseen circumstance happened on tugs during the

operation process. All tugs would be navigating in the same speed.

5. Not considering the priority of ship operation.
6. All operations would take a day as a period, which means if one period is initiated, tugs would be standing-by at berth base and readily available. Vessels would be all waiting for operation at anchorage or berth. At the end of one period, tugs would return to berth base and terminate all operations.
7. The mapping relationship between tugs and vessels is shown in Table 1.

PS1, PS2, PS3, PS4, PS5, and PS6 represent the following tug types: 1200HP, 2600HP, 3200HP, 3400HP, 4000HP, above 5000HP, respectively. S1, S2, S3, S4, and S5 represent the following tug lengths: less than 100m, above 100 but less than 200m, above 200m but less than 250m, and above 250m but less than or equal to 300m, respectively. Because this article only discusses mapping principle in general case, special need vessels are not taken into consideration.

2.2 Symbol definition

2.2.1 Parameters

J, l stands for stage number, j, $l \in J = \{1,2\}$ stands for berthing and unberthing, i, k stands for ship number, m is the tug number, M stands for the set of all tugs, ta_m is the type of tug m, $N:N = \{1,2, ..., n\}$ stands for the set of all ships, S_i is the type of NO.i ship, set_i is the set of tugs providing service for NO.i ship, O_{ij} is the task of ship i in phase j, and $M_{ij}:M_{ij} = \{m \mid ta_m = set_i\}$ stands for the set of tugs, which could not only execute task Oj but also satisfy tug and ship mapping principle. These tugs shall be berthing at berthing base. E_{jm} stands for the set of vessels that receive tug m's phase j assistance. LOS_{ij} stands for the starting position of task O_{ij}, when $j = 1$, LOS_{ij} represents the confluence point of ship and tug, when $j = 2$, LOS_{ij} represents the berth at which the ship would conduct loading and unloading and LOF_{ij} stands for the finish position of task O_{ij}, when $j = 1$, LOF_{ij} stands for the

berth at which the ship would conduct loading and unloading, when $j = 2$, LOF_{ij} stands for the confluence point of ship and tug, $ST(a, b)$ stands for the time cost when the tug is navigating from a to b, p_{ij} stands for the operating time of task p_{ij}, tb_i stands for the time cost when ship i is navigating from holding point to berthing location, $tb_i = ST (LOS_{ij}, LOF_{ij})$, te_i stands for the time a ship is berthing at berth i, to_i stands for loading and unloading time a ship would spend at berth i, tl_i stands for the time cost when ship i is navigating from unberthing location to departure location, $tl_i = ST(LOS_{i2}, LOF_{i2})$, S^m_{ijkl} stands for the switch time that tug m would spend when it is navigating from the finishing point of task O_{ij} to the starting point of task O_{kl}, bp stands for tug berthing base, and H stands for a sufficiently large positive number.

2.2.2 Decision variable

$$x_{ijm} = \begin{cases} 1 \text{ tug m execute task } O_{ij} \\ 0 \text{ otherwise} \end{cases}$$

$$y^m_{ijkl} = \begin{cases} 1 \text{ tug m execute both task } O_{ij} \text{ and task } O_{kl} \\ 0 \text{ otherwise} \end{cases}$$

$$u^m_{ijkl} = \begin{cases} 1 \text{ tug m execute task } O_{ij} \text{ before task } O_{kl} \\ \quad (O_{kl} \text{ shall not necessarily be the last task}) \\ 0 \text{ otherwise} \end{cases}$$

$$Z^m_{ijkl} = \begin{cases} 1 \text{ tug m execute task } O_{ij} \text{ before task } O_{kl} \\ \quad (O_{kl} \text{ shall be the last task}) \\ 0 \text{ otherwise} \end{cases}$$

2.2.3 State variable

TS_{ij} stands for the starting time of task O_{ij};
TF_{ij} stands for the finishing time of task O_{ij};
BT_m stands for the departure time of tug m from berth base during the planned period;
FT_m stands for the arriving time of tug m after finishing the last task within the planned period.

2.3 Objective function

This article would construct model so as to shorten the tug operation time. In single task, there are two important timings in a certain operation process, namely the starting time and finishing time of the task. The operation time function of a certain tug is the time when a tug has finished with its task and returned to berth base (FT_m) and the time when a tug is about to set out from berth base and execute the next task. Therefore, the whole time cost of tug operation calculation formula could be concluded as:

$$F = \sum_{m \in M} \left(FT_m - BT_m\right) \tag{1}$$

Table 1. Mapping relationship between tugboats and ships.

Ship type	Number of tugs required	Tug type
S1	1	PS1, PS2, PS3, PS4, PS5, PS6
S2	2	PS2, PS3, PS4, PS5, PS6
S3	2	PS3, PS4, PS5, PS6
S4	2	PS4, PS5, PS6
S5	2	PS5, PS6

FT_m and BT_m could be expressed as follows:

$$FT_m = TF_{ij} + ST\left(LOF_{ij}, bp\right)$$
$$\left\{(i,j,m)\middle|z_{ijkl}^m = 0, \forall k \in N, j \in J\right\} \quad (2)$$

$$BT_m = TS_{ij} - ST\left(bp, LOS_{ij}\right)$$
$$\left\{(i,j,m)\middle|z_{ijkl}^m = 0, \forall k \in N, j \in J\right\} \quad (3)$$

In order to set the shortest operation time as objective function and tally with the actual situation, the author defines the following objective function:

$$F = \sum_{m \in M}\left(FT_m - BT_m\right) \quad (4)$$

Constraint conditions are:

$$TS_{ij} \ge 0, \forall i \in N, \forall j \in J \quad (5)$$

$$TS_{i1} + p_{i1} + to_i \le TS_{i2}, \qquad \forall i \in N \quad (6)$$

$$\begin{cases} \sum_{m \in M} x_{ijm} = 1, S_i = S_1 \\ \sum_{m \in M} x_{ijm} = 2, \text{otherwise} \quad \forall_i \in N, \forall_j \in J \end{cases} \quad (7)$$

$$\begin{cases} set_i = \{PS1, PS2, PS3, PS4, PS5, PS6\} & S_i = S1 \\ set_i = \{PS2, PS3, PS4, PS5, PS6\} & S_i = S2 \\ set_i = \{PS3, PS4, PS5, PS6\} & S_i = S3 \quad \forall i \in N \\ set_i = \{PS4, PS5, PS6\} & S_i = S4 \\ set_i = \{PS5, PS6\} & S_i = S5 \end{cases} \quad (8)$$

$$y_{ijkl}^m \le 0.5\left(x_{ijm} + x_{klm}\right) \le y_{ijkl}^m + 0.5, \forall i, k \in E_{jm},$$
$$\forall m \in M_{ij}, \forall j, l \in J \quad (9)$$

$$y_{ijkl}^m = y_{klij}^m, \forall i, k \in E_{jm}, \forall m \in M_{ij}, \forall j, l \in J \quad (10)$$

$$u_{ijkl}^m + u_{klij}^m = y_{ijkl}^m, \forall i, k \in E_{jm}, \forall m \in M_{ij}, \forall j, l \in J \quad (11)$$

$$u_{ijkl}^m - Z_{ijkl}^m \ge 0, \forall i, k \in E_{jm}, \forall m \in M_{ij}, \forall j, l \in J \quad (12)$$

$$\sum_{k \in El_m} z_{ijkl}^m \le 1, \forall i \in E_{jm}, \forall m \in M_{ij}, \forall j, l \in J \quad (13)$$

$$\sum_{k \in El_m} z_{klij}^m \le 1, \forall i \in E_{jm}, \forall m \in M_{ij}, \forall j, l \in J \quad (14)$$

$$TS_{ij} + p_{ij} + s_{ijkl}^m \le TS_{kl} + H\left(1 - z_{ijkl}^m\right), \forall i, k \in N,$$
$$\forall m \in M_{ij}, \forall j \in J \quad (15)$$

$$TS_{kl} + p_{kl} + s_{klij}^m \le TS_{ij} + H\left(1 - z_{klij}^m\right), \forall i, k \in N,$$
$$\forall m \in M_{ij}, \forall j \in J \quad (16)$$

$$\begin{cases} p_{ij} = tb_i + te_i, & j = 1 \\ p_{ij} = tu_i + tl_i, & j = 2 \end{cases} \quad (17)$$

$$s_{ijkl}^m = ST\left(LOF_{ij}, LOS_{kl}\right) \cdot z_{ijkl}^m$$
$$\forall i, k \in N, \forall m \in M_{ij}, \forall j, l \in J \quad (18)$$

$$x_{ijm}, y_{ijkl}^m, u_{ijkl}^m, z_{ijkl}^m = 0 \text{ or } 1, \forall i, k \in N,$$
$$\forall m \in M_{ij}, \forall j, l \in J \quad (19)$$

Equation (5) indicates that the task starting time is over and above 0. Equation (6) indicates that ships could conduct unberthing or shifting berth operation only after it has finished with the berthing operation. Equation (7) indicates that one tug shall be only allocated to ship type 1. Equation (8) defines mapping principle in Table 1. Equations (9) and (10) indicate that when $x_{ijm} = x_{klm} = 1$, $y_{ijkl}^m = y_{klij}^m = 1$. Equation (11) indicates that one tug shall only help assist one ship at a time. Equations (13) and (14) indicate that after scheduling sequences are fixed, any tug shall only have one task before and after the adjacent task. Equations (15) and (16) indicate that the starting time of next task shall not be over and above the former task's finishing time. Equation (18) indicates the switch time between two tasks. Equation (19) indicates the value of decision variable.

3 DESIGN OF TUG SCHEDULING SCHEME BASED ON ARTIFICIAL BEE COLONY ALGORITHM

3.1 *Principle of ABC algorithm*

Suppose the solution domain is of D dimension. The number of employed bees and observation bees is equal to SN. The number of employed bees and observation bees is equal to the amount of nectar source. Then, the standard ABC algorithm would transform optimization solving process to search in dimension searching space. Every nectar source represents a possible solution. Nectar volume of each nectar source corresponds to the fitness of solution. One employed bee corresponds to one nectar source. Employed bee, which corresponds to NO.i nectar source, would search for new nectar according to the following function:

$$x_{id}' = x_{id} + \phi_{id} + \left(x_{id} - x_{kd}\right) \quad (20)$$

In this function, $i, k \in \{1, 2, ..., SN\}$, $d \in \{1, 2, ..., D\}$, Φ_{id} is a random number in the interval $[-1, 1]$. Standard ABC algorithm would compare new generated possible solutions $X_i' \in \{x_{i1}', x_{i2}', ... x_{id}'\}$ with original solution $X_i \in \{x_{i1}, x_{i2}, ... x_{id}\}$ and adapt greedy selection strategy to reserve better solution. Each observation bee would select a nectar source according to probability. The probability formula is as follows:

$$p_i = \frac{fit_i}{\sum_{i=1}^{SN} fit_i} \quad (21)$$

Where fit_i is the adaptive value of possible solution X_i. As for the selected nectar, the observation bee would search for a new possible solution according to the above-mentioned probability formula. When all employed bees and observation bees have finished their search in the entire searching space, on the condition that a nectar source's adaptive value has not been improved within given steps (defined as controlling parameter "limit"), then the nectar source shall be abandoned. Meanwhile, the corresponding employed bee would become a scout. The scout would search for new possible solution through the following formula:

$$x_{id} = x_d^{\min} + r\left(x_d^{\max} - x_d^{\min}\right) \qquad (22)$$

where r is a random number in the interval [0,1] and $x^{min}{}_d$ and $x^{max}{}_d$ are the lower and upper bonds of NO.d dimension, respectively.

3.2 Solve the model with ABC algorithm

In order to solve parameter selecting problem, test function is introduced in this article. Through combining ABC algorithm and trail function, adjusting initialization parameter of ABC algorithm, the pattern of initialization parameter and stable solution could be concluded. We could set initialization parameter according to the following pattern:

$$\min f\left(x_1, x_2\right) = 0.5 + \frac{\left(\sin\sqrt{x_1^2 + x_2^2}\right) - 0.5}{\left(1 + 0.001\left(x_1^2 + x_2^2\right)\right)^2} \qquad (23)$$

where $x_1, x_2 \in [-100,100]$. This function is a complicated two-dimensional function. There are countless minimum point in this function. The minimum reaches 0 at (0,0). Global optimum is difficult to find, as this function possesses a strong concussion form.

4 INSTANT ANALYSIS

4.1 Instant calculation

In order to verify the effectiveness of the algorithm, this article selects statistics from Tianjin port tug and barge corporation over a period of time on a certain day in August 2015. When initializing simulation, we assume that all conditions are in full accordance with model assumptions mentioned in Chapter 2. In the process of executing tasks, time consumed would differ from each other due to tugs and operation type. At present, 16 tugs exist, including the five types of tug presented in Table 1. In practice, the assumed time cost is shown

Table 2. Tugboat configuration and operation time.

Tug No.	Tug horsepower (HP)	Average task time (min)
1	1200	40
2	2145	
3	2230	
4	2230	
5	2230	
6	2600	48
7	2600	
8	2600	
9	2600	
10	2800	60
11	3200	
12	4000	75
13	5000	85
14	5000	
15	5200	
16	6000	

Table 3. Ships operate database.

Ship no.	Ship type	Task type	Tug required
1	S2	Berthing	2
2	S1	Unberthing	1
3	S1	Unberthing	1
4	S3	Berthing	2
5	S2	Shifting berth	2
6	S1	Berthing	1
7	S1	Unberthing	1
8	S2	Berthing	2
9	S1	Shifting berth	1
10	S1	Unberthing	1
11	S5	Shifting berth	2
12	S1	Berthing	1
13	S3	Berthing	2
14	S4	Shifting berth	2
15	S2	Shifting berth	2
16	S4	Unberthing	2
17	S1	Berthing	1
18	S2	Shifting berth	2
19	S5	Unberthing	2
20	S1	Unberthing	1

in Table 2. There are 20 stand-by ships, whose tasks are shown in Table 3.

First, substitute statistics into algorithm. Parameters are set as follows: S = 40, limit = 100, and maximum number of iteration = 400. As shown in Figure 1, time (min) is on vertical axis and iteration time (time) is on horizontal axis.

In early stage, the convergence is too fast to see if the result is the optimum solution. The author suggests increasing the number of maximum iteration times (as shown in Figure 2).

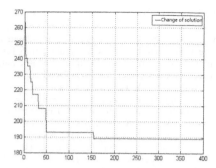

Figure 1.   The NO. 1 converge curve.

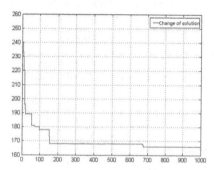

Figure 2.   The No. 2 converge curve.

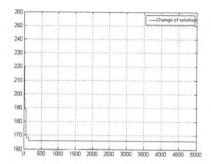

Figure 3.   The No. 3 converge curve.

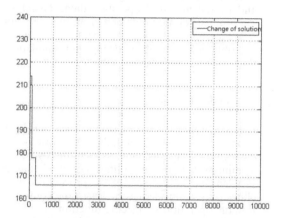

Figure 4.   The No. 4 converge curve.

Tug 1 assist ship 9. Tug 2 would first assist ship 12 and then ship 3. Tug 3,4 would help ship 18. Tug 5,6 would first assist ship 1 and then ship 5. Tug 7,8 would first assist ship 4 and then ship 8. Tug 9 would first assist ship 17, then ship 2, and finally ship 7. Tug 10,11 would first assist ship 13 and then ship 15. Tug 12 would first assist ship 6, then ship 10, and finally ship 20. Tug 13,14 would assist ship 14. Tug 15,16 would first assist ship 19, then ship 16, and finally ship 11.

In early stage, the convergence is too fast, later it would rapidly turn slow. The author suggest to increase the number of maximum iteration times to see if the optimal solution has been obtained (as shown in Figure 4). The corresponding scheduling sequences are as follows:

Tug 1 would help ship 12. Tug 2 would first assist ship 3 and then ship 2. Tug 3,5 would assist ship 8. Tug 4,8 would first assist ship 4 and then ship 5. Tug 6,7 would first assist ship 13 and then ship 15. Tug 9 would first assist ship 10, then ship 17, and finally help ship 9. Tug 10,11 would first assist ship 1 and then ship 18. Tug 12 would first assist ship 20, then ship 6, and finally help ship 7. Tug 13,14 would first assist ship 16 and then ship 11. Tug 15,16 would first assist ship 19 and then ship 14.

Comparing Figure 2 with Figure 3, the solution would become stable after 5000 iteration times.

The corresponding scheduling consequences are as follows:

Tug 1 would assist ship 6. Tug 2 would first assist ship 17, then ship 3, and finally help ship 9. Tug 4,5 would assist ship 8. Tug 3,8 would first assist ship 1 and then ship 15. Tug 6,7 would first assist ship 13 and then ship 5. Tug 9 would first assist ship 20, then ship 10, and finally ship 7. Tug 10,11 would first assist ship 4 and then ship18. Tug 12 would first assist ship 12 and then ship 2. Tug 13,14 would first assist ship 19 and then ship 14. Tug 15,16 would first assist ship 16 and then ship 11.

4.2   Result analysis

Comparing scheduling sequences of the above-mentioned tests, we could find that ships, which need collaborative operation, could not be estimated. The amount of assignments could not be balanced. Overall planning could only achieve a global short time scheduling. Since Tug 1 is short in horsepower, its work load is the lightest among other tugs in three solution procedures.

Because algorithm parameters have been discussed in the previous part of the article, final instant analysis could be simulated using algorithm parameters. From these simulation results, we could find that after S and the limit have been determined, when the number of maximum iteration reached 5000, solution would become stable. Minimum task executing time reached a stable level at 167 min. All optimum scheduling sequences are the solution when the number of maximum iterations is 5000.

Artificial bee colony algorithm could bring about good solving accuracy in the field of scheduling. In this article, collaborative task time of tugs could not be precisely determined. Statistics used in this article are gathered through empirical estimation and partial historical operation data. Therefore, the optimal solution obtained in this article is for reference only. However, the operation sequences are obtained through modeling and simulation, thus having some reference value.

5 CONCLUSION

Port scheduling problem has long been unsolved. From a practical standpoint, this article briefly discusses practical scheduling problem discovered in Tianjin port. It aims at optimizing tug scheduling problem in Tianjin port. In the process of modeling, we set considering the least time consumption tug operation as objective function to conduct modeling. Taken berthing, unberthing, and shifting berth into consideration, the author introduces ABC (artificial bee colony) algorithm to solve the model. During calculation, the ABC algorithm shows a great kinematic performance and solves the model in a high speed without running into local optimum. Finally, according to simulation results of relevant parameters, the author analyzes the influence of different iteration times posed on solution. A final reasonable scheduling scheme is concluded, through which the feasibility and practicability of the algorithm are verified.

ACKNOWLEDGMENT

This work is supported by "the Fundamental Research Funds for the Central Universities" (No. 3132014031).

REFERENCES

Chen, S.M., Sarosh, A. Dong, Y.F. Simulated annealing based artificial bee colony algorithm for global numerical optimization [J]. Applied Mathematics and Computations, 2012, 219:3575–3589.

Gao, W.F., Liu, S.Y. Huang, L.L. A novel artificial bee colony algorithm based on modified search equation and orthogonal learning [J]. IEEE Transaction on Systems, Man and Cybernetics, 2013, 43(3):1011–1024.

Kap Hwan Kim, Young-Man Park. A crane scheduling method for port container terminals [J]. European Journal of Operational Research. 2003 (3):752–768.

Kap Hwan Kim, Young-Man Park. A crane scheduling method for port container terminals [J]. European Journal of Operational Research. 2003, (3):752–768.

Karaboga, D. Akayb. A comparative study of artificial bee colony algorithm [J]. Applied Mathematics and Computation, 2009, 2(14):108–132.

Karaboga, D. An idea based on honey bee swarm for numerical optimization [R]. Computers Engineering Department, Engineering Faculty, Erciyes University, 2005.

Karaboga, D. Basturkb. On the performance of artificial bee colony (ABC) algorithm [J]. Applied Soft Computing, 2008, 8(1):687–697.

Liu, Y., Ling, X.X. Liang, Y. et al. Improved artificial bee colony algorithm with mutual learning [J]. Journal of Systems Engineering and Electronics, 2012, 23(2):265–275.

Mustafa, S.K., Mesut, G. A recombination-based hybridization of particle swarm optimization and artificial bee colony algorithm for continuous optimization problems [J]. Applied Soft Computing, 2013, 4(13):2188–2203.

Pietro Canonaco, Pasquale Legato, Rina M. Mazza, Roberto Musmanno. A queuing network model for the management of berth crane operations [J]. Computers and Operations Research. 2007 (8): 2432–2446.

Yan Zhichao. The Port Tugboat Scheduling Optimization Based on Artificial Be Colony [D]. Dalian Maritime University, 2015.

Advances in Energy Science and Equipment Engineering II – Zhou, Patty & Chen (Eds)
© *2017 Taylor & Francis Group, London, ISBN 978-1-138-71798-5*

Analysis of landscape and spatial patterns in land use of Pudong New Area, Shanghai

Guang Yang & Fangyuan Chen
College of Surveying and Geo-information, Tongji University, Shanghai, China

ABSTRACT: Landscape and spatial patterns can reflect land use condition of urban development, providing effective decision support for sustainable development of the region. The paper analyses spot area, spot density, landscape complexity and landscape isolation of landscape pattern and aggregation of spatial pattern using spot data of Pudong New Area, Shanghai in 2015. The result shows: The spots of agricultural land and industrial land of Pudong New Area, Shanghai in 2015 are dispersed, while the aggregation of residential land is high. The implementation of policies should be adjusted accordingly based on the characteristics of various types of land use.

1 INTRODUCTION

Land use is an important aspect of the research on geography, and one of the most frontier issues of global changes (Hao, 2007). Land use modes have a significant impact on Earth's climate, hydrology, soil, biodiversity and population bearing capacity. The form of land use has changed the natural landscape scene of Earth's surface, in addition, it has an important influence on material recycling and energy distribution. In recent years, the relationship studies among land use, landscape pattern and spatial pattern have gradually become an important way for human beings to understand nature (Zhang, 2006).

At present, though China is in a stage of rapid development, many problems still exist in the process of institutional reform. The landscape and spatial patterns of land use are to determine the rationality of land use forms, take fully use of land, and provide effective decision support in the implementation of policies for the sustainable development of the region (Zhang, 2012). From the perspective of landscape and spatial features of land use, the paper studies landscape index and spatial aggregation on the basis of spots, which can describe various types of land, find spot aggregation and spatial distribution rules and provide a theory foundation to promote the optimal utilization of land (Miao, 2013).

2 LANDSCAPE AND SPATIAL FEATURES OF LAND USE

2.1 Landscape features of land use

In landscape ecology, landscape feature refers to landscape index of shape, size, density, relative distance, interrelation of the spots at local scale. At present, domestic and foreign scholars have proposed a variety of landscape pattern metrics (Zheng, 2010; Zhang, 2009; Wang, 2005), by calculating the index in spot area, spot density, landscape complexity and landscape isolation, the paper describes the landscape patterns of land use situation quantitatively (Jin, 2003; Li, 2010). Calculation methods and the meaning of the relevant indicators are shown in Table 1.

2.2 Spatial features of land use

Spatial feature of land use refers to spatial distribution of spots at global scale. Spatial distributions (such as aggregated distribution, discrete distribution) and spatial relationships (such as spatial autocorrelation, spatial heterogeneity) reflect different spatial distribution patterns and spatial relationships of diverse land types, reveal the utilization of land, changes in industrial structure and the influence of national policies on land use patterns (Kimberly, 2002; Eklund, 1998).

Aggregation of land is one of the most significant features of spatial pattern. In ArcGIS10.3, common methods of spatial aggregation include Similarity Search, Local Spatial Autocorrelation (Moran's I), Subgroup Analysis (Grouping Analysis), Optimized Hotspot Analysis, Local Gi Statistics Index (Getis- ord Gi *).

In this paper, it calculates high value (hot spot) or low value (cold spot) by the means of Getis-ord Gi * at local area. In ArcGIS, the sample with red mark represents high value, showing aggregated distribution in space; the blue represents low value, presenting aggregated distribution in the same space, but the distribution of eigenvalue

Table 1. Landscape index of land use.

Landscape index	Equation	Meaning
Spot area	Constant	Refers to the area of a certain type of land, and the larger value indicates that the type of land in the region accounts for a larger proportion.
Spot density	$D = \sum_{i=1}^{m} \dfrac{n_i}{A}$	Refers to intensive degree of spots, used to measure the degree of fragmentation of spots in the study area. The smaller value means the spots are more regular, human factors are less, and nature attributes are more primitive. The larger value indicates the spots have serious fragmentation, the boundary is more complex, human factors are more, and the level of development is higher.
Landscape complexity	$L_{ij} = \dfrac{0.25 \cdot E_{ij}}{\sqrt{A_{ij}}}$	Refers to complexity of the spot shape. The greater value represents the complicated land shape, the greater intensity of land planning and development, and a high degree of development. The smaller value represents less human factors and land utilization degree.
Landscape isolation	$F_i = D_i / S_i$ $D_i = 1/(2\sqrt{n/A})$ $S_i = A_i / A$	Refers to discrete degree of spots in a certain landscape type. The greater value means poor correlation among the same land types and dispersed spots. The smaller value represents good connectivity.

belongs to smallness. The yellow sample is random without the characteristics of aggregated distribution (Yin, 2011).

3 LANDSCAPE AND SPATIAL PATTERNS OF PUDONG NEW AREA, SHANGHAI IN 2015

The paper calculates landscape and spatial index (Getis- ord Gi *) of Pudong New Area, Shanghai in 2015, analyses landscape features and spatial aggregated distribution modes in the course of land use. And it focuses on spatial distribution of agricultural, industrial and residential lands. At last, it analyses layout and development trend of the land (Geits, 2003).

3.1 General situation in the study area

Pudong New Area is located in the east of Shanghai and the eastern margin of the Yangtze River Delta. In 1990, the State Council made the decision that developing Pudong New Area. After twenty years of development, the fields in finance, cultures etc. have developed rapidly, which has an important role in Shanghai's urban development (Cen, 2011). Also, the ecological role of Pudong has been of great importance. Therefore, the study of land use in Pudong New Area is representative (Yin, 2007; Zeng, 2009).

3.2 Landscape features of the area

The total area of agricultural, industrial and residential lands in Pudong New Area is 1053.7 km² in

Table 2. Landscape index statistics of land use.

Types of land use	Agricultural land	Industrial land	Residential land
Spot area (km²)	496.6	167.0	281.6
Spot density (block/km²)	242.4	72.2	256.5
Landscape complexity	1.86	1.36	1.28
Landscape isolation	88.5	830.6	201.4

2015. Total number of spots and the average area is 214,459 and 4913.1 m² respectively. The paper calculates four landscape index, namely spot area, spot density, landscape complexity and landscape isolation, and analyses landscape pattern of land use. The relevant statistical results are shown in Table 2.

The value of landscape index in three different land use types can be seen from Table 2. Agricultural, industrial and residential lands account for 52.5%, 17.7% and 29.8% of total area respectively. Spot densities of agricultural and residential lands reach to 242.4 blocks per km² and 256.5 blocks per km², which means agricultural land and residential land are the primary land use ways in Pudong New Area. From the point of view of landscape complexity, the biggest agricultural land index, indicating that human factors on agricultural land are relatively large, and the degree of development is high. From the perspective of landscape isolation, a larger index of industrial land, explaining that the spots are scattered and have poor connectivity, while residential land and agricultural land have smaller index, showing a fine connection and aggregation of spots.

3.3 Spatial features of the area

The paper studies spatial patterns of the three different land types, namely agricultural land, industrial land and residential land in Pudong New Area. It analyses aggregation of these three types of land spots, according to the degree of aggregation, it can be classified as non-aggregation, low-aggregation, moderate-aggregation and high-aggregation, each aggregation type includes aggregation of low and high value. The aggregating spot number and proportion of each corresponding type of land are shown in Table 2, and the specific proportion of the case is shown in Figure 1.

3.3.1 Hotspot distribution of agricultural land

As Figure 1 shows, the non-aggregation area of agricultural land accounts for 58.38% of the total. The specific distribution is shown in Figure 2. The agricultural land of Pudong New Area concentrates mainly in the south (the original Nanhui District, Shanghai). The red part represents a high value aggregated distribution in Caolu Town, Zhuqiao Town, Laogang Town, Shuyuan Town, Shengang Town and Luchaogang Town. The agricultural area of these towns and streets is large and distributes densely. The blue part represents a low value aggregated distribution in Nicheng Town, Luzao Town, Zhoupu Town, Dingqiao Town, Datuan Town, spots of agricultural area in these towns distribute densely, while aggregation of other areas is not obvious.

3.3.2 Hotspot distribution of industrial land

As Figure 1 shows, the non-aggregation area of industrial land accounts for 63.90% of the total,

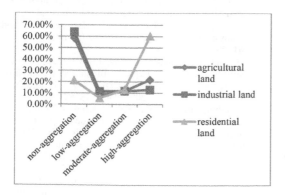

Figure 1. The aggregation of land use of Pudong New Area in 2015.

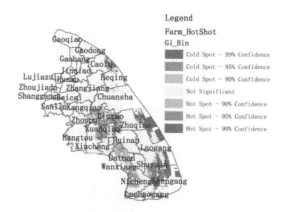

Figure 2. Spatial hotspot distribution of agricultural land.

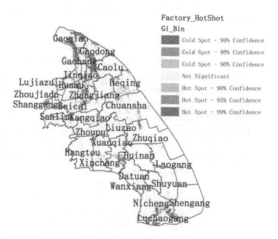

Figure 3. Spatial hotspot distribution of industrial land.

which is more than the sum of the other three types of aggregation, indicating that aggregation degree of industrial land is not high, and there is considerable potential for aggregation. The specific distribution is shown in Figure 3, the industrial land of Pudong New Area concentrates mainly in the northwest, including Gaoqiao Town, Gaohang Town, Gaodong Town, Jinqiao Town, Zhangjiang Town, LuChaogang Town (its southeastern region) and surrounding areas, the spots of these areas are large and dense. And low aggregated distribution areas are mainly in Heqing Town, Xinchang Town and Dongming Road Street, the spots of these areas are small and dense. These areas distribute mainly in Shanghai Biotechnology Industrial Park, Shanghai Wailianfa Industrial Park, Shanghai Jinling & Jinqiao Industrial Park, etc.

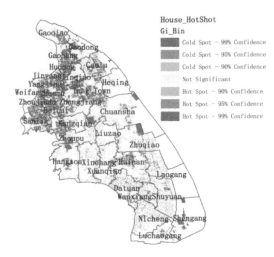

House_HotShot
Gi_Bin

■ Cold Spot - 99% Confidence
■ Cold Spot - 95% Confidence
□ Cold Spot - 90% Confidence
□ Not Significant
□ Hot Spot - 90% Confidence
■ Hot Spot - 95% Confidence
■ Hot Spot - 99% Confidence

Figure 4. Spatial hotspot distribution of residential land.

3.3.3 Hotspot distribution of residential land

As Figure 1 shows, the high-aggregation area of residential land accounts for 63.90% of the total, indicating that aggregation of residential land is relatively high, otherwise, the non-aggregation area is still up to 21.17%, indicating that the residential land still has further room for improvement. The specific distribution is shown in Figure 4, residential land distributes mainly in the west of Pudong New Area, the red part represents a high value aggregated distribution, including Gaoqiao Town, Gaohang Town, Hudong Street, Puxinglu Street, Jinyang Street, Yang Jing Street, Lujiazui Street, these regions are close to the main city of Shanghai, which are affected by migration obviously. The spots are large and regular, showing residential land in these areas is more mature. The blue small broken spots are mainly in the south of the Pudong New Area, including Liuzao Town, Xuanqiao Town, Xinchang Town, Datuan Town, Wanxiang Town, Laogang Town, Shuyuan Town and Nicheng Town, these areas are far away from the city center and the main types are homestead land, so the spatial distribution is broken, discrete and irregular.

4 CONCLUSIONS

Based on the hotspot analysis on three different land use types, agricultural, industrial and residential lands, the following conclusions can be made:

1. The aggregations of agricultural and industrial lands are low and random. On the contrary, the aggregation of residential land is higher.

2. The proportion of agricultural land in Pudong is relatively large, mainly in the southeastern part of the original Nanhui District. When the Government implements policies, it should focus on promoting intensive land use, scientsific resource allocation, a reasonable increase in agricultural materials and scalable labor inputs. To achieve greatest degree of improvement in land use pattern, it should adopt advanced agricultural technology and mine the potential of local land use.

3. Residential land distributes mainly in the west, which is close to more developed areas. In these areas, population is dense, so the government should strengthen the local road network construction, promote the development of infrastructure. To ensure the reasonable demand of farmers on the homestead, the government should focus on mining development potential and reform the distribution system and configuration.

This paper researches landscape features using four kinds of landscape index, and spatial characteristics of land use with spot aggregation index of Pudong New Area in 2015. Landscape index can study the basic features of landscape pattern, while GI statistical method is used for spatial aggregation partition. Both can reflect the distribution pattern of each kind of land types objectively, express the spatial proximity and correlation, reveal all kinds of characteristics and rules of spatial distribution effectively, in order to promote the optimization and provide the principle of the land use.

REFERENCES

Cen Yi. Pudong Construction Pattern in the Function, Characteristics and Value of National Strategic Practice [J]. Shanghai Studies on CCP history and Construction, 2011, 9: 39–41.

Eklund, P. W., S. D. Kirkby, A. Salim. Data mining and soil salinity analysis [J]. International Journal of Geographical Information Science, 1998, 12(3): 247–268.

Geits, A., J. Franklin. Second-order neighborhood analysis of mapped point patterns [J]. Ecology, 2003, 61: 99–105.

Hao Chengyuan, Wu Shaogang, Liu Chunguo. A Feasibility Analysis on Regional Integrated Research of Geography Science in China [J]. Geography and Geo-Information Science, 2007, 23(4): 48–52.

Jin Weibin, Hu Bingmin. Several Kinds of Landscape Separation Degree Evaluation Indexes [J]. Chinese Journal of Applied Ecology, 2003, 14(2): 314–316.

Kimberly, A.W., W.K. Anthony. Dispersal success in fractal landscapes: A consequence of lacunarity thresholds [J]. Landscape Ecol, 2002, 14: 73–82.

Li You, Wang Yanglin, Peng Jian, et al. Assessment of Urban Land Suitability for Construction in View of Landscape Ecology: A Case Study of Dandong City [J]. Acta Ecologica Sinica, 2010, 30(8): 2141–2148.

Miao Zuohua, Chen Yong, Zeng Xiangyang. Spatial Pattern Research on Urban Land-use Based on Patch Aggregation [J]. Geography and Geo-Information Science, 2013, 29(1): 56–59.

Wang Yongjun, Li Tuansheng, Liu Kang, et al. Landscape Patterns and Fragmentation in Yulin Prefecture [J]. Resources Science, 2005, 27(2): 161–166.

Yin Ke, Xiao Yi. Urban Function Zoning Based on Multivariate Spatial Analysis—Taking Tubei District of Chongqing City as Example [J]. Bulletin of Soil and Water Conservation, 2011, 31(6): 181–185.

Yin Zhan'e, Xu Shiyuan. Changes in Land Use and Land Cover and their Effect on Eco-Environment in Pudong New Area of Shanghai [J]. Resources and Environment in the Yangtze Basin, 2007, 16(4): 430–433.

Zeng Gang, Zhao Jianji. Study on the Shanghai Pudong Model [J]. Economic Geography, 2009, 29(3): 357–362.

Zhang Benyun, Shen Huaifei, Zheng Jinggang, et al. GIS-based Analysis on the Landscape Spatial Patterns of Land-use in Henan Province [J]. Resources Science, 2009, 31(2): 317–323.

Zhang Kan, Zhang Jianying, Chen Yingxu, et al. City Green Evaluation Model Based on the Value of Ecosystem Services in Hangzhoug [J]. China Journal of Applied Ecology, 2006, 17(10): 1918–1922.

Zhang Rongtian, Zhang Xiaolin, Li Chuanwu. Analysis on the Landscape Spatial Patterns of Land-use in Zhenjiang [J]. Economic Geography, 2012, 32(9): 132–134.

Zheng Xingqi, Fu Meichen. *Landscape Spatial Pattern Analysis Technology and Application* [M]. Beijing: Science Press, 2010: 34–38.

Advances in Energy Science and Equipment Engineering II – Zhou, Patty & Chen (Eds)
© 2017 Taylor & Francis Group, London, ISBN 978-1-138-71798-5

Optimal control of wind-induced vibrations of tall buildings using composite tuned mass dampers

Young-Moon Kim & Ki-Pyo You
Department of Architecture Engineering, Chonbuk National University, Jeonju, Korea
Long-Span steel Frame System Research Center, Jeonju, Korea

Jang-Youl You
Department of Architecture Engineering, Songwon University, Gwangju, Korea

ABSTRACT: Load identification is a major concept in the field of smart homes and smart grids. Nonintrusive Load Monitoring (NILM) method is applied to solve this problem, which is performed by analyzing the total current and voltage signal of the main distribution board to estimate the energy consumption of individual appliance and turning on/off or other operation. In this paper, we used the theory of NILM to identify household electric load. By analyzing the total current signal, extracting related features, and using Genetic Algorithm (GA) and Support Vector Machine (SVM), we identify different electric loads. In this paper, we used the BLUED data set (Anderson et al. 2012) as the experimental data. Finally, the rationality and effectiveness of the proposed method was verified by MATLAB simulation.

1 INTRODUCTION

Tuned Mass Damper (TMD) is a classical vibration control device consisting of a mass, a spring, and a damper supported at the primary vibrating system (Patil, 2011). The original idea of TMD was proposed by Frahm in 1909, who invented a vibration control device called a vibration absorber using a spring-supported mass without damper (Frahm, 1909). It was effective when absorber's natural frequency was close to the excitation frequency. This shortcoming was improved by introducing a damper in the spring-supported mass by Ormondroyd and Den Hartog (Ormondroyd, 1928). Later, Den Hartog derived optimum tuning frequency and damping ratio for the undamped primary structure under harmonic load (Den, 1985). While Den Hartog considered harmonic loading only, Warburton and Ayroinde derived optimum parameters of tuning frequency and damping ratio of TMD for the undamped primary system under harmonic and white noise random excitations (Ayoringde, 1980). Krenk derived the optimum parameters of TMD for the damped primary system subjected to random loads under the conditions that the mass ratio is small and the damping ratio of the primary system is less than that of the TMD (Krenk, 2008). A number of TMDs have been installed in tall buildings to suppress wind-induced vibrations (McNamara, 1977; Housner, 1997).

However, Nishimura and Wang pointed out that the disadvantage of a single TMD is its error in tuning the natural frequency of primary system to that of TMD and fitting the optimum damping ratio of TMD. The size restriction of TMD limits the vibration control effect (Nishimra, 1994; Nishimra, 1998; Wang, 1999). In overcoming such problem of a Passive Tuned Mass Damper (PTMD), the idea an active–passive tuned mass damper (composite tuned mass dampers, CTMD) and an Active Tuned Mass Damper (ATMD) be attached to the PTMD was proposed by Nishimura and Wang (Nishimra, 1994; Nishimra, 1998; Wang, 1999).

In this study, the performance of CTMD for suppressing across-wind-induced vibrations of tall building is investigated. The control force generated by the actuator of ATMD is estimated by Linear Quadratic Gaussian regulator (LQG) controller. Fluctuating across-wind load is simulated numerically using the across-wind load spectrum proposed by Kareem (1982). Dynamic across-wind responses of tall buildings with CTMD are estimated and compared to that of the original tall buildings without CTMD. The controlled rms responses with CTMD are reduced by about 17–30% rms response of the original tall building without CTMD. Therefore, CTMD is effective in suppressing excessive wind-induced vibrations of tall buildings.

2 EQUATIONS OF MOTION

Dynamic response analysis procedure can be simplified if the contribution of higher modes of tall buildings is ignored, so the response is represented by the motion of the first mode (Kareem, 1995). Therefore, tall building–CTMD system can be modeled as the first-mode generalized SDOF/CTMD system as shown in Figure 1.

The linear dynamic equations of motion of the CTMD system under fluctuating across-wind load f(t) and active control force u(t) of LQG controller can be written as:

$$M_S\ddot{X}_S + C_S\dot{X}_S + K_SX_S - C_P\dot{X}_P - K_PX_P = f(t) \quad (1)$$

$$M_P\ddot{X}_P + C_P\dot{X}_P + K_PX_P - C_A\dot{X}_A \\ - K_AX_A + M_P\ddot{X}_S = u(t) \quad (2)$$

$$M_A\ddot{X}_A + C_A\dot{X}_A + K_AX_A \\ + M_A\ddot{X}_S + M_A\ddot{X}_P = -u(t) \quad (3)$$

These equations of motion can be rewritten in the state space variable representation as:

$$\dot{X} = AX + Bu(t) + Ff(t) \quad (4)$$

where

$$X = \begin{bmatrix} X_S & X_P & X_A & \dot{X}_S & \dot{X}_P & \ddot{X}_A \end{bmatrix} \quad (5)$$

$$A = \begin{bmatrix} 0 & 0 & 0 & 1 & 0 & 0 \\ 0 & 0 & 0 & 0 & 1 & 0 \\ 0 & 0 & 0 & 0 & 0 & 1 \\ -\dfrac{K_S}{M_S} & \dfrac{K_P}{M_S} & 0 & -\dfrac{C_S}{M_S} & \dfrac{C_P}{M_S} & 0 \\ \dfrac{K_S}{M_S} & -\left(\dfrac{K_P}{M_P}+\dfrac{K_P}{M_S}\right) & \dfrac{K_A}{M_P} & \dfrac{C_S}{M_S} & -\left(\dfrac{C_P}{M_P}+\dfrac{C_P}{M_S}\right) & \dfrac{C_A}{M_P} \\ 0 & \dfrac{K_P}{M_P} & -\left(\dfrac{K_A}{M_A}+\dfrac{K_A}{M_P}\right) & 0 & \dfrac{C_P}{M_P} & -\left(\dfrac{C_A}{M_A}+\dfrac{C_A}{M_P}\right) \end{bmatrix} \quad (6)$$

$$B = \begin{bmatrix} 0 & 0 & 0 & 0 & \dfrac{1}{M_P} & -\left(\dfrac{1}{M_A}+\dfrac{1}{M_P}\right) \end{bmatrix}^T \quad (7)$$

$$F = \begin{bmatrix} 0 & 0 & 0 & \dfrac{1}{M_S} & -\dfrac{1}{M_S} & 0 \end{bmatrix}^T \quad (8)$$

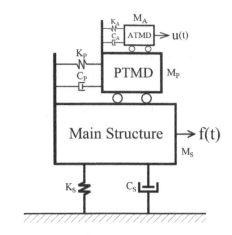

Figure 1. Tall building–CTMD model.

where M_S, C_S, and K_S are respectively the generalized mass, damping, and stiffness coefficients corresponding to the first mode of tall buildings;

M_P, C_P, and K_P are respectively the mass, damping, and stiffness coefficients of the PTMD;

M_A, C_A, and K_A are respectively the mass, damping, and stiffness coefficients of the ATMD;

X_S and $\dot{X}_S$ are respectively the displacement and velocity of main structure;

X_P and $\dot{X}_P$ are respectively the relative displacement and velocity of the PTMD with respect to the main structure;

X_A and $\dot{X}_A$ are respectively the relative displacement and velocity of the ATMD with respect to the PTMD;

u(t) is the active control force;

f(t) is generalized fluctuating across-wind load associated with the first mode.

3 OPTIMUM PARAMETERS OF CTMD

Krenk derived the optimum parameters of PTMD for the damped primary system subjected to random loads when the mass ratio is small and the damping ratio of the primary system is less than that of PTMD as follows (Krenk, 2008):

$$f_{opt} = \frac{1}{1+\mu} \qquad (9)$$

$$\xi_{opt} = \frac{\sqrt{\mu}}{2} \qquad (10)$$

where f_{opt} is the optimum tuning frequency ratio; ξ_{opt} is the optimum damping ratio; and μ is the mass ratio.

Determining the optimum parameters of CTMD for minimizing rms responses of the main structure is similar to that for PTMD. It was known that tuning the natural frequencies of PTMD to the fundamental natural frequency in the primary structure is more effective than tuning it to different natural frequencies (Kareem, 1995). Accordingly, the natural frequency and applied damping ratio of ATMD are tuned to the optimum parameters for PTMD as shown in Equations (9) and (10) with a small difference in the mass ratios of ATMD to PTMD. However, the total mass of ATMD and PTMD of CTMD is the same as that of the mass of a single PTMD.

4 LINEAR QUADRATIC GAUSSIAN CONTROLLER

Tall building under the fluctuating across-wind loads, which have a constant power spectral density function, can be considered as the linear dynamic system having a system noise with a constant power spectral density. Then, we can formulate a dynamic plant model with an LQG controller u(t) in modern optimal control theory as follows (Dorato, 1995; Lewis, 2012):

$$\dot{X}(t) = AX(t) + Bu(t) + \omega(t) \qquad (11)$$

$$Y(t) = CX(t) + Du(t) + v(t) \qquad (12)$$

where we let $D = 0$ for simplicity. $X(t)$ and $Y(t)$ are the state and output vectors, respectively, and $w(t)$ and $v(t)$ are the system and measurement noises, respectively, which are assumed to be uncorrelated zero mean Gaussian white noises with constant power spectral densities of W and V, respectively. That is:

$$E[\omega(t)\omega(t+\tau)^T] = W\,\delta(\tau) \qquad (13)$$

$$E[v(t)v(t+\tau)^T] = V\,\delta(\tau) \qquad (14)$$

and

$$E[\omega(t)v(t+\tau)^T] = 0, \ E[v(t)\omega(t+\tau)^T] = 0 \qquad (15)$$

where E is the expectation operator and $\delta(T-\tau)$ is a Dirac delta function.

It was known that the optimal controller u(t) in Equation (11) is obtained when all states X(t) of the system and the output Y(t) are a combination of all states. However, in practice, all states X(t) are not available and the system and output measurements are driven by stochastic disturbances called noises, which have constant power spectral densities of Gaussian white noise. In such situations, we need a state estimator or an observer to estimate all the states of the system. Then, the state observer could be designed using Kalman filter, which is an optimal state observer for the stochastic dynamic system (Dorato, 1995). Let $\hat{X}(t)$ be the state estimator and $Xe(t) = X(t) - \hat{x}(t)$ denote the estimation error. Then, the state is estimated using Kalman filter described as:

$$\dot{\hat{X}}(t) = A\hat{X}(t) + Bu(t) + L\left(Y(t) - C\hat{X}(t)\right) \qquad (16)$$

where the observer gain matrix L is given by:

$$L = \Gamma C^T V^{-1} \qquad (17)$$

where $\Gamma = E[Xe^T(t)\ Xe\ (t+\tau)]$
and Γ is the unique positive semi-definite solution of the algebraic Riccati equation (FARE) as below:

$$A\Gamma + \Gamma AT + W - \Gamma C\ V^{-1}C\Gamma = 0 \qquad (18)$$

In the LQG controller problem, the optimal controller $u(t)$ in Equation (11) that minimizes the cost functional J of Equation (19) is subjected to the constrained Equation (16) when determined separately as the deterministic Linear Quadratic Regulator (LQR) controller problem.

However, for a stochastic dynamic system, such as an optimal control problem of the wind-induced vibration of tall buildings, the cost functional of the deterministic LQR problem cannot be employed because of the stochastic nature of the state space formulation of the stochastic differential equation of (11). That is to say, the ensemble average over all possible realizations of the excitation is considered, so the cost functional J is given by:

$$J = E\left\{\lim_{T\to\infty} \frac{1}{T} \int_0^T [X^T(t)Qx(t) + u^T(t)Ru(t)]dt\right\} \qquad (19)$$

where Q and R are the positive semi-definite and positive definite weighting matrices, respectively.

The term $X^T(t)Qx(t)$ in Equation (19) is a measure of control accuracy and $u^T(t)Ru(t)$ is a measure of control effort. Minimizing J by keeping the system response and the control effort close to zero needs an appropriate choice of the weighting matrices Q and R (Lewis, 2012). If it is desirable that the system response be small, then large values for the elements of Q should be chosen by selecting the matrix to be diagonal and to make the values of diagonal elements large for any respective state variable to be small. If it is required that the control energy be small, then large values of the elements of R should be chosen (Suhardjo, 1992). The unique state-feedback optimal controller u(t) that minimizes the cost functional J of Equation (19) is determined as follows (Dorato, 1995; Lewis, 2012):

$$u(t) = -R^{-1}B^T P\hat{X}(t) \qquad (20)$$

where P is the unique symmetric, positive semi-definite solution to the Algebraic Riccati Equation (ARE) given by:

$$AP + PA^T - PBR^{-1}B^T P + Q = 0 \qquad (21)$$

It is noted that we can determine the LQR controller feedback gain matrix K and the observer gain Kalman filter matrix L independently. This is the so-called separation principle (Dorato, 1995).

5 NUMERICAL SIMULATION OF A FLUCTUATING ACROSS-WIND LOAD

The dynamic along-wind response of tall buildings can be estimated reasonably by a gust factor approach (Solari, 1993). However, dynamic across-wind response cannot be estimated by a gust factor approach. The complex nature of the across-wind loading, which are from an interaction of incident turbulence, flow separation, vortex-shedding, and unsteady wake development, has prevented theoretical prediction from estimation (Kareem, 1982). The fluctuating across-wind load can be treated as stationary random process, which can be simulated numerically in the time domain using the across-wind load power spectral density data. That is particularly useful for some response estimations that are more or less narrow-banded random processes, such as the across-wind response of tall buildings (Shinozuka, 1987). The numerical simulation procedure presented in this work is taken from Shinozuka (1987):

$$f(t) = \sum_{k=1}^{N} \sqrt{S_F(\omega_1)\Delta\omega}\cos(\omega_k t + \varphi_t) \qquad (22)$$

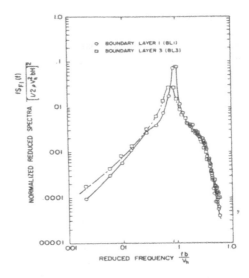

Figure 2. Normalized reduced spectra of across-wind load in suburban (BL1) and urban (BL3).

where $S_F(\omega_1)$ is the value of the spectral density of across-wind load corresponding to the first modal frequency:

$\Delta\omega = (\omega_u - \omega_1)/N$;

$\omega_k = \omega_1 + (k - 1/2)\Delta\omega$;

ω_u = upper frequency of $S(\omega)$;

ω_1 = lower frequency of $S(\omega)$;

Φ_t = uniformly distributed random numbers between 0 and 2π;

N = number of random numbers.

The across-wind load power spectral density used in Equation (22) is that given by Kareem (1982). Kareem used a 5 sq in. (127 mm²) square, 20-in. (508 mm²) tall prism model, and eight pressure transducers, integrating eight simultaneously monitored channels of pressure data on the surface of a building model, so obtained the normalized across-wind load spectra of the across-wind forcing function on a square tall building that is exposed to urban and suburban atmospheric flow conditions as shown in Figure 2.

6 NUMERICAL EXAMPLE

This numerical example is from "Numerical Example" of Kareem (1982). The tall building's height (H) is 180 m, width (B) is 30 m, depth (D) is 30 m, natural frequency (f_1) is 0.2Hz, critical damping ratio is 0.01, air density is 0.973 kg/m³, hourly mean wind speed at the building height V_h is 24.4 m/s, the reduced velocity (V_h/f_B) corresponding to the mean hourly wind speed is 4.0, generalized mass of a fundamental mode shape (m_j) is 10,942,500 kg, and $S_f(f_1)$ is 3.149*10^8 kg²/Hz. The optimum parameters of the PTMD are a mass ratio ($\mu = 0.01$), tuning frequency ($f_{opt} = 1.0$), and damping ratio ($\xi_{opt} = 0.05$).

6.1 Across-wind response without CTMD

The numerically simulated fluctuating across-wind loads and response without CTMD are shown in Figures 3 and 4. The rms displacement response without CTMD shown in Figure 4 is 0.0047 m, which is a good approximation to that of Kareem's closed-form response of 0.0040 m [12]. Therefore, numerically simulated across-wind loads in the time domain can be used for estimating optimally controlled responses of tall buildings using CTMD.

6.2 Across-wind responses with CTMD

For estimating LQG controller, the diagonal elements of weighting matrices Q and R are selected as:

$$Q = 1.0*10^8 * \begin{bmatrix} 100 & 0 & 0 & 0 & 0 & 0 \\ 0 & 1 & 0 & 0 & 0 & 0 \\ 0 & 0 & 1 & 0 & 0 & 0 \\ 0 & 0 & 0 & 100 & 0 & 0 \\ 0 & 0 & 0 & 0 & 1 & 0 \\ 0 & 0 & 0 & 0 & 0 & 1 \end{bmatrix},$$

$$R = [1.0*10^{-12}]$$

and the assumed value of V is 1.0*10^{-8}.

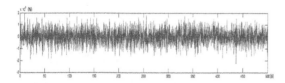

Figure 3. Simulated across-wind load in the time domain.

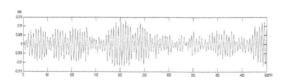

Figure 4. Across-wind response without CTMD (rms = 0.0047 m).

The dynamic across-wind responses of tall building with CTMD, which have different mass ratios of ATMD to PTMD, 0.01, 0.03, 0.05, 0.1, 0.3, and 0.5, are presented in Figures 5–10.

As shown above, comparing the controlled rms responses with CTMD to that of the original tall building, a 17–30% reduction effect is presented. The effectiveness of CTMD increases with the increase in the mass ratio of ATMD to

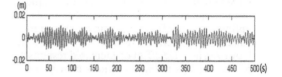

Figure 5. Estimated across-wind responses with CTMD (mass ratio = 0.01, rms = 0.0038 m).

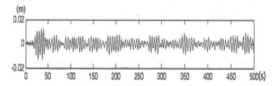

Figure 6. Estimated across-wind responses with CTMD (mass ratio = 0.03, rms = 0.0035 m).

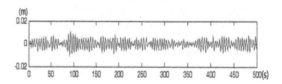

Figure 7. Estimated across-wind responses with CTMD (mass ratio = 0.05, rms = 0.0033 m).

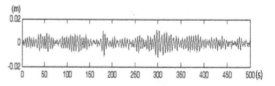

Figure 8. Estimated across-wind responses with CTMD (mass ratio = 0.1, rms = 0.0034 m).

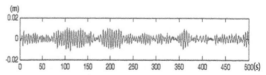

Figure 9. Estimated across-wind responses with CTMD (mass ratio = 0.3, rms = 0.0036 m).

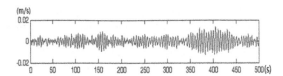

Figure 10. Estimated across-wind responses with CTMD (mass ratio = 0.5, rms = 0.0039 m).

PTMD within allowable limit. That is, as the mass ratio is increased from 0.01 to 0.1 with the same optimum tuning frequency and damping ratio, the maximum reduction effect of 30% is presented. However, adverse effect of increased responses can be obtained as the mass ratio is increased to 0.5. Therefore, CTMD is effective in reducing excessive wind-induced vibration of tall buildings.

CONCLUSIONS

The performance of Composite Tuned Mass Dampers (CTMD) for suppressing across-wind responses of tall buildings is investigated. Optimum parameters of tuning frequency, damping, and mass ratio for CTMD were used. Fluctuating across-wind load treated as a stationary Gaussian random process was simulated numerically using the across-wind load spectra. Comparing the controlled rms responses with CTMD to that of the original tall building, a reduction effect of 17–30% is presented. Therefore, CTMD is effective in reducing excessive wind-induced vibration of tall buildings.

ACKNOWLEDGMENT

This work was supported by the National Research Foundation of Korea (NRF) grant funded by the Korea government (MEST) (No. 2011-0028567).

REFERENCES

Ayoringde, E.O. Warburton, G.B, Earthquake Engineering and Structural Dynamics, **8**: 219–236. (1980).

Den Hartog, J.P. Mechanical Vibration, 4th edn. McGraw-Hill, New York, (1956). (Reprinted by Dover, NewYork, 1985).

Dorato, P. Abdallah, C. Cerone, V. Prentice Hall, New Jersey. (1995).

Frahm, H. U.S. Patent No. 989958. October 30. (1909).

Housner, G.W. Bergman, L.A. Caughey, T.K. Chassiakos, A.G. Claus, R.O. Masri, S.F. Skelton, R.E. Soong, T.T. Spencer, B.F. Yao, J.T.P. Journal of Engineering Mechanics (ASCE), **123**(9): 897–971. (1997).

Kareem, A. Journal of the SD, ASCE 108(ST4): 869–887. (1982).

Kareem, A. Kline, S. Journal of Structural Engineering ASCE, **121**(2): 348–361. (1995).

Krenk, S. Hogsberg, J. Probabilistic Engineering Mechanics, **23**: 408–415. (2008).

Lewis, F.L. Vrabie, D.L. Syrmos, V.L. John Wiley & Sons. (2012).

McNamara, R.J. Journal of the Structural Division. 103: 1785–1798. (1977).

Nishimra, I. Sakamoto, M. Yamada, T. Koshika, N. Kobori, T. Journal of Structural Control 1(1–2), 103–116. (1994).

Nishimra, I. Yamada. T. Sakamoto, M. Kobori, T. Journal of Smart Mater. Struct. 7(5), 637–753. (1998).

Ormondroyd, J. Den Hartog, J.P. Transactions of ASME, 1928; **50** (APM-50-7): 9–22. (1928).

Patil, V.B. Jangid, R.S. Journal of Civil Engineering and Management, **17**(4): 540–557. (2011).

Shinozuka, M. Stochastic fields and their digital simulation.93–133, edited by Schueller, G.I. Shinozuka, M. Martinus Nijhoff Publishers, (1987).

Solari, G. Journal of Structural Engineering ASCE, **119**(2): 383–398. (1993).

Suhardjo, J. Spencer, JR, B.F. Kareem, A., Jour. Wind Eng. Ind. Aerodyn. 41–44, 1985–1996. (1992).

Wang, C.M., Yan, N. Balendra, T. Journal of Vibration and Control. **5**, 475–489. (1999).

Advances in Energy Science and Equipment Engineering II – Zhou, Patty & Chen (Eds)
© 2017 Taylor & Francis Group, London, ISBN 978-1-138-71798-5

Study on the rainstorm-flood events and the reservoir engineering system of China under climate change—taking the Huai river basin as an example

S.M. Liu & H. Wang
Beijing Forestry University, Beijing, China

D.H. Yan
China Institute of Water Resources and Hydropower Research, Beijing, China

ABSTRACT: The impacts of climate change have caused rainstorm-flood events and other extreme weather events to occur with greater frequency, affecting increasing areas. This generates new challenges for water conservation projects to deal with these extreme events. For water conservation projects in China to better respond to climate change, it is important to study the scope of influence and trend of rainstorm-flood events. In this paper, we select the Huai River Basin as a typical, representative study area, and generate statistics using hourly rainfall events to analyze the distribution and evolution of those events. We also conduct a statistical analysis across the country as well as the study area on the distribution of water conservation engineering projects, exploring the problems and countermeasures of China's water conservation under climate change and storm-flood trends. The results show that: (1)The areas of coverage and disaster losses due to rainstorm—flood events in China had increasing trends; however, in the Huai River Basin the disaster losses showed a decreasing trend even though rainstorm-flood events increased at the same time. This phenomenon indicates that water conservation engineering to respond to rainstorm-flood events plays an important role in disaster response; (2) Due to climate change and the fact that the current reservoir system has reached its designed service period limit, the engineering system cannot meet the expected future flood disaster demands; and (3) The frequency of rainstorm events has increased year by year in the Huai River Basin and reached a mutation point around 2003, transitioning into a storm prone period. At the same time, China's urbanization rate is surging while the rain-island effect and urban flooding problems have gradually appeared, demonstrating the need for new requirements for urban construction of sponge type city and etc.in China.

1 INTRODUCTION

Excessive rainfall often leads to disaster flooding, which is mainly caused by extreme precipitation events as well as repeat numbers of rainstorms (Zhang 2008). With temperatures expected to rise in the future, extreme precipitation events are forecast to be more intense as well as more frequent (Zhang, 2009; Li, 2013) (Benistonet, 2007; IPCC, 2013). Climate change and variability have already affected global ecosystems, biodiversity, and the social economy (Kotir, 2011). In recent decades, significant changes in extreme climatic events have taken place that can have devastating impacts on human society and the environment (IPCC, 2007; Min, 2011), and the impacts are likely to become more pronounced in the future (Song, 2014; Jiang, 2007; Milly, 2002). Water conservation measures are an important way to cope with flooding from rainstorm events,

and play an important role in the protection of water resources, regional flood control, and other features affected by climate change (Jiao, 2009; People's Republic of China Ministry of water conservancy 2013). Related studies have indicated that climate change has adverse effects on the service environment and engineering materials for reservoirs in China (Zhang, 2009; Zhang, 2000; Zhang, 2014). For storm-flood events, during the past 50 years (1961–2011 years) the overall trend of heavy rainfall in China has been downward in the North, while the South and Western regions have seen increases. In addition, China's pattern of geomorphological evolution has a three-tiered ladder distribution. Analyzing how regional rainstorm events occurred over time and space under the backdrop of climate change will improve storm-flood control and the sustainable development of water conservation engineering projects to better cope with those changes.

2 RESEARCH SCOPE AND TYPICAL BASIN

China is one of the countries that is most affected by serious storm-flood disasters. On the one hand, China is located in the southeast of Eurasia and the Pacific Ocean, and the dramatic thermal difference between land and see creates a climate that is characteristically rainy in the summer and in the Southeast region, and drier in the winter and in the Northwest region (Figure 1). On the other hand, the topography is characterized by the West having higher elevations than the East, while the Midwest is the highest point of an anisomerous triangle, creating a three-layer distribution pattern across the country (Figure 2). The uplift of the Qinghai-Tibet Plateau plays a major role in the formation of the extreme climate. Simultaneously, as the winter and summer monsoons become less frequent, the intensity and number of typhoons that make landfall also change, resulting in interannual changes in precipitation. This fundamentally affects rainstorm-flood events that frequently occur in China.

The unique location of the Huai river basin, affected by multiple atmospheric circulation and weather systems, is also located in the key subtropical and sensitive area within the northward rain belt, causing more frequent storm-flood events. This can have an extremely negative impact on agriculture in the plains region, urban flooding, and so on. Therefore, the Huai river basin is highly representative of storm-flood events. This study selected the construction of the sponge city. Huai river basin as a typical basin, and used high precision measurements of hourly precipitation and statistics of rainstorm events to reveal the temporal and spatial variation of rainstorms under climate change. Statistical methodologies and theories

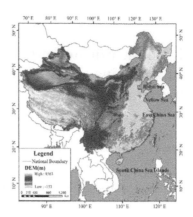

Figure 2. Geomorphological evolution pattern of a three-layer distribution in China.

of atmospheric dynamics and climatology were employed, and we analyzed a water conservation project to provide a scientific basis for its construction of water conservancy project and sponge city etc. In addition, although the Huai river basin was chosen to represent a typical area, reference values for the eastern coastal region and other areas that experience extreme precipitation were generated.

3 DATA SOURCES

The data used for this study were hourly precipitation records (1951 through 2012) from 229 meteorological stations in the Huai River basin. Stations near the basin boundary were also considered. The data for the study period were chosen based on their continuous, uninterrupted temporal record. The number of rainstorms and the locations of the stations are shown in Figure 3. Climate data were obtained from the China Meteorological Data Sharing Service System (http://cdc.cma.gov.cn/). Basic terrain data were acquired from the National Basic Geographic Information System for the Huai River basin (1:250,000). Social and economic data were obtained from the Statistical Yearbook of China, the Water Resources Bulletin, and the Emergency Disaster Database (EM-DAT), whereas data related to the water conservation project in the basin were taken from the latest Water Conservancy Survey.

4 ANALYSIS AND RESULTS

4.1 Rainfall characteristics

Statistics on the average annual number of rainstorm days and the total number of rainstorm

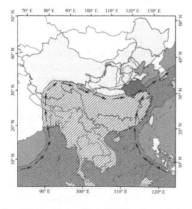

Figure 1. Water vapor flux distribution of the interaction between key plateau areas and the Asian monsoon.

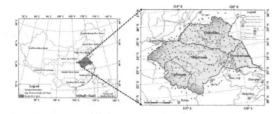

Figure 3. 229 meteorological stations in Huai river basin.

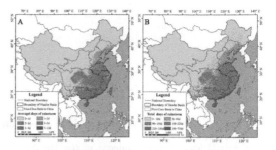

Figure 4. The spatial distribution of average and total days of rainstorm number in 1961–2011.

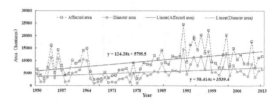

Figure 5. Changes in the 1950–2013 national flood event disaster rates and hazard rates.

days show that in the past 60 years, China's average annual number of rainstorm days was between 0~15, while the total number of rainstorm days ranged between 0~738 days, We choose the Inverse Distance Weighted (IDW) method to spatially interpolate these two statistics over the entire country, and Figure 4A shows the average annual rainstorm days, while the total number of rainstorm days is shown in Figure 4B. Areas with the highest values are found in the East and Southeast of China, which includes the area of the Huai river basin, the Yangtze River region, the Pearl River Basin, and others.

Rainstorm-flood disasters have a direct impact on crop yields, and statistics on heavy rainstorm events have been collected and analyzed in China for the past 60 years. They show that rainstorm-flood disaster and hazard rates are increasing every 10 years by 1.3% and 0.7%, respectively. The annual disaster and inundated areas are 9834.68 hectares and 5437.90 hectares, respectively, accounting for an 8.64% (hit rate) and a 4.76% (hazard rate) of China's agricultural regions. Since 1990, the scope of rainstorm and flooding impacts significantly increased. The average hit rate and hazard rate were 12.28% and 6.77% during the period of 1990–2013, which are respectively 1.90 and 1.89 times higher than during the period of 1950–1989. The years 1991 and 2003 were the most seriously affected by flood disasters, when the hit rate was more than 20% and disaster rate was higher than 13% (Figure 5).

We examined changes in the disaster and hazard rates for rainstorm and flood events in the Huai basin over the past 50 years. The affected area, disaster area, disaster rate, and hazard rate of flood events in Huai basin showed a decreasing trend, but the change is not significant with respective values of −2.33 million hectares/10a, −5.70 million hectares/10a, −0.12%/10a, and −0.95%/10a. From the perspective of temporal changes, the most serious period of flooding occurred in the 1960s in the Huai river basin, and after that, it was somewhat alleviated. Compared with the 1960s, the disaster rate greatly declined during 1970s, partly because

the affected area is relatively small; on the other hand, the construction of a large reservoir significantly reduced the losses caused by flood disasters during the 1970s. With the impacts of climate change, however, the affected area and affected rate have increased since the 1990s.

In summary, since the 1980s with economic development, a large number of water conservation projects have been created, which has improved flood disaster resilience. However, at the national scale, the ability to resist rainstorm-flood disasters is still inadequate.

4.2 Spatial and temporal evolution of rainstorm-flood characteristics in the region

4.2.1 Temporal variation of rainstorm processes in Huai river basin

The occurrence of rainstorm-flood disasters has obvious regional and seasonal characteristics. After statistical processing of the isolated rainstorm data from the Huai Basin, we found that rainstorm events occurred mainly between May and September. The duration of each event was generally one to three days, which accounted for 94.3% of the annual total of 24,536 instances. To analyze variation in the start times of isolated rainstorm events, we selected a representative period between May and September and used the ratio of the frequency of isolated rainstorm events in each 10-day interval divided by the total number of isolated rainstorm events in that interval to be the contribution rate.

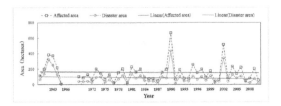

Figure 6. Inter annual variation of flood disaster area in Huaihe River Basin.
* partial loss of data sequence in the Huaihe River Basin.

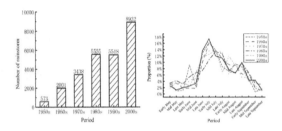

Figure 7. Variation of rainstorm events in ten-day intervals during 1950s–2000s.

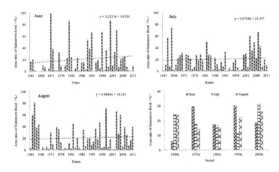

Figure 8. Changes of the flood covered area of 6~8 months in the Huai river basin.

The event start times showed a bimodal annual trend with clear differences between ten-day intervals (Figure 7). A major peak occurred mid to late July, with a secondary peak during the last 10 days of August. What's more, with the effects of climate change, the time of occurrence became earlier with each decade. The area affected by rainstorm and flood events of 6~8 in the Huai river basin is increasing (Figure 8). Among them, the increase in the area affected by rainstorm-flood events in June is relatively large, and its linear growth rate is 2.21%/10a. From the 1960s to the 1970s, the area affected by rainstorm-flood events increased from 17.89% to 20.43%; while during the 1980s the flood cover area was relatively low, alleviating the problem to some extent with a monthly

average of 16.43%. The 1990s and 2000s saw the most serious disaster problems in the Huai River Basin, with monthly average flood covered areas at 21.18% and 25.28%, respectively.

The results of the analysis of rainstorm-flood area in 6~8 months showed that the storm-flood area in Huai River Basin increased. From the view of inter annual variation, the greatest increases occurred in the 1970s, but the disaster rate decreased, showing that water conservation projects have played an important role in reducing the losses associated with rainstorm-flood events.

4.2.2 Spatial distribution and trend analysis in Huai river basin

We used the IDW method to convert the frequency of isolated rainstorm events observed at the 229 meteorological stations to a gridded surface (Figure 9). The spatial resolution for the interpolated hourly rainstorm event surface was 500×500 m². Data processing was done using the ArcInfo Workstation 10.2 platform. The ArcGIS Python language was also used for estimation. Based on our qualitative understanding of the spatial distribution and evolution of rainstorm events, we selected the MK mutation test to determine the mutation year.

According to the statistics from the period of 1950–2012, 26,017 isolated rainstorm events occurred across the Huai Basin. The spatial distributions of the average frequency of occurrence for these events were essentially the same across different time periods (Figure 9). They all showed a gradually increasing trend from the Northwest to the Southeast. Areas of high-frequency inter annual variability showed a gradually increasing trend in the Southwest, and the extent of these areas increased. The increasing trend of the events and frequency of occurrence over the entire basin were significantly affected by climate change beginning in the 1990s.

On inspection of Figure 10, we see that the intersection of the UF and UB curves occurred in 2003, after which there was an upward trend.

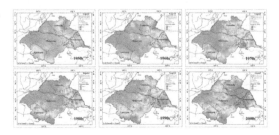

Figure 9. Spatial distribution of isolated rainstorm event frequency during the 1950s–2000s.

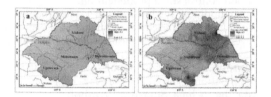

Figure 10. M-K mutation test results of rainstorm event frequency in the Huaihe Basin.

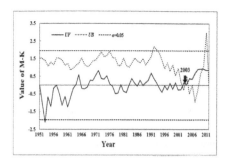

Figure 11. Rainstorm mutation and spatial distribution map in Huaihe Basin.

From both the China Statistical Yearbook and flood control plans for the Huai Basin, we know that there were persistent rainstorms in July 2003. Due to the El Niño event of 2002–2003, extreme precipitation changes across all of southwestern China were observed *(Chen, 2006)*. This implies that the MK mutation test for frequency was accurate. Relative to the years before 2003 (Figure 11a), the average frequency of occurrence increased by 25.5%. Areas of high-frequency inter annual variability showed a gradually increasing trend in the Southwest, with the extent of those areas also increasing (Figure 11b). Obviously, "the present day situation of isolated rainstorm events is undergoing a very strong trend of increase.

4.3 Construction of reservoir engineering system in China

Water conservation projects are an important measure to deal with flood and waterlogging disasters. Between 1950 and 1979, China completed construction of numerous water conservation projects as part of the "Great Leap Forward" to become the country with the most reservoirs in the world; today China has 98,002 reservoirs, most of which were built during the 1950s and 1960s (Figure 12). In 2015, 49 reservoir hub projects were built in seven major river basins in China (Figure 13). Compared to the fact that only 33 hub reservoir projects were completed prior to 2000, flood regulation storage increased by 1.76 times.

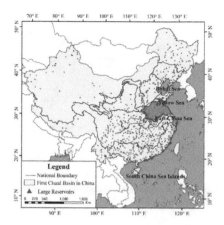

Figure 12. National large reservoir distribution map.

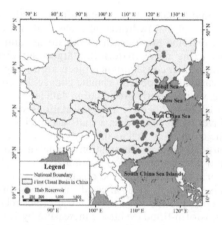

Figure 13. Distribution map of national hub reservoirs.

Based on the latest water census data and a map of the large reservoirs in China (Figure 14), we see that the distribution of the 44 large reservoirs in the basin is not proportional to the distribution of isolated rainstorm events. In the central midstream and along the entire downstream areas, there are almost no large reservoirs. This uneven distribution of large reservoirs will seriously affect flood control abilities. Of the large reservoirs in the Huai river basin, 59% were built during the 1950s and 1960s. These reservoirs have now reached the limits of their design service life of ~50 years. Because the frequency of rainstorms has significantly increased, the percentage for reservoir fill has also increased in the basin, so the threat of rainstorm disasters is much greater than before. Therefore, we recommend strengthening the capabilities of planning and management in water conservation projects, and improving the layout and distribution of those projects.

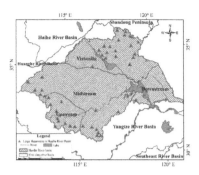

Figure 14.　Large reservoir distribution map of the Huai river basin.

5　CONCLUSION AND DISCUSSION

Climate change has deeply influenced the status of the global hydrological cycle, with hydrological extreme events such as rainstorms and flood occurring more frequently. Large reservoirs and other water conservation projects are an important measure to cope with climate change. In this article, we use hourly precipitation data (1951 through 2012) from 229 meteorological stations in the Huai River basin to conduct a statistical analysis across the country as well as the study area of the distribution of water conservation engineering projects, and discuss the spatio-temporal evolution of rainstorm-flood events. The results show that China has a broad area affected by rainstorm-flood events, and the disaster and hazard rates related to rainstorm-flood events are increasing by 1.3% and 0.7% every 10 years, respectively. Occurrence time of Huai river basin rainstorms concentrated in 6~8 months, the covered area is increasing. Because the disaster rate in the Huai river basin decreased, however, this indicates that the newly-built water conservation project has played an important role in reducing rainstorm-flood disaster losses. The event starting time showed a bimodal annual trend, with the primary peak occurring in mid to late July, and the secondary peak occurring during the last 10 days of August. Moreover, due to climate change, their time of occurrence is occurring earlier. The spatial distribution of the average occurrence frequency of rainstorms events showed a gradually increasing trend from the Northwest to the Southeast. Areas of high-frequency inter annual variability showed a gradually increasing trend in the Southwest, and the extent of these areas increased. The MK mutation test for frequency identified that the intersection of the UF and UB curves occurred in 2003, and after 2003, the basin entered a heavy rainfall period. However, a large number of older reservoirs have reached

Figure 15.　National urbanization rate change chart of 1949–2014.

the end of their service life, making the problem of flood control still a prominent one.

In addition, the influence of rapid urbanization and flooding caused by frequent rainstorm-flood events has also become a hot issue of public concern. The urbanization rate is defined as the proportion of the urban population out of the total population at the end of the year. Therefore, based on the "China Statistical Yearbook", Figure 15 shows that since 1949, the rate of urbanization has gradually been increasing, before the period of reform and then speeding up, with the average urbanization rate holding at around 20%. However, after the 1980s, China entered into its peak period of urbanization, population and wealth accumulating in the cities. From 2003 and onward urban growth has been booming, and in 2011 it exceeded 50% for the first time, reaching 54.77% in 2014. We expect population expansion to continue at the rate of about 20 million people per year, with city spaces also continuing to expand.

Under the influence of climate change, the mutation point of isolated rainstorm-flood event frequency occurred in 2003. Since 2003, China also entered its boom period for urbanization. This increase in urbanization will inevitably lead an increase in the impervious surface area in cities, with associated increases in rainfall runoff levels. Research shows that the rain island effect of cities increases the frequency of rainstorm events, and also influences rainfall distribution and duration, resulting in serious urban flooding (Gou, 2012; Xiao, 2012; Zhou, 2012; Yan, 2015). Therefore, based on the above conclusion, we have identified a "poor" combination: a period of high incidence of urban flood disasters that coincides with urbanization increases in the Huai river basin under climate change, potentially leading to significant increases in urban flooding-related problems.

ACKNOWLEDGMENTS

This work was supported by the Representative Achievements and Cultivation Project of State

Key Laboratory of Simulation and Regulation of Water Cycle in River Basin (No. 2016CG02) and National Key Research and Development Project (No. 2016YFA0601503).

REFERENCES

Beniston, M., Stephenson, D. B., Christensen, O. B., Ferro, C. A. T., Frei, C., Goyette, S., et al. (2007). Future extreme events in european climate: an exploration of regional climate model projections. *J. Climatic Change*, 81(1 Supplement):71–95.

Gou, A.N., Li,W.J., Huang, Y.G., et al. (2012). Analysis of the first heavy rain in hubei province during meiyu season. *J. Meteorological & Environmental Research.* (9):25–29.

IPCC. (2013). Contribution of Working Group I to the Fifth Assessment Report of the Intergovernmental Panel on Climate Change. In: Stocker T.F., Qin D., Plattner G.K., Tignor M., Allen S.K., Boschung J., Nauels A., Xia Y., Bex V., Midgley P.M. (eds) Climate change 2013: the physical science basis. Contribution of Working Group I to the Fifth Assessment Report of the Intergovernmental Panel on Climate Change. Cambridge University Press, Cambridge, United Kingdom and New York, NY, USA.

Jiang, T., Su, B., Hartmann, H. (2007). Temporal and spatial trends of precipitation and river flow in the yangtze river basin, 1961–2000. *J. Geomorphology,* 85(85):143–154.

Jiao, Y. (2009). Dam reservoir and harmonious development:Exploration and practice in China. *J. China Water Resources.* (12):1–3.

Kotir, J. H. (2011). Climate change and variability in sub-saharan africa: a review of current and future trends and impacts on agriculture and food security. *J. Environment Development & Sustainability*, 13(3): 587–605.

Li, F.P., Zhang G.X., Dong, L.Q. (2013). Studies for Impact of Climate Change on Hydrology and Water Resources. *J. Scientia Geographica Sinica.* 33(4):457–464.

Milly, P. C., Wetherald, R. T., Dunne, K. A., Delworth, T. L. (2002). Increasing risk of great floods in a changing climate. *J. Nature*, 415(6871):514–7.

Ministry of Water Resources of the People's Republic of China. (2013). Bulletin of First National Census for Water. *J. China Water Resources.* (7):1–3.

Nicholls, R.J. (2004). Coastal flooding and wetland loss in the 21st century: changes under the sres climate and socio-economic scenarios. *J. Global Environmental Change,* 14(1), 69–86.

Song, F., Qi., H. Wei, H., et al. (2014). Projected climate regime shift under future global warming from multi-model, multi-scenario CMIP5 simulations. *J. Glob Planet Chang.*112(1):41–52.

Xiao, H., Mei, L. (2012). Analysis on the Rainstorm Process from June 14 to 15 in 2011. *J. Meteorological and Environmental Research.* (11):15–21.

Yan, S. (2015). The Rainstorm Characteristics and Rain Island Effect Evolvement rule in Chinese Large Cities. *D. Donghua University.*

Zhang, J.M., Huang, Z.Y., Wu, J.D. (2000). Impacts of Climate Change on Risk in Running of the Three Gorges Reservoir. *J. Acta Geographica Sinica.* 55(s1): 26–33.

Zhang, J.Y. (2009). Climate change and safety of water conservancy projects. *J. Chinese Journal of Geotechnical Engineering.* 2009(6):41–46.

Zhang, J.Y., Lu, C.R., Wang, G.Q., et al. (2015). Impacts of and Adaptation to Climate Change for Water Conservancy Projects. *J. Progressus Inquisitiones DE Mutatione Climatis.* 11(5):301–307.

Zhang, J.Y., Wang, G.Q., Liu, J.F., et al. (2009). Review on worldwide studies for impact of climate change on water. *J. Yangtze River.* 40(8):39–41.

Zhou, Z.G., Jiang, Y.Q., et al. (2012). Numerical Simulation on a Heavy Rainfall Event over Jiangxi Province. *J. Meteorological & Environmental Research.* (12):8–12.

Advances in Energy Science and Equipment Engineering II – Zhou, Patty & Chen (Eds)
© 2017 Taylor & Francis Group, London, ISBN 978-1-138-71798-5

Analysis of the competition of aquatic products' trade between China and ASEAN: In the background of the 21st-century Maritime Silk Road

Xiao-fei Luo

Faculty of Logistics, Guangdong Mechanical and Electrical Technical College, Guangzhou, Guangdong, China

Yong-hui Han

Guangdong Institute for International Strategies, Guangdong University of Foreign Studies, Guangzhou, Guangdong, China

ABSTRACT: China and ASEAN are now facing a historic opportunity to accelerate their economic and trade cooperation under the constructions of the "21st century Maritime Silk Road." In this background, this paper thoroughly analyzes the competition of the bilateral trade on aquatic products between China and ASEAN based on the HS classification. By calculating the export similarity index and trade intensity index, it is found that the competition of aquatic products' export between China and ASEAN to the world and Japan tends to increase, but in different groups, trade imbalance appears obviously. Overall, the aquatic products' trade between China and ASEAN is not substantially competitive, indicating its great potential.

1 INTRODUCTION

1.1 *The 21st-century Maritime Silk Road*

The Silk Road is the main trade route combining the West World and China as an important network of economic, political, and cultural communication. Starting from the southeast coast of China, through Southeast Asia and South Asia to Africa, the Maritime Silk Road was also called Road of China, Tea Road, and Fragrant Road. In order to accelerate the construction of China's new system of open economy, the third Plenary Session of the 18th CPC Central Committee proposes a strategic plan to "Put forth the construction of Silk Road Economic Belt and the 21st-Century Maritime Silk Road, build a new pattern of all-dimensional opening-up". Therefore, the construction of the 21st-Century Maritime Silk Road will become another starting point for China to build a new pattern of all-dimensional opening-up.

1.2 *Trade of aquatic products between China and ASEAN*

Aquatic products are one of the world's most important agricultural products. As the world's leading country in the realm of aquaculture, aquatic production, and trading, China's development in the trade of aquatic products has great contribution to its national economy development (Shao, 2007).

Because the land and water of ASEAN and China are interlinked, their bilateral aquatic products' trade has been thriving and enduring. From 2002 to 2012, aquatic products trade volume between China and ASEAN increased rapidly from US $309.31 million to US $1754.19 million, with an average annual growth rate of 22%, and the trade surplus is also increasing every year (see Table 1). Especially, since 2010, after the completion of the China–ASEAN Free Trade Area, the tariffs on aquatic products between China and ASEAN have been reduced to zero and the growth of their aquatic products trade has been accelerating accordingly. In 2011, China's export of aquatic products to ASEAN, as reported from the database of China's General Administration of Customs, increased by nearly 66% over a year. With a series of policies supporting the trade of aquatic products, that between China and ASEAN will continue to develop rapidly, whereas ASEAN will become more and more important to China's export of aquatic products. In view of this, this paper makes a comprehensive and detailed analysis on the competition of aquatic products' trade between China and ASEAN based on the HS classification. We hope to provide a theoretical basis and empirical evidence for strengthening the trade cooperation of aquatic products between China and ASEAN in the background of the "Maritime Silk Road" in 21st century. Furthermore, we try to propose important policy guidance for upgrading

Table 1. China–ASEAN's trade situation of aquatic products (US $ million).

Year	Export	Proportion of total exports of Chinese aquatic products	Import	Proportion of total imports of Chinese aquatic products	Net export
2002	75.09	2.6%	234.22	15.0%	−159.13
2003	114.48	3.4%	91.42	4.9%	23.05
2004	161.25	4.0%	129.20	5.5%	32.05
2005	112.65	2.6%	154.54	5.4%	−41.89
2006	142.94	3.0%	158.21	5.0%	−15.27
2007	126.79	2.7%	214.56	6.2%	−87.76
2008	176.16	3.4%	291.93	8.0%	−115.77
2009	483.06	7.1%	277.71	7.7%	205.35
2010	680.84	7.7%	347.05	7.9%	333.79
2011	1129.19	10.3%	368.62	6.6%	760.57
2012	1277.18	11.3%	477.01	8.7%	800.17

* Data source: UN COMTRADE database statistics.

the competitiveness of China's aquatic products export, as well as the corresponding export enterprises.

The data in this paper were taken from the UN COMTRADE database (2002–2012). In many occasions of trade cooperation research, the academics will focus on the competitiveness of trade between the two places. This paper will follow the paradigm of two normative empirical measure indices including export similarity index, trade intensity index and revealed comparative advantages index to analyze the competitiveness of aquatic products' trade between China and ASEAN. In fact, the chosen indices are common empirical methods widely applied to study the trade relations between two countries or areas in recent years. For example, Cheng Rong and Cheng Hui-fang (2011) used the index system to study the bilateral trade cooperation between China and India; LV Hong-fen and Yu Cen (2012) used the index system to study bilateral trade between China and Brazil; Hou Min (2011) used the index system to study the agricultural products' trade between ASEAN and Australia; and Zheng Si-ning (2013) studied the aquatic products' trade between Fujian and Taiwan.

2 METHODS OF ANALYSIS

The aquatic products' trade between China and ASEAN probably seems competitive because China and most of the ASEAN countries are emerging market countries with similar development stage of marine industry economy. Now we use the following indices to examine the assumption.

2.1 Expor similarity index

This paper applied the export similarity index proposed by Finger and Krenin (1979) to measure the export similarity of aquatic production of China and ASEAN, defined by the following formula:

$$ESI_{ab} = \left\{ \sum_{i=0}^{n} \left[\left(\frac{X_{ak}^i / X_{ak} + X_{bk}^i / X_{bk}}{2} \right) \times \left(1 - \left| \frac{X_{ak}^i / X_{ak} + X_{bk}^i / X_{bk}}{X_{ak}^i / X_{ak} + X_{bk}^i / X_{bk}} \right| \right) \right] \right\} \times 100$$

The ESI measures the similarity of the export patterns of country "a" and "b" to market "k." Xaki/Xak is the share of commodity of sector i in a's exports of designated product group to market k while Xbki/Xbk is the share in b's exports to k. The index ranges from 0 to 100. The higher index value indicates that the export product group structure of the two places are more similar and of more intense competition. Considering Japan as the most important target market of both China and ASEAN, we calculate two groups of ESI, respectively, to the world market and Japan market to make comparative analysis (Table 2).

As Table 3 shows, from 2002 to 2012, the ESIs of aquatic products from China and ASEAN to the world market are higher than 65, which are much higher than that of Japan, which are mostly below 50. The trends to the two markets are basically the same from dynamic view with upward ESIs in the last 10 years. The data indicate the high similarity of export structure of aquatic products from China and ASEAN to that of the world market

Table 2. ESI of aquatic products of China and ASEAN.

Year	ESI	
	World market	Japan
2002	68.53	45.35
2003	69.77	47.00
2004	71.55	44.70
2005	67.26	41.34
2006	67.10	35.90
2007	67.81	39.16
2008	73.12	44.56
2009	75.34	46.52
2010	73.00	44.18
2011	74.66	48.16
2012	74.94	50.23

Table 3. China–ASEAN's bilateral trade intensity index of aquatic products and total products.

Year	Aquatic products		Total products	
	TII from China to ASEAN	TII from ASEAN to China	TII from China to ASEAN	TII from ASEAN to China
2002	0.727	0.803	1.228	1.583
2003	0.958	0.478	1.325	1.827
2004	1.268	0.395	1.313	1.800
2005	1.131	0.382	1.290	1.814
2006	1.120	0.412	1.320	1.765
2007	1.037	0.545	1.413	1.837
2008	1.364	0.503	1.360	1.691
2009	1.647	0.504	1.493	1.618
2010	1.685	0.591	1.370	1.575
2011	2.692	0.487	1.389	1.597
2012	2.122	0.517	1.427	1.495

but less competitive to that of Japan. These results come up a perfect match to the reality—the main aquatic products that China exports to Japan are live or fresh or frozen fish like eel, and fish eggs while those from ASEAN are live or fresh or processed crustaceans. However, it seems more and more competitive every year like the trend in the world market. Although in some year, like 2005, 2006, and 2010, the ESIs drop in some degree to the two target market (which may be because of the time lag effects from the RMB appreciation caused by China's exchange rate reform in 2005 and the Global Financial Crisis in 2008–2009), the export similarity of aquatic products of China and ASEAN is generally increasing. In other words, the competitiveness of aquatic products' export between China and ASEAN is increasing. This results from China's expansion of export scale of aquatic products, especially aquaculture products, which accounts for 61.69% in total global output by 2012 with a 5.5% growth rate annually from 2000 to 2012 (data from Food and Agriculture Organization of the United Nations). The development of domestic aquaculture directly improves the export of aquatic products in China and its international competiveness.

However, these results of ESI are in view of market angle. From the viewpoint of specific products, we find that China's export of aquatic products of HS classification was US $18.596 billion, which still lags behind that of ASEAN (i.e., US $ 20.127 billion) in 2012. The gap mainly lies on the differences in geographical distribution and deep processing technology of products: the ASEAN's exports of marine fisheries like products 0302 (whole fresh or chilled fish) and 0306 (crustaceans), and deep processed aquatic products like 1504

(not chemically modified fish, marine mammal fat or oil), 1604 (prepared or preserved fish, fish eggs, caviar), 230120 (flour or meal, pellet, fish, etc., for animal feed), and 391310 (alginic acid, its salts & esters, in primary forms), are significantly greater than those from China. This indicates that the competition of aquatic products' export from China and ASEAN lies in different groups of products. In other words, the product difference may conceal complementary relation between the export of aquatic products in China and ASEAN.

2.2 Trade intensity index

The concept of "trade intensity" was apparently first used by A. J. Brown (1949) to analyze the strength of bilateral trade ties between countries. Applied and developed by several scholars like Kojima (1964) and Kunimoto (1977), the Trade Intensity Index (TII) has been described to measure the level of trade interdependence between two countries (Edmonds, 2010). The index formula is:

$$TII_{ab} = (X_{ab}/X_a)/(M_b/M_w),$$

where TII_{ab} is the trade intensity degree from country a to country b, X_{ab} is the export from country a to country b, X_a is the total export of country a, M_b is the import of country b, and M_b is the import of country b from the world. Take 1 as the critical point, when the value of TII_{ab} is higher, the trade between two countries is expected to be more intensive. This paper computes the trade intensity degree of aquatic products' trade of China and ASEAN separately, to carry out comparative analysis on them (see Table 3).

In view of the total products in 2002–2012, both the TII from China and from ASEAN are above 1, which reveals the high trade intensive degree between the two areas. The all annual index from ASEAN to China are higher than that from China to ASEAN, but with a narrowing gap. In view of the aquatic products, the TII from China to ASEAN increases rapidly, which exceeded 1 from 2004 and peaked at 2.692 in 2011, and has been through a shift from lagging behind to moving dramatically beyond the index of total products. In comparison, the trade intensity degree from ASEAN to China goes much weaker, which has been around 0.5 or no more than 1 and much lower than that of total products.

We conclude our analysis of TII: with the establishment of CAFTA, the commodity trade connection between China and ASEAN has been continuously strengthening from 2002 to 2012, especially in the field of aquatic products, which reflects the reality that ASEAN has become China's important export market of aquatic products, as China has been increasing its export to ASEAN. Despite this, the export share of aquatic products from ASEAN to China seems hesitating to go forward, which comes in a realistic imbalance of bilateral trade—ASEAN's export of aquatic products to China is less intensive. It results from two aspects: (1) China's domestic freshwater aquaculture is flourishing with an area of nearly 6 million hectares, accounting for 73.04% of the total domestic aquaculture area, making adequate supply of freshwater fish to satisfy the domestic market, which mainly consumes low-valued freshwater fish; (2) Nearly 60% of China's import of aquatic products are fresh or chilled fish, mostly from Russia, Peru, and the United States, for processing into frozen fish or fresh, frozen fish fillets to export but not for domestic consumption. Therefore, China has a deficient import demand from ASEAN countries even their marine aquaculture and fishing is relatively developed.

3 CONCLUSIONS AND SUGGESTIONS

With the implementation of the "going out" strategy of opening up, China has deepened the economic and trade cooperation with ASEAN and broadened its field of international trade and economic cooperation. Under the new situation of building the 21st-century Maritime Silk Road, China and the ASEAN countries are facing a historical opportunity to accelerate their economic and trade cooperation. This paper thoroughly analyzes the competition of the bilateral trade on aquatic products between the two areas by calculating the export similarity index and trade intensity index. From the results, we found the truth as follows: by ESI, the competition of the export of aquatic products between China and ASEAN tends to increase, but it lies in different product groups; by TII, trade imbalance appears in the bilateral aquatic trade, as the contribution of the export of China's aquatic products to ASEAN is growing while the contribution of ASEAN's export of aquatic products to China stays low. In general, the trade of aquatic products between China and ASEAN is not highly competitive. This indicates that the potential of bilateral trade of aquatic products between China and ASEAN is still underestimated, and more cooperation will be useful to promote the product difference to reduce the degree of competition.

On the basis of such historical background, it is worth considering how to make advantage of resource endowment of China and ASEAN to find a feasible path to develop their bilateral trade of aquatic products. This paper attempts to propose the following suggestions.

3.1 Jointly build the Maritime Silk Road to promote the trade of aquatic products with ASEAN

The government should make efforts to continue to provide more policy supports to carry out the consensus, initiative, and overall planning on the economic cooperation of Maritime Silk Road between China and ASEAN. Specifically, it can set up a special leading group or an expert committee to make a top-level design of an overall planning for the construction of Maritime Silk Road. It can also put forward a development strategy to promote this construction by refining China's regional trade cooperation planning with ASEAN so as to advance multifield, multitier, and multiform trade cooperation with ASEAN. By jointly building a "Maritime Silk Road" marine economic circle with ASEAN, it would harmonize the standards for aquatic products' quality, health, and safety of the two areas with the international standards so as to resolve the technical trade barriers, simplify the customs clearance procedures, and improve the trade efficiency.

3.2 Make advantage of resource endowment on both sides to deepen the aquatic products' trade cooperation with ASEAN

On the one hand, the government shall support a "Multiple In, Produce Excellent" strategy. In other words, China should import more primary aquatic products of lower cost in ASEAN than in domestic market and produce those similar products with higher quality to meet the high-end market by technology upgrading. On the other hand, the government can strongly encourage the well-financed and large-scale processing enterprises to go out to collaborate with the local enterprises to make full use of the rich aquatic resources and lower-cost labor resource in ASEAN so that the Chinese "going out" enterprises can produce high-demand aquatic products like tuna, lobster, and other high-end seafood and primary processed products, which are difficult to produce or of high cost in domestic market, and then sell back to China.

ACKNOWLEDGMENTS

The authors acknowledges the support from the National Natural Science Foundation of China (71603060), the Ministry of education of Humanities and Social Science project (16YJC790023), the 2015 Philosophy and Social Science Planning Project of Guangdong, China (GD15YYJ01), the 2016 Philosophy and Social Science Development Planning Project of Guangzhou, China (2016GZYB04), the 2015 Young Creative Talents project from the Guangdong Department of Education, China (2015WQNCX029), and the Soft Science Research Program of Guangdong, China (2016A070706006).

*Corresponding author at: Guangdong Institute for International Strategies, Guangdong University of Foreign Studies, No. 2 Baiyun Avenue, Guangzhou, Guangdong, 510420, China, Tel: +86 20 36317679.

E-mail address: hanyonghui2006@foxmail.com.

REFERENCES

Brown, AJ. (1949). *Applied Economics: Aspects of World Economy in War and Peace*. George Allen and Unwin, London.

Cheng R & Cheng HF. (2011). China-India Trade Relation: Competition or Complementary. *Journal of International Trade*. 348 (Jun):85–94.

Edmonds C & Li Y. (2010). A new perspective on china trade growth: application of a new index of bilateral trade intensity.Working Papers. University of Hawaii at Manoa, Department of Economics.

Finger JM & Kreinin M E. (1979). A Measure of `Export Similarity' and Its Possible Uses. *The Economic Journal*, 89(356):905–912.

Hou M. (2011). Study on Complementary Relations of Agricultural Product Trade between ASEAN and Australia-New Zealand: an Analysis Based on RTA and OBC. *Journal of International Trade*. 348(Dec): 89–96

Kojima K. (1964). "The pattern of international trade among advanced countries," *Hitotsubashi Journal of Economics*. 5(1):16–36.

Kunimoto K. (1977). Typology of Trade Intensity Indices, *Hitotsubashi Journal of Economics*. 17(1):15–32

Lv HF & Yu C. (2012). Research on Competitiveness and Complementarity of Bilateral Trade between China and Brazil. *Journal of International Trade*. 360 (Feb):56–64.

Shao GL & Jiang H. (2007). Analysis and Comparison for the Nontraditional Barriers to Main Target Country for China's Seafood Export. *Issues in Agricultural Economy*. 28(7):81–84.

Zheng SN. (2013). Research on the Rivalrousness and Complementarity of Aquatic Products between Fujian and Taiwan. International, *Economics and Trade Research*. 29(1):103–112.

Advances in Energy Science and Equipment Engineering II – Zhou, Patty & Chen (Eds)
© 2017 Taylor & Francis Group, London, ISBN 978-1-138-71798-5

The selection of the leading industries in Shanghai Pudong airport economic zone

Liping Zhu & Xubiao Yang
Air Transport College, Shanghai University of Engineering Science, Shanghai, China

ABSTRACT: The paper clarifies the meaning, characteristics and importance of the leading industries in the airport economic zone. Then, it analyses the development level and opportunity of the Shanghai Pudong airport economic zone. By qualitative analysis and quantitative analysis, the paper points out that aviation logistics and warehousing industry, aviation maintenance and parts manufacturing industry should be selected as the leading industry in Shanghai Pudong airport economic zone at present.

1 INTRODUCTION

As the airport plays leading role in the regional economy apparently, the airport economic zone have appeared surrounding major domestic and international airports. It becomes regional growth pole and updates local industrial structure, which has advantages of airport location, convenient transportation and industry diversification. Shanghai Pudong airport economic zone has developed for more than ten years. It attracts many enterprises to enter, but the integral industry chains within the region has not formed. Industry is the key factor connecting regional macroeconomy and microeconomy, so we should make an objective and comprehensive selection of the leading industry in the airport economic zone. Therefore, this paper discusses the characters and selection of leading industry in the Shanghai Pudong airport economic zone, which would benefit for the upgrading of industrial structure and regional economic development.

2 LITERATURE REVIEW

Generally, the researches can be roughly divided into two types. One is discussed by professional researchers from the perspective of economics theory. Researchers focus on the relationship between the airport and the city, including direct and indirect economic impact, by model analysis and case studies. The other is airport advisory reports. The representative report is "The Social and Economic Impact of Airports in Europe", published by Airports Council International Europe Branch (ACI-Europe) and York aviation consulting firm.

China researches on airport economic zone start for a short time, which mainly summarize the development experience. Wei Jie (2004) points there are three criteria to determine the development of the airport economic zone: First, the airport's passenger and cargo traffic should reach a certain scale in order to form the airport economy. Second, the national economy and the cities economy surrounding the airport should achieve a certain level. Third, the airport is surrounded by a large number enterprises of tax revenue and employment opportunities, which is the most difficult to achieve in the three criteria. Cao Yunchun (2004) analyzes the relationship between regional economy level with airport economic zone development at its infancy stage, developing stage and mature stage. Sun Yanhai (2004) sums up the development experience of the area nearby domestic and foreign airport and points that airport area nearby developing to the airport economic zone is the inevitable trend, which is driven by the need of airport development, enterprise development, city and country economy development. Peng Jiyan (1997), Jiang Ling (1999), Jin Zhongmin (2004), Wang Hui (2005) Fang Zhongquan (2005), Jiang Xinsheng (2005), and many other scholars, from the aspects of urban planning, regional economy development, and airport economic zone management, have researches on the opportunity, industry development, and development strategies of Beijing Capital Airport, Nanjing Lukou Airport, Shanghai Pudong International Airport, Guangdong New Baiyun Airport, Tianjin Airport.

From above, we can see that there are few study on the development of the airport economic zone from the aspect of the leading industries. The study on the definition, function and selection ways of leading industries in airport economic zone would benefit for local government to support the industry development.

3 THE CHARACTERISTICS OF THE LEADING INDUSTRIES IN THE AIRPORT ECONOMIC ZONE

Airport economic zone is a special economic space with the core of airport relying on airport facilities and aviation manufacturing and transporting activity. It attracts technology, capital, information, trade, population and other production factors to the airport vicinity and forms industry clusters related to aviation.

The leading industry can prompt to the upgrading of the regional industrial structure different from other regions which would develop constantly. The essential characteristics of the leading industry are the function of promoting economic development and upgrading of industrial structure. Firstly, the leading industries can drive technological innovation and upgrade industrial structure. Secondly, the leading industries are the driving wheel of the regional economic growth. Thirdly, leading industries are associated with a large of industries and can lead other industries develop. Fourthly, leading industry can create new market demand.

The well-functioning industry clusters in airport economic zone need a long history. According to the size and quality of the airport economic zone, the situation airport economic zone can be divided into three stages: forming stage, growing stage and mature stage. At forming stage, the industry entering the airport economic zone is characterized with a strong connection with airport including aviation industry and aviation related industries. These industries such as aviation logistics industry and other high-tech products manufacturing industry, and aviation assembling manufacturing industry are gathering around the airport. At growing stage, the important features of the airport economic zone development are the extension of various manufacturing industries and the emergence of the modern service industry which improve the chain. At the mature stage, the impetus to the development of airport economic zone, mainly due to the endogeneity force, which is different from the situation at the growing stage. On the basis of the improvement of sound industrial chain and the formation of industrial clusters, innovation has become main force mechanism. The region would gain dynamic competitive advantage by continuously improving the enterprises learning ability.

4 EQUATIONS AND MATHEMATICS

The selection of leading industry in airport economic zone can be divided into two steps. The first step is initial selecting of industries around the airport industry through qualitative analysis. The second step is to select leading industry from the initial selecting scope via quantitative analysis, such as comprehensive evaluation method.

4.1 Qualitative analysis

Before quantitative analysis of selecting leading industries in airport economic zone, it is prerequisite to conduct a qualitative judgment in order to have a preliminary screening. Quantitative analysis consists of three aspects.

Firstly, we should analyze national and regional industry strategic layout. Because national and regional industry strategic layout is an important factor to affect the regional division of industries, and thus has an important impact on the development of the regional economy.

Secondly, we should analyze the internal and external environment of industry development. The factors such as regional comparative advantages, potential competitiveness and sustainable development capacity represents the industry's basic conditions, potential conditions and future development conditions, which can provide more accurate and more scientific reference for the selecting of leading industry.

Thirdly, we should analyse the stage of the airport economic zone development. The well-functioning industry clusters in airport economic zone need a long history including forming stage, growing stage and mature stage. At different development stage, it attracts different category of leading industry.

4.2 Quatitative analysis

The selection of leading industry is a multi-target, multi-object decision-making problems which generally conducted by comprehensive evaluation method. Comprehensive evaluation method determines the objects relative level by the establishment of a multi-level index system, and calculating weight of various factors affecting variable. Quantitative analysis has five following steps.

Firstly, according to the selection principles of the leading industries in airport economic zone, we establish an evaluation indicator system, as Table 1.

Secondly, the data should be nondimensionalized. This paper uses extremum processing method to process the original data in order to eliminate the effect of the indicators dimension.

Thirdly, we calculate weight of indicators on all levels by Analytic Hierarchy Process method. AHP requires questionnaires issued to a group of experts. After sorting out results of questionnaires, we can calculate weight of indicators on all levels.

Table 1. The indicator system of selecting the leading industries in the airport economic zone.

The indicator system of selecting the leading industries in airport economic zone (A)	Prompting regional economy principle (B₁)	The proportion of production value (X_{11})
		Cost margin ratio (X_{12})
		Added value of investment in fixed assets (X_{13})
	Industry potential development principle (B₂)	Income elasticity of demand (X_{21})
		Output growth ratio (X_{22})
	High correlation principle (B₃)	Sensitivity coefficient (X_{31})
		Influential coefficient (X_{32})
	Technical progress principle (B₄)	Technology diffusion ratio (X_{41})
		Number of R&D personnel (X_{42})
		Expenditure of R&D activities (X_{43})
	Relative comparative advantage principle (B₅)	Location Quotient (X_{51})
	Social benefits principle (B₆)	Scale of employment (X_{61})
		Jobs per million output value (X_{62})
		Jobs per million of fixed assets (X_{63})

Fourthly, we can get the index of the comprehensive evaluation of leading industries by calculating data into the formula using the weight of indicators.

Fifthly, we can determine the leading industries in airport economic zone on the basis of qualitative and quantitative analysis results.

5 EMPIRICAL ANALYSIS OF LEADING INDUSTRIES IN SHANGHAI PUDONG AIRPORT ECONOMIC ZONE

Relying on the Pudong International Airport, there has been built fast and secure transportation and aviation networks around the world. At present, the Shanghai Pudong International Airport accelerates the construction of aviation hub, creating good conditions for the development of airport economic zone. Reasonable planning and steady implementation will promote sustainable development of Pudong Airport and the surrounding areas.

5.1 Initial selecting of industries by qulatitative analysis

At present, there are a number of airport economic zones all over the country. But the development of the airport economic zones is not at the same level. Shanghai Pudong airport economic zone has developed for more than ten years and it belongs to growing stage according to some researches. At growing stage, the important features of the airport economic zone development are the extension of various manufacturing industries and the emergence of the modern service industry which improve the chain. Shanghai Pudong airport economic zone has good opportunities such as Commercial Aircraft project, construction of the

port area, the follow-up development of Expo and Shanghai Disneyland construction.

To sum up, according to the opportunities and the actual situation of the various types of industries, there are seven industries by initial selecting: (1) Leasing and Business Services industry (2) Information transmission, computer services and software industry (3) Accommodation and catering industry (4) Logistics and Warehousing industry (5) High-tech and equipment manufacturing industry (6) Exhibition industry (7) Tourism industry.

5.2 Comprehensive evaluation of the leading industries

5.2.1 Description of data

According to above indicators system, we collect data from "Shanghai Statistical Yearbook (2011)", "Shanghai Pudong New District Statistical Yearbook (2011)", "Shanghai Input&Output Table Analysis and Application (2007)" and the web of Shanghai Bureau of Statistics (http://www.stats-sh. gov.cn/). Then, the data are nondimensionalized, as Table 2. It should be noted that there are lack of statistical data of airport economic zone because data are collected mainly in the form of administrative regions. So, we take the data of Shanghai Pudong New District instead of Shanghai Pudong airport economic zone.

5.2.2 Calculating weight of each indicator by AHP

In this paper, we calculate weight of each indicator by AHP. We put out questionnaire to some experts working in government and colleges. Then, according to the scores by experts, we get the judgment matrix and test the consistency of the test matrix. Finally, we calculate the Weight of each indicator, as Table 3.

Table 2. Nondimensionalized results of data.

	X_{11}	X_{12}	X_{13}	X_{21}	X_{22}	X_{31}	X_{32}
(1)	0.026	0.195	0.478	0.355	0.846	0.002	0.758
(2)	0.063	0.337	0.0597	0.386	1.000	0.400	0.775
(3)	0.291	0.724	0.164	0.716	0.664	1.000	0.800
(4)	0.670	0.733	0.883	0.773	0.839	0.382	1.000
(5)	1.000	0.249	1.000	0.152	0.367	0.172	0.975
(6)	0.397	1.000	0.009	1.000	0.785	0.434	0.907
(7)	0.847	0.421	0.018	0.875	0.319	0.447	0.881

	X_{41}	X_{42}	X_{43}	X_{51}	X_{61}	X_{62}	X_{63}
(1)	0.070	0.109	0.095	0.134	0.075	0.154	0.037
(2)	1.000	0.712	0.583	0.195	0.229	0.572	0.041
(3)	0.151	0.012	0.101	0.397	0.315	0.296	0.257
(4)	0.231	0.187	0.358	0.603	0.789	0.783	1.000
(5)	0.776	1.000	1.000	0.235	1.000	0.815	0.891
(6)	0.132	0.043	0.121	0.432	0.463	0.231	0.011
(7)	0.015	0.065	0.119	1.000	0.196	0.467	0.015

Table 3. Weight of each indicator.

B_1	0.156	X_{11}	0.582
		X_{12}	0.334
		X_{13}	0.083
B_2	0.289	X_{21}	0.75
		X_{22}	0.25
B_3	0.301	X_{31}	0.667
		X_{32}	0.332
B_4	0.155	X_{41}	0.137
		X_{42}	0.286
		X_{43}	0.575
B_5	0.102	X_{51}	1
B_6	0.078	X_{61}	0.157
		X_{62}	0.297
		X_{63}	0.545

Table 4. Rank of comprehensive evaluation index of the leading industries in Shanghai Pudong airport economic zone.

	Comprehensive evaluation index	Rank
Leasing and Business Services industry	0.314	7
Information transmission, computer services and software industry	1.691	4
Accommodation and catering industry	1.118	6
Logistics and Warehousing industry	4.413	1
High-tech and equipment manufacturing industry	1.824	2
Exhibition industry	1.643	5
Tourism industry	1.762	3

5.2.3 *Calculating the comprehensive evaluation index*

According to the above results, we can get the comprehensive evaluation index by the formula:

$$Z = \sum B_i \times X_{ij} \left(i = 1,2,...6; j = 1,2,3 \right) \quad (1)$$

6 CONCLUSION

According to the results of the quantitative analysis, it can be seen that the rank of comprehensive evaluation index of the leading industries in Shanghai Pudong airport economic zone. In conclusion,

a. logistics and warehousing industry; (b) high-tech and equipment manufacturing industry are selected as the leading industries of the Shanghai Pudong airport economic zone. The reasons are as follows:

1. The logistics and warehousing industry as the leading industry is consistent with the results of quantitative analysis and it is in line with the general law of the development of airport economic zone.
2. Due to Shanghai Pudong airport economic zone is at growing stage, the expansion of the industry chain is an inherent requirement. Aviation maintenance and parts manufacturing industry is in line with the direction, which is an important driving force for regional industry upgrading. Therefore, the high-tech and equipment manufacturing industry is also selected as the regional leading industries.

The goal of selecting the leading industry in airport economic zone is to play the key role of the leading industries. Therefore, it is suggested that we should accelerate the expansion of the industry chain to gather industry clusters, create and develop brands in airport economic zone, improve the financing channels and strengthen the senior human resources.

REFERENCES

Brenner T., Greif S., The Dependence of Innovativeness on the Local Firm Population:An Emprical Study of German Patents, Industry & Innovation, 21–39 3(2006).

Jean Marc Callois, The two sides of proximity in industrial clusters: The trade-off between process and product innovation, Journal of Urban Economics, 1–17 1(2007).

John D Kasarda, From Airport City to Aerotropolist, Airport, 42–45 6(2001).

Liu Xueni, Ning Xuanxi, Zhangdongqing, The analysis of industry cluster around Beijing International Airport, Soft Science, 41–44 3(2008).

Yang Youxiao, Cheng Cheng, The study of airport economic development stage and government functions, International Economics and Trade Research, 69–73 10(2005).

Advances in Energy Science and Equipment Engineering II – Zhou, Patty & Chen (Eds)
© *2017 Taylor & Francis Group, London, ISBN 978-1-138-71798-5*

Model of the route choice of passengers in an airport terminal

Shuyi Hou & Yukun Liu

Logistics Engineering, School of Automation, Beijing University of Posts and Telecommunications, China

ABSTRACT: Nowadays, many people prefer to travel by air. To deal with problems about passenger flows in complex airport system, several simulation models have been built. However, most of them are macroscopic, ignoring the process that individuals make choice of their discretionary activities. In this paper, a model of route choice of passengers in airport terminal is presented. Then, a simulation model is built to demonstrate the practicality of the theoretical model.

1 INTRODUCTION

Modern airport terminal is composed of check-in counters, custom control counters, kinds of catering facilities, and many other submodules. With the development of economy, large number of people prefer air travel. The airport management has to deal with larger and larger number of passengers who use airport. Obviously, simulation is a reliable tool to deal with the complex airport system. At present, most research about airport terminal simulation concentrate on passengers' process, neglecting discretionary activities at other facilities. Furthermore, most passenger flow studies about airport terminal are macroscopic and do not refer to the process that individuals make route choices. However, standard processing facilities and discretionary facilities are always used by passengers simultaneously. Only making standard procedure into consideration is not comprehensive. In order to know the way that passengers behave in the real airport terminal environment, it is essential to understand the route selection method. In this paper, a model of route choice of passengers is proposed, and a simulation experiment is made to illustrate the application of this method.

2 PASSENGER FLOWS

Typically, on the basis of the passengers' process when they are in the terminal, passengers can be divided into three categories: departing passengers, arriving passengers, and transferring passengers. After passengers arrive at airport terminal, there are many possible facilities they may use and the ways they use are different in terms of different categories. Departing passengers are those who check in and who transfer aircraft. Arriving passengers are those who directly leave airport and

who transfer to another aircraft. Thus, transferring passengers' process can be included into arriving passengers and departing passengers' process. Departing passengers can go to duty-free shops, cafes, restaurants, toilets, and so on. Arriving passengers can go to cloakroom, take baggage cart, go to toilets, and so on. Passengers' route choice is influenced by time stress, stimulation of the environment, and mental needs. The following flowchart can illustrate the standard process and discretionary activities of passengers. The full line represents the standard process of a passenger. The dotted line represents the discretionary activities that passengers can take. Standard process of arriving passengers in domestic flight can leave out quarantine, frontier inspection, and custom controlas shown in Figure 1 and Figure 2.

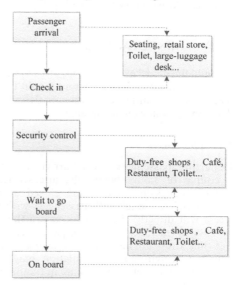

Figure 1.   Departing passengers flow.

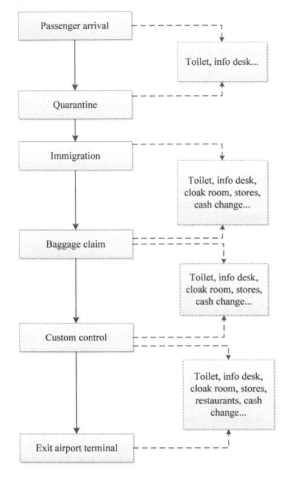

Figure 2. Arriving passengers flow.

3 INFLUENCE FACTORS

To build the model of route choice of passengers in airport terminal, influence factors should be analyzed first. The influence factors are described in detail in the following paragraphs.

3.1 Time stress

Departing passengers always arrive at airport a couple of hours before the aircraft leaves, because check-in counters are closed at least 30 min before the take-off. Passengers with time stress will directly check in to ensure they can go aboard on time. Otherwise, passengers without time stress are more likely to take additional activities. Environment stimulation may make passengers take activities that are not under their consideration.

A considerable time is needed for the arrived baggage to be transferred to the baggage claim area. Especially, the inspection mechanism of cross-border baggage is strict and may cause long delay. More time is needed when several planes arrive simultaneously, and this kind of situation happens a lot in large airport terminal. In real cases, the situation that passengers wait for baggage happens occasionally. During the waiting time, passengers have plenty of time to use other facilities in the terminal. The discretionary time is influenced by travel purpose of passengers.

3.2 Stimulation of the environment

Passengers interact with the surrounded environment. Thus, the environment has an impact on the behaviors of passengers. The congestion and long queue at some facilities may influence the route choice of passengers. For instance, when there are so many passengers at the baggage claim area waiting for their baggage, some passengers may go to vending machine first and claim their baggage until there is less people waiting. Moreover, passengers may be attracted by the promotion in shops and change their destination to the shops.

3.3 Mental needs

Mental needs of passengers at some degree dominate the behaviors of passengers. Passengers go to shops or ATM. Mental needs change over time. In the airport terminal, the activities and facilities have one-to-one correspondence. Every passenger has different dynamic mental needs to different facilities.

Mental needs can be abstracted as dynamic needs parameter to indicate the need degree of using facilities. The higher the dynamic needs parameter is, the more possibility of going to corresponding destination passengers have.

Table 1 shows the relationship between airport facilities and dynamic needs parameters of passengers. The first column describes the dynamic needs parameters, which influence the choice whether passengers use the facilities. The second column is the activities that passengers may take. The third column is the facilities that passengers may use. For instance, restaurants and café provide catering services and meet requirement of diet, and the parameter can be described by the level of hunger. The passengers with problems can go to information desks, which provide consulting service. The willingness to ask for assistance can be used as dynamic needs parameter.

Table 1. Relationship between airport facilities and dynamic needs parameter.

Dynamic needs parameter	Activities	Use of facilities
Willingness to ask for assistance	Inquiry	Information desk
Degree of needing cash	Encashment	ATM, money exchange agency
Level of hunger and thirst	Eating and drinking	Restaurant, café
Desire to shop	Shopping	Duty-free shops
Desire to relax	Seating	Check-in hall lounge
Comfort with technology	Self check-in	Self check-in kiosks
Need baggage cart?	Retrieving baggage cart	Baggage cart area
Desire to change clothes	Changing clothes	Cloakroom
Need to go to toilet?	Going to toilet	toilet
Need air insurance?	Buying air insurance	Air insurance
Have large luggage?	Retrieving large luggage	Large luggage claim area
Desire to charge phone	Going to charge phone	Phone charging station

4 THE MODEL OF DESTINATION CHOICE OF PASSENGERS IN AIRPORT TERMINAL

After analyzing the influence factors, the model of route choice of passengers in airport terminal is described as follows

1. Use time stress to determine whether the passengers can take additional activities such as eating and shopping. List every possible activity that the passengers without time stress can take. The passengers with time stress will directly take activities in the standard process. When an activity is finished, the passenger will estimate the time left. If the left time is enough, passengers can decide where to go. Otherwise, passengers should finish their procedure as soon as possible.
2. Compare every dynamic needs parameter of a passenger. The higher the parameter is, the more possibility that the passenger takes part in the corresponding activity is. The passenger will choose the facility with the highest parameter.
3. Stimulation of the environment can increase the dynamic parameter. The environment factors include shop promotion, congestion, and long queue at facilities.
4. After a passenger finishes an activity, time stress and dynamic needs parameter will change. It spends time for passengers to use facilities in terminal. The time stress is increased, urging passengers to finish their procedure. Meanwhile, the parameter will change to zero, which means the passenger would not choose the previous facility as next destination.

5 SIMULATION MODEL

AnyLogic, a simulation software, provides pedestrian library, which can be used to simulate

Table 2. Time and discretionary facilities.

Time	Discretionary facilities
Before check-in	Toilet, lounges, retail store, oversized & overweight baggage check, newsstand, restaurant, cafe
Before boarding and after check-in	Duty-free shops, cafe, restaurants, toilets

pedestrian flow in real environment. For example, pedestrian library can be used to create pedestrian activities in building such as subway station and railway station. It is a useful tool to build flexible simulation model, collect statistics, and show the model visually. In the simulation model made by AnyLogic, pedestrians travel in continuous space and react to barriers and other pedestrians. To demonstrate the practicality of the model presented above, a simulation model is built based on an airport in China using AnyLogic. The following table lists discretionary facilities in this airport terminal and the time that the activities can be done.

In general, there are three steps in building this simulation model.

1. First, set layout of the airport terminal according to the CAD drawing provided. Model the environment, including walls, barriers, galleries, service facilities, waiting area, and queue area.
2. Drag parameter in general palette into the agent type representing dynamic needs parameter and variable representing shopping promotion. When the value of shopping promotion variable is above zero, the shopping needs parameter will be increased.
3. Then, model the behaviors of passengers. Use the flowchart to define the passengers' process

from arriving to leaving the system. The following flowchart is made based on the model of route choice of passengers presented above. The simulation model is built in terms of Figure 3.

The Figure 4 is the simulation presentation.

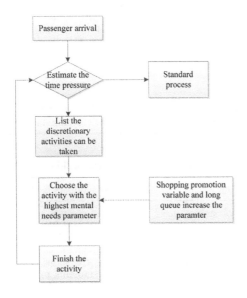

Figure 3. Logic in simulation model.

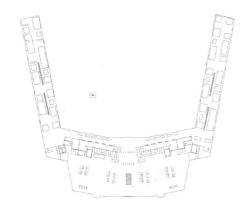

Figure 4. Simulation presentation.

6 SUMMARY

In this paper, passengers' procedure is given first, including standard processing procedure and discretionary activities. Then, the factors that influence the passengers' route choice are investigated. The influence factors include time stress, stimulation of the environment, and mental needs. To represent these influence factors, dynamic needs parameters are proposed. Then, the model of route choice of passengers in airport terminal is described. Finally, a simulation model is built using AnyLogic to prove the practicality of the theoretical model.

REFERENCES

Bovy, P.H.L., E. Stern, Kluwer Academic Publishers. Dordrecht (1990).
Gipps, P.G. Schriftenreihe des Instituts fuer Verkehrswesen 35, University of Karlsruhe (1986).
Helbing, D., P. Molnar, Physical Review E Statistical, Nonlinear and Soft Matter Physical (1995).
Henderson, L.F. Nature, 11, 381 (1971).
Hoogendoorn, S.P., P.H.L. Bovy, Transportation Research Part B, 38, 169 (2004).
Hughes, R. Transportation Research B 36, 6, 507 (2002).
Seth, B. Young, Transportation research record 1674, 99 (1999).
Wenbo, M. Queensland University of Technology (2012).
Xi, H., Y.J. Son, S. Lee, Simulation Conference (WSC), Proceedings of the 2010 Winter, 824 (2010).

Advances in Energy Science and Equipment Engineering II – Zhou, Patty & Chen (Eds)
© 2017 Taylor & Francis Group, London, ISBN 978-1-138-71798-5

Influence of fracture energy on the rotating model of fiber-reinforced concrete

Xiaoyue Zhang
Department of Hydraulic Engineering, Zhejiang University of Water Resources and Electric Power, Hangzhou, China

Dong Wang
Zhejiang Water Conservancy and Hydropower Engineering Quality and Safety Supervision and Management Center, Hangzhou, China

ABSTRACT: Ultra-High-Performance Fiber-Reinforced Concrete (UHPFRC) has been developed over the last century for its superior strength, high ductility, low permeability, and excellent durability. One of the most important features of UHPFRC is its superior ability to resist fracture and achieve higher loads after first crack to traditional concrete. As there is a special strain-hardening behavior, the simulation of UHPFRC is also different from that of traditional concrete. In this research, a smeared rotating crack constitutive model for the fracture behavior of UHPFRC with steel fibers was given. If the fracture energy already used is less than the total fracture energy, the old crack of the element closes and a new crack plane normal to the maximum principal stress forms. A three-point bending test was performed on single-edge notched prism, based on which the influence of fracture energy on load-resisting ability of the material was studied, also the simulation performances of fixed crack model and rotating crack model are compared.

1 INTRODUCTION

Fracture is one of the pressing issues in modeling civil engineering materials like concrete. The pursuit of genuine fracture properties has posed a continuous challenge to materials scientists, and to date, we have observed that the pool of data on properties such as fracture energy still extends (Nima, 2013). Ultra-High-Performance Fiber-Reinforced Concrete (UHPFRC) possesses a volume fraction of steel fibers of about 2%, compression strength higher than 150 MPa, and tension strength of 7–15 MPa, with the largest tensile strain of about 4%. Compared with conventional concrete, this material offers superior strength, high ductility, low permeability, and excellent durability (Zhang, 2012). By using UHPFRC, the amount of steel rebar needed in concrete construction can be reduced, and the thickness of concrete structure could be decreased. Because of these properties, UHPFRC is seen as a promising construction material to strengthen, retrofit, and repair current and future infrastructure. The material has already been used in large-scale applications, such as bridges, retaining walls, airport runways, and nuclear plants.

One of the most important features of UHPFRC is its superior ability to resist fracture to traditional concrete through the use of randomly dispersed fibers to the fiber matrix bond, and achieve higher loads after first crack. When conventional concrete is subjected to uniaxial tension load, cracking initiation is closely followed by the softening behavior (Simon, 2012). The tensile stress of UHPFRC goes on increasing after the linear–elastic stage because the steel fiber has much higher strength than concrete.

With a fixed crack method, the orientation of the crack is fixed during the entire computational process, whereas a rotating crack concept allows the axes of principal stress rotate after crack formation, which leads to an increasing discrepancy between the axes of principal stress and fixed crack. The model of fiber-reinforced concrete for fixed cracks has been widely accepted, but it begins to emerge that the rotating crack model may simulate the strength and stiffness of the material during the fracture process more precisely.

Strategies have been put forward providing solutions for multidirectional cracks or rotating cracks, but to date their performance for strain hardening and softening conditions are unsatisfactory. Raj Das gave a mesh-free method to simulate the concrete material under impact loading (Raj, 2015). S Muralidhara studied the size-independent fracture energy in plain concrete beams (Muralidhara, 2011). A. Benin studied the fracture energy of reinforced concrete bridge structures (VBenin, 2014). J. Bolander studied the fracture in concrete specimens of differing scale (EBolander, 2013).

In this paper, on the basis of a user-defined smeared rotating crack constitutive model of UHPFRC, in which the axes of principal stress rotate after crack formation, and a three-point bending test on single-edge notched prism, the influence of fracture energy on the load-resisting ability of UHPFRC was studied. Furthermore, the performances of the fixed crack model and rotating crack model are compared.

2 FRACTURE ENERGY OF UHPFRC

2.1 Constitutive model

The proposed UHPFRC material model under tensile loading builds up on Rots' smeared crack theory for conventional concrete fracture (Rots, 1998). Upon crack formation, an orthotropic stress–strain law is used, in which the total strain consists of elastic and cracking strains. Here we extend Rots' smeared crack theory of the strain softening stage to the whole fracture stage, including strain hardening process of the fiber-reinforced concrete. On the basis of Rots' theory, the relation between global stress and global strain can be written as:

$$\Delta\sigma = [D^{co} - D^{co}\hat{N}[\hat{D}^{cr} + \hat{N}^T D^{co}\hat{N}]^{-1}\hat{N}^T D^{co}]\Delta\varepsilon \quad (1)$$

where $\Delta\sigma$ is the increment of global stress; $\Delta\varepsilon$ is the increment of global strain; $[x, y, z]$ represents the global coordinate system:

$$\sigma = [\sigma_x\ \sigma_y\ \sigma_z\ \tau_{xy}\ \tau_{yz}\ \tau_{zx}]^T \quad (2)$$

$$\varepsilon = [\varepsilon_x\ \varepsilon_y\ \varepsilon_z\ \gamma_{xy}\ \gamma_{yz}\ \gamma_{zx}]^T \quad (3)$$

The increment of global strain $\Delta\varepsilon$ is composed of two parts: the cracking strain and the elastic strain:

$$\Delta\varepsilon = \Delta\varepsilon^{cr} + \Delta\varepsilon^{el} \quad (4)$$

$\hat{N}$ is the transformation matrix reflecting the orientation of the crack:

$$\hat{N} = \begin{bmatrix} l_x^2 & l_x l_y & l_z l_x \\ m_x^2 & m_x m_y & m_z m_x \\ n_x^2 & n_x n_y & n_z n_x \\ 2l_x m_x & l_x m_y + l_y m_x & l_z m_x + l_x m_z \\ 2m_x n_x & m_x n_y + m_y n_x & m_z n_x + m_x n_z \\ 2n_x l_x & n_x l_y + n_y l_x & n_z l_x + n_x l_z \end{bmatrix} \quad (5)$$

where l_x, m_x, and n_x form a vector, which indicates the direction of the crack plane in the global coordinate system, $l_x = \cos(n, x)$, $m_x = \cos(n, y)$, $n_x = \cos(n, z)$. In accordance with this convention, l_y, m_y, and n_y indicate the s-axis direction, and l_z, m_z and n_z indicate the t-axis direction.

$\hat{D}^{cr}$ is a 3×3 matrix incorporating the stiffness of the material in fracture mode I D^I, mode II D^{II}, and mode III D^{III} of the smeared crack:

$$\hat{D}^{cr} = \begin{bmatrix} D^I & 0 & 0 \\ 0 & D^{II} & 0 \\ 0 & 0 & D^{III} \end{bmatrix} \quad (6)$$

Mode I is defined as the opening mode, which occurs under tension loading. The other two modes are a result of shear of the material. Mode I D^I can be written as:

$$D^I = -\frac{1}{q}\frac{f_{cr}^2 h}{G_f} \quad (7)$$

where $q = 2$ reflects the shape of the softening diagram; f_{cr} is the cracking strength of the material; and h is the crack bandwidth.

Mode II leads to a sliding action between fracture surfaces. Mode III is a tearing action due to out-of-plane shearing. For simplicity, this study will only focus on mode I- and mode II-type fractures. In order to calculate more conveniently, use normal stress instead of shear stress to calculate mode II D^{II}, which indicates the relationship between shear stress and shear strain on the fracture surface:

$$D^{II} = \frac{\left(1 - \dfrac{\varepsilon_n^{cr}}{\varepsilon_u^{cr}}\right)^p}{1 - \left(1 - \dfrac{\varepsilon_n^{cr}}{\varepsilon_u^{cr}}\right)^p} G \quad (8)$$

where ε_n^{cr} is the crack normal strain at the beginning of the load increment, ε_u^{cr} is the ultimate crack normal strain, p is a constant, and G is the elastic shear modulus.

2.2 Fracture energy

Traditionally, a set of strength parameters has been introduced to control fracture initiation. The parameters are the fracture energy G_f, which is defined as the amount of energy required to create one unit of area of a mode I crack, and the shape of the tensile-softening diagram. These two parameters are assumed fixed material properties. The crack stiffness moduli must be expressed in terms of the strength parameter, the energy parameter, and the shape of the softening diagram. For a fixed single crack, the definition of G_f corresponds to the area under the softening curve:

$$G_f = \int \sigma_n^{cr} du_n^{cr} \qquad (9)$$

where σ_n^{cr} is the crack normal traction and u_n^{cr} is the crack normal displacement.

While the crack width increases to a certain value, the crack plane will fix. In every time step, whether the crack rotates or not should be judged before the calculation of global stresses. For more convenient calculation, fracture energy is used instead of crack width. Thus, if the fracture energy already used (which is calculated in last time step) is less than the total fracture energy, G_f, the old crack of the element closes and a new crack plane normal to the maximum principal stress forms. Otherwise, the crack will not rotate. For smeared cracks, the fracture is distributed over a crack bandwidth, h.

The fracture energy of strain-hardening material under tensile stress used in every time step can be calculated as:

$$\Delta G_f = (w_{cr,k+1} - w_{cr,k})(\sigma_{n,k+1} + \sigma_{n,k})/2 \qquad (10)$$
$$w_{cr} = \varepsilon_{cr} h \qquad (11)$$

where w_{cr} is the crack width; k represents the time step; h is the crack bandwidth; and $\varepsilon_{cr} = (\varepsilon_n - \varepsilon_{el})$ is the crack strain, equal to the crack normal strain minus the strain caused by elasticity.

3 THREE-POINT BENDING TEST

The purpose of the fracture test is to measure the influence of mode I fracture energy on the steel fiber-reinforced ultra-high-performance concrete, and to check the differences between fixed crack method and rotating crack method on the simulation of load versus crack opening width curves. The most common method for testing the fracture of concrete is a notched-beam three-point bending test. A typical fracture test involves notched specimen, which induces a crack to propagate at the tip of the notch. A specimen with a centralized notch was used to test the ability of the material to resist a load under different mode I-type fracture energies.

Simulate the test conducted by Eric L in laboratory, where a single-edge notch prism was tested using a crack mouth opening displacement-controlled three-point bend test (Eric, 2012). The notched prism with a width of 51 mm, depth of 51 mm, and span of 229 mm was used in this study. The size of the specimen, the support, and load condition are shown in Figure 1. Cement, silica fume, glass powder, water, superplasticizer, and steel fibers were used to produce the UHPFRC specimen. A Ductal brand

BS1000 UHPFRC material with high-strength steel fiber reinforcement was used. Fibers were 12.5 mm in length and 200 μm in diameter. The specimen was tested at 28 days. The percent of fibers by volume was 2%. The fracture energy is $G_f = 21 KJ/m^2$, elastic modulus is $E = 20000$ N/mm^2, cracking strength is 17 MPa, and the ultimate strength is 33 MPa, while the ultimate strain is 0.09.

The test was simulated by the user-defined constitutive model of UHPFRC in Ls-dyna, and fracture energy was used in the calculation and to judge whether the crack direction rotates or not. The mesh of the test is shown in Figure 2. The fracture energy was changed from $0.5G_f$, $1.0G_f$, to $1.5G_f$, and the resisting load of UHPFRC was checked in different fracture energy cases. In order to simulate the test by using the mid-span node displacement as input data, one elastic pad was added to the loading points. The load is applied in a displacement control mode in LS-DYNA. The simulation results are shown in Figure 3. Result

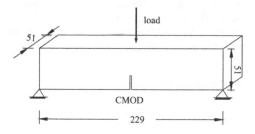

Figure 1. Size of the specimen (unit: mm).

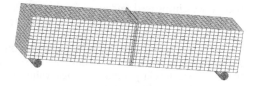

Figure 2. Mesh of the simulation.

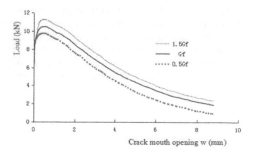

Figure 3. Load versus crack mouth opening curves of different fracture energies.

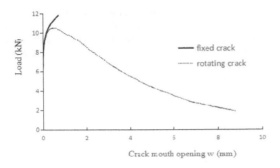

Figure 4. Load versus crack mouth opening curves of fixed and rotating methods.

shows that a 50% increase in fracture energy causes the maximum load to increase by about 8%.

Figure 4 shows the load versus the crack mouth opening by using the fixed crack method and the rotating crack method. It shows that it is unreasonable to use the fixed crack method in the simulation of UHPFRC constitutive model. The load goes on increasing with the increment of crack mouth opening while using the fixed crack method.

4 CONCLUSION

In this paper, a three-dimensional constitutive model considering strain hardening behavior was proposed for finite-element analysis of UHPFRC structures subjected to tensile loading. In this model, Rots' smeared rotating crack theory was improved and used in the simulation of strain-hardening and strain-softening stages of UHPFRC. The directions of the crack planes were assumed to coincide with the principal strain directions while the crack width is small and fixed and the crack width is larger than a critical value, which is determined by the fracture energy.

Fracture testing was performed on a single-edge notched prism loaded under a crack mouth opening displacement-controlled three-point bending test. By changing the fracture energy value in the numerical simulation, the load crack mouth opening curves are obtained, from which it is found that an increase of fracture energy of the material by about 50% can cause the maximum load to increase by about 8%, which obviously influence the simulation result. Furthermore, the simulation results of fixed crack method and rotating crack method are compared, which shows that rotating crack method should be proposed to simulate UHPFRC.

ACKNOWLEDGMENTS

This work was supported by Zhejiang provincial natural science foundation (Grant No. LQ15E090003) and 2016 college students' innovation training project of Zhejiang University of Water Resources and Electric Power (component optimization of hydraulic steel fiber concrete)

REFERENCES

EBolander J., Hikosaka H., etc. Fracture in concrete specimens of differing scale. Engineering Computations, 15(8): 1094–1116, 2013.

Eric L. Kreiger. A Model to describe the mode I fracture of steel fiber reinforced ultra-high performance concrete, Master Thesis, Michigan technological university, 2012.

Muralidhara S., RaghuPrasad B.K., etc. Size-independent fracture energy in plain concrete beams using tri-linear model. 7(25): 3051–3058, 2011.

Nima G. Corrosion control in underground concrete structures using double waterproofing shield system. Int J Min Sci Technol, 23(4): 603–11, 2013.

Raj Das, Paul WCleary. Application of a mesh-free method to modelling brittle fracture and fragmentation of a concrete column during projectile impact. Computers and Concrete, 6(16): 933–961, 2015.

Rots, J.G. Computational modeling of concrete fracture [D]. Delft University of Technology, 1998.

Simon Rouchier, Hans Janssen, etc. Characterization of fracture patterns and hygric properties for moisture flow modelling in cracked concrete. Construction and Building Materials, 34: 54–62, 2012.

VBenin A., Artem SSemenov, etc. Fracture Analysis of reinforced concrete bridge structures with account of concrete cracking under steel corrosion. Advanced Materials Research, 831: 364–369, 2014.

Zhang M.H. and Chen J.K. Analysis of interfacial fracture strength of an inclusion in a polymeric composite considering cohesive force. Computational Materials Science, 2012, 45: 334–343, 2012.

Advances in Energy Science and Equipment Engineering II – Zhou, Patty & Chen (Eds)
© *2017 Taylor & Francis Group, London, ISBN 978-1-138-71798-5*

Effect of high-speed railway on the accessibility and spatial pattern around cities

Gongding Wei, Xuemei Li & Yun Zhao
School of Economics and Management, Beijing Jiaotong University, Beijing, China

ABSTRACT: The most basic effect of High-Speed Railway (HSR) is that it changes the accessibility along its area and promotes the region's transportation advantages into economic advantages. In this paper, taking Shandong as an example, the weighted-average travel time and inverse distance-weighted interpolation method are used to analyze the evolution of accessibility and spatial pattern in cities along the HSR. Research shows that: (1) The effect of time-space convergence is remarkable; (2) The effect of HSR on the regional accessibility has significant corridor effect; and (3) The dynamic changes of urban accessibility accelerate the gradation and reconstruction of spatial pattern in Shandong Province.

1 INTRODUCTION

At present, China is constructing High-Speed Railway (HSR) in a large scale. By the end of 2014, China's high-speed railway, whose operating mileage is more than 16,000 km, has covered over 28 provinces in the country as well as over 160 prefecture-level cities. As a new important transportation mode, HSR construction and development will have a continuous important impact on the development of urban and regional space. The impact of HSR on urban and regional spatial development has been widely concerned by the academic community. There has been an extensive research in all aspects of theoretical and empirical methods.

Scholars found that the impact of HSR on regional accessibility is related to the accessibility's initial level, site distribution, transport network between sites, and other factors (Monzón, 2013). In normal condition, the changes of regional accessibility caused by HSR will lead to the transformation of "center-edge" and the formation of corridor or island. The accessibility layers will present the distance-decay effect (Gutiérrez, 1996). Accessibility changes caused by high-speed railway will accelerate the flow of population, capital, and other factors of production; cause the diffusion and radiation of population and regional economies; and promote spatial pattern (Murayama, 1994; Sasaki, 1997; Kim, 2000). The effect of HSR on regional economic development also depends on the improvement of the regional transportation system (Vickerman, 1997; Chen, 2011). The effect of HSR on regional development has time lag, emerged gradually, and will deepen over time (Blum, 1997; Kim, 2015).

In view of this, this paper analyzes the accessibility spatial pattern evolution of cities under the effect of high-speed rail in Shandong Province by referencing relational research of accessibility and combining the GIS space analysis function to analyze the feature of pattern evolution of accessibility by contrastive analysis of the accessibility of cities after the construction of HSR.

2 STUDY METHODOLOGY AND DATA

2.1 Methodology

The measurement of Weighted-Average Travel Time (WATT) measure emphasizes the relationship between regions. This indicator focuses on the shortest travel time rather than the shortest distance. WATT calculates travel time of a destination to all other destinations considering size of destinations. The size of the destination is used as weight in order to value the importance of the minimal travel time routes. The mathematical expression is as follows (Tang, 2012):

$$T_i = \sum_{j=1}^{n} \left(T_{ij} * M_j \right) / \sum_{j=1}^{n} M_j \qquad (1)$$

where T_i is the accessibility of destination i, T_{ij} is the railway travel time from destination i to destination j, and M_j is the size of destination j. The minimal railway travel time is used for T_{ij}, and the radication of the product of the number of population and gross product of each city is used for M_j. The interpretation of this indicator is simple: the reduced value of T_i after the open of the new HSR means a travel time saving of city i, and the lowest average travel time at the city is considered to have the highest accessibility to all other cities.

Inverse Distance-Weighted (IDW) interpolation is a common and simple spatial interpolation method. The mathematical expression is as follows:

$$z_0 = \left[\sum_{i=1}^{n} \frac{z_i}{d_i^2} \right] / \left[\sum_{i=1}^{n} \frac{1}{d_i^2} \right] \qquad (2)$$

where Z_0 is the estimated value of destination 0, Z_i is the value of destination i, d_i is the distance between destination 0 and i, n is the number of monitoring points, and k is the specified power.

2.2 Study area and data source

Considering the data availability, we will choose the following seven cities along the Beijing–Shanghai HSR in Shandong Province and the Qingdao–Jinan PDL: Jinan, Qingdao, Zibo, Weifang, Dezhou, Tai'an, and Zaozhuang. The study area covers 64030 square kilometers and has 44.8211 million resident population and 2.61559 trillion GDP in 2013. The study area is the core area of Shandong peninsular urban agglomerations, involving the core cities of the economic circle of capital city agglomerations, peninsular blue economic zone, and west booming economic belts, and has dense cities with large population, developed economy, top nationwide GDP, gross output value of industry and agriculture and income per capita, and considerable urbanization, modernization level, and economic opening rate.

In this paper, we will get the population and economic indicator data from <<Shandong Statistic Year Book 2011>> and <<Shandong Statistic Year Book 2014>>, get all shifts and time data of any two cities in the seven cities before and after the operation of the HSR from <<The national railway passenger train schedules (2011.01)>> and <<The national railway passenger train schedules (2015.01)>>, and then use the minimal time to calculate the weighted-average travel time and potential accessibility by the data. Then, through the IDW method, we analyze the evolution of accessibility and spatial pattern in cities along the HSR.

3 RESULTS AND DISCUSSION

In this paper, the time before and after the opening of HSR are designated as 2010 and 2013, respectively. The cities along the HSR refer to the seven cities in the area along the Shandong section of Beijing–Shanghai HSR and Qingdao–Jinan PDL, including Jinan, Qingdao, Zibo, Weifang, Dezhou, Tai'an, and Zaozhuang.

Ti average value of the effective average travel time of the cities along the HSR before and after opening to traffic are 147.37 and 82.48 min, respectively, and the reduction rate is as high as 39% (Table 1).

Before the open of HSR, the effective average travel time of the cities along is quite different. The numerical value of the effective average travel time of Jinan is minimum, that is, 64.36 min, so its accessibility is optimal. The following two cities are Zibo and Weifang, whose effective average travel times are 93.31 and 107.58 min, respectively. The effective average travel time of other cities is all above 2 h, and the numerical value of Zaozhuang's effective average travel time is maximum, as high as 250.71 min. After the opening of HSR, the effective average travel time of the cities along is greatly decreased and the accessibility value difference is significantly shrunk as well. The numerical value of the effective average travel time of Jinan is minimum, that is 60.62 min, while its accessibility is optimal. The following two cities are Zibo and Weifang, whose effective average travel times are 62.37 and 68.72 min, respectively. The effective average travel time of Tai'an, Dezhou, and Qingdao are all within 2 h. The numerical value of the effective average travel time of Zaozhuang is maximum, as high as 109.88 min, and its accessibility is significantly improved compared with that of 2010. The reduction degree of WATT of the cities along HSR is relatively close. After the opening of HSR, the accessibility increase range of Zaozhuang is maximum, and its reduction degree of the effective average travel time is 56%. The following two cities are Dezhou and Tai'an, whose reduction degrees of the effective average travel time are 51%. Compared with other cities, the accessibility increase range of Jinan is minimum, and its reduction degree is only 6%. In addition, the reduction degree of the effective average travel times of Qingdao City, Weifang City, and Zibo City are close: 38%, 36%, and 33% respectively.

Table 1. WATT of cities along the high-speed rail.

City	WATT (min)			
	2010	2013	Difference	Change
Jinan	64.36	60.62	3.74	6%
Qingdao	165.29	102.06	63.23	38%
Zibo	93.31	62.37	30.95	33%
Zaozhuang	250.71	109.88	140.83	56%
Weifang	107.58	68.72	38.86	36%
Tai'an	162.06	82.11	79.95	49%
Dezhou	188.25	91.58	96.67	51%
Average	147.37	82.48	64.89	39%

From the viewpoint of spatial distribution, the accessibility value presents an irregular ring distribution pattern, which gradually increases toward its peripheral area with Jinan as the core. Jinan's transportation hub position is outstanding. The accessibility layers present the distance–decay effect. The accessibilities of northwestern Shandong, southwestern Shandong plains, and peninsular hilly region are relatively low (Figure 1).

After the opening of HSR, an accessibility space pattern was initially formed along Beijing–Shanghai HSR outward southwestern Shandong and along Qingdao–Jinan PDL outward the peninsular hilly region. The accessibilities of southwestern Shandong plain and peninsular hilly region are relatively low, and accessibility layers take on distance attenuation effects (Figure 2).

From the viewpoint of spatial distribution of regional accessibility changes, the HSR produces significant corridor effects on the impact of regional accessibility. The regions along Beijing–Shanghai HSR are the largest beneficiary of the convergence time and also have highest accessibility changes. The regional distribution of high

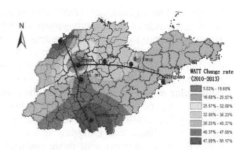

Figure 3. Spatial changing analysis of WATT of cities along the high-speed railway.

accessibility changes presents the characteristics of jumping and discontinuity. The regions take on a strip that runs along the direction of Beijing–Shanghai HSR. The accessibility change rate of the peripheral area of the HSR decreases with the increase in the distance to Beijing–Shanghai HSR (Figure 3).

4 CONCLUSION

The opening of HSR has deeply influenced the regional spatial structure in Shandong Province. In this paper, we measured the evolution of accessibility and spatial pattern in cities along the HSR in Shandong Province by using accessibility model and GIS spatial analysis technology. The results are essentially threefold:

1. The effect of time-space convergence is remarkable. The travel time has sharply reduced and regional accessibility has obviously improved in cities along the HSR. After the opening of HSR, the WATT of cities along the HSR has reduced 30%. The accessibility improvement of Zaozhuang, Dezhou, and Tai'an is outstanding, while Jinan, Zibo, and Weifang have small improvement. As the start and terminal of HSR, Qingdao has also gained considerable improved accessibility.

2. The effect of HSR on the regional accessibility has significant corridor effect. The regions along the HSR become the largest beneficiaries through the effect. Jinan will be more prominent as the HSR hub. The accessibility decreases from center to periphery in circular gradients. The distribution of high accessibility change area is noncontinuous, and suggests a belt be consistent with the same trend as that of the Beijing–Shanghai HSR and Qingdao–Jinan PDL.

3. Geographically, the dynamic changes of urban accessibility accelerate the gradation and reconstruction of spatial pattern in Shandong

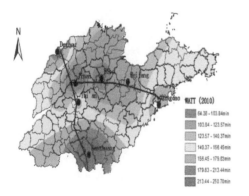

Figure 1. Spatial changing analysis of WATT of cities along the high-speed railway (2010).

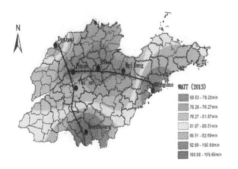

Figure 2. Spatial changing analysis of WATT of cities along the high-speed railway (2013).

Province. The opening of HSR reinforces the economic circle of capital city group, for which Jinan is the core city, and the peninsular blue economic zone, for which Qingdao is the core city that connects with the west uplift economic belts and promotes the integration of Shandong peninsular urban agglomeration. In the process of the integration of Shandong peninsular urban agglomeration, the radiation and driving function of Jinan and Qingdao as a core city will further strengthen, will more significantly promote the regional economy of Shandong Province, and the effect will be prolonged.

This paper offers a platform for further assessments of accessibility issues and spatial pattern for monitoring the long-term progress of HSR development, yet it has some limitations. This paper demonstrates the spatial economic effect of HSR opening from the perspectives of accessibility and spatial economic evolution. In fact, the opening of HSR has an important effect on the flow of production factors, growth of traffic economic belt, reconstruction of regional spatial structure, and other fields. Future research should consider these fields.

REFERENCES

Blum, U., K E Haynes, C Karlsson, The Annals of Regional Science, 31: 1–20. 1997.

Chen, C. L., H Peter, Journal of Transport Geography, 24: 89–110. 2011.

Gutiérrez, J., R González, G Gómez, Journal of Transport Geography, 4(4): 227–238. 1996.

Kim, H., S Sultana, Journal of Transport Geography, 45: 48–61, 2015.

Kim, K. S. Cities, 17(4): 251–262. 2000.

Monzón, A., E Ortega, E López, Cities, 30(1): 18–30, 2013.

Murayama, Journal of Transport Geography, 2(2): 87–100. 1994.

Sasaki, K., T Ohashi, A Ando, Annals of Regional Science, 31(1): 77–98. 1997.

Tang, G.A., X Yang, Beijing: Science Press, 2012.

Vickerman, R. The Annals of Regional Science, 31: 21–38. 1997.

Advances in Energy Science and Equipment Engineering II – Zhou, Patty & Chen (Eds)
© 2017 Taylor & Francis Group, London, ISBN 978-1-138-71798-5

A new safety culture index system for civil aviation organizations

Mingliang Chen, Mei Rong & Yanqiu Chen
Institute of Aviation Safety, China Academy of Civil Aviation Science and Technology, Beijing, China

ABSTRACT: Improving and maintaining safety has always been the most important task for civil aviation. As many other high-risk industries, safety culture has been recognized as playing a key role in achieving an effective level of safety in civil aviation. In this paper, an overview and review of the concept of safety culture is presented. Various elements of safety culture from different perspectives are introduced based on reviewing the literature. A new index system was proposed for assessing the safety culture of civil aviation by analyzing the safety management system of civil aviation and combining with some existing key elements in the literature. A civil aviation organization's safety culture can be assessed based on the proposed safety culture index system.

1 INTRODUCTION

Although civil aviation accidents are rare, they have caused fatal casualties in recent years worldwide, such as Yichun air crash in China in 2010 and FlyDubai air crash in Russia in 2016. In the last decades, researchers and practitioners in many high-hazard industries around the world had recognized some critical factors that result in catastrophic accidents. At the beginning, technical factors were recognized as important. Later, in the 1980s, attention was focused on human errors/factors. Later, it became clear that management plays a key role in achieving an effective level of safety, and the introduction of safety management systems followed (Mary, 2011). Finally yet importantly, the investigation result of many accidents revealed that weaknesses of safety culture were the root cause and they increase the severity of consequence. Then, the role of safety culture in maintaining and improving the safety level was recognized. The evolution of various contributing factors to safety is illustrated in Figure 1.

The concept of "safety culture" was first identified by the International Nuclear Safety Advisory Group (INSAG) from the analysis of the Chernobyl nuclear power plant accident in 1986. INSAG recognized that a lack of safety culture contributed to the incident. The increasing interest in safety culture leads to intensive research in high-risk industries. In the following years, safety culture is used across a broad spectrum of industries with different types of accident hazards, such as work site activities and risk scenarios (Baram, 2007). These areas include air traffic control (Gill, 2004), chemical industry (Frank, 2007), construction (Ling, 2009), shipping (Havold, 2009), mining

Figure 1. Various contributing factors to safety (Source: Mary Kay O'Connor Process Safety Center (Mary, 2011)).

(Sanmiquel, 2010; Lu, 2015), aircraft maintenance (Atak, 2011), and airport (Fu, 2014). Research results clearly identified a strong relationship between the level of safety culture and the number of accidents. The higher level of safety culture had a positive impact on the decrease of the number of accidents (Warszawska, 2016).

Safety is the most important factor for civil aviation. Improving and maintaining the safety culture is a goal for every civil aviation organization. Because of the recognized links between a positive safety culture and good safety performance based on the existing research, many civil aviation organizations have attempted to measure or assess their safety culture, thereby identifying strengths and weaknesses for improvement. For this purpose, it is first necessary to understand "what is safety culture" and identify the key elements of safety culture in civil aviation. Then, the index system for assessing the safety culture in their organization should be presented. The rest of this paper is organized as follows: Section 2 introduces the concept of safety culture and its component elements. The proposed safety culture index system for assessing safety culture of civil aviation organizations is illustrated in Section 3. Finally, conclusions are presented in the last section.

2 SAFETY CULTURE

As mentioned above, safety culture was first identified by INSAG in 1986. After 5 years, INSAG provided the following definition of safety culture: "Safety culture is that assembly of characteristics and attitudes in organisations and individuals which establishes that, as an overriding priority, nuclear plant safety issues receive the attention warranted by their significance." Despite clearly pointing out that lack of safety culture is contributed to the nuclear accident, there was little academic background. Thus, studies began to explore safety culture, how it could be measured and, if possible, how it could be used to improve safety (Edwards, 2014).

2.1 Definition of safety culture

Despite much research in this field, there were many different attempts to define "safety culture," and the concept had been widely debated among researchers

Table 1. Definitions of safety culture.

Source	Definition
INSAG (1991)	Safety culture is that assembly of characteristics and attitudes in organizations and individuals that establishes that, as an overriding priority, nuclear plant safety issues receive the attention warranted by their significance.
Cox & Cox (1991)	The attitudes, beliefs, perceptions, and values that employees share in relation to safety (Cox, 1991).
Pidgeon (1991)	The set of beliefs, norms, attitudes, roles, and social and technical practices that are concerned with minimizing the exposure of employees, managers, customers, and members of the public to conditions considered dangerous or injurious (Pidgeon, 1991).
Ostrom, Wilhelmsen, & Kaplan (1993)	The concept that the organization's beliefs and attitudes, manifested in actions, policies, and procedures, affects its safety performance (Ostrom, 1993).
Geller (1994)	In a Total Safety Culture (TSC), everyone feels responsible for safety and pursues it on a daily basis (Geller, 1994).
Ciavarelli & Figlock (1996)	Safety culture is defined as the shared values, beliefs, assumptions, and norms which may govern organizational decision making, as well as individual and group attitudes about safety (Ciavarelli, 1996).
Carroll (1998)	Safety culture refers to a high value (priority) placed on worker safety and public (nuclear) safety by everyone in every group and at every level of the plant. It also refers to expectations that people will act to preserve and enhance safety, take personal responsibility for safety, and be rewarded consistent with these values (Carroll, 1998).
Eiff (1999)	A safety culture exists within an organization where each individual employee, regardless of their position, assumes an active role in error prevention and that role is supported by the organization (Eiff, 1999).
Cooper (2000)	Safety culture is a sub-facet of organizational culture, which is thought to affect member's attitudes and behavior in relation to an organization's ongoing health and safety performance (Cooper, 2000).
Guldenmund (2000)	Those aspects of the organizational culture that will impact on attitudes and behavior related to increasing or decreasing risk (Guldenmund, 2000).
Hale (2000)	The attitudes, beliefs, and perceptions shared by natural groups as defining norms and values, which determine how they act and react in relation to risks and risk control systems (Hale, 2000).
Westrum (2004)	The organization's pattern of response to the problems and opportunities it encounters (Westrum, 2004).
Fernandez-Muniz, Montes-Peon, Vazquez-Ordas (2007)	A set of values, perceptions, attitudes and patterns of behaviour with regard to safety shared by members of the organization; as well as a set of policies, practices and procedures relating to the reduction of employees' exposure to occupational risks, implemented at every level of the organization, and reflecting a high level of concern and commitment to the prevention of accidents and illnesses (Fernandez, 2007).
Edwards (2013)	The assembly of underlying assumptions, beliefs, values and attitudes shared by members of an organisation, which interact with an organisation's structures and systems and the broader contextual setting to result in those external, readily-visible, practices that influence safety (Edwards, 2013).

and participators (Guldenmund, 2000; Hale, 2000; Reader, 2015). Guldenmund (Guldenmund, 2000) highlighted that this has led many researchers to re-define safety culture in relation to their specific area of interest. As a result, there are numerous definitions of safety culture proposed by different researchers and practitioners. Some representative definitions of safety culture are presented in Table 1.

Despite there are so many different definitions of safety culture, each conceptualization has its strengths and weaknesses. Nonetheless, there does appear to be several commonalities among these various definitions regardless of the particular industry being considered (Wiegmann, 2002). There is agreement that a safety culture, which positively influences safety, is an organizational culture that prioritizes safety-related beliefs, values, and attitudes (Edwards, 2014).

In civil aviation, the International Civil Aviation Organization (ICAO) has provided the following definition: Safety culture is the safety "personality" of the whole organization, that is, "it's the way we do things around here." Safety culture refers to the underlying (safety) values, beliefs, and practices that make a business what it is.

2.2 Key elements of safety culture

Researchers and practitioners have analyzed key elements (factors, or dimensions) of safety culture from the 1980s onward. There is little consensus in the key elements proposed in the literature that reflect an organization's safety culture. Some different key elements for assessing safety culture in recent years are presented in Table 2.

Table 2. Different key elements for assessing safety culture.

Source	Key elements
Kao, Lai, Chuang, & Lee (2008)	Organization's commitment: safety management system & organization, accident & emergency; managers' commitment: safety supervision & audit, safety commitment & support, rewards & benefits; individuals' commitment: safety training & competence, safety attitude & behavior, safety communication & involvement (Kao, 2008).
Wang, Xia, Pan, Zong (2012)	Decision-making level: belief and value, awareness and policy, learning and training; management level: awareness and attitude, commitment and communication, system and award or punish, learning and training; implementation level: responsibility and attitude, feedback and spreading, learning and training; external cause: policy environment, supervise environment, working environment, local environment; Persistence (Wang, 2012).
Wang & Sun (2012)	Organizational safety commitment, safety organization, safety regulation and rule, safety management behavior, safety operation behavior, safety education and training, safety information exchange, safety rewards and punishment (Wang, 2012).
Wang (2012)	Environment factor: safety environment, safety rules; organization factor: safety commitment, safety training, safety system, safety leadership, health activities, risk management, safety encouragement and punishment, performance measurement, contractor management, management of change, procurement management, safety communication; personal factor: safety knowledge worker participation; psychology factor: safety awareness and attitude, safe behavior (Wang, 2012).
Forest (2012)	Commit to process safety, understand hazards & risks, manage risk, learn from experience (Forest, 2012).
Frazier, Ludwig, Whitaker, & Roberts (2013)	Management concern: supervisor concern, senior management concern, work pressure; personal responsibility: supervisor/management blame, risky behavior; peer support: caution others, respectful feedback; safety management systems: safety policy, procedures, and rules, training, communication, incident reporting & analysis, safety audits and inspections, rewards and recognition, employee engagement, safety meetings/committees, suggestions/concerns, discipline (Frazier, 2013).
Fu & Chan (2014)	Person: attitude toward safety, rewards and recognition, sharing safety knowledge, safety values, incident investigation, respectful feedback safety management updates, organization type, job burnout; Situation: safety education and training, safety working environment, safety communication and commitment, employee engagement, resource improvement, safety policy/procedure/rules, injury rate, risk management, division of areas of responsibility; Behavior: system of incentives and penalties, safety behavior, safety supervision, employee participation, system integration (Fu, 2014).
Noort, Reader, Shorrock, & Kirwan (2016)	Management commitment to safety, collaborating for safety, incident reporting, communication, colleague commitment to safety, safety support (Noort, 2016).

3 SAFETY CULTURE INDEX SYSTEM

One of the major subjects of safety culture research is identification of the key elements of safety culture. Some key elements for assessing safety culture had been identified by various researchers. However, these studies were not consistent concerning which elements should be used for assessing safety culture. This research tried to develop a general safety culture index system to be suitable for civil aviation.

Safety culture of a civil aviation organization and its Safety Management System (SMS) are closely related. SMS is the formal, top–down, organization-wide approach to managing safety risk and assuring the effectiveness of safety risk controls. It includes systematic procedures, practices, and policies for the management of safety risk. The existence and understanding of safety culture is the prerequisite for the successful implementation and sustained performance of SMS. Assessing safety culture in civil aviation is always related to SMS, as safety culture is ultimately reflected by the way in which safety is managed.

The framework of SMS includes four components and 12 elements, representing the minimum requirements for SMS implementation. The framework of SMS is presented in Figure 2.

In order to establish a safety culture index system suitable for civil aviation, some indices were extracted from the relevant elements of SMS, and additional indices were selected from the key elements of safety culture identified in the literature by considering the characteristics of civil aviation. The safety culture index system is shown in Figure 3.

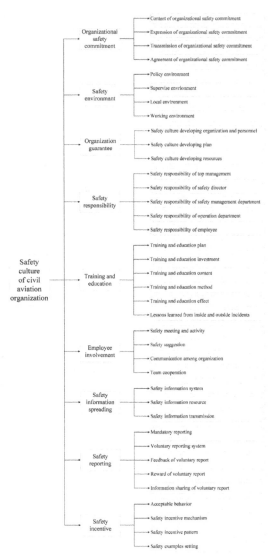

Figure 3. Safety culture index system for civil aviation.

Figure 2. Framework of safety management system (Note: MCR = management commitment and responsibility, SA = safety accountabilities, ASP = appointment of key safety personnel, ERP = coordination of emergency response plan, SD = SMS documentation, HI = hazard identification, RAM = safety risk assessment and mitigation, SPM = safety performance monitoring and measurement, MOC = management of change, CI = continuous improvement of the SMS, TE = training and education, SC = safety communication.).

In this safety culture index system, the following nine key elements are used to assess safety culture of civil aviation organization: organizational safety commitment, safety environment, organization guarantee, safety responsibility, training and education, employee involvement, safety information spreading, safety reporting, and safety incentive. For these key elements, there are some sub elements used as evaluation criteria. For example, for the element "training and education," there are six evaluation criteria: training and education plan, training and education investment, training and education investment, training and education content, training and education effect, and lessons learned from inside and outside incidents.

For assessing the safety culture of civil aviation organization, there are some assessment approaches, such as document analysis, workplace observation, safety culture survey, and interviews. The proposed safety culture index system in this paper can be used as the foundation for these assessment approaches.

4 CONCLUSION

Safety culture plays an important role in maintaining and improving the safety level of civil aviation organizations. Various definitions and key elements of safety culture were summarized for extracting elements that can be used for assessing safety culture of civil aviation organizations. A new safety culture index system containing nine key elements and 38 evaluation criteria was proposed. This index system can be used as the foundation for assessing safety culture. An important aspect of safety culture research is attempts to quantitatively evaluate the level of safety culture. A quantitative assessment model based on this safety culture index system should be studied in the future.

REFERENCES

Atak, A., S. Kingma, Safety culture in an aircraft maintenance organisation: a view from the inside, Safety Sci., 49: 268–278 (2011).

Baram, M., M. Schoebel, Safety culture and behavioural change at the workplace, Safety Sci., 45: 631–636 (2007).

Carroll, J.S. Safety culture as an ongoing process: Culture surveys as opportunities for enquiry and change, Work & Strss, 12: 272–284 (1998).

Ciavarelli, A., R. Figlock, Organizational factors in aviation accidents, *Proceedings of the Ninth Internatioinal Symposium on Aviation Psychology*, Columbus: 1033–1035 (1996).

Cooper, M.D. Towards a model of safety culture, Safety Sci., 36: 111–136 (2000).

Cox, S., T. Cox, The structure of employee attitudes to safety: A European example, Work & Stress, 5: 93–104 (1991).

Edwards, J., J. Freeman, D. Soole, B. Watson, A framework for conceptualising traffic safety culture, Transport. Res. F, 26: 293–302 (2014).

Edwards, J.R.D., J. Davey, K. Armstrong, Returning to the roots of culture: A review and re-conceptualisation of safety culture, Safety Sci., 55: 70–80 (2013).

Eiff, G. Organizational safety culture. *Proceedings of the Tenth International Symposium on Aviation Psychology*, Columbus: 1–14 (1999).

Fernandez-Muniz, B., J.M. Montes-Peon, C.J. Vazquez-Ordas, Safety culture: Analysis of the causal relationships between its key dimensions, J. Saf. Res., 38: 627–641 (2007).

Forest, J.J. How to evaluate process safety culture, Process Saf. Prog., 31: 195–197 (2012).

Frank, W.L., Process safety culture in the CCPS risk based process safety model, Process Saf. Prog., 26: 203–208 (2007).

Frazier, C.B., T.D. Ludwig, B. Whitaker, D.S. Roberts, A hierarchical factor analysis of a safety culture survey, J. Saf. Res., 45: 15–28 (2013).

Fu, Y.K., T.L. Chan, A conceptual evaluation framework for organisational safety culture: An empirical study of Taipei Songshan Airport, J. Air Trans. Manage., 34: 101–108 (2014).

Geller, E.S. Ten principles for achieving a total safety culture, Prof. Saf., 39: 18–24 (1994)

Gill, G.K., G.S. Shergill, Perceptions of safety management and safety culture in the aviation industry in New Zealand, J. Air Trans. Manage., 10: 231–237 (2004).

Guldenmund, F.W. The nature of safety culture: a review of theory and research, Safety Sci., 34: 131–150 (2000).

Hale, A.R. Culture's confusions, Safety Sci., 34: 1–14 (2000).

Havold, J.I., E. Nesset, From safety culture to safety orientation: validation and simplification of a safety orientation scale using a sample of seafarers working for Norwegian ship owners, Safety Sci., 47: 305–326 (2009).

Health & Safety Laboratory, Safety culture: A review of the literature, HSL/2002/25 (2002).

International Civil Aviation Organization, Safety culture and the future enhancement of ICAO provisions related to SMS implementation, Working Paper A38-WP/206 (2013).

International Nuclear Safety Advisory Group, Safety culture, 75-INSAG-4 Safety Series, Vienna: International Atomic Energy Agency (1991).

International Nuclear Safety Advisory Group, Summary report on the post-accident review meeting on the Chernobyl accident, International Nuclear Safety Advisory Group, Vienna: International Atomic Energy Agency (1986).

Kao, C., W. Lai, T. Chuang, J. Lee, Safety culture factors, group differences, and risk perceptioin in five petrochemical plants, Process Saf. Prog., 27: 145–152 (2008).

Ling, F.Y.Y., M. Liu, Y.C. Woo, Construction fatalities in Singapore, Int. J. Proj. Manage., 27: 717–726 (2009).

Lu, H., H. Chen, Does a people-oriented safety culture strengthen miners' rule-following behaviour? The role of mine supplies-miners' needs congruence, Safety Sci., 76: 121–132 (2015).

Mary Kay O'Connor Process Safety Center, Process safety research agenda for the 21st century, Texas: Mary Kay O'Connor Process Safety Center (2011).

Noort, M.C., T.W. Reader, S. Shorrock, B. Kirwan, The relationship between national culture and safety culture: Implications for international safety culture assessments, J. Occup. Organ. Psychol., 89: 515–538 (2016).

Ostrom, L., C. Wilhelmsen, B. Kaplan, Assessing safety culture, Nucl. Saf., 34: 163–172 (1993).

Pidgeon, N.F. Safety culture and risk management in organizations, J. Cross-Cultural Psychol., 22: 129–141 (1991).

Reader, T.W., M.C. Noort, S. Shorrock, B. Kirwan, Safety sans frontieres: An international safety culture model, Risk Anal., 35: 770–789 (2015).

Sanmiquel, L., M. Freijo, J. Edo, J.M. Rossell, Analysis of work related accidents in the Spanish mining sector from 1982–2006, J. Saf. Res., 41: 1–7 (2010).

Wang, C., Y. Liu, The dimensions and analysis of safety culture, Process Saf. Prog., 31: 193–194 (2012).

Wang, L., R. Sun, The development of a new safety culture evaluation index system, Procedia Eng., 43: 331–337 (2012).

Wang, Y., L. Xia, J. Pan, H. Zong, Application of factor restructuring analysis in enterprise safety culture evaluation, Phys. Procedia, 24: 1642–1648 (2012).

Warszawska, K., A. Kraslawski, Method for quantitative assessment of safety culture, J. Loss Prevent. Proc., 42: 27–34 (2016).

Westrum, R. A typology of organisational cultures, Qual. Saf. Health Car., 13: 22–27 (2004).

Wiegmann, D.A., H. Zhang, T. Thaden, G. Sharma, A. Mitchell, A synthesis of safety culture and safety climate research, Technical Report ARL-02–3/FAA-02–2 (2002).

Advances in Energy Science and Equipment Engineering II – Zhou, Patty & Chen (Eds)
© 2017 Taylor & Francis Group, London, ISBN 978-1-138-71798-5

Thinking about the legislation for multimodal transport of goods in China

Jingling Jiang
China Academy of Transportation Sciences, Beijing, China

ABSTRACT: Laws and regulations on multimodal transport of goods can provide foundation and guarantee for the development of multimodal transport. By reviewing the literature, conducting survey questionnaire among multimodal transport operators, and visiting many provinces to get knowledge of their legislation for multimodal transport of goods, this paper summarizes the status quo of legislation for multimodal transport of goods in China, dissects pertinent problems, describes the demands for laws and policies in this respect, and puts forward specific countermeasures and suggestions on accelerating the building of law and regulation system for multimodal transport of goods in China.

1 INTRODUCTION

Multimodal transport of goods began to develop in China in the 1980s, and after more than three decades of development, it has been used as a strategic tool to drive the development of logistics industry in the country. Planning of Logistics Industry Adjustment and Revitalization issued by the State Council in 2009, Guidance of the Ministry of Transport on Promoting the Sound Development of Logistics Industry issued by the Ministry of Transport in 2013, and Mid-Long Term Development Plan of the Logistics Industry (2014–2020) prepared by the State Council take multimodal transport as a key project to push the development of the logistics industry without exception. Compared with international advanced level, multimodal transport of goods is still in its infancy in China. Accelerating the building of a multimodal transport system with smoothly connected infrastructure, effectively arranged transportation, widely applied advanced equipment, interconnectedly shared information resource, and well improved regulations and standards should be taken as an important measure to expand effective supply and extend supply-side reform, which plays a significant role in propelling long-term stable and rapid development of China's economy and building a well-off society in an all-around way. Regulations and policies will provide strong foundation and support for the sound development of multimodal transport of goods. Therefore, in the context that both international and domestic multimodal transport of goods are witnessing continuous development, the formulation of laws and policies with regard to multimodal transport of goods will be conducive to promoting multimodal transport of goods in China.

2 STATUS QUO OF LEGISLATION FOR MULTIMODAL TRANSPORT OF GOODS IN CHINA

China has not enacted any law dedicated to multimodal transport of goods, and relevant provisions are scattered in Maritime Law of the People's Republic of China (hereinafter referred to as Maritime Law) and Contract Law of the People's Republic of China (hereinafter referred to as Contract Law). The former mainly regulates the legal relation of international multimodal transport of goods with shipping included (Zhu, 2013) while the latter regulates general multimodal transport contracts with shipping excluded. The two are complementary and basically constitute civil and commercial laws in this field. By analyzing foreign and domestic literature concerning multimodal transport of goods, conducting survey questionnaire among multimodal transport operators nationwide, and visiting Tianjin, Shandong, and Guangdong, some problems on the legislation for multimodal transport of goods in China are concluded as follows.

2.1 *No special law on multimodal transport of goods*

China remains in a state of relative lagging in multimodal transport legislation. The existing legal system of multimodal transport is subject to Maritime Law and Contract Law as the core (Zhao, 2015), and at the same time regulated by such laws and regulations on single mode of transport as Railway Law, Civil Aviation Law, and Postal Law. Due to restrictions by various factors, Integrated Transport Promotion Law and Multimodal Transport Law, one comprehensive law, and one administrative regulation to be formulated to promote the development of multimodal transport have just been brought into their legislative processes. In a word, there is no legal system for multimodal transport of goods, which exerts adverse effects on the healthy development of multimodal transport of goods in China.

2.2 No administrative regulations on multimodal transport of goods

Multimodal transport laws and regulations mainly include administrative laws between multimodal transport operation bodies and multimodal transport operators as well as civil and commercial laws between multimodal transport operators and consignors. Maritime Law and Contract Law mainly relate to the civil relation of multimodal transport, but there is still no administrative regulation on multimodal transport supervision by competent departments. Administrative regulations in this respect should specify which rules should be abided by in multimodal transport and set certain norms and restrictions on multimodal transport operations within the territory of China for market order maintenance and business practice regulation by competent departments as well as other public purposes. Violators should assume the corresponding legal liability.

2.3 Unclear division of responsibility for multimodal transport of goods

Maritime Law and Contract Law are primarily applicable to claim disputes of multimodal transport, with reference to laws on single mode of transport as well. For sections in which accidents often occur, such as highways, inland waters, and nonmarine sections, we can only refer to concerned departmental rules (Fang, 2010). On the one hand, these rules are less effective, and on the other hand, when it comes to liability limitation system, if goods are damaged or lost in a particular section of multimodal transport, rules relating to the adjustment of transport mode in such section should be referred to determine the compensation liability and liability limit of multimodal transport operators. Thus, damages to goods in certain multimodal transport section, especially in terms of liability limitation, should be dealt with in accordance with rules for such single mode of transport as railway, highway, waterway, and aviation (Pang, 2005). From here, we can see that China now has an unclear division of responsibility for multimodal transport of goods.

2.4 Lack of cooperative interactions among provinces in the country

In the context of accelerating the development of modern logistics at the national level, various cities and provinces nationwide have given positive response to national policies and begun to make exploration and carry out practice, with a series of supportive policies on multimodal transport of goods being drafted. However, because of different geographical environments and competitive transport organization forms, they are more inclined to invest limited resources into their competitive transport organization forms, so that they often compete for more goods supplies through fiscal subsidies and price cutting but ignore the cooperation between neighboring cities and provinces. As a result, insufficient connection of multimodal transport, geographic isolation, and regional protectionism seriously hampered the development of multimodal transport of goods.

3 LEGISLATIVE DEMANDS OF MULTIMODAL TRANSPORT OF GOODS IN CHINA

3.1 Questionnaire analysis

Questionnaires have been distributed to national authorities of the industry of multimodal transport of goods, industry associations, cargo transportation (logistics) customers and enterprises, transportation equipment manufacturers, and so on, based on the survey of the whole industry chain of multimodal transport development conducted by Ministry of Transport in 2016, with 629 ones received, including 309 effective ones. Among these questionnaires, 222 ones show demands for multimodal transport service specifications with respect to demands for multimodal transport regulations and rules, ranking the first and accounting for 72%. A total of 147 ones show demands for multimodal transport contracts, ranking the second and accounting for 48%; 121 ones show demands for claims and actions against multimodal transport operators, ranking the third and accounting for 39%; equally 120 ones show demands for multimodal transport documents and actual carriers' liability, both ranking the fourth and accounting for 38%; and 119 ones show demands for liability of multimodal transport operators, ranking the fifth and accounting for 37%. Only 73 ones show demands for relevant regulations on shippers' liability, which is the least and accounts for 24%. Survey findings are shown in Fig.1.

Moreover, with respect to deeper analysis of data, 146 enterprises checked all options, while 256 enterprises checked only multimodal transport

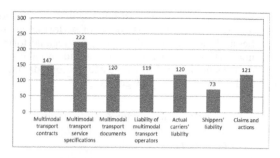

Figure 1. Findings of the questionnaire survey on laws and regulations among 309 enterprises.

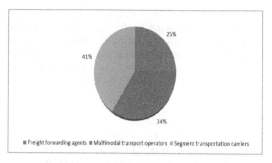

■ Freight forwarding agents ■ Multimodal transport operators ■ Segment transportation carriers

Figure 2. Findings of the questionnaire survey on laws and regulations among 309 enterprises.

service rules and multimodal transport documents. In view of respondents' understanding of the table, it can be seen that, at present, China lacks rules and regulations in the field of multimodal transport; under this premise, development of multimodal transport service rules and laws and regulations on multimodal transport documents is of utmost importance.

In respect of recognition of multimodal transport operators by 309 enterprises, although the principal status of multimodal transport operators is not yet recognized, in practical investigation, most enterprises position themselves as segment carriers and only one-third of them position themselves as multimodal transport carriers. See Fig.2 for details.

With the continuous development of practices of multimodal transport of goods, laws, regulations, and policies on multimodal transport of goods have been in urgent need. On the one hand, the law is required to regulate behaviors of parties to multimodal transport of goods and protect lawful rights and interests of different parties; on the other hand, the law is also required to ensure that multimodal transport of goods is smooth and provides a stable and orderly development environment.

3.2 Legislative claim

With constant development of practices of multimodal transport of goods, demands for regulations and policies on multimodal transport of goods are in an increasingly urgent need. A nationwide statistical analysis of 309 questionnaires and demands of China's multimodal transport development for laws and regulation can be summed up in two parts: (1) the law is required to regulate behaviors of parties to multimodal transport of goods and protect lawful rights and interests of different parties; and (2) the law is also required to ensure that multimodal transport of goods is smooth and provides a stable and orderly development environment.

3.2.1 Demands for administrative regulations on multimodal transport of goods

Legislation for multimodal transport of goods can regulate the market of multimodal transport of goods and preserve the operational order. Multimodal transport of goods accounts for a higher proportion in the transport structure, and the market of multimodal transport of goods is larger, triggering the demands for regulations on multimodal transport market (Yang, 2006). Necessary control of market access and exit with respect to multimodal transport of goods and essential regulation of market competition play a crucial role in preventing cutthroat competition and creating a good market environment for multimodal transport of goods. Meanwhile, legal means can guide market behaviors and help foster the market of multimodal transport of goods and form an incentive mechanism.

3.2.2 Demands for civil and commercial regulations on multimodal transport of goods

Legislation for multimodal transport of goods is required to regulate the legal relationship of the multimodal transport and safeguard lawful rights and interests of different parties. The constant expansion of multimodal transport in the form and the increasingly sound organizational structure have complicated the legal relationship of multimodal transport, as different parties are involved in the multimodal transport of goods, making the definition of the legal status of different parties to multimodal transport and their respective rights and obligations an urgent need (Hoeks, 2009). The basic legal relationship of multimodal transport of goods is the legal relationship set forth in multimodal transport contracts and the legal relationship set forth in separate subcontracts in accordance with master contracts, which observes the principle of freedom of contract in that the relationship of rights and obligations is agreed upon by both parties and will not be subject to the law unless it is not agreed or clearly agreed upon. However, in order to safeguard lawful rights and interests of different parties to multimodal transport of goods, the law is required to make mandatory adjustments to some special issues so as to balance various relations of interests concerning multimodal transport of goods.

4 COUNTERMEASURES AND SUGGESTIONS FOR PROMOTING THE LEGISLATION FOR MULTIMODAL TRANSPORT OF GOODS IN CHINA

4.1 To accelerate administrative legislation for multimodal transport of goods

At present, multimodal transport of goods in China is still in the stage of exploration and innovation, which is featured by the late start, few practices, less experience from other countries due to different stages, and immaturity. From perspectives such as seizing opportunities to promote the development of multimodal transport of goods, deepening the construction of an integrated transport system, establishing an efficient modern logistics operation platform, and optimizing systems and policy environment for the development of multimodal transport of goods, guiding opinions can be provided in advance to put into practice job requirements such as development planning for multimodal transport of goods, facilities construction, equipment and technology, organizations and services, incentive measures, and supervision and management. In the future, on the basis of successful experience in multimodal transport demonstration and comparatively clear market exploration, these guiding opinions will be gradually raised to rules or regulations.

4.2 To accelerate relevant legislation for multimodal transport of goods

The legislation of Integrated Transport Promotion Law should be accelerated to make the legal status and basic principles of China's promotion of multimodal transport construction clear and serve as the higher-level law and the basis of Multimodal Transport Law. The legislation of Multimodal Transport Law should be accelerated, which is a civil and commercial law adjusting the relationship between equal civil subjects, for example, multimodal transport operators and shippers, to regulate parties to multimodal transport contracts, conditions for conclusion and validity of such contracts, the relationship between such contracts and transport documents, and other issues; regulate period of responsibilities of multimodal transport operators,

their basis of liability and liability limitations and make clear liability for damages of performing parties and the principle of recovery between performing parties and operators; and provide regulations on lawsuits and arbitrations against loss, damage, and delay in delivery of goods.

4.3 To perfect the coordinated advancement mechanism and strengthen legislation propulsion

Laws and regulations on multimodal transport of goods are quite comprehensive, which tend to involve interest relationship in many respects and relate to work, systems, and mechanisms of different departments, so the legislation is very difficult. Meanwhile, diversification of interests and demands of parties to multimodal transport will be increasingly apparent. In the face of new situations and problems, it is required to practically improve the legislation quality, further give play the leading role of relevant organizations in legislation, strengthen organization and coordination for legislation so as to guarantee efficient utilization of limited legislative resources and ensure that legislation for multimodal transport of goods better serves the overall economic and social development in China.

4.4 To enhance propaganda and guidance of regulations on multimodal transport of goods and create a favorable atmosphere

Transport is closely related to people's work and life. With the rapid development of transport, transport work is increasingly concerned by society and propaganda, and guidance is therefore especially important. It is required to further give full play to the guiding role of public opinion, establish the concept of transport marketing, strengthen propaganda and guidance of regulations on multimodal transport of goods, as there is a severe shortage of such regulations, enhance the development of multimodal transport and the law consciousness of those engaged in multimodal transport business activities, and improve the influence of regulations on multimodal transport in the industry, all of which are conducive to creating market circumstances featured by legal operation, integrity, and fair competition for multimodal transport.

4.5 To strengthen monitoring of the development of multimodal transport of goods and enhance decision-making support

Efforts should be made to accelerate the establishment of a statistical system for the development of multimodal transport of goods and incorporate the system into the statistical working system of the transport industry. Transport authorities should establish a monitoring mechanism for the development and operation of multimodal transport; regularly organize market surveys on multimodal transport of goods, operation monitoring, and performance assessment; define index constitution, data connotation, evaluation method, and monitoring channel; release monitoring reports on multimodal transport market operation; and promote the establishment of a long-acting monitoring mechanism for the development of multimodal transport of goods.

REFERENCES

Fang, Y. Analysis of International Multimodal Transport Operator Liability Recovery and Defense. Journal of Chongqing Jiaotong University (Social Sciences Edition), 2010(01).

M.A.I.H Hoeks. Multimodal Transport Law. Haeue: Kluwer Law International, 2009.

Pang, Y. Study on the Legal Issues of Liability of Compensation in Respect of International Multimodal Transport: (Master Dissertation), Dalian: Shanghai Maritime University, 2005.

Yang, Y. Research on Legal Relationship in International Multimodal Transport of Goods: (Doctoral Dissertation), Beijing: University of International Business and Economics, 2006.

Zhao, Y. Multimodal Transport Practice and Rules, Shanghai: East China Normal University Press, 2015.

Zhu, Z. Review and Prospect on the Maritime Code of the People's Republic of China, Chinese Journal of Maritime Law, 2013(24).

Author index

Song, J.B. 1685
Song, J.J. 1297
Song, W.W. 1135, 1141
Song, X. 407
Song, Z.H. 1681
Su, C. 1781
Su, D.L. 1005
Su, K.Z. 521
Su, L. 881
Su, Y.S. 1261
Su, Z.J. 767
Sua, J. 1039
Sun, C. 1431
Sun, G.L. 709
Sun, G.Q. 1677
Sun, H. 467
Sun, H.Y. 919
Sun, J. 993
Sun, K.K. 511
Sun, K.L. 907
Sun, L. 1597
Sun, L. 1833
Sun, N. 381
Sun, P. 1661
Sun, S. 547
Sun, S.L. 1073
Sun, T. 1521
Sun, W. 269
Sun, W. 863
Sun, W. 1871
Sun, W.-H. 915
Sun, X.H. 965
Sun, X.Y. 349
Sun, Y. 687
Sun, Y.G. 235
Suo, Y.H. 31

Tan, B.-K. 595
Tan, D. 227
Tan, D.Y. 619
Tan, J. 891
Tan, J. 1817
Tan, T.L. 639
Tan, W. 187
Tan, Y.H. 525
Tan, Z.-Y. 447
Tang, C. 259
Tang, D.G. 441
Tang, H. 1891
Tang, K. 1351
Tang, M.X. 565
Tang, Y.F. 945
Tao, R. 461
Tao, W.H. 1231, 1243
Tao, Z.G. 659
Teng, W. 61
Tian, H. 451
Tian, K.W. 969

Tian, L. 151
Tian, Q.C. 295
Tong, S.K. 243
Tong, X. 977
Tu, C.Q. 853
Tu, X.-H. 45
Tu, X.H. 869
Tu, X.S. 1103
Tzou, G.-Y. 1307

Vikneswaran, M. 89

Wang, A.X. 859
Wang, B. 1427
Wang, B.-C. 301
Wang, C. 321, 1581
Wang, C.W. 499
Wang, D. 111
Wang, D. 249
Wang, D. 807
Wang, D. 1813
Wang, E.L. 1167
Wang, F. 311
Wang, F. 1283, 1291
Wang, F. 1613
Wang, G.H. 1301
Wang, G.Q. 255
Wang, G.S. 977
Wang, H. 129, 785
Wang, H. 687
Wang, H.M. 417
Wang, J. 259
Wang, J.C. 1451
Wang, J.J. 1457
Wang, J.L. 387
Wang, J.X. 1591
Wang, L. 315
Wang, L. 339
Wang, L. 1465
Wang, L. 1863
Wang, L. 1883
Wang, L.B. 863
Wang, L.M. 1681
Wang, L.R. 203
Wang, M. 1953
Wang, M.G. 837
Wang, M.-H. 559
Wang, M.J. 1789
Wang, M.M. 1677
Wang, M.N. 505
Wang, N. 307
Wang, N. 1891
Wang, P. 57
Wang, P. 977
Wang, P.L. 1367
Wang, Q. 259
Wang, Q. 881
Wang, Q.K. 667

Wang, Q.M. 767
Wang, R. 31
Wang, R.J. 901
Wang, R.J. 1129
Wang, S. 397
Wang, S. 609
Wang, S. 983
Wang, S.G. 1375
Wang, S.H. 863
Wang, S.M. 1681
Wang, S.Q. 1107
Wang, S.R. 255
Wang, S.W. 551
Wang, S.X. 235
Wang, S.Y. 1361
Wang, T. 53
Wang, T. 673
Wang, T.Y. 1253
Wang, W. 1427
Wang, W. 1591
Wang, W.H. 151, 159
Wang, W.H. 327
Wang, X. 1173
Wang, X.-B. 1887
Wang, X.D. 535
Wang, X.F. 1941
Wang, X.G. 1277
Wang, X.-J. 897
Wang, X.L. 1227
Wang, X.M. 13
Wang, X.S. 255
Wang, Y. 1113
Wang, Y. 1217
Wang, Y. 1487
Wang, Y. 1891
Wang, Y.C. 263
Wang, Y.D. 489
Wang, Y.H. 579
Wang, Y.H. 1457
Wang, Y.J. 959
Wang, Y.J. 1531
Wang, Y.J. 1575
Wang, Y.L. 1129
Wang, Y.N. 1565
Wang, Y.S. 315
Wang, Y.T. 929
Wang, Y.T. 1785
Wang, Y.W. 1775
Wang, Y.Y. 1173
Wang, Y.Y. 1653
Wang, Z.C. 159
Wang, Z.D. 863
Wang, Z.D. 1237, 1243
Wang, Z.G. 243
Wang, Z.H. 369
Wang, Z.J. 333, 1639
Wang, Z.Q. 1545
Wang, Z.W. 243

Printed and bound by CPI Group (UK) Ltd, Croydon, CR0 4YY

24/10/2024

01778293-0014